COGNITIVE NEUROSCIENCE

THE BIOLOGY OF THE MIND

SECOND EDITION

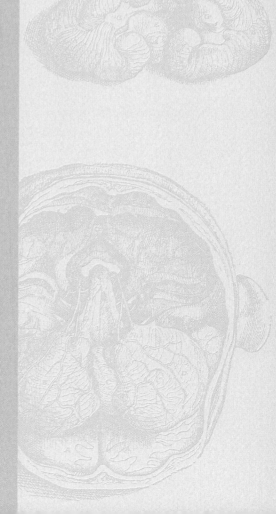

Michael S. Gazzaniga
DARTMOUTH COLLEGE

Richard B. Ivry
UNIVERSITY OF CALIFORNIA,
BERKELEY

George R. Mangun
DUKE UNIVERSITY

COGNITIVE NEUROSCIENCE

THE BIOLOGY OF THE MIND

SECOND EDITION

W · W · NORTON & COMPANY
NEW YORK · LONDON

PRINTED IN THE UNITED STATES OF AMERICA

*The text of this book is composed in Minion with the display set
in Myriad.*
Composition by TSI Graphics.
Manufacturing by R. R. Donnelley & Sons Company.
Editor: Jon W. Durbin.
Production manager: JoAnn C. Simony.
Copy editor: Mary Babcock.
Editorial assistant: Aaron Javsicas.
Production editor: Kim J. Yi.
Book design by Jack Meserole and Rubina Yeh.
Illustrations by Frank Forney.
*Cover image: Courtesy of Dr. Kevin Wilson, Laboratory for
Attention and Cognition, Center for Cognitive Neuroscience,
Duke University.*

Library of Congress Cataloging-in-Publication Data:

Gazzaniga, Michael S.
  Cognitive neuroscience : the biology of the mind / Michael S.
Gazzaniga, Richard B. Ivry, George R. Mangun.—2nd ed.
    p. cm.
  Includes bibliographical references and index.
  ISBN 0-393-97777-3
  1. Cognitive neuroscience. I. Ivry, Richard B. II. Mangun, G. R. (George
Ronald), 1956–   III. Title.
  QP360.5.G39 2002
  612.8'2—dc21   2001044542

W. W. Norton & Company, Inc., 500 Fifth Avenue, New York, NY 10110
www.wwnorton.com
W. W. Norton & Company Ltd., Castle House, 75/76 Wells Street, London W1T 3QT

4 5 6 7 8 9 0

FOR FRANCESCA AND ZACHARY
M.S.G.

FOR HENRY AND SAM
R.B.I.

FOR ALEXANDER
G.R.M.

CONTENTS

7 Selective Attention and Orienting 244

8 Learning and Memory 301

9 Language and the Brain (with Tamara Y. Swaab) 351

10 Cerebral Lateralization and Specialization 400

11 The Control of Action   445

12 Executive Functions and Frontal Lobes   499

13 Emotion (with Elizabeth A. Phelps)   537

BOXES

THE COGNITIVE NEUROSCIENTIST'S TOOLKIT

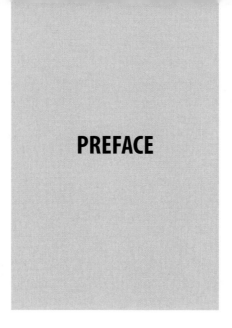

PREFACE

Welcome to the second edition! When new enterprises like cognitive neuroscience emerge, it is not always clear if they will survive. The youthful enthusiasm of a new field is not unlike a young doe standing up for the first time after birth. Can she make it? Will this new life develop and grow? Will true meaning emerge from the new life?

Five years later, we are confident that cognitive neuroscience has not only survived its infancy, it has blossomed in a spectacular fashion. This can be measured in many ways. Leading universities have undertaken major initiatives to develop dedicated cognitive neuroscience programs. The number of journals devoted to the field has increased exponentially. The annual meeting of the Society for Cognitive Neuroscience has burst its seams with attendance increasing 600% between 1994 and 2001.

What constitutes the first principles that make cognitive neuroscience distinct from physiological psychology, neuroscience, cognitive psychology, or neuropsychology? This question was our first challenge in laying the groundwork for our first edition, and constituted our defining point. We concluded that it is indeed a critical question—but paradoxically, not a question at all. Cognitive neuroscience certainly overlaps with and synthesizes these traditional approaches, but our book went beyond that function to define how cognitive neuroscientists will address the neural bases of cognition in the years ahead.

Our approach remains to balance cognitive theory with neuropsychological and neuroscientific evidence, plus add examples of the use of computational techniques to complete the story. We make liberal use of patient case studies, but this is to illustrate essential points, not to provide an exhaustive description of brain disorders. In every section, we strive to include the most current information and theoretical views, supported by cutting-edge technology that is such an important part of cognitive neuroscience. In contrast to purely cognitive or neuropsychological approaches, this text emphasizes the convergence of evidence that is a crucial aspect of any science, particularly studies of higher mental function.

Cognitive neuroscience takes on cognitive concepts and studies mind/brain matters with psychophysical and brain imaging techniques such as fMRI, MR, PET, and ERPs. The field requires one to become knowledgeable in each of these areas and to practice several different approaches when undertaking a single study. This book is intended to prepare students of cognitive neuroscience to do just that.

Since the first edition, there have been many major developments, both methodological and theoretical. There has been an explosion of brain imaging studies, almost 1,500 a year for the past four years. Other technologies such as transcranial magnetic stimulation have been added to the arsenal of the cognitive neuroscientist. New links have emerged with genetics, comparative anatomy, and robotics. All of the chapters in the second edition have been updated to capture these changes.

In addition, we have added two new chapters to the textbook. First, we now include a chapter on emotion. This information was interspersed across a number of chapters in the first edition. With the emergence of affective neuroscience as an important subfield, it became clear that the book needed a dedicated chapter. Second, the growth and importance of the fundamentals of neuroscience that are essential for any student of cognitive neuroscience led us to split this chapter into two, one focused on the basics of molecular and cellular neurobiology and the other on gross and functional anatomy.

Throughout this second edition we have updated each topic with new information. All of the chapters have undergone major revisions with the additions mostly devoted to incorporating results and ideas that have emerged from new studies. The second edition also allowed us to correct any errors in the first edition.

As usual, we are indebted to a number of people. Of special note are three colleagues who lent their expertise to specific chapters of the text, as noted in the table of contents. We thank Elizabeth Phelps for her contribution to Chapter 13, Emotion. Not only has Liz produced some of the most spectacular work in this field over the past few years, she provides a broad perspective to the topic and deep understanding of the issues. Leah Krubitzer wrote the first section of Chapter 14, Evolutionary Perspectives. She provides a convincing argument for why cognitive neuroscientists should understand comparative anatomy. As with the first edition of this textbook, Chapter 9, Language and the Brain, was coauthored by Tamara Swaab whose ability to lead the reader from current psycholinguistic theory to brain physiology and language dysfunction makes a complex field accessible.

As in the first edition, we make a special effort to bring cognitive neuroscience alive with color. Frank Forney is again the book's artist and is to be congratulated for his continued fine work. We also thank our many colleagues who have provided original artwork or scientific figures.

In sum, this book has been an interactive effort between ourselves, our colleagues, our students, and our reviewers! The product has benefited immeasurably from these interactions, but we will not rest here. We stand ready to modify and improve any and all of our work. In our first edition, we asked our readers to contact us with your suggestions and questions, and we do so again. We live in an age where interaction is swift and easy. We are to be found as follows: gazzaniga@dartmouth.edu; mangun@duke.edu; ivry@socrates.berkeley.edu. Good reading and learning.

ACKNOWLEDGMENTS

We are indebted to many scientists and personal friends. Writing a textbook is a major commitment of time, intellect, and affect! Those who have helped so much are noted below. Some reviewed our words and critiqued our thoughts. Others allowed us to interview them. We owe all our deep gratitude and thanks.

Linda P. Acredolo, University of California, Davis; Eyal Aharoni, University of California, Santa Barbara; David G. Amaral, University of California, Davis; Franklin R. Amthor, University of Alabama, Birmingham; Paul Aparicio, University of California, Berkeley; Ignacio Badiola, Florida International University; Gary Banker, Oregon Health Sciences University; Horace Barlow, Cambridge University; Kathleen Baynes, University of California, Davis; N. P. Bechtereva, Russian Academy of Science; Mark Beeman, University of Pennsylvania; Marlene Behrmann, Carnegie Mellon University; Ira B. Black, Robert Wood Johnson Medical School; Elizabeth Brannon, Duke University; Rainer Breitling, San Diego State University; Lindy A. Buck, Medical College of Ohio; Davina Chan, University of California, Berkeley; Valerie Clark, University of California, Davis; Clay Clayworth, VA Medical Center, Martinez, CA; Asher Cohen, Hebrew University; Jonathan Cohen, Princeton University; J. M. Coquery, Université des Sciences et Technologies de Lille; Michael Corballis, University Auckland; Paul Corballis, Dartmouth College; Anders Dale, Massachusetts General Hospital; Antonio Damasio, University of Iowa; Hanna Damasio, University of Iowa; Daniel C. Dennett, Tufts University; Mark D'Esposito, University of California, Berkeley; Joern Diedrichsen, University of California, Berkeley; Nina Dronkers, University of California, Davis; Martha Farah, University of Pennsylvania; Harlen Fichtenholtz, Duke University; Peter T. Fox, University of Texas; Karl Friston, Institute of Neurology, London; Rusty Gage, Salk Institute; Jack Gallant, University of California, Berkeley; Mitchell Glickstein, University College, London and Dartmouth College; Patricia S. Goldman-Rakic, Yale University School of Medicine; Gail Goodman, University of California, Davis; Elizabeth Gould, Princeton University; Jay E. Gould, University of West Florida; Scott Grafton, Dartmouth College; Charlie Gross, Princeton University; Nouchine Hadjikhani, Massachusetts General Hospital; Peter Hagoort, Max Planck Institute for Psycholinguistics; Todd Handy, Dartmouth College; Eliot Hazeltine, University of California, Berkeley; Hans-Jochen Heinze, University of Madgeberg; Steven A. Hillyard, University of California, San Diego; Hermann Hinrichs, University of Madgeberg; Joseph Hopfinger, University of California, Davis; Richard Howard, National University of Singapore; Drew Hudson, University of California, Berkeley; Akira Ishiguchi, Ochanomizu University; Lucy Jacobs, University of California, Berkeley; Amishi Jha, University of California, Davis; Cindy Jordan, Michigan State University; Tim Justus, University of California, Berkeley; Nancy Kanwisher, Massachusetts Institute of Technology; Larry Katz, Duke University; Steven Keele, University of Oregon; Leon Kenemans, University of Utrecht; Steve Kennerley, Oxford University; Alan Kingstone, University of British Columbia; Robert T. Knight, University of California, Berkeley; Stephen M. Kosslyn, Harvard University; Neal Kroll, University of California, Davis; Leah Krubitzer, University of California, Davis; Marta Kutas, University of California, San Diego; Joseph E. Le Doux, New York University; Steven J. Luck, University of Iowa; Nancy Martin, University of California, Davis; James L. McClelland, Carnegie Mellon University; Christina Middleton, University of California, Berkeley; George A. Miller, Princeton University; Teresa

Mitchell, Duke University; Amy Needham, Duke University; Ken A. Paller, Northwestern University; Jasmeet K. Pannu, University of Arizona; Steven E. Petersen, Washington University School of Medicine; Steven Pinker, Massachusetts Institute of Technology; Michael I. Posner, University of Oregon; David Presti, University of California, Berkeley; Robert Rafal, University of Wales, Bangor; Marcus Raichle, Washington University School of Medicine; Noam Sagiv, University of California, Berkeley; Mikko E. Sams, University of Tampere; Donatella Scabini, University of California, Berkeley; Daniel Schacter, Harvard University; Michael Scholz, University of Magdeberg; Art Shimamura, University of California, Berkeley; Michael Silverman, Oregon Health Sciences University; Allen W. Song, Duke University; Larry Squire, University of California, San Diego; Thomas M. Talavage, Massachusetts General Hospital; Keiji Tanaka, Riken Institute; Michael Tarr, Brown University; Ed Taylor; Sharon L. Thompson-Schill, University of Pennsylvania; Roger Tootell, Massachusetts General Hospital; Anne M. Treisman, Princeton University; Carrie Trutt, Duke University; Endel Tulving, Rotman Research Institute, Baycrest Center; John Vollrath; C. Mark Wessinger, Dartmouth College; Kevin Wilson, Duke University; Ginger Withers, Oregon Health Sciences University; Marty G. Woldorff, University of Texas at San Antonio; Andrew Yonelinas, University of California, Davis.

And finally, our special thanks to our four main reviewers: Jennifer Mangels, Columbia University; Chad Marsolek, University of Minnesota; Nancy Kim, University of Michigan; and Francoise Macar, CNRS Marseille.

1

A Brief History of Cognitive Neuroscience

What is the field of cognitive neuroscience all about? Where did it come from? Where is it going? We start this book with a brief history of the people and ideas that led to the new field of cognitive neuroscience, one that has roots in neurology, neuroscience, and cognitive science. Modern-day cognitive neuroscience represents a hybrid of disciplines, and therefore the student of the mind must become aware and knowledgeable in many areas to fully understand the issues studied in cognitive neuroscience. And the field changes rapidly. At the end of this chapter we introduce the short and very new history of brain imaging. Brain imaging has become central to the study of the mind in the last few years.

PONDERING THE BIG QUESTIONS

Do you wonder about big things like the meaning of life, or the meaning of meaning? Or are you the type who does not wonder about such evanescent questions? If you are the latter, do not read this book—even though you should. This book is for those who wonder what life, mind, sex, love, thinking, feeling, moving, attending, remembering, communicating, and being are all about. And better, it is about scientific approaches to these grand issues. So prepare yourself for learning about a fantastic story still in the making.

The scientific field of cognitive neuroscience received its name in the late 1970s in the back seat of a New York City taxi. One of us (M.S.G.) was riding with the great cognitive psychologist George A. Miller on the way to a dinner meeting at the Algonquin Hotel. The dinner was being held for scientists from Rockefeller University and Cornell University, who were joining forces to study how the brain enables the mind, a subject in need of a name. Out of that taxi ride came the term *cognitive neuroscience,* which took hold in the scientific community.

Now the question is, What does it mean? In answering this ponderous question, we need to step back and look at not only the history of human thought but also the history of the scientific disciplines of biology, psychology, and medicine.

To grasp the miraculous properties of brain function, one must bear in mind that Mother Nature built it, not a team of rational engineers. Although the earth formed approximately 5 billion years ago, and life first appeared around 3.5 billion years ago, human brains in their present form have been around only about 100,000 years. The primate brain appeared approximately 20 million years ago, and evolution took its course to build our present human brain, capable of all sorts of wondrous—and banal—feats.

During most of history, humans were too busy to think about thought. While there can be little doubt that human brains could engage in such activities, life was given over to more practical work such as surviving in tough environments, developing ways to live better by inventing agriculture or by domesticating

1

animals, and so forth. However, as soon as civilization developed to the point when day-to-day survival did not occupy every hour of every day, our ancestors began to spend time constructing complex theories about the motives of fellow humans. Examples of attempts to understand the world and our place in it include *Oedipus Rex,* the ancient Greek play that deals with the nature of the child-parent conflict, and Mesopotamian and Egyptian theories on the nature of religion and the universe. The brain mechanisms that enabled the generation of theories about the nature of human nature thrived inside the heads of ancient humans. Yet they had one big problem: They did not have the ability to systematically explore the mind through experimentation.

In a diary entry of 1846, the brilliant philosopher Søren Kierkegaard wrote:

> . . . That a man should simply and profoundly say that he cannot understand how consciousness comes into existence—is perfectly natural. But that a man should glue his eye to a microscope and stare and stare and stare—and still not be able to see how it happens— is ridiculous, and it is particularly ridiculous when it is supposed to be serious. . . . If the natural sciences had been developed in Socrates' day as they are now, all the sophists would have been scientists. One would have hung a microscope outside his shop in order to attract custom, and then would have had a sign painted saying: "Learn and see through a giant microscope how a man thinks (and on reading the advertisement Socrates would have said: 'That is how men who do not think behave')."

The Nobel laureate Max Delbrück (1986) began his fascinating account of the evolution of the cosmos in his book *Mind from Matter?* with the foregoing quote of Kierkegaard. Delbrück is part of the modern tradition that started in the nineteenth century. Observe, manipulate, measure, and start to determine how the brain gets its job done. Armchair thinking is a wonderful thing and has produced fascinating science such as theoretical physics and mathematics. But to understand how a biological system works, a laboratory is needed and experiments have to be performed. Ideas derived from introspection can be eloquent and fascinating, but are they true? Philosophy can add perspective, but is it right? Only scientific method can move a topic along on sure footing. And just think about the rich phenomena to study. Take the perception of faces. Some say that the brain has a special system for recognizing faces. This specialized system was revealed because patients with certain brain lesions had a hard time recognizing faces of all kinds. Scientists immediately debated whether there was a specialized system. No, some said, the impairment is with object perception in general, not faces in particular. They pointed to research which suggested that people who had a hard time recognizing faces also had a hard time seeing objects or faces of animals.

But then comes a new case. A patient has a terrible time seeing everyday objects but has no problem seeing faces! In fact, if the faces are composed of fruit arranged to look like a face, the patient says he sees the face but does not realize it is made up of fruit! Incredible but true. It appears as though a special system in the brain sees faces; it is triggered to produce the percept for our conscious lives by the configuration of elements. The special face processor does not know or care about what elements it is composed of; as long as they are in proper arrangement, a face is perceived. What could be more fascinating than to study how the brain does such things?

THE BRAIN STORY

You are given a problem to solve. A hunk of biological tissue is known to think, remember, attend, solve problems, want sex, play games, write novels, exhibit prejudice, and do a zillion other things. You are supposed to figure out how it works. Before starting, you might ask a few questions. Does the blob work as a unit with each part contributing to a whole? Or is the blob full of individual processing parts, each carrying out specific functions, with the result being something that looks like it is acting as a whole unit? After all, the blob of the city of New York looks like an integrated whole from a distance, but it is actually composed of millions of individual processors, which is to say people. Perhaps people, in turn, are made of smaller, more specialized units.

This central issue—whether the whole brain working in concert or parts of the brain working independently enable mind—is what fuels much of modern research. As we will see, the dominant view has changed over the past 100 years, and continues to do so today. It all started in the nineteenth century when **phrenologists,** led by Franz Joseph Gall and J.G. Spurzheim (1810–1819), declared that the brain was organized around some thirty-five specific functions (Figure 1.1). These functions,

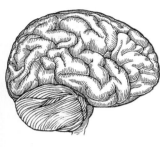

Figure 1.1 **Left:** Franz Joseph Gall. One of the founders of phrenology in the early nineteenth century. **Right:** The right hemisphere of the brain, from Gall and Spurzheim in 1810.

Figure 1.3 **Left:** Pierre Jean Marie Flourens (1794–1867), who supported the idea later termed the *aggregate field*. **Right:** The position of a pigeon deprived of its cerebral hemispheres described by Flourens.

which ranged from cognitive basics such as language and color perception to more ephemeral capacities such as hope and self-esteem, were thought to be supported by specific brain regions. Moreover, if a person used one of the faculties with greater frequency than the others, the part of the brain representing that function grew. According to the phrenologists, this increase in local brain size would cause a bump in the overlying skull. Logically, then, Gall and his colleagues believed a careful analysis of the skull could go a long way in describing the personality of the person inside the skull. He called this technique *anatomical personology* (Figure 1.2).

Gall, an Austrian physician and neuroanatomist, was not a scientist in the sense that he did not test his ideas.

The best part of his life's efforts was to direct attention to the cerebral cortex, particularly to its surface, and to emphasize the idea that different brain functions are localized to discrete brain regions.

The experimental physiologist Pierre Flourens challenged Gall's **localizationist** view (Figure 1.3). A large group of people rejected the idea that specific processes such as language and memory were localized within circumscribed brain regions, and Flourens became their champion. He studied animals, especially birds, and discovered that lesions in particular areas of the brain did not cause certain deficits in behavior. No matter where he made a lesion in the brain, the bird recovered. He

Figure 1.2 **Left:** An analysis of Presidents Washington, Jackson, Taylor, and McKinley by Jessie A. Fowler, from the *Phrenological Journal,* June 1898. **Center:** The phrenological map of personal characteristics on the skull, from the *American Phrenological Journal,* 1850. **Right:** Fowler and Wells Company publication on marriage compatibility in connection with phrenology, 1888.

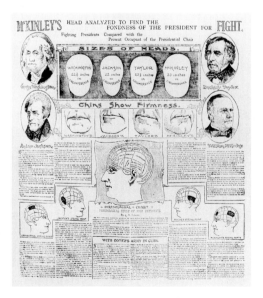
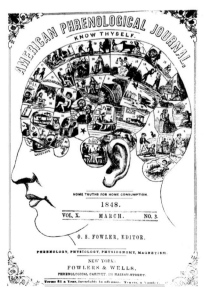
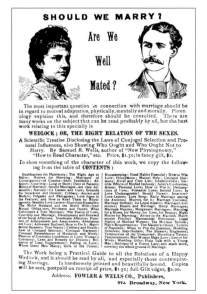

developed the notion that the whole brain participated in behavior, a view later known as the **aggregate field.** In 1824 Flourens wrote, "All sensations, all perceptions, and all volitions occupy the same seat in these (cerebral) organs. The faculty of sensation, percept and volition is then essentially one faculty."

Work on the European continent and in England helped to swing the notion back to the localizationist view. In England, for example, the neurologist John Hughlings Jackson (Figure 1.4) began to publish his observations on the behavior of persons with brain damage. One of the key features of Hughlings Jackson's writings was the incorporation of suggestions for experiments to test his observations. He noticed, for example, that during the start of their seizures, some epileptic patients moved in such characteristic ways that the seizure appeared to be stimulating a set map of the body in the brain; hence, clonic and tonic jerks in muscles, produced by the abnormal epileptic firings of neurons in the brain, progressed in an orderly way from one body part to another. This phenomenon led him to propose a *topographic* organization in the cerebral cortex: In this view, a map of the body was represented in a particular cortical area. Hughlings Jackson was one of the first to realize this essential feature of brain organization.

Although Hughlings Jackson was also the first to observe that lesions on the right side of the brain affect visual-spatial processes more than do lesions on the left side, he did not maintain that specific parts of the right side of the brain were solely committed to this important human cognitive function. Hughlings Jackson, being an

Figure 1.4 John Hughlings Jackson, an English neurologist who was one of the first to recognize the localizationist view.

observant clinical neurologist, noticed that it was rare for a patient to totally lose a function. For example, most people who lost their capacity to speak following a cerebral stroke could still say some words. Patients unable to direct their hands voluntarily to specific places on their bodies could still easily scratch those places if they itched. When Hughlings Jackson made these observations, he concluded that many regions of the brain contributed to a given behavior. Meanwhile, in France, perhaps the most famous neurological case in history was being reported by Paul Broca (Figure 1.5). In 1861 he treated a man who had suffered a stroke; the patient could understand language but could not speak. Consistent with Hughlings Jackson's observations though, the patient could utter something—the

Figure 1.5 **Left:** Pierre Paul Broca. **Right:** The connections between the speech centers, from Wernicke's article on aphasia. B = Broca's area of motor speech; A = the sensory speech center of Wernicke; Pc = area concerned with language.

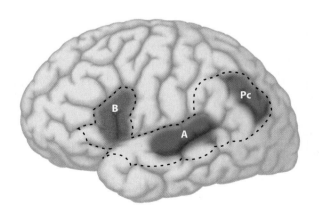

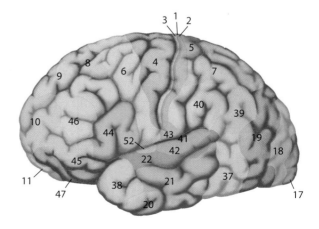

Figure 1.6 Left: Physiologist and anatomist Gustav Theodor Fritsch (1838–1907). **Center:** Professor of Psychiatry Eduard Hitzig (1838–1927). **Right:** The original illustration of the dog's cortex by Fritsch and Hitzig.

sound "tan." Such patients also can usually speak automatically, so while they might say, "Tan, tan, tan . . ." in response to the question, "Who are you?," they might easily count from one to ten in normal fashion.

The exact part of the brain that was damaged in Broca's patient was the left frontal lobe. It has come to be called *Broca's area*. The impact of this finding was huge. Here was a specific aspect of language that was impaired by a specific lesion. This theme was picked up by the German neurologist Carl Wernicke. In 1876, when he was only 26 years old, he reported a stroke victim who could talk quite freely, unlike Broca's patient, but what he said made little sense. Wernicke's patient also could not understand spoken or written language. He had a lesion in a more posterior region of the left hemisphere, an area in and around where the temporal and parietal lobes meet. Today, these differences in how the brain responds to focal disease are well known. Every neurologist in every community hospital knows these things. But a little over a hundred years ago, Broca's and Wernicke's discoveries were earth shattering. Philosophers, physicians, and early psychologists zeroed in on a startling point: Focal disease causes specific deficits. In those days investigators were limited in their ability to identify a patient's lesion. Physicians could observe the site of injury, for example, a penetrating head wound from a bullet, but they had to wait for the patient to die in order to determine the site of the brain lesion. The latter might take months or years, and usually was not possible: The physician would lose track of the patient after he or she recovered, and when the patient eventually died, the physician was not informed and thus could not examine the brain to correlate brain damage with the person's behavioral deficit(s). Today, the site of brain injury can be determined in a few minutes with imaging methods that scan the living brain and produce a picture. We learn about these techniques as we progress through this book. (As an interesting historical footnote, the brain of Broca's famous patient has been preserved. Recent scanning of the brain revealed the patient's lesion to be much larger than what Broca had originally described.)

As is so often the case, the study of humans leads to questions for those who work on animal models. Shortly after Broca's discovery, the German physiologists Gustav Fritsch and Eduard Hitzig electrically stimulated discrete parts of a dog brain and observed that this stimulation produced characteristic movements in the dog (Figure 1.6). This discovery led neuroanatomists to a closer analysis of the cerebral cortex and its cellular organization; they wanted support for their ideas about the importance of local regions. Because these regions performed different functions, it followed that they ought to look different at the cellular level.

Following this logic, German neuroanatomists began to analyze the brain by using microscopic methods to view the cell types in different brain regions. Perhaps the most famous of the group was Korbinian Brodmann, who analyzed the cellular organization of the cortex and characterized fifty-two distinct regions (Figure 1.7). Brodmann used tissue stains, such as the one developed by Franz Nissl, which permitted him to visualize the different cell types in different brain regions. How cells differed between

Figure 1.7 The fifty-two distinct areas described by Brodmann based on cell structure and arrangement. Adapted from Brodmann (1909).

brain regions was called **cytoarchitectonics,** or cellular architecture. Soon, many now-famous anatomists, including Oskar Vogt, Vladimir Betz, Theodor Meynert, Constantin von Economo, Gerhardt von Bonin, and Percival Bailey, contributed to this work, and several subdivided the cortex even further than Brodmann did. To a large extent, these investigators discovered that various cytoarchitectonically described brain areas do indeed represent functionally distinct brain regions. For example, Brodmann first distinguished between area 17 and area 18—a distinction that has proved correct in subsequent functional studies. The characterization of the primary visual area of the cortex, area 17, as distinct from surrounding area 18, remarkably demonstrates the power of the cytoarchitectonic approach, as we discover in Chapter 3.

Yet, the truly huge revolution in our understanding of the nervous system was happening down south, in Italy and Spain. There, an intense struggle was going on between two brilliant neuroanatomists. Oddly, it was the work of one that led to the insights of the other. Camillo Golgi, an Italian, developed a stain that impregnated individual neurons with silver (Figure 1.8). This stain permitted full visualization of single neurons. Using Golgi's method, Santiago Ramón y Cajal, a Spaniard, went on to find that, contrary to the view of Golgi and others, the neurons were discrete entities (Figure 1.9). Golgi had believed that the whole brain was a **syncytium,** or a continuous mass of tissue that shares a common cytoplasm! Cajal extended his findings and was the first to identify not only the unitary nature of neurons but also their transmission of electrical information in only one direction, from the dendrites down to the axonal tip (Figure 1.10).

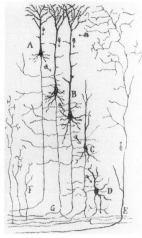

Figure 1.9 **Left:** Santiago Ramón y Cajal (1852–1934), co-winner of the Nobel Prize in 1906. **Right:** Cajal's drawing of the afferent inflow to the mammalian cortex.

Many truly gifted scientists were involved in the early history of the **neuron doctrine.** For example, Johannes Evangelista Purkinje, a Czech trained in German-controlled Prague who had to travel to Poland to get a university position (Figure 1.11), not only described the first nerve cell in the nervous system but also invented the stroboscope, described common visual phenomena, and made a host of other major discoveries.

Even Sigmund Freud got into the neuron act (Figure 1.12). As a young man he studied microscopic anatomy with the great German anatomist Ernst Brücke. Freud even wrote an essay about his subsequent and independent work with crayfish. Indeed, some of Freud's biographers suggest that he, too, had come up with the idea of the neuron as a separate and distinct physiological unit.

Hermann von Helmholtz, perhaps one of the most famous scientists of all time, also contributed to the early study of the nervous system (Figure 1.13). He was

Figure 1.8 **Left:** Camillo Golgi (1843–1926), co-winner of the Nobel Prize in 1906. **Right:** Golgi's drawings of different types of ganglion cells in dog and cat.

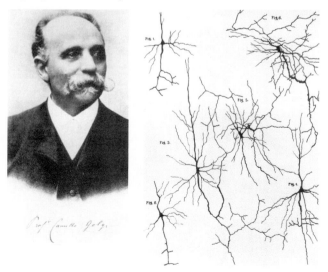

Figure 1.10 A bipolar retinal cell.

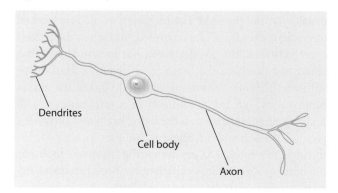

Dendrites

Cell body

Axon

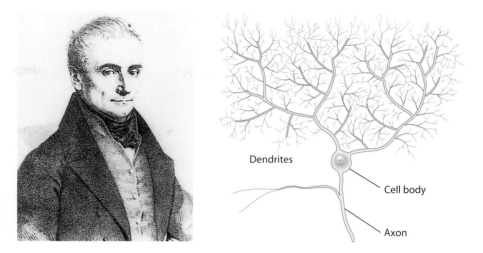

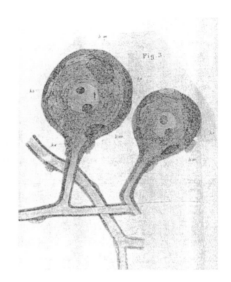

Figure 1.11   Left: Johannes Evangelista Purkinje, who described the first nerve cell in the nervous system. **Right:** A Purkinje cell of the cerebellum.

Figure 1.12   Left: Sigmund Freud (1856–1939). **Right:** From his work with crayfish, Freud published this illustration as an example of anastomosis of nerve fibers, a concept Cajal disproved.

Figure 1.13   Left: Hermann Ludwig von Helmholtz (1821–1894). **Right:** Helmholtz's apparatus for measuring the velocity of nerve conduction.

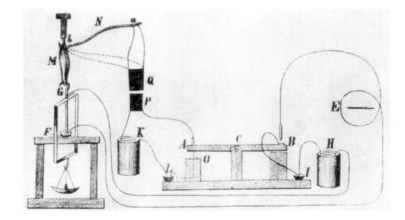

An Interview with Mitchell Glickstein, Ph.D.
Dr. Glickstein is a professor of neuroanatomy and neuroscience at University College, London. He has written extensively on the history of neuroscience.

Authors: You have taken on the task of studying a variety of issues in neuroscience from a historical point of view. What motivated this interest?

MG: Two things. One was my undergraduate education at the University of Chicago, a Baptist organization where a Jewish faculty taught Catholicism to an atheist student body. There were no textbooks (or almost none), only a syllabus of original readings. In history, for example, we read Lenin and Martov on the Russian revolution. In chemistry we read Lavoisier. I always thought that the people who made up new ideas could teach them more effectively than textbook writers. The second influence was Harry Patton's lectures to the medical students at the University of Washington. He taught reflexes brilliantly—experiment by experiment. After his lectures you understood that the shortest pathway for the knee jerk involved only a single synapse, but the evidence that led up to that conclusion was made brilliantly clear. Patton taught not just what we know, but more importantly, how we know.

Authors: You are now caught in the modern world of textbooks, TV, magazines, and frequent conferences. All of these give rise to an image of what the important issues are in neuroscience and how they came to be. Are you suggesting that investigation of original papers and the history of ideas gives rise to a different interpretation of the present than the one most people know? If so, could you give an example?

MG: There is so much to say.... For starters, consider Otto Loewi's experiment on "Vagusstoff." At the time Loewi did his critical experiment, there was the suspicion that nerves could activate muscles or other nerves by releasing a small amount of chemical substance—but no proof. Loewi set up a simple experiment in his own home after having an idea while sleeping. It was known at that time that vertebrate hearts, from frogs and even mammals, can continue to beat if removed from the body and placed in a suitable solution. Al-

though hearts have a nerve supply, they can beat even when the supply is cut off. The largest of the autonomic nerves, the vagus, provides the innervation to the heart and causes slowing of the heartbeat. If the vagus nerve is stimulated electrically, the heart rate slows and the force of the beat decreases. All of these facts were well known in Loewi's time. What was not known was how the vagus acted. Loewi stimulated a frog nerve-heart preparation at a high rate, causing the heart rate to slow. He collected the fluid from within the heart just after the electrical stimulation. He then injected it into the same heart—or a second heart—and it produced the same effect of slowing the heart rate and decreasing the pressure. When fluid was removed when the vagus had not been previously stimulated, there was no effect on the same heart or the second heart. Loewi concluded that something must have been released when the vagus nerve was stimulated, and it was that chemical that caused the slowing and decrease in force of the beat. He named it provisionally *vagusstoff*, meaning literally "vagus material." Some years later, Loewi and others demonstrated that the substance was the neurotransmitter acetylcholine. Loewi at first believed that the vagusstoff—acetylcholine—was specific. Sir Henry Dale, Walter Feldberg (a student of Loewi's), and Marthe Vogt (Cecile and Otto's daughter) demonstrated that acetylcholine is also a transmitter at voluntary nerve-muscle junctions.

Consider self-observation. Wollaston (1824) described his own transient hemianopia (partial visual loss) and the more permanent hemianopia of two acquaintances, when the concept did not exist, and his description was picked up by the *Boston Medical and Surgical Intelligencer* as a curiosity (after which they described a boy in Philadelphia who saw a candle flame upside down!). Wollaston was one of those nineteenth-century English geniuses, working at a time when people could make contributions to several sciences. He had invented a

the first to suggest that invertebrates would be good models for vertebrate brain mechanisms. This was the same Helmholtz who went on to make major contributions to physics, medicine, and psychology.

The phenomenon of a famous scientist making significant contributions to several distinct fields may be a thing of the past. Not that that sort of thing does not happen today; it is just not easily recognized. Science

kind of prism for optical work, and also developed a technique for pulling very fine wires for use in precision instruments. Prior to Wollaston's work, horse hairs had been used. Wollaston made the simplest of self-observations. He noted that he sometimes had an attack of partial loss of vision. The visual loss was the same in both eyes; half of the visual field could not be seen. Always it was the same half of the visual scene in both eyes. Wollaston gives this example: He went to see his friend—called Hughlings Jackson—and looked at the name plate on Hughlings Jackson's door. He said that he saw only the word "son." He was blind in the left half of the visual field in each eye. Wollaston's transient half-blindness (now called *hemianopia*, literally "half-not-seeing") was transitory, but he knew of two other individuals with the same half-blindness that was permanent. Both sorts of hemianopia are now well known to physicians. Sir Isaac Newton in his text on optics had speculated that each of the optic nerves divided in such a way that the optic nerve linking the left side of each retina went to the left side of the brain, and the optic nerve from the right went to the right side of the brain. The implications of Newton's anatomical suggestions had never been recognized, and Wollaston did not appear to be aware of them. His observation was thought to be a rare and unusual example of a visual disturbance. It was not for another 70 years that the two insights were fully integrated. Then Newton's speculations about the course of optic nerve fibers were verified, and physicians began to recognize the fact that hemianopia was all too common, and related to the underlying anatomy.

Consider the transcendent and international nature of science. It did not involve only Germans in the nineteenth century. It was a little guy, Cajal, from a backward country who set neuroanatomy on the right course for a hundred years.

These are a few of the examples of why I use historical sources. They help to get around the obscenity of 35-year-olds with five postdoctoral degrees who are brilliant at the last 5 years of research, and a bit hazy on the previous 5, and know nothing about how all of what we know came to be known. If you want to teach about science, it is not a bad choice to teach how we got here.

Authors: Surely all of that is true and more. For example, would you comment on Cajal's enunciation of the neuron doctrine? While he is largely credited as the founder, is it not really the case that the idea was in the air and many people were talking about it at the time? While Cajal was unquestionably a brilliant anatomist, perhaps the most brilliant to ever live, did he not simply crystallize and successfully market the idea of the time?

MG: Cajal was more than that. You appreciate Cajal more if you read his contemporaries (Golgi, Dogiel, Kölliker). When Cajal began his work, the dominant view was of a rather vague syncytium. Golgi thought there was an anastomosis between the axon collaterals and the brain was a fused network. I recently gave a lecture on Cajal at the Cajal Institute—a bold venture. I focused on Cajal's visit to England in 1894. I had all the documents relating to his invitation, his acceptance, and his honorary degree at Cambridge. (He was arrested briefly in Cambridge, a delightful side episode.) The reason I spent a day at the Institute was to look into a particular issue—the discovery of dendritic spines. Golgi certainly saw them but left them out of his figures. Kölliker solemnly reviewed the question in his (1896) textbook and concluded that the spines were an artifact. Cajal wrote a brilliant paper in which he raised the question of why should dendrites show them but not axons? Why could the Golgi method and the (mercurybased) variants show them? But to prove their validity, he said that they had to be revealed by a totally independent method. He fiddled with the methylene blue method and got it to impregnate (to stain) as well as the Golgi method did. I went to Madrid to look at his drawings. As Cajal said, "They are there with methylene blue staining; they are really there. The matter is over." There were others. Waldeyer, for example, coined the term *neuron* but contributed little else. I agree with Cajal who said of Waldeyer that what he had done was "...to publish my research in a weekly medical journal." How many textbooks written 100 years ago are still useful today?

Authors: Well, only time will tell if this one joins those ranks. Thank you.

today is an enormous enterprise, and each subdiscipline has its own cadre of heroes and villains. These guardians of a subject are loath to let a stranger enter into their debates.

Shepherd (1992) recounted the fascinating story of Camillo Golgi and contrasted it with Cajal's. Born near Milan, Golgi, the son of a physician, received his medical degree at the age of 22 from the University of Pavia in

Interlude

In textbook writing, authors use broad strokes to communicate milestones that have become important to our thinking over a long period of time. It would be folly, however, not to alert the reader to the complex and intriguing cultural, intellectual, and personal setting. The problems that besieged the world's first scientists remain today, in full glory. Issues of authorship, ego, funding, and credit are all integral to the fabric of intellectual life. Much as teenagers never imagine that their parents once had the same interests and desires as they do, novitiates in science believe they are tackling new issues for the first time in human history. Gordon Shepherd (1992), in his riveting account *Foundations of the Neuron Doctrine,* detailed the variety of forces at work on the figures we now feature in our brief history.

Shepherd noted how the explosion of research on the nervous system started in the eighteenth century as part of the intense activity swirling around the birth of modern science. As examples, Robert Fulton invented the steam engine in 1807, and Hans Chris-tian Ørsted discovered electromagnetism. Of more interest to our concerns, Leopoldo Nobili, an Italian physicist, invented a precursor to the galvanometer—a device that laid the foundation for studying electrical currents in living tissue. Many years before, in 1674, Anton van Leeuwenhoek in Holland had used a primitive microscope to view animal tissue (Figure A). One of his first observations was of a cross section of a cow's nerve in which he noted "very minute vessels." This observation was consistent with René Descartes's idea that nerves contained fluid or "spirits," and these spirits were responsible for the flow of sensory and motor information in the body (Figure B). To go further, however, this revolutionary work would have to overcome the problems with early microscopes, not the least of which was the quality of glass used in the lens. Chromatic aberrations made them useless at higher magnification. It was not until lens makers solved this problem that microscopic anatomy again took center stage in the history of biology.

Figure A **Left:** Anton van Leeuwenhoek. **Right:** One of the original microscopes used by Leeuwenhoek, which is composed of two brass plates holding the lens.

Figure B Portrait of René Descartes by Frans Hals.

Italy, yet he was outside the mainstream of research in Germany's great universities—a fact that haunted his career. Still, Pavia produced many spectacular scientists, like the physicist Alessandro Volta, Christopher Colum-bus, and a host of great biologists. Among this august group, Golgi was a well-trained physician and scientist.

While his career started off with great promise, Golgi was forced, out of financial necessity, to take a job outside of Pavia, in the town of Abbiategrasso, where he became the resident physician at the Home for the Incurables. In such circumstances, it was highly unlikely that Golgi would continue his scientific life. But he persevered, and working by the candlelight in his kitchen, he developed the most famous cell stain in the history of the world. Golgi discovered the silver method for staining neurons—*la reazione negra,* the "black reaction."

The Golgi stain became famous, as did Golgi himself. His intense interest in disease led him to several discoveries in pathobiology, and eventually he was recruited back to Pavia to be a professor. Still, his writings were not widely known outside of Italy, so he translated them, publishing them in the *Italian Archives of Biology*—which, oddly, was in French. Golgi's reputation grew, and in 1906 he was awarded the Nobel Prize jointly with Cajal.

Meanwhile, Cajal was the rambunctious son of another physician. It was not until Cajal's ambitious father started tutoring him at home that he developed an interest in biology. From a complex childhood, Cajal emerged as what some call the "father of modern neuroscience." He is credited as the first to articulate fully the neuron doctrine.

The irony in the story, as we already mentioned, is that Cajal made many of his discoveries as a result of the Golgi stain. He first saw the stain in the Madrid home of a colleague, Don Luis Simarro, who had learned the technique while attending a meeting in Paris. Cajal said it was there, in Simarro's home laboratory—not unlike the laboratory where Golgi invented the stain—that he saw for the first time "the famous sections of the brain impregnated by the silver method of the Savant of Pavia."

Cajal's response to seeing the stain remains fascinating (translated by Sherrington, 1935):

> Against a clear background stood black threadlets, some slender and smooth, some thick and thorny, in a pattern punctuated by small dense spots, stellate or fusiform. All was sharp as a sketch with Chinese ink on transparent Japanese paper. And to think that that was the same tissue which stained with carmine or logwood left the eye in a tangled thicket where sight may stare and grope for ever fruitlessly, baffled in its effort to unravel confusion and lost forever in a twilit doubt. Here, on the contrary, all was clear and plain as a diagram. A look was enough. Dumbfounded, I could not take my eye from the microscope.

And yet many years later, the scene at the Nobel Prize ceremony in Stockholm was ugly. Golgi came off as a huge egotist, set in his ways, and unwilling to acknowledge Cajal's discoveries, which by that time had established the neuron doctrine. Both were using the same stain, both were using the same microscopes, both were studying the same tissue. One saw the answer, one did not. Golgi continued to see his beloved syncytium of neurons as a single unit, whereas Cajal saw each neuron as the independent unit it has proved to be.

THE TWENTIETH CENTURY

Some scientists are chagrined to learn that the results of this early research of Cajal and others confused scientists in the first half of the twentieth century. All the work, especially when viewed in hindsight, argued for the importance of single neurons. Knowing how the nervous system works requires understanding how single neurons interact and behave, just like understanding proteins requires a knowledge of how the constituent amino acids are organized. The vagaries of syncytial processes, of nerve nets, and of **holistic** processes were not needed. The nervous system is not a big blob; it is built from discrete units. If we can figure out how these units work, and describe the laws and principles of their interaction, then the problem of how the brain enables mind can be addressed and eventually solved.

This is the ideal view, namely, that by knowing all the elements of a system, we can figure out the system. Yet, the human brain is composed of billions of neurons, and to think that we need to know the actions of all of them to figure out how the brain works would be preposterous. Indeed, it took a tremendous effort to figure out how the stomatogastric ganglion of the lobster, with eight neurons, produced rhythmic activity. Advances are made by working at different levels of organization, the backbone strategy in cognitive neuroscience. By knowing what behavior is actually produced, we need not know all the possible interactions that occur with the underlying elements. In this manner a problem becomes constrained and solvable. But that was not the dominant question of the early twentieth century. Even though the renowned British physiologist Sir Charles Sherrington pursued vigorously the neuron's behavior as a unit, and indeed coined the term *synapse* to describe the junction between two neurons, the scientists working on "larger" issues of brain and behavior remained committed to the belief in holistic processes. It took time for the new ideas to be widely accepted, especially when major figures of early brain science were so divided in their views. In addition, many early views were actually quite reasonable given the state of scientific knowledge at the time.

Take, for example, the ideas of Broca's contemporaries, like Pierre Marie. While Broca was selling the importance of localized function, Marie was demonstrating the variability in cortical localization. Marie reported that only half of his patients displayed speech impediments when lesions were localized to the third frontal convolution of the left cerebral hemisphere—Broca's area. He also noted that several patients with lesions sparing this same area had Broca-like aphasia. Marie was both right and wrong. There is great variation in the human brain, but by looking at the underlying structure, he may well have discovered that the critical brain area was simply shifted from one place to another during development. The notion of localization, therefore, is not really challenged with the sort of observation that Marie offered.

Still, some people simply refuse to see that understanding the function of single neurons or small areas of the brain can explain how the brain works. That view is as prevalent today as it was in the early twentieth century. Everybody concerned with the matter has a favorite example of the seemingly deep contradictions in this logic. During Broca's time, a German professor of physiology named Friedrich Goltz was prancing his dog around at scientific meetings (Figure 1.14). Goltz had removed large parts of the dog's cortex, and although some impairments were noted, the dog was remarkably functional. It turned out that the dog's lesion was smaller than claimed, but the example is not uncommon. With brain-damaged humans it is common to be surprised by a patient's lack of symptoms, given the extent of lesion depicted on a brain scan.

Still, one major advance had been made. By the beginning of the twentieth century, almost everyone was willing to grant that some localization of function occurs in the cerebral cortex. Even Goltz noticed substantive differences in his animals when the occipital lobe was removed compared with when the motor cortex was removed. The critics were now claiming that it was impossible to localize "higher cortical functions" like thinking and memory, a modification of the original view that no localization of function existed in the brain. This reservation combined with the insight first articulated by Hughlings Jackson; namely, one has to distinguish between evidence for localization of *symptoms* and the idea of localization of *function*. By this, Hughlings Jackson meant that while a brain lesion might produce a bizarre symptom, it did not follow that the injured area was specialized for only that function. The lesion might well affect other structures in the brain because the lesion might have damaged neurons connected to other regions. Hughlings Jackson's distinction was also an early warning that behaviors were constellations of independent activities, not a single whole unit. This distinction is crucial when analyzing modern brain imaging data, as we will see.

Stanley Finger (1994), in his historic account in *Origins of Neuroscience* of the events surrounding this key issue, provided telling quotes from antilocalizationists. At the turn of the century a broad-based movement was absorbed with gestalt processes, the idea that the whole is different from the sum of the parts. One of the movement's members, the great French biologist Claude Bernard, wrote in 1855:

Figure 1.14   **Left:** Friedrich Leopold Goltz (1834–1902). **Center:** The dog Goltz showed to the International Medical Congress in 1881. **Right:** The brain of the dog from which Goltz removed a section of cortex.

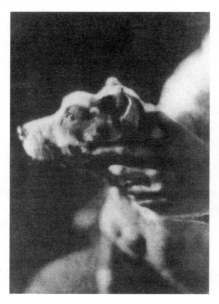
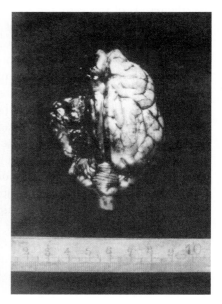

If it is possible to dissect all the parts of the body, to isolate them in order to study them in their structure, form and connections it is not the same in life, where all parts cooperate at the same time in a common aim. An organ does not live on its own, one could often say it did not exist anatomically, as the boundary established is sometimes purely arbitrary. What lives, what exists, is the whole, and if one studies all the parts of any mechanisms separately, one does not know the way they work. In the same way, anatomically, we take the organism apart, but we cannot grasp the whole. This whole can only be seen when the organs are in motion.

This sort of thinking motivated much of the later work of two neurologists, Constantin von Monakow and Sir Henry Head (Figure 1.15). Monakow is credited with the concept of **diaschisis,** the idea that damage to one part of the brain can create problems for another, a fact that has been demonstrated time and time again. Head, who worked in London, also saw the brain as a dynamic system, interconnected and mutable. When there was injury, Head believed that the behavior resulting from a lesion was due to the whole system being out of whack. He believed that a lesioned brain was like a new system, not an old system with one part missing. To quote Head: "So far as the loss of function or negative manifestations are concerned, this response does not reveal the elements out of which the original form of behavior was composed. . . . It is a new condition, the consequence of fresh readjustment of the organism as a whole to the factors at work at the particular functional level disturbed by the local lesions."

Despite their loss of this intellectual battle, these bright scientists did formulate transcendent arguments. The points raised by the holistic team and the reasons they put forth still have merit. With the appearance of Karl Lashley, the great experimental psychologist, the importance of single neurons and localized function was cast in doubt.

His studies and writings were based on a strong academic context, bolstered by his experimental data. Lashley's point was that lesions made throughout the brain did not appear to create problems in learning or performing a task. Lashley's animal of choice was the rat, and he used a maze-learning task almost exclusively. Since then, we have learned that Lashley's conclusions contain some weaknesses. For example, maze-learning tasks require so many modalities, which in turn involve so much of the brain, that no single lesion could produce a deficit in learning. If an animal had learned a maze task using proprioceptive and visual information, a lesion to the visual system or to the proprioceptive system might not be sufficient to create a deficit, as the other modality could compensate for the lesion. Following this logic, if the lesions made were very large and included all modalities, then deficits ought to be seen. Indeed, this is also what Lashley found.

Nonetheless, the message of the holistic school still holds some valid lessons. The pendulum gradually swung back to the localizationist view as neurophysiological research began to unearth certain regularities in the organization of the cerebral cortex. Starting in the 1930s, Clinton Woolsey, Philip Bard, and others began to discover motor and sensory "maps" in the brain. Indeed, it became clear that each modality had more than one of these maps. In the 1970s and 1980s, we learned that multiple maps exist in each sensory modality, reaching a pinnacle of complexity in the primate visual system. To date, more than thirty maps of visual information have been found in the primate brain. Even more spectacular are the discoveries that very localized areas in the brain, such as the middle temporal area, are highly specialized for the processing of visual motion information. In short, neuroscience is continuing to reveal the startling complexity and specialization of the cerebral cortex.

Figure 1.15 Left: Sir Henry Head and W.H.R. Rivers at St. John's College in Cambridge, 1903. Head sectioned a branch of his own radial nerve and had Rivers perform experiments on his sensory loss. **Right:** Quote from Sir Henry Head.

The charm of neurology, above all other branches of practical medicine, lies in the way it forces us into daily contact with principles. A knowledge of the structure and functions of the nervous system is necessary to explain the simplest phenomena of disease, and this can be only attained by thinking scientifically.
    SIR HENRY HEAD, *Some Principles of Neurology,* 1918.

Female Historical Figures in Neuroscience

As in most scientific fields, until relatively recent years few women have been recognized for their contributions to neuroscience. Although it is easy to draw the conclusion that this is because of the lack of participation by women, an all too frequent alternative is that women have been involved but have not been fully credited for their work. Consider the case of the microelectrode. The microelectrode is an important tool in electrophysiological studies today and is used to deliver discrete electrical or chemical stimulation to a cell and to record the electrical activity from within individual nerve and muscle cells. Ralph Gerard won the Nobel Prize in the 1950s, fully credited with discovery of the microelectrode among many other accomplishments. However, two notable women made significant contributions to the microelectrode's discovery, though they were not formally recognized.

In 1902, Ida Hyde (1854–1945) was the first woman to be elected to the American Physiological Society; she remained the only woman in the society until 1914. She was the first woman to be awarded a doctorate in physiology from a German university (University of Heidelberg) and to do research at Harvard Medical School. She invented the first microelectrode for intracellular work with lower organisms by combining instruments used by F.H. Pratt (1917) and M.A. Barber (1912). Hyde constructed a salt solution–filled electrode of very small diameter (3 microns or so) and connected it to a small column of mercury whose level could be altered by the introduction of varying amounts of positive or negative electrical current. This mercury in turn could force the salt solution toward or away from the tip of the capillary and also transmit electrical stimuli to the cell through the salt solu-

tion. In 1921, using this method, Hyde provided the first evidence that the recently discovered principle of all-or-nothing contraction was not universally true for all contractile cells. The microelectrode was subsequently lost during the war and had to be reinvented.

The first report that a microelectrode had been used to record resting potentials from the membranes of a frog muscle was published in 1942 by Judith Graham Pool (1917–1975), G.R. Carlson, and Ralph Gerard. Their electrodes (2–3 microns in diameter) were also bent capillary tubes drawn out to a tip and filled with an isotonic salt solution of potassium chloride that allowed Pool and Gerard to record resting potentials and action potentials evoked with their electrode. Although Gerard was fully credited with the microelectrode discovery because he decreased the diameter to 0.25 micron, it appears that at the least, Pool was instrumental in its development, and according to her, she actually invented it without any credit for her achievement.

In addition to being unrecognized for their work, women have also been overshadowed by similar work done by their male contemporaries, as is exemplified by single-nerve-cell recording studies. Though most neuroscience textbooks discuss the work of Hodgkin and Huxley on the giant axon of the squid, another scientist who is less recognized, Angelique Arvanitaki, made significant contributions to the study of the single neuron.

By the 1940s, neurobiologists had learned to penetrate electrically active cells with microelectrodes and wanted to study the properties of nerves from within a single cell. Although vertebrate nerves are minute and housed in a brain and nervous system, this is not true of the large nerve cells of certain invertebrates. Angelique

---

Stephen Kosslyn, one of the founders of cognitive neuroscience, tidily summarized the conflict between the localizationists and holists (Kosslyn and Andersen, 1992):

The mistake of early localizationists is that they tried to map behaviors and perceptions into single locations in the cortex. Any particular behavior or perception is produced by many areas, located in various parts of the brain. Thus, the key to resolving the debate is to realize that complex functions such as perception, memory, reasoning, and

movement are accomplished by a host of underlying processes that are carried out in a single region of the brain. Indeed, the abilities themselves typically can be accomplished in numerous different ways, which involve different combinations of processes. . . . Any given complex ability, then, is not accomplished by a single part of the brain. So in this sense, the globalists were right. The kinds of functions posited by the phrenologists are not localized to a single brain region. However, simple processes that are recruited to exercise such abilities are localized. So in this sense, the localizationists were right.

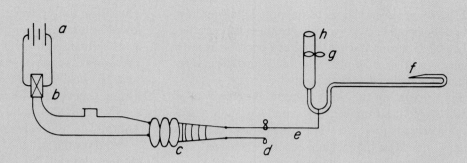

a, battery; *b*, commutator; *c*, induction coil; *d*, clamp; *e*, platinum wire; *f*, tip of pipette; *g*, clamp; *h*, rubber tubing

Left: Ida Hyde (1854–1945). The first woman elected to the American Physiological Society, 1902. **Right:** Ida Hyde's microelectrode (1921).

Arvanitaki (1939) developed the ganglion preparation of large identifiable nerves in the snails *Aplysia* (sea hare) and *Helix* (land snail). Arvanitaki also discovered that in low-calcium solutions, isolated nerve fibers of the cuttlefish *Sepia* (a relative of the octopus) produced regular electrical oscillations that periodically became larger and larger, until from time to time the nerve fired a series of action potentials. She was the first to demonstrate that spontaneous, rhythmically recurring activity could be an inherent property of a single nerve without the requirement of an entire neuronal circuit to generate it. Also, she found that when two or more nerves run close together, the activity in one nerve can entrain the activity in its neighbor. Hodgkin and Huxley won the 1963 Nobel Prize in physiology and medicine for analyzing the ionic basis of the action potential in the squid axon and are recognized in most neuroscience textbooks, overshadowing the significant contributions to this area of study by Arvanitaki.

Since Brenda Milner's memory work in the 1960s, many women in numerous areas of neuroscience have been recognized as being leading scientists in their field: Patricia Goldman-Rakic (neurophysiology and neuroanatomy of the frontal cortex)—past president of the Society for Neuroscience; Margaret Livingstone (visual neurophysiology); Leslie Ungerleider (cortical functional neuroimaging); Carol Colby (vision and the parietal cortex); Mary Hatten (developmental cellular neurophysiology); Carla Shatz (visual neurophysiology)—past president of the Society for Neuroscience; Christine Nussellin-Volhard (molecular neurophysiology); and perhaps most well-recognized, Rita Levi-Montalcini, the neurobiologist who shared the 1986 Nobel Prize in medicine for the discovery of nerve growth factor. Although the field of neuroscience still contains more male members than females, this inequality is quickly disappearing, as can be seen by the numbers of female graduate students pursuing degrees in neuroscience.

THE PSYCHOLOGICAL STORY

While the medical profession was pioneering most early studies of how the brain worked, psychologists began to claim that they could measure behavior and indeed study the mind. Until the start of experimental psychological science, the mind had been the province of philosophers, who wondered about the nature of knowledge, about how we come to know things. The philosophers had two main positions: **empiricism** and **rationalism.** Rationalism grew out of the Enlightenment period. It replaced religion, and among intellectuals and scientists it became the only way to think about the world. Through right thinking, rationalists would determine true beliefs. They would reject beliefs that while perhaps comforting, were unsupportable and even superstitious.

Although rationalism frequently is equated with logical thinking, it is different. Rationalism takes into account such issues as the meaning of life, whereas logic does not. Logic simply relies on inductive reasoning, statistics, probabilities, and the like. It does not concern such personal mental states as happiness, self-interest, and public good. Each person weighs these issues differently, and as a consequence a rational decision is more problematic than a simple logical decision. Clearly, rationalism is a complex mental activity.

Empiricism, on the other hand, is the idea that all knowledge comes from sensory experience. Direct sensory experience produces simple ideas and concepts. When simple ideas interact and become associated with each other, complex ideas and concepts are created in an individual's knowledge system. The British philosophers—from Thomas Hobbes in the seventeenth century, up through John Locke and David Hume to John Stuart Mill in the nineteenth century—all emphasized the role of experience. It is no surprise, then, that a major school of experimental psychology arose from this associationist view.

One of the first scientists to believe in **associationism** was Hermann Ebbinghaus. In the late 1800s he decided that complex processes like memory could be measured and analyzed. He had taken his lead from the great psychophysicists Gustav Fechner and E.H. Weber, who were hard at work relating the physical properties of things, like light and sound, to the psychological experiences they produce in the observer. These measurements were rigorous and reproducible. Ebbinghaus was one of the first to understand that one could also measure more internal mental processes like memory (see Chapter 8).

Even more influential was Edward Thorndike's classic 1911 monograph *Animal Intelligence: An Experimental Study of the Associative Processes in Animals* (Figure 1.16). In this volume, Thorndike articulated his law of

Figure 1.16 Edward L. Thorndike.

effect, which was the first general statement about the nature of associations. In many ways it was so simple. Thorndike simply observed that a response that was followed by a reward would be stamped into the organism as a habitual response. If there was no reward following a response, the response would disappear. Thus, rewards were responsible for providing a mechanism for establishing a more adaptive response. This idea sounds a little like Darwin's natural selection idea and indeed, Thorndike was deeply influenced by Darwin.

And yet the father of associationist thinking in psychology mixed his terminology. *Associationism* is hardly consonant with *nativism* (that is, the idea that many forms of knowledge are built into the organism from birth). Associationism is committed to an idea widely popularized by the American psychologist John B. Watson, who promoted the notion that he could take any baby and turn him or her into anything (Figure 1.17). Learning was the key, he proclaimed, and everybody had the same neural equipment on which learning could build. American psychology was giddy with this idea. Consumed with it, all strong psychology departments in the country were run by people who held this view.

All this hubbub in behaviorist psychology went on despite the well-established position—first articulated by Descartes, Leibniz, Kant, and others—that complexity is built into the organism. Sensory information is merely data on which preexisting mental structures act. This idea, which dominates psychology today, was blithely asserted in that golden age. As the associationists took over, they ran thousands upon thousands of experiments, and thus by volume of activity stole the issue as theirs.

The behaviorists' armor started to crack, however, when gestalt psychologists, working with perceptual phenomena, reported that percepts were best understood in relation to a stimulus's emergent properties. Apparent motion, for instance, was an emergent property of real-world stimuli. It existed only as a function of built-in properties of the brain. It was not learned. The gestalt psychologists developed hundreds of demonstrations making similar points.

The true end of the dominance of **behaviorism** and stimulus-response psychology did not come until the late 1950s. Almost overnight, psychologists began to think in terms of cognition, not just behavior. George Miller, who had been a confirmed behaviorist, offered what he calls his "very personal memory" of that event (Figure 1.18). Miller placed the revolution in the 1950s. In 1951 he wrote an influential book entitled *Language and Communication* and noted, "The bias is behavioristic . . . ." Eleven years later he wrote another book called

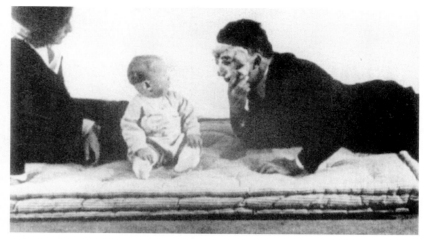

Figure 1.17   Left: John B. Watson. **Right:** John B. Watson and "Little Albert" during one of Watson's fear-conditioning experiments.

Psychology, the Science of Mental Life, a title that signals a complete rejection of the idea that psychology should study just behavior. As Miller put it, "My cognitive awakening must have occurred in the 1950s."

With a little searching, Miller put the exact date of his cognitive awakening at September 11, 1956, during the Second Symposium on Information Theory, held at the Massachusetts Institute of Technology (MIT). That year had been a rich one for several disciplines. In computer science, Allen Newell and Herbert Simon successfully ran Information Processing Language I, a powerful program that simulated the proof of logic theorems. The computer guru John von Neumann wrote the Silliman Lectures on neural organization. A famous meeting on artificial intelligence was held at Dartmouth College, with Marvin Minsky, Claude Shannon (known as the father of information theory), and many others in attendance.

Big things were also happening in psychology. As a result of World War II, new techniques were being applied to psychology. James Tanner and John Swets applied signal detection, servo theory, and computer technology to the study of perception. (These techniques had been developed, in large part, to help the Defense Department detect submarines.) Miller also wrote his classic paper, "The Magical Number Seven, Plus-or-Minus Two," in which he showed there was a limit on the amount of information that could be apprehended in a brief period of time. Also, the developmental psychologist Jerome Bruner was working on the problem of thinking. While he saw limited utility to associationist ideas in childhood learning, he believed in higher-level processes involved in thinking that were built on representations and mental maps. Perhaps the most important development, however, was from Noam Chomsky's work (Figure 1.19). A preliminary version of his ideas

Figure 1.18   George A. Miller.

Figure 1.19   Noam Chomsky.

An Interview with George A. Miller, Ph.D. Dr. Miller is professor emeritus at Princeton University and is one of the founders of modern cognitive science.

Authors: Psychology underwent a revolutionary change in the late fifties. After years of being dominated by behaviorists there was suddenly an interest in cognition. Some say it happened in September 1956. Could you tell us a little about that?

GAM: I once picked September 11, 1956, as an appropriate birthday for cognitive science. It was the date of a meeting at the Massachusetts Institute of Technology when leading cognitivists from computer science, linguistics, and psychology all came together for the first time and began to realize they shared their interest in the human mind. The interest in cognition had been growing for at least a decade, but this was the first time we realized that these separate fields were all parts of a larger whole—even before we knew what to call it.

Authors: Was there a particular set of scientific results that students of the mind were hard pressed to explain in terms of behaviorist principles? Or were people simply becoming bored with those types of explanations?

GAM: There were many such phenomena. For example, linguists found it impossible to describe the phrase structure of a grammatical sentence in terms of a linear sequence of stimuli-response reflexes. Bruner and his collaborators found clear evidence of problem-solving strategies that looked nothing like objective stimuli, responses, or reinforcements, and Simon and Newell were actually able to program a primitive computer to tackle problems the way people did, not by some blind trial-and-error procedure. I could not explain how people could "tune" themselves to discriminate optimally among a particular set of alternative stimuli without talking about subjective expectations. Ulric Neisser pulled together some of those phenomena a few years later in his book *Cognitive Psychology* (1967). But I remember that I was personally impressed by some experiments that seemed to show that a cat's attention could be monitored in terms of the response of the auditory nerve. I think the results later turned out to have been artifactual, but at the time it was exciting to think that something as subjective as attention might be reflected in something as objective as nerve impulses. During the fifties, it became increasingly clear that behavior is simply the evidence, not the subject matter of psychology.

Authors: Interestingly, the term *psychology* has largely been abandoned, even though most universities still have departments of psychology. One hears of *cognitive science*, even though psychology is defined as the "science of mental processes and behavior." Was this both necessary and deliberate for some sociological reason?

GAM: Psychology has been defined many ways. In the nineteenth century it was thought to have three branches, which would translate today into emotion, motivation, and cognition. Behaviorism could deal with emotion and motivation, but the refusal to admit mentalism in any form made it very difficult to give a plausible

on syntactic theories was published as *Three Models of Language.* Chomsky's effort transformed the study of language virtually overnight. The deep message was that learning theory—which is to say associationism, then heavily championed by B.F. Skinner—could in no way explain how language was learned. The complexity of language was built into the brain, and it ran on rules and principles that transcended all people and all languages. It was universal.

Herbert Simon and Allen Newell also publicly reported their efforts to simulate a cognitive process, presented in the context of linguistic work. A form of simple associationism was up and running once again. Remarkably, on the same day, there was an early attempt to test Donald Hebb's neuropsychological theory of cell assemblies, which suggested that any set of neurons can learn anything. Recently Hebb's ideas have been greatly advanced by computational neuroscientists. While the field gradually moves toward the importance of built-in and universal neural structures that govern cognitive and perceptual life, there are continuing efforts to demonstrate that the laws of simple association are responsible for our learning about the world. We shall return to this issue in the chapter on evolutionary perspectives.

behavioristic account of cognition. Some psychologists felt that a psychology without cognition was absurd, and so began a counterrevolution. We quickly discovered allies in other disciplines, particularly in linguistics and artificial intelligence, but also in philosophy and neuroscience. Personally, I was relieved to start collaborating with these new people and their new ideas, and to stop wasting my time explaining what was wrong with behaviorism. As collaboration grew, the need to name this new enterprise grew along with it. Hence, *cognitive science.*

Authors: And now cognitive science is a major force in the study of the mind. Sophisticated models of the mind have been built as the result of experimentation. One of the motivations for the development of cognitive neuroscience was to test these models in a biological system, to test their validity. Now people tend to come to problems with what might be called the *physicist perception* or the *biologist perception.* The physicists look for a few general principles to explain complex processes and that attitude would hope for a few principles for the mind. Biologists more or less give up on this and say the biological creature is a bag of tricks, a Swiss army knife with lots of specialized functions built in. Where do you stand in this debate?

GAM: They are both right. Detailed understanding of specialized functions probably has to come first, before the general principles can become clear. The fact that we cannot intuit those principles in advance doesn't mean that they will never be understood.

Authors: OK, to take a specific example from cognitive science, you are now working on an electronic dictionary which has become a model for how the human lexicon might work in the human brain. Could you tell us a little about that and how it might lead to insight on how our own lexicon works?

GAM: It is difficult to be concise when you are as close to a problem as I am to the problem of lexical knowledge, but let me try to characterize it with an example. Years ago the English philosopher Grice noted that the two sentences "I'm out of petrol" and "There's a garage around the corner" are immediately seen as related by normally intelligent persons. The question is, How do people fill in all the unstated information in such an exchange? One answer has been phrased in terms of spreading excitation in the brain. The first sentence activates the lexical node for "petrol" and the second activates the lexical node for "garage." These activations spread until eventually they intersect on some intermediate lexical node like "car" and so a bridging sentence can be constructed: Cars are supplied with petrol at garages.

When you try to build a system that will actually do this, however, you find that spreading activation results in far too many intersections. It has been estimated that only one-tenth of the intersections resulting from an unguided spread of activity will be appropriate. Clearly, if there is such spreading excitation, the spread is guided somehow. So, what kind of information would a system need—either a brain system or a computer system—in order to be able to constrain the spread of excitation into appropriate channels?

A group of us here at Princeton have been trying to propose an answer to that question in terms of a tightly interconnected network of lexical concepts. We don't think that a brain does it the same way our computer does it, but we are gaining a much deeper appreciation of the problem that the brain must solve. Once we understand that, a cognitive neuroscientist should have a much clearer idea of what to look for.

COGNITIVE NEUROSCIENCE

The term *cognitive neuroscience* was coined in that New York City taxi in the late 1970s because by that time a new mission was clearly required. Neuroscientists were discovering how the cerebral cortex was organized and how it functioned in response to simple stimuli. They were able to describe specific mechanisms such as those relating to visual perception. For example, David Hubel and Torsten Wiesel at Harvard were showing how single neurons in the visual cortex responded in a reliable way to particular forms of visual stimulation. The field of studying the mind had moved beyond the simple lesion method of assessing what perceptual or cognitive disorders might occur after brain damage. Neuroscientists were beginning to build models of how single cells interact to produce percepts. Most psychologists were no longer taking behaviorism seriously as a viable way to explain complex cognition. People like George Miller abandoned their earlier approach, which was totally behavioristic, and tried instead to articulate how language was represented.

No longer considered the product of simple learning and associationism, language came to be accepted as a complex construct delivered via the brain. Since Chomsky's breakthrough, it became clear that grammar is an instinct, whereas the lexicon is learned.

The extraordinarily talented David Marr at MIT made a major effort to bridge the gap between brain mechanisms and perception. Marr, who tragically died as a young man, provided a vision of what a cognitive neuroscience might look like. As Kosslyn and Andersen (1992) put it, "At the time, Marr's work was uniquely interdisciplinary and was particularly important because it provided the first rigorous examples of cognitive neuroscience theories . . . ."

Marr stressed the idea that neural computation can be understood at multiple levels by analysis. Philosophers of science had observed long ago that a single phenomenon can be examined at multiple levels of analysis. When considering psychology, philosophers such as Jerry Fodor distinguished between a functional and physical level; the functional level ascribed roles and purposes to events, and the physical level characterized the electrical and chemical characteristics of those events.

Marr took these earlier analyses several steps further. He posited a hierarchy of levels rooted in the idea that the brain computes. He divided the functional level into two levels, one that characterizes what is computed and another that characterizes how the computation is accomplished (i.e., algorithms), and he showed how these levels related to the lowest one, the level of implementation.

Although bold and fresh, Marr's approach was not quite accurate. Marr's ideas were embraced by cognitive theorists because he suggested that we could understand the cognitive level by reason alone. Theories that purported to explain a mental skill such as language, memory, or attention required deeper analysis, including algorithms to describe how neurophysiological processes produce the cognitive state.

But Marr's ideas have not worked out entirely. The distinction between the levels—that is, between the algorithms and the implementation mechanisms of the neurons themselves—has been vague. There is not one kind of neuron in the brain; there are dozens of kinds, each with different properties, each triggered by different neurotransmitters, and so on. Any computational theory, therefore, must be sensitive to the real biology of the nervous system, constrained by how the brain actually works—and it works differently for different functions.

Not that broad generalities on nervous system function do not exist. They do, and they enable scientists to search for specific mechanisms, which has given rise to the burgeoning field of neural network research. Here, scientists build models of how the brain might work and attempt to limit how their networks function by including information from neurophysiology and neuroanatomy.

THE SUDDEN RISE OF BRAIN IMAGING

It is mind boggling to consider the meteoric rise in brain imaging. When the first edition of this book was published, brain imaging had a foothold in cognitive neuroscience, but the vast majority of the work came from relatively few laboratories throughout the world. By the time of this edition, brain imaging has spread to literally dozens upon dozens of centers. Indeed, the traditional academic psychology department now finds itself with either a magnetic resonance imaging (MRI) scanner in its basement or plans to have one placed there. How did this all come about? How does studying blood flow by brain imaging help us understand such things as attention or reading?

Many good things start in Italy, and that includes the research by Angelo Mosso, a physiologist. Mosso worked in a neurosurgical ward and studied patients with skull defects. He noted that pulsations of the human cortex increased regionally during mental activity. With this work, he had established the correlation between blood flow and neuronal activity.

It was not until after World War II that the relationship between blood flow and neuronal function began to be quantified. Seymour Kety, Lou Sokoloff, and many others working at the National Institutes of Health (NIH) began to measure blood flow in the whole brain in animals. This work became foundational for the first brain imaging devices. First, Scandinavian researchers David Ingvar and Neils Lassen developed a helmet with scintillation counters that surrounded the whole head and allowed for crude regional measurements of changes in blood flow during mentation. This technique soon gave way to the much more powerful and spatially accurate technology called *positron emission tomography* (PET).

With PET, a technique developed at Washington University in St. Louis, both blood flow and metabolism could be measured. Using the procedures developed by Kety and Sokoloff, researchers could now similarly image the human brain. Quickly, however, measuring metabolism lost out to measuring blood flow. With the development of radiopharmaceuticals (e.g., $H_2^{15}0$) with a short half-life (123 seconds), blood flow could be measured quickly (<1 minute) and each subject could be studied several times, allowing for complex cognitive measurements in one individual.

Starting in the 1980s, there was great interest in how PET could illuminate human cognition. Cognitive psychology quickly became involved. Michael I. Posner and Steve Petersen established a major effort at Washington University and collaborated with Marcus Raichle. Their pioneering work launched an entire field of research. Over the next 10 years, researchers carried out study upon study, and by and large most used what is called the *method of subtraction.* Derived from the original work by the Dutch physiologist Franciscus Donders in 1868, this method involves subtracting one brain scan acquired during a particular behavioral state from another scan made during a different behavioral state. Thus, a scan taken while someone viewed a blank screen could be subtracted from a scan taken when the same person viewed the same screen with a word on it.

The subtracted scan isolated a process specifically associated with reading.

While these breakthrough studies were being conducted, yet another major advance in brain imaging was developing. It capitalized on yet another physical principle, the behavior of hydrogen atoms or protons in a magnetic field. Paul Lauterbur, now at the University of Illinois, saw how earlier work in physics could be used to make biological images, and his insights led to the development of MRI. At first the images dealt with the anatomy of the brain; these were called *structural images.* But Seiji Ogawa and colleagues quickly realized that functional states of the brain also could be imaged. Based on early chemical facts discovered by Linus Pauling and colleagues that the amount of oxygen carried by hemoglobin changes the degree to which hemoglobin disturbs a magnetic field, the idea of tracking blood flow using MRI became a reality. This signal became known as the blood oxygen level dependent, or BOLD, signal and is the basis for most brain imaging studies. We return to the details of imaging methods in Chapter 4. For a more extensive account of this fascinating history, read the superb review by one of the founding fathers of the field, Marcus Raichle (1998).

SUMMARY

We have learned how at least two rich and powerful academic fields have come together to produce yet another field of scientific research, cognitive neuroscience. Brain science emerged from the last century and gave us the knowledge that the brain is made up of discrete units—neurons. Cajal brought together the story about the importance of discrete entities, functioning neurons, and how they might interact to produce behavior. At a more general level, battle lines were drawn on how the brain as a whole was organized. Some researchers believed that functions were localized to discrete areas of the brain; others adamantly opposed this idea and maintained that functions were represented throughout the cerebral cortex.

As the brain localization debate continued into this century, psychologists began to think differently about their theories. Putting Freudian ideas aside, the major experimental scientists working on psychological issues believed in some form of associationism. Somehow, understanding how rewards and punishment influenced the organism would be all one needed to know to understand why and what people come to learn and remember. This belief that environmental contingencies

could explain all became welded into the very fabric of thought. After all, it reflected the American Dream. Anybody could become anything in the right environment.

All of this came crashing down in the late 1950s. Empiricism failed to explain complex mental functions such as language and other perceptual functions. Scientists began to consider representations of information as being almost built into the brain at birth. Hence, the advent of cognitive psychology, which fostered the notion that processing stages and cognitive activity could be analyzed with respect to their interlinked components.

All of this activity produced a new realization, however. If one wanted to understand how the brain enabled cognition, the thinking in neuroscience was not up to the job. Likewise, in psychology per se, models were being constructed and minds were being simulated—but without concern for how the brain did the job. The field became interested in how the mind might work or how it could work, but not how it does work. In this book, we explore how the brain actually does enable mind.

KEY TERMS

aggregate field

associationism

behaviorism

cytoarchitectonics

diaschisis

empiricism

holism

localization

neuron doctrine

phrenology

rationalism

syncytium

THOUGHT QUESTIONS

1. Can there be a study of how the mind works without studying the brain?
2. Will modern brain imaging experiments become the new phrenology?
3. What do cognitive psychologists mean by the term representation? What do neuroscientists mean by the term?
4. Can you imagine how the brain might be imaged in the future?

SUGGESTED READINGS

KASS-SIMON, G., and FARNES, P. (1990). *Women of Science: Righting the Record.* Bloomington, IN: Indiana University Press.

LINDZEY, G. (Ed.). (1936). *History of Psychology in Autobiography,* Vol. III. Worcester, MA: Clark University Press.

RAICHLE, M.E. (1998). Behind the scenes of functional brain imaging: A historical and physiological perspective. *Proc. Nat. Acad. Sci. U.S.A.* 95:765–772.

SHEPHERD, G.M. (1992). *Foundations of the Neuron Doctrine.* New York: Oxford University Press.

2

The Cellular and Molecular Basis of Cognition

G.W. was a 21-year-old student when her life hit a bump in the road. During her senior year at college she began to have difficulties managing the details of her life. Although she attended classes regularly, her grades suffered and she performed poorly on exams and even failed courses. Her parents worried that she might have become involved with illicit drugs or be suffering from depression, but neither one was the reason for her problems. Her behavior during holiday visits home caused her parents further unease. G.W. talked of plans they had never heard before. She spoke of traveling the world or of attending law school or medical school—each not an uncommon idea for a college student—but the organization of her plans had some clear flaws that she did not seem aware of. When her parents asked her why she was having trouble in school, she insisted that her instructors did not appreciate or understand her ideas, and hinted that they were "out to get her." Her family believed she was "stressed out" from too many demands in college, so they offered support and encouragement, and G.W. returned to school. Several months later she came home again, unable to continue with college; she was on a leave of absence at the suggestion of her college advisor. Over the course of the next few months, matters took a dramatic turn for the worse. G.W. began to express, sometimes aggressively, that she was being persecuted by friends and family, and later described voices that spoke warnings to her about dangers all around. She complained of difficulty with a tooth and seemed convinced that her dentist had done something terrible to her during a past visit, something that might have to do with her current "emotional" problems. A new dentist found no evidence of anything wrong with her teeth. Her family doctor referred her to a specialist. On the advice of that physician, she was hospitalized in a private psychiatric hospital for observation and treatment. After one month of treatment G.W. seemed much better and returned home, and the next semester she returned to school. She was given medication that her physician wanted her to continue taking for the foreseeable future. G.W. did not finish college, however, and over the next few years was in and out of the hospital several times. She lived at home, where her parents tried their best to help her. Although their support seemed to make a difference, she had sudden downturns characterized by bouts of paranoia, disorderly thinking, and emotional behavior. G.W. wrote long stories in letters to her grandmother. These notes became progressively disordered in both style and content, showing strange associations between topics and sometimes involving references to conspiracies against her and naming prominent individuals with whom she claimed some sort of connection. It seemed to her grandmother that G.W. could no longer distinguish between the real and the imaginary. G.W. suffered from a psychiatric disease, once called *dementia praecox,* but now named *schizophrenia,* a term coined by Swiss psychiatrist Eugen Bleuler (1857–1939) in his 1911 book on the subject.

What is schizophrenia? Simply put, it is a disorder of cognition, sometimes called a "thought disorder." Schizophrenia affects nearly 1% of the population worldwide. It has a genetic component: There is an increased likelihood that a person will develop schizophrenia if a family member has the disorder; the odds of developing the disorder are nearly 50% for the identical twin of a person with schizophrenia. These findings are clear evidence that schizophrenia has a biological component. But what is the neurobiological basis of the disorder? At present this question remains unanswered, but one of the major clues is the finding that certain drugs acting on major chemical transmitter systems in the brain can positively influence the symptoms of the disease. It is for this reason that we introduce Chapter 2 with a story about a psychiatric disorder. The delicate balance in communication between neurons in the brain yields cognition, and when this balance is disrupted by damage or disease, perception, movement, and thought itself may be at risk. In the rest of this chapter we will examine how neural cells communicate, returning later to the story of schizophrenia and chemical transmission.

In Chapter 1 we painted the historical canvas of modern cognitive neuroscience to present how psychology and biology have coalesced to create this new subdiscipline. In so doing, we introduced the theories, thoughts, and personalities of the history of psychology and neuroscience. Although thinking about the mind and brain can be traced to ancient civilizations, most of what we now consider to be the basic biology of the nervous system began from humble beginnings in the last half of the nineteenth century. Only in the last 100 years have the neurosciences laid the foundation for understanding how the human brain's cells and circuits enable behavior.

Neuroscientists now understand key aspects of neuronal signaling, synaptic transmission, and neuronal coding of information for storage and processing. Researchers have worked intensively to relate elementary molecular and cellular processes to the activity of individual neurons and to the activity of neuronal circuits. Although the precise manner by which biological events give rise to behaviors is only just beginning to be understood, there has been a vast accumulation of knowledge about events at the neuronal level. Some findings are sufficiently well understood to form a set of elementary principles of brain science. In this chapter, we briefly review these first principles of neuroscience, as they provide a necessary foundation for the chapters that follow. Here we focus on the cellular anatomy and physiology of neurons, whereas in Chapter 3 we present the gross organization and anatomy of the nervous system. To develop an understanding of the biology of information processing, we ask the following questions: How do neurons communicate? What are the chemical signals that mediate that communication? How do drugs modify these interactions, sometimes for the good and sometimes to the detriment of the individual?

CELLS OF THE NERVOUS SYSTEM

Scientists often adopt a reductionist strategy in the hope of understanding the whole by first identifying the parts. For example, in physics and chemistry the fundamental building blocks of matter were pursued with a vengeance over the past 400 years, a chase that led to the identification of the basic elements from which all molecules are made—and, of course, the principles of subatomic structure. This fascination with the essential building blocks of the world around us was also shared by the early biologists, who in the seventeenth and eighteenth centuries attempted to visualize the constituents of biological tissues by using a magnifying lens. Out of this effort, a fundamental principle, the cell theory, was derived. The idea was that the body was composed of elementary units, or cells. As we noted in Chapter 1, however, the idea that individual cells formed elementary functional units of the nervous system was controversial even as late as the turn of the twentieth century. In the end, Cajal's concept of the nervous system being composed of individual neurons prevailed.

Cajal observed that although neurons are close to each other, they are separated by small gaps. From these observations Cajal defined two main principles of neurons: connectional specificity and dynamic polarization. *Connectional specificity* incorporates the ideas that cells are separate because the cytoplasms of neurons are not in contact and that the connections between neurons are not random—circuits pass information through specific pathways. *Dynamic polarization* is the appreciation by Cajal that some parts of neurons are specialized for taking information in, while others are specialized for sending it out to other neurons or muscles. These two principles, and the overarching theory of the *neuron doctrine,* provide the focus for the first part of this chapter, which reviews the cellular structure and function of the nervous system.

There are two main classes of cells in the nervous system: **neurons** and **glial cells.** Neurons not only share similarities with all cells but also have unique morphological and physiological properties for special functions. Glial cells are a class of nonneural cells located in the nervous system that, in general, have a supportive function.

The Structure of Neurons

Neurons, the basic signaling units, are distinguished by their form, function, location, and interconnectivity within the nervous system (Figure 2.1). As Cajal and others of his time deduced, neurons take in information, make a "decision" about it following some relatively simple rules, and then, by changes in their activity

levels, pass it along to other neurons. These functions have close relations to the morphological specializations of neurons.

The neuron consists of a cell body, or **soma** from the Greek word meaning "body" (Figure 2.2). As with most cells, the cell body contains metabolic machinery that maintains the neuron. The components of the machinery include a nucleus, endoplasmic reticulum, ribosomes, mitochondria, Golgi apparatus, and other intracellular organelles common to most cells. These structures are surrounded by the neuronal membrane, which is composed of a lipid bilayer, and are suspended in cytoplasm, the intracellular fluid that is present inside all cells of the body.

In addition to the cell body, neurons also have specialized processes, the **dendrites** and **axon,** that extend away from the cell body. The role of these two types of structures reflects the principle of dynamic polarization. Dendrites are (usually) large treelike processes that receive inputs from other neurons at locations called **synapses.** As a result of being located after the synapse with respect to information flow, the dendrites of a neuron are said to be **postsynaptic** (after the synapse). The axon is said to be **presynaptic,** being before the synapse with respect to information flow. More generally then,

Figure 2.1 Mammalian neuron showing the cell body (soma) in the center surrounded by dendrites. The axon is visible as a hair-thin line emerging from the lower right quadrant of the soma and trailing off to the right of the image. This image shows a neuron expressing transferrin (TfR) receptor, a dendritic protein, tagged with green fluorescent protein (GFP) (green); TfR was produced in the cell body and transported to the dendrites. The red stain is anti-TfR, which distinguishes another source of TfR. Staining for MAP2, a dendrite-specific cytoskeletal protein, is in blue. Note the similarity to the idealized neuron in Figure 2.2. However, also note the relatively large size of the dendritic field in comparison to the soma. Courtesy of Michael Silverman and Gary Banker at Oregon Health Sciences University.

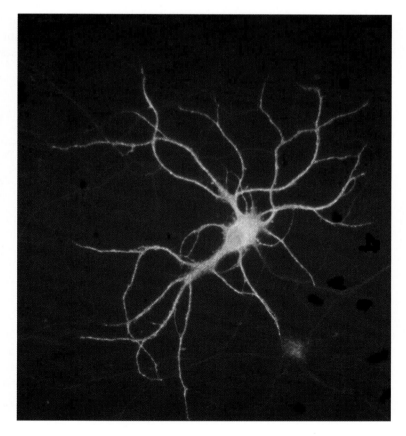

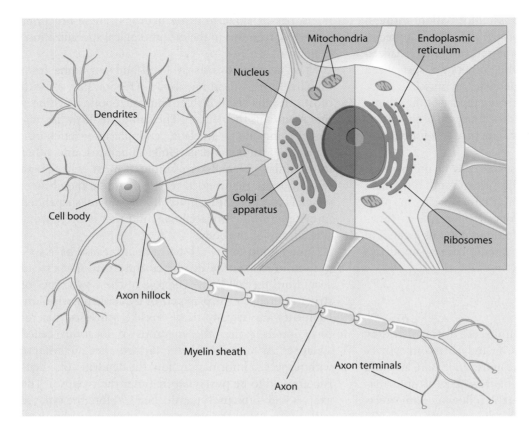

Figure 2.2 Idealized mammalian neuron. The cell body contains the cellular machinery for the production of proteins and other cellular macromolecules. Like other cells, the neuron contains a nucleus, endoplasmic reticulum, ribosomes, mitochondria, Golgi apparatus, and other intracellular organelles (inset). These structures are suspended in the intracellular fluid—cytoplasm—and are contained by a cell membrane composed of a lipid bilayer. Extending from the cell body are various processes that are extensions of the cell membrane and contain cytoplasm that is continuous with that in the cell body. These processes are the dendrites and axon.

we can refer to neurons as either presynaptic or postsynaptic with respect to any particular synapse, but most neurons are both presynaptic and postsynaptic—they are presynaptic when their axon makes a connection onto other neurons, and postsynaptic when other neurons make a connection onto their dendrites. It is useful to point out here that neurons signal electrically, but at synapses the signal between neurons is usually mediated

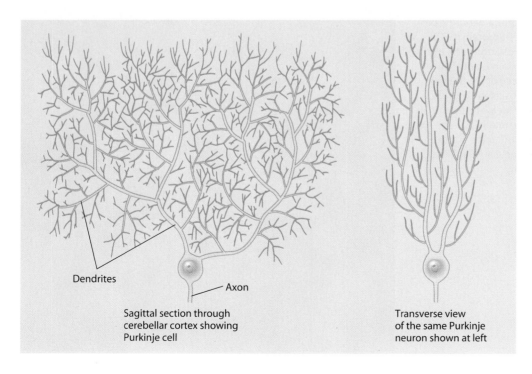

Figure 2.3 A diagrammatic representation of the soma and dendritic tree of a Purkinje cell from the cerebellum. The Purkinje cells are arrayed in rows in the cerebellum. They have a large dendritic tree that is wider in one direction than the other. Adapted from Carpenter (1976).

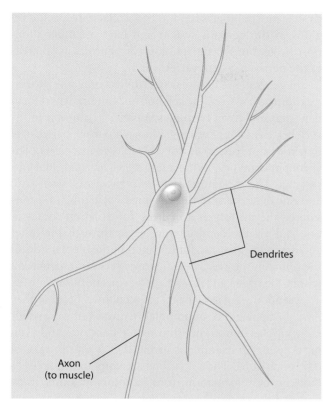

Figure 2.4 Diagrammatic ventral horn motor neuron. The multipolar neurons are located in the spinal cord and send their axons out the ventral root to make synapses on muscle fibers.

by chemical transmission, as Cajal surmised. However, in limited circumstances, some neurons signal between one another using electrical transmission at specialized electrical synapses, so it turns out that Golgi was correct too.

Dendrites take many varied and complex forms. They may appear as large arborizations like the branches and twigs of an old oak tree, as with the complex dendritic structures of the cerebellar Purkinje cells (Figure 2.3), or they may be somewhat simpler, as with the dendrites of spinal motor neurons or the neurons of the thalamus (Figure 2.4). Dendrites can also exhibit specialized processes called **spines,** little knobs attached by small necks to the surface of the dendrites; synapses are located on these spines (Figure 2.5). Synapses also are found elsewhere on neurons, including on cell bodies without spines.

The other type of process that extends away from the cell body is the axon. This structure represents the output side of the neuron, down which electrical signals can travel to the axon terminals where synapses are located. As noted, these axons, and axon terminals of the neuron, are presynaptic with respect to the synapses they form. The axon terminals have specialized morphologies and intracellular structures that enable communication via the release of **neurotransmitters,** the chemical substances that transmit the signal between neurons at chemical synapses. The dendrites and the axon of neurons are extensions of the cell body; their internal volume is filled with the same cytoplasm as the cell body's.

Figure 2.5 Cultured rat hippocampal neuron double labeled using immunofluorescent methods. Shown are presynaptic terminals (small green dots) making contact with spines located on the dendrites (red). The stain is an antibody that recognizes synuclein (sometimes called *synelfin*), a presynaptic protein. The dendrite is labeled with antibodies to MAP2, a dendrite-specific cytoskeletal protein. The large green area at the lower right is the soma. From Withers et al. (1997).

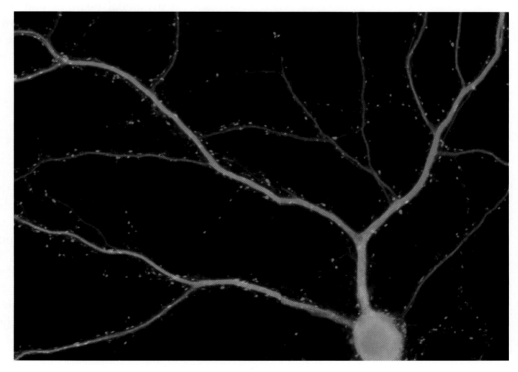

Hence, the cell body, dendrites, and axon are parts of one neuron. The continuity of the intracellular space between these neuronal components is necessary for the electrical signaling that neurons perform, as we will see.

The idealized neuron depicted in Figure 2.2 is modeled after a spinal motor neuron, but neurons have many forms. Indeed, the variation in the morphology of neurons is impressive; descriptions of the multitudes in the human nervous system would take a lot of space—more than we have here. Early anatomists described this morphological variance in great detail and grouped neurons into three or four broad categories according to their shapes. These distinctions have largely to do with how the cell's dendrites and axon are oriented with respect to each other and to the soma. Figure 2.6 presents four general morphological classes of neurons: unipolar, bipolar, multipolar, and pseudounipolar. Neurons with

Figure 2.6 Various forms that mammalian neurons may take. Some have few and others many processes extending from their cell bodies. These diagrammatic neurons are shown with short axons, which is not intended to be illustrative of all neurons. For example, motor neurons in the spinal cord have axons that are a meter or more in length, depending on the muscle innervated or the animal or human in question. Adapted from Kandel et al. (1991).

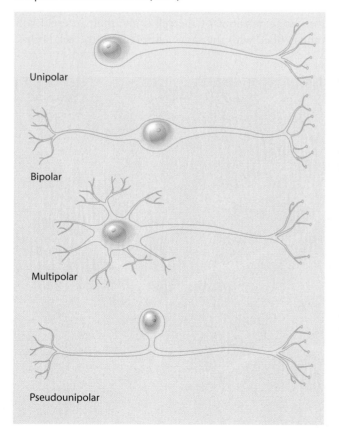

Unipolar

Bipolar

Multipolar

Pseudounipolar

similar morphologies tend to be localized in specific regions of the nervous system and have a similar functional role, but the morphology of neurons in all areas of the nervous system varies widely.

In general, the unipolar neuron has only one process extending away from the cell body; it can branch to form dendrites and axon terminals, a pattern common in invertebrate nervous systems. Bipolar neurons participate in sensory processes, as, for example, the neurons conveying information in the auditory, visual, and olfactory systems. These neurons have two processes, one axon and one dendrite, and thus in some sense might be considered the prototypical neuron: Information comes in one end via the dendrite and leaves through the other end down the axon, as, for instance, in the bipolar cells of the retina in the eye. These neurons process information within the retina and do not send projections outside it.

Pseudounipolar neurons are so named because they have the appearance of unipolar neurons but were originally bipolar sensory neurons whose dendrites and axon have fused. An example is in the dorsal root ganglia of the spinal cord. These neurons are somatosensory cells that convey information from receptors in joints, muscles, and skin into the central nervous system.

Finally, multipolar neurons exist in several areas of the nervous system and participate in motor and sensory processing. They have one axon but can have a few or many dendrites emerging from their cell bodies. The myriad of multipolar cells include spinal motor neurons, cortical sensory neurons such as stellate cells and pyramidal neurons, and some neurons of the autonomic nervous system. Figures 2.1 and 2.5 show stained multipolar neurons. For the most part, when we think about neurons in the brain, we are thinking about this *class* of neurons.

The Role of Glial Cells

The other type of cell in the nervous system is the glial cell (also called neuroglial cell or glia). These cells are more numerous than neurons, by perhaps ten times, and may account for more than half of the brain's volume. Neuroglial cells probably do not conduct signals themselves, but without them, the functionality of neurons would be severely diminished. The term *neuroglia* means literally "nerve glue," because anatomists in the nineteenth century believed that neuroglial cells had a role in the nervous system's structural support.

Glial cells are in the central nervous system (brain and spinal cord) and the peripheral nervous system (sensory and motor inputs and outputs to the brain and spinal cord). The types of glia in each are different,

Glia, Myelin, and Disease

J.C. was 32 years old when she stumbled while walking down her level, clean driveway. Embarrassed at her clumsiness, she laughed and told her friend that she should have the driveway repaired. Two weeks later she fell while walking across her living room. Six months after these seemingly trivial instances of clumsiness, some minor but annoying visual problems developed but disappeared over the next 6 months. During the next year she experienced some strange numbness in her hands and weakness in her left leg, which prompted her to seek medical treatment. Following a series of diagnostic tests, J.C. was diagnosed with a neuromuscular disease called *multiple sclerosis* (MS). What is this disease, and why did it lead to the varied symptoms, and why was she sometimes clumsy?

Moment to moment, on a millisecond scale, the human nervous system processes sensory information and executes motor responses. Some of these are under voluntary control; perhaps the vast majority are reflexive or automatic in one fashion or another. For example, for standing we not only decide voluntarily whether to stand or sit, but also use reflexive systems to maintain balance and posture. To accomplish this in real time takes speed and accuracy of timing. Thus, in many ways the nervous system is a high-speed machine. The timing of this machine is compromised in some disease states when the integrity of the neurons and their component parts are destroyed. MS is one disease among many that manifests itself, in part, as a loss of coordination among information transactions within specified neural systems. Specifically, MS is manifest as damage of the myelin sheaths surrounding axons in the central nervous system or peripheral nervous system, or both. Through mechanisms not completely understood (likely some form of autoimmune reaction of the body against the molecules in the myelin itself), the myelin is damaged, and in a spotty fashion may be broken down completely. The result for the patient can be mild or very severe. The symptoms of MS depend on which axons are affected by demyelination. If the affected axons are in the optic tract leading from an eye to the brain, then visual problems will be encountered. If demyelination happens to axons in peripheral nerves, losses of sensation or muscular control and strength may result. If the demyelination occurs in white matter tracts interconnecting regions of cerebral cortex involved in higher function, the damage may affect cognition or personality.

Why does damage to myelin lead to these problems? Damage to myelin can lead to slowing or complete disruption of neural signaling, and hence, a loss of function in the portion of the neural circuitry affected. In addition, however, inflammation that occurs as a result of the demyelination can lead to damage of the axons themselves and ultimately to their destruction. So the symptoms of MS may have two causes, one being the demyelination and the other neuronal damage. In the text, the role of myelin in axonal conduction is discussed. After reading about neuronal signaling, try to deduce how intermittent damage in myelin leads to the symptoms of MS.

though. The central nervous system has three main types: astrocytes, oligodendrocytes, and microglia (Figure 2.7). *Astrocytes* are large glial cells having round or radially symmetrical forms; they surround neurons and come in close contact with the brain's vasculature. Astrocytes actually make contact with blood vessels at specializations called *end-feet*, which permit the astrocyte to transport ions across the vascular wall and to create a barrier between the tissues of the central nervous system and the blood. This astrocytic barrier between neuronal tissue and blood is called the **blood-brain barrier (BBB)**. It plays a vital role in protecting the central nervous system from blood-borne agents or chemical compounds that might unduly affect neuronal activity. For example, many drugs cannot cross the BBB, and certain neuroactive agents, such as dopamine and norepinephrine, when placed in the blood, do not cross the BBB. This inability to pass through the BBB has significance for the treatment of disorders such as Parkinson's disease, in which the loss of the neurotransmitter dopamine in the basal ganglia of the brain (see Chapter 3) leads to a severe movement disorder; neurons located in the substantia nigra that produce dopamine and send their axons to the basal ganglia are lost. It is not possible to replenish the

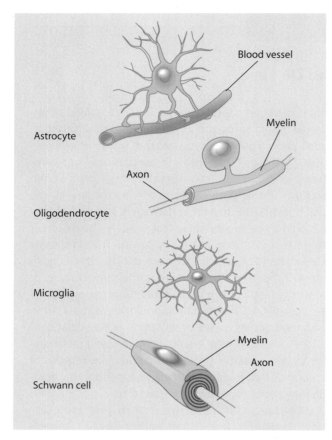

Figure 2.7 Various types of glial cells in the mammalian central and peripheral nervous systems. An astrocyte is shown with end-feet attached to a blood vessel. Oligodendrocytes and Schwann cells produce myelin around the axons of neurons. An oligodendrocyte is wrapped around an axon in the central nervous system. A Schwann cell is wrapped around an axon in the peripheral nervous system. A microglial cell is also shown.

The most obvious role of glial cells in the nervous system is in the formation of myelin. **Myelin** is a fatty substance that surrounds the axons of many neurons. In the central nervous system, *oligodendrocytes* act to produce this material; in the peripheral nervous system, the *Schwann cell* is involved in the myelination of axons. Both glial cell types produce myelin by wrapping their cell membranes around the axon in a concentric manner during development and maturation, and the cytoplasm in that portion of the glial cell is squeezed out. This squeezing out of the cytoplasm leaves primarily the lipid bilayer of the glial cell membrane—hence, the characteristic appearance of myelin as a fatty-looking material (it actually looks white and shiny under normal physiological conditions). But there are differences in how oligodendrocytes and Schwann cells produce myelin.

One oligodendrocyte of the central nervous system can form myelin sheaths around several axons. Figure 2.8 shows the wrappings of oligodendrocytes around the axons of several adjacent neurons in a white matter tract. The analogous situation is shown

Figure 2.8 Oligodendrocytes and Schwann cells produce myelin around axons. The oligodendrocytes in the central nervous system wrap around more than one axon to form myelin for many axons, but the Schwann cells in the peripheral nervous system wrap around a segment of only one axon. Whether in the central or peripheral nervous system, it takes many glial cells to produce myelin for the length of an axon because each wraps only a short segment of the axon. Adapted from Netter (1991).

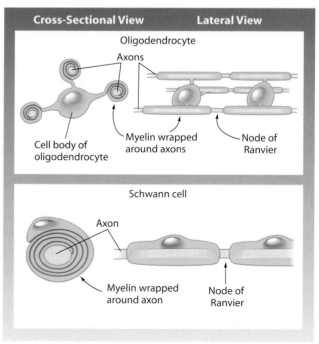

lost dopamine by getting it into the bloodstream because the BBB prevents it from entering the brain. But molecules like L-dopa that are precursors (i.e., molecules from which dopamine can be made) for the physiological synthesis of dopamine, when placed in the bloodstream, can cross the BBB, be taken up by neurons, and be converted to dopamine in the brain tissue.

Microglia, which are small, irregularly shaped cells, have a role to play once tissue is damaged (see Figure 2.7). Microscopic analysis of tissue that has been injured reveals that the damaged region is invaded by these glial cells. They also serve a phagocytic role; they literally devour and remove damaged cells. Microglia can proliferate even in adults (as do other glial cells), whereas neurons in the central nervous system are typically prevented from so doing (but see Chapter 15 for exciting new information on the birth of neurons in the adult brain).

for a Schwann cell in the peripheral nervous system, but the Schwann cell produces myelin only for a single axon in a peripheral nerve. The goal of each is similar: to provide electrical insulation around the axon that changes the way intracellular electrical currents flow in axons. In myelinated axons (many axons are unmyelinated), the myelin is interrupted at locations called *nodes*. These nodes, first described in the late nineteenth century by the French histologist and anatomist Louis Antoine Ranvier, are commonly referred to as the **nodes of Ranvier.** At the nodes, important membrane specializations permit the generation of action potentials that are conducted down the axon. Thus, both the portions of the axon where glial cells form myelin and the regions where this is interrupted at the nodes are significant for the way neurons are able to signal electrically. However, each of these segments has a different role in signaling.

NEURONAL SIGNALING

In reviewing the morphology of neurons, we noted that their function is to analyze and transmit information, which is the goal of nervous systems. Now we turn to how neurons accomplish the transfer of information from one to another; we refer to this globally as *neuronal signaling.*

Neuronal signaling has several requirements. As with most things in the physical world, energy is a primary requirement. How do neurons provide energy for neuronal signaling? How is this energy used to create a signal within a neuron? How is this intraneuronal signal transmitted to other neurons? In this next section, we will answer these questions, and learn how one can record the electrical activity of neurons to investigate their physical properties. First we begin with an overview, followed by the details of each step.

Overview of Neuronal Communication

Neurons are the integral units of the nervous system, and each neuron communicates with other cells at synapses, usually on dendrites or the cell body, or both. For outgoing information the synapses are at axon terminals. The goal of neuronal processing is to take in information, evaluate it, and pass a signal to other neurons, thereby forming local and long-distance circuits and networks of neurons. The process of signaling has several stages.

Neurons first receive a signal that is in either a chemical form (a neurotransmitter, or a chemical in the environment for sensations such as smell) or a physical form (such as touch in somatosensory receptors in the skin, or light in photoreceptors in the eye, or electrical signals at electrical synapses). These signals initiate changes in the membrane of the postsynaptic neuron, changes that make electrical currents flow in and around the neuron. The electrical currents act as signals within the neuron, potentially affecting the neuronal membrane at sites that are remote from where the input synapse was located. The current flow is mediated by ionic currents carried by electrically charged atoms (ions) such as sodium, potassium, and chloride that are dissolved in the fluid inside and outside of neurons. These ionic currents are related to but also different from the electrical currents that flow in metal wires, which are mediated by electrons rather than ions in solution. Long-distance signals (action potentials) can be generated in a region of the neuron that integrates the currents from many synaptic inputs, or from the stimulation of a sense receptor; this region of the neuron is generally referred to as the **spike triggering zone,** where *spike* refers to an action potential. What results in most cases is a signal that travels down the axon to its terminals, where it eventually causes the release of neurotransmitters at synapses.

Remember that this review highlights aspects of neuronal signaling that might be considered typical, but numerous variations on this pattern are present in the nervous system. For example, many neurons communicate without ever generating a long-range electrical signal. Such neurons are located in the retina and elsewhere where electrical synapses can be found.

Properties of the Neuronal Membrane and the Membrane Potential

The *neuronal membrane* is a bilayer of lipid molecules that separates intracellular space from extracellular space. Because the membrane is composed of lipids (fatty molecules), the membrane does not dissolve in the watery environments inside and outside the neuron (Figure 2.9). This is essentially how the membrane remains intact and controls the flow of water-soluble substances across its breadth. That is, the membrane does not allow some things to cross it; it is a barrier to ions, proteins, and other molecules dissolved in the intracellular and extracellular fluid. One way to think of it is that anything that dissolves in water does not dissolve

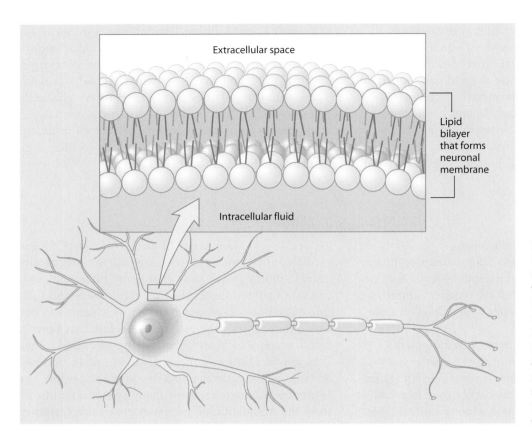

Extracellular space

Lipid bilayer that forms neuronal membrane

Intracellular fluid

Figure 2.9 An idealized neuron and the lipid bilayer (inset) that separates the intracellular from the extracellular space. The lipid bilayer makes up the cell membrane. The lipids (fatty material) are not water soluble and thus form a barrier to materials dissolved in watery solution such as sodium and potassium ions, proteins, and other molecules.

well in the membrane's lipids, and so it cannot readily cross into or out of the cell. This principle is a familiar one: Oil and water do not mix.

THE BASIS OF THE RESTING MEMBRANE POTENTIAL

The neuronal membrane, though, is not merely a lipid bilayer. It has numerous transmembrane proteins that form a variety of specialized structures in the membrane, including ion channels and active transporters or pumps. Other transmembrane proteins of interest include the class known as *receptor molecules*. We will discuss all of these channels in more detail later. For now, however, it is important to define a few special properties of ion channels and pumps in order to understand how the neuronal membrane creates the **resting membrane potential,** the difference in voltage across the neuronal membrane (Figure 2.10).

Ion channels are formed by transmembrane proteins that create pores, actual passageways through the membrane via which ions (charged atoms in solution) of sodium, potassium, and chloride (Na^+, K^+, and Cl^-, respectively) might pass (Figure 2.11a). Thousands of ion channels exist in the neuronal membrane. Some ion channels are passive (nongated) and are always in the same state (open to certain ions). Other channels

are gated—that is, they can be opened or closed by electrical, chemical, or physical stimuli. We will discuss gated channels later when we consider active membrane processes. The extent to which a channel permits ions to cross the membrane is referred to as **permeability.** Although the membrane itself is relatively impermeable to ions crossing it, some ion channels (nongated in this case) allow ions to move from inside to outside or vice versa. The membrane is more permeable to some ions (K^+) than others (Na^+, Cl^-), and therefore is said to be *selectively permeable*. The reason why the neuronal membrane is more permeable to K^+ than Na^+ is that it has many more nongated K^+-selective channels than nongated Na^+ channels (or other nongated channels).

Active transporters such as the Na^+/K^+ ATPase pump can move ions across the membrane. Energy-storing molecules called *adenosine triphosphate* (ATP) provide a form of fuel that the neuron uses to operate these small transmembrane pumps. The pumps are actually enzymes (proteins) located in the neuronal membrane; they can break a chemical bond in the ATP molecule and release energy that moves Na^+ out of the cell and K^+ into the cell. Every molecule of ATP can provide enough energy to move two K^+ ions inside for every three Na^+ ions extruded (see Figure 2.11a). Over time, pumping changes

Figure 2.10 Intracellular recordings are used to measure the resting membrane potential. One method is to use a small glass pipette electrode with a fine tip, to penetrate the neuronal cell membrane without damaging it too severely. Then recordings of the differences between the voltage measured by the electrode inside and that measured by the one outside the neuron (transmembrane differences) can be obtained. When the electrode enters the neuron, the difference in voltage between the two electrodes reflects the membrane potential, as shown in the idealized oscilloscope record.

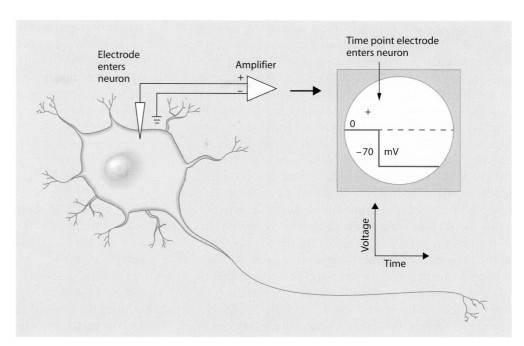

the internal-to-external neuronal concentrations of Na^+ and K^+ and creates ionic concentration gradients across the membrane (see Figure 2.11b). The relative impermeability of the neuronal membrane tends to keep these ions from traveling back across the membrane, even though the natural tendency is for diffusion to push ions down their concentration gradients (from areas of high concentration to areas of low concentration) to eliminate the differential in concentration. Hence, in the *resting state*, a higher concentration of Na^+ is outside the neuron and a higher concentration of K^+ is inside.

Figure 2.11 **(a)** Diagram of active transporters (Na^+/K^+ ATPase pump) and nongated ion channels in a segment of neuronal membrane. The ionic species that each ion channel and pump is related to are shown. The brackets around the letters representing the ions are symbols for concentration, and the print size of the letters in brackets corresponds to the relative concentration inside and outside the neuron. For example, there is a higher concentration of sodium ($[Na^+]$) outside than inside the cell. Properties of the membrane combined with the active pumping of ions across the neuronal membrane lead to ionic concentration gradients for sodium (Na^+), potassium (K^+), chloride (Cl^-), and large charged proteins (A^-) across the membrane. **(b)** The membrane's selective permeability to some ions and the concentration gradients formed by active pumping lead to a difference in electrical potential across the membrane called the *resting membrane potential*. The electrical potential across the membrane is shown in the figure as more positive charges outside the neuron along the membrane than inside, and is the basis for the transmembrane voltage difference shown in Figure 2.10.

The property of selective membrane permeability and the establishment of transmembrane ionic concentration gradients act together to create differences in

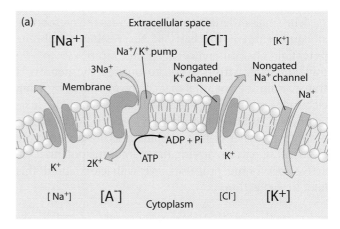

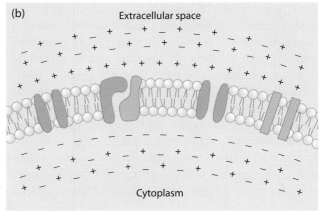

charge across the membrane. Here is how: As we just discussed, the pumps establish concentration gradients such that there is more Na^+ outside the neuron and more K^+ inside. These gradients create a force, the force of the unequal distribution of ions. This force wants to push Na^+ from an area of high to one of low concentration (from outside to inside) and K^+ from an area of high to an area of low concentration (from inside to outside). But because the membrane is more permeable (leaky) to K^+ than Na^+, the force of the concentration gradient pushes some K^+ out of the cell. In so doing, another force, an **electrical gradient,** begins to develop because the K^+ ions carry one unit of positive charge out of the neuron as they move across the membrane. That is, the environment outside becomes more positive than the environment inside the neuron. These two gradients (electrical and ionic concentration) are in opposition to one another with respect to K^+. As the positively charged K^+ leaves the cell, the increasingly negatively charged environment it creates makes it harder for the K^+ to leave (positive and negative charge attract). A point is reached where the force of the concentration gradient pushing K^+ out through the nongated K^+ channels is equivalent to the force of the electrical gradient acting to keep K^+ in, and this is where the opposing forces are said to reach *electrochemical equilibrium.*

The small difference in charge across the membrane at equilibrium is the basis of the resting membrane potential. This difference can range from −40 to −90 millivolts (mV) from inside to outside the neuron, a transmembrane potential that can be measured by using electrodes as shown in Figure 2.10. In essence, the neuron now is equivalent to a battery—that is, it has a form of potential energy that enables neurons to generate electrical signals by varying the membrane potential via active membrane processes. We will learn about this later in the section.

Modeling the Membrane Potential

The difference in the concentrations of ions across the selectively permeable membrane of the neuron leads to the resting membrane potential. The values for the resting membrane potential that are typically observed can be quantified, and the nineteenth-century German chemist Walter Nernst provided the means. His equation, the Nernst equation, is derived from principles of physical chemistry and thermodynamics and describes how unequal distributions of ions in solution will develop a difference in potential across a permeable membrane (an artificial one or one made of biological tissue). The equation provides for the calculation of that potential when one ionic species is involved. Julius Bernstein utilized the equation

at the turn of the twentieth century to resolve the question of how biological tissue such as muscle produces electrical potentials. (More than a century earlier Luigi Galvani had demonstrated electrical potentials in frog muscle.) The Nernst equation helped Bernstein argue that differences in intracellular and extracellular ion concentration led to a potential difference across the membranes of neurons. The Nernst equation for K^+ is

$$E_k = (RT/zF) \log_e([K^+]_o/[K^+]_i),$$

where E_k is the equilibrium potential of K^+, R is the universal gas constant, T is the absolute temperature in degrees Kelvin, z is the valence of the ion (for K^+ that is 1), F is Faraday's constant, $[K^+]_o$ is the concentration of K^+ outside, and $[K^+]_i$ is the concentration of K^+ inside the cell; $\log_e$ is the natural logarithm. By inserting values of 400 mM for the concentration of K^+ inside and 20 mM for the concentration outside the neuron (from studies in the squid axon), one can simplify the Nernst equation for K^+ at 25° C to

$$E_k = 59.8 \log_{10}(20/400) = -75 \text{ mV},$$

which is the **equilibrium potential** for K^+, that is, the membrane voltage at which there is no net flow of ions in or out. Bernstein performed measurements in muscle tissue that were in fair agreement with the results of his calculations using the Nernst equation. In neurons, then, the resting membrane potential is (roughly) equal to the equilibrium potential of K^+, except that the contribution of other ions also must be considered. K^+ is not the only ion to which the membrane is permeable; one must account for Na^+ and Cl^-. In the 1940s, David Goldman derived an equation that permits this calculation by taking into consideration the permeability of all relevant ionic species; we refer to the result as the *Goldman equation.*

The findings from simulations with the Nernst and Goldman equations were tested in the classic studies of Alan Hodgkin and Andrew Huxley and their students and colleagues in the 1930s, 1940s, and 1950s. They performed an ingenious series of studies that included artificial manipulation of extracellular and intracellular ion concentrations to observe the effects on the membrane potential. They also injected current while recording the membrane potential, to see how it changed, and to identify the associated transmembrane currents. They found that indeed, the equations did a good job of characterizing the changes in membrane potential they recorded. Their use of novel methods for recording is discussed in History of the Action Potential.

Electrical Conduction in Neurons

Neurons have two properties important for cellular signaling. One is that they are *volume conductors,* and hence currents can flow through them and across their membrane. Second, they generate a variety of electrical currents, called **receptor potentials**, **synaptic potentials**, and **action potentials**. Let us first look at how to measure the flow of current across membranes and within neurons before delving into passive electrical properties.

RECORDING THE MEMBRANE POTENTIAL

In the 1940s and early 1950s, techniques were developed to record the neuronal membrane potential with high accuracy. In principle the early techniques were not complex, but in practice, because of the small size of neurons (50–100 microns in diameter) and axons (1–3 microns in diameter), new techniques were needed. To record the transmembrane difference in potential, one must have a very small recording electrode (fine wire or micropipette electrode) inside a neuron and one outside; the difference in potential between these two electrodes is the value for the membrane potential (see Figure 2.10).

A micropipette electrode is created by heating and stretching a small glass tube to a fine point (<1 micron). When the microelectrode is filled with a conductive solution, it acts as both the recording tip and a kind of harpoon for puncturing the neurons under a micro-scope. The signal representing the difference between the voltage recorded from the electrode inside the cell and that recorded from the electrode outside the cell can be amplified and displayed on an oscilloscope for viewing. Typically, one arbitrarily defines the outside of the neuron to be at zero voltage; thus, the inside of resting neurons is observed to be a negative voltage.

This technique has been adapted so that current can be injected into the neuron, which is easily done by attaching a current source instead of an amplifier across the two electrodes. When electrodes are placed inside a neuron, one can inject current and record changes in the neuron's membrane potential (Figure 2.12). The action of the stimulating electrode approximates the role of synaptic potentials in a neuron, or of receptor potentials at a sensory receptor, and so we can learn about how tiny currents injected into a postsynaptic neuron affect it. Variations of this technique are now sophisticated enough to enable us to investigate the properties of ion channels.

PASSIVE ELECTRICAL PROPERTIES OF NEURONS

We now know (1) the gross morphology of neurons, (2) that they have a potential difference across their membranes that arises from the properties of the neuronal membrane and the distributions of ions across the membrane, and (3) that this electrical power source can be used by the neuron and measured by investigators. The next question we turn to is, How does the

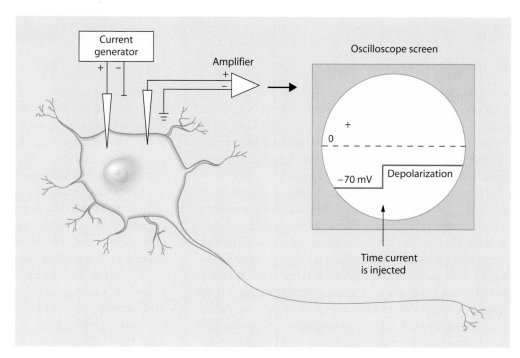

Figure 2.12 Intracellular recording and intracellular injection of current. Electrodes can be attached to current-generating equipment and used to pass current into the neuron. The effect this current has on the membrane potential can then be measured. Here, a depolarizing current is injected by making the tip of the electrode inside the neuron more positive. This depolarizes the membrane as the current flows out across the membrane to complete the electrical circuit with the other electrode outside the cell.

Building Neurons out of Batteries and Baling Wire

The components of the neuron have physical properties that determine how it conducts currents. These properties include the resistivities of the neuronal membrane, cytoplasm, and extracellular fluid, and the amplitude of the current generated by the power of the membrane potential. In order to model these components of the neuron to help understand the electrical behavior of dendrites, neurons, and axons, physiologists have developed the concept of the "equivalent circuit." The equivalent circuit uses terminology and principles from electronics to approximate the neuron, and thus to provide quantitative values for modeling the state of the neuron given a specific set of conditions. For example, as shown in the figure, we can assign specific resistances and capacitances (the ability to store charge) to the membrane, the intracellular fluid, and the extracellular fluid. Thus, diagrammatically the neuronal membrane becomes a set of resistors and capacitors in parallel, while the intracellular and extracellular fluids are represented as resistors in series.

The membrane of the neuron contains variable resistors because ion channels can open and close, thereby changing the resistance of the membrane to current flow (i.e., when ions move through the ion channels, conductance increases). The membrane potential can be modeled as a battery across the membrane because the membrane potential represents a steady power source that can direct current through the equivalent circuit.

Modeling the neuron as an electrical circuit permits quantitative predictions as to how the neuron should behave in the laboratory setting and under normal physiological conditions. By modeling the neuron in this fashion, physiologists have been able to attain theoretical explanations for the behavior of neurons from which they actually recorded data using microelectrodes. For example, the giant axon in squid has a diameter of about 0.5 mm, whereas the diameter of a muscle fiber in a frog may be as small as 0.1 mm. In these two cells, the resistance of the cytoplasmic fluid itself is approximately equivalent, but in the larger of the two cells, the resistance inside is lower because of the greater volume, and this can be modeled by decreasing the value of the resistors in series that represent the internal resistance of the intracellular space. Thereupon, it is possible to calculate the differential effect that a given change in current at one location would have on the membrane potential at any given distance away. Experimentally, this "length constant" (see text) is known to be different, and changes in the values of the resistors in an equivalent circuit model of the neuron can effectively describe this effect.

membrane potential permit neurons to communicate with one another? The first things to understand are the passive physical properties of neurons, and how these properties influence current flowing in their intracellular space; this is important to know because it is involved in all aspects of neuronal signaling.

Neurons are essentially sacks of electrically conductive fluid (cytoplasm) bounded by an electrical insulator (cell membrane), making them excellent volume conductors; that is, they are conductive volumes (as is your entire body, which is why sticking your finger into a light socket while standing in a puddle of water is an especially bad idea!). All this sits in a salty sea of extracellular fluid that has a conductivity similar to that of the cytoplasm. Thus, neurons and their environment can be broken down into conductors (cytoplasm and extracellular fluid) and insulators (the membranes)—the latter are structures with high but variable resistance, plus the ability to store charge briefly (capacitance). These components, plus the membrane potential and ion channels, can be modeled by using equivalent electrical circuits (see Building Neurons out of Batteries and Baling Wire).

We can also grasp the behavior of neurons in simple, qualitative terms. The first goal is to learn how electrical currents flow passively through neurons owing to sensory stimulation or synaptic activity. In so doing, we make a distinction between active membrane processes such as those that generate action potentials, synaptic potentials, or receptor potentials (all active currents), all of which involve opening of ion channels, and the passive currents that result. It is how well these passive currents flow through the neuron that sets the constraints

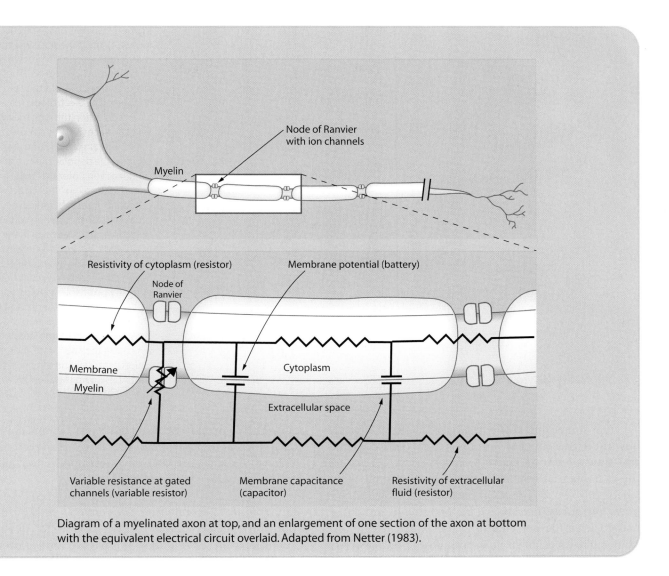

Diagram of a myelinated axon at top, and an enlargement of one section of the axon at bottom with the equivalent electrical circuit overlaid. Adapted from Netter (1983).

on how far a neuron can conduct, and whether and how the neuron will use action potentials to aid in communication. We will return to the action potential later when we discuss active membrane processes.

Overview of Signaling: Active Versus Passive Currents

Let's begin with an overview of some of the steps in neuronal signaling, in this case, between neurons. Figure 2.13 shows these steps from synaptic input to the generation of action potentials in the axon; the numbers in parentheses in this paragraph refer to those in the figure. When a synapse is activated (1), active electrical currents are generated across the cell membrane near the synapse and these generate synaptic potentials (2). This current flow across the postsynaptic membrane in a localized region results in current that is passively conducted throughout the neuron. This passive current conduction is called **electrotonic conduction.** These passive currents (conducted via volume conduction through the cytoplasm) pass through the dendrites and soma of the postsynaptic neuron (3). If these passive currents are strong enough, they can trigger action potentials in the spike triggering zone at the axon hillock of the neuron (4). The action potential is an active membrane process (because it involves changes in membrane conductances by opening and closing ion channels), but the result, as at the synapse, is to create passive currents that flow in the axon (5). These passive currents can depolarize nearby patches of the axon, causing them to generate a new action potential (6), permitting the process to continue down the axon to the terminals where neurotransmitters are released and the whole set of steps occur again in the next neuron.

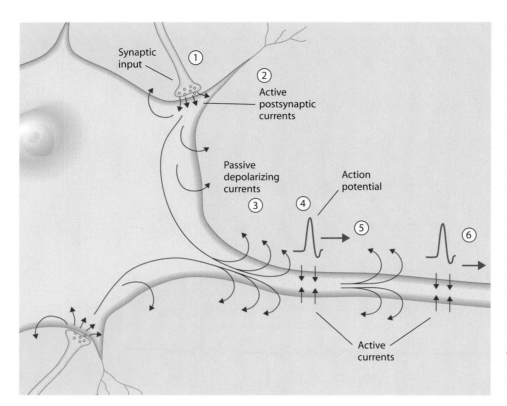

Figure 2.13 Overview of signaling between neurons. Synaptic inputs (1) affect the postsynaptic membrane, leading to postsynaptic currents at synapses (2). These currents are active, but they are conducted by electrotonic (passive) conduction through the intracellular space (3). They depolarize the membrane, and if large enough, the depolarization can trigger action potentials to be generated in the axon (4). Each action potential represents an active process that occurs when voltage-gated Na^+ channels open in the membrane. The inward current generated during this phase of the action potential is then conducted down the axon following principles of electrotonic conduction (5). This leads to depolarization of adjacent regions of membrane, which can then generate another action potential, and the process continues down the axon (6).

It is sometimes difficult to understand precisely how the active currents become passive, as there is no real transition from active to passive current. Rather the currents that flow in neurons are said to be active when they are in the neighborhood of active membrane processes that involve the opening of ion channels to let ionic currents flow (described later). The same currents can also flow through the cell passively by electrotonic conduction; we refer to these currents as *passive currents* because as they flow through the cell and across the membrane, they are not mediated by openings of ion channels but by the physical properties of the membrane and the nature of electrical circuits. Current must always flow so as to complete an electrical circuit; thus, the movement of ions to the inside of a neuron is accompanied by the return of currents to the outside of the neuron, to form the complete circuit (Figure 2.14).

As described earlier, electrical current in neurons is ionic; that is, it is carried by charged atoms (i.e., ions) in solution. With neurons, the charge carriers are primarily Na^+, K^+, and Cl^- ions, but there are also large charged proteins in solution. Since transmembrane currents are carried by the smaller ions, we usually speak of sodium currents, potassium currents, and chloride currents across the membrane and of ionic currents more generally inside or outside of the neu-

ron. These currents typically occur when ionic species such as Na^+ or K^+ cross the neuronal membrane through ion channels that open in response to chemical, electrical, or physical stimuli.

Electrotonic (Decremental) Conduction A key question for understanding signaling is, How far down a dendrite, axon, or cell body will passive ionic currents flow when not replenished by active processes (i.e., action potentials)? The distance is a function of three

Figure 2.14 Current flow generated by an action potential. Diagram of currents crossing the membrane, traveling down the neuron, exiting, and returning to the source. Adapted from Netter (1991).

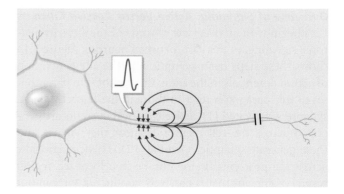

main physical properties of the neuron: the amplitude of the original current, the resistance (and capacitance) of the neuronal membrane, and the conductivity of the intracellular and extracellular fluid. We have come to understand these principles by modeling and by direct experimentation in neurons; we can record from inside neurons (if they are large enough) and can inject current into them to investigate the result (Figure 2.15). This method has greatly illuminated passive membrane properties and how they figure into the big picture of neuronal signaling.

First, consider the diagrammatic neuron in Figure 2.16a. A stimulating electrode placed inside the axon can inject current into the cell, mimicking the action of synaptic inputs. Recording electrodes are inserted into the axon at varying distances down the axon from the point at which a stimulating electrode is located. When the stimulating electrode injects positive current into the axon at that location (this means the electrode tip is made positive by a current generator), the current flows down the axon where it emerges through the axonal membrane and creates return currents that flow toward the other electrode located in the extracellular fluid, thereby completing the circuit; the example shown here is for when the depolarization induced by

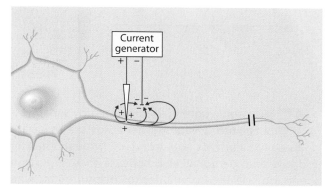

Figure 2.15 Current flow induced artificially. Currents can be made to flow in neurons using an intracellular electrode attached to a current generator.

the current injection is below the threshold for generating action potentials (see next section). Near the electrode inside the axon, the current is strongest, and the passive electronic currents that flow down the axon decrease with distance away from the source. If we measure the voltage across the membrane, we see a large voltage change near the stimulating electrode and progressively smaller voltage changes as we move farther away from it. If we assume that the membrane's

Figure 2.16 **(a)** A neuron with an electrode injecting current at one end of the axon and recording electrodes at varying distances down the axon away from the soma. The amplitude of the depolarization of the membrane that is induced by a current injection of below threshold amplitude drops systematically with increasing distance from the site of injection. **(b)** The decline in membrane depolarization induced by passive currents flowing down the axon with distance follows an exponential function, and a value known as the *length constant* (lc) can be defined as the distance from the origin where the change in membrane potential reaches 37% of its original value. Adapted from Kuffler and Nicholls (1976).

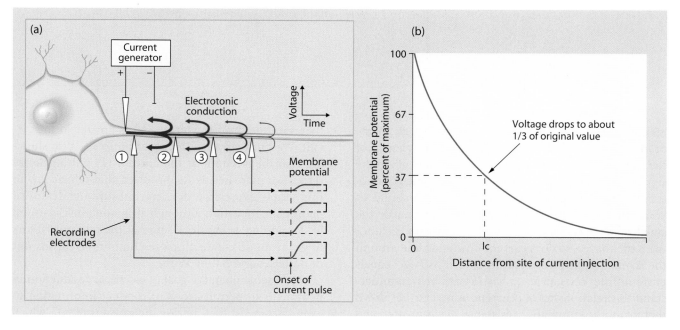

Hear Me Shout!

Electrotonic potentials generated by synaptic inputs onto a neuron's dendrites are associated with currents that conduct through the dendrites and neuronal cell body. If they are strong enough (via the summation of many inputs) when they reach the axon hillock region of the neuron (spike triggering zone), then an action potential may be initiated. But given that these synaptic inputs create electrotonic currents, and as described in the text, these currents are decremental, declining exponentially in strength from the site of generation at the synapse, how do synaptic inputs on distant (distal) dendrites have any influence on the behavior of the neuron? Would they not be so weak that they would be completely overcome by inputs that are closer to the axon hillock, for example, on nearby (proximal) portions of the dendrites near the cell body? This question has been a central mystery and point of controversy for many years. Recently, Jeffrey Magee at Louisiana State University Medical Center and Erik Cook at Baylor College of Medicine (2000) reported the results of an experiment in neurons from the hippocampus, a structure in the medial temporal lobe (see Chapter 3). They recorded and stimulated dendrites in slice preparations (see text) of rat hippocampal pyramidal cells. Their recordings were from apical dendrites near and far from the cell body as well as from the cell body of the neuron. Magee and Cook

found that inducing synaptic activity close to the cell body or far from it along the dendrite produced depolarization of the cell body of equivalent magnitude (based on the recordings in the cell body). Thus, even though some inputs were farther away, and the principles of decremental electrotonic conduction should have diminished the effects produced by the more distant synaptic inputs, they in fact produced equivalent effects at the cell body. Magee and Cook then asked how this was accomplished. Did the dendrites use some active mechanisms, perhaps the voltage-gated channels known to be present along the dendrite, to boost the strength of the distant signals? Or instead, was the equivalence of the effects of near and distant synaptic inputs the result of differences in the strengths of the synaptic currents for different regions along the dendrite? They found evidence that the latter mechanism is correct. When they recorded the currents along the dendrite produced by synaptic inputs, they found a systematic increase in the strength of these currents as they recorded farther and farther from the cell body. This increase in the more distant synaptic currents overcomes and compensates for the progressive decrement in the amplitude of the related electrotonic currents due to their more remote sites of generation! Or to put it another way, to be "heard," synapses, just like people, have to "shout more loudly" when farther away.

resistivity does not change, then the change in membrane voltage is due to the diminished amplitude of the current at more distant loci (i.e., decrements in the signal), and we therefore refer to electrotonic conduction as *decremental conduction*. Just as the sound of your voice, the strength of a flashlight beam, or the smell of your favorite pizza diminishes with distance from the source, so does current flowing in and around the axon. Application of Ohm's law (voltage equals current times resistance, or $V = IR$) can help to understand this relationship. As current moves farther down that axon, the cumulative resistance (R) is greater, and

therefore the drop in voltage (V) is greater in the circuit. We can rearrange the equation to $I = V/R$, showing that the strength of the current (I) diminishes as resistance increases. Note, however, that Ohm's law in this simple expression does not capture the details of neuronal properties, although it is nonetheless useful in developing an intuitive sense of the relevant factors in electrotonic conduction (see Building Neurons out of Batteries and Baling Wire).

Now remember, the goal of electrical conduction in neurons is to permit axons to conduct information to axon terminals that release neurotransmitter, which

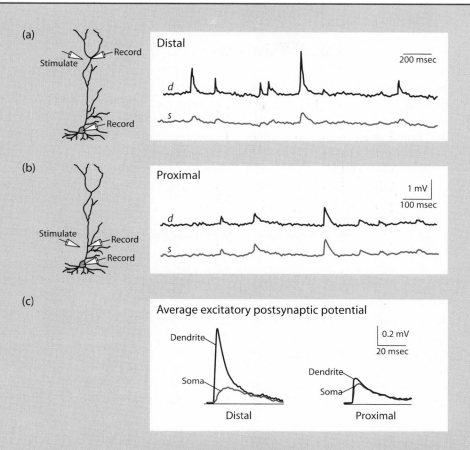

Recordings from **(a)** a distal dendrite and cell body and **(b)** a proximal dendrite and cell body in a rat hippocampal pyramidal neuron in slice preparation. The amplitudes of the potentials recorded in the cell body (traces labeled s for "soma") were similar for proximal and distal synaptic inputs (compare traces labeled s in (a) vs. (b)). However, the synaptic potentials were different for the proximal and distal inputs (compare traces labeled d, for "dendrite", in (a) vs. (b)). **(c)** The averaged excitatory postsynaptic potentials recorded from the dendrites and cell body (soma) for the distal and proximal synaptic inputs, showing the larger amplitude of synaptic currents for synapses located farther from the cell body, and demonstrating that the sizes of the resulting potentials at the cell body are similar. Adapted from Magee and Cook (2000).

results in synaptic transmission between neurons. The question of most interest now is whether passive electrotonic conduction is sufficient to allow this communication. The answer is a loud "No!"—with a qualifying "Yes" in many circumstances. Because the current's strength diminishes with distance from the source, such a signal is not appropriate for *long-distance* communication, but it can work well for short distances.

As noted, how far electrotonic currents can be effective for communication depends in part on the size of the original current. The greater the current, the farther it will conduct. In the laboratory, this can be manipulated by increasing the size of the current injected in the neuron. However, under normal physiological conditions the amplitude of the current is determined by physiological factors such as the intensity of a physical stimulus at a receptor, or the strength and number of synaptic inputs onto the neuron. Because these events may be associated with currents of different sizes, the changes in membrane potential they induce are said to be *graded*. We can also ask about the influence of the membrane's resistivity on the conduction of electrotonic potentials. Put simply, as the membrane resistivity increases, more current

HOW THE BRAIN WORKS

History of the Action Potential

As recently as about 65 years ago, we did not know much about the underlying mechanisms of the generation of action potentials. That is, humankind had built devices like airplanes and radios, both of which are sophisticated pieces of machinery for war and peace, but no living human had recorded an action potential from inside a neuron. But many wanted to, and two key figures in the story of the action potential are Alan Hodgkin and Andrew Huxley, who from 1938 to 1952 published a series of foundational articles on the action potential. In a classic 1938 report in the journal *Nature,* they published the first recording of an action potential from the giant axon of the squid. Note this was recorded from the giant axon of the squid (an animal about 20 cm long), not the axon of a "giant squid"; Hodgkin and Huxley may have been pioneers of the nervous system, but they left giant squids to the likes of Jacques-Yves Cousteau, preferring to challenge the natural world in the laboratory. The details of their work led to the "ionic hypothesis" and ulti-

mately to the Hodgkin-Huxley theory of the action potential. The bottom line of their model was that ionic currents, the Na^+ and K^+ currents, acted independently to create the action potential, and that these currents became active in a specified time period as the result of rapid changes in the conductances of the neuronal membrane. Hodgkin and Huxley made use of an ingenious technique invented by Kenneth Cole, called the *voltage clamp.* This technique involved an electrical circuit connected to the intracellular recording system. It permitted the voltage of the membrane to be maintained at a constant value ("clamped") even during changes in ionic concentration in the extracellular fluid and after injections of electrical current that caused currents to flow across the membrane through ion channels. Using these techniques, Hodgkin and Huxley and their colleagues demonstrated that separate Na^+ and K^+ currents contributed to the different phases of the action potential, via changes in conductances for Na^+ (g_{Na}) and K^+ (g_K).

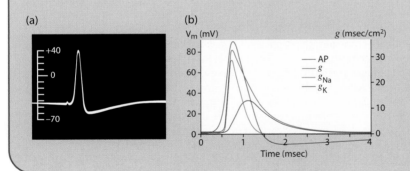

(a)

(b)

(a) First recorded axon potential, reported in 1938 by Hodgkin and Huxley. **(b)** The plot shows transmembrane voltage (membrane potential) and membrane conductance (g) as a function of time during an action potential (AP). From Haeusser (2000).

will be shunted down the axon and less will leak out. Consider the analogy of a garden hose. If the tubing of the hose is intact (i.e., has high resistance to water crossing it), then water put in one end is forced down the hose. Yet if the hose is full of holes (i.e., has low resistance to water), the water forced in one end does not move as far down the hose because it leaks out first. If the hose is leaky enough, the water may move only a short distance before the water pressure falls to the level where water no longer flows in the hose. Similarly, the resistivity of the neuronal membrane influences the distance down an axon that a current

can travel (this will be relevant later when we discuss myelination and conduction).

Finally, the conductivity (resistivity = 1/conductivity) of the intracellular space also affects how far through a neuron a current will flow. Usually the intracellular and extracellular fluids have high conductivities because as salt solutions, they are generally good conductors of electrical current (although they are vastly inferior to metal wire). Even so, the resistivity of dendrites, cell bodies, and axons changes as a function of their size. Returning to the garden hose analogy, we find that if the hose has a large diameter, water flows easily, but in a thin hose, water pressure, and thus resistivity, rises because of the restricted flow. In a similar way, if the axon is large, the current flow is greater. Thus, high-amplitude receptor or synaptic currents, high membrane resistance, and low-resistance intracellular pathways enhance electrotonic conduction of currents by permitting them to affect the neuronal membrane at more distant loci from their site of generation.

But under the best of conditions, how far can electrotonic conduction provide an effective means of electrical communication? Not far, generally about a millimeter. The decremental nature of passive electrical currents can be described by plotting the change in membrane potential as a function of distance from the current source to a recording site (see Figure 2.16b). What we find in this type of plot is that the recorded change in membrane potential drops exponentially. The *length constant* is the distance down the axon where the potential reaches about one-third of its original value. This parameter can be used to quantify the effectiveness of electrotonic conduction in different neurons or their parts.

A millimeter may seem too short to be effective for conducting electrical signals, but in a structure like the retina, a millimeter is enough to permit neuron-to-neuron communication (see also Electrical Transmission, later in this chapter). In the spinal cord, though, where axons may have cell bodies in the motor cortex of the brain, and axonal terminals on motor neurons in the spinal cord, a millimeter is too short. Under such circumstances, electrical signals down an axon might have to cover several meters; consider a giraffe, elephant, or whale! Another mechanism, an active regenerative mechanism, has evolved to conduct electrical signals long distances in the central and peripheral nervous systems.

ACTIVE ELECTRICAL PROPERTIES OF NEURONS

In the preceding section we described how the physical properties of neurons affect the way electrical currents passively flow through them, as a result of synaptic inputs or sensory stimulation at receptors. We showed that electrotonic conduction is good only for short-distance communication, not for long-distance communication. Long-distance communication requires active or regenerative electrical signals called action potentials.

Action Potentials   To understand action potentials, it is useful to appreciate that graded receptor and synaptic potentials, decremental electrotonic potentials, and action potentials involve changes in the membrane potential over time, but that these changes are fundamentally

different. Further, we have to remember to distinguish between the resting membrane potential and the membrane potential when current is flowing, as, for example, during an action potential. In addition, the membrane potential can become either more (hyperpolarized) or less (depolarized) negative with respect to the resting membrane potentials (Figure 2.17), and different types of inputs (excitatory and inhibitory) influence these differences. And finally, active processes such as action potentials are initiated by passive electrotonic currents that alter the local membrane potential, thereby triggering action potentials in specific regions of the neuronal membrane. Hence, the two types of electrical conduction—passive electrotonic and active regenerative—in neurons are intimately related (see Figure 2.13).

Stimulating neurons to depolarize them can generate action potentials in axons. These action potentials can be observed without intracellular recording, by placing recording electrodes on the outsides of axons or bundles of axons (i.e., nerves) (Figure 2.18). To find out about the mechanisms that generate action potentials, however, researchers use intracellular recording and stimulation. In axons, if stimulation is continued with successively higher and higher amplitudes of current, an action potential can be elicited. The **action potential** is a rapid depolarization and repolarization of the membrane in a localized area. Figure 2.19 shows the depolarizations of the axonal membrane that can ultimately lead to action potentials, if the depolarizations are large enough. The value of the membrane potential to which the axon must be depolarized to initiate an action potential is the **threshold.** Depolarizations that do not reach threshold will not elicit action potentials; those that do, lead to characteristic action potentials or "spikes" (the action potential can resemble a spike when viewed on an oscilloscope, and the term *spikes* can be used as shorthand for action potentials—e.g., the neuron fired at a rate of 10 spikes/sec means 10 action potentials/sec).

The amplitude of the action potential, once initiated, does not depend on the size of the initial depolarization. Thus, even if the depolarizing current is raised higher than what is needed to reach threshold, the action potential generated does not change size; hence, the action potential is said to be *all or none.* The reason is that the electrotonic currents that depolarize the membrane vary in amplitude (i.e., are graded), depending, for example, on the amplitude of the stimulus or synaptic input, but the action potential is generated by active processes unrelated to the amplitude of the triggering currents (as long as they are large enough to push the membrane potential to threshold).

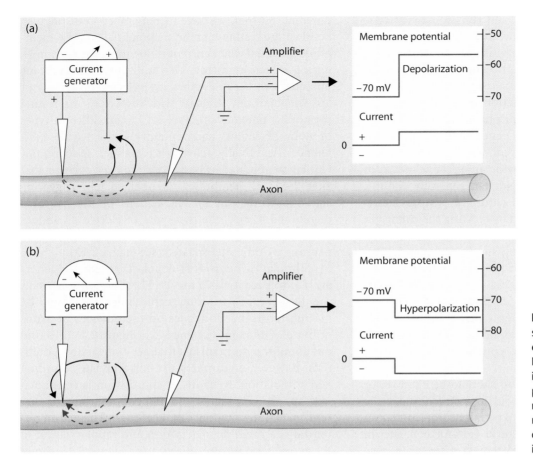

Figure 2.17 An axon with stimulating and recording electrodes placed inside. Injection of current by making the electrode tip more positive depolarizes the membrane **(a)**, whereas current flow in the opposite direction **(b)** can hyperpolarize the neuronal membrane.

As introduced at the beginning of this section, neuronal membranes contain ion channels that are structurally specialized to be selective to specific ionic species.

As well, ion channels are distinguished by whether or not they are gated. That is, their conductivity is affected by external influences such as voltage (e.g., from passive cur-

Figure 2.18 Extracellular stimulating and recording electrodes and recorded compound action potentials. A large electrode placed extracellularly can also depolarize the membrane and generate action potentials. In this figure, a peripheral nerve composed of many individual axons is being stimulated, and the compound action potential (the summation of the action potentials from all the axons) is recorded using an extracellular electrode. The electrodes pictured are hook electrodes, pieces of wire curled into hooks that the nerve lays against.

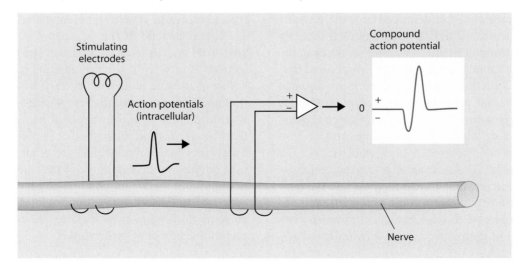

Figure 2.19 Injection of positive current into an axon leads to depolarization, which, if large enough, triggers an action potential. The changes in membrane potential induced by injected currents can be considered to be graded, that is, they can take on varying amplitudes depending on the size of the depolarizing current. Compare the initial portions of the membrane depolarizations in 1 through 3 at bottom as a function of the size of the injected current indicated in 1 through 3 above. However, when the depolarization reaches threshold, then the action potential that is generated reaches an amplitude that is not related to the size of the original depolarizing current. That is, the action potential is said to be all or none (compare 2 to 3 at bottom). Adapted from Kandel et al. (1991).

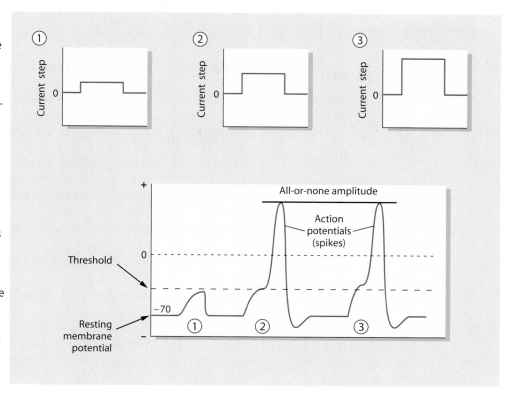

rents), chemicals (e.g., neurotransmitters), or physical stimulation (e.g., stretch or pressure such as at sensory receptors). The gated channels have different physical states; essentially they can be open or closed, like small gates in the membrane. If the channels are open, ions can move through them by being driven by electrical and ionic concentration gradients. In contrast, the nongated ion channels are essentially always in the same, open state. An example of the latter is the nongated K^+ channel that we described earlier; this channel establishes the membrane potential by letting K^+ leak out of neurons at rest. Nongated ion channels for Na^+ and Cl^- are also present in neuronal membranes. Voltage-gated ion channels are of prime importance in generating action potentials.

Voltage-gated ion channels open and close according to the membrane potential. They are closed at the resting membrane potential but open as the membrane is depolarized. When a passive current flows across the neuronal membrane following a synaptic or receptor potential, the membrane depolarizes and affects voltage-gated Na^+ channels. At this point, some of the channels begin to admit Na^+ into the neuron, which further depolarizes the neuron. This in turn leads to further opening of other voltage-gated Na^+ channels, and thus more depolarization, which continues the cycle by opening yet more Na^+ channels. The process, the Hodgkin-Huxley cycle, is depicted in Figure 2.20.

This rapid self-reinforcing cycle generates the large depolarization that is the first portion of the action

Figure 2.20 The Hodgkin-Huxley cycle. Depolarization of the axonal membrane leads to the opening of voltage-gated Na^+ channels. The inward Na^+ current then adds to the depolarization, thereby causing even more Na^+ channels to open. When this reaches threshold, the Na^+ channel currents dominate the membrane potential, leading to the action potential's positive-going initial phase (see Figure 2.21).

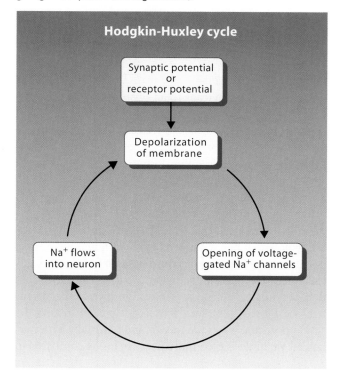

An Interview with Ira Black, M.D. Dr. Black is a professor and the chair of neuroscience and cell biology at the Robert Wood Johnson Medical School, of the University of Medicine and Dentistry of New Jersey. Dr. Black is an authority on cellular and molecular mechanisms of neuronal plasticity, and a former president of the Society for Neuroscience.

Authors: Many neuroscientists investigate the molecular events that underlie neural processing. How will an understanding of processes such as second-messenger systems help us conceive of memory, language, or consciousness? That is, if it turned out that second-messenger systems within neurons were accomplished in a manner entirely different from what is currently being discovered, would that affect the higher-level computations that neuronal systems and neurons perform to support cognition?

IB: A satisfactory mechanistic description of any well-framed cognitive process requires that we simultaneously explain it at multiple levels of analysis. Different levels provide complementary insights into characterization and causality that are unobtainable from any single line of analysis. Considering the molecular events associated with memory may take us beyond the foregoing pat generalizations. An account of the molecular basis of a memory subtype provides the starting point. With this molecular alphabet, critical genes are identified, molecular deficits leading to dementia can be characterized, and potential therapeutic targets may become apparent. But use of the molecular alphabet may provide insight into the nature of memory itself. For example, an understanding of the evolution of memories may help to define neural antecedents, relationships to underlying cell biology, and the selective pressures (or lack thereof) responsible. In turn, this information may indicate whether memory is a form of a widespread, cell biological phenomenon, or whether it represents a special case fundamentally different from other processes. Molecular analysis is particularly useful in this regard, since the component processes can be traced in a convenient, rigorous, and unambiguous fashion. Ideally, this type of evolutionary analysis may define the origins of memory and the forces driving its emergence. In addition, the same strategy may be used to define relations among different forms of extant memory. Such information could help us to understand the taxonomy of memory. In turn, we could begin to understand how memory relates to other cognitive phenomena. Surprises may well occur. For example, molecular analysis can reveal occult relations among cognitive processes that seem thoroughly dissimilar behaviorally. One not-so-modest goal entails constructing a genealogy of cognition that places processes

---

potential (Figure 2.21). Then, with a short delay, membrane depolarization leads to an opening of voltage-gated K^+ channels, which allow K^+ to flow out of the neuron and begin to repolarize it—and to reestablish the value of the resting membrane potential. The opening of the K^+ channels occurs and outlasts the closing of the Na^+ channels. The efflux of K^+ without the influx of Na^+ leads to the second repolarizing phase of the action potential, which brings the membrane potential back to the level of the resting membrane potential. The K^+ efflux drives the membrane potential below the level of the original membrane potential toward the equilibrium potential of K^+, which is more negative than the resting potential. As a result, the action potential undershoots the resting membrane potential (see Figure 2.21), and a transient period follows the action potential when the membrane is actually hyperpolarized; that is, it is even more negative inside compared to

outside than when at rest. This transient hyperpolarization lasts only a couple of milliseconds as the membrane gradually returns to the resting membrane potential.

The result of membrane hyperpolarization following the depolarization and repolarization of the action potential leaves the neuron in a state where it is temporarily more difficult to generate an action potential. This temporary state (known as the **refractory period**) occurs because when the membrane is hyperpolarized, the membrane potential is much lower than the threshold for triggering an action potential. However, one other factor also influences the neuron's ability to generate another immediate action potential. For a time immediately following the repolarizing phase of the action potential, the voltage-gated Na^+ channels are inactivated and unable to be opened, regardless of how much the membrane is depolarized. This inactivation of Na^+ channels produces the *absolute*

in a cognitive context and enables us to tentatively formulate a true cognitive structure. A skeletal example could indicate how different levels can and do cooperate in defining the nature of a cognitive process.

Authors: There are more glia in the brain than neurons. Can it really be the case that they have no computational role in brain processing?

IB: There are ten times as many glia in the brain as neurons. And there are about 100 billion neurons. Yet, for years we've known almost nothing about the function of these trillion or so glia. In fact, their name, which means "glue" in Greek, provides some sense of their low esteem. Early scientists thought they held the brain together. In the past few years, though, interest in glia has exploded and we know a lot more. We now know that glia produce growth factors and survival (or trophic) factors that serve as signals in the brain regulating growth, communication, and survival during maturity and development. Different glial types in the brain serve different functions. So, we are now beginning to appreciate the fact that many types of glia communicate with each other and with neurons. Instead of thinking of the neuron-neuron link as the unit of communication, we are now thinking of the neuron-glia-neuron loop. Consequently, while most scientists strongly suspect that glia are intimately involved in computations of the brain, their precise roles remain mysterious.

Authors: It has seemingly taken neuroscience many difficult years to uncover even the simplest of neural processes such as ion channel kinetics or transmitter system physiology, and one could argue that we do not understand these very well at all. Given this, what hope do we have for understanding higher behaviors?

IB: Is the glass half empty or half full? I would reframe the question to indicate that neuroscience has made astounding progress in a few short years. For perspective, recall that the neolithic agricultural revolution took place approximately 10,000 years ago, the Renaissance commenced about 500 years ago, and the Society for Neuroscience of North America began roughly 30 years ago. Yet, in a few short years, we already have an early understanding of the chemistry of vision, the physiology of hearing and touch, the physiology of muscle movement, the biochemistry of certain emotions, and a beginning molecular biology of learning, memory, and some behaviors. Not bad for a few short years. Viewed in this light, there is every reason to be optimistic about our progress. Researchers are now beginning to elucidate the genetics of schizophrenia, the biochemistry of Alzheimer's disease, and the brain system's basis of drug addiction. For the first time in the history of humankind, we have treatments for depression, epilepsy, and stroke. We are beginning to understand the genes that control brain function, the molecular messages they use, the nature of the neurons that comprise the brain and the systems that they form, and finally, the behavioral output of these critical brain systems. We now need more neuroscientists to put all these brain and mind pieces of the puzzle together. For all these reasons, there is unparalleled excitement in neuroscience, and unparalleled opportunity.

---

refractory period, which is followed by the *relative refractory period,* when the neuron can generate action potentials but only with larger-than-normal depolarizing currents. The consequence is that the neuron's speed in generating action potentials is limited. This limitation of the neuron is reflected in how it can be used for temporal coding. An example is the coding of sound frequency in the auditory system. Some sounds have frequencies in the tens of thousands of cycles per second (Hertz, abbreviated Hz), but the fastest neurons can follow frequencies one-to-one (spikes per cycle) up to about 1000 Hz (depending on the neuron). It is noteworthy, however, that not all Na^+ channels show a rapid deactivation. Some Na^+ channels in mammalian axons do not inactivate and therefore lead to longer-duration action potentials, although most show rapid inactivation. The classic picture we described here was derived from studies in the squid axon by Hodgkin and Huxley, but more recent molecular-genetic approaches have revealed that there are many different types of ion channels, even for one ionic species such as Na^+.

We now know enough about passive and active properties of neurons to understand how action potentials can propagate down the length of even very long axons, even if they are meters long. The key is simply to remember that passive electrotonic currents depolarize the membrane to threshold. This depolarization triggers an action potential which creates more passive currents that travel down the axon to depolarize new pieces of membrane to continue the process. The regeneration of action potentials is required because electrotonic currents die out quickly—regeneration has to keep the signal going. The **propagation** of action potentials is thus a continual interplay between electrotonic and active currents. Imagine the regeneration as firefighters in a bucket brigade; each person in the line is

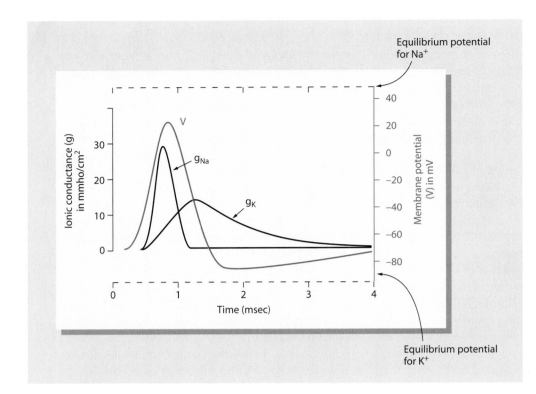

Figure 2.21 The relative time course of changes in membrane voltage during an action potential, and the underlying causative changes in membrane conductance to Na$^+$ (g_{Na}) and K$^+$ (g_K). The initial depolarizing phase of the action potential is mediated by Na$^+$ current, and the later repolarizing descending phase of the action potential is mediated by an increase in K$^+$ conductance that occurs when the K$^+$ channels open. The Na$^+$ channels have closed during the last part of the action potential, when repolarization by the K$^+$ current is taking place. The action potential has an undershoot where the membrane becomes more negative than the resting membrane potential. Adapted from Kuffler and Nicholls (1976).

analogous to the action potential, and the handing off of the bucket represents the electrotonic current flowing down the axon. The analogy would be better if the firefighters threw the water out of their buckets as far as they could to the next person, but analogies and firefighters have their limitations. Consider for a moment what happens if the firefighters in the bucket brigade are too far apart, then they could not get the water to the next firefighter in the chain, and the water could not be passed along. Something analogous to this would also occur in axons if the next segment of axon with voltage-dependent Na$^+$ and K$^+$ channels were too distant. Is this ever a concern? The answer is yes, when the axon is myelinated and conducts via saltatory conduction, as described next.

SALTATORY CONDUCTION AND THE ROLE OF MYELIN

A key aspect of neuronal communication is the speed of signaling between neurons, or between neurons and muscles. For example, as noted earlier, in animals as large as a giraffe or whale, the motor neurons located in the brain's motor cortex project axons down to the spinal cord, and therefore may have to send signals over many meters. As well, alpha motor neurons in the spinal cord that receive the signal from the brain must pass that along via peripheral nerves to the muscles they innervate, which might be located in the foot, for example. It is essential that these signals occur rapidly; otherwise coordination of motor ac-

tivity would be compromised. It is not enough that the signals can travel that distance; they also must do it quickly.

The physical properties of neurons affect how currents flow through them, and the resistances of the intracellular fluid and neuronal membrane are of prime importance. If the resistance of the membrane increases, or the resistance of the axon decreases, then currents flow down the axon more effectively, following the path of least resistance, and will therefore flow down the axon farther.

One way to lower the axon's internal resistance is to expand the axon's diameter; larger-diameter axons conduct axon potentials faster. In squid this is how the conduction speed needed to quickly contract muscles and avoid predators is achieved. The question is, How big would an axon's diameter have to be to achieve the speed required to communicate signals from a giraffe's brain to the motor neurons of the hind limbs? The answer is, Too big to fit all the needed axons into the spinal cord. Evolution had to solve this dilemma in another way before larger animals could evolve.

Myelination holds the key: Myelin wrapped around the axons of peripheral and central neurons increases membrane resistance. This increased resistance occurs because myelin is actually the concentric multiple wrappings of glial cell membrane around the axon; the cytoplasm in this portion of the glial cell is essentially squeezed out, leaving layer upon layer of cell membrane, which is the myelin (Figure 2.22). The increased electrical insulation produced by the myelin results in currents being shunted a

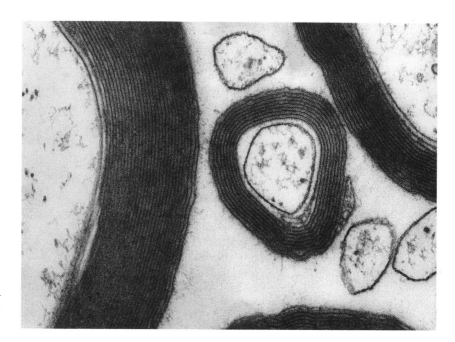

Figure 2.22 Cross section through peripheral nerve showing concentric wrappings of myelin around the axon.

greater distance down the axon. The result is that action potentials do not have to be generated as often, and they can be spread out along the axon at more distant intervals. Indeed, action potentials in myelinated axons need appear only at the nodes of Ranvier where myelination is interrupted. At the nodes, voltage-gated Na⁺ channels can trigger action potentials that regenerate fast electrotonic currents that flow to the next node, where another action potential is generated. So the action potential appears to jump along the axons from node to node—hence, the name **saltatory conduction,** which means "to jump" (Figure 2.23). By such conduction, mammalian nerves can transmit at roughly 120 m/sec (the length of a football field in 1 second—quite fast!).

Figure 2.23 Saltatory conduction in a myelinated nerve. Action potentials in a myelinated nerve occur only at the nodes of Ranvier where there is a gap in the myelin. The current generated at the nodes flows in the internode region under the constraints determined by the principles of electrotonic conduction. The distance between the nodes is determined by the length constant of the axon, which is affected by the myelination of the axon and the axon's diameter. Hence, the nodes are optimally spaced such that electrotonic currents from the action potential at the last node are still strong enough to depolarize the membrane to threshold at the present node, thus handing the signal down the axon and regenerating it at the nodes of Ranvier.

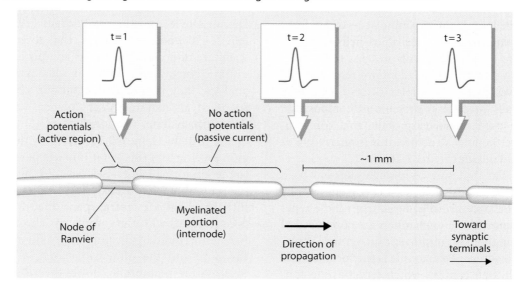

Transmembrane Proteins: Ion Channels and Pumps

In the preceding sections we discussed how the resting membrane potential arises and how changes in membrane permeability permit electrical signaling. We introduced the concept of transmembrane proteins, and defined two main types, the active transporters and the ion channels, the former using energy to pump ions across the membrane to create ionic concentration gradients and the latter contributing to the membrane permeability of ions. These molecules come in a variety of forms. For example, some ion channels are gated and others are not. Here we describe the structure and properties of ion channels in more detail.

ION CHANNELS

Ion channels are proteins, which are polypeptides (many peptides). Peptides in turn are formed from the molecular building blocks called *amino acids.* Any two amino acids can form a peptide, but many can come together to form small and large proteins. For example, human myoglobin contains 153 amino acid residues (individual amino acid molecules). There are twenty-one amino acids, including familiar ones like leucine, lysine, methionine, phenylalanine, threonine, and tryptophan that you can find on the labels of products in any health food store, and others such as glutamate that are important in their own right as neurotransmitters, as we will discover later.

Proteins are not only long strings of amino acids. They also have a three-dimensional structure that is important for their function. We can define several levels of structure in proteins. The *primary structure* is simply the order of the amino acids. The *secondary structure* refers to how the long chains of amino acids coil to form characteristic patterns such as the alpha-helix (a helical coiling). The *tertiary structure* of proteins reflects the fact that the long strands of coiled amino acids can fold upon themselves to form complex three-dimensional structures. Finally, many large proteins are composed of subunits, each itself a tertiary protein, that come together to create the final structure; this is referred to as the *quaternary structure* (Figure 2.24).

The molecular structure of ion channels has been investigated by methods such as x-ray crystallography in simple model systems. In general, from these types of studies we know that ion channels are composed of subunits, giving them an important quaternary three-dimensional structure. An example is the formation of a pore through their centers, which serves as the con-

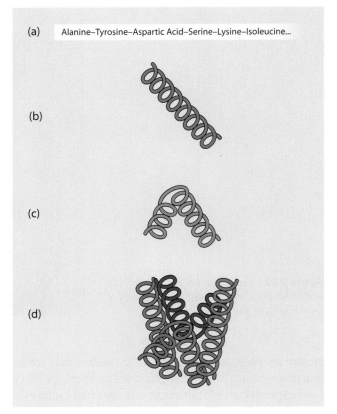

(a) Alanine–Tyrosine–Aspartic Acid–Serine–Lysine–Isoleucine...

(b)

(c)

(d)

Figure 2.24 General structure of proteins. **(a)** The amino acid sequence is the primary structure. **(b)** The strings of amino acids can coil into helical shapes. **(c)** Folding of the helix can give rise to globular proteins (many are enzymes, for example). **(d)** The quaternary structure is a complex three-dimensional structure (such as an ion channel) composed of globular proteins.

duit for ions to traverse the membrane. Recent work using the crystallographic approach in bacteria has led to a proposal for how K^+, Na^+, and Cl^- channels operate at the molecular level. Roderick MacKinnon, Declan Doyle, and their colleagues (1998) at Rockefeller University identified the three-dimensional structure of an ion channel as composed of four subunits (Figure 2.25). Based on their crystallographic analysis, they proposed that the ion channel has the shape of an inverted V or "teepee," created by the three-dimensional relationships of the four subunits of the protein (Figure 2.26). At its wider end is a region serving as a filter that imparts permeability only to the ion for which it is selective (in this case K^+); this region is about 12 angstroms (Å) wide. The selectivity filter opens into a central space filled with water that is continuous with the cytoplasm. The chemical environment created by the portions of the subunits lining the pore and the dimensions of the pore help K^+ to cross the membrane, down the K^+ concentration

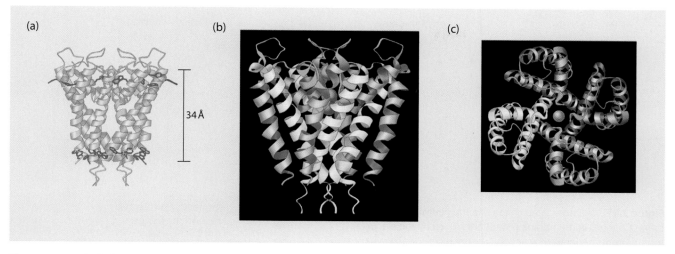

Figure 2.25 The helical structure of the transmembrane K^+ ion channel. **(a)** Cross-sectional diagram of the four subunits spanning the membrane from the extracellular (top) to the intracellular space, a length of about 34 Å. **(b)** The individual subunits in cross-sectional view as in (a) shown in different colors. **(c)** A "top" view of the ion channel showing the central pore. Adapted from Doyle et al. (1998).

gradient. Other ions such as Na^+ are not able to use K^+ channels to cross the membrane because the chemical environment inside the pore is not effective in enabling Na^+ to pass through. However, other ion channels that are specialized to be selective for Na^+, Cl^-, or Ca^{2+} probably act by similar principles as the K^+ ion channel.

Gated and Nongated Ion Channels Ion channels are either passive, called nongated, or active, being gated by electrical, chemical, or physical stimuli. The Na^+ and K^+ channels involved in the generation of the action poten-

Figure 2.26 Diagram of the K^+ channel in the membrane. In cross-sectional view, the channel is shown embedded in the membrane (the lipid bilayer is to the left and right of the channel). The extracellular space is at top. K^+ ions (green) are shown traversing the membrane via the pore. Adapted from Doyle et al. (1998).

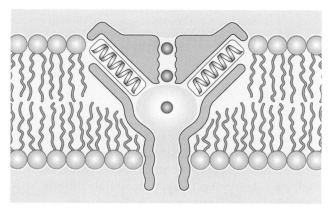

tial are voltage-gated channels. Electrotonically conducted currents that pass out across the neuronal membrane can depolarize it, causing the channels to open like gates. The opening permits the ions the channels are selective for to be driven through the pores by the concentration and electrical gradients across the membrane. Voltage-gated channels exist for Na^+, K^+, Cl^-, and Ca^{2+}. Voltage-gated Cl^- channels are involved in homeostatic processes, including stabilization of the membrane potential. Voltage-gated Ca^{2+} channels are, among other things, relevant for the release of neurotransmitters from presynaptic terminals. You might wonder how the voltage triggers the channels to open and close. Changes in the transmembrane potential influence the molecular structure of the proteins in the channels, changing the three-dimensional configuration of the pore region (Figure 2.27).

Chemically gated ion channels are those that serve as neurotransmitter receptors located chiefly on the postsynaptic membrane; examples are found on the dendrites of cortical neurons, or on muscle cells at the neuromuscular junction where motor neurons innervate muscle fibers. The chemicals (also called *ligands*) are typically neurotransmitters such as glycine, dopamine, and acetylcholine. So **receptors** are specialized ion channels that mediate signals at synapses; we will describe the process of synaptic transmission in the next section.

There are two general classes (and many subclasses) of postsynaptic receptors: directly and indirectly coupled receptors (Figure 2.28). The *directly coupled* receptors are those that bind (i.e., chemically

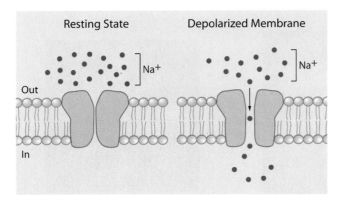

Figure 2.27 Voltage-gated channel in the neuronal membrane. Changes in membrane potential induced by the electrotonic currents generated by action, sensory receptor, and synaptic potentials alter the three-dimensional structure of the ion channel to permit certain ions to cross. The ions are pushed across the membrane by electrical and concentration gradients.

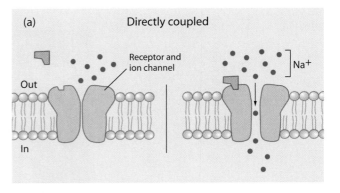

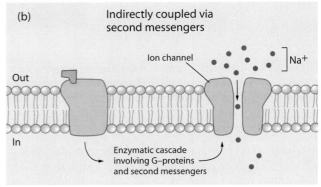

Figure 2.28 Mechanisms of neurotransmitter receptor molecules. **(a)** After binding with a neurotransmitter, the directly coupled receptor undergoes a conformational (structural) change that opens the pore to permit ions to cross the membrane, leading to (in this case) depolarization of the postsynaptic neuron. **(b)** An indirectly coupled receptor binds the neurotransmitter but is not itself an ion channel and hence forms no pore. Instead, the binding of the neurotransmitter leads to a series of intracellular biochemical events, beginning with the action of G-proteins and the subsequent generation of second messengers, ultimately leading to a signal that affects ion channels in the postsynaptic membrane, which then open and (in this case also) depolarize the postsynaptic neuron.

interact with) neurotransmitters on their extracellular surfaces. This binding leads to structural changes in the receptor that permit the passage of ions. The *indirectly coupled* receptors do not open themselves to create a pore. Instead, the binding of neurotransmitters leads to the generation of a signal inside the postsynaptic cell that then activates ion channels in the membrane, causing them to open. This latter mechanism involves so-called G-proteins (intracellular proteins that bind guanosine triphosphate, or GTP) and **second messengers** such as Ca^{2+}, cyclic nucleotides including cyclic guanosine monophosphate (c-GMP) and cyclic adenosine monophosphate (c-AMP), nitric oxide, and other substances. One advantage of second-messenger systems is they can permit signal amplification by triggering enzymatic cascades. These cascades exponentially increase the final signal via several intermediate biochemical steps, thereby allowing a relatively small neurotransmitter signal to lead to a large postsynaptic response. The second-messenger systems not only act to modulate ion channel conductances but also are essential for intracellular signaling that leads, for example, to changes in gene expression in response to neurotransmitters and hormones. There are a variety of receptor subtypes, and they play specific roles in neuronal communication, disease, and cognitive processes. An example is the role of the NMDA receptor in learning and memory, which we will investigate in Chapter 8.

SYNAPTIC TRANSMISSION

The ultimate act in neuronal signaling is for a neuron to communicate with other neurons or muscles. In order for neurons to do this, they must transmit signals—an action called *synaptic transmission*. Synaptic transmission (as its name implies) occurs at synapses, and there are two major kinds of synapses, chemical and electrical, each using very different mechanisms from the other. The essentials of synaptic transmission have been understood for decades, but the details continue to be unraveled with some recent exciting discoveries illuminating mechanisms that only a few years ago had been mysteries.

Chemical Transmission

Chemical transmission involves a series of fairly well-understood steps. First, in most cases an action potential must arrive at the axon terminals where the synapses are located. The invasion of the axon terminals by the action potential leads to the depolarization of the terminals, which initiates an influx of Ca^{2+} ions into the terminal region. This influx is mediated by voltage-gated Ca^{2+} channels. Ca^{2+} acts as an intracellular messenger and via mechanisms that remain poorly understood, triggers the next step in neural transmission. As a result of the increased intracellular Ca^{2+} concentration, small **vesicles** containing neurotransmitter fuse with the membrane at the synapse and release this transmitter into the synaptic cleft, the space between the presynaptic and postsynaptic neuronal membranes. The transmitter diffuses across the cleft and on reaching the postsynaptic membrane, binds with protein molecules (receptor molecules) embedded in the postsynaptic membrane, initiating changes in the postsynaptic membrane via the mechanisms reviewed in the last section (Figure 2.29).

The chemical interaction of the neurotransmitter and the postsynaptic receptor initiates events that lead to either depolarization (excitation) or hyperpolarization (inhibition) of the postsynaptic cell (Figure 2.30). If the postsynaptic cell is a neuron, then an excitatory postsynaptic potential (EPSP) might lead to the generation of action potentials, depending on a variety of factors. In most neurons, action potentials are initiated at the interface between the cell body and the axon (axon hillock), where synaptic currents are summated electrically. That is, the influences of many synaptic potentials sum together. This principle of summation is key to the potentially distant, axon hillock region of the postsynaptic membrane achieving sufficient depolarization to reach threshold for the generation of action potentials. If the postsynaptic cell is a muscle cell, the EPSP leads to action potentials in the muscle and to muscular contraction. If the neurotransmitter has an inhibitory action on the postsynaptic neuron, then hyperpolarization of the membrane potential might occur, causing an inhibitory postsynaptic potential (IPSP). Hyperpolarization would result in the postsynaptic neuron being less likely to generate an action potential because the membrane potential would become more negative and hence farther away from threshold. This hyperpolarization can result from opening of Cl^- channels on the postsynaptic membrane. The concentration of Cl^- is much higher outside the neuron than inside. Therefore, if the Cl^- channels are open, Cl^- will be forced into the neuron by the force of the concentration gradient, making the inside of the cell even more negative. Or, even if no hyperpolarization of the neuron occurs, the opening of Cl^- channels will minimize the effects of any simultaneous excitatory inputs that would open Na^+ channels. The effects are minimized because the open Cl^- channels and the open Na^+ channels have opposite influences on the membrane potential; hence, via summation, their effects tend to cancel each other. Thousands of inputs to a single neuron may try to influence its behavior, and summation helps determine the outcome.

Interestingly, inhibitory synaptic inputs might even lead to slight *depolarization* of the postsynaptic neuron and still act to inhibit the neuron from being depolarized to threshold by other (excitatory) inputs. How can this be? The answer is relatively straightforward. The equilibrium potential for Cl^- is around -60 mV; recall that the equilibrium potential, which is described by the Nernst equation, is the membrane potential that would yield no net flux of ions, in this case Cl^-, across the membrane. If the *resting membrane potential* is -70 mV, then opening Cl^- channels would tend to move the membrane potential toward the -60 mV equilibrium potential for Cl^-. However, further depolarization past -60 mV would be resisted, as the action of the Cl^- ions entering the neuron would cause the potential to move toward the -60 mV. Hence, if the *threshold* to generate an action potential was around -40 mV, the result of depolarizing to -60 mV would still be to inhibit the neuron. So, inhibitory inputs do not always hyperpolarize the postsynaptic neurons, but they do act to maintain the membrane potential in the direction of hyperpolarization with respect to threshold.

TRANSMITTER RELEASE

One of the mysteries of neuronal transmission has been how neurotransmitter is released. As recently as the 1980s, researchers still hotly debated whether or not neurotransmitter released from presynaptic terminals was the result of exocytosis of the vesicles in the terminals. A combination of methods was brought to bear to convince the scientific community that exocytosis was involved in most neuronal transmission. But until the mid 1990s or so, it was not clear exactly how the vesicles moved to the presynaptic membrane, fused with it, and extruded their contents into the synaptic cleft. Because of an amazing set of studies using molecular methods, we now have a clearer picture of the mechanisms producing this feat (Figure 2.31). The story has to do with a group of specialized intracellular proteins, some of which are attached to vesicles and some of which are attached to the neuronal membrane in the terminals where the

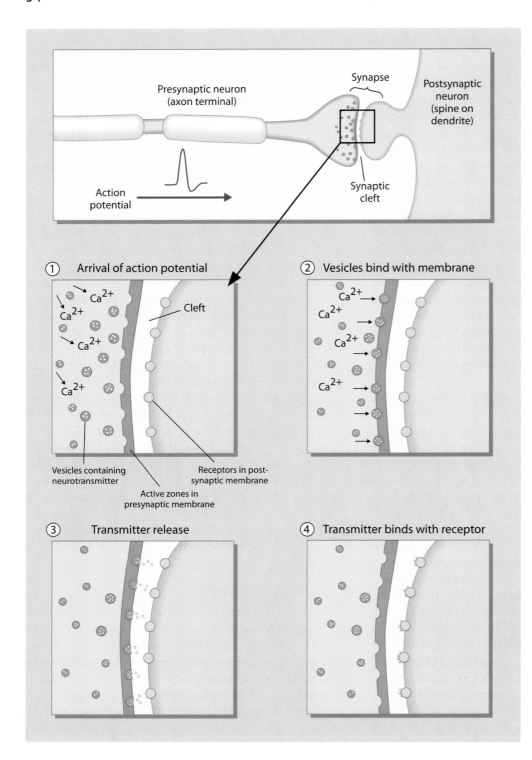

Figure 2.29 Neurotransmitter release at the synapse. The synapse consists of various specializations where the presynaptic and postsynaptic membranes are in close apposition. When the action potential invades the axon terminals, it causes voltage-gated Ca^{2+} channels to open (1), which triggers vesicles to bind to the presynaptic membrane (2). Neurotransmitter is released into the synaptic cleft by exocytosis and diffuses across the cleft (3). Binding of the neurotransmitter to receptor molecules in the postsynaptic membrane completes the process of transmission (4). Adapted from Kandel et al. (1991).

synapses are located. These proteins act as structural elements that can bring the vesicle up to the membrane and "dock" it there, like a ship next to a pier. Then via biochemical processes triggered by the influx of Ca^{2+}, the vesicle fuses with the neuronal membrane, permitting neurotransmitter to be extruded into the synaptic cleft.

Electrical Transmission

Some neurons communicate via electrical synapses. These synapses are very different, however, from chemical synapses because in electrical synapses, there is no synaptic cleft separating the neurons. Instead, the neuronal

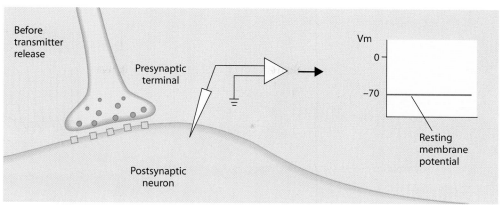

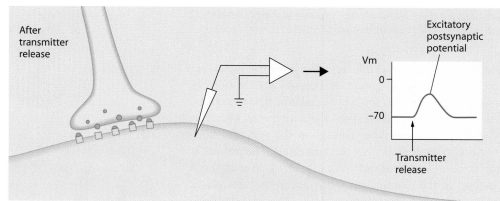

Figure 2.30 Neurotransmitter leading to postsynaptic potential. The binding of neurotransmitter to the postsynaptic membrane receptors changes the membrane potential. These postsynaptic potentials can be either excitatory as shown here (depolarize the membrane) or inhibitory (hyperpolarize the membrane).

membranes are touching, and the cytoplasms of the two neurons are essentially continuous. This continuity occurs via specialized transmembrane channels called *gap junctions* that create pores connecting the cytoplasms of the two neurons (Figure 2.32). As a result, the two neurons are isopotential (have the same potential), meaning that electrical changes in one are reflected in the other, virtually instantaneously. However, following the principles of electrotonic conduction, the passive currents that flow between the neurons when one of them is depolarized (or hyperpolarized) decrease and are therefore smaller in the postsynaptic neuron than the presynaptic neuron. Under most circumstances the communication is bidirectional, but so-called rectifying synapses limit current flow in one direction, as is typical in chemical synapses.

Electrical synapses are useful when information must be conducted rapidly, such as in the escape reflex of some invertebrates. Groups of neurons with these synapses can activate muscles fast, to get the animal out of harm's way quickly. For example, the well-known tail-flip reflex of the

Figure 2.31 Release of neurotransmitter (NT) from the presynaptic terminal. **(a)** Vesicles in the presynaptic terminal move toward and bind with the cell membrane, with the aid of several proteins including synaptogamin (SG), synaptobrevin (SB), syntaxin (ST), and SNAP-25 (S25). **(b)** These proteins act a bit like ropes on a ship and dock, and are used to bring the vesicle and presynaptic terminal membrane close to one another. **(c)** The influx of Ca^{2+} into the presynaptic terminal is believed to induce synaptogamin to fuse the two membranes, permitting neurotransmitter to be extruded into the synaptic cleft. Adapted from Purves et al. (2001).

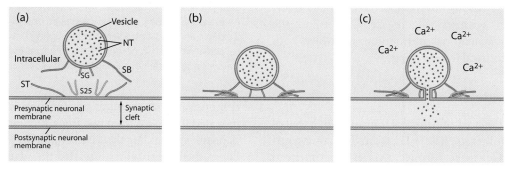

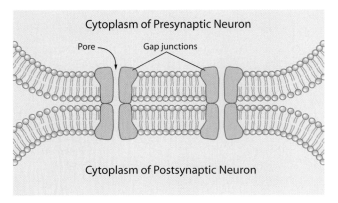

Figure 2.32 Electrical synapse between two neurons. Gap junctions are the transmembrane proteins in the presynaptic and postsynaptic neurons that are aligned to create a pore.

crayfish involves powerful rectifying electrical synapses. Another use is when groups of neurons should operate synchronously, as with some hypothalamic neurosecretory neurons. Are electrical synapses, then, merely an odd type of synapse found in invertebrates on the run, and hormonal systems that are just plain running? The answer is no; there is well-accepted evidence for a role of these synapses in the mammalian retina and some subcortical nuclei. Recently, Barry Connors and his colleagues Jay Gibson and Michael Beierlein at Brown University provided a powerful demonstration of electrically coupled neurons in the somatosensory cortex of the rat (Gibson et al., 1999). They investigated inhibitory interneurons (in cortical layers 4 and 6), neurons that make local circuit connections within the cerebral cortex, by recording from them in slice preparations using microelectrodes. The slice preparation is literally a slice of brain taken from an animal killed moments before. The slice is then placed in a dish with physiologically friendly solutions. With the use of a microscope and micromanipulators to position electrodes, electrical activity is recorded from the neurons within the slice where the local circuitry remains intact. When Connors and his colleagues recorded from neurons that were electrically but not chemically coupled, they were able to show high synchrony of electrical activity between two neurons when one or both were stimulated, and the delay between the activity in the two neurons was too short to be due to traditional chemical synapses (Figure 2.33). Thus, in the mammalian brain, neuronal communication involves both chemical and electrical transmission.

Neurotransmitters

A final topic in the story of neuronal transmission is that of the neurotransmitter itself. Otto Loewi, in the early twentieth century in Germany, designed a clever experiment to

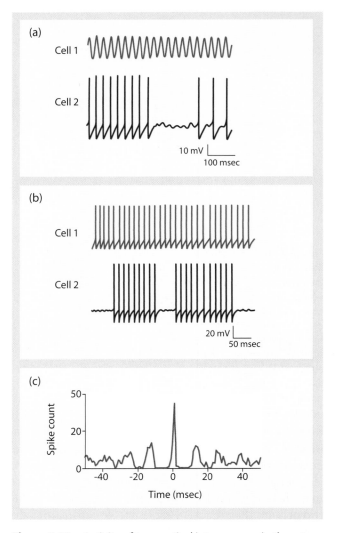

Figure 2.33 Activity of two cortical interneurons in the rat somatosensory cortex connected by electrical synapses. **(a)** Electrodes were placed in two coupled neurons, and a current whose amplitude varied at about 40 Hz was injected into cell 1 while cell 2 was maintained below spontaneous spike threshold with constant current. Cell 2 showed a tendency to spike at the peaks of the depolarizing phases in cell 1. **(b)** Depolarizing current was injected into two different neurons simultaneously, and the neurons were found to spike in synchrony. **(c)** To test the synchrony of the firing of the two neurons, the cross-correlogram method was used. The activity of cell 2 was cross-correlated with the firing of cell 1. If they were firing in synchrony, the cross correlogram should show a strong peak around 0 msec, and this was indeed observed. The peak is at 2 msec, demonstrating a very close coupling between the electrical activity in the two neurons. These neurons did not have chemical synaptic interactions. Adapted from Gibson et al. (1999).

demonstrate that chemical transmission was possible. He prepared the hearts from two frogs, keeping them as intact as possible, including their innervation by the vagus nerve. Loewi then electrically stimulated the vagus nerve of one frog, causing the heartbeat to slow, and collected the fluid perfusing (running through) that heart. He then transferred this perfusate from the stimulated heart to the second unstimulated heart, and observed a fascinating sight: The second heart, although not stimulated by Loewi, also slowed its rate of pumping. He surmised that a chemical agent mediated this effect, and named it *vagusstoff* ("vagus stuff" or "vagus substance" in English). Later, Loewi and his colleagues were able to isolate the substance, which turned out to be acetylcholine (ACh). An interesting aspect to this story is that Loewi claimed he had the idea for the experiment in a dream. He woke up, wrote it down, and went back to sleep. But upon waking in the morning, he could not decipher the writing on his note. He later (luckily) had the dream again, and this time remembered the idea long enough to do the history-making experiment. A few years later, British physiologist Henry Dale demonstrated that ACh was also the neurotransmitter for skeletal muscles, not merely cardiac muscles or other parts of the autonomic nervous system (see Chapter 3). The age of neurotransmitters was firmly established in neuroscience.

Today more than one hundred neurotransmitters are recognized. But precisely what are neurotransmitters? Well, they vary significantly in their molecular structures, so the best general definition is that they are molecules varying in size and chemical composition, that share some common properties: All are synthesized by the presynaptic neuron; transported to the axon terminals to be stored in vesicles that extrude the transmitter into the synaptic cleft; and after binding with the postsynaptic cell, are removed or degraded by enzymatic action. Several criteria have been established to classify a substance as a neurotransmitter. This is important because thousands of substances are present in the brain and nervous system, and determining which of these are involved in neural transmission is not easy. Candidates (putative neurotransmitters) have to meet the following criteria: (1) The putative neurotransmitter should be synthesized by and localized within the presynaptic neuron, and stored in the presynaptic terminal prior to release. (2) It must be released by the presynaptic neuron when action potentials invade and depolarize the terminal (mediated by Ca^{2+}). (3) The postsynaptic neuron must contain receptors that are specific for the substance. (4) When artificially applied to the postsynaptic cell, it should lead to the same response that stimulating the presynaptic neuron would lead to.

CLASSES OF NEUROTRANSMITTERS

There are several ways to classify neurotransmitters. They can be classified biochemically as particular substances like ACh; *amino acids* such as aspartate, gamma-aminobutyric acid (GABA), glutamate, and glycine; *biogenic amines,* including dopamine, norepinephrine, epinephrine (these three are known as *catecholamines*), serotonin (5-hydroxytryptamine), and histamine; and *neuropeptides,* of which there are more than one hundred.

Recall that peptides are strings of amino acids. Neuropeptides typically vary in length from three to more than thirty amino acid residues (the numbers are changing constantly as new ones are identified). In any case, neuropeptides are large molecules relative to the nonpeptide neurotransmitters. The peptides can be subdivided further into five major groups: (1) the *tachykinins* (brain/gut peptides) such as substance P, which affects vasoconstriction and is a spinal neurotransmitter involved in pain; (2) the *neurohypophyseal hormones* oxytocin and vasopressin, the former involved in mammary functions and the latter as an antidiuretic hormone; (3) the *hypothalamic releasing hormones* including corticotropin-releasing hormone and somatostatin; (4) the *opioid peptides* (so named for their similarity to opiate drugs, permitting them to bind to opiate receptors) such as the endorphins and enkephalins; and (5) the "other" peptides (ones that do not fit neatly into another category).

In general, each neuron typically produces one, two, or more neurotransmitters, and these may be released together or separately depending on the conditions of stimulation. For example, the rate of stimulation can induce the same neuron to release one or another neurotransmitter. In most cases, the action that a neurotransmitter has on the postsynaptic neuron is determined by that postsynaptic neuron rather than by the transmitter itself. That is, the same transmitter released from the same presynaptic neuron onto two different postsynaptic cells might cause one to increase firing and the other to decrease. Despite this fact, neurotransmitters can be classified by the typical effect they induce in the postsynaptic neuron. Excitatory neurotransmitters include ACh, the catecholamines, glutamate, histamine, serotonin, and some of the neuropeptides. Inhibitory neurotransmitters include GABA, glycine, and some of the peptides. In addition to acting directly to excite or inhibit a postsynaptic neuron, some neurotransmitters act only in concert with other factors and are sometimes referred to as *conditional neurotransmitters;* that is, their action is conditioned on the presence of another transmitter in the synaptic cleft or activity in the neuronal circuit. These types of mechanisms permit the nervous system to achieve complex modulations of information processing via modulation of the effectiveness of neurotransmission.

SYNTHESIS OF NEUROTRANSMITTERS

Large molecule transmitters (peptides) are synthesized in the cell body, whereas small molecule transmitters are synthesized in the synaptic terminals. The enzymes (i.e., biological catalysts) necessary for the synthesis (i.e., biochemical production of specific molecules) of small molecule transmitters are produced in the cell body. They then are transported down the axon via slow axonal transport, at a rate of about 1 mm/day, to the terminals where the small molecule transmitters are synthesized. In contrast, after synthesis in the cell body, peptides are placed in vesicles and transported down the axon via fast axonal transport, at a rate up to 400 mm/day. Intra-axonal microtubules transport the vesicles using energy from ATP.

Synthesis of the catecholamines is one of the most interesting chemical pathways, and it has important implications in the treatment of Parkinson's disease, a disease in which dopamine-containing neurons in the midbrain nucleus, the substantia nigra, lead to a loss of dopamine in the basal ganglia and a serious neurological disorder. The synthesis of dopamine, norepinephrine, and epinephrine begins with the amino acid tyrosine; the metabolic pathway for dopamine and norepinephrine is shown in Figure 2.34.

INACTIVATION OF NEUROTRANSMITTERS AFTER RELEASE

Following the release of neurotransmitter into the synaptic cleft and its binding with the postsynaptic membrane receptors, the remaining transmitter must be removed. This can occur (1) by active reuptake of the substance back into the presynaptic terminal, (2) by enzymatic breakdown of the transmitter in the synaptic cleft, or (3) merely by diffusion of the neurotransmitter away from the region of the synapse or site of action (e.g., in the case of hormones that act on target cells distant from the synaptic terminals).

Neurotransmitters that are removed from the synaptic cleft by reuptake mechanisms include the biogenic amines (dopamine, norepinephrine, epinephrine, histamine, and serotonin). The reuptake mechanism is mediated by active transporters, which are transmembrane proteins that pump the neurotransmitter back across the presynaptic membrane.

An example of a neurotransmitter that is eliminated from the synaptic cleft by enzymatic action is ACh. The enzyme acetylcholinesterase (AChE), which is located in the synaptic cleft, breaks down ACh after it has acted on the postsynaptic membrane. In fact, special AChE stains (chemicals that bind to AChE) can be used to label

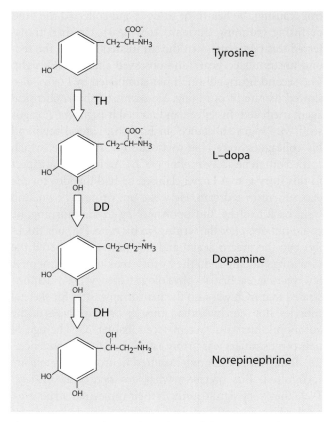

Figure 2.34 Biochemical synthesis of dopamine and norepinephrine from the amino acid tyrosine. The synthetic enzymes are tyrosine hydroxylase (TH), dopa decarboxylase (DD), and dopamine beta hydroxylase (DH). Norepinephrine can be converted to epinephrine in neurons containing phenylethanolamine *N*-methyltransferase (not shown).

AChE on muscle cells, thus revealing where motor neurons innervate the muscle.

To monitor the level of neurotransmitter in the synaptic cleft, presynaptic neurons have autoreceptors that provide feedback about the amount. These receptors, which are located on the presynaptic terminal, bind with the released neurotransmitter and permit the presynaptic neuron to regulate its synthesis and release.

ANATOMICAL PATHWAYS OF THE BIOGENIC AMINES

Neurotransmitters are both localized and widely distributed in the brain. Glutamate is located almost everywhere in the brain. In contrast, the biogenic amines are more specifically localized. Indeed, the biogenic amines form major neurotransmitter systems from the nuclei (groups of cells) containing the cell bodies in the midbrain (or in the hypothalamus in the case of histamine). Their axons project to widespread regions of the cortex and can play key modulatory roles. An example is the norepinephrine system (also called the *noradrenergic system*). The widespread projections of the norepinephrine neurotransmitter system to the cortex are involved in arousal and attentive behavior. The major projection pathways of the biogenic amines are shown in Figure 2.35.

Drugs, Neurotransmission, and Disease Many classes of drugs act to modify neurotransmission and are useful in treating psychiatric and neurological disorders. Other drugs actually induce depression, psychosis, and other symptoms of psychiatric disorders. (*Psychosis* is a disorder characterized by a constellation of symptoms such as irrational thought and is a component of schizophrenia and affective disorders.) For example, abuse of amphetamines, which increases the release of dopamine and norepinephrine, leads to amphetamine psychosis. These observations on the effects of drugs led to major hypotheses about the biochemical basis of mental disorders. The

Figure 2.35 Major projection pathways of the biogenic amine neurotransmitter systems in the human brain. Shown are the projections of the **(a)** dopamine, **(b)** norepinephrine, **(c)** histamine, and **(d)** serotonin systems. The views are midsagittal cuts through the human brain, showing the medial surface of the right hemisphere; the frontal pole is at left (see Chapter 3 for anatomical information). Adapted from Purves et al. (2001).

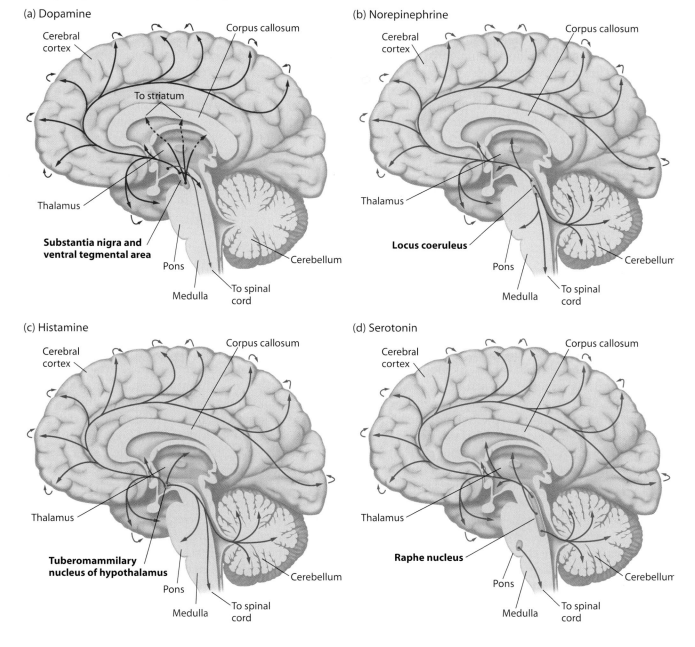

hypotheses propose that these disorders might reflect an improper functioning of specific brain neurotransmitter systems. If they are correct, then it is logical that drug therapies targeting neurotransmitter action might be useful in treating the disorder.

Drugs can influence neurotransmission by directly mimicking the action of the neurotransmitter at postsynaptic receptors, increasing the release of neurotransmitter, or preventing its reuptake (agonists). Alternatively, a drug might act by reducing the ability of the neurotransmitter to bind to the postsynaptic receptors (antagonists). In neurotransmitter systems implicated in psychiatric and neurological diseases, conditional transmitters are often involved (see above), and here the drugs may fine-tune neural transmission via subtle modulatory influences on transmitter function.

Drugs that manipulate reuptake mechanisms, such as the selective serotonin reuptake inhibitors, are used to treat depression, anxiety, and obsessive-compulsive disorder. Other drugs such as the monoamine oxidase (MAO) inhibitors act to prevent the breakdown of dopamine, norepinephrine, and serotonin by the enzyme MAO, thereby leading to more transmitter in the synaptic cleft for longer periods of time.

We began this chapter with the story of G.W., a patient who developed schizophrenia as a young adult. In passing, we noted that part of her treatment was drug therapy. Here, with a knowledge of neurotransmission and its biochemical bases in hand, you should be able to infer what sorts of drugs her psychiatrist prescribed. Let's put the story together. In the preceeding text we noted that psychosis, a major component of schizophrenia, can be induced by amphetamine abuse. Moreover, amphetamines act to stimulate the release of some neurotransmitters, such as dopamine. This and other related evidence led to the "dopamine theory of schizophrenia." The idea behind this theory is that too much dopamine may contribute to the symptoms of the disorder. Neuroleptic drugs (*lepsis* means "to take hold of"), also referred to as *antipsychotic drugs*, are those that relieve the paranoia and hallucinations associated with psychosis in schizophrenia, as well as psychosis induced by amphetamines. Some common neuroleptic drugs are antagonists of dopamine. More recently, however, researchers have found that drugs that are agonists of glutamate, an excitatory transmitter, also may be effective in the treatment of schizophrenia. It is now actually quite common for physicians to prescribe "drug cocktails," or combinations of drugs that affect different transmitter systems, to treat schizophrenia. So, we can conclude that patient G.W. has a disease, schizophrenia, that causes a disruption of normal neurotransmitter function in the brain, perhaps in the dopamine or glutamate systems, or both!

SUMMARY

Cells are the elementary units of the body, and this is also true in the nervous system where nerve cells or neurons provide the mechanism for information processing. When at rest, the neuronal membrane has properties that allow some materials (primarily ions) dissolved in intracellular and extracellular fluids to pass through better than others. In addition, active transport processes pump ions across the membrane to separate different species of ions, thereby setting the stage for electrical potential differences inside and outside the neuron. These electrical differences are a form of energy that can be used to generate electrical currents, which, via action potentials, can travel great distances down axons away from the neuron's cell body. When the action potential reaches an axon terminal, it prompts the release of chemicals at a specialized region, the synapse, where the neuron contacts another neuron. These chemicals (neurotransmitters) diffuse across the synaptic cleft between the neurons and contact receptor molecules in the next (postsynaptic) neuron. This chemical transmission of signal leads to the generation of currents in the postsynaptic neuron and the continuation of the signal through the system of neurons that comprise a neuronal circuit. Ion channels are the specialized mediators of neuronal membrane potential. They are large transmembrane proteins that create pores through the membrane. Transmembrane proteins also form receptors on postsynaptic neurons. These are the receptors that bind with neurotransmitters, leading to changes in the membrane potential. Neurotransmitters come in a large variety of forms. Small molecule transmitters include amino acids, biogenic amines, and substances like ACh, while large molecule transmitters are the neuropeptides. All neurotransmitters share the following characteristics: They are synthesized, transported to the terminal (or are synthesized in the terminal), stored in vesicles, released, and removed. Neurological and psychiatric diseases can involve disturbances in neurotransmission, and major therapies rely on drugs that modulate neurotransmitter function with the hope of reestablishing the delicate balance of function that characterizes normal neuronal communication.

KEY TERMS

action potentials	ion channels	propagation	spike-triggering zone
axon	myelin	receptor	spine
blood-brain barrier (BBB)	neuron	receptor potentials	synapse
dendrite	neurotransmitter	refractory period	synaptic potential
electrical gradient	nodes of Ranvier	resting membrane potential	threshold
electrotonic conduction	permeability	saltatory conduction	vesicle
equilibrium potential	postsynaptic	second messenger	
glial cell	presynaptic	soma	

THOUGHT QUESTIONS

1. If action potentials are all or none, how does the nervous system code differences in sensory stimulus amplitudes?
2. What property (or properties) of ion channels makes them selective to only one ion such as K^+, and not another such as Na^+? Is it the size of the channel, other factors, or a combination?
3. Synaptic currents produce electrotonic potentials that are decremental. Given this, how do inputs located distantly on a neuron's dendrites have any influence on the firing of the cell?
4. What would be the consequence for the activity of a postsynaptic neuron if reuptake or degradation systems for neurotransmitters were damaged?
5. Why does the brain have receptors for products produced in plants such as the opium poppy?

SUGGESTED READINGS

BLOOM, F., NELSON, C.A., and LAZERSON, A. (2001). *Brain, Mind and Behavior,* 3rd edition. New York: Worth Publishers.

DOYLE, D., CABRAL, J., PFUETZNER, R., KUO, A., GULBIS, J., COHEN, S., CHAIT, B., and MACKINNON, R. (1998). The structure of the potassium channel: Molecular basis of K^+ conduction and selectivity. *Science,* 280: 69–77.

HAEUSSER, M. (2000). The Hodgkin-Huxley theory of the action potential. *Nature Rev. Neurosci.,* 3:1165.

KATZ, B. (1966). *Nerve, Muscle and Synapse.* New York: McGraw-Hill.

3

Gross and Functional Anatomy of Cognition

There has been a long history of interest in the relationship between brain anatomy and cognition. For example, many studies have attempted to relate aspects of brain organization to intelligence in humans. Consider the story of one of the greatest brains in human history, that of Albert Einstein. He was clearly a person of exceptional ability. During his life he mentally tiptoed among atoms and stars and brought his incredible insights about the physics of the universe to humankind in the form of concrete mathematical descriptions. He said of his scientific thinking that "... words do not seem to play any role," yet there is an "associative play" involving "more or less clear images" of a "visual and muscular type."

Numerous efforts have been made to identify unique anatomical features of Einstein's brain that might explain his genius. When Einstein died in 1955 at the age of 76, his brain was extracted and saved. It was weighed, perfused with a solution of the chemical preservative formalin, measured, and photographed. The brain was then sectioned and embedded in a material that permitted it to be thinly sliced to make histological slides for microscopic analysis. These specimens and materials documenting their original three-dimensional relationships in the whole brain were stored for future scientific study. Sandra Witelson and her colleagues (1999) at McMaster University in Canada performed one such study. They investigated the dimensions and gyral morphology of Einstein's brain based on the original measurements, their own new measurements of the brain, and measurements calculated from the original calibrated photographs taken shortly after his death. The characteristics of Einstein's brain were compared to those of brains from a few dozen normal persons that had been donated for scientific research. These researchers found two prominent features of Einstein's brain that were very different from the normal control group of brains. The first observation was that Einstein's brain

had a **Sylvian fissure,** which separates the temporal from the frontal and parietal lobes, with an unusual anatomical organization (see description of gross anatomy later in the chapter to appreciate the anatomical features described here). Unlike the control brains, Einstein's brain showed a strange confluence of the Sylvian fissure with the central sulcus on the brain's lateral surface; most brains have a Sylvian fissure that projects posteriorly to end in an area surrounded by the supramarginal gyrus. The second unusual feature resulted from this anatomy: Einstein's inferior parietal lobe was actually larger, and indeed thicker (15%) in a lateral to medial extent. Witelson and her colleagues hypothesized that the increased size of Einstein's inferior parietal cortex might have been related to his intellectual capacity, although they pointed out that it is difficult to make any conclusions about causal relationships from their data. Nonetheless, it is remarkable that they found gross anatomical features that were so very unique to Einstein's brain in comparison to the brains in the rest of us, who for the most part are merely trying to understand what Einstein said, never mind attempt to generate new insights ourselves. From such investigations hypotheses can be generated and then tested using brains removed at autopsy or, in the modern age of neuroimaging, brains in the living human. Perhaps the next Einstein, like the last one, who permitted electroencephalograms to be recorded from electrodes on his head, will lay inside a scanner so young cognitive neuroscientists can measure his or her living, thinking brain and its function.

From form is derived function—a central tenet in biology. Throughout the history of neuroscience, researchers have probed the organization of the nervous system, hoping that by doing so they would unlock its secrets. In Chapter 1 we saw how the great neuroanatomists Santiago Ramón y Cajal of Spain and Camillo Golgi of Italy used information from cellular neuroanatomy to argue for different theories of information processing in the nervous system. This work continues to the present, with a march toward new and higher-resolution techniques for probing the biological foundations of the mind. In Chapter 4 we will learn about some of these new techniques for revealing the functional anatomy of the brain, techniques that are based on understanding the relationships between form and function. Then in later chapters of this text we will see myriad examples of the anatomical correlates of behavioral and physiological (functional) findings about cognition. It will become clear that gross anatomy must be interpreted in light of the functional interactions between structures, circuits, and systems in the nervous system because similar structures may perform very different computations, and the same structure may participate in different functions at different times. Understanding how the form of the nervous system supports the myriad functions of the mind is one of the major challenges of cognitive neuroscience.

In this chapter, we will review the gross anatomy and microanatomy of the brain and nervous system. We will describe the major anatomical structures of the brain and make note of the vasculature and ventricular systems. Some discussion of the functional organization of the mental functions that the brain supports begins here; subsequent chapters will describe in more detail the many functions of the human brain in perception and cognition. Hence, the aim of this chapter is to provide a reference for the material to be covered later, and to introduce the neuroanatomical terms that will be important throughout the book. So, let this chapter be a guide to your first introduction to the brain and nervous system and a companion to the chapters that follow. Refer back to these contents as needed to interpret the complexities about the form and function of the nervous system that are introduced later.

NEUROANATOMY

Neuroanatomy is the study of the nervous system's structure. It is concerned with identifying the parts of the nervous system and describing how the parts are connected. As with all of anatomy, descriptions can be made at many levels. For the neuroanatomist, investigations occur at one of two levels: *gross neuroanatomy,* in which the focus is on general structures and connections visible to the naked eye, and *fine neuroanatomy* (also referred to as *microscopic anatomy*), in which the main task is to describe the organization of neurons and their connections and subcellular structure. In Chapter 2, in order to provide the background necessary to consider the physiology of the nervous system in signaling, we introduced cellular neuroanatomy. Refer back to the first portion of Chapter 2 for an overview of cells of the nervous system. Here, we begin with a short description of the methods that help reveal the gross, microscopic, and functional neuroanatomy of the nervous system, the latter being the anatomical organization of cells, circuits, and systems that support a particular function such as object perception.

Methods in Neuroanatomy

Different approaches are used to investigate the nervous system at different levels of description, and indeed, these methods are highly specialized such that they represent different subdisciplines of neuroanatomy. Their development follows roughly the historical story laid out in Chapter 1, as the field of neuroanatomy developed in step with the development of new techniques.

GROSS DISSECTION

For identification of gross anatomy, the first challenge is to figure out how to view the brain. Until recently, when neuroimaging made it possible to view the living brain within the heads of animals and humans in vivo (see Chapter 4), the only method of doing so was the tried and true method of removing the brain from the head and putting it on the table, or more accurately, in a container filled with a preservative such as formalin. A neuroanatomist undertaking this procedure immediately is made aware of the specialized defenses that protect the brain. Not only is

Cortical Topography

Early insights into human cortical organization were made possible by studies that involved direct stimulation of the cortex of awake humans undergoing brain surgery. Wilder Penfield and Herbert Jasper (1954) at the Montreal Neurological Institute carried out such pioneering work in the 1940s. These surgeons took advantage of the fact that the cortex is exposed during surgery to remove damaged brain tissue, and systematically explored the effects of small levels of electrical current applied to the cortical surface. Panel **(a)** shows the exposed cortex of an epileptic patient and panel **(b),** a gridwork of electrodes laid over the surface for stimulation and recording; stimulation during surgery can be done with a single electrode rather than a grid. Because there are no pain receptors in the central nervous system (CNS), patients do not experience any discomfort from stimulation. Thus, stimulation can be applied even when they are awake and fully conscious, enabling researchers to gather the patients' subjective experiences—a relative impossibility in animal studies.

In their studies, Penfield and his associates found a topographical correspondence between cortical regions and body surface with respect to somatosensory and motor processes. This correspondence is represented in panel **(c)** by overlaying drawings of body parts on drawings of coronal sections of the motor and somatosensory cortex. These coronal sections are from the regions indicated by the color codes in the lateral view of the whole brain at the top of panel (c) (only the left hemisphere is shown here, representing the right body surface). The resulting map of the body surface on the cortex is sometimes called a *homunculus,* referring to the fact that there is an organized representation of the body across a given cortical area. Note that there is an indirect relation between the actual size of body parts and the cortical representation of the body's parts. For example, areas within the motor homunculus that activate muscles in the fingers, mouth, and tongue are much larger than would be expected from proportional representation. The large drawings of the fingers and mouth indicate that large areas of cortex are involved in the fine coordination required when we manipulate objects or speak.

Is the representation of the homunculus in the figure correct? Recent evidence from brain imaging studies using a technique described in Chapter 4, functional magnetic resonance imaging (fMRI), suggests that it may not be. Ravi Menon and his colleagues (Servos et al., 1999) in Canada stimulated the foreheads and chins of healthy volunteers while their brains were being scanned. In contrast to the results of the electrical stimulation studies, they found that stimulating the forehead produced activity in a region that was below (inferior) the region for activity related to chin stimulation, the reverse of the drawing in the figure based on the work of Penfield and his colleagues. If the latter pattern from neuroimaging turns out to be accurate, this will constitute a dramatic example of scientific revisionism.

(c) Adapted from Penfield and Jaspers, 1954, and Ramachandran, 1993.

the brain enclosed within the skull's protective bony structure, but also it is surrounded by the tough **dura mater,** which is dense layers of collagenous fibers.

After the dura mater is removed, superficial examination reveals many prominent structures (Figure 3.1). The **gyri** (the protruding rounded surfaces) and primary **sulci** (the enfolded regions that appear as lines and creases) of the cerebrum, the gradual narrowing of the brainstem, and the elaborate folding of the cerebellar cortex can be identified without the aid of a microscope. Further dissections expose internal structures and reveal organizational principles. For example, slicing through the brain reveals the dichotomy of **gray matter** and **white matter:** The gray matter forms a continuous cortical sheath enshrouding a seemingly homogeneous mass of white matter. Gray matter is so named

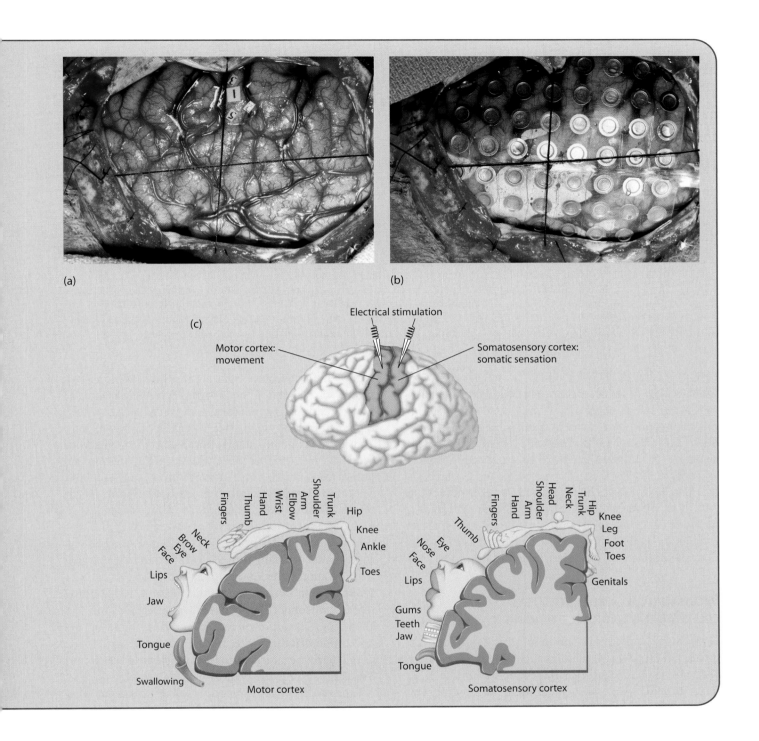

(a)

(b)

(c)

Electrical stimulation

Motor cortex: movement

Somatosensory cortex: somatic sensation

Fingers Thumb Hand Wrist Elbow Arm Shoulder Trunk Hip

Neck Brow Eye Face Lips Jaw Tongue Swallowing

Knee Ankle Toes

Motor cortex

Fingers Hand Arm Shoulder Head Neck Trunk Hip

Thumb Eye Nose Face Lips Gums Teeth Jaw Tongue

Knee Leg Foot Toes Genitals

Somatosensory cortex

because it appears darker and even a bit grayish in preserved brains, although it is more typically pinkish to reddish in the living brain, owing to its vascularization. The gray matter contains cell bodies of neurons and glia. The white matter is so termed because it looks lighter than the gray matter in preserved tissues; it actually appears quite milky white in the living brain. This coloring is due to the fatty myelin surrounding the axons.

Neuroanatomical dissection of the brain can provide a relatively detailed description of the organization of the major systems, structures, and connections of the brain. For example, gross dissection will reveal the inputs and outputs from peripheral sensory structures or motor effectors (muscles), via nerve bundles. It is also possible to follow the trajectory of axons grouped in large fiber **tracts** (bundles of axons) as they course through the

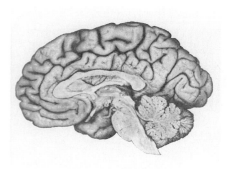 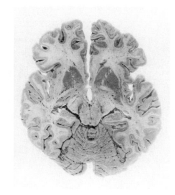 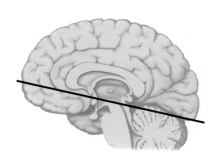

(a)   (b)

Figure 3.1 **(a)** Photograph of a midsagittal section through the human cerebral cortex. This medial view of the right hemisphere of a postmortem brain reveals prominent features of the gross anatomy, including cortical, subcortical, cerebellar, and brainstem regions. The frontal pole is on the left. On the surface of the cortex (top portions mostly), the crowns of the various cortical gyri can be seen; the enfolded regions of the cortex, the sulci, appear as lines or creases. **(b)** A photograph of a horizontal (also called axial) section through the human brain at the level diagrammed in the drawing at the right (dark line). The left and right cerebral hemispheres (frontal pole now at the top), the underlying subcortical structures, and portions of the cerebellum (bottom) can be seen in the horizontal section. The gray matter (darker regions) and white matter (lighter regions) are clearly visible.

brain and connect with different regions. Such dissection will show the geniculostriate projections from the thalamus to the visual cortex that we will touch upon later in this chapter and learn about in detail in Chapter 5. In another good example, dissection will reveal the arcuate fasciculus, a prominent fiber tract interconnecting Broca's and Wernicke's areas in support of language functions (see Chapter 9). These interconnections observable at the gross level are revealed in greater detail with tract tracing methods using chemical substances in animals and degeneration methods in animals and humans. Some of these methods are reviewed in the next section.

MICROANATOMY AND HISTOLOGY: CELL STAINING AND TRACT TRACING

If we simply looked at the brain in gross dissections, we would not know that structures such as the white matter were neural elements rather than supportive tissues. To make this inference, neuroanatomists must probe for finer detail with high-power microscopes. For example, microscopic examination reveals that the white matter is actually composed of millions of individual fibers, each surrounded by myelin. This level of analysis is the domain of the histologist. *Histology* is the study of tissue structure through microscopic techniques like those introduced in Chapter 1 in the work of the early neuroanatomists Cajal, Golgi, and Purkinje.

A primary concern for neuroanatomists is to identify the patterns of connectivity in the nervous system in order to lay out the neural "highways" that allow information to get from one place to another. This problem is made com-

plex by the fact that neurons are not wired together in a simple, serial circuit. A single cortical neuron is likely to be innervated by many neurons, and the axons from these input neurons can originate in widely distributed regions. That is, there is tremendous convergence in the nervous system as well as divergence, where a single neuron can project to multiple target neurons in different regions. Most axons are short projections from neighboring cortical cells. Others can be quite long, having originated in more distant cortical regions and only reaching their target after descending below the cortical sheath into the white matter, traveling through long fiber tracts, and then entering another region of cortex, subcortical nucleus, or spinal layer to synapse on another neuron. Neighboring and distant connections between two cortical regions are referred to as *cortico-cortical connections*, following the convention that the first term identifies the source and the second term the target. Inputs that originate in subcortical structures such as the thalamus would be referred to as *thalamo-cortical connections;* the reverse are *cortico-thalamic,* or more generally, *cortifugal projections* (projection from the cortex outward toward the periphery).

Much of the progress in neuroanatomy has been prompted by the development and refinement of new *stains,* chemicals of various sorts that are selectively absorbed by specific neural elements (see Chapter 1). These staining methods were discovered by trial and error or by intention, such as when the interaction of a chemical with tissues yielded an interesting pattern of staining that then became useful. There are different levels of analysis that one can pursue using cell staining. One can stain pieces of nervous tissue to reveal, for example, the laminar organization of the retina (Figure 3.2a), or zoom in on individual cells that have either taken up or been injected with intracellular stains that show the neural architecture (Figure 3.2b); these can be examined using light microscopic methods. At a higher level of resolution, one can examine subcellular organelles such as synaptic vesicles within an axon terminal using electron microscopy, which as the name suggests, uses a beam of electrons rather than light to "illuminate" neural tissues (Figure 3.2c). Another approach is to use fluorescent stains and ultraviolet light to visualize the structures that take up the stains (Figure 3.3).

Tract tracing methods permit the connections between different neurons and brain regions to be identified. An old but still effective technique is the degeneration method used to trace axonal pathways that are degenerating following damage or disease. At the simplest level, one can look for missing myelinated axons in damaged brains. At a more advanced level, the Marchi stain can be applied. It selectively stains dark the myelin in degenerating axons, a process that can result when the cell bodies die or the axons are cut off from the cell body. A more modern but now common method involves horseradish peroxidase (HRP), an enzyme. HRP can be used as a retrograde tracer in that it is taken up by the axons when it is injected at axon terminals and transported back to their cell bodies; it also can be used as an anterograde tracer, lightly taken up by cell bodies to stain axons. Thus, it provides a tool to visualize where the input to a particular neural region originates or where a region projects its axons (Figure 3.4). Suppose that a researcher wants to know which subcortical structures project to the primary visual cortex; she could use HRP as a retrograde transport tool. She injects the visual cortex's input layers with HRP. HRP is absorbed through the same axonal channels that allow neural transduction via the inflow and outflow of sodium, potassium, and calcium ions. Once inside, the HRP diffuses up the axon to the cell body. The animal is then killed and its brain removed. The HRP-injected tissue can then be treated with chemicals that turn the HRP a variety of colors from black, to blue or brown. Examining thin slices of brain tissue cut with a sharp mechanical knife called a *microtome*, the researcher can identify regions containing cell bodies labeled by HRP.

HRP is just one of many retrograde tracers. Other chemicals serve as anterograde tracers, in that they are absorbed at the dendrites or soma and then diffuse along the axons. One popular approach is to use radioactively labeled materials as traces, for example, radiolabeled amino acids. Autoradiography then can be used to visualize the pattern of staining. This technique involves slicing the labeled tissue and placing it on a photographic material to let the radioactivity expose the photographic emulsion, producing a picture of the distribution of label in the tissue. In combination, retro-

(a)

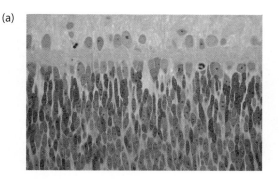

(b)

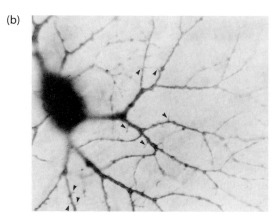

(c)

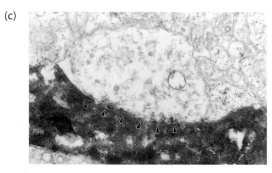

Figure 3.2 Neuroanatomical analysis at different scales. **(a)** The retina from a young ferret, stained and magnified by a light microscope. This section of the retina spans approximately 1.5 mm. **(b)** A ganglion cell from the cat retina that has been injected with stain to highlight its dendritic arborization. The magnification by light microscopy here is much greater than in (a). The arrowheads highlight places where short dendritic extensions, or spines can be seen. Processes along one dendritic branch can be visualized with the electron microscope. **(c)** An electron micrograph. The dark region is one of the stained dendrites of the cell shown in (b). The light regions are axons forming synapses along the dendrite. At this magnification, it is possible to see the synaptic vesicles in the axons. These contain neurotransmitter as described in Chapter 2.

grade and anterograde tracers allow researchers to identify the inputs to a specific region and determine where the axons from a particular region terminate. In this manner, the neuroanatomist can construct projection

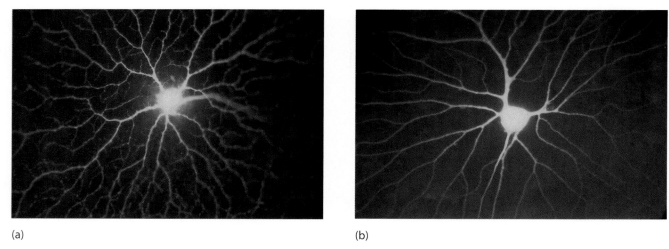

(a)

(b)

Figure 3.3 Fluorescent staining method. **(a)** The soma and dendritic arbor of a retinal neuron in the prenatal cat, 8 days before birth. **(b)** The soma and dendritic arbor of a neuron from this same structure in the adult cat. Many dendritic branches are eliminated during development and staining permits this developmental change to be visualized and quantified. Fluorescent stains come in a variety of types and may include adding a fluorescent molecule to an antibody or other molecule that binds with material in a neuron, or injecting a fluorescent material into a neuron and allowing it to diffuse throughout the cell.

maps of the patterns of connectivity for small and large circuits and systems.

Neuroanatomists have used histological techniques to classify neurons into groups based on their morphology (form). On close examination it is clear that despite the commonalities across neurons (most having cell bodies, axons, and dendrites), they are very heterogeneous, varying in size and shape. For some neurons such as the giant pyramidal cells in the cerebral cortex, the dendritic arbor is relatively small. In contrast, the Purkinje cells of the cerebellar cortex have vast arbors, allowing for more than 200,000 synaptic contacts (Figure 3.5—see also Chapter 2).

This level of detail was available early on to neuroanatomists. Recall that the silver impregnation stain introduced by Golgi allows an entire cell to be visualized, providing observers with the kind of details seen in the drawings in Figure 3.5. This mysterious procedure only stains, however, about 1% of the cells exposed to the stain. Such selectivity is advantageous in that it becomes possible to visualize individual cells without visual interference from their anatomical neighbors. With the Golgi technique and a host of other methods such as those using the Nissl stain, which stains rough endoplasmic reticulum (an intra-

Figure 3.4 Staining techniques reveal connections between distant neural structures. Horseradish peroxidase (HRP) is a commonly used tracer. When used as a retrograde label, HRP is taken up by axonal terminals and is transported back to the cell bodies. The animal is then killed and histological methods are used to identify the location of the stain. **(a)** Injection site in the lateral geniculate nucleus of the thalamus of a rat. **(b)** The flattened retina shown under very low-level magnification. **(c)** Individual neurons (the black spots, which are cell bodies of neurons filled with stain) under high magnification. If you look closely, you can see the shapes of many of these neurons and their dendrites.

(a)

(b)

(c)

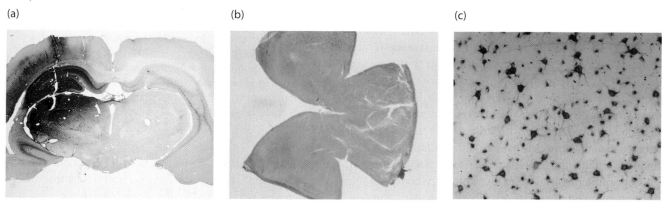

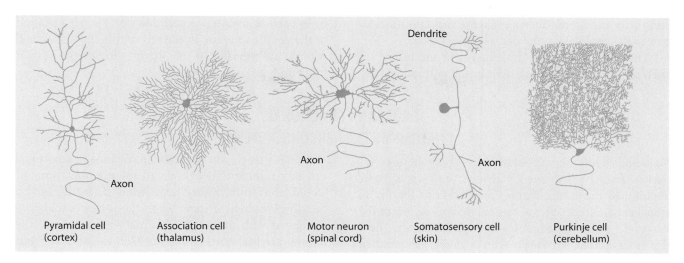

Figure 3.5 Five different neurons in the central and peripheral nervous systems. These neurons vary greatly in size (not drawn to scale): The axon of an association cell in the thalamus may extend less than 1 mm, whereas the axon of the pyramidal cell may traverse the length of the spinal cord.

cellular organelle, also referred to as *Nissl substance*), revealing the distribution of cell bodies (soma), or the Weigert stain, which selectively stains myelin, revealing the axons of neurons, a careful researcher can (painstakingly) catalogue the cell architecture and layering patterns of different brain regions. Indeed, as noted in Chapter 1, and described in more detail later in this chapter, Korbinian Brodmann used this technique to devise his **cytoarchitectonic map** of the brain. He methodically applied stains to samples from the entire surface of the cortex and obtained a picture of regional variations in cell architecture. Brodmann partitioned cortical areas according to these differences in cell morphology, density, and layering (Figure 3.6)

Figure 3.6 The gray matter of the cerebral cortex is composed of unmyelinated cell bodies that give a layered appearance as a function of the different cell types and their groupings in cross-sectional views (cortical surface at top). As shown in these examples from the macaque monkey, across different cortical areas, the density and layering of the cell types vary **(top)**. Brodmann used these variations in density to define the boundaries between different cortical areas. An example shows a cross-section through the visual cortex **(bottom)**. The arrows show the border between two cytoarchitectonically defined Brodmann areas: Note the change in the pattern of layering. Adapted from McClelland and Rummelhart (1986).

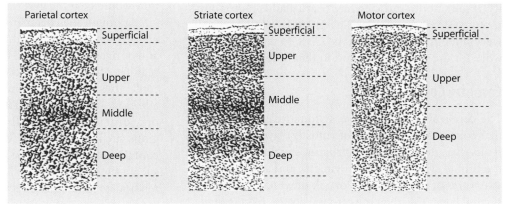

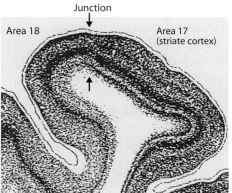

and introduced a numbering scheme that continues to be widely employed in the neurosciences today.

The methods that we described here and others related to these general neuroanatomical approaches have been used to investigate and describe the organization of the nervous system. In the rest of this chapter we will provide an overview of the organization of the human brain and nervous system.

GROSS AND FUNCTIONAL ANATOMY OF THE NERVOUS SYSTEM

Signaling in the nervous system occurs along well-defined pathways, and via specific anatomical relays to circumscribed areas of the brain, spinal cord, and peripheral musculature. A review of the nervous system's anatomical organization clarifies how this is organized. We begin with a global view and in later chapters focus on specific anatomical and physiological systems relevant to specific sensory, cognitive, and motor processes.

The nervous system has two major subdivisions: the **central nervous system (CNS)** consisting of the brain and spinal cord, and the **peripheral nervous system (PNS),** consisting of everything outside the CNS. The CNS can be thought of as the command-and-control portion of the nervous system. The PNS represents a courier network that delivers sensory information to the CNS and then carries the motor commands of the CNS to the muscles, to control the voluntary muscles of the body (somatic motor system) and the involuntary activity of the smooth muscles, heart, and glands (autonomic motor system). In most of the remainder of this chapter, we will focus on describing the CNS, to lay the groundwork for the studies of cognition that follow.

Cerebral Cortex

The **cerebral cortex** has two symmetrical hemispheres that consist of large sheets of (mostly) layered neurons. It sits over the top of core structures including parts of the limbic system and basal ganglia and surrounds the structures of the diencephalon, all of which will be considered later: Together, the cerebral cortex, basal ganglia, and diencephalon form the forebrain. The term *cortex* means "bark," as in tree bark, and in higher mammals and humans it contains many infoldings or convolutions (Figure 3.7). As noted earlier, the infoldings of the cortical sheet are further defined as sulci (the enfolded regions) and gyri (the crowns of the folded tissue that one observes when viewing the surface). Many mammal species, such as the rat or even New World monkeys like the owl monkey, have smoother, less folded cortices with few sulci and gyri.

The folds of the human cortex serve a functional purpose: to pack more cortical surface into the skull. If the human cortex were smoothed out to resemble that of the rat, for example, humans would need to have very large heads. There is about a one-third savings in space when the cortex is folded as compared with unfolded. The total surface area of the human cerebral cortex is about 2200 to 2400 cm^2, but because of the folding, about two-thirds of this area is confined within the depths of the sulci. Another advantage of having a highly folded cortex is that neurons are brought into closer three-dimensional relationships to one another, saving axonal distance and hence neuronal conduction time between different areas. This savings occurs because the axons that make long-distance cortico-cortical connections run under the cortex through the white matter, and do not follow the foldings of the cortical surface in their paths to distant cortical areas. In addition, by folding, the cortex brings some nearby regions closer together; for example, the opposing layers of cortex in each gyrus are in closer linear proximity than they would be if the gyri were flattened.

Although the cortex is composed of several cell layers in most regions, its thickness averages only 3 mm and ranges from 1.5 to 4.5 mm in different cortical regions. The cortex itself contains the cell bodies of neurons, their dendrites, and some of their axons. In addition, the cortex includes axons and axon terminals of neurons projecting to the cortex from other brain regions such as the subcortical thalamus. The cortex also contains blood vessels. Because the cerebral cortex has such a high density of cell bodies, it appears grayish in relation to underlying regions that are composed primarily of axons of neurons and appear slightly paler or even white. As described earlier, for this reason anatomists used the terms *gray matter* and *white matter* when referring to areas of cell bodies and axon tracts, respectively. The latter tracts represent the billions of axons that connect the neurons of the cerebral cortex to other locations in the brain (Figure 3.8).

ANATOMICAL SUBDIVISIONS

The cerebral hemispheres have four main divisions or lobes, and a fifth if you consider that the limbic system is sometimes referred to as the *limbic lobe,* as described

(a)

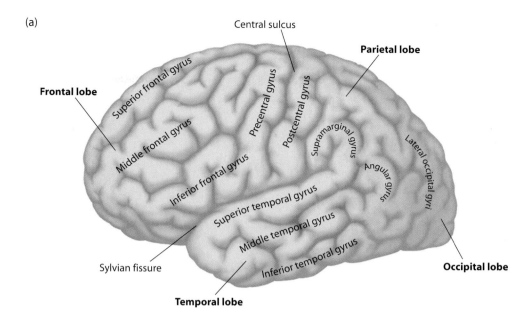

(b)

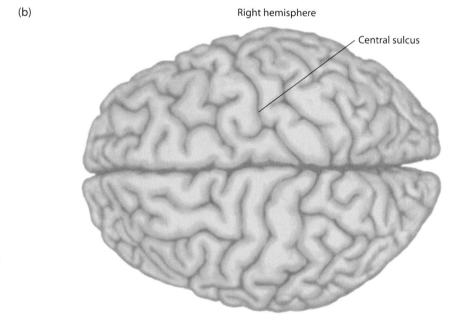

Figure 3.7 Lateral view of the left hemisphere **(a)** and dorsal view of the cerebral cortex **(b)** in humans. The major features of the cortex include the four cortical lobes and various key gyri. Gyri (singular is *gyrus*) are separated by sulci (singular is *sulcus*) and result from the folding of the cerebral cortex that occurs during development of the nervous system, to achieve an economy of size.

later. These regions have different functional properties and can be distinguished from one another by anatomical landmarks, principally sulci. The names of the brain areas were derived from names originally given to the overlying skull bones; for example, the temporal lobe lies underneath the temporal bone. The temporal bone derived its name from the graying of hair overlying the temporal bone—a sign of passing time if there ever was one.

The four lobes are the frontal, parietal, temporal, and occipital lobes (Figure 3.9). The central sulcus (singular of sulci) divides the **frontal lobe** from the **parietal lobe,** and the lateral fissure (sometimes referred to as the *Sylvian fissure*) separates the **temporal lobe** from the frontal and parietal lobes. The **occipital lobe** is demarcated from the parietal and temporal lobes by the parieto-occipital sulcus on the brain's dorsal surface and the preoccipital notch located on the ventral-lateral surface. The left and right cerebral hemispheres are separated by the interhemispheric fissure (also called the *longitudinal fissure*) that runs from the rostral to the

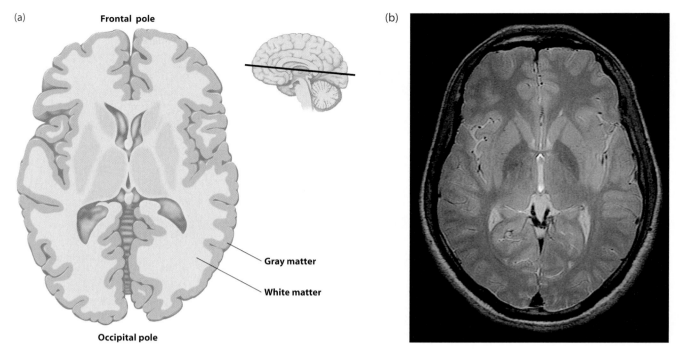

Figure 3.8 **(a)** Horizontal section through the cerebral hemispheres at the level indicated at upper right. White matter is composed of myelinated axons and gray matter is composed primarily of neurons. From this diagram you can see that the gray matter on the surface of the cerebral hemispheres forms a continuous sheet that is heavily enfolded. **(b)** High-resolution structural magnetic resonance image in a similar plane of section in a living human. This T2 image was obtained on a 4-tesla scanner (a high-magnetic-field scanner) using a 512 × 512 square matrix for acquisition. Note that on T2 images the white matter appears darker than the grey matter. The skull and scalp can be seen here but are not shown in (a). (a) Adapted from DeArmond et al. (1976). (b) Courtesy of Dr. Allen Song, Duke University.

caudal end of the forebrain. Interconnections between the cerebral hemispheres are accomplished by axons from cortical neurons that travel through the **corpus callosum,** which represents the largest white matter

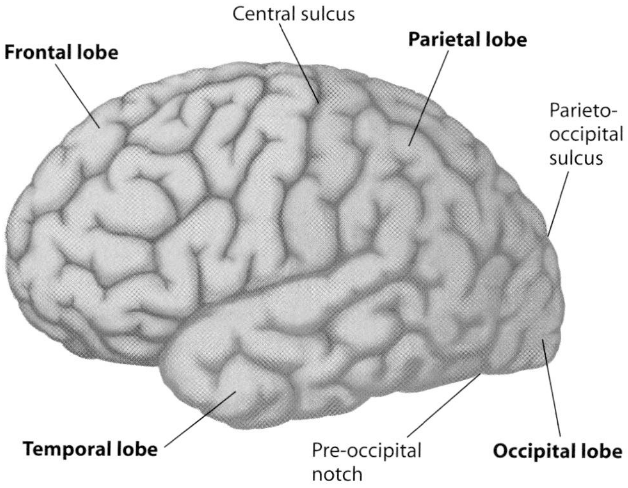

Figure 3.9 Four lobes of the cerebral cortex, in lateral view of the left hemisphere. See text for details.

commissure in the nervous system (*commissure* is a special term for the white matter tracts that cross from the left to the right side, or vice versa, of the CNS). The term *corpus callosum* means "hard body," so named because of its tough consistency. Indeed, very early anatomists believed that the corpus callosum served a structural function in supporting the cerebral hemispheres because it prevented them from collapsing onto structures below; this is not correct. As we discuss later in the book, the corpus callosum carries out valuable integrative functions for the two hemispheres.

CYTOARCHITECTURE

The cerebral cortex can be divided more finely than the four or five main lobes, in various ways. For example, it can be divided according to functional subdivisions of the cortex. But there are other, more purely anatomical criteria for subdividing the cortex: One is by the microanatomy of cell types and their organization. This is generally referred to as *cytoarchitectonics*—*cyto* means "cell" and *architectonics* means "architecture"—and has to do with how cells in a region appear morphologically and are arranged with respect to each other. Cytoarchitectonic investigations entail performing detailed histological analysis of the tissue from different regions of the cerebral cortex as described earlier in the chapter. The goal is to define the extent of regions wherein the cellular architecture looks similar and therefore might indicate a homogeneous area of the cortex that represents, perhaps, a functional area. This work began in earnest with Korbinian Brodmann at the turn of this century (see Chapter 1).

Brodmann (1909) identified approximately fifty-two regions of the cerebral cortex. These areas, categorized according to differences in cellular morphology and organization, were numbered (Figure 3.10). Other anatomists further subdivided the cortex into almost 200 cytoarchitectonically defined areas, but many classified transition zones as separate areas when perhaps they should not be considered so. A combination of cytoarchitectonic and functional descriptions of the cortex is probably most effective in dividing the cerebral cortex into meaningful units; this type of work will likely continue into the foreseeable future because we are only beginning to learn the cerebral cortex's functional organization. In the sections that follow, we use Brodmann's numbering system to describe the cerebral cortex as well as anatomical names (e.g., superior temporal gyrus).

The Brodmann system often seems unsystematic. Indeed, the numbering has more to do with the order in which Brodmann sampled a region than with any meaningful relation between areas. Nonetheless, in some regions the numbering system has a rough correspondence with the relations between areas that carry out similar functions, such as vision (i.e., areas 17, 18, and 19). It is worth noting here that the nomenclature of cortex (and indeed the nervous system) is not uniform. Hence, a region might be referred to by its Brodmann name, a cytoarchitectonic name, a gross anatomical name, or a functional name, and the latter changes rapidly as new information is gathered. For example, let's consider the naming of the first area in the cortex to receive visual inputs from the thalamus—the primary sensory cortex for vision. The Brodmann name is *area 17,* the gross anatomical name is *calcarine cortex* (the cortex surrounding the calcarine fissure), the cytoarchitectonic name is *striate cortex* (owing to its striated appearance under the microscope), and its functional name is *primary visual cortex,* which has been labeled area *V1* (for visual area 1) based on studies of the primate visual system, primarily in monkeys. The choice here of primary visual cortex as an example was fortuitous because all these different terms refer to the

Figure 3.10 **(a)** Brodmann's original cytoarchitectonic map from his work around the turn of the twentieth century. Different regions of cortex have been demarcated by histological examination of the cellular microanatomy. Brodmann divided the cortex into about fifty-two areas. **(b)** Lateral view of the left hemisphere showing Brodmann areas. Over the years, the map has been modified, and the standard version no longer includes some areas. **(c)** Medial view of the right hemisphere showing Brodmann's areas. Brodmann's areas are mostly symmetrical in the two hemispheres.

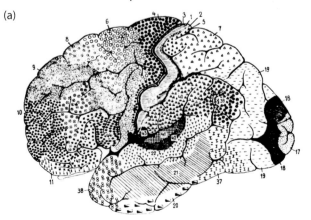

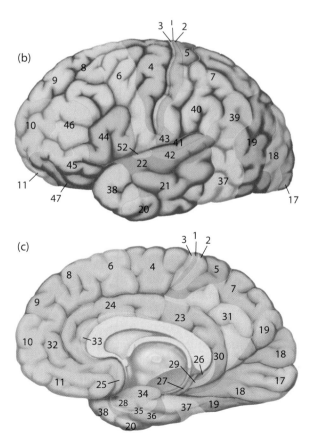

The Chambers of the Mind

We have understood for many decades that neurons in the brain are functional units, and that how they are interconnected yields specific circuits for the support of particular behaviors. Centuries ago, early anatomists, believing that the head contained the seat of behavior, examined the brain to see where the conscious self (soul, if you wish) was located. They found a likely candidate: Some chambers in the brain seemed to be empty (except for some fluid) and thus were possible containers for higher functions. These chambers are called *ventricles* (see Figure 3.8). What is the function of these chambers within the brain?

The brain weighs a considerable amount but has little or no structural support—there is no skeletal system for the brain. To overcome this potential difficulty, the brain is immersed in a fluid called *cerebrospinal fluid* (CSF). This fluid allows the brain to float to help offset the pressure that would be present if the brain were merely sitting on the base of the skull. CSF also reduces shock to the brain and spinal cord during rapid acceler-

ations or decelerations, such as when we fall or are struck on the head.

The ventricles inside the brain are continuous with the CSF surrounding the brain. The largest of these chambers are the lateral ventricles, which are connected to the third ventricle in the brain's midline. The cerebral aqueduct joins the third to the fourth ventricle in the brainstem below the cerebellum. The CSF is produced in the lateral ventricles and in the third ventricle by the choroid plexus, an outpouching of blood vessels from the ventricular wall. Hence, CSF is similar to blood, being formed by the transport of what resembles an ultrafiltrate of blood plasma; essentially, CSF is a clear fluid containing proteins, glucose, and ions, especially potassium, sodium, and chloride. It slowly circulates from the lateral and third ventricles through the cerebral aqueduct to the fourth ventricle and on to the subarachnoid space surrounding the brain, to be reabsorbed by the arachnoid villi in the sagittal sinus (the large venous system located between the two hemispheres on the dorsal surface).

---

same cortical area. But for much of the cortex this is not the case, and different nomenclatures may not refer to precisely the same area with a one-to-one mapping. Area 18 of the visual system, for example, is not synonymous with V2 (for visual area 2).

It is also possible to subdivide the cerebral cortex based on the general patterns of layering (Figure 3.11). Most of cortex is composed of **neocortex,** which typically contains six main cortical layers with a high degree of specialization of neuronal organization. Neocortex is composed of areas like primary sensory and motor cortex and association cortex (areas not obviously primary sensory or motor), although some models would make distinctions between primary sensory and motor cortex and association cortex. For the purposes of this text, we will refer to them collectively as *neocortex. Mesocortex* is a term for the so-called paralimbic region, which includes the cingulate gyrus, parahippocampal gyrus, insular cortex, and the orbitofrontal cortex, all of which will be defined later. Mesocortex is interposed between neocortex and allocortex and has six layers. *Allocortex* typically has only one to four layers of neurons and in-

cludes the hippocampal complex (sometimes referred to as *archicortex*) and primary olfactory cortex (sometimes referred to as *paleocortex*). The take-home message here is that the cerebral cortex can be subdivided into major regions that differ based on the degree of complexity of the neuronal layering.

FUNCTIONAL DIVISIONS

The lobes of the cerebral cortex have a variety of functional roles in neural processing. Major identifiable systems can be localized within each lobe, but systems of the brain also cross different lobes. That is, these brain systems do not map one-to-one onto the lobe in which they primarily reside, but in part the gross anatomical subdivisions of the cerebral cortex can be related to different sensory-motor functions. Cognitive brain systems have as a principal organizational feature the fact they are often composed of networks whose individual elements are located in different lobes of the cortex. Finally, most functions in the brain rely on both cortical and subcortical components, whether they are sensory,

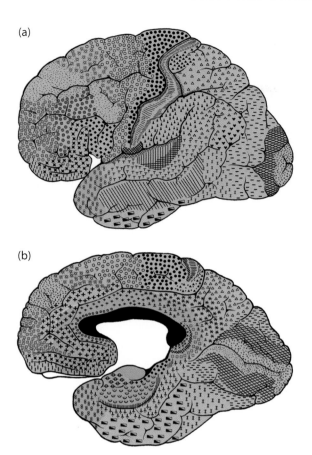

(a)

(b)

Figure 3.11 Cerebral cortex color coded to show the regional differences in cortical layering that specify different types of cortex. **(a)** The lateral surface of the left hemisphere. **(b)** The medial surface of the right hemisphere. Neocortex is shown in red, mesocortex in blue, and allocortex in green. Adapted from Mesulam (2000).

motor, or cognitive processes. Because one of the goals of this book is to review what we know about the functional localization of higher cognitive and perceptual processes, what follows is a beginner's guide to the cortex's functional anatomy.

Motor Areas of the Frontal Lobe   The frontal lobe plays a major role in the planning and execution of movements. There are two main subdivisions of the frontal lobe, the motor cortex and the prefrontal cortex. The motor cortex includes the precentral gyrus, which is also called the *motor strip* (Brodmann's area 4) and is located just anterior to the central sulcus. The precentral gyrus represents the primary motor cortex (motor area 1, or MI). Anterior to this area are two more main motor areas of cortex (within Brodmann's area 6), the premotor cortex on the lateral surface of the hemisphere and the supplementary motor cortex that lies dorsal to the premotor area and extends around to the hemi-

sphere's medial surface. These motor cortical areas contain motor neurons whose axons extend to the spinal cord and brainstem and synapse on motor neurons in the spinal cord. The motor neurons, located in the output layer 5 of the motor cortex, have fascinating specializations. In particular, layer 5 of the primary motor cortex contains large pyramidal neurons known as *Betz's cells,* named after Vladimir Aleksandrovich Betz who described them. They are the largest neurons in the cerebral cortex, reaching 60 to 80 microns in diameter at the cell body.

The most anterior region of the frontal lobe, the **prefrontal cortex,** takes part in the higher aspects of motor control and the planning and execution of behavior, tasks that require the integration of information over time. The prefrontal cortex has three or more main areas that are commonly referred to in descriptions of the gross anatomy of the frontal lobe (although different definitions can be used): the dorsolateral prefrontal cortex, which is found on the lateral surface of the frontal lobe anterior to the premotor regions and has been implicated in working memory functions; the orbitofrontal cortex (Figure 3.12); and the anterior cingulate and medial frontal regions (not visible in Figure 3.12). The orbitofrontal cortex is located on the frontal lobe's anterior-ventral surface and extends medially to limbic lobe structures, with which it maintains interconnectivity.

Somatosensory Areas of the Parietal Lobe   The somatosensory cortex is in the postcentral gyrus and adjacent areas (Brodmann's areas 1, 2, and 3). These cortical regions receive inputs from the somatosensory relays of the thalamus and represent information about touch, pain, temperature sense, and limb proprioception (limb position). The primary somatosensory cortex (or SI) is immediately caudal to the central sulcus, and a secondary somatosensory cortex (SII), receiving information via projections primarily from SI, is located ventrally to SI. Somatosensory inputs projecting to the posterior parietal cortex arise from SI and SII. Somatosensory information coming into the thalamus and then to the primary somatosensory cortex traverses two main pathways: the anterolateral system for pain and temperature sense, and the dorsal column–medial lemniscal system for information about touch, proprioception, and movement (Figure 3.13). Receptor cells in the periphery transduce physical stimuli into neuronal impulses conducted to the spinal cord and toward the brain, making synaptic connections at relay sites along the ascending pathway. The two systems for somatosensory information take slightly different paths in the spinal cord, brainstem, and midbrain on their route to the thalamus, and thence the cortex.

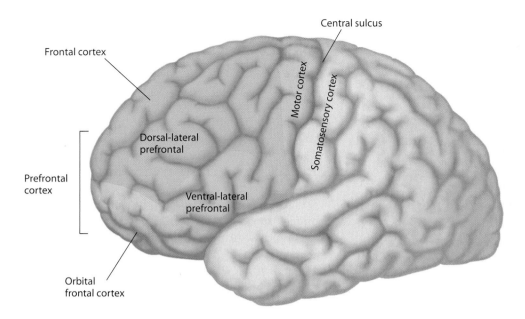

Figure 3.12 Divisions of the frontal cortex. The frontal lobe contains both motor and higher-order association areas. For example, the prefrontal cortex is involved in executive functions, memory, and other processes.

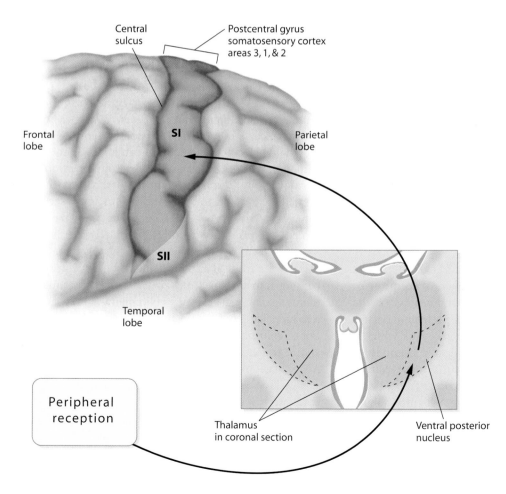

Figure 3.13 The somatosensory cortex, which is located in the postcentral gyrus. Inputs from peripheral receptors project via the thalamus (shown in cross section) to the primary somatosensory cortex (SI). Secondary somatosensory cortex (SII) is also shown.

Visual Processing Areas in the Occipital Lobe The primary visual cortex (also known as striate cortex, V1, or Brodmann's area 17) receives visual inputs relayed from the lateral geniculate nucleus of the thalamus (Figure 3.14). In humans, the primary visual cortex is primarily on the medial surface of the cerebral hemispheres, extending only slightly onto the posterior hemispheric pole. Thus, most of the primary visual cortex is effectively hidden from view, between the two hemispheres. The cortex in this area has six layers; it begins the cortical coding of visual features like color, luminance, spatial frequency, orientation, and movement.

Visual information from the outside world is processed by multiple layers of cells in the retina and transmitted via the optic nerve to the lateral geniculate nucleus of the thalamus, and thence to V1, a pathway often referred to as the *retino-geniculo-striate,* or *primary visual pathway.* Note that visual projections from the retina also reach other subcortical brain regions by way of secondary projection systems. The superior colliculus of the midbrain is the main target of the secondary pathway and participates in visuomotor functions such as eye movements. Later, in Chapter 7, we will review the role of the cortical and subcortical projection pathways in visual attention.

Surrounding the striate cortex is a large visual cortical region called the *extrastriate* ("outside the striate") visual cortex (sometimes referred to as the *prestriate cortex* in monkeys, to signify that it is anatomically anterior to the striate cortex). The extrastriate cortex includes Brodmann's areas 18 and 19. From physiological recordings and anatomical studies in monkeys, it is now known that there are roughly three-dozen distinct visual areas in the primate extrastriate cortex. These visual areas contain partially redundant spatial maps of the visual world, but each is specialized to analyze specific aspects of a scene, such as color, motion, location, and form. Two pathways from the striate cortex to extrastriate regions convey prominent streams of information (Figure 3.15). One pathway flows from V1 to the temporal lobe (ventral pathway

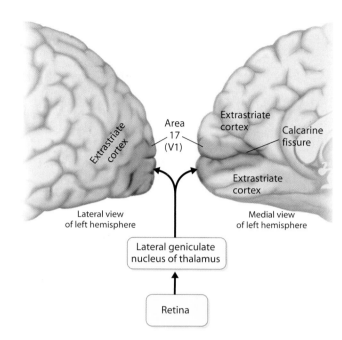

Figure 3.14 The visual cortex, which is located in the occipital lobe. Area 17 of Brodmann, also called the *primary visual cortex* (V1), is located at the occipital pole and extends onto the medial surface of the hemisphere, where it is largely buried within the calcarine fissure.

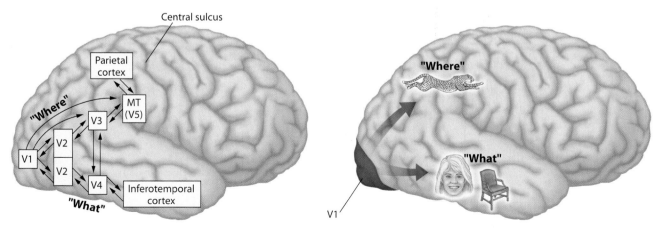

Figure 3.15 The two main projection routes from the primary visual cortex (V1) to visual areas in the extrastriate cortex: a dorsal "where" pathway that codes motion and location, and a ventral "what" pathway that processes detailed stimulus features, form, and object identity.

Navigating in the Brain

Because the brain is a complex three-dimensional object with numerous structures and pathways that are difficult to imagine in two-dimensional pictures, it is important to utilize conventions for describing the relations of regions. In general, the terms we use were derived from those used by anatomists to describe similar relations in the body as a whole, and therefore the brain's orientation with respect to the body determines the coordinate frame of reference that is used to describe anatomical relationships in the brain. But some confusing aspects of the terminology arise from differences in how the head and body are arranged in animals that walk on four legs versus humans, who are upright. Consider a dog's body surfaces. The front end is the rostral end, meaning "nose." The opposite end of this is the caudal end, the "tail." The back is the dorsal surface and the bottom surface is the ventral surface **(a, top).** We can now refer to the dog's nervous system by using the same coordinates **(a, bottom).** The part of the brain toward the front is the rostral end, toward the frontal lobes; the posterior end is the caudal end, toward the occipital lobe; and the top and bottom are the dorsal and ventral surfaces of the brain. This seems to be a reasonable set of conventions, or is it? Consider the human **(b).**

Humans are atypical, and thus create confusing problems with respect to anatomical nomenclature. The reason is simple: Humans stand upright and therefore tilt their heads down in order to be parallel with the ground. Thus, the dorsal surface of the body and brain are now at right angles to each other. But the conventions still apply, though there may be some confusion unless we remember that humans have tilted their heads. *Rostral* still means toward the frontal pole, while *caudal* still means toward the occipital pole as long as we are referring to the brain; however, when we discuss the spinal cord, the coordinate systems shift with respect to one another but not with respect to the local body axis. Thus, in the spinal cord, *rostral* means in the direction toward the brain, just as it does in the dog.

In humans, perhaps in part because of the differences in posture between humans and quadrupedal animals, we also use terms like *superior* and *inferior* to refer to the top of the brain and bottom, and *anterior* and *posterior* to refer to the front and back, respectively.

(a)

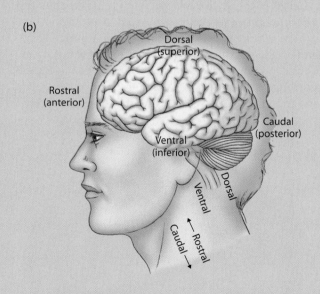

(b)

Anatomical terms for describing various views of anatomy and sections through brain and body. The relationship between a bipedal human and quadruped animal leads to some important considerations in describing the surfaces of the brain and spinal cord.

or "what" pathway) and conveys analysis of stimulus features and their conjunctions, and ultimately the information is used to carry out form discrimination and object identification. The other pathway projects from V1 toward the parietal lobe (dorsal or "where" pathway) and carries information about stimulus motion and localization within visual space, as described in detail in Chapters 5 and 6. Each of the multiple extrastriate areas maintains strong neural interconnec-

tivity with areas prior to it in the visual hierarchy (a reciprocal connectivity) and with other areas in the same processing stream. As well, many interconnections are between the dorsal and ventral visual processing streams. Indeed, interconnectivity of the visual cortex is complex but not random; that is, despite how it might appear on first look, not every area of the visual cortex is connected with every other (Figure 3.16).

Figure 3.16 Multiple representations of the visual world exist in extrastriate visual areas. Somewhere between thirty and thirty-five of these visual areas have been identified in monkeys, using a combination of anatomical and physiological methods. Each contains neurons performing specialized processing of the visual inputs. This representation shows how the cortex of a macaque monkey looks in normal perspective and when the cortex is flattened to aid in viewing the relationships of the various cortical areas. This flattened representation produced by David Van Essen and his colleagues includes areas **(right)** outside of the visual cortex as well as the visual areas **(left).**

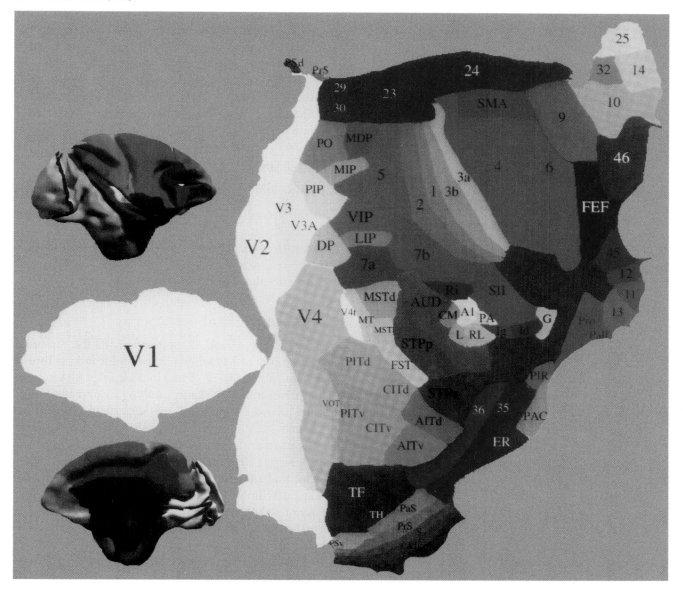

(a)

Auditory cortex

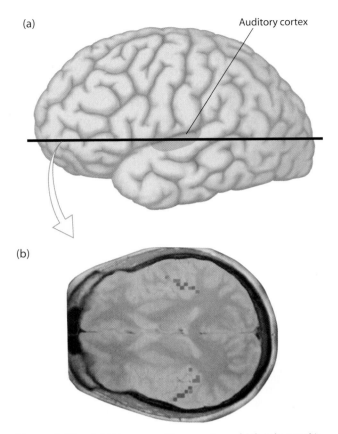

(b)

Figure 3.17 **(a)** Primary auditory cortex, which is located in the superior temporal lobe. The primary auditory cortex and surrounding association auditory areas contain representations of auditory stimuli and show a tonotopic organization. **(b)** The magnetic resonance image shows areas of the superior temporal region in horizontal section that have been stimulated by tones of different frequencies and showed increased blood flow as a result of neuronal activity. From Wessinger et al. (1997).

Auditory Processing Areas in the Temporal Lobe

The auditory cortex lies in the superior part of the temporal lobe and is buried within the sylvian fissure (also called *lateral fissure*) (Figure 3.17). The projection from the cochlea (the auditory sensory organ in the inner ear) proceeds through the subcortical relays to the medial geniculate of the thalamus, and then to the supratemporal cortex in a region known as *Heschl's gyri*. This region represents AI, the primary auditory cortex, and AII, the auditory association area surrounding it and posterior to the primary auditory cortex (Brodmann's areas 41 and 42). Area 22, which surrounds the auditory cortex, aids in the perception of auditory inputs; when this area is stimulated, sensations of sound are produced in humans. One can represent the sensory inputs to the auditory

cortex using a tonotopic map; the orderly representation of sound frequency within the auditory cortex can be determined with several tonotopic maps (see Chapters 4 and 5).

Association Cortex The volume of neocortex that is not sensory or motor has traditionally been termed the **association cortex,** which is composed of regions that receive inputs from one or more modalities. These regions have specific functional roles that are not exclusively sensory or motor. For example, take the visual association cortex. Though the primary visual cortex is necessary for the conscious sensation of vision, neither it nor the extrastriate cortex are the sole loci of visual perception. Regions of visual association cortex in the parietal and temporal lobes are important for correct perception of the visual world. As another example, the association areas of the parietal-temporal-occipital junction of the left hemisphere have a prominent role in language processing, while this region in the right hemisphere is implicated in attentional orienting. Thus, higher mental processes are the domain of the association cortical areas, in interaction with sensory and motor areas of cortex (Figure 3.18).

Limbic System, Basal Ganglia, Hippocampus, Hypothalamus, and Diencephalon

In the preceding sections we focused on the neocortex. Here we will consider the mesocortical and allocortical regions of the cerebrum. Then we will look at the subcortical structures of the basal ganglia and the diencephalon.

LIMBIC LOBE

We now take a look at the portions of the forebrain that are collectively known as the limbic lobe or **limbic system** (Figure 3.19). These include several structures that form a border (Latin, *limbus*) around the brainstem, named the *grand lobe limbique* by Paul Broca (see Chapter 1). Above the corpus callosum, a band of cortex known as the *cingulate gyrus* reaches from anterior to posterior. Together with the hypothalamus, the anterior thalamic nuclei and hippocampus constitute the "classical" limbic lobe (Figure 3.20). In the 1930s James Papez first suggested the idea that these structures were organized into a system for emotional behavior, which led to use of the term *Papez circuit*. Since his initial formulation, much has been learned about the structures

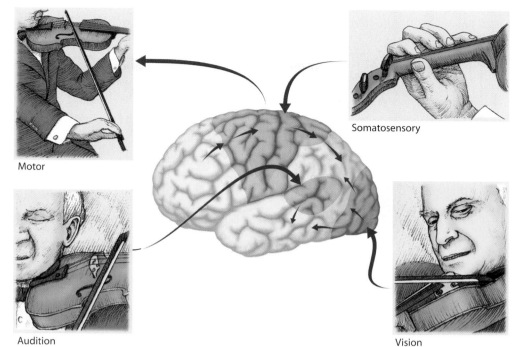

Figure 3.18 Primary sensory and motor cortex and surrounding association cortex. The blue regions show the primary cortical receiving areas of the ascending sensory pathways and the primary output region to the spinal cord. The secondary sensory and motor areas are colored green. The remainder is considered association cortex. Adapted from Kolb and Whishaw (1996).

participating in the limbic system, and today the **amygdala,** a group of neurons anterior to the hippocampus, is usually considered a key component, along with the orbitofrontal cortex and parts of the basal ganglia (described below, but not shown in Figure 3.20); in some descriptions, the medial dorsal nucleus of the thalamus is also included. The organization and role of the limbic system are described in more detail in Chapter 13.

The limbic lobe comes from a more primitive type of cortex (mostly), not the neocortex of the rest of the

cortical mantle. As described earlier, much of limbic lobe is considered mesocortex (e.g., cingulate gyrus) and allocortex (e.g., hippocampus). Limbic lobe structures are also phylogenetically older than the surrounding neocortex and are more prominent in the brains of nonmammalian species; for example, reptiles have little neocortex, whereas the primate brain is composed mostly of neocortex. The limbic system participates in emotional processing, learning, and memory. With each passing year we discover new functions of this system.

Figure 3.19 The limbic lobe as seen from a medial view of the right hemisphere. The structures that comprise the limbic system are in purple. These include the cingulate gyrus, the parahippocampal gyrus, and the subcallosal gyrus, as well as the dentate gyrus and hippocampal formation not visible in this view.

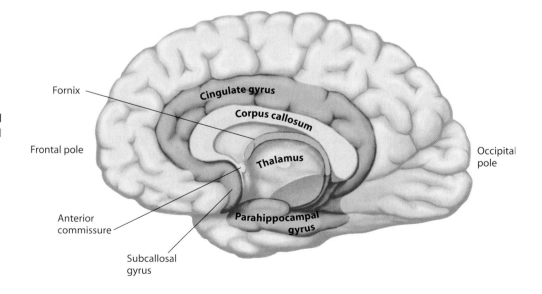

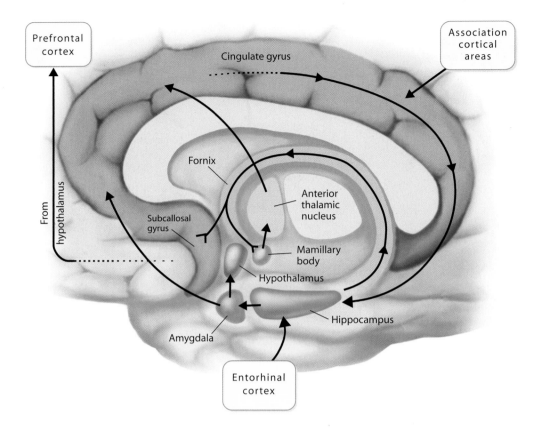

Figure 3.20 Major connections of the limbic system shown diagrammatically in a medial view of the right hemisphere. The figure zooms in on the region in purple in Figure 3.19. The basal ganglia are not represented in this figure, nor is the medial dorsal nucleus of the thalamus. There is more detail in this figure than needs to be committed to memory, but this figure provides a reference that will come in handy in later chapters. Adapted from Kandel et al. (1991).

THE BASAL GANGLIA

The **basal ganglia** are a collection of subcortical neuronal groups in the forebrain located beneath the anterior portion of the lateral ventricles (see below). The basal ganglia have a significant role in the control of movement. The three main subdivisions are the globus pallidus, caudate nucleus, and putamen (Figure 3.21). The caudate and putamen are referred to together as the *neostriatum* because they are phylogenetically the most recent of the basal ganglia to appear and are developmentally related.

Some anatomists have considered the amygdala and associated nuclei (amygdaloid complex) to be part of the basal ganglia, but this is not the convention agreed on by most neuroscientists today. Some say that both the subthalamic nucleus and the substantia nigra are part of the basal ganglia because of their similar neuronal structure and strong interconnectivity with the principal cell groups forming the basal ganglia. Yet the substantia nigra, at least, is generally considered part of the midbrain—actually located at the juncture of the midbrain and diencephalon—while the basal ganglia are in the forebrain. This gross anatomical distinction between the substantia nigra and the nuclear groups of the basal ganglia may not be as important as the mi-

croanatomical (cellular) and functional relationships, however. Perhaps, then, deciding how to classify these structures anatomically is less important than understanding their functional relations, an observation with general application in neural anatomy and physiology.

The basal ganglia, subthalamic nucleus, and substantia nigra participate in circuits with the cortex and thalamus to mediate aspects of motor control (both somatic motor and oculomotor systems), as well as cognitive functions such as the short-term memory processes of the dorsolateral prefrontal cortex, and some functions that can be called *executive functions* because they involve high-level control of behavior, something we will discuss later in this book.

The primary circuits projecting to the basal ganglia include a "cortico-striatal" projection that involves direct projections from all major cortical regions onto neurons in the caudate and putamen, which are the input structures of the basal ganglia. In addition, motor areas of cortex can project to the basal ganglia via the cell groups in the thalamus, and the subthalamic nucleus. The major outputs of the basal ganglia project from the globus pallidus to thalamic nuclei and thence to cortex, primarily motor and premotor cortex, as well as prefrontal cortex

(a)

Level of anterior commissure

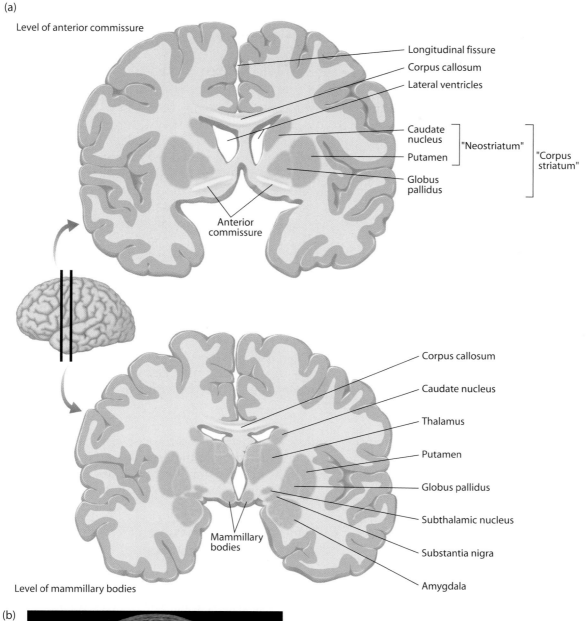

Longitudinal fissure
Corpus callosum
Lateral ventricles
Caudate nucleus
Putamen
Globus pallidus
"Neostriatum"
"Corpus striatum"
Anterior commissure

Corpus callosum
Caudate nucleus
Thalamus
Putamen
Globus pallidus
Subthalamic nucleus
Substantia nigra
Amygdala
Mammillary bodies

Level of mammillary bodies

(b)

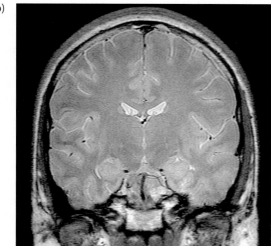

Figure 3.21 **(a)** Cross-sectional drawings through the brain at two anterior-posterior levels (as indicated) showing the basal ganglia. **(b)** Corresponding high-resolution, structural magnetic resonance image (4-tesla scanner) taken at the same level as the more posterior drawing. This image also shows the brainstem, and the skull and scalp, which are not shown in (a). (a) Adapted from Carpenter (1976). (b) Courtesy of Dr. Allen Song, Duke University.

Dancing to Death: Diseases of the Motor System

In 1982 Nancy Wexler and a team of scientists from the Huntington's Disease Foundation traveled to a small fishing village in Venezuela. There, Nancy observed a startling sight—a population full of people walking as though drunk, some with wildly swaying gaits, and grand hand gestures or simple repetitive movements of the body, limbs, and face. Many were unable to speak clearly, if at all. She recognized this disease because it had killed her mother, and it was her arch enemy—she too had a 50% chance of becoming like these people, as did her siblings. The disease is known as *Huntington's chorea* or *Huntington's disease*. Huntington's chorea (*chorea* means "dance") was first described by Huntington in the 1870s. He was a physician in a rural area of Long Island, New York, and he and his father, also a physician, had observed the disease over generations in certain families. It typically has an onset after the age of 40 (but can occur earlier, even in

childhood) and includes both motor and cognitive symptoms, which finally lead to death after progressive decline over a 10- to 20-year period. These symptoms are caused by cell losses in the basal ganglia, and probably also in part from damage to cortical neurons as a result of the basal ganglia cell loss and subsequent deterioration in the cortical-striatal circuitry. There is no cure. Thanks to research on the population living in Venezuela, however, there is now a genetic marker for the disease. Using molecular genetic techniques, we can determine whether a person carries the disease and will develop it in his or her lifetime. To date, although the Wexler family has dedicated itself to tracking down and curing this disease, none of the Wexler children has used the genetic techniques they helped to develop to look into their own futures. Nancy Wexler remains healthy and continues her phenomenal personal struggle against this terrible disease.

(Figure 3.22). Thus, the basal ganglia are not in a projection pathway from motor cortical areas to the spinal cord, to control muscular activity directly, but

instead are part of a cortical-subcortical motor loop that is thought to monitor aspects of how motor activity as well as nonmotoric functions are progressing (see Dancing to Death: Diseases of the Motor System).

Figure 3.22 Major inputs and outputs of the basal ganglia. The basal ganglia form a cortical-subcortical motor loop that monitors motor behavior. Adapted from Kandel et al. (1991).

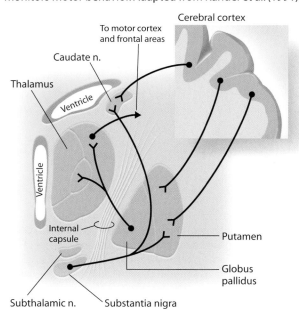

HIPPOCAMPAL FORMATION AND MEDIAL TEMPORAL LOBE

The region of the forebrain along the ventral medial surface of the temporal lobe contains the **hippocampus** and the associated areas of the dentate gyrus, parahippocampal gyrus, and entorhinal cortex—the latter being the anterior portion of the parahippocampal gyrus (Brodmann's area 28) (Figure 3.23). The hippocampus and dentate gyrus are composed of three- or four-layer cortex (archicortex), whereas entorhinal cortex and the parahippocampal gyrus is composed of a form of six-layer cortex in humans (although it is mesocortex, not neocortex).

The hippocampus has been subdivided into zones referred to as the *CA* (*cornu ammonis*) *fields*, divided into CA1, CA2, CA3, and CA4 based on differences in cellular morphology, connectivity, and development (Figure 3.24). The entorhinal cortex provides the main inputs to the dentate gyrus and hippocampus, via two projection pathways that terminate primarily on pyramidal neurons in the hippocampus. In turn, the entorhinal cortex receives

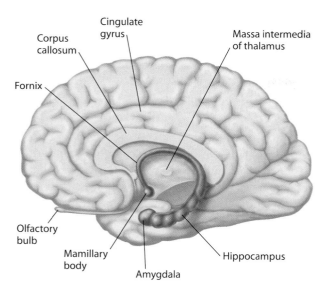

Figure 3.23 Anatomy of the hippocampal formation. The hippocampus is located on the inferior and medial aspect of the temporal lobe. Adapted from drawings provided courtesy of David Amaral.

Figure 3.24 Histological slide of a cross section through the hippocampus. The dentate gyrus (DG), entorhinal cortex (EC), and subiculum (S) can be seen, as well as cells of CA fields. The presubiculum (PrS) and parasubiculum (PaS) are also labeled. Courtesy of David Amaral.

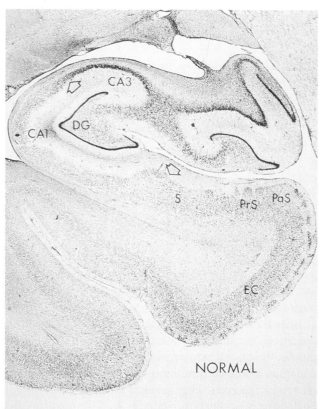

inputs from the cingulate cortex, and thus, a pathway to the hippocampus from the cingulate cortex appears to project via the entorhinal cortex. The hippocampal pyramidal neurons project out of the hippocampus via the fornix, a large white matter tract. Some of these fibers cross to the opposite hemisphere, but the majority make an arching projection, first posteriorly and then anteriorly running below the corpus callosum. Finally, this projection pathway dives ventrally and passes anterior to the thalamus, through the hypothalamus to make contacts with the mammillary bodies and some regions of the thalamus, and via a separate projection, to the medial septal area of the cortex. The hippocampus has been implicated in emotional processing (because of its interconnections with cingulate and mammillary bodies and participation in the limbic system) and memory.

DIENCEPHALON

The remaining portions of the forebrain to consider are the **thalamus** and **hypothalamus,** which together compose the diencephalon. These subcortical nuclei are composed of groups of specialized cells with interconnections to widespread brain areas.

Thalamus *Thalamus* is the Greek term for "inner chamber," even though it is not actually hollow. It lies at the most rostral end of the brainstem (Figure 3.25),

Figure 3.25 Gross anatomy of the thalamus. This diagram shows the thalamus of the left and right hemispheres in a see-through brain. The thalamus is egg shaped. It serves as the gateway to the cortex for the sensory systems and contains reciprocal loops with all cortical regions, organized according to subdivisions of the thalamus. The thalamus also is innervated by brainstem projection systems.

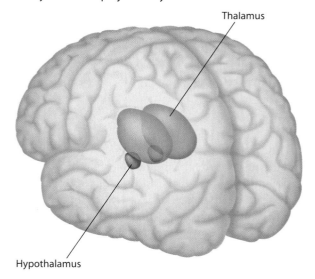

An Interview with David Amaral, Ph.D.
Dr. Amaral is a professor of psychiatry at the Center for Neuroscience, University of California, Davis, and the founding and current editor of the scientific journal *Hippocampus*.

Authors: You have done some of the pioneering work on the anatomy of the hippocampus and its function. How would you characterize how studying neuroanatomy informs our understanding of how the brain works?

DA: Understanding the intrinsic circuitry and the inputs and outputs of a brain region not only provides insight into how it functions, but also provides essential information on how it interacts with other brain regions. Understanding the unique neuroanatomical features of a brain region is particularly helpful in interpreting the unique neural computations that it is capable of carrying out.

Using the hippocampal formation as an example, we now know that virtually all of the sensory input that it receives arises from higher-order, multimodal cortical regions. This would indicate that whatever processing is done by the hippocampus in the service of forming long-term memories is accomplished with fairly abstract, gestalt-like representations of experience. It is also now clear that in addition to its subcortical connections, the hippocampal formation has massive return connections to the neocortex. This fits well with the emerging view that the hippocampal formation is not the final repository of long-term memories. Thus, a reasonable strategy for defining the storage sites would be to search the cortical areas that receive hippocampal inputs.

Authors: The intrinsic connections of the hippocampus have also been extensively studied. Why has this received so much attention?

DA: Well, the effects of hippocampal lesions on memory suggest the hippocampus ought to be a neural machine designed for forming associations. Indeed by studying the local connections, one can get a feel for how this structure goes about its business. It appears that by way of the extensive associational connections, hippocampal neurons can form essentially limitless networks to create representations of perceived experiences. We have learned that one of the cardinal features of the intrinsic anatomy of the hippocampal formation is the high degree of divergence and convergence of its stepwise connections. Unlike regions of primary sensory neocortex that demonstrate a point-to-point mapping of connections, connections originating in one portion of the hippocampal formation project to as much as half of the entire region of the next processing step. This is particularly true in the hippocampus proper where the vast majority of the inputs to a pyramidal cell in the area known as *CA3* originate from other CA3 neurons. Thus, the neuroanatomical fact of high levels of associational connections predicts that individual neurons can be addressed by myriad inputs; that is, their response properties are not hard wired as a neuron in V1 might be.

Authors: Lesions to most cortical structures are never as dramatically devastating as are those to the hippocampus. There always seems to be some sparing of function with cortical lesions. Why is that?

DA: A unique feature about the intrinsic hippocampal circuitry is that the major intrinsic connections are all unidirectional. This neuroanatomical feature suggests that information processing in the hippocampal formation follows an obligatory sequence of steps at which different computations are carried out. The

in the dorsal part of the diencephalon in each hemisphere, and is bordered medially by the third ventricle, dorsally by the fornix and corpus callosum, and laterally by the internal capsule—the projection fibers from the motor cortex to the brainstem and spinal cord—which separates the thalamus from the basal ganglia. In some people, the thalamus in the left hemisphere and the thalamus in the right hemisphere are connected by a bridge of gray matter called the *massa intermedia*.

The thalamus has been referred to as the "gateway to the cortex" because, with the exception of some olfactory inputs, all sensory modalities make synaptic relays in the thalamus before continuing to the primary cortical sensory receiving areas. The thalamus is divided into several nuclei that act as specific relays for incoming sensory in-practical ramification of this unique circuitry is that damaging any link in the chain can have devastating effects on function. An example of this is the amnesic syndrome that is produced in human patients who have suffered an ischemic loss confined to the CA1 field. These patients show a severe, anterograde amnesic syndrome despite the fact that the remainder of the hippocampal formation is intact.

Authors: Actually, the hippocampus was once viewed as the essential brain region involved with establishing memories. Now that view has undergone a significant change. Why has there been this change?

DA: The hippocampal formation was implicated in memory function because it was sensitive to a variety of neurological pathologies which resulted both in damage to the hippocampus and in impairment of memory (see Chapter 8). But the capricious nature of neuropathology couldn't be relied on to define all of the memory-related brain regions. So, again, one can come to appreciate the value of careful neuroanatomical studies. Such studies can define the total system of brain structures involved in a particular cognitive function. By starting at the hippocampal formation and examining its inputs and outputs, neuroanatomists have provided a much more comprehensive understanding of brain regions involved in memory. In the last 10 years, for example, studies carried out in the monkey have determined that the largest contributor of sensory information to the hippocampal formation comes from two regions of neocortex, the perirhinal and parahippocampal cortices, that lie adjacent to the hippocampal formation in the primate temporal lobe. These brain regions were ignored for many years by behavioral neuroscientists interested in the medial temporal lobe substrate of memory, and their damage during lesions of the hippocampus or amygdala was considered to be inconsequential to observed behavioral deficits. Spurred on by the newer neuroanatomical findings, however, behavioral and electrophysiological studies have demonstrated that the perirhinal and parahippocampal cortices not only play an important role in contributing sensory information to the hippocampal formation but also subserve some forms of memory function on their own, that is, independent of the hippocampal formation. Thus, neuroanatomical studies can provide clues as to which regions to study by more functionally oriented methodologies.

Authors: And what about the role neuroanatomy plays for those interested in more molecular approaches to the nervous system?

DA: The neuroanatomy of the last 20 years has been heavily involved in defining the chemical identity of neurons and pathways in the brain. The neuroanatomical fact that one of the highest brain densities of N-methyl-D-aspartate (NMDA) receptors is found in the CA1 field of the hippocampus has heightened interest in the role of this glutamate receptor in memory function. Chemical neuroanatomy is essential in defining the functional valence (excitatory versus inhibitory) of defined brain pathways and for determining potential modulators of cognitive function. Again, confirmation of the neuroanatomical predictions must rely on functional analyses such as electrophysiology and neuropharmacology. All in all, neuroanatomy provides essential information on the components and organization of brain systems that carry out particular cognitive functions. It can provide insight into the kinds of inputs the system is using and unique features of the information processing within the system. All of this information constrains hypotheses as to how the system functions. While I have used examples from studies of the hippocampal formation and memory, similar examples could be generated from the visual cortex, striatum, or even the spinal cord. In each case, neuroanatomy provides essential clues that facilitate the design and interpretation of functionally oriented studies.

---

formation. The lateral geniculate nucleus receives information from the ganglion cells of the retina and sends axons to the primary visual cortex, area 17 (Figure 3.26). Similarly, the medial geniculate nucleus receives information from the inner ear, via other brainstem nuclei in the ascending auditory pathway, and sends axons to the primary auditory cortex (AI). Somatosensory information projects via the ventral posterior (medial and lateral) nuclei of the thalamus and thence to primary somatosensory cortex in Brodmann's areas 1, 2, and 3.

Not only is the thalamus involved in relaying primary sensory information, but also it receives inputs from the basal ganglia, cerebellum, neocortex, and medial temporal lobe and sends projections back to these structures to create circuits involved in many different functions. One important structure within the thalamus

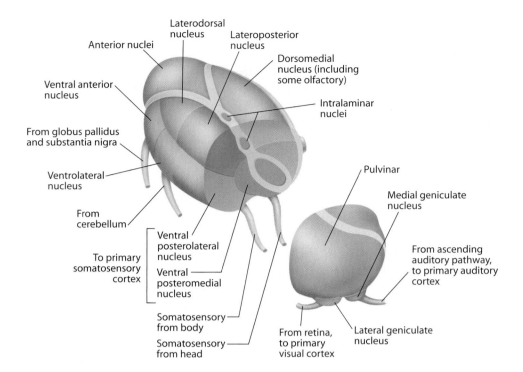

Figure 3.26 Diagram of the left thalamus showing inputs and outputs and major subdivisions. The various subdivisions of the thalamus serve different sensory systems and participate in various cortical-subcortical circuits. The posterior portion of the thalamus (at **lower right**) is cut away in cross section and separated from the rest of the thalamus to reveal the internal organization of the thalamic nuclei **(upper left)**. Adapted from Netter (1983).

is the pulvinar nucleus, located at the posterior pole of the thalamus. The pulvinar nucleus has a series of reciprocal connections with posterior cortical areas including the parietal lobe and areas in the visual cortex. The pulvinar also has several subdivisions that make different connections with the cortex and other subcortical areas. As described in Chapter 7, evidence implicated the pulvinar as a critical structure in attentional processing because of its heavy interconnectivity with regions of the cortex known to be involved with attentional control (posterior area of the parietal lobe) and the areas of visual cortex where feature analysis and object recognition are accomplished (ventral projection pathway).

A final note is that the sensory relay nuclei of the thalamus not only project axons to the cortex but also receive heavy descending projections back from the same cortical area they contact. This descending cortico-thalamic projection terminates in a thin layer of cells that surrounds the thalamic nuclei, known as the *thalamic reticular nucleus* (also called *nucleus reticularis thalami* and the *reticular nucleus of the thalamus*). Neurons in the thalamic reticular nucleus form a lateral inhibitory network of cells that may act in the modulation of thalamocortical outputs, perhaps to fine-tune sensory transmission or partially gate the flow of information to the cortex. This structure is not a single nucleus but rather can be conceived of as many nuclei that interact with the specific thalamic subdivisions they overlay. For example, that portion of the thalamic reticular nucleus that is ad-

jacent to the lateral geniculate nucleus (the visual relay nucleus) is also known as the *perigeniculate nucleus,* to reflect its association with the lateral geniculate.

Hypothalamus Below the thalamus is the hypothalamus, a small collection of nuclei and fiber tracts that lie on the floor of the third ventricle (Figure 3.27). The hypothalamus is important for the autonomic nervous system and the endocrine system, and controls functions necessary for the maintenance of homeostasis (i.e., maintaining the normal state of the body). The hypothalamus is also involved in emotional processes and in control of the pituitary gland, which is attached to the base of the hypothalamus.

The hormones produced by the hypothalamus control much of the endocrine system. For example, hypothalamic neurons in the region that surrounds the third ventricle send axonal projections to an area at the border of the hypothalamus and pituitary gland—the median eminence—where releasing factors (e.g., peptides) are released into the portal system that provides circulation to the anterior pituitary gland. In the anterior pituitary, these hypothalamic peptides trigger the release of (or inhibit the release of) a variety of hormones into the bloodstream; growth hormone, thyroid-stimulating hormone, adrenocorticotropic hormone, and the gonadotropic hormones are examples of those released by the cells of the anterior pituitary under hypothalamic control. Hypothalamic neurons in the anterior-medial region, including the supraoptic nucleus and paraventricular nuclei, send axonal

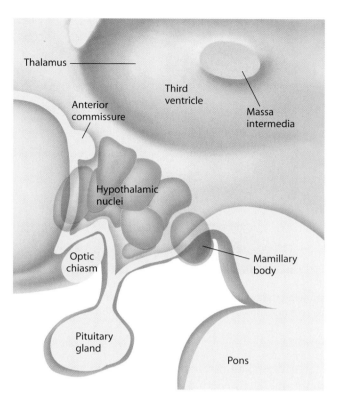

Figure 3.27 Hypothalamus shown in midsagittal section. Various nuclear groups are shown diagrammatically. The hypothalamus is the floor of the third ventricle, and as the name suggests, it sits below the thalamus. Anterior is to the left in this drawing. Adapted from Netter (1983).

projections into the posterior pituitary, where they stimulate the posterior pituitary to release the hormones vasopressin and oxytocin into the blood to regulate water retention in the kidneys and the production of milk and uterine contractility, respectively. The hypothalamus not only receives inputs from the limbic cortex but also receives inputs from other brain areas including the mesencephalic reticular formation, amygdala, and the retina to control circadian rhythms (light-dark cycles). Projections from the hypothalamus include a major projection to the prefrontal cortex, amygdala, and spinal cord. One of the most prominent projections is the one to the pituitary.

In addition to the direct neuronal projections of the hypothalamus, one important manner in which the hypothalamus influences the activity of other neurons is via neuromodulatory processes that involve the secretion of peptide hormones into the blood. These circulating peptide hormones can influence a wide range of behaviors by acting on distant sites through the bloodstream. In a similar fashion, the hypothalamus can be affected by hormones circulating in the blood, and thereby produce a neural response.

Brainstem

We usually think of the **brainstem** as having three main parts, the mesencephalon (midbrain), metencephalon (pons), and myelencephalon (medulla), which range from the level of the diencephalon to the spinal cord. Compared to the vast bulk of the forebrain, the brainstem is rather small (Figure 3.28). This region of the nervous system contains groups of motor and sensory nuclei, nuclei of widespread modulatory neurotransmitter systems, and white matter tracts of ascending sensory information and descending motor signals. The organization becomes more complex as it proceeds from the spinal cord through the medulla, pons, and midbrain to the diencephalon and cerebral cortex. This neuronal complexity reflects the increasingly complex behaviors that these regions enable. However, this does not mean that the brainstem is unimportant or simplistic in its processing, nor does it signify that the brainstem's functions are ancillary. Indeed, damage to the brainstem is highly life-threatening, in part because of the brainstem's size—being small means that a small lesion encompasses a large percentage of the tissue—and also because brainstem nuclei control respiration and even states of consciousness such as sleep and wakefulness. Therefore, damage to the brainstem can often be fatal, while damage to the cerebral cortex may have (relatively) minor consequences, depending on where and how much cortex is damaged.

MIDBRAIN

The mesencephalon or **midbrain** lies caudal to the diencephalon and is bounded posteriorly by the pons. It surrounds the cerebral aqueduct, which connects the third and fourth ventricles, and consists of the tectum (meaning "roof," and representing the dorsal portion of the mesencephalon), tegmentum (the main portion of the midbrain), and ventral regions occupied by large fiber tracts (crus cerebri) from the forebrain to the spinal cord (cortico-spinal tract), cerebellum, and brainstem (cortico-bulbar tract). The midbrain contains neurons that participate in visuomotor functions (e.g., superior colliculus, oculomotor nucleus, trochlear nucleus), visual reflexes (e.g., pretectal region), and auditory relays (inferior colliculus), and the mesencephalic tegmental nuclei involved in motor coordination (red nucleus) (Figure 3.29).

Much of the midbrain is occupied by the mesencephalic reticular formation, a rostral continuation of the pontine and medullary reticular formation. It is lateral and dorsal to the red nucleus and ventral to the cerebral aqueduct (the connection between the third and fourth ventricles). The reticular formation is best seen as a set of motor and sensory nuclei in the brainstem that

(a)

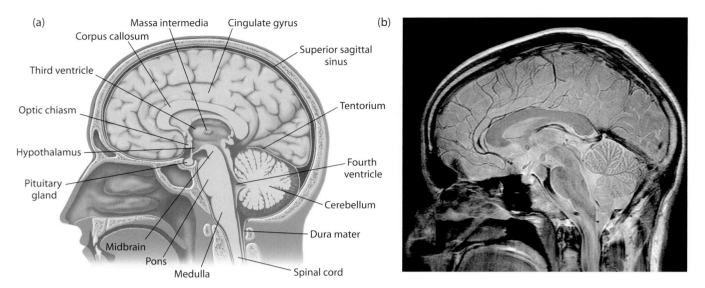

(b)

Figure 3.28 **(a)** Drawing of a midsagittal section through the head showing the brainstem, cerebellum, and spinal cord. **(b)** High-resolution structural magnetic resonance image obtained with a 4-tesla scanner, showing the same plane of section as in (a). Courtesy of Dr. Allen Song, Duke University.

participate in arousal, respiration, cardiac modulations, modulation of reflex muscular activity at the segmental level (i.e., in the limbs), and pain regulation.

Another major structure of the midbrain is the substantia nigra, located ventral to the reticular formation and red nucleus and dorsal to the white matter tracts on the ventral surface. The substantia nigra is so named because when prepared as fixed tissue for microscopic analysis, it can look sort of blackish. Neurons containing dopamine as their neurotransmitter have their cell bodies in a subdivision of the substantia nigra and project their axons into the basal ganglia, synapsing on neurons in the caudate and putamen.

As described in Chapter 2, projection systems arising in the locus coeruleus, a midbrain nucleus, project widely to the cortex, using norepinephrine as their neurotransmitter. Similar, the raphe nuclei in the midbrain project widely to the cortex and use serotonin as their transmitter. Finally, in addition to the projections of the substantia nigra noted above, the ventral tegmental area contains dopaminergic neurons that project throughout the cortex. These neurotransmitters (all biogenic amines) and their projection systems are involved in a variety of modulatory functions in the brain and spinal cord (see Figure 2.35).

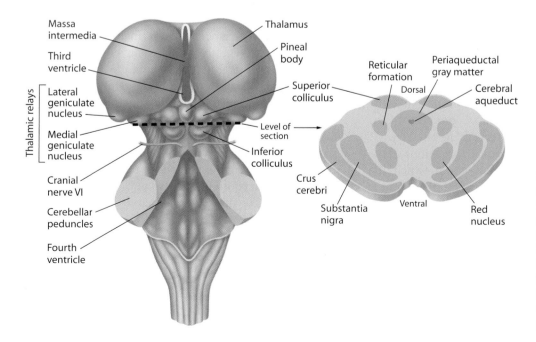

Figure 3.29 Anatomy of the midbrain. The dorsal surface of the brainstem is shown with the cerebral cortex and cerebellum removed. Cross section is through the midbrain at the level of the superior colliculus, the subcortical visuomotor nuclei. Adapted from Carpenter (1976).

PONS AND MEDULLA

The last areas of the brainstem to consider are the metencephalon (pons) and myelencephalon (medulla). Together they form the hindbrain. The **pons** includes the pontine tegmental regions on the floor of the fourth ventricle, and the pons itself, a vast system of fiber tracts interspersed with pontine nuclei. The fibers are continuations of the cortical projections to the spinal cord, brainstem, and cerebellar regions; initially they are compact fiber tracts on the ventral surface of the midbrain. At the pons, they explode into smaller tracts that continue to their final destinations as they course around the pontine nuclei, some terminating on neurons in this region (pons is labeled in Figure 3.28a).

The many nuclei at the pontine level have auditory and vestibular (balance) functions; primary CNS synapses of axons coming from the auditory and vestibular periphery are located in cell groups of the pontine tegmentum. As well, sensory and motor nuclei from the face and mouth are located here, as are visuomotor nuclei controlling some of the extraocular muscles. This level of the brainstem also contains a large portion of the reticular formation.

Finally, the brain's most caudal portion is the **medulla,** which is continuous with the spinal cord. Here, the fourth ventricle narrows and begins to shift more ventrally until at the level of the transition from the medulla to the spinal cord, it is a narrow tube running through the medulla to connect with the central canal of the spinal cord. The medulla has two prominent bilateral nuclear groups on the ventral surface (the gracile and cuneate nuclei) that are the primary relay nuclei for ascending somatosensory information entering the spinal cord. These projection systems continue through the brainstem to synapse in the thalamus en route to the somatosensory cortex. On the ventral surface of the medulla the continuations of the corticospinal motor projections are grouped once again as tight bundles of nerve fibers into the *pyramids,* bilateral bumps on the ventral medulla. At the level of the medulla these motor axons to the spinal cord cross (forming the pyramidal decussation) in order to project to the contralateral side of the spinal cord; that is, for example, the right-hemisphere motor systems control the left side of the body. At the rostral end of the medulla are large and characteristically formed nuclei of the olivary complex (inferior and medial accessory olivary nuclei). They appear in cross section as highly enfolded nuclei that are part of the cortical-cerebellar motor system (Figure 3.30). The olivary nuclei receive inputs from the cortex and red nucleus and project them to the cerebellum. Sensory nuclei that carry out vestibular processing (caudal portions of the vestibular nuclei) and some sensory inputs from the face, mouth, throat (including taste), and abdomen are in the medulla. There are also motor nuclei that innervate the heart and muscles of the neck, tongue, and throat.

In sum, the brainstem's neurons carry out numerous sensory and motor processes, especially visuomotor,

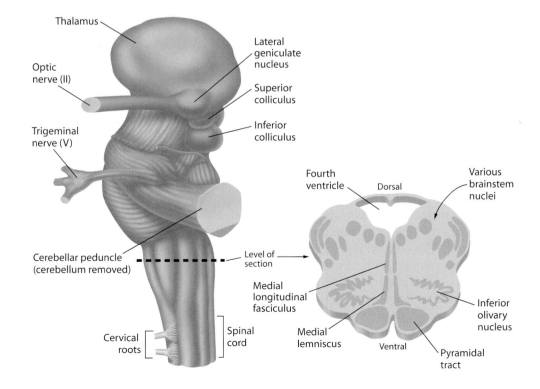

Figure 3.30 Lateral view of the brainstem showing the midbrain, medulla, and spinal cord. Section is through the medulla at the level of the inferior olivary nucleus. Adapted from Carpenter (1976).

Thalamus

Optic nerve (II)

Trigeminal nerve (V)

Lateral geniculate nucleus

Superior colliculus

Inferior colliculus

Cerebellar peduncle (cerebellum removed)

Level of section

Fourth ventricle

Dorsal

Various brainstem nuclei

Medial longitudinal fasciculus

Inferior olivary nucleus

Cervical roots

Spinal cord

Medial lemniscus

Ventral

Pyramidal tract

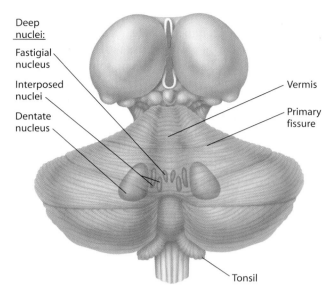

Figure 3.31 Gross anatomy of the cerebellum. Anterior in the brain is at top and the spinal cord is toward the bottom (not shown). This dorsal view of the cerebellum shows the underlying deep nuclei in a see-through projection. Adapted from Carpenter (1976).

auditory, and vestibular functions, and sensation and motor control of the face, mouth, throat, respiratory system, and heart. The brainstem houses fibers that pass from the cortex to the spinal cord and cerebellum, and sensory fibers from spinal levels to the thalamus and thence the cortex. Many neurochemical systems have nuclei in the brainstem that project widely to the cerebral cortex, limbic system, thalamus, and hypothalamus.

Cerebellum

The **cerebellum** (small cerebrum, or little brain) is actually a very large neuronal structure overlying the brainstem at the level of the pons (see Figure 3.28). It forms the roof of the fourth ventricle and sits on the cerebellar peduncles (meaning "feet"), which are massive input and output fiber tracts of the cerebellum (Figure 3.30, left). The cerebellum has several gross subdivisions, including the cerebellar cortex, the four pairs of deep nuclei, and the internal white matter (Figure 3.31). In this way the cerebellum resembles the forebrain's cerebral hemispheres.

The inputs to the cerebellum come into the deep nuclei, but some projections extend directly to the cerebellar cortex. Most outputs of the cerebellum are via the deep nuclei. Inputs to the cerebellum come from the parts of the brain that participate in motor and sensory processing; hence, they convey information about motor outputs, and sensory inputs describing body position. Inputs from vestibular projections involved in balance, and audi-

tory and visual inputs also project to the cerebellum from the brainstem. The cerebellum's output is to the thalamus and thence to the motor and premotor cortex. Cerebellar projections to the brainstem's nuclei ultimately influence descending projections to the spinal cord. The cerebellum is key to maintaining posture, walking, and performing coordinated movements. By itself, the cerebellum does not control movements directly; instead it integrates information about the body and motor commands and modifies motor outflow to effect smooth, coordinated movements. The cerebellum's role in motor control is given more attention in Chapter 11.

Spinal Cord

The last neural portion of the CNS to be reviewed is the spinal cord, which runs from the medulla to its termination in the cauda equina (meaning "horse's tail") at the base of the spine, where only nerve bundles remain and there is no longer a cord per se. The spinal cord primarily conducts the final motor signals to muscles, and takes in sensory information from the body's peripheral sensory receptors and relays it to the brain. In addition, at each level of the spinal cord, reflex pathways exist, as

Figure 3.32 Cross-sectional and three-dimensional representation of the spinal cord, showing the central butterfly-shaped gray matter, which contains neurons, and the surrounding white matter tracts, which convey information down the spinal cord from the brain to neurons in the cord and up the spinal cord from peripheral receptors to the brain. The dorsal and ventral nerve roots are shown exiting and entering the cord—they fuse to form peripheral nerves. The cell bodies of peripheral sensory inputs reside in the dorsal root ganglion and project their axons into the central nervous system via the dorsal root. The ventral horn of the spinal cord houses motor neurons that project their axons out the ventral roots to innervate peripheral muscles.

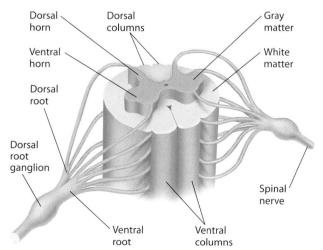

for example, that for the knee-jerk reflex. The gross anatomy of the spinal cord is simple: It consists of white matter tracts of ascending and descending, sensory and motor information, respectively (plus intraspinal projection fibers), and neuronal cell bodies organized in a more central gray matter (Figure 3.32). These include motor neurons, interneurons, and sensory neurons. The gray matter, when viewed in cross section, resembles a butterfly, with two separate sections, or horns, called *dorsal* and *ventral horns*. The ventral horn contains the large motor neurons that project to muscles, while the dorsal horn contains sensory neurons and interneurons. The latter project to motor neurons on the same and opposite sides of the spinal cord to aid in the coordination of limb movements.

The spinal cord is protected within the bone of the spine, but with the spine removed, one can view the bilateral pairs of spinal nerves that carry motor output (ventral root) and sensory information (dorsal root) into and out of the spinal cord. They pass through small gaps in the spine.

Autonomic Nervous System

The **autonomic nervous system** (also called the *autonomic* or *visceral motor system*) is part of the PNS and is involved in controlling the action of smooth muscles, the heart, and various glands. It has two subdivisions, the *sympathetic* and *parasympathetic* branches (Figure 3.33). These two systems innervate smooth muscles and glands and have antagonistic actions. The sympathetic system uses the neurotransmitter norepinephrine, while the parasympathetic fibers use acetylcholine as the transmitter. Hence, for example, activation of the sympathetic system increases heart rate, diverts blood from the digestive tract to the somatic musculature, and prepares the body for action (fight or flight) by stimulating the adrenal glands to release adrenaline. In contrast, activation of the parasympathetic system slows heart rate, stimulates digestion, and in general helps the body in normal functions germane to maintaining the body.

There is a great deal of specialization in the autonomic system that is beyond the scope of this chapter,

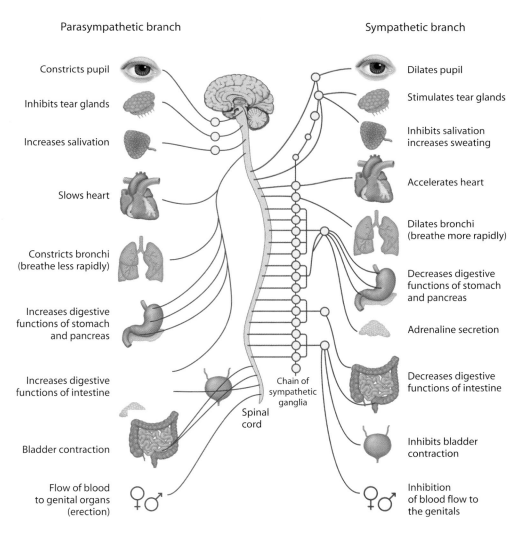

Figure 3.33 The organization of the autonomic nervous system, showing sympathetic and parasympathetic branches.

Parasympathetic branch

Constricts pupil

Inhibits tear glands

Increases salivation

Slows heart

Constricts bronchi (breathe less rapidly)

Increases digestive functions of stomach and pancreas

Increases digestive functions of intestine

Bladder contraction

Flow of blood to genital organs (erection)

Sympathetic branch

Dilates pupil

Stimulates tear glands

Inhibits salivation increases sweating

Accelerates heart

Dilates bronchi (breathe more rapidly)

Decreases digestive functions of stomach and pancreas

Adrenaline secretion

Decreases digestive functions of intestine

Inhibits bladder contraction

Inhibition of blood flow to the genitals

Chain of sympathetic ganglia

Spinal cord

Blood Supply and the Brain

Approximately 20% of the blood flowing from the heart is pumped to the brain. A constant flow of blood is necessary because the brain has no way of storing glucose or extracting energy without oxygen. Disruption in the flow of oxygenated blood to the brain lasting only a few minutes can produce unconsciousness, and finally death. Two sets of arteries bring blood to the brain: the vertebral artery, which supplies blood to the caudal portion of the brain, and the internal carotid artery, which supplies blood to the rostral portions. Although the major arteries sometimes join together and then separate again, there is actually little mixing of blood from the rostral and caudal arterial supplies or from the right and left sides of the rostral portion of the brain. As a safety measure, in the event of a blockage or ischemic attack, blood should be rerouted to reduce the probability of loss of blood supply, but in practice this backup system is relatively poor.

The blood flow in the brain is tightly coupled with metabolic demand of the local neurons. Hence, increases in neuronal activity lead to a coupled increase in regional cerebral blood flow. The increased blood flow is not primarily for increasing the delivery of oxygen and glucose to the active tissue, but rather to hasten the removal of the resultant metabolic by-products of the increased neuronal activity. The precise mechanisms, however, remain hotly debated. These local changes in blood flow permit regional cerebral blood flow to be used as a measure of local changes in neuronal activity. This is the principle on which some types of functional neuroimaging are based. Particular examples are positron emission tomography using techniques such as the ^{15}O-water method, and functional magnetic resonance imaging, which is sensitive to changes in the concentration of oxygenated versus deoxygenated blood in the region of active tissue.

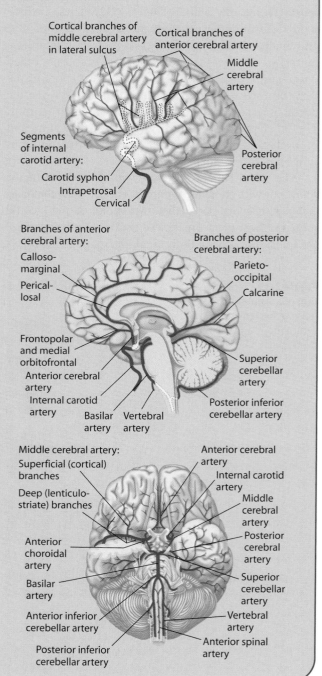

but understanding that the autonomic system is involved in a variety of reflex and involuntary behaviors is useful for interpreting information presented later in the book. In Chapter 13, we will discuss arousal of the autonomic nervous system and the way in which changes in a number of psychophysiological measures tap into emotion-related changes in the autonomic nervous system. One example is the change in skin conductance that is related to sweat gland activity—sweat glands are under the control of the autonomic nervous system.

SUMMARY

The nervous system is composed of cells—neurons—and their supportive counterparts, the glial cells. The neuron is the elementary unit of structure and function within the brain and spinal cord of the central nervous system and the peripheral nervous system, which includes the autonomic nervous system. These elementary units and their interconnections can be viewed directly using a variety of gross dissection methods, and analyzed in detail using microanatomical methods. Connections can be traced with special substances that are taken up by neurons and transported down their axons either in an anterograde (from soma to axon terminals) or a retrograde (from terminals to soma) direction. Neuronal circuits are organized to form highly specific interconnections between groups of neurons in subdivisions of the central nervous system. Different neuronal groups have different functional roles. The functions may be localized within discrete regions that contain a few or many subdivisions, identifiable either anatomically or functionally, but usually by a combination of both. Brain areas are also interconnected to form higher-level circuits or systems that are involved in complex behaviors such as motor control, visual perception, or cognitive processes such as memory, language, and attention.

KEY TERMS

amygdala

association cortex

autonomic nervous system

brainstem

basal ganglia

central nervous system (CNS)

cerebellum

cerebral cortex

commissure

corpus callosum

cytoarchitectonic map

dura mater

frontal lobe

gray matter

gyri

hippocampus

hypothalamus

limbic system

medulla

midbrain

neocortex

occipital lobe

parietal lobe

peripheral nervous system (PNS)

pons

prefrontal cortex

sulci

Sylvian fissure

temporal lobe

thalamus

tract

white matter

THOUGHT QUESTIONS

1. What is the evolutionary significance of the different types of cortex (e.g., neocortex versus allocortex)?
2. What is the functional advantage of organizing cortex as a sheet of neurons, rather than groups of nuclei such as are found in subcortical structures?
3. What regions of the cerebral cortex have increased in size the most across species during evolution? What does this brain region subserve in humans that is absent or reduced in animals?

4. Why are almost all sensory inputs routed through the thalamus on the way to cortex? Would it not be faster and therefore more efficient to project these inputs directly from sensory receptors to the primary sensory cortex?
5. Although the brainstem is relatively small in comparison with the forebrain, it contains some essential structures. Select one and describe how its anatomical organization supports its role in brain function.

SUGGESTED READINGS

GAN, W.B., GRUNTZENDLER, J., WONG, W.T., WONG, R.O., AND LICHTMAN, J.W. (2000). Multicolor "diolistic" labeling of the nervous system using lipophilic dye combinations. *Neuron* 27: 219–225.

KATZ, B. (1966). *Nerve, Muscle and Synapse.* New York: McGraw-Hill.

MESULAM, M.-M. (2000). Behavioral Neuroanatomy: Large-scale Networks, Association Cortex, Frontal Syndromes, the Limbic System, and Hemispheric Specialization. In M.-M. Mesulam, *Principals of Behavioral and Cognitive Neurology* (pp. 1–34). New York: Oxford University Press.

SHEPHERD, G.M. (1988). *Neurobiology,* 2nd edition. New York: Oxford University Press.

4

The Methods of Cognitive Neuroscience

The frontiers of scientific discovery are defined as much by the tools available for observation as by conceptual innovation. In the sixteenth century, the earth was considered the center of the solar system. Simple observation verified it: The sun rose each morning in the east and slowly moved across the sky to set in the west. But the invention of the telescope in 1608 changed astronomers' observational methods. With this new tool, astronomers suddenly found galactic entities that they could track as these entities moved across the night sky. These observations exposed geocentric theories as painfully wrong. Indeed, within 5 years, Galileo spoke out for a heliocentric universe—a heretical claim that even the powerful Roman Catholic Church could not suppress in the face of new technology.

Similar breakthroughs in theory can be linked in all scientific domains to the advent of new methods for observation. The invention of the bubble chamber allowed particle physicists to discover new and unexpected elementary particles such as mesons, discoveries that have totally transformed our understanding of the microscopic structure of matter. Gene cloning and sequencing techniques provided the tools for identifying new forms of proteins and for recognizing that these proteins formed previously unknown biological structures such as the neurotransmitter receptor that binds with tetrahydrocannabinol, the psychoactive ingredient in marijuana. Research in this area is now devoted to searching for endogenous substances that utilize these receptors rather than following the more traditional view that tetrahydrocannabinol produces its effects by binding to receptors linked to known transmitters.

The emergence of cognitive neuroscience has been similarly fueled by new methods, some of which utilize high-technology tools unavailable to scientists of previous generations (Sejnowski and Churchland, 1989). The positron emission tomography (PET) scanner, for instance, enables scientists to observe the brain's activity.

Brain lesions can be localized with amazing precision owing to methods such as magnetic resonance imaging (MRI). High-speed computers allow investigators to construct elaborate models to simulate patterns of connections and processing. Powerful electron microscopes bring previously unseen neural elements into view.

The real power of these tools, though, is still constrained by the types of problems one chooses to investigate. The dominant theory at any point in time defines the research paradigms and shapes the questions to be explored. The telescope helped Galileo to plot planets' positions with respect to the sun. But without an appreciation of the forces of gravity, he would have been at a loss to provide a causal account of planetary revolution. In an analogous manner, the problems investigated with the new tools of neuroscience are shaped by contemporary ideas of how the brain works in perception, thought, and action. Put simply, if well-formulated questions are not asked, even the most powerful tools will not provide a sensible answer.

In this chapter we examine the methods used in cognitive neuroscience. In the preceding chapter we discussed some of the tools neuroanatomists use to investigate the cellular and gross structure of the nervous system. Here we focus on methods for studying behavior, those employed by cognitive psychologists, computer modelers, neurophysiologists, and neurologists. While each of the areas represented by these professionals has blossomed in its own way, the interdisciplinary nature of cognitive neuroscience has depended on the clever ways scientists have integrated paradigms across these areas. The chapter concludes with examples of this integration.

WHAT IS COGNITIVE PSYCHOLOGY?

It would be naive to suppose that people have only recently sought to relate behavior to brain function. What marks cognitive neuroscience as a new field for this endeavor are the paradigms developed in **cognitive psychology,** the study of mental activity as an information-processing problem. Cognitive psychology rests on the assumption that we do not directly perceive and act in the world. Rather, our perceptions, thoughts, and actions depend on internal transformations or computations. Information is obtained by sense organs, but our ability to comprehend the information, to recognize it as something we have experienced before, and to choose an appropriate response depends on a complex interplay of processes.

Mental Representations and Transformations

Two key concepts underlie the cognitive approach. The first idea, that information processing depends on internal representations, we usually take for granted. Consider the concept "ball." If we met someone from a planet composed of straight lines, we could try to convey what this concept means in many ways. We could draw a picture of a sphere, we could provide a verbal definition indicating that such a three-dimensional object is circular along any circumference, or we could write a mathematical definition. Each instance is an alternative form of representing the "circle" concept. Whether one form of representation is better than another depends on our visitor. To understand the picture, our visitor would need a visual system and the ability to comprehend the spatial arrangement of a curved drawing. To understand the mathematical definition, our visitor must comprehend geometrical and algebraic relations. Assuming our visitor has these capabilities, the task will help dictate which representational format is most useful. For example, if we want to show that the "ball" rolls down a hill, a pictorial representation is likely to be much more useful than an algebraic formula.

The second critical notion of cognitive psychology is that mental representations undergo transformations. The need to transform mental representations is most obvious when we consider how sensory signals are connected with stored knowledge in memory. Perceptual representations must be translated into action representations if we wish to achieve a goal. Moreover, information processing is not simply a sequential process from sensation to perception to memory to action. Memory may alter how we perceive something, and the manner in which information is processed is subject to attentional constraints. Cognitive psychology is all about how we manipulate representations.

Consider the categorization experiment introduced by Michael Posner (1986) at the University of Oregon that is schematized in Figure 4.1. Two letters are presented simultaneously in each trial. The subject's task is to evaluate whether they are both vowels, both consonants, or one vowel and one consonant. If the letters are from the same category, the subject presses one button. If they are from different categories, then he or she should respond with the other button.

One version of this experiment includes five conditions. In the physical identity condition, the two letters are exactly the same. In the phonetic identity condition, the two letters have the same identity, but one letter is a capital and the other is lowercase. There are two types of same-category conditions, conditions in which the two letters are different members of the same category. In one, both letters are vowels; in the other, both letters are consonants. Finally, in the different condition, the two letters are from different categories and can be either of the same type size or of a different one. Note that the first four conditions—physical identity, phonetic identity, and the two same-category conditions—require the same response. On all three types of trials, a correct response is yes, the two letters are from the same category. Nonetheless, as shown in Figure 4.1, response latencies differ significantly. Subjects respond fastest to the physical identity condition, next fastest to the phonetic identity condition, and slowest to the same-category condition, especially when the two letters are both consonants.

The results of Posner's experiment suggest that we derive multiple representations of stimuli. One representation is based on the physical aspects of the stimulus. In this experiment, it is a visually derived representation of the shape presented on the screen. A second representation corresponds to the letter's identity. This representation

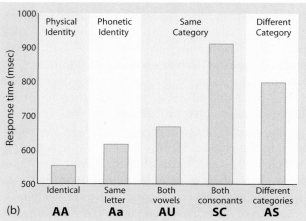

Figure 4.1 **(a)** In this version of the letter-matching task, the subject responds "SAME" when both letters are either vowels or consonants and "DIFFERENT" when they are from different categories. **(b)** The reaction times for different conditions. (b) Adapted from Posner (1986).

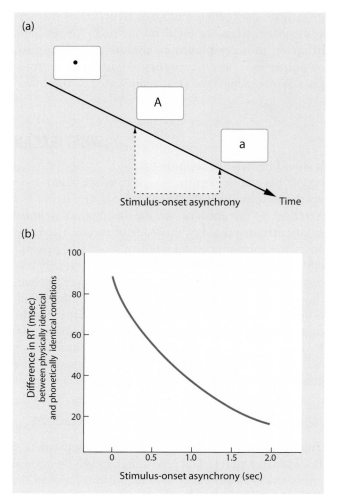

Figure 4.2 **(a)** The same letter-matching task as in Figure 4.1 except that an interval, defined as the stimulus-onset asynchrony, separates the presentation of the two letters. **(b)** As this interval is lengthened, the difference in the reaction time (RT) to the physical identity and phonetic identity condition becomes reduced, suggesting a transformation of the representation into a more abstract code. Data from Posner (1986).

proves that many stimuli can correspond to the same letter. For example, we can recognize that *A*, *a*, and *a* all represent the same letter. A third level of abstraction represents the category a letter belongs to. At this level, the letters *A* and *E* activate our internal representation of the category "vowel." Posner maintains that different response latencies reflect the degrees of processing required to do the letter-matching task. By this logic, we infer that physical representations are activated first, phonetic representations next, and category representations last.

This experiment provides a powerful demonstration that even with simple stimuli, the mind derives multiple representations. Other manipulations with this task have

explored how representations are transformed from one form to another. In a follow-up study, Posner and his colleagues used a sequential mode of presentation. Two letters were presented again, but a brief interval (referred to as the *stimulus-onset asynchrony*, the time between the two stimuli) separated the presentations for the letters. As shown in Figure 4.2, the difference in response time to the physical identity and phonetic identity conditions was reduced as the stimulus-onset asynchrony became longer. Hence, the internal representation of the first letter is transformed during the interval. The representation of the physical stimulus gives way to the more abstract representation of the letter's phonetic identity.

As you may have experienced personally, experiments such as these elicit as many questions as answers. Why are subjects slower to judge that two letters are consonants in comparison to two letters that are vowels? Would the same advantage for identical stimuli exist if the letters were spoken? What about if one letter were visual and the other were auditory? Suppose that the task is to judge whether two letters are physically identical. Would manipulating the stimulus-onset asynchrony affect reaction times on this version? Cognitive psychologists address these questions and then devise methods for inferring the mind's machinery from observable behaviors.

In the preceding example, the primary dependent variable has always been reaction time, the speed with which the subjects make their judgments. Reaction time experiments utilize the chronometric methodology. *Chrono* comes from the Greek word for "time," and *metric* for "measure." The chronometric study of mind is essential for cognitive psychologists because mental events occur rapidly and efficiently. If we only consider whether a person is correct or incorrect on a task, we miss subtle differences in performance. Measuring reaction time permits a finer analysis of internal processes. In addition to measuring processing time as a dependent variable, chronometric manipulations can be applied to independent variables, as with the letter-matching experiment in which the stimulus-onset asynchrony was varied.

Characterizing Mental Operations

Suppose you arrive at the grocery store and discover that you forgot to bring your shopping list. As you wander up and down the aisles, you gaze upon the thousands of items lining the shelves, hoping that they will help prompt your memory. You can cruise through the pet food section, but hesitate when you come to the dairy section: Was there a carton of eggs in the refrigerator? Was the milk supply low? Were there any cheeses not covered by a 6-month rind of mold?

This memory retrieval task draws on a number of cognitive capabilities. A fundamental goal of cognitive psychology is to identify the different mental operations that are required to perform tasks such as these. Not only are cognitive psychologists interested in describing human performance—the observable behavior of humans and other animals—but also they seek to identify the internal processing that underlies this performance. A basic assumption of cognitive psychology is that tasks are composed of a set of mental operations. Mental operations involve taking a representation as an input, performing some sort of process on the input, and then producing a new representation, or output. Thus, mental operations are processes that generate, elaborate upon, or manipulate mental representations. Cognitive psychologists design experiments to test hypotheses about mental operations.

Consider an experimental task introduced by Saul Sternberg (1975) when he was working at Bell Laboratories. The task bears some similarity to the problem faced by our absent-minded shopper, except that in Sternberg's task, the difficulty is not so much in terms of forgetting items in memory, but rather of how well people can compare sensory information with representations that are active in memory. On each trial, the subject is first presented with a set of letters to memorize (Figure 4.3). The memory set could consist of one, two, or four letters. Following this, a single letter is presented and the subject has to decide if this letter was part of the memorized set. One button is pressed if the subject thinks the target was part of the memory set ("yes" response) and a second button if the target was not part of the set ("no" response). The primary dependent variable is reaction time.

Sternberg postulated that to respond on this task, the subject must engage in four primary mental operations. First, the target must be encoded. That is, the subject must identify the visible target. Second, the mental representation of the target must be compared to the representations of the items in memory. Third, a decision must be made as to whether the target matches one of the memorized items. Finally, based on this decision, the appropriate response must be generated. Note that each of these operations—encode, compare, decide, and respond—is likely to be composed of additional operations. For example, responding might be further fractionated into processes involved in selecting the appropriate finger and processes involved in activating the muscles that make the finger move. Nonetheless, by postulating a set of mental operations, experiments can be devised to explore the operation of putative mental operations.

A basic question for Sternberg was how to characterize the efficiency of recognition memory. Assuming that there are active representations of all of the items in the memory set, one could conceive of (at least) two different ways the recognition process might work. A highly efficient system might compare a representation of the target with all of the items in the memory set simultaneously. On the other hand, the recognition operation might be limited in terms of how much information it can handle at any point in time. For example, it might require that the input be compared successively to each item in memory. Sternberg realized that the reaction time data could distinguish between these two alternatives. If the comparison process could occur for all

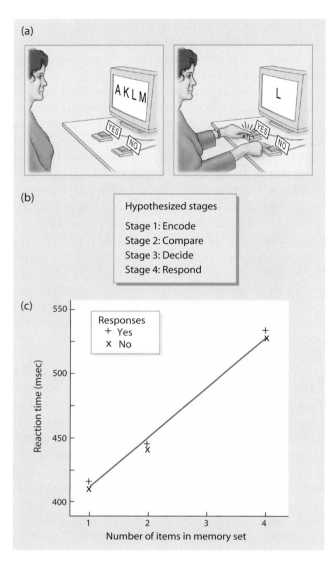

Figure 4.3 The memory compare task. **(a)** The subject is shown a set of letters (either one, two, or four) and is asked to memorize them. After a delay, a single probe letter appears and the subject indicates whether it was a member of the memory set. **(b)** Hypothesized mental operations required to perform this task. **(c)** Reaction time increases with set size, indicating that the comparison of the target letter with the memory set must be performed sequentially rather than in parallel. Adapted from Sternberg (1966).

items simultaneously, what would be called a *parallel process*, then reaction time should be independent of the number of items in the memory set. But if the comparison process operates in a sequential or serial manner, then reaction time would be expected to slow down as the memory set became larger. It would take more time to compare an item with a large memory list than with a small memory list. Sternberg's results convincingly supported the serial hypothesis. In fact, reaction time increased in a constant, or linear, manner with set size, and the functions for the "yes" and "no" trials were essentially identical.

The parallel, linear functions allowed Sternberg to make two inferences about the mental operations associated with this task. First, the linear increase in reaction time as the set size increased implied that the memory comparison operation took a fixed amount of internal processing time. In the initial study, the slope of the function was approximately 40 msec/item, implying that it takes about 40 msec for each successive comparison of the target to the items in the memory set. This does not mean that this value represents a fixed property of memory comparison. It is likely to be affected by factors such as task difficulty (e.g., whether the nontarget items in the memory set are similar or dissimilar to the target item) and experience. Nonetheless, the experiment demonstrates how both qualitative and quantitative characterizations of mental operations can be made from simple behavioral tasks.

Second, the fact that the two functions were parallel implied that subjects compared all of the memory items to the target before making a response. If subjects had terminated the comparison as soon as a match was found, then the slope of the "no" function should have been twice as steep as the slope of the "yes" function. This follows because in "no" trials all of the items have to be checked. With "yes" trials, on average only half the items need to be checked before a match is found. The fact that the functions were parallel implies that comparisons were carried out on all items, and that comparison was serial and exhaustive (as opposed to serial and self-terminating). An exhaustive process seems illogical, though. Why continue to compare the target to the memory set once a match is detected? One possible answer is that it is easier to store the result of each comparison for later evaluation than to monitor "on-line" the results of successive comparisons.

While memory comparison appears to involve a serial process, many other tasks demonstrate cognitive operations that operate in parallel. A classic demonstration of this is the word superiority effect (Reicher, 1969). In this experiment, a stimulus is shown briefly and the subjects are asked which of two target letters (e.g., A or E) was present. The stimuli can be composed of either words, nonsense letter strings, or letter strings in which all of the letters are *X*'s except for the target letter (Figure 4.4). Brief presentation times are used so that errors will be observed, with the critical question centering on whether the context affects performance. The *word superiority effect* refers to the fact that subjects are most accurate when the stimuli are words. Somewhat counterintuitively, this

Does the stimulus contain an "A" or "E"?

Condition	Stimulus	Accuracy
Word	**RACK**	90%
Nonsense string	**KARC**	80%
X's	**XAXX**	80%

Figure 4.4 The word superiority effect. Subjects are more accurate in identifying the target vowel when it is embedded in a word. This result suggests that both letter and word levels of representation are activated in parallel.

suggests that we do not need to identify all of the letters of a word before we recognize the word. Rather, when we are processing the word lists, parallel activation occurs for representations corresponding to the individual letters and to the entire word. Performance is facilitated because both representations can provide information as to whether the target letter is present. A word-level representation is not possible with nonsense words and letter strings, and thus judgments must be based solely on letter-level representation.

Constraints on Information Processing

In Sternberg's memory search experiment, information processing operates in a certain manner because the memory comparison process is limited. The subjects cannot compare simultaneously the target item to all of the items in the memory set. An important question is whether this limitation reflects properties that are specific to memory or a more general processing limitation. Perhaps there is a limitation to how much internal processing people can do at any one time, regardless of the task. An alternative explanation is that processing limitations are task-specific. Processing constraints are defined only by the particular set of mental operations associated with a particular task. For example, while the comparison of a probe item to the memory set might require a serial operation, encoding might occur in parallel such that it would not matter whether the probe was presented by itself or among a noisy array of competing stimuli.

Exploring the limitations in task performance is a central concern for cognitive psychologists. Consider a

Color matches word	Random colors	Color doesn't match word
RED	XXXXX	GREEN
GREEN	XXXXX	BLUE
RED	XXXXX	RED
BLUE	XXXXX	BLUE
BLUE	XXXXX	GREEN
GREEN	XXXXX	RED
BLUE	XXXXX	GREEN
RED	XXXXX	BLUE

Figure 4.5 The Stroop task. Time yourself as you work through each column, naming the color of the ink of each stimulus as fast as possible. Assuming you do not squint to blur the words, it should be easy to read the first and second columns, but quite difficult to read the third.

simple color-naming task that was devised in the early 1930s by an aspiring doctoral student, J.R. Stroop (1935; for a recent review, see MacLeod, 1991), and that has become one of the most widely employed tasks in all of cognitive psychology. In this task, a list of words is presented and the subject is asked to name the color of each stimulus as fast as possible. As you can experience from Figure 4.5, it is much easier to do this task when the words match the ink colors. The Stroop effect powerfully demonstrates the multiplicity of mental representations. The stimuli in this task appear to activate at least two separable representations. One representation corresponds to each stimulus's color; it is what allows the subject to perform the task. The second representation corresponds to the color concept associated with the words. The fact that you are slower to name the colors when the ink color and words are mismatched indicates that this representation is activated even though it is irrelevant to the task. Indeed, the activation of a representation based on the words rather than the colors of the words appears to be automatic. The Stroop effect persists even after thousands of trials of practice, reflecting the fact that skilled readers have years of practice in analyzing letter strings for their symbolic meaning. On the other hand, the interference from the words is markedly reduced if the response requires a speeded key press rather than a vocal response. Thus, the word-based representations are closely linked to the vocal response system and have little effect when the responses are produced manually.

A second method used to examine constraints on information processing involves dual tasks. For these studies, performance on a primary task alone is compared to performance on that task concurrently with a secondary task. The decrement in primary-task performance during the dual-task situation helps elucidate the limits in cognition. Sophisticated use of dual-task methodology also can identify the exact source of interference. For example, the Stroop effect is not reduced when the color-naming task is performed simultaneously with a secondary task in which the subject must judge the pitch of an auditory tone. However, if the auditory stimuli for the secondary task are a list of words and the subject must monitor this list for a particular target, the Stroop effect is attenuated. It appears that the verbal demands of the secondary task interfere with the automatic activation of the word-based representations in the Stroop task, thus leaving the color-based representations relatively free from interference.

The efficiency of our mental abilities and the way mental operations interact can change with experience. The beginning driver has her hands rigidly locked to the steering wheel; within a few months, she is unfazed to steer with her left hand while using the right hand to scan for a good radio station and maintain a conversation with the person in the passenger seat. Even more impressive is that with extensive practice, people can become proficient in simultaneously performing two tasks that were originally quite incompatible. Elizabeth Spelke and her colleagues at Cornell University studied how well college students read for comprehension while taking dictation (Spelke et al., 1976). Prior to any training, their subjects could read about 400 words/min when faced with difficult reading material such as modern short stories. As we would expect, this rate fell to 280 words/min when the subjects were required to simultaneously take dictation, and their comprehension of the stories was also impaired. Remarkably, after 85 hours of training spread over a 17-week period, the students' proficiency in reading while taking dictation was essentially as good as when reading alone. The results offer an elixir for all college students. Imagine finishing the reading for an upcoming psychology examination while taking notes during a history lecture!

COMPUTER MODELING

The computer is a powerful metaphor for cognitive neuroscience. Both the brain and the computer chip are impressive processing machines, capable of representing and transforming large amounts of information. While there are vast differences in how these machines process information, cognitive scientists use computers to simulate cognitive processes. **Simulation** means to imitate, to reproduce behavior in an alternative medium. The simulated cognitive processes are commonly referred to as *artificial intelligence*—artificial in the sense that they are artifacts, man-made creations, and intelligent in that the computers perform complex functions. Computer programs control robots on factory production lines, assist physicians in making differential diagnoses or in detecting breast cancer, and create models of the universe in the first nanoseconds after the "big bang."

Many commercial computer applications are developed without reference to how brains think. More relevant to our present concerns are the efforts of cognitive scientists to create models of cognition (Rummelhart et al., 1986). In these investigations, simulations are designed to mimic behavior and the cognitive processes that support that behavior. The computer is given input and then must perform internal operations to create a behavior. By observing the behavior, the researcher can assess how well it matches behavior produced by a real mind. Of course, to get the computer to succeed, the modeler must specify how information is represented and transformed within the program. To do this, concrete hypotheses regarding the "mental" operations needed for the machine must be generated. As such, computer simulations provide a useful tool for testing theories of cognition. Successes and failures of models give valuable insights to a theory's strengths and weaknesses.

Models Are Explicit

Computer models of cognition are useful because they can be analyzed in detail. In creating a simulation the researcher has to be completely explicit; the way the computer represents and processes information must be totally specified. This does not mean that a computer's operation is always completely predictable and that the outcome of a simulation is known in advance. Computer simulations can incorporate random events or be on such a large scale that analytic tools do not reveal the solution. But the internal operations, the way information is computed, must be specified. Computer simulations are especially helpful to cognitive neuroscientists in recognizing problems the brain must solve to produce coherent behavior.

Braitenberg (1984) gave elegant examples of how modeling brings insights to information processing. Imagine observing the two creatures shown in Figure 4.6, as they move about a minimalist world consisting of a single heat source such as a sun. From the outside, the creatures look identical: They both have two sensors and four wheels. Despite this similarity, their behavior is distinct. One creature moves away from the sun and the other homes in on it. Why the difference? As an outsider with no access to the internal operations of these creatures, we might conjecture that they have had different experiences and so the same input activates different representations. Perhaps one was burned at an early age and fears the sun, and maybe the other likes the warmth.

But, as can be seen from their internal wiring, the behavioral differences depend on how the creatures are wired. The uncrossed connections make the creature on the left turn away from the sun; the crossed connections force the creature on the right to orient toward it. Thus, the two creatures' behavioral differences arise from a slight variation in how sensory information is mapped onto motor processes.

These creatures are exceedingly simple—and inflexible in their actions. At best, they offer only the crudest model of how an invertebrate might move in response to a phototropic sensor. The point of Braitenberg's example is not to model a behavior; rather, it represents how a single computational change—from crossed to uncrossed wiring—can yield a major behavioral change. When interpreting such a behavioral difference, we might postulate extensive internal operations and representations. However, when we

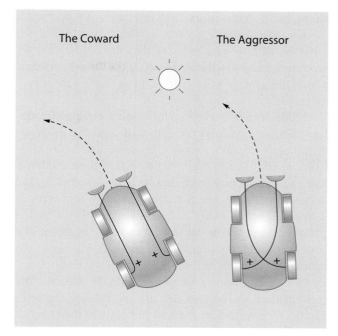

Figure 4.6   Two very simple vehicles, each equipped with two sensors that excite motors on the rear wheels. The wheel linked to the sensor closest to the sun will turn faster than the other wheel, thus causing the vehicle to turn. By simply changing the wiring scheme from uncrossed to crossed, the behavior of the vehicles is radically altered. The "coward" will always avoid the source, whereas the "aggressor" will relentlessly pursue it. Adapted from Braitenberg (1984).

look inside Braitenberg's models, we see that there is no difference in how the two models process information, only in their patterns of connectivity.

Representations in Computer Models

Computer models differ widely in their representations. Symbolic models include, as we might expect, units that represent symbolic entities. A model for object recognition might have units that represent visual features like corners or volumetric shapes. Over the past decade or so, an alternative architecture has gained popularity: neural networks. In neural networks, processing is distributed over innumerable units whose input and output can represent specific features. For example, they may indicate whether a stimulus contains a visual feature such as a vertical or a horizontal line. Of critical importance in many of these models, however, is that so-called hidden units are connected with input and output units. Hidden units provide intermediate processing steps between the input and output units. They allow the model to extract the information that allows for the best mapping between the input and desired output by changing the strengths of connections between units. To do this, a modeler must specify a learning rule, a quantitative description of how processing within the model changes according to how well it performs. If the model performs poorly, the change is likely to be large. If the model performs well, the change is small.

Models can be very powerful in solving complex problems. Simulations cover the gamut of cognitive processes including perception, memory, language, and motor control. One of the most appealing aspects of these models is that the architecture resembles, at least superficially, the nervous system. In neural network models, processing is distributed across many units, similar to the way neural structures depend on the activity of many neurons. The contribution of any unit may be small in relation to the system's total output, but complex behaviors can be generated by the aggregate action of all units. In addition, the computations in these models are simulated to occur in parallel. The activation levels of the units in the network are updated simultaneously.

Computational models can vary widely in the level of explanation they seek to provide. Some models simulate behavior at the systems level, seeking to show how cognitive operations such as motion perception or skilled movements can be generated from a network of interconnected processing units. In other cases, the simulations operate at a cellular or even molecular level. For example, neural network models have been used to investigate how the varation in transmitter uptake is a function of dendrite geometry (Volfovsky et al., 1999). The amount of detail that must be incorporated into the model will be dictated to a large extent by the type of question being investigated. Many of these problems are difficult to evaluate without simulations, either experimentally because the available experimental methods are insufficient or mathematically because the solutions become too complicated given the many interactions between the processing elements.

An appealing aspect of neural network models, especially for people interested in cognitive neuroscience, is that "lesion" techniques demonstrate how a model's performance changes when its parts are altered. Unlike strictly serial computer models that collapse if a circuit is broken, neural network models degrade gracefully. The model may continue to perform appropriately after some units are removed, because each unit plays only a small part in the processing. "Artificial lesioning" is thus a fascinating way to test a model's validity. At the first level, a model is constructed to see if it adequately simulates normal behavior. Then "lesions" are made to see if the breakdown in the model's performance resembles the behavioral deficits observed in neurological patients.

Models Lead to Testable Predictions

The contribution of computer modeling usually goes beyond the assessment of whether a model succeeds in mimicking a cognitive process. Models can generate novel predictions that can be tested with real brains. An example of the predictive power of computer modeling comes from the work of John Desmond and John Moore (1991) at the University of Massachusetts. One of the best-studied animal models of learning is the classic conditioning of a rabbit's nictitating membrane response. An unconditioned, aversive stimulus—a puff of air to the eye—is preceded by a tone. The rabbit blinks after the air puff, which is referred to as an *unconditioned response* because it does not have to be learned; the blink occurs spontaneously. With repeated pairings of the tone and air puff, the animal begins to blink in response to the tone, which is a *conditioned response* because the animal must learn that the tone predicts the air puff.

Rabbits are sensitive to the temporal relation between the tone and the air puff. The conditioned response is timed so it peaks prior to the air puff, which is what makes this type of learning adaptive. By timing the response appropriately, the animal can minimize the air puff's impact. Animals can learn to associate these two stimuli even when a silent interval separates the tone's offset and the air puff's onset. This is referred to as a *trace learning task* because the rabbit must learn to associate the air puff with a stimulus that no longer is present. Only a trace remains upon presentation of the air puff.

Desmond and Moore were interested in how animals timed their responses in a trace learning task. Two hypotheses appeared tenable. The conditioned response could be time locked either to the tone's onset or to its offset. If the former, we assume that the animal has a way to sustain a stimulus's activation beyond its actual duration. Alternatively, the offset of the tone might be the salient signal that triggers the response. Desmond and Moore simulated learning with either type of representation. And they also found that a model incorporating both representations not only simulated the behavior but also led to a novel prediction. Consider a condition in which the tone and trace interval last 200 and 200 msec, respectively. With these parameters, the rabbit learns to close its eye just short of 400 msec. Learning that is time locked to the tone's onset requires the animal to represent an interval of 400 msec; the tone's offset (or trace interval) requires only 200 msec.

These observations led Desmond and Moore to explore a series of simulations with a previously untested transfer task. After training a "model rabbit," the tone's duration was extended to 400 msec while the trace interval remained constant at 200 msec (Figure 4.7). Models with either an onset or an offset representation yielded unsurprising predictions: When the response was time locked to the tone's offset, the response was delayed by 200 msec, resulting in an optimally timed behavior. When the response was time locked to the tone's onset, the latency of the response was unchanged, and it now preceded the air puff by 200 msec. The model that incorporated two timing mechanisms, on the other hand, revealed a surprising prediction: During transfer, this model produced two conditioned responses, one that was timed to the tone's onset and a second that was timed to the tone's offset. When researchers returned to real rabbits, they confirmed the prediction!

With hindsight, it may seem like Desmond and Moore did not need computer modeling to derive this prediction. The problem could have been revealed by a careful analysis of the information that rabbits must represent. But this analysis eluded researchers who had worked on this problem for many decades. The modeling work led Desmond and Moore to be explicit in deriving mechanisms for representing temporal information, and with an explicit model, new predictions could be generated.

Limitations with Computer Models

Computer modeling is limited as a method for studying the operation of living nervous systems. For one thing, there are almost always radical simplifications in how the nervous system is modeled. While the units in a typical neural network model bear some similarity to neurons—for example, nonlinear activation rules produce spike-like behavior—the models are limited in scope, usually consisting of just a few hundred or so elements, and it is not always clear whether the elements correspond to single neurons or ensembles of neurons. Second, some requirements and problems arise in modeling work, particularly on learning, and are at odds with what we know occurs in biological organisms. Many network models require a homunculus-like teacher who "knows" the right answer and can be used to correct the behavior of the internal elements. And these models can suffer "catastrophic interference"—the loss of old information when new material is presented.

Third, most modeling efforts are restricted to relatively narrow problems such as demonstrating how the Stroop effect can be simulated by postulating separate word-name and word-color representations, under the control of a common attentional system. As such, they provide useful computational tests of the viability of a particular hypothesis but are typically less useful in generating new predictions. Moreover, as some critics have argued, unlike experimental work that by its nature is cumulative, modeling research tends to occur in isolation. There may be lots of ways

Figure 4.7 **(a)** Basic paradigm for eyeblink conditioning. A 200-msec tone is followed by a 50-msec air puff. Early in training, the animal blinks in response to the air puff. With repeated pairings, the animal makes a conditioned response to the tone, minimizing the aversive effects of the air puff. **(b)** In the transfer test, the duration of the tone is extended to 400 msec. Divergent predictions are derived from models in which the timing between the tone and air puff is triggered by either the onset or the offset of the tone. The experimental results showed that the rabbits actually made two conditioned responses in the transfer phase, one linked to the tone onset and one linked to the tone offset.

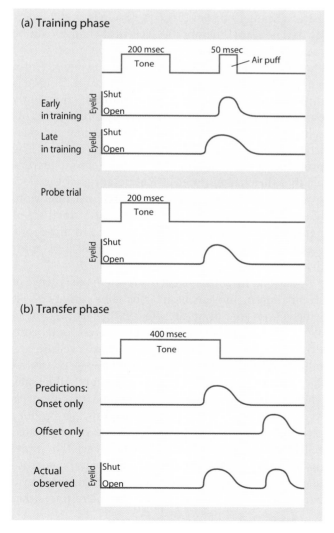

to model a particular phenomenon, but less effort has been devoted to devising critical tests that pit one theory against another.

These limitations are by no means insurmountable, and we should expect the contribution of computer simulations to continue to grow in the cognitive neurosciences. Indeed, the trend in the field is for modeling work to be more constrained by neuroscience, with researchers replacing generic processing units with elements that embody the biophysics of the brain. In a reciprocal manner, computer simulations provide a useful way to develop theory, which may then aid experimentalists in designing experiments and interpreting results.

EXPERIMENTAL TECHNIQUES USED WITH ANIMALS

The use of animals for experimental procedures has played a critical role in the medical and biological sciences. Although many insights can be gleaned from careful observations of people with neurological disorders, as we will see throughout this book, such methods are in essence correlational. We can observe how behavior is disturbed following a neurological insult, but it can be difficult to pinpoint the exact cause of the disorder. For one thing, insults such as stroke or tumor tend to be quite large, with the damage extending across many neural structures. Moreover, damage in one part of the brain may disturb function in other parts of the brain that are spared. There is also increasing evidence that the brain is a plastic device: Neural function is constantly being reshaped as a function of our experiences, and such reorganization can be quite remarkable following neurological damage.

The use of animals in scientific research allows researchers to adopt a more experimental approach. Because neural function depends on electrochemical processes, neurophysiologists have developed techniques that can be used to measure and manipulate neuron activity. Some measure and record cell activity, either in passive or in active conditions. Others manipulate activity by creating lesions through the destruction or temporary inactivation of targeted brain regions. Lesion studies in animals face the same limitations associated with the study of human neurological dysfunction. However, modern techniques allow the researcher to be highly selective in creating these lesions, and the effects of the damage can be monitored carefully following the surgery.

Single-Cell Recording

The most important technological advance in **neurophysiology**—perhaps all of neuroscience—was the development of methods to record the activity of single neurons in laboratory animals. With this method, the understanding of neural activity jumped a quantum leap. No longer did the neuroscientist have to be content with describing nervous system action in terms of functional regions. **Single-cell recording** enabled researchers to describe response characteristics of individual elements.

In single-cell recording, a thin electrode is inserted into an animal's brain. If the electrode is in the vicinity of a neuronal membrane, electrical changes can be measured (see Chapter 2). Although the surest way to guarantee that the electrode records the activity of a single cell is to record intracellularly, this technique is difficult and penetrating the membrane frequently damages the cell. Thus, single-cell recording is typically done extracellularly. With this method, the electrode is situated on the outside of the neuron. The problem with this approach is that there is no guarantee that the changes in electrical potential at the electrode tip reflect the activity of a single neuron. More likely, the tip will record the activity of a small set of neurons. Computer algorithms are used to differentiate this pooled activity into the contributions from individual neurons.

Neurons are constantly active, even in the absence of stimulation or movement. This baseline activity varies widely from one brain area to another. For example, some cells within the basal ganglia have spontaneous firing rates of over 100 spikes/sec, whereas cells in another basal ganglia region have a baseline rate of around 1 spike/sec. These spontaneous firing levels fluctuate. The primary goal of single-cell recording experiments is to determine experimental manipulations that produce a consistent change in the response rate of an isolated cell. Does the cell increase its firing rate when the animal moves its arm? Is this specific to movements in a particular direction? When is the movement terminated by a particular stimulus (e.g., food)? As interesting, what makes the cell decrease its response rate? The neurophysiologist is interested in what causes change in the synaptic activity of a neuron.

As such, single-cell recording is essentially a correlational approach. The experimenter seeks to determine the response characteristics of individual neurons by correlating their activity with a given stimulus pattern or behavior. The technique has been used in almost all regions of the brain in a wide range of nonhuman species. For sensory neurons, the experimenter might manipulate the type of stimulus presented to the animal. For motor neurons, recordings can be made as the animal performs a task or moves about the cage. Single-cell recordings also have been made in higher brain centers to examine changes in cellular activity related to emotion and learning.

In the typical neurophysiological experiment, recordings are obtained from a series of cells in a targeted area of interest. In this manner, a functional map can describe similarities and differences between neurons in a specified cortical region. One area where the single-cell method has been used extensively is in the study of the visual system of primates. In a typical experiment, the researcher will target the electrode to a cortical area that contains cells thought to respond to visual stimulation. Once a cell is identified, the researcher tries to characterize its response properties.

A single cell is not responsive to all visual stimuli. There are a number of stimulus parameters that might correlate with the variation in the cell's firing rate such as the shape of the stimulus, its color, and whether or not it is moving (see Chapter 5). An important factor is the location of the stimulus. As shown in Figure 4.8, all visually sensitive cells only respond to stimuli in a limited region of space. This region of space is referred to as that cell's **receptive field**. For example, some neurons will respond when the stimulus is located in the lower left portion of the visible field. For other neurons, the stimulus may have to be in the upper left region of the visible field. The size of the receptive fields of visual cells varies, with a general rule being that they are smallest in primary visual cortex and become larger toward association visual areas. Thus, a stimulus will alter only the response of a cell in primary visual cortex when it is positioned in a very restricted region of the visible world. If the stimulus is moved outside this region of space, the cell will return to its spontaneous level of activity. In contrast, displacing a stimulus over a large distance may produce a similar increase in the firing rate of visually sensitive cells in the temporal lobe.

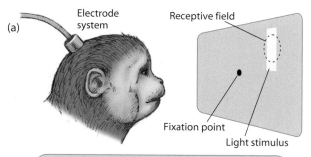

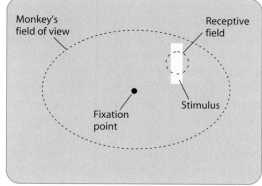

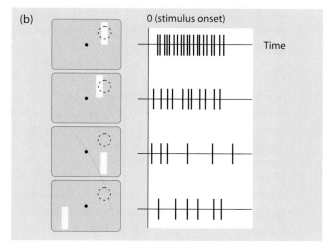

Figure 4.8 Electrophysiological methods are used to identify the response characteristics of cells in the visual cortex. **(a)** While the activity of a single cell is monitored, the monkey is required to maintain fixation, and stimuli are presented at various positions in the visual field. **(b)** The vertical lines to the right of each stimulus correspond to individual action potentials. The cell fires vigorously when the stimulus is presented in the upper right quadrant, thus defining the receptive field for this cell.

Neighboring cells have at least partially overlapping receptive fields. As a region of visually responsive cells is traversed, there is an orderly relation between the receptive field properties of these cells and the external world. A representation of external space is

The Ethics and Practice of Animal Research

Students on most university campuses are familiar with the annual protests during Animal Liberation Week of groups opposed to the use of animals for research purposes. These groups hand out inflammatory pamphlets graphically depicting how researchers callously exploit laboratory rats, cats, and especially monkeys in their pursuit of knowledge. While many of the protests are peaceful and law-abiding, there have been numerous incidents of violence against people and property: Windows have been smashed, equipment destroyed, animals kidnapped, and professors harassed at their homes. The protesters argue that university administrators have become dependent on the money generated by public and private grants supporting "frivolous and unnecessary" research, and thus are forced to resort to more dramatic measures to raise the public's consciousness.

Scientists have taken the offense, eager to educate the public about the importance of these forms of research. Almost all that we know about the structure and physiology of the nervous system has depended on invasive studies of animals. Human autopsies can provide only the crudest understanding of the anatomy of the nervous system. The fine structure that is revealed by sophisticated staining procedures is not possible unless the stains are injected into live animals and allowed to propagate along metabolically active tissue. All that we know about the operation of single neurons has been made possible only by the use of laboratory animals.

This basic research does not exist in a vacuum, sought after as part of an ephemeral quest for knowledge. Scientists and public policy makers have long recognized that advances in medicine depend on the ability to carry out both basic research and clinically inspired research. Human studies of brain metabolites may reveal that excessive dopamine levels are associated with schizophrenia, but animal research is essential for understanding where dopamine is produced and the metabolic processes that regulate the production and uptake of this transmitter. Only rarely have medical treatments been discovered in the absence of animal models. In most cases, new medical treatments only become available after years of careful experimental work with animals, first involving years of basic research, then followed by the careful development of clinical measures involving animal studies.

The scientific research community has served as a visible target for animal rights groups. One reason has been a few well-publicized cases in which videotapes or photographs were used to demonstrate the extreme experimental manipulations being practiced on laboratory animals. Two of the more well-known cases involved monkeys. In one case the animals were subjected to severe blows to the head as part of a study of head trauma in car accidents. In the other case, the animals were in poor health after undergoing fetal surgery to destroy sensory fibers. The latter case resulted in the only conviction recorded in this country of a scientist on animal cruelty charges.

While abuses most certainly exist, research with laboratory animals is closely regulated by both research institutions and federal agencies. All research protocols must be approved by institutional animal care and use committees that include not only scientific peers but also lay members of the community. Three basic principles must be met for any research project to be certified (Rowan and Rollin, 1983). First, the goals of the research must be articulated clearly, showing that the studies are worthwhile and will advance our knowledge of the nervous system. Second, experiments must be conducted so as to minimize pain and distress through the use of anesthetics and analgesics. Any animals that show evidence of suffering must be humanely destroyed. Third, alternative methods that might yield similar knowledge must be considered.

The evidence suggests that these codes are strictly followed. In almost all studies involving surgical procedures, animals are provided with anesthesia. The rare exceptions are those in which the focus is on how the brain reacts to pain, an important problem faced by people recovering from surgery. The argument that computer models can serve as realistic substitutes for animal research is specious. Current models can simulate only the simplest of neural functions, and even these models are only useful in terms of how well they converge with the results of studies on living organisms.

Nonetheless, it is important that the public debate the fundamental question regarding whether it is ethical for humans to exploit another species for their own benefit. A researcher in 1898 was quoted as saying, "Animals have no more rights than inanimate objects, and it is no worse from an ethical point of view to flay the forearm of an ape or lacerate the leg of a dog than to rip open the sleeve or rend a pair of pantaloons." It is clear that few would take this stance in our modern society. We must understand both the benefits and the costs of animal research in order to form educated opinions on these ethical matters.

The morality of using animals for research purposes has been debated for centuries. People on both sides of the debate have recognized the importance of engaging the public, a point underscored by the fact that members of the U.S. Congress receive more letters on this issue than any other.

reflected in a continuous manner across the cortical surface; neighboring cells have receptive fields of neighboring regions of external space (Figure 4.9). As such, cells form a **topographic representation,** an orderly mapping between an external dimension such as spatial location and the neural representation of that dimension. In vision, topographic representations are frequently referred to as being **retinotopic.** The retina is composed of a continuous sheet of photoreceptors, neurons that respond to visible light passing through the lens of the eye. Visual cells in subcortical and cortical areas maintain retinotopic information. Thus, if light falls on one spot of the retina, cells with receptive fields spanning this area will be activated. If the light moves and falls on a different region of the retina, activity ceases in these cells and begins in other cells whose receptive fields encompass the new region of stimulation. In this manner, visual areas provide a representation of the location of the stimulus. Cell activity

within a retinotopic map correlates with (i.e., predicts) the location of the stimulus.

There are other types of topographic maps. In Chapter 3 we reviewed the motor and somatosensory maps along the central sulcus that provide topographic representations of the body surface. In a similar sense, auditory areas in the subcortex and cortex contain tonotopic maps, in which the physical dimension reflected in neural organization is a stimulus's sound frequency. With a tonotopic map, some cells are maximally activated by a 1000-Hz tone, and others by a 5000-Hz tone. In addition, neighboring cells tend to be tuned to similar frequencies. As such, sound frequencies are reflected in cells that are activated upon the presentation of a sound. Tonotopic maps are sometimes referred to as *cochleotopic* because the cochlea, the sensory apparatus in the ear, contains hair cells tuned to distinct regions of the auditory spectrum.

Single-cell recordings have limitations. As with any correlational technique, it is hard to establish cause and

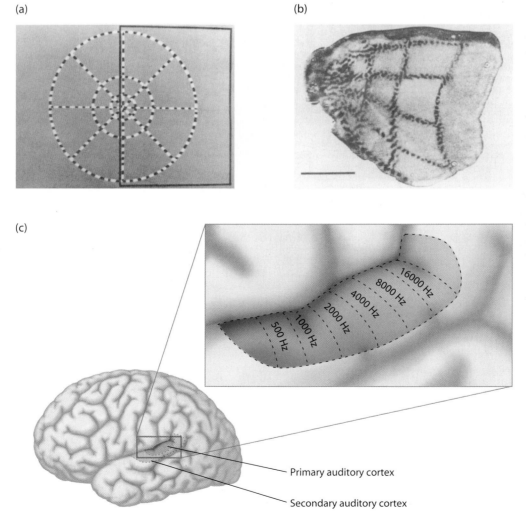

(a)

(b)

(c)

Primary auditory cortex

Secondary auditory cortex

Figure 4.9 Topographic maps of the visual and auditory cortex. In the visual cortex, the receptive fields of the cells define a retinotopic map. While viewing the stimulus **(a),** the monkey was injected with a radioactive agent. **(b)** Metabolically active cells in the visual cortex absorb the agent, revealing how the topography of the retina is preserved across the striate cortex (right). **(c)** In the auditory cortex, the frequency tuning properties of the cells define a tonotopic map. Topographic maps are also seen in the somatosensory and motor cortex. (a,b) From Tootell et al. (1982). (c) Adapted from Bear et al. (1996).

effect. Suppose that we find two areas, areas X and Y, with visually responsive cells. We could consider three distinct possibilities: Cells in area X might be elaborating on information provided by cells in area Y, cells in area Y might be activated by cells in area X, or areas X and Y might reflect independent processing of input from other visual sensory cells.

Some problems can be overcome by combining physiological and neuroanatomical methods. If the anatomist can establish that the primary visual cortex is the main recipient of output from the thalamus's lateral geniculate nucleus, it is likely that what gets processed in the primary visual cortex is limited by what is provided by the lateral geniculate nucleus. Yet reciprocal connections happen between many brain regions: Area X is innervated by area Y and in turn sends axonal projections back to area Y. This makes it difficult to determine if one area is downstream of another, especially for cortical areas that are not part of the primary sensory zones.

When the single-cell method was first introduced, neuroscientists were optimistic that the mysteries of brain function would be solved. All they needed was a catalogue of different cells' contributions. Yet it soon became clear that with neurons, the aggregate behavior of cells might be more than the sum of its parts. The function of an area might be better understood by identifying correlations in the firing patterns of groups of neurons rather than by identifying the response properties of each individual neuron. This inspired single-cell physiologists to develop new techniques that allow recordings to be made in many neurons simultaneously. Bruce McNaughton at the University of Arizona studied how the rat hippocampus represents spatial information by simultaneously recording from 150 cells (Wilson and McNaughton, 1994)! Other researchers use multiple-cell recordings to ask whether temporal correlations between neurons might provide significant processing information. Multiple single-cell recording may bring about the next revolution in neurophysiology.

Lesions

The brain is a complicated structure, composed of many structures including subcortical nuclei and distinct cortical areas. It seems evident that any task a person performs requires the successful operation of the brain's components. A long-standing method of the neurophysiologist has been to study how behavior is altered by selectively removing one or more of these parts. The logic of this approach is straightforward. If a neural structure contributes to a task, then rendering the structure dysfunctional should impair the performance of that task.

Humans obviously cannot be subjected to **brain lesions** to investigate their nervous system's function. Human neuropsychology requires patients with naturally occurring lesions. But animal researchers have not been constrained in this way. They share a long tradition of studying brain function by comparing the effects of different brain lesions. Around the turn of the twentieth century, the English physiologist Charles Sherrington employed the lesion method to investigate the importance of feedback in limb movement in the dog (see Chapter 11). By severing the nerve fibers carrying sensory information into the spinal cord, he observed that the animals stopped walking.

Lesioning a neural structure will eliminate that structure's contribution. But the lesion also might force the animal to change its normal behavior and alter the operation of intact structures. One cannot be confident that the effect of a lesion eliminates only the contribution of a single structure. The function of neural regions that are connected to the lesioned area might be altered, either because they are deprived of their normal neural input or because their axons fail to make normal synaptic connections. The lesion might also cause the animal to develop a compensatory strategy to minimize the consequences of the lesion. For example, when monkeys are deprived of sensory feedback to one arm, they stop using the limb. However, if the sensory feedback to the other arm is eliminated at a later date, the animals begin to use both limbs (Taub and Berman, 1968). The monkeys prefer to use a limb that has normal sensation. But the second surgery shows that they could indeed use the other limb.

So with this methodology we should remember that a lesion may do more than eliminate the function provided by the lesioned structure. Nonetheless, the method has been critical for neurophysiologists. Over the years, lesioning techniques have been refined, allowing for much greater precision. Most lesions were originally made by aspirating neural tissue. In aspiration experiments, a suction device is used to remove the targeted structures. Other methods for destroying tissue involved applying electrical charges strong enough to destroy tissue. One problem with these methods is the difficulty of being selective. Any tissue within range of the voltage generated by the electrode tip would be destroyed. For example, a researcher might want to observe the effects of a lesion to a certain cortical area, but if the electrolytic lesion extends into underlying white matter, these fibers also will be destroyed. This might render a distant structure dysfunctional because it will be deprived of some input.

Newer methods allow for more control over the extent of lesions. Most notable are neurochemical lesions. Sometimes a drug will selectively destroy cells that use a certain transmitter. For instance, systemic injection of 1-methyl-4-phenyl-1,2,3,6-tetrahydropyridine (MPTP) destroys dopaminergic cells in the substantia nigra, producing an animal version of Parkinson's disease (see Chapter 11). Other neurochemical lesions require applying the drug to the targeted region. Kainic acid is used in many studies because its toxic effects are limited to cell bodies. Application to an area will destroy the neurons whose cell bodies are near the site of the injection, but will spare any axonal fibers passing through this area. Other researchers choose to make reversible lesions using chemicals that produce a transient disruption in nerve conductivity. So long as the drug is active, the exposed neurons will not function. When the drug wears off, function gradually returns. The appeal of this method is that each animal can serve as its own control. Performance can be compared during the "lesion" and "nonlesion" periods. A different form of reversible lesion involves cooling neural tissue by injecting a chemical that induces a low temperature. When the tissue is cooled, metabolic activity is disrupted, thereby creating a temporary lesion. When the coolant is removed, metabolic activities resume and the tissue becomes functional again.

Pharmacological manipulations also can be used to produce transient functional lesions. For example, the acetylcholine antagonist scopolamine produces temporary amnesia such that the recipient fails to remember much of what he or she was doing during the period when the drug was active. Since there are no adverse consequences from the low doses required to produce the amnesia, scopolamine provides a tool with which researchers can study the kinds of memory problems that plague patients with hippocampal damage (Nissen et al., 1987). However, systemic administration of this drug produces widespread changes in brain function, and thus limits its utility as a model of hippocampal dysfunction.

Genetic Manipulations

The turn of the millennium witnessed the climax of one of the great scientific challenges, the mapping of the human genome. Scientists now have a complete record of the genetic sequence on our chromosomes. At present, the utility of this knowledge is limited; we have only begun to understand how these genes code for all aspects of human structure and function. In essence, what we have is a map containing the secrets to many treasures: What causes people to grow old? Why are some people more susceptible to certain cancers than other people? What dictates whether embryonic tissue will become a skin cell or a brain cell? But deciphering this map is an imposing task that is likely to take many years of intensive study.

Genetic disorders are manifest in all aspects of life including brain function. Certain diseases such as Huntington's disease are clearly heritable (see Chapter 3 and Degenerative and Infectious Disorders later in this chapter). Indeed, scientists can now predict whether or not individuals will develop this debilitating disorder by analyzing their genetic code. This predictability was made possible by analyzing the genetic code of individuals who developed Huntington's disease, and comparing their genes with relatives who remained disease free. In this particular disease, the differences were restricted to a single chromosomal abnormality. This discovery not only allows for predictive diagnosis, but also is expected to lead to new treatments that will prevent the onset of Huntington's. Scientists hope to devise techniques to alter the aberrant genes, either by modifying them or by figuring out a way to prevent them from being expressed.

In a similar way, scientists have sought to understand other aspects of normal and abnormal brain function through the study of genetics. Behavioral geneticists have long known that many aspects of cognitive function are heritable. For example, by controlling mating patterns based on spatial learning performance, "maze bright" and "maze dull" strains of rats can be developed. Rats who are quick to learn to navigate mazes are likely to have offspring with similar abilities, even if the offspring are raised by rats who are slow to navigate the same mazes. Such correlations also are observed across a range of human behaviors including spatial reasoning, reading speed, and even preferences in watching television (Plomin et al., 1990)! This should not be taken to mean that our intelligence or behavior is genetically determined. "Maze bright" rats perform quite poorly if raised in an impoverished environment. The truth surely reflects complex interactions between the environment and genetics.

To understand the genetic component of this equation, neuroscientists are now working with many species, seeking to identify the genetic mechanisms of both brain structure and function. Dramatic advances have been made in studies with the fruit fly and mouse, two species with reproductive propensities that allow many generations to be spawned over a relatively short period of time. As with humans, the genome sequence for these species has been mapped out. More important, the functional role of many genes can be explored. A key methodology involves the development of genetically

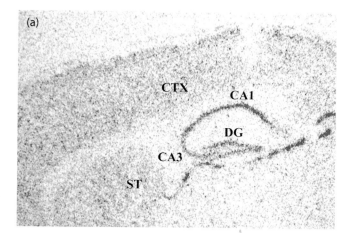

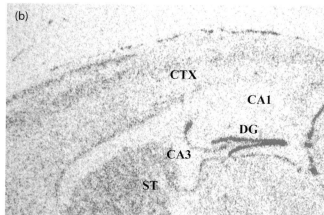

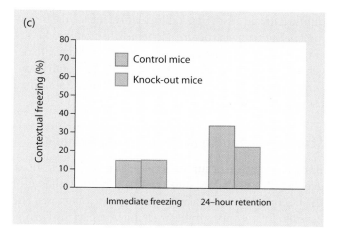

altered animals, or what are referred to as knock-out subspecies. The term *knock-out* comes from the fact that specific genes have been manipulated so that they are no longer present or expressed. Scientists can then study the new species to explore the consequences of these changes. For example, weaver mice are a knock-out strain in which Purkinje cells, the prominent cell type in the cerebellum, fail to develop. As the name implies, these mice exhibit coordination problems.

At an even more focal level, **knock-out procedures** have been used to create strains that lack single types of postsynaptic receptors in specific brain regions while leaving intact other types of receptors. Susumu Tonegawa at the Massachusetts Institute of Technology (MIT) and his colleagues developed a mouse strain in which they altered cells within a subregion of the hippocampus that typically contain a receptor for *N*-methyl-D-aspartate, or NMDA (Wilson and Tonegawa, 1997) (see Chapter 8). Knock-out strains lacking the NMDA receptor in the hippocampus exhibit poor learning on a variety of memory tasks, providing a novel approach for linking this complex behavior with its molecular substrate (Figure 4.10). In a sense, this approach constitutes a lesion method, but at a microscopic level.

Figure 4.10 Brain slices through the hippocampus, showing the absence of a particular receptor in genetically altered mice. **(a)** Cells containing the gene associated with the receptor are stained in black. **(b)** The gene is absent in the CA1 region of the slice from the knock-out mouse. **(c)** Fear conditioning is impaired in knock-out mice. After receiving a shock, the mice freeze. When normal mice are placed in the same context 24 hours later, they show strong learning by the large increase in the percentage of freezing responses. This increase is reduced in the knock-out mice. (a, b) From Rampon et al. (2000). (c) Adapted from Rampon et al. (2000).

NEUROLOGY

Human pathology has long provided key insights to the relation between brain and behavior. Observers of neurological dysfunction have certainly contributed much to our understanding of cognition—long before the advent of cognitive neuroscience. Discoveries concerning the contralateral wiring of sensory and motor systems were made by physicians in ancient societies attending to warriors with open head injuries. Postmortem studies by early neurologists, such as Broca and Wernicke, were instrumental in linking the left hemisphere with language functions (see Chapter 1). Many other disorders of cognition were described in the first decades of the twentieth century, when neurology became a specialty in medicine.

Even so, there is now an upsurge in testing neurological patients to elucidate issues related to normal and aberrant cognitive function. As with other subfields of cognitive neuroscience, this enthusiasm has been inspired partly by advances in the technologies for diagnosing neurological disorders. As important, studies of patients with brain damage have benefited from the use of experimental tasks derived from research with healthy people.

Examples of the merging of cognitive psychology and **neurology** are presented at the end of this chapter; in this section, we focus on the causes of neurological disorders and the tools that neurologists use to localize neural pathology. We also take a brief look at treatments for ameliorating neurological disorders.

Basic research questions, such as those attempting to link cognitive processes to neural structures, can be best addressed by selecting patients with a single neurological disturbance whose pathology is well circumscribed. Patients who have suffered trauma or infections will frequently have diffuse damage, rendering it difficult to associate the behavioral deficit with a structure. Nonetheless, extensive clinical and basic research studies have focused on patients with degenerative disorders such as Alzheimer's disease, both to understand the disease processes and to characterize abnormal cognitive function.

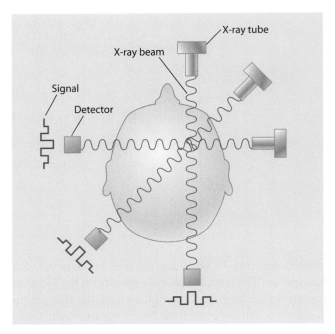

(a)

Structural Imaging of Neurological Damage

Brain damage can occur from vascular problems, tumors, degenerative disorders, and trauma. The first charge of the neurologist is to make the appropriate diagnosis. She needs to follow appropriate procedures, especially if a disorder is life-threatening, and to work toward stabilizing the patient's condition. While diagnosis frequently can be made on the basis of a clinical examination, most hospitals in the Western world are equipped with tools that help neurologists visualize brain structure.

Computed tomography (**CT** or **CAT** scanning), introduced commercially in 1983, has been an extremely important medical tool for structural imaging of neurological damage in living people. This method is an advanced version of the conventional x-ray study; whereas the conventional x-ray study compresses three-dimensional objects into two dimensions, CT allows for the reconstruction of three-dimensional space from the compressed two-dimensional images. Figure 4.11a depicts the method, showing how x-ray beams are passed through the head and a two-dimensional image is generated using sophisticated computer software.

To undergo CT, a patient lies supine in a scanning machine. The machine has two main parts: an x-ray source and a set of radiation detectors. The source and detectors are located on opposite sides of the scanner. These sides can rotate, allowing the radiologist to project x-ray beams from all possible directions. Starting at one position, an x-ray beam passes through the head. Some radiation in the x-ray is absorbed by intervening

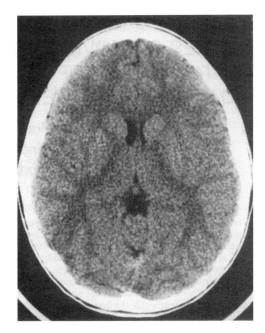

(b)

Figure 4.11 Computed tomography (CT) provides an important tool for imaging neurological pathology. **(a)** The CT process is based on the same principles as x-rays. An x-ray is projected through the head and the recorded image provides a measurement of the density of the intervening tissue. By projection of the x-ray from multiple angles and with the use of computer algorithms, a three-dimensional image based on tissue density is obtained. **(b)** A transverse CT image. The dark regions along the midline are the ventricles, the reservoirs of cerebrospinal fluid.

tissue. The remainder passes through and is picked up by the radiation detectors located on the opposite side of the head. The x-ray source and detectors are then rotated and a new beam is projected. This process is repeated until x-rays have been projected over 180 degrees. At this point, recordings made by the detectors are fed into a computer that reconstructs the images.

The key principle underlying CT is that the density of biological material varies, and that the absorption of x-ray radiation is correlated with tissue density. High-density material such as bone will absorb a lot of radiation. Low-density material such as air or blood will absorb little radiation. The absorption capacity of neural tissue lies between these extremes. Thus, the software for making CT scans really provides an image of the differential absorption of intervening tissue. The reconstructed images are usually contrast reversed: High-density regions show up as light colored and low-density regions are dark colored.

Figure 4.11b shows a CT scan from a healthy individual. Most of the cortex and white matter appear as homogeneous gray areas. The typical spatial resolution for CT scanners at most hospitals is approximately 0.5 to 1.0 cm in all directions. Each point on the image reflects an average density of that point and the surrounding 1.0 mm of tissue. Thus, it is not possible to discriminate two objects that are closer than approximately 5 mm. Since the cortex is only 4 mm thick, it is very difficult to see the boundary between white and gray matter on a CT scan. The white and gray matter are also of very similar density, further limiting the ability of this technique to distinguish them. But larger structures can be easily identified. The surrounding skull and eye sockets appear white because of the high density of bone. The ventricles are black owing to the cerebrospinal fluid's low density.

While CT machines are still widely used, most hospitals in the Western world have now added a second important imaging tool, the **magnetic resonance imaging (MRI)** scanner. In contrast to the x-rays used for CT, the MRI process exploits the magnetic properties of organic tissue. Based on the number of the protons and neutrons in their nuclei, certain atoms are especially sensitized to magnetic forces. One such atom that is pervasive in the brain, and indeed in all organic tissue, is hydrogen. The protons that form the nucleus of the hydrogen atom are in constant motion, spinning about their principal axis. This motion creates a tiny magnetic field. In their normal state, the orientation of these protons is randomly distributed, unaffected by the weak magnetic field created by the earth's gravity (Figure 4.12).

The MRI machine creates a powerful magnetic field, measured in tesla units. Whereas gravitational forces on the earth create a magnetic field of about 1/1000 tesla,

the scanners in most hospitals produce a magnetic field that ranges from 0.5 to 1.5 tesla. When a person is placed within the magnetic field of the MRI machine, the protons become oriented in the direction parallel to the magnetic force. Radio waves are then passed through the magnetized regions, and as the protons absorb the energy in these waves, their orientation is perturbed in a predictable direction. When the radio waves are turned off, the absorbed energy is dissipated and the protons rebound toward the orientation of the magnetic field. This synchronized rebound produces energy signals that are picked up by detectors surrounding the head. By systematically measuring the signals throughout the three-dimensional volume of the head, the MRI system can then construct the actual image, reflecting the distribution of the protons and other magnetic agents in the tissue.

As can be seen in Figure 4.12, MRI scans give a much clearer image of the brain than is possible with CT scans. This improvement reflects the fact that the density of protons is much greater in gray matter compared to white matter. With MRI, it is easy to see the individual sulci and gyri of the cerebral cortex. A sagittal section at the midline reveals the impressive size of the corpus callosum. The MRI scans can resolve structures that are less than 1 mm, allowing elegant views of small, subcortical structures such as the mamillary bodies or superior colliculus.

An imaging method also commonly used in neurology is angiography. As shown in Figure 4.13, this method helps visualize the distribution of blood by highlighting major arteries and veins. A dye is injected into the vertebral or carotid artery and then the person undergoes an x-ray study. This method is particularly useful for diagnosing disorders related to vascular abnormalities. Some people are born with an arteriovenous malformation, an irregularity in the shape of an arterial branch or wall. Such a malformation can leak, causing an ischemic disorder, a temporary loss of blood, or hemorrhage, resulting in a massive disruption in blood flow.

Causes of Neurological Disorders

Nature has sought to ensure that the brain remains healthy. Structurally, the skull provides a thick, protective encasement. The distribution of arteries is extensive and even redundant for much of the brain. Even so, the brain is subject to numerous disorders, and their rapid treatment is frequently essential to avoid chronic, debilitating problems or death.

VASCULAR DISORDERS

As with all tissue, neurons need a steady supply of oxygen and glucose. These substances are essential for the cells to produce energy and make transmitters for neural communication. The brain uses 20% of all the oxygen we breathe, an extraordinary amount considering that it accounts for only 2% of the total body mass. The continuous supply of oxygen is essential: A loss of oxygen for as briefly as 10 minutes can result in neural death.

Oxygen and glucose are distributed to the brain from four primary arteries: the two internal carotid and two vertebral arteries. Each carotid artery branches into two major arteries, the anterior cerebral artery and the middle cere-

bral artery, as well as several smaller ones. Together, these arteries act as a network to supply the anterior and middle portions of the cortex with blood. The vertebral arteries join together to form the basilar artery. Inferior branches from the basilar artery irrigate the cerebellum and posterior part of the brainstem. More superiorly, this system branches into two posterior cerebral arteries to provide blood to the occipital lobe and medial temporal lobe. The major cerebral arteries partially overlap in their distribution, in areas referred to as *borderzones* or *watershed areas*.

Cerebral vascular accidents, or strokes, occur when blood flow to the brain is suddenly disrupted. The most frequent cause of stroke is when a foreign substance occludes the normal passage of blood. Over years, arteriosclerosis, the buildup of fatty tissue, occurs in the heart. This tissue can break free, becoming an embolus that is carried off in the bloodstream. An embolism that enters the cranium may easily pass through the large carotid or vertebral arteries. But as the arteries and capillaries reach the end of their distribution, their size decreases. Eventually, the embolism becomes stuck, or infarcted, blocking the flow of blood and depriving all downstream tissue of oxygen and glucose. Within a short period of time, this tis-

(a)

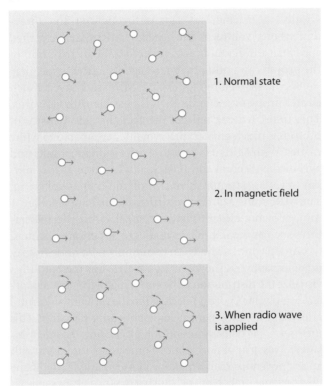

1. Normal state

2. In magnetic field

3. When radio wave is applied

Figure 4.12 Magnetic resonance imaging (MRI) exploits the fact that many organic elements such as hydrogen are magnetic. **(a)** In their normal state, the orientation of these elements is random. When an external magnetic field is applied, the elements become aligned and can be perturbed in a systematic fashion by the introduction of radio waves. The MRI scanner measures the endogenous magnetic fields generated by these elements as they spin. The density of hydrogen atoms is different in white and gray matter, making it easy to visualize these regions. **(b)** Transverse, coronal, and sagittal images. The finer resolution offered by MRI can be seen by comparing the transverse slice in this figure with the CT image in the previous figure. Both are from about the same level.

(b)

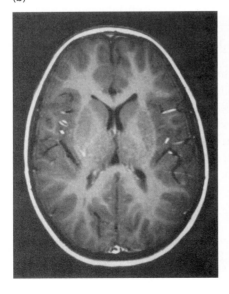
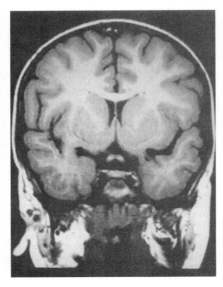
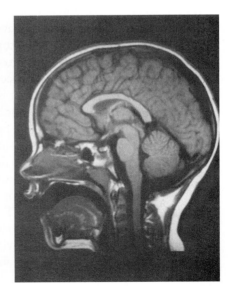

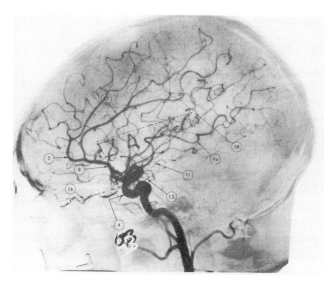

Figure 4.13 The angiogram provides an image of the arteries in the brain.

sue will become dysfunctional. If the blood flow is not rapidly restored, the cells will die (Figure 4.14a).

The onset of stroke can be quite varied, depending on the afflicted area. Sometimes the person may lose consciousness and die within minutes. Here the infarct is usually in the vicinity of the brainstem. When the infarct is cortical, the presenting symptoms may be striking, such as the sudden loss of speech and comprehension. In other cases, the onset can be innocuous. The person reports a mild headache or finds himself or herself unable to use a hand in an appropriate manner. The vascular system is fairly consistent between individuals; thus, stroke of a particular artery will typically lead to de-

struction of tissue in a consistent anatomical location. For example, occlusion of the posterior cerebral artery will invariably lead to deficits in visual perception.

There are many other types of cerebral vascular disorders. Ischemia can be caused by partial occlusion of an artery or capillary due to an embolus, or it can arise from a sudden drop in blood pressure that prevents blood from reaching the brain. A sudden rise in blood pressure can lead to cerebral hemorrhage (Figure 4.14b), or bleeding over a wide area of the brain due to the breakage of blood vessels. Spasms in the vessels can result in irregular blood flow and have been associated with migraine headaches.

Other disorders are due to problems in arterial structures. Cerebral arteriosclerosis is a chronic condition in which cerebral blood vessels become narrow due to the thickening and hardening of the arteries. This can result in persistent ischemia. More acute situations can arise if a person has an aneurysm, a weak spot or distention in a blood vessel. An aneurysm may suddenly expand or even burst, causing a rapid disruption of the blood circulation. While some aneurysms appear to develop spontaneously, others can develop in people born with an arteriovenous malformation. An arteriovenous malformation that has remained innocuous for many years may suddenly weaken.

Cerebral vascular accidents require immediate attention. Often a neurological examination and CT can reveal the problem. Arteriovenous malformations or aneurysms may require the more precise MRI or angiography. In these instances, surgery is frequently required to prevent further bleeding. Occlusive strokes, on the other hand, do not generally require surgery. Either the occluded tissue is already completely infarcted, the clot is too small to remove, or the

Figure 4.14 Vascular disorders. **(a)** Strokes occur when the blood flow to the brain is disrupted. This brain is from a person who had an occlusion of the middle cerebral artery. The person survived the stroke. Postmortem analysis showed that almost all of the tissue supplied by this artery had died and been absorbed. **(b)** This brain is from a person who died following a cerebral hemorrhage. The hemorrhage destroyed the dorsomedial region of the left hemisphere. The effects of a cerebrovascular accident 2 years prior to death can be seen in the temporal region of the right hemisphere.

(a)

(b)

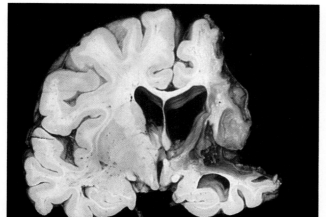

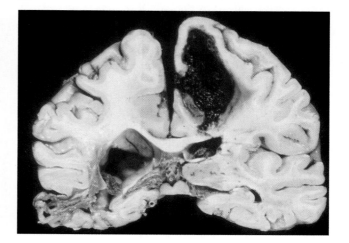

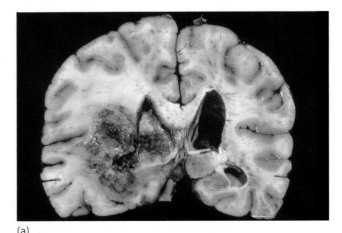

(a)

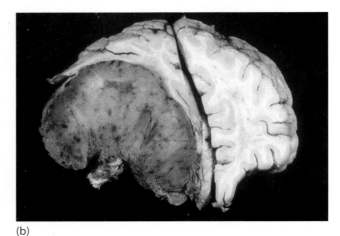

(b)

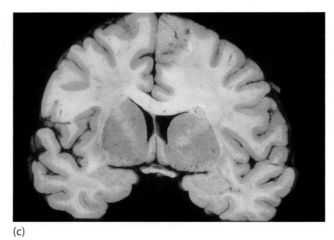

(c)

Figure 4.15 Postmortem views of three types of brain tumors. **(a)** A malignant glioma has infiltrated the white matter of the parietal lobe in the right hemisphere. **(b)** A large meningioma led to massive compression of the right frontal lobe. This patient had been hospitalized at age 41 for psychotic behavior, quite likely due to the effects of this slow-growing tumor. The tumor was not detected until autopsy. **(c)** A metastatic tumor is seen in the dorsomedial tip of the left hemisphere. This woman died 5 years after undergoing a mastectomy for breast cancer.

embolus has been absorbed into the surrounding tissue by the time of surgery. Treatment here usually involves the administration of drugs to dissolve the clot and restore circulation prior to permanent tissue injury.

TUMORS

Brain lesions also can result from tumors. A *tumor,* or *neoplasm,* is a mass of tissue that grows abnormally and has no physiological function. Brain tumors are relatively common, with most originating in the glia and other supporting white matter tissues. Tumors also can develop from gray matter or neurons, but these are much less common, particularly in adults. Tumors are classified as benign when they do not recur after removal and tend to remain in the area of their germination (although they can become quite large). Malignant, or cancerous, tumors are likely to recur after removal and are often distributed over a number of different areas. With brain tumors, the first concern is not usually whether the tumor is benign or malignant, but rather its location and prognosis. Concern is greatest when the tumor threatens critical neural structures. Neurons can be destroyed by an infiltrating tumor or become dysfunctional due to displacement by the tumor.

Three major types of brain tumors are distinguished according to where they originate (Figure 4.15). Gliomas are brain tumors that begin with the abnormal reproduction of glial cells. The rate of growth of different subtypes of gliomas can vary widely: Some escape detection for years; others expand rapidly and are malignant, with a poor prognosis because a lot of tissue is quickly disturbed. The second type of tumor, the meningioma, originates in the meninges, the protective membrane that surrounds the brain. Although meningiomas do not invade the brain, they can create severe neurological problems by producing abnormal pressure. Metastatic tumors, the third type, originate in a noncerebral structure such as the lungs, skin, and breasts. The malignant tissue invades the bloodstream or lymphatics and is ultimately carried to the brain. It is common for metastatic tumors to be widely distributed, affecting many structures.

DEGENERATIVE AND INFECTIOUS DISORDERS

Many neurological disorders result from progressive disease. Table 4.1 lists some of the more prominent degenerative and infectious disorders. In later chapters we will review some of these disorders in detail, exploring the cognitive problems associated with them and how these problems relate to underlying neuropathologies. Here, we focus on the etiology and clinical diagnosis of degenerative disorders.

Table 4.1 Prominent Degenerative and Infectious Disorders of the Central Nervous System

Disorder	Type	Most Common Pathology
Alzheimer's disease	Degenerative	Tangles and plaques in limbic and temporal-parietal cortex
Parkinson's disease	Degenerative	Loss of dopaminergic neurons
Huntington's disease	Degenerative	Atrophy of interneurons in caudate and putamen nuclei of basal ganglia
Pick's disease	Degenerative	Frontal-temporal atrophy
Progressive supranuclear palsy (PSP)	Degenerative	Brainstem atrophy including colliculus
Multiple sclerosis	Possibly infectious	Demyelination, especially of fibers near ventricles
AIDS dementia	Viral infection	Diffuse white matter lesions
Herpes simplex	Viral infection	Destruction of encephalitis neurons in temporal and limbic regions
Korsakoff's disease	Nutritional deficiency	Destruction of neurons in diencephalon and temporal lobes

Degenerative disorders have been associated with both genetic aberrations and environmental agents. A prime example of a degenerative disorder that is genetic in origin is Huntington's disease. The genetic link in degenerative disorders such as Parkinson's disease and Alzheimer's disease is weaker. Environmental factors are suspected to be important, perhaps in combination with genetic dispositions. The causes of Parkinson's disease are unknown, but it is suspected that the cell death in dopaminergic neurons may be accelerated by unknown toxins accumulating in the environment. Unlike many neurological disorders, there are no descriptions of people with parkinsonian symptoms in the pre–Industrial Age medical literature. The causes of Alzheimer's disease also remain a mystery, despite intense research efforts. The disease in about 5% of patients is clearly linked to a genetic deficiency; in the rest, there is no identifiable genetic component. Many hypotheses regarding the cause of Alzheimer's disease have been proposed: The disorder has been linked to exposure to aluminum silicates (e.g., an active ingredient in most antacids) and to overactivity in cortical neurons. More recently, it was suggested that the production of amyloid, a ubiquitous protein in organic tissue, goes awry and leads to the characteristic plaques found in the brains of patients with the disease (Figure 4.16).

Progressive neurological disorders can be caused by a virus. The human immunodeficiency virus (HIV) that causes acquired immunodeficiency syndrome (AIDS)–related dementia has a tendency to lodge in subcortical regions of the brain, producing diffuse lesions of the white matter through the destruction of axonal fibers. The herpes simplex virus, on the other hand, destroys neurons in cortical and limbic structures if it migrates to the brain. Viral infection is also suspected in multiple sclerosis, although evidence for this is indirect, coming from epidemiological studies. For example, the incidence of multiple sclerosis is highest in temperate climates, and a number of isolated tropical islands had not experienced multiple sclerosis until the population came in contact with Western visitors.

Degenerative and infectious disorders are usually characterized by a gradual onset of symptoms. Often the patient may not notice the deterioration of motor or cognitive abilities. Rather, the changes may be more apparent to a spouse or other family members. The neurological examination is especially critical for reliable diagnosis. The first signs of multiple sclerosis may be slight disturbances in sensation or double vision. The insidious onset of the memory problems associated with Alzheimer's disease may be difficult to detect given our expectations of cognitive changes in normal aging. A patient developing Parkinson's disease may first note difficulty standing up or initiating movement. The experienced clinician will recognize the implications of these signs. CT and MRI may confirm a diagnosis, but usually the scans do not reveal any pathology in the early phases of these disorders. As the diseases progress, evidence of neural atrophy or degeneration becomes obvious (see Figure 4.16).

The pace of deterioration varies enormously according to the degenerative disorder. Whereas Huntington's disease invariably results in progressive deterioration and death within 5 to 15 years, demyelinating disorders such as multiple sclerosis may go into remission for many years. Part of this variation is due to differences in the underlying mechanisms producing the neural pathology. For example, in multiple sclerosis, the disease process affects white matter, whereas Alzheimer's disease involves widespread atrophy of cell bodies. Another important factor affecting a patient's outcome is the use of medication to treat symptoms. Just 30 years ago, patients with Parkinson's disease were generally bedridden within a few years of diagnosis, and

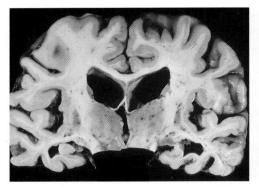
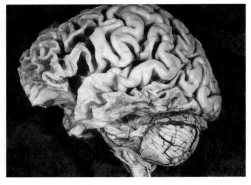

(a)

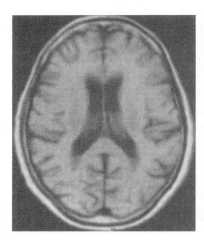
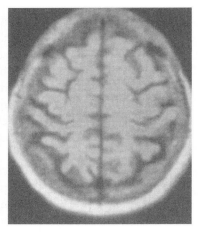

(b)

Figure 4.16 Degenerative disorders of the brain. **(a)** On the left is a coronal section from a patient with Alzheimer's disease who died at age 67, eight years after the first reports of memory problems. There is severe cortical atrophy; at death, her brain weighed only 750 gm, less than half that of a normal brain. The right side shows the brain of a patient who died of Pick's disease. The atrophy here is limited to frontal and temporal lobe regions. **(b)** Transverse MRI scans from a patient with Alzheimer's disease. Atrophy has led to the enlargement of the sulci and ventricles.

the disorder would be listed as the cause of death. With the introduction of drugs categorized as dopamine agonists, the symptoms are greatly minimized for many patients.

TRAUMA

More than any natural cause such as stroke or tumor, most patients arrive on neurology wards after a traumatic event such as a car accident, a gunshot wound, or an ill-advised dive into a shallow swimming hole. The traumatic event can lead to a closed- or an open-head injury. In closed-head injuries, the skull remains intact, but the brain is damaged by the mechanical forces generated by a blow to the head. The most common cause of closed-head injury is a car accident in which a person's head slams against the windshield. The damage may be at the site of the blow, for example, just below the forehead—damage referred to as a *coup*. In addition, reactive forces may bounce the brain against the skull on the opposite side of the head, resulting in a *countercoup*. Occipital deficits are sometimes observed in car accident victims who suffer a countercoup. Certain regions are especially sensitive to the effects of coups and countercoups. The inside sur-

face of the skull is markedly jagged above the eye sockets. As can be seen in Figure 4.17, this rough surface can produce extensive tearing of brain tissue in the orbitofrontal region. Open-head injuries happen when the skull is penetrated by an object like a bullet or shrapnel. With these injuries, tissue may be directly damaged by the penetrating object. The impact of the object can also be expected to create reactive forces producing coup and countercoup.

Additional damage can follow a traumatic event as a result of vascular problems and increased risk of infection. Trauma can disrupt blood flow by severing vessels or it can change intracranial pressure as a result of bleeding. Swelling after trauma might generate further brain damage, and seizures, common after trauma, can originate in scarred tissue.

Early work on localizing cognitive function often involved patients with traumatic injuries. The eminent British neurologist Sir Gordon Holmes (1919) provided some of the classic descriptions of cerebellar and occipital lobe function based on his observations of World War I soldiers who had open-head injuries. Indeed, prior to the invention of CT, open-head injuries offered the best way to localize brain damage while the patient

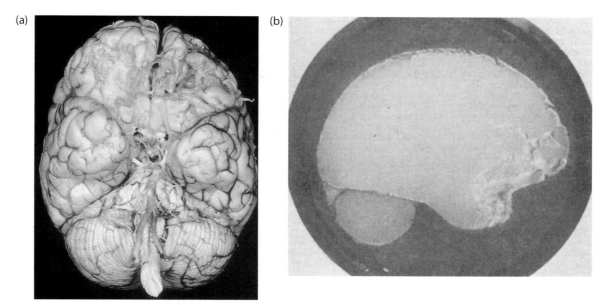

(a)   (b)

Figure 4.17 Trauma can produce extensive destruction of neural tissue. Damage can arise from the collision of the brain with the solid internal surface of the skull, especially along the jagged surface over the orbital region. In addition, accelerative forces created by the impact can cause extensive shearing of dendritic arbors. **(a)** The brain of a 54-year-old man who had sustained a severe head injury 24 years prior to death. Tissue damage is evident in the orbitofrontal regions, and was associated with intellectual deterioration subsequent to the injury. **(b)** The susceptibility of this region to trauma was made clear by A. Holbourn of Oxford in 1943, who filled a skull with gelatin and then violently rotated the skull. While most of the brain retains its smooth appearance, the orbitofrontal region has been chewed up.

was still alive. Researchers now avoid studying trauma patients to see how brain regions relate to cognitive functions; such patients often have multiple neuropsychological problems, perhaps because neurological damage is generally extensive and diffuse.

EPILEPSIES

Epilepsy is a condition characterized by excessive and abnormally patterned activity in the brain. The cardinal symptom is a seizure, a transient loss of consciousness. The extent of other disturbances varies. Some epileptics will shake violently and lose their balance. For others, the seizure may be perceptible only to the most attentive friends and family. Seizures are confirmed by performing electroencephalography (EEG) to record the patient's brain waves (see Electrical and Magnetic Signals later in this chapter). During the seizure, the EEG profile is marked by large-amplitude oscillations (Figure 4.18).

The frequency of seizures is also variable. The most severely affected patients can have hundreds of seizures each day, with each seizure disrupting function for a few minutes. Other epileptics will suffer only an occasional seizure, but it may incapacitate the person for a couple of hours. Furthermore, simply

having a seizure is not diagnostic of epilepsy. While 0.5% of the general population has epilepsy, it is estimated that one in twenty people will have a seizure at some point in their life. Often the seizure is triggered by an acute event such as trauma, exposure to toxic chemicals, or high fever. Approximately 50% to 70% of epileptics respond well when treated with antiseizure medication. For the remaining patients, surgery may be an option if the disorder is chronic and severely debilitating.

Neuropsychologists are generally not interested in the cognitive deficits associated with epilepsy as they relate to normal function. Since the seizures disrupt neural activity across large sections of the brain, it is difficult to link behavioral deficits with structural abnormalities.

Functional Neurosurgery

Surgical interventions for treating neurological disorders provide a unique opportunity to investigate the link between brain and behavior. The best example of this comes from research involving patients who have undergone surgical treatment for the control of intractable epilepsy. The extent of tissue removal is always well documented, enabling researchers to investigate correlations between

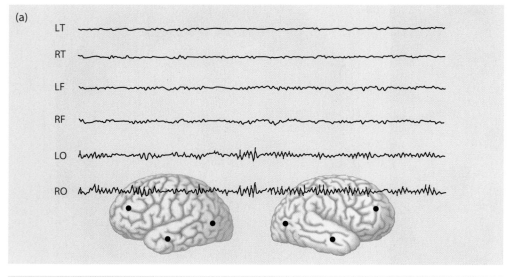

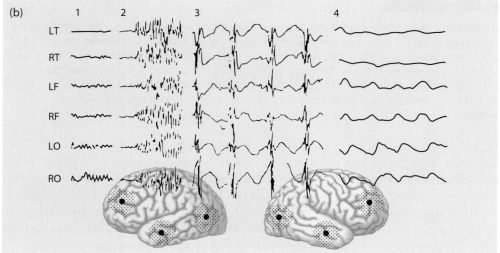

Figure 4.18 Electroencephalographic (EEG) recordings from six electrodes, positioned over the frontal, temporal, and occipital cortex on both the left and the right side. **(a)** Activity during normal cerebral activity. **(b)** Activity during a grand mal seizure. From Kolb and Whishaw (1996).

lesion site and cognitive deficits. But caution must be exercised in attributing cognitive deficits to surgically induced lesions. It is possible that other, structurally intact tissue is dysfunctional owing to the chronic effects of epilepsy because the seizures spread beyond the epileptogenic tissue. One method used with epilepsy patients compares their performance before and after surgery. The researcher can differentiate changes associated with the surgery from those associated with the epilepsy.

An especially fruitful paradigm for cognitive neuroscience has involved the study of patients who have had the fibers of the corpus callosum severed (corpus callosotomy). In these patients, the two hemispheres have been disconnected—the so-called split-brain procedure. While few patients have had this procedure, they have been extensively studied. Indeed, over the past 10 years,

one hundred studies published in scientific journals have been based on just the five most tested patients from the Dartmouth group. Insights to the role of the two hemispheres on a wide range of cognitive tasks have been made possible by studying these patients. Split-brain patients also offer the opportunity to assess the unity of consciousness. We return to these issues in more detail in subsequent chapters.

The logic of corpus callosotomy stems from the idea that a surgeon's knife can reduce the number or frequency of seizures and thereby eliminate physiological abnormalities that interfere with normal function. This idea, that surgery can eliminate abnormal brain function, has a long and sometimes troubled history in neurology. This reasoning motivated the notorious frontal lobotomy operation. Though the procedure enjoyed widespread popularity in the middle of this century as a treatment for depression, schizophrenia, and other psychiatric disorders, its theoretical motivation was weak—perhaps best captured in the *Life* magazine cartoon shown in Figure 4.19. A hyperactive superego, localized in the frontal lobes, was assumed to exert excessive control over posterior brain regions. Slicing the white matter in the frontal cortex (with crude techniques such as inserting an ice pick behind the eye sockets!) was thought to restore balance by reducing the connections between frontal and posterior brain regions.

Whereas in the preceding examples neurosurgery was eliminative in nature, it has also been used as an attempt to restore normal function. Examples of this are found in current treatments for Parkinson's disease, a movement disorder resulting from basal ganglia dysfunction (see Chapter 11). While the standard treatment involves the use of medication, the efficacy of the drugs can change over time and even produce debilitating side effects. Many of these patients are now treated surgically. The most popular technique involves the implantation of deep brain stimulators in the basal ganglia. These devices produce continuous electrical signals that trigger neural activity. Dramatic and sustained improvements have been reported with this procedure (Krack et al., 1998).

An even more radical approach for treating severe Parkinson's disease is the use of fetal brain transplants. Neurochemically, Parkinson's disease results from a loss of cells in the substantia nigra, a nucleus that is the source of dopaminergic inputs to the basal ganglia and frontal cortex. To restore function to the basal ganglia, cells from aborted fetuses are placed in structures that receive inputs from the nigra. The behavioral results from this procedure are encouraging, although the procedure is considerably riskier than the implantation of deep brain stimulators. Patients' performance on motor tests improves substantially following surgery, and metabolic imaging reveals activity in the vicinity of the implants. It appears that the fetal tissue is able to develop into dopamine-sensitive neurons (Kordower et al., 1995).

Transplantation techniques offer a promising method for clinical neurology and open up areas of research for

Figure 4.19 A *Life* magazine cartoon from 1947 sketches the mechanisms underlying the supposed benefits of the frontal lobotomy. Freudian theory was in its heyday at the time, as reflected in the accompanying caption: "In agitated depression (top drawing), the superego becomes overbearing and unreasonable, unbalancing the whole mind. . . . The surgeon's blade, slicing through the connections between the prefrontal areas (the location of the superego) and the rest of the brain, frees the tortured mind from its tyrannical ruler (bottom drawing). . . . Lobotomy, however, should be performed only on those patients whose intelligence is sufficient to take control of behavior when the moral authority is gone."

cognitive neuroscience. In animals, fetal grafts have been placed in the spinal cord to promote reconnectivity of descending motor fibers following spinal resection (see Chapter 15). Rather than simply focus on removing aberrant tissue, these procedures raise the possibility that neurosurgical techniques can improve function by restoring damaged tissue.

CONVERGING METHODS

So far, we have taken a brief look at the basic methodologies of cognitive psychology, computer modeling, neurosciences, and neurology. Each discipline has unique methodologies for learning about the nature of the mind and the relation between brain and mind. But the real strength of cognitive neuroscience comes from the way in which these diverse methodologies are integrated, which is the subject of this final section.

Cognitive Deficits Following Brain Damage

Perhaps the best-established paradigm of cognitive neuroscience involves the effects of brain injury on behavior. For many centuries lesions have been extensively studied in animals and humans, and as noted earlier in this chapter, the lesion model has laid an empirical foundation for learning about brain organization. Fundamental concepts, such as the left hemisphere's dominant role in language or the dependence of visual functions on posterior cortical regions, were developed by observing the effects of brain injury.

For two reasons, research on neurological patients has been booming over the past two decades. First, with neuroimaging methods such as CT and MRI, we can precisely localize brain injury in vivo. Second, the paradigms of cognitive psychology have provided the tools for making more sophisticated analyses of the behavioral deficits observed after brain injury. Early neuropsychological work focused on localizing complex tasks: language, vision, executive control, motor programming. The essence of the cognitive revolution has been that these complex tasks require the integrative activity of many component operations. Cognitive neuropsychologists have extended this to research on brain-injured patients. Indeed, the excitement about neuropsychological research is not restricted to its potential to link mental activities to brain structures. Equally important, many researchers recognize that the study of dysfunctional behavior can help identify the component operations that underlie normal cognitive performance.

The logic of this approach is straightforward. If a behavior depends on processing within a certain brain structure, then damage to this structure should disrupt the behavior. As such, this approach assumes that brain injury is eliminative—that brain injury disturbs or eliminates the processing ability of the affected structure.

Consider the following example. Suppose that damage to brain region A results in impaired performance on task X. One conclusion is that region A contributes to the processing required for task X. For example, if task X is reading, we might conclude that region A is critical for reading. But from cognitive psychology we know that a complex task such as reading has many component operations: Fonts must be perceived; letters and letter strings activate representations of their corresponding meanings; and syntactic operations link individual words into a coherent stream. By merely testing reading ability, we will not know which component operation or operations are impaired when there are lesions to region A. What the cognitive neuropsychologist wants to do is design tasks that diagnose the function of

specific operations. If a reading problem stems from a general perceptual problem, then comparable deficits should be seen on a range of tests of visual perception. If the problem reflects the loss of semantic knowledge, then the deficit should only be limited to tasks that require some form of object identification or recognition.

Associating neural structures with specific processing operations calls for appropriate control conditions. The most basic form of control is to compare the performance of a patient or group of patients with that of healthy subjects. Poorer performance by the patients might be taken as evidence that the affected brain regions are involved in the task. Thus, if we had a group of patients with lesions in the frontal cortex who showed impairment on our reading task, we might suppose that this region of the brain was critical for reading. However, it is important to keep in mind that brain injury can produce widespread changes in cognitive abilities. The frontal lobe patients not only might have trouble in reading but also might demonstrate impairment on just about any task we give them such as problem solving, memory, or motor planning. Thus, the challenge for the cognitive neuroscientist is to determine whether the observed behavioral problem results from damage to a particular mental operation or whether it is secondary to a more general disturbance. For example, many patients are depressed after a neurological disturbance such as a stroke, and depression is known to affect performance on a wide range of tasks.

SINGLE AND DOUBLE DISSOCIATIONS

More typically, cognitive neuropsychologists design experiments that have at least two tasks, an experimental task and a control task. The best experiments are those in which the two tasks are similar in most respects but differ in requiring one hypothetical mental operation. Suppose that a researcher is interested in the association between two aspects of memory. One aspect of memory is knowledge about when we learned a particular fact or piece of information. For example, people who were alive in 1963 can recall not only that President Kennedy was killed in Dallas but also where they were when they first heard about the tragedy. A second aspect relates to the familiarity we have with that fact or piece of information. We recognize that our memory of Kennedy's death is not simply the result of that initial experience, but also due to the fact that the event has been recalled in countless news documentaries, books, and movies.

Our researcher hypothesizes that these two aspects of memory are separable. One way to test this hypothesis would be to examine patients with memory disorders. For example, if the researcher hypothesizes that familiarity is associated with the temporal lobe, then he might test patients with temporal lobe lesions on two memory tests, one designed to look at memory of when information was acquired and the second designed to look at familiarity. To test this, patients with memory problems would be required to perform two tasks. For each task, the stimuli would be identical—a series of abstract drawings in which some items are shown once, others twice, and others three times. To test how well the patients remember when they learned something, a temporal-order task would be used. The subjects are presented with a pair of drawings and judge which was presented first. To test for familiarity, a frequency-judgment task can be used. Here, the subjects would again be presented with a pair of drawings, but now they must decide which drawing was seen more often. If temporal lobe lesions disrupt familiarity but not the ability to remember when something was learned, then the patients should demonstrate selective impairment on the frequency task. To detect the impairment, it would be necessary to include a control group such as people without any neurological problems.

Such a result would constitute a **single dissociation** (Figure 4.20a). Two groups are tested on two tasks and a between-group difference is apparent in only one task. Two groups are necessary to compare the patients' performance with that of a control group. Two tasks are necessary to examine whether a deficit is specific to a particular task or reflects a more general impairment. Many conclusions in neuropsychology are based on single dissociations: When compared to control subjects, patients with hippocampal lesions cannot develop long-term memories despite their short-term memory being intact. Patients with Broca's aphasia have intact comprehension but struggle to speak fluently.

Single dissociations have unavoidable problems. In particular, the two tasks are assumed to be equally sensitive to differences between the control and experimental groups. However, often this is not the case. One task may be more sensitive than the other because of differences in task difficulty or sensitivity problems in how the measurements are obtained. For example, the frequency-judgment task might be more demanding than the temporal-order task, requiring a greater degree of concentration. If the brain injury produced a generalized problem in concentration, then the patients might have difficulty with this task, but the problem would not be due to a specific problem in memory for familiarity. As an analogy, consider a comparison between two six-cylinder cars, one with an engine in mint condition and a second that is running on only five cylinders. It might be difficult to tell the difference between the two cars when driving through the city because the speed must be kept low and frequent stops are necessary. However, when the cars are taken out on the highway, it would quickly become apparent that one car drives rougher than the other. However, we would not want to conclude that this car has a selective deficit in highway driving. Rather, our city driving test was not sufficiently sensitive to detect the persistent problem.

Double dissociations avoid these problems. As with single dissociations, double dissociations require two groups and two tasks. What defines double dissociation is that group 1 shows impairment on task X and group 2, impairment on task Y (Figure 4.20b). The two groups' performances can be compared to each other. Or, each group's performance can be compared with a control group that shows no impairment. With a double dissociation, it is no longer reasonable to argue that a difference in performance merely results from the unequal sensitivity of the two tasks. In our memory example, the claim that temporal lobe patients have a selective problem with familiarity would be greatly strengthened if it were shown that a second group of patients (e.g., frontal lobe patients) showed selective impairment on the temporal-order task.

Figure 4.20 Hypothetical series of results conforming to either a single **(a)** or double **(b)** dissociation. With the single dissociation, the patient group shows impairment on one task and not on the other. With the double dissociation, one patient group shows impairment on one task and a second patient group shows impairment on the other task. Double dissociations provide much stronger evidence for a selective impairment.

(a) Single dissociation

Group	Tasks (% correct)	
	Recency memory	Familiarity memory
Temporal lobe damage	90%	70%
Controls	90%	95%

(b) Double dissociation

Group	Tasks (% correct)	
	Recency memory	Familiarity memory
Temporal lobe	90%	70%
Frontal lobe	60%	95%
Controls	90%	95%

Double dissociations offer the strongest neuropsychological evidence that a patient or patient group has a selective deficit in a certain cognitive operation.

The inferential power of double dissociations has been exploited in many settings beyond the neuropsychology laboratory. Lesion studies in animals have been most convincing when the conclusions were based on double dissociations. Cognitive research on healthy subjects has also benefited from the logic of double dissociations. Evidence of separable cognitive operations can be gained by demonstrating that one task is affected by one type of manipulation whereas a second task is selectively affected by a different manipulation. For example, it might be found in normal subjects that the rate at which the stimuli appear affects performance on the temporal-order task whereas the similarity between the stimuli affects familiarity performance. Here, the same subjects serve as their own controls. The double dissociation arises because the two manipulations—rate and similarity—differentially affect performance on the two tasks. Such a result would suggest that the two tasks involve nonoverlapping component operations (Figure 4.21).

GROUPS VERSUS INDIVIDUALS

In many neuropsychological studies, groups are defined according to whether patients have a common neurological diagnosis (e.g., Alzheimer's disease) or have pathology in a common neural region (e.g., frontal strokes). Group studies have been criticized as inappropriate for human neuropsychology because of the variability among

patients assigned to the same groups. No two strokes or tumors are exactly alike. In a similar sense, neurological and cognitive deficits found in degenerative diseases such as Alzheimer's disease vary from patient to patient. Human neuropsychology will never approximate the type of control that is possible in animal research, where the experimenter can control the size and location of the lesion. Even powerful MRI machines provide a relatively crude resolution compared to the histological procedures available to researchers working with animals, and some types of lesions go undetected with either CT or MRI. Given this anatomical variability, the utility of lumping patients into a single group has been questioned, as we should expect a similar lack of correspondence at the behavioral level. Instead, it has been argued that insights to cognitive processes can best be achieved by comprehensively documenting the performance of individual patients and making comparisons across case studies (Caramazza, 1992).

In general, proponents of the case study approach want to use patient studies to develop models of cognitive architecture. Double dissociations are especially prominent. Individuals with unique deficits help to isolate the component operations for a task. Yet the case study approach is more limited for linking neural structures to cognition operations. Lesions from strokes or tumors encompass a wide area and affect several disparate structures. It is difficult to know which affected area correlates with a deficit. Group studies offer hope. Though the extent and location of damage may be heterogeneous, reconstruction software can identify regions of overlap, as shown in Figure 4.22. While individual differences might occur because lesions extend into nonidentical regions, one hopes that the common site of pathology produces a consistent pattern of deficits on a task being studied.

The group-versus-case study debate reflects the difference between the root words in cognitive neuroscience and cognitive neuropsychology. The case study method affords powerful insights into the functional components of cognition. For example, case studies have been essential for demonstrating that brain lesions can selectively disrupt restricted semantic classes. One patient studied by Alfonzo Caramazza and his colleagues at Johns Hopkins University showed a peculiar anomia, an inability to name things (Hart et al., 1985). For this patient, the problem was restricted to certain classes of objects. He was unable to generate the names of fruits and vegetables. In contrast, he showed no impairment when asked to name objects such as tools or furniture. If a study had been conducted with a group of patients with anomia, this patient's selective problem

Figure 4.21 Identifying the mental operations required for a particular task can be accomplished by showing that different manipulations selectively influence different aspects of performance. In this example, it is hypothesized that recognition memory depends on both recency and familiarity, and that these two processes can be influenced by separate manipulations.

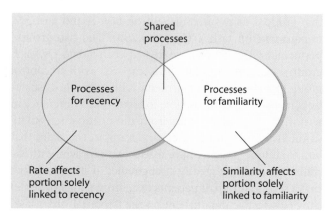

Left Prefrontal

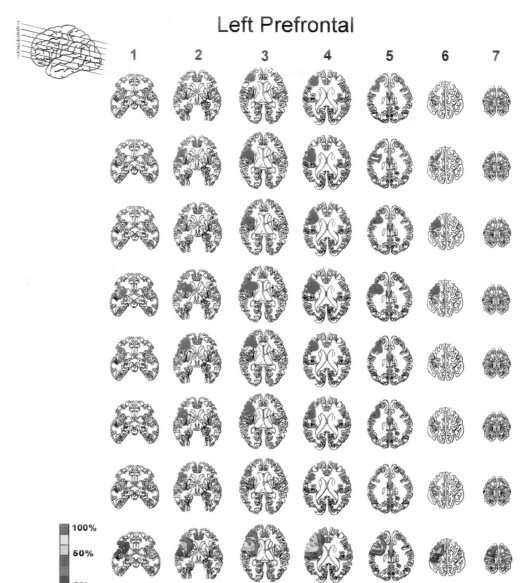

Figure 4.22 Drawing inferences from the study of humans with brain damage is difficult since naturally occurring brain lesions are never identical. Group studies can facilitate the functional analysis of brain structures by identifying regions of lesion overlap. Shown here are sketches of the extent of lesions in the left frontal cortex in a series of patients. The individual patients are represented in each row, with the transverse slices going from inferior to superior (as shown in the diagram in the upper left corner). The bottom row shows the extent of damage for the group in composite form.

might have been attributed to normal between-subject variability. When he was treated as an isolated case, the problem stood out and inspired researchers to look for patients with similar problems as well as patients with other category-specific anomias. This work has led to sophisticated models of the functional organization of semantic knowledge.

On the other hand, group studies have proved useful for relating cognitive processes to underlying brain structures (Robertson et al., 1993). If a brain structure is hypothesized to perform a particular mental operation, then lesions to this structure should be associated with deficits on tasks that depend on this putative operation. This does not mean that all patients will be similarly affected. For some patients, the pathology may not be as

extensive or encroach upon the critical tissue. Moreover, as with healthy subjects, individuals will differ in how they perform a particular task or may have developed idiosyncratic strategies. Nonetheless, group studies allow the researcher to look for similarities across patients with related lesions as well as make systematic comparisons between the effects of lesions centered in different brain structures.

Virtual Lesions: Transcranial Magnetic Stimulation

Lesion methods have been an important tool for both human and animal studies of the relationship between the brain and behavior. Observations of the performance of neurologically impaired individuals have tended to serve as the starting point for many theories. The impairments of patients who had suffered gunshot and shrapnel wounds during the First World War were instrumental in identifying the wide assortment of visual deficits that could accompany lesions of the occipital and parietal cortex. But with such human studies, the experimenter is limited by the vagaries of nature (or the types of damage caused by military technology). Lesion studies in animals have the advantage that the experimenter can control the site and size of the damage. Such work was important in identifying the complex network of processing systems in the posterior cortex that support visual perception. In these situations, a specific hypothesis can be tested by comparing the effects of lesions to one region versus another.

Transcranial magnetic stimulation (TMS) offers a new way to apply the experimental approach associated with animal lesion studies to human research. With this method, an experimenter can disrupt neural function in a selected region of the cortex, and as with other lesion studies, the consequences of the stimulation on behavior can shed light on the normal function of the disrupted tissue. What makes this method appealing is that the technique, when properly conducted, is safe and noninvasive. Since the principle use is with neurologically healthy individuals, the method avoids the concerns in traditional neuropsychology that brain function as a whole has been altered by the neurological disorder. In essence, TMS allows for the creation of "virtual" lesions (Pascual-Leone et al., 1999).

The TMS device consists of a tightly wrapped wire coil that is encased in an insulated sheath and connected to a source of powerful electrical capacitors. When triggered, the capacitors send a large electrical current through the coil, resulting in the generation of a magnetic field. When the coil is placed on the surface of the skull, the magnetic field passes through the skin and scalp and induces a physiological current that results in the firing of neurons (Figure 4.23). The exact mechanism causing the

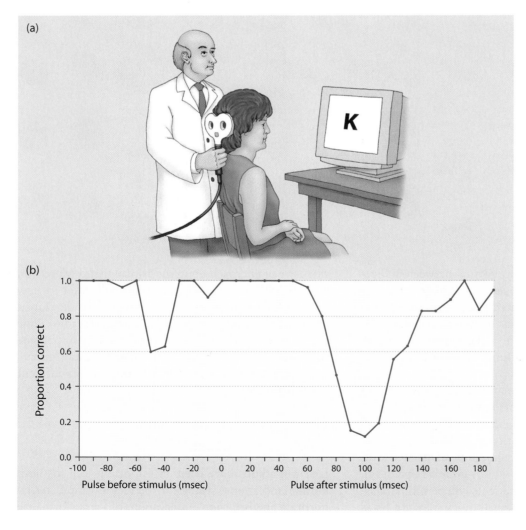

(a)

(b)

Proportion correct

Pulse before stimulus (msec)   Pulse after stimulus (msec)

Figure 4.23 Transcranial magnetic stimuluation (TMS) over the occipital lobe. **(a)** The center of the figure-8 coil is placed over the targeted area. When a large electrical field is passed through the coil, a magnetic pulse is generated and passes through the skull. This pulse activates neurons in the underlying cortex. **(b)** The time between the onset of the letter stimulus and the pulse is varied. When the pulse follows the stimulus by 70 to 130 msec, the subject fails to identify the stimulus on all trials. Note the failures when the pulse precedes the letter. These occur because the subject briefly blinks on some trials after the pulse. (b) Adapted from Corthout et al. (1999).

neural discharge is not well understood. It may be that the current leads to the generation of action potentials in the soma; alternatively, the current may directly stimulate axons. The area of neural activation will depend on the shape and positioning of the coil. With currently available coils, the primary activation can be restricted to an area of about 1.0 to 1.5 cm^2.

TMS has been used to explore the role of many different brain areas. When the coil is placed over the hand area of the motor cortex, stimulation will produce activation of the muscles of the wrist and fingers. The sensation can be rather bizarre—the hand will visibly twitch, yet the subject is aware that the movement is completely involuntary! Like many research tools, TMS was originally developed for clinical purposes. The direct stimulation of motor cortex provides a relatively simple way to assess the integrity of motor pathways since muscle activity in the periphery can be detected in about 20 msec following stimulation over motor cortex.

With other stimulation sites, the person usually is not aware of any effects from the stimulation. Yet measurable changes in performance can be detected easily if the timing between the stimulation and behaviorally relevant events is controlled. For example, stimulation over visual cortex can interfere with a person's ability to identify a letter (Corthout et al., 1999). The synchronized discharge of the underlying visual neurons interferes with their normal operation. The timing between the onset of the TMS pulse and the onset of the stimulus (e.g., presentation of a letter) can be manipulated to plot the time course of processing. In the letter identification task, the person will only err if the stimulation occurs between 70 and 170 msec after presentation of the letter. If the TMS is given before this interval, the neurons have time to recover; if the TMS is given after this interval, the visual neurons have already responded to the stimulus.

There are some notable limitations with TMS. As shown in the last example, the effects of TMS are generally quite brief. The method tends to work best with tasks in which the stimulation is closely linked to either stimulus events or movement. It remains to be seen if more complex tasks can be disrupted by brief volleys of externally induced stimulation. The fact that stimulation results in activation of a restricted area of the cortex is both a plus and a minus. The experimenter can restrict stimulation to a specific area, especially if the coordinates are based on MRI scans. But TMS will be of limited value in exploring the function of cortical areas that are not on the superficial surface of the brain. Despite these limitations, TMS offers the potential of providing the cognitive neuroscientist with a safe method for momentarily disrupting the activity of the human brain. Almost all other methods rely on correlational procedures, either through the study of naturally occurring lesions or, as we will see in the next section, through the observation of brain function with various neuroimaging tools.

Functional Imaging

We already mentioned that patient research rests on the assumption that brain injury is an eliminative process: The lesion disrupts certain mental operations and has little or no impact on others. But the brain is massively interconnected, and damage in one area might have widespread consequences. It is not always easy to analyze the function of a missing part by looking at the operation of the remaining system. Allowing the spark plugs to decay or cutting the line distributing the gas to the pistons will cause an automobile to stop running. This does not mean that spark plugs and distributors do the same thing; rather, their removal has similar functional consequences.

Concerns such as these point to the need for methods that measure activity in the normal brain. Along this front have occurred remarkable technological breakthroughs during the past decade. Indeed, new tools and methods of analysis develop at such an astonishing pace that new journals and scientific organizations have been created to rapidly disseminate this information. In the following section, we review some of the technologies that allow researchers to observe the electrical and metabolic activity of the healthy human brain in vivo.

ELECTRICAL AND MAGNETIC SIGNALS

Neural activity is an electrochemical process. Although the electrical potential produced by a single neuron is minute, when large populations of neurons are active together, they produce electrical potentials large enough to be measured by electrodes placed on the scalp. These surface electrodes are much larger than those used for single-cell recordings, but involve the same principles: A change in voltage corresponding to the difference in potential between the signal at a recording electrode and that at a reference electrode is measured. This potential can be recorded at the scalp because the tissues of the brain, skull, and scalp passively conduct the electrical currents produced by synaptic activity. The record of the signals is referred to as the *electroencephalogram.*

Electroencephalography, or **EEG,** provides a continuous recording of overall brain activity and has

An Interview with Robert T. Knight, M.D. Dr. Knight is a professor in the Department of Psychology at the University of California, Berkeley. His research provides an elegant demonstration of how measurements of evoked potentials in neurological patients can reveal interactions between different cortical regions.

Authors: Cognitive neuroscience is practiced by many people who are not medically trained. For those working on patients with lesions, what is it they should always know and keep in mind when they are considering their results?

RTK: The "golden rule" of lesion studies in cognitive neuroscience is that the neuroanatomy of the lesion, albeit structural (i.e., stroke or resection) or neurochemical (i.e., Parkinson's), drives interpretations of all results that make inferences about brain-behavior relationships. People often make the mistake that a large number of patients is by definition better than a smaller one. However, as in all experimental work, the variance of the group under investigation is paramount. Your interpretations are only as reliable as the variance in your group of interest. Confusion in this area has led to some of the divisive interactions in the literature over single-case versus group studies.

Variance raises its ugly head in many forms in experimental research employing clinical populations. For instance, medication effects in Parkinson's (i.e., How long since the last dopamine treatment were the data recorded in your Parkinson's patient?), seizure control in postlobectomy patients, and anatomy of damage in stroke or trauma patients are all major issues in data gathering and analysis. The nonmedical cognitive neuroscientist must rely on a collaborator to help address these issues. As in all forms of research, the more interactive the collaboration, the more likely fruitful results will emerge.

The tools (i.e., fMRI, PET, high-density EEG, and MEG) to address the issue of diaschisis or remote effects of lesions are also emerging for implementation in lesion, neuropsychological, and cognitive neuroscience research. In the future, clarification of the anatomy of a focal lesion may include the distributed cortical or subcortical network affected by the area of damage both in the resting and in the task conditions. This likely will have profound implications for brain-behavior theoretical formulations.

Authors: There are a growing number of instances where brain imaging studies using PET or fMRI (blood flow techniques) do not coincide with lesion data. Do you think this is a serious problem?

RTK: The divergence between PET and fMRI in comparison to lesion studies raises several interesting issues. One basic experimental design problem is the type of subjects studied by each method. Except for a few aging and lesion studies the metabolic techniques are limited to young, normal subjects. Conversely, lesion studies typically involve older populations who now have a superimposed brain lesion. Thus, you have a double confound of aging effects and postlesion brain reorganization when trying to compare metabolic physiological findings in young subjects to results in lesioned populations. Obviously, the correct design is between fMRI and PET results in older normal controls in comparison to lesioned groups. Hopefully, this approach coupled with examination by PET and fMRI in lesioned populations will tease out true differences and convergences between blood flow and lesion approaches.

A more fundamental issue is determining exactly what blood flow techniques actually measure. The well-known

proved to have many important clinical applications. The reasons for this stem from the fact that there are predictable EEG signatures associated with different behavioral states (Figure 4.24). For example, in deep sleep, the EEG is characterized by slow, high-amplitude oscillations, presumably resulting from rhythmic changes in the activity states of large groups of neurons. In other phases of sleep and during various wakeful states, this pattern changes, but in a predictable manner.

Since the normal EEG patterns are well established and consistent among individuals, EEG recordings can detect abnormalities in brain function. As noted earlier, EEG provides valuable information in the assessment and treatment of epilepsy (see Figure 4.18). Of the many forms of epileptic seizures, generalized seizures have no known locus of origin and appear bilaterally symmetrical in the EEG record. Focal seizures, in contrast, begin in a restricted area and spread

coupling between blood flow, glucose utilization, and neuronal firing supports an important role of fMRI and PET techniques in cognitive research. However, it is not clear what a 2% to 10% change in blood flow translates into in terms of neural processing. Most likely, fMRI and PET are measuring the most robust aspects of neural processing in any given task, whereas electrophysiological techniques may be measuring more distributed activity in many situations.

One frequently cited example involves imaging studies using tasks that lesion studies have shown to involve mesial temporal structures such as the hippocampus. When healthy people perform these same tasks, PET and fMRI studies consistently fail to reveal activity in these regions, a classic example of "technique divergence." However, the problem may not be as serious as first glance would suggest. Electrophysiological recordings in humans reveal strong hippocampal activation in a range of memory tasks. However, different regions of the hippocampal formation are also activated during a variety of tasks that might serve as the control task in a blood flow memory experiment. The subtraction technique would show no net differential activation since mesial temporal structures are active in both the control and the memory tasks. Obviously, no technique is wrong or right; each has its own strengths and weaknesses. One should note, however, that only the lesion approach clearly documents the critical role of the hippocampal formation in memory. The subtraction approach of blood flow techniques shows no or minimal activation, and EEG techniques show activation during many nonmemory tasks. Thus, neither would give an investigator an inkling about the paramount role of the hippocampus in memory. This strongly argues for the use of combined approaches in the field of cognitive neuroscience.

Authors: When doing developmental studies, is there a rule of thumb for when a cortical lesion most likely causes a reorganization of function? Is the pattern of deficits in a 6-year-old with a focal or diffuse brain lesion different from that in a 20-year-old with the same lesion?

RTK: Brain reorganization after central nervous system insult is a poorly understood phenomenon; the experimental data available are limited despite the great clinical and theoretical significance of the topic. As a practical rule, recovery of function is thought to decrease dramatically somewhere between the ages of 6 and 10 years. Support for this contention derives mainly from research in the field of neurolinguistics on aphasic children with focal lesions and children requiring hemispherectomy. Interestingly, "language plasticity" decreases at about the time the corpus callosum undergoes myelination connecting the two hemispheres, providing support for interhemispheric inhibition theories. However, even the evidence for this connection is largely indirect.

A more critical point may be the effects of age on the temporal parameters of reorganization of function. For instance, the notion that adults with aphasia do not recover substantial function over time is incorrect. What is true is that they may take a substantially longer time to recover than children. An example of this would be aphasia with a left-hemisphere lesion in a child at age 4, and aphasia with a lesion of comparable size and location in an adult at the age of 40 and in an adult at age 70. Major recovery might occur over 1 year in the child, over 10 years in the 40-year-old, and over the same 10 years in the 70-year-old. However, the 70-year-old is likely to be dead by age 75. Thus, the older adult would be noted to have limited recovery. Since the literature is biased toward deficits in older adults, it incorrectly concludes that recovery is weak or nonexistent in adults. However, the 40-year-old at age 50 might look as good language-wise as the 5-year-old 1 year after the insult. It appears that reorganization of function is available at all ages. A key issue may be understanding the influence of age on the time constant of recovery.

throughout the brain. Focal seizures frequently provide the first hint of a neurological abnormality. They can result from congenital abnormalities such as a vascular malformation or can develop as a result of a local infection, enlargement of a tumor, or residual damage from a stroke or traumatic event. Surface EEG can crudely localize focal seizures, as some electrodes detect the onset earlier and with higher amplitude than other electrodes.

EEG is limited in providing insight to cognitive processes because the recording tends to reflect the brain's global electrical activity. A more powerful approach used by many cognitive neuroscientists focuses on how brain activity is modulated in response to a particular task. The method requires extracting an evoked response from the global EEG signal.

The logic of this approach is straightforward. EEG traces from a series of trials are averaged together by

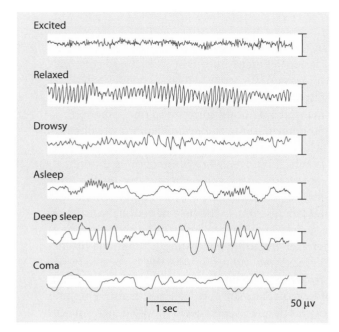

Figure 4.24 EEG profiles obtained during various states of consciousness. From Kolb and Whishaw (1986) after Penfield and Jasper (1954).

aligning the records according to an external event, such as the onset of a stimulus or the onset of a response. This alignment washes out variations in the brain's electrical activity that are unrelated to the events of interest. The evoked response, or **event-related potential (ERP),** is a tiny signal embedded in the ongoing EEG. By averaging the traces, investigators can extract this signal, which reflects neural activity that is specifically related to sensory,

motor, or cognitive events, hence, the name event-related potential (Figure 4.25). A significant feature of evoked responses is that they provide a precise temporal record of underlying neural activity. The evoked response gives a picture of how neural activity changes over time as information is being processed in the human brain.

ERPs have proved to be an important tool for both clinicians and researchers. Sensory evoked responses offer a useful window for identifying the level of disturbance in patients with neurological disorders. For example, the visual evoked potential can be very useful in the diagnosis of multiple sclerosis, a disorder that leads to demyelination. When demyelination occurs in the optic nerve, the early peaks of the visual evoked response are delayed in their time of appearance. Similarly, in the auditory system, tumors that compromise hearing by compressing or damaging auditory processing areas can be localized using auditory evoked potentials (AEPs) because characteristic peaks and troughs in the AEP are known to arise from neuronal activity in anatomically defined areas of the ascending auditory system. The earliest of these AEPs indexes activity in the auditory nerve, occurring within just a few milliseconds of the sound. Within the first 20 to 30 msec, there are a series of responses that index, in sequence, neural firing in the brainstem, midbrain, thalamus, and cortex (Figure 4.26). These stereotyped responses allow the neurologist to pinpoint the level at which the pathology has occurred. Thus, by looking at the sensory evoked responses in patients with hearing problems, the clinician can determine if the problem is due to poor sensory processing, and, if so, at what level the deficit becomes apparent.

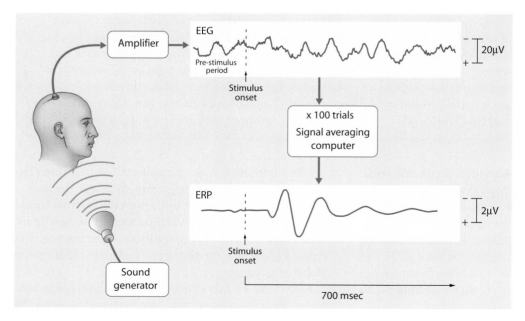

Figure 4.25 The relatively small electrical responses to specific events can only be observed by averaging the EEG traces over a series of trials. The large background oscillations of the EEG trace make it impossible to detect the evoked response to the sensory stimulus from a single trial. However, by averaging across tens or hundreds of trials, the background EEG is removed, leaving the event-related potential (ERP). Note the difference in scale between the EEG and ERP waveforms.

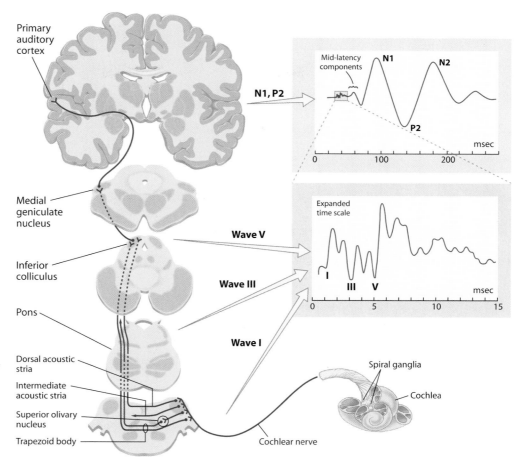

Figure 4.26 The evoked potential shows a series of positive and negative peaks at predictable points in time. In this auditory evoked response potential, the early peaks are invariant and have been linked to neural activity in specific brain structures. Later peaks are task dependent, and localization of their source has been a subject of much investigation and debate.

In this example, we specified the neural structures associated with the early components of the ERP. It is important to note that these localization claims are based on indirect methods since the electrical recordings are made on the surface of the scalp. For early components related to the transmission of signals along the sensory pathways, the neural generators are inferred from the findings of other studies that used direct recording techniques as well as considerations of the time required for peripheral pathways to transmit neural signals. This is not possible when we look at evoked responses generated by cortical structures. The auditory cortex relays its message to many cortical areas; all contribute to the measured evoked response. Thus, the problem of localization becomes much harder once we look at these latter components of the ERP.

For this reason, at present, ERPs are best suited for addressing questions about the time course of cognition rather than elucidating the brain structures that produce the electrical events. For example, as we will see in Chapter 7, evoked responses can tell us when attention affects how a stimulus is processed. ERPs also provide physiological indices of when a person decides to respond, or when an error is detected.

In the meantime, researchers have made much progress in developing analytic tools to localize the sources of ERPs recorded at the scalp. This localization problem has a long history: In the late nineteenth century, the German physicist Hermann von Helmholtz showed that an electrical event located within a spherical volume of homogeneously conducting material (approximated by the brain) produced one unique pattern of electrical activity on the surface of the sphere. This is called the *forward solution* (Figure 4.27). However, he also determined that given a particular pattern of electrical charge on the surface of the sphere, it was impossible to determine the distribution of charge within the sphere that caused it. This is called the *inverse problem*. The problem arises because an infinite number of possible charge distributions in the sphere could lead to the same pattern on the surface. ERP researchers unfortunately face the inverse problem, given that all of their measurements are made at the scalp. The challenge is to determine what areas of the brain must have been active to produce the recorded pattern. In other words, where are the generators of a particular event in the ERP?

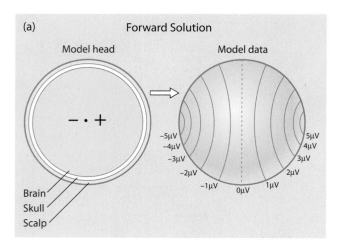

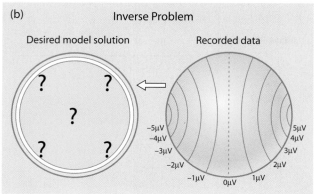

Figure 4.27 **(a)** In the forward solution, a model head is created based on known conductivities of various tissues of the brain, skull, and scalp. The pattern results from the location and orientation of a single dipolar charge, used to simulate an active neuronal population. The dipolar charge creates electrical currents that flow to the surface of the sphere, creating a distinct pattern of electrical voltages in the surface—this is the forward solution. **(b)** The inverse problem arises because a given pattern observed on the surface of the scalp can result from many possible locations of underlying neural generators.

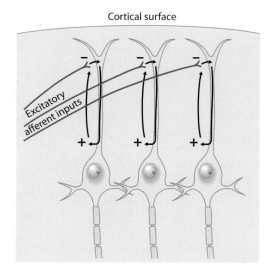

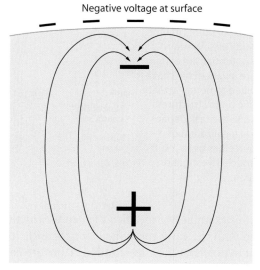

Figure 4.28 Inverse dipole modeling. Drawings of three cortical pyramidal neurons whose long axes are oriented perpendicular to the cortical surface (top of figure). Each is receiving an excitatory synaptic input at its apical dendrite. The postsynaptic potential that is created causes current to flow into the neuron near the cell body. This pattern of current flow produces an electrical field with one pole being the negative (–) electrical potential at the apical dendrite, and the other pole being the positive (+) electrical potential near the neuronal cell body. The electrical fields of the pyramidal cells sum and can be represented by a single current dipole (shown on the right). Inverse dipole modeling involves estimating the source of the neural activity from the electrical activity that is recorded at the scalp.

To solve this problem, researchers have turned to sophisticated modeling techniques. They have been able to do so by simplifying assumptions about the physics of the brain and head tissues, as well as the electrical nature of the active neurons. Of critical importance is the assumption that neural generators can be modeled as electrical dipoles, conductors with one positive end and one negative end, as shown in Figure 4.28. For example, the excitatory postsynaptic potential generated at the synapse of a cortical pyramidal cell can be viewed as a dipole.

Inverse dipole modeling is relatively straightforward. Using a high-speed computer, one creates a model of a spherical head and places a dipole at some location within the sphere. The forward solution is then calculated to determine the distribution of voltages that this dipole would create on the surface of the sphere. This predicted pattern is then compared to the data actually recorded. If the difference between the predicted and obtained results is small, then the model is supported; if the difference is large, then the model is rejected and another solution is tested by shifting the location of the dipole. In this manner, the location of the dipole is moved about the inside of the sphere until the best match between predicted and actual results is obtained. In many cases, it is necessary to use more than one dipole to obtain a good match. But, this should not be surprising: It is likely that many ERPs are the result of processing in multiple brain areas!

Unfortunately, as more dipoles are added, it becomes harder to identify a unique solution—the inverse problem returns. Researchers are exploring two new ways to overcome this problem. First, by using anatomical MRI, investigators can study precise three-dimensional models of the head instead of generic spherical models. Second, results from anatomically based neuroimaging techniques (see below) can be used to constrain the locations of the dipoles. In this way, the set of possible solutions can be made much smaller.

A related technique to ERP is **magnetoencephalography,** or **MEG.** In addition to the electrical events associated with synaptic activity, active neurons also produce small magnetic fields (see Chapter 2). Just as with EEG, MEG traces can be averaged over a series of trials to obtain event-related fields (ERFs). MEG provides the same temporal resolution as with ERPs and also has an advantage in terms of localizing the source of the signal. This advantage stems from the fact that unlike electrical signals, magnetic fields are not distorted as they pass through the brain, skull, and scalp. Inverse modeling techniques similar to those used in EEG are necessary, but the solutions are more accurate.

Indeed, the reliability of spatial resolution with MEG has made it a useful tool in neurosurgery (Figure 4.29). Consider a situation where an MRI scan reveals a large tumor near the central sulcus. Such tumors present a surgical dilemma. If the tumor extends into the precentral sulcus, surgery might be avoided or delayed

Figure 4.29 Use of magnetoencephalography (MEG) as a non-invasive presurgical mapping procedure. **(a)** MRI showing a large tumor in the vicinity of the central sulcus. **(b)** The event-related fields (ERFs) produced following repeated tactile stimulation of the index finger. Each trace shows the magnetic signal recorded from an array of detectors placed over the scalp. **(c)** Inverse modeling indicated that the dipole producing the surface recordings in (b) was anterior to the lesion. **(d)** Three-dimensional reconstruction showing stimulation of the fingers and toes on the left side of the body in red and the tumor outlined in green. From Roberts et al. (1998).

(a)
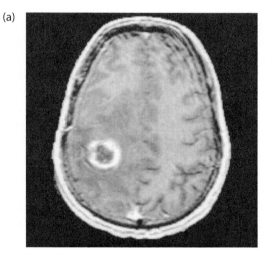

(b)
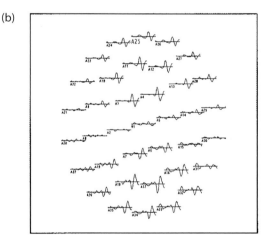

(c)
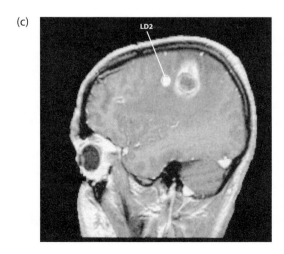

(d)
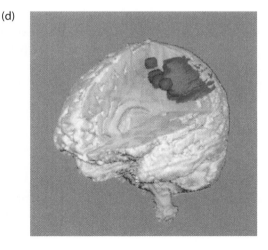

as the procedure is likely to damage motor cortex and leave the person with partial paralysis. However, if the tumor does not extend into the motor cortex, surgery is usually warranted. MEG provides a noninvasive procedure to identify somatosensory cortex. From the ERFs produced following repeated stimulation of the fingers, arm, and foot, inverse modeling techniques are used to determine if the underlying neural generators are anterior to the lesion. In the case shown in Figure 4.29, the surgeon can proceed to excise the tumor without fear of producing paralysis. The tumor borders on the posterior part of the postcentral sulcus, clearly sparing the motor cortex.

There are, however, disadvantages with MEG, at least in its present form. First, it is only possible to detect current flow that is oriented parallel to the surface of the skull. Most cortical MEG signals are produced by intracellular current flowing within the apical dendrites of pyramidal neurons. Because of this, the neurons that can be recorded with MEG tend to be located within sulci where the long axis of each apical dendrite tends to be oriented parallel to the skull surface (see Chapter 2). Second, the cost of a MEG recording system is quite prohibitive compared to the equipment needed for ERPs. A system that covers the entire head, one with 150 to 300 sensors, will cost more than a million dollars. Thus, at present, relatively few MEG studies have appeared in the cognitive neuroscience literature.

METABOLIC SIGNALS

The most exciting methodological advances for cognitive neuroscience have been provided by new imaging techniques that identify anatomical correlates of cognitive processes (Raichle, 1994). The two prominent methods are **positron emission tomography,** commonly referred to as **PET,** and functional MRI, or fMRI. These methods detect changes in metabolism or blood flow in the brain while the subject is engaged in cognitive tasks. As such, they enable researchers to identify brain regions that are activated during these tasks, and to test hypotheses about functional anatomy.

Unlike EEG and MEG, PET and fMRI do not directly measure neural events. Rather, they measure metabolic changes correlated with neural activity. Neurons are no different from other cells of the human body. They require energy in the form of oxygen and glucose, both to sustain their cellular integrity and to perform their specialized functions. As with all parts of the body, oxygen and glucose are distributed to the brain by the circulatory system. The brain is an extremely metabolically demanding organ. As noted previously, the central nervous system uses approximately 20% of all the oxygen we breathe. Yet the amount of blood supplied to the brain varies only a little between the time when the brain is most active and when it is quiet (perhaps because what we regard as active and quiet in relation to behavior does not correlate with active and quiet in the context of neural activity). Thus, the brain must regulate itself. When a brain area is active, more oxygen and glucose are made available by increased blood flow.

PET activation studies measure local variations in cerebral blood flow that are correlated with mental activity (Figure 4.30). To do this, a tracer must be introduced into the bloodstream. For PET, radioactive elements, or isotopes, are used as tracers. Owing to their unstable state, these isotopes rapidly decay by emitting a positron from their atomic nucleus. When a positron collides with an electron, two photons, or gamma rays are created. Not only do the two photons move at the speed of light, passing unimpeded through all tissue, but also they move in opposite directions from one another. The PET scanner—essentially a gamma ray detector—can determine where the collision took place. Because these tracers are in the blood, a reconstructed image can show the distribution of blood flow: Where there is more blood flow, there will be more radiation.

The most common isotope used in cognitive studies is ^{15}O, an unstable form of oxygen with a half-life of 123 seconds. This isotope, in the form of water (H_2O), is injected in the bloodstream while a person is engaged in a cognitive task. Although all areas of the body will absorb some radioactive oxygen, the fundamental assumption of PET is that there will be increased blood flow to the brain regions that have heightened neural activity. Thus, PET activation studies do not measure absolute metabolic activity, but rather relative activity. In the typical PET experiment, the injection is administered at least twice: during a control condition and during an ex-perimental condition. The results are usually reported in terms of a change in **regional cerebral blood flow (rCBF)** between the two conditions.

Consider, for example, a PET study designed to identify brain areas involved in visual perception: In the experimental condition the subject views a circular checkerboard surrounding a small fixation point (to keep subjects from moving their eyes); in the control condition, only the fixation point is presented. With PET analysis, researchers subtract the radiation counts measured during the control condition from those measured during the experimental condition. Areas that were active when the subject was viewing the checkerboard stimulus will have higher counts, reflecting increased blood flow. This subtractive procedure ignores

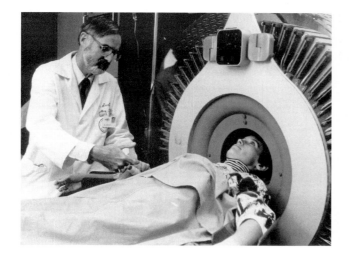

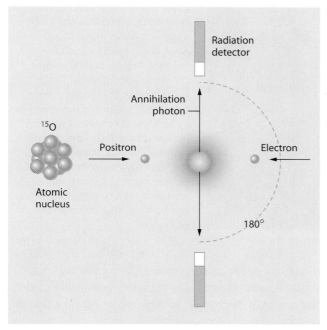

Figure 4.30 Positron emission tomography (PET) scanning. This technique allows metabolic activity to be measured in the human brain. In the most common form of PET, water labeled with radioactive oxygen, ^{15}O, is injected into the subject. As positrons break off from this unstable isotope, they collide with electrons. A by-product of this collision is the generation of two gamma rays, or photons, that move in opposite directions. The PET scanner measures these photons and calculates their source. Regions of the brain that are most active will increase their demand for oxygen. From Posner and Raichle (1994).

variations in absolute blood flow between the brain's areas. The difference image identifies areas that show changes in metabolic activity as a function of experimental manipulation (Figure 4.31).

The control condition need not be a simple resting, or fixation-only condition; it could include a second experi-

mental task. The difference image would then show the areas that are more active during the first task in comparison to the second task, and also the areas that are more active during the second task. Some PET studies compare groups of subjects rather than experimental conditions, for example, how brain activity in schizophrenics compares with that in healthy individuals.

Currently, PET scanners are capable of resolving metabolic activity to regions that are approximately 5 to 10 mm³ in volume. While this size includes thousands of neurons, it is sufficient to identify cortical and subcortical areas and can even show functional variation within a given cortical area. The panels in Figure 4.31 show a shift in activation within the visual cortex as the stimulus's location moves from being adjacent to the fixation point to more eccentric places.

As with PET, fMRI exploits the fact that local blood flow increases in active parts of the brain. The procedure is essentially identical to the one used in traditional MRI: Radio waves make the protons in hydrogen atoms oscillate, and a detector measures local energy fields that are emitted as the protons return to the orientation of the external magnetic field. With fMRI, however, imaging is focused on the magnetic properties of hemoglobin. Hemoglobin carries oxygen in the bloodstream, and when the oxygen is absorbed, the hemoglobin becomes deoxygenated. Deoxygenated hemoglobin is more sensitive, or paramagnetic than oxygenated hemoglobin. The fMRI detectors measure the ratio of oxygenated to deoxygenated hemoglobin. This ratio is referred to as the **blood oxygenation level dependent effect,** or **BOLD** effect.

Intuitively, one would expect that the proportion of deoxygenated tissue would be greater in active tissue given the intensive metabolic costs associated with neural function. However, fMRI results are generally reported in terms of an increase in the ratio of oxygenated to deoxygenated hemoglobin. This change occurs because as a brain area becomes active, the amount of blood being directed to that area increases. The neural tissue is unable to absorb all of the excess oxygen. The time course of this regulatory process is what is measured in fMRI studies. While neural events occur on a scale measured in milliseconds, blood flow modulation takes place much more slowly, with the initial rise not evident for at least a couple of seconds and peaking 6 to 10 seconds later. This delay suggests that right after a neural region is engaged, there should be a small drop in the ratio of oxygenated to deoxygenated hemoglobin. In fact, the newest generation of MRI scanners, reaching strengths of 4 tesla and above, are able to detect the initial drop (Figure 4.32). This decrease is small, representing no more than 1% of the total hemoglobin signal. The subsequent increase in the oxygenated

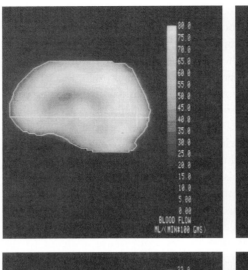

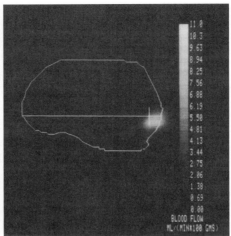

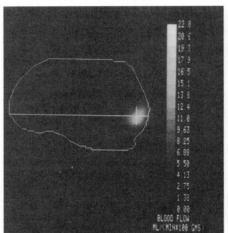

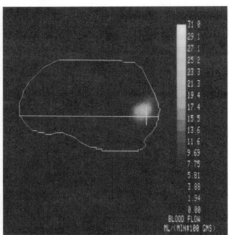

Figure 4.31 Measurements of cerebral blood flow using PET. The upper left panel shows blood flow when the subject fixated on a central spot. Activity in this baseline condition was subtracted from that in three other conditions in which the central spot was surrounded by a checkboard, either in the center of view **(top right)**, more toward the periphery **(bottom left)**, or in the far periphery **(bottom right)**. A retinotopographic map can be identified, with central vision represented more inferiorly than peripheral vision.

Figure 4.32 Functional MRI (fMRI) signal observed from visual cortex in the cat with 4.7-tesla scanner. The black line indicates the duration of a visual stimulus. Initially, there is a dip in the blood oxygenation level dependent (BOLD) signal, reflecting the depletion of oxygen from the activated cells. Over time, the BOLD signal increases, reflecting the increased hemodynamic response to the activated area. Scanners of this strength are now being used with human subjects. Adapted from Duong et al. (2000).

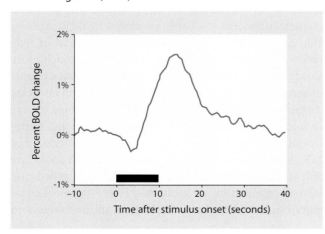

blood can produce a signal as large as 5%. By continuously measuring the fMRI signal, it is possible to construct a map of changes in regional blood flow that are coupled to local neuronal activity.

While PET scanning provided a breakthrough for cognitive neuroscience, fMRI has led to revolutionary changes. Less than a decade after the first fMRI papers appeared (in the early 1990s), fMRI imaging studies now fill the pages of neuroscience journals and proceedings of conferences. There are a number of reasons spurring the popularity of fMRI. For one thing, compared to PET, fMRI is a more practical option for most cognitive neuroscientists. MRI scanners are present in almost all hospitals in technologically advanced countries, and with modest hardware modifications, most of them can be used for functional imaging. In contrast, PET scanners are present in only a handful of major medical facilities and require a large technical staff to run the scanner and the cyclotron used to produce the radioactive tracers.

In addition, there are important methodological advantages favoring fMRI over PET. The spatial resolution is superior with fMRI, with current scanners able to resolve volumetric areas of around 3 mm^3 and the potential to become even finer as high-powered magnets become more available. Because fMRI does not involve the injection of radioactive tracers, the same individual can be tested repeatedly, either in a single session or over multiple sessions. With these multiple observations it becomes possible to perform a complete statistical analysis on the data from a single subject. This advantage is important given the individual differences in brain anatomy. With PET, computer algorithms must be used to average the data and superimpose them on a "standardized" brain, since each person can only be given a limited number of injections. Even with the newest generation of high-resolution PET scanners, subjects can only receive between twelve and sixteen injections.

The localization process is also improved with fMRI because high-resolution anatomical images are obtained during the same session as when the functional scanning occurs. With PET, not only is anatomical precision compromised by averaging across individuals, but precise localization requires that structural MRIs be obtained from the subjects. Error will be introduced when trying to align anatomical markers between the PET and MRI scans.

Temporal resolution is also much better with fMRI. It takes time to collect sufficient "counts" of radioactivity in order to create images of adequate quality with PET. The person must be engaged continuously in a given experimental task for at least 40 seconds, and metabolic activity is averaged over this epoch. The signal changes in fMRI also require averaging over successive observations, and many fMRI studies utilize a block design similar to that of PET in which activation is compared between experimental and control scanning phases (Figure 4.33).

However, the BOLD effect in fMRI can be time locked to specific events to allow a picture of the time course of neural activity. This method is called *event-related fMRI* and follows the same logic as is used in ERP studies. The BOLD signal can be measured in response to single events such as the presentation of a stimulus or onset of a movement. While metabolic changes to any single event are likely to be hard to detect among background fluctuations in the brain's hemodynamic response, a clear signal can be obtained by averaging over repetitions of these events. Event-related fMRI allows for improved experimental designs since the experimental and control trials can be presented in a random fashion. With this method, the researcher can be more confident that the subjects are in a similar attentional state during both types of trials, thus increasing the likelihood that the observed differences reflect the hypothesized processing demands rather than more generic factors such as a change in overall arousal.

A powerful advantage of event-related fMRI is that the experimenter can choose to combine the data in many different ways once the scanning is completed. As

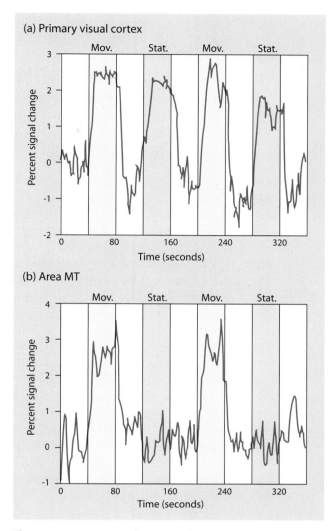

Figure 4.33 Functional MRI measures time-dependent fluctuations in oxygenation with excellent spatial resolution. The subject viewed a field of randomly positioned white dots on a black background. The dots would either remain stationary or move along the radial axis. The 40-second epochs of stimulation alternated with 40-second epochs during which the screen was blank. **(a)** Measurements from primary visual cortex (V1) showed consistent increases during the stimulation epochs compared to the blank epochs. **(b)** In area MT, a visual region associated with motion perception (see Chapter 5), the increase was only observed when the dots were moving.

an example, consider memory failure. We all have experienced the frustration of being introduced to someone at a party and then being unable to remember the person's name just 2 minutes later. Is this because we failed to listen carefully during the original introduction and thus the information never really entered memory? Or did the information enter our memory stores but after 2 minutes of distraction, we are unable to access the information? The former would constitute a problem with memory encoding; the latter would reflect a problem with memory

An Interview with Marcus E. Raichle, M.D. Dr. Raichle works in the Department of Neurology at Washington University School of Medicine. Dr. Raichle's PET group provided the seminal cognitive neuroscience imaging studies in the mid-1980s.

Authors: As one of the world's pioneers and authorities on brain imaging, and in particular PET, how would you characterize its short history? Are the kinds of problems you think about now the kinds you guessed you would be thinking about 10 years ago?

MER: The history of modern functional brain imaging, now exemplified by a combination of PET, fMRI, and ERPs, in my mind represents a remarkably successful merging of developments in imaging technology, neuroscience, and behavior. The pieces of the puzzle had developed quite separately until about 10 years ago, when cognitive science joined neuroscience in using the newly developed PET techniques to measure changes in brain blood flow in relation to changes in normal human behavior. The power of this combined approach became apparent almost immediately, although the learning curve remains steep.

I certainly didn't envision my current scientific agenda when I began working in the late sixties on issues of brain metabolism and blood flow. I was intrigued by the unique properties of positron-emitting radionuclides for measuring regional brain metabolism and blood flow in humans—little did I know at the time how regional and how unique. Luckily, I was in the right place at the right time, as events unfolded rather quickly with the introduction of x-ray CT in about 1972 and the invention of PET in our laboratory over the ensuing 2 years.

Authors: PET was initially built to deal with medical issues, perhaps looking at cerebral stroke per se or studying chemotherapeutic agents for brain tumor, or looking at neurotransmitters in psychiatric and degenerative diseases. PET today seems mostly committed to the study of functional correlates of cognitive function. Is this true, and if so, why?

MER: Actually, at its inception PET had a very varied agenda in the minds of the people who created it. The physics and engineering people who developed the imaging devices themselves had what I would describe as a clinical nuclear medicine orientation. In my estimation, they saw PET as the means by which clinical nuclear medicine could maintain a position in the clinical area along with x-ray CT, which was clearly getting all of the attention at the time. Nuclear brain scans, which had been staples of the practice of nuclear medicine, were quickly replaced by x-ray CT. Imaging had clearly captured everyone's imagination.

With PET, we suddenly had a tool that could give us measures of blood flow, blood volume, oxygen consumption, glucose utilization, tissue pH, and receptor pharmacology, among other things. These measurements in the brain had never been a part of the clinical practice of medicine. We had to develop our existent methods to develop an understanding of how to use this new information. That process is still very much ongoing, and in many areas such as brain pharmacology, it is a slow and tedious process. I'm still optimistic that it will provide important information in a variety of areas ranging from movement disorders and psychiatric diseases to certain types of brain injury. PET is, however, "competing," so to speak, with many other approaches in these areas. These approaches range from cellular and molecular techniques to various animal models.

You're absolutely right that PET and, more recently, fMRI have established a preeminent position in the study of the functional anatomical correlates of cognitive function in humans. The wonderful relationship between blood flow and neuronal activity, and the accuracy and simplicity of the technique, allowed for the design of an elegant paradigm, and thanks to input from my good friend Mike Posner, functional imaging with PET was off and running. The final ingredient was, certainly, that the questions we could address were immensely interesting and important. This was not a technique in search of a question.

Authors: Could you elaborate on the principle of image subtraction that is the standard in PET studies of cognition?

MER: The image subtraction methodology represents the wedding of objectives from the cognitive and imaging sciences. From an imaging perspective, the objective was to identify areas of the brain active during the performance of a particular task. Prior to the advent of the subtraction methodology, investigators using brain imaging techniques, as well as their predecessors who used simpler regional blood flow techniques, made a priori decisions about where in the brain they would look for changes. This was the so-called region-of-interest, or ROI, approach. The brain was arbitrarily divided, according to various schemes, into regions that would be analyzed for a change in blood flow or metabolism. This approach was particularly problematic when it came to the human cerebral cortex, where uncertainty was the rule, rather than the exception, in the areas much beyond primary motor and sensory cortices.

The subtraction methodology changed our perspective completely. In this approach, images obtained in two states (what we have come to refer to as a *task state* and a *control state*) are subtracted from one another to create a difference image. This image identifies for us those areas of the human brain that differ between the task and control states. There are no a priori assumptions about where such regions lie within the cerebral cortex or elsewhere. The subtraction images define the location and shape of the regions and also allow us to quantify the magnitude of the change. In one sense, this is a hypothesis-generating exercise— we're letting the human brain tell us how it is organized.

Authors: What concerns arise with the subtractive procedure? It has certainly had its critics.

MER: One of the most common criticisms is that the assumption of "pure insertion" is an incorrect assumption, and therefore, the subtraction methodology is invalid. The idea of pure insertion assumes that when a task state is compared to a control state, the difference represents the addition of processing components unique to the task state without affecting processing components in the control state. The issue is, How do we know this to be true? In imaging, we can also ask whether this is a serious concern.

Consider two scenarios. In the first, the control state and the task state are different only by the addition of brain processing components unique to the task state. Everything used in the control state remains unchanged. A subtraction image, under such circumstances, will predictably reveal areas of increased brain activity unique to the task state without changes in the areas known to be used in the control state.

Now, let us consider a second scenario, in which the notion of pure insertion is violated. Under these circumstances, areas of the brain that are active in the control state are not active in the task state. Now the subtraction image reveals not only areas of increased activity relative to the task state but also areas of decreased activity, reflecting areas that are used in the control state but not in the task state. Far from presenting us with a frustrating dilemma of interpretation, such data provide us with an even richer and less ambiguous understanding of human brain functional organization.

A second major criticism of the subtraction method has centered on the issue of averaging. Averaging, of course, is used to enhance the signal-to-noise properties of the images and is common to both PET and fMRI. Initially, the naysayers suggested, despite considerable empirical imaging data to the contrary, that subtraction-image averaging wouldn't work because of "obvious" individual differences among subjects. I'm just glad we didn't hear this criticism before we got started, or we might never have gotten into this work! If individual differences had been the methodological limitation portrayed by some, this entire enterprise would never have gotten off the ground.

So, does this mean that individual differences don't exist? Hardly. One has only to inspect individual human brains to appreciate that they do differ. However, general organizing principles emerge that transcend these differences. Such principles, coupled with our increasing ability to anatomically warp images to match one another anatomically in the averaging process, will further reduce the effect of individual differences. I find it amusing to reflect on the fact that our initial work was aided by the relative crude resolution of PET scanners. The blurring of data brought responses common across individuals together and allowed us to "see" them. Early on, even robust responses could be caused to "disappear" when one attempted to go to too high a resolution.

retrieval. Distinguishing between these two possibilities has proved very difficult, as witnessed by the thousands of articles on this topic that have appeared in cognitive psychology journals over the past 100 years.

Anthony Wagner and his colleagues at Harvard University used event-related fMRI to take a fresh look at the encoding-retrieval question (Wagner et al., 1998). They obtained fMRI scans while the subjects were studying a list of words, with one word appearing every 2 seconds. About 20 minutes after the scanning session was completed, the subjects were given a recognition memory test. On average, the subjects correctly recognized 88% of the words. The researchers then separated the trials based on whether a word had been remembered or forgotten. If the

memory failure was due to retrieval difficulties, no differences should be detected in the fMRI response to these two trials, since the scans were obtained only while the subjects were reading the words. However, if the memory failure was due to poor encoding, then one would expect to see a different fMRI pattern following presentation of the words that were later remembered compared to those that were forgotten. The results clearly favored the encoding-failure hypothesis (Figure 4.34). The BOLD signal recorded from two areas, the prefrontal cortex and the hippocampus, was stronger following the presentation of words that were later remembered. As we shall see in Chapters 8 and 12, these two areas play a critical role in memory formation. This type of study would not be possible with a

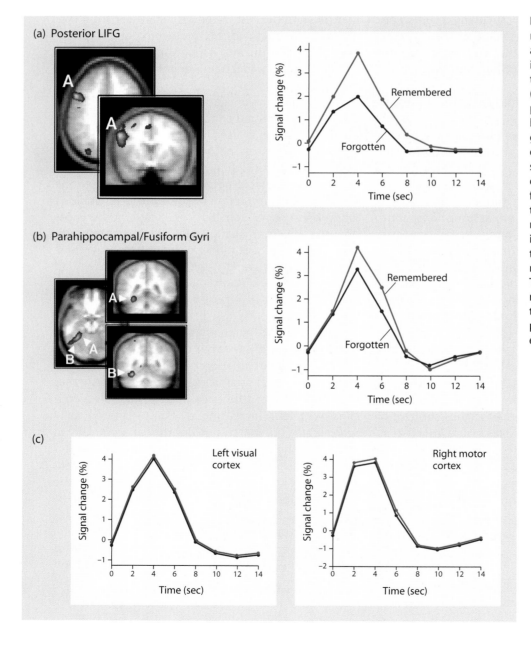

Figure 4.34 Use of event-related fMRI to identify areas associated with failures during memory encoding. Both the inferior frontal gyrus (LIFG) **(a)** and the parahippocampal region **(b)** in the left hemisphere exhibit greater activity during encoding for words that are subsequently remembered compared to those that are forgotten. **(c)** Activity over the left visual cortex and right motor cortex is identical following words that subsequently are either remembered or forgotten. This demonstrates that the memory effect is specific to the frontal and hippocampal regions. From Wagner et al. (1998).

block design method since the signal is averaged over all of the events within each scanning phase.

The limitations of imaging techniques such as PET and fMRI must be kept in mind. Even if we discover that the metabolic activity in a particular area correlates with an experimental variation, we still need to make inferences about the area's functional contribution. Correlation does not imply causation. For example, an area may be activated during a task but not play a critical role in the task's performance. The area simply might be "listening" to other brain areas that provide the critical computations. In this respect, imaging studies frequently are guided by other methodologies. For example, single-cell recording studies in primates can be used to identify regions of interest in a PET study with humans. Or, imaging studies can be used to isolate a component operation that is thought to be linked to a particular brain region based on the performance of patients with injuries to that area.

In turn, imaging studies can be used to generate hypotheses that are tested with alternative methodologies. For example, fMRI was used to identify neural areas that become activated when people recognize objects through touch alone (Figure 4.35). Suprisingly, tactile object recognition led to pronounced activation of the visual cortex even though the subjects' eyes were shut during the entire experiment (Deibert et al., 1999). One possible reason for the pronounced activation is that the subjects identified the objects through touch and then generated visual images of them. Alternatively, the subjects might have constructed visual images during tactile exploration and then used the images to identify the objects. A follow-up study with TMS was used to pit these hypotheses against one another (Zangaladze et al., 1999). TMS stimulation over the visual cortex impaired tactile object recognition. The disruption was only observed when the TMS pulses were delivered 180 msec after the hand touched the object; no effects were seen with earlier or later stimulation. Thus, the results indicate that the visual representations generated during tactile exploration were essential for inferring object shape from touch. These studies demonstrate how the combination of fMRI and TMS test causal accounts of neural function as well as make inferences about the time course of processing. Obtaining converging evidence from various methodologies enables us to make the strongest conclusions possible.

Another limitation of PET and fMRI is that both methods have poor temporal resolution in comparison to techniques such as single-cell recording or ERPs. PET is constrained by the decay rate of the radioactive agent. Even the fastest isotopes, such as ^{15}O, require measurements for 40 seconds to obtain stable radiation counts. While fMRI can operate much faster, it still lacks synchrony between changes in the stimulation and changes in the measured signal. This lack of synchrony occurs because alterations in blood flow do not happen immediately on stimulation but take a few seconds. Thus, PET and fMRI cannot give a temporal picture of the "online" operation of mental operations. Even complex tasks such as deciding whether the sum of the square roots of 16 and 25 is an even or odd number can be performed in seconds. Researchers have opted to combine the temporal resolution of evoked potentials with the spatial resolution of PET or fMRI for a better picture of the physiology and anatomy of cognition.

Figure 4.35 Combined use of fMRI and TMS to demonstrate the role of the visual cortex in tactile perception. **(a)** Functional MRI showing areas of activation in nine people during tactile exploration with the eyes closed. All of the subjects show some activation in striate and extrastriate cortex. **(b)** Accuracy in making tactile orientation judgments when a textured object is vibrated against the right index finger. Performance is impaired when the TMS pulse is applied 180 msec after the vibration. This effect is only observed when the pulse is applied over the central occipital cortex or over the contralateral occipital lobe. (a) From Deibert et al. (1999). (b) Adapted from Zangaladze et al. (1999).

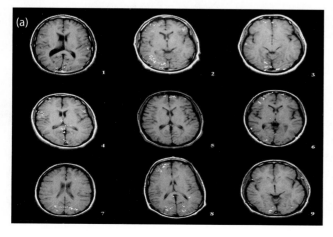

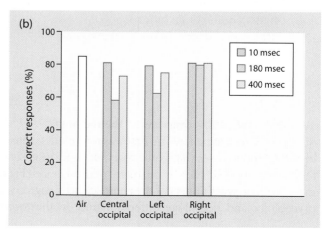

An Interview with Michael I. Posner, Ph.D.

Dr. Posner works in the Department of Psychology at the University of Oregon. His research on attention has encompassed many of the major methodologies of cognitive neuroscience and has been extended to populations ranging from newborn infants to patients with schizophrenia.

Authors: Your work in cognitive psychology is singular and established a whole field of research. You then became interested in cognitive neuroscience and have employed a variety of new techniques to measure brain function, including PET and ERPs. Do you think the traditional behavioral methods of cognitive psychology are antiquated and no longer sufficient for studying how the brain enables mind?

MIP: By traditional methods, I assume you mean accuracy and reaction time including the many variants such as signal detection theory, additive factor theory, etc. In my view, an impressive aspect of the anatomical methods such as PET and fMRI is how much they have supported the view that cognitive measures can be used to suggest separate neural structures. One impressive example is the paper in the *Journal of Cognitive Neuroscience* by Stan Dehaene (1996). He applied additive factors theory to determine separate stages in simple numerical judgments. Additive factor theory assumes separate serial stages, and tests this by showing independence between the variables that influence the time for each stage. Dehaene found that each stage, as determined from an analysis of independence in reaction time, was generated by a separate brain area. Another example is the close correspondence we have found between the conditions required for executive function based on cognitive models and the conditions in which PET studies have found activation of the anterior cingulate. This is not to argue that cognitive and anatomical methods will always converge, but to support the importance of their joint contribution to understanding brain function.

Authors: Would it be fair to say your center has changed? By that we mean 25 years ago you read the cognitive psychology literature which dealt solely with behavioral measures and from those one inferred cognitive states. Do you now read mostly in the area of cognitive neuroscience?

MIP: Yes, it would be fair to say that my reading habits have changed, but I still like to read the cognitive literature to help understand the theoretical issues involved. In the end we have to interpret the anatomy and circuitry found in imaging studies in terms of the functions they serve. The methods of cognitive neuroscience have made the dream of a deep understanding of how the brain carries out thought seem so much closer now than 25 years ago. It is almost exactly 25 years ago that I began work with parietal patients in hopes of connecting the cellular studies with cognitive studies of normal human beings. Now most neuroimaging studies illustrate the close connection between local neuronal activity on the one hand and cognitive operations on the other.

Authors: Suppose PET studies reveal the circuitry for a particular cognitive process. How do we then go on to seek an explanation of the mechanism of the process? Obviously this is a difficult, perhaps impossible question. But how are you beginning to think about the job of the next generation of scientists?

MIP: It's probably best to answer by example. In the 1970s Roger Shepard showed that reaction time to rotate something in your head was exquisitely related to the angle through which the rotation had to occur. Recent PET studies have provided a nice treatment of the anatomy involved in this act. More recently, Georgopoulos provided a

SUMMARY

Two goals guided this overview of the methods of cognitive neuroscience. The first was to provide a sense of the methodologies that come together to form an interdisciplinary field such as cognitive neuroscience (Figure 4.36). The practitioners of the neurosciences, cognitive psychology, and neurology differ not only in the tools

they use but also in the questions they seek to answer. The neurologist may request a CT scan of an aged boxer to find out if the patient's confusional state is reflected in atrophy of the frontal lobes. The neuroscientist may want a blood sample from the patient to search for metabolic markers indicating a reduction in a transmitter

neuronal model of how a changing population vector of cells could produce a rotation. The population vector orientation shifted during the reaction time in a way that fit with the idea of a rotation. Similar efforts to relate the cellular, anatomical, and cognitive level are now emerging in areas such as motion perception, visual search, and shifts of attention. Some of these areas involve accounts of the transmitters involved as well as the cellular activity. I don't necessarily want to claim that the population vector is a complete explanation of mental rotation, but just a few years ago it would have been hard to imagine even a good start in the direction of answering your question.

Authors: For new students wanting to study those lofty issues of cognition and indeed the very nature of consciousness itself, what do you recommend they do? Put differently, what does the research scientist of tomorrow working on these questions have to be trained in today?

MIP: A good background in cognition that might involve courses in cognitive psychology, linguistics/anthropology, and philosophy would be important. A strong background in computational methods involved in model building is also important. The neuroscience background should include studying systems and cellular levels, and a good knowledge of genetics is becoming important.

Authors: What do you expect a cognitive neuroscience lab, or conference, or journal will look like in 2005?

MIP: Here is my idea of what the Table of Contents of an issue of the *Journal of Cognitive Neuroscience* in 2005 might look like:

- How communication between brain areas involved in first and second language comprehension changes with mastery of the new language
- At what age are genes coding for extroversion expressed?
- Laser images of neuronal activity in parietal cortex during mental rotation

- Function of monkey cortex homologous to the human visual word-form system
- A pharmacological study designed to reduce loss of brain plasticity with age
- Size of brain areas devoted to the semantic category "animal" as a function of expertise: A functional MRI study
- Change in blood flow and dopamine uptake in auditory areas following pharmacological treatment for auditory hallucinations in first break schizophrenics

Addendum: Prior to sending this second edition off to press, we contacted Dr. Posner and asked him to reflect back on his predictions from 1995. His response is given below.

MIP: In the light of the time that has passed, my comments seem rather conservative. Much of what I predicted in 1995 might be rejected in 2005 as already having been done. I had always regarded myself as an optimist, but the swift pace of advance in cognitive neuroscience and related fields has caught me by surprise. Examples are the developments in fMRI that have produced event-related methods and the ability to image white matter. Nor did I anticipate the astounding advance that the human genome project has made possible, in the potential for both direct genetic studies of humans and indirect studies by conditional mouse knock-outs. Even less did I think about how the amazing ability to isolate stem cells and to repair damaged tissue would bring issues of rehabilitation to the forefront. These findings and others have led to my commitment to the study of neural plasticity as found in the developing control systems of infants and young children. We believe our studies will show how genes and experience shape the systems that allow for our voluntary control of feelings, thoughts, and actions. Of course, I always hope to understand attention in all its forms, but I am most excited about seeing the fascinating changes in natural behaviors as these control circuits become organized.

---

system. The cognitive psychologist may design a reaction time experiment to test whether a component of a decision-making model is selectively impaired. Cognitive neuroscience endeavors to answer these questions by taking advantage of the insights that each approach has to offer and using them together.

The second goal was to introduce methods we will encounter in subsequent chapters. These chapters focus on content domains such as perception, language, and memory, and on how the tools are being applied to understand the brain and behavior. Each chapter draws on research that uses the diverse methods of cognitive neuroscience. Often the convergence of results yielded by different methodologies offers the most complete theories. A single method cannot bring about a complete understanding of the complex processes of cognition.

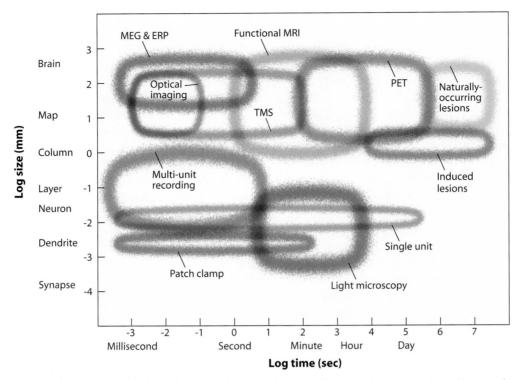

Figure 4.36 Spatial and temporal resolution of the prominent methods used in cognitive neuroscience. Temporal sensitivity, plotted on the x axis, refers to the time scale over which a particular measurement is obtained. It can range from the millisecond activity of single cells to the behavioral changes observed over years in patients who have had strokes. Spatial sensitivity, plotted on the y axis, refers to the localization capability of the methods. For example, real-time changes in the membrane potential of isolated dendritic regions can be detected with the patch clamp method, providing excellent temporal and spatial resolution. In contrast, naturally occurring lesions damage large regions of the cortex and are detectable with MRI.

We have reviewed many methods, but the review is incomplete, in part because new methodologies for investigating the relation of the brain and behavior spring to life each year. Neuroscientists are continually refining techniques for measuring and manipulating neural processes at a finer and finer level. Patch clamp techniques isolate restricted regions on the neuron, enabling studies of the membrane changes that underlie the inflow of neurotransmitters. Laser surgery can be used to restrict lesions to just a few neurons in simple organisms, providing a means to study specific neural interactions. The use of genetic techniques such as knock-out procedures has exploded over the past decade, promising to reveal the mechanisms involved in normal and pathological brain function.

Technological change is also the driving force in our understanding of the human mind. Our current imaging tools are undergoing constant refinement. Each year witnesses the development of more sensitive equipment to measure the electrophysiological signals of the brain or the metabolic correlates of neural activity, and the mathematical tools for analyzing these data are constantly becoming more sophisticated. In addition, entire new classes of imaging techniques are just beginning to gain prominence. One

such noninvasive method is optical imaging (Gratton and Fabiani, 1998). Beams of near-infrared light are projected at the head. The light diffuses through the tissues, and sensors placed on the skull detect the light as it exits the head. Brain areas that are active scatter the light more than areas that are inactive, allowing a direct measure of neural activity. Noninvasive optical imaging offers excellent temporal resolution. Its spatial resolution is comparable to that of current high-field MRI systems, although the technique at present is limited to measuring structures near the cortical surface. Another plus is that the method is relatively inexpensive. Whereas an MRI system might cost $5 million, optical imaging systems can be built with less than $100,000.

We began this chapter by pointing out that paradigmatic changes in science are often fueled by technological developments. In a symbiotic way, the maturation of a scientific field such as cognitive neuroscience provides a tremendous impetus for the development of new methods. The questions we ask are constrained by the available tools, but new research tools are promoted by the questions we ask. It would be foolish to imagine that current methodologies will become the status quo for the field, which makes it an exciting time to study brain and behavior.

KEY TERMS

blood oxygenation level dependent (BOLD) effect

brain lesions

cognitive psychology

computed tomography (CT or CAT)

double dissociations

electroencephalography (EEG)

event-related potential (ERP)

knock-out procedures

magnetic resonance imaging (MRI)

magnetoencephalography (MEG)

neurology

neurophysiology

positron emission tomography (PET)

receptive field

regional cerebral blood flow (rCBF)

retinotopic

simulation

single-cell recording

single dissociation

topographic representation

transcranial magnetic stimulation (TMS)

THOUGHT QUESTIONS

1. To a large extent, progress in all scientific fields depends on the development of new technologies and methodologies. What technological and methodological developments have advanced the field of cognitive neuroscience the most?

2. Cognitive neuroscience is an interdisciplinary field that incorporates aspects of neuroanatomy, neurophysiology, neurology, cognitive psychology, and computer science. What do you consider the core feature of each discipline that allows it to contribute to cognitive neuroscience? What are the limits of each discipline in addressing questions related to the brain and mind?

3. Describe the requirements for establishing single and double dissociations, and explain why double dissociations provide stronger evidence for claims about the brain and behavior. Choose a task that interests you, and generate an example of each type of dissociation you might find if you were to study neurological patients whose ability to perfom this task is impaired. Be sure to note what we learn about the brain *and* mind from these examples.

4. A skeptic wrote in the *New York Times* Science section in the spring of 2000 that despite all of the excitement surrounding new imaging techniques like PET and fMRI, these methods were unlikely to lead to profound insights into brain function. The author's focus was that these tools were limited in terms of their resolution and were unlikely to provide more insight than simply showing which parts of the brain are activated during particular tasks. Make explicit the concerns that the writer was expressing, comparing these imaging tools to other methods that might not have similar problems. Discuss whether or not you share these concerns.

5. In anticipation of the next chapter, consider how you might study a problem such as color perception using the multidisciplinary techniques of cognitive neuroscience. Can you predict the questions that one might ask about this topic, and outline the types of studies that cognitive psychologists, neurophysiologists, and neurologists might consider?

SUGGESTED READINGS

CHURCHLAND, P.S. (1986). *Neurophilosophy: Toward a Unified Science of the Mind/Brain.* Cambridge, MA: MIT Press.

D'ESPOSITO, M., ZARAHN, E., and AGUIRRE, G.K. (1999). Event-related functional MRI: Implications for cognitive psychology. *Psychol. Bull.* 125:155–164.

HILLYARD, S.A. (1993). Electrical and magnetic brain recordings: Contributions to cognitive neuroscience. *Curr. Opin. Neurobiol.* 3:710–717.

KANDEL, E.R., SCHWARTZ, J.H., and JESSELL, T.M. (1995). *Essentials of Neural Science and Behavior.* Norwalk, CT: Appleton and Lange.

POSNER, M.I., and RAICHLE, M.E. (1994). *Images of Mind.* New York: W.H. Freeman.

RAPP, B. (2001). *The Handbook of Cognitive Neuropsychology: What Deficits Reveal about the Human Mind.* Philadelphia: Psychology Press.

5

Perception and Encoding

How does the brain convert sensory signals into a perception of a coherent world? Our phenomenal experience suggests that this is an effortless process: We absorb the familiar sights of the neighborhood, recognize the sounds of children playing at the nearby playground, and take delight in the smells of a backyard barbecue. Considering the complexity of the nervous system, it is quite amazing how easily we take in all of this information. Our sensory apparatus cannot make multisensory snapshots of the world; the receptors in our eyes respond to the photons of light in a manner clearly distinct from the fine filaments in the inner ear that sense the changes in air pressure corresponding to sounds. Moreover, when we focus on a single sensory modality, we can appreciate the complexity of the information-processing task we face. The cars along the street are of different shapes and colors. Some are parked; some are moving.

In the next few chapters, we explore the cognitive neuroscience of perception. In this chapter, the focus is on the initial stages of perception—how sensory information is represented and processed to form integrated percepts. In Chapter 6, we will examine how we recognize these percepts as meaningful entities: objects that we can manipulate or navigate about, or other organisms like our friends and family. A full appreciation of perception also requires consideration of how we attend to certain stimuli at the expense of others, and how we connect this information to our stored knowledge of the world. In Chapters 7 and 8 we will address the problems of attention and memory.

To begin, let us consider a person for whom perception has become a challenge. By studying the damaged brain, we can gain insight to the processes required for perceiving the world.

DISORDERS OF PERCEPTION: A CASE STUDY

Patient P.T. was presented at the Neurology Grand Rounds in Portland, Oregon. The Grand Rounds, a weekly event, is when staff neurologists, internists, and residents gather to review the most puzzling and unusual cases being treated on the ward.

The cause of P.T.'s neurological disorders was not a mystery; he had had a stroke. Four months previously,

P.T. had awakened and experienced acute dizziness and weakness in his left hand and leg. On visiting his family physician that morning, he was told that he had probably had a cerebrovascular accident, a diagnosis based on his history of chronic hypertension. Indeed, P.T. had suffered a left-hemisphere stroke 6 years previously, and so the physician was confident in his diagnosis, especially since multiple strokes are not uncommon. P.T. was referred to the hospital, where the stroke, localized in the right hemisphere this time, was confirmed by computed tomography (CT).

What was unusual about P.T. was the collection of symptoms he continued to experience 4 months after the stroke occurred. The dizziness had ended by the second day after the stroke, and the left-sided weakness had mostly subsided during the first month. The only remaining sign of these initial symptoms was that P.T. continued to drag his left leg slightly, although he was not aware of it.

Yet P.T. experienced some problems as he tried to resume the daily routines required on his small family farm, his home for 66 years. He had particular difficulty recognizing places and objects. He would be working on a stretch of fence, look out over the hills, and suddenly realize that he did not know the landscape. It was hard for him to pick out individual dairy cows, a matter of concern lest he attempt to milk a bull! And, most troubling of all, P.T. no longer recognized the people around him, including his wife. The woman who had served him breakfast each morning when he sat down at the table was a stranger. He had no trouble seeing her standing over the kitchen stove, he could describe her actions when she served his bacon and eggs, and he noticed when she walked. But, in simply looking at her, he failed to identify her as his wife. He knew that her parts—body, legs, arms, and head—formed a person. But P.T. failed to see these parts as belonging to a specific individual. His deficit was not limited to his wife; he had the same problem with other members of his family and friends from his small town.

A striking feature was that P.T.'s inability to recognize objects and people was limited to the visual modality. As soon as his wife spoke, he immediately recognized her voice. Indeed, he claimed that on hearing her voice, the visual percept of her would "fall into place." The shape in front of him would suddenly metamorphose into his wife. In a similar fashion, he could recognize specific objects by touching them.

P.T. is not the first person to have difficulty visually recognizing objects and people. Indeed, neurologists have a term for this type of disorder, *visual agnosia*. Although visual agnosia is rare, patients such as P.T. have been described and studied extensively. In presenting the case during Grand Rounds, the chief neurologist assessed P.T.'s processing abilities to pinpoint the problems underlying his deficit.

During the examination, P.T. demonstrated a striking dissociation. He was shown two paintings, one by Monet depicting a subdued nineteenth-century countryman dressed in his Sunday suit, and the other by Picasso of a woman with a terrified expression (Figure 5.1). P.T. was asked to describe what he saw in each painting. When shown the Monet, he looked puzzled.

(a)

(b)

Figure 5.1 Two portraits. **(a)** Detail from "Luncheon on the Grass," painted in 1886 by the French Impressionist Claude Monet. **(b)** Pablo Picasso's "Weeping Woman," painted in 1937 during his Cubist period.

He saw no definable forms, just an abstract blend of colors and shapes. His problem in interpreting the painting was consonant with the deficits he experienced at home. Yet he readily identified the figure in Picasso's painting and pointed out that it was a woman, or perhaps a young girl. This dissociation is even more compelling, as most would readily agree that the Monet is more realistic.

The two paintings differ radically from each other. A psychoanalyst might choose to focus on the different emotional responses evoked by the images. Picasso's painting is clearly unsettling, and perhaps this stirring of emotions facilitates recognition. But this is an unlikely connection to P.T.'s problems at home, which are manifest across a range of objects. Moreover, he does not have difficulty recognizing objects through modalities other than vision.

In Grand Rounds, attention was paid to how the visual information in the two paintings differs. For example, facial features are well marked in the Picasso. The oval eyes are clearly demarcated by black contours encircling white irises. The teeth are individually drawn, as are the hairs of each eyebrow. In Monet's portrait, these features either are absent or slowly emerge from the background. A slight change of shading cues to the transition from forehead to eyebrows. In a similar way, the nose has only faint detailing. These differences in contours are matched by the level of contrast, or brightness, used by the two artists. Picasso, following his cubist tradition, used bold colors and sharp contrasts. Each part has a different color; even in the facial skin yellow switches rapidly to white, a transition emphasized by a black contour. Monet, in contrast, selected colors that are fairly equal in brightness. Using a variety of colors, he blended each region into the next.

This example demonstrates at least three prominent differences. Picasso painted the parts as separate units. He used sharp contrasts in brightness and vivid colors to highlight facial regions. Monet opted for a softer approach, in which parts are best seen in a continuous whole, with gradual changes in contrast and color. Can any of these factors account for P.T.'s performance in identifying the figures in Picasso and Monet? Do the differences explain his problems with recognizing familiar objects? Is his problem related to one of these factors or a combination of them? To answer these questions, we need to know more about how visual information is processed and represented. After considering these issues, we will return to P.T.'s deficits and formulate hypotheses about their cause.

OVERVIEW OF NEURAL PATHWAYS

Humans, like most diurnal creatures, depend on the sense of vision. While other senses such as hearing and touch are essential, visual information dominates our perceptions and frames the way we think. We use visually derived metaphors such as "I see" and "Your hypothesis is murky" to describe cognitive states.

Consider the vast amount of neuroanatomical tissue involved in visual perception. Not only are there millions of neurons in the eye, but also these sensors project countless bits of data to the subcortex and cortex. Indeed, over 50% of the cortex in the macaque monkey is devoted to visual perception. Beware, though: Estimates such as these may be biased by researchers' reliance on visual stimulation. If auditory or somatosensory stimuli were used as readily, many brain regions might be found to be responsive to these modalities.

The Eye, Retina, and Receptors

One reason why vision is so important is that it enables us to perceive information at a distance, to engage in what is called *remote sensing* or *exteroceptive perception*. We need not be in immediate contact with a stimulus to process it. Contrast this ability with the sense of touch. For this sense, we must be in direct contact with the stimulus. The advantages to engaging in remote sensing are obvious. An organism surely can avoid a predator better when it can detect the predator at a distance. It is probably too late to flee once shark teeth pierce the skin.

Visual information is contained in the light reflected from objects. To perceive objects, we need sensory detectors that respond to the reflected light. As light passes through the lens of the eye, the image is inverted and focused to project on the back surface of the eye (Figure 5.2), the **retina.** The deepest layer of the retina is composed of millions of **photoreceptors** (see Figure 5.2), each containing light-sensitive molecules, or *photopigments*. When exposed to light, the photopigments become unstable and split apart. Their decomposition alters the flow of the electrical current around the photoreceptors. This light-induced change

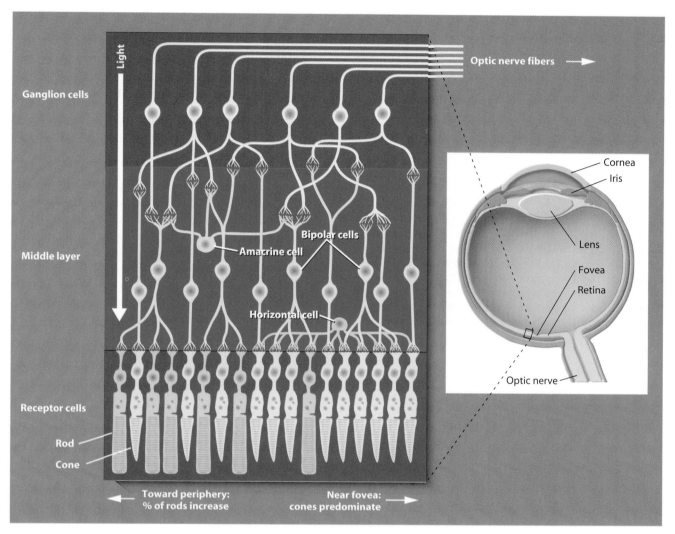

Figure 5.2 Anatomy of the eye **(right)** and retina **(left)**. Light enters through the cornea and activates the receptor cells of the retina located along the rear surface. There are two types of receptor cells, rods and cones. The output of the receptor cells is processed in the middle layer of the retina and then relayed to the central nervous system via the optic nerve, the axons of the ganglion cells. Adapted from Sekuler and Blake (1990).

triggers action potentials in downstream neurons. Thus, photoreceptors provide for translation from an external stimulus, light, into an internal neural signal, the detection of that stimulus.

The two types of photoreceptors are rods and cones. *Rods* are sensitive to low levels of stimulation. They are most useful at night when light energy is reduced. They will also respond to a bright light, but because replenishing the photopigment in rods takes time, they are of little use during daytime. *Cones* require more intense levels of light and use photopigments that can be generated rapidly. Thus, cones are most active during daytime vision. Cones are essential for color vision, and we commonly speak of them as

being one of three types: red, green, or blue. These names are somewhat misleading. Cones do not respond to colors per se; as shown in Figure 5.3, they differ in the sensitivity of their photopigments to different wavelengths of visible light.

Rods and cones are not distributed equally across the retina. Cones are densely packed near the center of the retina, in a region called the *fovea*. Few cones are in the more eccentric regions of the retina. In contrast, rods are distributed throughout the retina. An easy demonstration of the differential distribution of rods and cones can be made by having a friend slowly bring a colored marker into your view from one side of your head. Notice that you see the marker

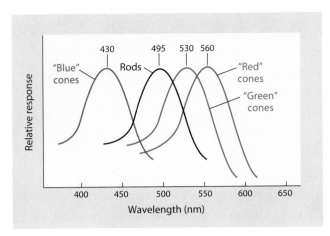

Figure 5.3 Spectral sensitivity functions for rods and the three types of cones. The short, or "blue" cones are maximally responsive to light with a wavelength of 430 nm. The peak sensitivities of the medium, "green" and long, "red" cones are shifted to longer wavelengths. White light such as daylight will activate all three receptors because it contains all wavelengths.

and its shape well before you identify its color because of the sparse distribution of cones in the retina's peripheral regions.

From the Eye to the Central Nervous System

Extensive signal processing of visual information is performed within the retina. The output from the photoreceptors is first processed in the bipolar cells and from there to ganglion cells. Extensive convergence of information happens within these processing layers of the retina. Indeed, while humans have an estimated 260 million photoreceptors, there are only 2 million ganglion cells, the eye's sole output source. This compression of information suggests that higher-level visual centers should be efficient processors to recover the details of the visual world. Axons of the ganglion cells form a bundle, the *optic nerve*. By way of this nerve, visual information is transmitted to the central nervous system.

Figure 5.4 diagrams how visual information is conveyed from the eyes to the central nervous system.

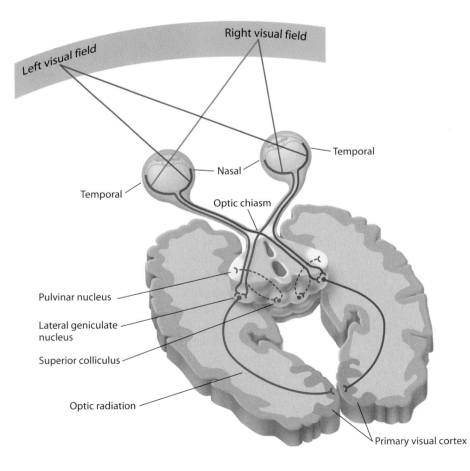

Figure 5.4 The primary projection pathways of the visual system. The optic fibers from the temporal half of the retina project ipsilaterally, while the nasal fibers cross over at the optic chiasm. In this way, the input from each visual field is projected to the primary visual cortex in the contralateral hemisphere after the fibers synapse in the lateral geniculate nucleus (geniculo-cortical pathway). A small percentage of visual fibers of the optic nerve terminate in the superior colliculus and pulvinar nucleus.

Before entering the brain, each optic nerve splits into two parts. The temporal, or lateral, branch continues to traverse along the same side. The nasal, or medial, branch crosses over to project to the opposite side; this crossover place is called the *optic chiasm*. Given the eye's optics, the crossover of nasal fibers ensures that visual information from each side of external space will be projected to contralateral brain structures. For example, because of the retina's curvature, the temporal half of the right retina is stimulated by objects in the left visual field. In the same fashion the nasal hemiretina of the left eye is stimulated by this same region of external space. Since fibers from each nasal hemiretina cross, all information from the left visual field is projected to the right hemisphere and information from the right visual field is projected to the left hemisphere.

Once inside the brain, each optic nerve divides into pathways that differ with respect to where they terminate within the subcortex. Figure 5.4 focuses on the retino-geniculate pathway, the projection from the retina to the **lateral geniculate nuclei (LGN)** of the thalamus. This pathway contains more than 90%

of the axons in the optic nerve and provides input to the cortex via the geniculo-cortical projections. The remaining 10% of the fibers innervate other subcortical structures including the pulvinar nucleus of the thalamus and the **superior colliculus** of the midbrain. However, the fact that these other receiving nuclei are innervated by only 10% of the fibers does not mean these pathways are unimportant. The human optic nerve is so large that 10% of the optic nerve constitutes more fibers than are found in the entire auditory pathway. The superior colliculus and pulvinar nucleus play a big role in visual attention, and the retino-collicular pathway is sometimes viewed as a more primitive visual system. We return to possible functions of this pathway later in this chapter.

The final projection to the visual cortex is via the geniculo-cortical pathway. This bundle of axons exits from the LGN and ascends to the cortex, with almost all of the fibers terminating in the primary visual area of the occipital lobe. Thus, visual information in the cortex has been processed by at least five distinct neurons: photoreceptors, bipolar cells, ganglion cells, LGN cells, and cortical cells.

PARALLEL PROCESSING IN THE VISUAL SYSTEM

We have described the pathways enabling the transmission of visual information from the eye to the cortex. We have focused on anatomical patterns of connectivity, the wiring that permits neural regions to communicate with other neural regions. What we also need is a description of the information carried in these tracts: What do neural signals in the visual pathway represent?

A central hypothesis in visual perception is that visual information is distributed across distinct subsystems. By this view, perception is analytic. Early processes are devoted to analyzing attributes of a stimulus: Some processes represent shape, other processes focus on color, and others provide information about the dynamics or movement in the visual scene.

In some ways, this hypothesis is counterintuitive. Our introspection is that objects are perceived as unified wholes. If a blue Volkswagon passes you on the highway, you do not have the impression that your final percept was produced in a piecemeal manner. For example, introspection would not suggest that the analysis of the Volkswagon-like shape and the color associated with that shape occurred separately. Rather, from the mo-

ment you are aware of the car, its color and shape appear as an integrated whole.

Nonetheless, converging evidence provides compelling support for the idea that perception operates in an analytic manner. Indeed, the feature-extraction hypothesis is one of the best examples of how different branches of cognitive neuroscience can provide complementary evidence.

Organization of the Lateral Geniculate Nucleus

We noted that 90% of the fibers in the optic nerve terminate in the LGN. The projection of these fibers is not random. Rather, the architecture of the LGN is highly organized, and this is manifest at several levels of analysis. At a macroscopic level, the LGN contains six well-defined layers (Figure 5.5). These layers should not be confused with the six-layer structure of the cerebral cortex. In the cortex, the layering reflects functional differences such as whether cells receive input from subcortical areas, project to subcortical areas, or are involved in intracortical processing. In contrast, each layer within the LGN receives inputs

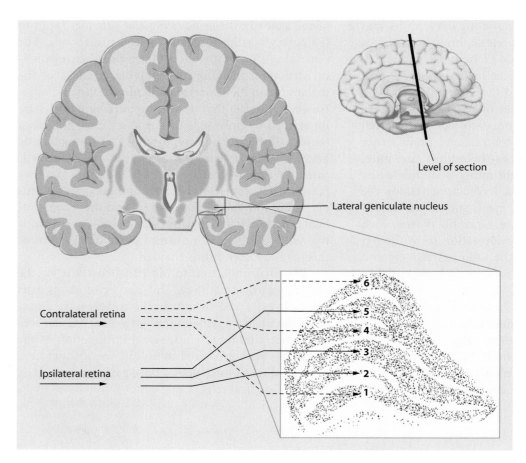

Level of section

Lateral geniculate nucleus

Contralateral retina

Ipsilateral retina

Figure 5.5 The lateral geniculate nucleus, located in the most lateral, inferior region of the thalamus. It is composed of six layers, with the ipsilateral eye projecting to layers 2, 3, and 5 and the contralateral eye projecting to layers 1, 4, and 6. Layers 3 through 6 contain the smaller neurons of the parvocellular system; layers 1 and 2 contain the larger neurons of the magnocellular system.

from axons of the optic tract and sends outputs that terminate in the cortex.

At a more microscopic level, additional organizational principles can be seen in the structure and connectivity of individual cells within the LGN. In particular, the LGN are characterized by three organizational properties. First, three of the layers receive input from one retina while the other three receive input from the other retina. Ganglion cell axons from the right temporal hemiretina terminate in layers 2, 3, and 5 of the right LGN, and ganglion cell axons from the left nasal hemiretina terminate in layers 1, 4, and 6. The opposite pattern is seen for the left LGN. It is important to keep in mind that the ipsilateral temporal hemiretina and contralateral nasal hemiretina are responsive to stimuli in the same visual field (see Figure 5.4). This ensures that visual information from a region in space is projected to the same LGN.

The second organizational principle pertains to the specificity of projections from the visual field to the LGN. Each layer of the LGN contains a topographic map of the retina (and thus of external space) that is in tight register. An object at a certain position in space will activate cells within each layer that fall along a line perpendicular to the LGN's surface.

Consider a stimulus located in the upper portion of the right visual field. This stimulus activates the nasal hemiretina in the right eye and the temporal hemiretina in the left eye. Ganglion cells with receptive fields that encompass this region of space activate cells in the left LGN. Activation occurs in all six layers, and the selected cells are located at comparable positions within the LGN. So each LGN has six distinct retinotopic representations, one within each layer. The evidence for this relies not only on anatomical tracing techniques that detail the patterns of connectivity between the retina and the LGN, but also on cellular recordings from the LGN that verify the functional relevance of these connections. If a stimulus is presented at a location, cellular activity is found within each layer as the electrode successively penetrates the six layers.

The third organizational principle of the LGN shows that the multilayered system is not simply redundant. With regard to cytoarchitecture, the cell types of each layer have clear distinctions. As can be

Pioneers in the Visual Cortex

Akin to the voyages of the fifteenth-century European explorers, the initial investigations of the cerebral cortex's neurophysiology required a willingness to sail in uncharted waters. The two admirals in this enterprise were David Hubel and Torsten Wiesel. Hubel and Wiesel arrived at Johns Hopkins University in the late 1950s, hoping to extend the pioneering work of Steve Kuffler. His research had elegantly described the receptive field organization of ganglion cells in the cat retina, laying out the mechanisms that allowed cells to detect edges that defined objects in the visual world. Rather than focus on the lateral geniculate nucleus (LGN), the next relay in the system, Hubel and Wiesel (1977) set their sights on the primary visual cortex. Vernon Mountcastle, another Hopkins researcher, was just completing his seminal work in which he laid out the complex topographic organization of the somatosensory cortex. Hubel and Wiesel were inspired to look for similar principles in vision.

During the first few weeks of their recordings, the research duo was puzzled by what they observed. While they had little difficulty identifying individual cortical cells, the cells failed to respond to the kinds of stimuli that had proved so effective in Kuffler's studies—small spots of light positioned within a cell's receptive fields. Indeed, the lack of consistent responses made it difficult to determine where the receptive field was situated. Their breakthrough came when they switched to dark spots, created by placing an opaque disk on a glass slide. While the cell did not respond to the dark spot, Hubel and Wiesel noticed a burst in activity as the edge of the glass moved across part of the retina. After hours of playing with this stimulus, the first organizational principle of primary visual cortex neurons became clear: Unlike the circular receptive fields of ganglion cells, cortical neurons were responsive to edges.

Subsequent work revealed that LGN cells behaved similarly to ganglion cells, being maximally
continued on the following page

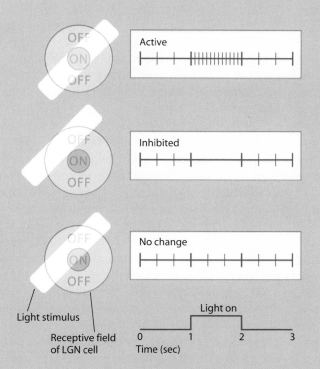

Figure A Characteristic response of a lateral geniculate cell. Cells in the lateral geniculate nucleus (LGN) have concentric receptive fields with either an on center–off surround or off center–on surround organization. This on center–off surround cell fires rapidly when the light encompasses the center region **(top)** and is inhibited when the light is positioned over the surround **(middle)**. A stimulus that spans both the center and the surround produces little change in activity **(bottom)**. As such, LGN cells are ideal for signaling changes in illumination such as those that arise from stimulus edges.

continued from previous page

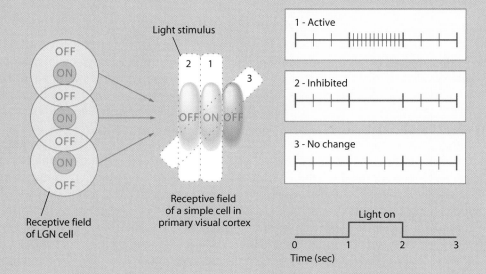

Figure B Simple cells in the primary visual cortex can be formed by linking the outputs from concentric LGN cells with adjacent receptive fields. In addition to signaling the presence of an edge, simple cells are selective for orientation. This simple cell is either excited or inhibited by an edge that follows its preferred orientation. It shows no change in activity to an edge at a perpendicular orientation.

excited by small spots of light. Such cells are best characterized as exhibiting a concentric center-surround organization. Figure A shows the receptive field of an LGN cell. When the spot of light falls within the excitatory center region, the cell is activated. If the same spot is moved into the surrounding region, the activity is inhibited. A stimulus that encompasses both the center and the surrounding will fail to activate the cell, as the activity from the excitatory and inhibitory regions will cancel each other. This clarifies a fundamental principle of perception: The nervous system is most interested in change. We recognize an elephant not by the homogeneous gray surface of its body, but by the contrast between the gray edge of its shape against the background.

In Figure B, outputs from three LGN cells with receptive fields centered at slightly different positions are linked to a single cortical neuron. This cortical neuron would continue to have a center-surround organization, but for this cell the optimal stimulus would have to be an edge. In addition, the cell would be selective for edges in a certain orientation. As the same stimulus is rotated within the receptive field, the cell would cease to respond because the edge would now span

excitatory and inhibitory regions of the cell. Hubel and Wiesel called these cells *simple cells,* to connote the fact that their simple organization would extract a fundamental feature for shape perception: the border of an object. The same linking principle can yield more complex cells, cells with a receptive field organization that makes them sensitive to other features such as corners or edge terminations.

Orientation selectivity has proved to be a hallmark of neurons in the primary visual cortex. Across a 2-×-2-mm chunk of cortex, the receptive fields of neurons are centered on a similar region of space (Figure C). Within the chunk, the cells vary in terms of their preferred orientation, and alternate between columns that are responsive to inputs from the right and left eyes. A series of such chunks allows for the full representation of external space, providing the visual system with a means for extracting the visible edges in a scene.

Hubel and Wiesel's studies established how a few organizational principles can serve as building blocks of perception derived from simple sensory neurons. The importance of their pioneering studies was acknowledged in 1981, when they shared the Nobel Prize in physiology or medicine.

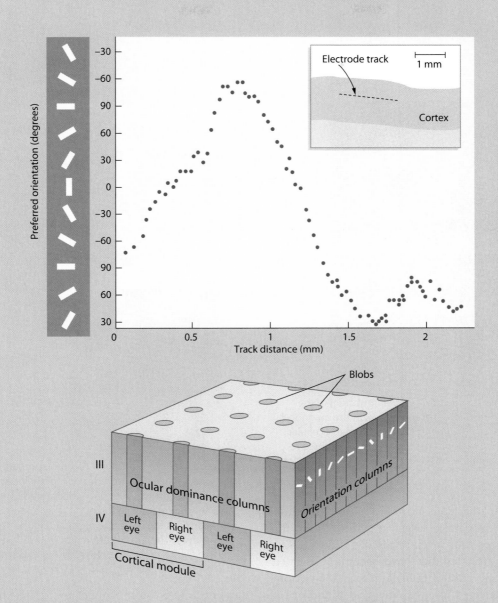

Figure C Feature representation within the primary visual cortex. **Top:** As the recording electrode is moved along the cortex, the preferred orientation of the cells varies in a continuous manner. The preferred orientation is plotted as a function of the location of the electrode. **Bottom:** The orientation columns are crossed with ocular dominance columns to form a cortical module. Within a module, the cells have similar receptive fields (location sensitivity), but vary in terms of input source (left or right eye), orientation sensitivity, color sensitivity, and size sensitivity (not shown). This organization is repeated for each location. Adapted from Bear et al. (1996). Top panel after Hubel and Wiesel (1968).

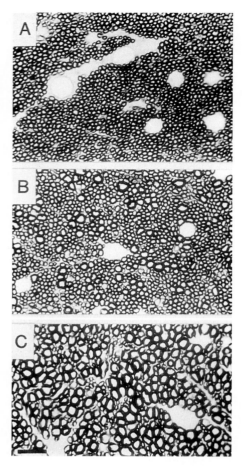

Figure 5.6 A cross section through the optic tract of the macaque, showing the large-diameter fibers of the magnocellular layer and the small-diameter fibers of the parvocellular layer. The large white regions are blood vessels. The three regions **(A–C)** are magnifications of different sections from dorsal to ventral.

seen in Figure 5.6, the axons of the cells within the lower two layers are considerably larger in diameter than the cells of the upper four layers. This difference in size gives rise to the names that have become associated with the layers. Because of their large size, the bottom two layers are referred to as the **magnocellular, or M, system,** while smaller cells in the upper four layers constitute the **parvocellular, or P, system.** In the macaque, 80% of the LGN neurons are part of the P system. We discuss functional differences between the M and P systems after we describe their cortical projections.

Multiple Pathways in the Visual Cortex

The first cortical synapses for neurons carrying visual information are in the medial portion of the occipital lobe, area 17 in Brodmann's map. This receiving area is located medially and buried below the superficial surface of the cortex along the calcarine sulcus. The cyto-

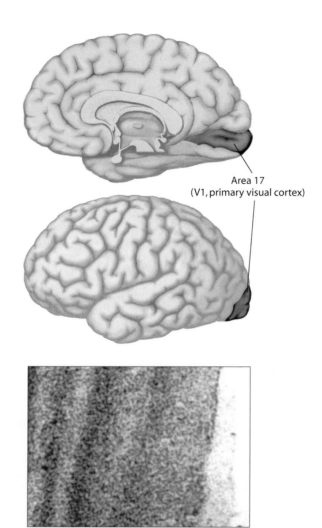

Figure 5.7 Brodmann's area 17, or the primary visual cortex. It extends to the most posterior pole of the occipital lobe and, in humans, extends along the medial surface. The bottom panel shows a magnification of the gray matter of the primary visual cortex in the macaque, highlighting the striated, or striped, layers. This area is referred to as *V1* by physiologists. Adapted from Bear et al. (1996).

architecture is quite regular and stippled, thus giving rise to the name *striate cortex.* Area 17 is also referred to as the **primary visual cortex** and, especially among physiologists, *V1* (Figure 5.7). This latter nomenclature refers to the hypothesis that this region is the first visual processing area in the cortex.

The segregation of M and P pathways is maintained in the cortex. Axons from both regions terminate in layer 4 of the striate cortex. But the terminal zones of the axons within this layer are offset from one another. And within the striate cortex, the P pathway involves a second synapse. An interlaminar projection carries information in the P pathway from layer 4 to the more superficial layers 2 and 3 (Figure 5.8).

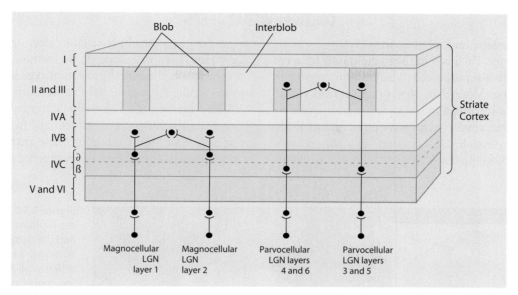

Figure 5.8 The terminal projections in area 17 of the geniculo-cortical pathway. While all of the inputs terminate in layer 4, the parvocellular inputs synapse on intracortical neurons that terminate in layers 2 and 3. Adapted from Bear et al. (1996).

Even more striking anatomical evidence for multiple pathways within the visual stream comes from a staining technique called the *cytochrome oxidase method,* which stains for the enzyme of the same name. Cytochrome oxidase is concentrated in areas of high metabolic activity. When applied to the surface of the primary visual cortex, this stain reveals a beautiful mosaic in which regions of low saturation surround regions of high saturation in the top layers (Figure 5.9). In essence, the stain indicates that metabolic activity does not vary randomly across the surface but in a systematic way. Darker regions associated with high metabolic rates are referred to by the highly technical name *blobs;* lighter regions are called *interblobs.* Thus, data derived from the cytochrome oxidase method suggest that the P pathway has at least two branches.

Visual information is segregated into distinct pathways in the adjacent cortical area. This area is referred to as the *prestriate cortex* to indicate its location in front of the striate cortex; physiologists know it as *V2,* indicating that this is the second physiological visual area. When cytochrome oxidase is applied to the cortical surface of this visual region, the stain reveals three subregions—thick stripes, thin stripes, and interstripes—which tracing techniques show as the continuation of the M, P-blob, and P-interblob pathways, respectively.

Whether these pathways are completely independent of one another is arguable. Many researchers maintain that there is cross talk between the M and the two P pathways. More important, a lot of convergence and divergence takes place after the first few synapses in the cortex.

Figure 5.9 Cytochrome oxidase stain revealing the blob/interblob regions in the primary visual cortex (V1) of the macaque. The blob areas are metabolically more active and absorb more of the stain. **(a)** A section taken parallel to the cortical surface through layers 2 and 3 shows the islands of blobs. **(b)** A radial section through the cortex. The white circles are blood vessels.

(a)   (b)

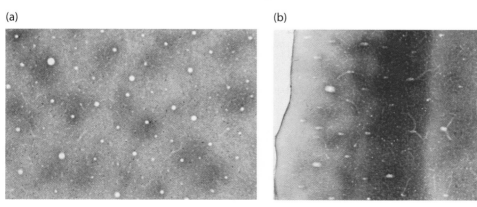

CORTICAL VISUAL AREAS

Figure 5.10 shows a map of the visual areas of the cortex. Each box in the figure stands for a region of cortex that is purported to be a distinct region of visual processing. More than 30 distinct **cortical visual areas** have been identified in the monkey, an increase of about 200% since similar maps were published in 1983. Some say that physiologists "discover" new visual areas faster than rabbits reproduce. Note that this figure follows the nomenclature developed by physiologists for functional maps. As we mentioned, striate cortex, or V1, is the initial projection region of geniculate axons. While other areas have names such as V2, V3, and V4, this numbering scheme should not be taken to mean that the synapses proceed sequentially from one area to the next. The lines connecting these **extrastriate** visual areas demonstrate extensive convergence and divergence

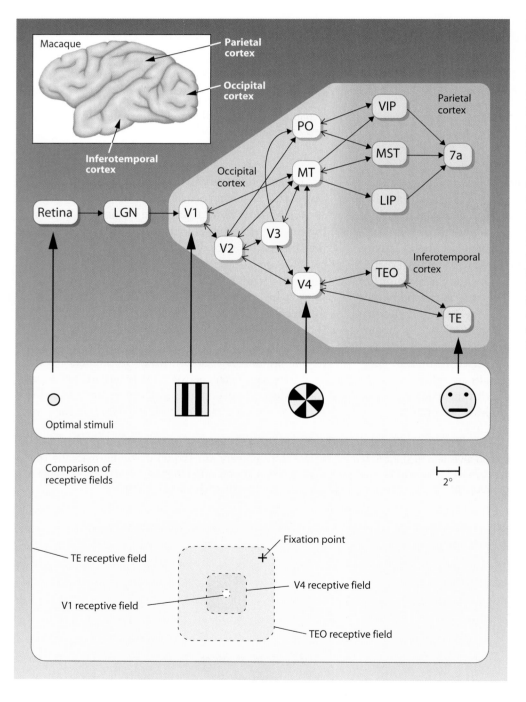

Figure 5.10 Summary of the prominent visual areas and the pattern of connectivity in the macaque. Whereas all cortical processing begins in V1, there are two processing streams that extend either dorsally to the parietal lobe or ventrally to the temporal lobe (see Chapter 6). The stimulus required to produce optimal activation of a cell becomes more complex along the ventral stream. In addition, the size of the receptive fields of these cells increases, ranging from the 0.5-degree span of a V1 cell to the 40-degree span of a cell in area TE. Adapted from art provided courtesy of Steven Luck.

across visual areas. Also, connections between many areas are reciprocal; areas frequently receive input from an area to which they project.

How a visual area is defined depends on the criteria used. An obvious criterion is that cells within the area respond to visual stimuli; however, if this were the sole criterion, it would be difficult to tell where one visual area begins and another ends. Sometimes neuroanatomy might help. For example, the border between V1 and V2 corresponds to the boundary between Brodmann's areas 17 and 18. But boundaries often cannot be identified with anatomical methods. For the physiologist, area 19 has many distinct visual areas. Physiologists depend on criteria different from the ones used by anatomists.

A primary physiological method for establishing visual areas is to measure how spatial information is represented across a region of cortex. Each visual area has a topographic representation of external space in the contralateral hemifield, and the boundaries between anatomically adjacent visual areas are marked by topographic discontinuities (Figure 5.11). In the same way that each LGN has six representations of the retina, the cortex has at least as many retinotopic maps as visual areas. The replication of topography within the cortex does not result from independent inputs to each area. As one area projects to another, topography is preserved. Precise spatial information is preserved by these multiple retinotopic maps, at least in early visual areas, reflecting the fact that the system has to link features that emanate from a common location.

Cellular Correlates of Visual Features

Why would it be useful for the primate brain to have evolved so many visual areas? One possibility is that the areas form a hierarchy in which each area successively elaborates on the representation derived by processing in earlier areas, representing the stimulus in a specific way. The simple cells of the primary visual cortex calculate edges used by more complex cells to detect corners and edge terminations used by higher-order neurons to represent shapes. Successive elaboration culminates in formatting the representation of the stimulus so it matches (or not) information in memory. As shown in Figure 5.10 though, there is not a simple hierarchy; extensive patterns of convergence and divergence result in multiple pathways.

An alternative hypothesis relates to the idea of visual perception as an analytic process. Although each visual area provides a map of external space, the maps differ with regard to the type of information they represent. For instance, neurons in some areas are highly sensitive

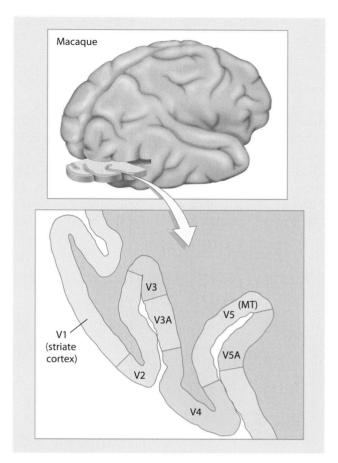

Figure 5.11 Physiological methods are used to identify different visual areas. An area is defined by a discontinuity in the retinotopic representation of the cells. Along this continuous ribbon of cortex, seven different visual areas can be identified. However, processing is not restricted to proceeding from one area to the next in a sequential order. For example, axons from V2 project to V3, V4, and V5/MT. Adapted from Zeki (1993).

to color variation. In other areas, the neurons may be movement sensitive but color insensitive. By this hypothesis, neurons within an area not only code where an object is located in visual space but also provide information about the object's attributes. Visual perception is a divide-and-conquer strategy. Rather than each visual area representing all attributes of an object, each provides its own limited analysis. Processing is distributed and specialized. As we advance through the visual system, different areas elaborate on the initial information in V1 and begin to integrate this information across dimensions to form recognizable percepts.

Extensive physiological evidence supports the specialization hypothesis. For instance, single-cell recordings in the M pathway reveal that these neurons do not show specificity in terms of the color of the stimulus. These cells, extending from the magnocellular layer of the LGN, through layer 4b of V1 and the thick stripes of V2, will respond similarly to either a green or red circle on a white background. Even more striking, these neurons respond weakly when presented with an alternating pattern of red and green stripes whose colors are of equal brightness.

In contrast, these neurons are quite sensitive to movement and direction, as shown in Figure 5.12 (Maunsell and Van Essen, 1983). The neuron shown in Figure 5.12 was located in **area MT,** a visual area in the middle temporal region of the macaque monkey. This area is a prominent recipient of information in the M pathway. The stimulus, a rectangular bar, was passed through the receptive field in varying directions. The cell's response was greatest when the stimulus was moved downward and left. In contrast, this cell was essentially silent when the stimulus was moved upward or to the right. Thus, the cell's activity correlates with two attributes of the stimulus. First, it is active only when the stimulus falls within its receptive field. Second, the response is greatest when the stimulus moves in a certain direction. This specificity is even more remarkable. Activity in MT cells also correlates with the speed of motion. The cell in Figure 5.12 responded maximally when the bar was moved rapidly. At slower speeds, the bar's movement in the same direction failed to raise the response rate above baseline.

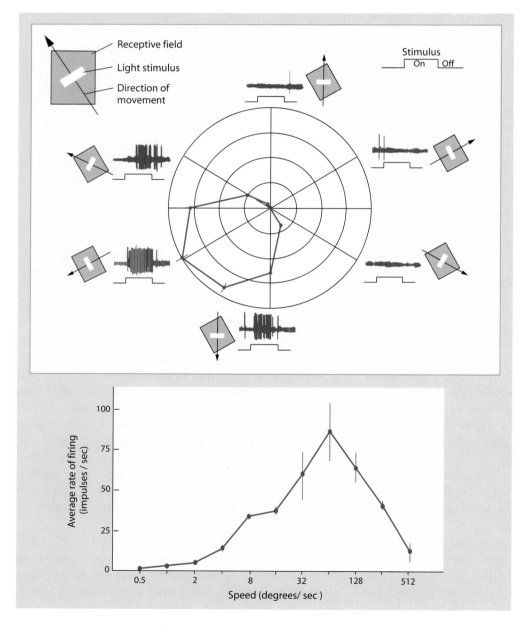

Figure 5.12 Directional and speed tuning of a neuron from area MT. **Top:** A rectangle was moved through the receptive field of this cell in various directions. The red traces beside the stimulus cartoons indicate the response of the cell to these stimuli. In the polar graph, the firing rates are plotted, with the angular direction of each point indicating the stimulus direction and the distance from the center indicating the firing rate as a percentage of the maximum firing rate. The polygon formed by connecting the points indicates that the cell was maximally responsive to stimuli moved down and to the left; the cell responded minimally when the stimulus moved in the opposite direction. **Bottom:** The graph shows speed tuning for a cell in MT. In all conditions, the motion was in the optimal direction. This cell responded most vigorously when the stimulus moved at 64 degrees/sec. Adapted from Maunsell and Van Essen (1983).

Figure 5.13 summarizes stimulus variations used in neurophysiological studies to identify the representational characteristics of cells in the M, P-blob, and P-interblob pathways. As we noted, neurons in the M pathway are movement sensitive and color (or, more precisely, wavelength) insensitive. In contrast, neurons in the P-blob pathway are highly selective to color and are minimally responsive to movement or changes in orientation. A cell within the blobs of V1 that responds to a red stimulus will respond regardless of whether the stimulus is oriented vertically or horizontally. Orientation information, on the other hand, is well represented by cells in the P-interblob pathway and is poorly represented by neurons in the P-blob pathway and weakly represented by M pathway neurons.

Imaging Visual Areas in Humans

Single-cell recording studies have provided physiologists with a powerful tool to map out the visual areas in the monkey brain and characterize the functional

Figure 5.13 Summary of the responsiveness of cells in the three pathways to different stimulus properties. Cells in the magnocellular pathway are very responsive to motion and are very sensitive to contrast (brightness) differences. Cells in the blob, thin-stripe regions of the parvocellular pathway also have high contrast sensitivity. While they show little sensitivity to motion, they are sharply tuned for wavelength, or color. Cells in the interblob, interstripe regions of the parvocellular pathway are sensitive to location and orientation, and show some variation in firing rates as motion and color parameters of a stimulus are varied.

Parallel Pathways in Visual Perception			
Neural Structure	**Cell Types**		
Thalamus (LGN)	**Magnocellular**	**Parvocellular**	
Area 17 (VI)	Layer 4b	Blobs	Interblobs
Area 18 (V2)	Thick stripes	Thin stripes	Interstripes
Cellular Correlates			
Contrast (brightness)	high	high	low
Location	low	low	high
Motion	high	low	middle
Color	low	high	middle
Orientation	middle	low	high

properties of the neurons within these areas. This work has provided strong evidence that different visual areas are specialized to represent distinct attributes of the visual scene. Inspired by these results, researchers have employed neuroimaging techniques to ask whether a similar architecture can be discerned in the human brain.

Semir Zeki (1993) of University College in London and his colleagues at Hammersmith Hospital used position emission tomography (PET) to verify that different visual areas are activated when subjects are processing color or motion information. Subtractive logic was used by factoring out the activation in a control condition from the activation in an experimental one. Consider first the color experiment. For the control condition, subjects passively viewed a collage of achromatic rectangles. Various shades of gray, spanning a wide range of luminances, were chosen. The control stimulus would be expected to activate neural regions with cells that are contrast sensitive (e.g., sensitive to differences in luminance).

For the experimental condition, the gray patches were replaced by a variety of colors (Figure 5.14). Each color patch was matched in luminance to its corresponding gray patch. Thus, neurons sensitive to luminance information should be equally activated in control and experimental conditions. However, the colored stimulus should produce more activity in neural regions sensitive to chromatic information. These regions should be detected if the metabolic activity recorded when subjects viewed the gray stimulus is subtracted from the one recorded when subjects viewed the colored stimulus.

The same logic was used to design the motion experiment. For this study, the control stimulus consisted of a complex black-and-white collage of squares. The same stimulus was used in the experimental condition, except that the squares were set in motion. They would begin to move in one direction for 5 seconds and then the reverse direction for the next 5 seconds.

The results of the two studies provided clear evidence that the two tasks activated distinct brain regions (Figure 5.15). After subtracting activation during the viewing of the achromatic collage, investigators found numerous residual foci of activation when subjects viewed the colored collage. These foci were bilateral and located in the most anterior and inferior regions of the occipital lobe. Although the spatial resolution of PET is coarse, these areas were determined to be in front of the striate (V1) and prestriate cortex (V2). In contrast, after the appropriate subtraction in the motion experiment, the residual

(a) Mondrian display

(b) Pattern of moving squares

Figure 5.14 Schematic of the stimuli used in a PET experiment to identify regions involved in color and motion perception. **(a)** For the color experiment, the stimuli were composed of an arrangement of rectangles that were either shades of gray (control) or various colors (experimental). **(b)** For the motion experiment, a random pattern of black and white regions was either stationary (control) or moving (experimental). Adapted from Zeki (1993).

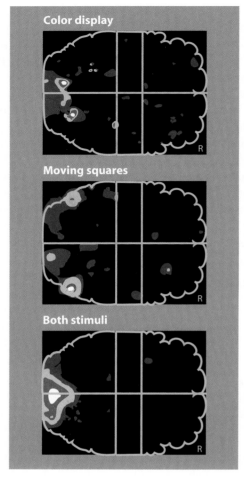

Figure 5.15 Regions of activation when the control conditions were subtracted from the experimental conditions. In the color condition, the prominent activation is medial, in areas corresponding to human V4. In the motion condition, the activation is more lateral, in areas corresponding to human MT. The foci also differ along the dorsal-ventral axis: The slice showing MT is superior to that showing V4. Both stimuli also produced significant activation in primary visual cortex when compared to a control condition in which there was no visual stimulation. Adapted from Zeki (1993).

foci were bilateral but near the junction of the temporal, parietal, and occipital cortices. These foci were more superior and much more lateral than the color foci. Zeki and his colleagues were so taken with this dissociation that they proposed that the nomenclature developed by primate researchers should be applied here. They labeled the area activated in the color foci as *human V4* and the area activated in the motion task as *V5* (most researchers now refer to the latter as *human MT* even though the area is not in the temporal lobe in the human brain). Of course, with PET

data we cannot be sure that the foci of activation really consist of just one visual area.

It is important to note that there are also striking between-species differences in terms of the relative position of the color and motion areas (compare Figures 5.11 and 5.15). Such differences likely result from the fact that the surface area of the human brain is substantially larger, and this expansion required additional folding of the continuous cortical sheet. Thus, we can ask questions about functional homology (see Chapter 14), the correspondence in structure and function between species, but we need to be aware that the mapping is unlikely to be straightforward. Indeed, it is quite possible that humans have visual areas that do not correspond to any region in our close primate relatives.

The activation maps in this PET study are rather crude. More recently, sophisticated functional magnetic resonance imaging (fMRI) techniques have been applied to study the organization of human visual cortex. In these studies, a stimulus is systematically moved across the visual field (Figure 5.16). For example, a semicircular checkerboard pattern is slowly rotated about the center of view. In this way, the blood oxygen level dependent (BOLD) response for areas representing the superior quadrant will increase at a different time than the response for areas representing the inferior quadrant, and in fact, the representation of the entire visual field can be continuously tracked. To compare areas that respond to foveal stimulation and those that respond to peripheral stimulation, a dilating and contracting ring stimulus is used. By combining these different stimuli, researchers can measure the cortical representation of the contralateral visual field.

Because of the convoluted nature of the human visual cortex, the results from such an experiment would be indecipherable if we were to plot the data on the anatomical maps found in a brain atlas. To avoid this problem, a flat representation is constructed. High-resolution anatomical MRI scans are obtained and computer algorithms are employed to transform the folded, cortical surface into a two-dimensional map by tracing the gray matter. The activation signals from the fMRI study are then plotted on

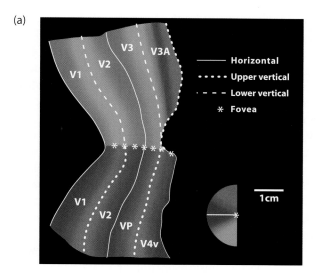

(a)

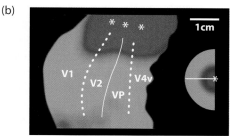

(b)

Figure 5.17 Retinotopic mapping in the human. **(a)** The correspondence between the position of the rotating wedge and activation across the visual cortex is coded by color on the flattened map. Boundaries between visual areas are defined by reversals in the retinotopic representation within a quadrant. Note that V3 in the ventral portion is referred to as *VP.* **(b)** The expanding/contracting stimulus is used to map eccentricity. Only the ventral visual cortex is shown. The foveal regions are marked by the asterisks. Adapted from Hadjikhani et al. (1998).

Figure 5.16 Mapping visual fields with functional magnetic resonance imaging (fMRI). The subject views a rotating circular wedge while fMRI scans are obtained. The wedge passes from one visual quadrant to the next, and the blood oxygenation level dependent (BOLD) response in visual cortex is measured continuously to map out how the regions of activation change in a corresponding manner.

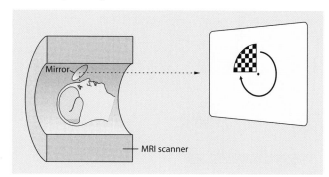

the flattened map, with color coding used to indicate areas that were activated at similar times.

Researchers at the Massachusetts General Hospital have used this procedure to reveal the organization of the human visual system in exquisite detail (Hadjikhani et al., 1998). As can be seen in Figure 5.17a, the physical world is inverted in the visual cortex. Areas above the calcarine sulcus, along which lies the primary visual cortex, are active when the rotating stimulus is in the lower quadrant; the reverse is true when the stimulus is in the upper quadrant. In addition, the activation patterns show a series of repetitions across the visual cortex. The green and blue repetitions in the superior portion of visual cortex indicate that the lower quadrant of the visual field is represented multiple times. Similarly, the red and purple alternations in the lower portion indicate the upper quadrant of the visual field is represented multiple times. These repetitions indicate separate topographic maps. Following the conventions adopted in the single-cell studies, the visual areas are numbered in increasing order, with primary visual cortex (V1) most posterior and secondary visual areas (V2, V3/VP, V4) more anterior. Figure 5.17b shows how eccentricity is represented in the ventral part of the visual cortex. The cortical representation of the fovea, the regions shown in red, is quite large. Visual acuity is much greater at the fovea owing to the disproportionate amount of cortex that encodes information from this part of space.

Imaging studies have opened a new window for studying visual illusion phenomena that have puzzled philosophers and psychologists alike for many years. Patterns of brain activation during illusory states can be compared with those observed during visual stimulation. In this way, areas of overlap can provide insight into the level of processing at which illusions arise, as well as indicate how information is represented in different visual areas.

Stare at the Enigma pattern shown in Figure 5.18a. After a few seconds, you should begin to see scintillating motion within the blue circles, an illusion created by their opposed orientation to the radial black and white lines. What are the neural correlates of this illusion? We know that moving patterns produce a strong hemodynamic response in human MT. Is this same area also activated during illusory motion? Both PET and fMRI have been used to show that indeed, viewing displays such as the Enigma pattern leads to pronounced activity in human MT. This activation is selective: V1 is silent during illusory motion. A different illusion is demonstrated in Figure 5.18b. After staring at the bright green circle for about 30 seconds, shift your gaze to the neighboring gray circle. You should perceive the gray circle as tinged with purple, the complementary color to green. Scanning of subjects while they perceive this illusion reveals a high level of activation in an inferior visual area anterior to V4 for an extended period of time after a saturated color turns to gray. In a control condition, the color patch alternates between two saturated colors before turning gray, thus negating the illusion. Here, activity quickly returns to baseline when the gray patch is presented. As with the motion illusion, such aftereffects are not seen in V1.

These results suggest that perception involves higher visual areas rather than the primary visual cortex. Indeed, even knowledge concerning the perceptual

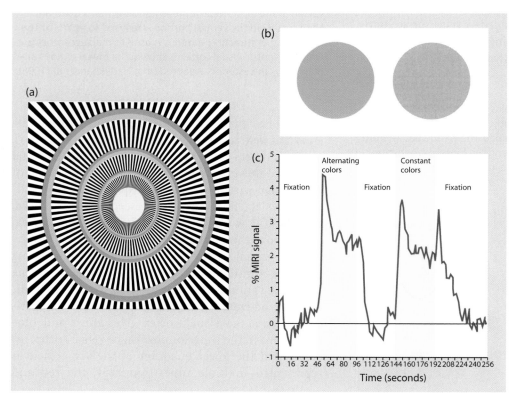

Figure 5.18 The neural basis of visual illusions. **(a)** Illusory motion is perceived when one is viewing the Enigma pattern, and activation is observed in area MT. **(b)** Color aftereffects are produced after staring at the bright green patch. **(c)** Activation in a visual region anterior to V4 is shown graphically (see text for details). (a) From Zeki (1993). (c) From Hadjikhani et al. (1998).

properties of objects can be sufficient to produce measurable activity in higher visual areas. Viewing a static picture of a dynamic scene such as the rolling surf or an athlete in motion boosted the BOLD signal in human MT (Kourtzi and Kanwisher, 2000). Likewise, accessing knowledge about the colors of familiar objects activated the area anterior to V4 in PET studies (Chao and Martin, 1999). For example, recalling that a wagon is red or a banana is yellow activated this higher-level visual area and not V1 or V4.

Analysis and Representation of Visual Features

Anatomical and physiological results suggest that the visual system is characterized by multiple pathways, each with their individual specializations. Thus, information is processed in such a way that analysis across the different pathways proceeds concurrently. While single-cell recording methods can elucidate what makes a cell fire and imaging studies can identify correlations between stimulus parameters and neural activation patterns, it is important to ask whether perception is constrained by an architecture based on concurrent processing pathways.

Figure 5.19 Sample displays in a visual search task. In each panel, find the target, the green T as quickly as possible. The target is readily apparent in both panels on the left (feature search) despite the variation in the number of distracters, because it is the only green object. In contrast, the target is harder to find in the conjunction search, especially when there are many distracters.

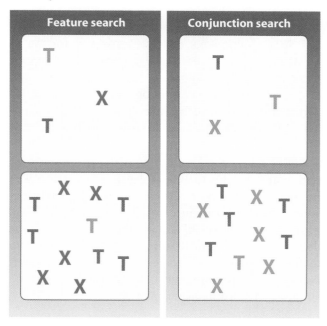

VISUAL SEARCH

Before the modern techniques of neuroanatomy, neurophysiology, and neuroimaging were targeted to examine the organization of the visual cortex, experimental psychologists were tackling the question of whether perception is an analytic process. Anne Treisman (1988), working at the University of British Columbia in the 1970s, introduced the **visual search task** as a model of the perceptual problems we may encounter in the world. Sherlock Holmes, for instance, describes looking for his red-headed friend at a train station in London by rapidly scanning the crowd and examining only the faces of people with red hair. Treisman asked if we are equally capable of searching by any arbitrary criteria, or whether certain targets are easier than others.

In the visual search task, subjects are presented with displays containing multidimensional objects. Two types of conditions can be defined. For the feature condition, the target object differs from all of the distracter objects by a single dimension, for example, color. For the conjunction condition, the target cannot be distinguished by a single dimension; instead, it is defined by the conjunction of information from two or more dimensions, for example, color and shape. Thus, while the target in all four panels of Figure 5.19 is always a green T, its color is unique in the feature condition (the left panels), whereas it shares its color with half of the distracters and its shape with the other half of the distracters in the conjunction condition (the right panels). As you can readily see, the target is much easier to detect in the feature condition compared to the conjunction condition. Indeed, reaction time in the feature condition is not only fast but remains constant even when more distracters are added to the display. In contrast, search time in the conjunction condition increases in a linear manner as the set size increases, a result taken to imply that attention must be deployed in a sequential manner to link together information from distinct feature maps (see Chapter 7).

Targets that yield horizontal search functions are assumed to reflect visual primitives, the basic building blocks of perception. The dimensions that neurophysiologists use to describe activity in early visual areas generally correspond well with the stimulus conditions that produce horizontal search functions. This correspondence holds not only for features that differ in shape and color but also when the target moves in a different direction (Nakayama and Silverman, 1986) or at a significantly different speed from the distracters (Ivry and Cohen, 1992). Horizontal search functions have also been found when the

An Interview with Anne Treisman, Ph.D.

Dr. Treisman is associated with the Department of Psychology at Princeton University. Based on behavioral methods, her work anticipated the neuroanatomical and neurophysiological work on the analytic aspects of visual perception.

Authors: How did you get on to the importance of studying features in your research on attention?

AT: I'm not sure I can accurately reconstruct the past, but certainly the distinction between parallel processing of simple physical features and limited attention to more complex items goes back to the development of the filter model of attention, and played a central role in the selective listening research on the "cocktail party problem." [The *cocktail party problem* refers to the problems that arise when we selectively attend to one input among a host of competing signals. At a cocktail party, we may wish to filter out the buzz from neighboring conversations to hone in on an intimate discussion. Alternatively, if we are conversing with a boring acquaintance, we may find ourselves picking up bits and pieces of chatter being shared across the room.] That was the subject of my Ph.D. dissertation. I was also influenced by research on discrimination learning in animals, particularly the thought that the more they learn about one property, the less they learn about others. My interest in applying the distinction to vision (rather than hearing) goes back to a paper I wrote in 1966, when I suggested that different properties are analyzed in parallel in separate modules, and that perhaps we could attend selectively either to one property, by switching out the analyzers for other properties, or to one object (or source of stimuli), by selecting which sensory inputs to gate before they reach certain analyzers. Looking at my flow diagrams with separate boxes for color, shape, motion, and so on started me wondering how

the properties get reassembled, and whether we might sometimes make the wrong bundles and experience illusory conjunctions.

I remember sitting on the lawn with my children in the early 1970s, drawing displays of mixed red *X*'s and green *O*'s on scraps of paper, and asking them to find either a blue *O* or a red *O*. I noticed that they took much longer to find the conjunction targets. This led to the idea that focused attention to each element would be one way of solving what is now known as the *binding problem*. Either space or time can be used to individuate objects; these two fundamental attributes also provide media within which we can attend selectively to a subset of the features present in the field. A crucial test (which I put off for a while because I didn't really believe it would work) was to see whether illusory conjunctions (binding errors) actually would be experienced when attention was overloaded or directed elsewhere. One of the high points of my research was when my research assistant told me that, contrary to my firm impression, I was making many of those errors in reporting colored letters that I believed I was seeing quite clearly.

Authors: How does one decide what constitutes an elementary feature of visual perception?

AT: Defining what counts as a feature for the visual system is an empirical question. I suggested some criteria for finding out: parallel processing and pop out with divided attention, minimal interference when responding along one dimension from varia-

target differs in depth from the distracters or is uniquely illuminated, creating a distinct shadowing pattern for the target in comparison to the distracters (Figure 5.20). The effect of illumination is especially striking, as it depends on the orientation of the stimulus display. If the display in Figure 5.20b is rotated 90 degrees, the target becomes easier to find. Visual detectors can be sensitive to the direction of illumi-

nation and use this to extract depth information. The human visual system evolved in an environment where the primary source of illumination, the sun, projects from overhead and so illumination-sensitive detectors are limited. Results such as these can help the neurophysiologist decide on appropriate stimulus parameters to manipulate in their time-consuming single-cell recording experiments.

tions along other dimensions, the occurrence of illusory recombinations of features or binding errors, selective adaptation. The hope was that these would converge on a limited number of candidates. I also found an interesting asymmetry on visual search tasks between many pairs of contrasting features, suggesting that one of the two counts (for the visual system) as the presence of something and one as its absence, just as the presence of a slash on the *Q* distinguishes it from an *O*, which lacks the slash. In all these pairs (*Q* versus *O*, tilted versus vertical line, curved versus straight line, gap versus closed curve, ellipse versus circle, converging versus parallel lines) the search for the first of the pair in a background of the second was much easier than the reverse. What is shared by all but the *Q/O* pair is that the second is a standard or reference value and the first can be thought of as a deviation from it. By analogy with the *Q/O* pair, it is as if the brain assumes the standard value and codes deviations from it as the presence of something to be signaled. This signal makes the target that has it pop out from a background of elements in which it is absent.

Authors: These are behavioral criteria, yet in your interpretation you make inferences about brain function. What evidence supports the idea that we can expect a convergence between what we define as a feature in tasks such as visual search and how the brain extracts information about the visual world?

AT: It would be tempting to look for single units coding the features that pop out in behavioral studies. While this might be illuminating, I suspect that the activity of a single unit will not always correspond directly to what the visual system as a whole is doing. It's more likely that some form of coarse coding is used—activities in different cell populations with overlapping sensitivities carry the relevant information. I recently generated illusory conjunctions of slightly tilted purple

bars by presenting brief exposures of vertical blue bars and more tilted red bars. Thus, the observer's perception can reflect a blending of the stimulus information rather than simply a miscombination of the available features. This blending is likely a result of the crude coding by individual neurons.

Some neural evidence supports the idea that the search for feature-defined targets differs from the search for targets defined only by conjunctions of features. By looking at the early P1 component of event-related potentials, researchers have found attentional suppression at distracter locations in a visual search task only when participants needed to conjoin features and not when they were asked simply to detect the presence of particular features. PET studies have shown similar patterns of activation in the parietal lobes both when participants shift attention in space and when they search for a conjunction target, suggesting that the same spatial selection mechanisms are used in both. This is consistent with my account of the need for a serial scan with focused attention when features must be bound. Neurophysiologists have begun to look also at attention effects on single-unit responses in V1, V2, and V4 in relation to the risk of binding errors: Attention effects were seen only when the monkey had to respond to one of two objects presented within the same receptive field, where presumably their features would otherwise risk being exchanged or pooled.

Exciting research is taking place to explore the possibility that synchronized firing is used as a temporal code to link the neural representations of the features from one object—an intriguing solution for how the brain might solve the binding problem. In this context, attention limits might reflect limits on the number of different synchronies that can be maintained at one time, and illusory conjunctions might reflect accidental synchronies that result when too many objects are coded at once.

The visual search task also enables us to ask questions about what makes a feature (Beck, 1982). Consider shape. An S is easily seen among a display of X's and T's. Does this mean that each letter should be viewed as a primitive of visual perception? It is unlikely that specialized detectors would be tuned to letters since alphanumeric characters in the English language are arbitrary. Indeed, as can be seen in Figure

5.20c, the letter L does not pop out when embedded among a set of T's. It is much easier to find the upright L among a display of tilted L's than among upright T's. So the critical feature is not the letter as a whole but its parts. When the target L is among tilted L's, the orientation of both lines forming the target is unique compared to the orientation of the lines forming the distracters. When the distracters are upright T's, the

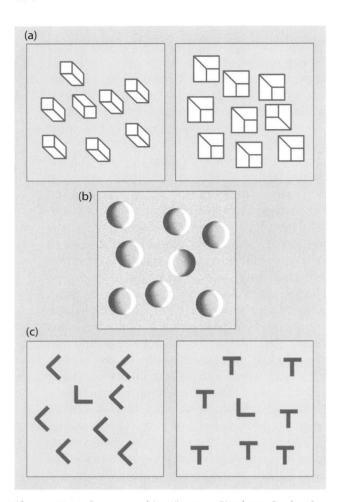

Figure 5.20 Does one object "pop out" in these displays? The ability to rapidly identify a unique stimulus property has been used as one defining criterion for identifying the basic building blocks of perception. **(a)** The deviant object is created by varying the end at which the arrowhead is placed. This difference is much easier to detect when this attribute changes the orientation of a three-dimensional box rather than a flat, abstract shape. **(b)** Variations in lightness can also provide a cue to depth. Turn your book sideways and examine (b) again. The deviant object will now pop out, literally! **(c)** Despite our familiarity with letters, it is easier to find the upright L among tilted L's than among the T's. The pop-out phenomenon here is likely due to the difference in orientation of the component parts of the letters rather than their overall shapes. (a) Adapted from Enns and Rensick (1990). (b) From Ramachandran (1988). (c) From Beck (1982).

target's components are shared by all distracters. Again, this finding meshes nicely with the results of neurophysiology. Within the early processing regions of the P pathway, cells are highly selective to orientation; more complex representations emerge at later stages (Croner and Albright, 1999).

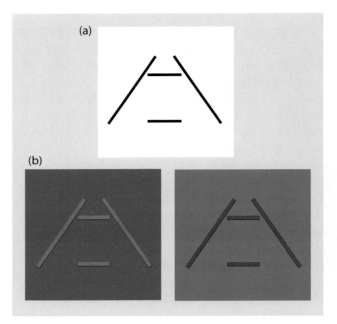

Figure 5.21 The Ponzo illusion. **(a)** Which horizontal line is longer, the one on the top or the one on the bottom? People almost always perceive the top line as longer, even when they are aware of the illusion. The diagonal lines give the impression of train tracks receding into the horizon, and thus the top line is judged as farther away. **(b)** The Ponzo illusion is also effective with color **(left)**, except when the stimulus is isoluminant with the background **(right)** and thus reduces the perception of depth.

ILLUSIONS AND FEATURE INDEPENDENCE

Examine the black and white drawing in Figure 5.21a. Compare the length of the two horizontal lines. Is the upper line longer than the lower line? Or are they equal? Your ruler can confirm that the two lines are equal, yet this does not match our percept, at least not for the stimulus in Figure 5.21a. For this black-on-white stimulus, we have a striking illusion that the upper line is longer. This *Ponzo illusion* arises because the diagonal lines appear to recede in depth. While the two horizontal lines produce retinal images of equal size, the diagonal lines make the upper line look farther away, which leads us to perceive it to be much longer than the lower line.

Now consider the red-on-green versions of the illusion in Figure 5.21b. Most people readily see the illusion in the left version. But, surprisingly, the illusion in the right version is reduced. To understand this discrepancy, we need to know how the borders between the background and the lines are created. For the black-on-white version, the brightness between the white background and the black lines is quite different. This is not so for the

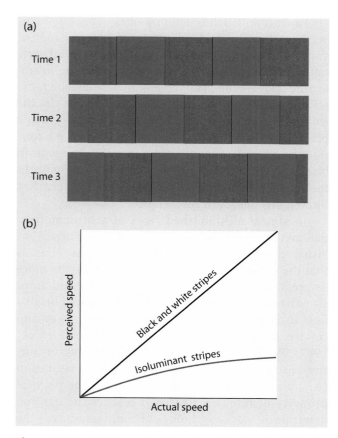

(a)

Time 1

Time 2

Time 3

(b)

Perceived speed

Black and white stripes

Isoluminant stripes

Actual speed

Figure 5.22 **(a)** Schematic of a series of displays in which the striped pattern moves across the page similar to a rotating barber pole. **(b)** The perceived speed of the stimulus is veridical when the striped regions are black and white. However, when the stripes are isoluminant, the perceived speed is much slower, suggesting that color differences are, at best, weakly signaled in the motion system.

red-on-green version shown on the right. Here, the intensities of the red lines and green background are identical or isoluminant. *Luminance* refers to the perceived brightness of a stimulus. For isoluminant stimuli, the brightness of each region is set so that each part is of equal brightness. Thus, for the stimulus on the left, brightness and color distinguish the lines from the background; for the stimulus on the right, color is the only cue.

The loss of the Ponzo illusion at isoluminance provides further evidence for a segregation of function within the visual pathways. As shown in Figure 5.13, color-sensitive cells in the P-blob pathway are depth insensitive. Since only these cells are activated by the isoluminant stimulus, depth information is lost. Without the impression of depth, the two lines appear to be equidistant from the viewer and hence equal in length.

Isoluminant stimuli have been used in many other illusions to demonstrate the functional implications of the concurrent pathway architecture (Livingstone and Hubel, 1988). Especially dramatic is the loss of motion information with isoluminant stimuli. Figure 5.22 presents a computer-generated motion display in which a striped pattern is continuously shifted over time, creating the impression of rightward motion. If the striped regions are isoluminant, though, the perceived speed of the moving display slows down. Indeed, one almost has the impression that the display is stopped. This percept is paradoxical. We can clearly tell that a region has moved from one location to another. With isoluminant stimuli, the perception of motion becomes inferential, as if we conclude that because something is not where it was, it must have moved. Yet the percept of the movement is attenuated.

DEFICITS IN FEATURE PERCEPTION

Within the realm of neurology, the idea of concurrent processing has a long and tumultuous history. In 1888 a Swiss ophthalmologist, Louis Verrey (cited in Zeki, 1993), described a patient who had lost the ability to perceive colors in her right visual field. Verrey reported that the patient had problems with acuity within restricted portions of this right visual field. But the color deficit was uniform and complete. We can guess that this patient's world looked similar to the drawing in Figure 5.23: On one side of space, the world was multicolored; on the other, it was a montage of grays.

Verrey speculated that if a person could selectively lose color perception, then multiple specializations of function within the visual cortex should be presumed.

The idea of a specialized color center, let alone multiple visual centers, met with great resistance over the next century. The anatomy of the cortex was just beginning to be known in detail, and so the focus was not on specializations within a domain. Rather, scientists were debating about which cortical regions were critical for visual perception. They sought to find the "cortical retina." War victims provided compelling evidence that losing a retina led to total loss of vision in that eye. In a corresponding manner, lesions to the cortical retina were assumed to disrupt all aspects of vision. More recent studies of patients with neurological disorders provided convincing evidence that brain lesions can produce selective impairments.

Figure 5.23 In achromatopsia, the world is seen as devoid of color. Because color differences are usually correlated with brightness differences, the objects in a scene might be distinguishable and appear as different shades of gray. This figure shows how the world might appear to a person with hemiachromatopsia. In most cases, there is some residual color perception, although the person cannot distinguish between subtle color variations.

Deficits in Color Perception: Achromatopsia

When we speak of someone who is color blind, we are usually describing a person who has inherited a gene that produces an abnormality in the photoreceptor system. *Dichromats,* or people with only two photopigments, can be classified as red-green color blind if they are missing the photopigment sensitive to either medium or long wavelengths, or blue-yellow color blind if they are missing the short-wavelength photopigment. *Anomalous trichromats,* in contrast, have all three photopigments but one has abnormal sensitivity. The incidence of these genetic disorders is high in males—about 8% of the population. The incidence rate is much lower in females: less than 1%.

Much rarer are disorders of color perception that arise from disturbances of the central nervous system. These disorders are called **achromatopsia,** taken from the prefix *a* or "without" and the stem *chroma* or "hue." As evident in Verrey's description, these patients see the world without color. J.C. Meadows (1974) of the National Hospital in London described one such patient as follows: "Everything looked black or grey. He had difficulty distinguishing British postage stamps of different value which look alike, but are of different colors. He was a keen gardener, but found that he pruned live rather than dead vines. He had difficulty distinguishing certain foods on his plate where color was the distinguishing mark."

Patients with achromatopsia often report that colors have become a bland palate of "dirty shades of gray." The shading reflects variations in brightness rather than hue. Other aspects of vision like depth and texture perception also remain intact, enabling the achromatopsic to see and recognize objects in the world. Indeed, color is not a necessary cue for shape perception. Its subtlety is underscored when we consider how people often do not notice the change from black and white to color when Dorothy lands in Oz in the movie *The Wizard of Oz.* Nonetheless, when lost forever, this subtlety is sorely missed.

In almost all published cases of achromatopsia, patients exhibit abnormalities in other areas of visual perception as well. What remains clear is that the deficit in color perception is markedly more severe. Consider the performance of an achromatopsia patient on tests of hue discrimination and brightness discrimination conducted by Alan Cowey and colleagues at Oxford University (Heywood et al., 1987). While we think of colors as differing in hue, they also vary in other properties such as saturation (e.g., pink vs. deep red) and reflectance, the physical property of reflected light that reaches the eye and determines our psychological impression of the brightness of a visual scene. Although there is no objective way to equate a difference in hue with a difference in reflectance, we can determine the psychological equivalence of differences across the dimensions (Figure 5.24).

In the study, on each trial the subject viewed three color patches. Two patches were identical; the third varied in hue or reflectance. The subject was asked to identify the stimulus that was different. As evident in Figure 5.24, the task was much more difficult when stimuli varied in hue. Reflectance differences, for example, were almost always correctly identified with a four-step difference. For hue differences, the patient still scored only 70% correct even with a ten-step difference, which corresponds to a transition from a yellowish green to a deep orange! Though this hue perception deficit is striking, the patient's performance on reflectance differences was also impaired. Healthy subjects rarely made errors with a one-step difference on any of the dimensions. Moreover, this patient had lost visual acuity after his stroke. He could read, but only if the print was enlarged.

We should not be surprised that people with achromatopsia exhibit visual deficits that extend beyond color perception. Neurological disorders such as strokes and tumors do not respect the borders of

Figure 5.24 **(a)** Psychological scaling techniques can be used with healthy individuals to create stimulus sets in which the similarity of all neighboring pairs is judged as equal. These techniques are used to create norms for the similarity across the different dimensions of a color (hue, saturation, and reflectance, the physical measure that underlies our perception of brightness). **(b)** Pairs of color chips were presented and the achromatopsia patient was asked to judge whether they were the same or different. This patient's ability to make such judgments was severely impaired when the pairs differed in hue, even when the stimuli differed by 10 units. His ability to discriminate brightness, although not normal, was much better. Here, he almost always labeled the stimuli as different when they were separated by at least 4 units. (b) From Heywood et al. (1987).

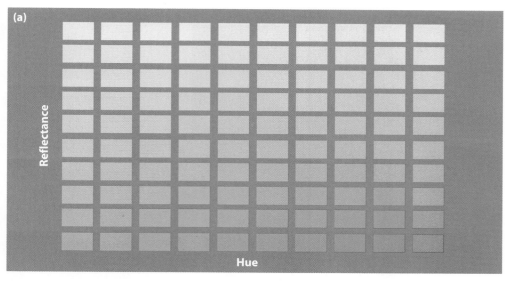

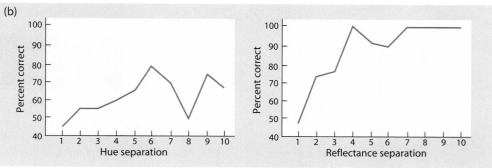

visual areas. Achromatopsia has consistently been associated with lesions that encompass V4 and the area anterior to V4, but the lesions typically extend to neighboring regions of the visual cortex. Color-sensitive neurons are also orientation sensitive, so we would expect that losing these neurons would affect form perception.

This hypothesis was carefully explored in a case study of a patient who suffered a stroke resulting in a small lesion near the temporal-occipital border in the right hemisphere. The damage was centered in area V4 and anterior parts of the visual cortex (Figure 5.25). To assess the patient's achromatopsia, a hue-matching experiment was performed in which a sample color was presented at the fovea, followed by a test color in one of the four quadrants of space. The patient's task was to judge if the two colors were the same or not. The difference between the sample and test color was adjusted until the patient was performing correctly on 80% of the trials, and this difference was measured separately for each quadrant. Regard-

less of the sample hue, the patient was severely impaired on the hue-matching task when the test color was presented in the upper left visual field. The fact that the deficit was only found in the upper contralesional visual field is consistent with previous reports of achromatopsia.

The main focus of the study, however, was to examine form perception. Would the patient show similar deficits in form perception in this quadrant? If so, what types of form perception tasks would reveal the impairment? To answer these questions, a variety of tasks were administered. The stimuli are shown in Figure 5.26. On the basic visual discriminations of contrast, orientation, and motion, the patient's performance was similar for all four quadrants and comparable to the performance of control participants. However, he showed impairment on tests of higher-order shape perception, and again, this impairment was restricted to the upper left quadrant. For these tasks, shape information requires combining information from neurons that might detect simple properties such as line orientation. For example, the orientation of the line

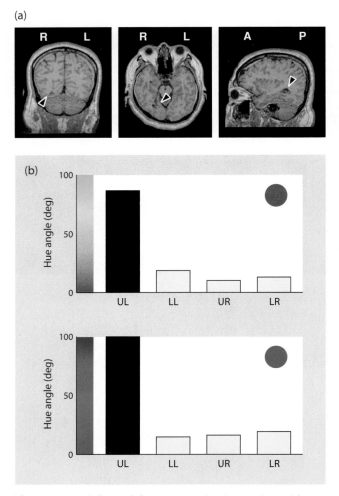

Figure 5.25 Color and shape perception in a patient with a unilateral lesion of V4. **(a)** MRI scans showing a small lesion encompassing V4 in the right hemisphere. **(b)** Color perception thresholds in each visual quadrant. The y axis indicates the color required to detect a difference between a patch shown in each visual quadrant (upper right [UR], upper left [UL], lower left [LL], lower right [LR]) and the target color shown at the fovea. The target color was red for the results shown in the top panel and green for the results shown in the bottom panel. (a) From Gallant et al. (2000). (b) Adapted from Gallant et al. (2000).

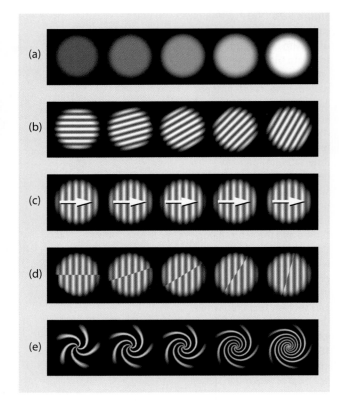

Figure 5.26 Stimuli used to assess form perception in the patient with damage to area V4. On basic tests of luminance **(a)**, orientation **(b)**, and motion **(c)**, the patient's perceptual threshold was similar in all four quadrants. Thresholds for illusory contours **(d)** and complex shapes **(e)** were elevated in the upper left quadrant. Adapted from Gallant et al. (2000).

separating the two semicircles is only defined by the combination of the lengths of the individual stripes and their offset. Thus, characterizing area V4 as a "color" area is too simplistic. This area is part of secondary visual areas devoted to shape perception. Color can provide an important cue about an object's shape.

A selective deficit of color perception is even less likely to occur if the lesion is in regions upstream to V4 such as the blobs of V1 or the thin stripes of V2. A vascular lesion or trauma that affects these areas would not be confined to cells in the P-blob pathway. Yet the puzzling case reported by Zeki (1993) of a patient with transient achromatopsia may reflect a temporary loss of function within these cells. Over several weeks, this patient experienced brief episodes when the world was suddenly drained of color. This loss of color perception occurred despite the fact that his vision remained intact and he could still identify objects. After a minute or so, color perception slowly returned. The disorder's transient nature suggests that blood flow to the brain temporarily diminishes. This decrease would be most disruptive to the most metabolically active cells. Cytochrome oxidase–sensitive cells within blobs and thin stripes are candidates for being disproportionately affected by reduced blood flow.

Animal lesions might provide the strongest clues that color perception depends on the P-blob pathway. Lesions of the parvocellular layers of the LGN produce a severe deficit in color discrimination (Schiller and Logothetis, 1990). As expected, these lesions also disturb form perception. However, when cortical lesions are made in primate V4, the results are more problematic. The lesions do not severely hamper hue discrimination; the monkeys can still discriminate colored regions from an isoluminant background. What is disturbed is their ability to maintain color constancy, to recognize that a region of space might have the same surface color but vary in how that surface reflects color to the eye because of variations in the light source or shadowing. Perhaps color discrimination can be sustained by hue-sensitive neurons in upstream visual areas. Again, we see that V4 may be invaluable in using color information to define regions of visual space and partition the objects within that space.

Deficits in Motion Perception: Akinetopsia

Researchers at the Max Planck Institute in Munich reported in 1983 a striking case of a woman who had incurred a selective loss of motion perception, or **akinetopsia** (Zihl et al., 1983). For this woman, whom we call M.P., perception was akin to viewing the world as snapshots. Rather than seeing things move continuously in space, she would see moving objects appear in one position and then another. When pouring a cup of tea, M.P. would see the liquid frozen in air, like a glacier. She would fail to notice the tea rising in her cup and would be surprised when the cup overflowed (Figure 5.27). The loss of motion perception also made M.P. hesitant about crossing the street. As she noted, "When I'm looking at the car first, it seems far away. But then when I want to cross the road, suddenly the car is very near."

Examination revealed M.P.'s color and form perception to be intact. Her ability to perceive briefly presented objects and letters, for example, was within the normal range. But her ability to judge the direction and speed of moving objects was severely impaired. This deficit was most apparent with stimuli moving at high speeds. At speeds faster than 20 degrees/sec, M.P. never reported detecting the motion. She could see that a dot's position had changed and hence could infer motion. But her percept was of two static images; there was no continuity from one image to the other. Even when presented with stimuli moving at lower speeds, M.P. was hesitant to report a clear impression of motion.

CT scans of M.P. revealed large, bilateral lesions that involved the temporoparietal cortices. On each side, the lesions included posterior and lateral portions of the middle temporal gyrus. These lesions roughly corresponded to areas that participate in motion perception. Furthermore, the lesions were lateral and superior to human V4, including the area identified as the human equivalent of area MT. The importance of MT was confirmed in primate studies in which motion perception deficits were induced by focal lesions restricted to this area. Additionally, the

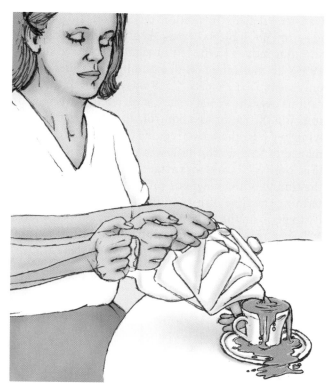

Figure 5.27 For the patient with motion blindness, the world appears as if viewed through a strobe light. Rather than see the liquid rise continuously in the teacup, the patient reports seeing the liquid jump from one level to the next.

contribution of the M system was demonstrated by the fact that even more striking deficits were found when the lesions were targeted to the magnocellular layer of the LGN. Such lesions caused no deficits in hue discrimination.

While the case of M.P. has been cited widely for many years, the fact that similar patients have not been identified suggests that severe forms of akinetopsia may result only from bilateral lesions. With unilateral lesions, the motion perception deficits are much more subtle (Plant et al., 1993). Perhaps people can perceive motion as long as human MT is intact in at least one hemisphere. Motion, by definition, is a dynamic percept, one that typically unfolds over an extended period of time. As the viewing time is extended, there is an opportunity for signals from early visual areas in the impaired hemisphere to reach secondary visual areas in the unimpaired hemisphere. The receptive fields in primate area MT are huge and have cells that can be activated by stimuli presented in either visual field.

Nonetheless, the application of transcranial magnetic stimulation (TMS) over human MT can produce transient deficits in motion perception. In one such experiment, subjects were asked to judge whether a stimulus moved to the left or right (Beckers and Zeki, 1995). To make the task demanding, the stimulus was only visible for 25 msec. TMS was applied to three locations over the visual cortex, targeted to activate neurons in MT, V1, or as a control, an extrastriate region lying between these two regions (Figure 5.28). Performance of the motion task was disrupted by stimulation over MT. One advantage of this method is that the timing of the magnetic pulses can be varied to determine the time of maximum disruption in the ability to perform the task. When TMS was applied to area MT just prior to the onset of the stimulus, performance dropped to one of chance. When the TMS was delayed until 40 msec after stimulus onset, the performance was perfect.

These results are puzzling, especially when one considers that only at much longer delays did stimulation over V1 disrupt performance. If V1 provides the gateway to all information processing in the visual cortex, one would expect the optimal time of stimulation over V1 to precede the optimal time of stimulation for human MT. Alternatively, it is possible that there may be at least two ways for signals to reach MT. The first is a cortical route, the extension of the M pathway from the LGN to the primary visual cortex and eventually on to area MT. The second route bypasses V1. It may involve direct projections to area MT from the magnocellular layers of the LGN or projections from other subcortical

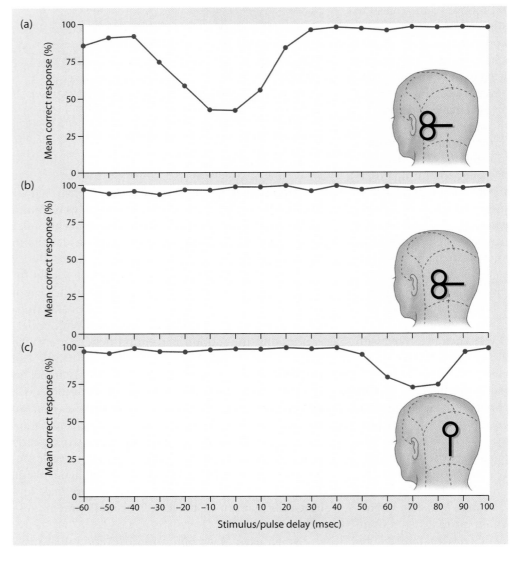

Figure 5.28 Perception of motion during transcranial magnetic stimulation (TMS) over the visual cortex. Results are rates of accuracy in determining the direction of motion as a function of the time between stimulus onset and the TMS pulses applied over area MT **(a)**, V1 **(c)**, and an extrastriate region between these two **(b)**. Adapted from Beckers and Zeki (1995).

structures that are sensitive to moving stimuli, such as the superior colliculus or pulvinar. Physiological and anatomical studies in primates will likely be needed to address these hypotheses.

Deficits in Other Aspects of Visual Perception

Physiological evidence indicates that the strongest segregation of function is in motion and color perception. This hypothesis is supported by double dissociations found in the neurological literature. As with patient M.P., motion perception can be impaired without losing color perception. Conversely, deficits in color perception arise without losing the ability to perceive movement.

The case for specific deficits in other visual features is more contentious. A report from early in this century (Riddoch, 1917) described a patient to whom the world appeared essentially flat. This depth perception impairment was said to exist despite the patient's ability to perceive variations in color and shading. However, it is not possible to evaluate the selectivity of the deficit because careful assessments of other visual functions were not performed. In a similar manner, patients who have lost visual acuity after suffering cortical lesions generally have widespread problems in visual perception. Thus, there are no unambiguous reports of selective deficits in form or depth perception.

The visual system is likely to represent form and depth redundantly. Depth perception arises from a multitude of cues. One potent cue, binocular disparity, arises when each eye has a slightly different view of the world. Inputs from the two eyes project to the magnocellular and parvocellular layers of the LGN and converge on common cortical neurons. Cells in all three cortical streams are sensitive to depth.

Depth also can be inferred from motion. A moving object will obscure a more distant object or be hidden by a nearer object. In a similar sense, as we move, the change in the retinal image is greater for near objects than for ones far away. All these cues can be used in normal depth perception. This redundancy would be expected to help preserve depth perception even if certain modules were lesioned.

Form perception also stems from multiple sources of information. Indeed, form perception is an essential goal of vision, and all visual processing is devoted to determining what objects are in the visual field and where they are located. Motion perception may be important for a predator anticipating the location of moving prey. Nonetheless, movement also can help to identify prey, to separate it from the background, and to determine if its motion is characteristic of an animal worthy of pursuit. A hungry frog does not want to flick its tongue at any moving object. Rather, the frog can recognize a potential snack by its specific motion—for example, that produced by a fly. Color perception is important only in that it facilitates form perception. Colors do not exist without forms. Color is a potent cue for discriminating an object or part of one from other regions of space. Unlike borders, which can change as a function of illumination conditions—consider the effects of a shadow falling across an object—color remains invariant as long as sufficient light is available for the cones to fire.

Therefore, it is not surprising that we have no clinical reports of patients who are "form blind." The P-interblob pathway may have heightened sensitivity to features that correspond to shape in comparison to the M and P-blob pathways. But the latter two also contribute to form perception, in ways that the P-interblob pathway cannot. Indeed, evolutionary theorists have proposed that color perception is a recent adaptation that refines the recognition capabilities of the visual system. It is reasonable to suppose that concurrent pathways provide multiple inputs to systems devoted to object recognition. In the next chapter we will explore how these higher-level aspects of perception operate.

INDEPENDENT OR CONVERGENT PATHWAYS

In this chapter we have emphasized that the visual system contains multiple pathways, each specialized to abstract specific information. Yet the outputs from these systems are designed to complement each other. This paradox has led to a lively debate concerning the independence of pathways. Advocates favoring a segregationist viewpoint have focused on anatomical, physi-ological, and behavioral data indicating that the processing of features such as motion and color involves separable mechanisms from the first synapses in the central nervous system all the way to the extrastriate cortex. To these theorists, neurological dissociations are the crowning piece of evidence. On the other side are those who emphasize the lack of perfect segregation.

Tactile "Seeing" in the Visual Cortex

What happens in the visual cortex of people who are blind because of damage to the peripheral visual apparatus? Unlike patients with cortical lesions, the central nervous system is intact—there is no widespread loss of tissue and the cells continue to fire. But what causes cells in the visual cortex to respond under such conditions? Is it just noise, random firing that is unrelated to any external events? Or does the brain become reorganized and exploit this tissue in other functions?

This issue was explored in a recent PET study (Sadato et al., 1996). The results suggest a remarkable degree of functional reorganization, or what neuroscientists refer to as *cortical plasticity*. The subjects for this study included people with normal vision and people who were blind, either congenitally or from a young age (average age at onset of blindness < 5 years). In the first experiment the subjects were scanned under two experimental conditions. In one condition the subjects were simply required to sweep their fingers back and forth over a rough surface covered with dots. In the second condition they were given tactile discrimination tasks such as deciding whether two grooves in the surface were the same or different. Blood flow in the visual cortex during each of these tasks was compared to that during a rest condition when the subjects were scanned while keeping their hands still.

Amazingly, changes in blood flow in the visual cortex were in opposite directions for the two groups of subjects. For the sighted subjects, a significant drop in blood flow was found in the primary visual cortex during the tactile discrimination tasks. Analogous decreases in the auditory or somatosensory cortex occurred during visual tasks given to normal subjects; this means that as attention was directed to one modality, blood flow decreased in other sensory systems. In contrast, in blind subjects the blood flow increased—but only during discrimination tasks and not when they swept their fingers over the surface without having to use tactile information. As with normal subjects, blood flow changes were most apparent when the sensory information required a response. For blind subjects, the enhanced activity was seen in what would be the visual cortex of normal subjects.

The second experiment explored the same issue, but with a task of great importance to blind subjects: reading Braille. Here blind subjects explored strings of eight Braille letters and had to decide whether the strings formed a word or nonword. In accord with the results of the first study, blood flow in the primary and secondary visual cortex rose during Braille reading in comparison with the rest state. It is unlikely that this rise was associated with the motor demands of the experimental task. Subjects responded only on the nonword trials, which accounted for less than 10% of all trials.

Cells in the M and P-blob pathways exhibit some orientation selectivity. Color sensitivity is not the exclusive domain of the P-blob pathway; it is also found in cells of the P-interblob pathway. Even the cells in the magnocellular layers of the thalamus respond to a stimulus border defined solely by variation in hue, although the cells may respond similarly to either a green or a red border on a blue background (Croner and Albright, 1999). Lesion studies of animals also have brought into question the notion that the processing of features like depth, color, and orientation depends solely on a single pathway.

For these reasons we prefer the term **concurrent processing,** coined by David Van Essen of Washington University in St. Louis (Van Essen and DeYoe, 1995). This phrase emphasizes the analytic characteristic of visual perception. There is strong and converging evidence that the visual system does not analyze the input as a unified whole but engages in a strategy of divide and conquer. It has evolved mechanisms that enable visual information to be distributed over specialized subsystems. Even with this strategy, though, there is a benefit to letting systems interact. Concurrent processing emphasizes the analytic nature of perception devoted to common goals.

With this in mind, we can reconsider the case of the Oregon farmer, P.T. For him, the most distressing symptom was his inability to recognize faces, especially his wife's. But the patient's deficit was not limited to face perception. He had difficulty discriminating among his cows and identifying the landscape around his farm. He demonstrated a puzzling dissociation in which he readily identified the portrait in Picasso's painting while failing to discern the face in Monet's.

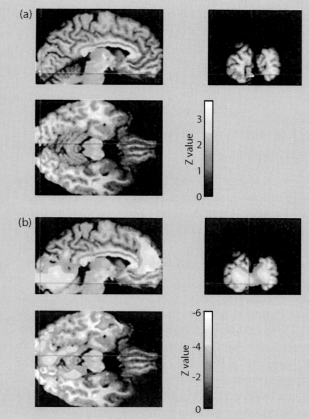

(a)

(b)

Z value

3
2
1
0

Z value

-6
-4
-2
0

Activation loci when blind and sighted subjects performed a tactile discrimination task compared with a control condition in which subjects simply rested. For the blind subjects **(a)**, activation increased in the primary visual cortex during the discrimination task. For the sighted subjects **(b)**, activation decreased in the primary visual cortex during this condition.

At present, it is unclear how tactile information ends up activating visual cortex neurons in blind people. One possibility is that somatosensory projections to thalamic relays spread into the nearby LGN, exploiting the geniculo-striate pathway. But this hypothesis is unlikely since the blood flow changes in the blind subjects were bilateral. Somatosensory inputs to the thalamus are strictly lateralized. Because the subjects performed the tactile tasks with their right hand, the blood flow changes should have been restricted to the left hemisphere. A more viable hypothesis is that a massive reorganization of cortico-cortical connections follows peripheral blindness. The sensory-deprived visual cortex is taken over, perhaps through back projections originating in polymodal association cortical areas.

While this study provided a dramatic demonstration of cortical plasticity, the results also suggest a neurobiological mechanism for the greater nonvisual perceptual acuity exhibited by blind people. Indeed, Louis Braille's motivation to develop his tactile reading system was spurred by his belief that vision loss was offset by heightened sensitivity in the fingertips. One account of this compensation focuses on nonperceptual mechanisms. While the sensory representation of somatosensory information is similar for blind and sighted subjects, the former group is not distracted by vision (or visual imagery) and thus can use somatosensation more efficiently. These new PET data, though, suggest a perceptual-based mechanism. Sensitivity increases because more cortical tissue is devoted to representing nonvisual information.

We can see how difficult it is to make an unambiguous diagnosis. In his conclusion, the neurologist evaluating P.T. emphasized that the primary problem stemmed from a deficit in color perception. This hypothesis is in accord with one of the primary differences between the Monet and the Picasso. In the Monet painting, the boundaries between the face and the background are blended: Gradual variations in color demarcate the facial regions and separate them from the background landscape. A deficit in color perception provided a parsimonious account of the patient's problems in recognizing faces and landscapes. The rolling green hills of an Oregon farm can blur into a homogeneous mass if one cannot discern fine variations in color. In a similar way, each face has its characteristic coloration.

Yet it seems equally plausible that the problem may have stemmed from a deficit in contrast or contour perception. These features are salient in the Picasso and absent in the Monet. What is clear is that the patient's stroke primarily affected the cortical projections of the P pathways, the pathways essential for color and form perception. In contrast, the cortical projections of the M pathway were intact. The patient did not have any trouble recognizing his wife as she moved from the stove to the kitchen table; indeed, P.T. commented that her idiosyncratic movement enabled him to recognize her.

Differentiating between a color-based hypothesis and a contour-based one requires more details than were offered at Grand Rounds. It would have been informative to test P.T. on Treisman's visual search tasks to compare his ability to detect features based on color and form differences. Given the neurologist's hypothesis, we would expect the patient to have difficulty when the target is defined by a unique color. For P.T., like other achromatopsia patients, the ability to find a red target among isoluminant green distracters might be similar to what a normal person experiences when searching for a red target among orange distracters. The critical question centers on whether P.T. would demonstrate difficulty with a visual search task that included shape differences such as a tilted line among vertical lines. The analytic tools of cognitive psychology could provide the key to making a much more precise diagnosis of P.T.'s deficits.

DISSOCIATIONS OF CORTICAL AND SUBCORTICAL VISUAL PATHWAYS

We know that the retino-geniculate-cortical tract contains 90% of the fibers in the optic tract, and we know how subdivisions of this pathway provide building blocks for form, color, and motion perception. From these building blocks, we can recognize and identify complex visual scenes.

The importance of this pathway is underscored by the dramatic deficits that develop when this pathway is damaged. While lesions of the LGN are rare, it is not uncommon to come across patients who have had a stroke that encompasses portions of the primary visual cortex. Such lesions are devastating for visual processing. The patients become blind to stimuli falling within the receptive fields of the affected area. This deficit, referred to as a **scotoma,** or field cut, renders patients incapable of detecting an object presented within the scotoma, let alone reporting the object's color or shape (Figure 5.29). When the entire visual cortex is lesioned, the patient is unable to see anything in the contralesional hemifield. When the lesion is more circumscribed, affecting merely a portion of the visual cortex, the scotoma is more limited. For instance, if a lesion is restricted to the lower bank of the calcarine fissure, the scotoma is limited to the upper quadrant of the contralesional hemifield. Correspondingly, a lesion in the upper bank produces a scotoma in the lower visual field on the side opposite the lesion.

Clinicians can easily map the extent of a scotoma. The standard technique, called *perimetry,* involves presenting a small spot of light in various places in the visual field and asking the patient to report whether she can detect the light. When this stimulus falls outside the scotoma, detection is immediate. Within the scotoma, the patient fails to see the stimulus. The patient is, in essence, cortically blind. While rods and cones continue to function in a normal manner, the cells fail to produce any sensation or percept.

The phenomenology of cortical blindness led neurologists to assume that visual processing in humans depends entirely on the geniculo-striate pathway. Anatomists could point out that numerous optic nerve fibers terminate in subcortical structures other than the LGN, but that the function of these pathways was vestigial—they had become dormant. This hypothesis has come into question owing to surprising findings in animal and human research.

Spatial Orientation and Object Perception in the Hamster

Lesion studies in animals provided the first line of evidence. Many nongeniculate fibers terminate in the superior colliculus. This structure plays a critical role in producing eye movements. If this midbrain structure becomes atrophied, as with supranucleur palsy, the patient will lose the ability to generate eye movements. It is as if the eyes have become paralyzed. Stimulation studies in primates further demonstrated the role of the superior colliculus in eye movement. When this area was stimulated, the eyes moved, and their direction depended on the stimulation site.

Gerald Schneider (1969), working at the Massachusetts Institute of Technology (MIT), provided initial evidence of the importance of the colliculus in studies of hamsters. These animals were trained to do the two tasks schematized in Figure 5.30. In one task, the hamsters were trained to turn their head in the direction of a sunflower seed held in an experimenter's hand. The task was easy for hamsters since they have a strong propensity to find sunflower seeds and put them in their cheek.

The second task presented more of a challenge. Here, the animals were trained to run down a two-arm maze and to enter the door behind which a sunflower seed was hidden. The task required the animals to make simple visual discriminations such as distinguish between black and white doors or doors with vertical or horizontal stripes. With normal hamsters, the discriminations are not taxing. Within a few trials, they became proficient in selecting the right door in almost all trials.

After training hamsters for both tasks, Schneider divided them into two experimental groups. One group received bilateral lesions of the visual cortex, including all of

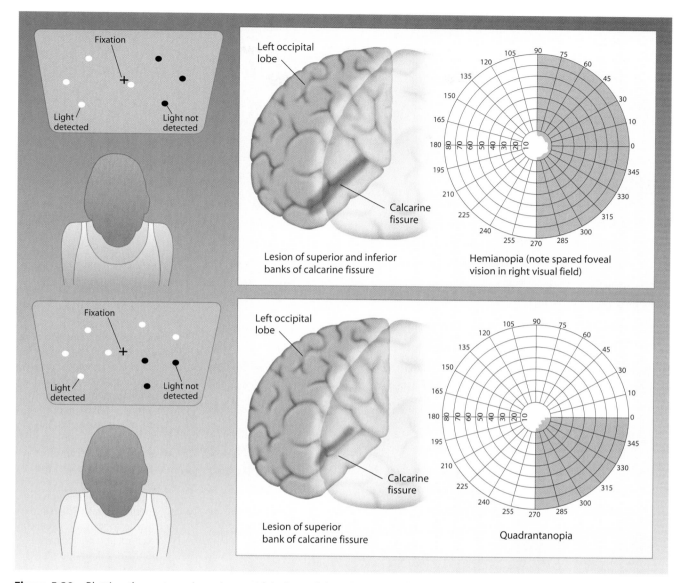

Figure 5.29 Plotting the scotoma in patients with lesions of the primary visual cortex. While the patient fixates on a central marker, a small light is flashed at various locations. The patient is asked to report when she sees the stimulus. If the lesion includes tissue on both the upper and lower banks of the calcarine fissure, the scotoma will include the entire contralesional hemifield. This is referred to as a *hemianopia*. If the lesion is restricted to the upper bank, the patient will have a *quadrantanopia*, referring to the fact that she only misses targets in one quadrant, the lower region of the contralesional hemifield. Although each eye is tested separately, the scotoma will be essentially identical because the lesions are in the cortex.

areas 17 and 18. For the second group, the superior colliculus was rendered nonfunctional by ablating all input fibers. This strategy was necessary because direct lesions to the colliculus were likely to kill the animals; the structure borders numerous brainstem nuclei that are essential for life.

The two lesions yielded a double dissociation. Cortical lesions severely impaired the animals' performance on the visual identification tasks. The animals could run down the maze and had sufficient motor capabilities to enter one of the doors, but they could not discriminate

black from white or horizontal from vertical stripes. In contrast, the animals with collicular lesions demonstrated no impairment.

The deficits were reversed on the sunflower localization task. Animals with cortical lesions were perfect at this task once they had recovered from the surgery. Yet animals with collicular lesions acted as though they were blind. They made no attempt to orient toward the seeds, and not because they were unmotivated or had a motor problem. If the seed brushed

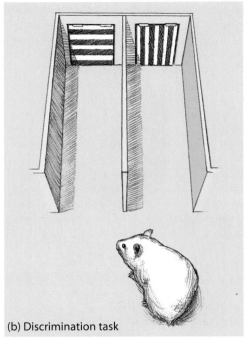

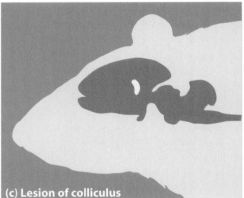

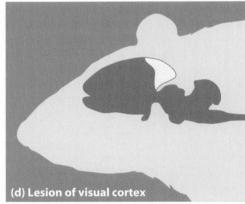

(a) Localization task

(b) Discrimination task

(c) Lesion of colliculus

(d) Lesion of visual cortex

Figure 5.30 Schneider's tests to demonstrate the double dissociation between lesions of the superior colliculus and visual cortex. **(a)** In the localization task, the animals were trained to collect sunflower seeds that were held at various positions in space. **(b)** In the discrimination task, the animals were trained to run down one of two alleys that differed in terms of whether the stripes were horizontal or vertical. **(c)** Lesions of the colliculus selectively disrupted performance on the orientation task. **(d)** In contrast, lesions of the visual cortex impaired performance on the discrimination task.

against a whisker, the animal rapidly turned in its direction and gobbled it up.

These data provide compelling evidence for dissociable functions of the hamsters' superior colliculus and visual cortex. The collicular lesions impaired their ability to orient toward the position of a stimulus, while the cortical lesions disrupted visual acuity. For the hamster, one might think of this double dissociation as reflecting two systems: One devoted for spatial orientation and the other devoted to object identification.

Blindsight: Evidence of Residual Visual Function Following Cortical Blindness

Collicular function in the hamster cannot be extrapolated directly to humans. It might be that when elabo-

rate visual systems evolved in humans, cortical areas subsumed functions that had depended on subcortical areas in our ancestors. As we will see in Chapter 6, within the cortical pathways, humans and other primates exhibit a functional division between areas devoted to representing spatial information, or "where" processing, and areas devoted to object recognition, or "what" processing. Nonetheless, the subcortical visual pathways in humans do appear to play an important role in spatial orientation. Evidence for this has come from intriguing studies involving patients who have suffered lesions producing cortical blindness.

Inspired by Schneider's findings that cortically blind hamsters could still locate the sunflower seeds, Lawrence Weiskrantz (1986) at Oxford University tested a hemianopic patient, D.B., to determine whether he could still detect the location of objects presented within his scotoma. Weiskrantz was not satisfied with the usual method of self-report to test for visual function. Rather, he designed a clever behavioral assay for residual function. After presenting a spot of light, the experimenters sounded a tone and D.B. was asked to move his eyes to the stimulus's location. This task was easy when stimuli were presented in his intact right visual field. But when stimuli were presented within the scotoma, the task struck D.B. as utter nonsense. He could not understand how he should know where

to move his eyes when he failed to see anything. Nonetheless, the experimenters encouraged D.B. to guess. On control trials, a tone was sounded, requiring D.B. to move his eyes, but it was not preceded by a light. D.B., of course, was not aware of this difference between the control and experimental trials. To him, all of the trials seemed bizarre in that he was being asked to look at stimuli that he had not been aware of (Figure 5.31).

The results could hardly have been more dramatic. As expected, D.B.'s eye movements on the phantom

Figure 5.31 Experimental conditions to demonstrate blindsight. In the experimental trials, a light beam is projected at one of seven locations within D.B.'s scotoma. Upon hearing a tone, D.B. was required to move his eyes to the target location. In the control conditions, a target was selected but not illuminated prior to the tone. To D.B., the two conditions seemed identical, and he had to be encouraged to move his eyes since he reported not seeing anything in either condition. Nonetheless, D.B.'s eye movements showed a systematic relationship to the target position in the experimental condition, at least for the targets within 20 degrees of fixation. Adapted from Weiskrantz (1986).

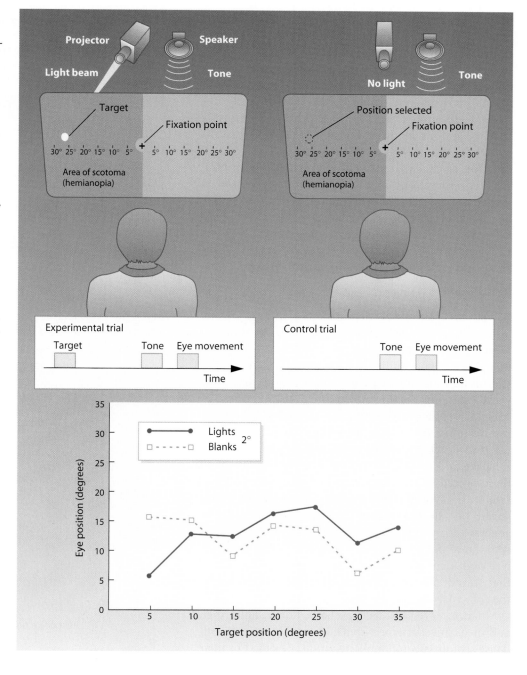

trials were random. But in the experimental trials, D.B. did much better than chance at localizing the stimuli. When the stimuli were within 20 degrees of the fovea, D.B.'s eye movements were highly correlated with the position of the stimuli. Weiskrantz named this paradoxical phenomenon *blindsight*. The patient acts and feels as if he is blind, yet shows a residual ability to localize stimuli. While the blindsight phenomenon has been reported in other patients over the past 20 years, the interpretation of the effect remains controversial. It would be premature to conclude that this phenomenon solely reflects processing in subcortical pathways. It is possible that information can reach extrastriate visual areas in the cortex, either through direct geniculate projections or via projections from other subcortical structures. Using PET, researchers have shown that extrastriate regions such as human MT can be activated by moving stimuli, even when the ipsilateral striate cortex is completely destroyed (Barbur et al., 1993). Another possibility is that the lesions in the primary visual cortex are incomplete and that blindsight results from residual function in the spared tissue (Fendrich et al., 1992). The representations in the damaged region may be sufficient to guide eye movements, even though they fail to achieve awareness. We will return to this issue is Chapter 16 when we discuss consciousness.

Functions of the Retino-collicular Pathway in Humans

It may well turn out that different aspects of the blindsight phenomenon are attributable to different mechanisms. Nonetheless, an important feature of many blindsight studies is that patients can localize stimuli they are not aware of. This ability to orient toward a stimulus fits the known functions of the retino-collicular pathway. The superior and inferior colliculi receive input from visual and auditory pathways and use it to develop a representation of where objects are in space and to generate eye movements to attend to these objects. Subcortical visual systems may collaborate with cortical visual systems. As an object is detected on the periphery, the colliculi may help to generate an eye movement that would bring the object into the center of view so the cortical system can analyze and identify it. From this view, we might expect that stimuli falling within the scotoma would activate the collicular orienting system.

In a clever test of this hypothesis, Robert Rafal, then at Brown University, measured how quickly patients with dense scotomas could look at stimuli presented in their intact visual field (Rafal et al., 1990). Since the patients were conscious of the stimuli, the task seemed quite reasonable—unlike the traditional blindsight studies. The key manipulation, though, was that simultaneously with the stimulus's presentation, an irrelevant stimulus was presented within the scotoma (Figure 5.32). The pa-

Figure 5.32 Subcortical pathways remain active in the presence of lesions of the primary visual cortex. Hemianopic patients were asked to look at targets in the ipsilesional (intact) hemifield as quickly as possible. In the experimental condition, the onset of the target was preceded by a distracter presented in the scotoma. Despite their lack of awareness of the distracters, the patients were slower to look at the targets in comparison to a control condition in which the distracter was not presented. The slower response times are attributed to competing eye movement signals being generated by the superior colliculus following the onset of the distracter.

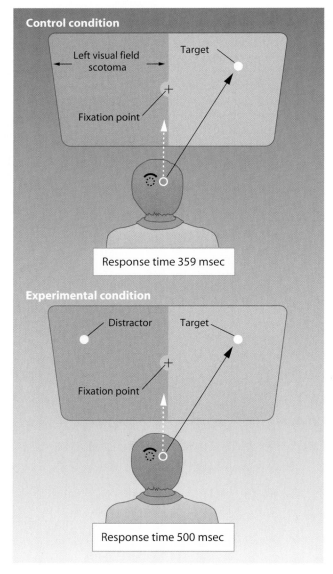

tients, of course, did not report seeing it. Nonetheless, the irrelevant stimulus led to a significant boost in reaction time in comparison to a control condition where the irrelevant stimulus did not appear until after the response. Through this implicit measure of interference, then, we see evidence of residual processing within scotomas.

Rafal and his colleagues argued that this interference arises because the irrelevant stimulus provides competing activation between the distracter and target in the intact retino-collicular pathway. This competi-

tion slows eye movement toward the target. Further evidence that this interference is related to the collicular system comes from a second experiment where subjects were asked to press a button after detecting the target rather than look at the target. Now the competing stimulus had no effect on reaction time. Thus, the interference was present only when the response required an eye movement, a result that meshes with the functions of the superior colliculus. As with Schneider's hamsters, it appears that a collicular orienting system remains intact in hemianopic patients.

AUDITORY PERCEPTION

Compared to what we know about visual perception, our knowledge of the functional organization of other sensory systems is sparse. We have a clear understanding of the anatomy of the senses of audition, touch, position, olfaction, and taste. But the physiology and functional analysis of these systems have not been developed with the degree of sophistication applied to vision. In this section we present an

organization scheme for auditory perception similar to the one for visual perception.

Overview of the Auditory Pathways

An overview of the auditory pathways is presented in Figure 5.33. The complex structures of the inner ear provide the mechanisms for transforming sounds, or

Figure 5.33 An overview of the auditory pathway. The hair cells of the cochlea are the primary receptors. The output from the cochlear nerve projects to the cochlear nuclei in the brainstem. Ascending fibers reach the auditory cortex following synapses in the inferior colliculus and medial geniculate nucleus. Adapted from Bear et al. (1996).

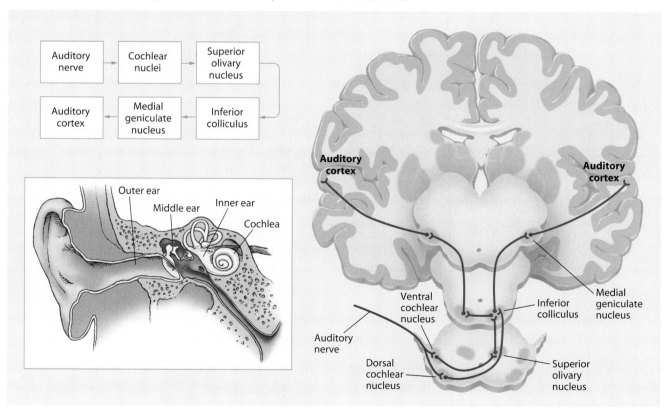

HOW THE BRAIN WORKS

Follow Your Nose

Sight, sound, and touch are the senses we are most aware of. And yet, the more primitive senses of taste and smell are in many ways equally essential to our survival. While the baleen whale probably does not taste or smell the tons of plankton it ingests, these two senses are essential for terrestrial mammals, helping recognize foods that are palatable, nutritional, and safe. Spending a few minutes to observe dogs meeting in the park or a pet rodent wandering around the living room provides a quick reminder of the importance of smell. While olfaction might have evolved as a mechanism for evaluating whether a potential food is edible, it also has come to serve important social functions. Just consider the endless display of perfumes and colognes that line the first-floor counters of every department store!

Anatomical and lesion studies indicate that the primary olfactory cortex (POC) is located in the ventral region of the anterior temporal lobe. This area is occasionally destroyed during temporal lobe surgery for the treatment of severe epilepsy; after surgery, these patients have difficulty recognizing familiar odors (Jones-Gotman et al., 1997). Interestingly, unlike the crossed pattern of connectivity that is found for the rest of the cortex, the olfactory nerve innervates ipsilateral cortex. Thus, a left-sided lesion will produce a more severe deficit when the odor is sniffed with the left nostril.

Functional magnetic resonance imaging (fMRI) has been used to study the human olfactory system (Sobel et al., 1998). Conducting such a study is technically quite challenging. First, there is the problem of delivering odors to the subject in a controlled manner. Nonmagnetic systems must be constructed to allow the odorized air to be directed at the subjects' nostrils. Second, it is hard to determine when an odor is no longer present. The chemicals that carry the odor can linger in the air for a long time. Third, while some odors overwhelm our senses, most are quite subtle and require exploration through the act of sniffing. Whereas it is almost impossible to ignore a sound, we can exert some control over the intensity of our olfactory experience.

Indeed, the results of the imaging studies suggest an intimate relationship between smelling and sniffing. Subjects were scanned while they were exposed to either nonodorized, clean air or one of two chemicals, vanillin or decanoic acid. The former smells like vanilla; the latter like crayons. The odor-absent and odor-present conditions alternated every 40 seconds. Throughout the entire scanning session, the instruction, "Sniff and respond, is there an odor?" was presented every 8 seconds. In this manner, the researchers sought to identify those areas in which brain activity was correlated with sniffing and those areas in which activity was correlated with smelling.

Surprisingly, smelling failed to produce consistent activation in the POC. Instead, the presence of the odor produced a consistent increase in the BOLD effect in lateral parts of the orbitofrontal cortex, a region that is thought to constitute a secondary, olfactory area. Activity in the POC was closely linked to the rate of sniffing. Each time the person took a sniff, the BOLD signal would increase regardless of whether or not the odor was present. These results seem quite puzzling, suggesting that the POC might be more part of the motor system for olfaction. Upon further study, however, the lack of activation in the POC became clear. Neurophysiological studies of the POC in the rat had shown that these neurons habituate quickly. Perhaps the lack of a smell-related response in the POC is due to the fact that the hemodynamic response exhibits a similar habituation. To test this idea, the BOLD signal was modeled by assuming a sharp increase followed by an extended drop during the odor-present epoch, an elegant example of how single-cell results can be used to interpret imaging data. When analyzed in this manner, the hemodynamic response in the POC was found to be related not only to sniffing but also to smell (Sobel et al., 2000). However, the POC role may be restricted to detecting a change in the external odor, with the secondary olfactory cortex playing the critical role in identifying the smell itself. Each sniff represents an active sample of the olfactory environ-

variations in sound pressure, into neural signals. Sound waves arriving at the ear make the eardrum vibrate. These vibrations produce tiny waves within the inner ear's fluid that stimulate tiny *hair cells* located along the surface of the basilar membrane. Analogous to the eye's photoreceptors, hair cells are primary auditory receptors. Oscillations of the basilar membrane prompt hair cells to generate action potentials. In this way, a me-

(a)

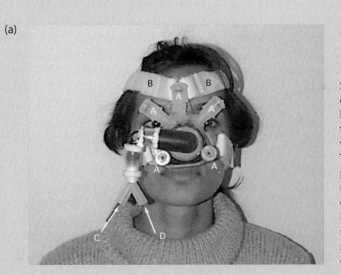

Sniffing and smelling. **(a)** Special device constructed to deliver controlled odors during fMRI scanning. **(b) Top:** Activated regions during sniffing. The circled region includes the primary olfactory cortex and a medial-posterior region of the orbitofrontal cortex. **Bottom:** Regions more active during sniffing when an odor was present compared to when the odor was absent.

(b)

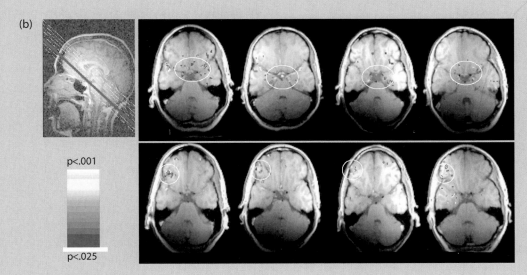

p<.001

p<.025

ment, and that POC plays a critical role in determining if a new smell is present.

The importance of sniffing for olfactory perception is also underscored by the fact that our ability to smell is continually being modulated by changes in the size of the nasal passages. In fact, the two nostrils appear to switch back and forth, with one larger than the other for a few hours and then the reverse. Why might the nose behave this way? One hypothesis is that the olfactory percept depends not only on the intensity of the odor but also on the efficiency with which we sample the odor. By having two nostrils of slightly different sizes, the brain is provided with slightly different images of the olfactory environment. As we will see in Chapter 10, asymmetrical representations are the rule in human cognition, perhaps providing a more efficient manner of processing complex information. With the ancient sense of olfaction, it appears that this asymmetry is introduced at the peripheral level.

chanical signal, the fluid oscillations, is converted to a neural signal—the output from hair cells.

The basilar membrane and hair cells are located within a spiral structure known as the *cochlea*. Even at this early stage of the auditory system, much information about the sound source can be discerned. Hair cells have receptive fields analogous to the retina's ganglion cells. While receptive fields of ganglion cells refer to a coding of locations in space, receptive fields of hair cells refer to a coding of sound frequency. Human auditory sensitivity ranges from a low of about 20 Hz to a high of 20,000 Hz. Hair cells at the thick end, or base of the cochlea, are activated by high-frequency sounds; cells at the opposite end, or apex, are activated by low-frequency sounds. These receptive fields have extensive overlap. Further, natural sounds like music or speech are made up of complex frequencies; thus, sounds activate a broad range of hair cells.

Important subcortical relays are in the auditory system. The output from the cochlea is projected to two midbrain structures, the **cochlear nucleus** and the inferior colliculus. From there, information is sent to the medial geniculate nucleus. As with vision, the geniculate nucleus serves as the final relay point to the auditory cortex. Area 41 is considered the primary auditory cortex in humans. Areas 42 and 43 are secondary cortical areas, although they too receive direct projections from the medial geniculate nucleus.

Neurons throughout the auditory pathway continue to have frequency tuning. As seen in Figure 5.34, the tuning curves for auditory cells can be quite broad; single cells respond to sounds ranging over a couple of octaves, where an octave corresponds to a doubling in frequency. The fact that individual cells do not give pre-

cise frequency information but provide coarse coding indicates that our perception must depend on the integrated activity of many neurons. Such integration is probably facilitated by tonotopic maps found in most auditory areas; within such maps, there is an orderly correspondence between the location of the neurons and their specific frequency tuning. For example, cells in one region of an auditory area will respond to low-frequency stimuli; cells in another region will respond to middle or high frequencies. This organization can be seen in humans with the resolution provided by fMRI (Figure 5.35).

Electrophysiological studies of the cat revealed a second general principle of auditory processing. The tuning specificity of auditory receptive fields becomes more refined as the stimulus proceeds through the system. A neuron in the cochlear nucleus that responds maximally to a pure tone of 5000 Hz may also respond to tones ranging from 2000 to 10,000 Hz. A comparable neuron in the auditory cortex is likely to respond over a much narrower range.

Computational Goals in Audition

Frequency data are essential for deciphering a sound. Sound-producing objects have unique resonant properties that provide a characteristic signature. The same note played on a clarinet and a trumpet will sound different. Though they may share the same base frequency, the resonant properties of each instrument will produce great differences in the note's harmonic structure. In a similar way we produce our range of speech sounds by varying the resonant properties of the vocal tract. Movements of our lips, tongue, and jaw change the frequency content of the acoustic stream produced during speech. Frequency variation is essential for a listener to identify words or music.

Auditory perception does not merely identify the content of an acoustic stimulus. A second important function of audition is to localize sounds in space. Consider a bat, which hunts by echo location. The echoes create an auditory image of an object, preferably a tasty moth. But knowing a moth is present will not lead to a successful hunt. The bat also has to determine the moth's precise location. As demonstrated in the comparison of collicular and cortical vision, vision researchers have intensively studied "what" (i.e., object recognition) and "where" (i.e., spatial cognition) problems. In contrast, the cognitive neuroscience of audition has focused primarily on "where" problems. In solving the "where" problem, the auditory system relies on concurrent processing.

Figure 5.34 Tuning curves for a cell in the auditory nerve of the squirrel monkey. This cell is maximally sensitive to a sound of 1600 Hz, and the firing rate falls off rapidly for either lower- or higher-frequency sounds. The cell is also sensitive to intensity differences, although this does not cause a shift in its tuning profile. Other cells in the auditory nerve would be tuned for different frequencies. Adapted from Bear et al. (1996), after Rose et al. (1971).

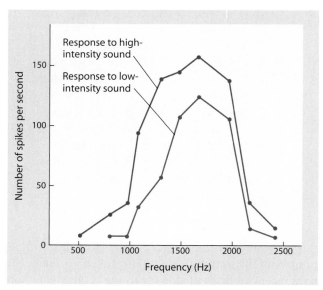

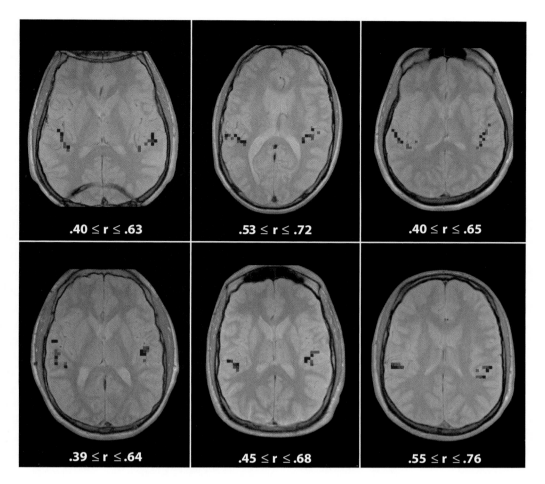

Figure 5.35 Tonotopic representation revealed with fMRI. In most cases, the region responding to the low tones (blue) is more posterior and medial to the region responding to the high tones (red). Adapted from Wessinger et al. (1997).

.40 ≤ r ≤ .63

.53 ≤ r ≤ .72

.40 ≤ r ≤ .65

.39 ≤ r ≤ .64

.45 ≤ r ≤ .68

.55 ≤ r ≤ .76

Concurrent Processing for Sound Localization

In developing animal models to study auditory perception, we select an animal with well-developed hearing. A favorite species for this work has been the barn owl, a nocturnal creature. Barn owls have excellent *scotopia*, or night vision that guides them to their prey. But barn owls also must utilize an exquisitely tuned sense of hearing to locate their prey because visual information is unreliable at night. The low levels of illumination provided by the moon fluctuate over the lunar cycle; this and stellar sources can be obscured by dense cloud cover. Sound, such as the patter of a mouse scurrying across a field, offers a more reliable stimulus. Indeed, barn owls have little trouble finding prey in a totally dark laboratory.

Barn owls rely on two cues to localize sounds: the difference in time between when a sound reaches the two ears, the **interaural time,** and the difference in the sound's intensity at the two ears. Both cues result from the fact that the sound reaching two ears is not identical. Unless the sound source is directly parallel to the head's orientation, the sound will reach one ear before the other. Moreover, since the intensity of a sound wave be-

comes attenuated over time, the magnitude of the signal at the two ears will not be identical. The time and intensity differences are quite small in magnitude. For example, if the stimulus is located at a 45-degree angle to the line of sight, the interaural time difference will be approximately 1/10,000th of a second.

The intensity differences resulting from sound attenuation are even smaller—indistinguishable from variations due to noise. These small differences are amplified by a unique asymmetry of owl anatomy, however. The left ear is higher than eye level and points downward, while the right ear is lower than eye level and points upward. Because of this asymmetry, sounds coming from below are louder in the left ear than the right. Humans do not have this asymmetry, but the complex structure of the human outer ear, or *pinna*, may amplify the intensity difference between a sound at the two ears.

Interaural time and intensity differences provide independent cues for sound localization. To show this, stimuli are presented over headphones and the owl is trained to turn its head in the sound's perceived direction. The headphones allow the experimenter to manipulate each cue separately. When amplitude is held constant, asynchronies in presentation times prompt the owl to shift its head in the horizontal plane. Variations in amplitude produce vertical head movements. Combining the two cues by way of binaural fusion provides the owl with a complete representation of three-dimensional space. If one ear is plugged, the owl's response indicates that a sound has been detected, but it cannot localize the source.

These two cues are processed by independent neural pathways. The auditory nerve synapses on the cochlear nucleus. Each cochlear nucleus is composed of two parts, the magnocellular nucleus and the angular nucleus (Figure 5.36). The auditory nerve fibers innervating the cochlear nucleus bifurcate, sending one axonal branch to the magnocellular nucleus and a second branch to the angular nucleus. This parallelism is maintained in the ascending projections to the midbrain lemniscal nuclei. The magnocellular and angular nuclei project to the anterior and posterior regions of the lemniscal nucleus, respectively.

Mark Konishi (1993) of the California Institute of Technology provided a well-specified neural model of how the owl brain codes interaural time and intensity differences. For detecting time differences, Konishi assumed that the first site of convergence from the two ears is in the anterior lemniscal neurons. These neurons operate as coincidence detectors. To activate an anterior lemniscal neuron, it must simultaneously receive an input from each ear. In computer science terms, these neurons act as AND operators. An input from either ear alone or in succession is not sufficient; they will fire only if an input is received at the same time from both ears. Simultaneous activation from a sound source is restricted to only some of these neurons because the magnocellular axons from each half of the brain converge from opposite directions on the coincidence detectors.

Figure 5.37 explains how this model works. In the left half of the figure, the sound source is directly in front of the animal. In this situation, the coincidence detector in the middle is activated because the stimulus is received at each ear at the same time. In the right half of the figure, the sound source is to the animal's right, which gives the magnocellular axon from the right ear a slight head start. Simultaneous activation now occurs in a coincidence detector to the left of center. This simple arrangement provides the owl with a complete representation of the sound source's horizontal position. Physiological studies have confirmed that neurons of the anterior lemniscal nucleus act as coincidence detectors.

A different coding scheme represents interaural intensities. The lemniscal nucleus's posterior region is again

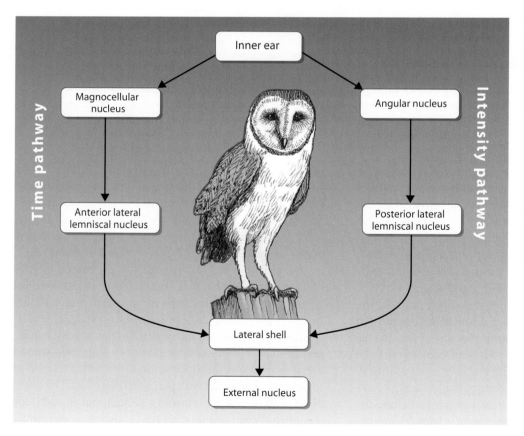

Figure 5.36 Parallel pathways in the auditory system of the barn owl. The cochlear nucleus is actually composed of two parts, the magnocellular nucleus and the angular nucleus. The magnocellular pathway is specialized to compute interaural time differences, information essential for locating the lateral position of a stimulus. The angular pathway is specialized to compute intensity differences, information essential for locating the distance to a stimulus.

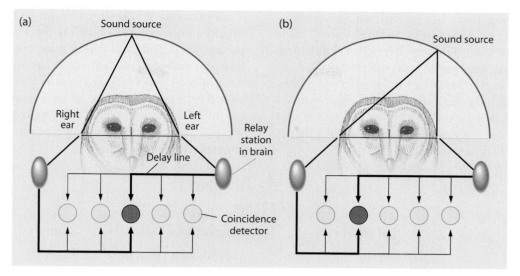

Figure 5.37 Slight asymmetries in the arrival times at the two ears can be used to locate the lateral position of a stimulus. **(a)** When the sound source is located directly in front of the owl, the stimulus will reach the two ears at the same time. As activation is transmitted across the delay lines, the coincidence detector representing the central location will be activated simultaneously from both ears. **(b)** When the sound source is located to the left, the sound reaches the left ear first. Now a coincidence detector offset to the opposite side receives simultaneous activation from the two ears. Adapted from Konishi (1993).

the first point of convergence. But the neural code for intensity is based on the input's firing rate. The louder the stimulus, the more action potentials come into play. Neurons in the posterior region use combined signals from both ears to pinpoint the source's vertical position.

Localization is incomplete at the level of the lemniscal nucleus; horizontal and vertical positions must be combined. Outputs from each region of the lemniscal nucleus converge on another brainstem nucleus, the external nucleus. Neurons within this nucleus are considered space-specific. They activate only when sound comes from a certain location.

Konishi provided an elegant account of how the barn owl can pinpoint a sound in three-dimensional space. The theory stands as an excellent example of the power of cognitive neuroscience. We know from an owl's behavior that it can localize sounds precisely. In vision, neural representation of space is more straightforward. Sensory detectors in the retina provide a topographic map that is maintained throughout the processing system. In audition, though, sensors do not code spatial in-

formation. The location of a sound source must be computed by integrating signals received from two ears. Anatomy and physiology led Konishi to realize that temporal information and intensity information were processed independently. Computational models of coincidence detectors and neural summation then guided their search for the derivation of location information.

In Konishi's model, the problem of sound localization by the barn owl is solved at the level of the brainstem. To date, this theory has not explained higher stages of processing such as the auditory cortex. Perhaps cortical processing is essential for converting location information into action. The owl does not want to simply attack every sound; it must decide if the sound is generated by a potential prey. Another way of thinking about this is to reconsider the issues surrounding the computational goals of audition. Konishi's brainstem system provides the owl with a way to solve "where" problems but has not addressed the "what" question. The owl may need other auditory processing to discriminate whether a sound is the movement of a mouse or a deer.

SUMMARY

This chapter provided an overview of the organization of the pathways involved in visual and auditory perception. This review is not complete: Not only have we ignored the senses of somatosensation, taste, and smell, but also we have described only briefly the opera-

tion of peripheral mechanisms that transduce external information into neural signals.

A point emphasized in this chapter is that specialized mechanisms for solving different computational problems have evolved in the brain. In auditory and visual perception, there is a segregation between systems that focus on determining "where" something is and "what" it is. Within these two functional domains, the sensory systems apply an analytic strategy. The auditory system exploits two distinct cues contained in the sounds reaching the ears and uses this information to localize the source. With the visual system, specialized mechanisms are devoted to processing different attributes of a complex stimulus.

An account of how sensory information is processed provides only the initial building blocks of a theory of perception. A complete theory must take into account how this information is used: What are the ultimate goals of having sophisticated sensory machinery? To answer this question, we need to know how we recognize the information that reaches our eyes and ears, and how we select which information is relevant. The following chapters tackle these issues.

KEY TERMS

achromatopsia

akinetopsia

area MT

cochlear nucleus

concurrent processing

cortical visual areas

extrastriate

interaural time

lateral geniculate nuclei (LGN)

magnocellular (M) system

parvocellular (P) system

photoreceptors

primary visual cortex

retina

scotoma

superior colliculus

visual search task

THOUGHT QUESTIONS

1. You watch a short video segment in which a large purple dinosaur appears briefly in the left visual field. Trace the flow of information about this stimulus and its separate features (color, shape, luminance, motion, position) from the eye through the secondary visual areas.

2. Compare and contrast the functional organization of the visual and auditory systems. What are the computational problems that each system must solve, and how are these solutions achieved in the nervous system?

3. A person arrives at the hospital in a confused state and appears to have some impairments in visual perception. As the attending neurologist, you suspect the person may have had a stroke. How would you go about examining the patient to determine at what level in the visual pathways the damage has occurred? You should emphasize the behavioral tests you would administer, although you should feel free to make predictions about what you expect to see on MRI scans.

4. Define the physiological concepts of the receptive field and visual area. How is the receptive field of a cell established? How are the boundaries between visual areas identified? Can either receptive fields or visual areas be studied noninvasively in humans?

5. Much of the focus in this chapter has been on salient visual properties such as color, shape, and motion. In looking around the environment, do these properties seem to reflect the most important cues for a highly skilled visual creature? What other sources of information might an adaptive visual system exploit?

SUGGESTED READINGS

CRONER, L.J., and ALBRIGHT, T.D. (1999). Seeing the big picture: Integration of image cues in the primate visual system. *Neuron* 24: 777–789.

LIVINGSTONE, M., and HUBEL, D. (1988). Segregation of form, color, movement, and depth: Anatomy, physiology, and perception. *Science* 240:740–749.

PALMER, S.E. (1999). *Vision Science: Photons to Phenomenology.* Cambridge, MA: MIT Press.

TREISMAN, A. (1988). Features and objects: The Fourteenth Bartlett Memorial Lecture. *Q. J. Exp. Psychol. A* 40:201–237.

ZEKI, S. (1993). *A Vision of the Brain.* Oxford, UK: Blackwell Scientific.

6

Higher Perceptual Functions

With analytic perception, visual and auditory systems use a divide-and-conquer strategy. Features like color, shape, and motion are processed along distinct neural pathways. But perception requires more than simply perceiving the features of objects. When gazing at the coastline of San Francisco, we do not have the impression of blurs of color floating among a sea of various shapes (Figure 6.1). We perceive unified objects: the deep-blue water of the bay, the peaked towers of the Golden Gate Bridge, the silver skyscrapers, and the fog lingering over the city.

What is more, perceptual capabilities are enormously flexible and robust. The city vista looks the same whether we view it with both eyes or with only the left or the right eye. A change in position may reveal Golden Gate Park in the distance, but we readily recognize that we are looking at the same city. The percept remains stable even if we stand on our head. The retinal image may be inverted, but we readily attribute this change to our inverted viewing position. We do not see the world as upside-down.

The product of perception is intimately interwoven with memory. Object recognition is more than linking features to form a coherent whole. That whole triggers memories. For those of us who have spent many hours roaming the hilly streets of San Francisco, we recognize that the pictures in Figure 6.1 were taken from the Marin headlands to the north of the city. Even if we have never been to San Francisco, there is an interplay of perception and memory. For the traveler arriving from Australia, the first view of San Francisco is likely to evoke comparisons to Sydney; for the traveler from Kansas, the vista may be so unusual that she recognizes it as such: a place unlike any other she has seen.

This chapter delves into the cognitive neuroscience of object recognition. We consider the problems inherent in a computational system that not only processes sensory information but also links this information to memory.

To begin with, we describe a patient who lost the ability to recognize visually presented objects. Surprisingly, this deficit cannot be attributed to any sort of sensory problem. Patient G.S. retained all the fundamental capabilities for identifying shapes and colors. He was able to use this information to navigate through busy city streets, use silverware to feed himself, or copy complex drawings. But these abilities obscured a severe limitation in G.S.'s use of visual information. He frequently failed to recognize the very objects that he was drawing! Moreover, G.S.'s performance on other visual tasks was not impaired. For example, he had no difficulty recognizing his friends. Paradoxes such as these raise many intriguing questions about how the brain represents and stores knowledge of the world. Are there separate representational systems for different types of information such as objects and faces? Do the sensory modalities have their own memory systems, or do they access a modality-independent knowledge base? At an even more fundamental level, the case of patient G.S. forces us to be precise when we use terms like *perceive* or *recognize*. Perception and recognition do not appear to be unitary phenomena but are manifest in many guises. This problem is one of the core issues in cognitive neuroscience. Not only will this issue be central in this chapter, but also it will reappear when we turn to problems of attention, memory, and consciousness.

Figure 6.1 Our view of the world depends on our vantage point. These two photographs are taken of the same scene, but from two different positions and under two different conditions. Each vantage point reveals new views of the scene, including objects that were obscured from the other vantage point. Moreover, the colors change, depending on the time of day and weather. Despite this variability, we easily recognize that both of the photographs are of San Francisco and the Golden Gate Bridge.

AGNOSIA: A CASE STUDY

Failures of perception can happen even when processes such as the analysis of color, shape, and motion are intact. This disorder is referred to as **agnosia.** The label was coined by Sigmund Freud and is derived from the Greek word *gnostic,* meaning "to know." To be agnosic means to experience a failure of knowledge, or recognition. When the disorder is limited to the visual modality, the syndrome is referred to as **visual agnosia.** Patient G.S. had one form of agnosia. He had had a stroke while still in his thirties. Though the stroke was initially life-threatening, he eventually recovered most of his cognitive functions. On examination a few years after the stroke occurred, his sensory abilities were intact. Language function was normal and there were no problems in coordination, yet G.S. had severe problems recognizing objects. When shown household objects such as a candle or a salad bowl, he was unable to name them. His deficits were even more marked when he viewed photographs of the objects.

G.S.'s problems did not seem to result from a loss of visual acuity. When presented with two lines, he had no difficulty judging which was longer. Indeed, he could describe the color and general shape of stimuli without hesitation. The candle was described as elongated, the salad bowl as curved. His deficit also did not reflect an inability to retrieve verbal labels of objects, as was demonstrated in two ways: First, if the experimenter asked G.S. the name of a round, wooden object in which lettuce, tomatoes, and cucumbers are mixed, he would respond "salad bowl;" second, G.S. could identify objects by using other senses such as touch or smell. When presented with a candle, he reported that it was a "long object." Upon touching it, he labeled it a crayon, but he corrected himself after smelling it and came up with the correct answer of "candle." Thus, G.S.'s deficit was modality-specific. His agnosia was limited to the visual mode.

As with many neuropsychological labels, the term *visual agnosia* has been applied to a number of distinct disorders. As we see in this chapter, by focusing on these subtypes of agnosia, cognitive neuroscientists have been able to develop detailed models of the processes involved in object recognition. In some patients, the problem is one of developing a coherent percept; in others, the agnosia reflects an inability to access conceptual knowledge of perceived objects and use this knowledge to identify them. G.S.'s problem seems to be more closely related to this latter type of agnosia. However, despite his relatively uniform difficulty in identifying visually presented objects, other aspects of his performance indicated that in some situations, this knowledge was being accessed.

Consider what happened when G.S. was shown a picture of a combination lock. At first, he failed to respond. He then noted the round shape but did not describe any of the details of the lock's face. When prompted by experimenters to make a guess, G.S. reported that the picture was of a telephone. Indeed, he remained adamant, insisting that the picture was of a telephone, even after the experimenter emphatically informed him that the picture did not contain a telephone. G.S.'s choice of a telephone was not random. He had perceived the numeric markings around the lock's circumference, and this was enough to make him believe that it was a telephone.

G.S.'s actions also indicated that his knowledge of the stimulus went beyond his erroneous report of a telephone. While viewing the picture, he kept twirling his fingers, pantomiming the actions one would make when opening a combination lock (Figure 6.2). When asked about this, G.S. reported that it was a nervous habit, something to do to keep his hands busy so he would not reach out and touch the picture. Finally, the experimenter asked G.S. if the picture was of a "telephone, lock, or radio." By this time, G.S. was convinced from the experimenter's responses that the picture was not of a telephone. After a moment's hesitation, he responded "clock." And then, after a look at his fingers, he proudly announced, "It's a lock, a combination lock." Truly a case where he had let his fingers do the talking!

A similar thing happened when he saw a picture of a clarinet. Partial retrieval was evident by his initial report that the picture was of a "flute." But again his actions indicated otherwise. Rather than adopt the posture for playing a flute, G.S. configured his hands as if he were playing a clarinet. When told that "flute" was incorrect, G.S. quickly changed his response to "clarinet." We will return to this interesting dissociation of awareness and indirect measures of processing in later chapters; for now we focus on the role of the cortex in making sense of the visible world.

Figure 6.2 When asked to name the depicted object, the patient failed to come up with the correct answer, "combination lock," but rather continued to insist that the object is a telephone. The similar components of a lock and telephone—the numbers and dial—indicate that some information is being processed. In addition, the patient's hands mimed the actions that would be required to open a combination lock.

TWO CORTICAL PATHWAYS FOR VISUAL PERCEPTION

The pathways carrying visual information from the retina to the first few synapses in the cortex clearly segregate into multiple processing streams. Early on there is partitioning into magnocellular and parvocellular pathways, followed by differential projection of the latter to blob and interblob zones in the primary visual cortex. But once past the initial cortical regions, convergence and divergence become the anatomical rules (see Figure 5.10). This complex wiring diagram presents a daunting challenge to theorists interested in the functional organizational principles of visual perception. One hypothesis, though, is inspired by the fact that the output from the occipital lobe is primarily contained in two major fiber bundles, or *fasciculi*. As shown in Figure 6.3, the inferior longitudinal fasciculus follows a ventral route into the temporal lobe. The superior longitudinal fasciculus takes a more dorsal path, with most of its terminations in the posterior regions of the parietal lobe.

Two researchers at the National Institutes of Health in Washington, D.C., Leslie Ungerleider and Mortimer Mishkin (1982), proposed that processing along these two pathways is designed to extract fundamentally different types of information. The **ventral** or **occipito-temporal pathway** is specialized for object perception and recognition, for determining what it is we are looking at. The **dorsal** or **occipito-parietal pathway** is specialized for spatial perception, for determining where an object is, and for analyzing the spatial configuration between different objects in a scene. "What?" and "Where?" are the two basic questions to be answered in visual perception. Not only must we recognize what we are looking at, but also we need to know where it is in order to respond appropriately.

Simple discrimination learning tasks have been used in primate research to demonstrate functional dissociations of the ventral and dorsal processing pathways. Walter Pohl (1973) of the University of

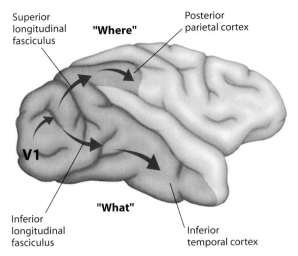

Figure 6.3 The what-where pathways for object recognition. The outputs from the primary visual cortex (V1) follow two general pathways. The superior longitudinal fasciculus includes axons terminating in the posterior parietal cortex, a region associated with identifying the location of the object (where). The inferior longitudinal fasciculus contains axons terminating in the inferior temporal cortex, a region implicated in object recognition (what).

Wisconsin trained rhesus monkeys to do two spatial learning tasks. The landmark discrimination task, the first of Pohl's tasks, was to retrieve food from two food wells, recessed into a tabletop. In this task, shown in Figure 6.4, the well located near a small cylinder was always baited and the other was empty. The position of the cylinder was randomized: Sometimes it was next to the well on the right and sometimes next to the well on the left. Each day the animals were given thirty trials, a simple task for humans but difficult for other primates. The animals had a strong tendency to probe the well that contained the reward in preceding trials. Control monkeys averaged 150 errors over successive sessions before reaching a criterion of twenty-eight out of thirty correct responses.

Once the discrimination had been learned well enough to meet the criterion, the reward was reversed: The food was placed in the well farthest from the cylinder. Learning would then contain this new contingency until the animals reached criterion, at which point the contingency would reverse again. The experiment continued until the animal successfully completed seven reversals.

The second task, the object discrimination task, was quite similar, with one critical difference. Rather than have a single object near one of the food wells,

the displays now contained two objects, a cylinder and a cube. Each object was located near a food well, and the left-right position of the objects was randomly varied from trial to trial. Working with different animals, training began with the well near the cylinder being baited. After the animal reached criterion, the contingency was reversed. Thus, if the cube is not taken into consideration, the animals' training is essentially identical in the two tasks: First, they must learn to respond to the food well closest to the cylinder and then they must learn to respond to the food well farthest from the cylinder.

Despite such similarities, Pohl found a double dissociation between the effects of temporal lobe and parietal lobe lesions. Bilateral lesions of the parietal lobe selectively disrupted performance on the landmark discrimination task. The animals learned the initial contingency but failed to improve with each reversal. In contrast, monkeys with bilateral lesions of the temporal lobes demonstrated impairment on the object discrimination task. They were slow to learn the initial contingency and took more trials than did the animals with parietal lobe lesions to switch once the contingency was reversed.

In Ungerleider and Mishkin's terminology, the temporal lobe lesions disrupted the "what" pathway. These animals had great difficulty discriminating between the cylinder and the cube. Recognizing "what" the object is, though, is not needed in the landmark discrimination task. Here the task depends on a "where" computation. The animal needs only to perceive the single object's position in the display and to respond on the basis of its position relative to the food wells. Animals with parietal lobe lesions could not learn the relative position contingencies: The "where" pathway was malfunctioning.

Subsequent lesion work confirmed the what-where dissociation between ventral and dorsal pathways. This research also showed that at least for the ventral pathway, object discrimination deficits are restricted to the visual modality. Moreover, lesion studies indicated a dissociation between posterior and anterior regions of the temporal lobe. Whereas damage to the posterior region disrupts performance on visual discrimination tasks, damage to the anterior region impairs visual memory.

Bear in mind that deficits in these experiments are apparent only in the presence of bilateral lesions, especially ones in the temporal lobe. One explanation is that while information is segregated within each hemisphere, the monkeys scan the entire visual field; thus, stimuli fall within the left and right visual fields.

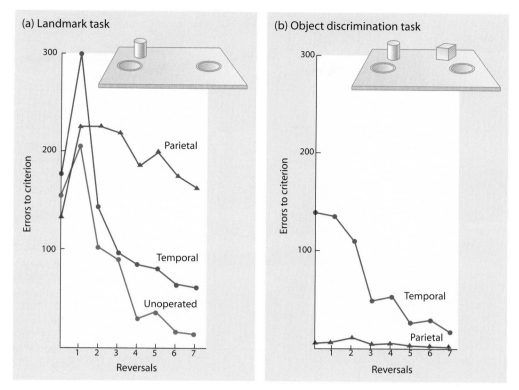

Figure 6.4 Double dissociation in support of the what–where dichotomy. **(a)** In the landmark task, the monkey initially finds a reward in the food well closest to the cylinder. Once this association is learned, the rule is reversed so that now the food is always placed in the well farthest from the cylinder. While all the animals have difficulty following the first few reversals, the control animals as well as those with bilateral temporal lobe lesions show significant improvements with subsequent reversals. Animals with bilateral parietal lobe lesions fail to improve. **(b)** In the object discrimination task, the location of the food is associated with one of the objects. Now the animals with temporal lobe damage show more impairment than those with parietal lobe lesions. Control animals were not tested on the object discrimination task. Adapted from Pohl (1973).

Alternatively, cells in the temporal cortex can respond to stimuli in either visual field as long as those stimuli are not presented far off in the periphery. Bilateral responsiveness happens because visual areas in the temporal lobe are innervated by ipsilateral projections from occipital areas such as V2 and V4, and by callosal projections from the contralateral hemisphere.

Experiments with combination lesions provided convincing evidence that the "what" pathway involves bilateral inputs. For example, animals were tested with an object discrimination task after lesions were created in the right striate cortex and left inferior temporal cortex. The striate lesions rendered the animals cortically blind to any stimuli falling within the left visual field. Thus, the pathway providing intrahemispheric input to the right temporal lobe was silenced. With a lesioned left temporal lobe, these animals are deprived of all temporal lobe function if all the connections are intrahemispheric. But this combination lesion produced little or no deficit in performance.

When the callosal fibers were cut, though, the animals' performance immediately fell to chance (Figure 6.5). After combined right striate–left temporal lobe lesions were made, the intact right temporal lobe processed information received from the contralateral left striate cortex. This input was blocked after the callosal resection.

Combination lesions of the dorsal pathway revealed a slightly different situation. Unilateral parietal lobe lesions produced only mild deficits in the landmark reversal task. When these lesions were followed by a second lesion in the contralateral striate cortex, performance was greatly disrupted. The animals relearned the task but needed extensive training. Sectioning the corpus callosum disrupted performance again, yet the deficit was not as marked as after striate lesions were made. Thus, while callosal fibers are essential for discriminating objects in the setting of opposite-side striate and temporal lobe lesions, these fibers are less important in transmitting location information from the contralateral hemisphere—results

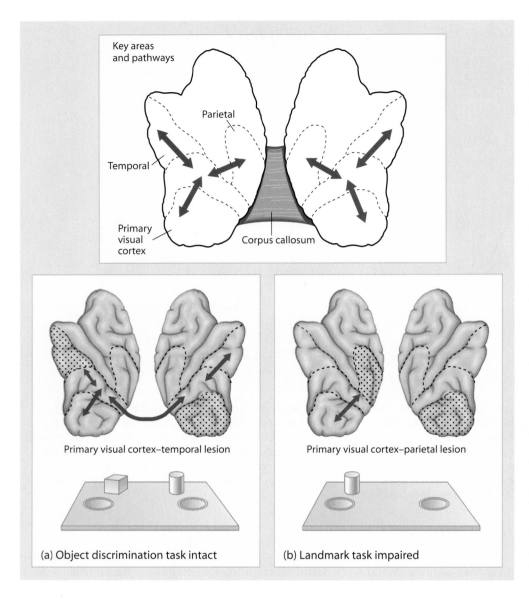

Key areas and pathways

Parietal

Temporal

Primary visual cortex

Corpus callosum

Primary visual cortex–temporal lesion

Primary visual cortex–parietal lesion

(a) Object discrimination task intact

(b) Landmark task impaired

Figure 6.5 Combination lesion studies indicated that connections over the corpus callosum are more important for transmitting information related to "what" processing. **(a)** Monkeys were given a primary visual cortex lesion on one side and a temporal lobe lesion on the other side. If each hemisphere only received input from the ipsilateral visual cortex, then the animal should demonstrate severe impairment on the object discrimination task since the intact temporal lobe would be deprived of all input. However, animals with this combination of lesions showed little deficit. Further evidence of the importance of callosal fibers for interhemispheric "what" processing is given by the fact that cutting the corpus callosum in these animals severely disrupted performance. **(b)** A combined parietal lobe and visual cortex lesion led to a severe deficit on the landmark task, suggesting that "where" processing is primarily intrahemispheric. Adapted from Ungerleider and Mishkin (1982).

that agree with physiological studies showing that the majority of parietal lobe neurons have unilateral receptive fields. Processing within the "where" pathway appears to be more segregated within each hemisphere.

Representational Differences Between the Dorsal and Ventral Pathways

The physiological properties of neurons within the temporal and parietal lobes are quite distinct. Neurons in both lobes have large receptive fields. Neurons in the parietal lobe, however, have an interesting property in that they can respond in a nonselective way (Robinson et al., 1978). For example, a parietal neuron, recorded in an awake monkey, might be activated when a stimulus such as a spot of light is restricted to a small region of space or when the stimulus is a large object that encom-

passes much of the hemifield (Figure 6.6). In addition, many parietal neurons are responsive to stimuli presented in the more eccentric parts of the visual field. While 40% of these neurons have receptive fields near the central region of vision, the fovea, the remaining cells have receptive fields that exclude the foveal region.

These eccentrically tuned cells are ideally suited for detecting the presence of a stimulus, especially one that has just entered the field of view. Remember that in examining subcortical vision, we suggested a similar role for the superior colliculus. Chapter 7 reviews the important role of the fovea and superior colliculus in visual attention.

The response of neurons in the temporal lobe is quite different (Ito et al., 1995). The receptive fields for these neurons always encompass the fovea, and the majority of these neurons can be activated by a stimulus

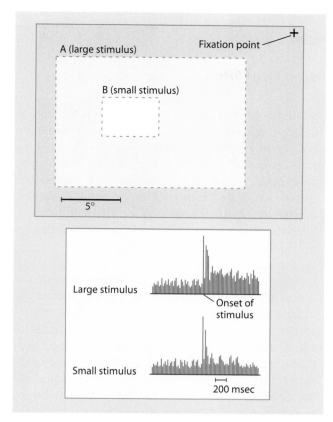

Figure 6.6 Single-cell recordings from a neuron in the posterior parietal cortex of the monkey. The receptive field of the cell was centered in the lower quadrant of the left visual field. While fixating, the animal was presented with either a large or a small stimulus. The cell responds to both stimuli, although the magnitude of activity is correlated with the size of the stimulus.

that falls within either the left or the right visual field. The disproportionate representation of central vision appears to be ideal for a system devoted to object recognition. We usually look directly at things we wish to identify, thereby taking advantage of the greater acuity of foveal vision.

Cells within the visual areas of the temporal lobe have a diverse pattern of selectivity. Robert Desimone (1991), also working at the National Institutes of Health, studied these cells in detail. One study employed a wide range of stimuli. Some stimuli were simple, involving edges or bars at orientations that varied in color or brightness. Others were complex and included photographs and three-dimensional models of objects like a human head, hand, apple, flower, and snake. Of the 151 cells sampled, 110 consistently responded to at least one of the stimuli. A large minority (41%) of these were similar to parietal neurons: They were activated by any of the stimuli, and their firing rates were similar across the set of stimuli. The remaining 59% of the cells exhibited some selectivity and responded more vigorously when presented with the complex stimuli.

Recording from such a cell, located in the inferior temporal cortex, are shown in Figure 6.7. This cell is most highly activated by a model of the human hand. The top row shows the response of the cell to views of a hand. Activity is high regardless of the hand's orientation and is only slightly reduced when the hand is considerably

Figure 6.7 Single-cell recordings from a neuron in the inferior temporal cortex. These cells rarely respond to simple stimuli such as lines or spots of light. Rather, they respond to more complex objects such as the hand drawings shown in the top row. Note that the cell shows only a weak response to the mitten, indicating that its activity is not associated with the general shape of a hand. This cell does not respond to the comblike objects that contain the series of parallel lines as in the hand stimuli. Adapted from Desimone et al. (1984).

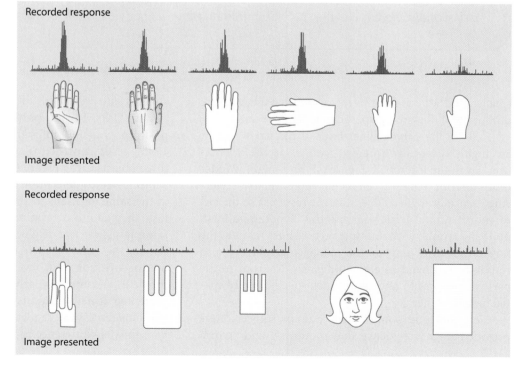

Now You See It, Now You Don't

Gaze at the picture in Figure A for a couple of minutes. If you are like most people, you initially saw a vase. But, surprise! After a while, the vase changed to a picture of two human profiles staring at each other. With continued viewing, your perception changes back and forth, content with one interpretation until suddenly, the other asserts itself and refuses to yield the floor. This is an example of *multistable perception*.

How do multistable percepts get resolved in the brain? The stimulus information does not change at the points of transition. Rather, the interpretation of the pictorial cues changes. When staring at the white region, you see the vase. If you shift attention to the black regions, then you see the profiles. But here we run into a chicken and egg question. Did the representation of individual features change first and thus cause the percept to change? Or did the percept change and lead to a reinterpretation of the features?

To explore these questions, Nikos Logothetis of the Max Planck Institute in Tuebingen, Germany, has turned to a different form of multistable perception, binocular rivalry (Sheinberg and Logothetis, 1997). The exquisite focusing capability of our eyes (perhaps assisted by an optometrist) makes us forget that they provide two separate snapshots of the world. These snapshots are only slightly different, and they provide important cues for depth perception. However, with some technological tricks, it is possible to present radically different inputs to the two eyes. To accomplish this while keeping the eyes fixated at a common location, researchers employ special glasses that alternately block the input to one eye and then the other at very rapid rates. Varying the stimulus in synchrony with the alternate blocking by these glasses allows a different stimulus to be presented to each eye simultaneously.

But do we see two things simultaneously at the same location? The answer is no. As with the ambiguous vase/face profiles picture, only one object or the other is seen at any one point in time, although at transitions there is sometimes a period of fuzziness in which neither object is clearly perceived. Logothetis trained his monkeys to press one of two levers to indicate which object was being perceived. To make sure the animals were not responding randomly, he included nonrivalrous trials in which only one of the objects was presented. Recordings were then made from single cells in various areas of the visual cortex. Within each area, he would select two objects, only one of which was effective in driving the cell. In this way, he could correlate the activity of the cell with the animal's perceptual experience.

As his recordings moved up the ventral pathway, Logothetis found an increase in the percentage of ac-

smaller. The last panel in this row shows that the response diminishes if the same shape lacks defining fingers.

Cells in the parvocellular layer of the striate cortex are highly responsive to edges with a specific orientation. From the top row in Figure 6.7, you might wonder if this inferior temporal cell is behaving similarly, preferring vertically aligned edges. But the response of the cell to other stimuli argues against this hypothesis. First, nonhand stimuli with multiple vertical lines (akin to the fingers) failed to activate the cell. Second, the cell also responds to a hand rotated 90 degrees. Neurons in the parietal lobe would not exhibit such specificity and spatial independence.

Neuroimaging studies with human subjects have provided further evidence that the dorsal and ventral streams are activated differentially by "where" and "what" tasks. In one particularly elegant study using positron emission tomography (PET) (Kohler et al., 1995), trials consisted of pairs of displays containing three objects each (Figure 6.8). In the position task, the subjects had to determine if the objects were presented at the same locations in the two displays. In the object task, they had to determine if the objects remained the same across the two displays. The irrelevant factor could remain the same or change: The objects might change on the position task, even though the locations remained the same; similarly, the same objects might be presented at new locations in the object task. Thus, the stimulus displays were identical for the two conditions, with the only difference being the task instructions.

tive cells, with activity mirroring the animals' perception rather than the stimulus conditions. In V1, the responses of less than 20% of the cells fluctuated as a function of whether the animal perceived the effective or ineffective stimulus. In V4, this percentage increased to over 33%. In contrast, the activity of all the cells in the visual areas of the temporal lobe was tightly correlated with the animal's perception. Here, the cells would respond only when the effective stimulus, the monkey face, was perceived (Figure B). When the animal pressed the lever indicating that it perceived the ineffective stimulus, the starburst, under rivalrous conditions, the cells were essentially silent. In both V4 and the temporal lobe, the change in cell activity occurred in advance of the animal's response that the percept had changed. Thus, even when the stimulus did not change, an increase in activity was observed prior to the transition from a perception of the ineffective stimulus to that of the effective stimulus.

These results suggest a competition during the early stages of cortical processing between the two possible percepts. The activity of the cells in V1 and in V4 can be thought of as perceptual hypotheses, with the patterns across an ensemble of cells reflecting the strength of the different hypotheses. Interactions between these cells ensure that by the time the information reaches the inferior temporal lobe, one of these hypotheses has coalesced into a stable percept. Reflecting the properties of the real world, the brain is not fooled into believing two objects exist at the same place at the same time.

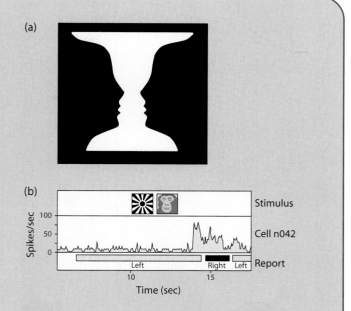

Figure A   Does your perception change over time as you continue to stare at this drawing?
Figure B   When the starburst or monkey face is presented alone, the cell in the temporal cortex responds vigorously to the monkey face but not to the starburst. In the rivalrous condition, the two stimuli are presented simultaneously, one to the left eye and one to the right eye. The bottom bar shows the monkey's perception, indicated by a lever press. About 1 second after the onset of the rivalrous stimulus, the animal perceives the starburst. The cell is silent during this period. About 7 seconds later, the cell shows a large increase in activity and, correspondingly, indicates that its perception has changed to the monkey face shortly thereafter. Two seconds later the percept flips back to the starburst and the cell's activity is again reduced.

The PET data for the two tasks were compared directly to identify neural regions that were selectively activated by one task or the other. In this way, areas that are engaged similarly for both tasks, either because of similar perceptual, decision, or response requirements, are masked. During the position task, regional cerebral blood flow was higher in the parietal lobe in the right hemisphere. In contrast, the object task led to increased regional cerebral blood flow bilaterally at the junction of the occipital and temporal lobes.

A number of other imaging studies have confirmed the dorsal-ventral distinction, although there was considerable variability across the studies in the exact neural loci that were activated within the parietal and occipital-temporal cortex. Two sources likely produced this variability. First, the methods of analysis varied across the experiments. Many studies used a passive viewing condition for their baseline rather than direct comparisons between two experimental conditions. Comparisons of this type generally lead to larger areas of bilateral activation. For example, both inferior and superior regions of the parietal lobe were activated bilaterally when subjects were required to do a spatial matching task that requires mental rotation (Haxby et al., 1994).

Second, human subjects are clever creatures and can easily adopt a variety of strategies to solve a task. Consider the object task in Figure 6.8. Do subjects make their decisions by comparing the objects in the second display with an imagined, perceptually based

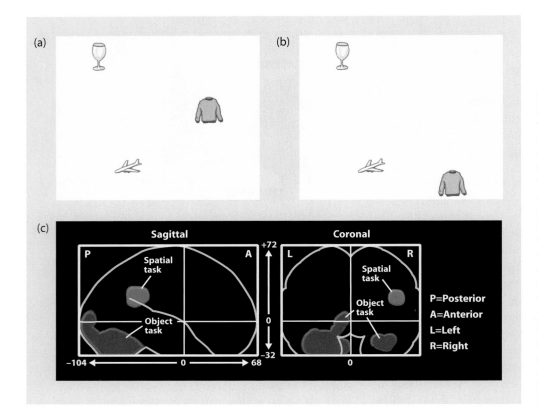

Figure 6.8 Same-different matching task used to contrast spatial and object discrimination. **(a)** Sample stimulus. **(b)** Test stimulus in which a same response would be required in the object task and a different response would be required in the spatial task. **(c)** Sagittal section showing activation centers for the object task (red) and spatial task (green). While both activations are superimposed on the same slice, activation during the spatial task was restricted to the right hemisphere whereas activation during the object task was bilateral. Adapted from Kohler et al. (1995).

representation from the first display? Or do they generate the names of the objects and make their comparison based on these verbal labels? Isolating specific mental operations can be a tricky business in imaging studies.

The design of one study that attempted to isolate specific mental operations (Kanwisher et al., 1997) is outlined in Figure 6.9. While in the PET scanner, subjects were presented with three types of line drawings. To eliminate activations related to decision and response processes, testing always involved only passive viewing conditions. In one condition, the lines were scrambled. In the other two conditions, the line drawings formed objects that were either familiar or novel. All three conditions should engage processes involved in initial visual perception, or what is labeled *feature extraction*. To identify areas involved in object perception, activation during presentation of the scrambled condition was subtracted from that measured during the novel and familiar object conditions. If passive object perception also engaged memory retrieval processes, then additional areas should be activated during the familiar object condition compared to the novel object condition. In fact, no such differences were found. However, both the novel and familiar object conditions led to increases in regional cerebral

blood flow bilaterally in the inferior occipitotemporal areas. The center of activation was lateral and inferior to V4 and likely constituted upstream projections of the parvocellular pathway.

Perception for Identification Versus Perception for Action

Patient studies offer even more support for a dissociation of what-where processing, although there are discrepancies between the findings from human and primate lesion studies. The parietal cortex is central to spatial attention. Lesions of this lobe can produce severe disturbances in the ability to represent the world's spatial layout and the spatial relations of objects in it. Lesions associated with agnosia do not have a clear pattern that corresponds to findings in animal studies. The lesions often fall along the ventral pathway, especially in patients with impaired face perception, or **prosopagnosia,** but they frequently extend into parietal structures. Unlike examples from animal research, patients with severe deficits in object recognition can have unilateral lesions of either the right or the left hemisphere.

More revealing have been functional dissociations in the performance of patients with agnosia. Mel

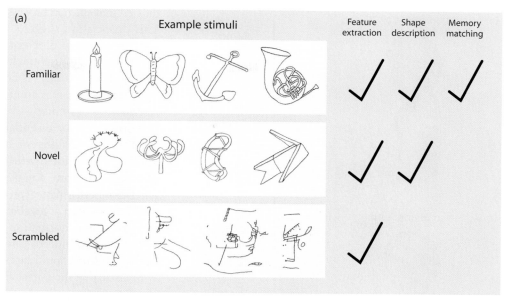

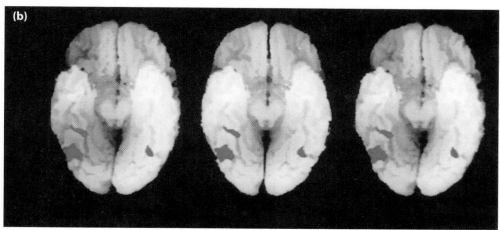

Figure 6.9 Component analysis of object recognition. **(a)** Stimuli for the three conditions and the list of required mental operations. Novel objects are hypothesized to engage processes involved in perception even when verbal labels do not exist. **(b)** Activation for both novel and familiar objects was bilateral along the ventral surface of the occipitotemporal cortex. From Kanwisher et al. (1997).

Goodale and David Milner (1992) at the University of Western Ontario described a 34-year-old patient, D.F., who suffered carbon monoxide intoxication due to a leaky propane gas heater. Although D.F.'s computed tomography (CT) scans appeared normal, magnetic resonance imaging (MRI) revealed bilateral lesions in the occipital lobes. D.F. had a severe disorder of object recognition. When asked to name household items, she made errors such as labeling a cup an ashtray or a fork a knife. She usually gave crude descriptions of a displayed object; for example, a screwdriver was "long, black, and thin." Picture recognition was even more disrupted. When shown drawings of common objects, she could not identify a single one. Her deficit could not be attributed to *anomia,* a problem with naming objects; whenever an object was placed in her hand, she identified it. Sensory testing further indicated that D.F.'s agnosia could not be attributed to a loss of visual acuity.

She could detect small gray targets displayed against a black background. While her ability to discriminate small differences in hue was abnormal, she correctly identified primary colors.

Most relevant to our concerns is the dissociation of her performance on two tasks designed to assess her ability to perceive the orientation of a three-dimensional object. For these tasks, D.F. was asked to view a circular block into which a slot had been cut. The orientation of the slot could be varied by rotating the block. In the recognition task, D.F. was given a card and asked to orient her hand so the card would fit into the slot. As can be seen in Figure 6.10, she failed miserably, orienting the card vertically even when the slot was horizontal. When asked to insert the card into the slot, D.F. quickly reached forward and inserted the card. Her performance on this visuomotor version did not depend on tactile feedback that would result when the

(a) Perception condition

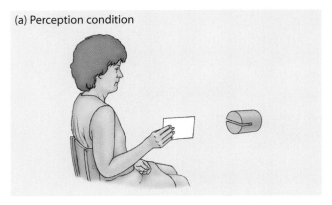

(b) Action condition

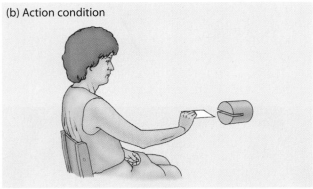

(c) Memory (recall)

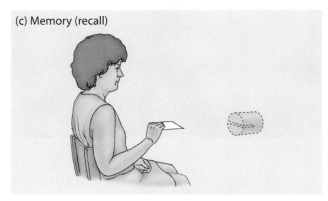

Figure 6.10 A dissociation of perception linked to aware-ness and perception linked to action. **(a)** When asked to match the orientation of the card to that of the slot, the patient demonstrated severe impairment. **(b)** When instructed to in-sert the card in the slot, the patient produced the correct ac-tion without hesitation. **(c)** The patient also did not show any impairment in the memory condition, verifying that her knowledge of orientation was intact.

card contacted the slot. She simply oriented her hand prior to touching the block.

D.F.'s performance shows that processing systems make use of different sources of perceptual information. From the first task, it was clear that D.F. could not recog-nize the orientation of a three-dimensional object. This deficit is indicative of her severe agnosia. Yet when D.F. was asked to insert the card, her performance clearly in-

dicated that she had processed the orientation of the slot. This dissociation suggests that the "what" and "where" systems support different aspects of cognition. The "what" system is essential for determining the identity of an object. If the object is familiar, we will recognize it as such; if novel, we may compare the percept to stored rep-resentations of similarly shaped objects. The "where" sys-tem appears to be essential for more than determining the location of different objects; it is also critical for guiding interactions with these objects. Similar to the blindsight phenomenon described in the previous chap-ter, D.F.'s performance provides a another example of how information accessible to action systems can be dis-sociated from information accessible to knowledge and consciousness. Indeed, Goodale and Milner argued that the dichotomy should be between "what" and "how," to emphasize that the dorsal visual system provides a strong input to motor systems to compute how a movement should be produced.

The opposite dissociation can be found in the clinical literature. Patients with **optic ataxia** can recognize ob-jects yet cannot use visual information to guide their action. When reaching for an object, they fail to move di-rectly toward it, but they grope about like a person trying to find a light switch in the dark. Their eye movements present a similar loss of spatial knowledge; *saccades*, or directed eye movements, may be directed inappropri-ately and fail to bring the object within the fovea. Such patients have the inverse problem of D.F.'s: They can re-port the orientation of a visual slot even though they cannot use this information for moving their hand to-ward the slot. In accord with what we expect based on dorsal-ventral dichotomy, optic ataxia is associated with lesions of the parietal cortex.

In sum, the what-where, or what-how dichotomy offers a functional account of two computational goals for higher visual processing. This distinction is best viewed as a heuristic one rather than reflecting an ab-solute distinction. The dorsal and ventral pathways are not isolated from one another but communicate exten-sively. Processing within the parietal lobe, the termina-tion of the "where" pathway, serves many purposes. We focused here on its guiding of action; in the next chapter we will see that the parietal lobe plays a critical role in *selective attention*, the enhancement of processing at some locations instead of others. Moreover, spatial in-formation also can be useful for solving "what" prob-lems. For example, depth cues help to segregate a complex scene into its component objects. The remain-der of this chapter concentrates on object recognition: how the visual system has an assortment of strategies for perceiving and recognizing the world.

COMPUTATIONAL PROBLEMS
IN OBJECT RECOGNITION

Object perception depends primarily on the analysis of the shape of a visual stimulus, although cues such as color, texture, and motion certainly contribute to normal perception. For example, when we look at the surf breaking on the shore, our acuity is not sufficient to see grains of sand, and water is essentially amorphous, lacking any definable shape. Yet the texture of the sand's surface and the water's edge, and their differences in color, enable us to distinguish between the two regions. The water's motion is important, too. But even if surface features like texture and color are absent or applied inappropriately, recognition is minimally affected: We can readily identify pink elephants, striped apples, and stick figure drawings in Figure 6.11. Here object recognition is derived from a perceptual ability to match an analysis of shape and form.

To account for shape-based recognition, consider two things. The first has to do with shape encoding. How is a shape internally represented? What salient features enable us to recognize differences between a triangle and a square or a monkey and a person? The second centers on how shape is processed when the perceiver's viewing position is rarely constant. We can recognize shapes from an infinite array of positions and orientations, and our recognition system is not hampered by scale changes in the retinal image as we move close to or far away from an object.

Variability in Sensory Information

Questions about shape encoding and processing arise because our perceptual capabilities must be maintained in the face of a variable world. **Object constancy** refers to the amazing ability to recognize an object in count-less situations. Figure 6.12 shows four drawings of an automobile, each having little in common with respect to sensory information reaching the eye. And yet we have no problem identifying the car in each picture and discerning that all four cars are the same model. Object constancy is thus essential for perception. Imagine how difficult life would be if we could not recognize familiar things or people unless we gazed at them from a specific point of view.

Variability in the visual information emanating from an object arises for several reasons. First, sensory information depends highly on viewing position. Changes in viewpoint arise not only as the perceiver moves about the world but also as objects change their orientation. If we command a dog to roll over, our interpretation of the object (the dog) remains the same despite the change in our perception. The retinal projection of shape, and even the visible components of an object, can change dramatically as the object's and viewer's positions alter. For instance, when a car is viewed from the front, the retinal projection of its width is larger than the projection of its length. Yet our percept does not reflect this distortion. We do not have the impression of a squashed car. Our perceptual system is adept at separating changes caused by shifts in viewpoint from those intrinsic to the objects themselves.

Changes in the illumination of an object introduce a second source of sensory variability (Figure 6.13). The visible parts of an object may differ depending on whether the object is illuminated from above or from the side. Variations in shadowing can also alter the illumination of an object's parts. Because so many visual cells are sensitive to brightness, we might expect them to have an impact on perception, but recognition is mainly insensitive to changes in illumination.

Figure 6.11 Despite the irregularities in how these objects are depicted, we have little problem in recognizing them. We may never have seen pink elephants or plaid apples, but our object recognition system can still discern the essential features.

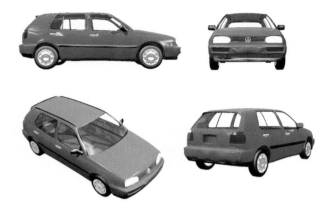

Figure 6.12 The image on the retina is vastly different for these four drawings of a car. Despite this sensory variability, our phenomenology is that we rapidly recognize that the drawings are of the same car.

A third source of variability arises because objects are rarely seen in isolation. While cars can be drawn on a blank page, they are usually seen on streets with pedestrians, buildings, and traffic signs—and these other objects frequently occlude part of the car. We have no trouble separating other objects from a car; our perceptual system quickly partitions the scene into components.

While object recognition must overcome these three sources of variability, it also must accommodate the fact that changes in perceived shape can reflect changes in the object. Not only must object recognition be general enough to support object constancy, but also it must be specific enough to pick out slight differences between members of a category or class.

View-Dependent or View-Invariant Recognition?

A central debate in object recognition has to do with defining the frame of reference where recognition oc-

Figure 6.13 Object constancy must be achieved in spite of the many sources of variation in the sensory input, including shadows (**left**) and occlusion (**right**).

curs (Perrett et al., 1994). Two general approaches have been proposed. In **view-dependent theories,** perception is assumed to depend on recognizing an object from a certain viewpoint. According to this theory, perception proceeds from analyzing a viewpoint's information (Figure 6.14). When a bicycle is seen from the side, we may recognize it by the shape of its frame, wheels, and seat. We may also recognize a bicycle when viewed from an aerial perspective, even though this vantage point obscures the wheels. Perhaps from this orientation the handlebars and elongated shape provide the defining components. View-dependent theories posit that we have a cornucopia of specific representations in memory; we simply need to match a stimulus to a stored representation. The key idea is that the stored representation for recognizing a bicycle from the side is different from the one for recognizing a bicycle from above. Hence, our ability to recognize that two stimuli are depicting the same object is assumed to arise at a later stage of processing.

Figure 6.14 View-dependent theories of object recognition posit that recognition processes are dependent on the vantage point. Recognizing that these drawings depict bicycles, one from a side view and the other from an aerial view, requires matching the distinct sensory inputs to view-dependent representations. For the side view, we may include the features of the seat, chain, and spokes to match the drawing with a stored representation of a bicycle seen from this perspective. For the aerial view, the representation would have encompass the axis, seat, handlebars, and pedals.

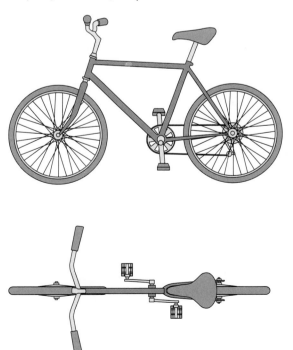

One shortcoming with view-dependent theories is that they would seem to place a heavy burden on perceptual memory. Each object would require multiple representations in memory, each associated with a different vantage point. This problem, however, is less daunting if we assume that recognition processes are able to match the input to stored representations through an interpolation process. Recognition of an object seen from a novel viewpoint occurs by comparing the stimulus information to the stored representations and choosing the best match. This idea is supported by experiments using novel objects, an approach that minimizes the contribution of past experience and the possibility of verbal strategies. The time needed to decide if two objects are the same or different increases as the viewpoints diverge, even when each member of the object set contains a unique feature (Tarr et al., 1997).

An alternative scheme is to posit that recognition occurs in a **view-invariant frame of reference**. With this approach, recognition does not happen by simply analyzing the stimulus information. Rather, sensory input defines basic properties; the object's other properties are defined with respect to these properties. David Marr of MIT, in his book titled *Vision: A Computational Investigation into the Human Representation and Processing of Visual Information* (1982), articulated a comprehensive and influential theory of recognition based on view-invariant representations. In Marr's theory, a critical property for recognition is establishing the major and minor axes inherent to the object. The major axis of a bicycle runs along its length. The handlebars can be represented as the minor axis, two appendages arranged perpendicularly to the primary axis. These properties will generally hold across different vantage points. It is true that if viewed head on, the minor axis produces a larger retinal image than the foreshortened major axis. But Marr would consider this a degenerate case, one that may pose a challenge for our usual effortless recognition systems. In such situations, recognition may depend on an inferential process based on a few salient features.

Shape Encoding

View-dependent and view-independent theories of recognition propose that recognition depends on breaking a scene or object into its component parts. In Chapter 5 we introduced the idea that recognition may involve hierarchical representations in which each successive stage adds more complexity. Features such as lines can be combined into edges, corners, and intersections to parse a visual array into component regions. These, in turn, are grouped into parts and the parts into objects. Recognition of a triangle entails perceiving three seg-

ments that join together to form three corners, which define an enclosed region. Another object could be defined if these same features were combined differently, for example, an arrow derived from three lines connected at their endpoints. But there is only one point of intersection, and the lines do not enclose an area. The triangle and arrow might activate similar representations at the lowest levels of the hierarchy. Yet the combination of these features produces distinct representations at higher levels of the processing hierarchy.

EXPLOITING SALIENT SOURCES OF INFORMATION

What information is coded at these higher levels? We might expect that the representations would include the most salient aspects of the stimulus. Each of the abstract shapes in Figure 6.15 is composed of a single, continuous curved line. In comparing the two shapes, however, it is apparent that some regions of curvature provide more information than others. We are not sensitive to the slight differences in curvature along the right side, but the sharp inflection point along the bottom of one of the figures is very salient. We would describe the shape on top as having a flat bottom; the shape on the bottom

Figure 6.15 Study these two drawings for 10 seconds and then attempt to draw them from memory. Your drawings are likely to reflect the fact that points of greatest curvature provide the most salient sources of information. The knob on the left edge of the top figure is more distinctive than the wavy edge on the right edge. Reproductions of the bottom figure will surely include the gap separating the two appendages.

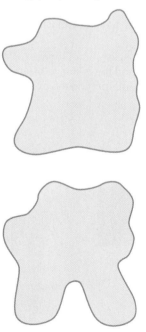

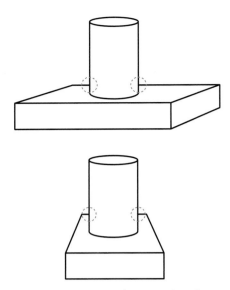

Figure 6.16 Certain perceptual cues are invariant across vantage points. The parallel edges of the rectangle will remain parallel when viewed from any position (given slight distortions due to depth). In contrast, the border between two objects is likely to depend on the viewpoint. In the top drawing, T-junctions are formed by the long edge of the rectangle and the sides of the cylinder. These junctions shift to the short edge of the rectangle when the objects are viewed from the end as in the lower drawing.

resembles an object with two bases of support, or legs. We might use this information to guess that the figure on the bottom is more mobile than the object on top.

Sensory information can change from one viewpoint to another. Nonetheless, the world has regularities that the visual system exploits—sources of information that remain constant across vantage points. Theories of object recognition have emphasized the significance of

these invariant properties. Apples can be red, green, or yellow, but they all share a basic shape. Moreover, invariant cues are often retained in the two-dimensional retinal projection of the three-dimensional world.

Consider the stimulus in the top panel of Figure 6.16. This percept completely depends on analyzing the objects' shapes: No color or texture distinguishes the two parts. Rather, the stimulus is simply the composition of lines at various orientations. Even so, from a glance we see that the scene is composed of two objects, a cylinder perched on a rectangle.

We are able to infer invariant properties on the basis of the depicted information. In this drawing, the simplest invariance has to do with the line segments themselves. In all viewing positions, straight edges on a three-dimensional object will appear as straight in a two-dimensional projection. In a similar manner, curved surfaces are seen as such over a wide range of viewpoints.

Other invariants like parallelism and symmetry bring into play comparisons of features in a stimulus. In Figure 6.16, parallelism is violated in the two-dimensional projection because of perspective. Even so, the visual system infers parallelism, which reflects the ubiquity of this feature in natural objects like trees and riverbanks and in human creations like streets and buildings. Indeed, psychologists take advantage of our bias to infer parallelism in creating compelling illusions (Figure 6.17).

RECOGNITION BY PARTS ANALYSIS

To recognize the cylinder and rectangle in Figure 6.16, it is necessary to resolve the ambiguity introduced at points where the two objects intersect. Our percept is that the cylinder sits on the rectangle and occludes its

Figure 6.17 The illusory Ames room exploits the fact that, given our knowledge of rooms, we expect the back wall to be parallel with the front wall and the ceiling to be at a constant height. Our perceptual system assumes the two people are at the same distance, and ends up interpreting the closer one as a giant.

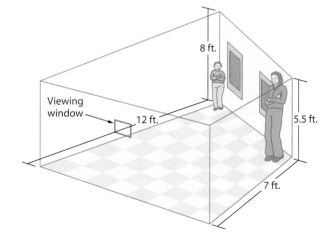

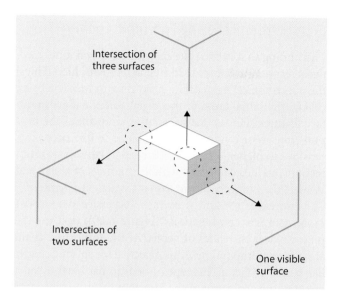

are found in the two-dimensional projection of the rectangle; indeed, the junctions here uniquely define the three-dimensional shape of the object (Figure 6.18).

These properties are invariant. The two-dimensional projection of collinearity and T-junctions will persist as an object rotates, allowing the stimulus to be parsed into its component parts. Recognition is then possible by determining that the stimulus contains defining component parts. A bicycle contains an elongated axis, two wheels, and handlebars. A third wheel and compression of the difference between the major and minor axes would correspond to a tricycle. Perhaps the difficulty in recognizing a bicycle from an aerial vantage point is that one of the critical features, the wheels, is obscured.

But what defines a part? Is it a list of higher-level features such as corners, parallelism, symmetry, and enclosure—invariant properties that can be extracted from the two-dimensional projection of three-dimensional objects? Or, do we use these properties to build representations of defining parts?

Irv Biederman (1990) of the University of Southern California proposed a theory of object recognition that emphasizes a part-to-whole analysis. The central tenet of his theory is that any object can be described as a configuration of limited parts. Indeed, these parts form a perceptual alphabet, a set of simple geometric volumes referred to as *geons,* an acronym for "geometric ions." In the way ions form the building blocks of chemical elements, visual geons are hypothesized to constitute fundamental shapes for object recognition (Figure 6.19). In

Figure 6.18   Invariant cues to surface perception. The intersection of three visible surfaces creates a Y-junction. An arrow and shaft indicate the intersection of two visible surfaces whereas an arrow alone indicates the corner of one visible surface.

rear edge, an interpretation reflecting two cues. First, the fact that the upper borders of the two rear segments are collinear, forming a discontinuous line, is a powerful cue that these borders are part of the same object. Second, the edges of the rectangle and cylinder form a T-junction. Such intersections almost always define two objects (or two parts of a single object such as the base and the top of a table). In contrast, no T-junctions

Figure 6.19   The geon theory posits that object recognition is based on identifying the defining geons, or geometrical "ions," that constitute an object. **(a)** Twelve geon shapes, defined by the shape of the cross section (circular, straight, or varying) and parallelism (long or short axes either parallel or not parallel). **(b)** A set of common objects with their component geons marked. The numbers indicate the geons in the left panel. Adapted from Biedermann (1990).

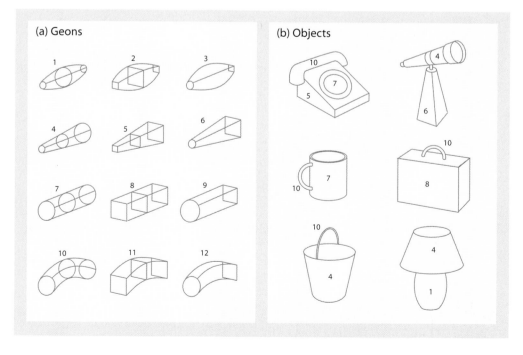

Biederman's theory, twenty-four geons uniquely capture the full set of view-invariant volumes. Objects are defined by their unique set of constituent geons and the spatial relationships between these geons. For example, a cup is composed of two geons, a cylinder with a handle "attached to the side."

This last example also underscores a limitation of geon theory. We have little difficulty recognizing the similarity of two coffee cups, even if one doesn't have a handle. Yet their geon-based descriptions would be quite different. On the other hand, we are able to make fine discriminations between objects that would have near-identical geon-based descriptions. It isn't obvious how geon theory could account for the ease with which we identify a household pet as a dog or cat while still recognizing the perceptual similarity across the various species of dogs. Geon theory emphasizes that an analysis of component parts can be useful for making rough distinctions between object classes. A complete theory of object recognition must also be able to account for our ability to recognize specific objects.

Grandmother Cells and Ensemble Coding

The finding that cells in the inferior temporal lobe selectively respond to complex stimuli agrees with hierarchical theories of object perception. According to these theories, cells in the initial areas of the visual cortex code elementary features such as line orientation and color. The outputs from these cells are then combined to form detectors sensitive to higher-order features such as corners or intersections, an idea consistent with the findings of Hubel and Wiesel (see Pioneers in the Visual Cortex, in Chapter 5). The process is continued as each successive stage codes more complex combinations (Figure 6.20), perhaps the physiological correlates of geons. At the top of the chain are inferior temporal neurons, selective for specific shapes like hands or faces. This type of neuron has been called a **gnostic unit,** referring to the idea that the cells can signal the presence of a known stimulus, an object, place, or animal that has been encountered in the past.

It is tempting to conclude that the cell represented by the recordings in Figure 6.7 signals the presence of a hand, independent of viewpoint. Other cells in the inferior tem-

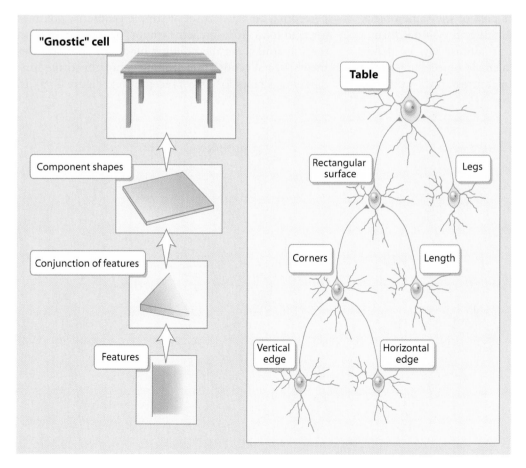

Figure 6.20 Hierarchical coding hypothesis in which elementary features are combined to create gnostic units that recognize complex objects. At the lowest level of the hierarchy are edge detectors, units that operate similar to the simple cells discussed in Chapter 5. These feature units combine to form corner detectors, which in turn are combined to create cells that respond to even more complex stimuli such as surfaces. The left side of the figure shows the hypothesized computational stages for hierarchical coding; the right side is a neural implementation based on the idea of grandmother cells.

poral area respond preferentially to complex stimuli such as jagged contours or fuzzy textures. The latter might be useful for a monkey, to identify that an object has a fur-covered surface, perhaps the backside of another member of its group. Even more intriguing has been the discovery of cells in the inferior temporal gyrus and the floor of the superior temporal sulcus that are selectively activated by faces. In a tongue-in-cheek manner, the term *grandmother cell* has been coined to refer to the notion that there might be gnostic cells that become excited only when one's grandmother comes into view. Other gnostic cells would be specialized to recognize grandpa, or a blue Volkswagon, or the Golden Gate Bridge.

These results raise the question of how specific the responsiveness of an individual cell is. Does recognition depend on the specificity of a few neurons or the aggregate behavior of large groups of them? A physiologist might find that a particular cell exhibits a strong preference for a hand, and be tempted to conclude that this cell is the gnostic unit for that percept. However, it is only possible to test a limited subset of all possible objects when one is recording from any single cell, and the physiologist's explorations are usually limited to a relatively small number of cells, usually on the order of a couple of hundred.

With these limitations in mind, it is useful to consider two troubling problems for the grandmother cell hypothesis. First, the idea of grandmother cells rests on the assumption that the final percept of an object is coded by a single cell. Since cells are in a constant state of spontaneous firing and refractoriness, a coding scheme of this nature would be highly susceptible to error. If a gnostic unit were to die, we would expect to experience a sudden loss for an object. Second, the grandmother cell hypothesis cannot adequately account for the fact that we perceive novel objects, a perception whose mechanism is unexplained.

One alternative to the grandmother cell hypothesis is to regard object recognition as resulting from activation across complex feature detectors (Figure 6.21). Granny, then, is perceived when higher-order neurons are activated; some respond to her shape, others to the

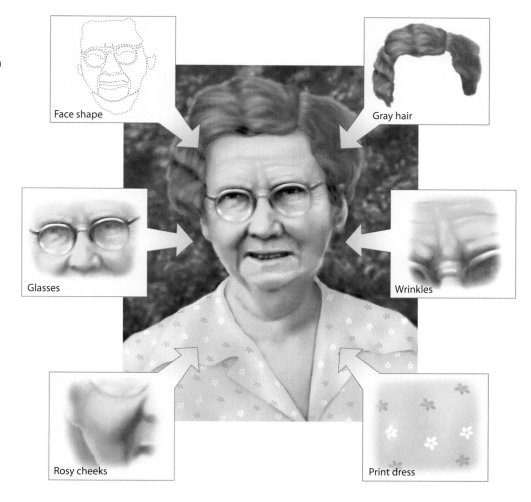

Figure 6.21 Ensemble coding hypothesis in which an object is defined by the simultaneous activation of a set of defining properties. "Granny" is recognized here by the co-occurrence of her glasses, facial shape, hair color, and so on.

Face shape

Gray hair

Glasses

Wrinkles

Rosy cheeks

Print dress

An Interview with Horace Barlow, Ph.D. Dr. Barlow is affiliated with the Department of Physiology at Cambridge University. While neurophysiology has made great inroads in describing the activity of single cells, Dr. Barlow's theoretical discussions of "grandmother" cells have ensured that researchers in this field keep in mind the larger-scale problems faced by a system seeking to recognize objects.

Authors: Do you still believe in the grandmother cell?

HB: I don't believe in grandmother cells and never have, but there are several important associated ideas. I think grandmother cells are the elements of a mutually exclusive representation—a representation in which only a single element is active at any one time—of the whole sensory input. One good reason for not believing in them is the one usually advanced, namely, that you would require an impossibly large number of such elements. Another equally good reason is that they don't achieve the kind of representation one needs, namely, classifying similar things together: In a mutually exclusive representation you need a separate grandmother cell for each view of grandmother, for example.

Now, for some of the good ideas lurking behind grandmother cells, consider the softer idea, that of mother cells. These respond to any view of mother, and only to views of mother. Thus, they generalize in the appropriate way and are selective in the appropriate way; furthermore, mothers are behaviorally important not only for humans but also for our ancient ancestors. It makes sense to have cells that respond to all views of behaviorally important objects, and only to views of these objects.

I still think gnostic cells and their implied sparse representations are good ideas. *Gnostic cells* are representational elements that are infrequently active so they carry a lot of information when they do fire. Their firing corresponds to a complex feature or event in the sensory input. Thus, a whole scene can be represented by a small number of them, and they can classify or categorize parts of the scene in sensible ways. Such cells might account for the "unreasonable effectiveness" or real perceptual mechanisms, compared with standard machine vision mechanisms.

Finally, the idea of grandmother cells emphasizes that individual neurons are the brain's computational elements. They are the only things we know of that can do the logical operations required for survival in a complex environment. The network can connect them to each other, but it is the elements themselves that must make the logical decisions underlying our behavior.

So although I don't believe in grandmother cells, I think we owe a lot to them.

Authors: Surely the neuron doctrine—that each neuronal element has an all-or-none character to its

---

color of her hair, and still others to the markings on her face. With this population, or ensemble hypothesis, recognition is not due to one unit but to collective activation. Ensemble theories readily account for why we can confuse one visually similar object with another; both objects activate many of the same neurons. Losing some units might degrade our ability to recognize an object, but the remaining units might suffice. Ensemble theories also account for our ability to recognize novel objects. Such objects bear a similarity to familiar things, and our percept results from activating units that represent their features.

The results of single-cell studies of temporal lobe neurons are in accord with ensemble theories of object recognition. While it is striking that some cells are selective for complex objects, the selectivity is almost always relative, not absolute. The cells in the inferior temporal cortex prefer certain stimuli over others, but they are also activated by visually similar stimuli. The cell represented in Figure 6.7, for instance, increases its activity when presented with a mitten-like stimulus. No cells respond to a particular individual's hand; the hand-selective cell responds equally to just about any hand. In contrast, our perceptual abilities demonstrate that we make much finer discriminations.

Summary of Computational Issues

We have considered several computational issues that must be solved by an object recognition system. Information is represented on multiple scales. While early visual

discharge—is not threatened by the idea that whole neuronal networks are the fundamental building blocks of object perception. Given that an object has a spatial component, a size component, an angular component, a color component, and so on, is it not mandated that a whole network is essential to its being perceived and acted on?

HB: You generally don't react to the whole of a sensory scene but just a part of it. If it contains mother, or hungry tiger, you have to ascertain that, and only then can you react appropriately. The trouble is that the mother or hungry tiger can occupy any parts of the sensory input, and it is only by detecting the pattern among these widely dispersed parts that you can decide whether either is present or which it is. You have to do some analysis, or computation, on the whole sensory input, and you will certainly need all the network's parts connected to your mother or tiger cells, which might be quite a large part of the whole network.

The tiger or mother cells, and those detecting more primitive features in the hierarchy, collect the information required to make efficient decisions at each level. The connectivity of the network is invaluable in this task, but only because the connections consolidate all the information so a decision can be made. Again, cells are the only elements we know of that can make these decisions. If you think you know other ways of making them, tell me!

Authors: OK, both of these ideas—sparse and coarse coding schemes—have been around for a long time. Lots of smart people have done clever experiments, both in physiology and in perception. Why

aren't we any closer to resolving the dilemma now than we were 30 years ago? Or phrased another way, what do you think we need to distinguish the two extremes?

HB: Our perceptual system performs a horrendously difficult and complex task when it gives us knowledge of the world. It all seems so easy—we just look and see a world full of objects that we instantly know a lot about. But doing this implies that our visual system has accumulated extraordinarily deep knowledge of the associative structure of sensory stimuli we have received in the past, and that it continues to do so all the time.

I think the ideas we have been talking about do have the potential to explain how all this is done, but they are not like the glorious general laws of physics, which state absolute limits that cannot be broken in just a few words or even shorter mathematical expressions. Ideas such as sparse coding require fleshing out with a lot of detail. We do not know yet what these steps are, but we shall probably soon find out because machines that use images and other data effectively will have to perform analogous tasks.

It's the complexity of the task that has held things up, and although I would not be a bit surprised at new principles emerging over the next few years, I believe the ones we already have will take us quite a distance. The view I'm proposing is that perception performs tasks that are genuinely difficult, in the information-processing sense. When we adequately define these tasks, we shall discover that single neurons perform them well, that is, well compared with the whole brain's performance, and well compared with current computer vision methods of doing the same task.

input can specify simple features, object perception involves intermediate stages of representation where features are assembled into parts. Objects are not determined solely by their parts; they are defined by relations between parts. An arrow and the letter *Y* contain the same parts but differ in their arrangement. For object recognition to be flexible and robust, the perceived spatial relations among parts should not vary over viewing conditions.

FAILURES OF OBJECT RECOGNITION

Where does the syndrome of visual agnosia fit in with these considerations? To repeat, visual agnosia is when patients have difficulty recognizing visually presented objects. A key word here is *visual:* Many people who have incurred a traumatic neurological insult or who have a degenerative disease such as Alzheimer's disease may experience problems recognizing things. But these problems will be generic: If the same object is placed in their hands or described verbally, patients with Alzheimer's disease will still have problems recognizing it. For visual agnosics, the deficit is restricted to the visual domain. Recognition can occur through other sensory modalities like touch or audition. They may fail to recognize the fork, but as soon as

HOW THE BRAIN WORKS

Auditory Agnosia

Other sensory modalities besides visual perception surely contribute to object recognition. Distinctive odors in a grocery store enable us to figure out which bunch of greens is thyme and which is basil. Touch can differentiate between a cheap polyester and a fine silk garment. And we depend on sounds, both natural and man-made, to cue our actions. A siren prompts us to search for a nearby police car or ambulance. Or an anxious parent immediately recognizes the cries of his or her infant and rushes to the baby's aid. Indeed, we often overlook our exquisite auditory capabilities for object recognition. Have a friend rap on a wooden tabletop, or metal filing cabinet, or glass window. You will easily distinguish between these objects.

Numerous studies have documented failures of object recognition in other sensory modalities. As with visual agnosia, a patient has to meet two criteria to be labeled agnosic. First, a deficit in object recognition cannot be secondary to a problem with perceptual processes. For example, to be classified as having auditory agnosia, patients must perform within normal limits on tests of tone detection: How loud must a sound be for the person to detect it? Second, the deficit in recognizing objects must be restricted to a single modality. A patient who cannot identify environmental sounds such as the ones made by flowing water or jet engines must be able to recognize a picture of a waterfall or an airplane.

Consider a patient, C.N., reported by Isabelle Peretz and her colleagues (1994) at the University of Montreal. C.N., a 35-year-old nurse, had suffered two aneurysms, one in the right middle cerebral artery and the other in the left middle cerebral artery, over a 3-month period. Both required surgery. Postoperatively, her ability to detect tones and comprehend and produce speech was not impaired. But she immediately complained that her perception of music was deranged. Her *amusia,* or impairment in music abilities, was verified by tests. For example, she could not recognize melodies taken from her personal record collection, nor could she recall the names of 140 popular tunes including the Canadian national anthem. C.N.'s deficit could not be attributed to a problem with long-term memory. She also failed when asked to decide if two melodies were the same or different. That the problem was selective to auditory perception was evidenced by her excellent ability to identify these same songs when shown the lyrics. Or when given the title of a musical piece such as *The Four Seasons,* C.N. would respond that the composer was Vivaldi and even recalled when she had first heard the piece.

As interesting as C.N.'s amusia is her absence of problems with other auditory recognition tests. C.N. understood speech, and she was able to identify environmental sounds such as animal cries, transportation noises, and human voices. Even within the musical domain, C.N. did not have a generalized problem with all aspects of music comprehension. She performed as well as normal subjects when asked to judge if two tone sequences had the same rhythm. But her performance fell to a level of near chance when she had to decide if the two sequences were the same melody. This dissociation makes it less surprising that, despite her inability to recognize songs, she still enjoyed dancing!

Other cases of domain-specific auditory agnosia have been reported. Many patients have an impaired ability to recognize environmental sounds, and as with amusia, this deficit is independent of language-comprehension problems. In contrast, patients with pure word deafness cannot recognize oral speech even though they exhibit normal auditory perception for other types of sounds and have normal reading abilities. Such category specificity suggests that auditory object recognition involves several distinct processing systems. Whether the operation of these processes should be defined by content—for example, verbal versus nonverbal input—or by computations—for example, words and melodies may vary with regard to the need for part versus whole analysis—remains to be seen ... or, rather, heard.

the object is placed in their hand, they will immediately recognize it (Figure 6.22). Indeed, after touching the object, an agnosia patient may report seeing the object clearly. Because the patient can recognize the object through other modalities, and through vision with supplementary support, we know that the problem does not reflect a general loss of knowledge. Rather, it must reflect a loss of knowledge that is intimately linked to visual perception or an inability for the products of visual perception to access modality-independent knowledge stores.

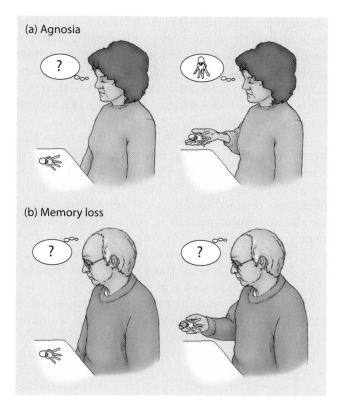

(a) Agnosia

(b) Memory loss

Figure 6.22 To diagnose an agnosic disorder, it is essential to rule out general memory problems. **(a)** The patient with agnosia is unable to recognize the keys by vision alone, but immediately recognizes the keys when she picks them up. **(b)** The patient with a memory disorder is unable to recognize the keys even when he picks them up.

Subtypes of Agnosia

Visual agnosia is differentiated from visual deficits caused by an impairment in sensory abilities. A patient who is completely blind will be unable to recognize a visually presented object. But a blind patient does not have a problem linking visual information to stored knowledge about the world; rather, a blind patient lacks the perceptual input needed to activate this internal knowledge. The label *visual agnosia* is restricted to individuals who demonstrate object recognition problems despite the fact that visual information continues to be registered at the cortical level.

The nineteenth-century German neurologist H. Lissauer first suggested the existence of two types of visual agnosia. Lissauer (1890) distinguished between recognition deficits that were primarily sensory based and those that reflected an inability to visually access memory, a disorder that he melodramatically referred to as *Seelenblindheit,* or "soul blindness." In the current literature, the general distinction is made between apper-

ceptive agnosia and associative agnosia. **Apperceptive agnosia** describes failures in object recognition linked to problems in perceptual processing. **Associative agnosia**, in contrast, is reserved for patients who derive normal visual representations but cannot use this information to recognize things. While research has led to more refined taxonomies, the distinction between apperceptive and associative agnosias continues to provide a useful classification scheme.

APPERCEPTIVE AGNOSIA

The inability to recognize objects can arise from a host of perceptual disorders. A severe case was apparent in a young man who had suffered carbon monoxide poisoning, an injury that produced widespread bilateral cortical damage. The poisoning had not produced any scotomas, and the patient could distinguish small differences in brightness and color. Yet his ability to discriminate between even the simplest shapes was essentially nonexistent. For example, he was unable to distinguish between two rectangles of equal area, even when they were oriented in opposite directions. He could not read letters, except for those composed of straight segments (e.g., *I*), nor could he copy drawings. Face perception was impossible for him. He even failed to recognize his own face in a mirror; he thought it might be his doctor's.

In patients with apperceptive agnosia, the perceptual problems are subtle. Indeed, a standard clinical evaluation may fail to reveal any visual problems, and the patient will have to insist that he or she is having difficulties in recognizing objects to receive a more detailed examination. The patient may perform normally on shape discrimination tasks yet make many mistakes when asked to recognize line drawings or photographs of objects. To demonstrate that the agnosia is truly of the apperceptive subtype and not an associative agnosia, it is necessary to devise refined tests of perceptual acuity. For example, tasks like the Gollin Picture Task and Incomplete Letters Task examine whether patients can recognize objects in a degraded format (Figure 6.23).

Working out of the National Hospital in London, Elizabeth Warrington (1985) has carefully investigated perceptual disabilities in patients over the past 25 years. One of her first studies included more than eighty patients with right- or left-hemisphere lesions. Inclusion was not dependent on whether the patients showed signs of agnosia; indeed, it is unlikely that many patients would have presented any clinical signs of object recognition problems. Instead, the main criterion for inclusion was evidence of a unilateral cerebral lesion (stroke or tumor). In addition, all the patients were right-handed with

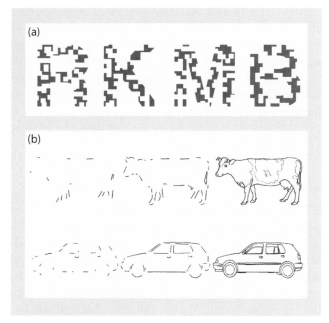

Figure 6.23 **(a)** Patients with agnosia following right-hemisphere lesions have more difficulty than patients with left-hemisphere lesions, despite the more severe language problems for the left-hemisphere group. **(b)** In the Gollin picture test, the subjects are presented with a series of drawings of an object, each successive drawing being more complete than the previous one. Patients with right-hemisphere lesions required more complete drawings in order to correctly identify the objects. Adapted from Warrington (1985).

normal visual acuity, either with or without glasses. This last criterion excluded patients with object recognition deficits caused by obvious sensory problems.

In the Gollin Picture Task and the Incomplete Letters Task, patients with right-sided lesions performed more poorly than did either control subjects or patients with left-sided lesions. Left-sided damage had little effect on performance, a result made even more impressive by the fact that many of these patients had language problems. This laterality effect, with apperceptive agnosia being more apparent after right-hemisphere damage, has been observed during a variety of object recognition tasks. How are we to interpret these results? Do they implicate a primary role for the right hemisphere in object recognition? Or does the right hemisphere perform a special cognition operation essential for tasks like the Gollin Picture Task? Can we interpret this deficit within the framework of the computational problems described in the previous section?

Warrington maintained that such an analysis is possible and, indeed, provided a most cogent account of the deficit in patients with right parietal lesions. She contended that the problem is not one of contour discrimi-

nation. If patients with right parietal lesions are asked to make fine discriminations between the shapes of simple geometric figures, their performance is only slightly poorer than that of healthy subjects. The critical problem, then, is that perceptual categorization in these patients is impaired—the patient's ability to achieve object constancy is disrupted.

To test this hypothesis, Warrington devised an Unusual Views Object Test. For this test, photographs were taken of twenty objects, each from two distinct views, as shown in the example in Figure 6.24. For one view, the object was oriented so the photograph would depict a standard or prototypical view. For example, a cat was photographed with its head facing forward. In the other view, the photograph depicted an unusual or atypical view. For example, the cat was photographed from behind, without its face or feet in the picture.

Subjects were shown the photographs and asked to name them. While normal subjects made few, if any, errors, patients with right posterior lesions had difficulty identifying objects photographed from unusual orientations. But they could name the objects when photographed in the prototypical orientation, further confirming that their problem is not a deficit of lost visual knowledge.

A second task verified the claim that the pairs of photographs were not recognized as depicting the same object. Here pairs of photographs were presented in each trial, one from the prototypical set and one from the unusual set. The two photographs were either of the same or of different objects, and the subjects' task was to judge if the objects were the same. As expected, patients with right posterior lesions showed more impairment than did healthy subjects and patient control groups. Similar results were obtained from other photographic transformations that presumably tax the object constancy mechanism. The patients had problems matching an object photographed from the exact same position but in two different lighting conditions.

Warrington's analysis thus specified the critical deficit underlying apperceptive agnosia. Her evaluation did not simply restate the problem—that patients with right-hemisphere lesions have difficulty with object recognition because the right hemisphere is so prominent in object perception. Instead, she postulated a cognitive process associated with the posterior part of the right hemisphere, a principal player in perceptual categorization. People can normally recognize an object under nearly infinite orientations, distances, and illuminations. Although we see these percepts as distinct, we can identify the similarities. A cat will be identified whether it is sleeping on the couch, rolling on the floor, or prowling through the gar-

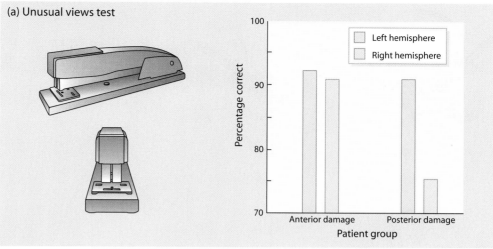

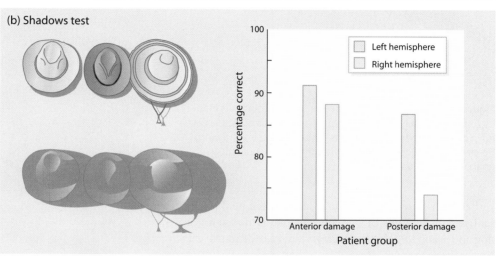

Figure 6.24 Tests used to identify apperceptive agnosia. **(a)** The unusual views test. The subject must judge if the two pictures are of the same object. **(b)** The shadows test. Subjects must identify the object(s) when seen under normal or shadowed illumination. In both tests, patients with right-hemisphere lesions, especially in the posterior area, performed much more poorly than did control subjects (not shown) or patients with left-hemisphere lesions. Adapted from Warrington (1982).

den. This implies that we can readily achieve object constancy; from the infinity of percepts, we extract critical features to identify the object. Patients with right parietal lobe lesions are less able to recognize objects. When presented so the perceptual input highlights the most salient features, the object can be recognized. If these features must be inferred or extracted from a limited perceptual input, apperceptive agnosics have difficulties.

ASSOCIATIVE AGNOSIA

Associative agnosia is, by definition, a failure of visual object recognition that cannot be attributed to perceptual abilities. These patients rarely perform normally on perceptual tests, but their perceptual deficiencies are not proportional to their recognition problems. One patient, F.R.A., awoke one morning and discovered that he could not read his newspaper, a condition known as *alexia,* or *acquired dyslexia* (McCarthy and Warrington, 1986). When the problem did not subside, he visited his local hospital. A CT scan revealed an infarct of the left posterior cerebral artery. The lesioned area was primarily in the occipital region of the left hemisphere, although the damage probably extended into the posterior temporal cortex. This patient's story is not atypical: Strokes often go unnoticed until the patient discovers an inability to perform a task.

After F.R.A.'s condition had stabilized, perceptual tests were administered. He copied geometric shapes with ease and could point to objects when they were named. Most impressive was his ability to segment a complex drawing into its parts (Figure 6.25). Apperceptive agnosia patients fail miserably when instructed to color each object differently. In contrast, F.R.A. performed the task effortlessly. Despite this ability, though, he could not name the objects he had colored.

Further testing revealed the extent of F.R.A.'s deficits. When shown line drawings of common objects, he could name or describe the function of only half of them. If he was given the name verbally, he could readily generate a

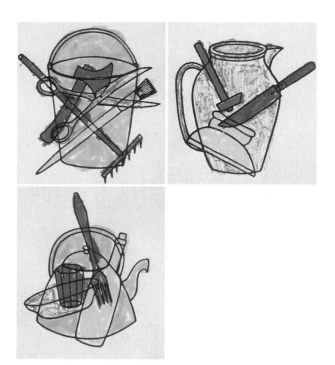

Figure 6.25 The drawings of patient F.R.A. Despite his inability to name visually presented objects, F.R.A. was quite successful in coloring in the components of these complex drawings. He clearly had succeeded in parsing the stimuli but still was unable to identify the objects.

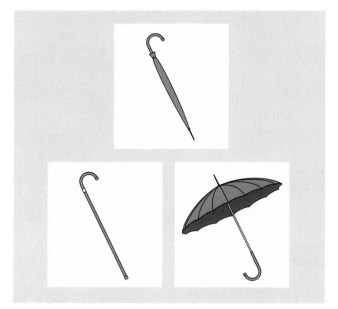

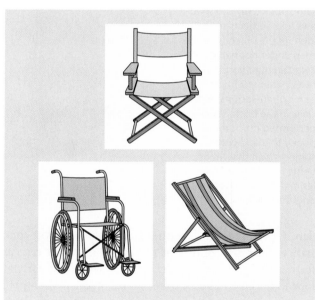

Figure 6.26 The Matching-by-Function test. The subject is asked to choose the two objects that are most similar in terms of function. In both examples, the correct match is with the object on the right despite the greater physical similarity of the objects on the left. Agnosic patients with left-hemisphere lesions demonstrate impairment on this task. (Top) Adapted from Warrington (1982). (Bottom) Adapted from Warrington and Taylor (1978).

verbal description. In a similar manner, when he was presented with pictures of two animals like a mouse and a dog and asked to point to the largest, his performance was barely above that which would occur by chance. If the two animal names were said aloud, F.R.A. could do the task perfectly. His problems were clearly restricted to the visual modality. The ability to recognize the meaning of visually presented objects was compromised by the stroke.

This hypothesis was explored in another test used by Warrington in her group studies. Recall that in the Unusual Object Views Test, subjects are required to judge if two pictures depict the same object from different orientations. This task requires that the subjects categorize information according to their perceptual qualities. In an alternative task, the Matching-by-Function test, subjects are shown three pictures and asked to point to the two that are functionally similar. In Figure 6.26, the correct response would be to match the closed umbrella to the open umbrella, even though the former is physically more similar to the cane. Thus, the Matching-by-Function test requires subjects to categorize stimuli on the basis of their semantic properties, that is, in relation to how they are used.

Patients with posterior lesions in either the right or the left hemisphere showed impairment on this task. Yet Warrington maintained that the problems in the two

groups happen for different reasons. Patients with right-sided lesions cannot do the task because they fail to recognize many objects, especially those depicted in an unconventional manner, such as the closed umbrella. Patients with left-sided lesions can frequently recognize objects in isolation, but they cannot make the functional connection between the two visual percepts. They lack the semantic representations needed to link the functional association between the open and closed umbrella.

Integrating Parts into Wholes

Warrington (1985) proposed an anatomical model of the cognitive operations required for object recognition (Figure 6.27). The central feature of the model is two categorical stages of object recognition. Initial visual processing is assumed to involve both occipital cortices. Subsequently, the first categorical operation, *perceptual categorization,* is invoked. Perceptual inputs are matched with stored representations of visual objects, a stage of processing essential for dealing with the variability of sensory information. The visual system must distinguish between idiosyncratic sources of information, such as shadowing patterns, and those that provide invariant sources of information. This stage, associated with processing in the right hemisphere, can be characterized as presemantic in the sense that we may recognize two pictures that depict the same object without being able to name the object or describe its function. To do this, a second categorization stage, *semantic categorization,* is needed, one that depends on the left hemisphere. Here, visual input is linked with knowledge in long-term memory concerning the name and functions of that input.

The Warrington model is a simple look at how we go about recognizing objects. The model requires elaboration, though. First, neuropathological findings have not always proved a correspondence between associative agnosia and left-hemisphere lesions. More typically, these patients usually have bilateral lesions that affect occipitotemporal regions. In addition, unilateral right-hemisphere lesions restricted to this region can produce an agnosia more similar to the associative subtype than to the apperceptive subtype.

Detailed analysis of a few case studies also pointed to the need for more refined analysis at the cognitive level of what happens in perceptual categorization. Glyn Humphreys and Jane Riddoch and colleagues (1994) of the University of Birmingham in England described a patient, H.J.A., who suffered a stroke while undergoing an appendectomy. The stroke resulted in bilateral lesions of the occipital lobe that extended anteriorly into the ventral part of the temporal lobe. Over 15 years of testing revealed an extensive and stable agnosia. When initially asked to name common household objects, H.J.A. succeeded on about 80% of the trials; his performance fell to 40% when line drawings were used as stimuli. As with all visual agnosics, his ability to identify objects tactilely was normal and he demonstrated good long-term knowledge concerning the visual properties of objects (e.g., does the prime minister of England have a moustache?).

An assessment of H.J.A.'s performance on standard measures of agnosia led to a classification of associative agnosia. He had no problem with shape-matching tasks and was able to copy accurately. He was also successful in matching photographs of objects from unusual views, Warrington's test for a perceptual categorization deficit.

But it was apparent that H.J.A.'s perceptual abilities were quite abnormal. When presented with overlapping figures, his object recognition ability failed completely; he recognized objects in isolation but had great problems when the contours of two objects overlapped. Further evidence of his limited capabilities was found in his drawings. His renditions were produced in a piecemeal, slavish fashion. He drew each segment individually, repeatedly examining the model before progressing to the next part.

Humphreys and Riddoch proposed that H.J.A. had difficulty in integrating the parts of an object into a coherent whole, a deficit they labeled **integrative agnosia.** He was unable to perceive an object "at a glance." Instead, his ability to identify objects depended on recognizing salient

Figure 6.27 Warrington's two-stage model of object recognition. Visual analysis occurs in both hemispheres, at least when we look directly at an object. The first stage of object categorization is perceptual, the processes required to overcome the perceptual variability in the stimulus. This stage is dependent on the right hemisphere. The second stage involves semantic categorization in which the perceptual representation is linked to semantic knowledge. This stage is dependent on the left hemisphere.

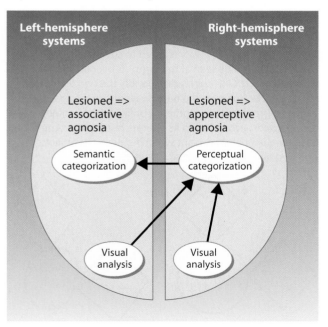

features or parts. For H.J.A. to recognize a dog, he must independently perceive each of the legs and the characteristic shape of the body and head. These part representations are then used to identify the whole. Such a strategy would run into problems with overlapping figures where it is necessary not only to identify the parts but also to make the correct assignment of parts to objects.

Many tests can provide evidence that H.J.A. has a deficit in grouping and integrating features. Consider the task in Figure 6.28, where subjects must determine whether a display contains an upside-down T among distracters that are upright or tilted T's. This task can be considered a conjunction-search task: The upside-down target contains the same features (a horizontal and a vertical line)

as the upright distracters. Normal subjects are much faster when the distracters form a homogeneous set (e.g., all upright T's) than when they are drawn from a heterogeneous set (e.g., upright and tilted T's). It appears that we can rapidly group homogeneous elements and detect a deviation from this consistent texture. H.J.A., though, did not have this advantage with organized displays. His reaction times were comparable in organized and random conditions, which accords with the hypothesis that H.J.A. must treat each object as an independent entity, examining each pair of lines to determine whether the T is upright or upside-down. H.J.A. also failed to identify objects when contours were well structured as opposed to being fragmented.

H.J.A. does not represent a special case of integrative agnosia. Indeed, an inability to integrate features into a coherent whole may be the hallmark of many agnosia patients. A telling example of this deficit is provided by the drawings of patient C.K., a young man who suffered a head injury in an automobile accident (Behrmann et al., 1994). C.K. was asked to draw three geometric shapes in a specific configuration. We might be impressed if we simply look at the final product of his drawing, shown in Figure 6.29. What is abnormal is the order in which he produced the segments. For example, after drawing the left-hand segments of the upper diamond, C.K. proceeded to draw the upper left-hand arc of the circle. Rather than continue with the circle, he branched off to draw the lower diamond before returning to complete the upper diamond and the rest of the circle. For C.K., each intersection defined the segments of different parts. He failed to link these parts in recognizable wholes.

Another dramatic case involved an artist who became agnosic after an occipital stroke. The patient retained a

Figure 6.28 Patients with integrative agnosia have difficulty grouping common elements together. Normal subjects find the upside-down T much faster when all of the distracters are upright T's. When the distracters are heterogeneous, their reaction times (RT) are much slower. The benefit is likely due to the fact that the distracter elements group together in the homogeneous condition. Patient H.J.A.'s reaction times are comparable in the two conditions, indicating a failure to group. Adapted from Humphreys et al. (1994).

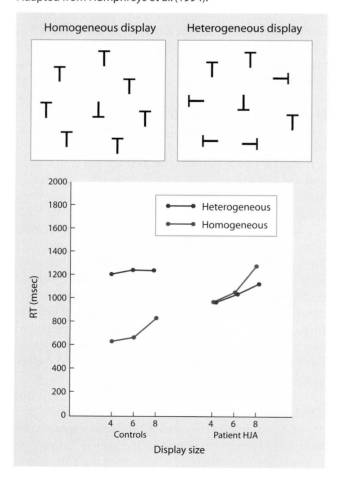

Figure 6.29 Patient C.K. was asked to copy the figure shown at the left. As can be seen at the right, his overall performance was quite good: One can readily identify the two diamonds and the circle. However, the numbers indicate the order in which he produced the segments. Unlike healthy subjects, C.K. did not draw each object in its entirety but followed the outer boundary, even if this meant switching between objects. Adapted from Behrmann et al. (1994).

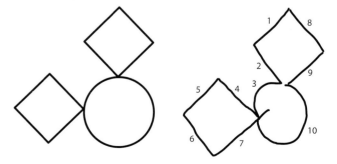

keen sense of drawing, as can be seen in his copy of a silk screen in Figure 6.30. Even after drawing the picture, though, he could not recognize the objects; he failed to identify the stove, and only after prodding was he willing to admit that the animal might be a "domestic bird."

Warrington's analysis of perceptual and semantic categorization fails to capture the integration problems faced by these patients, the inability to synthesize parts into a coherent whole. An isolated part may provide the perceiver with a hypothesis about the identity of a stimulus, perhaps even proving sufficient with unique objects (e.g., the frame of a bicycle). But, in most situations, parts must be integrated into wholes. Patient G.S. was fixated on the belief that the depicted combination lock was a telephone because of the circular array of numbers, a salient feature on the rotary phones of his time. He was unable to integrate this part with the other components of the combination lock. Object perception is truly a situation where the whole is greater than the sum of its parts.

Category Specificity in Agnosia

Associative agnosia results from the loss of semantic knowledge regarding the visual structure or properties of objects. Early perceptual analyses proceed normally, but the long-term knowledge of visual information is lost, and thus the object cannot be recognized. Further support for the idea of a loss of visual semantics comes from bizarre reports of patients whose object recognition deficit appears to be selective for specific categories of objects (Warrington and Shallice, 1984).

Patient J.B.R. was diagnosed as having herpes simplex encephalitis in 1980. His illness left him with a complicated array of deficits including a dense amnesia, or memory loss, and word-finding difficulties. While his performance on tests of apperceptive agnosia was normal, J.B.R. had a severe associative agnosia. Most notable about his agnosia was that it was disproportionately worse for living objects than inanimate ones. For drawings of common objects such as scissors, clocks, and chairs, his success rate was about 90%. In contrast, he could correctly identify only 6% of the pictures of living things. A similar dissociation of agnosia for living and nonliving things has been reported for other patients (Satori and Job, 1988).

How are we to interpret such puzzling deficits? Warrington's visual semantics hypothesis provides an answer. If we assume that associative agnosia represents a loss of knowledge about visual properties, we might suppose that a category-specific disorder results from the selective loss within this knowledge system.

Figure 6.30 Patient J.R., a skilled artist, copied this drawing. While his reproduction is quite faithful to the original, he was not able to identify the stove or the rooster.

Semantic knowledge is structured: We recognize that birds, dogs, and dinosaurs are animals because they share common features. In a similar way, scissors, saws, and knives share characteristics. Some might be physical (e.g., they all have an elongated shape) and others functional (e.g., they all are used for cutting). Brain injuries that produce agnosia in humans do not completely destroy the connections to semantic knowledge. Even the most severely affected patient will recognize some objects. Because the damage is not total, it seems reasonable that circumscribed lesions might destroy tissue that is devoted to processing similar types of information. **Category-specific deficits** support this form of organization. The lesions in such patients have affected regions associated with processing information about living things.

If this interpretation were valid, we might expect to find opposite results: patients whose recognition of nonliving things is disproportionately impaired. Indeed, there have been a few; for example, one patient showed much more impairment in recognizing common objects than animals. Each trial consisted of presenting an array of five objects. A target, presented either visually or auditorily, was then given, and the subject's task was to choose the object in the array that was from the same category (Figure 6.31). For both

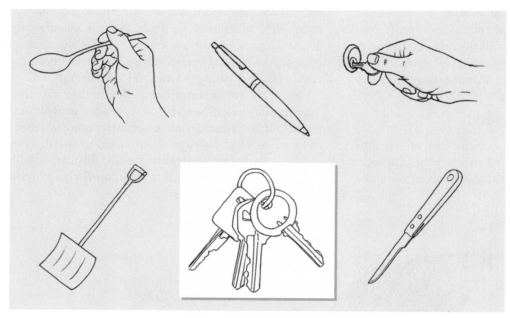

Figure 6.31 Tests used to demonstrate category-specific agnosia. In each condition, the subject is asked to choose from the array of five drawings the one that goes best with the probe item in the box. A patient with a selective deficit for common objects would perform poorly in the top example but perform normally in the bottom example. The reverse would be expected in a patient with a category-specific deficit for living things. Adapted from Warrington and McCarthy (1994).

modalities, the patient was much slower and made more errors with the objects.

So we do have evidence of a double dissociation between agnosia for living things and that for nonliving things. But we should be cautious: No studies have shown this double dissociation by using the exact same stimulus materials. In the study just reviewed, no patients showed the opposite pattern of results, nor were there any control data. It may be, then, that the nonliving-object trials were harder than the living-object trials for these stimuli. Given that reports of patients with selective deficits in recognizing nonliving objects are much rarer, we should consider alternative accounts for this pattern of category-specific agnosias.

One idea is that many nonliving things evoke representations not elicited by living things (Damasio, 1990). In particular, manufactured objects can be manipulated. As such, they are associated with kinesthetic and motoric representations. When viewing an object, we can activate a sense of how it feels or of the actions required to manipulate the object (Figure 6.32). Corresponding representations will not exist for living objects. While we

Figure 6.32 The spared ability to recognize common objects has been attributed to the fact that our visual knowledge is supplemented by kinesthetic codes developed through our interactions with these objects. When a picture of the scissors is presented, the visual code may not be sufficient for recognition. However, when supplemented with priming of kinesthetic codes, the person is able to name the object. Kinesthetic codes are unlikely to exist for most living things.

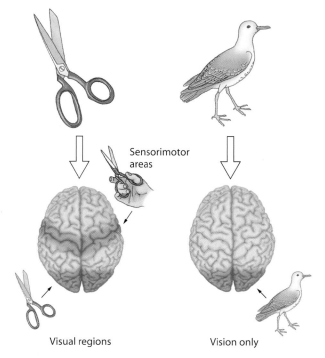

Sensorimotor areas

Visual regions   Vision only

may have a kinesthetic sense of how a cat's fur feels, few of us have ever stroked an elephant. And we certainly have no sense of what it feels like to pounce like a cat or fly like a bird.

According to this hypothesis, manufactured objects are easier to recognize because they activate additional forms of representation. While brain injury can produce a common processing deficit for all categories of stimuli, these extra representations may be sufficient to recognize nonliving objects. The viability of this hypothesis can be seen in the behavior of patient G.S., described at the beginning of this chapter. Remember that when he was shown the picture of the combination lock, his first response was to call it a telephone. But even when verbalizing "telephone," his hands began to move as if they were opening a combination lock. Indeed, he was able to name the object after he looked at his hands in wonder and realized what they were trying to tell him.

A less glamorous interpretation of category-specific agnosia for living things must also be considered. David Gaffan and Charles Heywood (1993) of Oxford University created 260 pictures, some of which depicted living things and the rest nonliving things. Healthy observers were able to name all the objects. To make the task more difficult, the pictures were presented for only 20 msec. This manipulation was highly effective: Error rates were higher than 35%, thereby creating a transient pseudo-agnosia in the normal subjects. Of greatest interest, all five normal subjects made many more errors on pictures of living things than on those of nonliving things. In a second experiment, similar results were obtained with six monkeys.

To account for these findings, Gaffan and Heywood argued that living things are inherently more difficult to discriminate than are nonliving things. Members of categories such as mammals and fruits share more salient and distinctive features than do categories such as tools. For example, most of the animals in Figure 6.33 have appendages or legs; it is hard to identify a commonality across the nonliving objects. The greater similarity between the pictures of living things made it harder for normal subjects to determine the correct label under limited viewing conditions.

If this interpretation is correct, agnosia patients' selective deficit for living things may not reflect differences in the types of representations; it may simply reflect differences in object similarity. A fundamental tenet of testing brain-injured patients is that the more difficult a task is, the more apparent their deficits are. Agnosia patients may fail more often at recognizing living things simply because these objects are less discriminable than are nonliving things.

Figure 6.33 Line drawings of objects used to compare recognition of living and nonliving things by normal subjects. Notice the greater similarity (and thus confusability) of the living things: they tend to have rounded bodies and appendages of some sort. There is little similarity among the set of nonliving things. Adapted from Snodgrass and Vanderwart (1980).

Computational Account of Category-Specific Deficits

Martha Farah and Jay McClelland (1991) used a series of computer simulations to integrate a number of these ideas. Their study was designed to contrast two ways of conceptualizing the organization of semantic memory (Figure 6.34). One hypothesis is that semantic memory is organized by category membership. According to this hypothesis, there are distinct representational systems for living and nonliving things, and perhaps further subdivisions within these two broad categories. An alternative hypothesis is that semantic memory reflects an organization based on object properties. The idea that nonliving things are more likely to entail kinesthetic and motor representations is one

variant of this view. The computer simulations were designed to demonstrate that category-specific deficits can result from lesions to a semantic memory system organized by object properties.

The architecture of their model involved a simple connectionist network, designed to simulate performance when people are asked to associate names for objects with visual representations of the objects. As with standard connectionist networks, information was distributed across a number of processing units. One set of units corresponded to peripheral input systems, divided into a verbal and visual system. Each of these was composed of twenty-four input units. The visual representation of an object involved a unique pattern of activation across the twenty-four visual units. Similarly, the name of an object involved a

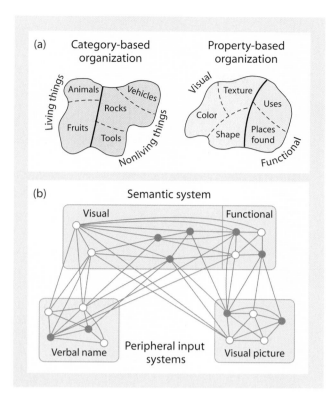

Figure 6.34 **(a)** Two hypotheses concerning the organization of semantic knowledge. A category-based hypothesis proposes that semantic knowledge is organized according to our categories of the world. For example, one prominent split would be between living and nonliving things. A property-based hypothesis is that semantic knowledge is organized according to the properties of objects. These properties may be visual or functional. **(b)** The architecture of Farah and McClelland's (1991) connectionist model of a property-based semantic system. The initial activation for each object is represented by a unique pattern of activation in two input systems and the semantic system. In this example, the darkened units would correspond to the pattern for one object. The final activation would be determined by the initial pattern and the connection weights between the units. There are no connections between the two input systems. The names and pictures are linked through the semantic system.

unique pattern of activation across the twenty-four verbal units. In the simulations, the model was presented with twenty unique patterns representing twenty objects, half of them living objects and the other half nonliving.

Each object was also linked to a unique pattern of activation across the second type of unit in the model, the semantic memory. Within the semantic system, there were two types of units, visual and functional. While these units did not correspond to specific types of information (e.g., colors or shapes), the idea here is that

semantic knowledge consists of at least two types of information. One type of semantic knowledge is visually based, akin to Warrington's visual semantics. For example, visual semantics would embody facts, such as a tiger has stripes or a chair has legs. The other type of semantic memory corresponds to our functional knowledge of objects. For example, functional semantics would include our knowledge that tigers are dangerous or that a chair is a type of office furniture.

The researchers imposed two constraints on the semantic memory system, intended to capture psychological differences in how visual and semantic information might be stored. The first constraint was that of the eighty semantic units, sixty were visual and twenty were semantic. This 3 : 1 ratio was based on a preliminary study in which human subjects were asked to read the dictionary definitions of living and nonliving objects and indicate whether a descriptor was visual or functional. On average, three times as many descriptors were classified as visual. Second, the preliminary study indicated that the ratio of visual to functional descriptors differed for the two classes of objects. For living objects, the ratio was 7.7 : 1; for nonliving objects this ratio dropped to 1.4 : 1. Thus, as discussed previously, our knowledge of living objects is much more dependent on visual information than is our knowledge of nonliving objects. This constraint was incorporated into the model by varying the number of visual and functional semantic units used for the living and nonliving objects.

The model was trained to link the verbal and visual representations of a set of twenty objects, half of them living and the other half nonliving. Note that the verbal and visual units were not directly linked but can only interact by way of their connections with the semantic system. The strength of these connections was adjusted in a training procedure. This procedure was not intended to simulate how people acquire semantic knowledge. Rather, the experimenters would set all of the units, both input and semantic, to their values for a particular object and then allow the activation of each unit to change depending on both its initial activation and the input it received from other units. The connection weights were then adjusted to minimize the difference between the resulting pattern and the original pattern. The model's object recognition capabilities can be tested by measuring the probability of correctly associating the names and pictures.

This model proved to be extremely adept. After forty training trials, it was perfect when tested with both the living and the nonliving stimuli. The key question centered on how well the model did after receiving "lesions" to its semantic memory, lesions assumed to correspond to what happens in patients with visual associative agnosia. Lesions in a model consist of deactivating a certain percentage of the semantic units. As can be seen in Figure 6.35, selective lesions in either the visual or the functional semantics system produced category-specific deficits. When the damage was restricted to visual semantic memory, the model had great difficulty in correctly associating the names and pictures for living objects. In contrast, when the damage was restricted to functional semantic memory, failures were limited to nonliving objects. Note that the resulting "deficit" is much more dramatic in the former simulation, the visual semantic lesion's effect on identifying living things. This result meshes nicely with reports in the neuropsychological literature of many more instances of patients with a category-specific agnosia for living things. Even when functional semantic memory is damaged, the model remains proficient in identifying the nonliving objects, presumably because knowledge of these objects is distributed across both the visual and the functional memory units.

These simulations demonstrated how category-specific deficits might reflect the organization of semantic knowledge. The modeling work makes the important point that we need not postulate that our knowledge of objects is organized along categories such as living and nonliving. The double dissociation between living and nonliving things has been taken to suggest that we have specialized systems sensitive to these categorical distinctions. In contrast, in Farah and McClelland's model, the same system is used to recognize both living and nonliving things. Rather than assume a partitioning of representational systems in terms of the type of object, they proposed that semantic memory is organized according to the properties that define the objects. Category-specific deficits are best viewed as an emergent property of the fact that different sources of information are needed to recognize living and nonliving objects. Our knowledge of living things is highly dependent on visual information, whereas this dependency is lessened for nonliving objects. While stripes are the defining property of a tiger, chairs come in all shapes and sizes but have the consistent property that they are to be sat on.

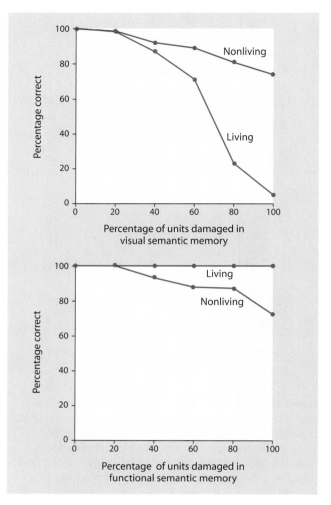

Figure 6.35 Lesions in the semantic units resulted in a double dissociation between the recognition of living and nonliving objects. After a percentage of the semantic units were eliminated, two measurements were made. In the picture-naming task, the visual input pattern for each object was activated and the resultant pattern on the verbal units was evaluated as either correct or incorrect. In the matching-to-sample test, the name pattern was activated and the resulting pattern on the picture units was evaluated. When the lesion was restricted to the visual semantic memory units, the model showed much more impairment in correctly associating the name and picture patterns for the living things. When the lesion was restricted to the functional semantic memory units, the model showed impairment only in associating the input patterns for nonliving things. Adapted from Farah and McClelland (1991).

PROSOPAGNOSIA

It is hard to deny that the most important things we must recognize are faces. Though people may have characteristic physiques and mannerisms, facial features provide the strongest distinctions of one person from another. Think about what you say when describing a new acquaintance to a friend. You may begin by providing information about the person's shape (e.g., large or small), but beyond that, most of the description will center on facial features: Does the person have a long or round face? Are the eyes spaced far apart or close together? Do the lips turn upright into a perpetual smile, or are they drawn down into a pout?

The importance of face perception is reflected in our extraordinary ability to remember faces. In browsing through an old yearbook, we readily recognize the faces of people we have not seen for over 30 years. Unfortunately, our other memory abilities are not as keen. While we may recall that the person in this photograph was a lab partner in biology, her name may remain elusive. Of course, it does not take 30 years to experience this frustration. On a regular basis we run into an acquaintance whose face is familiar, but we cannot remember the name or where and when a previous encounter had taken place.

Given the importance of face perception, prosopagnosia is one of the most fascinating disorders of object recognition. Prosopagnosia is a deficit in the ability to recognize faces that cannot be directly attributed to a deterioration in intellectual function. As with all visual agnosias, prosopagnosia requires that the deficit be specific to the visual modality. Like the patient described in the beginning of Chapter 5, patients with prosopagnosia can recognize a person upon hearing that person's voice.

Prosopagnosiacs can have difficulty recognizing the faces of familiar and unfamiliar people. A patient with bilateral occipital lesions failed to identify not only his wife but also an even more familiar person, himself (Pallis, 1955). As he reported, "At the club I saw someone strange staring at me, and asked the steward who it was. You'll laugh at me. I'd been looking at myself in the mirror." Not surprisingly, this patient was also unable to recognize pictures of famous individuals of his time including Churchill, Hitler, Stalin, Marilyn Monroe, and Groucho Marx. This inability was particularly striking in that the patient had an excellent memory, recognized common objects with no hesitation, and could read and recognize line drawings—all tests that agnosia patients often fail.

While this patient had "normal" object perception, most patients who have been studied in laboratories show abnormalities on other tests of visual perception. Nonetheless, it is clear that their recognition problems for faces are disproportionate to their ability to recognize other objects. For example, patients with severe prosopagnosia may perform perfectly on War-

rington's standard perceptual categorization tasks (e.g., Unusual Views Object Test) yet fail miserably when face stimuli are substituted for the objects. Results such as these suggest that prosopagnosia is not related in a simple way to problems with recognizing nonfacial stimuli.

Are Faces Special?

Face perception, some say, may not use the same processing mechanisms as the ones for object recognition, but this hypothesis is counterintuitive. It is more reasonable and certainly more parsimonious to assume a single, general-purpose system for recognizing all sorts of visual inputs. Why should faces be treated differently from other objects?

One argument in favor of a specialized face-processing module is based on evolutionary arguments. When we meet other individuals, we usually look at their faces rather than their bodies—a behavior that is not a newly acquired cultural convention. The tendency to focus on faces reflects behaviors deeply engraved in our evolutionary history. Across cultures, facial expressions provide the most salient cues regarding affective states. Facial gestures help to discriminate between pleasure and displeasure, friendship and antagonism, agreement and confusion.

While evolutionary arguments can motivate a hypothesis, it is essential to develop empirical tests. Three criteria can be useful in considering whether face and other forms of object perception utilize distinct processing systems. First, do the processes involve physically distinct mechanisms? Do face perception and other forms of object perception depend on different regions of the brain? Second, are the systems functionally independent? Can each operate without the other? The logic of this criterion is essentially the same as that underlying the idea of double dissociations. Third, do the two systems process information differently?

Neural Mechanisms for Face Perception

Some patients show impairment only on tests of face perception. More often, though, the patients' performance on other object recognition tasks is below normal. This result is in itself inconclusive with regard to the existence of specialized brain mechanisms for face perception. Do not forget that brain injury in humans is an uncontrolled experiment! Prosopagnosia rarely occurs with single, well-circumscribed lesions. This syndrome is frequently associated with

An Interview with Nancy Kanwisher, Ph.D.

Professor Kanwisher is a member of the Department of Cognitive and Brain Sciences at the Massachusetts Institute of Technology. Trained as a cognitive psychologist, she has turned to neuroimaging in her research on the building blocks of visual perception.

Authors: You have recently stated that "a primary goal in scientific research is to carve nature at its joints." How does an aspiring scientist decide where the joints are?

NK: By invoking Plato's famous phrase, I was alluding to the hope shared, I think, by all scientists that the categories we find in the world are not completely arbitrary inventions, but that they at least partly reflect real properties of the world itself. The problem, of course, is that there are many different ways to categorize any set of phenomena, so we have to make choices about the properties we deem important. In cognitive science, as in biology, a concept that is at the very core of most theories is *function*. So in cognitive neuroscience, carving nature at its "joints" means discovering the functional components of the brain and mind. It is therefore a great stroke of luck that the brain really does seem to have joints, with different bits of brain apparently carrying out very different functions. One of the things that makes cognitive neuroscience so much fun is the large number of different kinds of evidence we can draw on to try to discover these bits, from computation to monkey physiology to behavioral studies of children, neuropsychological patients, and normal adults. In the work in my lab, we try to use clues from all of these fields when we are generating hypotheses for brain imaging experiments.

Authors: But is there a risk that if brain imaging research is too hypothesis-driven, it will lead researchers to find whatever they want in their data?

NK: Yes, there is a risk, but it is an avoidable one. The enormous richness of brain imaging data puts us at risk of always being able to find something in our data that will fit with the preconceptions we bring to the experiment. If we don't take measures to avoid this problem, brain imaging research just becomes one big projective test. Images of brain activations often end up looking like Rorschach ink blots! But this problem only arises when the hypotheses that guide the research are too open-ended. The way to avoid it is to state our hypotheses explicitly enough that we give them an opportunity to fail. So in a sense, the problem is not that brain imaging research is too hypothesis-driven, but rather that it isn't hypothesis-driven enough.

A related problem is that the field has yet to converge on a consensus about the criteria for a solid and interesting imaging result. Here are two things that I think are important in analyzing and reporting fMRI data. First, look at your raw time-course data. It is surprising how often this simple step will reveal that even a nice-looking activation blob is artifactual. Second, report not just significance levels but also means. Imagine trying to publish a behavioral study that reported only *P*-levels and not means—you would get laughed out of town! Yet this is widespread in the brain imaging literature.

---

bilateral lesions caused by multiple strokes, head trauma, encephalitis, or poisoning. It is likely that even if there were a specialized face perception module, other visual areas would be affected. As Nancy Kanwisher (2000) writes, "The chance that a stroke or head trauma . . . will obliterate all of the hypothesized face-processing region of cortex without affecting nearby cortical areas is similar to the chance that an asteroid hitting New England would obliterate all of the state of Rhode Island without affecting Massachusetts or Connecticut."

With this caveat in mind, we can still evaluate whether patients with prosopagnosia have a common foci of lesions. In her comprehensive monograph, *Visual Agnosia: Disorders of Object Recognition and What They Tell Us about Normal Vision*, Martha Farah (1990) iden-

Authors: You've taken a different tack than most imagers in your work, choosing to focus solely on a single neural area rather than describe all of the areas activated by a particular task. For example, your face perception studies only report activity in the fusiform face area. It makes for a simpler paper. But couldn't this be misleading?

NK: I think both region-of-interest approaches and whole-brain approaches to imaging have value. In part, the choice of one approach over the other reflects one's hunch about just how modular the mind/brain is. If it does have natural joints and parts, then it seems a sensible approach to try to get to the bottom of what each of these parts does. Repeated testing of the same region of interest is a good way—if not the only way—to do that. If instead you think that the brain is more like porridge, with most functions distributed over many different cortical regions, then it might make more sense to look at the whole brain. The worry in the latter case is that you may end up with a laundry list of activated regions, about which all you can conclude is that the function you have studied is carried out in a "distributed cortical network."

One of the things I love about the visual system is that it really does appear to have joints and parts, perhaps to a greater degree than other parts of the brain. That's why we focus on region-of-interest approaches in my lab. But we also do whole-brain analyses. In fact, you have to look at the whole brain (or much of it) to discover regions of interest in the first place. And even in studies designed for region-of-interest analyses, we look to see what is going on in the rest of the brain. Indeed, it is clear from studies in many different labs that the fusiform face area is not the only face-selective region of the cortex. Sometimes you see a face-selective region in the superior temporal sulcus and sometimes you see one in the posterior lateral occipital region. But we see these regions clearly in only about half of subjects, making it difficult to reach strong conclusions about their function. Ultimately, a full account of face perception will require an understanding of the contributions of these and probably other areas.

When all is said and done, it will probably turn out that the mind/brain is like a porridge, one with some tempting raisins. Right now we're having a lot of fun picking out the raisins. If other scientists are brave enough to try to say something about the goo in between, more power to them.

Authors: To stick with your food metaphor, do you envision a different sort of recipe will be required for understanding the goo? Your studies involve looking at brain activity when subjects passively view different stimulus sets such as faces and places. Must we require that the subjects engage in an active task—seek particular types of information—in order to make sense of the activation patterns in these other areas?

NK: Of course! It is an unusual luxury that much of the visual system functions quite automatically, so you can crank the activity up and down in different areas just by changing the images the subjects are looking at. But much of the rest of cognition (outside of perception and language understanding) is less automatic. You can't get subjects to carry out a difficult arithmetic problem simply by flashing it on their retinae. So for less automatic mental activities, varying the task makes more sense than varying the stimulus. What is generally *not* a good idea, though it is surprisingly common, is varying both the stimulus and the task at the same time.

Authors: Thanks, and bon apetit.

---

tified 81 patients reported to be prosopagnosic. Table 6.1 summarizes the general location of the pathology in 71 of these patients. What is most notable is that the lesions were bilateral in 46 patients (65%). The number would probably increase if the brains were examined at autopsy, as CT and MRI frequently do not detect lesions, especially with head trauma. For the remaining 25 patients (35%) with unilateral lesions, the incidence was much higher for right-sided lesions than left-sided ones. For bilateral and unilateral situations, the lesions generally involved occipital and temporal cortices.

Given the messiness of human neuropsychology, it is also important to look for converging evidence with the physiological tools of cognitive neuroscience. Neurophysiologists have recorded from the temporal lobe of primates to see if cells in this region respond

Table 6.1	Summary of Lesion Foci in Patients Described as Prosopagnosic*	
Bilateral (n = 46)		65%
Temporal		61%
Parietal		9%
Occipital		91%
Left only (n = 4)		6%
Temporal		75%
Parietal		25%
Occipital		50%
Right only (n = 21)		29%
Temporal		67%
Parietal		28%
Occipital		95%

*Within each subcategory, the percentages indicate how the lesions were distributed across the temporal, parietal, and occipital lobes. The sum of these percentages is greater than 100% because many of the lesions spanned more than one lobe. The majority of the patients had bilateral lesions. Adapted from Farah (1990).

specifically to faces. In one study (Baylis et al., 1985), recordings were made from cells in the superior temporal sulcus while a monkey was presented with stimuli like the ones in Figure 6.36. Five of these stimuli were faces, four of other monkeys and one of an experimenter. The other five stimuli ranged in complexity but included the most prominent features in the facial stimuli. For example, the grating reflected the symmetry of faces, and the circle was similar to eyes. Some cells were highly selective: They responded only to the clear frontal profile of another monkey. Other cells raised their firing rate for all facial stimuli. Nonfacial stimuli hardly activated the cells. In fact, compared to spontaneous firing rates, activity decreased for some nonfacial stimuli. The behavior of these cells closely resembles what would be expected of a grandmother cell in almost a literal sense, although no one has ever compared pictures of grandma and grandpa.

Two decades of research confirmed that cells in two distinct regions of the temporal lobe are preferentially activated by faces, one in the superior temporal sulcus and the other in the inferior temporal gyrus (Rolls, 1992). We cannot conclude that cells like these respond only to faces. It is impossible to test all stimuli, and many cells in this cortical region are triggered by nonfacial stimuli. Various alternative ideas, though, have

been ruled out. For example, one might suppose that facial stimuli evoke emotional responses and it is this property that causes a cell to respond strongly to a face and not to other, equally complex stimuli. However, the same cells are not activated by stimuli that produce a fear response in monkeys.

The debate concerning a dedicated face perception area has sprung full bloom in the functional magnetic resonance imaging (fMRI) literature. This methodology is well suited to address this problem as the spatial resolution of fMRI can yield a much more precise picture of face-specific areas than can be deduced from lesion studies. Two questions can be asked by comparing conditions in which subjects view different classes of stimuli: First, what neural regions show differential activation patterns when the subject is shown faces compared to the other stimulus conditions? Second, do these "face" regions also respond when the nonfacial stimuli are presented?

In one such study (McCarthy et al., 1997), subjects were presented with pictures of faces, inanimate objects, or random texture patterns (Figure 6.37). Compared to passive viewing of the random patterns, faces led to a stronger BOLD response along the ventral surface of the temporal lobe in the **fusiform gyrus**. When the assorted-objects condition was used as the comparison, the response in the fusiform gyrus of the right hemisphere remained significant. These results are consistent with the lesion data in that they indicate an important role for the fusiform gyrus in face perception, especially in the right hemisphere. Indeed, the fMRI data from this study suggest that a dedicated face module is only found in the right hemisphere, given that the left-hemisphere fusiform area responded similarly to faces and the inanimate objects. A second study in which the pictures of faces were replaced by pictures of flowers confirmed the general nature of the left-hemisphere response. As with faces, the left-hemisphere fusiform region was activated when these stimuli were compared to the random texture patterns; the right-hemisphere region was silent.

Numerous imaging studies have confirmed engagement of the fusiform gyrus during a range of face perception tasks. Indeed, this region has come to be referred to as the *fusiform face area*, or *FFA*, in the literature. However, even in the right hemisphere, the FFA does not appear to be exclusively activated by faces. Tasks requiring judgments about birds or cars also produce greater activation in the FFA, compared to a baseline condition in which no stimulus is presented (Gauthier et al., 2000). The increase, though,

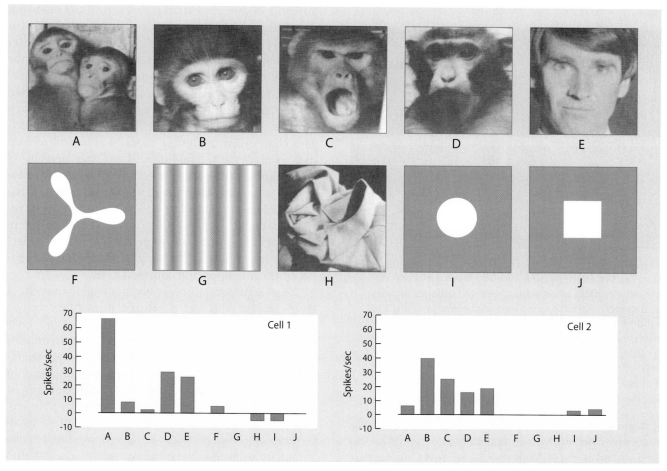

Figure 6.36 Face cells in the superior temporal sulcus of the macaque monkey. The graphs show the response of two cells to the ten stimuli (labeled A–J). Both cells responded vigorously to many of the facial stimuli. Either there was no change in activity when the animal looked at the objects, or in some cases, the cells were actually inhibited relative to baseline. The firing rate data are plotted as a change from baseline activity for that cell when no stimulus was presented. (Bottom) Adapted from Baylis et al. (1985).

is considerably smaller than with faces. One object class that does not produce significant activation in the FFA is letter strings. For these stimuli, the activation is primarily in the left hemisphere, including the inferior occipital cortex and a lateral region along the occipital-temporal border.

Dissociations of Face and Object Perception

The provocative physiological results just described support our first criterion: Face perception appears to utilize distinct physical processing systems. In turning to the second criterion—whether face and object perception can be dissociated—we must keep in mind the problems with single dissociations. As we discovered, many case reports recount patients who have a selective disorder in face perception; they cannot recognize faces but have little problem recognizing

objects. Even so, this evidence does not mandate a specialized processor for faces. Perhaps the tests that assess face perception are more sensitive to the effects of brain damage than the ones that evaluate object recognition.

Michael Tarr of Brown University and Isabel Gauthier (2000) of Vanderbilt University have elegantly articulated this point. One concern is that neuropsychological studies tend to focus on simple measures such as overall accuracy—a person either correctly identifies an object or face or fails to do so. As emphasized in Chapter 4, reaction time measures can provide additional insight, revealing the time course of the underlying processes. When such measures are used in face and object recognition tasks, prosopagnosia patients demonstrate impairment on both tasks; if accuracy alone had been measured, their deficit would have been selective for the face task. Why might this be? One

(a)   (b)

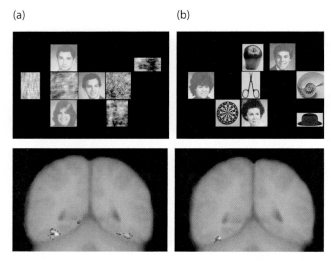

Figure 6.37 Isolating neural regions during face perception. **(a)** Bilateral activation in the fusiform gyrus was observed with fMRI when subjects viewed collages of faces and random patterns compared with collages of just the random patterns. **(b)** The activation was restricted to the right hemisphere in the comparison of faces plus objects to objects alone. Note that following neuroradiological conventions, the right hemisphere is on the left. Adapted from McCarthy et al. (1997).

intriguing hypothesis comes from the Web page comments of a person with congenital prosopagnosia, who writes, "People will have come and gone long before you can identify them. So you never do. . . . People in your presence who don't know of your face blindness will be offended at your failure to recognize them" (quoted in Gauthier et al., 1999). As this quote suggests, experience may have taught prosopagnosics to shy away from face perception tasks. Objects, on the other hand, never get offended.

Nonetheless, there are some striking cases of the reverse situation—patients with severe object recognition problems without any evidence of prosopagnosia—to support the second criterion. A dramatic example is C.K., the patient who could only reproduce pictures by laboriously following the outer boundaries (see Figure 6.29). Consider his perception of the still life shown in Figure 6.38, a painting by the quirky sixteenth-century Italian Giuseppe Arcimbaldo. This scene stumped C.K.; he reported a mishmash of colors and shapes that failed to gel into a coherent percept. But when the painting was turned upside down, C.K.'s perception of the face, even in caricature form, was immediate.

When we consider the kinds of tasks typically used to assess face and object perception, a different concern arises. Face perception tests are qualitatively

different in one important respect from the ones that test recognition of common objects. The stimuli for assessing face perception are all from the same category: faces. The subjects' task may be to decide whether two faces are the same or different, or they may be asked to identify specific individuals. When object perception is tested, the stimuli cover a much broader range. Here, subjects are asked to discriminate chairs from tables, or identify common objects such as clocks and telephones. Thus, face perception tasks involve within-category discriminations, whereas object perception tasks involve between-category discriminations. Perhaps the deficits seen in prosopagnosia patients do not reflect a selective problem in face perception but a more general problem in perceiving the subtle differences that distinguish between the members of a common category.

This argument is probably not valid, though. A sheep farmer who had developed prosopagnosia was tested on a set of within-category identification tasks (Figure 6.39), one involving people and the second involving sheep (McNeil and Warrington, 1993). In the familiar faces test, the farmer performed at the level of chance. In contrast, he was able to pick out photographs of sheep from his own flock. In a second experiment, recognition memory was tested. After viewing a set of pictures of sheep or faces, he was shown these same stimuli intermixed with new photographs. The patient's performance in recognizing the sheep was higher than that of other control subjects, including other sheep farmers! For faces, though, the patient's performance was at the level of chance, while the control subjects' performances were close to perfect.

Farah (1994) also argued that prosopagnosia cannot be attributed to the fact that face recognition entails a difficult within-category discrimination. Her experiments demonstrated that face perception does not involve the same processes and representations as object perception. A central figure in this work was a prosopagnosia patient, L.H., who had diffuse brain damage from a car accident. L.H. was tested with two recognition memory tests, one with face stimuli and the other with eyeglasses. Using only eyeglasses for the second set mitigates the criticism that nonfacial object recognition tests, compared to facial recognition tests, involve stimuli much more distinct from one another. In the study phase, half the stimuli from each set were presented. Then, in the memory test, all the stimuli were presented and the patient had to judge if the stimulus was old or new. Control subjects correctly recognized 85% of the faces and 69% of the eyeglasses. L.H. recognized only 64% of the faces correctly in

Figure 6.38 Keep an eye on the turnip when you turn the book upside down. From Moscovitch et al. (1997).

comparison to 63% of the eyeglasses. A patient with integrative agnosia had an opposite pattern of deficits: an accuracy of 98% for recognizing faces but only at the level of chance for eyeglasses.

An unusual phenomenon, the face-inversion effect has also been offered in support of the faces-are-special argument. Look at the face in Figure 6.40. Try recognizing the face in its upside-down orientation and then turn your book upside down. The effect is striking. Recognition is immediate when the stimulus is viewed in its proper orientation, but difficult with inverted faces. The effect is not limited to humans. Chimpanzees also are better at recognizing other chimp faces when presented upright (Parr et al., 1998). Interestingly, they do not show the inversion effect when shown the faces of another primate species, the capuchin, or a set of inanimate objects, automobiles.

One interpretation of the inversion effect is that we no longer can use a specialized face-processing system but must revert to a more analytic, analysis-by-parts mode of processing for an upside-down face. These faces constitute the perfect control stimulus for assessing the special status of face perception. While stimuli in the normal orientation and upside-down orientation contain identical information, only the normal condition should be disrupted in patients with prosopagnosia. This is exactly what was found. In fact, L.H. had the opposite of an inversion effect. For the upright and upside-down conditions, control subjects were correct on 94% and 82% of the trials, respectively. L.H.'s performance was better with upside-down faces (72%) in comparison with upright faces (58%). One surprising finding was that when viewing upright faces, L.H. could not bypass his damaged face recognition system. If this

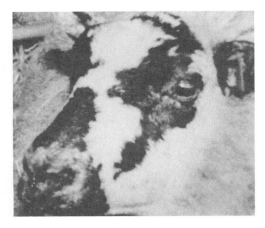

Figure 6.39 Face-specific recognition deficits cannot be attributed to the fact that faces represent a difficult within-category discrimination. A sheep farmer, W.J., with prosopagnosia demonstrated impairment in recognizing famous faces yet was more accurate in recognizing individual sheep in comparison to control subjects. In one of the face discrimination tasks, the subjects were shown three faces and asked to point to the one familiar face. In this example, the familiar face is Norman Tebbit, a well-known (to the British population) politician of the Thatcher era. The faces on the left and right were unfamiliar to the subjects. The sheep were either familiar or unfamiliar to the subjects. From McNeil and Warrington (1993).

were so, he would have performed well with both stimuli. But he continued to utilize his face recognition system, even though it failed to produce the representations required for the recognition memory task.

Research with healthy people reinforced the notion that face perception requires a representation that is

Figure 6.40 Who is this person? Is there anything unusual about the picture? Recognition can be quite difficult when faces are viewed upside down. Even more surprising, we fail to note a severe distortion created by inverting the eyes and mouth—something that would be immediately apparent when viewed right side up. The person is former British Prime Minister Margaret Thatcher, a face that would be widely recognized even in the United States in the late 1970s.

not simply a concatenation of individual parts. In one study, subjects were asked to recognize line drawings of faces and houses (Tanaka and Farah, 1993). Each stimulus was constructed of limited parts. For faces, the parts were eyes, nose, and mouth; for houses, the parts were doors, living room windows, and bedroom windows. In a learning phase, subjects saw a name and either a face or a house (Figure 6.41). For the face, they were instructed to associate the name with the face. For example, Larry had round eyes, a pointed nose, and a narrow mouth. For the house, they were instructed to learn the name of the person who lived in the house. For example, Larry lived in a house with a three-paneled door, shuttered living room window, and rounded bedroom windows.

After learning, a recognition memory test was given. The critical manipulation was whether the probe item was presented in isolation or in context, embedded in the whole object. For example, when asked whether the stimulus matched Larry's nose, the nose was presented either by itself or in the context of Larry's eyes and mouth. As predicted, house perception did not depend on whether the test items were presented in isolation or as an entire object. But face perception did. The subjects were much better at identifying the person when shown a face containing all three features.

Two Systems for Object Recognition

If we accept that there is a specialized system for face perception, a natural question is whether other forms of object perception also entail specialized systems. To answer this question, Farah (1990) focused on another subtype of visual agnosia, acquired **alexia.** *Acquired alexia* refers to reading problems after a patient has had a stroke or head trauma. They understand spoken speech and can speak without problems. Moreover, many patients retain the ability to write, a syndrome known as *alexia without agraphia.* But reading is painstakingly difficult. Errors usually reflect visual confusions. The word *ball* may be misread as *doll* or *snake* as *stale.* Like prosopagnosia, alexia involves a within-category deficit, the failure to discriminate between items that are a lot alike.

Anatomically, alexia and prosopagnosia are associated with very different neural regions. Alexia is seen with left-hemisphere lesions, particularly lesions that encompass an extrastriate occipital "word-form" area or the angular gyrus at the junction of the occipital, temporal, and parietal lobes. For prosopagnosia, the damage is either bilateral or unilateral in the right hemisphere (De Renzi et al., 1994) and includes ventral regions of the occipital and temporal lobes. Also demonstrating an anatomical dissociation, fMRI scans show different regions being activated when letter strings are presented compared to when faces are presented.

Prosopagnosia and alexia rarely occur in isolation. Both types of patients usually have problems with other types of object recognition. Importantly, a double dissociation between prosopagnosia and acquired alexia becomes evident when we consider the patterns of correlation between three types of agnosia: agnosia for faces, agnosia for objects, and agnosia for words. Table 6.2 lists the pattern of co-occurrence reported by Farah (1990) in a thorough review of the literature on visual associative agnosia. She did not tally the number of patients with a selective deficit in reading because they were so numerous. She did find several patients whose ability to recognize all three types of materials was impaired. The lesions were probably extensive and affected multiple processes. More telling was the dissociation of face and word perception. Either deficit frequently occurred along with a disorder in object recognition. But, most critically, Farah could not find a single unequivocal patient in whom face perception and word perception were impaired but object perception was not. Thus, one could be prosopagnosic and agnosic without being alexic. Or, one could be agnosic and alexic without being prosopagnosic.

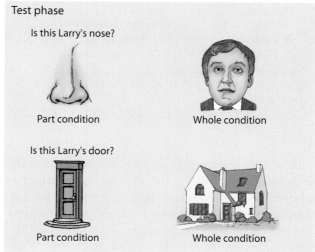

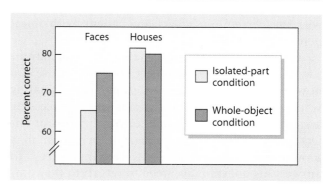

Figure 6.41 Facial features are poorly recognized in isolation. Subjects first learned the names that correspond with a set of faces and houses. During the recognition test, subjects were presented with either a face, a house, or a single feature from the face or house. They were asked if a particular feature belonged to an individual. When presented with the entire face, subjects were much better in identifying the facial features. Recognition of the house features was the same in both conditions. Adapted from Farah (1994).

A second significant result can be gleaned from Table 6.2. While alexia and prosopagnosia do not occur together, it is equally telling that agnosia for objects is always accompanied by a deficit in either word or face perception, or both. Because patients

Table 6.2	Patterns of Co-occurrence of Prosopagnosia, Visual Agnosia, and Alexia*	
Deficits in all three		21 patients
Selective deficits		
Face and objects		14 patients
Words and objects		15 patients
Faces and words		1 patient (possibly)
Faces alone		35 patients
Words alone		Many patients described in literature
Objects only		1 patient (possibly)

*Note that there was only a single case in the literature reporting a patient who showed impairment in recognizing both faces and words, but not objects. Adapted from Farah (1990).

with deficits in object perception also have a problem with one of the other types of stimuli, we might be tempted to conclude that object recognition involves two independent processes. It would be unparsimonious to postulate three processing subsystems. If this were so, we would expect to find three sets of patients: those with deficits in word perception, those with face perception deficits, and patients with object perception deficits.

Given that the neuropsychological dissociations suggest two systems for object recognition, we can now examine the third criterion for evaluating whether face perception depends on a processing system distinct from the one for other forms of object perception: Do we process information in a unique way when attempting to recognize faces? To answer this question, we need to return to the computational issues surrounding the perception of facial and nonfacial stimuli. Are there differences in the way information is represented when we recognize faces in comparison to when we recognize common objects and words? Farah maintained that face perception is unique in one special way: Whereas object recognition decomposes a stimulus into its parts, face perception is more holistic. We recognize an individual according to the facial configuration. A person is not recognized by his or her idiosyncratic nose or eyes or chin structure. The individual parts are not sufficient for face recognition; analyzing the structure and configuration of these features is what counts—hence, integrative agnosia and prosopagnosia represent two extremes.

The patient with integrative agnosia cannot identify the critical parts that form an object. The prosopagnosic patient may succeed in extracting these parts but cannot derive the holistic representation necessary for face perception.

In Chapter 10 we will see that the distinction between **analytic processing** and **holistic processing** has been important in theories of hemispheric specialization, with the former associated with the left hemisphere and the latter associated with the right hemisphere. In that chapter, we will review computational hypotheses that might underlie the difference between analytic and holistic processing. For present purposes, it is useful to note that alexia and prosopagnosia are in accord with this laterality hypothesis.

Theodor Landis and his colleagues at the University of Munich provide a nice example of how the distinction between holistic and analytic processing can be useful for understanding the representational deficits associated with prosopagnosia and alexia, respectively (Rentschler et al., 1994). When presented with various handwriting examples obtained from the patients' friends, the prosopagnosic patient could read the words but was unable to recognize the handwriting. The alexic patient showed the opposite pattern: She was unable to read the words despite being able to identify the handwriting. Recognizing the unique handwriting style of an individual would depend on the overall pattern formed by the letter strings, rather than the shapes of individual letters. Thus, the impaired holistic processing ability seen in prosopagnosia is not limited to faces. Holistic processing has also been assumed to increase with expertise. Interestingly, perceptual expertise is associated with increased activation in the FFA for various classes of stimuli (Tarr and Gauthier, 2000).

When viewed in this way, the question of whether face perception is special has been changed in a subtle, yet important way. Farah's model emphasizes that higher-level perception reflects the operation of two, distinct representational systems. The relative contribution of the analysis-by-parts and holistic systems will depend on the task (Figure 6.42). Face perception is at one extreme. Here, the critical information requires a holistic representation to capture the configuration of the defining parts. For these stimuli, discerning the parts is of little importance. Think how hard it is to notice that a casual acquaintance has shaved his mustache. Rather, recognition requires that we perceive a familiar arrangement of the parts. Faces are special, in the sense that the representation derived from an analysis by parts is not sufficient.

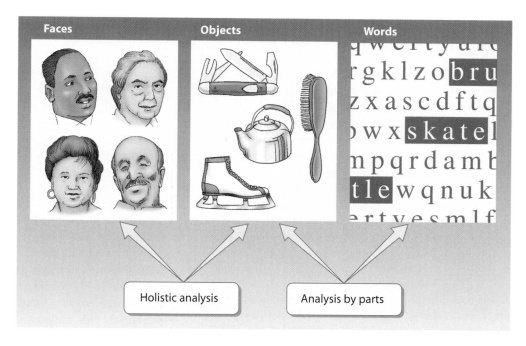

Figure 6.42 Farah's two-process model for object recognition. Recognition can be based on two forms of analysis, holistic and part based. The contribution of these two systems varies for different classes of stimuli. Analysis by parts is essential for reading and is central for recognizing objects. A unique aspect of face recognition is its dependence on holistic analysis. This process also contributes to object recognition.

Words represent another special class of objects, but at the other extreme. Reading requires that the letter strings be successfully decomposed into their constituent parts. We benefit little from noting general features such as word length or handwriting. We have to recognize the individual letters in order to differentiate one word from another.

Recognizing objects falls somewhere between the two extremes of words and faces. Defining features such as the number pad and receiver can identify a telephone, but recognition is also possible when we perceive the overall shape of this familiar object. If either the analytic or holistic system is damaged, object recognition may still be possible through the operation of the intact system. But performance is likely to be suboptimal. Thus, agnosia for objects can occur with either alexia or prosopagnosia.

THE RELATIONSHIP BETWEEN VISUAL PERCEPTION, IMAGERY, AND MEMORY

Imagine walking along the beach at sunset. The glowing orange sun has kissed the water's edge and the cloud-streaked sky bursts into a palette of reds, violets, and blues. Screeching seagulls flutter over the remains of a fish washed up in the afternoon tide. The smell of the rotting carcass is pungent, adding to the saltiness of the ocean mist.

Our images can be vivid. Most likely our images are of a specific place where we once enjoyed an ocean sunset. Some details may be quite salient; others may require further reflection. Were any boats passing by on the horizon in the image? Was the surf calm or were the waves rolling in, perhaps topped by a couple of surfers trying to catch one last ride for the day?

The relation between perception and imagery has been a point of heated debate in psychology for more than two decades (Kosslyn, 1988). At the center of this debate has been the question of whether imagery uses the same neural machinery as perception uses. When we imagine our beachside sunset, are we activating the same neural pathways and performing the same internal operations as when we gaze upon such a scene with our eyes?

Introspection provides few answers to this debate. If asked to imagine a ripe banana, we can easily describe the oblong shape, the bright yellow color, and perhaps the splotches of black where the fruit has been bruised. Have we really activated neurons in the shape and color regions of visual areas? Or is this image created from the knowledge of what bananas look like without having to "see with the mind's eye"?

Experimental psychologists have devised many ways to answer these questions. Several studies demonstrated similarities in how we process images and percepts. Roger Shepard of Stanford University developed several tasks to explore the dynamics of mental processing (Podgorny and Shepard, 1978). In one task, subjects view a grid of twenty-five squares that contains either a block letter or one that they are to imagine as a designated letter (Figure 6.43). The display is then turned off, and after a short delay, the grid reappears, containing a dot in one of the squares. The subjects must decide if the dot falls on or off the previously seen or imagined letter. Responses are considerably slower when the dot is near an edge of the letter. Most important, this effect is as strong in the imagery condition, as if the subjects must inspect their image in detail.

Figure 6.43 Recall processes appear to be similar when remembering either a seen or imaged object. Subjects are shown a grid that contains a letter **(left)** or upon which they are to imagine the letter. After a delay, a probe dot is presented **(right)** and the subjects must judge whether the probe was on or off the seen or imagined letter. The time required to make these judgments is quite similar in the two conditions. For example, subjects are slower to respond "NO" in both conditions when the probe dot is close to the edge of the seen or imagined letter. Adapted from Podgorny and Shepard (1978).

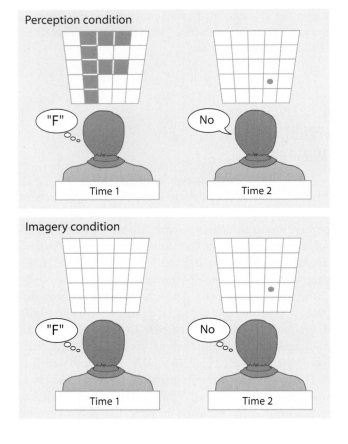

Neuropsychological research may provide the most compelling evidence of shared processing for imagery and perception. Martha Farah (1988) carefully reviewed the literature, identifying patients in whom brain lesions that caused perceptual deficits also led to corresponding deficits in imagery. Strokes may isolate visual information from areas that represent more abstract knowledge, causing difficulty in both perception and imagery tasks. One patient was able to sort objects according to color, but when asked to name a color or point to a named color, her performance was impaired. With imagery tasks, the patient could not answer questions about the colors of objects. She could say that a banana is a fruit that grows in southern climates but could not name its color. Even more surprising, the patient answered metaphorical questions about colors. For example, she could answer the question, "What is the color of envy?" Questions like these cannot be answered through imagery.

Patients with higher-order visual deficits have related deficits in imagery. Consider two patients, one with bilateral lesions of the temporal cortex and left-sided occipital involvement and another with bilateral lesions of the parieto-occipital cortex. The patient with the more ventral, occipitotemporal lesions had difficulty imaging faces or animals. He described an elephant as having long legs and a neck that could reach the ground to pick things up, and Abraham Lincoln as having a short, round face. Despite these difficulties, this patient could readily draw a floor plan of his house and locate major cities on a map of the United States. In contrast, the patient with damage to the dorsal pathways produced vivid descriptions when he was asked to image objects, but failed spatial imagery tasks. He described the elephant as being big, with a long nose, big floppy ears, a little tail, and four thick legs. Yet this patient was unable to describe how to get around his own neighborhood or provide knowledge of more extensive geographic regions. Chicago was described as being north of Boston. Together, these patients provide evidence of a dissociation in imagery of what-where processing that closely parallels that observed in perception.

Another source of neuropsychological evidence comes from physiological studies of normal subjects. Farah measured evoked potentials while subjects read a list of words, or read the words and simultaneously imaged their referents. The early component of the event-related potential was identical for the two conditions. But in the imaging condition, the waveform representing the occipital electrodes increased. Although the stimuli were the same in both conditions, the imagery condition selectively enhanced activation in visual areas.

Similar findings have been reported in PET studies (Kosslyn et al., 1993). In fact, the evidence suggests that imagery not only activates visual association areas but also produces increases in regional cerebral blood flow in the primary visual cortex. As with all modern imaging studies, it is difficult to know if this activity is essential for imagery. It may be that mental images are generated in secondary visual areas and that the activation in primary visual cortex results from feedback connections. Steven Kosslyn and his associates at Harvard University used a novel transcranial magnetic stimulation (TMS) method to assess the functional importance of visual cortex in imagery. The TMS coil was placed over primary visual cortex, and subjects received a pulse every second for 10 minutes. This procedure was hypothesized to disrupt neural activity in the target structure for about 10 minutes after the stimulation ends. As a control condition, the TMS procedure was repeated, but here the coil was reoriented such that the magnetic pulses did not reach the brain. In comparison to the sham condition, subjects were slower on making imagery-based judgments following TMS to the visual cortex. These results are consistent with the idea that representations in primary visual cortex are invoked during mental imagery, although it is possible that secondary visual areas were also disrupted by the stimulation via the feedforward connections from the primary visual cortex.

The evidence provides a compelling case that mental imagery uses many of the same processes critical for perception. The sights in an image are likely to activate visual areas of the brain; the sounds, auditory areas; and the smells, olfactory areas. Indeed, the imagery results demonstrate that memory for perceptual information is not independent of perceptual processes. We need not think of perceptual processing and the memory of that processing as distinct neural entities. Perceptual memory might simply reactivate perceptual pathways.

Despite the similarities between perception and imagery, it would be premature to conclude that the two are identical. Patients frequently have corresponding deficits in perception and imagery, but there are notable exceptions. Patients with no evidence of agnosia may have imagery deficits, being unable to recall whether animals have long or short tails, or whether the letter *P* has a curved segment. By contrast, intact imagery has been observed in patients with severe agnosia. Similarly, there are patients with severe achromatopsia without corresponding deficits on tasks that would appear to require the use of color imagery. One such patient not only could recall the colors of familiar objects (e.g., the U.S. Postal Service's Mailboxes), but also could make more subtle distinctions such as gauge that a plum contains more red than an eggplant (Shuren et al., 1996).

We can try to account for the dissociations between imagery and perception by supposing that while imagery and perception share forms of representation, they may achieve them in distinct ways. Consider two versions of a task that requires an observer to decide if a donkey's ears are pointed or floppy. In the imagery version, the observer must generate an internal representation. It is unlikely that he or she could do this by imaging the donkey's ears. Instead, an image of the entire animal will probably be created, and the observer will zoom in for a closer examination of the head. Imagery as such requires that the parts be generated and arranged appropriately. Now consider the perceptual version where an observer is shown a donkey's picture and, as with imagery, is asked to identify parts and make comparisons. In perception, the parts need not be generated but must be extracted from the representation of the stimulus. The critical process is to partition the percept into its component parts—to segment the ears from the head.

Dissociations between imagery and perception have led to models in which access to visual knowledge can occur through multiple pathways. Consider again patient C.K., the patient with severe integrative agnosia who can recognize vegetable faces. C.K. had no problems with imagery tasks. When read the names of objects, C.K. produced excellent drawings (Figure 6.44). When shown the same drawings during a subsequent visit, he failed to identify any of them. In a similar way, while C.K. had difficulty recognizing letters, he was successful at imagery tasks that required multiple transformations. He readily provided the answer to the question, "What letter do you get when you take the letter *L*, flip it from top to bottom and add a horizontal line in the middle." In these tasks, C.K. demonstrated that he can represent objects visually and has knowledge of visual attributes. His representations are not simply of whole objects: The letter transformation task demonstrated that he can generate and manipulate the parts. C.K. failed when he had to extract parts from the perceptual input. Without this capability, his object recognition capacity is limited.

One provocative issue that has received relatively little attention is how visual memory changes over time following damage to systems involved in visual perception. When deprived of a consistent input, it would seem reasonable to expect that our knowledge base would be reorganized. In his essay "The Case of the Colorblind Painter," Oliver Sacks (1995) described Mr. I, a successful artist who suffered complete achromatopsia following a car accident. For Mr. I, a colorless world was initially "awful, disgusting." He was horrified upon returning to his studio to discover all of his brilliant abstract paintings, done in the most vivid colors, now appeared a morass of grays, blacks, and whites. His abhorrence for a colorless world extended outside of the studio. He lost interest in food given its uniform, "grayish, dead appearance," and even sex became repugnant after he viewed his wife's

flesh and indeed, his own flesh as a "rat-colored" gray. Interestingly, his shock underscores the fact that his color knowledge was still intact. Mr. I could remember with great detail the colors he expected to see in his paintings. It was the mismatch between his expectation and what he saw that was so depressing. He shunned museums because the familiar pictures just looked wrong.

However, over the subsequent year, a transition occurred. Mr. I's memory for colors started to slip away. He no longer despaired when gazing at a tomato devoid of red or a sunset drained of all color. He knew that something wasn't quite right, but his sense of the missing colors was much more vague. Indeed, he began to appreciate the subtleties of a black and white world. Overwhelmed by the brightness of the day, Mr. I became a night person in which he could appreciate forms in the purity, "uncluttered by color." This change can be seen in his art (Figure 6.45). Prior to the accident, Mr. I relied on color to create subtle boundaries, to evoke movement across the canvas. In his black and white world, geometric patterns delineated sharp boundaries.

H.J.A., the patient described previously as having integrative agnosia, also experienced a similar loss of visual memory over time (Riddoch et al., 1999). Figure 6.46 shows two sets of drawings, one set drawn when he was first tested 4 years after his accident and the other set drawn 10 years later. While the changes might be attributed to the effects of aging, H.J.A. also showed a change over time in how he described common objects. When he was tested 15 years after his stroke, his descriptions were primarily verbal, focusing on the functions of the objects. He mentioned visual attributes less often than in his earlier test sessions. Interestingly, H.J.A.'s recognition of real objects had improved to near-normal levels by the time of the follow-up session, although he still showed impairment when he was tested on line drawings. As he reported, "My actual recognition has never got any worse or better. My ways of getting around it have improved—but not my recognition, at least that's what I believe." "Getting around" here would seem to mean a greater reliance on noting distinctive features such as the handle on a coffee cup or using depth and texture information. But his ability to access visual memory by integrating parts into a coherent percept has not improved. Correspondingly, the perceptual deficit has led to a decline in visual knowledge itself. We should not be surprised that a system that is constructed through perceptual experience would require that experience to remain intact, an issue that we will return to in Chapter 15.

Figure 6.44 The ability to draw objects from memory may be spared in agnosia. Patient C.K.'s drawings of a map of the United Kingdom and an electric guitar are shown. His ability to generate an internal visual representation would appear to be intact. However, C.K. was unable to recognize the objects in his own drawings on a subsequent visit.

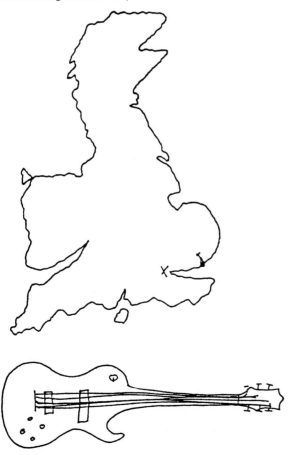

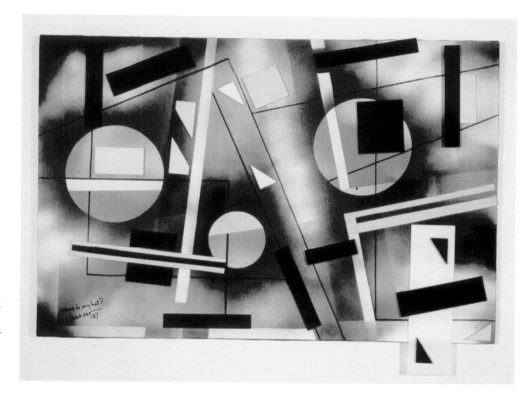

Figure 6.45 An abstract painting by Mr. I, produced 2 years after his accident. Mr. I was experimenting with colors at this time, although he was still unable to see them. From Sacks (1995).

Figure 6.46 Drawings produced by H.J.A 4 **(a)** and 14 **(b)** years after his stroke. His renditions of celery are shown on the left and an owl, on the right. The later drawings are much poorer, suggesting a loss of visual knowledge associated with his persistent agnosia. Age-matched control subjects show little deterioration in their drawing skills over a similar period. From Riddoch et al. (1999).

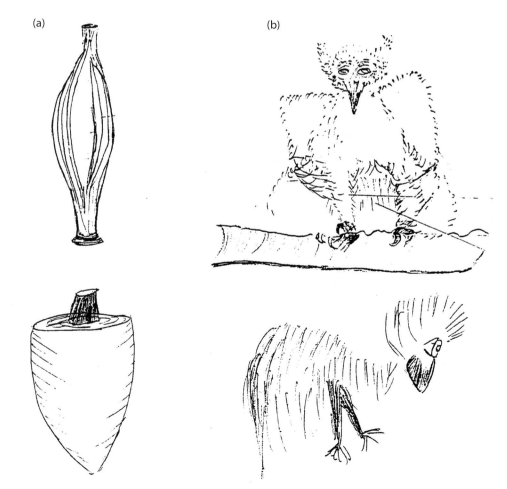

SUMMARY

This chapter provided an overview of the higher-level processes involved in visual perception. People, like most mammals, are visual creatures: Most of us rely on our eyes to identify not only *what* it is we are looking at, but also *where* to look, to guide our actions. These processes are surely interactive. To accomplish a skilled behavior such as catching a thrown object, we have to determine the object's size and shape and to track its path through space so we can anticipate where to place our hands.

Object recognition can be achieved in a multiplicity of ways and involves many levels of representation. To recognize an object, our visual system must overcome the variability inherent in the sensory input by extracting the critical information that distinguishes one shape from another. Perceptual categorization solves only part of the recognition problem. For this information to be useful, the contents of current processing must be connected to our stored knowledge about visual objects. Semantic categorization allows us to see the similarities between percepts and recognize the unique features of objects. We do not see a meaningless array of shapes and forms. Rather, visual perception is an efficient avenue for recognizing and interacting with the world (e.g. determining what path to take across a cluttered room or which tools make our actions more efficient).

Moreover, vision provides a salient means for one of the most essential goals of perception: recognizing conspecifics. One might suppose that the importance of face perception led to the evolution of an alternative form of representation, one that analyzed the global configuration of a stimulus rather than its parts. On the other hand, multiple forms of representation may have evolved, and face perception may be highly dependent on the configural form of representation. As seen in Farah's model, the evidence leads us to regard the dichotomy between part-based and configural-based representations as absolute. In object recognition, these two forms of representation interact, and their relative contribution varies according to the task and the objects being perceived.

The goals-oriented aspects of perception—that we use vision to guide our movements, to manipulate tools, or to recognize faces—also underscore the selective aspects of cognition. We are not passive processors of information. We can select from the dazzling array that impinges on our senses at any one time. Depending on our goals, the relative importance of different sources of information constantly changes.

KEY TERMS

agnosia

alexia

analytic processing

apperceptive agnosia

associative agnosia

category-specific deficits

dorsal (occipito-parietal) pathway

fusiform gyrus

gnostic unit

holistic processing

integrative agnosia

object constancy

optic ataxia

prosopagnosia

visual agnosia

ventral (occipito-temporal) pathway

view-dependent theories

view-invariant frame of reference

THOUGHT QUESTIONS

1. What are some of the differences between processing in the dorsal and ventral visual pathways? In what ways are these differences useful? In what ways is it misleading to imply a functional dichotomy of two distinct visual pathways?

2. Mrs. S recently suffered a brain injury. She claims to have difficulty in "seeing" as a result of her injury. Her neurologist has made a preliminary diagnosis of "agnosia," but nothing more specific is noted. In order to determine the exact nature of her agnosia, a cognitive neuropsychologist is called in. What behavioral tests could the neuropsychologist use to make a more specific diagnosis? Remember that it is also important to conduct tests to determine if her deficit reflects a more general problem in visual perception or memory.

3. A scientist working with the MRI system at the hospital hears about the case. Which anatomical and functional neuroimaging tests would the scientist

recommend, and what specific results would support each of the possible diagnoses?

4. Review different hypotheses concerning why brain injury may produce the puzzling symptom of disproportionate impairment in recognizing living things. What sorts of evidence would support one hypothesis over another?

5. As part of a debating team, you are assigned the task of defending the hypothesis that the brain has evolved a specialized system for perceiving faces. What arguments will you use to make your case? Now change sides. Defend the argument that face perception reflects the operation of a highly experienced system that is good at making fine discriminations.

SUGGESTED READINGS

DESIMONE, R. (1991). Face-selective cells in the temporal cortex of monkeys. *J. Cogn. Neurosci.* 3:1–8.

FARAH, M.J. (1990). *Visual Agnosia: Disorders of Object Recognition and What They Tell Us about Normal Vision.* Cambridge, MA: MIT Press.

KANWISHER, N. (2000). Domain specificity in face perception. *Nature Neurosci.* 3:759–763.

MILNER, A.D., and GOODALE, M.A. (1995). *The Visual Brain in Action.* New York: Oxford University Press.

RIDDOCH, M.J., and HUMPHREYS, G.W. (2001). Object recognition. In B. Repp (Ed.), *The Handbook of Cognitive Neuropsychology: What Deficits Reveal about the Human Mind,* pp. 45–74. Philadelphia: Psychology Press.

7

Selective Attention and Orienting

Do we perceive everything that illuminates our retinas, rattles our cochleas, caresses our skin, and wafts up our nostrils? How much do our internal desires, beliefs, momentary necessities, and intent affect our perceptual experiences? The answers to these two questions are "no" and "greatly." What we choose to attend to, and what grabs our attention can dominate our experience. Precisely what neuronal mechanisms are involved when we concentrate our attention on external events or when they capture our awareness is one of the most intriguing issues in cognitive neuroscience. Consider the following encounter.

A man in his early sixties is sitting comfortably in his doctor's office. Several weeks before, he had suffered a stroke but appeared to recover well, having no lasting speech or motor problems of note. However, his wife was very concerned that his vision seemed to be impaired, that he could not "see" normally, making it difficult for him to function in his daily life. His doctor, a neurologist, began to examine the man, and indeed, he could see that the patient was recovering well but did display some difficulty seeing objects. The doctor took a comb from his pocket and held it in front of the man (Figure 7.1a). "What do you see?" he asked. "Well, I'm not sure, but . . . oh . . . it's a comb, a pocket comb." Fine, said the doctor. Then, he held up a spoon whisked from the recesses of his white coat and asked the same question (Figure 7.1b). After a moment the patient replied, "I see a spoon." Fine again. The doctor then held up the spoon and the comb together, asking the man again to tell him what he saw. "I guess . . . I see a spoon," said the man. So the doctor held up the comb right behind the spoon in a crossed fashion so they would be overlapping in the same location but still visible, and asked again what the man saw (Figure 7.1c). He said he saw the comb. When the doctor asked about the spoon,

the patient said he did not see it, but then suddenly he said, "Yes, there it is, I see the spoon now." "And what else?" asked the doctor. "Nothing," replied the man. "You don't see anything else, nothing at all?" asked the doctor as he shook the spoon and the now-unseen comb vigorously in front of the man's face, hoping to attract his attention. The man stared straight ahead, looking intently, and finally said, "Yes, I . . . yes, I see them now . . . I see some numbers." "What?" said the amazed doctor, "Numbers, there are no numbers. What do you see?" The man squinted and appeared to strain his vision, moving his head ever so slightly, and replied, "Yes, I see numbers." The doctor then noticed that the man's direction of gaze seemed far away, and turned to glance over his own shoulder to spot a large clock on the wall behind him. Turning back toward the man, he said in a brisk tone, "Er . . . yes . . . numbers, on the face of that clock on the wall . . . fine. Thank you."

What is interesting about this scene is that the doctor was holding both objects in front of the man at the same time, and even overlapping them in space, in plain view and in good lighting. But the man only saw one item at a time, or saw yet some other item in the direction of his gaze, even when

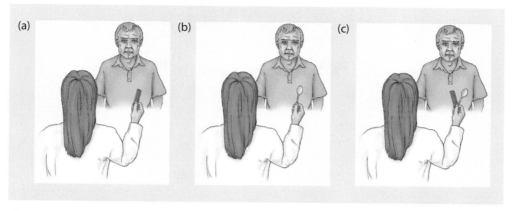

Figure 7.1   A patient is examined after recovering from a stroke affecting the cortex. **(a)** The doctor holds up a pocket comb and asks the man what he sees. The man says he sees the comb. **(b)** The doctor then holds up a spoon, and the man says he sees this too. **(c)** But when the doctor holds up both the spoon and the comb at the same time, the man says he can only see one object at a time. The patient has Balint's syndrome.

closer items were being presented for his viewing right in front of his nose. This patient has **Balint's syndrome,** which we will discuss later in the text. What form of brain damage could lead to the deficits observed in this patient, and what does this tell us about how vision and attention are organized in the brain? Consider the following question: Do you attend to things or to regions of space, or both? Perhaps the answer is somewhere here. Remember the discussion in Chapter 6 (see Figure 6.3) about the two dominant streams of visual cortical analysis? How might this organization of the visual system into spatial (where) and object (what) pathways interact with attention to create the deficits observed in Balint's syndrome patients?

In this chapter we will learn that attention involves both top-down voluntary processes and bottom-up reflexive or stimulus-driven mechanisms, and that they are in dynamic competition for control of the momentary focus of attention. These effects of attention influence the way information is processed in the brain, and can occur early during sensory processing, but perhaps not too early. Widespread brain networks interact to enable us to attend to relevant events and importantly, to ignore the irrelevant. These attentional networks involve structures in all the lobes of the cortex, as well as subcortical structures. However, these networks are highly specific for certain aspects of attention, and damage to different brain regions can lead to a variety of attention disorders, including Balint's syndrome and disorders of spatial attention such as unilateral neglect, which we will review at the end of the chapter.

We first consider the cognitive psychology of attention. Then we delve into the neurobiology and neurology of attention, using evidence from studies in animals and humans to understand why some things gain access to awareness while others are excluded, how the brain controls this key ability, and what happens when it is lost.

THEORETICAL MODELS OF ATTENTION

The concept of "attention" is at once intuitive and enigmatic. It involves everyday concepts like paying attention to your mother when she instructs you in how to behave, but also carries with it the elusive nature of human awareness, as when your attention is so focused on a thought that you do not hear your mother's admonitions—at your own peril. At first pass it may appear that attention is identical to seeing or perceiving, but on deeper reflection it is clear that attention involves something more than sensation and perception, though it undoubtedly interacts with them. For example, we can attend to things other than sensory inputs. Attention can be directed to internal mental processes such as thinking about memories or adding numbers in our head.

HOW THE BRAIN WORKS

Behavioral Arousal and Selective Attention

Attention has many meanings. It involves consciousness, awareness, and attentiveness, not to mention related deficits. Scientists concentrated primarily on global states of attention in the late 1950s and early 1960s, prior to the beginnings of the information-processing revolution. The emphasis then was on distinctions between conscious and unconscious states, such as sleep versus wakefulness. In the late 1940s, Giuseppi Moruzzi and Horace Magoun (1949) at the University of California at Los Angeles showed that stimulating a cat's brainstem changed its global behavioral state. In some regions of the brainstem, sleep was induced by stimulation, while in other regions wakefulness resulted.

These changes in global state or arousal can be related to specific neurons in the brain and are reflected by changes in the electroencephalogram (EEG), the small voltage fluctuations that accompany neuronal activity in the brains of all animals. The EEG is a clear indicator of the state of global arousal and sleep in normal subjects; one can pinpoint the change from wakefulness to sleep and to different stages of sleep by viewing the EEG alone. Such changes must, however, be distinguished from the changes in brain activity that form the basis for *selective* attention and selective perception.

Attentive behaviors have a hierarchical structure. At the most global level are states of alertness such as sleep and wakefulness. Wakefulness includes more and less attentive and more and less selective states: drowsiness, alertness, and hyperalertness such as when a life-threatening situation arises. At a finer level of description are levels within each awake, global state of awareness; here we reach levels of description that are appropriate for considering selectivity, as with the *cocktail party effect* described in this chapter.

The cocktail party effect is not just a global change in behavioral arousal or attentive state, but a reflection of a selective attention because the listener's goal is to attend to one source of sound while simultaneously ignoring potentially distracting inputs. This distinction between selective processes and global state changes is essential for its implications in selective attention research, and for the implications that nonselective behavioral changes can have for experimental design. For instance, finding that neurons are more active when an animal is excitedly performing a task, as compared with drowsily watching stimuli, might be an example only of a nonspecific state difference induced by the arousal of the task, and not evidence of neuronal mechanisms of selective processing. These design considerations plagued the interpretation of many early studies of selective attention, as described in the text. The problems were overcome by controls that cognitive neuroscientists used for studying auditory attention in humans. This victory is largely due to the appreciation of a theoretical framework in the cognitive psychology literature that distinguished between nonspecific global state changes and selective processes—just one example of how hard-fought theories fuel scientific advances.

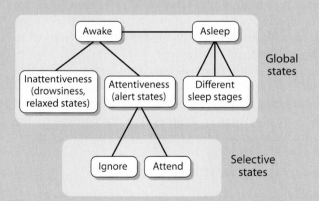

Hierarchical relationships between states of arousal, attentiveness, and selective attention.

Much of the introspective flavor of attention was captured by early psychologists like William James from Harvard University (Figure 7.2), brother of Henry, the novelist and playwright. In the nineteeth century, William James (1890) wrote

> Everyone knows what attention is. It is the taking possession by the mind, in clear and vivid form, of one out of what seem several simultaneously possible objects or trains of thought. Focalization, concentration of consciousness are of its essence. It implies withdrawal from some things in order to deal effectively with others, and is a condition which has a real opposite in the confused, dazed, scatterbrain state. . . .

In this insightful quote, James captures characteristics of attentional phenomena that are under investigation today. For example, his statement that "it is the taking possession by the mind" comments on the voluntary aspects of attention; we can control the focus of our attention. And his mention of "one out of what seem several simultaneously possible objects or trains of thought" refers to the inability to attend to many things at once, and hence the selective aspects of attention. James reinforces the idea of limited capacity in attention by noting that "it implies withdrawal from some things in order to deal effectively with others. . . ."

As clear and articulate as James's writings were, very little of the computational mechanisms and virtually none of the neurophysiological implementation of these processes were understood during his lifetime. Even after more than 100 years of investigation, the study of attentional mechanisms continues to accelerate as we seek to understand the neural substrates of attention and selective perception.

Let's define *attention* for the purpose of the first part of this chapter. Attention is a cognitive brain mechanism that enables one to process relevant inputs, thoughts, or actions while ignoring irrelevant or distracting ones (here we define **selective attention**—see Behavioral Arousal and Selective Attention). Attention is a covert mechanism, meaning that it can operate without overt adjustments of external sensory structures (e.g., one can direct visual attention to locations or objects in the world without moving the eyes). We can break attention down into two broad categories: (i) **voluntary attention** or "endogenous" attention, and (ii) **reflexive attention** (automatic) or "exogenous" attention. *Voluntary attention* refers to our ability to intentionally attend to something, while *reflexive attention* describes the phenomena where something, such as a sensory event, captures our attention. These two forms of attention differ in their properties and neural mechanisms, as we will see later in this chapter.

Studies of the cognitive neuroscience of attention have three principal goals: (1) to understand how attention enables and influences the detection, perception, and encoding of stimulus events, as well as the generation of actions based on the stimuli; (2) to describe what computational algorithms enable these effects; and (3) to uncover how these algorithms are implemented in the human brain's neuronal circuits and neural systems. Understanding these processes will ultimately permit us to comprehend how damage or disease in these systems leads to attentional deficits and cognitive problems.

We will begin our consideration of attention with a look at early studies of sensory attention. In 1894, Hermann von Helmholtz performed a fascinating experiment in visual perception (Figure 7.3). In one set of studies, he constructed a screen on which letters were painted at various distances from the center of the screen. He hung the screen at one end of his lab, which he completely darkened to prevent any light from illuminating the screen. Helmholtz then used an electrical spark to briefly illuminate the screen, essentially just a flash of light, like a camera flash. Helmholtz wanted to investigate aspects of visual processing when stimuli were briefly perceived, but he also came upon an interesting phenomenon. The screen was too large to view without moving the eyes. But if Helmholtz held his eyes still in the center of the screen, he could decide in

Figure 7.2 William James, the great American psychologist (1842–1910).

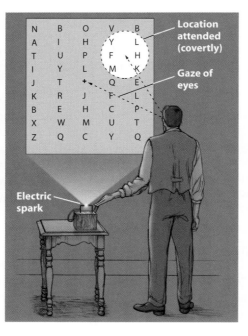

Figure 7.3 Hermann von Helmholtz (1821–1894). Experimental setup by Helmholtz to study visual attention. Helmholtz observed that while keeping his eyes fixated in the center of the screen, during a very brief illumination of the screen he could "covertly" attend to any location on the screen and perceive the letters located within this region but had difficulty perceiving the letters at other locations. He described this phenomenon as due to attention.

advance where he would pay attention; that is, he made use of something we now refer to as **covert attention,** wherein the visual field location to which gaze is directed can be different from the location to which one chooses to pay attention. In so doing, he observed that during the brief period of illumination, he could perceive the letters located within the region of the screen to which he had directed his attention in advance (without moving his eyes) but could not discriminate the letters elsewhere on the screen. Helmholtz noted that "these experiments demonstrated, so it seems to me, that by a voluntary kind of intention, even without eye movements, and without changes of accommodation, one can concentrate attention on the sensation from a particular part of our peripheral nervous system and at the same time exclude attention from all other parts."

From the work of James and Helmholtz, much of what we understood from introspective examination of attentional phenomena was laid out by the turn of the twentieth century. Not until the last half of the twentieth century, though, have psychologists been able to provide empirical support for any of the concepts put forth by James and Helmholtz. In the 1950s, psychologists asked about the mechanisms of attention, and the scientific study of Helmholtz's covert attention began.

The Cocktail Party Effect

British psychologist E.C. Cherry (1953) examined the so-called cocktail party effect. How is it that in the noisy, confusing environment of a cocktail party, people can focus on a single conversation? Is the conversation more salient because of the speaker's and listener's proximity? No, this explanation is inadequate because our everyday experience indicates that the loudest inputs are not always the best perceived, although this has an influence. Indeed, one goal of the listener at the cocktail party is to overcome the louder inputs of the environment (e.g., the music or nearby boisterous conversations) and to attend to the conversation of interest. Imagine that you have just met an interesting person at a party and you are desperately trying to hear what the person is saying so you can respond intelligently and with panache. The desired response to comments is not "Huh? . . . Excuse me? . . . What did you say?" Rather, you want to reply with style, and so you pay close attention to what is said and try to avoid interference from other inputs.

Selective auditory attention is the mechanism by which you achieve your goal—perception of a weak speech signal in a noisy environment—and avoid the necessity of cupping your ear, requesting the band to play more quietly, or insulting other party goers by telling them to pipe down. By selectively attending, you can perceive the signal of interest amidst the louder noise, and remain charming in a difficult social context. But if you think the person is boring, you may choose (or be unable to avoid) to attend to another conversation using covert attention to do so (Figure 7.4).

This cocktail party effect clearly demonstrates the kinds of phenomena that interest cognitive psychologists studying attention. Cherry investigated the effect by providing competing speech inputs to the two ears of

Figure 7.4 The cocktail party effect of Cherry (1953), illustrating how in the noisy confusing environment of the cocktail party, people are able to focus attention on a single conversation.

a normal subject through headphones (**dichotic listening**). In different conditions he asked people to attend to and verbally "shadow" (immediately repeat each word) a train of speech coming into one ear, while simultaneously ignoring similar inputs to the other ear.

Cherry discovered that when a different train of speech was played into each ear, and the subjects were asked to shadow what was played into only one ear at a time, they could not report any of the details of the speech in the unattended ear (Figure 7.5). This led

Figure 7.5 Cherry's shadowing experimental setup that presents auditory information to both ears of a subject. The subject is asked to "shadow" (immediately repeat) the auditory stimuli from one input.

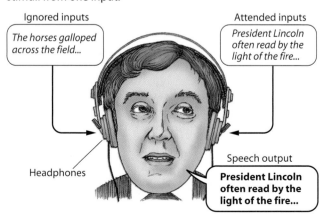

Cherry and others to propose that attention to one ear resulted in better encoding of the inputs to the attended ear, and perhaps loss or degradation of the unattended inputs to the other ear.

British psychologist Donald Broadbent (1958) at Cambridge University proposed a model of attention to explain data from researchers like Cherry. He conceptualized the *information-processing system*, a term that encompasses all aspect of the brain's processing of data including sensory inputs, as having a limited-capacity channel through which only a certain amount of information could pass (Figure 7.6). Hence, the many sensory inputs capable of entering higher levels of the brain for processing had to be screened to let the most important (attended) events through. Broadbent described this mechanism as a gate that could be opened for attended information and closed for ignored information.

Some features of Broadbent's model fit the data well and explained the processing limitations revealed in experiments like those of Cherry; but not all data could be understood by a strict gating mechanism. Cherry and his contemporaries such as Neville Moray (1959) also noted that even though information coming into the unattended ear was not noticed in most circumstances, if it contained information of high-enough priority to the subjects—for example, if their own name was presented to the unattended channel of inputs—then they often would be able to discriminate and orient to the higher-priority information. This concept of *intrusion of the unattended* inputs led many to believe that all information was actually analyzed equivalently, regardless of whether it was attended or ignored later during processing. This

Figure 7.6 Broadbent's model of selective attention. In this model, a gating mechanism determines what limited information is passed on for higher analysis. The gating mechanism shown here takes the form of top-down influences on early neural processing, under the control of voluntary, executive processes. Adapted from Broadbent (1958).

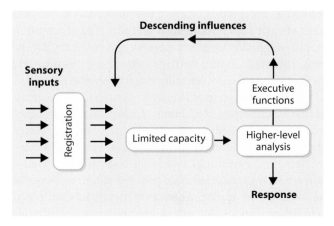

view set the stage for a debate that has persisted to the present, although we now have significant evidence from physiology to guide current models. The questions are: Can we completely close the gate on some inputs and eliminate them from further processing? If so, where does this gate reside in the information-processing system, and does this affect all subsequent encoding and memory processes that might normally act on those inputs? These questions have been addressed by *selection theories,* as described next.

Early- Versus Late-Selection Theories

A central issue raised by the cocktail party experiments and the gating model of Broadbent has to do with the stage(s) of stimulus processing where incoming signals can be selected or rejected by internal attentional processes. This is the "early- versus late-selection" idea in which cognitive psychologists asked about the extent of processing that supposedly ignored signals might actually attain. That is, if we are unaware of ignored conversations, perhaps we simply close the gate on irrelevant inputs before they reach conscious awareness. But the key question is, How much before?

Early selection is the idea that a stimulus need not be completely perceptually analyzed and encoded as semantic or categorical information before it can be either selected for further processing or rejected as irrelevant. Using the experiments and observations of Helmholtz (see Figure 7.3), we can describe early selection as the idea that during visual processing, it may be possible to process only the letters from the to-be-attended spatial position by gating information from irrelevant positions before that information is encoded as letters of the alphabet (i.e., letters are the nonperceptual meaning representations) within the brain. In early-selection models, the idea is that it may even be possible to select inputs prior to a full perceptual analysis of the stimuli's elementary features—for example, before analyses of luminance, form, and color are complete. Thus, the concept of early selection suggests that this type of attention could potentially alter our perceptions by changing the way ignored (and perhaps attended) inputs are processed at an early point in visual analysis.

Early-selection models can be contrasted with an alternative concept called **late selection,** which hypothesizes that attended and ignored inputs are processed equivalently by the perceptual system, reaching a stage of semantic (meaning) encoding and analysis. Thereafter, selection for additional processing and subsequent representation in conscious awareness can take place. This view implies that attentional processes cannot af-

fect our perceptions of stimuli by changing the way they are processed by the sensory-perceptual processing system. Instead, all selection takes place at higher stages of information processing that involve internal decisions about whether the stimuli should gain complete access to awareness, be encoded in memory, or initiate a response—the term *decisions* in this context refers to nonconscious processes, not a conscious decision on the part of the observer. Differential stages of early versus late selection are presented in Figure 7.7.

The strict versions of early-selection models proposed by Broadbent did not permit semantic encoding or analysis of unattended information. Given that high-priority signals could break through from supposedly unattended sources of sensory input (as in the case of subjects hearing their own name in dichotic listening studies), this original view of Broadbent was in need of modification. A simple gating mechanism which assumed that ignored information was lost could not accommodate the fact that sometimes unattended information was not perceived but at other times it was. Anne Treisman (1969), now at Princeton University, proposed that perhaps unattended channel information was not completely gated from higher analysis but was merely degraded or attenuated. Broadbent (1970) agreed with this view, and hence early versus late selection was modified to make room for the possibility that information on the unattended channel could pass the gate but in a greatly attenuated fashion. Because such information could reach the level of semantic analysis, if it was important enough, it might lead to an attentional switch to the unattended ear inputs (selection) after the level of semantic encoding, and thereby reach awareness.

Figure 7.7 Diagram of early versus late selection of perceptual processing. This conceptualization is concerned with the extent of processing an input signal might attain before it can be selected or rejected by internal attentional processes.

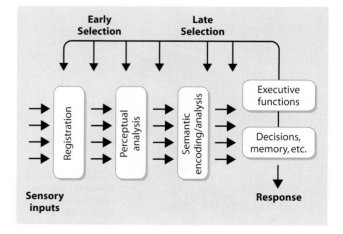

LIMITED-CAPACITY PROCESSES

One aspect of the early- versus late-selection debate is the concept of **limited capacity,** a concept that naturally flows from the observation that human performance suffers when overloaded by multiple inputs. The early- and late-selection views assume that the human (or animal) information-processing system cannot simultaneously process multiple inputs if there is a high information load, so the system must make hard "decisions" about what to process next—the core of selection. When a stage of processing has a lower capacity than the previous stage, then a processing **bottleneck** occurs where all inputs cannot gain access to or pass that stage of processing; this is hypothesized to happen at key way stations during information processing. Presumably the information-processing system evolved selection mechanisms to control information flow at these bottlenecks and to establish priorities.

The difference between early- and late-selection theories has sometimes been framed by the following questions: Where does the bottleneck reside, and hence, where must selection take place? Is it at an early perceptual level or at later encoding, decision, or response levels? Let us examine the evidence from normal subjects who were given detection and discrimination tasks that manipulated attention.

Quantifying Attention in Perception

One way of measuring the effect of attention on information processing is to examine how subjects respond to target stimuli. This has been done in various experiments. For example, one can measure the time it takes subjects to discriminate letters at a target location in the presence of distracting letters next to the target location. The target letter requires one response (e.g. press the button with the right hand) when, for example, it is the letter *A*, but another response (e.g. press the button with the left hand) when it is the letter *E*. The reaction time to make these responses can be compared when the nearby distracter letters are the same (congruent condition, e.g., *AAA*) as the target letter on that trial, versus when the nearby distracter letters are different from the target letter (incongruent condition, e.g., *(EAE)* or are neutral and are not mapped to any response (e.g., *XAX*). In this type of study, Charles Eriksen and his colleagues (e.g., Eriksen and Eriksen, 1974) at the University of Illinois at Urbana-Champaign showed that reaction times to the target letters were slowed when the distracters were incongruent, as compared to congruent or neutral. These types of designs are referred to as *interference tasks* or *flanker tasks.*

Directing attention to some location or locations while ignoring others is called **spatial attention**. In the laboratory, spatial attention can be manipulated as in the flanker task or by inducing subjects to believe that targets are more likely to be in one place than another (voluntary attentional orienting). Attracting visual attention to locations by using sensory stimuli (automatic attentional orienting) can also be used to manipulate spatial attention. Attention is manipulated by an attention-directing cue, which is presented prior to the delivery of each target stimulus, and hence these are referred to as *cuing tasks.* The speed of responses to attended- or unattended-location stimuli is an index of attention-related processing.

VOLUNTARY ORIENTING

In reaction time studies of voluntary spatial attention using cuing tasks, participants are asked to respond as fast as they can to targets presented on a computer screen. They are instructed that the most likely location for the next target is the one indicated by a prior cue, such as an arrow, pointing to that location. When this happens, we refer to these as *valid trials* because the cue validly predicted the target location on that trial. If the relation between cue and target is strong—that is, the cue usually predicts the target location (e.g., 80% of the time)—then subjects learn to use the cue to predict the next target's location. Sometimes, though, the target is presented to locations not indicated by the cue on that trial, and these are referred to as *invalid trials*. Finally, on yet other trials, the cue might not give any information about the most likely location of the impending target. In voluntary cuing studies, the time between the presentation of the attention-directing cue and the presentation of the subsequent target might be 200 msec to a second or more. These trial types are illustrated in Figure 7.8. The result is that subjects respond faster to the target stimulus when the cue correctly predicts the target's location, even when they do not move their eyes to the cued spot. That is, if the target appears where subjects expect it, they are faster to respond to it; they are also slower to respond to targets at unexpected locations (Figure 7.9).

Differences in reaction time as a function of positional expectancy have been called *benefits* (speeding of reaction time) and *costs* (slowing of reaction time) with respect to the neutral situation in which the subject does not expect the target at one location more than another. These effects have been attributed to the influence of covert attention on the efficiency of information processing. According to some theories, such effects result when

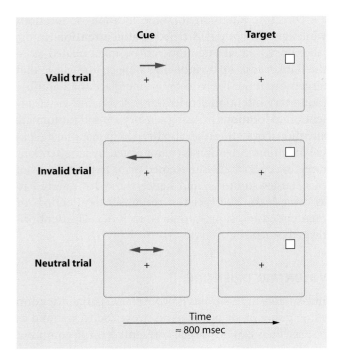

Figure 7.8 The spatial cuing paradigm of Posner and colleagues. A subject sits in front of a computer screen and fixates on the central cross. An arrow cue indicates to which visual hemifield the subject is to covertly attend. The cue is then followed by a target in either the correctly or the incorrectly cued location.

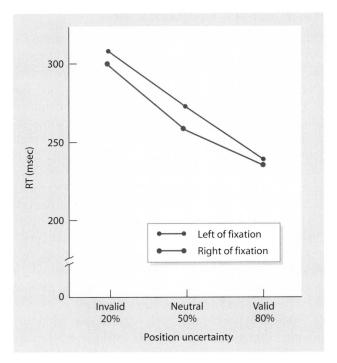

Figure 7.9 Results of the study by Posner and colleagues, as shown by reaction times (RT) to unexpected, neutral, and expected targets for the right and left visual hemifields. Reaction times for expected locations are significantly faster than those for neutral or unexpected targets. Adapted from Posner et al. (1980).

the predictiveness of the cue induces the subjects to internally direct covert attention—a sort of mental "spotlight" of attention—to the cued visual field location. The spotlight is merely a metaphor for changes in brain processes that accompany attending to a spatial location. Because subjects are typically required to keep their eyes on a central fixation spot on the viewing screen, internal or covert mechanisms must be at work. Michael Posner, of the University of Oregon, and his colleagues (1980) suggested that this attentional spotlight may affect reaction times by influencing sensory and perceptual processing; hence, representations of attended-location stimuli are enhanced with respect to unattended-location stimuli. This formulation, one early-selection model, proposes that changes in perceptual processing can happen when the subject is attending a stimulus location. However, these measures of reaction time, or behavioral measures more generally, are not measures of specific stages of neural processing, but rather are indirect measures, requiring interpretation to be related to hypothesized stages of neural analysis. Later we will see how cognitive neuroscience measures have been combined with the voluntary cuing paradigm described here to test early- and late-selection models more directly.

REFLEXIVE ORIENTING

Introspection and experimental data converge to support the idea that attention can be directed voluntarily to locations or objects in the sensory world—and to internal mental events. In addition, as described earlier, things in the environment sometimes attract our attention without our cooperation, as any student sitting in the back of a classroom can attest. What happens when the lecturer leaves the door to the hallway open and, in the middle of lecture, someone walks down the hallway passing the open door? Heads turn toward the sounds and sights in the hall and then wag back to the lecturer a moment or two later. This sometimes happens before we can stop ourselves, and is a sign of reflexive (automatic) attention leading to overt orienting to the sensory stimulus—"overt" because heads and eyes turn toward the event in the hallway. But even in the absence of overt signs of orienting, covert attention is still attracted to the sensory stimulus, which can be demonstrated experimentally by using a variant of cuing methods.

As with voluntary attention, reaction times can be faster for targets when automatic attention is invoked. One way to demonstrate this phenomenon is to examine the effects of presenting an irrelevant flash of light somewhere in the visual field and then measuring the speed of responses to subsequent task-relevant target stimuli. This method is referred to as reflexive cuing or **exogenous cuing** because attention is controlled by external stimuli and not by internal voluntary control. The typical reflexive cuing design uses cues (light flashes) that do not predict the location of subsequent targets; nonetheless, responses are faster to targets at the cued location, but only for a short time after the light flash (usually within about 50–200 msec). When more time passes between the task-irrelevant cuing light flash and the target (> about 300 msec), the pattern of reaction time effects is reversed: Subjects respond more slowly to these stimuli. This slowing of responses is called the *inhibitory aftereffect* or, more commonly, **inhibition of return;** as the names suggest, the recently attended location becomes inhibited over time such that responses to stimuli occurring there are slowed.

Why would automatic attentional orienting have such profound variations in its effect over time after a sensory cue? The answer may be rather intuitively simple: If sensory events in the environment caused automatic orienting that lasted for many seconds or tens of seconds, it would be difficult to function normally in the world. We would be continually distracted by things happening around us. Imagine the potential disaster for someone driving a car. It is no surprise, then, that the automatic orienting system has built-in mechanisms to prevent this; so, over time (a few tens or a couple of hundred milliseconds) the automatic capturing of attention subsides and the likelihood that attention will be attracted back to that location is slightly reduced. If the event is important, we merely invoke our voluntary mechanisms to sustain attention longer. Thus, the nervous system has evolved clever, complementary mechanisms to control attention so we can function in this hectic sensory world.

SEARCHING THE VISUAL SCENE

As introduced in Chapter 5, a natural aspect of everyday perception is the search for items in busy scenes. Imagine that you are at the airport and lose track of your companion in the crowded terminal. You begin a visual search of the scene to find him or her. Or, if you have misplaced your car keys in your home, you search the nooks and crannies of your abode to locate them. What happens during the visual search for a target (i.e., your companion or your keys) among distractors (i.e., the hundreds of people in the airport terminal, or the dozens of things in your home)? What role do attentional processes play in this search, and how does your knowledge of the appearance of the lost items affect the way you hunt for them? If your airport companion was last seen wearing a bright-red sweater, you may use the color information to constrain your search. This method might prove an easy way of finding your friend if nobody else in the crowd is dressed in red. But if many people are wearing red, you may have to use a conjunction of many pieces of information like hair color, height, and facial features in addition to the color red.

These mechanisms of visual search were investigated and modeled by Anne Treisman and her colleagues (Treisman and Gelade, 1980—see Chapter 5 for additional discussion). They presented arrays of multiple simple stimuli (distracters) on a video screen that sometimes contained a target item requiring a speeded motor response. The complexity of the search task could be manipulated by having the target item differ from distracters by a unique feature (e.g., red O among O's and X's of a different color, as in Figure 7.10a) or by having the target and distracters share some features such that the target was defined only by a conjunction of two or more relevant stimulus features (e.g., red O as a target among distracters that could be O's and X's of a different color and red X's, as in Figure 7.10b). In the former, simpler case, the target could be identified simply by color. These targets popped out of the visual scene; that is, they were immediately obvious to the observer as soon as the array was shown. Under these so-called pop-out conditions, the average time it took the observers to respond when the target was presented was not affected by how many distracter stimuli were included in the display. Displays could have two, four, eight, sixteen, thirty-two, and more distracters. Hence a plot, called a *search function,* of reaction time as a function of distracter number was flat for pop-out targets (Figure 7.10c). The finding that some stimuli can pop out of the background has been interpreted as evidence that attention is not required to detect the target, and has been attributed to a **preattentive mechanism**. Preattentive processes are based on perceptual analyses that can simply signal, for example, the presence of a difference in the visual scene, rather than on the action of focused attention.

In contrast, when the target is defined by a conjunction of features shared with distracters, the search functions have positive slopes (see Figure 7.10c). The positive slope for conjunction search can be interpreted as evidence that the stimulus items are searched one at a time, in a serial, self-terminating search—that is, when the target is detected, the search can stop. This process occurs independently of eye movements; that is, if subjects are not permitted to move their eyes during the

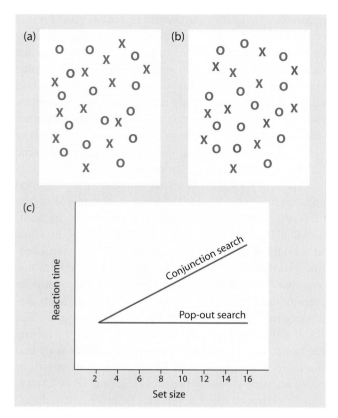

Figure 7.10 Searching for the targets among distracters. **(a)** Example of a search array with a pop-out target (red O). Stimuli are said to pop out when they differ from distracters by simple single features and the observer does not need to search the entire array to find the target. **(b)** Example of a search array where the target is defined by a conjunction of features shared with the distracters. **(c)** Idealized plot of reaction times as a function of set size (the number of items in the array) during visual search for pop-out stimuli versus feature conjunction stimuli. In pop-out searches, where an item differs from distracters by a single feature, the subjects' reaction times do not increase as much as a function of set size as they do in conjunction searches.

search, the results are still obtained, which means that the search is covert. Treisman defined this as an attentional mechanism that must move from item to item to identify individual stimuli, sort of an automatic spotlight that searches the visual scene serially, albeit rapidly (only a few tens of milliseconds per item), in order to analyze the conjunction of features and compare this to the predefined target characteristics.

Evidence about whether or not the spotlight of attention employed in visual search tasks is automatic or under top-down control was recently provided by Jeremy Wolfe and his colleagues (2000) at Harvard University. They employed a visual search task to show that when deliberate movements of attention are required,

they are slower than when these deliberate movements of attention are not required, and search is permitted to proceed automatically. To demonstrate this, they compared two conditions where subjects either knew in advance or did not know in advance where to focus their attention. Wolfe's team presented subjects with a series of stimulus arrays. The subjects viewed each of eight frames for 53 msec, followed by a masking stimulus (random patterns) that varied in duration from trial to trial. A letter target (*Y* or *N*) was present on each trial and required a response, but in one condition (command condition) the target letter could appear only at location *one* (i.e., the 12 o'clock position) in the first frame, location *two* (halfway between the 12 o'clock and 3 o'clock positions) in the second frame, and so on, throughout the eight frames. The subjects were instructed to expect the target and therefore covertly attend to the next position in the clockwise direction as each frame appeared, following a warning tone; the eyes had to remain fixated on the center of the frame (Figure 7.11a–d). In a second condition (random anarchy condition), the location of the *Y* or *N* target letter varied randomly from frame to frame, and thus the subjects could not anticipate the location and were required to search each frame.

The investigators then determined the fastest rate of frame presentation that yielded correct button press responses from the subjects 70% of the time—this told them how fast the subjects could find the target under two different conditions (command versus random anarchy). They found that the average minimum rate for movements of voluntary attention was 217 msec/frame (command condition), which can be interpreted as 217 msec/item. However, surprisingly perhaps, in the random anarchy condition, for which subjects did not know where the target would actually be on the next trial, the rate was faster, between 105 and 187 msec/item (Figure 7.11e). This remarkable demonstration shows that as attentional theories would predict, automatic processes are faster than voluntary ones with respect to attentional orienting. The striking finding here, however, is that visual search for conjunction targets in tasks such as that shown in Figure 7.10 can proceed most rapidly when the subject permits the attention system to control the search based on visual sensory information, rather than executing a slow, voluntarily controlled search of the items. Or to put it another way, the brain automatically scans the visual world with a fast, automatic spotlight of attention. As Wolfe and colleagues put it, "Anarchy is faster than order in this case."

Under some circumstances, the rate of search in standard search arrays similar to those in Figure 7.10 is much faster for conjunction targets than what the Treisman model predicts. This faster search occurs when a subset of

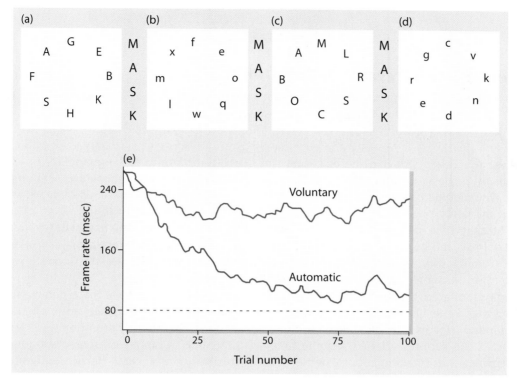

Figure 7.11 Evidence that visual search can proceed automatically. **(a–d)** In this study, successive frames were presented one at a time for 53 msec each, with a mask presented in between of variable duration. Target letter *Y* or *N* was presented in the arrays under two different conditions. In one condition (command condition) subjects knew where the target would occur for each frame, and therefore needed to only covertly attend to that location to determine whether the letter was *Y* or *N*. Subjects were instructed to search in a clockwise direction at the sound of a warning tone. In the other condition (random anarchy condition) the target letter would appear anywhere on each frame in a random fashion. Note: Because the figure is used to describe the two different conditions, no targets are shown in most arrays here. **(e)** The fastest speed at which the subjects could detect the targets at 70% accuracy was then determined, and this is plotted for the two conditions. The time per frame was more for the command (voluntary) condition than for the random anarchy (automatic) condition. Adapted from Wolfe et al. (2000).

the distracters in the search array share a common feature with each other but not with the target, and this feature can be used to reject these distracters without searching them. Let's take as an example the conjunction search array in Figure 7.10b. If the target is a red O and the distracters are black O's and red X's and green O's and X's, then the green distracters share some features with the target (i.e., letter O) but can be rejected based simply on color. But can the attention system use this fact to reject green distracters or must each be searched in turn? The empirical finding is that when a significant part of a search array (e.g., 30%) consists of these green distracters, the search rate to find the conjunction target (red O) is faster than would be predicted by the total number of distracters alone. The idea here is something called *guided search*, a term Jeremy Wolfe coined. One way to conceive of this result is that the attention system rejects the irrelevant distracters based on the feature green (i.e., the green O's and X's), perhaps by searching them in parallel with the search among nongreen items, and therefore the search among the red and black X's and O's to find the red O target is more efficient and the target discovered more quickly.

NEURAL SYSTEMS IN ATTENTION AND SELECTIVE PERCEPTION

So far, we have focused on selective attention in visual and auditory perception, and on how we select some visual inputs for further high-priority analysis while rejecting or attenuating others. The methods we described are those of experimental and cognitive psychology. They are essential for

An Interview with Steven A. Hillyard, Ph.D.

Dr. Hillyard is a professor of neuroscience at the University of California School of Medicine at San Diego. He is the world's foremost authority on the electrophysiology of human attention.

Authors: You began studying the electrophysiology of sensory processing during attention some 25 years ago. Why did you choose to work with humans rather than animals?

SAH: At that time people were excited about studies in animals and humans showing that evoked potentials from sensory pathways became enlarged when attention was paid to the evoking stimuli. These findings offered the first real possibility of identifying specific levels of the sensory pathways where attended signals are enhanced and unattended inputs are suppressed. At the same time, a lively controversy was brewing about whether these electrophysiological changes were associated with selective sensory processing or with changes in general arousal or alertness.

We chose to study the neural bases of attention in humans rather than in animals for several reasons. First, computer technology enabled us to make noninvasive scalp recordings of evoked activity in a person's sensory pathways as he or she performed a demanding attentional task. Since humans can be trained to carry out complex sensory discrimination and decision tasks much more rapidly than any animal, we could implement rigorous experimental designs that revealed brain activity specifically related to selective attention rather than to general arousal states. Second, we wanted to test a hypothesis about attention—that is, early versus late levels of selection—that had emerged from the psychological literature and for which experimental designs had already been well established in humans. Third, we didn't really know to what extent studies in animals, including those in our fellow primates, could serve as valid models for human attention. Of course, we still don't know this.

The intellectual repertoire of humans is so much more elaborate than our nearest relatives' it would be presumptuous to assume that all the rich manifestations of human attention have their equivalents in animals. Perhaps animal models may be suitable for studying simpler aspects of attention, but the ability to attend to language, for example, has evidently been evolving separately in humans for tens if not hundreds of thousands of years. Thus, if we are ever to understand the full range of human attentional abilities and the neural bases thereof, I believe it is necessary to direct a major research effort toward humans, despite the limitations inherent in noninvasive studies.

Authors: What have these noninvasive studies revealed about attention mechanisms in humans?

SAH: For auditory attention we examined evoked potentials to rapidly presented sounds coming from speakers in a cocktail party situation; this was designed to rule out nonspecific arousal and alertness effects. We found that short-latency evoked potentials, beginning at 50 to 60 msec after stimulus onset, became markedly enlarged in response to attended sounds in one ear versus unattended sounds in the other ear. In follow-up studies by Marty Woldorff—my colleague who is a professor at Duke University—using neuromagnetic recordings in conjunction with MRI-based source localization, evoked sensory activity was enhanced by attention as early as 20 to 50 msec after stimulus onset in the primary auditory cortex. These results strongly confirm early selection theories of attention, which assert that attended sensory inputs are boosted and unattended inputs are suppressed at an early stage of processing, before more complex features of the stimuli are analyzed in higher cortical areas.

Authors: What about visual attention?

SAH: Analogous effects were demonstrated in visual attention, most notably in spatial attention tasks where a subject was cued to attend to stimuli at one location in the visual fields while ignoring stimuli at others. In this situation, evoked potentials in the visual cortex were enlarged for attended versus unattended stimuli starting at 70 to 80 msec after stimulus onset. By combining techniques for localizing the sources of this attention-related activity, including parallel experiments using PET, Hans-Jochen Heinze and one of the authors of this text (G.R.M.), my colleagues of many years, determined that incoming sensory signals are modulated by attention most strongly in extrastriate visual areas throughout the interval 80 to 200 msec. In contrast, earlier activity evoked in the primary visual cortex in the 50- to 80-msec range is less influenced by spatial attention, although attention has a role in primary sensory cortex as well.

Authors: Tell us more about this amplification mechanism. For example, what does it mean when a brain potential—or for that matter a PET or fMRI signal or single-unit discharge—is enlarged in response to an attended signal? What is the behavioral significance of a bigger evoked response and how has this link been established?

SAH: What you're asking about is the nature of the neural codes for sensory/perceptual information. There is a general assumption that more activity—greater firing rates, etc.—within a nerve cell population means that more information is being represented or processed by those cells. Conversely, a suppressed response reflects diminished processing. But you're right, this is something that needs to be established in each case rather than taken for granted.

In simple, low-level sensory signals, greater physical stimulus energy produces systematic increases in neural response amplitudes that are paralleled by elevations in the perceived stimulus magnitude. Also, signal detection experiments in humans and monkeys have shown that moment-to-moment fluctuations in neural response amplitudes can predict precisely whether the person or animal detects a faint signal. Thus, for low to moderately intense sensory inputs, there is good evidence that heightened neural responses often represent enhanced perceptual information. This relationship has also been observed in attention experiments which showed that increased evoked potential amplitudes to stimuli at attended locations were associated with improved accuracy at discriminating the features of those stimuli. It's unlikely, though, that such simple relationships would hold in general. The neural codes for representing complex perceptual or cognitive events undoubtedly involve elaborate patterns of excitation and inhibition in neuronal populations that cannot be simply characterized by increases or decreases in response amplitude.

Authors: Very well, you have identified evoked responses that vary with attention. What about animal studies of detailed anatomical circuits and cellular-level events?

SAH: I think we need a combined approach that brings together human and animal studies in common conceptual and experimental frameworks. Invasive animal studies are certainly necessary to reveal the fine-grained neurophysiology of the brain's attentional systems, and also the neurotransmitters and microcircuitry involved. Numerous monkey experiments demonstrated attentional modulation of single-cell activity patterns in several well-defined cortical areas. With few exceptions, however, these monkey experiments used attention tasks that differed from the ones employed in behavioral and physiological studies of humans. It is usually difficult to know whether the attentional phenomena were truly equivalent or homologous to those displayed by humans, but that hasn't stopped researchers such as my colleague Steve Luck, now at the University of Iowa, or our colleague, neuroscientist Robert Desimone at the National Institute of Mental Health, and others. Their promising studies have great potential for bridging the gap between the human and animal literature.

I think the most effective research strategy would be to carry out animal and human experiments in parallel and to use equivalent or identical experimental tasks. Then let the results in one species guide the experiments in the other. For example, with current technologies, noninvasive human studies can reveal the brain areas activated by a task with a resolution of about 1 cm or better (using PET and fMRI) and the timing of neural population dynamics with a millisecond level of resolution (using ERPs and MEG). These macrolevel activity patterns can be studied in relation to precisely defined cognitive processes in humans and can point the way to brain systems that should be explored in parallel with animal experiments. Moreover, if similar macrolevel brain activity patterns were observed in humans and monkeys performing equivalent tasks, this would greatly strengthen the validity of the animal model for the cognitive process under investigation, which can then be investigated further by using invasive techniques to get at the microlevel structure of the underlying neural events. Unless we have such interplay between human and animal studies, we are whistling in the dark with regard to the applicability of animal findings to human selective attention.

investigating attentional mechanisms, but they cannot fully describe these mechanisms because they do not inform us about neural substrates. For example, where in the human brain does attention influence signal processing? If we can respond more quickly to a target appearing at an attended location, is this because it was more efficiently processed in our visual cortex or because motor systems in the frontal lobe were biased to generate faster responses to it? What brain systems control attention during voluntary and reflexive attention, and are they the same, different, or overlapping?

The limitations in behavioral and psychophysical measures for elucidating neural mechanisms should not be taken to mean that these measures are inferior or uninformative—far from it. Rather, they have helped to describe attentional phenomena in terms of the observer's behavior and have generated testable hypotheses that cognitive neuroscientists can address. In their search for answers to the question of how attention works, many investigators have turned to physiological methods in humans and animals to determine how intermediate neural events give rise to changes in perceptual and cognitive experience, and task performance during attention. Physiological recording studies and data from patients with neurological damage or disease provide a wealth of information about the neural mechanisms of attentional processes. In this section we will look at studies that demonstrated how selective attention works in modulating sensory and perceptual processing. We begin with research that makes use of physiological recording methods that provide measures of human brain activity during attentional performance.

Neurophysiology of Human Attention

To describe the mechanisms of attentional selection at the neural level in humans, researchers use several tools, including electrical and magnetic recordings of brain activity and neuroimaging techniques (see Chapter 4 for a discussion of these methods). Each method measures human brain activity to provide a window into the sensory, perceptual, cognitive, and motor processes, but each does so using different parameters of neural activity. Electrical and magnetic recordings (electroencephalography, event-related potentials, magnetoencephalography, and event-related magnetic fields) directly measure neuronal activity as millisecond-to-millisecond fluctuations in electrical and magnetic fields. In humans, these small signals can be recorded noninvasively by placing electrodes or sensors on the head (and inside it during certain neurological procedures—see Human Brain Activity During Attention). How well the noninvasive electrical and mag-

netic recordings can localize a signal to a specific brain structure is limited, though. It is difficult to infer where inside the brain the signals are coming from based on patterns recorded outside the head. To facilitate the localization of neural activity while the brain senses, acts, and thinks, functional neuroimaging methods such as positron emission tomography (PET) and functional magnetic resonance imaging (fMRI) are employed to detect and measure local changes in blood flow or metabolic activity related to neuronal activity. When considered together, electromagnetic and neuroimaging methods can outline the time course and neuroanatomy of higher mental processes such as attention.

Studies using electrical, magnetic, or functional neuroimaging methods to address questions about the neural mechanisms of attention do so in the context of the cognitive models and paradigms we described in the foregoing section. From this integrative approach, which is the core of cognitive neuroscience, the aim is to identify the functional architecture and neuronal mechanisms of brain attention systems.

ELECTRICAL AND MAGNETIC RECORDINGS OF THE HUMAN BRAIN

In humans, attentional selection in the brain was first demonstrated electrophysiologically. Electrophysiological studies of attention were initially tried with animals but failed because the experimental designs employed in the early studies did not carefully control for behavioral and physiological factors that could confound the interpretation of the results. Yet the history of the early neurophysiology studies of attention makes crystal clear some important aspects of the search for evidence of selective attention in the nervous system. For this reason, let us consider some research on cats that was performed more than four decades ago.

Auditory Selective Attention Raul Hernandez-Peon and his colleagues (1956) attempted to determine whether phenomena like the cocktail party effect might result from the gating of auditory inputs in the ascending auditory-sensory pathways. They tested their theory on cats by presenting sounds and recording from subcortical stages of the auditory pathway when the animals attended (as opposed to ignored) the sounds. They reported attention-related modulations in the neuronal activity of the cochlear nucleus and auditory nerve, a finding that was interpreted as evidence that voluntary attention can affect the earliest stages of sensory processing—strong evidence for early-selection theories. Unfortunately, these findings were later shown to result from the animals differentially overtly orienting their ears with respect to the sound-emitting speaker, rather than from the covert orienting of auditory attention. As anyone familiar with cats knows, they can rotate the external ear toward sounds of potential interest, such as the sound of an electric can opener, without moving any other part of their body (Figure 7.12). The result of this change in the relationship between the external ear and the sound source is certainly one mechanism for improving hearing, and does indeed involve attentional orienting in some fashion, but it also presents a significant confound when one is searching for any signs of modulation of early sensory processing by covert attentional mechanisms.

Given this type of experimental failure, one may well ask, Why start a section on the neurophysiology of human attention with a negative result from animal work? The answer is that it points to a more fruitful approach and highlights significant controls germane to the physiological measurement of selective attention.

Hernandez-Peon and his colleagues' principal error related to the control over stimuli. The altered positions of a cat's external ear with respect to the loudspeaker led to varied amplitude signals reaching the eardrum in the attended versus ignored conditions—hence, differing activity levels were recorded from the auditory structures. We already know that changes in stimulus amplitude at the eardrum will alter the firing rates of neurons. The question was whether such changes in neuronal firing could occur as the result of selective internal neural control over sensory processing. One lesson that emerges from this work is that experimental controls must pre-

vent any adjustment of the peripheral organs of sensory reception, a lesson that applies to all sensory modalities when one is performing attention research. The solution is rigorous monitoring of ear, head, and eye positions, and controls such as placing small headphones in the external ear canal, a method similar to the one used to reveal that Hernandez-Peon's results were incorrect and why. Hernandez-Peon was a pioneer, and like many working at the cutting edge of science, his findings in the field of attention research were not what he believed at the time. However, the approach he developed led to successful studies many years later, in part owing to the lessons learned by Hernandez-Peon and his colleagues and students (who, by the way, were the ones responsible for correcting and thereby overturning the results of their mentor).

In the late 1960s and early 1970s, the first properly controlled experiments on the neurophysiology of attention were conducted in humans. These studies were made possible by the development of methods to perform signal averaging of the electroencephalogram (EEG). Signal averaging revealed tiny brain waves that were directly related to the processing of a stimulus but were hidden in the larger EEG rhythms. These tiny brain waves are referred to as *event-related potentials* (ERPs), a term stemming from the fact that they are "potentials" in the electrical sense of the word and that they are related to some "event," namely, the stimulus and its processing in the brain (see Chapter 4 for a description of the methods). By designing clever experiments that controlled the subject's global state of arousal and how the stimuli impinged on the receptors across conditions of attention, researchers demonstrated that brain wave responses (i.e., ERPs) elicited by stimuli when they were attended differed from those elicited by the same physical stimuli when they were ignored.

These experiments were first performed by Steven Hillyard and his colleagues (1973) at the University of California at San Diego. They developed a selective listening method that involved recording ERPs from the scalps of humans in response to auditory stimuli. Streams of sounds that differed in pitch were presented to the two ears of volunteers wearing headphones. During one condition they were asked to attend the sounds in one ear while ignoring those in the other (e.g., attend right-ear sounds and ignore left-ear sounds). Then in a second condition, they were asked to pay attention to the stimuli in the other ear (e.g., attend left-ear sounds and ignore right-ear sounds). In this manner, the researchers separately obtained auditory ERPs to stimuli entering one ear when that ear's input was attended and when it was ignored. The significant design feature of the experiment

Figure 7.12 Early work by Hernandez-Peon and colleagues attempted to relate single-cell recordings in cats to auditory attention. By recording from subcortical structures while the cat attended or ignored sounds, Hernandez-Peon and colleagues initially observed modulations at the subcortical level of processing. This was later shown to be due to differential orienting of the cat's ears rather than auditory attention.

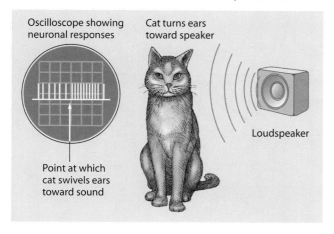

Oscilloscope showing neuronal responses

Cat turns ears toward speaker

Loudspeaker

Point at which cat swivels ears toward sound

Human Brain Activity During Attention

On rare occasions it is necessary to record the human brain's electrical activity with electrodes located in the brain rather than outside on the scalp. This medical procedure is performed when physicians want to localize a defective region of brain tissue, such as occurs in epilepsy. The goal is to locate the spot in the brain where abnormal neural tissue (the seizure focus) touches off uncontrolled neural activity that spreads to otherwise normal tissue, producing a seizure. Seizures can be so severe as to render the victim unconscious. If a confined region of the brain is responsible for starting epileptic seizures, it is possible to remove that neuronal tissue and perhaps alleviate the attacks.

Often when it is necessary to implant electrodes in the brain surgically, the electrodes are in place for several days. Under these circumstances, if the patient permits, it might be possible to record from the electrodes while the patient performs cognitive tasks such as selective attention.

Gregory McCarthy (now at Duke University), Truet Allison, and their colleagues at Yale Medical School have been performing such experiments for years. These researchers asked patients who had had grid electrodes (eight-by-eight array of electrodes with 1-cm spacing) placed on the cortical surface of the occipital cortex (top figure) to attend some stimuli while ignoring others, just as with scalp recording in healthy volunteers. In this manner, the researchers obtained event-related potentials (ERPs) directly from within the cerebral cortex of humans. The result was clear evidence for modulations of activity in the extrastriate cortex as a function of spatial attention. This evidence is represented in the graph, where the ERPs from two left-hemisphere occipital electrodes located a scant 1.4 cm apart are shown. The task required subjects to attend the flashed black-and-white checkerboard stimuli in one visual field quadrant (e.g., upper right = RU) while ignoring those in the other field (e.g., upper left = LU) and to detect target checkerboards that contained red checks, which were presented infrequently (about 10% of the time). As seen in the graph, large responses to the upper-right stimuli were obtained when they were attended (RU attend trace), as compared to when the same stimuli were ignored in other blocks (RU unattend trace). Moreover, the exquisite localizing ability of this method can be seen by the fact that these responses reversed in polarity over the 1.4 cm separating the two cortical electrode sites. As well, at these sites, stimuli in the upper-left visual field quadrant (LU attend trace) elicited virtually no responses. These dramatic medical circumstances support the idea that the amplitude of scalp-recorded activity generated in modality-specific cortical areas is modulated during selective attention.

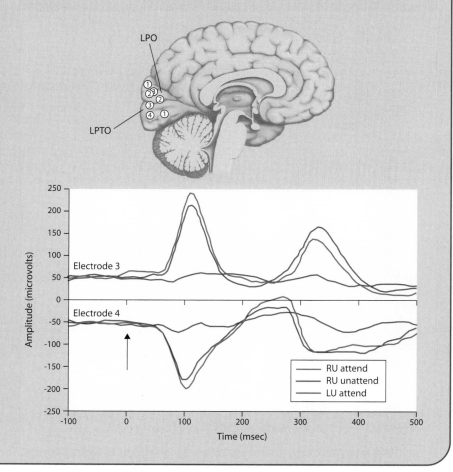

was that during the two conditions of attention, the subjects were always engaged in a difficult attention task: All that varied was the direction of covert attention, that is, which ear the subjects directed their attention to. In addition, the researchers controlled the orientation of the subject's head with respect to the source of sound by using headphones to deliver the sounds. Thus, they controlled for the confounding factors that had led to failures in previous animal research.

Hillyard and his colleagues discovered that auditory ERPs were enlarged in amplitude when stimuli were attended compared with when they were ignored. This effect in the ERPs appeared as a deflection in the ERP waveform known as the *auditory N1 potential,* so named because it is the first of the large, negative polarity deflections in the signal-averaged waveform (Figure 7.13). The N1 component is believed to be a sensory evoked wave with a latency of less than 90 msec to the peak after stimulus onset. Given that this sensory response was labile to manipulation by covert attention, these data provided evidence in support of early-selection models of attention. However, one might argue that during auditory processing something occurring at 90 msec is rather late, especially given that the auditory cortex can be activated as early as 20 to 25 msec after the sound onset. Can attention not influence auditory processing earlier than the N1 wave?

To answer this question Marty Woldorff and Steven Hillyard (1991) extended these earlier studies by increasing the signal-to-noise ratios of their averaged auditory ERPs by including greater numbers of trials, and modifying their paradigm so as to force subjects to attend selectively by increasing the difficulty of the auditory discrimination task. They then used methods designed to enable them to simultaneously record responses from as early in auditory processing as the auditory nerve, and on throughout the ascending sensory pathways to the level of the cortex; although recordings of auditory responses from all levels of the ascending auditory pathway can be done routinely in the clinic to diagnose hearing problems, it presents a considerable challenge when designing a carefully controlled attention study, because the stimulus parameters are not typically optimal for revealing the smallest, early auditory responses.

Woldorff and Hillyard replicated the prior N1 attention effect—larger N1 responses to tones presented to the attended ear. In addition, they observed a new finding. Between 20 and 50 msec after stimulus onset, much earlier than the time range of the N1, the waveforms were also different as a function of selective auditory attention. This finding, the *P20-50 effect,* named for its positive polarity and latency of occurrence, provided extremely compelling evidence in support of early-selection models. Support for such models was based on the fact that 20 msec is about the time it takes auditory inputs to reach the human cortex, and therefore, the P20-50 attention effect was likely the result of modulations of information processing prior to complete analysis of the inputs by the perceptual system.

To interpret the auditory P20-50 and N1 attention effects, it is important to understand what brain regions generate them. Where do the P20-50 effects and N1 potential come from inside the brain? If these effects represent early selection of stimulus inputs—a sort of filtering of perceptual signals—they should be generated in sensory brain structures. If we could show that the P20-50 effect or the N1 potential were generated in the medial geniculate nucleus of the thalamus (the auditory thalamic relay) and the primary auditory cortex in Heschl's gyrus, then we would know that both sensory brain structures are involved in the initial analysis of auditory processing or sound inputs, and this might bolster the argument for early-selection models. How can we learn what brain regions generate these effects in humans, for whom recording directly from the brain is not (usually) possible?

Powerful new methods come to the rescue. The magnetic counterpart of the EEG is called the *magnetoencephalogram* (MEG), and MEG signals can be averaged to derive event-related magnetic fields (ERFs). By combining ERP recordings in selective listening paradigms with recordings of magnetic field fluxes generated by active neurons, we can find out where the auditory attention effects are generated and what they might mean for information processing.

Figure 7.13 Event-related potentials (ERPs) in a dichotic listening task. The solid line represents the average voltage response to an attended input over time, and the dashed line to an unattended input. Hillyard and colleagues found that the amplitude of the N1 component was enhanced with attention as compared to when ignored. Adapted from Hillyard et al. (1973).

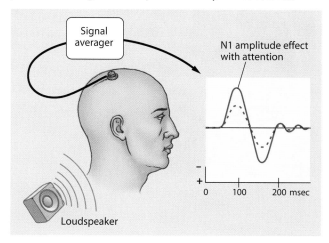

This is because localizing the intracranial sources of MEG signals is more simple than for their electrical counterparts (see Chapter 4).

In 1993, Marty Woldorff and Steven Hillyard and their colleagues combined recordings of ERPs and ERFs to replicate and extend their original P20-50 data. They identified a magnetic correlate of the electrical P20-50, the so-called M20-50 attention effect (*M* for magnetic). The external magnetic field patterns that reflected the M20-50 suggested the signals arose in the auditory cortex (Figure 7.14a). By estimating the best dipolar model source within the head that would explain the attention-related responses recorded on the surface of the head, and mapping those onto structural MRI scans of the subject's brains, the researchers localized the M20-50 effect to the auditory cortex in Heschl's gyri (Figure 7.14b). Similar magnetic effects can be observed in the time range of the N1 attention effect (the so-called M1 attention effect), and these effects also appear to be generated within the auditory cortex, but at a later stage of neural processing than the shorter-latency M20-50. Attention, therefore, can affect stimulus processing in the auditory cortex. From these data, we do not know for sure whether in humans the attention effects in auditory cortex represent modulations in the primary auditory cortex or in immediately surrounding secondary auditory areas. Even so, attention clearly affects sensory-specific stages of auditory analysis at latencies strongly indicating that auditory sensory analysis has not been completed. We know this based in part on the observation that at this early stage of processing, the auditory system cannot select information based on higher-order aspects of auditory inputs such as speech cues; rather, the ear of entry is the cue for selection. Longer latency changes in an ERP or ERF waveform reflect attentional selection involving the stimuli's higher-order properties such as a speaker's voice or semantic meaning.

Could the attention effects at the auditory cortex be the result of gating that occurred earlier in auditory processing, such as in the thalamus, brainstem, or even the cochlea? For the auditory system we know that there exist neuroanatomical systems that, at least in principle, could exert top-down control over auditory inputs at the earliest processing stages. Each neural relay in the ascending auditory pathway sends return axons back down to the preceding processing stage, even out to the cochlea in the form of the olivocochlear bundle (OCB). We also know that activity in the cortex can activate the descending OCB pathway all the way to the cochlea, as demonstrated by direct electrical stimulation of the cortex (even secondary auditory cortex). Perhaps this system can be used in the service of early attentional gating of auditory inputs. This idea has been tested in recordings of ERPs.

Figure 7.14 **(a)** Topographic map of magnetic event-related fields (ERFs) that are associated with auditory attention in a selective listening paradigm. The field map was created by subtracting the field elicited by unattended tones (in the 20–50-msec time range after stimulus onset) from the field elicited by the same tones when they were attended. The arrow indicates the location and orientation of a model dipolar neural generator that best explains the surface field. **(b)** The localization of the P20-50 attention effect in the brain. When the three-dimensional location of this model dipolar generator (red asterisk) was mapped onto a structural magnetic resonance imaging (MRI) scan (after coregistration of the data sets), the activity was found to originate in the primary auditory cortex, in Heschl's gyri in the supratemporal plane. Adapted from Woldorff et al. (1993).

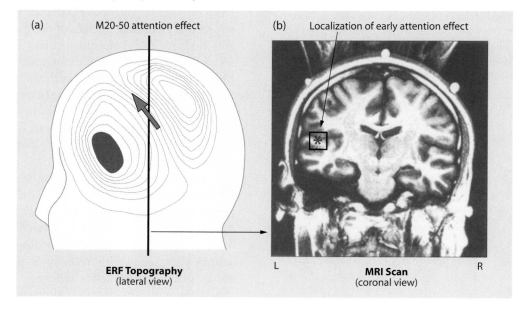

(a) M20-50 attention effect

(b) Localization of early attention effect

ERF Topography
(lateral view)

L   **MRI Scan**   R
(coronal view)

Electrical recordings of human auditory processing from scalp electrodes can pick up sensory activity as early as the brainstem relays, and even the compound action potential of the auditory nerve itself. This activity is called the *auditory brainstem response* (ABR) and is manifest as voltage deflections in the auditory ERPs with latencies between 1 and 10 msec after stimulus onset. Can these ERPs be modulated by selective attention? The answer is that voluntary selective attention in the auditory system does not appear to involve modulations of subcortical auditory neurons as measured by the ABR recordings, even under conditions where attention must be very highly focused on inputs from one channel (i.e., ear) such as those employed by Woldorff and Hillyard. The evidence from these scalp recordings have been confirmed in other studies in humans that used an even more sensitive measure of the auditory nerve response—electrical recordings from a needle electrode inside the ear. Such recordings similarly failed to find modulations of auditory nerve activity with attention.

In part as a result of the negative findings for an effect of attention on the very earliest stages of subcortical auditory processing, research on the function of the descending OCB pathway to the cochlea has tended to focus on its possible role as part of a system for tuning auditory nerve responses, rather than as a mechanism for attentional gating. Nevertheless, the issue of brainstem-level gating as a mechanism of auditory selective attention is not yet resolved. Recent evidence from Marie-Helene Giard and colleagues (2000) in France provided evidence that under conditions very similar to those that revealed the P20-50 attention effect, otoacoustic emissions emanating from the ear canal are modulated by auditory selective attention. Otoacoustic emissions are minute sounds produced by mechanical changes in the cochlea during hearing, and they can be recorded with a sensitive microphone placed in the ear canal. Changes in these sounds with attention could reflect attention-related modulations of the mechanical properties of the cochlea, which might represent a form of attentional control over auditory processing. Although one might expect such effects in the cochlea to influence subsequent neural processing in the ascending auditory pathways, it remains possible that such influence does not occur, in line with negative findings from recordings of the ABR during selective attention. Suffice it to say, however, that modulation of subcortical auditory inputs, if present, is not a robust phenomenon.

Visual Selective Attention   Neural mechanisms of visual selective attention have been investigated using similar methods. As with auditory studies, visual studies using ERPs directly measure the electrical activity generated by visually responsive neurons in the brain to indicate stages of processing that can be affected by selective attention. The first successful investigations, by psychologist Robert Eason and his colleagues in 1969, found that during visual-spatial attention (selection by location), certain visual ERPs showed changes in their amplitudes. Many studies replicated this effect and showed that such modulation can begin about 70 msec after the onset of the visual stimulus.

In these studies, subjects were given verbal instructions to covertly attend to stimuli presented to one location (e.g., right-field location) and ignore those presented to another (e.g., left-field location). The pattern of attention effects in the visual system is now well established. Directing voluntary covert attention toward rather than away from a stimulus (Figure 7.15) leads to modulations in the amplitude of ERPs as early as 70 to 90 msec after stimulus onset. These effects are first observed in an ERP wave known as *P1* (first major positive wave), which can be recorded over lateral occipital regions of the scalp. When a visual stimulus appears at a location a subject is attending, P1 is larger in amplitude than when the same stimulus appears at the same location but attention is focused elsewhere (Figure 7.16). Such a pattern is therefore consistent with filtering mechanisms like the one demonstrated for auditory attentional processing.

These early so-called P1 attention effects (latency of 70–90 msec) have been obtained only during spatial attention, not reliably during selection based on other stimulus features like color, spatial frequency, orientation, or conjunctions of these features. As well, selection of visual inputs based on higher-order properties such as what object it is (e.g., attend chairs but ignore tables) does not lead to P1 attention effects; rather, attention effects for these more complex tasks appear later in the ERPs (> 120-msec latency).

We can ask the same questions about spatial attention effects as we did about auditory ones: Do attention-related modulations of visual ERPs during spatial attention reflect processing in cortical or subcortical stages? And can we say whether they reflect selection occurring before or after semantic encoding? The answers to these questions are remarkably similar to those for the auditory system.

Spatial selective attention occurs in part via the modulation of sensory processes in the visual cortex. These effects happen primarily in extrastriate cortical areas but also may involve mechanisms in the striate cortex. We can be certain that attention is acting on visual cortical processing for several reasons: First, the early P1 spatial attention effect has properties indicative of a visual sensory stage of processing. For example, the sensory

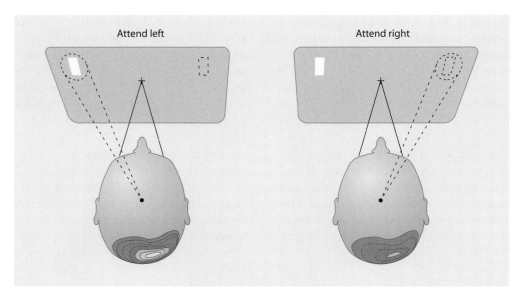

Figure 7.15 Stimulus display used to investigate sustained, spatial selective attention. For periods of seconds to minutes, the subject fixates the eyes on the central cross-hairs (+), while stimuli are flashed in random order to the left- and right-field locations. During some blocks the subject is instructed to covertly attend left stimuli **(left picture)** in order to detect infrequent targets, and to ignore those flashing on the right. During other blocks the subject is instructed to attend the right location, ignoring the left **(right picture).** Then a comparison is made between responses to the same physical stimuli, like the white rectangle being flashed in the figure, when they are attended versus when they are ignored (i.e., during the attend-right condition). Over the contralateral posterior scalp, beginning around 70 msec after the onset of the stimulus, an increased positive voltage response can be observed for the left stimulus when it is attended (left picture) versus when it is ignored (right picture). Larger-amplitude positive responses to the stimuli over the occipital scalp are colored more brightly as yellow. The activity is represented in a topographic map where the voltages at a time point or narrow range of times are recorded simultaneously from many electrodes. Then, the voltage across the scalp in between the electrode sites are estimated by interpolation, and isovoltage contours are drawn as outlines and filled with colors corresponding to different voltage values. The response to the stimulus produces a sort of "mountain peak" of voltage over contralateral scalp, and the peak is higher, if you will, for attended as opposed to ignored stimuli. The dotted lines from the head to the screen indicate the direction of covert attention for the different attention conditions.

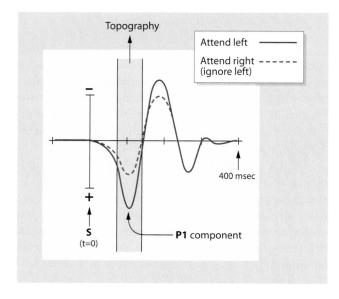

Figure 7.16 Event-related potentials (ERPs) recorded from a single location over contralateral occipital scalp (from the region of the maximal responses represented in the topographic maps of Figure 7.17) during visual spatial–selective attention tasks. Attended stimuli (red trace) elicit greater amplitude ERPs than unattended stimuli (dashed blue trace). The shaded area on the ERP shows the difference in amplitude between attended and unattended events in the P1 latency range. The label "topography" indicates that the differences in amplitude of the P1 wave in this time range (from this and adjacent electrodes) were used to generate the topographic maps of Figure 7.17. Time zero (t=0) represents the onset of the stimuli evoking these ERPs. Positive voltage is plotted downward here, and the time scale is from 100 msec prior to stimulus onset to 400 msec after stimulus onset.

evoked P1 wave and the attention-related modulations of P1 have highly similar distributions on the scalp surface, and these topographies follow a retinotopic pattern (Figure 7.17). That is, initially the P1 attention effect is maximal over the contralateral occipital region of the scalp, as would be expected from the contralateral projection of the visual pathways (compare the top two images to the bottom two in Figure 7.17). Moreover, upper-field stimuli produce a scalp maxima that is different from the maxima for lower-field stimuli (compare the first and second images from top, and the third and fourth in Figure 7.17). Second, its latency (70 msec) is consistent with neuronal firing latencies in early extrastriate areas, as we know from research in monkeys. Finally, functional neuroimaging performed in these paradigms localized attentional modulations of sensory processing to regions of extrastriate cortex, as we will describe later in the chapter (see Neuroimaging of Spatial Attention and Integration with Electrophysiology).

To what extent can we assume that spatial attention effects in ERPs might be reflections of the neural processes underlying the behavioral effects of spatial attention? Can

Figure 7.17 Topographic maps showing the scalp distribution of attention-related brain responses to stimuli presented in each of the four quadrants of a computer screen while subjects had their eyes fixated on the cross in the middle. These maps represent the differences in the P1 waves in Figure 7.16. The differences are computed by subtracting the activity (ERPs) representing the unattended stimulus response from that representing the attended stimulus response, and the topographic maps are created as described in the caption of Figure 7.15. In this figure, the data are rendered as scalp current density rather than voltage. This transformation of the data was done by taking (mathematically) the second spatial derivative of the voltage values at each point on the scalp. This method highlights regions where the voltage values are changing rapidly across the scalp, and provides a better view of the location of the brain activity on the scalp surface. Small amplitudes are represented as whites and blues and larger amplitudes, as yellows and reds. The key feature of these data is that the activity "moves" over the occipital scalp as a function of the location of the stimulus in the visual field, reflecting attentional modulations of sensory responses that follow the retinotopic mapping of the visual field onto the visual cortex. That is, left-field stimuli project to contralateral right-hemisphere regions (**top two images**) and right-field stimuli project to left-hemisphere regions (**bottom two images**). Adapted from Mangun et al. (1993).

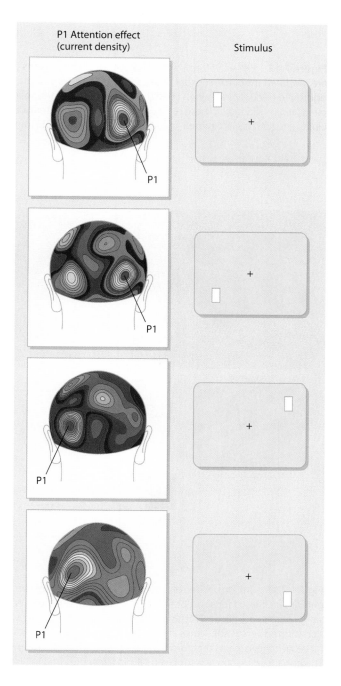

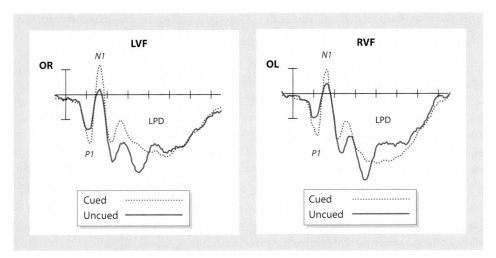

Figure 7.18 ERP waveforms averaged over fourteen persons performing a spatial cuing task like that depicted in Figure 7.8. Shown are modulations of the sensory evoked P1 and N1 ERPs to visual stimuli. The dotted traces show the responses to left-field (LVF) and right-field (RVF) targets when cued (i.e., valid trials), and the solid red traces show the responses elicited by the same stimuli when uncued (i.e., invalid trials). These recordings were taken at lateral occipital scalp sites contralateral to the stimulus—that is, at a right occipital site (OR) for left stimuli and at a left occipital site (OL) for right stimuli. Adapted from Mangun and Hillyard (1991).

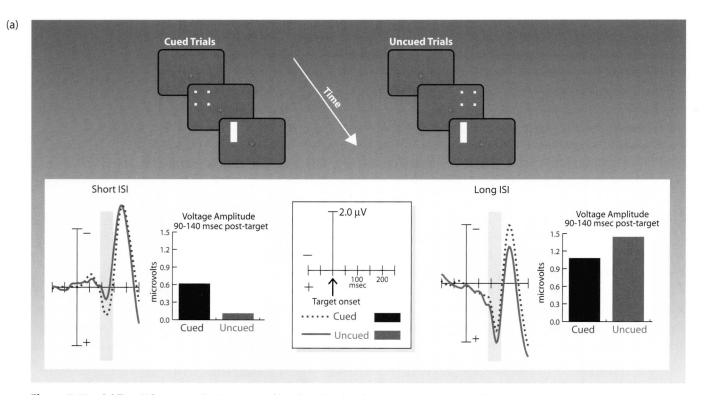

Figure 7.19 **(a) Top:** When attention is attracted to a location by abrupt onset of a visual stimulus, reaction times to subsequent targets are facilitated for short periods of time, as described earlier in this chapter. **Bottom:** When ERPs are measured to these targets, the same extrastriate cortical response (P1 wave) that is affected by voluntary spatial attention appears to be enhanced by reflexive attention (dotted versus solid lines at left) at short cue-to-target interstimulus intervals (ISIs). The time course of this reflexive attention effect is not the same as that for voluntary cuing, but rather is similar to the pattern observed in reaction times during reflexive cuing. The enhanced response is replaced within a few hundred milliseconds by a relative inhibition of the response (at right).

we conclude, for example, that better performance in spatial cuing paradigms is brought about by processing changes in the extrastriate visual cortex? To address this difficult question, researchers recorded ERPs using the same paradigms as prior cognitive studies (see Figure 7.8). If the improved performance—faster response times and more accurate responses—that was observed for targets presented to precued locations was due to better processing in the visual cortex, this should be reflected in the ERPs. Indeed, the target stimuli flashed to cued (valid) locations elicited larger sensory ERPs than did target stimuli presented to uncued (invalid) locations (Figure 7.18). It is unlikely that these modulations in early sensory processing are the sole neural basis of the subsequently recorded reaction benefits and costs, but a finding such as this bolsters the idea of Posner and colleagues described earlier (see prior section Voluntary Orienting). They proposed that when subjects expect that a relevant target is about to

occur at a particular location in visual space, they prepare for this eventuality by covertly orienting visual attention to that location, leading to changes in sensory/perceptual processing and reaction times.

Now that we know that voluntarily focusing attention on a location in response to verbal instructions or visual precues enhances visual ERPs to stimuli occurring at that location, we might ask whether similar neural mechanisms participate when attention is attracted to locations automatically, via reflexive mechanisms. That is, when our attention is reflexively attracted to a location in the visual field by a sensory event, does this heighten neural processing in the visual cortex? The answer is yes. Remember that reflexive cuing can be observed when a sensory event immediately precedes a target at the same location (Figure 7.19a, top). Under such conditions, response times are faster at the cued location than at the uncued location, but only when the time between the cuing and target stimuli is

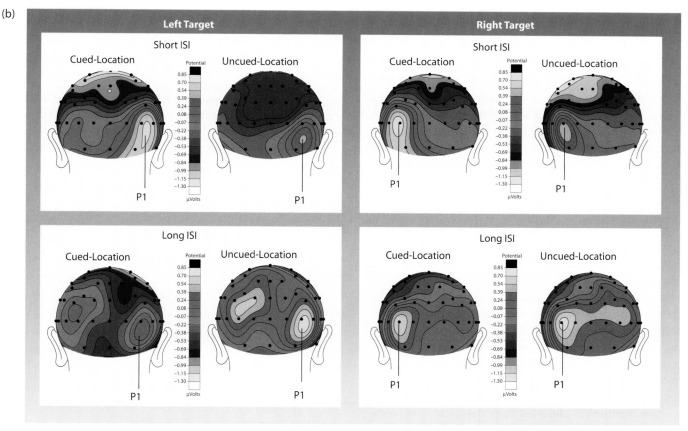

Figure 7.19 con't   **(b) Top:** These reflexive attention effects show the same scalp distribution as the P1 attention effect to voluntary cues. The topographic maps show the enhanced responses over contralateral occipital scalp sites for cued-location targets versus uncued-location targets at short cue-to-target intervals. The increased peak amplitudes (labeled P1) are indicated in the topographic maps and are represented by areas of yellow. **Bottom:** At longer cue-to-target ISIs, this facilitated response declines and may be replaced by slight inhibition such that cued-location responses are now smaller than uncued-location responses (especially noticeable for right-field targets in the long ISI condition—bottom right of figure). This physiological finding parallels that observed with reaction time and provides a neuronal mechanism for the behavioral findings. After Hopfinger and Mangun (1998).

short (< about 250 msec); with longer periods, this effect reverses, producing inhibition of return.

Joseph Hopfinger and colleagues (1998, 2001) showed that when ERPs are recorded in response to target stimuli in this type of design, the result is that the early occipital P1 wave is enlarged for targets that quickly follow a sensory cue at the same location versus when the sensory cue and target occur at different locations. But as the time after cuing grows longer, this effect reverses and the P1 response diminishes, and may even be inhibited, just as in reaction time measures (see Figure 7.19). Therefore, these data indicate that both reflexive (bottom-up) attention and voluntary (top-down) attention to locations involve a common mechanism with regard to the effect on early cortical stimulus processing. Presumably, however, the neural networks implementing these attentional modulations of sensory analysis are different, reflecting the differing triggers for the two forms of attention.

The idea that has emerged from studies like the ones reviewed in Figures 7.15 through 7.19 is that during both voluntary attention and reflexive attention, a "spotlight" of enhanced perceptual processing is focused on a discrete spatial location. Is this hypothetical spotlight similar to the one that visually searches for a target among multiple distracters? This hypothesis was investigated electrophysiologically by Steven Luck and colleagues, who presented subjects with arrays of targets and distracters in a conjunction search paradigm, like that in Figure 7.10b. At brief time intervals after the search array was presented, a solitary probe stimulus was flashed on the screen to elicit a visual ERP (Figure 7.20). The probe elicited larger early visual responses (i.e., P1) at the location of a conjunction target rather than in regions where nothing other than distracters was present. Hence, the idea that focal spatial attention is required to analyze conjunction targets found electrophysiological support. However, does this mean that this spotlight of attention moves rapidly around visual space? The data in Figure 7.20 do not address this question, but subsequent work by Steven Luck and his colleagues has addressed this issue well.

Behavioral research into the mechanisms of visual search has not resolved this question because it uses indirect measures of the action of the attention system, and therefore does not provide a moment-by-moment index of how attention is allocated in visual space. Such measures infer mechanisms based on the pattern of results (see Figure 7.11 and related discussion in the text). There is, however, an attention-related component of the ERP that can provide such an index. This is the ERP component with the pithy name *N2pc*, first identified by Steven Luck and Steven Hillyard (1994) in studies of visual search. The N2pc is observed 200 to 300 msec after the onset of a visual search array. It is largest over areas of vi-

sual cortex in the hemisphere contralateral to the location of an attended item within a search array (hence, the name N2pc—for second large negative posterior contralateral component). Experimental work has established that the N2pc component is a sign of the covert orienting of visual attention, and that it represents a stage of processing that occurs before the completion of object recognition. Interestingly, the N2pc component is believed to index a process of spatial filtering that resembles one that was observed in recordings of single cortical neurons in monkeys performing similar tasks.

Luck and colleagues reasoned that if visual search involves the rapid, serial shifting of attention from location to location, then the N2pc component should shift rapidly between the left and right hemispheres when attention is covertly shifted between the right and left sides of search arrays that span the vertical meridian. This is indeed what

Figure 7.20 ERP waveforms at frontal, parietal, and posterior temporal scalp sites show the different responses to probes that were flashed to locations where relevant and irrelevant items were located in the conjunction search array. The probe followed the onset of the arrays by 250 msec. The irrelevant probe elicited a larger sensory evoked P1 wave when it occurred at the location of a conjunction target than when it was presented to regions of the array where no target was present, supporting the idea that focal spatial attention is required in order to conjoin features of targets. Adapted from Luck et al. (1993).

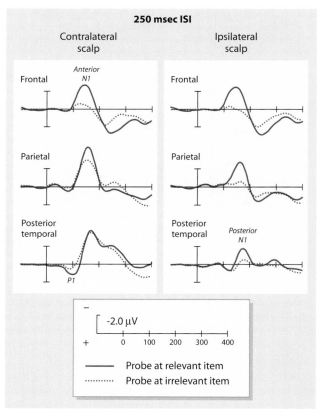

they found. They accomplished this by modifying the standard search task to include color as a means of biasing the subjects' search. Each array contained four colored stimuli (red, green, blue, and violet) and twenty-one black distracters. All stimuli had gaps on their surface: The target was a shape with a gap on its left side; distracters had gaps in other locations (Figure 7.21). For each subject, the target was one color (e.g., red) 75% of the time (designated C75) and another color (e.g., green) the remaining 25% of the time (designated C25). The researchers recorded the N2pc in response to arrays that had no targets, because under these conditions the subjects would search the red and green stimuli, for example, in that order. When the red and green stimuli are in the opposite hemifield, then the N2pc component first should be observed over the scalp contralateral to the red target (C75) and then after a few

tens of milliseconds delay, should appear over the opposite side of the scalp, given that that was contralateral to the green (C25) target that would be searched next. This is what Luck and colleagues observed (Figure 7.21b and c): The N2pc component was first contralateral to the C75 stimulus and then within about 80 msec or so, shifted to be contralateral to the C25 stimulus. These data provide significant electrophysiological support for serial models of attention search for conjunction targets.

To summarize what we know now, human electrophysiological and neuromagnetic recordings prove the concept of early selection and extend it with evidence on the neural stages of perceptual processing that are likely involved. The general model supported by these data is that incoming sensory signals can be altered within the sensory-specific cortex when stimuli having relevant

Figure 7.21 ERP study of visual search. **(a)** The search array consisted of four colored stimuli, one in each quadrant, and twenty-one black distracters. Targets were stimuli with a gap on the left (see text). For each subject, one color was designated as predicting the target 75% of the time (C75) and the other 25% of the time (C25) to bias the subject's search. The ERPs recorded from lateral occipital sites showing the N2pc when the two relevant colors were in the opposite hemifield **(b)** and the same hemifield **(c)**. The polarity reversal of the N2pc can be seen in (b) being first contralateral to the C75 stimulus and then later reversing to be more contralateral to the C25 stimulus. From Woodman and Luck (1999).

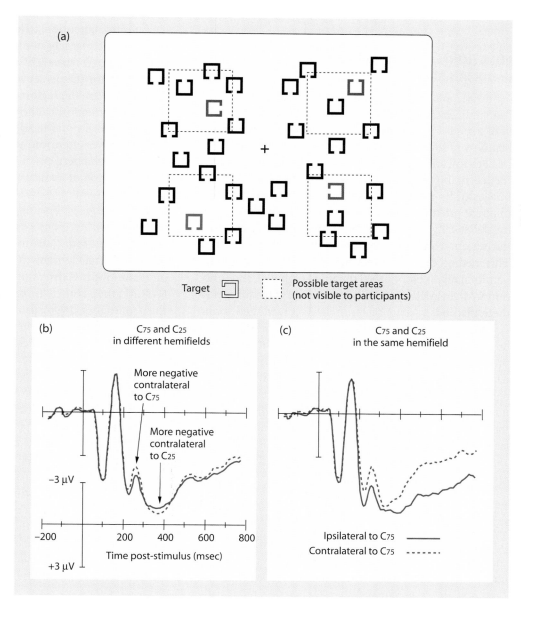

physical features are encountered. The earliest form of this selection can be defined by location in auditory and visual systems. The implication of these data is that descending projections from attentional control systems affect the excitability of cortical neurons coding the features of the stimuli to be attended or ignored. These data inform us about the sites (cortical versus subcortical, sensory cortex versus association cortex) where attention effects are manifest during perceptual processing. However, we also want to know precisely which brain structures are being modulated, as well as what systems and circuits are involved in the voluntary and reflexive control of these perceptual modulations. These issues can be investigated in animals using single-neuron recordings and lesion methods, in neurological patients who have lesions to specific brain structures, and in healthy humans by using functional neuroimaging methods. We examine evidence from the latter next.

FUNCTIONAL NEUROIMAGING IN HUMAN ATTENTION

Modern tools that measure physiological correlates of human brain activity in the normal living organism provide a powerful means for investigating the functional anatomy of higher mental functions. In neuroimaging, as in electrical and magnetic recording, the effects of experimental manipulations are observed as differences in the images of brain activity obtained as volunteers perform different attention tasks. These differences are visualized by statistically contrasting or subtracting images made during one condition from those made during another. This method can be used for images of regional hemodynamic changes (blood flow) obtained with PET and functional MRI (fMRI) (see Chapter 4 for an introduction to these methods). Many scientists have contributed to the development of neuroimaging methods for studies of human cognition, but it is fair to say that these approaches were significantly refined by classic cognitive neuroscience collaboration between cognitive psychologist Mike Posner, neuroscientist Steven Petersen, and neurologists Maurizio Corbetta and Marcus Raichle, and their physics, computer science, and engineering colleagues at Washington University in St. Louis in the late 1980s—work that marked the beginning of a new era of research into human brain function.

Neuroimaging of Feature Attention   In PET studies, Corbetta and colleagues (1991) found that attention to stimulus features like color, form, and movement leads to characteristic increases in blood flow (also referred to as *activation*) in the extrastriate visual cortex. While blood

flow in the brain was monitored by using radioactive water as a tracer, the volunteers were shown pairs of visual displays that had arrays of stimulus elements. The first display of each trial was a reference stimulus; the second was a test stimulus. The subjects' task was to view the first array and compare it with a second array to detect changes in prespecified aspects of the stimulus. In different selective-attention blocks (groups of test trials), the subjects were required to discriminate whether the arrays differed in color, shape, or motion. In other blocks, subjects were required to detect changes in any of the features (divided attention) or to passively view the stimuli or only a fixation point. Separate PET scans were made during each type of experimental block. To uncover which brain areas were activated during selective attention to color versus shape or motion, PET scans corresponding to each condition were subtracted from one another, thereby showing only the regions that differed between the conditions.

Selective attention to color, form, or motion activated distinct, largely nonoverlapping regions of extrastriate cortex (Figure 7.22). Extrastriate cortical regions specialized for the perceptual processing of color, form, or motion were the regions found to be modulated during visual attention to the individual stimulus features. These findings provide additional support for the idea that selective attention in modality-specific cortical areas alters the perceptual processing of inputs prior to the completion of feature analysis.

Neuroimaging of Spatial Attention and Integration with Electrophysiology   In studies investigating spatial selective attention, Hans-Jochen Heinze of the University of Magdeburg, Germany (then from the Medical School of Hannover, Germany), and one of the authors (G.R.M., then at the University of California, Davis) and their colleagues (1994) also performed PET studies. They presented subjects with bilateral stimulus arrays of nonsense symbols (two in each hemifield) flashed at a varying rate averaging about two per second. In separate conditions, the subjects were required to pay attention to the left half of the array and ignore the right, or vice versa, while keeping their eyes carefully fixated on a central spot and thereby employing covert spatial attention. Then the hemodynamic responses (changes in blood flow) in the brain could be compared for attend-left versus attend-right conditions (or to passive viewing conditions) to reveal attention-related PET activations. The researchers found that spatial attention led to activations in extrastriate cortex in the hemisphere contralateral to the attended side of the bilateral stimulus array, especially in the posterior fusiform gyrus on the ventral cortical surface (see summary data in Figure 7.22 and data

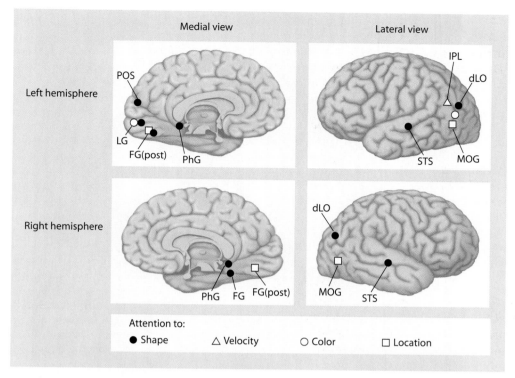

Figure 7.22 Summary of position emission tomography (PET) studies by Corbetta and colleagues (1991), Heinze and colleagues (1994), and Mangun and colleagues (1997). The figure illustrates that regions of extrastriate cortex specialized for the processing of color, form, or motion (from the work of Corbetta) are selectively modulated during visual attention to these stimulus features (selective feature attention). Selective attention to color activated areas in bilateral extrastriate cortex in the lingual gyrus (LG) and dorsal lateral occipital (dLO) area, but the activations were generally larger for left-hemisphere structures. Attention to shape led to bilateral activations in the fusiform gyrus (FG) and parahippocampal gyri (PhG), collateral sulcus of the left hemisphere, right superior temporal sulcus (STS), and a region between the calcarine sulcus and parieto-occipital sulcus (POS) in the right hemisphere. Selective attention to the motion (velocity) of the stimuli activated the left parahippocampal gyrus, the right superior temporal sulcus (in a region different from that activated for attention to shape), and left inferior parietal lobule (IPL). The PET study by Heinze and colleagues showed that attention to spatial location (selective spatial attention) produced activations in the posterior fusiform gyrus, and Mangun and colleagues replicated and extended this to show that the middle occipital gyri (HOG), was also activated during spatial attention. Adapted from Corbetta et al. (1991), Heinze et al. (1994), and Mangun et al. (1997).

in Figure 7.23 from a follow-up study reviewed below). Such a finding fit extremely well with evidence provided in prior ERP studies by these researchers and others (see Figure 7.17). But it is difficult to make inferences about the relationship between neuroimaging results and the prior ERP findings, because even when the same experimental paradigm is used, the two approaches provide very different, albeit complementary information about brain activity. To address this issue, Heinze and colleagues attempted a new approach: They included ERP recordings from the same subjects using the same stimuli, parameters, and tasks as in the PET study. This approach had never been attempted for a cognitive process.

The idea is, in some regards, simple: Acquire the PET data to learn where in the brain visual spatial attention affected visual processing, as just described. Then in parallel sessions, record ERPs from the same subjects to learn

about the time course of attentional processing on a millisecond level of temporal resolution, something not possible with the PET measures. Take the two data sets and compare them to establish both the time course and the functional anatomy of attentional selection based on location. An example of how this was accomplished comes from a follow-up study by one of us (G.R.M., now at Duke University) and his colleagues (1997) at the University of California, Davis; the study follows the earlier study with Heinze, and here we focus on the main findings of interest to illustrate the logic of the multimethodological approach.

To establish the relationship between activity in the ERPs and PET activations, dipole modeling of the ERPs was performed (see Chapter 4 for an overview of the method). The PET method had revealed two areas of attention-related activation in each hemisphere (Figure 7.23a), and the aim was to learn whether these could serve

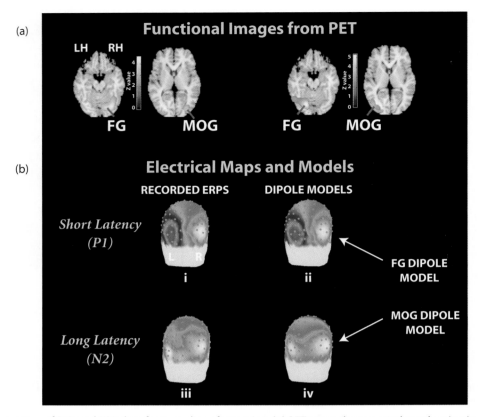

Figure 7.23 Integration of PET and ERP data from studies of attention. **(a)** PET scans show contralateral activations with spatial attention in the posterior fusiform gyrus (FG) and the middle occipital gyrus (MOG). **(b)** Attention effects (attend left minus attend right) as recorded by ERPs and models of the ERPs in realistic head simulations. The relationship between the recorded and modeled data is illustrated in **i** and **ii** for the P1 latency range, and in **iii** and **iv** for the longer latency N2 range. The recorded and model data are shown as topographic voltage maps on the surface of the realistic head model viewed from the rear (left side of brain and head on left side of figure). Topographies of the model data are much more similar to those of the recorded data in the P1 time range when the dipoles were seeded to (placed in) the areas of the FG activation (compare i to ii) than when placed in the MOG. In contrast, for the longer latency N2 effects (260–300 msec) dipoles placed in MOG produced a better fit for the 260- to 300-msec ERP effects (compare iii to iv) than when they were placed in the more medial, fusiform location. Adapted from Mangun et al. (1997).

as possible generators of the P1 attention effect (and a longer-latency N2 attention effect) in the ERPs that had been revealed in the same study. One activation was ventral medial in the posterior fusiform gyrus, and the other was more lateral and slightly ventral occipital in the middle occipital gyrus. To determine which of these PET activations might be related to which attention effect in the ERPs (P1 versus N2), each locus (fusiform gyrus and middle occipital gyrus) was used in turn to create the simulations. The details of the brain, skull, and scalp surfaces were derived from anatomical MRI scans of the subjects in the experiment, and were coregistered with the PET data in three-dimensional space. When model dipolar sources (simple point dipoles that were equivalent to the electrical field produced when cortical pyramidal cells are active) were placed in the fusiform gyrus in the head model, the model data provided a better explanation (lower residual variance, or residual difference) of the recorded data in the time range of the P1 attention effect than did placement of the dipoles

in the more lateral area of PET activation in the middle occipital gyrus (see Figure 7.23b, i and ii). Conversely, dipoles in the middle occipital gyrus in the model simulation were better at explaining the longer-latency ERP attention effects in the N2 range (260–300 msec after onset of stimulus) (see Figure 7.23b, iii and iv). Thus, in addition to localizing *where* the effects of spatial attention took place in the brain, this type of integrative approach can also tell *when* those effects occurred, yielding a spatiotemporal view of cognition—something not possible with either method alone.

These neuroimaging findings on modulations of the visual cortex during attention provide powerful evidence about the stages of processing affected by attention, and fit well with findings in animals that we will describe later. However, as we learned in Chapter 5, the visual system in primate occipital cortex is composed of multiple representations of the visual world called *retinotopic maps* that define successive stages, or visual areas in extrastriate cortex. How do these neuroimaging and ERP findings in humans map onto this complexity of visual areas in the brain? Although PET has proved a powerful tool in studies of cognition, for most purposes, as we have already seen, fMRI provides a higher-resolution view of brain activity.

In an fMRI study of spatial attention, Roger Tootell and Anders Dale and their colleagues (1998) at Massachusetts General Hospital have addressed questions about the details of the organization of attentional modulations in the visual cortex. They made use of the novel method of cortical unfolding and retinotopic mapping of human visual cortex (see Chapter 5) developed by Steve Engel and his colleagues (1994) at Stanford and Marty Sereno and Anders Dale and their colleagues (1995) at the University of California, San Diego, and Massachusetts General Hospital. Using knowledge gained about the probable organizational principles of the primate visual cortex from prior studies in monkeys, this method permits the borders of the early vi-

sual areas to be defined in humans. Tootell and his group implemented a simple spatial attention paradigm designed after that shown in Figure 7.17; it involved attending to bar stimuli in either the right upper, left upper, right lower, or left lower visual quadrant in different blocks and recording the attention-related changes in visual processing with fMRI, much as was done in the earlier ERP and PET studies just described. They then performed careful retinotopic mapping of the subjects to reveal the borders of the early visual areas from V1 (primary visual cortex) through V4 and beyond. The attentional activations were then mapped onto the flattened representations of the visual cortex, permitting the attention effects to be related directly to the multiple visual areas of human visual cortex. These researchers found that spatial attention was represented by robust modulatations of activity in multiple early extrastriate visual areas, including small modulations in V1 (Figure 7.24).

Figure 7.24 Spatial attention effects in multiple visual cortical areas as demonstrated by functional MRI (fMRI). Activations with spatial attention to left-field stimuli shown in the flattened right visual cortices of two subjects (one in the left column and the other on the right). The white lines (dotted and solid) indicate the borders of the visual areas as defined by representations of the horizontal and vertical meridians; each area is labeled from V1 (striate cortex) through V7, a retinotopic area adjacent to V3A. The solid black line is the representation of the horizontal meridian in V3A. Panels A and B show the retinotopic mappings of the left visual field for each subject, corresponding to the polar angles shown at right (which represents the left visual field). Panels C and D show the attention-related modulations (attended vs. unattended) of sensory responses to a target in the upper left quadrant (the quadrant of the stimuli is shown at right). Panels E and F show the same for stimuli in the lower left quadrant. In C through F, the yellow to red colors indicate areas where activity was greater when the stimulus was attended to than when ignored; the bluish colors represent the opposite, where the activity was greater when the stimulus was ignored than when attended. The attention effects in C through F can be compared to the pure sensory responses to the target bars when passively viewed. Note the retinotopic pattern of the attention effects in C through F: The attention effects to targets in the lower left quadrant produced activity in several lower field representations, which included the appropriate half of V3A (inferior V3A labeled with an i) in both subjects and V3 and V2 in one subject. In contrast, attention to the upper left quadrant produced activity in the upper field representation of V3A (S) and in the adjacent upper field representation of area V7. From Tootell et al. (1998).

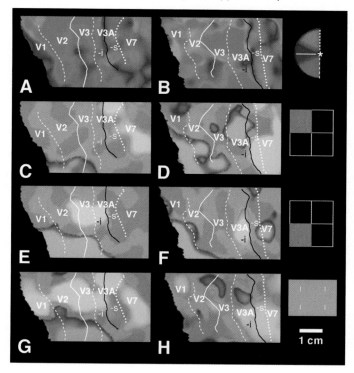

Moreover, these effects were retinotopically mapped within retinotopic visual cortex (V1–V8). This work provides a high-resolution view of the functional anatomy of multiple areas of extrastriate and striate cortex during sustained spatial attention in human visual cortex.

Neuroimaging of Attention to Object Representations

We have focused much of the discussion of attention on studies of spatial attention, as far more is known about the mechanisms of spatial attention than nonspatial attention. Yet, models of attention do account for spatial, feature, and **object-based attention.** What are the mechanisms supporting attention to color, form, or speed (see Figure 7.22) versus attention to locations (spatial attention) (see Figure 7.23)? Behavioral work has demonstrated evidence for object-based attentional mechanisms. In a clever study John Duncan (1984) contrasted attention to location (spatial attention) with attention to objects (object-based attention). Holding spatial distance constant, he discovered that two perceptual judgments concerning the same object can be made simultaneously without loss of accuracy, whereas the same two judgments about different objects cannot. This processing limitation in attending to two objects implicates an object-based attention system in addition to a space-based system. In line with this view, in a more recent behavioral study, Rob Egly, Jon Driver, and Robert Rafal (1994) demonstrated that the *costs* and *benefits* of spatially valid versus invalid cuing were greater *between* two objects than they were *within* one object. Their data suggest that the presence of otherwise irrelevant object forms facilitated either the spread of attention within the confines of each object or the switching of attention within an object, whereas crossing between objects hindered spatial attention. Thus, object representations can modulate spatial attention, but are these representations labile to the effects of attention in the absence of spatial factors?

In a recent study, Kathleen O'Craven, Paul Downing, and Nancy Kanwisher at MIT and Massachusetts General Hospital (1999) provided a fascinating demonstration of the direct action of attention on object representations. They made use of the facts that faces activate a region of the fusiform gyrus (in a zone anterior to the area of spatial attention activation shown in Figure 7.23—see discussion in Chapter 6) that is less active in response to images of other objects such as houses (known as the fusiform face area or FFA), and that a region of parahippocampal cortex is more active in response to images of houses than faces (the so-called parahippocampal place areas or PPA). The researchers then superimposed the face and house stimuli transparently over one another so they occupied the same region of space (Figure 7.25); one of the objects was moving

Figure 7.25 Example of stimuli in an fMRI study of object-based attention. Houses and faces were superimposed transparently to create stimuli that could not be attended to using spatial mechanisms. From O'Craven et al. (1999).

back and forth, while the other was stationary. The motion of the moving stimulus activated cortical motion areas MT/MST (sometimes referred to as *V5*). Which stimulus moved and which was stationary varied in different blocks. In these different blocks, subjects were told to attend selectively either to the face, to the house, or to the motion. The activity in the FFA, the PPA, or MT/MST provided relatively pure measures of the responses to each stimulus, respectively. The researchers found that when subjects attended to faces, activity in the FFA increased but activity in the PPA did not, and vice versa for when the subjects attended to houses. Interestingly, when they selectively attended to the motion, activity in the MT/MST increased, as did activity in the region (FFA or PPA) that corresponded to the object that was moving (face or house, respectively).

These results demonstrate how attention acts on object representations. Attention facilitates processing of all the features of the attended object. For example, face processing was facilitated when the attended moving stimulus was a face, even though the task did not require identification or attention to the face it-

self. These findings do not indicate that attention does not work in spatial coordinates or does not modulate activity as a function of spatial locations. It clearly does. Rather these findings show that when spatial attention is not involved, object representations can be the level of perceptual analysis affected by top-down attentional control.

What are the effects of attention on the visual cortex when it modulates stimulus processing? Leslie Ungerleider and Robert Desimone and their colleagues (Kastner et al., 1998) proposed the biased competition model. They suggested that stimulus features compete for control of neuronal resources, and that this competition can be influenced by directed attention. In a series of fMRI studies, they showed that nearby stimuli (*X* and *Y*) when presented simultaneously can interfere with one another, reducing the neural response in comparison to when only one stimulus (e.g., *X*) is presented alone. However, when attention is introduced and directed to one stimulus (*X*), then simultaneous presentation of the competing stimulus (*Y*) no longer interferes (Figure 7.26).

Figure 7.26 Task and fMRI effects for attention to competing stimuli. Panels **(a)** and **(b)** show the task design. Competing stimuli were presented either simultaneously or sequentially. During the attention condition, covert attention was directed to the stimulus closest to the point of fixation, and the others were merely distracters.

(a) **Sequential Condition (SEQ)**

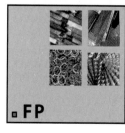 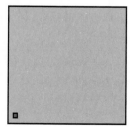  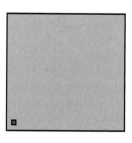

(b) **Simultaneous Condition (SIM)**

   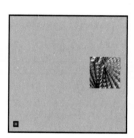

250 ms   250 ms   250 ms   250 ms

Time

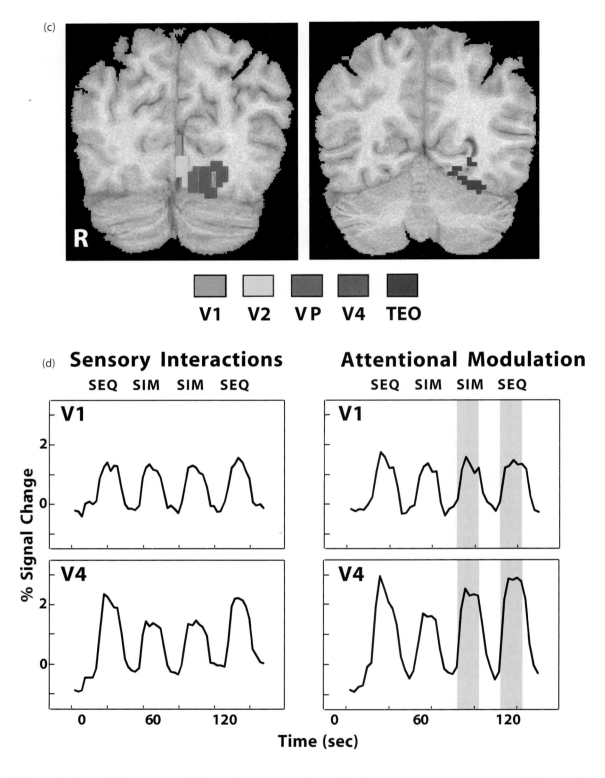

Figure 7.26 con't Panel **(c)** shows coronal MRI sections in one subject, with the pure sensory responses in multiple visual areas mapped using meridian mapping (similar to that used in Figure 7.23, but without cortical unfolding). Panel **(d)** shows the percentage of signal changes over time in areas V1 and V4 as a function of whether the stimuli were presented in the simultaneous (SIM) or sequential (SEQ) condition, and as a function of whether they were unattended (left side) or whether attention was directed to the target stimulus (far right side shaded blue). In V4 especially, the amplitudes during the SEQ and SIM conditions were more similar when attention was directed to the target stimulus (shaded blue areas at right) than when it was not (unshaded areas). Adapted from Kastner et al. (1999).

Neuroimaging Studies of Attentional Control Networks

In addition to determining the activation of sensory-specific cortical areas, PET and fMRI studies of selective attention have revealed attention-related activity in several other brain areas: the thalamus (probably exclusively the pulvinar nucleus), the basal ganglia, the insular cortex, the frontal cortex, the anterior cingulate nucleus, the posterior parietal cortex, and the temporal lobe. Some regions (frontal cortex, parietal cortex, and pulvinar nucleus) are implicated in directing attention according to a task's requirements. For instance, Corbetta and colleagues (1993, 1995) showed that the superior parietal cortex is activated when attention is switched from one location to another to detect a target (Figure 7.27); it is

interesting that this occurs whether attention is voluntarily switched in a controlled manner to expected target locations, or moves through a field of potential targets in a visual search task. These findings suggest that movements of attention during visual search are subserved by the same neuronal machinery as are voluntary movements in response to cues, a finding consistent with the ERP studies described earlier (see Figure 7.20).

How do activations in nonsensory brain regions and in the pulvinar nucleus of the thalamus explain the modulations in extrastriate visual areas observed in PET and ERP studies? Presumably, neuronal projections from executive control systems contact neurons in sensory-specific cortical areas in order to alter their excitability. Therefore, when a stimulus excites those neurons, the response is enlarged. A network including the pulvinar nucleus of the thalamus, the posterior parietal cortex, and the dorsolateral prefrontal cortex may mediate cortical excitability in the extrastriate cortex as a function of selective attention (Figure 7.28). The pulvinar nucleus is not part of the projection pathway from the retina to the primary visual cortex; instead, it maintains reciprocal connections with cortical and subcortical areas. Hence, the pulvinar nucleus is ideally suited to afford the frontal and parietal cortices a way to influence extrastriate visual cortical processing. In doing so, it presumably uses information from location maps in the parietal cortex. But how can we measure the activity of the different brain regions during the execution of complex tasks, and decompose and identify the distinct neuronal networks associated with the variety of elementary mental operations hypothesized to be involved in attentional orienting and selection?

One method that has been exploited for this purpose is event-related fMRI. As described in Chapter 4, this approach is analogous to the ERP methods of selectively

Figure 7.27 Corbetta and colleagues found that regional cerebral blood flow in the posterior parietal cortex increased when attention was switched from one location to another, in order to detect a relevant target. This same area was activated when attention was moved through the visual field during visual search. RVF = right visual field; LVF = left visual field. Adapted from Corbetta et al. (1993).

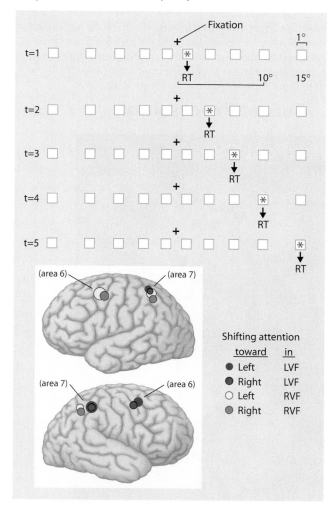

Figure 7.28 Model of executive control systems and the way in which extrastriate cortex processing is affected by a network of brain areas.

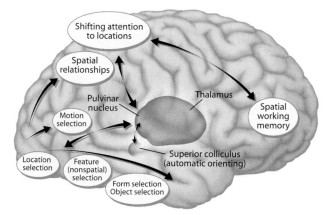

averaging responses to different intermingled trials to differentiate, in this case, the time-locked hemodynamic responses to a specific class of stimuli from others that are temporally adjacent. The approach provides more detailed information than is possible using blocked designs in PET (which is the only way such analyses can be done) or fMRI (in which either blocked or event-related analyses are possible) (see Chapter 4). Event-related fMRI has been used to understand attentional control with a specificity not previously available in neuroimaging studies of attention.

Joseph Hopfinger and his colleagues (2000) investigated the brain network involved in controlling spatial attention. To do this, they used a spatial cuing paradigm similar to that in Figure 7.8, but modified it to make use of bilateral stimuli. An arrow cued the subjects as to

which hemifield to attend on each trial, and then 8 seconds later, the bilateral target array (flickering black and white checkerboards) appeared for 500 msec; the subjects had to press a button if some checks were gray rather than white on the cued side only. This covert attention task required the subjects to fixate their eyes on a central spot. The hemodynamic responses to the attention-directing cues were extracted separately from the responses to the later targets. This permitted the brain's responses to the cuing instruction to be separated from attention-related processing of the target. In response to the cue, a top-down attentional control network was isolated. It included the superior frontal cortex, inferior parietal cortex, superior temporal cortex, and portions of the posterior cingulate cortex and insula (Figure 7.29; compare

Figure 7.29 Attentional control network activated in response to an attention-directing spatial cue and identified using event-related fMRI in a cuing paradigm. Activations time locked to the cue, indicating attentional control networks, are shown on the left. Those time locked to the targets, indicating target processing and motor responses, are on the right. The activated areas are shown in different colors to reflect the statistical contrasts that revealed the activations: bluish for brain regions that were more activated to cues than targets, and reddish-yellow to indicate regions that were more active to targets than cues. The activity is collapsed from the data gathered from six normal subjects and is presented on a canonical brain rendered in three dimensions derived from structural MRI scans. The top row shows the lateral view of the left hemisphere, while the bottom row shows its medial surface; the activity was the same for the right hemisphere, which is not shown. The attentional control network included dorsolateral prefrontal cortex (DFL), inferior parietal lobe (IPL), superior temporal sulcus (STS), medial frontal cortex (MF), and the posterior cingulate gyrus. Target processing activated the visual cortex (VC), superior parietal lobe (SPL), anterior cingulate cortex (AC), supplementary motor area (SMA), the ventrolateral frontal lobe (VLF), and the precentral and postcentral gyri (Pre/Post CS). Adapted from Hopfinger et al. (2000).

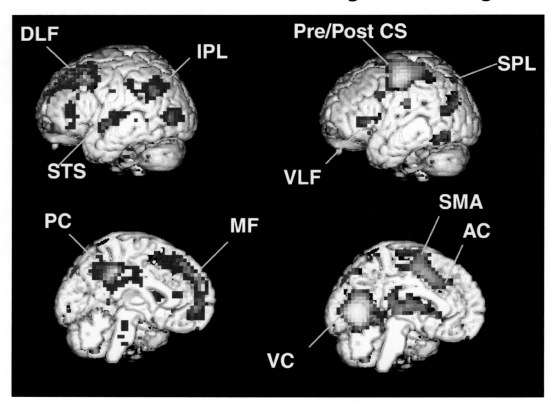

Control Network   **Target Processing**

to model in Figure 7.28). In contrast, activity time locked to the targets reflected motor processes (the subjects responded to the targets) and visual analysis. Thus, one striking finding here is that activity reflecting different components of a complex task such as cued spatial attention can be decomposed to reveal the distinct brain networks subserving attentional control versus target processing. What effects did attention control have on target processing? To answer this question, the researchers compared attend-left to attend-right conditions as described in earlier studies. Figure 7.30 shows these effects of attention in the visual cortex. As in related research, attention to one hemifield activated the visual cortex in the opposite (contralateral) hemisphere, and this activity occurred in multiple extrastriate cortical areas.

Interestingly, this study also revealed something about how top-down attention affects visual cortex prior to the presentation of task-relevant targets. After the cue was presented but before the target stimuli were presented, activations were observed in the regions of cortex that would later process the incoming target (Figure 7.31). This sort of attentional priming of sensory cortex may form the basis for later selective processing of target inputs. These data bear similarity to the single-neuron data Steve Luck and Robert Desimone and their colleagues (1997) obtained from attending monkeys. These researchers showed that neuronal firing rates increased after attention was focused on the receptive field corresponding to the recorded neuron, but prior to the appearance of the task-relevant target. How might these effects lead to changes in the processing of later targets? In a computational model, Dave Chawla and Karl Friston and their colleagues (1999a) in London have described this type of mechanism as a means for increasing the gain of sensory processes with attention, and related this model to empirical data from fMRI studies of color attention (1999b). Their idea is that attention may act to synchronize neuronal firing, and this may be visible with fMRI as priming of the sensory cortex. Presumably, the

Figure 7.30 Effects of spatial attention in the visual cortex as demonstrated by event-related fMRI. **(a)** Areas of activation in a single subject were overlaid onto a coronal section through the visual cortex (left hemisphere on the left) obtained by structural MRI. The statistical contrasts reveal where attention to the left hemifield produced more activity than attention to the right (reddish to yellow colors—on left) and the reverse, where attention to the right elicited more activity than attention to the left (bluish colors—on right). As demonstrated in prior studies, the effects of spatial attention were activations in the visual cortex contralateral to the attended hemifield. **(b)** Meridian mapping allowed the borders of early visual areas to be defined (on the left) and the effects of attending the left and right fields to be overlaid (outlines at right). Adapted from Hopfinger et al. (2000).

Selective Attention (Targets)
(Single Subject)

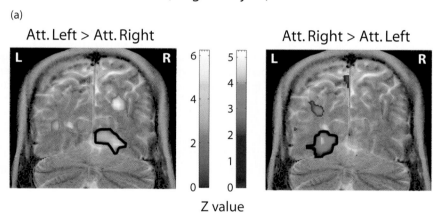

(a)

Att. Left > Att. Right   Att. Right > Att. Left

Z value

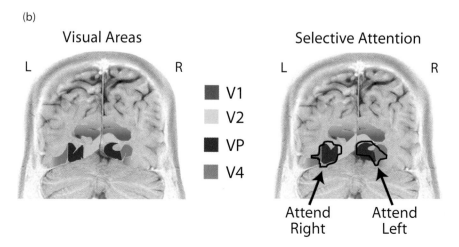

(b)

Visual Areas   Selective Attention

V1
V2
VP
V4

Attend Right   Attend Left

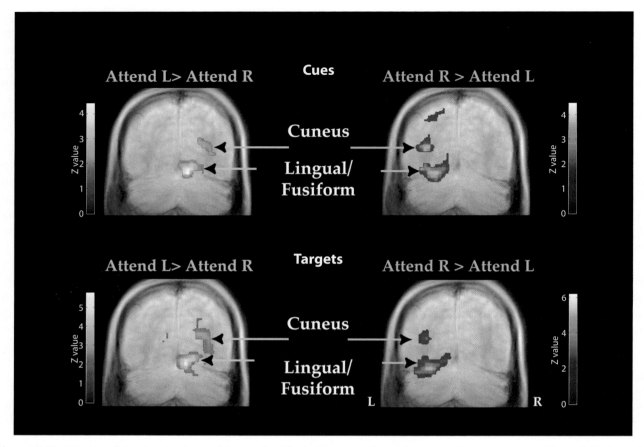

Figure 7.31 Priming of visual cortex by spatial attention. The **bottom row** shows the same effects to targets as Figure 7.30a but collapsed over a group of six subjects. When these same regions of visual cortex are investigated before the targets actually appear but after the cue was presented **(top row)**, a preparatory priming of these areas can be observed as increased activity, and these regions of increased activity precisely overlap with the regions that will later receive the target stimuli shown at bottom. Adapted from Hopfinger et al. (2000).

synchronization permits subsequent stimulus processing to be more efficient, leading to increased processing of the stimuli that activate the synchronized neurons at later points in time. In the next section, on single neuronal responses during attention in animals, we consider converging evidence about brain attention mechanisms.

Animal Studies of Attentional Mechanisms

Neurophysiological studies of humans that employ electrical, magnetic, and functional imaging techniques have revealed some of the inner workings of attentional selection and control. We now see that complex interactions take place within neuronal circuits to achieve selective processing. What are the underlying cellular-level events, and what can we learn from animal studies that confirm, amplify, or perhaps contradict what we have found in humans?

We expressed earlier that animal studies preceded neurophysiological recording in humans (see Figure

7.12), but the first experiments were flawed and provided few real answers about whether selective sensory-level gating of inputs occurred during attention. The first well-controlled studies that demonstrated such effects involved recordings of ERPs in humans in the early 1970s (see Figure 7.13). Nevertheless, single-cell recording methods provide details on the attentional mechanisms at the synaptic, cellular, and circuit levels—a difficult thing to do in humans (but also see Human Brain Activity During Attention).

Activity in the visual pathways of monkeys has been investigated during selective attention. A microelectrode is lowered into a candidate structure in the visual pathways (as described in Chapter 4), and the extracellular activity of cortical or subcortical neurons is measured in different conditions of attention. In principle, the methods are similar to those for recording human ERPs or ERFs, except that typically, individual neuronal responses are measured by the former rather than the

HOW THE BRAIN WORKS

Cell-to-Cell Firing in the Visual System

If attentional selectivity leads to changes in the firing rates of neurons in extrastriate cortical maps, what happens at the neuronal level? We do not really know. Neurons can be modulated by numerous hypothetical mechanisms, but at present it remains unknown which mechanism might produce the effects observed in monkeys. Robert Desimone and his colleagues (1990) at the National Institute of Mental Health described two possible ways that attention might affect neuronal firing rates in a visual area like V4.

One is the input-gating model, where the receptive fields of hypothesized intermediate neurons correspond to the attended or ignored regions of the visual field. The primary purpose of these neurons is to provide an inhibitory signal to V4 neurons for any regions of the V4 cell's receptive field to which attention is not currently directed. Although in this case it is

the V4 neurons that show attention-related modulations of firing, the site of action of the attention effect is on the inputs to V4, not the V4 neurons themselves. The main alternative is a neuron-gating model; the effect here is hypothesized to be on the actual V4 neurons showing enhancement and suppression with attention. At present, too little is known about cellular-level interactions that produce attention effects in extrastriate neurons, such as those in V4, to choose between the competing models. Nor can we be certain what the source of the attention-related command may be. In principle, it is possible to obtain the information necessary to precisely specify the neuronal mechanisms underlying selective attention, but it is technically very challenging. There is little doubt, however, that in the not-too-distant future, these mechanisms will be revealed.

Two models for how attentional gating of extrastriate neurons is represented at the neuronal level. **(a)** In the input-gating model, intermediate neurons provide an inhibitory signal to regions that are outside of the to-be-attended location. **(b)** In the neuron-gating model, the V4 neurons of interest show both direct enhancement and suppression with attention, with mutual inhibition present between cells coding different stimulus features in the same general region of space.

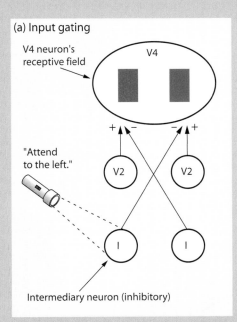

(a) Input gating

V4 neuron's receptive field

V4

"Attend to the left."

V2

V2

I

I

Intermediary neuron (inhibitory)

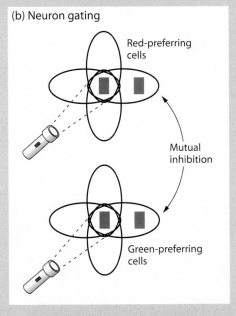

(b) Neuron gating

Red-preferring cells

Mutual inhibition

Green-preferring cells

ensemble responses of tens or thousands of activated neurons measured by the latter.

Single neurons in area V4 of a monkey's extrastriate cortex lie in the ventral, or "what" pathway from the primary visual cortex to the inferior temporal cortex, the pathway that carries out feature analysis and object discrimination (see Chapter 5). It is plausible, then, that the firing of neurons in this cortical stream of visual processing might be modulated during spatial attention, thereby providing a way to gate relevant feature information for perception. Indeed, Jeff Moran and Robert Desimone (1985) of the National Institute of Mental Health found that the firing rates of neurons in visual area V4 are significantly modulated during selective spatial attention. When a stimulus is attended, it elicits a stronger response than when it is ignored; yet this effect manifests itself differently from what was seen in the ERP studies described earlier. Interestingly, with V4 neurons, attentional modulation of neuronal firing is most robust when attended and ignored locations are located within the V4 neuron's receptive field.

In these studies, Moran and Desimone first identified the V4 neuron's classical receptive field and then determined which stimulus was good for exciting a response from the neuron (e.g., perhaps a vertical red bar) and which was a poor one (e.g., perhaps a horizontal green bar); that is, the one stimulus (red bar) makes the cell fire whereas another (the green bar) might not. Both stimuli could fit simultaneously within the neuron's receptive field, which for V4 neurons may extend over several degrees of visual angle. The task involved attending covertly to a cued visual field location, and thus, prior to each testing block, the monkey was cued to the location where a discrimination task would be required—similar to what is done with humans in spatial cuing paradigms.

At the beginning of each stimulus trial, the monkey viewed a sample stimulus and then simultaneously the pair of target stimuli, one within the attended location and one at another location, but both within the large V4 receptive field. If the target stimulus at the attended location matched the sample stimulus, the monkey had to make a motor response, for which a food or drink re-

Figure 7.32 Desimone and associates studied the effect of selective attention on the responses of a neuron in area V4 of macaque monkeys. The areas that are circled in broken lines indicate the attended locations for each trial. Bars in blue are effective sensory stimuli and bars without color are ineffective sensory stimuli for this neuron. When the animal attended to effective sensory stimuli, the V4 neuron gave a good response, whereas a poor response was generated when the animal attended to the ineffective sensory stimuli. Since both stimuli were presented simultaneously, this effect can be attributed only to the change in the focus of attention from one stimulus to the other; the animal was precued where to attend. The neuronal firing rates are shown to the right of each monkey head. The first burst of activity is to the cue, but the second burst in each image is to the target array, and here the reduction can be observed when the animals attend the ineffective stimulus (right). Adapted from Moran and Desimone (1985).

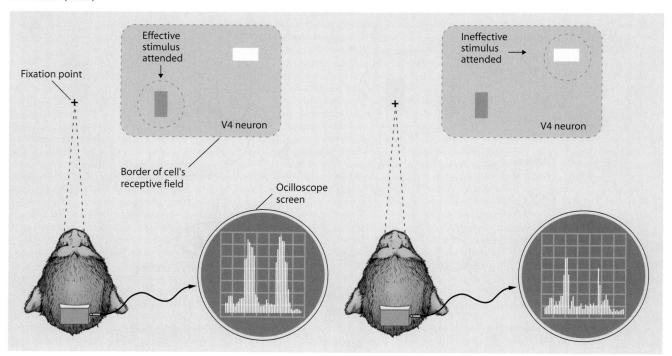

ward was given. When the optimal stimulus for eliciting responses from the V4 neuron was within the attended location, the responses were enhanced in comparison to when the optimal stimulus fell within the receptive field but outside the attended region. The changes in neuronal firing were related to the location of attention and not to the stimulus features (Figure 7.32). A fascinating aspect of this neuronal attention effect was that it directed attention to or away from stimuli that were inside the cell's receptive field. In V4 neurons, when attention was directed to locations outside the receptive field, modulations of responses to the optimal stimuli presented in the receptive field were not found.

The pattern for V4 neurons is markedly different from that of neurons in subsequent stages of visual analysis in the ventral pathway. In the inferior temporal cortex, attention can modulate neural activity even when the to-be-ignored stimulus is far from the to-be-attended stimulus. This happens partly because the receptive fields of inferior temporal neurons are huge, encompassing the entire central visual field. Processing in visual areas prior to V4 appears to be less successful in modulating cortical visual processing even when proper controls are utilized, like when the conditions are equally well matched for arousal. Thus, spatial attention effects of the type evident in V4 and inferior temporal cortex have been difficult to obtain, for example, in V1, which led to the idea that perhaps attentional modulations happen only in extrastriate cortical areas—a conclusion supported by some electrophysiological studies in humans. But changes in V1 firing rates may occur with attention when perceptual demands require filtering of sensory inputs on a spatial scale appropriate for the size of the V1 receptive fields, which are very small (< 1 degree of visual angle at the fovea). For example, Brad Motter (1993) showed modulations of V1 neurons when competing stimuli were presented simultaneously in the visual field. More recent work by Carrie McAdams and John Maunsell (1999) demonstrated similar effects in V1 and related them to activity in extrastriate cortex to show that the effects in V1 were smaller but nonetheless present. These findings fit well with those from numerous recent neuroimaging studies in humans showing V1 modulations with selective attention (see Figure 7.24). What might this mean about where attention can modulate visual inputs?

If there are attentional modulations of the primary visual cortex, could they reflect earlier gating in the thalamus or even in the retina? Unlike the cochlea, no neural projections to the retinas of humans could serve as a substrate for modulating retinal activity. But in pri-

mates massive neuronal projections extend from the visual cortex back to the thalamus. These projections synapse on neurons in what is known as the **reticular nucleus** *of the thalamus* or, more specifically for the visual system, on the *perigeniculate nucleus,* which is the portion of the reticular nucleus that surrounds the lateral geniculate nucleus (Figure 7.33).

These neurons maintain complex interconnections with neurons in the thalamic relays and could, in principle, gate information flow from the thalamus to the cortex. Indeed, this can happen during intermodal (visual-auditory) attention, as demonstrated in cats by Charles Yingling and James Skinner in 1976. Such a mechanism might also select the visual field location for the current spotlight of attention in perception, according to theories by Nobel laureate Francis Crick (1992). Effects in the thalamus do not appear to participate in selection for locations during voluntary visual attention, but may have

Figure 7.33 Diagram of the thalamus, perigeniculate nucleus, and projections to and from the thalamus and visual cortex.

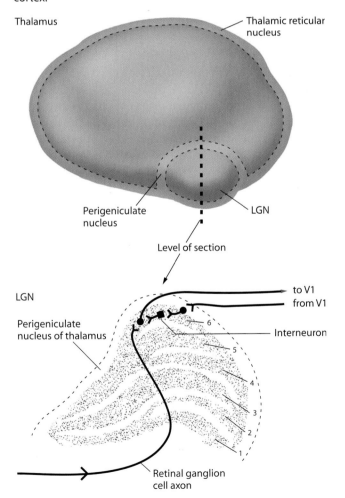

a role in reflexive attention and nonspecific arousal. Therefore, although we can say with confidence that modulations of retinal processing with spatial attention do not occur (given that no neural substrate exists to accomplish this), we must leave open the possibility that the thalamus might be gated during some forms of visual attention. By the level of the extrastriate visual cortex, though, robust changes in neuronal firing rates are involved in selective spatial attention and likely for nonspatial attention as well (see Cell-to-Cell Firing in the Visual System).

ATTENTIONAL MODULATIONS IN SUBCORTICAL STRUCTURES AND PARIETAL CORTEX

In the early 1970s, Robert Wurtz and his colleagues at the National Eye Institute began to investigate the superior colliculus in the midbrain for its role in attentional processes. They discovered visually responsive neurons that were affected by how monkeys responded to stimuli. When a monkey's eyes made a rapid movement from one location to another (saccadic eye movement) where a target stimulus was presented, the firing rates of the cells whose receptive fields included this region increased. Eye movement did not enhance cell firing, because without sensory stimuli, eye movements would not elicit neuronal responses (Figure 7.34). Eye movements to other locations also did not enhance responses to the stimulus. These cells responded as though they were involved in attentional mechanisms having to do with the stimulus. In subsequent experiments, these investigators found that when the stimulus was task relevant but not the target for saccadic eye movement, the cell did not become more responsive. These neurons, then, were facilitated not just by attending to the stimulus's location, as were inferior temporal neurons in the studies of Desimone and colleagues; they also needed the eyes to move to the target. The conclusion was that cells in the superior colliculus do not participate in voluntary visual selective attention per se, but are part of an eye movement system and may therefore have a role in overt rather than covert aspects of attention. However, fortuitous studies in humans by Ayelet Sapir and Avishai Henik and their colleagues (1999) in Israel provide some additional information about the role of the colliculus in attention. In a rare patient with a unilateral bleed that damaged only one of the superior colliculi, they demonstrated that inhibition of return (described earlier) was reduced for inputs to the lesioned colliculus, indicating that the superior

colliculus may have an important role in the inhibitory component of reflexive attention.

Scientists have demonstrated that local deactivation of **superior colliculus** neurons lowers performance in discriminating targets when more than one is present in the visual field (one inside and another outside the neuron's receptive field). When there is a distracter stimulus, there is no decline in performance, but if the target location is inhibited, by injection of a gamma-aminobutyric acid (GABA) agonist (see Chapter 2), then the presence of a distracter leads to poor performance (Figure 7.35). This pattern is in line with the speculation that the superior colliculus may indeed participate in attentional processes that do not involve eye movements, though we do not yet know how.

Another major subcortical structure that has been implicated in attentional processes is the **pulvinar nucleus** of the thalamus (Figure 7.36). We know that the pulvinar is activated during attentional filtering tasks, and later we will consider human lesions that suggest an orienting role for this nucleus. Here we discuss what has been learned from animal studies about the role of the pulvinar in visual attention.

The pulvinar nucleus has visually responsive neurons that show color, motion, and orientation selectivity. In addition, it has subdivisions containing retinotopic maps of the visual world and interconnections with frontal, parietal, occipital, and temporal cortical areas. Recordings in awake monkeys have demonstrated longer response latencies in neurons in the dorsal medial region of the lateral pulvinar (Pdm), than in neurons in other regions of the pulvinar nucleus. In addition, responses in these neurons are enhanced when the stimulus either is the target of a saccadic eye movement or is attended without eye movements to the target.

Steve Petersen, David Lee Robinson, and their colleagues (1987, 1992) examined the effects of unilateral Pdm deactivations on spatial cuing tasks similar to the ones used in humans. They injected drugs that inhibited or excited Pdm neurons and examined how these drugs influenced the monkey's behavior. The drugs either mimic (agonists) or oppose (antagonists) the actions of neurotransmitters such as the inhibitory transmitter GABA. When muscimol, a GABA agonist that inhibits neuronal activity, was injected, the monkey could not easily orient attention to targets in the contralateral visual field. In contrast, when a GABA antagonist, which prevents the normal inhibition caused by GABA, was administered, the monkey readily directed its attention

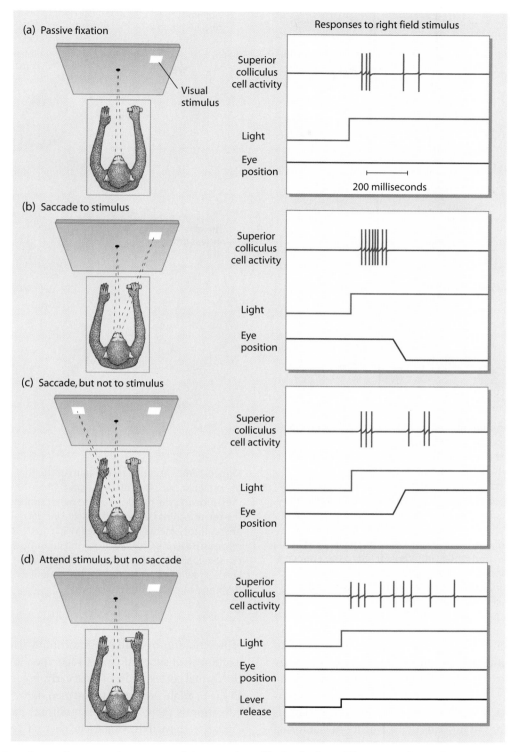

Figure 7.34 Experimental setup and responses of the superior colliculus for four different conditions from Wurtz and colleagues. **(a)** The monkey kept its eyes fixated on the central point and ignored the stimulus that was flashed in the right visual field. A few bursts of activity were found in the superior colliculus. **(b)** The monkey made a rapid saccadic eye movement to the location of a nonfoveal stimulus, and the firing rates of the cells with receptive fields coding that area of space greatly increased. **(c)** The monkey moved its eyes to a location outside the cell's receptive field and the cells in the superior colliculus did not respond. **(d)** The monkey attended to the location in the field but did not make a saccadic eye movement, and only a few bursts of activity were found. Adapted from Wurtz et al. (1982).

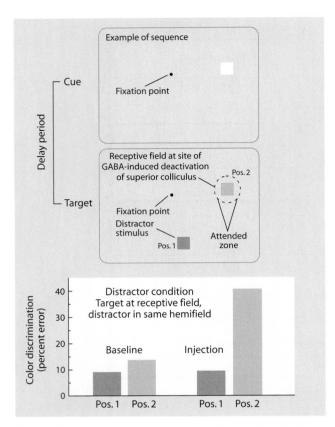

Figure 7.35 Experimental setup and effects of deactivation of a site in the superior colliculus on the discrimination of a target in the same hemifield. **Top:** A cue (white box) indicating to the monkey where to attend. **Middle:** The cued target (green), which is coded to the region of the superior colliculus where a GABA agonist is injected via a cannula surgically implanted in that region. The other stimulus (red) is a distracter stimulus located elsewhere in the same hemifield. **Bottom:** The percentage of errors in the task when no GABA agonist is injected (Baseline) and when GABA is injected to the attended location (Injection). The percentage of errors for the targets at the site of GABA injection increased in the presence of distracters. From Desimone et al. (1990).

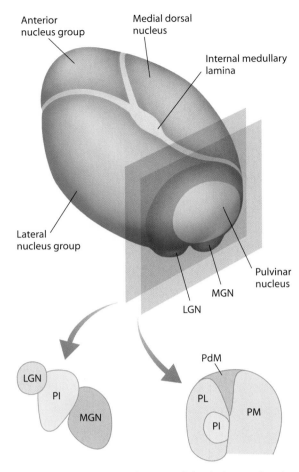

Figure 7.36 Anatomical diagram of the thalamus showing the pulvinar nucleus. **Top:** A diagram of the entire left thalamus showing the divisions of the major groups of nuclei, and the relationships between the visual lateral geniculate (LGN) and auditory medial geniculate (MGN) nuclei, and the pulvinar nucleus. **Bottom:** Cross sections through the pulvinar at anterior levels showing the LGN and MGN, and at more posterior levels showing the lateral (PL), dorsal-medial (PdM), medial (PM), and inferior (PI) subdivisions.

to contralesional targets. Hence, the pulvinar is central to covert spatial attention (Figure 7.37). The pulvinar may also filter distracting information. Injections of muscimol led to deficits in color or form discrimination when competing distracter stimuli were in the visual field. This is similar to findings in humans by University of California at Irvine cognitive psychologist David LaBerge (1990). He employed a task in which filtering of distracter stimuli was required for successful task performance, and found PET activations in the pulvinar in humans.

The parietal cortex is a key region for attentional control, as we know from human neurological lesion evidence (described later) and also its activation in neuroimaging studies (see Figures 7.27 to 7.29), and participates in the representation of spatial relationships. It maintains connection with subcortical areas like the pulvinar as well as the frontal cortex. Indeed, most early research on attention in the primate brain was performed in the parietal cortex. Attention is correlated with significant increases and decreases in the activity of parietal cortical neurons.

Some time ago, Vernon Mountcastle (1976) from Johns Hopkins University discovered that attentive fixation to visual stimuli led to higher firing rates in parietal neurons. The main question is how these neuronal enhancements occur. Studies by Robert Wurtz and his colleagues (1982) at the National Eye Institute showed that the firing rates in neurons of the parietal cortex increase in response to a target stimulus when an animal is using the stimulus as a target for a saccade or when covertly discriminating its features. But if a monkey is merely waiting attentively for the next trial in a sequence of trials, the parietal neurons do not usually have enhanced responses to visual stimuli in their receptive fields (Figure 7.38).

The parietal cortex's response pattern is different from that of the superior colliculus in that parietal neurons become enhanced when stimuli either are the target for saccades or are covertly attended (Figure 7.38c). These neurons are participants in visual selec-

tive attention—but how? What is the parietal cortex's role? To answer the question, we can turn to evidence that parietal lesions in monkeys cause deficits in tasks that require them to discern spatial relations between objects. Also, the parietal cortex is activated when covert attention shifts from location to location, and when subjects have to analyze the spatial relations between targets or shift attention, as found in PET and fMRI studies in humans. Finally, damage to the human parietal cortex leads to deficits in attention, which are

Figure 7.37 Effects on behavior when the pulvinar nucleus is injected with GABA agonists and antagonists. The trial types (predictive peripheral cue in left column and target in middle column) correspond to the data presented to the right. The measure is reaction time to target detection as a function of cue-to-target interval (msec) (see text for description of results). Adapted from Robinson and Petersen (1992).

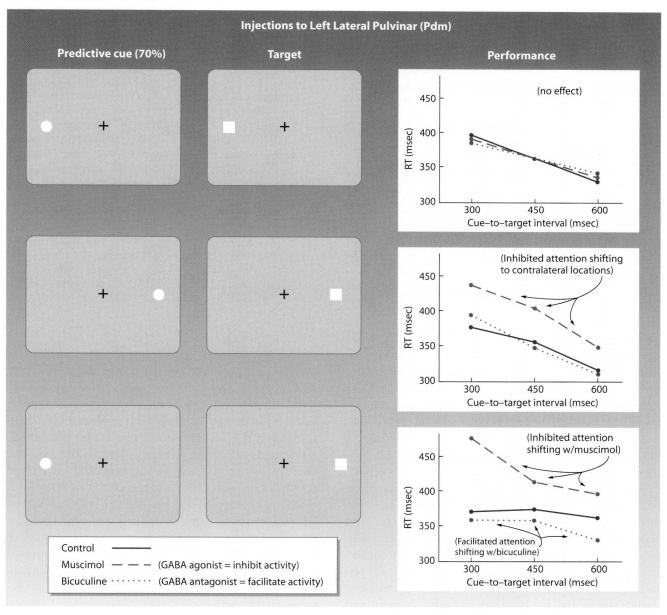

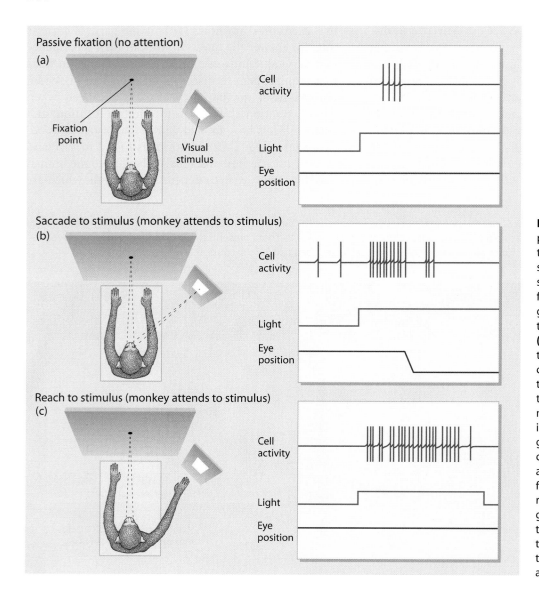

Passive fixation (no attention)
(a)

Fixation point

Visual stimulus

Cell activity

Light

Eye position

Saccade to stimulus (monkey attends to stimulus)
(b)

Cell activity

Light

Eye position

Reach to stimulus (monkey attends to stimulus)
(c)

Cell activity

Light

Eye position

Figure 7.38 Properties of parietal neurons in visual attention. Three conditions are shown. **(a)** The monkey passively fixates while a lateral field stimulus is presented, generating some action potentials from the neuron **(right)**. **(b)** The monkey has the task of making a saccadic eye movement to the target when it appears, and this increases the rate of neuron firing. **(c)** The neuron increases its firing rate to targets that are presented and covertly attended, when the animal must keep its eyes fixated straight ahead but is required to reach to the target. This demonstrates that the neuron is spatially selective, a sign of covert attention. Adapted from Wurtz et al. (1982).

described below. Thus, the parietal lobe appears to function both to represent spatial locations and to control voluntary orienting to locations. Presumably, however, different regions of the parietal lobe are involved in each of these processes, and evidence not presented here is accumulating to support this idea.

In summary, single-unit recordings and lesion studies in animals have shown that subcortical and cortical brain areas are involved in various aspects of attentive behavior and orienting. However, concerning selective attention, the evidence that gating in the subcortical relays to the visual cortex plays a role is weak. The first clear evidence for gating or filtering of inputs occurs in the visual cortex. In primary visual cortex, effects of selective attentional are rather small, but in extrastriate cortex attentional modulations of neural activity are often large. Neurons in other cortical areas such as the parietal cortex are also sensitive to attentional processes, and appear to play a role in coding the locations of behaviorally relevant stimuli and perhaps in coordinating shifts of attention. Together with evidence from neuroimaging, the take-home message is that the parietal cortex and subcortical structures like the pulvinar nucleus are key parts of a system for orienting attention to relevant locations. The consequence of this attentional orienting during perception is seen in the visual cortex, where analyses of stimulus features and forms are carried out. Convergent evidence for the role of the parietal and subcortical structures in orienting comes from observations of what happens to these abilities in humans with neurological damage. These observations are reviewed in the next section.

NEUROLOGY AND NEUROPSYCHOLOGY OF ATTENTION

One of our most informative sources of data on brain attention systems is also one of the oldest: the human brain damaged by stroke, tumor, trauma, or disease. Early neurologists appreciated that brain damage led to characteristic deficits, including problems in attention and orienting. We now investigate these deficits from the perspective of cognitive neuroscience theory and with the use of cognitive experimental approaches; further, we can now more precisely describe damage in an individual. Anatomical neuroimaging such as computed tomography (CT) and MRI enables us to look with high anatomical resolution at intracranial damage associated with an external behavioral symptom (Figure 7.39). With these advanced tools, a new era of cognitive neuroscience investigation of neurological patients has begun.

Extinction and Neglect

In neurological patients, attention deficits can result from damage to various brain regions. One type of damage, unilateral parietal damage, often leads to symptoms

Figure 7.39 Imaging damage to the brain. **(a)** Brodmann's representation of the human brain showing architectonic areas. **(b)** Three-dimensional reconstruction of the brain of a patient with a cortical lesion. The lesion was produced by an infarct of parietal and temporal cortex that was caused by occlusion of the posterior trunk of the middle cerebral artery. The patient has a conduction aphasia. **(c)** The lesion traced onto templates of horizontal sections through lower **(top left)** and upper **(lower right)** regions of the brain. Courtesy of Dr. Robert T. Knight.

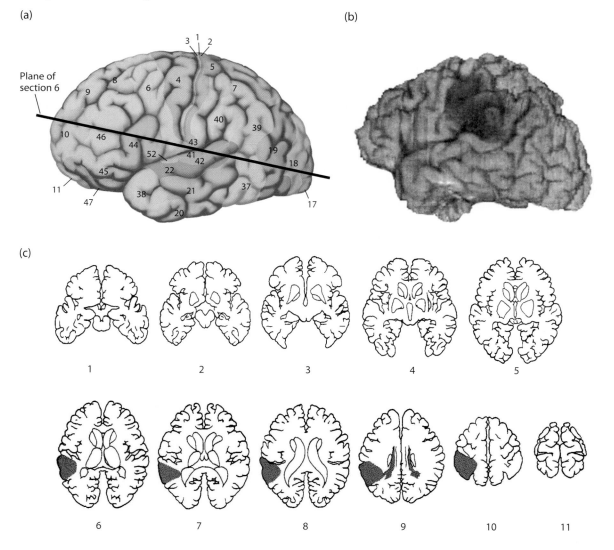

THE COGNITIVE NEUROSCIENTIST'S TOOLKIT

Two Brains Are Better Than One

The human visual system is remarkably adept at searching for target items in complex scenes. But when a target differs from other items by a conjunction of simple features like brightness and shape, search time grows linearly with the number of distracters. What is happening, then, is a serial process in which a unitary focus of attention searches the representation of the visual scene in the normal brain. But how would a unitary process such as this function after the cerebral hemispheres in split-brain patients are disconnected by surgically sectioning the corpus callosum?

Because split-brain patients have had their corpus callosum cut to treat epilepsy, there is little or no perceptual or cognitive interaction between the hemispheres. And so, specializations of the two halves of the brain can be investigated separately. From studying visual search in these patients, we know that they, unlike subjects with an intact corpus callosum, can search at *twice* the rate when items are split between hemifields rather than concentrated within a single field. It is clear that each disconnected hemisphere possesses its own attentional scanning mechanism. So the old adage that two brains are better than one has real meaning for this cognitive mechanism in split-brain patients.

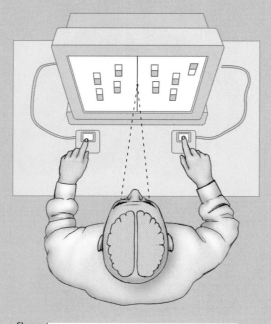

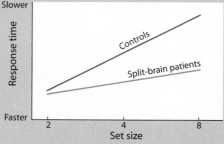

A split-brain patient seated in front of a computer monitor and performing a visual search task for conjunction targets **(top).** Graphs of the patient's reaction times in comparison to healthy control subjects for bilateral stimulus arrays of various set sizes **(bottom).** Control subjects take longer to perform the task (red line) than does the split-brain patient (green line) as set size increases. The split-brain patient has a search rate that is twice as fast, indicating that he is searching the array for the target separately with each hemisphere.

of the **neglect syndrome,** wherein patients fail to acknowledge that objects or events exist in the hemispace opposite their lesion. A prominent feature of the neglect syndrome is **extinction,** the failure to perceive or respond to a stimulus contralateral to the lesion (contralesional) when presented simultaneously with a stimulus ipsilateral to the lesion (ipsilesional). Neglect of contralesional space can be diagnosed by neuropsychological tests such as line cancellation. Patients suffering from neglect are given a sheet of paper containing many horizontal lines and are asked under free-viewing conditions to bisect the lines precisely in the middle by placing a vertical line. Patients with lesions of the right hemisphere tend to bisect the lines to the right of the midline. They may also miss lines altogether in the direction opposite the lesioned hemisphere (Figure 7.40).

The syndrome of neglect is unfortunate for those afflicted, and yet fascinating to contemplate. What would it mean to have a deficit that makes part of the world inaccessible to us, and we were completely unaware of the fact? A graphic example comes from paintings by the late German artist Anton Raederscheidt. The artist had a se-

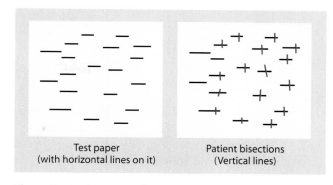

Test paper
(with horizontal lines on it)

Patient bisections
(Vertical lines)

Figure 7.40 Patients suffering from neglect are given a sheet of paper containing many horizontal lines and asked under free-viewing conditions to bisect the lines precisely in the middle with a vertical line. They tend to bisect the lines to the right (for a right-hemisphere lesion) of midline, owing to neglect for contralesional space.

vere stroke that left him with neglect; the pictures in Figure 7.41 are self-portraits done at different times after the stroke occurred and during his partial recovery. They show his failure to represent a portion of contralateral space, including, remarkably, portions of his own face!

Figure 7.41 The late German artist Anton Raederscheidt's self-portraits painted at different times following a severe stroke, which left him with neglect to contralesional space.

When Attention Is Lost

Patient N.R. is a 61-year-old man who had a stroke in the right hemisphere one week prior to a neurological examination. He experienced weakness, clumsiness, and some loss of sensation in his left hand and leg. His speech was fine, and he comprehended what was said to him. But when approached by his neurologist from the left side, N.R. was unaware of her presence. This is called *neglect* because the patient appears not to notice or pay attention to some stimuli. Yet when the neurologist spoke, the patient moved his upper torso and head toward her and greeted her, showing surprise to find her there. This did not happen when N.R. was approached from the right side; he detected and recognized the physician normally. When the neurologist showed N.R. visual stimuli in the left and right hemifields of his visual field, he detected and responded to them if presented one at a time. Therefore, N.R. was not blind in any portion of the visual field.

When stimuli were presented simultaneously in the left and right visual fields, however, the patient reported seeing only the one in the right visual field. This effect is called *extinction* because the presence of the stimulus in the right field leads to the stimulus on the left being extinguished from awareness. When asked how he felt, N.R. complained of being ill, but when describing his problems he did not mention any difficulty in noticing things on his left side. MRI studies of his brain verified that much of the inferior parietal lobe in the right hemisphere had been damaged. There is no treatment for the symptoms of the neglect syndrome, and the patient partially recovers over time. Six months after the incident, N.R. was vastly improved and rarely failed to notice when someone approached from the left; yet for weak stimuli, he still manifested some extinction and was slower to respond to stimuli on the left side of visual and body-centered space.

Neurologist testing a patient with a right parietal lesion for signs of extinction. When confronted with simultaneous stimuli in the left and right visual fields, the patient only responds to the one in his unaffected (right) visual field, extinguishing the stimulus in the contralesional (left) visual field.

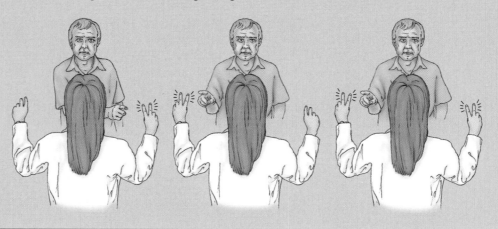

We might speculate that neglect patients have disorders of sensation and perception, but this is not really so. It can be clearly demonstrated that neglect happens even in the absence of damage to the visual system and can have nonvisual components such as motor and representational problems—like the ones in Raederscheidt's self-portraits. To relate the clinical syndrome of neglect to attention in normal volunteers, researchers have tried to determine

which components of normal attentional orienting and selection are lost or damaged after parietal damage.

Michael Posner, Robert Rafal, and their colleagues (1984) studied patients with unilateral lesions of the parietal and posterior temporal cortex (i.e., temporal-parietal junction) in a trial-by-trial cuing design like the one shown in Figure 7.8. Even though under normal conditions these patients may neglect contralesional stimuli, neurologists have noted that they can often be instructed to attend to the neglected field. Indeed, in the cuing paradigm, when they were cued to attend to locations in the contralesional hemifield, they responded relatively quickly to target stimuli there. But if they were cued to expect the target stimulus in the ipsilesional field—for example, the right visual field for a right temporal-parietal junction lesion—they were unusually slow to respond to the target when it unexpectedly appeared in the contralesional field. These reaction times to uncued contralesional targets were much slower even than those to uncued trials presented to the field ipsilesional to the lesion (Figure 7.42). This reaction time pattern has been described as an "extinction-like reaction time pattern," to indicate its similarity to the clinical finding of extinction in bedside neurological testing of these patients (see also When Attention Is Lost).

These results were interpreted according to a hypothetical three-stage model of attention and orienting having the following components: (1) disengagement of attention from its current focus, (2) movement or shifting of attention to a new location or object, and (3) engagement of attention on the new location or object to facilitate perceptual processing of the stimuli (Figure 7.43). If patients had exceptionally long reaction times when the cue was in the ipsilesional field and the target was in the

contralateral field, it signified a deficit in the disengage operation. One role for the temporal-parietal region, then, was to disengage attention from its spatial location. Disengagement probably causes the deficit, as subjects can shift attention to contralesional space and engage targets if cued there, but they show particular impairment when their attention is first attracted somewhere else in the visual field and then moves toward the contralesional hemispace.

In a striking recent event-related fMRI study, this model for the role of temporal-parietal junction was supported in normal volunteers. Maurizio Corbetta and his colleagues (2000) used a trial-by-trial cuing paradigm like that in Figure 7.8, where the cue predicted the most likely location of a subsequent visual target, but sometimes the cue was invalid and the target would occur elsewhere. By time locking the event-related analysis to the onset of the invalid targets, the precise experimental situation where neglect patients showed their disengage deficit, they found neural activation in the region of the right temporal-parietal junction. Thus, this region does in fact support something like attentional disengagement or reorienting to relevant targets appearing at unexpected locations.

Figure 7.43 The three-stage model of attention of Posner and colleagues (1984). A sensory event generates a signal that produces localization of the event spatially. Thereupon attention disengages from the current focus, moves to the location of the event, and engages the stimulus. Attentional engagement may be followed by inhibition if this is purely reflexive orienting to the sensory event—the phenomena of inhibition of return.

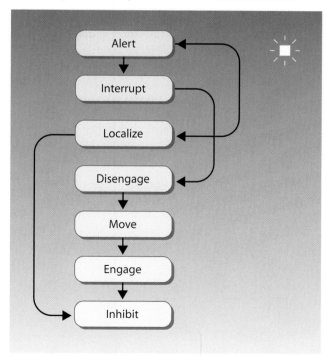

Figure 7.42 Diagrammatic representation of the extinction-like reaction time pattern in patients with unilateral lesions of the parietal cortex. Reaction times to precued (valid) targets contralateral to the lesion were almost "normal"; that is, while being slower than reaction times produced by healthy control subjects, they were not much slower than the patients' reaction times to targets that occurred in the ipsilesional hemifield when that field was cued. When the patients were cued to expect the target stimulus in the field ipsilateral to the lesion (e.g., right visual field for a right parietal lesion), they were unusually slow to respond to the target when it occurred in the opposite field (invalid trials). Adapted from Posner et al. (1984).

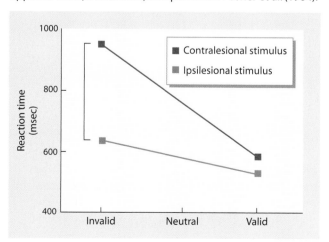

In patients having lesions in other brain regions, Rafal and Posner (1987) suggested that the move and engage operations are associated with subcortical structures. They targeted the thalamic area, perhaps the pulvinar nucleus, as being involved in the hypothetical engage operation. In patients with lesions to the thalamic regions, reaction times were slowed for validly cued and invalidly cued targets that appeared in the contralesional space, but only a minor disengage deficit was demonstrated; that is, the pattern of reaction times was different for patients with parietal and thalamic lesions. In contrast, patients with lesions of the midbrain, in the area of the superior colliculus, resulting from a neurological disease, progressive supranuclear palsy (PSP), had a different pattern of reaction times in the task. They were slowed in responding to cued targets in the direction of their orienting deficit, and thus displayed a deficit in the hypothetical move operation. These patterns of reaction time deficits in the different patients are interpretable within the predictions of the three-stage model of attention (see Figure 7.43), and therefore relate specific brain regions with specific components of the model: the parietal cortex (temporal-parietal junction) with attentional disengagement, the midbrain with the hypothetical move operation, and the thalamus with attentional engagement.

Although these data provide a compelling story and elaborate a nice model, an alternative model has been proposed to explain reaction time patterns in neglect patients. This alternative is a computational model developed by Jonathan Cohen and his colleagues (1994) when at the University of Pittsburgh. It is based on how systems in a spatial localization and attention network might interact when components are damaged. The computational model, based on a connectionist architecture, is diagrammed in Figure 7.44. It was able to predict the behavioral performance of normal subjects and patients with lesions to the temporal-parietal area without having to identify an explicit mechanism for disengaging attention. This model is a competitive interaction one where excitatory and inhibitory interactions between modules in the attentional orienting system are in equilibrium in the normal state; however, when part of the system is damaged, various behavioral manifestations, including the disengage pattern, can arise. Computer simulations are quite useful for investigating alternative theories. However, a limitation with such models is that alone they do not tell us how a process may

Figure 7.45 The effect of damage to parietal and temporal cortex on visual search for bilateral versus unilateral arrays. **(a)** Visual stimulus arrays in the left (LVF search) or right hemifield (RVF search) and arrays spanning both hemifields (Bilateral search). **(b)** The graph summarizes the changes in reaction times (RT) for targets as a function of the numbers of distracters present in the hemifield opposite to that in which the targets appear (see text for details). Adapted from Eglin et al. (1989).

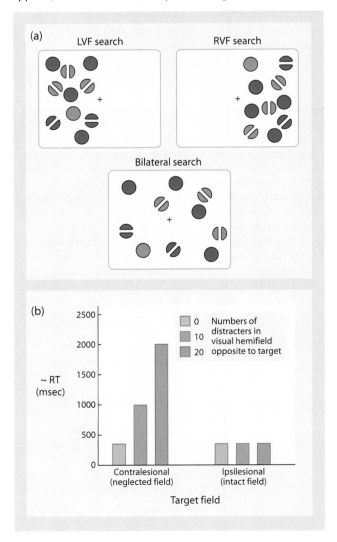

Figure 7.44 The attentional model of Cohen and colleagues. This model represents various processing stages as nodes in a computer connectionist network. Biases that were intended to simulate the damage in neglect were imposed on certain nodes by altering their weighting such that they contributed more or less to activity in the entire network. Adapted from Cohen et al. (1994).

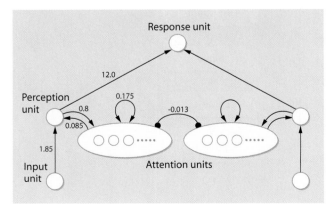

actually be manifest in the biology of the brain. Nonetheless, they permit theories of brain processing to be formally modeled and tested against data gathered in direct experiments, including those involving physiological methods.

The idea from either the disengage or the competitive interaction model is still that attention to the intact visual hemifield causes poorer performance for items presented to the damaged hemifield; thus, items in the intact field affect processing in the contralesional field. Support for this notion comes from studies of visual search in patients with unilateral parietal and temporal-parietal lesions causing neglect. When visual search arrays were presented to only the contralesional or ipsilesional hemifield, the patients' performance in target detection did not differ between the two hemifields. But when search arrays were presented to both hemifields simultaneously, their performance did differ between the two hemifields (Figure 7.45). Search for targets in the intact hemifield was not affected by the presence of stimuli in the contralesional hemifield. Yet the reverse was not true: When patients searched for targets in the contralesional field, the presence of stimuli in the intact field did affect their performance. They were slower at finding the targets on the contralesional side, and this worsened when more stimuli were on the intact side. Thus, in spatial cuing and visual search paradigms, patients with unilateral lesions of the temporal-parietal regions perform poorly at target detection on the contralesional side, and this performance

worsens when stimuli are presented to the ipsilesional side or when they are cued to expect a target to be in the ipsilesional field (as demonstrated earlier in Figure 7.42).

This pattern of interference in the good versus the neglected field raises a fascinating question: How are items processed in the neglected field, and to what extent? Is information in the neglected field not available because it is neglected behaviorally? Bruce Volpe and one of the authors (M.S.G.) and their colleagues (1979) demonstrated that neglect patients could make accurate same-different judgments between items in the intact versus the neglected field, even when they could not identify items on the neglected side of visual hemispace. These provocative data suggest that information on the neglected side can be processed to a high level of analysis, even though it may not be accessible to the patient's awareness.

Information processed in the neglected field can sometimes positively affect the processing of stimuli also in the neglected hemifield. In search tasks, the presence of distracter stimuli in the peripheral regions of the neglected, contralesional hemifield actually helps improve the detection of targets closer to the vertical midline but still in the neglected field (Figure 7.46). This effect was observed as a reduction in deficits induced by the presence of distracters ipsilateral to the lesion. The logic is that normally the patients are biased toward the right (for right-hemisphere lesions) and therefore neglect the left. But if distracter stimuli

Figure 7.46 The bias toward the field ipsilateral to the lesion in unilateral neglect patients can be reduced by the presentation of target or distracter items in the neglected hemifield. This reduction in bias is indicated by the improvement in the detection of the target near the midline (i.e., in the direction of the neglected field) in the good hemifield. The logic for this effect is described in the text. Adapted from Grabowecky et al. (1993).

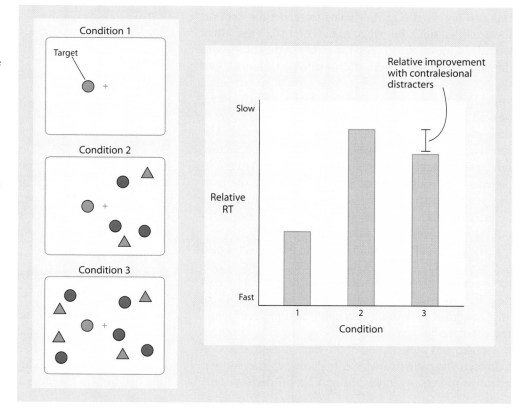

are presented bilaterally, the stimuli contralateral to the lesion partially attract attention to that field or, rather, reduce the asymmetrical contralateral-ipsilateral attentional bias caused by the unilateral temporal-parietal damage. Targets near the visual field's midline are now more likely to be quickly detected even when located in the neglected field (or more generally, the neglected direction). One interpretation for this finding is that even though the targets are in the neglected field, the center of mass of the stimulus array has shifted to the direction contralateral to the lesion and this helped patients direct attention leftward.

To what level is information in the neglected field processed? Patients may display extinction for simultaneous bilateral stimulus presentations when the stimuli are the same. For example, in testing visual extinction, when a physician holds up two items, one in each visual field, and they are different objects, the extinction is less than when the two items are the same, like two forks versus a fork and a key (Figure 7.47) (see also When Attention Is Lost). This tells us that neglect patients process information in the neglected field at least to the extent that they can detect it is different from information in the intact field, just as Volpe and his colleagues showed. For instance, in one study, colored letters were presented to both fields. The letters were either different but of the same color, the same but of a different color, or different with different colors. Patients were required to say where the stimulus occurred, and what letter or color it was. If the color was the same when it was the relevant dimension to be reported, then extinction was observed (i.e., they failed to report the item on the neglected side). But when in the color trials the letters were the same on the left and the right but the colors were different, there was no effect on extinction. Hence, both the neglected and intact fields show normal extraction of visual feature information even if it does not reach awareness all the time in the neglected regions of space. Yet, a difference signal can still be calculated between contralesional and ipsilesional visual fields using the visual feature information that is unconsciously represented. As a result, patients both exhibit greater extinction for similar items in the two fields and can make same-different judgments about items in the intact versus neglected field even without knowing what exactly was in the neglected field.

The general idea that information in the neglected field is processed to a high level has been extended to include the meanings of items. Words presented in the neglected field are processed to a semantic level even when patients are unaware of the stimuli because of their neglect. Hence, for instance, the word *doctor* in the neglected field primes the subsequent presentation of the word *nurse* in the good field such that patients are faster to decide whether nurse is a real word versus a nonword in a lexical decision task.

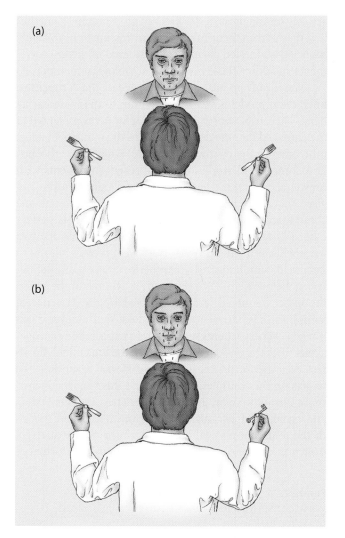

Figure 7.47 Extinction test in a patient with neglect. **(a)** A neurologist holding up two identical items in two hemifields (and getting extinction of the item on the neglected side). **(b)** A neurologist holding up different items in the two fields (resulting in less extinction).

These types of findings represent one form of evidence for unconscious processing resulting from damage to the cerebral cortex, *unconscious* because the patients are unaware of it (are not conscious of it), and *processing* because these studies show that the information can influence behavior.

COORDINATE SYSTEMS OF ATTENTION AND NEGLECT

Does neglect affect only external scenes and action to external objects? This major question about neglect was illustrated in a clever study by Eduordo Bisiach and Claudio Luzzatti (1978) in Italy. They studied patients with neglect who had lived in the same town for most of their lives and

Figure 7.48 Diagram of a neglect patient recollecting visual memory from two ends of a piazza. The neglected side in visual memory (shaded gray) was contralateral to the side with cortical damage. From Bisiach and Luzzatti (1978).

were familiar with the city's major landmarks, especially the central piazza where there is a church, restaurants, shops, and buildings known to all the townspeople. They asked patients with neglect following stroke, but who had recovered from the acute symptoms, to imagine themselves in the piazza standing in a certain location, like on the church steps and to describe the piazza and the buildings surrounding it. Amazingly, in their description from memory, they neglected things on the side of the piazza contralateral to their lesion as defined from their perspective looking out from the church steps, just as if they had actually been standing there surveying the scene. But the clincher is, when the researchers asked the patients to now imagine themselves standing at the other end of the piazza, facing the other way *toward* the church, they reported items they had previously neglected and now neglected the side of the piazza they previously reported! Thus, in this striking demonstration, neglect was not merely found for items in the external sensory world but also for items in visual memory during remembrance of a known scene (Figure 7.48). The key point is that the piazza was well known to the patients prior to the damage; thus, their neglect of buildings and objects on the side contralateral to the lesion could not be attributed to not having memories of previously neglected scenes, but rather reflected that attention to parts of the recalled images was biased by parietal damage.

Patients with lesions of the parietal and posterior temporal cortex leading to extinction and neglect sometimes have disturbed attention in object-based coordinates and not merely eye-, head-, or body-centered coordinates. For example, in tests of line bisections, patients neglect the left half of the page (scene-based neglect) and each object (each line) on the page (object-based neglect), and they

often do so simultaneously. Such effects are still spatially defined since the side of each object (the line or the page) contralateral to the lesion is neglected. But these results show that object-based coordinates for neglect can exist even when the objects themselves are not in the neglected field. This interesting observation was put to the test by Marlene Behrmann and Steve Tipper (1994) in a study using moving dumbbells (Figure 7.49). When dumbbell-shaped

Figure 7.49 Behrmann and Tipper's rotating dumbbell experiment showing that neglect can be object based (see text for details). After Behrmann and Tipper (1994).

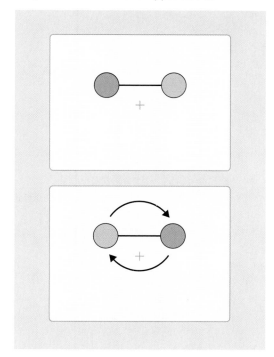

stimuli had targets flashed within the confines of the two circular ends, parietal lesion patients manifested a deficit in responding to the side contralateral to the lesion, displaying spatial neglect. Yet, when the dumbbell was rotated around its center in the visual field, the targets subsequently presented on the end of the dumbbell that was originally on the neglected side continued to be neglected, even when this end was rotated into the ipsilesional field. These data demonstrate that neglect can be distributed within objects; neglect can then follow the object's movements and be represented in its new coordinate frame of reference.

Balint's Syndrome That attention can be object based is supported by a remarkable condition known as *Balint's syndrome*, which was introduced at the beginning of this chapter. Patients with this malady have less ability to attend to multiple objects. The typical lesion in a Balint's

syndrome patient is bilateral, affecting posterior parietal and lateral occipital areas (Figure 7.50), unlike the hemispatial neglect lesion, which is typically unilateral.

Patients with Balint's syndrome have the peculiar deficit of only being able to perceive objects one at a time; recall the stroke patient at the opening of the chapter and his unusual "visual" problems. The patients correctly identify objects but have difficulty relating objects to one another. For example, when shown two or more objects such as a comb, brush, and pen, they may report seeing only one of the objects. Even when the objects are close together or overlapping, they may still report only one of them. Which object is perceived may vary over time, and the patients cannot control their attention to the objects. For example, patients shown a person wearing glasses may report the face or glasses, but perhaps not a person wearing glasses. The patient's attention is drawn to one object to the exclusion of others.

These clinical observations of Balint's syndrome patients were supported in an engaging study by Glyn Humphreys and Jane Riddoch (1992). They asked Balint's syndrome patients to report what colors were presented in an array of colored circles. The circles were all green, or all red, or half were green and half were red (Figure 7.51).

Figure 7.50 Diagrams showing unilateral parietal lesions **(a)** typical of neglect, and bilateral posterior pariental/lateral occipital lesions **(b)** typical of Balint's syndrome.

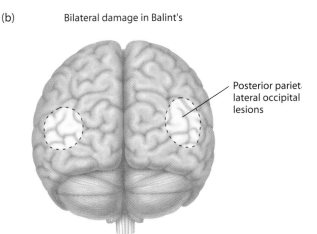

(a)   Unilateral parietal damage

Parietal lesion

(b)   Bilateral damage in Balint's

Posterior pariet lateral occipital lesions

Figure 7.51 Stimuli used to test object attention in Balint's patients. Colored circles were presented in two conditions: with and without connecting lines. When the circles were not connected, these patients would only report the color of one object at a time, but when the circles were connected to form one object, the patients could report both colors of the new object. After Humphreys and Riddoch (1992).

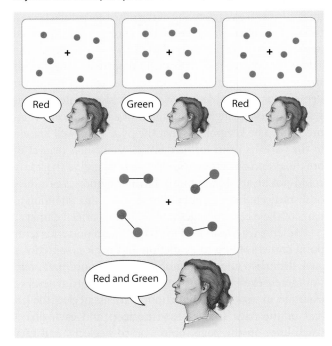

The patients were not very good at saying they saw red and green in the two-color arrays, presumably because they would perceive only one object at a time and each object was either red or green. But when the red and green circles were connected by lines, the patients improved markedly, saying that two colors existed. This improvement in performance happened because connecting the circles of colors formed a single object (a dumbbell) that the patients could attend to. They could perceive the object in its entirety and therefore could perceive the two colors.

SUMMARY

Animal and human studies that address theoretical issues about the brain's attention systems have been contemplated for more than 100 years. If Helmholtz were alive today, he would marvel at how much behavioral and physiological data we can provide to answer his questions about selective attention. Although we have not addressed the whole of attention—it is too great a task—we have looked at selective aspects of perception that are affected by attentional mechanisms, and we have examined the top-down executive systems and bottom-up stimulus-driven mechanisms that engender orienting and selection within the sensory pathways.

The picture we have formed is of distributed but highly specific brain systems participating in attentional control. The roles and limits of these systems in attention are becoming more clearly defined as we combine attentional theory, experimental and cognitive psychological findings, and neurophysiological approaches. Systems for controlling attention include the parietal lobe, temporal cortex, frontal cortex, and subcortical structures—these comprise the source of attentional selection (i.e., the control processes). The result, in visual processing, for example, is that in the cortex we observe modulations in the activity of neurons as they analyze and encode perceptual information as a function of their relevance; these areas of modulation are the sites of attentional selection. We no longer wonder whether early or late selection is the mechanism for selective attention because we now know that both are part of the act. The fascinating fact is that physical stimuli that impinge on the retina or cochlea may not be expressed in our conscious awareness, either when they occur or later via our recollections. Attentional phenomena are diverse and entail many brain computations and mechanisms, and when these are damaged by insult or disease, the result can be devastating for the individual. Cognitive neuroscience is vigorously carving away at the physiological and computational underpinnings of these phenomena, not only to provide a complete account of the functioning of the healthy brain but also to shed light on how to ameliorate attentional deficits in all their forms.

KEY TERMS

Balint's syndrome	exogenous cuing	neglect syndrome	reticular nucleus
bottleneck	extinction	object-based attention	selective attention
covert attention	inhibition of return	preattentive mechanism	spatial attention
dichotic listening	late selection	pulvinar nucleus	superior colliculus
early selection	limited capacity	reflexive attention	voluntary attention

THOUGHT QUESTIONS

1. Do we perceive everything that strikes our retinas? If not, does something interfere with vision? What might be the fate of stimuli that we do not perceive but that nonetheless stimulate our sensory receptors?

2. Are the same brain mechanisms involved when we focus our intention voluntarily as opposed to when our attention is captured by a sensory event, such as a flash of light?

3. Does attention act on inputs from locations or object representations or both? If both, how are the two levels of representation related during selective attention?

4. To what extent does evidence from neurological patients with brain damage enlighten us about the organization of spatial and object attention?

5. Do all neurons sensitive to attention in the monkey brain behave the same in any given attentional task? If not, why not?

SUGGESTED READINGS

HOPFINGER, J.B., BUONOCORE, M.H., and MANGUN, G.R. (2000). The neural mechanisms of top-down attentional control. *Nature Neurosci.* 3:284–291.

KANWISHER, N., and WOJCIULIK, E. (2000). Visual attention: Insights from brain imaging. *Nature Rev. Neurosci.* 1: 91–100.

KASTNER, S., and UNGERLEIDER, L. (2000). Mechanisms of visual attention in the human cortex.

Annu. Rev. Neurosci. 23:315–341.

LUCK, S.J., CHELAZZI, L., HILLYARD, S.A., and DESIMONE, R. (1997). Neural mechanisms of spatial selective attention in areas V1, V2 and V4 of macaque visual cortex. *J. Neurophysiol.* 77: 24–42.

SAPIR, A., SOROKER, N., BERGER, A., and HENIK, A. (1999). Inhibition of return in spatial attention: Direct evidence for collicular generation. *Nature Neurosci.* 2:1053–1054.

8

Learning and Memory

H.M. was a young man who suffered from a difficult-to-treat form of epilepsy that progressed in severity during his teen years. He was, however, successful in completing high school, after which H.M. worked in a factory winding wire in electrical motors. Over the years his physicians had treated him with the available drugs to minimize his seizures, but H.M. was not lucky in that these drugs, which worked so well for some, were largely ineffective for him. As his seizures worsened in his twenties, they prevented him from working, and he decided to try a then-radical new therapy that involved surgery; H.M.'s story began in the 1950s. At that time neurologists knew that many seizures originated in the medial portions of the temporal lobe and from there spread to other areas of the brain, leading to violent seizures and often loss of consciousness. It was also becoming increasingly clear that surgically removing the brain region in which the seizure activity originated, the so-called seizure focus, could help patients with epilepsy. The decision in H.M.'s case was to remove his medial temporal lobe bilaterally, in a procedure called *temporal lobectomy.*

Following recovery from this major neurosurgical procedure, H.M.'s epilepsy did improve. In fact, it improved enough to permit a reduction in the dosage of his anticonvulsant medications. Thus, the surgery was a success, both with regard to his surviving the risks associated with any surgery of the brain and with regard to the epilepsy. However, physicians, family, and friends began to realize that H.M. was experiencing new difficulties. For example, a year and a half after the surgery, which was performed in September 1953, H.M. displayed clear problems with his memory. Although it was April 1955 and H.M. was 29 years old, he reported the date as March 1955 and his age to be 27. H.M. would say he did not remember ever meeting certain individuals, even when he actually spoke to them a few minutes earlier and they merely left the room, returning after a short delay! H.M. was profoundly amnesic—that is, he suffered from disorders of memory. However, H.M. did not have the kind of amnesia one sees depicted in television shows or movies,

where the character has a total loss of all prior memories. Indeed, H.M. knew who he was and could remember things about his life—that is, up until a period prior to his surgery. However, it became increasingly clear that H.M. could not form new long-term memories.

Formal neuropsychological tests were performed on H.M. to establish the nature of his cognitive deficits. These tests showed that H.M.'s intelligence was well above normal after the surgery; in fact, he actually scored higher after the surgery than before. H.M. also had no perceptual or language problems and seemed generally fine, with no changes in his personality or motivation. However, when memory tests were administered, H.M. scored well below normal. The bilateral removal of H.M.'s medial temporal lobe produced a highly selective deficit in his memory ability, leaving other cognitive functions intact.

Obviously, the surgeons who performed the surgery on H.M. did not know that bilateral removal of the

301

medial temporal lobes would lead to dense amnesia, or they would not have performed the surgery. Indeed, ever since H.M.'s case was revealed, surgeons have taken great care to avoid removing both medial temporal lobes, and even one medial temporal lobe if the other is compromised in any way. So surgery is still used with success today for certain patients suffering from epilepsy, but nobody would consider doing the surgery the way it was done on H.M.

What does H.M.'s case reveal about the brain's organization for long-term memory? The answer to this question, and the controversies regarding the role of medial temporal lobe structures in memory, are addressed later in this chapter.

The ability to learn and remember information about the world around us and our experiences in it is a key cognitive ability that we all share. Remarkably, we retain millions of pieces of information, sometimes with ease and sometimes with tremendous effort. Most people have no difficulty remembering a birthday party during their youth. What may stand out vividly are the taste and color of the cake, the friends who attended, and some gifts. If we pick a salient event such as a birthday party, we will be amazed by how much we can recall about that special day. But what about the birthday party from the year before or the year after? Or what we did the next day?

Memory has the peculiar quality of being incomplete, and yet the totality of life's accumulated experiences is immense. Despite the vast stores of information contained in our brains, we continuously acquire new information and form new memories. How is it possible that we can recall the first day of school, the hair color of our best friend in first grade, or the smell of our grandmother's apple strudel as we entered her home for one of those memorable weekend visits?

As the case of H.M. reveals, not all memories are the same. That is, not all forms of information have the same essential quality. A dominant trend in the cognitive neuroscience of memory has been to appreciate that multiple memory systems exist and to search in earnest for their properties and substrates in the brain. Today scientists ask questions such as, Are all memories the same? Are learning and remembering how to ride a bicycle the same as learning and remembering the relationship between the sides of a right triangle?

We search for neural correlates of learning and memory in many ways: by developing animal models of memory in simple (invertebrates) and complex systems (nonhuman primates); through case studies like the one of H.M., which reveal what is and is not lost in amnesia; and with brain imaging to investigate normal encoding, retrieval, and recall in healthy humans. In this chapter we will explore learning and memory.

THEORIES OF MEMORY

What is the relationship between learning and memory? **Learning** is the process of acquiring new information, whereas **memory** refers to the persistence of learning in a state that can be revealed at a later time (Squire, 1987). Learning, then, has an outcome, and we refer to that as *memory*. To put it another way, learning happens when a memory is created or is strengthened by repetition. This need not involve the conscious attempt to learn. Learning can occur and performance can improve simply from more exposure to information or to a task. For example, we remember the details of a person's face better by seeing it more, without having to try to consciously memorize facial features.

Learning and memory can be subdivided into major hypothetical stages: encoding, storage, and retrieval. **Encoding** refers to the processing of incoming information to be stored. The encoding stage has two separate steps: acquisition and consolidation. **Acquisition** registers inputs in sensory buffers and sensory analysis stages, while **consolidation** creates a stronger representation over time. **Storage,** the result of acquisition and consolida-

tion, creates and maintains a permanent record. Finally, **retrieval** utilizes stored information to create a conscious representation or to execute a learned behavior like a motor act.

Sensory and Short-Term Memory Mechanisms

Memory, as just defined, implies a temporal component. We remember things over time, short periods and long periods. This inherent characteristic of memory led Endel Tulving (1995), one of the leaders in research into memory, to describe some forms of memory as "mental time travel." By this, Tulving meant that the act of remembering something that happened in the past is to re-experience that past in the present. Hence, it is meaningful to characterize memory by how long we retain the information of interest.

Models of memory include the distinctions between sensory memory, short-term memory, and long-term memory, based on how long information is retained. **Sensory memory** has a lifetime measurable in milliseconds

HOW THE BRAIN WORKS

The Memorist

Do you think you have a pretty good memory for numbers? Can you remember your license plate number? Social Security number? Telephone numbers of friends? How about the digits of pi? What was that again, 3.1415...? How good can people be at remembering numbers?

Rajan Mahadevan recalled being very good as a child in memorizing license plates—lots of them. To test the extent of his memory for numbers, he spent a few months of his adult life memorizing the digits of pi and won his place in the *Guinness Book of World Records* in 1981 for reciting the first 31,811 digits. Not bad considering there is nothing too meaningful in strings and strings of numbers. Rajan is considered a "memorist," one with an extraordinary skill for learning and reciting information—most commonly, a genius for memorizing numbers, as contrasted with "prodigious savants" who are mentally and often physically impaired but have one phenomenal talent. Unlike the character played by Dustin Hoffman in the movie *Rain Man,* Rajan is a perfectly normal individual of above-average intelligence.

How is it that Rajan memorized over 30,000 digits? According to Rajan, he used a method of pairing matrix locations with individual digits, which consisted of a matrix of ten columns by *N* rows, each place holding a ten-digit number sequence (similar to the table below).

Amazingly, Rajan now knows the first 5000 digits of pi so well that when given a location number (counting from left to right and top to bottom in the matrix), he can tell the digit in that location, or he can fill in the remaining five digits within a few seconds of being read the first five digits of any ten-digit sequence.

Throughout the years, many memorists have been written about. The most famous was Shereshevskii, who was studied for 30 years by Alexander Luria, the great Russian neuropsychologist (1968). Instead of the location-digit matching used by Rajan, Shereshevskii claimed that he used rich visual images akin to photographic memories to reproduce number matrices and mathematical formulas. Memorists have varied in their specific abilities of memorization; Rajan's specialty is restricted to digit strings. If you are feeling like your memory is not quite up to par, you might be encouraged to know that he can perform no better than control subjects on tests of spatial memory and word lists.

Three general principles for skilled memory have been proposed: (1) meaningful encoding, the use of preexisting knowledge as a tool in storing new information in memory; (2) retrieval structure, the attachment of cues to new material for later retrieval of information; and (3) speed-up, the effect of practice on the ability to learn material. According to some experts on this phenomenon, anyone can demonstrate the skills of memorists such as Rajan if they follow these general principles. Yet, Rajan does not follow the first principle, as he does not relate the numbers to anything significant and already known, as some memorists have done. The memorist Arnould substituted consonants for numbers and made sentences to aid in memorization. Based on extensive comparisons between Rajan and control subjects, Rajan's ability is now attributed to a combination of practice and some underlying biological component. The idea of a genetic explanation for his unusual skill is interesting in light of the fact that he is named after a distant cousin, one of the most famous mathematicians in India, Srinivasa Ramanujan, who claimed that "every number was a personal friend."

How has Rajan's place as a world record holder withstood the test of time? In 1987, Hideaki Tomoyori claimed Rajan's place in the *Guinness Book of World Records* by reciting the first 40,000 digits of pi from memory.

pi = 3. +

1415926535	8979323846	2643383279	5028841971	6939937510
5820974944	5923078164	0628620899	8628034825	3421170679
8214808651	3282306647	0938446095	5058223172	5359408128
4811174502	8410270193	8521105559	6446229489	5493038196
4428810975	6659334461	2847564823	3786783165	2712019091
4564856692	3460348610	4543266482	1339360726	249141273...

to seconds, as when we recover what someone just said to us although we were not paying close attention. **Short-term memory** is associated with retention over seconds to minutes. This may include remembering a phone number provided by a telephone operator, as we frantically try to dial it. **Long-term memory** is measured in days or years—an event from childhood or last week.

SENSORY MEMORY

While you are watching the final game of the World Cup, with the score tied and seconds to go, your father enters the room and begins a soliloquy that you are not really paying attention to. Suddenly, you detect an increase in the volume of his voice and hear the words, "You haven't heard a word I said!" Wisely, your response is not to admit it, but instead, in the nick of time to avoid repercussions, you metaphorically reach back and retrieve the most recent sentence with enough accuracy to say, "Sure I did, you said that the neighbor's cat is in our yard again and you want me to call her to complain." Almost everyone you ask about this phenomenon knows what you mean. The auditory verbal information just presented to you seems to persist as a sort of "echo" in

---

HOW THE BRAIN WORKS

Flashbulb Memories

What were you doing when you heard the news of the Challenger shuttle explosion? Questions like this typically are asked when citing examples of "flashbulb" memories, the vivid memories of the circumstances surrounding shocking or emotionally charged news. Chances are that you can recall the details of your whereabouts, the source of the news, and who was with you at the time. Since the term was first coined by Roger Brown and James Kulik in 1977, flashbulb memories have been the topic of debate for many psychologists. The concept is much older, however. In the last century, the renowned psychologist William James wrote that "an impression may be so exciting emotionally as almost to leave a scar upon the cerebral tissues." This view has some similarity to Brown and Kulik's "Now Print!" mechanism as a possible basis of flashbulb memories. Under this proposed explanation, the mechanism for such vivid recollection is neurophysiological in nature. Researchers into this fascinating instance of long-term memory have asked, How do these memories compare to other recollections? Is there a special mechanism for flashbulb memories? Are flashbulb memories especially accurate or long-lasting?

The Challenger shuttle explosion in 1986 presented a unique opportunity to study flashbulb memories over a range of persons' ages, and at various periods of time after the event. This has provided interesting insights. Ulric Neisser and Nicole Harsch (1992) queried students at Emory University about hearing the news of the Challenger explosion at various times after the incident. They began as early as 24 hours after the occurrence, and then followed up 2½ years later.

Two and a half years after the incident, one student wrote, "When I first heard about the explosion I was sitting in my freshman dorm room with my roommate and we were watching TV. It came on a news flash and we were both totally shocked. I was really upset and I went upstairs to talk to a friend of mine and then I called my parents."

Along with this recollection, the student gave the highest score possible in confidence ratings on the accuracy of the memory. How close was this recollection to his previous report? Two and a half years earlier, 24 hours after the explosion, he had written, "I was in religion class and people walked in and started talking about it. I didn't know any details except that it had exploded and the schoolteacher's students had all been watching, which I thought was so sad. Then after class I went to my room and watched a TV program about it and I got all the details."

Like this student, over 40% of the participants were quite inconsistent in their descriptions of the incident; nonetheless, they gave high confidence ratings to their later recollections. How sure are you now of your answer to the first question in this box?

Alternative explanations for flashbulb-like memories have been offered by Neisser (1982) and others who postulated that ordinary aspects of memory can explain the phenomena of flashbulbs and apparent "phantom flashbulbs." So, are there flashbulb memories? The answer is that emotionally charged events have vivid "tags" in memory; therefore, confidence in the recollection, even years later, is high. However, flashbulb memories are no more accurate than other memories of everyday experiences.

your head, even when you are not really paying attention to it. But if you attempt to retrieve it quickly enough, you find it is still there, and you can repeat it out loud to assuage your interrogator. We refer to this as *sensory memory* or the *sensory memory trace* (sometimes as the *sensory registers*). Specifically for audition, we call it *echoic memory*. In vision, we refer to *iconic memory* (sometimes the *iconic store*). What is the nature of this effect? We can begin to answer this question by considering the time course of the sensory traces.

The key feature of the sensory memory traces is that they decay relatively swiftly. In general, most models hold that these sensory traces are not directly accessible to conscious awareness, although information can be sort of read out of sensory memory, analyzed, and brought into awareness if done promptly. Sensory memory also appears to have a relatively large capacity, in comparison to short-term memory, which we will discuss in the next section. In addition, sensory memories contain a sensory-based representation of information as opposed to a semantic (i.e., meaning-based) representation.

Let's consider the evidence that supports these characterizations of sensory memory. We do so first for the visual sensory trace, iconic memory, by asking how much information can be briefly held in the iconic store, and how long can this information be retained? The answer to the first question, how much can be stored, has been known for a long time: We can "see" much more than we can remember. As psychologist

George Sperling (1960) wrote a few decades ago, "When stimuli consisting of a number of items are shown briefly to an observer, only a limited number of the items can be correctly reported. The fact that observers commonly assert that they can see more than they can report suggests that memory sets a limit on a process that is otherwise rich in information."

How can this be demonstrated? Using the method of "partial report," investigators can assess how much is picked up by sensory systems immediately after a stimulus is briefly presented (Figure 8.1). This information cannot be readily indexed by merely asking the observer to state what she saw, because of the short duration of the iconic trace and the limits in short-term memory (described later). In the partial report method, subjects are presented briefly (for 50 msec) with stimuli like letters arranged in rows. If, as in Figure 8.1, subjects are presented with twelve letters and are asked to report as many items as possible, they typically cannot report all the letters that were presented. In this sort of design, on average subjects report between five and nine letters correctly, but never all twelve. However, one can demonstrate that all the letters were available in sensory memory very briefly by asking the subjects to report only a subset of the items, and this subset can be the letters in a particular row. Which row the subject is to report is indicated immediately after the stimulus is flashed. One method to indicate which row to report is to use an auditory tone. A high-frequency tone would indicate that the top row of letters

Figure 8.1 Partial report experiment of sensory memory. Subjects are presented with brief (50-msec) glimpses of stimuli like letter arrays that consist of three rows of four letters each. If asked to report as many items as possible, they would show the typical limitation in short-term memory and report about seven (plus or minus two) items. In the partial report method, they see the entire array briefly and then report on only a subset of the array. Which subset to report is indicated by an auditory cue presented with or just after the visual array (e.g., high-frequency tone indicates top row). The cue can tell the subjects to report the first, second, or third row of letters; the key here is that the subjects do not know prior to the stimulus which row of letters they will be asked to report. Nonetheless, the subjects are accurate at reporting the four letters from any row cued. This ability makes sense because the four letters in the row to be reported are well within the limits of the span of immediate memory. Given that the subjects do not know in advance which letters to retain in memory, this pattern of results indicates that for some brief period after the presentation of the visual stimulus, all twelve letters are retained, and so the letters in any row can be reported.

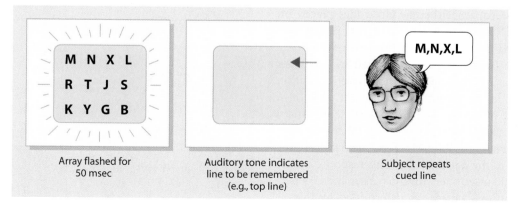

Array flashed for 50 msec

Auditory tone indicates line to be remembered (e.g., top line)

Subject repeats cued line

should be reported; a medium-frequency tone, that the middle row should be reported; and a low tone, that the bottom row should be reported. When this is done, regardless of which row is cued by the tone, the subjects accurately report all the items in that row. Since the subjects cannot predict which row will be cued and have to be reported, we can conclude that briefly, all the information was available to the subjects.

These sorts of experiments tell us that observers actually perceive much more information than they can report in a traditional verbal report test. However, this information is only present for a few hundred milliseconds and is then rapidly "forgotten." The decay time is about 1 second, but most of the decay occurs by 300 to 500 msec in the visual system, indicating that the visual sensory trace might be gone as soon as 500 msec. This time frame was determined by studies that varied the interval between the brief presentation of the visual stimulus, and the cue that indicates which information is to be reported, and by studies that used brief sensory masking stimuli that are believed to erase the contents of sensory memory.

How does the information about the short time course of the iconic trace fit with the phenomenon described at the beginning of this section, where auditory information may be accessible for longer than a few hundred milliseconds? Is there a difference between the temporal characteristics of the visual and auditory sensory memory systems? The answer is yes, but cognitive scientists disagree on exactly how long the echoic trace may last. Some estimates have suggested that the echoic trace may last as long as 20 seconds!

Cognitive neuroscience approaches have been employed to investigate the time course of auditory sensory memory and to establish whether it occurs in sensory-specific cortex in humans. The persistence of the auditory sensory memory trace has been measured by recording a human event-related brain potential known as the *electrical mismatch negativity* (MMN), or its magnetic counterpart, the *mismatch field* (MMF). The brain responses are recorded either by electrodes placed on the head (MMN) or with magnetic sensors placed near the scalp (MMF) (see Chapter 4 for a review of these methods). The brain response is elicited by a deviant stimulus such as a high tone frequency presented within a sequence of identical stimuli of a different pitch (e.g., low tones). The MMN is a negative polarity component in the event-related potential elicited by the deviant tone, and the corresponding MMF is a deviation in the magnetic field to the deviant tone. This response occurs about 150 to 200 msec after stimulus onset and is generated in the auditory cortex, as determined by inverse

modeling of the magnetic fields (see Chapter 4). These mismatch responses have been interpreted as representing sensory memory processes that hold recent auditory experience in echoic memory for comparison to new inputs. When these inputs differ, the MMN and MMF are generated. Hence, these brain responses could index how long the echoic memory trace persists.

Mikko Sams, Ritta Hari, and their colleagues (1993) at the Helsinki University of Technology in Finland did precisely this. They varied the interstimulus intervals between standard and deviant tones and found that the MMF could still be elicited by the deviant tone at interstimulus intervals of 9 to 10 seconds (Figure 8.2). After about 10 seconds the amplitude of the MMF declined to the point where it could no longer be distinguished reliably from noise. Some behavioral studies have yielded results that also support about 10 seconds as the duration of the echoic trace. Because inverse modeling stud-

Figure 8.2 The magnetic responses known as the *mismatch field* (MMF) elicited by deviant tones (blue trace) in comparison to the magnetic responses elicited by standard tones (red traces). The amplitude of the MMF (indicated by the shaded difference between the blue and red traces) declines as the time between the preceding standard tone and the deviant tone increases to 12 seconds. This result can be interpreted as evidence for an automatic process in sensory memory that has a time course on the order of approximately 10 seconds. Adapted from Sams et al. (1993).

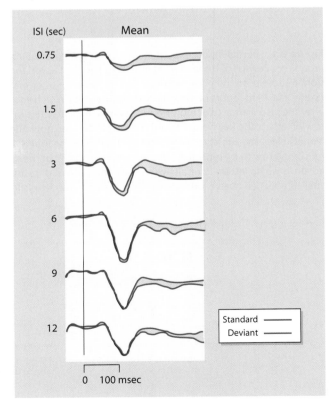

ies have localized the MMN/MMF to the auditory cortex on the supratemporal plane, these physiological studies also provide information about where sensory memories are stored; they are stored in sensory-specific cortex as a short-lived neural trace.

SHORT-TERM MEMORY: FORGETTING OVER THE COURSE OF SECONDS

In contrast to sensory memory, which is a high-capacity but short-lived sensory trace that is not itself considered to be available to conscious awareness, *short-term memory* is severely limited in capacity, has a time course counted in seconds to minutes, and is readily available to our conscious awareness. Here we will first review the evidence about the nature of short-term memory and then present competing models of these forms of memory.

A central debate in the 1950s and 1960s was whether forgetting was the result of interference from newly learned items or was the result of decay of previously learned material. The dominant thinking was that loss of information from memory was due to interference of newly learned material rather than decay of information. However, the studies supporting this view mostly involved long-term memory. In the late 1950s, Petersen and Petersen (1959) and others developed an experimental paradigm to measure how long it takes to forget information over the course of seconds—short-term memory and forgetting. Volunteers were presented with three-letter consonant strings, and after varying intervals of time, a light would cue the person to recall and say the letters aloud. First, Petersen and Petersen demonstrated that even after 30 seconds subjects remembered the consonant strings when they were permitted to rehearse (silently) the information during the delay interval. However, in another condition of the experiment, during the interval between the presentation of the letters and the cue, which lasted from 3 to 18 seconds, the subjects

---

HOW THE BRAIN WORKS

Short-Term Memory Capacity and Codes

Short-term memory is limited, but how limited? Precisely how much information a healthy individual can retain in short-term memory varies among individuals (see also The Memorist). However, experiments have demonstrated an interesting characteristic of human memory. In the 1950s, George Miller (see Chapter 1, Figure 1.18, and Suggested Readings—Miller, 1994) investigated how much information individuals can process. Although the initial work centered on perception, the research has been extended to memory for the retention of items.

Volunteers were presented with items to be remembered, in groups of varying size. The results were quite amazing: Regardless of the content of the items (e.g., digits, letters, or words), the number of items that were retained typically proved to be around seven. When more than seven items were presented, volunteers were less successful at recalling all of them. Miller referred to this characteristic feature of human memory as the *span of immediate memory*, or in the terminology we have been using up until now, the *span of short-term memory*. When digits are used, this feature is referred to as *digit span* and it is commonly measured in neuropsychological tests.

The memory limits discovered in these studies are defined by the number of items, not the content of each item, and therefore tell us about the way information is coded in short-term stores. This distinction has sometimes been cast as the difference between a *bit* of information and a *chunk*—a *bit* being the elementary piece of information and a *chunk* being a unit composed of bits. The use of words allows individual letters to be chunked into one meaningful piece of information. The word *cerebellum* is either ten letters, or one word. If ten letters have to be remembered, the short-term memory system is taxed, but if the letters can be chunked as one word (*cerebellum*), then about seven of these chunks (or words) can be remembered. The consequence of this chunking is that during recall of the material, the chunked information can be essentially unpacked (unchunked) to yield more bits of information than could normally be retained. That is, if we can retain in our memory seven words of ten letters each, we can unpack them into seventy bits of information by using knowledge about word spelling. This evidence points to the ability of humans to recode information in manageable packets, packets that can be handled within the constraints of short-term memory.

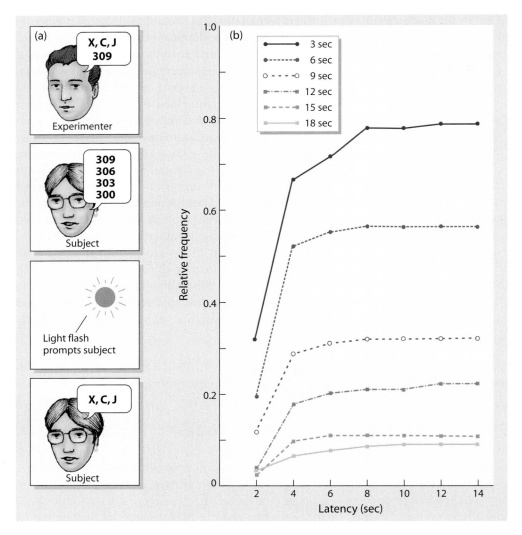

Figure 8.3 **(a)** To test a person's short-term retention of verbal items, an experimenter presents the subject with a string of three consonants such as *XCJ*, and then a number such as *309*. The subject listens to the consonant string and then starts counting backward by threes from the number given by the experimenter. After 3 to 18 seconds, a red light signals the subject to recall and repeat aloud the consonant string. **(b)** Correct recall as a function of response latency and retention delay. In the study by Petersen and Petersen, by 18 seconds the subjects could recall the consonant string less than 10% of the time. Adapted from Petersen and Petersen (1959).

were distracted from rehearsing the letters by being asked to perform mental arithmetic such as backward subtraction (Figure 8.3a). The effect of distraction on recalling the letter strings was clear; by 18 seconds the percentage of correct responses dropped below 10%. The curves that describe retention of the letter string as a function of response latency for the different delay intervals (3–18 sec) are shown in Figure 8.3b.

How should these findings be interpreted? Contemporary thought about forgetting things from memory held that new information interfered with older memories (interference models). This explanation was, however, inconsistent with the data from the short-term forgetting experiments of Petersen and Petersen, which showed that forgetting could happen when no *new* information was actually being learned. Therefore, interference from newly learned information could not have caused the forgetting. By this logic, decay of information from short-term memory must be the mechanism of forgetting. These new findings opened up the possi-

bility that there were indeed two distinct types of memory, or two memory systems—short-term and long-term—and that they could be distinguished in part by whether the loss of information resulted from decay (short-term memory) as opposed to interference (long-term memory). One point of interest about this view of memory is that rehearsal of information was considered critical for keeping information from decaying in short-term memory, and therefore was considered important for being able to transfer that information to a long-term memory store. Was there any information to support this general idea? The answer is yes, and it comes in part from the so-called serial position effect.

What is the serial position effect, and how might it help clarify the idea that short-term memory and long-term memory may be separate mechanisms, with rehearsal mediating the transfer of items from one to the other? When we make a shopping list but forget to take the list to the market, we often remember some items better than others. The pattern for what we typically re-

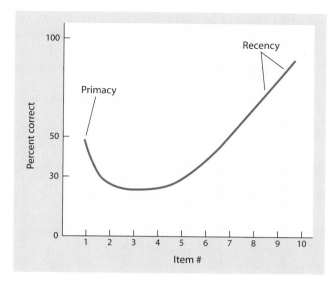

Figure 8.4 Serial position effect. The percentage of items recalled is plotted as a function of the item's position in the list. Primacy and recency effects are represented by the better recall of items presented at the beginning and end of the list, respectively.

member is called the **serial position effect.** We are better at recalling items at the beginning and end of a list, known as *primacy* and *recency effects,* respectively. Figure 8.4 shows the serial position effect, a U-shaped curve when accuracy of recall is plotted against the item's position on the list. The list might be of ten words presented one after another at 1-second intervals. Following presentation of the list, subjects are asked to repeat the list in free recall.

Two separate mechanisms are proposed for primacy and recency effects. One interpretation of the primacy effect is that it reflects the efficiency of transferring items from short-term memory to long-term storage. The logic of this view is as follows: At the beginning of the list, the memory system has enough capacity to transfer successfully from short- to long-term memory by rehearsal of the items. (Remember that when rehearsal was prevented in Figure 8.3b, the items were rapidly forgotten.) In contrast, the recency effect happens because items at the end of the list are available in short-term memory, having recently been seen, and therefore have not yet had a chance to decay.

Studies supporting these proposed mechanisms for primacy and recency employed a distracting task after a list of items was presented. The distracting task eliminated the recency effect, as one would predict if recency effects represented storage in short-term memory; the distracting task prevented rehearsal. However, the primacy effect was unchanged by adding a distracting task at the *end* of the list.

In contrast, the primacy effect could be eliminated if the list items were presented more quickly, a manipulation that did not affect the recency effect. Presumably, the primacy effect was reduced at faster rates because the rehearsal of early items was hindered by the large information load generated by the rapid presentation of subsequent items, and rehearsal was needed for successful transfer to long-term memory. The experiments, then, supported the idea that primacy and recency effects resulted from separate memory processes; the primacy effect reflects transfer from short- to long-term memory through rehearsal, but the recency effect reflects retention in short-term memory. These distinctions between primacy and recency effects highlight the distinction between short- and long-term storage and emphasize the concept of transfer from short-term memory to long-term memory. However, as we will see in the next section, the concept of a distinct short-term memory system that can pass rehearsed information into long-term memory is not supported by all available information.

Models of Short-Term Memory

The early data on short-term memory led to some influential models that proposed discrete stages of information processing during learning and memory. One model, the "modal model," represents a way of viewing data generated by studies on sensory memory, short-term memory, and long-term memory.

Cognitive psychologists Richard Atkinson (the current president of the University of California system) and Richard Shiffrin (1968) at Indiana University elaborated the details of the modal model (Figure 8.5). Information

Figure 8.5 The Atkinson and Shiffrin modal model of memory. Sensory information enters the information-processing system and is first stored in a sensory register. Items that are selected via attentional processes are then moved into short-term storage. With rehearsal, the item can move from short-term to long-term storage. Adapted from Atkinson and Shiffrin (1968).

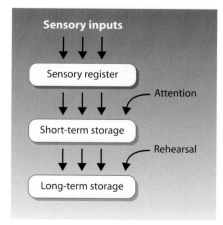

is first stored in sensory memory. Items selected by attentional processes can move into short-term storage. Once in short-term memory, if the item is rehearsed, it can be moved into long-term memory. The Atkinson and Shiffrin modal model suggests that at each stage, information can be lost by decay, interference, or a combination of both. This model formalizes the idea that discrete stages of memory exist and that they have different characteristics. In addition, this model has a strong serial structure: Information coming into the sensory register can be passed to short-term storage and only then onto long-term memory. This view is not supported by other theoretical and experimental evidence, however.

LEVELS OF PROCESSING MODELS

New formulations about memory mechanisms developed partly from challenges to the modal model shown in Figure 8.5. For example, other factors besides simply holding information in short-term memory seemed to influence long-term memory. One notion was that the "deeper" (i.e., more meaningfully) an item was processed, the more it was consolidated and stored in long-term memory. This view is known as the *levels of processing model.*

To illustrate the levels of processing view, let's consider experiments by Fergus Craik and Robert Lockhart (1972). Written words were presented to subjects in three conditions. In one condition, the subjects were told to identify whether the words were composed of uppercase or lowercase letters—this condition was considered a superficial processing task since it involved only consideration of the gross form of the letters and not the meaning of the words themselves. In a second condition, the subjects were asked to determine whether a word rhymed with another word—this condition was considered an intermediate level of processing because although the words were being considered as words, the judgment had to do with whether the words sounded like they rhymed and had nothing to do with meaning. Finally, in the third condition, the subjects were asked to actually make a judgment about the meaning of the word (e.g., does it fly?)—this one is considered a deep level of processing because the task involved semantic (meaning) analysis. Subjects showed better subsequent memory for items more deeply processed during learning. Thus, the manner in which new information is processed affects how it is remembered.

Deep or elaborate rehearsal and encoding create meaning-based codes that relate information directly to previously acquired knowledge. The result is better learning compared to when information is merely repeated and stored as simple visual or phonological codes. This feature of memory is inconsistent with the concept of a short-term store because it distinguishes between merely holding information in short-term memory long enough to get it into long-term storage and the type (superficial or deep) of encoding necessary to accomplish this process.

NEUROPSYCHOLOGICAL EVIDENCE AGAINST THE MODAL MODEL OF MEMORY

A key piece of information from studies of patients with brain damage also led to skepticism about the hierarchically structured modal model of memory. As introduced earlier in the chapter in the story of H.M.—which will be described in more detail shortly—the effect of brain damage on memory function can help to place constraints on cognitive models. In 1969, neuropsychologists Tim Shallice and Elizabeth Warrington of Great Britain reported a patient with damage to the left perisylvian cortex. The patient displayed reduced digit span ability (about two items, as opposed to five to nine items for healthy persons). This meant that he had a deficit in his ability to recall lists of items when the delay intervals between the presentation of the list and the test were short, only a few seconds. Remarkably, however, this patient still retained the ability to form new long-term memories.

A more recent example of a similar patient comes from the work of Hans Markowitsch and colleagues (1999). Their patient, E.E., had a tumor centered in the left angular gyrus, affecting the inferior parietal cortex and posterior superior temporal cortex (Figure 8.6). After the tumor was removed surgically, E.E. showed below-normal short-term memory ability but preserved long-term memory, just like the patient of Shallice and Warrington. E.E. showed normal speech production and comprehension and normal reading comprehension. But he had a minor deficit in the Token Test, which requires subjects to move plastic pieces (tokens) according to auditory instructions (e.g., "Place the green square behind the yellow circle after touching the black triangle"); this deficit was attributable to his difficulty in retaining the information long enough to complete the task. So, E.E.'s primary deficit was in short-term memory.

The pattern of behavior displayed by these patients demonstrates a selective loss of so-called short-term memory abilities but preservation of long-term memory. The implication of this work is that short-term memory cannot be the gateway to long-term memory in the manner laid out in the modal model. Rather, infor-

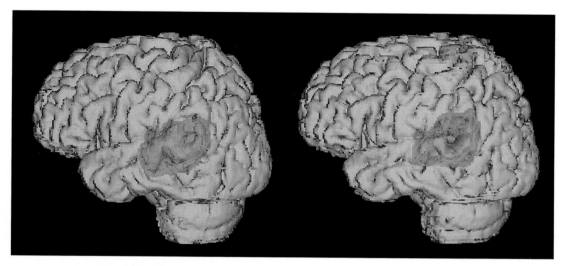

Figure 8.6 Magnetic resonance imaging (MRI) scans reconstructed to provide a three-dimensional rendering of patient E.E.'s left hemisphere. The reconstructed MRI scan taken prior to surgery is on the **left,** and the one taken after surgery is on the **right.** The area of the tumor is indicated by shading. Positron emission tomography (PET) with a radiolabeled methionine tracer was used to identify the tumor based on its increased metabolic profile in comparison to that of surrounding brain tissue. Adapted from Markowitsch et al. (1999).

mation from sensory memory registers can be encoded into long-term memory directly. However, these data do demonstrate a clear dissociation between long-term memory ability and short-term retention of information. In contrast, patient H.M. had preserved short-term memory but a loss in the ability to form new long-term memories—this provides a double dissociation for retention of information over the short and long term. What then is the nature of our ability to retain information over the short term? That is, if short-term stores are not the gateway to long-term storage, how shall we conceptualize our ability to remember a phone number long enough to dial it? Or to remember the numbers *34, 56,* and *21* long enough to add them together?

WORKING MEMORY MODELS

The concept of working memory was developed to address various shortcomings in the short-term memory concept as expressed in the modal model (see Figure 8.5), and to elaborate the kinds of mental processes that can occur when information is retained over a period of seconds to minutes. **Working memory** represents a limited-capacity store for retaining information over the short term *and* for performing mental operations on the contents of this store. The contents of working memory might originate from sensory inputs by way of sensory memory (as in the modal model) but also might be retrieved from long-term memory. In each case, working memory contains in-

formation that can be acted on and processed, not merely maintained by rehearsal, although this is one aspect of working memory.

In their formulation of working memory in the 1970s, Alan Baddeley and his colleagues (e.g., Baddeley and Hitch, 1974) proposed that a unitary short-term memory was insufficient to explain the maintenance and processing of information over short periods. They constructed a three-part working memory system containing a central executive mechanism for controlling two subordinate systems involved in rehearsal. These subordinate systems reflected the types of codes used by each in rehearsal; they are the phonological loop and the visuospatial sketch pad. What do these hypothesized subsystems do, and what evidence supported this formulation?

The *central executive mechanism* is a command-and-control center that presides over the interactions between the two subordinate systems and long-term memory. It is a modality nonspecific cognitive system that coordinates processes in working memory, and controls actions. Donald Norman and Tim Shallice (1980) thought of this concept of a central executive as a supervisory attentional system (SAS). The SAS overrides routine execution of learned behaviors when novel circumstances require modified actions, and it also coordinates and plans activities.

The *phonological loop* is a hypothesized mechanism for acoustically coding information in working memory. The evidence for this first came from studies that asked subjects to recall strings of consonants. The letters were presented visually, but the pattern of recall errors indicated that perhaps the letters were not coded visually over the short term. The subjects were apparently using an acoustic code because during recall they were more likely to replace a presented letter with an erroneous letter having a similar sound (e.g., *T* for *G*) rather than a similar shape (e.g., *Q* for *G*). The idea is that *tee* and *gee* are acoustically more similar that *kyoo* and *gee*, even though the round-shaped letter *Q* is more similar to *G* in appearance. This was the first insight that an acoustic code might participate in rehearsal. Immediate recall of lists of words also is poorer when many words on the list sound similar compared to when they sound dissimilar, even when the latter are semantically related. This finding indicates that an acoustic but not a semantic code is used in working memory, because words that sound similar interfere with one another, while words related by meaning do not.

The phonological loop may have two parts: a short-lived acoustic store for sound inputs and an articulatory component that plays a part in the subvocal rehearsal of items to be remembered over the short term. This latter part also would code visually presented information in working memory.

The *visuospatial sketchpad* is based short-term representation that parallels the phonological loop and permits information storage in either purely visual or visuospatial codes. Evidence for this system came from dissociations between verbal and visuospatial codes. The idea is that if two codes are separate, they should not interfere with one another in working memory. If some information is coded acoustically, the introduction of a secondary visuospatial task during retention should not disrupt performance. Indeed, acoustic and visuospatial codes are found to be separate; what emerges from these studies is a working memory with a central executive and two independent subsystems (Figure 8.7).

The advantage of the working memory conceptualization over the short-term memory idea represented in the modal model is that it fills in details about the relation between short-term and long-term memory. Further, the working memory concept does not assume a unitary short-term store. Instead, it accounts for short-term forgetting and for processing new information in a context that describes the codes used.

Neuropsychological Evidence and Working Memory Mechanisms Deficits in short-term memory abilities, such as remembering items on a digit span test, can be correlated with damage to subcomponents of

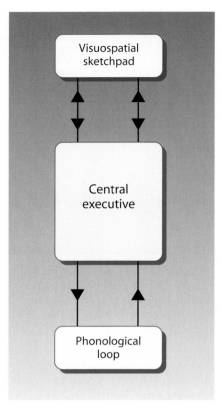

Figure 8.7 Simplified representation of the working memory model by Baddeley and Hitch. This three-part working memory system has a *central executive* that controls two subordinate systems: the *phonological loop*, by which information can be acoustically coded in working memory, and the *visuospatial sketchpad*, by which information is visually coded in working memory. Adapted from Baddeley (1995).

the working memory system. Remember that the working memory model of Baddeley and colleagues consists of three parts: a central executive (attentional) system and two slave systems, the phonological loop and the visuospatial sketch pad (see Figure 8.7). Evidence about the distinct nature of these subsystems and their anatomical substrates in the human brain first came from studies of patients with specific brain lesions. Indeed, each system can be damaged selectively by different brain lesions.

Lesions of the left supramarginal gyrus (Brodmann's area 40) lead to deficits in phonological working memory (Figure 8.8, see also Figure 8.6). Patients with lesions to this area have reduced auditory-verbal memory spans; they cannot hold strings of words in working memory. The rehearsal process of the phonological loop involves a region in the left premotor region (area 44). Thus, a left-hemisphere network consisting of the lateral frontal and inferior parietal lobes subserves phonological working memory. These

(a)

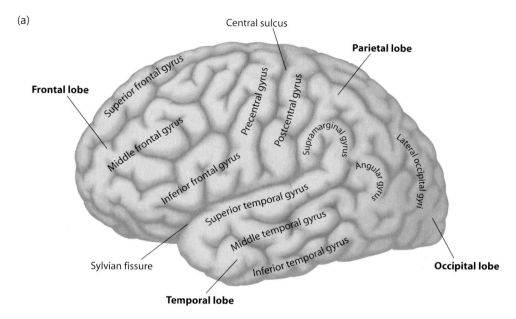

(b)

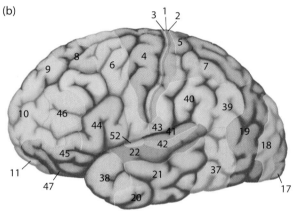

Figure 8.8 **(a)** Diagrammatic lateral view of the left hemisphere indicating major sulci and gyri. **(b)** Representation of Brodmann's areas in the left hemisphere.

deficits in working memory for auditory-verbal material (digits, letters, words) have not been found to be associated with deficits in speech perception or production. This distinction between aphasia—language deficits following brain damage (see Chapter 9)—and deficits in auditory-verbal short-term memory is important to keep in mind.

The visuospatial sketchpad is compromised by damage to the parieto-occipital region of both hemispheres, but damage to the right hemisphere produces more severe deficits in visuospatial short-term memory. Patients with lesions in the right parieto-occipital region have difficulty with nonverbal visuospatial working memory tasks like retaining and repeating the sequence of blocks touched by another person. For example, if an investigator touches blocks on a table in sequences that the patient has to repeat, and gradually increases the number of blocks touched, patients with parieto-occipital lesions would show below-normal performance, even when

their vision is otherwise normal. Similar lesions in the left hemisphere can lead to impairments in short-term memory for visually presented linguistic material.

Working memory impairments, then, are associated with discrete lesions in distinct brain areas. Further, as we will learn in the following sections, these areas are largely distinct from the areas associated with deficits in long-term memory. We will return to a discussion of the cognitive neuroscience of working memory in detail in Chapter 12, when we review executive functions.

Models of Long-Term Memory

Information maintained for a significant time is referred to as *long-term memory*. Much of memory research has focused on distinctions in long-term memory function. Theorists have tended to split long-term memory into two major divisions that reflect the characteristics of the information that is stored, and that

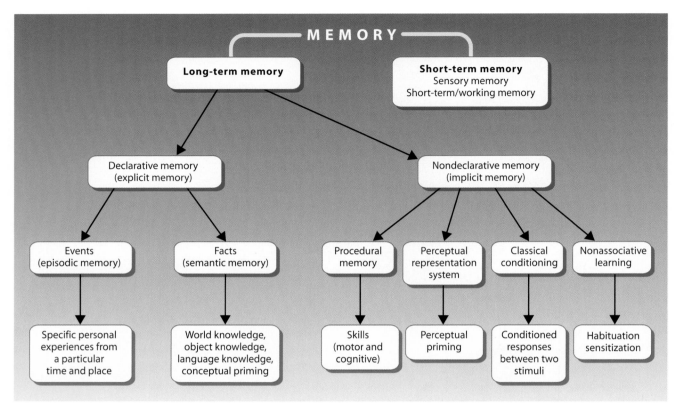

Figure 8.9 The hypothesized structure of human memory diagramming the relationship among different forms of memory.

take into account the observable fact that not all stored knowledge is the same. The key distinction is between declarative and nondeclarative memories.

Declarative memory refers to knowledge that we have conscious access to, including personal and world knowledge. In contrast, **nondeclarative memory** refers to knowledge we have no conscious access to, such as motor and cognitive skills (procedural knowledge), perceptual priming, and simple learned behaviors that derive from conditioning, habituation, or sensitization. The essential relations between these forms of long-term memory are summarized in Figure 8.9.

DECLARATIVE MEMORY

Declarative memory can be broken down further into things that we recall about our own lives (episodic memory) and world knowledge (semantic memory) that does not relate to events in our lives. Endel Tulving introduced this distinction between episodic and semantic memory. Examples of episodic memory include events in our personal history, such as meeting a friend for the first time, our eighteenth birthday party, or the time we fell off a new bicycle, badly skinning an elbow and scratching some of that beautiful red paint off of

the bike. *Episodic memory* involves conscious awareness of past events; it is our personal, autobiographical memory.

Semantic memory, in contrast, reflects knowing things such as how to tell time, who the twenty-first president was, how to cook an omelet, how to subtract two numbers, that rocks and animals are fundamentally different, that things you can sit on may (or may not) be chairs even if they look very different, and other similar facts and concepts. Semantic memory is world knowledge that we remember in the absence of any recollection of the specific circumstances surrounding its learning. That is, no episodic memory is associated with the semantic information. This distinction holds even though sometimes we might remember the episode for the acquisition of some facts. For example, you might have an episodic memory of your father explaining that paying $100 for a simple toy is expensive, whereas paying $100 for an automobile would be quite inexpensive (depending on the car!). In this example, the episodic memory is the memory of the specific interaction with your father, while the semantic memory involves the general world knowledge that cost and value are relative concepts—although the acquisition of such world knowledge presumably also involves many other "forgotten" episodes during which this knowledge of cost and value were refined in numerous ways. The take-home message is that world knowledge is fundamentally different from our recollection of events in our own life.

NONDECLARATIVE MEMORY

One of the most exciting areas of memory research in the past decade or so has been the area of nondeclarative memory. This memory is revealed when previous experiences facilitate performance on a task that does not require intentional recollection of the experiences.

Nondeclarative memory encompasses several forms of knowledge that we can observe on a day-to-day basis and that can be revealed with appropriately designed experimental testing. In Figure 8.9, which presents a taxonomy of memory, nondeclarative memory includes forms of memory that are learned and retained even when explicit memory for that knowledge does not exist. **Procedural memory** is one form of nondeclarative memory that involves the learning of a variety of motor (e.g., knowledge of how to ride a bike) and cognitive skills (e.g., the acquisition of reading skills). One form of nondeclarative memory that acts within the perceptual system is known as the **perceptual representation system** (PRS). In the PRS, the structure and form of objects and words can be primed by prior experience. *Priming* refers to a change in the response to or ability to identify a stimulus as the result of prior exposure to that stimulus.

Two other domains of nondeclarative memory include *classical conditioning* and *nonassociative learning.* Classical conditioning, sometimes referred to as *Pavlovian conditioning,* occurs when a conditioned stimulus (an otherwise neutral stimulus to the organism) is paired with an unconditioned stimulus (one that elicits some response from the organism). Following this pairing, the conditioned stimulus can evoke a response similar to that typically evoked by the unconditioned stimulus, as in the experiments of Pavlov with his dogs. Nonassociative learning, as its name implies, does not involve the association of two stimuli to elicit a behavioral change. Rather it

involves forms of simple learning such as habituation (the *decrease* in a response with repeated presentations of the stimulus) and sensitization (the *increase* in a response with repeated presentations of the stimulus). We will not consider classical conditioning or nonassociative memory further in this chapter but instead will focus on the neural substrates of declarative (episodic and semantic memory) and nondeclarative memory (procedural memory and the perceptual representation system).

Summary of Theories of Memory

Memory theories include two main distinctions about how we learn and retain knowledge. The first is that memory can be defined by retention time. Thus, we have identified sensory memory, short-term/working memory, and long-term memory. The second main concept involves the ideas that memories can be characterized by their content and that different types of information can be retained in partially or wholly distinct memory systems. For example, we have identified several subtypes of long-term memory. The main distinction is the declarative versus nondeclarative dichotomy. Knowledge is sometimes in the form of declarative memories, which are memories about events from our lives or even world knowledge that we have conscious access to and can make declarations about. Other forms of knowledge can be said to be nondeclarative, which involves procedural knowledge (e.g., how to ride a bike), perceptual priming, conditioned responses, and nonassociative learning. The information stored as nondeclarative memory is typically out of the reach of our conscious awareness. The question that remains is whether these different memory systems are supported by different neural circuits and systems in the brain. If they are, we would have converging evidence for the idea that these qualitatively different forms of learning and memory are reflections of distinct memory mechanisms. Hence, in the next section, we focus on the cognitive neuroscience of memory with a look at amnesia, including a detailed look at patient H.M., who we introduced briefly at the opening of this chapter.

MEMORY AND BRAIN

The field of the cognitive and experimental psychology of memory is rich with theory and data and has produced a consistent set of concepts about the organization of human memory. In the foregoing we crystallized a vast literature into a few general concepts. In the next section we will turn to neuroscientific stud-

ies of memory, both to understand how they have contributed to general theories of memory, and to investigate how specific neural circuits and systems enable the learning and retention of specific forms of knowledge. The story is unfolding as we write, and is one of the great challenges of cognitive neuroscience.

Eyewitness Testimony

Every prosecutor in a criminal case knows that an eyewitness account is among the most compelling evidence for establishing guilt. But is this type of testimony to be trusted? Elizabeth Loftus and her colleagues (1978, 1980) illustrated the difficulty of relying on witness's recall by showing subjects color slides detailing the time of an accident and asking people what they saw in a later test session. One of the slides showed a car at an intersection before it turned and hit a pedestrian. Half the subjects viewed a red stop sign and half a red yield sign. Subjects then answered questions about the slides: Half were presented with questions referring to the correct sign and half with the incorrect sign. For example, if a subject had been shown a yield sign, she might have been asked, "When the car came to the stop sign, did the driver stop?" During subsequent recognition tests for whether a certain slide was what they had previously seen, 75% of the subjects correctly recognized a previously seen slide if the correct sign had also been mentioned in the questioning session. But when subjects previously had been questioned with the wrong sign being mentioned, only 41% had correct recognition for the slides. These findings indicate that recollections of an event can be influenced by misleading statements made during questioning.

Misinformation about things as obvious as hair color and the presence of a mustache can lead subjects to wrongly identify people they had seen previously. What does this say about witness's suggestibility and the influence of misinformation on recall? Do witnesses really know the correct information but later fail to distinguish between their own memories and information supplied by another person? One line of thinking is that perhaps the information was not encoded initially, and when forced to guess, the subject provides the information given by someone else.

Not just adults are eyewitnesses in court cases. Children often are asked to testify as witnesses. Since adults with fully developed memories have difficulty recalling what they have seen, how do young children with potentially underdeveloped memory systems behave under the pressures of authorities and courtrooms? This is a controversial issue because of situations where children are eyewitnesses to crimes against themselves, such as child abuse or sexual abuse in which they may be the only witness.

The question of how well children remember and report things they have experienced is of special concern when the events may be traumatic. One way to effectively study such conditions is to use events traumatic for children and to involve contact between them and others that can be verified. Physicians sometimes must perform genital examinations on children, which may include painful medical procedures. The children's memories of these events can be examined systematically. In one study, half of seventy-two girls between the ages of 5 and 7 years were given a genital examination as part of necessary medical care, and half were not. Children who received the examination were unlikely to report anything about the event during free recall or when using anatomically detailed dolls. Only when asked leading questions did they reveal that they had been examined. The control group made no false reports during free recall or with the dolls, but with leading questions, three children made false reports. Psychologist Gail Goodman and her colleagues (1994) at the University of California at Davis emphasized that one of the most important predictors of accurate memory performance is age. Memory performance for traumatic events is significantly worse in children 3 to 5 years old than in older children. Dr. Goodman also noted, however, that other factors influence memory accuracy in children. These include how well the traumatic episode is actually understood, the degree of parental emotional support and communication, and the children's emotional (positive versus negative) feelings. The goal of this research is to determine the validity of children's reports on events—including negative ones such as abuse—that may have happened to them, and how they might invent stories. Of special interest is whether leading questions can induce children to fabricate testimony. We need to know this when interpreting their testimony that involves other persons such as therapists or members of the legal system.

Temporary Losses of Memory

A common theme in novels and movies is the temporary loss of memory after a bump on the head or psychological trauma. Such things do happen outside of Hollywood and might have many causes, including head injury or epilepsy. However, sometimes temporary memory loss occurs for other reasons, and one such syndrome is known as *transient global amnesia*, or TGA. Most clinicians consider TGA to be distinct from the temporary memory loss caused by head injury, epileptic activity, or stroke. Vascular effects are suspected as the main cause of TGA, but they are limited to transient ischemia (reduced blood flow to a brain area) that does not lead to cerebral infarct (permanent damage to brain tissue). The tissue recovers, and so does the patient's memory. The prime candidates for the brain areas affected are the medial temporal lobe and diencephalic areas. The patients have symptoms similar to those of persons with permanent damage to the medial temporal lobe, such as H.M.

A simple scenario common for TGA is the following: An otherwise healthy person over 50 years old is brought (or goes) to the hospital after experiencing some problems with memory. During questioning by a physician, the person is not sure why he is there, and may not be sure how he got there or where he is. He does know his name, birth date, job, and perhaps address, but has retrograde amnesia pertaining to a period of weeks or months—perhaps even years. Thus, if he had recently moved, he would supply his past address and circumstances. He has normal short-term memory and thus can repeat lists of words told to him. But his ability to form new episodic memories is impaired, and when told to remember a list of words, he forgets it within a couple of minutes if he is prevented from rehearsing it. He will continually ask who the physician is and why he is there. He does show an awareness that he *should* know the answer to some questions, which worries him. He performs normally on most neuropsychological tests, except for the ones indexing memory. He manifests a loss of time sense, and so responds incorrectly to questions asking how long he has been in the hospital. He also might have reduced verbal fluency and show impairment in generating words that begin with a certain letter, like *a*, when asked to do so. However, he does not have deficits in producing category names, for example, general names for fruits. This latter pattern is typical of frontal lobe problems.

Over the hours after the amnesia-inducing event occurred, distant memories return, and his anterograde memory deficit resolves. Within 24 to 48 hours he is essentially back to normal, although mild deficits may persist for days or even weeks.

Persons having one such attack are more likely to have TGA again, but it does not usually occur often. Some people may never have another episode. The recurrence rate is about 18% within a 20-year period. Men are more likely to experience TGA.

The physiological cause is thought to be ischemia due to a disruption of normal blood flow in the vertebral-basilar artery system, which supplies blood to the medial temporal lobe and diencephalon. TGA most often is triggered by physical exertion. Unlike patient R.B. described in the text, who had such a severe ischemic attack that it caused neurons in his hippocampus to die, TGA patients only suffer from a transient deficit in the medial temporal lobe or in thalamic functioning. PET and single-photon emission computed tomography studies in a few patients have demonstrated this transient deficit: Blood flow in the medial temporal lobe decreases during a TGA attack but later returns to normal. So far, we do not know whether these patients have normal implicit learning or memory, in part because their impairment does not last long enough for researchers to adequately index things like procedural learning. This would be informative in understanding human memory, and in learning more about an amnesia that could happen to any of us later in life.

Human Memory, Brain Damage, and Amnesia

Deficits in memory as a function of brain damage, disease, or psychological trauma are known as *amnesia*. Amnesia can involve either the inability to learn new things or a loss of previous knowledge, or both. It can differentially affect short-term/working memory and long-term memory abilities. Moreover, there is evidence to suggest that amnesia may selectively manifest as deficits in episodic memory, perceptual priming, and procedural knowledge. In at least one patient, there is evidence to distinguish episodic memory from semantic memory by the pattern of memory deficits. Thus, by examining amnesia in conjunction with cognitive theories derived from experiments on normal subjects, we can understand the organization of memory at a functional and a neural level. Much compelling information about the organization of human memory during amnesia was first derived from medical treatments that left patients amnesic. The history is fascinating, and so let's begin by turning back the clock more than fifty years.

BRAIN SURGERY AND MEMORY LOSS

In the late 1940s and early 1950s, surgeons attempted to treat neurological and psychiatric disease using a variety of neurosurgical procedures, including prefrontal lobotomy (removing or disconnecting the prefrontal lobe), corpus callosotomy (surgically sectioning the corpus callosum), amygdalotomies (removing the amygdala), and temporal lobe resection (removal of the temporal lobe). These surgical procedures opened a new window on human brain function as they revealed, usually quite by accident, fundamentally important principles of the organization of human cognition. One surgical procedure relevant to memory was removal of the medial portion of the temporal lobe, including the hippocampal formation.

In 1953 at a medical conference, the neurosurgeon William Beecher Scoville from the Montreal Neurological Institute reported on bilateral removal of the medial temporal lobe in one epileptic patient and several schizophrenic patients. Shortly thereafter he wrote:

> Bilateral resection of the uncus, and amygdalum alone, or in conjunction with the entire pyriform amygdaloid hippocampal complex, has resulted in no marked physiologic or behavioral changes with the one exception of a very grave, recent memory loss, so severe as to prevent the patient from remembering the locations of the rooms in which he lives, the names of his close associates, or even the way to the toilet. . . . (Scoville, 1954)

At the time, prefrontal lobotomy and orbitofrontal undercutting—a less radical method than lobotomy for treating severe mental disorders—had shown promise for treating severely psychotic patients. These procedures, confined to the prefrontal cortex, showed no other side effects—patients did not lose memory, speech, motor skills, or other abilities. Because of the links between the orbitofrontal cortex and the medial temporal lobe, surgeons reasoned that further resections of the medial temporal lobe might ameliorate certain mental illnesses. However, as indicated in the foregoing quote,

the epileptic patient that Scoville treated underwent similar surgery (without the orbitofrontal undercutting), removal of the presumably abnormal brain tissue in the medial temporal lobe, to control his severe epilepsy. This patient was not entirely unscathed following his surgery.

When Scoville introduced his patients to a young psychologist, Brenda Milner, the neuropsychological examination of them began in earnest. The findings in ten patients were reported by Scoville and Milner in 1957; two of them were reported in Scoville's earlier publication. What Milner revealed was that of the patients having medial temporal lobe resections as part of their treatment, those manifesting memory impairments did so in relation to how much of the medial temporal lobe was removed. The farther posterior along the medial temporal lobe the resection was, the worse the amnesia. Strikingly, however, only *bilateral* resection of the hippocampus resulted in severe amnesia. In comparison, in one patient whose entire right medial temporal lobe (hippocampus and hippocampal gyrus) was removed, no residual memory deficit was found. But the interesting patient was the young man who had had bilateral medial temporal resection.

The Story of Patient H.M. Patients who underwent bilateral medial temporal lobe resection developed dense amnesia. The most famous patient was H.M. The case of H.M. holds a prominent position in the history of memory research because, as we introduced at the beginning of the chapter, although he had epilepsy, he was of normal intelligence, had no psychological or mental illness, and as a result of the surgery developed a severe and permanent inability to acquire new information (Scoville and Milner, 1957; Milner et al., 1968).

After the surgery, H.M. had normal short-term memory (sensory registers and working memory). Like many other amnesics (Figure 8.10), H.M. had normal digit span abilities. He did well at holding strings of digits in working memory, but unlike normal subjects, he performed poorly on digit span tests that required the acquisition of new long-term memories—that is, his digit span ability hardly improved with practice. What appeared to be disrupted was the transfer of information from short-term storage to long-term memory. Recall the case of Shallice and Warrington's patient described in the section on short-term memory earlier in the chapter? These case studies of patients like H.M. and others provide a strong double dissociation of short-term/working memory and long-term memory. Shallice and Warrington's patient had below-normal digit span ability (short-term/working memory) but normal long-term memory ability, while H.M. showed the inverse pattern.

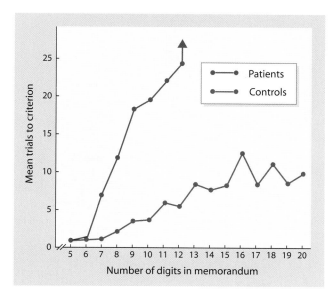

Figure 8.10 Digit span for amnesic patients and control subjects. A sequence of five digits was read to the subjects, who were then asked to repeat the digits back to the experimenter. If the digits were repeated correctly, one more digit was added to the next sequence presented. If the digits in a sequence were reported incorrectly, that sequence was repeated until the subject reported it correctly. Amnesic patients had relatively normal digit span ability, but required more trials to learn strings of digits. Adapted from Drachman and Arbit (1966).

AMNESIA AND THE MEDIAL TEMPORAL LOBE

Which region or regions of the medial temporal lobe were critical for supporting the long-term memory ability lost in H.M.? The medial temporal area includes the amygdala, the hippocampus, the entorhinal cortex, and the surrounding parahippocampal and perirhinal cortical areas (see Chapter 3). We will return to a more detailed analysis of these areas based on work in animals in a later section of this chapter. Here we focus on the story as it has emerged based on work in amnesic patients.

For the past 40 years, scientists studying H.M. used surgical reports of his lesions to guide theories of memory and amnesia and their neural bases. Original reports by Scoville, who performed the surgery, indicated that all of H.M.'s hippocampus in each hemisphere had been removed. Scoville indicated that this amounted to removal of about 8 cm of tissue, measuring from the anterior tip of the hippocampus toward the posterior portions of the medial temporal lobe (Figure 8.11). Although it may not change the conclusions that have been drawn from studies of H.M., the evidence suggests that less was removed than previously thought.

With high-resolution neuroimaging methods such as magnetic resonance imaging (MRI), H.M.'s brain and

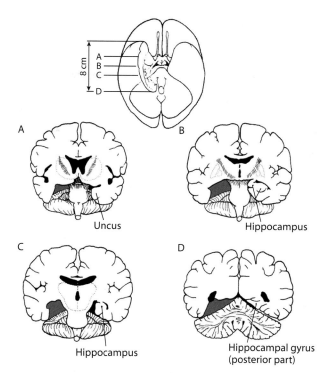

Figure 8.11 Anatomy of the medial temporal lobe and areas believed to have been removed from H.M. (in red) as reported by the surgeon (Note that the resection is shown here on one side only, to permit a comparison of the resected region with an intact brain at the same level. H.M.'s actual lesion was bilateral.) At the top is a ventral view of the brain showing both hemispheres and the details of the right medial temporal area (shown on left of figure). The four anterior to posterior levels (A–D) shown on this ventral view correspond to the coronal sections at the bottom of the figure. Adapted from Corkin et al. (1997).

the surgical lesions were re-evaluated with improved accuracy. For many years H.M. could not be scanned by MRI because of the metal clips that had been placed in his brain to prevent major vessels from bleeding during surgery—a normal procedure. The intense magnetic field of the scanner might have caused the indwelling metal objects to move, which could have led to injury. This restriction applies only to ferromagnetic items affected by magnetic fields; many metals are not ferromagnetic. With H.M. the problem was not knowing whether his clips were ferromagnetic. His surgery was performed long before MRI had been developed, and clips developed for surgery at that time might or might not have been ferromagnetic. Thus, it was possible that MRI scanning would endanger H.M.

In an unusual detective story, Sue Corkin of MIT and journalist and author Philip Hilts (1995) tracked the history of H.M.'s surgery. They determined that the clips used were not ferromagnetic, and that it would be safe to scan him using MRI. Hence, more than 30 years after his surgery, H.M.'s surgical lesion was investigated using modern neuroimaging techniques (Figure 8.12).

Data gathered by Corkin and her colleagues were analyzed by neuroanatomist David Amaral of the University of California at Davis (Corkin et al., 1997). This analysis, shown in Figure 8.13, is an example of remarkable scientific and historical revisionism. In contrast to Scoville's reports, approximately half of the posterior region of H.M.'s hippocampus was intact. And 5 cm, not 8 cm, of the medial temporal lobe was found to be removed. However, the remaining portions of H.M.'s hippocampi were atrophied, probably because of the loss of inputs from the surrounding perihippocampal cortex that had been removed. Thus, despite the original error in our knowledge about H.M.'s lesion, it may be that no functional hippocampal tissue actually remains. Bear in mind that in addition to the hippocampus, some of H.M.'s surrounding cortex was also removed.

Is damage to the hippocampus sufficient to block the formation of new long-term memories? Consider another patient, R.B., who lost his memory after an ischemic episode (reduction of blood to the brain) during heart bypass surgery. R.B. developed dense anterograde amnesia similar to H.M.'s: He could not form new long-term memories. He also had retrograde amnesia that extended back about 1 to 2 years, slightly less severe than H.M.'s retrograde loss. R.B.'s case is remarkable because his lesions were found to be restricted to his hippocampus. This was learned when he died a few years after surgery. After his death, his brain was donated for study, permitting a detailed analysis of the extent of his neuroanatomical damage. Because R.B.'s memory was tested extensively after his amnesia began and until his death,

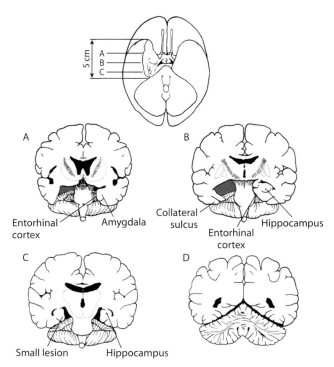

Figure 8.13 Modern reconstruction by Amaral and colleagues showing that portions of H.M.'s posterior hippocampus were not removed during surgery. However, this tissue does show signs of atrophy and may no longer be functioning normally. Adapted from Corkin et al. (1997).

much that was known about his memory capacities could then be related to the precise information about the extent of his lesions from postmortem analysis.

On gross examination of the medial temporal lobe of R.B., the hippocampus appeared to be intact (Figure 8.14a). But histological analysis revealed that within

Figure 8.12 Coronal MRI scans of H.M.'s brain. On the **left** is an anterior slice, showing that the hippocampus has been removed bilaterally. On the **right,** however, is a more posterior slice. Here, the hippocampus is still intact in both hemispheres! This finding is in marked contrast to the belief that H.M. has no hippocampus, a view held by the scientific community for the prior 40 years that was based on the surgeon's report. Adapted from Corkin et al. (1997).

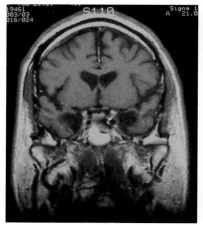 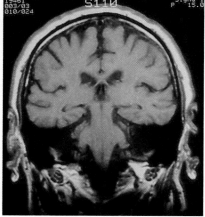

(a)

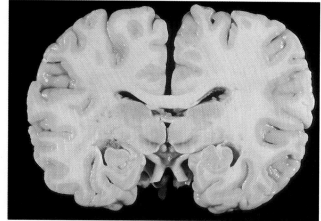

(b)

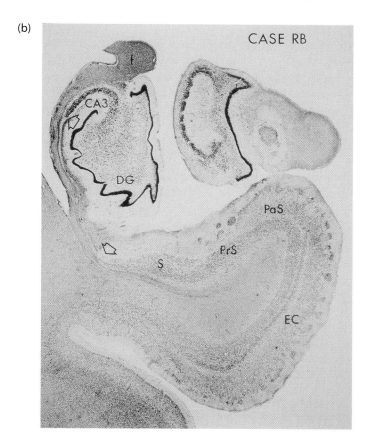

(c)

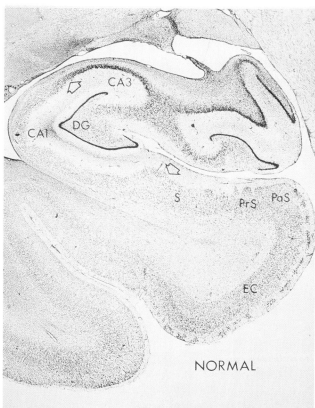

Figure 8.14 **(a)** Section of brain from patient R.B. following his death. In contrast to MRI sections from H.M. in Figure 8.12, showing absence of the anterior and middle portions of the hippocampus, R.B.'s medial temporal lobe appeared intact on gross examination. **(b)** However, upon careful histological examination, it was clear that cells in the CA1 region of the hippocampus were absent (see region between the arrows). The absence of cells was the result of an ischemic episode following surgery. Cells of the CA1 region are particularly sensitive to transient ischemia (temporary loss of blood supply to a brain region). **(c)** Histological section from brain of a normal subject showing an intact CA1 region (labeled CA1 and delimited as region between the arrows). Courtesy of David Amaral.

each hippocampus, R.B. had sustained a specific lesion restricted to the CA1 pyramidal cells (Figure 8.14b). Compare his hippocampus (Figure 8.14b) with that of a normal person after death (Figure 8.14c).

These findings in patient R.B. support the idea that the hippocampus is crucial in forming new long-term memories. R.B.'s case also supports the distinction between areas that store long-term memories and the role of the hippocampus in forming new memories. Even though retrograde amnesia is associated with medial temporal lobe damage, it is temporally limited and does not affect long-term memories of events that happened more than a few years prior to the amnesia-inducing event.

MEMORY CONSOLIDATION AND THE HIPPOCAMPUS

Memories are solidified in long-term stores over days, weeks, months, and years. This process is referred to as consolidation, an old concept that refers to how long-term memory develops over time after initial acquisition. This concept originally was developed to explain the behavioral consequences of the interference effects

MILESTONES IN COGNITIVE NEUROSCIENCE

An Interview with Endel Tulving, Ph.D. Dr. Tulving is an emeritus professor of psychology at the University of Toronto and a scientist at the Rotman Research Institute. He is considered the world's foremost authority on cognitive theories of memory.

Authors: When beginning to think about the problem of human memory, what are some of the essential principles a student should know?

ET: There are many such, of course, but let me tell you about one that I believe is the most important at the present stage of cognitive neuroscience of memory. Actually it is one that applies to all science: You must know exactly what it is that you want to study, or know about. There are as many ways (an infinite number of ways?) of thinking about human memory as there are ways of sightseeing in, say, Europe. Therefore, the most essential principle is to make sure that you study the most important problems around at any given time that can be studied. Timing is as important in science as it is in real life, and it is easy to do right things but at the wrong time, too early or too late, and therefore miss it all.

Authors: Well, what we are after is making the distinction between observations about memory as opposed to how memory works, why there are things like episodic memory, which are different from semantic memory, and so on. How does one go past merely describing the simple observations about a phenomenon, to actually elucidating the mechanisms of the phenomenon?

ET: Yes, this is what everybody wants to know in cognitive neuroscience—what are the mechanisms of cognition or, as in the case we are talking about, memory? I must admit I have misgivings about the questions about "mechanisms" applied to memory, or just about "neural correlates" of memory. I think they are a bad habit uncritically taken over from more mature sciences, or even more mature parts of cognitive neuroscience, where the questions do make good sense. In the case of memory, the timing is wrong. The questions about mechanisms are premature, for the simple reason that we do not yet know what it is that we want to know the mechanisms of. To try to specify the mechanisms of memory would be no easier than to specify the mechanisms of life. We must be more precise in our thinking about memory, as biologists have been about life.

No one in his or her right mind would pose questions about neural correlates of, say, perception. The question always is, neural correlates of what kind of perception, and then, what aspects of the chosen kind. But in memory, it is much more tempting to do just the wrong thing, assume that memory is memory is memory, and ask for its molecular or cellular underpinnings, or neuroanatomical correlates. Now, your textbook does not commit this error, but I would not be surprised to find that even you make general references to "memory" rather than its subdivisions.

In neuroscience, one of the most difficult tasks is to build satisfactory explanatory bridges from one level of

---

of newly learned items on previously learned ones as a function of their temporal proximity.

How do we know that long-term memories must be consolidated over time? One line of evidence for temporal consolidation comes from patients who have undergone electroconvulsive therapy to treat psychological disorders. Electroconvulsive therapy involves passing an electrical current through the brain by electrodes placed on the scalp, a useful treatment for conditions such as severe depression. In patients so treated, a retrograde amnesia is more likely to affect items that were learned close to the time of the treatment (Figure 8.15). A similar pattern is observed with severe head trauma that leads to closed-head injury. Retrograde amnesia is more likely for recent events, and even as the amnesia

Figure 8.15 Effects of electroconvulsive therapy (ECT) on memory performance. Memory apparently changes for a long time after initial learning, with some material being forgotten and the material that remains becoming more resistant to disruption. Adapted from Squire and Slater (1983).

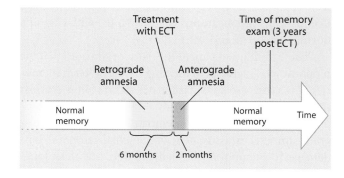

analysis to another, from molecules to cells, or from brain activations to phenomenal experience. But the daunting task becomes even more so if one is not sure about the units, or features, or subdivisions that one is trying to bring into a meaningful relation with one another at different levels. Cognitive psychologists, of whom I am one, ought to know how to analyze things such as perception, or attention, or memory, or thought, into their "component processes." We have made definite progress on that front over the last 30 or 40 years or so, in many cases decisively reaching beyond folk psychology. But now that we have powerful new techniques to connect the mind to the brain, and vice versa, we realize how puny our efforts actually have turned out to be. Yes, we can talk about memory systems and memory processes, and we can name them, but we have little idea how "real" these systems and processes are. And we do not spend enough systematic effort in discussing these issues.

The point I am trying to make is that what you refer to as "merely describing the simple observations about a phenomenon" is the crux of the matter. Unless we get that description right, we can forget about everything else, including any hope of elucidating the mechanisms of any phenomena, because we may end up elucidating the mechanisms of something that exists only in our minds but not in nature, that is, not in the evolved brain/mind. Describing phenomena of memory is a much harder problem than is usually realized, but we must address it and thus clear the bush for building the next stretch of the road.

Authors: In building the next stretch, what kind of equipment would you invest in? One might think enough has been learned already about how focal brain lesions produce one or another memory deficit. Now we need to understand why. Shouldn't we abandon behavioral measures and move on to physiology?

ET: I would not give up on brain lesions any more than I would give up on any existing technique. And I would certainly not abandon behavioral measures, because it is these measures, or their eventual future equivalents, that we need to understand in terms of what the brain does. We need every ruse and device and piece of equipment we can think of as possibly relevant to our mission of understanding how the mind arises from, and influences, the brain. The important thing is to use all the techniques intelligently and wisely.

Those of us who have access to the wonderful new toys—PET and fMRI and MEG—are placing their bets on functional neuroimaging techniques, all of which are promising and some of which are downright exciting. They do have shortcomings, of course, but there is every reason to believe that improvements are on their way. And they have opened windows on the brain/mind that were unimaginable only a generation ago.

The good news for all researchers, however, is that the single most critical piece of equipment is still the researcher's own brain. All the equipment in the world will not help us if we do not know how to use it properly, which requires more than merely knowing how to operate it. Aristotle would not necessarily have been more profound had he owned a laptop and known how to program. What is badly needed now, with all those scanners whirring away, is an understanding of exactly what we are observing, and seeing, and measuring, and wondering about. Thinking about scanning memory is much harder than scanning memory.

remits over time, the most recent events are affected longest, and sometimes permanently.

From a cognitive neuroscience perspective, consolidation is conceived of as biological changes that underlie the long-term retention of learned information, and we can ask what brain structures and systems support this process. Because damage to the medial temporal lobe does not wipe out most of the declarative memories formed over a lifetime, we know that the hippocampus is not the repository of stored knowledge. Rather, the medial temporal lobe appears to support the process of forming new memories; that is, the hippocampal region is critical for the consolidation of information in long-term memory. The strongest evidence that the hippocampus is involved in consolidation comes from the fact that amnesics have retrograde amnesia for memories from one to a few years prior to the damage to the medial temporal lobe or diencephalon, a pattern that does not support a storage role but rather a role in consolidation.

What might consolidation entail at the neural level? One idea is that consolidation strengthens the associations between multiple stimulus inputs and activations of previously stored information. The hippocampus is hypothesized to coordinate this strengthening, but the effects are believed to take place in the neocortex. The idea is that once consolidation is complete, the hippocampus is no longer required for storage or retrieval. Nonetheless, keep in mind that although the memories are stored in the neocortex, the hippocampus is crucial for consolidation.

Alcoholic Korsakoff's Syndrome and Diencephalic Amnesia The medial temporal lobe is not the only area of interest in human memory. Amnesia emerges from brain damage in other regions too. For example, damage to midline structures of the diencephalon of the brain causes amnesia. The prime structures are the dorsomedial nucleus of the thalamus and the mammillary bodies (Figure 8.16). Damage to these midline subcortical regions can be caused by stroke, tumors, and metabolic problems like those brought on by chronic alcoholism as well as by trauma.

In the last half of the nineteenth century, the Russian psychiatrist Sergei Korsakoff reported an anterograde and retrograde amnesia associated with alcoholism. Long-term alcohol abuse can lead to vitamin deficiencies that cause brain damage. Patients suffering from alcoholic Korsakoff's syndrome have degeneration in the diencephalon, especially the dorsomedial nucleus of the thalamus and the mammillary bodies. It remains unclear whether the dorsomedial thalamic nucleus, the mammillary bodies, or both are necessary for the patients' amnesia. Nonetheless, damage to the diencephalon can produce amnesia. The symptoms of amnesia resulting from Korsakoff's syndrome resemble those that accompany damage to the medial temporal lobe. Both types of amnesics lose declarative memory and preserve nondeclarative memory. Since patients with Korsakoff's syndrome have no damage to the medial temporal lobe, this region cannot be solely responsible for forming declarative memory. Rather, some researchers argue that both regions (medial temporal lobe and midline diencephalon) participate in this memory formation, and that we should consider them part of one system for acquiring new declarative memories; this view is not shared by all.

ANTERIOR AND LATERAL TEMPORAL LOBES AND MEMORY

If, as suggested earlier, the neocortex is crucial for the storage of memories, then it should be possible to demonstrate retrograde amnesia with cortical damage, even though most amnesias are anterograde. In line with this proposal, amnesia can be caused by damage to regions of the neocortex.

One region of special interest is the temporal neocortex outside the medial temporal lobe. Lesions that damage the lateral cortex of the anterior temporal lobe, near the anterior pole, lead to a dense amnesia that includes severe *retrograde* amnesia; in such cases the entorhinal cortex and perihippocampal cortex may be involved. The retrograde amnesia may be severe, extending back many decades before the amnesia-inducing event occurred or encompassing the patient's entire life. Various forms of damage can lead to this condition. Progressive neurological diseases like Alzheimer's, and herpes simplex encephalitis involving viral infection of the brain are two such conditions.

Some patients with dense retrograde amnesia might still form new long-term memories. This type of amnesia is called *isolated* retrograde amnesia. It is particularly related to damage of the anterior temporal lobe. This portion of the temporal lobe is not, therefore, essential for acquiring new information.

Are these lateral and anterior regions of the temporal lobe the sites of storage of long-term declarative memories? The answer is maybe, but another view is that these regions may be particularly important for the retrieval of information from long-term stores. Where then are memories stored? More recent evidence from neuroimaging studies suggests that memories are stored as distributed representations throughout neocortex (see Connectionist Modeling of Memory), involving the regions that originally encoded the perceptual information and regions representing information that was associated with this incoming information (as noted in the last section, the medial temporal lobe may coordinate the consolidation of this information over time). Why do we not believe that the lateral and anterior areas of the temporal lobe are where the brain stores all long-term memories? If these sites were the sole repository of all declarative memories, we would not find patients who lost all previously acquired memories but still acquired new ones; such patients have been identified.

Figure 8.16 A diagrammatic coronal section of the brain at the level of the anterior nucleus of the thalamus, mammillo-thalamic tract, and the rostral portion of the crus cerebri (see Chapter 3). The dorsomedial nucleus of the thalamus and mammillary bodies are damaged in Korsakoff's syndrome.

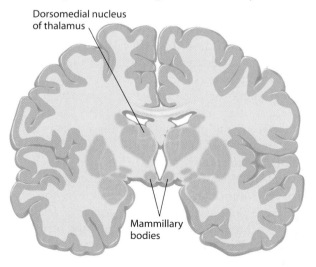

Dorsomedial nucleus
of thalamus

Mammillary
bodies

THE COGNITIVE NEUROSCIENTIST'S TOOLKIT

Connectionist Modeling of Memory

How are memories stored in the brain? We do not really know the answer to this question, but there are a lot of ideas. Some models consider memories to be stored as items—once created, they are filed in neat order somewhere in the brain. Other models propose that networks consisting of discrete nodes are formed. The nodes are interconnected via associative links that relate information together for storage, the basic organizing principles being that symbolic nodes and associated nodes become more tightly interconnected during learning. In contrast, connectionist models hold that memories are stored as changes in the instructions that neurons send to one another. New incoming information induces a specific pattern of activity over a population of neurons, and this pattern is the representation of that information. Such models, then, embody the concept of a **distributed representation,** and bear resemblance to concepts such as sparse coding, which has been observed in sensory systems and in the hippocampus in neuronal recordings in animals.

Connectionist models also rely on concepts such as hebbian learning (see text), where the interconnections between units in the population can change their strength (weight) to reflect the changing patterns of instructions that the units send to one another. The changes in these weights occur as learning occurs, and connectionist models use training algorithms such as "back propagation" to accomplish this learning and adjustments of the weights in the network. The figure demonstrates a simple model in which the visual inputs and olfactory inputs from a rose can be stored in the same network by the pattern of weights between the units. The four-by-four matrixes at the bottom show the pattern of weights from the visual inputs (left matrix) summed with those from the olfactory inputs (middle matrix) to yield the final matrix, which reflects the stored representation of the multimodal features of the rose. The key point is that the representation of the visual and olfactory features need not be stored separately but can be represented across the same network of units. Hence, it is the pattern of activity across the same units that accounts for the storage of visual or olfactory information, not activity in distinct groups of units (neurons).

Connectionist models vary from very simple ones, like that shown in the figure, to complex ones that are built from detailed knowledge of the neuronal connectivity from well-studied neural systems. These models, like all models, are formalizations of concepts that were derived from experimental work. They provide a powerful method for testing cognitive and biological theories because they can be implemented in computer simulations. In addition, such models might reveal new ways of thinking about old data, by demonstrating the consequences of one or another model through direct tests in the simulations and by making explicit predictions that can be tested experimentally.

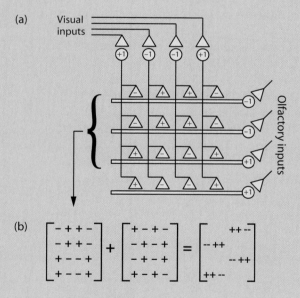

(a) Associative network that demonstrates how information from two sources can be stored as different sets of weights across the same population of neurons within the network. **(b)** The left matrix represents the weights in the network induced by the visual inputs (i.e., those shown in (a)), and the middle matrix represents the olfactory inputs. These can be summed to yield the right matrix, representing the network of combined visual and olfactory inputs. Note the middle matrix is not represented in connections in (a). In both (a) and (b), + = +.25 and – = –.25, except where noted (i.e., +1 or –1); ++ = +.50 and – – = –.50 in the summed matrix at lower right. Adapted from McClelland (2000).

AMNESICS CAN LEARN NEW INFORMATION

The insights derived from H.M. and other amnesic patients with medial temporal lobe or diencephalic damage indicate that a core problem with this form of amnesia is the inability to learn and retain new information. That is, as noted previously, medial temporal lobe damage generally is associated with the inability to learn new information, as opposed to losses of memories obtained prior to the damage (except for a time-limited retrograde amnesia). The inability to learn new information often was considered to be equally true for episodic and semantic information; we will return to consider procedural learning shortly. But is it true that damage to the medial temporal lobe (and diencephalic) memory systems affects both episodic and semantic memory? Let's consider whether new learning of semantic information (world knowledge) can take place in amnesia patients even though the ability to form new episodic memories is clearly lost.

H.M. had severe anterograde amnesia; nonetheless, it turns out that he was still able to learn some verbal information. Indeed, amnesics can retain new information, albeit to varying extents. Amnesics' learning of new information includes semantic knowledge, such as details related to other people, target words in sentences, new vocabulary for technical terms, and novel, one-word descriptions of ambiguous circumstances. This acquisition of new knowledge typically happens in amnesics who do not remember the source of the knowledge, however; that is, they do not know about the episode in which the information was learned, a condition sometimes termed *source amnesia* (we will return to this in Chapter 12). Endel Tulving and his colleagues (1991) examined this phenomenon in one very interesting amnesic patient, K.C.

K.C. developed amnesia following a motorcycle accident at the age of 30. He suffered severe head injury, and a subdural hematoma (a pool of blood under the sheath covering his brain) was surgically removed. Following a long convalescence and rehabilitation, K.C. exhibited several residual neurological complications. Neuroimaging revealed damage in several brain areas, including the medial temporal lobe and the frontal, parietal, and occipital cortices (Figure 8.17a). The damage was greater in the left hemisphere. Nonetheless, neuropsychological testing revealed normal intelligence (IQ was 94 on the revised version of the Wechsler Adult Intelligence Scale), although tests of frontal lobe function (see Chapter 12) and verbal fluency showed some deficits. He had normal short-term memory, remembering eight or nine digits in memory span tests. But, as is typical with amnesics, K.C.'s score on the Wechsler Memory Scale was below normal for delayed recall of information. Of particular note is that K.C. had severe anterograde *and* retrograde amnesia. Curiously, his retrograde amnesia involved episodic information from his entire life! He did not remember a single event in his personal history, although he had factual knowledge of things that pertained to his life, for example, that he went to high school at a particular place, that he had certain talents, that his family consisted of a particular number of people and who they were, and so on.

Despite K.C.'s dense source amnesia, he also had an intact semantic memory for general world knowledge acquired prior to his accident. K.C. had not forgotten how to use implements common to daily life or knowledge about the structure of the world around him. He was able to learn new information but could not remember how he acquired it. For example, in tests of K.C.'s memory skills, he was presented with three-word sentences together with a related picture (Figure 8.17b). In each sentence the final words were the critical ones that he was to be tested on later.

During the subsequent test he was presented with word fragments (perceptual cues) or conceptually based cues consisting of the two other words of the sentences or the associated picture. These latter cues were conceptual because they did not provide any perceptual information about the target word to be remembered, that is, any of its letters, or length, and so on. K.C. demonstrated perceptual priming effects with the word fragments that lasted up to 12 months; *priming* here means that K.C. was better at generating words from fragments if he had seen the word previously, in comparison to words he had not seen previously. This is a key pattern of results because it demonstrates that a form of nondeclarative memory—perceptual priming (see Figure 8.9)—was intact in K.C. even though he had no episodic memory of having seen the words before in the test setting.

Remarkably, K.C. also displayed an ability to learn new facts (semantic knowledge). He showed a greater likelihood of generating the correct target word when he saw the associated words or picture. Although it took K.C. longer to learn semantic information than it did a control person without amnesia, once it was learned, forgetting it occurred at the same rate in K.C. as it did in control subjects—and, again, K.C. had no inkling of how this information was acquired. Thus, some amnesics can learn semantic information, even though it might be more difficult to acquire and even when such learning is not possible for episodic knowledge. These sorts of findings demonstrate that episodic and semantic memory may not be subserved by the same underlying memory system.

Psychologists Andrew Yonelinas and Neal Kroll and their colleagues (1998) at the University of California at Davis believe this ability for some amnesics to learn new information can be conceptualized as a preservation of their ability to *recognize* prior events based on a kind of

(a)

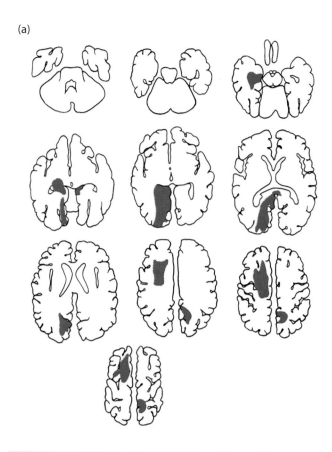

Figure 8.17 **(a)** K.C.'s lesions are represented on outline drawings of horizontal sections of his brain traced from computed tomography (CT) scans. The red areas indicate the sites of lesions due to trauma. **(b)** To test K.C.'s memory skills, Tulving and colleagues presented him with three-word sentences together with a related picture. In each sentence, the final word was the critical word tested for later. During subsequent test sessions, K.C. was presented with word fragments (a few letters forming a perceptual cue) or conceptually based cues consisting of the other two words of the sentences or the associated picture. K.C. showed priming effects (improved performance) with the word fragments for words he had seen previously, and these priming effects remained 12 months later. Adapted from Tulving et al. (1991)

strength signal. In their studies of amnesics who had damage in the left medial temporal lobe following stroke, they quantified the patients' ability to recognize words they had heard previously. A strength score was assigned that indicated how well the amnesics could distinguish a true signal (the words presented to them an hour before) from noise (distractor words not presented previously). The amnesics reliably indicated that they had heard a word before, even though they had no episodic memory for its prior presentation (Figure 8.18). Similar approaches are difficult to apply in persons with normal memory because the memory for items is aided by their intact episodic knowledge of the prior events.

Procedural Learning in Amnesia Studies of amnesia have revealed that fundamental distinctions between long- and short-term memory have biological correlates,

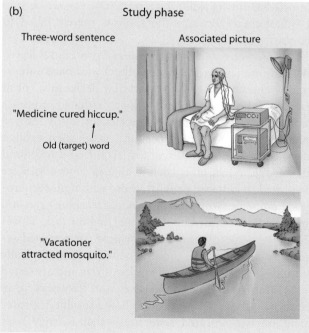

(b)

Study phase

Three-word sentence | Associated picture

"Medicine cured hiccup."
↑
Old (target) word

"Vacationer attracted mosquito."

Test phase (12 months later)

Perceptual cue: e.g., _ i c _ _ p

Conceptual cues: e.g., "Medicine cured _____"

Figure 8.18 Investigators read lists of words to amnesic patients. Then 1 hour later, the words were read again together with new words that were not in the previous reading. The patients were asked to indicate their confidence that the words had been presented previously. Their answers were used to compute quantitative metrics of their memory ability using signal-detection theory, a method borrowed from perceptual sciences and adapted to studies of memory.

"Sheep"

Tell us whether or not you think this was presented before.

Well, that seems very familiar, it is likely that I heard this from you before.

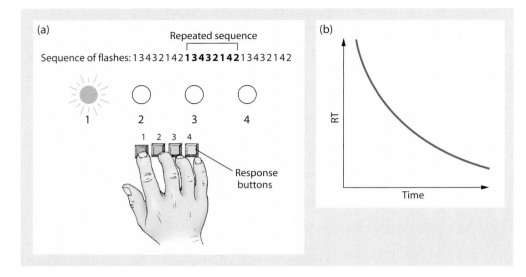

Figure 8.19 **(a)** Sequence learning paradigm. Using their fingers, subjects are asked to push buttons corresponding to flashes of lights in a complex sequence that is repeated. **(b)** Over time, subjects' reaction time (RT) to the repeating sequence becomes faster as compared to a sequence that is totally random, although they apparently have no knowledge that any pattern exists.

and that episodic and semantic memory can be distinguished from one another by their relative sensitivity to brain damage. However, learning with amnesia is not limited to semantic knowledge. Amnesics learn other nondeclarative information such as that for procedures.

One test of procedural learning is the "serial reaction time task." Subjects press buttons that correspond to locations of stimuli in front of them. For example, in such a task the subject is presented with four buttons located under each of the fingers on one hand. Each button corresponds to one of four lights—the mapping between button and light can be based on their spatial relationships (i.e., the left light maps to the left button). The task given the subjects is to press the button with the finger that corresponds to the light that is illuminated (Figure 8.19a). The lights can be flashed in different sequences: A totally random sequence can be flashed, or a pseudorandom sequence might be presented in which the lights seem to be flashing randomly to the participant, but in reality they are flashing in a complex repetitive sequence.

Over time, in normal subjects the speed of response to the repeating sequence, as compared with a totally random sequence, decreases. Thus, learning the sequence can be characterized by improved performance. The key point, however, is that when the subjects are questioned about whether or not the sequences were random, they report that they were completely random, and express no knowledge that any pattern existed (if the sequence is constructed properly, of course). Yet they learn the skill, which is procedural learning requiring no explicit knowledge about what was learned. This kind of evidence has been used to argue for the distinction between declarative and procedural knowledge, because subjects appear to acquire one (procedural knowledge) in the absence of the other (declarative knowledge).

Some investigators have challenged the idea that normal subjects learn in the absence of any explicit knowledge of what was learned. For example, sometimes the investigators ask normal volunteers about the sequences and find that in fact they can describe explicitly the learned material. So, perhaps the subjects who deny any such knowledge have low confidence and therefore do not state the declarative memory they have about the repeating sequence. Given this possibility in normal subjects, if we do not find evidence for explicit knowledge during skill acquisition, how can we be sure it is not there? Perhaps the subjects merely failed to demonstrate it.

An answer comes from studies of procedural learning in amnesics like H.M. and others who have anterograde amnesia and cannot form new declarative (or at least episodic) memories. The logic is that if amnesics cannot form new episodic memories for newly learned material but can learn procedural tasks, then explicit knowledge is not required for acquiring procedural knowledge. When tasks such as that in Figure 8.19 are presented to amnesic patients, those with dense anterograde amnesia (with loss of episodic learning) can still improve their performance by repeating the task. This is shown as a speeding of reaction time for repeated sequences compared to random ones over time. Typically amnesic patients practice the task one day and are tested on separate days. Even though they state they have never performed the task before, they show a similar improvement on the repeated sequences—that is, they have learned the procedure. Therefore, procedural learning can proceed independently of the brain systems required for episodic memory.

This type of dissociation of declarative and nondeclarative learning and memory for procedural learning also happens in patients with diencephalic amnesia such as Korsakoff's syndrome. As well, normal subjects taking drugs like benzodiazepines (tranquilizers) and scopolamine, an anticholinergic agent that acts on the muscarinic acetylcholine receptor (used to prevent motion sickness), also show this dissociation. These drugs cause deficits in explicit memory but do not affect procedural learning.

What brain systems support procedural memory? For motor skill learning, the basal ganglia have been implicated. When patients with disorders of the basal ganglia or inputs to these subcortical structures are tested on a variety of procedural learning tasks, they perform poorly. These disorders include Parkinson's disease, in which cell death in the substantia nigra disrupts dopaminergic projections into the basal ganglia, and Huntington's disease, in which neurons in the basal ganglia degenerate. Affected patients, who are not amnesic per se, have impairments in their ability to acquire and retain motor skills as assessed by a variety of tests involving motor skill learning.

Learning of Conceptual Information   Nondeclarative memory encompasses more than procedural learning and processes like priming. The learning of more abstract knowledge, such as artificial grammar, also exemplifies a nondeclarative (implicit) form of learning and memory.

Artificial grammar involves rules that govern how groups of letter strings can be formed. In artificial grammar tasks, subjects see groups of letter strings that follow the grammatical rules of an artificial system devised by the investigators, and about which the subjects have no knowledge. After studying the strings, the subjects are presented new letter strings and asked to judge whether the new strings follow the same rules as the original strings. Amazingly, healthy subjects demonstrate above-chance performance even when they report that they cannot describe the rule system. The argument is that they have implicitly learned the artificial grammar.

As with implicit learning of repeated motor sequences, a lingering concern about acquiring artificial grammar is that the normal subjects may have deduced the rule and have explicit knowledge about it but are not confident in reporting it. This possibility can be ruled out in amnesic patients because those with dense anterograde amnesia affecting episodic memories also can learn the rules of artificial grammars. Again, because amnesics with damaged medial temporal lobe–diencephalic structures have no ability to acquire new episodic information but still acquire implicit knowledge, there is a clear dichotomy between declarative memory and the nondeclarative memory that supports the learning of artificial grammar.

DEFICITS IN FORMING NEW NONDECLARATIVE MEMORIES

Amnesic patients provide evidence for a dissociation between declarative and nondeclarative memory. One strong interpretation of this evidence is that the dissociations in what amnesics can learn indicate that distinct neural systems mediate different forms of memory.

One alternative interpretation, however, has to do with the differential demands that different types of knowledge place on a single cognitive-neural memory system. If such were true, a single memory system might subserve declarative and nondeclarative memory, with deficits being more obvious for the more demanding type of memory. Here the argument would be that declarative memory is more cognitively demanding and thus is first affected by partial damage to a hypothesized unitary memory system. An analogy might be useful in conveying the point. Take locomotion, and consider the effects of a knee injury on walking versus running. One would not conclude that walking and running were subserved by different anatomical systems merely because damage to the knee caused severe impairments in running but left walking relatively intact. Rather, walking and running rely on the same anatomical systems, but running is more demanding on the now weakened locomotor system.

As introduced in Chapter 4, double dissociations are the strongest evidence for two separate systems. Patients like H.M. demonstrate only a single dissociation—preserved nondeclarative learning and memory but damaged declarative memory. To provide a double dissociation, it would be necessary to demonstrate that it is possible to have a loss in nondeclarative memory while preserving declarative memory. One form of nondeclarative memory that we defined earlier is the perceptual representation system (PRS) that is involved in perceptual priming. For example, in the studies of patient K.C. (see Figure 8.17b), word fragment completion was used as a test for perceptual priming, a nondeclarative form of learning and memory.

The Perceptual Representation System   One of the most exciting areas of memory research in the past decade or so has been the area of *implicit learning and memory,* a term Daniel Schacter (1987) coined. In most formulations, such as that in Figure 8.9, implicit memory and nondeclarative memory are essentially the same, and for simplicity, we will consider them so here. In this section, we will focus on the PRS and perceptual priming studies.

Priming tasks have been used extensively to demonstrate information storage in the PRS. *Priming* is the improvement in identifying or processing a stimulus as the result of its having been observed previously. For example, subjects can be presented with lists of words and then in a test phase, their memory of them can be evaluated using either an explicit or an implicit test. In explicit memory tests, subjects consciously attempt to recognize previously seen material. To achieve this, subjects are presented with a second word list (after a delay of hours or days) composed of new words plus ones in the previous list (old words). The subjects are asked to identify words that they had been shown in the initial presentation. In contrast, an implicit memory test might use a word fragment completion task. Here, during the later test phase, subjects view only some letters; for example, t_ou_h_s for *thoughts*. The fragments can be from either new words or old words. The task is to complete the fragment, as in the studies with K.C. Subjects are not told that some words might have been in the initial list; thus, there is no attempt to recall whether the words were part of the initial list, and anyway it is not relevant to the task.

These priming experiments found that in fragment completion tests (implicit memory tests), subjects are significantly better at correctly completing fragments for words presented in the initial list (Figure 8.20). Subjects benefit from having seen the words before, even if they are not told that the words were in the previous list! Priming for fragment completion does not lessen over time (hours versus days). In contrast, performance in the recognition task (explicit test) does decrease over time. The evidence we have that priming effects do not depend on explicit memory is simple: There is not a relation between the ability to recognize old words (e.g., "Did we show you this word before?") and the ability to complete fragments of those same words. However, are they independent in all ways?

Information is remembered better on recall or recognition tests if it is processed more deeply during encoding, as described earlier in the chapter. However, this was found not to be true for tests of implicit memory performance such as fragment completion. Subjects are *not* more likely to complete a fragment of a word if it was viewed previously under conditions requiring deep versus shallow encoding. Thus, explicit memory is independent from implicit memory regardless of how this is assessed.

When subjects complete fragments of words and perform better with words previously viewed even without explicitly recognizing them, does this reflect anything more than priming of the perceptual features of the written word? That is, is there implicit learning of the word's meaning or not? The answers to these questions can be determined in different ways. One method is to present the initial words auditorily and then perform the implicit tests visually. When this is done, priming from auditory to visual is reduced for the implicit tests, which suggests that perceptual information drives the priming phenomenon. Implicit memory in priming reflects a PRS that subserves structural, visual, and auditory word form representations.

Figure 8.20 A typical word-priming study. Subjects are shown lists of words and then wait varying amounts of time to perform a fragment completion task. Tulving and Schacter (1990) showed that subjects are better at completing fragments of words that were previously shown to them.

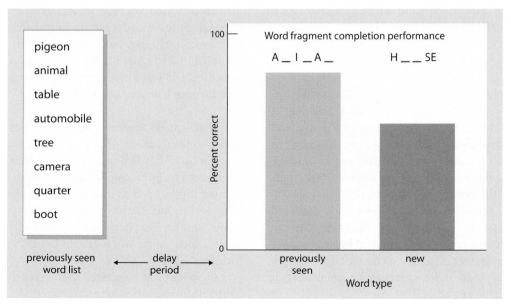

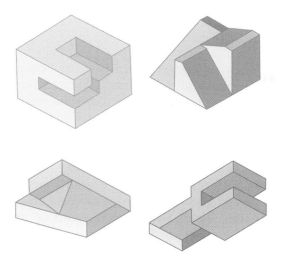

Figure 8.21 Possible and impossible objects used in experiments on object priming. During tests, the subjects were asked to make decisions about whether the stimuli were possible (top figures) or impossible objects (bottom figures). Schacter, Cooper, and Delaney (1990) found that subjects were better making the decision for possible but not impossible objects that were viewed previously if they had been instructed to pay attention to the global form of the stimulus at the time of encoding.

Additional evidence that priming involves a PRS comes from studies of nonverbal stimuli like pictures, shapes, and faces. In one study (Schachter et al., 1990), subjects viewed drawings of objects that were either possible forms or impossible forms (Figure 8.21). Impossible forms are those that cannot really exist in three-dimensional space, as in M.C. Escher's drawings of

Figure 8.22 Drawing by artist M.C. Escher that shows water circulating endlessly through waterwheels and troughs by apparently flowing downhill.

impossible scenes (Figure 8.22). During later tests, subjects had to decide whether stimuli were possible or impossible objects. They showed priming for object forms; that is, they were better at deciding on objects previously viewed. But this occurred only for possible objects and only when subjects were instructed to pay attention to the global form of the stimulus at the time of encoding. This effect did not increase when subjects elaborated during encoding by trying to match the object with a real-world object from memory.

In summary, the PRS apparently mediates word and nonword forms of priming, and so implicit memory of this type is not based on conceptual systems but rather is perceptual in nature. Additional evidence suggests that the PRS develops early in life and is preferentially maintained during aging. Further, it is insensitive to drug manipulations that affect declarative memory. As we have seen with patient K.C., implicit memory also can be dissociated from declarative memory in patients with amnesia. That is, K.C. shows normal implicit priming in word fragment completion tasks, while having decidedly abnormal episodic memory. But this is merely a single dissociation. Is there any evidence that brain lesions can affect the PRS system while leaving declarative memory intact?

The answer to this question is yes! John Gabrieli and his colleagues (1995) at Stanford University tested a patient, M.S., who had a right occipital lobe lesion, and they compared him to amnesics having anterograde amnesia for episodic memory. M.S. was an epileptic who had undergone surgery at the age of 14 to treat intractable seizures. The surgery removed most of areas 18 and 19 of his right occipital lobe, leaving him blind in the left visual field (Figure 8.23). M.S. demonstrated above-average intelligence and memory. Explicit tests of memory (recognition and cued recall) and implicit memory (perceptual priming) were administered to other amnesics and to M.S. The test materials were words presented briefly and then read aloud. During the implicit memory test, the words were presented and then masked with rows of X's. The duration of presentation increased from 16 msec to a time when the subject could read the word. Any reductions in duration of

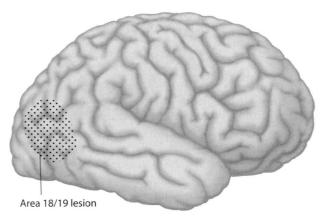

Area 18/19 lesion

Figure 8.23 A representation of the brain lesion of patient M.S. studied by Gabrieli and colleagues. Most of areas 18 and 19 (extrastriate cortex) of the right hemisphere were removed surgically to treat epilepsy. Patient M.S. showed intact declarative memory but had a deficit in perceptual priming, an implicit memory process. Adapted from Gabrieli et al. (1995).

word presentation that was required to read the word when the word was previously seen would be evidence for implicit perceptual priming. In a separate explicit recognition test, subjects saw old and new words and had to judge whether they had seen them before.

The amnesic patients displayed the expected impairments of explicit word recognition, but they did not show impairment in the implicit perceptual priming test. In contrast, M.S. had the novel pattern of normal performance on explicit recognition but impairment in the implicit perceptual priming test! This deficit was not due to his partial blindness because his explicit memory for word recognition and recall indicated that he perceived them normally by using the intact portions of his visual field.

M.S. showed a pattern opposite to that typical of amnesics like H.M. These data show that perceptual priming can be damaged in the absence of impairments in explicit memory, thereby completing a double dissociation for declarative and nondeclarative memory systems. The anatomical data indicate that perceptual priming depends on the perceptual system because M.S. had lesions to the visual cortex leading to deficits in perceptual priming. It is likely that this right occipital PRS is not specific to words but is more generally involved in priming of visual forms. This idea is supported by evidence that other patients with similar lesions show impairment in priming for nonword stimuli like fragmented pictures. The key point is that declarative and nondeclarative memory systems can be doubly dissociated by brain lesions; hence, separate areas mediate explicit and implicit learning.

Summary of Amnesia and Long-Term Memory Systems

The special cases we have reviewed have implicated specific brain regions in distinct forms of memory impairment. The learning and retention of new information about one's autobiographical history (episodic memory) require an intact medial temporal lobe (primarily hippocampus) and midline diencephalon (dorsomedial nucleus of the thalamus). Damage to these areas impedes the formation of new declarative memories (anterograde amnesia). Damage to these areas also leads to difficulties in remembering events in the years immediately prior to the injury (time-limited retrograde amnesia) but leaves intact most previous episodic and semantic memories acquired during life. Therefore, these structures are not storage sites of information in long-term memory but appear to be essential for the consolidation of new information in long-term stores. In contrast, damage to regions of the temporal lobe outside the hippocampus can produce dense retrograde amnesia, an apparent loss of episodic memories, even while the ability to acquire new ones may be intact. Patients with amnesia can learn new information, particularly nondeclarative knowledge (e.g., procedural knowledge) and in some cases new world knowledge (semantic knowledge), but those with dense anterograde amnesia do not remember the episodes during which they learned or observed the information previously. Using implicit memory tests such as fragment completion, researchers have observed impairments in nondeclarative memory in some patients with brain damage, and importantly, this has been observed in patients in whom declarative memory is intact. This pattern follows brain damage to regions outside the medial temporal lobe and midline diencephalon, and so declarative and nondeclarative memory processes probably rely on partially or wholly distinct brain systems. The PRS mediates forms of nondeclarative memory such as perceptual priming that rely on the structure and form of objects and words to create lasting memory traces which influence later perceptually based performance for previously seen or heard information.

Animal Models of Memory

Studies in monkeys with lesions to the hippocampus and surrounding cortex have been invaluable in learning about the contributions of the medial temporal lobe to primate memory systems. Hence, we will turn for a moment to a review of studies in animals aimed at elucidating the role of the medial temporal lobe in long-term memory. In general, the goal of such research is to develop animal models of human memory and amnesia. Through research, such models are providing key information on relations between specific memory and brain structures. Several animal species, ranging from invertebrates to monkeys, have been investigated for clues to human memory and its functional neuroanatomy and neurobiology; it is likely that monkeys will contribute the most directly applicable knowledge about human processes at the systems level given the similarity among primate brains. We must always keep in mind, however, that the gross organization and functional capabilities of the brains of monkeys and humans differ significantly. Thus, animal models of cognitive processes like memory are perhaps most informative when linked with studies in humans.

One of the key questions in memory research was how much the hippocampus alone, as compared with surrounding structures in the medial temporal lobe, participated in the memory deficits of patients like H.M. In other words, what structures of the medial temporal lobe system are involved in episodic memory? For example, does the amygdala influence memory deficits in amnesics (Figure 8.24)? Data from amnesics indicate that the amygdala is not part of the brain's episodic memory system, although as we will learn in Chapter 13, it has a role in emotion and emotional memories.

To verify this, surgical lesions were created in the medial temporal lobe and amygdala of monkeys, to cause memory impairment. In classic work by Mortimer

Mishkin (1978) at the National Institute of Mental Health (NIMH), the hippocampus or the amygdala, or both the hippocampus and the amygdala, of monkeys were removed surgically. He found that the amount of impairment, as measured on tests, varied according to what was lesioned.

The brain-lesioned monkeys were tested with a popular behavioral task known as the *delayed nonmatching to sample task* that Mishkin developed. A monkey is placed in a box with a retractable door in the front (Figure 8.25). When the door is closed, the monkey cannot see out. With the door closed, a food reward is placed under an object. The door is opened and the monkey is allowed to pick up the object to get the food. The door is closed again and the same object plus a new object are put in position. The *new* object now covers the food reward, and after a delay that can be varied, the door is reopened and the monkey must pick up the new object to get the food reward. If the monkey picks up the old object, there is no reward. With training, the monkey picks the new, or nonmatching, object—hence, learning and memory are measured by looking at the monkey's performance.

In his early work, Mishkin found that in the monkey, memory was impaired only if the lesion included the hippocampus and amygdala. This led to the idea that the amygdala was a key structure in memory. The idea does not fit well with data from amnesics like R.B., who had anterograde amnesia caused by a lesion restricted to CA1 neurons of the hippocampus and no damage to the amygdala. Stuart Zola, Larry Squire, and colleagues (Zola-Morgan et al.,1993) at the University of California at San Diego investigated this dilemma. They performed more selective lesions of the brains of monkeys by distinguishing between the amygdala and hippocampus, as well as the surrounding cortex near each structure. They surgically created lesions of the amygdala, the entorhinal cortex, or the surrounding neocortex of the parahippocampal gyrus and the perirhinal cortex (Brodmann's areas 35 and 36) (Figure 8.26). They wanted to extend Mishkin's work, which always involved lesions of the neocortex surrounding the amygdala or hippocampus, owing to the way the surgery was performed.

They found that lesions of the hippocampus and amygdala produced the most severe memory deficits only when the cortex surrounding these regions was also lesioned. When lesions of the hippocampus and amygdala were made but the surrounding cortex was spared, the presence or absence of the amygdala lesion did not affect the monkey's memory. The amygdala, then, could not be part of the system that supported the acquisition of long-term memory.

Figure 8.24 Drawing of the human right hemisphere from a medial view. The medial temporal lobe structures including the hippocampus and amygdala are shown in red.

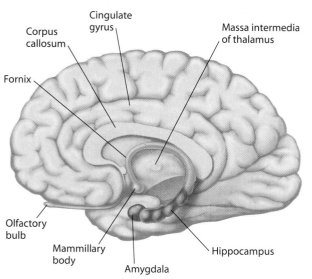

Cingulate gyrus

Corpus callosum

Massa intermedia of thalamus

Fornix

Olfactory bulb

Mammillary body

Amygdala

Hippocampus

(a)

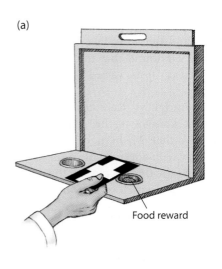

Food reward

(b)

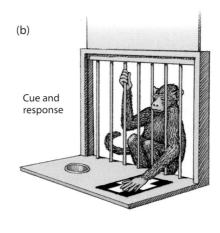

Cue and response

(c)

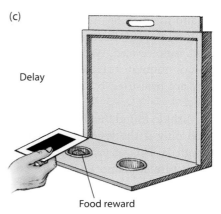

Delay

Food reward

(d)

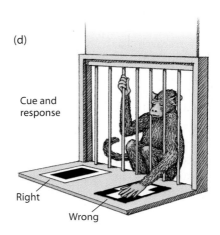

Cue and response

Right

Wrong

Figure 8.25 A monkey is presented with a memory task. **(a)** The correct response has a food reward located under it. **(b)** The monkey is shown the correct response, which will result in the monkey getting the reward. **(c)** The screen is drawn and the reward is placed under the additional response option. **(d)** The monkey is then shown two options and is to pick the correct response (the nonmatch to the original sample item) to get the reward. Here the monkey is shown making an error.

(a)

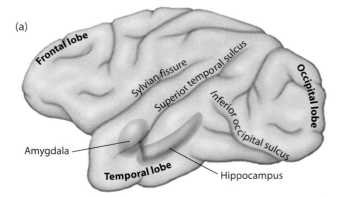

Frontal lobe
Sylvian fissure
Superior temporal sulcus
Inferior occipital sulcus
Occipital lobe
Amygdala
Temporal lobe
Hippocampus

(b)

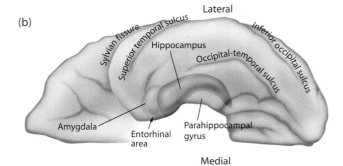

Lateral
Sylvian fissure
Superior temporal sulcus
Hippocampus
Inferior occipital sulcus
Occipital-temporal sulcus
Amygdala
Entorhinal area
Parahippocampal gyrus
Medial

Figure 8.26 Diagrammatic gross anatomy of the medial temporal lobe with the hippocampus, amygdala, and entorhinal area and the parahippocampal gyrus of the monkey. **(a)** Lateral, see-through view of the left hemisphere, with the amygdala and hippocampus seen within temporal lobe. **(b)** View from the ventral surface of the same hemisphere. Adapted from materials provided by David Amaral.

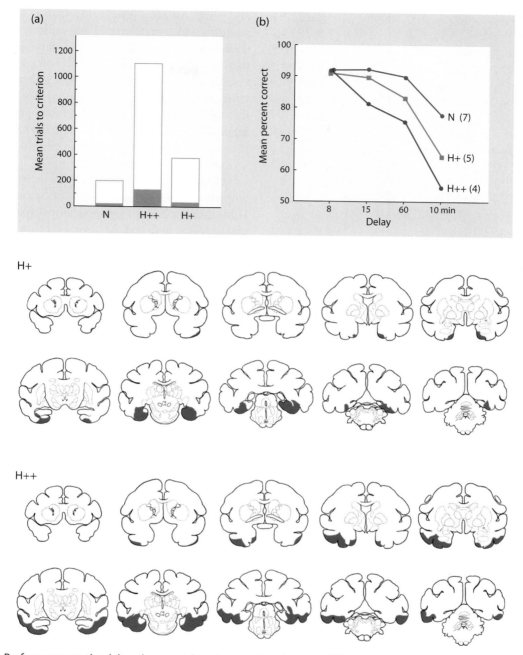

Figure 8.27 Performance on the delayed nonmatching to sample task at two different test sessions for normal monkeys (N); monkeys with lesions of the hippocampal formation, parahippocampal cortex, and perirhinal cortex (H++); and monkeys with lesions of the hippocampal formation and the parahippocampal cortex (H+). **(a)** Initial learning of the task with a delay of 8 seconds: Open bars = test 1; filled bars = test 2. **(b)** Performance across delays for the same groups. Lesions (red) in H+ **(top)** and H++ **(bottom)** are shown in coronal sections. Adapted from Zola-Morgan et al. (1993).

In subsequent investigations, Zola and his colleagues selectively created lesions of the surrounding cortex in the perirhinal, entorhinal, and parahippocampal regions. This worsened memory performance in delayed nonmatching to sample tests (Figure 8.27). Follow-up work showed that lesions of only the parahippocampal

and perirhinal cortices also produced significant memory deficits.

How does this make sense in relation to R.B.'s profound anterograde amnesia with damage limited to the hippocampus and not involving the surrounding parahippocampal or perirhinal cortex? The parahippocampal

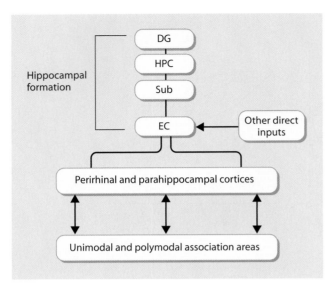

Figure 8.28 Diagrammatic flow of information between the neocortex and the hippocampal system. DG = dentate gyrus; HPC = hippocampal cortex; Sub = subiculum; EC = entorhinal cortex. From Cohen and Eichenbaum (1993)

and perirhinal areas receive information from the visual, auditory, and somatosensory association cortex and send these inputs to the hippocampus (Figure 8.28), and from there to other cortical regions (Figure 8.29). The hippocampus cannot function properly if these vital connections are damaged. But more than this, we now also know that these regions are involved in much processing themselves, and hence lesions restricted to the hippocampus do not produce as severe a form of amnesia as do lesions that include surrounding cortex.

In summary, the data from animals are highly consistent with evidence from amnesic patients such as R.B. and H.M. that implicates the hippocampal system in the medial temporal lobe and the associated cortex as critical for forming long-term memories. Lesions that damage the hippocampus directly, or damage the input-output relations of the hippocampus with the neocortex, produce severe memory impairments. The amygdala is not a crucial part of the system for episodic memory but is important for emotional memory (see Chapter 13). Moreover, the

Figure 8.29 The anatomy of the hippocampal memory system in monkeys and rats. Most areas of cortex send information to the hippocampus. Different neocortical zones (blue) project to one or more subdivision of the parahippocampal area. These subdivisions are the perirhinal cortex (purple), the parahippocampal cortex (darker purple), and the entorhinal cortex (lighter purple). These latter areas are interconnected and project to different regions of the hippocampus (green), including the dentate gyrus, CA3 and CA1 fields of the hippocampus, and the subiculum. As a result, the parahippocampal region serves as a site for the convergence of various cortical inputs. As well, the parahippocampal region passes this information from the cortex onto the hippocampus. Following processing in the hippocampus, information can be fed back via the parahippocampal region to the same areas of the cortex from which the original inputs came. Adapted from Eichenbaum (2000).

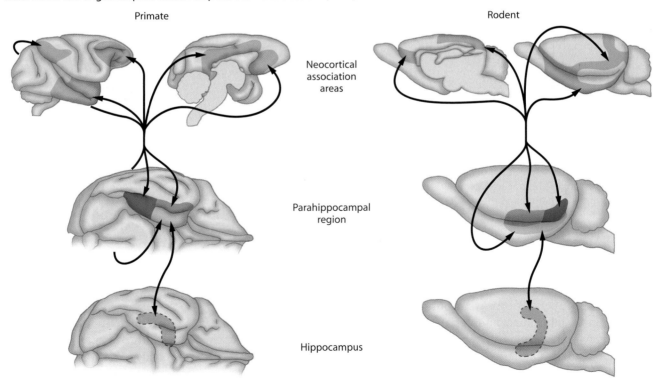

animal data match well with those from amnesics with regard to the preservation of short-term memory processes after the medial temporal lobe has been damaged; monkeys' memory deficits in the delay nonmatching to sample task became more pronounced as the interval between the sample and test increased. The medial temporal lobe, then, is not essential for short-term or working memory processes. As we noted earlier, the medial temporal lobe is not the locus of long-term storage because retrograde amnesia is not total after damage to this area; rather, the medial temporal lobe is a key component in organizing and consolidating long-term memory that is permanently stored in a distributed fashion in the neocortex.

Imaging the Human Brain and Memory

The work described so far has dealt with evidence from humans and animals with brain damage. These data are consistent with regard to the role of the medial temporal lobe in memory, and to the independence of procedural memory and perceptual priming (as well as conditioning and nonassociative learning) from the medial temporal lobe system. Over the past decade, there has been an exponential increase in studies of normal subjects using functional brain imaging methods. The results are quite provocative and are confirming and extending the findings from lesion studies. In the following, we review recent studies of the brain organization of episodic memory, semantic memory, procedural memory, and the PRS.

EPISODIC ENCODING AND RETRIEVAL

Given the purported role of the hippocampal system in encoding memory in long-term stores, researchers have eagerly addressed this issue using positron emission tomography (PET) and functional magnetic resonance imaging (fMRI). One such line of work involved face encoding and recognition. The question was whether the hippocampus becomes active when encoding new information. James Haxby, Leslie Ungerleider, and their colleagues (1996) at the NIMH presented subjects with pictures of either faces or nonsense patterns, and, using PET, investigated memory performance. In different conditions subjects were required to remember (encode) the face, recognize the face, and perceptually analyze the face by comparing two faces (Figure 8.30). During these periods, PET scans recorded changes in regional cerebral blood flow triggered by local neuronal activity.

These investigators observed that the right hippocampal region was activated during encoding of the face but not during recognition, where retrieval processes should have been engaged (Figure 8.31). These data are consistent

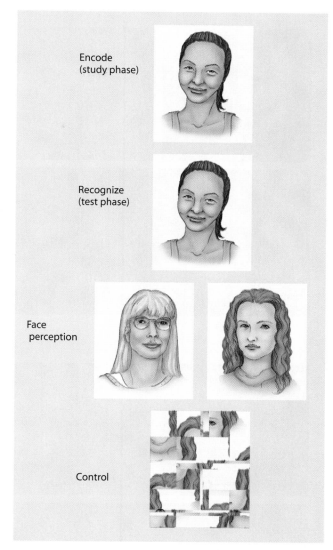

Figure 8.30 Examples of some of the face stimuli used in a PET study of encoding and retrieval of face information. During a face-encoding task, the subjects tried to memorize the faces for a later test. During a face recognition task, they were required to determine whether the face had been previously seen. During a face perception control task, they were supposed to match the two faces. Finally during a sensorimotor control task, nonsense patterns were presented and the subjects pressed a button.

with those from amnesic patients who had medial temporal lobe damage that led to anterograde amnesia but preserved distant retrograde memories. Encoding also activated the left prefrontal cortex, whereas recognition activated the right prefrontal cortex. Thus, we have more support for the hippocampus's role in memory, as well as possible support for hemispheric asymmetries in memory functions. Are hemispheric asymmetries in memory function specific to memory for faces or more general?

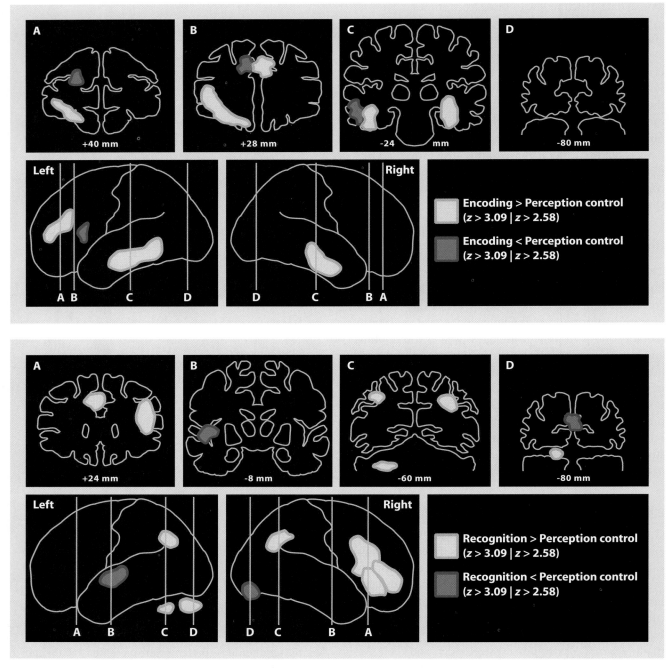

Figure 8.31 PET data show activation of the right hippocampal region and left prefrontal cortex during the encoding of face information (Panels A, B, and C, **top**). In contrast, when the face is recognized later, the activity is primarily in the right prefrontal cortex but not the left; no hippocampal activity is observed during the recognition task (Panels A, B, and C, **bottom**). Additional regions of significantly increased and decreased activity also can be observed. Adapted from Haxby et al. (1996).

Endel Tulving and his colleagues Lars Nyberg and Roberto Cabeza investigated encoding and retrieval of episodic information using PET (Nyberg et al., 1996). When subjects were asked to perform a retrieval task for episodic memories, the right, not the left, hemisphere was activated. In these PET studies, normal subjects were given phrases in a learning session on one day. The phrases were short definitions for a word that was presented at the end of the phrase (e.g., recreation for the jumpy—trampoline). These phrases were similar to ones used for new semantic learning in patient K.C.; however, normal subjects can readily remember the episode of

THE COGNITIVE NEUROSCIENTIST'S TOOLKIT

Monitoring Recollection Using Human Brain Electrophysiology

An invaluable criterion for ascertaining whether explicit and implicit processes are subserved by separate brain systems is to show that they are independent. What one wants to show is that performance on implicit tests is not correlated with explicit recollections about test items. One problem in memory research has been the difficulty of revealing how much recollective experience patients might have for previously viewed information, even though they are not required to recollect anything about that information. It would be tremendously helpful to distinguish between normal subjects who do have and those who do not have recollective experiences. Recently, Ken Paller, Marta Kutas, and their colleagues (1995) provided a means of doing this through electrical recordings (event-related potentials, or ERPs) of brain activity in healthy persons.

They used a method in which they varied the depth of encoding words by having subjects generate a mental image corresponding to the word or merely indicate whether the word contained more than one syllable. Later, the subjects were asked to make a lexical decision (is it a word?) about a list of words and nonword letter strings. The words were a mixture of those from the prior image task list, those from the prior syllable task list, and new words that they had not seen. In a second experiment, the researchers asked subjects to perform the lexical decision task and give a recognition judgment about whether they had seen the words before. In each experiment, the researchers performed separate recognition posttests to assess the extent to which subjects could recollect the words.

The behavioral measures revealed priming in the lexical decision task (previously seen words were classified as words faster than were new words). Priming was not affected by whether subjects had studied the words in the image versus the syllable condition. Recognition scores from the posttest, though, differed significantly according to the study task—subjects were better at recognizing previously seen words if at the time of the study they had generated an image corresponding to the word. Thus, the behavioral data differed in how the task affected implicit (lexical decision) versus explicit (recognition) processes. Were there any electrophysiological signs of recollection? Yes, the voltage of ERPs elicited by words was more positive 500 to 900 msec after the onset of words that had been seen previously in the image condition as compared with the ones in the syllable condition. Because this ERP mirrored the effect of the study task on recognition performance, it is a physiological sign of recollection. This is especially true since recollection was not necessary for the implicit memory test. The researchers noted that the ERPs permitted a view of human subjective experience that was not contaminated by introspective reports, which are notoriously unreliable.

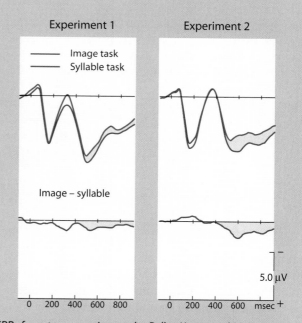

ERPs from two experiments by Paller, Kutas, and McIsaac (1995) showing responses for words that had been seen previously in one of two conditions: when the words were used to generate images (solid line) and when the subjects decided how many syllables were present in the word (dashed line). At bottom, the difference between these two conditions is plotted as the subtraction waveform (image minus syllable condition). Words that had been encoded during the image task elicited more positivity between latencies of 500 to 900 msec, and this effect was larger when on the test the subjects were required to include recognition judgments (bottom right).

False Memories and the Medial Temporal Lobes

When our memory fails, we usually forget events that happened in the past. Sometimes, however, something more surprising occurs: We remember events that *never* happened. Whereas forgetting has been a topic of research for more than a century, memory researchers did not have a good method to investigate false memories in the laboratory until Henry Roediger and Kathleen McDermott at Washington University rediscovered an old technique in 1995. The technique involves presenting subjects with a list of words (e.g., *thread, pin, eye, sewing, sharp, point, haystack, pain, injection,* etc.) in which all the words are highly associated to a word that is not presented (e.g., *needle*—did you have to go back and recheck the list?). When subjects are asked subsequently to recall or recognize the words in the list, they show a strong tendency to falsely remember the associated word that was not presented. The memory illusion is so powerful that participants are willing to claim that they vividly remember seeing the nonpresented critical word in the study list.

However, when participants are interrogated carefully about the conscious experience associated with remembering items from the list (true items) and the critical nonpresented words (false items), they tend to rate true items higher than false items in terms of sensory details (Mather et al., 1997; Norman and Schacter, 1997). This finding introduced a conundrum in false memory research: How can human participants believe in their illusory recollections, and at the same time be able to differentiate them from veridical recollections in terms of sensory detail?

Roberto Cabeza at Duke University and a team of collaborators (Cabeza et al., 2001) recently provided a possible answer to this conundrum. In their study, subjects watched a videotape segment in which two speakers alternatively presented lists of associated words. The subjects then were required to perform an old/new recognition test that included true items, false items, and unrelated new words (new items), while their brains were scanned by functional magnetic resonance imaging. Changes in blood flow in the brain that indicated changing patterns of neural activity were measured separately for each kind of item. Memory performance showed the same pattern as in previous studies: Participants were able to reject new items but showed a strong tendency to falsely recognize false items. The researchers found a dissociation between two medial temporal lobe regions, as depicted in the

learning new definitions, whereas K.C. learned them without any episodic memory of the learning experience.

On the first day of the study, 120 of the definitions were presented to normal volunteer subjects. Then on the next day, when scanning was performed, 120 new definitions were added to the 120 old ones. During some scanning sessions, new and old definitions were given in such a way that mostly old definitions were presented; during other sessions, mostly new ones were presented. Blood flow was compared for when subjects received mostly new versus mostly old definitions. The subjects' performance was better than 95% in identifying old versus new items; they correctly recognized the definitions they had been presented with previously. In conjunction with this, cerebral blood flow rose in the right dorsolateral prefrontal cortex. Some activation occurred in the left dorsolateral prefrontal cortex, but not in areas symmetrical to those in the right, and to a lesser degree. In addition, activation was found in the parietal lobe.

Encoding and retrieval processes were lateralized in the left and right hemispheres, respectively, giving rise to a model with the acronym of HERA, which stands for "hemispheric encoding-retrieval asymmetry." This model represents the idea that encoding involves the left hemisphere more than the right, and retrieval involves the right hemisphere more than the left. Both processes predominantly involve the dorsolateral prefrontal cortex. In encoding and retrieving information from long-term memory, neocortical areas were the most activated.

In a comprehensive survey of dozens of studies using PET and fMRI, Roberto Cabeza and Lars Nyberg (2000) assessed the generalizability of prefrontal, temporal, and

figure. In the hippocampus bilaterally, false items elicited more neural activity than did new items and as much activity as true items, whereas in the left parahippocampal gyrus—a region surrounding the hippocampus—false items elicited about the same amount of activity as new items and significantly less activity than true items. In other words, the hippocampus responded similarly to true and false items, while the parahippocampal gyrus responded more strongly to true than to false items. Since true and false items were similar in terms of semantic content but differed in terms of sensory content, these results suggest that the hippocampus is involved in the retrieval of semantic information, whereas the parahippocampal gyrus is involved in the retrieval of sensory information.

This dissociation provides a possible solution for the aforementioned conundrum: the memory system in the medial temporal lobes can generate two different types of messages when information is presented. Whereas anterior hippocampal activity suggests that false items are like true items, posterior parahippocampal activity suggests that false items are like new items. These two messages are not contradictory: False items are like true items in terms of their semantic properties, but they are like new items in terms of their sensory properties.

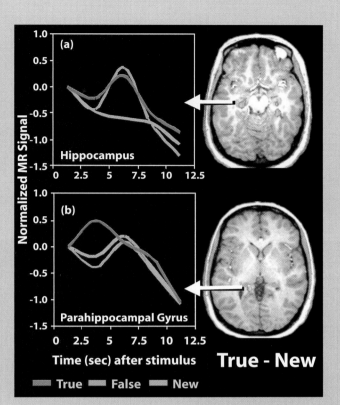

Significant increases in blood flow in the medial temporal lobes, and their corresponding hemodynamic response functions. **(a)** Bilateral hippocampal regions were more activated for true and false than for new items, with no difference between activations for true and false items. **(b)** A left posterior parahippocampal region that was more activated for true than for false and new items, with no difference between activations for false and new items.

medial temporal activity during a wide variety of tasks involving episodic encoding and retrieval. They found that across studies, the left prefrontal region and hippocampal region were active during the intentional encoding of information, whereas the right prefrontal region was activated during episodic retrieval (Figure 8.32).

Some more recent studies made use of event-related fMRI methods to track the processing of individual items as a function of the success of the encoding, as indexed by later memory performance. Anthony Wagner and his colleagues at MIT, Harvard University, and Massachusetts General Hospital (1998), and John Gabrieli and his colleagues at Stanford (Brewer et al., 1998) conducted such studies. They presented subjects with items and scanned their brains using fMRI while they were encoding the information, and then later tested them for their memory of the items. Wagner and colleagues used words, while Gabrieli and colleagues used complex color photographs. The event-related fMRI method permitted the original data gathered during encoding to be sorted and analyzed as a function of whether the items were later remembered or forgotten (as indexed by, for example, a recognition test). That is, data for the brain responses to the items that were later remembered were collapsed together, and the same was done separately for data for the responses to the items that were forgotten. What do you predict they observed?

Each research group found that event-related responses were larger in prefrontal and medial temporal regions (parahippocampal cortex) during encoding of words or pictures that were later remembered. Interestingly, though, whether the prefrontal activations were in the right or left hemisphere was material specific. Wagner

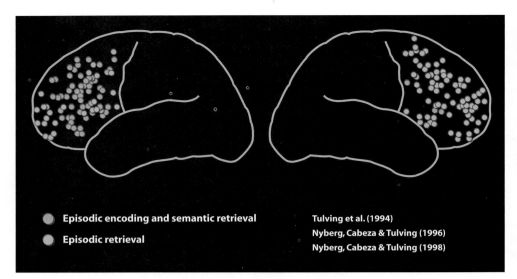

Figure 8.32 Summary of regions in the prefrontal cortex that showed activation for episodic encoding and semantic retrieval (blue), or episodic retrieval (red). The data are from many studies. Courtesy of Roberto Cabeza.

Figure 8.33 Increased activation during encoding for words that were remembered later versus those that were forgotten. Event-related activity and time course data are from various brain regions including the anterior and posterior left inferior frontal gyrus (LIFG) and the frontal operculum. Adapted from Wagner et al. (1998).

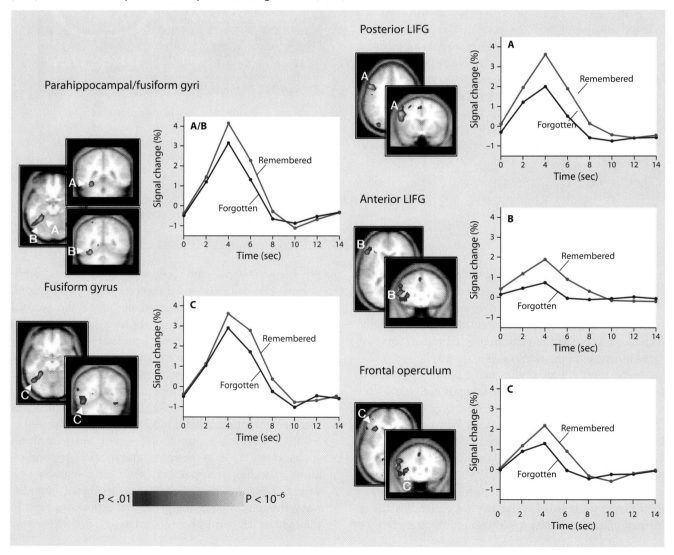

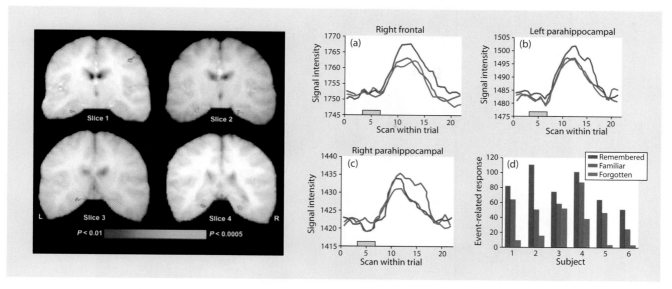

Figure 8.34 Activations and event-related (hemodynamic) time courses as determined by functional MRI during encoding of photographs that later were remembered or not remembered. Adapted from Brewer et al. (1998).

and colleagues used words and found effects primarily in the left hemisphere (Figure 8.33), whereas Gabrieli and colleagues used color photographs and found effects primarily in the right frontal cortex and bilaterally in the medial temporal lobe (Figure 8.34).

Although we have concentrated on activations in the medial temporal lobe and prefrontal cortex during memory encoding and retrieval, a few studies reported activation of the medial parietal cortex, near the precuneus, during episodic retrieval. These medial parietal effects and the right prefrontal cortical involvement in episodic retrieval are examples of new discoveries driven by functional neuroimaging studies; that is, these regions were not considered to be involved in episodic retrieval based on studies in either amnesic patients or animals.

SEMANTIC ENCODING AND RETRIEVAL

The encoding and retrieval of semantic knowledge also have been studied using functional neuroimaging, and significant new findings have been uncovered. In particular, evidence for domain-specific organization (i.e., knowledge of animate and inanimate objects is localized in different cortical regions) has proved to be a fascinating story; we will review the latest findings from this work in Chapter 9. Here, we will briefly consider semantic retrieval. As shown in Figure 8.32, unlike episodic retrieval that activates the right prefrontal cortex, semantic retrieval involves the left prefrontal cortex. The region includes Broca's area (Brodmann's area 44 extending into area 46) and the ventral lateral region (Brodmann's areas 44 and 45). This lateralization to the left hemisphere remains regardless of whether the memories being retrieved are of objects or words.

PROCEDURAL MEMORY ENCODING AND RETRIEVAL

Earlier we learned that amnesics demonstrate implicit learning of motor sequences (procedural knowledge) even when they cannot form explicit memories about the stimulus sequence (see Figure 8.19). Amnesics provide powerful evidence that implicit learning need not be mediated by explicit knowledge about the material.

Scott Grafton, Eliot Hazeltine, and one of the authors (R.B.I.) (1995) investigated the brain basis of procedural motor learning in normal subjects. They compared conditions in which the subjects learned motor sequences implicitly during dual-task conditions, which helped to prevent subjects from explicitly noticing and learning the sequence. The subjects responded to stimulus sequences by pushing a button with their right hand while monitoring a sequence of tones and counting those of lower frequency. The investigators compared this to the condition where subjects only performed the sequence learning task and thus might become explicitly aware of the sequence.

PET conducted during the dual-task condition demonstrated activation of the motor cortex and the supplementary motor area of the left hemisphere, and the putamen in the basal ganglia bilaterally. Also activated were the rostral prefrontal cortex and parietal cortex. Therefore, when subjects were implicitly learning the task, brain areas that control limb movements were activated. When the distracting auditory task was removed, the right dorsolateral prefrontal cortex, right premotor cortex, right

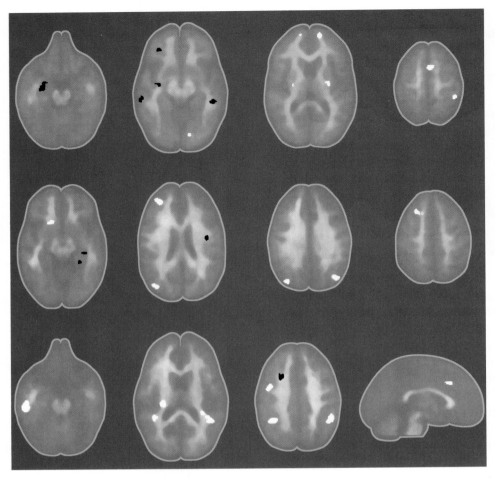

Figure 8.35 Activations corresponding to implicit learning of motor sequences in normal subjects. Scans showing the areas of activation (white areas) are overlaid onto MRI scans. Different brain areas were activated when subjects performed the task under dual-task conditions designed to emphasize implicit learning **(top row)** than when they performed the task under single-task conditions, during which the subject might develop explicit awareness **(middle row)**. The differences between the subjects who reported awareness of the sequence (N = 7) and those who did not (N = 5) in the single-task condition are shown **(bottom row)**. Right hemisphere on left. From Grafton et al. (1995).

putamen, and parieto-occipital cortex were activated bilaterally (Figure 8.35). Under these conditions, seven of the twelve subjects became aware of the sequence. The differences in activations for the seven aware and five unaware subjects are presented in Figure 8.35 (bottom). The awareness effects were correlated with activity in the right temporal lobe, left and right inferior parietal cortex, right premotor cortex, and anterior cingulate cortex.

These PET data, and convergent evidence from transcranial magnetic stimulation and animal studies, indicate that the motor cortex is critical for implicit procedural learning of movement patterns. Activations in the putamen would be expected given that patients with Huntington's disease (disease of the basal ganglia) have deficits in sequence learning tasks. Because the supplementary motor cortex was also activated during implicit learning, it may be part of a network of subcortical and cortical areas

collectively known as the *cortical-subcortical motor loop*, which regulates voluntary movements. Explicit learning and awareness of the sequences required increased activity in the right premotor cortex, the dorsolateral prefrontal cortex associated with working memory, the anterior cingulate, areas in the parietal cortex concerned with voluntary attention, and the lateral temporal cortical areas that store explicit memories. Demonstrated here are clear dissociations between brain systems involved in explicit and implicit learning and memory in healthy humans.

PERCEPTUAL PRIMING AND IMPLICIT AND EXPLICIT MEMORY

Daniel Schacter and his colleagues (1996) at Harvard University investigated the neural bases of perceptual priming (implicit learning) in a PET study. They presented the

Low recall minus baseline   High recall minus baseline

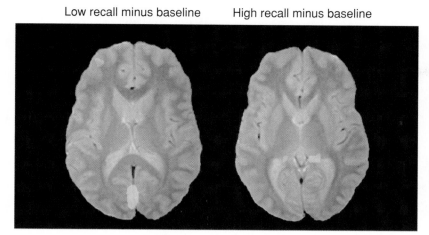

Figure 8.36 PET scans during recall of words that were previously seen under two conditions of encoding. In one condition, which produced low recall, the task at encoding was to view the words and make a perceptual judgment (detect T-junctions in the letters). In the other condition, which produced high recall, the words were shown repeatedly at encoding and semantic judgments were required. When recall was low (**left**), owing to the perceptual nature of the task, brain activity increased in the occipital cortex. But when meaning was relevant during encoding, and hence recall was high, then the hippocampal region was active. Adapted from Schacter et al. (1996).

subjects with words but attempted to eliminate the possibility that the subjects would have later explicit recollections of seeing a list of test words. Schacter and colleagues showed the subjects the words but required them to process the words only for surface perceptual features (the number of T junctions—any perpendicular lines that intersect in the shape of a T—in the letters of the word). Then they were assigned an implicit stem-completion task in which some experimental blocks had stems (the first few letters) from previously presented words and other blocks had stems of new words. The task was to say a word that completed the stem—for example, saying "house" when presented the stem "hou__." Scanning was performed only during the task. Subjects manifested implicit priming behaviorally; they completed more stems when they were from words previously seen. No activations or deactivations were noted in the hippocampus, but blood flow in the bilateral occipital cortex (area 19) decreased. The hippocampus was not activated, then, even though implicit perceptual priming was obtained.

In a separate experiment, encoding was manipulated to produce high recall during one test (words shown four times for semantic judgments) versus low recall during another test (one presentation of words and judgment about perceptual qualities—T-junction detection). The idea was that when the words were recalled, the subjects would have more explicit recollections for the list studied more deeply (i.e., the high-recall condition). Subjects did have better recall, and significant activations were now evoked bilaterally in the hippocampus (Figure 8.36).

The conclusions from this and other studies are that implicit and explicit retrieval of information is subserved by separate brain systems. Schacter's study indicated that the hippocampus is activated most when explicit retrieval includes significant experience with the material, not merely an attempt to retrieve information. Together with the face encoding data Haxby and colleagues obtained by PET (see Figure 8.31), and animal and human lesion data, a reasonable conclusion is that the hippocampus encodes new information but also retrieves *recent* information when explicit recollection is involved. Perhaps more interestingly, *deactivation* of the visual cortex for previously seen words is a correlate of perceptual priming.

In summary, neuroimaging studies have demonstrated patterns of neuronal activation that are consistent with memory systems derived from cognitive research, studies in human amnesics, and animal models. Neuroimaging also has provided some notable new findings in the cognitive neuroscience of memory, including, for example, the hemispheric asymmetries in encoding and retrieval. Functional neuroimaging clearly will continue to provide invaluable information about human memory and its neural substrates in the healthy human in the years to come.

CELLULAR BASES OF LEARNING AND MEMORY

Most models of the cellular bases of memory hold that it is the result of changes in the strength of synaptic interactions among neurons in neural networks. How would synaptic strength be altered to enable learning and memory? Hebb (1949) proposed one possibility. Hebb's law states that if a synapse is active when a postsynaptic neuron is active, the synapse will be strengthened—this is known as **Hebbian learning**.

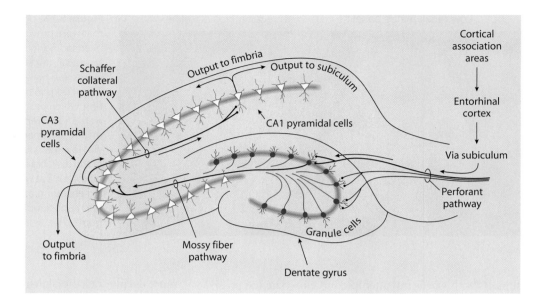

Figure 8.37 Diagram of the synaptic organization of the rat hippocampus.

Long-Term Potentiation and the Hippocampus

Because of the role of the hippocampal formation in memory, it has long been hypothesized that neurons in the hippocampus must be plastic, that is, able to change their synaptic interactions. Since the late 1960s, researchers have sought the mechanisms behind this type of plasticity in learning and memory storage. Although it is now clear that storage itself is not in the hippocampus, this was not understood when work on hippocampal cell physiology first began. This fact does not invalidate the hippocampal models we will examine, because the same cellular mechanisms can operate in various cortical and subcortical areas.

First, let's quickly review the three major excitatory neural components of the hippocampus (Figure 8.37): (1) the *perforant pathway* that forms excitatory connections between the parahippocampal cortex and the granule cells of the dentate gyrus, (2) the *mossy fibers* that connect the granule cells of the dentate gyrus to the CA3 pyramidal cells (on dendritic spines), and (3) the *Schaffer collaterals* that connect the CA3 pyramidal cells to the CA1 pyramidal cells. This system provides an opportunity for researchers to examine synaptic plasticity as the mechanism of learning at the cellular level.

In studies by Bliss and Lømo (1973), stimulation of axons of the perforant pathway of the rabbit resulted in a long-term increase in the magnitude of excitatory postsynaptic potentials (EPSPs). That is, the stimulation led to greater synaptic strength in the perforant pathway such that later stimulation created larger postsynaptic responses in the granule cells of the dentate gyrus. This phenomenon, named *long-term potentiation* (LTP) (*potentiate*

means "to strengthen or make more potent"), was later extended to the other two excitatory projection pathways of the hippocampus. The changes could last for hours in isolated slices of hippocampal tissue placed in dishes, where recording was easier. LTPs can even last for days or weeks in living animals. It has since been found that the LTPs in the three pathways vary. Nonetheless, Hebb's (1949) law is confirmed physiologically by the discovery of LTP.

One way that LTP can be recorded is by placing a stimulating electrode on the perforant pathway and a recording electrode in a granule cell of the dentate gyrus (Figure 8.38). First, a single pulse is presented, and the resulting EPSP is measured. The size of this first recording is the strength of the connection before the LTP is induced. Then the perforant pathway is stimulated with a burst of pulses; early studies used approximately 100 pulses/sec but more recent ones used as few as 5 pulses/sec. After LTP is induced, a single pulse is sent again, and the magnitude of the EPSP in the postsynaptic cell is measured. The magnitude of the EPSP grows after LTP is induced, signaling the greater strength of the synaptic effect (Figure 8.38a). A fascinating finding is that when the pulses are presented at a slow rate, the opposite effect, long-term depression (LTD), develops (Figure 8.38b).

HEBBIAN LEARNING

Associative LTP is an extension of Hebb's law and asserts that if a weak and a strong input act on a cell at the same time, the weak synapse becomes stronger. This has been tested directly by manipulating LTP in the CA1 neurons of the hippocampus. When two weak inputs (W1 and W2) and one strong input (S1) are given to the same

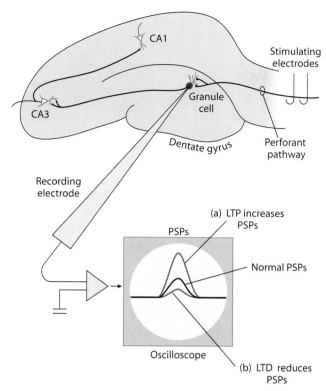

Figure 8.38 Stimulus and recording setup for the study of long-term potentiation (LTP) in perforant pathways. **(a)** The pattern of responses before and after inducing LTP is shown (microvolts). **(b)** The pattern of responses in long-term depression (LTD) is shown. PSPs = postsynaptic potentials.

cell, and when W1 and S1 are active together, W1 is strengthened whereas W2 is not. Subsequently, if W2 and S1 are active together, W1 is not affected by the LTP induced from W2 and S1. From this finding, three rules for associative LTP can be stated: More than one input must be active at the same time (cooperativity), weak inputs are potentiated when co-occurring with stronger inputs (associativity), and only the stimulated synapse shows potentiation (specificity).

For LTP to be produced, the postsynaptic cells must be depolarized in addition to receiving excitatory inputs; in fact, LTP is reduced by inhibitory inputs to postsynaptic cells. As well, when postsynaptic cells are hyperpolarized, LTP is prevented. Conversely, when postsynaptic inhibition is prevented, LTP is facilitated. If an input that is normally not strong enough to induce LTP is paired with a depolarizing current to the postsynaptic cell, LTP can be induced.

THE NMDA RECEPTOR

That an excitatory input and postsynaptic depolarization are needed to produce LTP is explained by the properties of the doubly gated N-methyl-D-aspartate (NMDA) receptor located on the dendritic spines of postsynaptic neurons that show LTP. Glutamate is the major excitatory transmitter in the hippocampus, and it can bind with NMDA and non-NMDA receptors. When 2-amino-5-phosphonopentanoate (AP5) is introduced to CA1 neurons, NMDA receptors are chemically blocked and LTP induction is prevented. But the AP5 treatment does not produce any effect on previously established LTP in these cells. Therefore, NMDA receptors are central to producing LTP but not maintaining it. It turns out that maintenance of LTP may depend on the non-NMDA receptors.

What is the cellular mechanism that permits LTP to develop via the NMDA receptors, and why does blocking them with AP5 prevent LTP? NMDA receptors are normally blocked by magnesium ions (Mg^{2+}). The Mg^{2+} ions can be ejected from the NMDA receptors only when the neurotransmitter glutamate binds to the receptors and when the membrane is depolarized; that is, the NMDA receptors are transmitter and voltage dependent (gated). When these two conditions are met, Mg^{2+} is ejected and calcium (Ca^{2+}) can enter the cell (Figure 8.39).

The effect of Ca^{2+} influx via the NMDA receptor is critical in forming LTP. Ca^{2+} acts as an intracellular messenger conveying the signal which changes enzyme activities that influence synaptic strength. Despite rapid advances in understanding the mechanisms of LTP at physiological and biochemical levels, the molecular mechanisms of synaptic strengthening in LTP are still subject to extensive debate.

The synaptic changes that create a stronger synapse after LTP induction likely include presynaptic and postsynaptic mechanisms. One hypothesis is that LTP raises the sensitivity of postsynaptic non-NMDA glutamate receptors and prompts more glutamate to be released presynaptically. Or perhaps changes in the physical characteristics of the dendritic spines transmit EPSPs more effectively to the dendrites. Finally, via a postsynaptic to presynaptic cell message, the efficiency of presynaptic neurotransmitter release is increased.

LONG-TERM POTENTIATION AND MEMORY PERFORMANCE

Having identified a candidate cellular mechanism for long-term plastic changes in synaptic strength, it should be possible to produce deficits in learning and memory by eliminating LTP. Chemically blocking LTP in the hippocampus of normal mice impairs their ability to demonstrate normal place learning; thus, blocking LTP

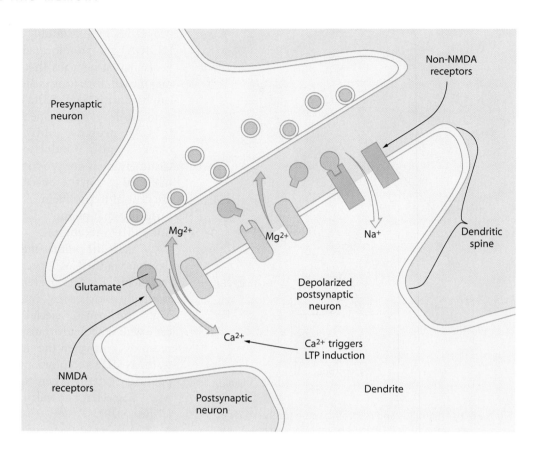

Figure 8.39 The role of Mg^{2+} and Ca^{2+} in the functioning of the NMDA receptor.

prevents normal spatial memory. In a similar way, genetic manipulations that block the cascade of molecular triggers for LTP also impair spatial learning. These experiments provide strong evidence of impairing spatial memory by blocking NMDA receptors and preventing LTP.

Two studies with different results cast doubt on the role of LTP in learning and memory, however (Bannerman et al., 1995; Saucier and Cain, 1995). Both experiments found that pharmacological NMDA receptor blockers did not stop rodents from learning how to navigate in a water maze; the animals were able to develop a new spatial map even when LTP was prevented. Unlike previous studies, these studies pretrained the rodents to swim to a platform, which prevented the impairment of new spatial learning when the NMDA receptors were blocked. When mice were pretrained by either a water maze task or a nonspatial task, the introduction of AP5 (the NMDA receptor blocker) in the hippocampus prevented new learning in mice pretrained for the nonspatial task, but it did not affect mice pretrained for the spatial task. The conclusion is that NMDA receptors may be needed to learn a spatial strategy but not to encode a new map. Another experiment also revealed that

blocking LTP did not affect behavior, but the pattern was slightly different. When mice were pretrained with a nonspatial task, spatial memory was not interrupted by the introduction of an NMDA antagonist. The conclusion is that the pretraining merely allowed the motor-related side effects of NMDA receptor blockage to be avoided. Although these two studies did not exclude the possibility that new spatial learning involves NMDA receptors, they do point to the possibility that at least two memory systems could utilize NMDA receptors. These systems participate in the water maze task but possibly could be consolidated by pretraining.

The role of LTP in memory on the cellular and behavioral levels is still being unraveled. There is great debate over whether the maintenance of LTP is located presynaptically or postsynaptically and even whether LTP is necessary for spatial memory. Two points of agreement are that LTP does exist at the cellular level and that NMDA receptors play a crucial role in LTP induction in many pathways of the brain. Because LTP is also in brain areas outside the hippocampal system, the possibility that LTP forms the basis for long-term modification within synaptic networks remains promising.

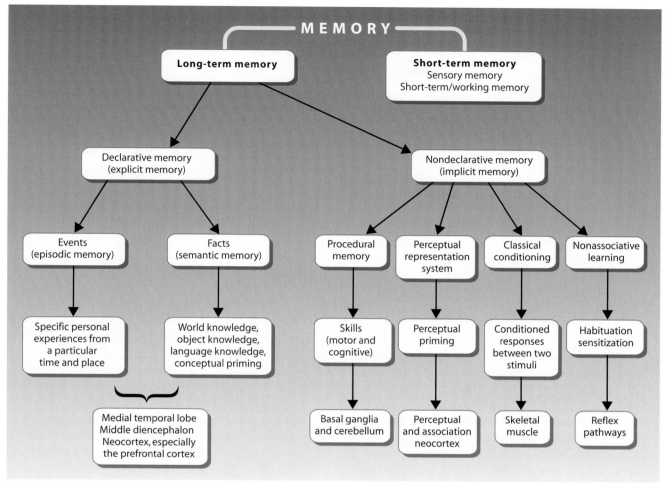

Figure 8.40 A generalized diagram of the relationships of long-term memory systems to the underlying brain systems. This figure is an elaboration on Figure 8.9 to include candidate brain areas.

SUMMARY

The ability to acquire new information and retain it over time defines learning and memory. Cognitive theory and neuroscientific evidence argue that memory is supported by multiple cognitive and neural systems. These systems support different aspects of memory, and their distinctions in quality can be readily identified. Sensory registration, perceptual representation, working memory, procedural memory, semantic memory, and episodic memory all represent systems or subsystems for learning and memory. The brain structures that support various memory processes differ, depending on the type of information to be retained and how it is encoded and retrieved.

The biological memory system includes the medial temporal lobe, which forms and consolidates new episodic and perhaps semantic memories; the prefrontal cortex, which is involved in encoding and retrieval of in-

formation; the temporal cortex, which stores episodic and semantic knowledge; and the association sensory cortices for the effects of perceptual priming. Other cortical and subcortical structures participate in learning skills and habits, especially those with implicit motor learning. The data from studies in human amnesic patients, in animals, and in normal volunteers using electrophysiological and neuroimaging methods permit us to elaborate on the cognitive model first presented in Figure 8.9, by including our best current estimates of the neural systems that support the memory functions listed, and this is shown in Figure 8.40.

The brain is not equipotential in the storage of information, and although widespread brain areas cooperate in learning and memory, the individual structures form systems that support and enable rather specific memory processes. At the cellular level, changes in the synaptic strengths between neurons in neural networks in the medial temporal lobe, neocortex, and elsewhere are the most likely mechanisms for learning and memory. We are rapidly developing a very clear understanding of the molecular processes that support synaptic plasticity, and thus learning and memory in the brain.

KEY TERMS

acquisition

consolidation

declarative memory

distributed representation

encoding

Hebbian learning

learning

long-term memory

memory

nondeclarative memory

perceptual representation system

procedural memory

retrieval

sensory memory

serial position effect

short-term memory

storage

working memory

THOUGHT QUESTIONS

1. Compare and contrast the different forms of memory based on their time course. Does the fact that some memories last seconds while some last a lifetime necessarily imply that different neural systems mediate the two forms of memory?

2. In the short-term memory studies of Petersen and Petersen from the 1950s, subjects forgot information within 18 seconds after it was presented if during the interval they also performed a distracting task such as backward subtraction. This finding was used to argue that information from short-term stores decayed over time, in contrast to the view that newly learned information interfered with the items in the short-term store. Could one not argue that the backward counting task introduced interference that led to the loss of information? Discuss.

3. Patient H.M. and others with damage to the medial temporal lobe develop amnesia. What form of amnesia do they develop, and what information can they retain, what can they learn, and what does this tell us about how memories are encoded in the brain?

4. Can you ride a bike? Do you remember learning to ride a two-wheeler? Can you describe to others the principles of riding a bike? Do you think that if you gave a detailed set of instructions to another person who has never ridden a bike, he or she could carefully study your instructions and then hop on a bike and ride happily off into the sunset? If not, why not?

5. Relate models of long-term potentiation (LTP) to changing weights in connectionist networks. What constraints do cognitive neuroscience findings place on connectionist models of memory?

SUGGESTED READINGS

COLLINGRIDGE, G.L., and BLISS, T.V.P. (1995). Memories of NMDA receptors and LTP. *Trends Neurosci.* 18:54–56.

GABRIELI, J., FLEISCHMAN, D., KEANE, M., REMINGER, S., and MORELL, F. (1995). Double dissociation between memory systems underlying explicit and implicit memory in the human brain. *Psychol. Sci.* 6:76–82.

MARKOWITSCH, H.J. (2000). Neuroanatomy of Memory. In E. Tulving and F.I.M. Craik (Eds.), *The Oxford Handbook of Memory* (pp. 465–484). New York: Oxford University Press.

McCLELLAND, J.L. (2000). Connectionist Models of Memory. In E. Tulving and F.I.M. Craik (Eds.), *The Oxford Handbook of Memory* (pp. 583–596). New York: Oxford University Press.

MILLER, G. (1994). The magical number seven, plus or minus two: Some limits on our capacity for processing information. *Psychol. Rev.* 101:343–352.

PALLER, K., KUTAS, M., and McISAAC, H. (1995). Monitoring conscious recollection via the electrical activity of the brain. *Psychol. Sci.* 6:107–111.

9
Language and the Brain

It had been an otherwise uneventful day at the major U.S. airport where Flight 301 was preparing to depart. The cabin crew was completing all preflight preparations. In the meantime, the flight attendants and gate personnel had finished boarding all the passengers, most having already stowed their oversized bags and parcels in every conceivable nook and cranny and taken their seats. Crammed into their "luxury" seats, belted in for their own safety, they were waiting for the plane to leave so that all manner of electronic devices could be retrieved, activated, and set into the usual frenzy of beeping, chirping, and flashing. Soon the passengers were rewarded as the plane was pushed back from the gate and began to taxi toward the runway. Suddenly the plane turned around and started to head back to the gate, naturally accompanied by the groans and sighs of now wary and unhappy passengers: "Yet another delay? What's wrong? Hopefully it's just another indicator light in the cockpit that burned out."

Back in the control tower, air traffic controllers looked tense and were frantically searching through airport procedures, following every guideline for the emergency facing them. Prepared and on alert, the airport security was rushing toward Flight 301 from every direction, while the Federal Aviation Commission, the FBI, and local law enforcement were mobilizing their resources to aid the passengers and crew of the airplane. What had happened? What was wrong with the airplane? What imminent danger elicited this massive response?

The controller in the tower who was in charge of guiding Flight 301 to a safe takeoff had been communicating with the pilots over the radio, as is normal during the preflight period. A checklist of details had to be dealt with before the plane could be cleared to depart the gate and begin takeoff. As the plane was pushing away from the gate, the controller in the tower heard the fateful words from the cockpit that everybody in aviation fears: "Hijack!" This simple word of course called for immedi-

ate action, and the controller engaged a well-prepared plan developed for just such a situation. He notified authorities and then ordered the plane back to the gate. The pilots did as they were instructed and turned the plane around and headed to the gate, only to be met by a swarm of armed and nervous officers.

After some initial tense moments and a thorough investigation, the security officers decided that Flight 301 was not being hijacked, and gave the green light for departure; none of the passengers even had to leave the plane. What had happened? Because of a heightened sensitivity for such situations and a need for extreme caution, a simple matter of human language had gotten in the way. One of the last passengers to board the plane had recognized his friend the pilot, Jack, and had greeted him with the carefree words, "Hi Jack!," which the tower personnel heard over the radio and took to mean something entirely different.

The misinterpretation of the simple phrase "Hi Jack" by the air traffic controller in the tower illustrates one of the many complexities that we encounter in everyday language. Language is indeed one of the most complex of the human brain's feats. The meanings of words, their organization into sentences, how they are produced as speech or in written form, and perhaps most importantly, how they are understood by the listener or reader comprise one of the most fascinating ongoing detective stories in cognitive neuroscience.

Communication and brain complexity took a monumental leap from the most intelligent of the nonhuman primates to humans, for whom communication includes language. Speech and symbolic language, being uniquely human, mark a dramatic shift between monkey brains and human brains. It has been surprisingly difficult, though, to identify human brain systems and

specializations that support language. Some brain regions have long been known to be critical for normal speech and comprehension, but these 100-year-old clues from neurology, although foundational, have provided only modest insights into the neural organization of language. Today, though, a growing body of work on the brain and language processes, guided by careful psycholinguistic models and aided by modern tools, is providing remarkable new data for cognitive neuroscientists to ponder. This chapter introduces a cognitive neuroscience perspective on the human language system, a system that is so complex that the smallest misinterpretation, such as the one in the story of Flight 301, can have sometimes dire or, at the very least, interesting consequences.

THEORIES OF LANGUAGE

How does our brain cope with spoken and written input to derive meaning? To answer this question we have to know how words are represented in the brain. It turns out this is a difficult question to answer, but cognitive neuroscience has elucidated a number of key principles. We will begin by tackling the question of how the brain stores words by reviewing data from the perspective of theoretical-experimental models and neurophysiological data.

The Storage of Words and Concepts: The Mental Lexicon

One of the central concepts in word representation is that of the **mental lexicon**—a mental store of information about words that includes *semantic* information (what is the word's meaning?), *syntactic* information (how are the words combined to form a sentence?), and the details of *word forms* (how are they spelled and what is their sound pattern?). Most psycholinguistic theories agree on the central role for a mental lexicon in language. But some theories propose one mental lexicon for both language comprehension and production, whereas other models distinguish between input and output lexica. In addition, the representation of *orthographic* (vision-based) and *phonological* (sound-based) forms must be considered in any model. The principal concept, though, is that a store (or stores) of information about words exists in the brain, and we have some, albeit limited, ideas about how it must be organized conceptually.

A normal adult speaker has passive knowledge of about 50,000 words and yet can recognize and produce about three words per second without any difficulty.

Given this speed and the size of the database, the mental lexicon must be organized in a highly efficient manner. It cannot be merely the equivalent of a dictionary. If, for example, the mental lexicon were organized in simple alphabetical order, it might take longer to find words in the middle of the alphabet, such as the ones starting with *K, L, O,* or *U,* than to find a word starting with an *A* or *Z.* Fortunately, this is not the case.

The mental lexicon differs in other respects from the dictionary on our shelves. For one thing, it has no fixed content: Words can be forgotten and new words can be learned—when was the last time you picked up a dictionary to see that new words had miraculously appeared (new editions excused)? Another significant difference is that in the mental lexicon, more frequently used words are accessed more quickly; for instance, the word *table* is more readily available than the word *snail*. In addition, the process of accessing lexical (word) representations in the mental lexicon is influenced by the so-called neighborhood effect. Let's consider this phenomenon from the perspective of understanding spoken words.

The "auditory neighborhood" of a word is defined as the number of words that differ from the target word by only one phoneme. A *phoneme* is the smallest unit of sound that makes a difference to meaning. In English, the sounds for "l" and "r" are two phonemes (the words *late* and *rate* mean different things), but in the Japanese language, no meaningful distinction is made between "l" and "r," and they are therefore represented by only one phoneme. Auditory neighbors for the word *hate* are *late, rate,* and *eight*. Words having more neighbors are identified more slowly. The idea is that there may be competition between activations of different words during recognition of speech, and this phenomenon tells us something about the organization of our mental lexicon. The mental lexicon, then, is quite dissimilar from the average college dictionary: How is it organized? The mental lexicon is thought to be organized as information-specific networks (Figure 9.1). The example in Figure 9.1 is taken from a model proposed by Willem Levelt from the Max Planck Institute for Psycholinguistics in the Netherlands (this model will be discussed in more detail later in this chapter in the section on speech production). As you can see in this fragment of a lexical network, information-specific networks exist for the word forms at the so-called *lexeme* level, and for the grammatical properties of words at the *lemma* level. At the lemma level the semantic specifications of words are represented as well. This semantic information defines the conceptual conditions under which it is appropriate to use a certain word, for example, whether the word represents an animate object (living thing) or an inanimate object (nonliving thing). These specifications are communicated by the " sense" connections between the lemma level and the conceptual level—the semantic knowledge of words is represented at the conceptual

level. As you can also see in Figure 9.1, the organization of the representations in the mental lexicon involves the relations between words, such that words that are related in meaning are connected and tend to be close together in the network (*sheep–goat*).

Support for the idea that representations in the mental lexicon are organized according to meaningful relationships between words comes from semantic priming studies that utilize a lexical decision task. In a semantic priming task, subjects are presented with pairs of words. The first member of the pair, the prime, is a word, while the second member, the target, can be a real word, or a nonword (like *sfhsi*), or a pseudoword (like *fisch*). If the target is a real word, it can be related or unrelated in meaning to the prime. For the lexical decision task in such a design, the subjects must decide as quickly and accurately as possible whether the target is a word, pressing a button indicating their decision. Subjects are faster and more accurate at making the lexical decisions when the target is preceded by a related prime (e.g., *car–truck*) than an unrelated prime (e.g., *tulip–truck*). Related patterns are found when the subject is asked to pronounce the target (under conditions when it is

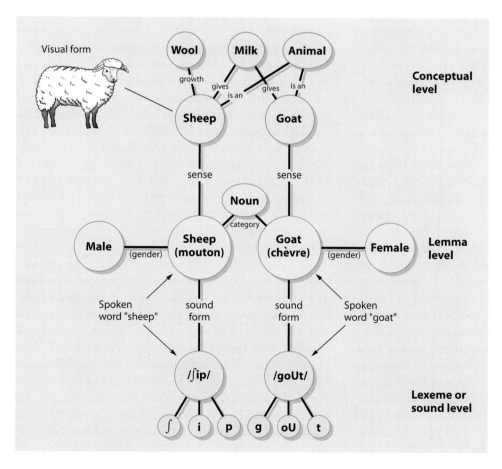

Figure 9.1 Fragment of a lexical network according to the Levelt model. See the text for a description. This model describes the processes for spoken word input, but a similar model applies to written word input. Adapted from Levelt (1994).

always a word, but is either related or unrelated to the prime). Here, naming latencies are faster for related words than for unrelated ones. What does this pattern of facilitated response speed tell us about the organization of the mental lexicon?

Initially, just like with other conceptual priming effects in memory (see Chapter 8), these effects of priming were thought to result exclusively from an automatic spread of activation (ASA) between the nodes in the word network. But over many years of research on priming with lexical decision tasks, it has become clear that priming can result from more than these implicit or automatic types of processes. *Expectancy-induced priming* might occur if the time between the presentation of primes and targets is long (e.g., > 500 msec) and the proportion of related word pairs in a list is large (e.g., when 50% or more of the pairs are like *car–truck, cat–dog,* etc.). Subjects may then expect some possible target words after hearing the prime. So if they hear the word *cat* for example, they may generate an expectancy set consisting of words that are related, such as *dog* and *mouse*. If the target word matches one of the words in the expectancy set (*dog* or *mouse*), then reaction times to this word will be faster or facilitated. In contrast, when the target word does not match the expectancy set, the response will be slower or inhibited. Another process, *semantic matching,* happens when subjects actively try to match the meaning of the target word with the meaning of the prime. In a lexical decision task this leads to faster "yes" responses to words that match than to words that do not match in meaning with the preceding context word. This process of semantic matching is postlexical; it happens after the representations in the mental lexicon have been accessed.

Let's now consider a little bit more about the conceptual level as represented in Figure 9.1. The conceptual level goes beyond our linguistic knowledge of words. Semantic knowledge of words can be distinguished from pure linguistic knowledge. This distinction becomes clear when we consider words that have one form representation but two or more unrelated meanings, for example, the word *bank*. In order to arrive at the meaning of the word *bank,* contextual information is needed to decide whether the "side of the river" (is sandy, trees can grow there) or the "money institution" (is a building, people can work there) is meant. As we saw in Chapter 8 on the organization of memory, Endel Tulving had proposed in the early 1970s that a distinction be made between episodic memory, our memory for personal events, and semantic memory, our memory for facts and general knowledge. Semantic memory is important for language comprehension and

production and is therefore clearly connected to our mental lexicon, as can be seen in Figure 9.1, but semantic memory and mental lexicon are not necessarily the same thing. In general, we can say that conceptual or semantic representations reflect our knowledge of the real world. These representations can become activated through our own thoughts and intentions or through our perceptions of words and sentences, of pictures and photographs, and of events, objects, and states in the real world. There have been many investigations into the nature of conceptual or semantic representations, but the issues of how and where in the brain concepts are represented are still very much disputed.

First, there is the matter of how many conceptual or semantic systems there are. Some investigators have argued for a unitary semantic system that utilizes a verbal- or proposition-based format, or a unitary perceptual-based format. Others favor the idea that different types of information can be stored on the basis of perceptual- *and* verbal-based codes. The question of unitary or multiple conceptual systems is related but not identical to the question of whether our conceptual or semantic system is amodal or whether modality-specific and independent meaning systems exist. That is to say, is the same conceptual representation of a robin activated regardless of whether one hears the word *robin* or sees one flying?

This question highlights some of the ongoing uncertainty about how conceptual information is represented, and indeed, many organizational structures such as feature lists, schemata, exemplars, and connectionist networks have been proposed. Collins and Loftus (1975) proposed one very influential model. In this model, word meanings are represented in a semantic network in which words, represented by conceptual nodes, are connected with each other. An example of a semantic network is presented in Figure 9.2. The strength of the connection and the distance between the nodes are determined by the semantic relations or associative relations between the words. For example, the node that represents the word *car* will be close to and have a strong connection with the node that represents the word *truck*. A major component of this model is the assumption that activation spreads from one conceptual node to others, and nodes that are closer together will benefit more from this spreading activation than will distant nodes. If we hear the word *car,* then the node that represents *car* in the semantic network will be activated. In addition, words like *truck* and *bus* that are closely related to the meaning of *car,* and are therefore nearby and well connected in the semantic network, will receive a considerable amount of activation. In contrast,

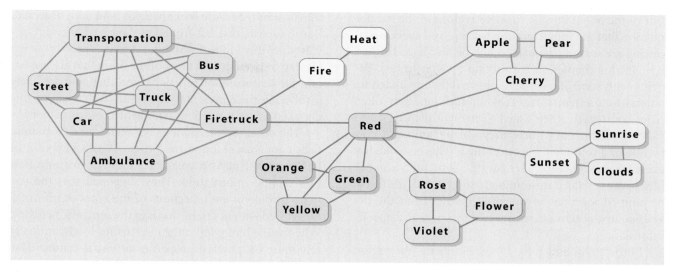

Figure 9.2. An example of a semantic network. Note that words that have strong associative or semantic relations are closer together in the network (e.g., *Car–Truck*) than are words that have no such relation (e.g., *Car–Clouds*). Semantically related words are colored similarly in the figure, and associatively related word (e.g., *Firetruck-Fire*) are closely connected.

a word like *tulip* most likely will not receive any activation at all upon hearing "car." This model would predict that hearing the word *car* should facilitate recognition of the word *truck* but not *tulip*.

Although the semantic network model that Collins and Loftus proposed has been extremely influential, it is still a matter of dispute as to how word meanings are organized. There are many other models and ideas on how conceptual knowledge is represented. Some models propose that concepts are represented by their semantic features or semantic properties. For example, the word *dog* has several semantic features, such as "is animate," "has four legs," and "barks," and they are assumed to be represented in the conceptual network. Such models are confronted with the problem of activation: How many features have to be activated in order for a person to recognize a dog? For example, it is possible to train our dogs not to bark (well, not any we own), yet we can recognize a dog even when it does not bark, and we can identify a barking dog that we cannot see. Furthermore, it is not exactly clear how many features would have to be stored. For example, a table could be made of wood or glass, and in both cases we would recognize it as a table. Does this mean that we have to store the features "is of wood/glass" with the "table" concept? In addition, some words are more "prototypical" examples of a semantic category than others, as reflected in our recognition and production of these words. When we are asked to generate bird names, for example, the word *robin* will come to mind as one of the first examples, while a word like *ostrich* might not come up at all, depending on where we grew up or have lived.

In sum, it is a matter of debate how word meanings are represented. No matter how they are represented, everyone agrees that a mental store of word meanings is crucial to normal language comprehension and production. Evidence from brain-lesioned patients and from functional brain imaging studies is revealing how the mental lexicon and conceptual knowledge may be organized.

NEURAL SUBSTRATES OF THE MENTAL LEXICON AND CONCEPTUAL KNOWLEDGE

Through observations of deficits in patients' language ability, we can infer a number of things about the functional organization of the mental lexicon. Different types of neurological problems create deficits in understanding and producing the appropriate meaning of a word or concept. Patients with **Wernicke's aphasia** (a language deficit usually due to brain lesions in the posterior parts of the left hemisphere—we will discuss aphasia in detail later) make errors in speech production that are known as **semantic paraphasias.** For example, they could use the word *horse* when they mean *cow*. Similar errors in reading are made by patients with *deep dyslexia:* They might read the word *horse* where "cow" is written (see From Written Text to Word Representation: One or Two Pathways?). Patients with *progressive semantic dementia* initially show impairments in the conceptual system, while other mental and language abilities are spared. For example, these patients can still understand and produce the syntactic structure of sentences. This impairment has been associated with progressive damage to the temporal lobes, mostly on the left side of the brain. But the superior regions of the temporal lobe that are important for hearing and speech processing are spared (these areas will be discussed further, later in this chapter, in the section Spoken Input). Patients with semantic dementia have difficulty assigning objects to a semantic category. In addition, they often name a category when asked to name a picture; when viewing a picture of a horse, they will say "animal," and a picture of a robin will produce "bird." This neurological evidence provides support for the semantic network idea because related meanings are substituted, confused, or lumped together, as we could predict from the degrading of a system interconnected by nodes that specify information.

In the 1970s and early 1980s, Elizabeth Warrington and her colleagues performed groundbreaking studies on the organization of conceptual knowledge in the brain. Some of these studies were done on patients that would now be classified as suffering from semantic dementia, as just described. These researchers found that semantic problems can be localized specifically to certain semantic categories, such as animals versus objects. They reported studies of patients who had great difficulties *pointing* to pictures of food or living things when presented with a word, whereas their performance with man-made objects like tools was much better. These difficulties were also found when the patients were asked to *name* foods or living things when presented with a picture, whereas their naming of man-made things was spared. The reverse pattern was found in another group of patients: Their ability to identify foods and living things was preserved while their ability to identify man-made objects was impaired. Since these original observations by Warrington, many cases of patients with category-specific deficits have been reported, and there appears to be a striking correspondence between the sites of lesions and the type of semantic deficit. The patients whose impairment involved living things had lesions that included the inferior and medial temporal cortex, and often these lesions were located anteriorly. The anterior inferior temporal cortex is located close to areas of the brain that are crucial for visual object perception, and the medial temporal lobe contains important relay projections from association cortex to the hippocampus, a structure that, as you might remember from Chapter 8, has an important function in the encoding of information in long-term stores. Furthermore, the inferior temporal lobe is the end station for the so-called what-information or object recognition stream in vision (see Chapter 6).

Less is known about the localization of the lesions in patients who show greater impairment for man-made things, simply because fewer patients have been observed. But it appears that left frontal and parietal areas are involved in this kind of semantic deficit. These areas are close to or overlap with areas of the brain that are important for sensorimotor functions, and therefore are likely involved in the representation of actions that can be undertaken when using man-made artifacts such as tools.

The observed correlations between type of semantic deficit and area of lesion are consistent with a hypothesis by Warrington and her colleagues about the organization of semantic information. They suggested that the patients' problems are reflections of the types of information stored with different words in the semantic network. Whereas the biological categories (fruits, foods, animals) rely more on physical properties or visual features (for example, what is the color of an apple?), man-made objects are identified by their functional properties (for example, how do we use a hammer?). This idea has been both tested and challenged. An example of evidence in favor of the idea of a modality-specific organization of the semantic network comes from a computational model by Martha Farah and James McClelland (1991). Computational models can provide important tests of the plausibility of particular hypotheses. In this specific case, Farah and McClelland modeled semantic memory as consisting of separate visual and functional subsystems, and further, they included the idea that living things must include representations based on visual attributes whereas nonliving things include information about functional attributes. They found, as predicted by the hypothesis of Warrington, that if they "lesioned" visual properties in their model (by removing nodes in their network in the computer simulation) the model was especially "impaired" in dealing with living things, whereas "lesioning" of functional properties led to "impairments" with nonliving things.

The challenge for Warrington's proposal comes from observations by Alfonso Caramazza and others (e.g., Caramazza and Shelton, 1998) that the studies in patients did not always use well-controlled linguistic materials. For example, when comparing living things versus man-made things, some studies did not control the stimulus materials so that the objects tested in either category were matched on visual complexity, visual similarity across objects, the frequency of use, and the familiarity of objects. If these variables vary widely between the categories, then clear-cut conclusions about differences in their representation in a semantic network cannot be drawn. Caramazza has proposed an alternative theory in which the semantic network is organized along lines of the conceptual categories of animacy and inanimacy. He argues that the selective damage that has been observed in brain-damaged patients, as in the studies of Warrington and others, genuinely reflects "evolutionarily adapted domain-specific knowledge systems that are subserved by distinct neural mechanisms."

Recent studies using imaging techniques in neurologically unimpaired human participants have looked further into the problem of the organization of semantic representations. Alex Martin and his colleagues (1996) at the National Institute of Mental Health (NIMH) conducted studies using positron emission tomography (PET) and functional magnetic resonance imaging (fMRI). Their findings reveal how the intriguing dissociations in neurological patients we just described can be identified in neurologically normal brains. When subjects read the names of or answered questions about animals, or when they named pictures of animals, the more lateral aspect of the fusiform gyrus (on the brain's ventral surface) and the superior temporal sulcus were found to be active. But naming of animals also activated a brain area associated with the early stages of visual processing, namely, the left medial occipital lobe. In contrast, identifying and naming tools were associated with activation in the more medial aspect of the fusiform gyrus, the left middle temporal gyrus, and the left premotor area, a region that is also activated by imagining hand movements. These findings are consistent with the idea that conceptual representations in our brains of living things versus man-made tools rely on separable neuronal circuits engaged in processing perceptual versus functional information.

One of the most compelling pieces of evidence for category-specific deficits comes from Hannah Damasio and her colleagues (1996) at the University of Iowa, who investigated a large population of patients with brain lesions. These patients had to perform a naming task in three different conditions: (1) naming famous faces, (2) naming animals, and (3) naming tools. To dissociate conceptual problems (damage to the conceptual representations of objects themselves) from problems at the level of word retrieval (accessing the name of an object), the following procedure was used: If a subject was able to describe many of the features of the picture but was unable to name it, then that was scored as a naming error. For example, when the subject was asked to name the picture of a skunk, the following answer showed that the subject recognized it but could not name it: "Oh, that animal makes a terrible smell if you get too close to it; it is black and white, and gets squashed on the road by cars sometimes." But if the subject was not specific in describing the picture, the answer was not included in the naming score (for example, "Some kind of animal; I don't know what . . . just an animal."). The naming re-sponse was scored correct when it was the same as that of normal subjects.

Thirty patients, twenty-nine of whom had a lesion in the left hemisphere, showed impairments on this task. The naming deficit was very selective in nineteen patients: Seven demonstrated impairment in naming faces; five patients, in naming animals; and seven, in naming tools. The remaining eleven patients had a combination of problems in word retrieval for faces, animals, and tools, or faces and animals, or animals and tools, but never for the combination of faces and tools together without also a deficit in animals. When the researchers examined where the patients had their brain lesions, they found that they could correlate naming deficits with specific regions (Figure 9.3).

Damasio discovered that brain damage in the left temporal pole (TP) correlated with problems in retrieving the names of persons, whereas lesions in the anterior part of the left inferior temporal (IT) lobe correlated with problems in naming animals, and finally, damage to the posterolateral part of the left inferior temporal lobe, along with the lateral temporo-occipito-parietal junction (IT+), was correlated with problems in retrieving the names of tools. In a coordinated study with neurologically normal subjects that used PET scanning, these same brain areas were activated when subjects engaged in naming persons (TP), animals (IT), and tools (IT+), as illustrated in Figure 9.4.

Damasio and her colleagues concluded that since they made sure that patients with problems retrieving the names of words could still activate many of the conceptual properties relevant to the word (i.e., saying in response to a picture of a skunk, "Oh, that animal makes a terrible smell if you get too close to it; it is black and white, and gets squashed on the road by cars sometimes"), the correlated brain areas that were lesioned in these patients must play a role in word retrieval. They therefore did not assume that their findings reflect the organization of a conceptual network in the brain, but rather proposed that their findings reflect the organization at the word (lexical) level. They argued that their results indicate that the brain has three levels of representation for word knowledge, as illustrated in Figure 9.5. These representational levels were predicted by cognitive models of word production.

The top level in Figure 9.5 represents the conceptual level, a preverbal level that contains the semantic features about words (four legs, furry, tail). At the lexical level the word form that matches the concept is represented (cat). Finally, at the phonological level the sound information that corresponds to the word is represented. (Note that this model of word representation is different from what

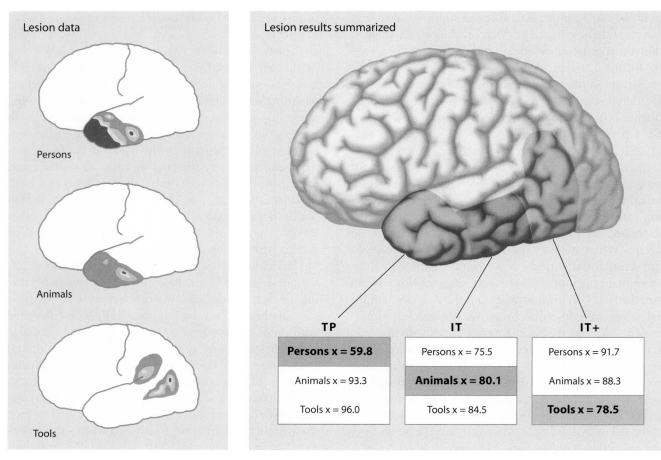

Figure 9.3 Locations of brain lesions that are correlated with selective deficits in naming persons, animals, or tools. On the left, the actual averaged lesion data are displayed for patients that had person-naming **(top)**, animal-naming **(middle)**, or tool-naming **(bottom)** deficits. The colors indicate the percentage of patients with a given deficit whose lesion is located in the indicated area. Red indicates that most patients had a lesion in that area, whereas purple indicates that few had a lesion in that area. On the right, the lesion results are summarized. The blue area corresponds to the temporal pole (TP); the red area, to the inferotemporal region (IT); and the green area, to the posterior part of the inferotemporal lobe extending to the anterior part of the lateral occipital region (IT+). Scores in the boxes indicate the percentage of recognized items that were correctly named. Patients with TP lesions scored lowest on naming persons (59.8%), patients with IT lesions scored lowest on naming animals (80.1%), and patients with IT+ lesions scored lowest on naming tools (78.5%). Adapted from Damasio et al. (1996).

we have discussed thus far, because there is no representation of the "lemma" level—see earlier text.) Damasio and her colleagues proposed that the brain's conceptual networks involve several neuronal structures in the left and the right hemisphere. These conceptual networks are connected to the lexical networks in the left temporal lobe and might contain specialized information for persons, animals, or tools. These areas in turn activate the phonological network that touches off the sound patterns needed to pronounce the word.

Perceptual Analyses of the Linguistic Input

All models of normal language comprehension have to begin with the problem of how words are represented.

The next step is to identify what enables understanding of the linguistic input; Figure 9.6 represents the processes and representations that are important to normal language comprehension.

As we will see, understanding spoken language and understanding written language share some processes, but there are also some striking differences in how spoken and written inputs are analyzed. When attempting to understand spoken words (see Figure 9.6), the listener has to decode the acoustic input. The result of this acoustic analysis is translated into a phonological code, because that is how the lexical representations of auditory word forms are stored in the mental lexicon. Both of these processing steps are prelexical and they do not involve the mental lexicon.

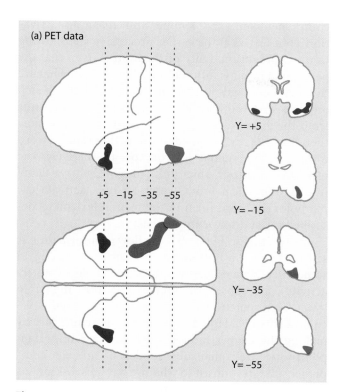
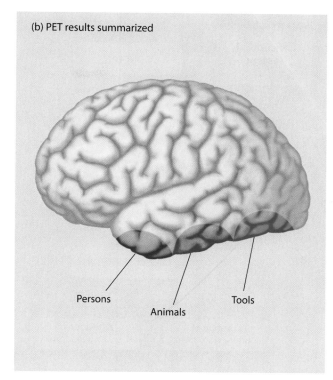

Figure 9.4 Activations in neurologically unimpaired subjects during naming of persons, animals, or tools as determined by positron emission tomography (PET). **(a)** The PET activations in lateral and ventral views (left), and in four coronal sections at the levels indicated by the dashed lines. The values correspond to millimeters in anterior and posterior directions from a zero point in the brain defined by a stereotactic coordinate system. **(b)** The summarized PET results. Naming persons mostly activated the temporal pole, naming animals mostly activated the middle portion of the inferior temporal gyri, and naming tools mostly activated the posterior portions of the inferior temporal gyrus. Adapted from Damasio et al. (1996).

Figure 9.5 Three levels of representation that are needed in speech production: semantic features, lexical nodes, and phonological segments. **(a)** The semantic features of the word *cat* (four legs, furry) activate the lexical node of the word *cat*, which in turn activates the phonological segments of that word. **(b)** A model that fits the data of Damasio and colleagues shown in Figures 9.3 and 9.4. The information at the lexical level is organized according to specific semantic categories (e.g., animals versus tools). Adapted from Caramazza (1996).

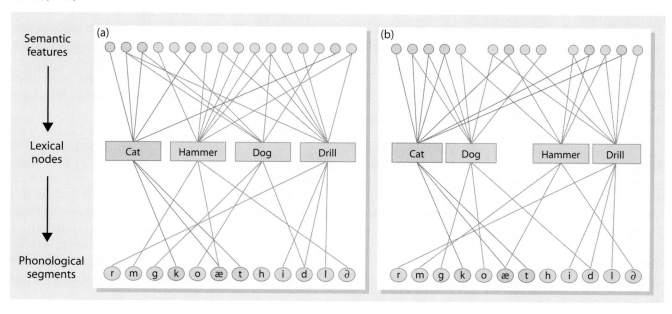

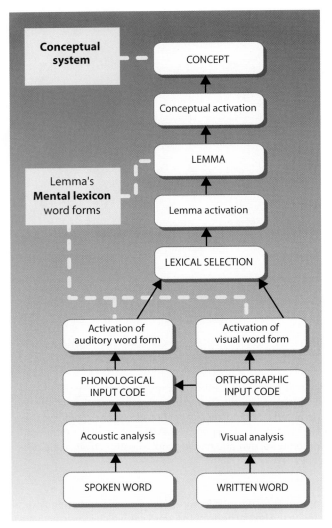

Figure 9.6 Schematic representation of the components that are involved in spoken and written language comprehension. Inputs can enter via either auditory (spoken word) or visual (written word) modalities. Notice that the flow of information is bottom up, from perceptual identification to "higher-level" word and lemma activation. Interactive models of language understanding would predict top-down influence to play a role as well. For example, activation at the word form level in this case would be able to influence earlier perceptual processes. This type of feedback could be introduced into this schematic representation by making the arrows bidirectional.

After the acoustic input has been translated into a phonological format, the lexical representation in the mental lexicon that best matches the auditory input can be selected—this is known as **lexical selection.** The selected word form in turn activates the lemma (store of grammatical information), and then the words' meaning.

The process of reading words shares at least the last two steps of linguistic analysis with auditory comprehension (i.e., lemma and meaning activation), but differs at the earlier processing steps, owing to the different modality of the input. Given that the perceptual input is different (see Figure 9.6), what are these earlier stages in reading? The first analysis step requires that the reader identify orthographic units from the visual input. These orthographic units may then be directly mapped onto orthographic word forms in the mental lexicon, or alternatively, the identified orthographic units might be translated into phonological units, which in turn activate the phonological word form in the mental lexicon as described for auditory comprehension.

Notice that in Figure 9.6 the flow of information is strictly from the bottom up. However, according to some interactive models (see Modularity Versus Interactivity), a feedback of information from higher to lower representational levels is possible. Figure 9.6 could easily be adapted to this type of architecture by making the arrows bidirectional. Another important point to realize is that some models of language comprehension, in contrast to what we have discussed thus far, do not assume that words are represented permanently in the brain (that means that there would be no word form or lemma representations), but rather that they emerge as a pattern of activation in a distributed network. Which type of model fits better will have to be determined in future experiments by clever scientists (this could be you!).

In the next few sections we will delve into the processes involved in the understanding of spoken and written inputs of words. Then we will consider the understanding of sentences. Let's now first have a closer look at the perceptual analyses of the linguistic input. The first task that faces a listener or reader is to identify individual words in a spoken utterance or in a written text. One has to perceptually analyze the input, which is a prelexical process that does not involve the mental lexicon. Naturally, the perceptual analyses required for spoken and written input are quite different.

SPOKEN INPUT

The input signal in spoken language is very different from that in written language. Whereas for a reader it is immediately clear that the letters on a page are the physical signals of importance, a listener is confronted with a variety of sounds in the environment and has to identify and distinguish the relevant speech signals from other "noise." Let's begin with some of the basic units that constitute the speech signal.

HOW THE BRAIN WORKS

Modularity Versus Interactivity

Two major questions that have puzzled cognitive psychologists and cognitive neuroscientists are, Do the different component processes involved in remembering, or paying attention, or understanding and producing language interact? If so, at what level of processing? In reading a text, for example, we need to analyze the features that comprise the letters, and the letters have to be grouped such that they form a word that we can understand. As we have seen in the text, scientists differ in their view of how these processes take place. Some (like Selfridge, see Figure 9.11) argue that the analysis process works purely from bottom up, from sensory input to higher-level representations of words. Others (like McClelland and Rumelhart, see Figure 9.12) think that higher-level processes can influence lower-level perceptual analyses. These different views can be recognized in the literature as modular or interactive. The most well-known proposal for modularity of cognitive processes comes from Jerry Fodor in his book *The Modularity of Mind* (1983). His proposal consists of two central ideas: (1) Language is an input system rather than part of a central system, where a central system can influence different knowledge domains, and (2) input systems have a modular architecture.

Fodor argued that input systems with a modular architecture should have the following characteristics:

a. *Domain specificity.* The input system receives information from several sensory systems, but processes the information with codes that are specific to the system. For example, the language input system translates visual input into a phonological or speech sound representation.

b. *Information encapsulation.* The processing strictly goes in one direction. The processing of the earliest module has to be finished before it can send its information onto the next module and so on. There can be no transmission of partial information. This direction of information flow is strictly from bottom up. In Fodor's model there can be no influence from higher cognitive modules on lower cognitive modules. In other words, there is no top-down influence in language processing.

c. *Localization of function.* Each module is implemented in a particular brain region.

Interactive views challenge all these premises. But the most important objection to the modular architecture is the idea that various subsystems can only communicate with each other in a bottom-up way. Interactive views maintain that higher cognitive processes can influence lower-level cognitive processes through systems of feedback.

In many aspects of language comprehension and production, the question of modularity versus interactivity is relevant, and debates on this issue pertain to all of the component processes discussed in this chapter.

As introduced earlier, important building blocks of spoken language are *phonemes*. These are the smallest units of sound that make a difference to meaning (for example "cap" versus "tap," where the first phoneme is different). In order to be able to describe the phonemes of our language, we make use of the phonetic alphabet, which is not the same as our orthographic or letter alphabet. For example, "i" is the phonetic symbol for the phoneme *ee* as in m*ee*t, but "I" is the phonetic symbol for *i* as in "*it.*" The English language uses forty phonemes, but other languages may use more or less.

To get a feeling for what a listener has to deal with when he or she is listening to spoken input, it is useful to know how different speech sounds are produced. We can produce different speech sounds by making use of different features of our vocal apparatus. Three main factors determine the actual speech sound that we produce. The first one is *voicing*, or whether or not we use our vocal cords to produce a speech sound. All vowels are voiced (*a, e, i, o, u, y*), but not all consonants are. Examples of voiced consonants are *b* and *m*, and of unvoiced consonants are *s* and *t*. For consonants there are two other ways to modify the speech sound. One has to do with the part of the speech apparatus we use to produce the sound (*point of articulation*), for example, whether we use both lips ("b"), or the lips and the mouth ("f"). The other has to do with the manner in which the airstream is changed (*manner of articulation*); for example, a "p" is produced by blocking the airstream, whereas "l" hardly obstructs the airflow at all. For vowels there is another way to modify speech sounds: via placement of the tongue in the mouth. The tongue can be placed high or low, and front or back.

Having identified that phonemes are important building blocks to spoken language by no means simplifies the task of the listener. There are a number of additional difficulties with the speech signal that his or her brain has to resolve; some of these have to do with the variability of the signal. For example, speech sounds vary on the basis of the context in which they are spoken (lack of invariance). So even though they start with the same phoneme, the speech signal of the syllable "bo" is very different from the speech signal of the syllable "bi," whereas "bi" greatly resembles "di," even though they start with a different phoneme. So pronunciation of a phoneme is modified dramatically as a function of whether it is followed by an "o" or an "i." And yet a listener will still be able to perceive a "b" and distinguish it from a "d." Another problem a listener faces is that variability in the production of the *same* speech sound is introduced when it is pronounced by different speakers, for example, by a male or a female.

Further problems for the listener emerge because the phonemes often do not appear as separated little chunks of information; in other words, they lack segmentation. Whereas physical boundaries divide words and sentences in written language, these boundaries are mostly absent in speech. As you read this text, you see each word separated by a blank space and each sentence terminated by a period. These physical cues help you to discriminate between words and sentences. By contrast, the word boundaries of speech are murky, as illustrated in Figure 9.7, where the speech waveform of the word

captain looks as though the signal represents two words because of a "silence" within the word.

In addition to silences within words, spoken sentences also typically lack a clear silence between words because they become *coarticulated*. Their ends and beginnings are united, as in "I dunno" instead of "I don't know." As shown in Figure 9.8, no clear boundaries fall between the words *do you mean* when they are uttered at normal conversational speed (i.e., as *connected speech*). This problem of unclear boundaries within and between words in spoken language has been labeled the segmentation problem.

We have now established that the listener is confronted with enormous variability in input. In spoken language, the perceptual analysis of auditory input must account for all these variables, and there cannot be a one-to-one relation between a physical signal and representations in memory. This has puzzled many researchers and has led to the question of what the abstract units of representation might be for spoken inputs.

Some researchers have proposed that these representations are built on spectral properties (frequency content) of the incoming signal. These spectral properties vary according to different sounds, as shown in Figure 9.9. The features derived from spectral analysis might form a phonetic representation, and this might be the access code to phonological representations. But other units of representation have been proposed, such as the phoneme itself, syllables, and the way the speaker has intended to pronounce the phoneme. Other theo-

Figure 9.8 Speech waveform for the question "What do you mean?" Note that the words *do you mean* are not physically separated. Even though the physical signal provides little cues to where the spoken words begin and end, the language system is able to parse them into the individual words for comprehension. Courtesy of Tamara Swaab.

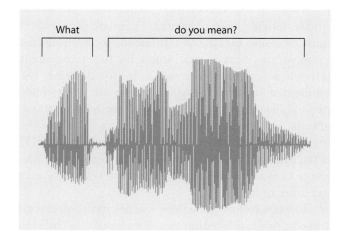

Figure 9.7 Speech waveform for the word *captain*. Note the silence within the word. Time progresses from left to right, and amplitude is registered in the vertical dimension. Courtesy of Tamara Swaab.

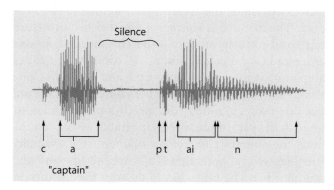

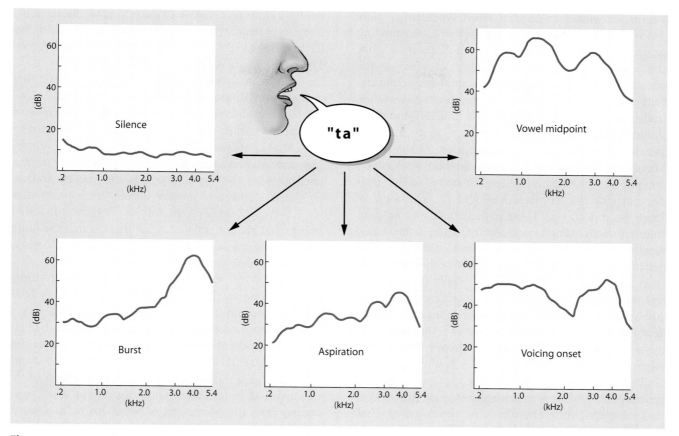

Figure 9.9 Spectral properties vary according to sounds. The sound *ta* can be analyzed to yield five different critical-band spectra. Each critical band has a particular frequency content (in kHz, horizontal axis). For example, the "burst" portions of the sound have a lot of higher-frequency content, with power (in dB, vertical axis) in the 4- to 5-kHz range. Adapted from Klatt (1989).

rists have altogether rejected the idea that there are discrete units of representation for the speech signal.

So how do we break up the spoken input? Fortunately there are some other clues as to how to divide the speech stream into meaningful segments, namely, through *prosodic* information, which is what the listener derives from the speech rhythm and the pitch of the speaker's voice. The speech rhythm is introduced by varying the duration of words and by placing pauses between them. Prosody is apparent in all spoken utterances, but it is perhaps most clearly illustrated when a speaker asks a question or emphasizes something. When asking a question, a speaker will raise the frequency of his or her voice toward the end of the question, and when emphasizing a part of speech, a speaker will raise the loudness of his or her voice and include a pause after the critical part of the sentence.

In their psycholinguistic research, Ann Cutler, now at the Max Planck Institute for Psycholinguistics in the Netherlands, and her colleagues have revealed other cues that can be used to segment the continuous speech stream. These researchers showed that English listeners use syllables that carry an accent or stress (strong syllables) to establish word boundaries. For example, a word like *lettuce*, with stress on the first syllable, is usually heard as a single word and not as two words ("let us"). In contrast, words such as *invests*, with stress on the last syllable, are usually heard as two words ("in vests") and not as one word (although we know it is one word). Interestingly, in other languages such as Dutch and French, syllables themselves seem more important than stress in helping the listener to segment the continuous speech stream. When listeners in these languages are asked to detect a string of letters in a word, they are faster when the string exactly corresponds to a syllable in a heard word, and have more difficulty when the string is longer or shorter than a syllable. For example, Dutch listeners would be very fast to detect the first syllable "bak" in "bakker" (*baker*) but slow when they hear the word "baken" (*beacon*) where "ba" and not "bak" is the first syllable.

Neural Substrates of Spoken Word Processing Now that we have identified the initial steps in understanding speech from the perspective of psycholinguistic theory and research, we turn to the question of where in the brain these processes may take place and what neural circuits and systems support them. From animal studies, studies in patients with brain lesions, and PET and fMRI studies in humans, we know that the superior temporal cortex is important to sound perception. At the beginning of the last century, it was already well understood that patients with bilateral lesions restricted to the superior parts of the temporal lobe had the syndrome of "pure word deafness." These patients had specific difficulties in recognizing speech sounds, whereas they could process other sounds relatively normally. Because these patients did not have any difficulty in other aspects of language processing, their problem really seemed to be restricted primarily to auditory or phonemic deficits, and hence the term *pure word deafness*. But with the evidence from more recent studies in hand, we can start to make some finer distinctions as to where speech sounds might be distinguished from nonspeech sounds.

When the speech signal hits the ear, it first is processed by pathways in the brain that are not specialized for speech but that are used for hearing in general. Heschl's gyri, which are located on the supratemporal plane, superior and medial to the superior temporal gyrus (STG) in each hemisphere, contain the primary auditory cortex, or the area of cortex that processes the auditory input first (see Chapter 3). The areas that surround Heschl's gyri and extend into the STG are auditory association cortex. PET and fMRI studies in humans have shown that Heschl's gyri and the STG of both hemispheres are activated by speech and nonspeech sounds (e.g., tones) alike, but that this activation is not particularly selective for one of the two. This pattern of activity to sounds suggests that the STG is important for sound perception but not necessarily specialized for linguistic processes. As we will see later in this chapter (in Aphasia), this finding is a big surprise. More than a century ago, the famous neurologist Karl Wernicke found that patients with lesions in the left temporal-parietal region that included the STG had difficulty with understanding spoken and written language. This led to the now-century-old notion that this area—which was christened *Wernicke's area*—is crucial to word comprehension. But even in Wernicke's original observations, the lesions were not restricted to the STG. And we can now conclude that the STG alone is probably not the seat of word comprehension.

So if not in the STG, where else in the brain do we distinguish speech from other sounds? There are some strong indications that the midsection of the superior temporal sulcus, of both hemispheres (but mostly of the left hemisphere), plays an important role in this process. These regions are ventral to the STG, which is important for acoustic analyses, as we just described. In one fMRI study, Jeffrey Binder and colleagues (2000) at the Medical College of Wisconsin compared different types of nonspeech sounds. The nonspeech sounds consisted of noise that contained no systematic frequency or amplitude modulations, or tones that were frequency modulated between 50 and 2400 Hz; the speech sounds consisted of reversed speech sounds (real words that were played backward), pseudowords (pronounceable strings that are not real words but that contain the same letters as the real word, for example "sked" from "desk"), and real words.

Figure 9.10 shows the results of the Binder study. Relative to noise, the frequency-modulated tones activated posterior portions of the STG bilaterally. Areas that were more sensitive to the speech sounds than to tones were located more ventrolaterally, in or near the superior temporal sulcus. In the same study, Binder and colleagues also showed that these areas are most likely not involved in lexical-semantic aspects of word processing (i.e., the processing of word forms and word meaning), because they were equally activated for words, pseudowords, and reversed speech. Recall that the results of the studies we discussed earlier (in Neural Substrates of the Mental Lexicon and Conceptual Knowledge) showed that the areas implicated in the processing of lexical-semantic information are lateralized much more to the left side and are mostly ventral to the superior temporal areas associated with speech perception per se.

Researchers also have tried to identify areas in the brain that are particularly important for the processing of phonemes. This has proved difficult because it is very hard to find ways to only "activate" phonemes, since hearing a word also will automatically activate its meaning. So the definitive answer is not in yet, but some PET and fMRI studies have looked at reading of words or of listening to spoken words and pseudowords and have found middle temporal gyrus activations, sometimes lateralized to the left hemisphere—evidence, perhaps, that reading can, and often does activate phonological information (see From Written Text to Word Representation: One or Two Pathways?).

WRITTEN INPUT

For written input, readers must recognize a visual pattern. These visual patterns vary across different writing

Lateral slices ←——————————————————→ Medial slices

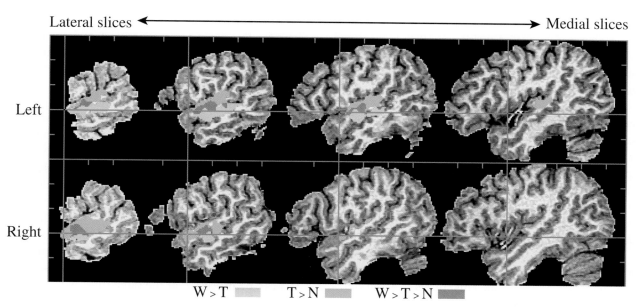

Left

Right

W > T ▦   T > N ▦   W > T > N ▦

Figure 9.10 Superior temporal activations to speech and nonspeech sounds. Four sagittal slices are shown for each hemisphere. The posterior areas of the superior temporal gyrus are more active bilaterally for frequency-modulated sounds than for simple noise (in blue), whereas areas that are more sensitive to speech sounds are located ventrolateral to this area (in yellow), in or near the superior temporal sulcus. This latter activation is somewhat lateralized to the left. Areas that are more active for words (W) and tones (T) than for noise (N) are also indicated (in red). Adapted from Binder et al. (2000).

systems. There are three different ways in which words can be symbolized in writing. Many Western languages (e.g., English) use the *alphabetic system* where the symbols approximate the *phonemes*. However, languages that use the alphabetic system differ in how close the correspondence between letters and sounds is. Some languages such as Finnish and Spanish have a close correspondence (shallow orthography). In contrast, English often lacks a correspondence between letter and sound, which means that English has a relatively *deep orthography*. Compare, for example, the differences in the pronunciation of the following three words, even though they start with the same four letters: *sign, signal,* and *signing.* The lack of correspondence between letter and sound makes it harder to learn to spell (and it might also have consequences for reading—see From Written Text to Word Representation: One or Two Pathways?). Luckily, at the next level of organization is the *morpheme,* the smallest unit of language that has meaning—that is, a word and the existing variations on a word (e.g., with prefixes or suffixes added, as in /sign/, /sign//ing/). There tends to be slightly more correspondence in the spelling of morphologically related words.

Other languages such as Chinese and Japanese have different writing systems. The Japanese writing system *Kana* uses the *syllabic system,* where each symbol reflects a syllable. This is possible because Japanese only has about 100 unique syllables, whereas English has close to 1000. In the third writing system, the *logographic system,* a unique symbol is used for each word or each morpheme. An example of a living language that uses something close to a logographic system is Chinese, where characters can symbolize whole morphemes. However, the characters also represent phonemes, so Chinese is not a pure logographic system. The reason for this representation system in writing is that Chinese is a tonal language. The same words can mean different things depending on the rise and fall in the pitch or tone of the vowels. This would be difficult to represent in a system that only symbolizes speech sounds or phonemes.

The three writing systems symbolize different parts of speech (phonemes, syllables, and morphemes or words), but they all use arbitrary symbols. In other words, the symbols used are abstract representations that do not resemble what they represent. So the written word *dog* in English does not resemble in any way the real-life Fluffy who wags his tail and barks when he sees the mailman.

Regardless of the writing system used, readers must be able to analyze the primitive features, or the shape of the symbols. In the alphabetic system, on which we will concentrate here, this process would amount to the visual analysis of horizontal lines, vertical lines, closed curves, open curves, intersections, and other elementary shapes.

A model for letter recognition, the *pandemonium model* proposed by O.G. Selfridge in 1959, is displayed in Figure 9.11. In this model, the sensory input ("R") is temporarily stored as an iconic memory by the so-called *image demon*. Selfridge used the term *demon* to refer to a discrete stage or substage of information processing. Then twenty-eight *feature demons* sensitive to features like curves, horizontal lines, and so forth start to decode features in the iconic representation of the sensory input. Figure 9.11 indicates where all the feature demons are located. In the next step, all the representations of letters with these features are activated by *cognitive demons*. Finally, the representation that best matches the input is selected by the *decision demon*.

In 1981 James McClelland and David Rumelhart proposed another model that has been very important for visual letter recognition. This model is a computational model that assumes three levels of representation: a layer for the features of the letters of words, a layer for letters, and a layer for the representation of words. A very important characteristic of this model is that it allows for top-down information (i.e., information from the higher cognitive levels, such as the word layer) to influence earlier processes that happen at lower levels of representation (the letter layer and/or the feature layer). This contrasts sharply with the model of Selfridge, where the flow of information is strictly bottom up (from the image demon, to the feature demons, to the cognitive demons, and finally to the decision demon). This difference between these models provides an example of a crucial theoretical distinction between modularity and interactivity of which we will see more examples later in this chapter (see Modularity Versus Interactivity). Another important difference between the two models is that in the McClelland and Rumelhart model, processes can take place in parallel such that several letters can be processed at the same time, whereas in Selfridge's model, one letter is processed at a time in a serial manner. Figure 9.12 shows that the model of McClelland and Rumelhart permits both excitatory and in-

Figure 9.11 The pandemonium model of letter recognition of Selfridge (1959). For written input, the reader must recognize a pattern that starts with the analysis of the sensory input. The sensory input is stored temporarily in iconic memory by the image demon, and a set of twenty-eight feature demons decode the iconic representations. The cognitive demons are activated by the representations of letters with these features, and the representation that best matches the input is then selected by the decision demon. Adapted from Coren et al. (1994).

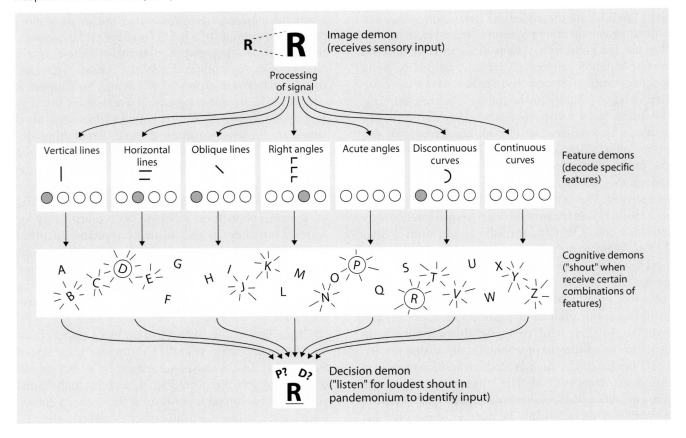

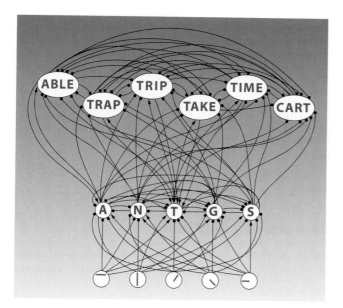

Figure 9.12 Fragment of a connectionist network for letter recognition. Nodes at three different layers represent the features of letters, letters, and words. Nodes in each layer can influence the activational status of the nodes in the other layers by excitatory (arrows) or inhibitory (lines) connections. Adapted from McClelland and Rumelhart (1981).

hibitory links between all the layers. If a reader reads the word *trip*, for example, then activations will ensue in all representational layers that match with the features and letters of the word *trip* and with the word *trip* itself. But when the word node *trip* is activated, it will send out inhibitory signals to lower layers, and letters and features that do not match the word *trip* will be inhibited.

Computational models are very important in cognitive neuroscience because they enable scientists to precisely formalize which components are essential to a given cognitive process and also how these components might interact. The empirical validity of a model can be tested on real-life behavioral phenomena or against physiological data. The connectionist model of McClelland and Rumelhart does an excellent job of mimicking reality for a phenomenon that has been labeled the *word superiority effect*. Experimentally, this effect can be observed when three types of visual stimuli are briefly flashed to the subjects. The stimulus can be a word (*trip*), a nonword (*pirt*), or a letter (*t*). The subjects are then asked to say whether they had seen a *t* or a *k*. Subjects perform better in this identification task when the letter is presented within a real word relative to a nonword, but interestingly, they also are better at identifying the letter when it is presented in the word than when it is presented as the individual letter. This remarkable result indicates that words are probably not perceived on a

letter-by-letter basis. The word superiority effect can be explained in terms of the McClelland and Rumelhart model, because the model proposes that top-down information of the words can either activate or inhibit letter activations, thereby helping the recognition of letters

As we learned in earlier chapters, the brain is well equipped for pattern recognition, so let's fill in the lower levels of the letter recognition models with physiological details. Single-cell recording techniques, for example, have enlightened us about some of the basics of visual feature analysis. We know how the brain analyzes edges, curves, and so on. But unresolved questions remain because letter and word recognition are not really understood at the cellular level, and recordings in monkeys are not likely to enlighten us about letter and word recognition in humans. Recent studies using PET and fMRI in humans have started to shed some light on where humans may process letters in the brain.

If subjects are presented with letters or letter-like symbols, the initial perceptual analysis of written symbols bilaterally activates areas of the brain that are specialized for visual processing, such as striate cortex and adjacent areas of the extrastriate visual cortex. Interestingly, these activations are not specific to letter-like symbols. The lack of specialization may have to do with the fact that humans invented written language only about 5500 years ago. In contrast, auditory processing of spoken language is much older, and may have been affected by evolutionary pressures, thereby creating neural specialization. The fact that written language is relatively young has some implications for visual word recognition: It is unlikely that reading is represented by a specialized input system. Rather, some aspects of reading will rely on some general visual processes. For example, to process the visual input, the brain has to analyze the featural aspects of the letters. This process can make use of the system that performs feature analysis in recognizing visual objects in general.

Neural Substrates of Written Word Processing The actual identification of orthographic units may take place in occipital-temporal regions of the left hemisphere. In studies using fMRI in normal subjects, Gregory McCarthy at Duke University and his colleagues (Puce et al., 1996) contrasted brain activation in response to letters with activation in response to faces and visual textures. They found that regions of the occipital-temporal cortex activated preferentially in response to unpronounceable letter strings (Figure 9.13). Interestingly this finding confirmed results from an earlier study by the same group (Nobre et al., 1994) in which intracranial electrical recordings were made

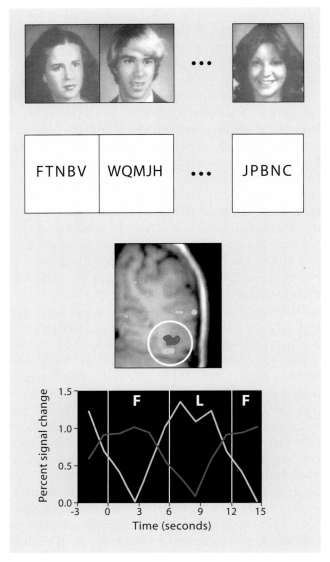

Figure 9.13 Activation of letters (L) compared to activation of faces (F) in the ventral part of the left hemisphere (white circle in brain scan). The slice is in the coronal plane and shows only the left hemisphere at the level of the anterior occipital cortex. The scan shows activated voxels superimposed on the corresponding anatomical image. Faces activated a region of the lateral fusiform gyrus (yellow), whereas letter strings activated a region of the occipitotemporal sulcus (pink). The graph shows the corresponding time course of activations averaged over all alternation cycles for faces (yellow lines) and letter strings (pink lines). The y axis shows the percentage of signal change with vertical ticks of 0.5%. Time (in seconds) is shown on the x axis. Note that hemodynamic responses are slow, peaking between 6 seconds and 9 seconds after stimulus onset (t=0). From Puce et al. (1996).

from this brain region in patients who later underwent surgery for intractable epilepsy. In this study they found a large negative polarity potential around 200 msec (N200) in occipital-temporal regions, in response to the visual presentation of letter strings. This area was not sensitive to other visual stimuli such as faces, and importantly, it also appeared to be insensitive to lexical or semantic features of words. Finally, lesions of this area can give rise to "pure alexia." *Pure alexia* is a condition in which patients cannot read words even though other aspects of language are normal. This convergent neurological and neuroimaging evidence gives clues to how the human brain solves the perceptual problems of letter recognition.

The Recognition of Words

Word or lexical processing is a well-investigated phenomenon in psycholinguistics. Most scientists agree on its main components: lexical access, lexical selection, and lexical integration. The output of perceptual analysis is probably projected onto word form representations in the mental lexicon. Labeled **lexical access,** this results when representations are activated and spread to semantic and syntactic attributes of the word forms.

Lexical access differs, in most cases, for the visual and auditory modalities, and decoding the input signal such that it can make contact with word form representations in the mental lexicon presents the language comprehender with a distinct set of challenges in either case. For the written input there is the question of how we can read not only words with a spelling that does not directly translate into sound, as, for example, the word *colonel,* but also pseudowords that have no matching word form as, for example, *lonocel.* Pseudowords cannot be read by directly mapping the orthographic output onto a word form because there is none. Therefore, researchers have suggested that we need to translate the letters into their corresponding phonemes to be able to read *loconel.* On the other hand, if we want to read aloud the word *colonel,* we would mispronounce it if we directly translated it into the corresponding phonemes, and to prevent such errors we might use a direct route from the orthographic units to the word form representation. This has led researchers to propose dual-route reading models, with one direct route from orthography to word form, and one indirect route (or assembled route) where the written input is translated into phonology before it is mapped onto the word form (for more details, see From Written Text to Word Representation: One or Two Pathways?).

As noted earlier, the continuity of the speech signal in spoken input is different from the clear boundaries in written input. A listener is challenged to segment speech and to control the speed of its input. When we read a text

From Written Text to Word Representation: One or Two Pathways?

Languages such as English pose difficulties in translating letters (graphemes) to sounds (phonemes), called *grapheme-to-phoneme conversion*. For example, the combination "ph" is pronounced differently in *physiology* and in *uphill*, and "c" is pronounced differently in *cop* and *cerebellum*. So, how do we know how to pronounce *physiology* when we read it? We might be relying on rules based on the language's regularities. For example, the combination "mb" signals separate pronunciation of the "m" and the "b" when it is in the middle of a word (*ambulance*) but not when it is at the end (*bomb*). But it is difficult to come up with rules that will predict all combinations of letters and sounds—consider *bomber* and *bombard*. A rule-based system alone will not do the job. This problem led Max Coltheart and colleagues (1993) at Macquarie University in Australia to propose that there must be a direct route from reading text to word representation: from the whole-word orthographic input to representations in the mental lexicon. Thus, getting from written text to word representations in the mental lexicon might be accomplished in two ways, or by dual routes: grapheme-to-phoneme translation (the so-called *assembled route*), or written input directly to the mental lexicon (or the *direct route*).

Evidence for a dual route comes from patients with acquired dyslexia. These patients' reading problems are due to brain damage. The deficit is modality specific; that is, the patients can comprehend spoken language and may even be able to produce written language (alexia without agraphia). Two types of acquired dyslexia have been found. Patients with deep or phonological dyslexia are unable to read aloud pseudowords like *grimp*, but

they do not have any problem with reading irregular words like *broad* or difficult ones like *chrysanthemum*. Thus, they cannot read aloud words that do not have a representation in the mental lexicon, but have no difficulties with words that are in the lexicon. In addition, the patients also make semantic errors (*rose* for *iris*). Such problems cannot be explained as a loss of phonological representations, because patients with deep dyslexia usually perform well on a rhyming task if the stimuli are presented auditorily. Instead, the patients rely exclusively on the direct lexical route when reading. In contrast, patients with surface dyslexia rely only on rules based on regularity. They overregularize the pronunciation of irregular words (e.g., they will read "heed" for *head*) and thereby show a propensity for a direct translation from the letter to the sound representations. These forms of dyslexia provide a powerful double dissociation that tells us there might be two routes by which text that we read can be converted into verbal output.

However, a computational model attempts to explain these data by assuming just a single route that always makes use of phonological information—it just uses more in some cases than in others (Seidenberg and McClelland, 1989). In this model there is a continuous interaction between written input units and phonological units, and feedback allows the model to learn about correct pronunciations of words. This model has been successful in reproducing some of the phenomena that we just described, but the model is not very good at reading pseudowords (words that sound and look like real words but are not, e.g., *fisch*), something that humans have no difficulty with.

in a book, we can go back and reread it. But when we are trying to understand what someone is saying, we can lose track of the conversational flow. Spoken language, then, has a temporal dimension that should be factored into any model of spoken language comprehension.

An influential model in this respect is the cohort model of British psychologist William Marslen-Wilson (Marslen-Wilson and Tyler, 1980). This model (Figure 9.14) assumes that processing in speech starts with the very first sound or phoneme that the listener has identified as the onset of a word. It is clear that initially, when not all the perceptual information is yet available, more

than one representation will be activated because more than one representation will fit the first part of the output of the perceptual system. This means that of the activated word form representations, the one that best matches the sensory input has to be selected—a process labeled *lexical selection*, mentioned earlier. Here is an example of how lexical selection plays a role in understanding words: The word initial sound "ca" of the word *captain* fits *captain*, but also fits words like *cap, capital, capitalist, caption*, and *capitol*. When we hear "ca," all the word forms that match this sound will become activated in what is called a *word initial cohort*. When more

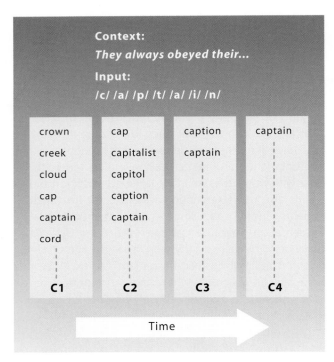

Figure 9.14 Cohort model of spoken word recognition. Initially all words that start with the same initial sound are activated (C1). As time progresses, fewer competitors match with the actual speech signal and are eliminated from the cohort, until only the actual candidate remains (see text for further explanation).

perceptual information becomes available, the number of activated representations narrows to the element that best matches the input, resulting in selection of the word *captain,* and deactivation of the other nonmatching word forms. In this model, selecting the appropriate word form depends on the incoming sensory information and the number of competitors in the word initial cohort. A word is selected at its uniqueness point, the time when a word is uniquely distinguishable from all its competitors.

This model clearly takes into account the temporal aspects of speech perception. However, there is another difficulty that faces the listener: We only have thirty to forty phonemes to represent words in our language, but our mental lexicon contains tens of thousands of words. This necessarily means that many words resemble each other, and that they have other words embedded within them. For example, a word such as "strange" contains the words "strain" and "range," and it resembles the words "straight" and "change." The listener has the difficult job of identifying what the actual words in the input stream are. As we noted previously, he or she can use sublexical cues such as stress patterns in the words to do this.

In addition, to address the problem of identifying words, recent models for auditory word recognition propose that the competition between word candidates is not limited to words that have the same word initial cohort (as in the cohort model just described). Instead, all lexical forms that partially overlap with the speech input become activated. Let's illustrate this with our previous example. If we hear the word "strange," all word forms that partially match this input are activated, such as "strain," "range," "straight," and "change." Because only the word form "strange" completely matches the actual input signal, it will win the competition against all the word forms that do not match the input signal completely, and will do so by means of inhibiting the competitors.

An example of a study from Anne Cutler's group (e.g., Norris et al., 1995) provides evidence for the idea that competition plays a role in word recognition. These researchers showed that subjects were *slower* to recognize real words embedded in nonsense strings that can activate competing words (e.g., "mess" in "domess," which could activate the real word "domestic"), than to recognize real words that are embedded in nonsense strings that have no overlap with real words (e.g., "mess" in "bemess," which could not activate any real words).

We have now sketched the ingredients that are important to understanding written and spoken words. We started out at the beginning of the chapter with a description of the mental representations of words and their meanings, and then surveyed the perception of speech and written input. In this section, we described how this perceptual information translates into the activation of word forms in the mental lexicon. Throughout we have given you some hints of where these processes and representations may be localized in the brain. This functional neuroanatomical story is based on convergent evidence from specific functional losses in some patients, the relationship between patient brain lesion data and their language deficits, and some imaging and intracranial recording data. More recently, some cognitive neuroscience studies have been tackling this organization directly using neuroimaging tools to study the brain circuits and systems involved in reading and listening to words.

BRAIN SYSTEMS FOR WORD RECOGNITION

For auditory word perception Jeffrey Binder and colleagues (2000) proposed a hierarchical model of speech processing in the brain that is illustrated in Figure 9.15. This model is based on studies that used fMRI to identify brain regions that become activated in relation to

Lateral slices ◀───────────────────────────────────▶ Medial slices

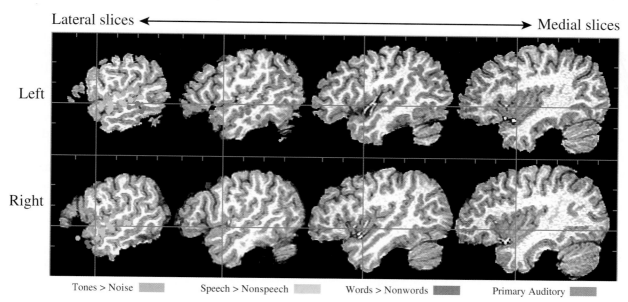

Left

Right

Tones > Noise �   Speech > Nonspeech ▒   Words > Nonwords ▒   Primary Auditory ▒

Figure 9.15 A hierarchical processing stream for speech processing (see text for explanation). Four slices are shown for the left and right hemispheres. Heschl's gyri, the site of primary auditory cortex, are indicated in purple. Indicated in blue are areas of the dorsal superior temporal gyri that are activated more by frequency-modulated tones than by random noise. Yellow areas are clustered in superior temporal sulcus and are speech specific; that is, they show more activation for speech sounds (words, pseudowords, or reversed speech) than for nonspeech sounds. Finally, red areas include regions of the middle temporal gyrus, inferior temporal gyrus, angular gyrus, and temporal pole and are more active for words than for pseudowords or nonwords. Note that these "word" areas mostly are lateralized to the left. The latter areas were identified in a number of studies (Démonet et al., 1992, 1994; Perani et al., 1996; Binder et al., 1999, 2000). From Binder et al. (2000).

subcomponents of speech processing. First, the stream of auditory information proceeds from auditory cortex in Heschl's gyri to the superior temporal gyrus. In these parts of the brain, no distinction is made between speech and nonspeech sounds, as noted earlier. The first evidence of such a distinction is in the adjacent superior temporal sulcus, but no lexical-semantic information is processed in this area. From the superior temporal sulcus, the information proceeds to the middle temporal gyrus and the inferior temporal gyrus, where phonological and lexical-semantic aspects of words may be processed. The next stage involves analysis in the angular gyrus, posterior to the temporal areas just described (see Chapter 3), but then in more anterior regions in the temporal pole. Only these latter four areas seem to be lateralized more to the left hemisphere.

For written input we have seen that the initial perceptual analysis of letters takes place in the primary and secondary visual cortex of both hemispheres, areas of the brain that are specialized for visual processing but not specific to letters. Occipital-temporal regions of the left hemisphere may be specialized for the identification of orthographic units (see Figure 9.13). Reading studies that compared areas of the brain that become activated during

the processing of words and pseudowords also identified the middle temporal gyrus as involved in phonological processing. Interestingly, the middle temporal gyrus is more interested in words than pseudowords, which indicates that these areas also might be involved in more semantic aspects of word processing. Finally, when we need to translate the orthographic input into phonological information that can be used to pronounce the word, the left inferior frontal gyrus, including the ventral part of Broca's area, seems to play a role.

Usually words are not processed in isolation but in the context of other words (sentences, stories, etc.). To understand words in their context, we have to integrate syntactic and semantic properties of the recognized word into a representation of the whole utterance.

THE ROLE OF CONTEXT IN WORD RECOGNITION

One question that puzzles psycholinguistic researchers is at what point during language comprehension does linguistic context influence word processing. More specifically, does context influence word processing before or after lexical access and lexical selection are complete? This question has to do with the issue of modularity versus interactivity (see Modularity Versus Interactivity). Consider the following sentence, which ends in an ambiguous word having more than one meaning: "The tall man planted a tree on the bank." *Bank* can mean both "money institution" and "side of the river." Semantic integration of the meaning of the final word *bank* into the context of the sentence allows us to interpret *bank* as the "side of the river" and not as a "money institution." The question that is of relevance here is, When does the sentence's context influence the activation of the multiple meanings of the word *bank?* Do both the contextually appropriate meaning of *bank* (in this case, "side of the river") and the contextually inappropriate meaning (in this case, "monetary institution") become briefly activated regardless of the context of the sentence? Or does the sentence context immediately constrain the activation to the contextually appropriate meaning of the word *bank?* From this example we can already see that two types of representations play a role in word processing in the context of other words: lower-level representations, those constructed from sensory input (in our example, the word *bank* itself), and higher-level representations, those constructed from the context preceding the word to be processed (in our example, the sentence preceding the word *bank*). Contextual representations are crucial to determine in what sense or what grammatical form a word should be used. But without sensory analysis, no message representation can take place. The information has to interact at some point. When this occurs differs in competing models.

As alluded to previously models of word recognition also vary in the amount of interactivity they propose. In general, three types of models attempt to explain word comprehension. *Modular models* (also called *autonomous models*) claim that normal language comprehension is executed within separate and independent modules. Thus, higher-level representations cannot influence lower-level ones, and therefore the flow is strictly data driven or bottom up. In modular models the representation of the context information cannot affect lexical access and lexical selection processes. For our *bank* example, an autonomous model of word recognition would predict activation of both the "side of the river" meaning *and* the "monetary institution" meaning, even if we have a context such as "The tall man planted a tree on the bank." Only after lexical access, selection of the word form, and activation of both meanings can context play a role in the selection of the contextually appropriate meaning of the word, in our example, the "side of the river" meaning.

In contrast, *interactive models* maintain that all types of information can participate in word recognition. In

these models, context can have its influence even before the sensory information is available, by changing the activational status of the word form representations in the mental lexicon. McClelland and colleagues (1989) have proposed this type of interactivity model. Many of these models also do not suggest that we actually have representation of word forms as whole units. A certain word is recognized when particular nodes in a distributed network become activated together. For our "bank" example, this would mean that the context of the sentence can guide the selection of the contextually appropriate meaning, and therefore we would find no activation of the "monetary institution" meaning of the word *bank* in our example.

Between these two extreme views is the notion that lexical access is autonomous and not influenced by higher-level information, but that lexical selection can be influenced by sensory and higher-level contextual information. In these *hybrid models* information is provided about word forms that are possible given the preceding context, thereby reducing the number of activated candidates. The version of Marslen-Wilson's model of auditory word recognition that we discussed earlier is an example of such a hybrid model.

A nice example of a study that addressed the question of modularity versus interactivity in word processing is one that Pienie Zwitserlood of the University of Münster in Germany performed while at the Max Planck Institute for Psycholinguistics in the Netherlands (Zwitserlood et al., 1989). She asked subjects to listen to short texts such as: "With dampened spirits the men stood around the grave. They mourned the loss of their *captain.*" At different points during the auditory presentation of the word "captain" (e.g., when only "c" or only "ca" or only "cap," etc., could be heard) a visual target stimulus was presented. This target stimulus could be related to the actual word *captain,* or to an auditory competitor, for example, *capital.* In this example, target words could be words like *ship* (related) or *money* (unrelated). In other cases a pseudoword would be presented. The task was to decide whether the target stimulus was a word or not (lexical decision task). The results of this study showed that subjects were faster to decide that *ship* was a word in the context of the story about the men mourning their fellow, and slower to decide that *money* was a word, even when only partial sensory information of the stimulus word "captain" was available (i.e., before the whole word was spoken). Apparently, the lexical selection process was influenced by the contextual information that was available from the text that the subjects had heard before the whole word "captain" was spoken. However, some sensory information of the word "captain" had to be heard, or the reaction time benefit was not found. This result is consistent with the ideas that the process of lexical access can be guided by the sensory input and not much by the higher-order contextual information alone, and importantly, that lexical selection can be influenced by sentence context. We do not know for certain which type of model fits word comprehension the best, but there is growing evidence from studies like that of Zwitserlood and others that at least lexical selection is influenced by higher-level contextual information.

Integration of Words in Sentences

In the foregoing, we examined processes dedicated to recognizing a word in the linguistic input. But normal language comprehension requires more than just recognizing individual words. To understand the message conveyed by a speaker or a writer, we have to integrate the syntactic and semantic properties of the recognized word into a representation of the whole sentence or utterance. Let's consider again the following sentence: "The tall man planted a tree on the bank." Why do we read *bank* to mean "side of the river" instead of "monetary institution"? We do so because the rest of the sentence has created a context that is compatible with one meaning and not the other. Let's consider another example: "The pianist rose to the applause of the audience." Once again, there is more than one way to interpret the sentence, and via integration processes, we interpret *rose* as a verb ("to stand") and not a noun ("a flower"). This integration process has to be executed in real time—as soon as we are confronted with the linguistic input. So if we come upon a word like *bank* in a sentence, we are usually not aware that this word has an alternative meaning because the appropriate meaning of this word has been rapidly integrated into the context.

Understanding a word with respect to the higher-order representation of sentence meaning involves semantic and syntactic integration processes. Higher-order semantic processing is important to determine the right sense or meaning of words in the context of a sentence, as with ambiguous words such as *bank*, which have the same form but more than one meaning. Semantic analysis is also needed for making sense of nonambiguous words such as *piano*, whose meanings can be emphasized in different contexts (as in "the piano is a heavy instrument" versus "the piano is a beautiful instrument").

For a full interpretation of linguistic input, though, we also have to assign a grammatical structure to the input. Semantic information in words alone is not enough to understand the message, as is clear from the following sentence: "The little old lady bites the gigantic dog." Syntactic analysis of this sentence reveals its structure: Who was the actor, what was the theme or action, and what was the subject? The syntax of the sentence demands that we imagine an implausible situation where an old lady is biting and not being bitten.

Syntactic analysis goes on even in the absence of real meaning. Normal subjects are faster at detecting a target word in a sentence when it does not make any sense but is grammatically correct, than when the grammar is locally disrupted. Typically, subjects in such studies are asked to listen to sentences in different conditions and are to press a button as soon as they hear a target word such as *kitchen*. In a baseline condition, they would be presented with normal sentences. A normal sentence might be, "The maid was carefully peeling the potatoes in the garden because during the summer the very hot kitchen is unbearable to work in." In this baseline condition, subjects take about 300 msec to press a button (from the onset of the target word "kitchen"). These reaction times can be compared with reaction times to sentences that are semantically absurd but grammatically normal, such as, "An orange dream was loudly watching the house during smelly nights because within these signs a very slow kitchen snored with crashing leaves." Here the subjects' mean response to the target word is slowed by about 60 msec. Yet when the syntax of the sentence is also disrupted, the response times are even slower (by another 45 msec). An example of such a sentence is, "An orange dream was loudly watching the house during smelly nights because within these signs a slow very kitchen snored with crashing leaves." The word order of the phrase *a slow very kitchen* is grammatically incorrect. These types of findings inform us that syntactic analysis proceeds even when sentences are meaningless.

From a psycholinguistic point of view, how do we process the structure of sentences? When we hear or read sentences, we activate word forms (lexemes) that in turn activate the lemma information. As we discussed earlier, the lemma is hypothesized to contain not only information about syntactic properties of the word (e.g., whether it is a noun or a verb) but also information about the possible sentence structures that can be generated given a certain verb. For example, a verb such as *eat* requires a subject and an object, like in "The cat (subject; the eater) *eats* the food (object; the eatee)." As soon as the appropriate lemma information is retrieved, structure building can start. This means that words in a sentence are assigned their syntactic roles and clustered into groups or syntactic phrases. For example, "The cat eats the food" consists of a noun phrase (*The cat*), a verb phrase (*eats*), and another noun phrase (*the food*). This process is incremental: As soon as the information for a lemma is retrieved, it is inserted into the constituent structure that has already been built. This immediate assignment of syntactic structure to incoming words has been labeled **parsing,** and the processor that does the parsing is called the *parser*. Unlike the representation of words and their syntactic properties (lexemes and lemmas) that are stored in our mental lexicon, representations of whole sentences are not stored in our brain. It just is not feasible for our brain to store the incredible number of different sentences that can be written and produced. Parsing is therefore a building process that does not and cannot rely on the retrieval of representations of sentences.

How does parsing work? A very influential theory on how syntactic analysis takes place is the *garden-path model*, championed by Lynn Frazier (1987). The essence of the model is that sentences have a preferred interpretation. This preference could lead to garden-path effects, which means being led to believe something that seems correct at first but is not. To understand this model, we must introduce the idea of sentence structure.

A sentence consists of a linear arrangement of phrases and words, which can be represented by a hierarchical tree that reflects the sentence's structure. An example of a hierarchical tree for the sentence "The spy saw the cop with binoculars" is presented in Figure 9.16, where the sentence consists of two large phrases, a noun phrase and a verb phrase. The noun phrase has an article (*the*) and a noun (*spy*). The verb phrase is divided into the main verb (*saw*); a noun phrase (*the cop*), which has an article (*the*) and a noun (*cop*); and a prepositional phrase (*with binoculars*), which has a preposition (*with*) and a noun (*binoculars*). Mental representations of the tree's components are labeled *syntactic nodes.*

The formation of constituent structures plays a central role in the garden-path model (but also in other models of parsing). The garden-path model assumes that we process syntactic information in a way that minimizes what we have to do to meet the demanding time pressure of normal comprehension. Two mechanisms that would help this economy principle are postulated: *minimal attachment* and *late closure*. The minimal attachment mechanism makes sure that syntactic analysis is done in such a way that the minimum number of additional syntactic nodes must be computed. The late closure mechanism tries to assign incoming words to the syntactic phrase or clause currently being processed.

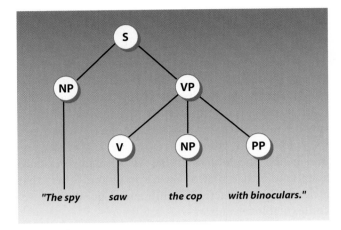

Figure 9.16 Constituent structure of a sentence. As explained in the text, this structure is based on the principles of minimal attachment.

Let's consider the same version of the example sentence represented by the hierarchical tree shown in Figure 9.16. In the example in Figure 9.17, the meaning of the sentence is changed because a different syntactic structure is assigned. This tree leads to the interpretation that the cop rather than the spy was equipped with binoculars. The reason why the interpretation in the tree in Figure 9.16 is preferred over that in Figure 9.17 is because it has fewer and less complex nodes. The minimal attachment and late closure principles were proposed for reasons of economy. Since the language processor must work under an enormous time pressure, the less time required for syntactic analysis, the better. Sometimes the minimal attachment principle leads to mistakes, as, for example, in "Ron loves Holland and his

Figure 9.17 Constituent structure of the same sentence as in Figure 9.16. The hierarchical tree in the present figure presents an example of nonminimal attachment. See the text for a further explanation.

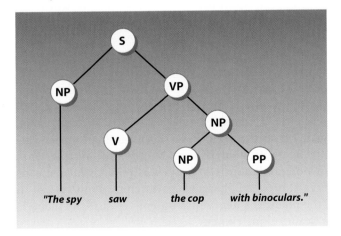

mother enjoyed her trip to Amsterdam." This is a garden-path sentence whose preferred interpretation leads to a wrong solution and hence must be reanalyzed.

Frazier's model of sentence processing is an example of a modular model (see Modularity Versus Interactivity). Syntactic processing is encapsulated, and the first analysis of the sentence is based on its syntactic structure alone and is not influenced by other sources of information such as the meaning of words or more general knowledge of the world. Let's illustrate this with the following example: "The manager sold the couch and the employee sold the desk." In this example, the meaning of the verb *sold* would make it easier to construct a syntactic structure where the noun *employee* starts a second clause, because employees usually are not sold by managers. In linguistic terms, *sold* can go together easily with an inanimate object (e.g., *couch*), but not very easily with an animate object (e.g., *employee*). However, Frazier's modular model claims that independent of this type of contextual information, in the first instance a structure is assigned on the basis of syntactic structure and the principles of minimal attachment and late closure alone. So even in this example, we would be led down the garden path, and we would structure the sentence first as if the manager sold the couch and the employee. Only in a later stage, when semantic information is evaluated by a so-called semantic interpreter, can we revise the structure of the sentence to reflect its actual meaning.

But there are more interactive views of sentence processing, and they propose that other sources of information can immediately influence syntactic processing. According to these models, the semantic information of the verb *sold* in our example would prevent us from being led down the garden path. Many studies indicate that we can immediately use meaning information, our knowledge of the world, and syntactic information to determine the correct structure of a sentence, which supports a more interactive view of sentence processing. To illustrate this we will concentrate here on one study by Thomas Münte and colleagues (1998) at the University of Magdeburg in Germany. These researchers asked the question, Can the way we process sentences be influenced immediately by our conceptual knowledge of how events are temporally organized? To test this, subjects were asked to read sentences that differed only in the first word, such as, for example, "After/Before the scientist submitted the paper, the journal changed its policy." According to our knowledge of the world, a more "logical" sequence of events would be that the journal changes its policy *after* the scientist had submitted the paper, because this sequence follows the actual order of

events. In contrast, the *before* sentences signal a reverse order of events and are therefore more difficult to understand. Event-related potentials (ERPs) were recorded while the subjects read the sentences. The issues the researchers addressed were if and when there would be any difference in the brain's response to the words *after* and *before* and if that would influence the processing of the rest of the sentence. Interactive models would predict that our conceptual knowledge of order of events can immediately influence the processing of the sentence, whereas modular models would predict that this influence does not occur until later. The results are more in line with an interactive view. As can be seen in Figure 9.18, beginning almost immediately, the ERPs to the word *before* are more negative in polarity than they are to the word *after*, and this effect lasts over the course of the sentence. These findings indicate that conceptual knowledge of the temporal order of events can indeed influence the processing of sentences almost from the onset, and moreover, this has lasting consequences for the processing of the rest of the sentence.

NEURAL SUBSTRATES OF SYNTACTIC PROCESSING

What do we know about the brain circuitry involved in syntactic processing? Let's consider evidence from brain-damaged patients and brain imaging studies to try and answer this question. Some brain-damaged patients have severe difficulties in producing or understanding the structure of sentences. These deficits are apparent in patients with **agrammatic aphasia,** who generally produce two- or three-word sentences that consist exclusively of content words and hardly any function words (words such as *and, then, the, a,* etc.), or grammatical or morphological markers and inflections (such as *was pushing*). An example of such an impoverished sentence would be, "Son . . . university" instead of "My son is at the university."

In language comprehension, agrammatic aphasic patients often have great difficulties understanding complex syntactic structures. So when they hear the sentence, "The gigantic dog was bitten by the little old lady," they would most likely understand this sentence to mean that the little old lady was bitten by the gigantic dog. This problem in assigning syntactic structures to sentences traditionally has been associated with lesions that include Broca's area of the left hemisphere, a region we will describe later (please refer ahead to Figure 9.24). But a lot of variability has been observed and some agrammatic aphasic patients do not have lesions in Broca's area as it is classically defined.

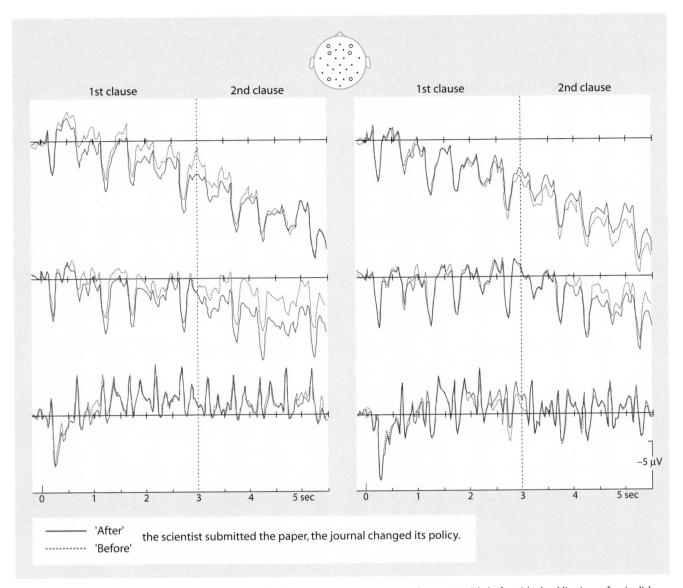

| | 1st clause | 2nd clause | | 1st clause | 2nd clause |

———— 'After'   the scientist submitted the paper, the journal changed its policy.
·············· 'Before'

Figure 9.18 Event-related potentials (ERPs) recorded in response to sentences that start with *before* (dashed line) or *after* (solid line). These ERPs are recorded from prefrontal **(top),** frontal, and occipital **(bottom)** scalp sites of the left (left side) and right (right side) hemispheres. At the left prefrontal site, the waveforms for *before* and *after* start to diverge almost immediately (at 300 msec). At prefrontal and frontal sites, this effect lasts throughout the sentence and is largest during the second clause. See text for a further explanation. From Münte et al. (1998).

Other evidence that Broca's area may be important for processing syntactic information comes from PET studies by David Caplan and colleagues (2000) at Harvard Medical School. In these studies PET scans were made while subjects read sentences that varied in syntactic complexity. For example, sentences with a relatively simple syntactic structure such as, "The child enjoyed the juice that stained the rug," were compared to sentences with a more complex syntactic structure, such as "The juice that the child enjoyed stained the rug." The second sentence is more complex because the main sentence (*the juice stained the rug*) is interrupted by a *relative clause*

(*that the child enjoyed*). Therefore, the thematic role (subject? object?) of the first noun (*the juice*) cannot be assigned until the verb that follows the relative clause (*stained*) is encountered. The first noun phrase (*the juice*) has to be kept in some kind of working memory buffer. This type of working memory load is not taxed as much in the first sentence because there the thematic roles can be assigned almost immediately. Caplan and colleagues found increased activation in Broca's area for the more complex syntactic structures (Figure 9.19).

However, the specificity of Broca's area for syntactic comprehension has been questioned. Sentence com-

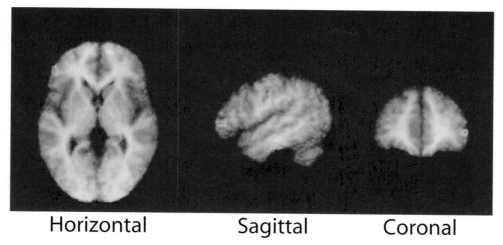

Horizontal   Sagittal   Coronal

Figure 9.19 Increase in blood flow in Broca's area when subjects are processing complex relative to simple syntactic structures. See text for a further explanation. Adapted from Caplan et al. (2000).

plexity manipulations in other studies led to activations of more than Broca's area alone. Marcel Just and colleagues (1996) reported activation in Broca's and Wernicke's areas (lateral frontal and posterior-superior temporal cortex, respectively—see later sections) and in the homologous areas in the right hemisphere. In addition, activations of Broca's area have been found in sentence production studies and in studies that did not look at sentence processing at all, but, for example, at memorizing, maintaining, and retrieving lists of items. In general then, the question of whether Broca's area is specific to syntactic processing remains a matter of debate. Some

have argued that this area has a more general role in working memory operations that are not specific to syntax.

Another candidate for syntactic processing in the brain has been identified by PET as being in the anterior portions of the STG, in the vicinity of area 22 (Figure 9.20a). Nina Dronkers at the University of California at Davis and colleagues (1994) also implicated this area in aphasics' syntactic processing deficits (Figure 9.20b). Since parsing is such a complex process, it is not at all surprising that more than one area in the brain would subserve it. Future research will have to elucidate further which aspects of the parsing process are subserved by

(a)

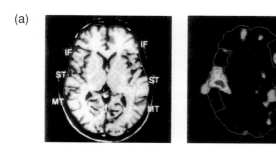

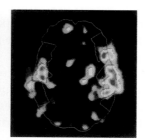

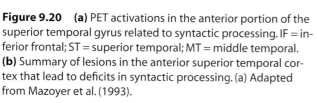

Figure 9.20 **(a)** PET activations in the anterior portion of the superior temporal gyrus related to syntactic processing. IF = inferior frontal; ST = superior temporal; MT = middle temporal. **(b)** Summary of lesions in the anterior superior temporal cortex that lead to deficits in syntactic processing. (a) Adapted from Mazoyer et al. (1993).

(b)

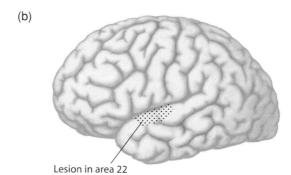

Lesion in area 22

which areas in the brain, but scientists are narrowing in on the answer using a variety of methods.

Speech Production

We have focused mainly on language comprehension, and now we turn our attention to language production. To provide a framework for discussion, we will concentrate mostly on one influential model for language production proposed by Willem Levelt (1989) of the Max Planck Institute for Psycholinguistics in the Netherlands. Figure 9.21 illustrates this model.

A seemingly trivial, but nonetheless important difference between comprehension and production is our starting point. Whereas language comprehension starts with spoken or written input that has to be transformed into a concept, language production starts with a concept for which we have to find the appropriate words.

The first step in speech production is to prepare the message. Levelt maintains that there are two crucial aspects to message preparation: *macroplanning* and *microplanning*. The speaker must determine what she wants to express in her message to the listener. The formulation of the message will be different when we direct someone to our home than when we want someone to close the door. This communicative intention is planned in goals and subgoals expressed in an order that best serves the communicative plan. This aspect of message planning is macroplanning. Microplanning, in contrast, proposes how the information is expressed, which means taking perspective. If we describe a situation where there is a house and a park, we must decide whether to say that "the park is next to the house" or "the house is next to the park." The microplan determines word choice and the grammatical role the words play (e.g., subject, object, theme).

The output of the macroplanning and microplanning is a conceptual message that constitutes the input for the hypothetical *formulator*, which puts the message in a grammatically and phonologically correct form. During grammatical encoding, a message's surface structure—its syntactic but not the conceptual representation, including information such as "is subject of," "is object of," the grammatically correct word order, and so on—is computed. The lowest-level elements of surface structure are the lemmas. These are stored in the mental lexicon and, as introduced earlier, contain information about a word's syntactic properties (e.g., whether the word is a noun or a verb, gender information, and other grammatical features) and its semantic specifications, and/or the conceptual conditions where it is appropriate to use a certain word. Lemmas in the mental lexicon are organized in a

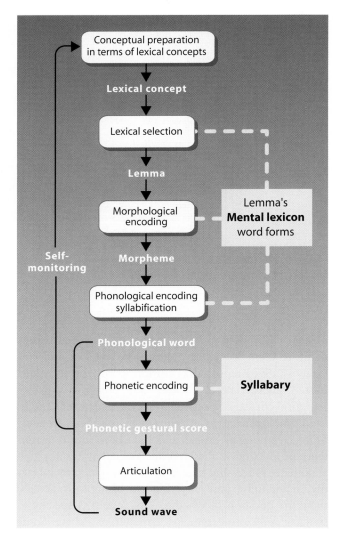

Figure 9.21 Outline of the theory of speech production developed by Willem Levelt. The processing components in language production are displayed schematically. Word production proceeds through stages of conceptual preparation, lexical selection, morphological and phonological encoding, and articulation. Speakers monitor their own speech by making use of their comprehension system. Adapted from Levelt (1999) and Levelt et al. (1999).

network that links them by meaning; as you might remember from the beginning of this chapter and is illustrated in Figure 9.1, which is a fragment of a lexical network.

Imagine that a subject is presented with a picture of a sheep, and her task is to name the picture. This is what will happen according to Levelt's model: First, the concept that represents sheep is activated, but also concepts related to the meaning of sheep are activated: goat, wool, milk. Evidence for the idea that related concepts are activated on seeing the picture representing a word like *sheep* comes from the picture-word interference paradigm, where subjects are asked to name pictures as fast and as accurately as they can. Soon after the picture is presented, an interfering auditory word is presented. The naming latency for the word *sheep* will increase if the interfering stimulus is the word *goat*, but not when the interfering stimulus is the word *house*.

Activated concepts in turn activate representations in the mental lexicon, starting with nodes at the lemma level to access syntactic information such as word class (in our example, *sheep* is a noun, not a verb). At this point, the lemma appropriate to the presented picture must be retrieved, which is called *lexical selection* in word production. The selected lemma (or the lemma that is most highly activated, in our example, "sheep") activates the lexeme or word form. This lexeme contains both phonological information and *metrical information,* which is information about the number of syllables in the word and the stress pattern (in our example, *sheep* consists of one syllable that is stressed). The process of phonological encoding makes sure that the phonological information is mapped onto the metrical information. Sometimes we cannot activate the sound form of a word: This is known as the "tip of the tongue" (TOT) state. You most likely have had a TOT state: You know a lot about the word, you can say it has four legs and white curly hair, you visualize it in your mind, you reject other words that do not match the concept (e.g., *goat*), and if someone tells you the word's first letter you probably say, "Oh yes, 'sheep.'"

In addition to mental blocking on a word, speech errors might also happen during the transition from the lemma level to the lexeme level. Sometimes we mix up speech sounds or exchange words in a sentence. But if all goes well, the appropriate word form is selected and phonetic and articulatory programs are matched. In the last phase of speech production, we plan our articulation: The word's syllables are mapped onto motor patterns that move the tongue, mouth, and vocal apparatus to generate the word. At this stage, we can repair any errors in our speech, for example, by saying "um" and gaining more time to generate the appropriate term. Not only do production errors happen in normal speech, but also brain damage can affect each of the processing stages. Some anomic patients (deficit in naming) are afflicted with an extreme TOT state. When asked to name a picture, they often can give you a fairly accurate description but cannot name the word. Their problem is not one of articulation because they can readily repeat the word aloud; these patients' problems are on the lexeme level. Patients with Wernicke's aphasia produce *semantic paraphasias;* they produce words related in meaning to the intended word. This problem can be one

of inappropriate selection of concepts or lemmas or lexemes. Patients with Wernicke's aphasia also might make errors at the phoneme level by incorrectly substituting one sound for another. Finally, Broca's aphasia is often accompanied by dysarthria, which hinders articulation and results in effortful speech, because the muscles that articulate the utterance cannot be controlled.

So it appears that the distinctions that are made in Levelt's model with respect to the different levels in speech production are adequate, and in the literature there is general agreement on the distinction between conceptual, syntactic, and phonological processing levels. However, there is discussion about the claim that the selection of the lemma of a word has to precede phonological encoding. This means that *only* the word form that matches the selected lemma information will become activated. According to this view, phonological information cannot influence the selection of lemma information.

In contrast to this modular view, interactive models such as the one proposed by Gary Dell (1986) at the University of Illinois suggest that phonological activation begins shortly after the semantic and syntactic information of words has been activated, and actually overlaps in time. Another big difference between modular and interactive models for speech production is that interactive models allow for feedback from the phonological activation to the semantic and syntactic properties of the word, thereby enhancing the activation of certain syntactic and semantic properties.

One study by Miranda van Turennout and her colleagues (1999) at the Max Planck Institute for Psycholinguistics in the Netherlands tested whether lemma selection indeed precedes activation of the appropriate lexeme (as in Levelt's model), or whether phonological information can feed back and change the activation levels of the lemma nodes (as in Dell's model). These researchers presented subjects with pictures of animals and objects whose names started with either an "s" or a "b" sound (e.g., pictures of a sheep, bear, shoe, and book). The study was done in Dutch, where in addition to word class, lemmas also represent the syntactic gender of a word, which can be either common gender or neuter gender. The syntactic gender determines which article to use in Dutch (so *the* has two translations from English to Dutch: *de* is used for common gender, and *het* for neuter gender), and it also modifies the adjectives that precede the common or neuter gender noun. In our example, *sheep* and *book* are neuter gender words, whereas *bear* and *shoe* are common gender. In one task (Figure 9.22), subjects were asked to press a button *only* when the picture represented a word that started with a "b." Furthermore, they had to respond with their left

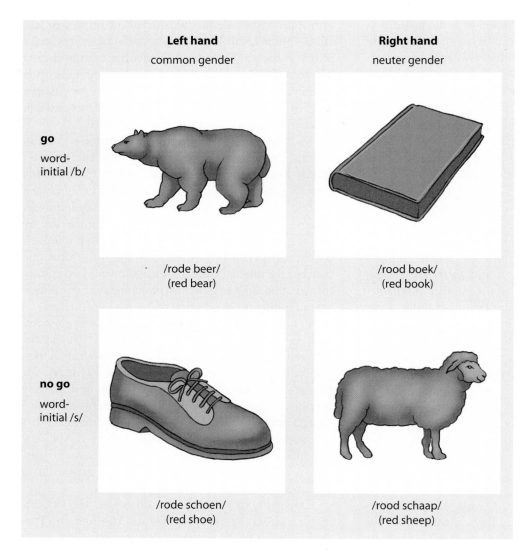

Left hand
common gender

Right hand
neuter gender

go
word-initial /b/

/rode beer/
(red bear)

/rood boek/
(red book)

no go
word-initial /s/

/rode schoen/
(red shoe)

/rood schaap/
(red sheep)

Figure 9.22 Examples of stimuli that were used in an ERP study of word production in Dutch. Below each picture is the Dutch name for it, and the adjective *red* inflected for common (*rode*) or neuter (*rood*) gender. Subjects were asked to make a response only when the words representing the picture started with a "b" (go trials), and to withhold their response when the word started with an "s" (no-go trials). On go trials, they were instructed to respond with their left hand for common gender words and with their right hand for neuter gender words. In addition, subjects also had to make a naming response for all trials. See text for further explanation. Adapted from van Turennout et al. (1999).

hand when the picture represented a common gender word that started with a "b" but with their right hand when it represented a neuter gender word that started with a "b." In this experimental design the decision of whether to respond or not was based on the phonological information at the lexeme level ("s" or "b"), and choice of response hand was based on lemma information (syntactic gender).

The researchers in this study made use of an ERP component that is sensitive to response preparation, the *lateralized readiness potential,* or LRP. The LRP is a brain wave recorded from the scalp over motor cortex and starts to appear when a person plans to make a movement, and hence, before the actual movement is carried out. This potential is a sensitive measure of motor activation because even when the button ultimately is not pressed, the LRP can signal that a response was being prepared (sort of like spying on the subjects' momentary intent). What van Turennout and her colleagues found was that the correct LRP started to develop even when

the subjects were not supposed to respond, that is, when the picture represented a word that started with an "s" (Figure 9.23). What does this mean? The subjects must have had the lemma information available before they had access to the phonological information that could tell them not to respond. These findings therefore indicate that lemma selection might occur before the phonological information at the lexeme level is activated.

NEURAL SUBSTRATES OF SPEECH PRODUCTION

Let's now turn briefly to what we know about language production in the brain. Imaging studies of the brain during picture naming and word generation found activation in the basal temporal regions of the left hemisphere and in the left frontal operculum (Broca's area). In addition, cortical stimulation of the basal temporal region of the left hemisphere in epileptic patients revealed a temporary inability to produce words. The activation in the frontal operculum might be specific to

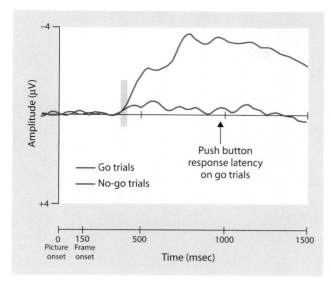

Figure 9.23 Lateralized readiness potential response to go trials (blue line) and no-go trials (red line). The shaded area indicates the time period in which the brains' response to no-go trials is the same as to go trials. This period represents preparation of the response for no-go trials, even when no overt response is given. See text for further explanation. From van Turennout et al. (1999).

phonological encoding in speech production. The articulation of words might involve the posterior parts of Broca's area (area 44), but in addition, studies showed bilateral activation of motor cortex, the supplementary motor area (SMA), and the insula. A lesion in the insula leads to **apraxia** of speech (a difficulty in pronouncing words) in patients with Broca's aphasia.

To summarize, language processing involves representing, comprehending, and communicating symbolic information, either written or spoken. During reading and listening, the most important issues are how the meanings of words are stored in the brain and how they are accessed by visual or auditory inputs. Specializations in the brain code language inputs and produce outputs, but the richness of linguistic capacity escapes simple analysis of language's organization. The most significant features are its amazingly rapid time course and vast store. Where is language hiding in the vast uncharted reaches of the human brain? We have reviewed some tentative answers from research in patients with brain lesions and also from imaging studies in neurologically unimpaired subjects. In the next part of this chapter, we will examine further which complex cortical networks may be involved in language comprehension and production.

NEUROPSYCHOLOGY OF LANGUAGE AND LANGUAGE DISORDERS

Investigators of brain function have struggled with the concept of localization, as we noted at the beginning of this book. The opportunity for research on brain structure-function relations that arises in brain-damaged or brain-diseased subjects has provided numerous insights to perception, attention, and memory, and the same holds for language. The complexity of the language system, as well as the absence of an animal model of language, however, hampers the elucidation of its structure and neural mechanisms, and so the struggle to clarify the neural mechanisms of language remains one of the great scientific challenges facing cognitive neuroscience.

We are unable to describe how the detailed psycholinguistically defined functions of the language system map directly onto the brain's complex anatomical structures. However, as we have seen in earlier sections of this chapter, we do have clues to the language-brain puzzle, many coming from patients with aphasia—language problems following brain damage or disease. In the next section we will begin by reviewing the contributions of aphasia research to our knowledge of language and the brain. We will start with the history of

aphasia and brain-language research, presenting the classic models that have dominated brain and language research for more than a century. Even though current evidence about the brain organization of language functions has gone much beyond the classic models, these models still influence modern-day thinking to a great degree, and the terms and concepts developed in the history of aphasia research are crucial to our understanding of current ideas and concepts about representation of language in the brain. So, we will take a short step back to tell the whole tale. Then, after considering classic models, we will return to ponder what the current studies of aphasia are revealing about the organization of language in the brain.

Aphasia

Brain injury can lead to language disorders called **aphasia,** which refers to the collective deficits in language comprehension and production that accompany neurological damage. Aphasia is extremely common. Approximately 40% of all strokes produce some aphasia, at least

HOW THE BRAIN WORKS

Does the Right Hemisphere Understand Language?

Although the left hemisphere is dominant for language processing, this does not mean that the right hemisphere is completely without language function or is unable to understand language. One bit of evidence that the right hemisphere has some ability to understand language comes from patients with surgically isolated hemispheres. These split-brain patients have had a resection of the fiber tracts that connect the right and the left hemisphere (the corpus callosum), to relieve severe epilepsy. In other words, the right and the left hemispheres can no longer communicate to each other at the cortical level; therefore, when visual information is presented to the left visual hemifield, it goes exclusively to the right hemisphere. Such patients open the possibility to study the language capabilities of the isolated right hemisphere because there can be no transfer of information from the perceptual areas of the right hemisphere to the language areas in the left.

Split-brain patients with a disconnected right hemisphere can make simple semantic judgments and can read material presented to the right hemisphere. However, only grammatically simple sentences can be managed well by the isolated right hemisphere; thus, such patients can misunderstand sentences like "The boy that was hit by the girl cried."

Other evidence for the right hemisphere's role in linguistic processing comes from patients with lesions in the right hemisphere. Although these patients are generally nonaphasic, they do have subtle language deficits. Peter Hagoort and his colleagues (1996) found that patients with right-hemisphere lesions have normal priming effects for words that are associatively related (*cottage cheese*). In contrast, these patients do not show priming effects for words that are not associatively related but from the same semantic category (*dog–horse*). This could mean that the left hemisphere is not very good at processing distant semantic relations but that the right hemisphere is—an idea supported by experimental research. Christine Chiarello

(1991) found a left-visual-field/right-hemisphere advantage for the processing of words that come from the same semantic category, but that have no associative relation like, for example, *dog–horse*. In short, the right hemisphere does have language abilities that might play an important role in the processing of meaning.

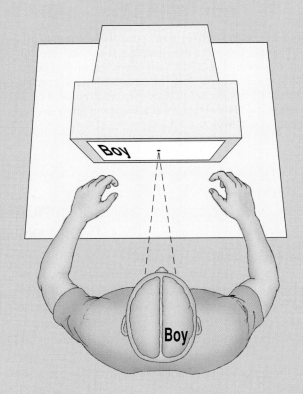

Words presented to the disconnected right hemisphere of a split-brain patient. Stimuli entering the left visual field are transmitted to the right hemisphere. In split-brain patients, the absence of the corpus callosum means that the information (*Boy*) does not get to the language-dominant left hemisphere. This permits the language capacity of the right hemisphere to be investigated.

in the acute period during the first few months after the stroke occurred. However, in many patients the aphasic symptoms persist, and they are confronted with lasting problems in understanding or producing spoken and written language. Primary and secondary aphasic impairments are distinct. Primary aphasia is due to problems with the language-processing mechanisms themselves. Secondary aphasia results from memory impairments, attention disorders, or perceptual problems. Some investigators only classify patients as aphasic when their problems are caused by impairment to the language system (i.e., primary aphasia).

HISTORICAL FINDINGS IN APHASIA

The most useful way to understand classification schemes of aphasia is to review the history of brain and language that emerged by studying aphasic patients. Nineteenth-century investigators held that localized brain lesions should lead to specific functional losses. One such person, the Frenchman Jean-Baptiste Bouillaud, collected evidence from hundreds of brain-injured patients who displayed language problems. Based on the side (left or right hemisphere) and position of the damage, he concluded that language resided in the frontal lobe. This line of thinking was dominant during the period when Paul Broca began his observations of language and brain, and indeed, it was not entirely incorrect.

Broca treated a patient with a leg infection. The patient had already been hospitalized for many years, having lost speech more than twenty years earlier. Ten years prior to coming to Broca's clinic, he had lost the use of his right arm. His name was Leborgne, but other patients called him "Tan" because for many years he had been unable to utter anything but the nonsense word *tan* ("Tan tan tan, tan tan, tan tan tan . . .") and an occasional curse word. The patient died only a few days after being transferred to Broca's clinic. Upon autopsy, Broca observed that the patient had a brain lesion in the posterior portion of the left inferior frontal gyrus. This region, which includes the subdivisions of the inferior frontal gyrus known as the *pars triangularis* and *pars opercularis*, is referred to as *Broca's area* (Figure 9.24). During the autopsy, only a superficial anatomical analysis was performed, and the brain was not sectioned or studied with microanatomical techniques. However, the patient's brain tissue was tested for signs of softening, a clue to damage. Broca found softening from the frontal lobe, where the lesion was noticeable on visual inspection, to the operculum, which included the most inferior portions of the precentral gyrus and adjoining regions.

After this patient, Broca studied other patients with language deficits associated with brain damage. They generally had right hemiparesis (weakness of the right arm and leg) in conjunction with language disorders. Since these persons were right-handed, he concluded that brain areas that produce speech were localized in the inferior frontal lobe of the left hemisphere. This deduction was based on the fact that damage to the right hemisphere led to the most severe deficits in sensation and motor control for the left side of the body, and vice versa. Thus, if the right side is hemiparetic, the left hemisphere is damaged. If language disorders are present, they must also result from left-hemisphere damage.

A second brain region involved in language was found in posterior areas, as opposed to the frontal region described by Broca. In Germany in the 1870s, another physician, Carl Wernicke, described two patients who had problems understanding spoken language after having a stroke. Unlike the aphasics described by Broca, these patients had fluent speech but spoke nonsensical sounds, words, and sentences. Moreover, they had severe deficits in comprehending what was spoken to them. Later, Wernicke performed an autopsy on one patient's brain and discovered damage in the posterior regions of the superior temporal gyrus. Since auditory processing occurred nearby (anteriorly) in the superior temporal

Figure 9.24 **(a)** The preserved brain of Leborgne, Broca's patient "Tan," which is maintained in a Paris museum. **(b)** The area in the left hemisphere lesioned in Leborgne's brain and now known as *Broca's area* (in red). The dotted lines indicate the location of Wernicke's area.

(a)

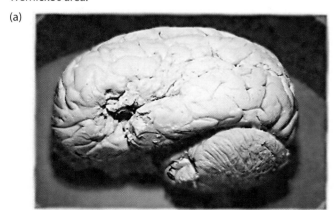

(b)

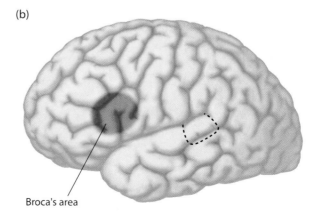

Broca's area

cortex within Heschl's gyri, Wernicke deduced that this more posterior region participated in the auditory storage for words, that is, as an auditory memory area for words. This area later became known as *Wernicke's area* (Figure 9.25). According to Wernicke, damage to this area produced poor language comprehension because it had lost word-related memories, whereas nonsense speech, he posited, resulted from patients' inability to monitor their own verbal output.

Although he was not entirely correct with respect to the nature of the functional loss he observed in language, Wernicke's discovery was the second key piece of information derived from observing language deficits following brain injury. It established the dominant view of brain and language for almost 100 years: Damage to Broca's area of the inferior-lateral left frontal lobe created difficulties in producing speech (*expressive aphasia*), and damage to the posterior inferior-lateral left parietal and supratemporal areas (including the supramarginal gyrus, angular gyrus, and posterior regions of the superior temporal gyrus) hampered the comprehension of language (*receptive aphasia*). During Broca's time, most emphasis was on word-level analysis; hence, little consideration was given to processing losses at the sentence level. This was reflected in the prevailing view of language in which word memory was taken to be the key: Broca's area was considered the locus of motor memory for words; Wernicke's area was the region concerned with the sensory memory for words. These ideas led to a view of language in which three brain centers interacted as the foundations of language: a production area, a comprehension area, and a conceptual area.

An Early Model of Language Organization

Wernicke, Broca, and their contemporaries fueled the idea that language was localized in structures interconnected anatomically to create the brain's total language system. Sometimes referred to as the *classical localizationist view* or the *connectionist model of language,* it dominated through the 1970s, having been revived in the 1960s by the American neuropsychologist Norman Geschwind (1967). You should note that the "connectionist" model of Geschwind is not the same as the interactive or connectionist models that were developed later by researchers such as McClelland and Rumelhart and implemented in computer simulations. In the latter models, the interactive nature of the processes plays a very important role, and in contrast to the Geschwind model, representations of functions are assumed to be distributed instead of localized; to avoid confusion we will refer to Geschwind's model as the *classical localizationist model.* Figure 9.26 presents a version of the model first described by Lichtheim in 1885. The three main centers for auditory or oral language processing in this model

Figure 9.26 Lichtheim's classic model of language processing. The area that stores permanent information about word sounds is represented by A. The speech planning and programming area is represented by M. Conceptual information is stored in area B. The arrows indicate in which direction the information flows. From this model it was predicted that lesions in the three main areas, or in the connections between the areas, or the inputs to or outputs from these areas, could account for seven main aphasic syndromes. The locations of possible lesions are indicated by the line segments transecting the connections between A, B, and M. Adapted from Caplan (1994).

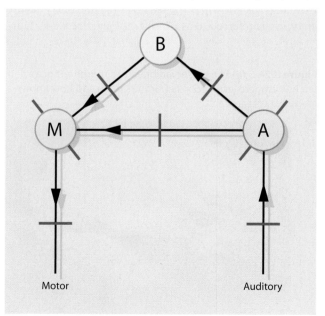

Figure 9.25 Lateral view of Wernicke's area. The arcuate fasciculus is the bundle of axons that connect Wernicke's and Broca's areas. It originates in Wernicke's area, goes through the angular gyrus, and terminates on neurons in Broca's area.

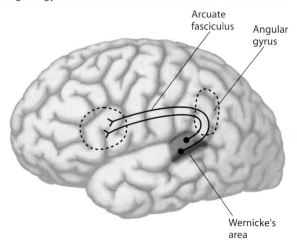

are labeled A, B, and M in the figure. Wernicke's area, A, represents the phonological lexicon—the area that stores permanent information about word sounds. Broca's area, M, is the speech planning and programming area. Concepts are stored in area B. The nineteenth-century language models had concepts distributed widely in the brain, but the relatively newer Wernicke-Lichtheim-Geschwind model localized concepts in more discrete areas. As an example, in this model the region containing the supramarginal gyrus and angular gyrus was proposed as the region where incoming sensory properties (auditory, visual, and tactile) or features of words were processed.

This classic localizationist model of language represents linguistic information as being localized to discrete brain regions interconnected by white matter tracts. Language processing was thought to activate these linguistic representations and involve their transfer between language areas. The idea is simple. According to the classical localizationist model, in auditory language, the flow of information is as follows: Auditory inputs are transduced in the auditory sensory system, which then passes the information to the parieto-temporo-occipital association cortex (centered on the angular gyrus), and then to Wernicke's area, where representations of words can be accessed from phonological information. From Wernicke's area, information flows through the arcuate fasciculus (a white matter tract) to Broca's area, where the grammatical properties are stored and phrase structure can be assigned. Word representations in turn activate related concepts in the concept centers, and, voilà, auditory comprehension occurs. In speech production, a similar process occurs, except concepts activated in the concept areas generate phonological representations of words in Wernicke's area and then are sent to Broca's area to program motor actions for speech articulation.

In Figure 9.26, lines connecting areas A, B, and M are transected. These lines represent white matter fibers in the brain that interconnect Wernicke's area, Broca's area, and the conceptual centers; lesions of these fibers, hypothetically, would disconnect the areas. Lesions in centers A, B, and M themselves would reflect damage to specific language areas. Thus, if the Wernicke-Lichtheim-Geschwind model were correct, we would expect forms of language deficits from brain damage that had symptoms and signs corresponding to the ones predicted by the model. Indeed, various aphasias correspond to what would be predicted by the model, and so the model was pretty good, actually.

Lichtheim reported cases of the aphasias predicted by the model in pure form, but it remains unclear how dis-

tinct they really were. Some extant data support the basic tenets of the Wernicke-Lichtheim-Geschwind model, yet the model still has significant shortcomings. For one thing, prior to the advent of neuroimaging with computed tomography (CT) and magnetic resonance imaging (MRI), lesion localization was poor and relied on sometimes inaccessible autopsy information, or guesses based on co-occurrence of other more well-defined symptoms (e.g., hemiparesis). Second, there is great variability in how lesions are defined in autopsy studies (as well as in neuroimaging data). And third, there is great variability in the lesions themselves; as an example, sometimes anterior brain lesions produce Wernicke's aphasia! Finally, when classified, patients often fall into more than one diagnostic category. Broca's aphasia has several components, for instance. Given this, it is worth reviewing the major aphasic syndromes and considering the interpretational difficulties in the literature on brain and language. The goal is to provide insights to aphasic syndromes and their relation to brain structures in classic models and modern ones. Prior to reviewing the syndromes, though, we should briefly consider how aphasia is diagnosed and classified in the clinic.

CLASSIFICATION OF APHASIA

Speech pathologists and aphasiologists have elaborate criteria for diagnosing and characterizing language disorders such as aphasia. Diagnostic classifications might exclude neuropathologies that may be treatable or require special management, care, and rehabilitation of aphasic patients. The schemes also vary in their ability to distinguish language deficits. This variability is partly because a brain lesion in one patient rarely, if ever, has the same pattern as that in another patient. Nonetheless, the classification of aphasic patients into subtypes is successful given its goal clinically. The three main test parameters for language disorders are spontaneous speech, auditory comprehension, and verbal repetition. The patients' performance permits a trained clinician to classify aphasics into general groups.

Classifying Broca's Aphasia **Broca's aphasia** is the oldest and perhaps most well-studied form of aphasia. Characterized by speech difficulties, it includes a wide range of symptoms (Figure 9.27). In the most severe forms of Broca's aphasia, single utterance patterns of speech such as that of Broca's original patient are often observed. But the variability is large and may include unintelligible mutterings, single syllables or words ("tan," "yes," "no"), short simple phrases or sentences that mostly lack function words or grammatical markers, or idioms such as

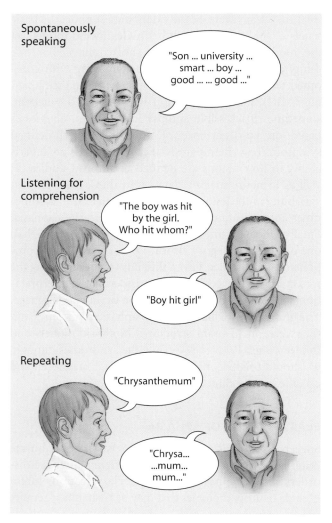

Spontaneously speaking

"Son ... university ... smart ... boy ... good ... ... good ..."

Listening for comprehension

"The boy was hit by the girl. Who hit whom?"

"Boy hit girl"

Repeating

"Chrysanthemum"

"Chrysa... ...mum... mum..."

Figure 9.27 Speech problems of Broca's aphasics. Broca's aphasics can have various problems when they speak or when they try to comprehend or repeat the linguistic input provided by the clinician (woman at left). The speech output of this patient is slow and effortful, and it lacks function words: It resembles a telegram **(top).** Broca's aphasics also may have a hard time understanding reversible sentences, where a full understanding of the sentence depends on correct syntactic assignment of the thematic roles (who hit whom?) **(middle).** Finally, these patients will sometimes have problems with speech articulation because of deficits in the regulation of the articulatory apparatus (e.g., muscles of the tongue) **(bottom).**

"Fit as a fiddle and ready for love." Sometimes the ability to sing normally is undisturbed, as might be the ability to recite phrases and prose, or to count.

The speech of Broca's aphasics is often telegraphic and very effortful, coming in uneven bursts. The ability to find the appropriate word or combination of words and to execute pronunciation is compromised in Broca's aphasics. Some problems are derived from speech deficits such as dysarthria (loss of control over articulatory muscles) and speech apraxia (deficits in the ability to program articulations).

The extent to which Broca's aphasia is limited to language production, especially spoken language, depends on the extent of any comprehension deficits that might also be present. These deficits are easiest to observe when the meaning of a sentence requires precise interpretation of its grammatical structure and cannot be derived simply from the sentence's individual words. Consider the following examples: "The boy ate the cookie" and "The boy was kicked by the girl." The second sentence is more complex and could be misunderstood by the listener except that grammatical rules determine who kicked whom; such rules are not needed for the aphasic to understand that the boy ate the cookie and not vice versa (i.e., "the cookie ate the boy" can be rejected based on world knowledge that cookies are eaten and that boys eat). The notion that Broca's aphasics have only an expressive disorder is not correct; they also can have comprehension deficits. Broca's aphasics are not devoid of grammatical knowledge, though; rather, they have deficits in processing grammatical aspects of language, as described by the term *agrammatism.*

Broca's Area When Broca first described the disorder now known as Broca's aphasia, he called it *aphemia* and related it to damage in the left ventral-lateral frontal cortex, in the anatomical regions known as the *pars triangularis* and *pars opercularis,* later defined cytoarchitectonically as Brodmann's areas 44 and 45 and commonly referred to now as *Broca's area* (see Figure 9.24b). He limited his conclusions to cortical damage in the inferior frontal lobe, which he believed was a participant in language production, but as we described, the symptoms may include comprehension deficits too, especially those related to certain grammatical forms. In any case, and somewhat unfortunately, it is difficult to correlate the symptoms of Broca's aphasia (a functionally defined syndrome) with the classically defined structures in Broca's area (an anatomically defined location in the brain).

Challenges to the idea that Broca's area was responsible for speech deficits in aphasia were laid down in Broca's time. Some reports noted damage to Broca's area without deficits in speech; others noted speech deficits when damage was limited to more posterior brain regions, outside Broca's area. Researchers continue to question the relation between Broca's area and Broca's aphasia. For example, aphasiologist Nina Dronkers (1996) at the University of California at Davis reported twenty-two patients with lesions in Broca's area, as defined by neuroimaging, with only ten having Broca's aphasia.

The regions classically defined as Broca's area are generally limited to the gray matter in the ventral-lateral frontal cortex of the left hemisphere. Broca did not consider the underlying white matter, deep cortex, and subcortical structures as potential contributors to the syndrome he observed. However, by the turn of the twentieth century, scientists were proposing that structures deep to Broca's area were involved in the deficits of Broca's aphasia. Brain areas including the insular cortex, the lenticular nucleus of the basal ganglia, and fibers of passage have been implicated in Broca's aphasia. Research by Dronkers (1996) found that all patients with Broca's aphasia and apraxia of speech (difficulty in pronunciation of words), including those studied by autopsy or neuroimaging, have damage in the insula.

This was the case even for Broca's original patient. His brain is now housed in a French museum, and recent CT scans showed that his lesions extended into regions underlying the superficial cortical zone of Broca's area, and thus included the insular cortex and portions of the basal ganglia. Damage to the classic regions of the frontal cortex known as Broca's area is not, therefore, solely responsible for the deficits of Broca's aphasics.

Classifying Wernicke's Aphasia Wernicke's aphasia is primarily a disorder of language comprehension. Patients with this syndrome have problems understanding spoken or written language, and sometimes cannot understand at all. They are cut off from communication with others by their comprehension problems, and they cannot speak meaningful sentences. Wernicke's patients can produce fluent-sounding speech, in contrast to the broken speech patterns of Broca's aphasics, but the speech is meaningless. To provide insight to what it means for a person to be fluent and to have Wernicke's aphasia, imagine you are listening to someone speaking a language you do not know. You would probably not detect any deficits. And as might be expected, the speech sounds fine but is composed of meaningless strings of words, sounds, and jargon. Figure 9.28 shows the speech of two patients with Wernicke's aphasia, one in English and the other in Dutch. Unless you speak Dutch, it is unlikely that you can discriminate this passage from that of normal Dutch prose.

Wernicke's Area The syndrome known as *Wernicke's aphasia* was first related to what is now called *Wernicke's area* in classic studies of the nineteenth century. Wernicke's area includes the posterior third of the STG. However, language comprehension deficits also arise

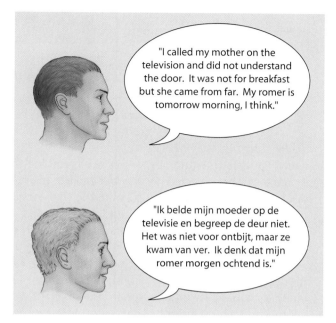

Figure 9.28 Wernicke's aphasics can speak fluently. If one is not a native speaker of a language, one might not detect that there was anything wrong with the speech of a patient with Wernicke's aphasia. For example, to most people in the United States, the message in Dutch in the lower part of the figure might sound perfectly normal (unless you speak Dutch) because it flows smoothly. However, as the English translation at the top indicates, the speech output of Wernicke's aphasics is often meaningless. They also will make semantic errors. For example, they will say "television" for telephone. Sometimes these patients will produce utterances that do not exist in their language (jargon), for example, "romer."

from damage to the junction between the parietal and temporal lobes, including the supramarginal and angular gyri. How well do language comprehension deficits relate to Wernicke's area, and how much can be attributed to damage in the surrounding cortex and white matter?

As with Broca's aphasia and Broca's area, sometimes lesions that spare Wernicke's area still lead to comprehension deficits. In one study of seventy patients with Wernicke's aphasia, seven had damage confined to regions outside of Wernicke's area. In contrast, some patients who have Wernicke's aphasia with damage in Wernicke's area may be able to comprehend more as their aphasia improves over time, leaving a milder aphasia or even only an anomic syndrome (see The Man Without Nouns).

Recent studies support this picture. Dense and persistent Wernicke's aphasia is ensured only if there is damage in Wernicke's area and in the surrounding cortex of the posterior temporal lobe, or with damage to the underlying white matter that connects temporal lobe language areas to other brain regions. Thus, Wernicke's area

HOW THE BRAIN WORKS

The Man Without Nouns

Damage to Wernicke's area or surrounding cortex can sometimes produce a disorder called *anomia*, the inability to name things in the world. Patients with anomia have difficulty finding the words that label objects. This disorder can be strikingly discrete; the language deficit is limited to the inability to name objects, but comprehension is intact and speech is unaffected. One such patient is H.W., an intelligent businessman who headed his own company. He was studied by cognitive neuroscientist Kathleen Baynes of the University of California at Davis. After suffering a stroke in his left hemisphere, H.W. was left with anomia and almost no other deficits of note except a slight right-side hemiparesis (which was actually quite minor; he was a physically robust individual) and a slight deficit in face recognition. But his anomia was extraordinary. Here is a passage of H.W.'s speech as he described the contents of a picture of a boy falling off a stool while reaching for cookies in a jar on the shelf and handing one to his sister:

First of all this is falling down, just about, and is gonna fall down and they're both getting something to eat . . . but the trouble is this is gonna let go and they're both gonna fall down . . . I can't see well enough but I believe that either she will have some food that's not good for you and she's to get some for her, too . . . and that you get it there because they shouldn't go up there and get it unless you tell them that they could have it. And so this is falling down and for sure there's one they're going to have for food and, and this didn't come out right, the, uh, the stuff that's uh, good for, it's not good for you but it, but you love, um mum mum [H.W. intentionally smacks lips] . . . and so they've . . . see that, I can't see whether it's in there or not . . . I think she's saying, I want two or three, I want one, I think, I think so, and so, so she's gonna get this one for sure it's gonna fall down there or whatever, she's gonna get that one and, and there, he's gonna get one himself or more, it all depends with this when they fall down . . . and when it falls down

there's no problem, all they got to do is fix it and go right back up and get some more.

H.W. was able to describe aspects of the scene, but without the ability to say nouns, the information was minimal. Nonetheless, H.W. used proper grammatical structures, such as noun phrases; they simply missed a noun. For example, H.W. would say "this" instead of "the chair." He substituted generic nouns such as "food" for "cookie" although he knew it was a cookie, that it tasted good, and that adults consider this food bad for children. The location of his lesion is enlightening and disappointing: He had a large left-hemisphere lesion that included large regions of the posterior language areas. Yet his functional deficit was discrete—he had no nouns.

Picture similar to that shown to H.W. by cognitive neuroscientist Kathleen Baynes. The patient's description of this scene is provided in the text.

remains in the center of a posterior region of the brain whose functioning is required for normal comprehension. Lesions confined to Wernicke's area lead to only temporary Wernicke's aphasia because the damage to this area does not actually cause the syndrome; instead,

secondary damage due to tissue swelling contributes to the most severe problems. When swelling around the lesioned cortex goes away, comprehension improves. Some researchers such as Dronkers have suggested that the white matter underlying Wernicke's area may hold the key. Cortical damage that temporarily compromises white matter tracts below the cortex may explain much of the "here and gone again" characteristics of Wernicke's aphasia. More functional analyses are clearly needed.

DAMAGE TO CONNECTIONS BETWEEN LANGUAGE AREAS

In describing the Wernicke-Lichtheim-Geschwind model, we alluded to strong links between anterior and posterior regions of the brain involved in speech production and comprehension, respectively. In the context of the model, as presented in Figure 9.26, systematic syndromes were implied to result from damage to these white matter tracts. Wernicke predicted that a certain type of aphasia should result from damage to fibers projecting from Wernicke's to Broca's areas. Indeed, a disconnection syndrome, now known as **conduction aphasia,** can occur when the **arcuate fasciculus,** the pathway from Wernicke's to Broca's area, is damaged. Conduction aphasics have problems producing spontaneous speech as well as repeating speech, and sometimes they use words incorrectly. They can understand words that they hear or see and can hear their own speech errors, but they cannot repair them. Hence, conduction aphasia arises from damage to connections between posterior and anterior language areas. But similar symptoms happen with lesions to the insula and portions of the auditory cortex. One explanation for this similarity may be that damage to other nerve fibers is not detected, or that connections between Wernicke's area and Broca's area are not as strong as connections between the more widely spread anterior and posterior language areas. Indeed, the emphasis should not really be on Broca's or Wernicke's areas but on brain regions better correlated with the syndromes of Broca's and Wernicke's aphasia. Considered in this way, a lesion to the area surrounding the insula could disconnect comprehension from production areas.

The Wernicke-Lichtheim-Geschwind model also predicted that damage to the connections between conceptual representation areas (area B in Figure 9.26, perhaps in the supramarginal and angular gyri) and Wernicke's area would harm the ability to comprehend spoken inputs but not the ability to repeat what was heard (*transcortical sensory aphasia*). Indeed, patients with lesions in the supramarginal and angular gyri regions display such problems. Further, they have the unique ability to repeat what they heard and to correct grammatical errors in what they heard when they repeat it. These findings have been interpreted as evidence that

this aphasia may come from losing the ability to access semantic information without losing syntactic or phonological abilities.

The picture that emerges from these classifications of major aphasic syndromes is not unlike the one that was put forward by the classical Wernicke-Lichtheim-Geschwind model. However, as we have seen, studies in patients with specific aphasic syndromes have revealed that the assumption of the classical model that only Broca's and Wernicke's areas are associated with the syndromes of Broca's and Wernicke's aphasia, respectively, is most likely incorrect. Part of the problem is that the original lesion localizations were not very sophisticated. But another part of the problem lies in the classification of the syndromes themselves, because Broca's and Wernicke's aphasias are associated with a conglomerate of symptoms that together constitute each syndrome. As you now know, Broca's aphasics, for example, may have apraxia of speech and agrammatic comprehension. From our discussion of psycholinguistic models of normal language comprehension and production, it should be clear that these are very different linguistic processes, and it is not at all surprising that this variety of language functions is supported by more than Broca's area alone.

MECHANISMS OF APHASIC DEFICITS

We have reviewed numerous language deficits associated with major classes of aphasia and implied that these must represent problems in language faculties. But what form would this take? Is the deficit the result of true losses of parts of the language system, like selective damage to the second gear in an automobile transmission while the other gears are intact? Or is the problem in aphasia more like losing the ability to use the linguistic information properly, like having a perfectly good set of gears but a burned-out clutch?

These questions reflect a pivotal concern in aphasia research: Do comprehension deficits in aphasic patients result from losses of stored linguistic information, or from disruption of computational language processes that act on representations of the linguistic inputs? Aphasic deficits have classically been attributed to losses of semantic or syntactic knowledge structures. In recent years, however, many deficits in aphasia have been attributed to processing impairments rather than true losses of knowledge. Two results are central to this change in perspective.

Broca's aphasics with severe problems in syntactic understanding were tested in a sentence-picture matching task. They often were unable to point to a picture that matched the meaning of the sentence they had to read, especially when more complex sentence constructions were used. For example, Broca's aphasics might not understand who is kicking whom in the sentence "The boy was kicked by the girl." If they have to choose from two pictures, one picture where a girl is kicking a boy and one where a boy is kicking a girl, they often select the wrong picture. But in one seminal study (Linebarger et al., 1983), Broca's aphasics who were impaired in their ability to match pictures to reversible sentences, as in the example above, performed considerably above chance when they had to distinguish syntactically well-formed sentences from incorrect ones in a grammar judgment task, even when they had to cope with numerous syntactic structures.

The task that one uses, then, directly influences the aphasic patient's performance. Moreover, sometimes Broca's aphasics also can access and exploit structural knowledge. Variable performance across linguistic tasks supports the notion that agrammatic comprehension results from a processing deficit rather than a representational one; patients still have syntactic knowledge but sometimes they cannot use it while at other times they can.

The classic notion of loss of semantic knowledge in Wernicke's aphasics also has been challenged by experiments in which aphasic patients were required to make speedy decisions about target words. In this lexical decision task, aphasics had to decide as quickly as possible whether a letter string or a sound sequence was a word. Neurologically normal persons made faster lexical decisions on words primed by a semantically or associatively related word than when preceding words were unrelated. Wernicke's aphasics consistently displayed the normal pattern of priming for semantically or associatively related word pairs in this task. They showed this effect even though they demonstrated severe impairment when asked to make judgments on word meaning. Thus, despite problems in comprehension, when lexical-semantic processing is tested in an implicit test (such as in the lexical decision task), patients do not show impairment. This finding suggests that lexical-semantic knowledge might be preserved but aphasics sometimes cannot access or exploit it to achieve normal comprehension. How might this work?

An alternative to the idea that linguistic information is lost is the idea that real-time mental processes that act on linguistic information might be compromised in aphasia. One view is that these patients' brains cannot keep up with the time challenges of language comprehension or production because accessing and exploiting stored linguistic information no longer operates at normal speeds. Consider the following analogy: A soccer player kicks the ball to his teammate who is charging the goal. In the nick of time, with perfect coordination, the teammate receives the ball and with a punishing kick, sends it past the goalkeeper for a score! In this normal situation all players are on the field and are healthy, and their interactions are in temporal coordination in a fast-paced setting. What would happen if there were no player charging toward the goal to receive and relay the ball into the net? No goal, of course! This situation would be analogous to the case where stored linguistic information in the brain is actually lost owing to aphasia and therefore could not be used in the service of language comprehension and production. But consider another situation, one where the player receiving the ball is merely injured and slow, making it hard to manage the pass from the teammate and leading to the ball being lost, and therefore, no goal. In both situations the outcome is the same—no goal—but the underlying causes for the failure are very different. In one situation a player is absent, whereas in the latter case the player is still on the field, still knows what to do, and can still receive passes and kick the ball, but simply cannot do so with the facility necessary to compete in a rapidly moving soccer match. Language processing is also a rapidly moving game, and deficits that undermine normal real-time processing could lead to aphasia. We can refer to this general idea as the *processing impairment model* of comprehension deficits in aphasia.

A central question about the processing impairment model of comprehension deficits in aphasics is, When is the impairment manifested during language processing? Two suggestions have been put forward: One is that the dynamics of lexical access are affected in aphasic comprehenders; an alternative proposal is that automatic lexical access might be largely intact but problems arise later, during lexical integration.

The claim for impaired lexical access has been derived, for example, from word priming experiments. Some studies found that Broca's aphasics lack semantic priming in lexical decision tasks, even though normal control subjects and Wernicke's aphasics show faster reaction times to words related to the prime. Yet support for this claim is not decisive, for two reasons. First, in other semantic priming studies, Broca's aphasics actually demonstrated an effect of semantic priming. Second, to make claims about automatic access, the time between the onset of the priming word and that of the target (stimulus onset asynchrony, or SOA) should be short (approximately 200 msec). In normal subjects, semantic priming at long SOAs does not result only from automatic lexical activation but may include other postlexical processes. In line with this, studies using long SOAs between prime and target words in aphasics failed to find priming, whereas those using shorter SOAs revealed semantic priming.

What emerges is more compatible with the idea that lexical integration rather than lexical access is impaired in these patients. The patients have difficulty integrating the meaning of words into the current context established by the preceding words in the sentence; that is,

they cannot use the contextual information to select the appropriate meaning of a word quickly enough for normal comprehension. Thus, in the sentence "The man planted a tree on the bank," under normal conditions the context tells us that *bank* refers to the "side of a river" and not a "monetary institution." One can readily appreciate the comprehension problems that would ensue if such processes were disrupted as with aphasia (see Aphasia and Electrophysiology).

NEUROPHYSIOLOGY OF LANGUAGE

The neuropsychological lesion approach epitomized by research on aphasia and other language disorders represents only one way to investigate the biological bases of language. In this section, we briefly review some additional evidence from functional neuroimaging not already presented, studies involving stimulation of the human cortex during neurosurgical procedures, and recordings of language-related brain potentials (ERPs) from healthy and aphasic persons. These approaches yield information about the normal language system that may not be obvious from observations of the performance of damaged brains alone. Further, the neurophysiological approach can provide measures of brain activity and functional processing in persons with language disorders that are not possible to infer from behavior or anatomical imaging alone.

Functional Neuroimaging of Language

As we have reviewed throughout the chapter, PET and fMRI can be used to identify areas in the brain that are active during specific psycholinguistic processes, or to identify linguistic and conceptual representations in normal, neurologically unimpaired volunteers. But these methods also can be used passively to investigate the metabolic correlates of language disorders such as Broca's aphasia during rest. The goal is to identify the extent of the functional lesion beyond the structural lesion on which most neuropsychological investigation is based. Wallerian (anterograde) degeneration and retrograde degeneration of neurons occur when part of the neuron is damaged, and this can lead to transneuronal degeneration for neurons projecting to or receiving inputs from the lesioned region. As a result, sites interconnected with a lesioned area also may be affected adversely, and these effects cannot be seen using structural imaging with CT or MRI, because they may manifest only as regions of lowered metabolism.

The ^{18}F-deoxyglucose PET method measures neuronal metabolism by using radioactively labeled glucose

analogues that are taken up by active cells and trapped within them. The radioactivity can then be imaged using a PET scanner. The advantage of this technique is that we can learn about the functional consequences of lesions that extend beyond the structurally damaged region.

METABOLIC CORRELATES OF APHASIA

Research on resting brain metabolism has revealed a complex picture of the effects of aphasia. Focal brain lesions from stroke can lead to widespread changes in metabolism; that is, the effects of stroke extend to regions outside the lesioned areas (Figure 9.29). Remote metabolic changes typically are observed in stroke victims who have observable symptoms, but not when they do not show neurological symptoms.

In 100% of the aphasic patients studied, PET measures obtained during a resting state revealed hypometabolism (lower glucose utilization) in the temporoparietal region, regardless of the type of aphasia. Of these patients, 97% had metabolic changes in the angular gyrus and 89% had such changes in the supramarginal gyrus; 87% showed hypometabolism in the STG. These PET-defined hypometabolic regions were then compared to the anatomically defined lesions. Such comparisons showed that 67% of the patients had parietal lesions, 67% had damage in Wernicke's area, and 58% had damage in the posterior middle temporal region. Even when the anatomically defined damage did not include the supramarginal and angular gyri, these brain areas had reduced neuronal metabolism during a resting state. The lesson here is that correlating the behavioral deficits in aphasia with only the visible anatomical lesions may not provide a complete picture and may even lead to confusion.

Some researchers maintain that correlations between aphasic symptoms and metabolism as defined by PET are better than those with anatomical lesion data. Broca's aphasics with anatomical damage to Broca's area and perhaps surrounding tissue have marked hypometabolism

Stimulation Mapping of the Human Brain

The scene is out of a Walt Disney or Steven Spielberg film. A young man lies on a table covered with clean light-green sheets. He is lying on his side and is awake. His head is partially covered by a sheet of cloth, so we can see his face if we wish. On the other side of the cloth is a woman wearing a surgical gown and mask. The man is a patient; the woman is his surgeon. His skull has been cut through and his left hemisphere is exposed. Yes, he is awake; no, this is not another time or another place. This is here and now at the University of Washington Medical School where George Ojemann and his colleagues (1989) have been using direct cortical stimulation and recording methods to map the brain's language areas.

The patient suffers from epilepsy and is about to undergo a surgical procedure to remove the epileptic tissue. But first, because this epileptic focus is in the left language-dominant hemisphere, it is essential to find where language processes are localized in the patient's brain. This can be done by electrical stimulation mapping. Electrodes are used to pass a small electrical current through the cortex, momentarily disrupting activity; thus, electrical stimulation can probe where a language process is localized. The patient has to be awake for this test. Language-related areas vary among patients and so these areas must be mapped carefully.

During surgery, it is essential to leave the critical language areas intact.

One benefit of this work is that we can learn more about the organization of the human language system. Patients are shown line drawings of everyday objects and are asked to name those objects. During naming, regions of the left perisylvian cortex are stimulated with low amounts of electricity. When the patient makes an error in naming or is unable to name the object, the deficit is correlated with the region being stimulated during that trial, and so that area of cortex is assumed to be critical for language production and comprehension.

Stimulation of between 100 and 200 patients revealed that aspects of language representation in the brain are organized in mosaic-like areas of 1 to 2 cm^2. These mosaics usually include regions in the frontal and posterior temporal areas. However, in some patients, only frontal or posterior areas were observed. The correlation between these effects in either Broca's or Wernicke's areas was weak; some patients had naming disruption in the classic areas and others did not. Perhaps the single most intriguing fact is how much the anatomical localizations vary across patients, a point that has implications for how across-subject averaging methods, such as PET activation studies, reveal significant effects.

---

in the prefrontal cortex. This effect is reduced in patients with Wernicke's aphasia and even further diminished in those with conduction aphasia—but half of these patients display hypometabolism in Broca's areas.

Anatomically defined correlates of aphasia, therefore, may not capture the physiological brain changes that accompany the aphasic syndrome. These findings remind us of the shortcomings of the anatomical correlation method for determining the functional anatomy of human language function. One certainty is the need for data from multiple methods in trying to pin down the brain organization of language.

Electrophysiology of Language

In Chapter 4 we introduced the ERP method and described how these ERPs are obtained for different stimuli and mental processes. ERPs also can be used to investigate human language. For many reasons, they provide a powerful tool to study language comprehen-

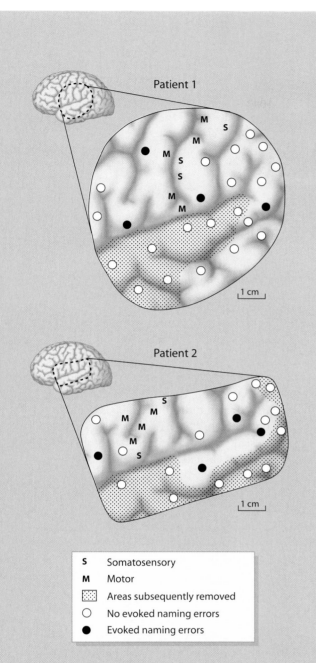

Patient 1

Patient 2

S	Somatosensory
M	Motor
░	Areas subsequently removed
○	No evoked naming errors
●	Evoked naming errors

Regions of the brain of two patients studied with cortical stimulation mapping. During surgery, with the patient awake and lightly anesthetized, the somatosensory and motor areas are mapped by stimulating the cortex and observing the responses. Then patients are shown pictures and asked to verbally name them. Discrete regions of the cortex are stimulated with electrical current during the task. Areas that induce errors in naming when they are stimulated are mapped, and those regions are implicated as being involved in language. The surgeon uses this mapping to avoid removing any brain tissue associated with language. The surgery not only treats brain tumors or epilepsy but also enlightens us about the cortical organization of language functions. Adapted from Ojeman et al. (1989).

sion. To appreciate this, let's look at some brain waves or ERP components that index aspects of semantic and syntactic processing during language comprehension.

SEMANTIC PROCESSING AND THE N400 WAVE

The N400 is a brain wave related to linguistic processes. It was named N400 because it is a negative polarity voltage peak in brain waves for words that usually reaches maximum amplitude around 400 msec after onset of the word stimulus. This brain wave is especially sensitive to semantic aspects of linguistic input.

Marta Kutas and Steven Hillyard (1980) at the University of California, San Diego, discovered the N400 wave. They compared the processing of the last word of sentences in three conditions: normal sentences that ended with a word congruent with the preceding context, like "It was his first day at work"; sentences that ended with a word anomalous to the preceding context, like "He spread the warm bread with socks"; and sentences that

(a) Slice drawn from postmortem tissue   (b) 18-FDG PET

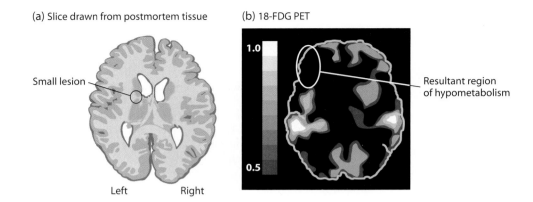

Figure 9.29 **(a)** Diagram of lesions and **(b)** PET scan showing regions of lowered metabolism in the brain of a stroke patient. The regions showing decreased metabolic activity include those that were damaged by the stroke and have cell loss. The regions connected to the damaged areas experience changes in activity and hence possible function, but are not themselves damaged by the stroke. These areas can be thought of as functionally lesioned and may contribute to deficits seen in the patients, thereby making it difficult to infer a structure-function correlation from computed tomography or structural magnetic resonance imaging scans, which only reveal the areas of physical damage.

ended with a word semantically congruent with the preceding context but physically deviant, like "She put on her high-heeled **shoes**." The sentences were presented on a computer screen, one word at a time. The subjects were to read the sentences attentively, knowing that questions about the sentences would be asked at the end of the experiment. The electroencephalograms (EEGs) were averaged for the sentences in each condition, and the ERPs were extracted by averaging data for the last word of the sentences separately for each sentence type.

The amplitude of the N400 to the anomalous words ending the sentence was increased when compared to that of the N400 to congruent words (Figure 9.30). This difference in the amplitude of the N400 is called the *N400 effect*. In contrast, words that were semantically congruent with the sentence but were merely physically deviant (larger letters, etc.) elicited a positive potential rather than

Figure 9.30 ERP waveforms differentiate between congruent words at the end of sentences *(work)* and anomalous last words that do not fit the semantic specifications of the preceding context *(socks)*. The anomalous words elicit a large negative deflection (plotted upward) in the ERP called the N400. Words that fit into the context but are printed with a larger font *(shoes)* elicit a positive wave (P560) and not the N400, indicating that the N400 is not generated only by surprises at the end of the sentence. Adapted from Kutas and Hillyard (1980).

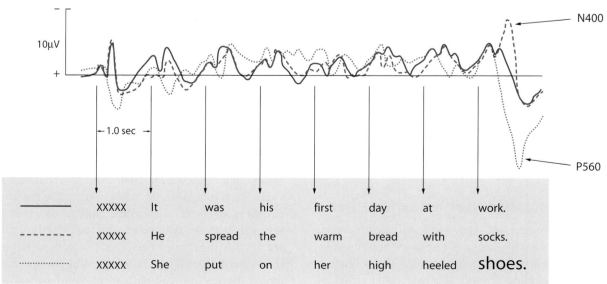

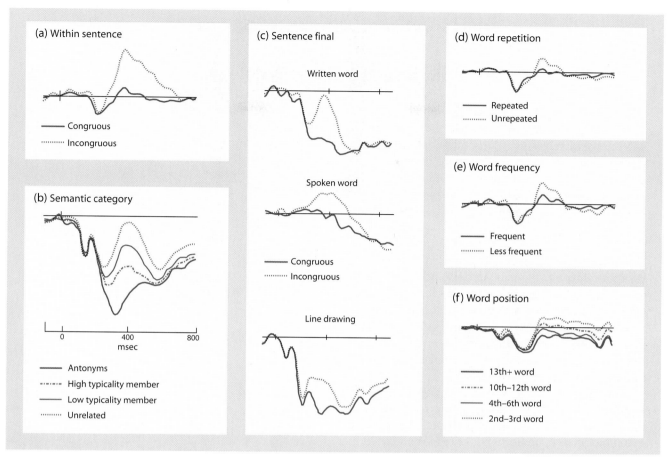

Figure 9.31 The N400 response to various language manipulations. The N400 is modulated as a function of congruency with the context **(a)**, degree of semantic relations between words **(b)**, word repetition **(d)**, word frequency **(e)**, and position of the word in a sentence **(f)**. These effects are modality independent **(c)**. From Kutas and Federmeier (2000).

an N400. Subsequent experiments showed that nonsemantic deviations like musical or grammatical violations also fail to elicit the N400 effect. Thus, the N400 is specific to semantic analysis. Figure 9.31 presents a summary of some other important findings with respect to the N400. As you can see in this figure, the N400 amplitude is modulated as a function of a number of different manipulations, including semantic relations between words, word repetition, word frequency, and the position of the word in a sentence. The figure also demonstrates that N400 effects are modality independent. They happen when subjects read sentences; when subjects are presented with auditory input, including languages like English, Dutch, French, and Japanese (for speakers of those languages); and even when congenitally deaf subjects are presented with American Sign Language (not shown in the figure). With respect to linguistic-neural processing, the N400 reflects primarily postlexical processes involved in lexical integration. Hence, an N400 effect will happen when lexical integration is hampered by a mismatch between semantic specifications of a word and semantic specifications of its preceding word or sentence context.

SYNTACTIC PROCESSING AND EVENT-RELATED POTENTIALS

Whereas work with the N400 had started in the early 1980s, only in the last decade have researchers started to study the electrophysiological correlates of other aspects of linguistic analysis such as syntax. One ERP component that has been identified has been labeled the *P600* (because of its polarity and latency of onset) or the *syntactic positive shift* (SPS, because of its assumed sensitivity to syntactic processes). From now on we will refer to this ERP component as the P600/SPS. The P600/SPS is a large positive component elicited by words after a syntactic violation. Lee Osterhout and Phil Holcomb in the United States (1992) and the Dutch psychologists Peter Hagoort, Colin Brown, and their colleagues (1993) at the Max Planck Institute for Psycholinguistics in the Netherlands discovered this component. To illustrate the sensitivity of the SPS/P600 to syntactic violations, we will concentrate here on a study in which Hagoort and Brown asked subjects to silently read sentences

Aphasia and Electrophysiology

Since investigators have been struggling with the question of whether aphasic symptoms reflect processing versus representational losses, interest in on-line measures of language processing has grown. Such measures include the ERPs elicited by language processing. The idea is to investigate the processing of spoken language and observe how the patient's brain responds to linguistic inputs, and to compare these responses to those in healthy control subjects. One study used the N400 component of the ERP to investigate spoken-sentence understanding in Broca's and Wernicke's aphasics. Tamara Swaab (now at Duke University), Colin Brown, and Peter Hagoort at the Max Planck Institute for Psycholinguistics in Holland (1997) tried to determine whether spoken-sentence comprehension might be hampered by a deficit in the online integration of lexical information.

The patients listened to sentences spoken at a normal rate. In half of the sentences, the meaning of the final word of the sentence matched the semantic meaning building up from the sentence context. In the other half of the sentences, the final word was anomalous with respect to the preceding context. As in Kutas and Hillyard's study (1980), the N400 amplitude should be larger to the final words that are anomalous than to those that are congruent. This result was obtained for normal age-matched control subjects. In comparison to the controls, nonaphasic brain-damaged patients (right hemisphere–damaged controls) and aphasic patients with a light comprehension deficit (high comprehenders) had an N400 effect comparable to that of neurologically unimpaired subjects. In aphasics with moderate to severe comprehension deficits (low comprehenders), the N400 effect was reduced and delayed. The results are compatible with the idea that aphasics with moderate to severe comprehension problems have an impaired ability to integrate lexical information into a higher-order representation of the sentence context because the N400 component indexes the process of lexical integration. The incorporation of electrical recordings into studies of neurological patients with behavioral deficits such as aphasia permits scientists to track the processing of information in real time as it occurs in the brain. This can be combined with analysis using traditional approaches such as reaction time measures in, for example, lexical decision tasks. But importantly, ERPs can also provide measures of processing in patients whose neurobehavioral deficit is too severe to use behavior alone because their comprehension is too low to understand the task instructions.

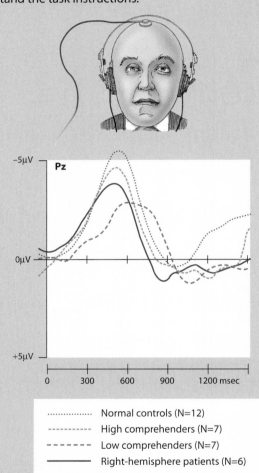

Normal controls (N=12)
High comprehenders (N=7)
Low comprehenders (N=7)
Right-hemisphere patients (N=6)

The N400 effect to different anomalous words at the end of a sentence in different groups of patients and healthy control subjects. The recording is from a single electrode located over centroparietal scalp regions in elderly healthy control subjects, aphasics with high comprehension scores, aphasics with low comprehension scores, and patients with right-hemisphere lesions (control patients). The waveform for the low comprehenders is clearly delayed and somewhat reduced compared to that for the other groups. The waveforms for the normal control subjects, the high comprehenders, and the patients with right-hemisphere lesions are comparable in size and do not differ in latency. This pattern implies a delay in time course of language processing in the patients with low comprehension. Adapted from Swaab et al. (1997).

that were presented one word at a time on a video monitor. Brain responses to normal sentences were compared with those to sentences containing a grammatical violation. The results are shown in Figure 9.32. As you can see in this figure, there is a large positive shift to the syntactic violation in the sentence (labeled SPS), and the onset of this effect is approxi-

mately 600 msec after the violating word (*throw* in the example). The P600/SPS shows up in response to a number of other syntactic violations as well, and it occurs both when subjects have to read sentences and when they have to listen to them. As with the N400, the P600/SPS has now been reported for a number of different languages.

Figure 9.32 ERPs from frontal (Fz), central (Cz), and parietal (Pz) scalp recording sites elicited in response to each word of sentences that are syntactically anomalous (dashed waveforms) versus those that are syntactically correct (solid waveforms). In the violated sentence, a positive shift emerges in the ERP waveform at about 600 msec after the syntactic violation (shaded). It is called the syntactic positive shift (SPS). Adapted from Hagoort et al. (1993).

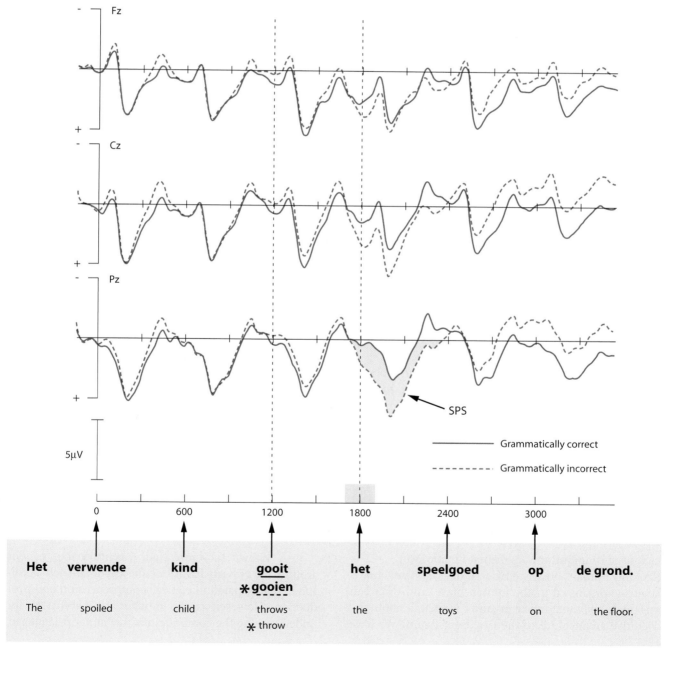

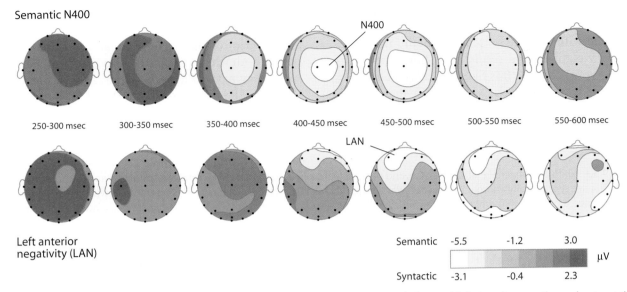

Figure 9.33 ERPs related to semantic and syntactic processing. The voltage recorded at multiple locations on the scalp at specific time periods can be displayed as a topographic voltage map. These maps show views of the topographies of the N400 to semantic violations (see Figure 9.30 for equivalent waveforms), and a left anterior negativity (LAN) to syntactic violations. The maps are read in a similar manner as are elevation maps of mountain ranges, except here the topography shows "mountains" and "valleys" of voltage. The N400 and LAN have different scalp topographies, which implies that they are generated in different neural structures in the brain. Adapted from Münte et al. (1993).

But the P600/SPS is sensitive to more than syntactic violations alone. Researchers have found that the P600/SPS also shows up when subjects read syntactically correct but dispreferred continuations or garden-path sentences. As you might remember from the section on syntactic processing earlier in this chapter, garden-path sentences are sentences that are misleading and require revision, after a first incorrect structural assignment has been made (for an example, see the section Integration of Words in Sentences). Recent studies also showed that the P600/SPS is sensitive to this revision process.

Syntactic processing is reflected in other types of brain waves as well. Neurologist Thomas Münte and colleagues (1993), and psycholinguist Angela Friederici and colleagues (1993) in Germany, described a negative wave over the left frontal areas of the brain. This brainwave has been labeled the *left anterior negativity* (LAN) and has been observed when words violate the required word category in a sentence (e.g., as in "the red eats," where a noun instead of a verb lemma is required), or when morphosyntactic features are violated (e.g., as in "he mow"). The LAN has about the same latency as the N400 but a different voltage distribution over the scalp. This difference in distribution is illustrated in Figure 9.33.

One issue that has long intrigued psycholinguistic researchers is whether semantic and syntactic processes are modular or interactive. The distinct electrophysiological signatures of these processes enable us to address this question physiologically. As well, now that clearly defined electrical activity has been linked to semantic and syntactic processing, these waves can be used as tools to probe human language in persons with deficits due to neurological disease and damage (see Aphasia and Electrophysiology). Such approaches for looking into the human language system are powerful tools for understanding language structure.

SUMMARY

Language is unique among mental functions in that only humans possess a true language system. How is language organized in the human brain, and what can this functional-anatomical organization tell us about the cognitive architecture of the language system? We have known for more than a century that regions around the Sylvian fissure of the dominant left hemisphere participate in language comprehension and production. However, classic models are insufficient for understanding the computations that support language.

Newer formulations based on detailed analysis of the effects of neurological lesions (supported by improvements in structural imaging), functional neuroimaging, human electrophysiology, and computational modeling now provide some surprising modifications of older models. But the human language system is complex, and much remains to be learned about how the biology of the brain enables the rich speech and language comprehension that characterize our daily lives. The future of language research is promising as psycholinguistic models combine with neuroscience to elucidate the neural code for this uniquely human mental faculty.

KEY TERMS

agrammatic aphasia

aphasia

apraxia of speech

arcuate fasciculus

Broca's aphasia

conduction aphasia

lexical access

lexical selection

mental lexicon

parsing

semantic paraphasias

Wernicke's aphasia

THOUGHT QUESTIONS

1. How might the mental lexicon be organized in the brain? Would one expect to find it localized in a particular spot in cortex? If not, why not?

2. At what stage of input processing are the comprehension of spoken and of written language the same, and where must they be different? Are there any exceptions to this rule?

3. Describe the route that an auditory speech signal might take in the cortex, from perceptual analysis to comprehension.

4. What evidence exists for the role of the right hemisphere in language processing? If the right hemisphere has a role in language, what might that be?

5. What role, if any, does working memory play in comprehension? Give an example of how disruptions in working memory might affect the understanding of spoken or written inputs.

SUGGESTED READINGS

BROWN, C.M., and HAGOORT, P. (1999). *The Neurocognition of Language.* Oxford: Oxford University Press.

CARAMAZZA, A. (1996). The brain's dictionary. *Nature* 380:485–486.

LEVELT, W.J.M. (1989). *Speaking: From Intention to Articulation.* Cambridge, MA: MIT Press.

MARTIN, A., HAXBY, J.V., LALONDE, F.M., WIGGS, C.L., and UNGERLEIDER, L.G. (1995). Discrete cortical regions associated with knowledge of color and knowledge of action. *Science* 270:102–105.

PINKER, S. (1994). *The Language Instinct.* New York: William Morrow.

SIGNORET, J.-L., CASTAINE, P., LEHRMITTE, F., ABELANET, R., and LAVOREL, P. (1984). Rediscovery of Leborgne's brain: Anatomical description with CT scan. *Brain Lang.* 22:303–319.

10

Cerebral Lateralization and Specialization

We live in a society that treasures the individual. Unique twists and interpretations of life's events are a marvel, and friends who offer other ways of looking at the world are a delight. So imagine how intriguing the world might be if our brain's two hemispheres were disconnected. Surgeons have done so, mainly in an effort to limit epilepsy, and their patients have been studied extensively. Each hemisphere is evaluated as to whether it and only it possesses specific functions. The big question is, Could each half of the brain provide a different view of the world?

DIVIDING THE MIND

W.J. was the first patient to have his brain split in recent years. A charismatic war veteran, W.J. appeared perfectly normal, possessed a sharp sense of humor, and always charmed all those whom he met in life. Following his surgery in 1961, his delightful personality remained unchanged but a most remarkable phenomenon happened. He was able to name and describe visual information presented to his left dominant hemisphere, but when the same information was presented to the right hemisphere, W.J. claimed he saw nothing. How could that be?

Such questions gave birth to human **split-brain research.** The phenomenon seen in W.J. has been observed repeatedly in other patients. Even so, much of the split-brain story was revealed in this one study. What was so remarkable was that this patient's right hemisphere could do things the left could not do. For example, there were striking differences between the performance of the two hemispheres on the block design task shown in Figure 10.1. Previously he could write dictated sentences and could carry out any kind of command such as making a fist or drawing geometrical shapes with his right hand. But after surgery, he

could not arrange four red and white blocks in a simple pattern with his right hand. The surgery had disconnected specialized systems in the right hemisphere from motor apparatus in the left hemisphere, which in turn controls the right hand. Even though W.J. was given as much time as needed, he was unable to perform tasks with his right hand because motor commands specific to the tasks could not be communicated from the isolated left hemisphere.

At the same time W.J.'s right hemisphere was a whiz. When blocks were presented to his left hand, he quickly and adeptly arranged them into the correct pattern. This simple observation gave birth to the idea that Mind Left did different things from Mind Right, supporting the idea that the central nervous system is laterally specialized. Each of the two cerebral hemispheres has processes the other does not have—hence, W.J.'s unique perspective on the world.

After the first testing session revealed this so clearly, investigators arranged to film W.J. carrying out tasks. The scientists knew a young fashion photographer, Baron Wolman, who dabbled in filmmaking (and would later help found *Rolling Stone* magazine); he was invited

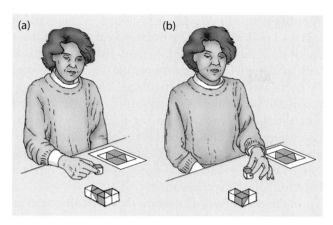

Figure 10.1 The pattern in red to the right is the shape that the patient is trying to create with the blocks given to her. **(a)** With her right hand (left hemisphere), she is unable to duplicate the pattern, whereas **(b)** with her left hand (right hemisphere), she is able to perform the task correctly.

to come to a session during which the whole test was carried out again. Wolman could not believe his eyes. During filming, W.J.'s right hand attempted to arrange the blocks and his left hand kept trying to intervene. Mind Right saw the problem, knew the solution, and tried to help out just like a good friend. W.J. had to sit on his left hand so the inadequate but dominant right hand could at least try.

It was all to no avail. For the film's final scene, everyone decided to see what would happen if both hands were allowed to arrange the blocks. Here we witnessed the beginning of the idea that Mind Left can have its own view of the world with its own desires and aspirations, and Mind Right can have another view. As soon as Mind Right, working through the left hand, began to arrange the blocks correctly, Mind Left would undo the good work. The hands were in competition. The specializations of each hemisphere were different, and growing out of that were the behaviors of each half of the brain.

From the perspective of natural selection, by which organisms acquire specific adaptations, one would imagine that the two hemispheres do not function identically. After all, one does not need two speech production systems or two places to store the memory of faces. Once a brain region has evolved a functional specialization, there would seem to be no need to duplicate it in a second region. Be that as it may, the cerebral cortex's organization suggests that duplication, rather than unilateral specialization, is the rule. The two hemispheres are much more similar to one another in function than you might think; differences surface at a more subtle level of analysis. Each hemisphere's visual areas are devoted to representing the shapes and colors of objects, and their primary specialization relates to the objects' position: Is the object on the left side of space or the right side? What is more, the motor cortices are roughly mirror images of one another, both in structure and in function. While each cortex can be represented by a motor homunculus of the body, it is dominated by fibers that project to muscles on the opposite side of the body.

Evolution could have utilized the available cerebral space in a completely different way. Visual functions could have been isolated in the left hemisphere, and motor projections to both limbs could have originated in the right hemisphere. Yet selective pressures appear to have favored a cerebral organization that reflects structural properties of the world and the organism. The world's spatial structure is reflected in our biology.

Hemispheric specializations are best conceived as superimposed on this fundamental symmetry. In some instances, specializations may have evolved because there was an advantage in having a single system devoted to a certain process. For example, one hypothesis postulates that speech production became strongly lateralized because of the need to communicate at rapid rates. **Transcortical** processing and integration take time and might slow down complicated articulatory gestures. Indeed, lateralization may underlie stuttering because of the two parallel systems that compete for control of speech output.

Others argue that hemispheric specializations evolved because of the inherent advantages in having nonidentical forms of representation. Homologous visual areas perform related operations, but differently enough that the resultant nonidentical representations are imbued with unique advantages in performing certain tasks. This does not mean that these tasks are strictly localized—that language functions are restricted to the left hemisphere or that spatial behavior emanates from the right hemisphere. Not only does the normal performance of these tasks require distributed operations that might span both hemispheres, but also usually both hemispheres contain the essential machinery for performing the task. This organization helps explain why patients with large unilateral lesions that damage 25% of the cortical tissue on one side still have an amazing capacity for recovery. Even more dramatic evidence comes from patients with isolated cerebral hemispheres, a condition that arises after a split-brain operation. Because this radical operation remarkably preserves function, such patients are invaluable for giving us clues to subtle **functional asymmetries** and basic capabilities of each cerebral hemisphere.

The study of **cerebral specialization** has a long and storied tradition in the behavioral and biomedical sciences. Today, left-brain/right-brain claims are everywhere. They are popular, and on the surface they are entertaining accounts of how the human brain might be organized. How much of it is true? Where did these stories come from? What are the facts about the left and right sides of the brain? These and other issues concerning a broad range of human cognitive processes, from perception and attention to language and emotions, are considered in this chapter.

We will find that research on laterality has provided extensive insights into the organization of the human brain. And that the simplistic left-brain/right-brain claims that permeate popular culture distort the complex mosaic of mental processes that contribute to cognition.

Split-brain studies provide a profound demonstration that the two hemispheres do not represent information in an identical manner. Complementary studies on patients with focal brain lesions underscore the crucial role played by lateralized processes in cognition. This research, and recent computational investigations of lateralization and specialization, have advanced the field far beyond the popular interpretations of left-brain/right-brain processes, and they provide the scientific basis for future explorations of many fascinating issues concerning cerebral lateralization and specialization.

In this chapter we will examine the differences between the right and left cerebral hemispheres. Research on the human brain has revealed marked differences between the two halves. We begin at the beginning: the anatomy and physiology of the two halves and their interconnections.

PRINCIPLES OF
CEREBRAL ORGANIZATION

The two cerebral cortices of the human brain are of equal size and surface area. The major lobes (occipital, parietal, temporal, and frontal) appear, at least superficially, to be symmetrical. Nonetheless, for centuries the effects of unilateral brain damage have revealed major functional differences. Most dramatic has been the effect of left-hemisphere damage on language functions. The dominant role of the left hemisphere in language has been confirmed by various behavioral methods in normal humans, including stimulation techniques in which the input is directed to one hemisphere or the other and injections of **amobarbital** (Amytal) into the carotid artery to produce a rapid and brief anesthesia of the ipsilateral hemisphere (i.e., the hemisphere that is on the same side as the injection) (Figure 10.2). This latter approach is called the *Wada test* and was pioneered by Juhn A. Wada and Theodore Rasmussen in the late 1950s as a way of determining, prior to elective surgery for the treatment of disorders such as epilepsy, which hemisphere possesses language. A patient is given an injection of amobarbital into the two hemispheres on separate days. When the injection is delivered to the hemisphere that possesses language, the ability of the patient to speak, or to comprehend speech, is disrupted for up to several minutes. Thus, this procedure unequivocally localizes language, and like the stimulation techniques, it revealed an extremely strong bias for language to be lateralized to the left hemisphere. Similarly, recent functional neuroimaging techniques, such as positron emission tomography (PET) and functional magnetic resonance imaging (fMRI), both of which measure signals associated with increases in blood flow, revealed that language processing is preferentially biased to the left hemisphere (Binder and Price, 2001).

Anatomical Correlates of Hemispheric Specialization

The dominant role of the left hemisphere in language is not strongly correlated with **handedness.** Approximately 50% of all left-handers also have left-hemisphere dominance for language, despite the fact that they comprise only 7% to 8% of the total population. Thus, taken together, over 96% of humans have left-hemisphere specialization for language. Given this dramatic functional asymmetry, most anatomical studies of hemispheric specialization have looked for structural asymmetries in regions of the brain associated with language functions.

Anatomists in the nineteenth century could see that the **lateral fissure**—the large sulcus that defines the superior border of the temporal lobe—has a more prominent upward curl in the right hemisphere in comparison to the left hemisphere (Figure 10.3). In more recent times, the 1960s, Norman Geschwind of the Harvard Medical School followed this up by examining brains obtained from 100 people known to be right-handed (Geschwind and Levitsky, 1968). After slicing through the lateral fissure, they measured the temporal lobe's surface area and discovered that the **planum temporale,** which encom-

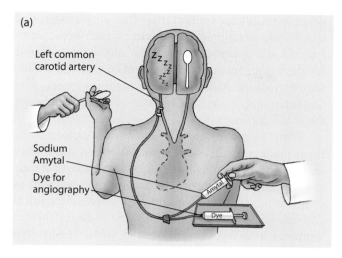

(a)

Left common carotid artery

Sodium Amytal

Dye for angiography

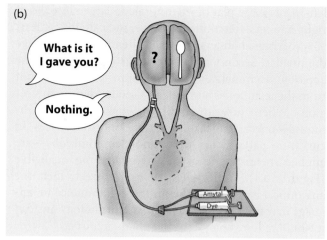

(b)

What is it I gave you?

Nothing.

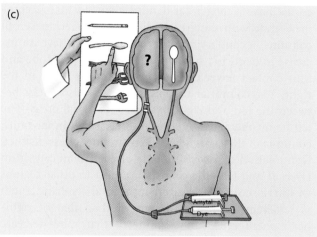

(c)

Figure 10.2 The methods used in amobarbital (Amytal) testing. **(a)** Subsequent to angiography, amobarbital is administered into the left hemisphere, anesthetizing the language and speech systems. A spoon is placed in the left hand, and the right hemisphere takes note. **(b)** When the left hemisphere regains consciousness, the subject is asked what was placed in his left hand, and he responds, "Nothing." **(c)** When a board with a variety of objects pinned to it is held up, it is discovered that the patient can easily point to the appropriate object due to the right hemisphere directing the left hand during the matching to sample task.

passes Wernicke's area, was larger in the left hemisphere, a pattern found in 65% of the brains. Of the remaining brains, 11% had a larger surface area in the right hemisphere and 24% had no asymmetry. The asymmetry in this region of the temporal lobe also extends to subcortical structures that are connected to these areas. For example, the lateral posterior nucleus of the thalamus also tends to be larger on the left.

Other investigations explored whether this asymmetry is absent in people with developmental language disorders. MRI studies revealed that the area of the planum temporale is approximately symmetrical in children with dyslexia, a clue that their language difficulties may stem from the lack of a specialized left hemisphere.

The planum temporale's asymmetry is one of the few examples in which an anatomical index is correlated with a well-defined functional asymmetry. The complex functions of language comprehension presumably require more cortical surface. However, a number of questions remain concerning both the validity and the explanatory power of this asymmetry. First, while the

Figure 10.3 Geschwind and Galaburda (1987) examined brains obtained from 100 people known to be right-handed and measured the temporal lobe's surface area. They discovered that the planum temporale was generally larger in the left hemisphere and that the asymmetry in this region of the temporal lobe also extended to subcortical structures that are connected to these areas.

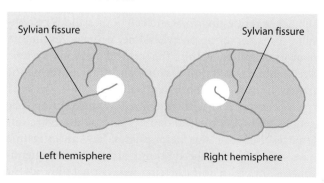

Sylvian fissure

Sylvian fissure

Left hemisphere

Right hemisphere

left-hemisphere planum temporale is larger in 65% of right-handers, functional measures indicate that at least 95% of right-handers show left-hemisphere language dominance. Second, recent work has suggested that the apparent asymmetries in the planum temporale result from the techniques and criteria used to identify this region. When new three-dimensional imaging techniques—ones that take into account differences in curvature patterns in the superior temporal lobe—are applied, hemispheric asymmetries become negligible (Figure 10.4). If this new view is correct, then the anatomical basis for left-hemisphere dominance in language may not be reflected in gross morphology; instead it would require analysis of the microcircuitry.

Microanatomical Investigations of Anatomical Asymmetries

The cellular basis of hemispheric specialization is an exciting new field of investigation. Although many investigators have examined gross size differences between the two hemispheres, relatively few have questioned whether differences in neural connectivity or organization un-

derlie hemispheric asymmetries. One possibility is that specific organizational characteristics, such as the number of local synaptic connections, may be responsible for the unique functions of different cortical areas. Current knowledge of local cortical organization is quite limited, mainly because it is based mostly on studies of the visual cortex.

With language, cortical differences between the hemispheres exist in both Wernicke's region and Broca's region. The cortex can be pictured as a sheet of tightly spaced columns, each of which comprises a circuit of cells that is repeated over and over across the cortical surface. These columns not only are wider in the left temporal lobe but also are spaced farther from each other. Cells within a column of the left primary auditory cortex have a tangential dendritic spread that is commensurate with the greater distance between these cell columns; secondary auditory areas do not have an increase in dendritic spread in the left temporal lobe and thus contact fewer cell columns than do those in the right. The columnar organization itself also varies between the left and right posterior temporal areas. The left hemisphere is organized into clear columnar units; columns in the right hemisphere are much less distinct. Examination of a small sample of pyramidal cells from layers II and III—the cells that project to other regions within the cortex—reveals that the total dendritic length of left-hemisphere cells is greater than that of right-hemisphere cells—an asymmetry that may decrease with age. Structural differences also have been documented in the anterior speech regions in Broca's area. Asymmetries include cell size differences between the hemispheres such as those shown in Figure 10.5, and a possible difference between the fine dendritic structure of each hemisphere.

Most of these comparisons between the hemispheres focus on the gross appearance of individual cells or their position within the cortex. Little effort has been expended on subgroups of cortical cells that can be identified by their neurotransmitters and related chemicals. Chemical assays have documented higher levels of choline acetyltransferase in the left hemisphere. This substance resides in axons and cell bodies and is responsible for assembling acetylcholine, a major excitatory neurotransmitter within the central nervous system. Acetylcholine-containing axons and their postsynaptic targets are similar in both hemispheres.

In comparison with the vast literature on aphasic deficits following cortical damage and putative size differences between the hemispheres, few investigators have homed in on the question of whether the hemispheres' microanatomical circuitry contributes to the

Figure 10.4 Three-dimensional computer reconstructions such as this one can be used to model the cortical surface as a folded mesh composed of many small polygons. These models, which are derived from magnetic resonance images (MRIs), permit measurement of specific regions of the cortex such as the planum temporale located on the dorsal surface of the temporal lobe. Measurements in both hemispheres can be used to assess left-right anatomical asymmetry in living subjects. Investigations of the planum temporale based on these models have revealed that it may not be as asymmetric as was previously thought.

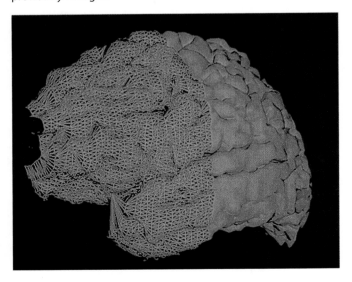

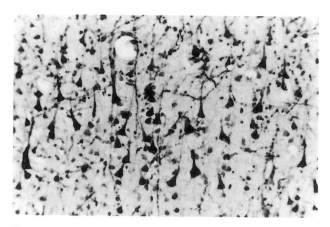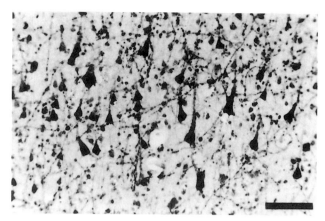

Figure 10.5 Visual examination reveals a subtle difference in the sizes of the largest subgroups of layer III pyramidal cells in the left hemisphere (pyramidal cells stained with acetylthiocholinesterase).

specialized functions of cortical regions. And even fewer have addressed this question from the standpoint of examining subgroups of cells within the cortex by using modern microanatomical labeling techniques.

A thorough understanding of the anatomy and physiology of language-associated cortices could shed considerable light on the cortical mechanisms that facilitate linguistic analysis and production. Since cortical areas have a basic similar organization, documenting cortical locations involved in certain functions should distinguish between the form and variety of neural structures common to all regions and the structures that are critical for a region to carry out particular cognitive functions. These questions hold importance not only for the greater understanding of a species-specific adaptation such as language, but also for understanding how evolution might build functional specialization into the framework of cortical organization.

HOW THE TWO HEMISPHERES COMMUNICATE

The two cerebral cortices are interconnected by the largest fiber system in the brain, the **corpus callosum.** In humans, this bundle of white matter includes more than 200 million axons. As shown in Figure 10.6, many of the callosal projections link together **homotopic areas,** areas in corresponding locations in the two hemispheres. For example, regions in the left prefrontal cortex project to homotopic regions in the right prefrontal cortex. Although this pattern holds for most areas of the association cortex, it is not always seen in primary cortices. Few callosal projections from the visual cortex represent the most eccentric regions of space, and homotopic callosal projections are sparse in the primary motor and the somatosensory cortices.

Callosal fibers also project to heterotopic areas. These projections generally mirror the ones found within a hemisphere. A prefrontal area sending projections to premotor areas in the same hemisphere is also likely to send projections to the same premotor area in the contralateral hemisphere. Yet, heterotopic projections are usually less extensive than comparable projections within the same hemisphere.

The functional role of callosal projections remains unclear. Some researchers point out that in the visual association cortex, receptive fields can span both visual fields. Communication across the callosum enables information from both visual fields to contribute to the activity of these cells. Indeed, the callosal connections could play a role in synchronizing oscillatory activity in cortical neurons as an object passes through these receptive fields (Figure 10.7). In this view, callosal connections facilitate processing by pooling together diverse inputs. Other researchers take a more competitive view of callosal function. If the callosal fibers are inhibitory, then they would provide a means for each hemisphere to compete for control of current processing. For example, multiple movements might be activated, all geared to a common goal; later processing selects one of these candidate movements (see Chapter 11). Inhibitory connections across the corpus callosum might be one contributor to this selection.

In developing animals, callosal projections are diffuse, being more evenly distributed across the cortical surface. Cats and monkeys lose approximately 70% of

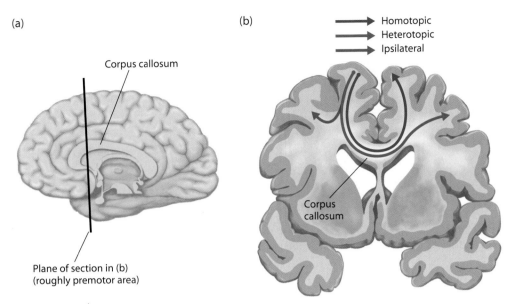

Figure 10.6 **(a)** Midsagittal view of the right cerebral hemisphere, with the large corpus callosum labeled. **(b)** The caudal surface of a coronal section of brain roughly through the premotor cortical area. Homotopic callosal fibers (blue) connect corresponding sections of the two hemispheres via the corpus callosum, whereas heterotopic connections (green) link different areas of the two hemispheres of the brain. In primates, both types of contralateral connections (blue and green) as well as ipsilateral connections (red) start and finish at the same layer of neocortex. Adapted from Gazzaniga and Le Doux (1978).

their callosal axons during development; some of these transient projections are between areas (portions of the primary sensory cortex) that in adults are not connected by the callosum. Yet, this loss of axons does not produce cell death in each cortical hemisphere. Recall that a single cell body can send out more than one axon terminal, one to cortical areas on the same side of the brain and one to the other side of the brain. Thus, loss of a callosal axon may well leave its cell body alive with its secondary collateral connection to the same hemisphere intact. The refinement of connections is a hallmark of callosal development, just as such refinement characterizes intrahemispheric development (see Chapter 15).

In general terms, hemispheric specialization must have been influenced and constrained by callosal evolution. The appearance of new cortical areas might be expected to require more connections across the callosum (expansion). In contrast, lateralization might have been facilitated by a lack of callosal connections. The resultant isolation would promote divergence among the functional capabilities of homotopic regions.

As with the cerebral hemispheres, researchers have investigated functional correlates of anatomical differences in the corpus callosum. Usually investigators measure gross aspects like the cross-sectional area of the callosum. Variations in this measure are linked to gender, handedness, mental retardation, autism, and schizophrenia. Interpretation of these data is complicated by methodological disagreements and by contradictory results. The underlying logic of measuring the corpus callosum's cross-sectional area relies on the relation of area to structural organization. Callosal size could be related to the number and diameter of axons, the proportion of myelinated axons, the thickness of myelin sheaths, and measures of nonneural structures such as the size of blood vessels or the volume of extracellular space. With large samples of callosal measures from age-matched control subjects, sex-based differences are seen in the shape of the callosum but not in the size. Handedness may be associated with the size of the callosum's subregions, but geometrical parsing of the callosum is artificial, at best. The interpretation of a smaller callosum, such as that found in autistic patients, remains unknown.

Cortical Disconnection

Because the corpus callosum is the primary means of communication between the two cerebral hemispheres, we learn a lot when we sever the fibers. This approach was most successfully used in the pioneering animal studies of Ronald Myers and Roger Sperry (Gazzaniga and Sperry, 1967) at the California Institute of Technology. They showed how crucial the corpus callosum is for unified cortical function. Myers and Sperry demonstrated that visual discrimination learned by one hemisphere did not transfer to the other. For example, cats were trained to choose a "plus" stimulus versus a "circle" stimulus randomly alternated between two doors. If the cats chose correctly, they were rewarded with food. Myers and Sperry made the startling discovery that such visual discriminations trained to one half of the brain were not known to the other half when the callosum and anterior commissure was sectioned.

This important research laid the groundwork for comparable human studies initiated by Sperry and Michael Gazzaniga in the 1960s (Sperry et al., 1969). Unlike lesion studies, the split-brain operation does not destroy any cortical tissue; instead, it eliminates the connections between the two hemispheres. With split-brain patients, functional inferences need not be based on how behavior might change after the cortical areas are eliminated. Rather, it becomes possible to see how the two hemispheres operate in relative isolation.

Patients who undergo split-brain surgery are those suffering from intractable epilepsy. As with all forms of neurosurgery, the procedure is applied only when other forms of treatment, such as medication, fail. By severing the transcortical connections, surgeons think that the epileptogenic activity is held in check. Indeed, the operation is almost always successful. Seizures generally subside immediately, even in patients who prior to the

operation experienced up to fifteen seizures per day. The operation enables them to resume a normal life with no obvious side effects.

Many methodological issues arise when evaluating the performance of split-brain patients. First, we must bear in mind that these patients were not normal prior to the operation: They were all chronic epileptics. Therefore, it is reasonable to ask whether they provide an appropriate barometer of normal hemispheric function after the operation. There is no easy answer to this question. Several patients do display abnormal performance on neuropsychological assessments, and they may even be mentally retarded. But in some patients the cognitive impairments are negligible; these are the patients studied in closest detail.

Second, it is important to consider whether the transcortical connections were completely sectioned, or whether some fibers remained intact. In the original California operations, one had to rely on surgical notes to determine the completeness of the surgical sections. In recent years, MRIs, such as in Figure 10.8, and electrical brain mapping techniques have provided a more accurate representation of the extent of surgical sections. Accurate documentation of a callosal section is crucial for learning about the organization of the cerebral commissure.

Figure 10.7 **(a)** When receptive fields (1 and 2) on either side of fixation are stimulated by two separate light bars moving in different directions, the firing rates of the two cells are not correlated. **(b)** In animals with an intact corpus callosum, cells with spatially separate receptive fields fire synchronously when they are stimulated by a common object, such as a long light bar spanning both fields. **(c)** In animals whose corpus callosum has been severed, synchrony is rarely observed. Adapted from Engel et al. (1991) and Gray et al. (1989).

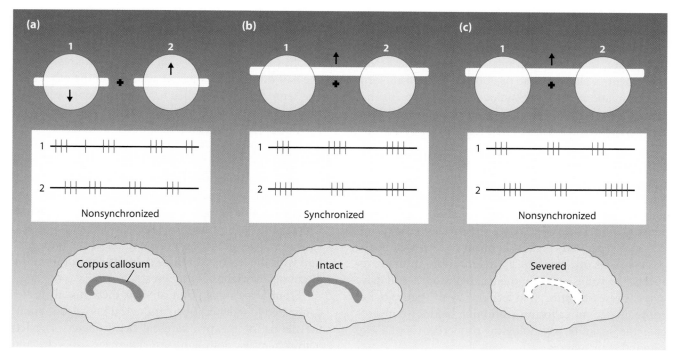

(a)

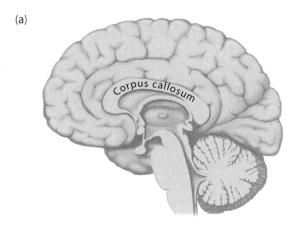

(b)

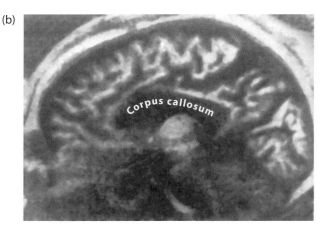

Figure 10.8 **(a)** A sagittal view of the human brain. The corpus callosum is the large fiber tract that is sectioned by a neurosurgeon seeking to control otherwise intractable epilepsy. **(b)** MRI of a corpus callosum that has been entirely sectioned, resulting in no transfer of information between the cerebral hemispheres.

The main methods of testing the perceptual and cognitive functions of each hemisphere have changed little over the past 30 years. Researchers mainly use visual stimulation, not only because of the preeminent status of this modality for humans but also because the visual system is more strictly lateralized than other sensory systems such as the auditory and olfactory systems. The visual stimulus is restricted to a single hemisphere by quickly flashing the stimulus in one visual field or the other (Figure 10.9). Prior to stimulation, the patient is required to fixate on a point in space. The brevity of stimulation is necessary to prevent eye movements, which would redirect the information into the unwanted hemisphere. Eye movements take roughly 200 msec, so if the stimulus is presented for a briefer period of time, the experimenter can be confident that the stimulus was lateralized. More recent image stabilization tools—ones that move in correspondence with the subject's eye movements—allow a more prolonged, naturalistic form of stimulation. This technological development has opened the way for new discoveries in the neurological and psychological aspects of hemisphere disconnection.

Functional Consequences of the Split-Brain Procedure

Reports on the California patients dealt with several fundamental issues concerning the psychological properties of separated cerebral hemispheres, as well as the basic aspects of neurological organization. In many respects, the issues raised in the original studies still drive current research efforts, so let's review the neurological consequences of callosum sectioning before we investigate the separate psychological properties of the two cerebral hemispheres.

Humans were studied in the context of new animal evidence that a division of the cerebral commissure produces a profound deficit in the interhemispheric transfer of sensory and motor information. After operations on cats, monkeys, and chimpanzees that followed the midline sectioning of the corpus callosum and anterior commissure, researchers discovered that visual and tactile information that was lateralized to one hemisphere did not transfer to the opposite hemisphere—hence, the split-brain animal. This startling discovery was completely contrary to earlier reports on the effects of human commissure section reported by A.J. Akelaitis (1941) of the University of Rochester. He claimed that no significant neurological and psychological effects were seen after the callosum was sectioned.

Careful testing with the California patients, however, revealed behavioral changes similar to those seen in split-brain primates. Visual information presented to one half of the brain was not available to the other half. The same principle applied to touch. Patients were able to name and describe objects placed in the right hand but not objects presented in the left hand. Sensory information restricted to one hemisphere was not available to accurately guide movements with the ipsilateral hand. For example, when a picture of a hand portraying the "ok sign" was presented to the left hemisphere, the patient was able to make the gesture with the right hand, which gains its control from the left half of the brain. However, the patient was unable to make the same gesture with the left hand, which gains its control from the disconnected right hemisphere.

From a cognitive point of view, these initial studies confirmed long-standing neurological knowledge about the nature of the two cerebral hemispheres. The left hemisphere is dominant for language, speech, and major problem solving, while the right one appears specialized for visuospatial tasks such as drawing cubes and other three-dimensional patterns. This means that split-brain patients cannot name or describe visual and tactile stimuli presented to the right hemisphere because the sensory information is disconnected from the dominant left (speech) hemisphere. But this does not mean that knowledge about the stimuli is absent in the right hemisphere. Nonverbal response techniques are required to demonstrate the competence of the right hemisphere. For example, the left hand can be used to point to named objects or to demonstrate the function of depicted objects presented in the left visual field.

Specificity of Callosal Function

When the corpus callosum is fully sectioned, little or no perceptual or cognitive interaction occurs between the hemispheres. Surgical cases where callosal section is limited or part of the callosum is inadvertently spared have enabled investigators to examine specific functions of the callosum by region. For example, when the **splenium,** the posterior area of the callosum that interconnects the occipital lobe, is spared, visual information is transferred normally between the two cerebral hemispheres (Figure 10.10). In these instances, pattern, color, and linguistic information presented anywhere in either visual field can be matched with information presented

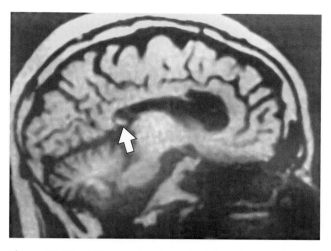

Figure 10.10 MRI scan showing splenial sparing, resulting in the ability of visual information to be transferred between the cerebral hemispheres.

to the other half of the brain. Even so, the patients show no evidence of interhemispheric transfer of tactile information from felt objects. These observations are consistent with other human and animal data showing that major callosum subdivisions are organized into functional zones whose posterior regions are more concerned with visual information and whose anterior regions transfer auditory and tactile information.

The anterior part of the callosum is involved in higher-order transfer of semantic information. Surgeons sometimes prefer to perform the split-brain procedure in stages, restricting the initial operation to the front (anterior) or back (posterior) half of the callosum. The remaining fibers are sectioned in a second operation only if the seizures continue to persist. This two-stage procedure offers a unique glimpse into what the anterior callosal regions transfer between the cerebral hemisphere. When the posterior half of the callosum is sectioned, transfer of visual, tactile, and auditory sensory information is severely disrupted. Yet the remaining intact anterior region of the callosum is able to transfer higher-order information. For example, one patient was able to name stimuli presented in the left visual field following a resection limited to the posterior callosal region. Close examination revealed that the left hemisphere was receiving higher-order cues about the stimulus without having access to the sensory information about the stimulus itself (Figure 10.11). In short, the anterior part of the callosum transfers semantic information about the stimulus but not the stimulus itself. After the anterior callosal region was sectioned in this patient, this capacity ceased.

Figure 10.9 The split-brain patient reports through the speaking hemisphere only the items flashed to the right half of the screen and denies seeing left-field stimuli or recognizing objects presented to the left hand. The left hand correctly retrieves objects presented in the left visual field of which the subject verbally denies having any knowledge.

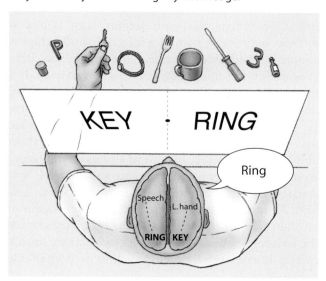

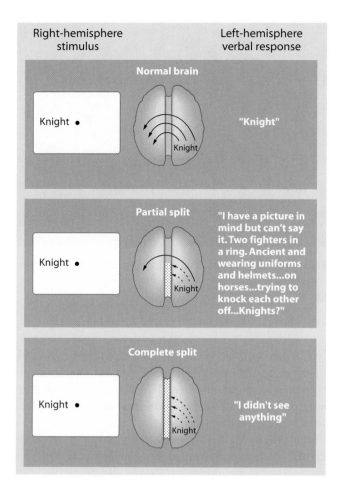

Right-hemisphere stimulus

Left-hemisphere verbal response

Normal brain

Knight ●

"Knight"

Partial split

Knight ●

"I have a picture in mind but can't say it. Two fighters in a ring. Ancient and wearing uniforms and helmets...on horses...trying to knock each other off...Knights?"

Complete split

Knight ●

"I didn't see anything"

Figure 10.11 Schematic representation of J.W.'s left-visual field-naming ability at each operative stage. J.W. was a split-brain patient. Adapted from Sidtis et al. (1981).

HEMISPHERIC SPECIALIZATION

One of the cardinal features of behavioral studies of human and animal brains has been the realization that specific brain areas appear to be involved with specific perceptual and cognitive functions. The split-brain approach to this issue has been straightforward. By testing each disconnected hemisphere, one can assess the different capacities each might possess. While some claims are exaggerated, there are marked differences between the two halves of the brain. The most prominent lateralized function in the human brain is the left hemisphere's capacity for language and speech.

Language and Speech

A useful dichotomy when trying to understand the neural bases of language is the distinction between grammatical and lexical functions. The grammar-lexicon distinction is different from the more traditional syntax-semantics distinction commonly invoked to understand differential effects of brain lesions on language processes

(see Chapter 9). In general terms, *grammar* is the rule-based system humans have for ordering words to facilitate communication. The *lexicon*, by contrast, is the mind's dictionary, where words are associated with specific meanings. The reason for the distinction is that it takes into account such factors as memory in the sense that, with memory, word strings as idioms can be learned by rote. For example, the cliche "You can't teach an old dog new tricks" is most likely a single lexical entry. Though it is clear that the lexicon cannot encompass most phrases and sentences—there are endless unique sentences, such as the one you are now reading—memory does play a role in many short phrases. When uttered, such word strings do not reflect an underlying interaction of syntax and semantic systems; they are, instead, essentially an entry from the lexicon. A modern view would predict that there ought to be brain areas wholly responsible for grammar, whereas the lexicon's location ought to be more elusive, since it reflects learned information and thus is part of the brain's general memory-knowledge systems. The grammar system, then, ought to be discrete and hence localizable, and the lexicon should be distributed and hence more difficult to completely damage.

Language and speech processes are rarely present in both hemispheres. While the separated left hemisphere normally comprehends all aspects of language, the right hemisphere is not always devoid of linguistic capabilities. Indeed, out of dozens of split-brain patients who have been carefully examined, only six showed clear evidence of residual linguistic functions in the right hemisphere. However, even in these patients, the extent of right-hemisphere language functions is severely limited and restricted to the lexical aspects of comprehension.

The left and right lexicons of these special patients can be near equal in their capacity, but are organized quite differently. For example, both hemispheres show a phenomenon called the *word superiority effect* (WSE) (see Chapter 4). Normal readers are better able to identify letters (e.g., *L*) in the context of real English words (e.g., *BELT*) than when the same letters appear in pseudowords (e.g., *KELT*) or nonsense letter strings (e.g., *KTLE*). Since pseudowords and nonwords do not have lexical entries, letters occurring in such strings do not receive the additional processing benefit bestowed on words. Thus, the WSE emerges.

The WSE may be a useful measure of the integrity of the visual lexicon. When these effects are assessed for each hemisphere, the patients with right-hemisphere language exhibit a visual lexicon, as evidenced by a WSE. Yet it may be that each hemisphere accesses this lexicon in a different way. To resolve this issue, investigators evaluated the possibility by using a letter priming task. The task was simply to indicate whether a briefly flashed uppercase letter was an *H* or a *T*. On each trial the uppercase letter was preceded by a lowercase letter that was either an *h* or a *t*. Normally subjects are significantly faster, or primed, when an uppercase *H* is preceded by a lowercase *h* than when it is preceded by a lowercase *t*.

The difference between response latency on compatible (*h–H*) versus incompatible (*t–H*) trials is taken to be a measure of letter priming. J.W., a split-brain participant, performed a lateralized version of this task in which the prime was exposed for 100 msec to either the right or the left visual field, and 400 msec later the target letter appeared in either the right or the left visual field. The results, shown in Figure 10.12, provide no evidence of letter priming for left-visual-field trials but clear evidence of priming for right-visual-field trials. Thus, the lack of a priming phenomenon in the disconnected

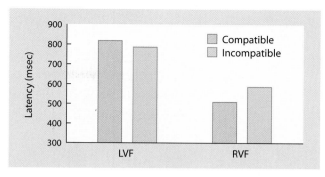

Figure 10.12 Letter priming as a function of visual field in split-brain patients. The graph shows the response latencies for compatible and incompatible words in the right (RVF) and left visual fields (LVF). The latencies for both types of words are much longer for the left visual field (right hemisphere).

right hemisphere suggests a deficit in letter recognition, prohibiting access to parallel processing mechanisms. There were a variety of other deficiencies in J.W.'s right-hemisphere functions. For example, he was unable to judge whether one word was superordinate to another (e.g., *furniture–chair*), or whether two words were antonyms (e.g., *love–hate*).

In sum, there can be two lexicons, one in each hemisphere—but this lexical organization is rare. When present, the right hemisphere's lexicon seems organized differently from the left hemisphere's lexicon, or at least is accessed in different ways. These observations are consistent with the view that lexicons reflect learning processes and as such are more widely distributed in the cerebral cortex. Still, it would be folly to ignore the fact that in the general population the lexicon appears to be in the left hemisphere. A right-hemisphere lexicon is rarely present and stores information in a more random fashion.

Generative syntax is present in only one hemisphere. While the right hemisphere of some patients clearly has a lexicon, the hemispheres perform erratically on other aspects of language such as understanding verbs, pluralizations, the possessive, or active-passive differences. In patients who possess some language, the right hemisphere also fails to use word order to disambiguate stimuli for correct meaning. Yet these right hemispheres can indicate when a sentence ends with a semantically odd word! What is more, right hemispheres with language capacities can make grammar judgments. For some peculiar reason, they cannot use syntax to disambiguate stimuli, but they can judge that one set of utterances is grammatical while another set is not. This startling

An Interview with Michael Corballis, Ph.D.   Dr. Corballis is associated with the Psychology Department at the University of Auckland, New Zealand. His writings have provided an important evolutionary perspective to the study of hemispheric specialization.

Authors: The human brain is full of lateralized functions. Are we the only species with such cerebral organization?

MC: There is, of course, growing evidence for cerebral asymmetries in other species, but I have yet to be convinced that they are of the same degree and consistency as our own. The most conspicuous asymmetries of the human brain, moreover, seem to apply to functions that are themselves characteristically human, namely, language and manual praxis, in which I include such things as writing, accurate throwing, manufacture of tools, carving, painting, etc. I think these activities themselves constitute something of a discontinuity in primate evolution, although I don't want to adopt an extreme Chomskian stance about language, or to deny the considerable manual dexterity seen in the gorilla, for example. I rather like Marian Annett's idea of a laterality gene, specific to humans, that imposed a left-hemispheric dominance for language and manual praxis. I think it may have been imposed on a structure that was already lateralized to some extent, specifically with respect to spatial and emotional functions, but it gave humans a distinctive lopsidedness.

Authors: But cognitive functions, save perhaps for grammar systems, are not wholly lateralized in humans. Disconnected left and right hemispheres each can process facial information, although the right hemisphere is much better at it. Children injured during development can learn grammar with their nondominant right hemisphere and so on. How do you think of these phenomena in the context of the laterality gene?

MC: I think that the laterality gene may control growth gradients. There is evidence for a spurt of left-hemispheric growth between the ages of about 2 and 4, and this coincides with the development of grammar. If this is disrupted, then grammar may be acquired by the right hemisphere in a later growth spurt, though at the expense of the usual right-hemisphere functions. For this mechanism to work, you need to postulate some sort of interhemispheric inhibition, so that if grammar is acquired in a normal fashion in the left hemisphere, the right hemisphere does not acquire it during its later growth phase. All of this is fairly hypothetical, of course, but may go some distance toward explaining the equipotentiality (or near-equipotentiality) of the two hemispheres for grammar. I should point out that equipotentiality is not the same as equality—the right hemisphere does appear to have a potential for grammar but is normally denied the opportunity to exploit it, so there must be some kind of inhibitory process.

I think that growth gradients provide an elegant explanation for hemispheric asymmetries—much simpler than supposing that the brain was somehow rewired in the course of evolution—yet powerful because they allow the environment to do much of the work. The other conspicuous feature of the human brain is the long postnatal period of growth, which maximizes environmental influences while the brain is at its most plastic. If you suppose that growth is programmed to occur asymmetrically during this period, then I think you have a recipe for much that is unique about the human mind.

To get back to the question of right-hemisphere language, I think that language can be quite sophisticated in the absence of grammar (and perhaps

finding suggests that patterns of speech are learned by rote. Yet, recognizing the pattern of acceptable utterances does not mean a neural system can use this to understand word strings (Figure 10.13).

One of the hallmarks of most split-brain patients is that their speech production is from the left hemisphere and not the right. This observation, consistent with the literature and with amobarbital studies, shows that the left hemisphere is the dominant hemisphere for speech production. Nonetheless, there are now a handful of documented cases of split-brain patients who can produce speech from both hemispheres. While speech is

phonology). However, grammarless language can't be stretched to generative, propositional speech, but may be able to get by on cliches and learned phrases. (The speech of L.B., the split-brain man well known in the literature, is most of the time like a long-playing record of cliches, endlessly repeated—I have often wondered if his right hemisphere plays a role in this.) I must say I am quite impressed by the apparent ability of the disconnected right hemisphere to follow verbal instructions, although I am not sure how general this is among the split-brain population. I am nevertheless still enough of a Chomskian to think that there is something special about normal, human left-hemispheric language, and that grammar is the key.

Authors: The general rule seems to be that only one hemisphere has a grammar but both can have a lexicon which would include cliches and phrases. But more generally, do you feel that all specialized systems as revealed through brain damage to adults occur or crystallize sometime during development? Are you saying the blueprint for the brain is symmetrical until some process stamps out the lateralization pattern?

MC: I think that most specialized systems are crystallized during development, and that some become lateralized in the process. Of course, we can obtain specialized knowledge in adulthood, but this is not a "system" in the way that grammar, face recognition, etc., are. By the way, I have yet to discover evidence for a right-hemispheric specialization that is anywhere near as pronounced as the left-hemispheric specialization for grammar and propositional language. I still think that much of what is called right-hemispheric specialization is really so by default—a slight weakness due to the presence of left-hemispheric functions rather than a programmed specialization. I think that goes for face recognition, but I may be wrong.

Is the blueprint for the brain symmetrical? Depends what you mean by "blueprint," I suppose. I think that there may be programmed gradients that are asymmetrical, and to the extent that these are part of the

blueprint, the answer is no. But I do think that millions of years of evolution have selected for symmetry, largely on the grounds that, in the absence of consistent asymmetry in the natural environment, locomotion is best accomplished by a symmetrical system. Symmetry of sensory systems would naturally follow, since to a freely moving organism there is no systematic sensory bias. Asymmetry could be a disadvantage—as Martin Gardner once put it, a predator "could sneak up on the right." However, once you get to a level of sophistication beyond mere sensorimotor function, toward a "computational brain," then there could be advantages to asymmetry. I like to distinguish between actions that are reactions to the environment, which are by and large symmetrical, and operations on the environment, which are much more likely to be asymmetrical. Skilled cricketers can usually catch equally well with either hand, but can only throw accurately with one.

Authors: What is the future for research on cerebral lateralization? Where is it going?

MC: I have thought for years that it was coming to an end, but it never seems to. Nevertheless, I think we should all remember that there was a good deal of fanciful speculation about cerebral lateralization in the late nineteenth century, but it pretty well dried up in the early twentieth. It may well happen again as we now cross into the twenty-first. I don't think we are getting any further in the search for some "fundamental" dichotomy that characterizes the difference between the hemispheres, and I'm not sure either that a multi-component view is very satisfying, although it may be nearer the truth. My guess is that further insights are likely to come from a more evolutionary approach, rather than yet more dichotic-listening and visual-hemifield studies, although imaging techniques may provide new information. It would be really great if we could discover the gene or genes that control cerebral lateralization. If such a gene exists, its discovery should not be too far away.

restricted to the left hemisphere following callosal bisection, in these rare patients the capacity to make one-word utterances from the disconnected right hemisphere has emerged over time.

This intriguing development raises the question of whether information is somehow transferring to the

dominant hemisphere for speech output or whether the right hemisphere itself is capable of developing speech production. After extensive testing it became apparent that the latter hypothesis is correct. For example, the patients were able to name an object presented in the left field and in the right field but were not able to judge

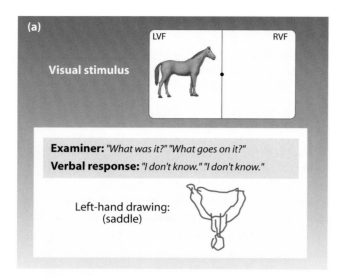

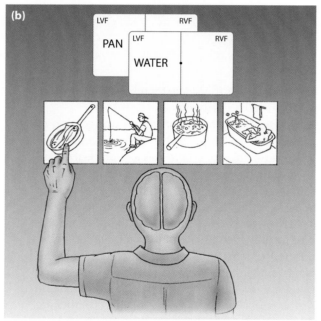

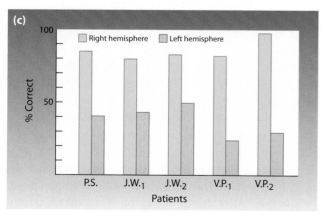

Figure 10.13 **(a)** The right hemisphere is capable of understanding language but not syntax. When presented with a horse stimulus to the left visual field (right hemisphere), the subjects maintain through the left hemisphere that they saw nothing. When asked to draw what goes on the object, the left hand (right hemisphere) is able to draw a saddle. **(b)** The capacity of the right hemisphere to make inferences is extremely limited. Two words are presented in serial order, and the right hemisphere (left hand) is simply required to point to a picture that best depicts what happens when the words are causally related. The left hemisphere finds these tasks trivial while the right cannot perform the task. **(c)** Data from three patients show how the right hemisphere is more accurate than the left in recognizing unfamiliar faces. From Gazzaniga and Smylie (1983).

whether they were the same objects! Or when words like *father* were presented such that the fixation point fell between the *t* and the *h*, the patients said either "fat" or "her," depending on in which hemisphere speech production dominated. These findings illustrate that an extraordinary plasticity lasts sometimes as long as 10 years after callosal surgery. In fact in one patient, the right hemisphere had no speech production capability for approximately 13 years before it "spoke."

Visuospatial Processing

The second primary task domain that has been studied in split-brain patients is visuospatial processing. As shown in Figure 10.1, the isolated right hemisphere is frequently found to be superior on neuropsychological tests such as the block design task, a subtest of the Wechsler Adult Intelligence Scale. Here, the simple task of arranging some red and white blocks to match those of a given pattern can find the left hemisphere performing poorly while the right triumphs. However, asymmetries with tasks such as these have proved to be inconsistent. In some patients, performance is impaired with either hand; in still others, the left hemisphere is quite adept at this task.

This inconsistency is, at least in part, due to the fact that the component operations required for the block design task have not been identified. Patients who demonstrate a right-hemisphere superiority for this task evince no superiority on the perceptual aspects of the task. If a picture of the block design pattern is lateralized, either hemisphere can easily find the match from a series of pictures. And since each hand is sufficiently dexterous, the crucial link must be in the mapping of the sensory message onto the capable motor system.

It is also true that monitoring and producing facial expressions are managed by different hemispheres. In the perceptual domain, it appears that the right hemisphere has special processes devoted to the efficient detection of upright faces (Kingstone et al., 2000).

Although the left hemisphere can also perceive and recognize faces and can reveal superior capacities when the faces are familiar, the right hemisphere appears specialized for unfamiliar facial stimuli. This pattern of asymmetry also has been shown for the rhesus monkey.

As the right hemisphere is superior for perception of faces, it would be reasonable to suppose it is also specialized for the management of facial expressions. Recent evidence, however, shows that while both hemispheres can generate spontaneous facial expressions, only the dominant left hemisphere can generate voluntary facial expressions.

Attention and Perception

The attentional and perceptual abilities of split-brain patients have been explored extensively. Visual perception is the easiest to study. After cortical disconnection, perceptual information does not interact between the two cerebral hemispheres, but the supporting cognitive processes of attentional mechanisms do sometimes interact.

Split-brain patients cannot cross-integrate visual information between the two visual fields. When visual information is lateralized to either the left or the right disconnected hemisphere, the unstimulated hemisphere cannot use the information for perceptual analysis. This is also true for stereognostic information presented to each hand. Although touching any part of the body is noted by either hemisphere, patterned somatosensory information is lateralized. Thus, when a split-brain patient is holding an object in the left hand, he is unable to find an identical object with the right hand. Although some investigators argue that higher-order perceptual information is integrated by way of subcortical structures, others have not replicated these results.

For example, in one study, split-brain patients sometimes drew pictures that combine word information presented to the two hemispheres (e.g., *ten + clock =* a picture of a clock set at 10). Although this outcome initially seemed to imply the subcortical transfer of higher-order information between the hemispheres, subsequent observations (Kingstone and Gazzaniga, 1995) suggested that it actually reflects dual hemispheric control of the drawing hand (with control biased to the left hemisphere). Conceptually ambiguous word pairs, such as *hot dog*, were always depicted literally (e.g., a dog

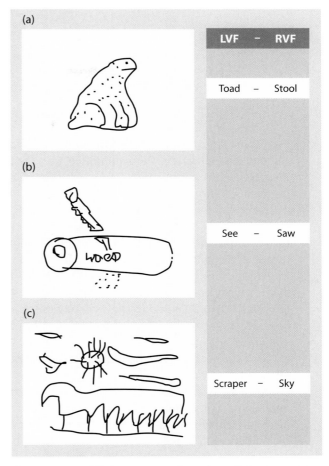

Figure 10.14 All pictures were drawn by split-brain participant J.W.'s left hand. **(a)** Drawing of the left-visual-field word *Toad* (ipsilateral to the drawing hand). **(b)** Drawing of the right-visual-field word *Saw* (contralateral to the drawing hand). **(c)** Drawing combines both words *Scraper* and *Sky* (ipsilateral + contralateral). Adapted from Kingstone and Gazzaniga (1995).

panting in the heat) and never as emergent objects (e.g., a frankfurter). Moreover, right- and left-hand drawings often depicted only the words presented to the left hemisphere (Figure 10.14).

Whereas the previous work showed that object identification processes work in isolation in the split-brain patient, the evidence does suggest that crude information concerning spatial locations can be cross-integrated. In one set of experiments, a four-point grid was presented to each visual field. On a given trial, one of the positions on the grid was highlighted, and one condition of the task required the subject to move her eyes to the highlighted point within the visual field stimulated (Figure 10.15). In the second condition, the subject was required to move her eyes to the relative point in the opposite visual field. Split-brain subjects did this easily, which raised the possibility of crude cross-integration of

Interhemispheric Communication: Cooperation or Competition?

Theories of callosal function generally have focused on the idea that this massive bundle of axonal fibers provides the primary pathway for interhemispheric transfer. For example, in Chapter 5 we discussed Warrington's model of object recognition. In her view, the right hemisphere performs a specialized operation essential for perceptual categorization. This operation is followed by a left-hemisphere operation for semantic categorization. Interhemispheric communication is essential in this model for shuttling the information through these two processing stages. Even in less serially oriented models, interhemispheric transfer frequently is assumed to allow each hemisphere to have access to the information that might have been lateralized initially.

On the other hand, interhemispheric communication need not be conceptualized as a cooperative process. Connections across the corpus callosum may actually underlie a competition between the hemispheres. Indeed, the primary mode of callosal communication may be inhibitory rather than excitatory. As activation builds in a region within a hemisphere, the homologous, corresponding region in the other hemisphere would be inhibited. By this view, we need not assume that interhemispheric communication is designed to share information processing within the two hemispheres to facilitate concurrent, and roughly identical, activity in homologous regions. Similar to the way in which split-brain behavior is assumed to reflect the independent operation of the

two hemispheres, behavior produced by intact brains may also reflect the (fluctuating) dominance of one or the other hemisphere.

Why might the brain have evolved such that the two hemispheres act as competitors rather than cooperators? One challenge for a cooperative system is that there must be a means to ensure that the two hemispheres are operating on roughly the same information. Such coordination might be difficult given that both the perceptual input and the focus of our attention are constantly changing. While computers can perform their operations at lightning speed, neural activity is a relatively slow process. The processing delays inherent in transcallosal communication may limit the extent to which the two hemispheres can cooperate.

There are a number of factors limiting the rate of neural activity. First, to generate an action potential, activity within the receiving dendritic branches must integrate tiny inputs across both space and time in order to reach threshold. Second, the rate at which individual neurons can fire is limited, owing to intrinsic differences in membrane properties, tonic sources of excitation and inhibition, and refractory periods between each spike-generating event. Third, and most important, neural signals need to be propagated along axons. These conduction times can be quite substantial, especially for the relatively long fibers of the corpus callosum.

James Ringo and his colleagues (1994) at the University of Rochester provided an interesting analysis of

spatial information. This was true even when the grid was randomly positioned in the test field.

The finding that some type of spatial information remains integrated between the two halves of the brain raises the question, Are the attentional processes associated with spatial information affected by cortical disconnection? Surprisingly, split-brain patients can use either hemisphere to direct attention to positions in either the left or the right visual field. This conclusion was based on studies using a modified version of the spatial cuing task (see Chapter 7). To review, in this task, subjects respond as quickly as possible upon detecting a target that appears at one of several possible locations. The target is preceded by a cue, either at the target location

(i.e., a valid cue) or at another location (i.e., an invalid cue). Responses are faster on valid trials, indicating spatial orienting to the cued location. In split-brain patients, as with normal subjects, a cue to direct attention to a particular point in the visual field was honored no matter which half of the brain was presented with the critical stimulus. These results suggest that the two hemispheres rely on a common orienting system to maintain a single focus of attention.

The discovery that spatial attention can be directed with ease to either visual field raised the question of whether each separate cognitive system in the split-brain patient, if instructed to do so, can independently direct attention to a part of its own visual field. Can the

this problem. They began by calculating estimates of transcallosal conduction delays. Two essential numbers were needed: the distance to be traveled and the speed at which the signal will be transmitted. If the distances were direct, the average distance of the callosal fibers would be short. However, most axons follow a circuitous route. Taking this into consideration, a value of 175 mm was used as representative of the average length of a callosal fiber in humans. The speed at which myelinated neural impulses travel is a function of the diameter of the fibers. Using the limited data available from humans, in combination with more thorough measures in the monkey, an estimate of the average conduction speed was calculated to be around 6.5 m/sec. Thus, to travel a distance of 175 mm would take almost 30 msec. Single-cell studies in primates have confirmed that interhemispheric processing entails relatively substantial delays.

Ringo used a neural network to demonstrate the consequences of slow interhemispheric conduction times. The network consisted of two identical sets of processing modules, each representing a cerebral hemisphere. It included both intrahemispheric and interhemispheric connections, with the latter much more sparse to reflect the known anatomy of the human brain. This network was trained to perform a pattern-recognition task. After it had learned to classify all of the patterns correctly, the interhemispheric connections were disconnected. Thus, performance could now be assessed when each hemisphere had to operate in isolation. The critical comparison was between networks in which the interhemispheric conduction times during learning had been either slow or fast. The results showed that for the network trained with fast interhemispheric connections, the disconnection procedure led to a substantial deterioration in performance. Thus, object recognition was dependent on cooperative processing for the network with fast interhemispheric connections. In contrast, for the network trained with slow interhemispheric connections, performance was minimally affected by the disconnection procedure. For this network, recognition was essentially dependent only on intrahemispheric processing. These results led Ringo to conclude that a system with slow interhemispheric conduction delays, for example, the human brain, ends up with each hemisphere operating in a relatively independent manner.

Interestingly, these delays could be reduced if the callosal fibers were larger, as this would increase conduction speed. However, larger fibers would require a corresponding increase in brain volume. For example, to reduce the conduction delay by a factor of two would lead to a 50% increase in brain volume. Such an increase would have severe consequences in terms of metabolic demands and for childbirth. The brain appears to have evolved such that each hemisphere can have rapid access to information from either side of space, but with limited capability for tasks that would require extensive communication back and forth across the corpus callosum. The delays associated with transcallosal communication not only might limit the degree of cooperation between two hemispheres but also might have provided an impetus for the development of hemispheric specialization. Independent processing systems would be more likely to evolve nonidentical computational capabilities.

right hemisphere direct attention to a point in the left visual field, while the left one simultaneously attends to a point in the right visual field? Normal subjects cannot so divide their attention. Perhaps the split-brain operation frees the two hemispheres from this constraint.

Alas, the split-brain patient cannot divide spatial attention between the two halves. There appears to be only one integrated spatial attention system that remains intact following cortical disconnection. Thus, like neurologically intact observers, the attentional system of split-brain patients is unifocal. They are unable to prepare for events in two spatially disparate locations.

The dramatic effects of disconnecting the cerebral hemispheres on perception and cognition might suggest that each half has its own attentional resources (Kinsbourne, 1982). If true, the cognitive operations of one half, no matter what the difficulty, would have scant influence on the cognitive activities of the other. The competing view is that the brain has limited resources that manage such processes; if they are being applied to task A, fewer are available for task B. According to this model, the harder hemisphere A worked on a task, the worse hemisphere B would do on a task of constant complexity.

Many have studied this phenomenon, and all confirmed that the central resources are limited. Consider the following experiment, diagrammed in Figure 10.16. In what was called the *mixed* or *hard condition*, two series of three different geometric shapes were displayed

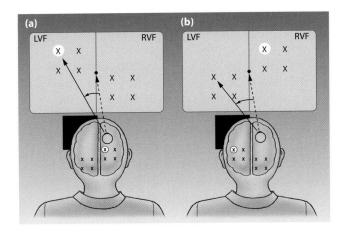

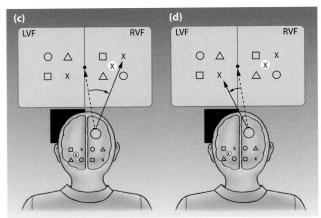

Figure 10.15 The upper example depicts the spatial tests. **(a)** On within-field trials, the eye moved to the stimulus that was surrounded by the probe. **(b)** On between-field trials, the eye moved to the corresponding stimulus in the other hemifield. The lower example depicts the perceptual tests. **(c)** On within-field trials, the probe appeared centered in one of the arrays, and the eye moved to the corresponding stimulus in the field in which the probe appeared. **(d)** On between-field trials, the eye moved to the corresponding stimulus in the opposite field.

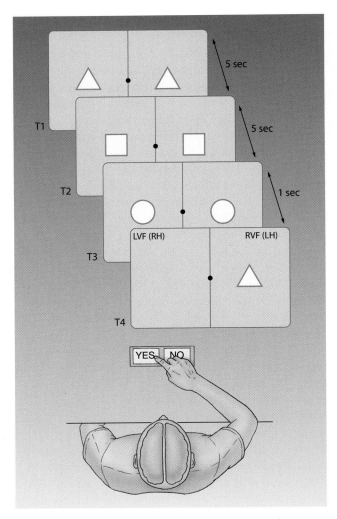

Figure 10.16 Sequence of events for a redundant three-condition trial. Stimuli were selected from a set of seven geometric forms. Each stimulus appeared for 150 msec. A delay followed the presentation of the last stimulus, and the unilateral probe stimulus was presented for 150 msec. The subject then made a decision as to whether or not the stimulus had been viewed previously in the corresponding visual field. Adapted from Holtzman and Gazzaniga (1982).

concurrently to the left and right of the central fixation point and thus were lateralized to the right and left hemispheres, respectively. In the *redundant* or *easy condition*, the three geometric shapes were presented to only one hemisphere, while three identical geometric shapes were presented to the other. To test recognition memory, a single shape was then presented in either the left or the right visual field. The observer indicated with a forced-choice key press whether this probe matched any of the shapes that had been presented in the same hemifield. When the shapes had all been identical in the other hemifield (redundant condition), performance was better than when three different shapes had been presented in the other hemifield (mixed condition). Indeed, in the mixed condition, where six distinct shapes had been presented, performance was poor with either hemisphere.

This concept of resources should be distinguished from other properties of sensory systems associated with information processing. The limits to information processing captured by the concept of resource limitations are more general than the limits and mechanisms now studied in such phenomena as searching a visual scene for information. For example, while the overall resources a brain commits to a task appear constant, the method by which they are deployed can vary. The time needed to detect a complex object increases as more items are added to the display. For example, normal control subjects require about an extra 70 msec to detect the target when two extra items are added to the display, and another 70 msec for each additional pair of items. In split-brain patients, when the items are distributed across the midline of the visual field, as opposed to all being in one visual field, the reaction time to added stimuli is cut almost in half.

This finding was recently qualified; the strategy differed in how each hemisphere examined the contents of its visual field. The left dominant hemisphere utilized a "guided" or "smart" strategy and the right hemisphere did not. Hence, the left hemisphere adopts a helpful cognitive strategy in problem solving, but the right hemisphere does not have extra cognitive skills (Kingstone et al., 1995).

The concept of resources in many ways refers to processes engaged when attention to an information-processing task is allocated voluntarily. Searching a visual scene, however, calls upon more automatic processes that may well be built-in properties of the visual system itself. That these systems are distinct is reflected in the discovery that splitting brains has a different effect on the processes.

CONVERGING EVIDENCE OF HEMISPHERIC SPECIALIZATION

Studies on split-brain patients have given us the most dramatic evidence of hemispheric specialization (Gazzaniga, 1995a). Many cognitive neuroscientists concentrate on identifying the capabilities of the two hemispheres when forced to act in isolation: Can the right hemisphere produce speech? Do the two hemispheres share common attentional resources? Can information be integrated at the subcortical level?

Functional Asymmetries in Patients with Unilateral Cortical Lesions

Research on hemispheric specialization has not been limited to split brains. Two other prominent methodologies come into play. First, continuing the tradition initiated by Broca, many researchers have examined the performance of patients with unilateral, focal brain lesions. The basic idea has been to compare how patients perform when their lesions are restricted to either the right or the left hemisphere. An appealing feature of this approach is that there is no need to lateralize the stimuli to one side or the other: Stimuli can be presented without restriction. Laterality effects are assumed to arise because of the unilateral lesions. If lesions to the left hemisphere result in more disruption in reading tasks, the deficit is attributed to the hemisphere's specialization in reading processes.

Of course, it is necessary to show that similar lesions to the right hemisphere do not produce a similar deficit. This approach, called the *double dissociation method*, provides strong evidence that many cognitive functions are lateralized to the left or right hemisphere (see Chapter 4). The logic is straightforward: If damage to one region affects process A (e.g., language) without affecting process B (e.g., spatial orientation), and a similar lesion to a different area affects process B without affecting process A, evidence for functional separation receives strong support. For instance, it has been demonstrated consistently that lesions in the left hemisphere can produce deficits in language function (such as speaking and reading) that do not follow comparable lesions to the right hemisphere. Similarly, lesions to the right hemisphere can disrupt spatial orientation (such as the ability to accurately locate visually presented items) that do not follow from comparable lesions to the left hemisphere.

To illustrate the nature of lateralized functions in patients suffering from unilateral brain lesions, consider the stimulus in Figure 10.17. If asked to describe the stimulus, your first response might be to say it is a house. When prodded for more description, you would note the intricate detailing on the front door, the windows running across the front facade, and the roof made of asphalt shingles. In recounting the picture, you would have provided a hierarchical description. The house can

Figure 10.17 We represent information at multiple scales. At its most global scale, this drawing is of a house. We also can recognize and focus on the component parts of the house.

be described on multiple levels: In terms of its shape and attributes, it is a house. But it is also a specific house, with a certain configuration of doors, windows, and materials. This description is hierarchical in that the finer levels of description are embedded in the higher levels. The house's shape evolves from the configuration of its component parts, an idea developed in Chapter 5.

David Navon (1977) of the University of Haifa introduced a model task for studying **hierarchical structure.** He created stimuli that could be identified on one of two levels, an example of which is given in Figure 10.18. At each level, the stimulus contains an identifiable letter. The critical feature is that the letter defined by the global shape is composed of smaller letters. In Figure 10.18a, the global *H* is composed of *F*'s.

Navon was interested in how we perceive hierarchical stimuli. He initially found that the perceptual system first extracted the global shape. Not only were reaction times slower when subjects had to identify local elements, but considerable interference happened when the global shape and local elements were discordant. The interference was not symmetrical. The time required to identify the global letter was independent of the identity of the constituent elements. Subsequent research qualified these conclusions, though. Global precedence is not always found but depends on factors like object size and number of local elements. Although Navon had assumed that a single system processes hierarchical stimuli, the lack of invariance suggested otherwise. Perhaps different processing systems were essential for representing local and global information.

Experiments with patients who have suffered a unilateral lesion to the left or right hemisphere support this proposal. After carefully determining the focus of patients' lesions, Lynn Robertson and her colleagues (1988) at the Veterans' Administration Medical Center in Martinez, California, presented the patients with local and global stimuli to the center of view (the critical laterality factor was the side of the lesion). Patients with left-sided lesions were slow to identify local targets, and patients with right-sided lesions were slow with global targets, demonstrating that the left hemisphere is more adept at representing local information and the right hemisphere is better with global information.

It is important to keep in mind, however, that both hemispheres can abstract either level of representation. How they differ is in the efficiency with which they represent local or global information. Thus, patients with left-hemisphere lesions are still able to analyze the local structure of a hierarchical stimulus, but they must rely on an intact right hemisphere, and this hemisphere is inefficient in abstracting local information. Further confirmation of this idea comes from local-global studies with split-brain patients (Robertson et al., 1993). Here, too, patients generally identify targets at either level regardless of the side of presentation. But, as with normal subjects and patients with unilateral lesions, split-brain patients are faster at identifying local targets presented to the right visual field and global targets presented to the left visual field.

In extreme cases, the hemispheric biases for one level of representation can completely override other levels. In the case study at the beginning of this chapter, W.J. was unable to manipulate blocks into their global configuration when he was restricted to using his right hand. Similar dramatic things happen with stroke victims. Figure 10.19. displays drawings made by patients who recently had strokes in either the right or the left hemisphere. They were shown a hierarchical stimulus and asked to reproduce it from memory. Drawings from patients with left-hemisphere lesions faithfully followed the contour, but without any hints of local elements. In contrast, patients with right-hemisphere lesions produced only local elements. Note that this pattern was consistent with both linguistic and nonlinguistic stimuli; hence, the representational deficits were not restricted to certain stimuli.

Functional Asymmetries in the Normal Brain

The second important alternative methodology tests people with intact brains. Here the emphasis is on designing experiments that differentially require processing

Figure 10.18 Local-global stimuli used to investigate hierarchical representation **(a-e)**. Each stimulus is composed of a series of identical letters whose global arrangement forms a larger letter. The subjects' task is to indicate whether the stimulus contained an *H* or *L*. When the stimulus set included competing targets at both levels (b), the subjects would be instructed to respond either only to local targets or only to global targets. Neither target is present in (e).

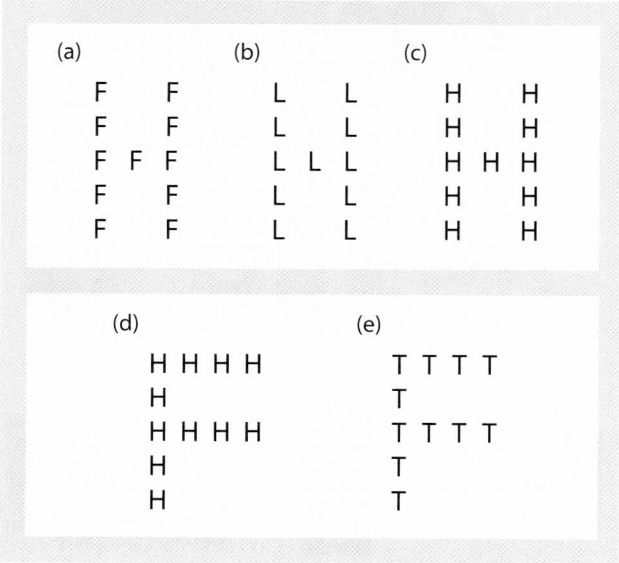

WHAT IS LATERALIZED?

Laterality researchers continually grapple with appropriate ways to describe asymmetries in the function of the two hemispheres (Allen, 1983; Bradshaw and Nettleton, 1981; Bryden, 1982). Early hypotheses fixed on the stimuli's properties and the tasks employed. For example, the left hemisphere was described as verbal and the right hemisphere as spatial. By this view, one might suppose that the left hemisphere has sole province over all language functions. In the dichotic listening task, words presented to the left ear are recognized only when they succeed in being transferred to the left hemisphere. But, as seen in split-brain and stroke research, such absolutes are rare. Many tasks can be performed successfully by either hemisphere, although they may differ in efficiency.

An alternative approach for characterizing hemispheric specialization is to look for differences in processing style. Here, the left hemisphere has been described as analytic and sequential, whereas the right hemisphere is viewed as holistic and parallel. The popular press has promoted such descriptions, triggering industries devoted to helping people learn to "think" with the right hemisphere or improve the "power" of the left hemisphere (Figure 10.22).

Hypotheses about modes of processing are not restricted to specific task domains—they describe general processing styles. Hemispheric specializations may emerge because certain tasks benefit from one style or another. Language, for example, is seen as sequential: We hear speech as a continuous stream that requires rapid analysis of its component parts; an accurate representation of space calls for not just perceiving the component parts but seeing them as a coherent whole.

A problem here is knowing whether a task requires analytic or holistic processing. The theoretical interpretation frequently disintegrates into a circular redescription of results. Though a right-ear advantage exists in perceiving consonants, no asymmetry is found for vowels; consonants require the sequential, analytic processors of the left hemisphere, while vowel perception entails a more holistic form of processing.

With verbal-spatial and analytic-holistic hypotheses, we assume that a single fundamental dichotomy can characterize the differences in function between the two hemispheres. The appeal of "dichotomania" is one of parsimony: The simplest account of hemispheric specialization rests on a single difference. Current dichotomies all have their limitations, but perhaps a new one will capture a unitary distinction.

Yet it is also reasonable to suppose that the fundamental dichotomy is a fiction. Hemispheric asymmetries have been observed in many task domains: language, motor control, attention, object recognition. It may be that specializations are specific to particular task domains. There need not be a causal connection between hemispheric specialization in motor control (e.g., why people are right- or left-handed) and why the hemispheres differ in representing language or visuospatial information. Maybe the commonality across task domains is their evolution: As the two hemispheres became segregated, there was a common impetus for the

Figure 10.22 Pop psychology books have tended to trivialize the differences between the two cerebral hemispheres. The right hemisphere has been suggested to be the creative, intuitive side and the left hemisphere, the analytic, logical side.

Left-hemisphere painting

Right-hemisphere painting

An Interview with Stephen M. Kosslyn, Ph.D.

Dr. Kosslyn is affiliated with the Department of Psychology at Harvard University. In addition to his study of hemispheric specialization, Dr. Kosslyn has used a variety of behavioral and imaging methods to examine the relationship between perception and imagery.

Authors: How do you think the two hemispheres of the human brain are different?

SMK: As a first stab, consider the possibility that the hemispheres differ in three ways. First, there are neuroanatomical differences. For example, the visual cortex is larger in the right hemisphere. I presume these differences have implications for function, but it isn't entirely clear yet what the implications are. Second, the hemispheres differ in how efficiently they can perform specific "elementary" operations, such as specifying an object's location in space, iterating through a mental list, or detecting basic geometric properties in a shape. In my view, relatively simple, mechanical functions are implemented in local regions of cortex, and any complex task requires combinations of such basic functions. Local cerebral activation detected using positron emission tomography, say, indicates that such an operation is used, and patterns of activation reflect combinations of operations that are used to carry out a task. Third, combinations of basic operations correspond to a specific "strategy." The hemispheres may differ in the types of strategies they use best. For example, perhaps the left hemisphere is better at analyzing things into basic units (shapes into parts, stories into phrases, etc.), as some researchers have suggested, whereas the right is better at using overall patterns (global shape, general theme, etc.). Such strategy differences may arise in part because of differences in the relative efficiency of specific operations, but may also reflect "strategy biases" that arise for other reasons.

Authors: What are the implications of your approach?

SMK: If you want to know whether to slice through the left or right hemisphere to remove a subcortical tumor, it would be nice to know what the two sides are doing. If the patient is an architect, I would recommend avoiding regions that deal with spatial relations, whereas that might not be quite so important if the patient is an English teacher.

Authors: We will keep this in mind next time we need neurosurgery. But are there deeper implications of this approach for our understanding of "human nature"?

SMK: The basic idea is that people are different, but not in boring ways; rather than trying to sort people into categories—such as "analyzers" versus "synthesizers," or "visuals" versus "audiles"—we need to think about the efficiency of different processing systems, and how these differences in efficiency can have widespread consequences. People are not going to be pigeonholed very easily.

Authors: What about cerebral lateralization more specifically?

SMK: This is an area where dichotomies have run amok. The hemispheres have been characterized in terms of many sorts of dichotomies, such as verbal versus perceptual, analytic versus holistic, logical versus intuitive, cooked versus raw, and so on. On my view, such dichotomies won't get us very far. For some tasks in some contexts, a particular dichotomy might help summarize the performance of the parts of systems that are more efficient in one or the other hemisphere. But these dichotomies are basically descriptive, and when

evolution of systems that were nonidentical. Asymmetry in how information is processed, represented, and responded to may be a more efficient and flexible design principle than redundancy. With a growing demand for cortical space, perhaps the forces of natural selection began to modify one hemisphere but not the other. Because the corpus callosum exchanges information between the hemispheres, mutational events could occur in one lateralized cortical area and leave the other mutation-free, thus continuing to provide the cortical function from the homologous area to the entire cognitive system. As these new functions develop, cortical regions that had been dedicated to other functions are likely to be co-opted. Given that these functions are still supported by the other hemisphere, there is no overall loss of function. In short, asymmetrical development al-we look at brain function more closely—for example, as revealed by current brain-scanning techniques—we see many and varied patterns of activation in each half-brain. In my view, we won't understand the functions of the hemispheres unless we think about the whole brain; the two halves work together, after all, and only by considering how they function in a single system will we be able to work out what each contributes.

Authors: But how does this "mechanical" view you have espoused tell us about mental life?

SMK: Well, depending on what question you ask, different things count as answers. I am interested in mental functions, such as memory, perception, language, imagery, and so on. This approach seems appropriate for answering questions about how such functions operate. Now, if you ask me why they are as they are, or why they are used in specific circumstances, this approach is less useful. These questions get us into talking about how specific types of information, specific contents ("meanings"), engage the system. But even for these kinds of questions, knowing that the structure of the system is a certain way will—at the least—help one to couch the question better and will give one a framework for formulating answers.

Authors: It is now widely agreed that most complex perceptual and cognitive skills such as language or the detection of upright faces are specific adaptations that have been established through natural selection. That such adaptations would be lateralized makes sense since there isn't an obvious reason why these circuits should be laid down twice. Any comments?

SMK: Hmmm, seems to me that the same argument would also apply to lungs, kidneys, and so forth. . . which are, of course, duplicated. Nothing like a little redundancy to increase survival value; in contemplating the possibility of a stroke, I rather like the idea of having all my cognitive functions duplicated in different parts of the brain. In general, it seems clear that the structure

of the brain is dictated by the genes, and hence is a product of evolution (not just natural selection, but the entire process), but it is not clear to me that any particular kind of "content" has been built into the brain by such mechanisms.

Authors: Do I detect a trace of "environmentalist" sentiment in that last remark?

SMK: Yes, indeed. I am impressed as much by the variability in human cognition and behavior as by the constancy. It is true that some of the variability could be due to noise around the mean (and evolutionary processes definitely produce such variation), but at least some of it is probably a consequence of one's interactions with the environment. My own view is that the genes program some parts of the brain precisely, but most of it is rather loosely prespecified—and these parts are subsequently configured by one's interaction with the environment. The fact that so many "extra" connections are present at birth, and are then "pruned" away after the animal has some interaction with the environment, is consistent with this view. Indeed, I spent a good chunk of time developing computer simulation models of how interaction with the environment could cause different patterns of laterality.

Authors: But research with split-brain patients has not revealed such variability. Perhaps the work with normal subjects has nothing to do with laterality, but instead reflects attentional variations or something like that.

SMK: Perhaps. But recall that one doesn't get to be a split-brain patient for no reason. Those people are very sick people before the surgery. So they aren't exactly a random, representative sample; maybe their epilepsy forces a greater consistency than would otherwise exist. In addition, there is a pretty small number of such patients, which again argues for caution in generalizing from them to the population at large. Finally, I believe that there is considerable variability even among split-brain patients, if one chooses to focus on it.

lowed for no-cost extensions; cortical capacity could expand by reducing redundancy and extending its space for new cortical zones. Support for this idea is provided by the fascinating work of Galuske and colleagues, which has revealed that differences in the neuronal organization of the left and right brain areas are related to the processing of auditory signals associated with human speech (Galuske et al., 2000; Gazzaniga, 2000).

The idea of asymmetrical processing also underscores an important point in modern conceptualizations of hemispheric specialization; namely, the two hemispheres may work in concert to perform a task, even though their contributions may vary. There is no need to suppose some sort of master director who decides which hemisphere is needed for a task. Language is not the exclusive domain of the left hemisphere. The right hemisphere also might contribute, although the types of representations it derives may not be efficient for certain tasks. And the left hemisphere does not defer to the right hemisphere on visuospatial tasks but processes this information in a different way. Seeing the brain organized in this way, we begin to realize that much of what we learn from clinical tests of hemispheric specialization tells us more about our tasks rather than the computations performed by each hemisphere. This point is also evident in split-brain research. With the notable exception of speech production, each hemisphere has competence in every cognitive domain.

Asymmetries in Perceptual Representations

Objects that are attended to are rarely restricted to a single visual field. If we notice a fast car approaching from the left, we quickly shift our gaze to its direction. Given that foveal input is projected bilaterally, coupled with small eye movements, a stimulus is projected to both hemispheres. But this does not mean that the two hemispheres will derive identical representations from a common input. Indeed, the two hemispheres will amplify different sources of information, which in turn vary in importance for operations devoted to identifying a stimulus and deciding whether it requires a response.

An excellent illustration of this point was provided in the preceding section when we considered how hierarchical (local/global) stimuli are represented by the human brain. Inspired by laterality dichotomies in analytic-holistic processing, Justine Sergent (1982) of the Montreal Neurological Institute modified the global-local task developed by Navon to look for hemispheric asymmetries. On each trial, a hierarchical pattern was presented for 150 msec in either the left or the right visual field (or at a central viewing location, a condition we will not consider here). As with split-brain studies, brief exposures ensured that subjects could not move their eyes and look directly at the peripheral stimuli, thus allowing the information to be projected to both hemispheres. The subjects were asked to press one button if the stimulus contained an *H* or *L*, and the other button if neither of these targets was present. The target could be either at the local and global level (one *H* and one *L*), at one level only, or on neither level. The letters *F* and *T* were shown at levels that did not contain a target.

Figure 10.23 shows the intriguing interaction between the target level and the side of presentation. When both the global and local levels contained a target, subjects were faster to respond when the stimulus was presented in the left visual field. Even though this might reflect heightened arousal for the right hemisphere or a

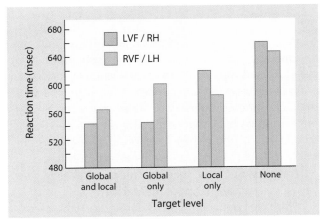

Figure 10.23 Reaction times to hierarchical letter stimuli, presented in either the left or the right visual field. Subjects judged whether the stimulus contained (at least) one of two letters, *H* or *L*. Reaction times were faster for left-visual-field (right-hemisphere) stimuli when the target was defined at the global level or at both levels. In contrast, reaction times were faster for the right-visual-field (left-hemisphere) stimuli when the target was defined at the local level. Adapted from Sergent (1982).

generalized advantage on visual tasks, the local-only results suggest otherwise; reaction times were fastest when the stimulus was in the right visual field. The asymmetries between the two target conditions were tied to the type of analysis required. The left hemisphere was more adept at representing local information and the right hemisphere was better with global information. And as we have discovered, experiments with neurologically impaired patients confirmed that this interaction stems from the asymmetrical functions of the two hemispheres.

COMPUTATIONAL BASIS OF LOCAL AND GLOBAL ASYMMETRIES

What defines the local and global structure of a stimulus? An obvious answer is that they differ in size: Global structure is larger than local structure. But how does the visual system process differences in size? To answer this, consider the receptive field properties of neurons involved in the analysis of shape.

A notable feature of visual receptive fields is that they are tuned to detect changes in contrast, or lightness. Most visual neurons respond weakly to a uniform field of light. Instead, as we saw in Chapter 4 (see Pioneers in the Visual Cortex), they are most sensitive to changes in lightness because the neurons have a center-surround structure in which one region is excited by light and flanking regions are inhibited by it. In an on-off cell, the center region is excited by light; in an off-on cell, the center is inhibited by it (Figure 10.24). Moreover, the sizes of the center-surround regions vary. Such a computational device is sensitive to an object's size. The narrow receptive field fails to respond to the large bar because the bar spans the excitatory and inhibitory regions; this yields a net effect in which the excitation is canceled by the inhibition. In contrast, the same receptive field is maximally activated by the narrow bar.

The center-surround structure has led some theorists to conceptualize visual neurons as spatial frequency fil-

ters, which refers to how the brightness of a visual pattern varies as a function of space. Consider the simple patterns in Figure 10.25 in which brightness varies in a sinusoidal manner along the horizontal axis. In the top pattern, the variation is slow. The transition from light to dark and back to light again spans most of the page. The variation is much faster in the bottom pattern; hence, there is more rapid transition between the contrasting stripes. The transition rate can be described by its periodicity with respect to a unit of space. For example, the top pattern has a frequency of 2 cycles/page and the bottom one has a frequency of 8 cycles/page. In visual perception, units are expressed as cycles per degree, where

Figure 10.24 Stimulus size is encoded by the spatial extent of a neuron's receptive field. **(a)** Striate cortex neurons not only have a preferred orientation but also are sensitive to the size of a stimulus. Size sensitivity is determined by the width of the excitatory and inhibitory regions of the cell's receptive field. An on-off cell responds when a light stimulus falls in the excitatory region and a dark stimulus falls in the inhibitory region. **(b)** A small stimulus produces a high response rate in the neuron with the smaller receptive field. This stimulus produces little activity in the neuron with the large receptive field because the excitatory center region is also stimulated by the dark background. The opposite occurs with the large stimulus. The larger receptive field is maximally responsive to this stimulus. The smaller receptive field does not respond because the stimulus spans both the excitatory and inhibitory regions.

Figure 10.25 Three sinusoidal gratings. Intensity varies in a sinusoidal manner across the horizontal dimension. The frequency of the sinusoids varies, being lowest for the stimulus on top and highest for the stimulus on the bottom. The firing rate of a visual cortex neuron to these stimuli will vary, depending on the width of the neuron's receptive field. In this way, visual neurons operate as spatial frequency filters.

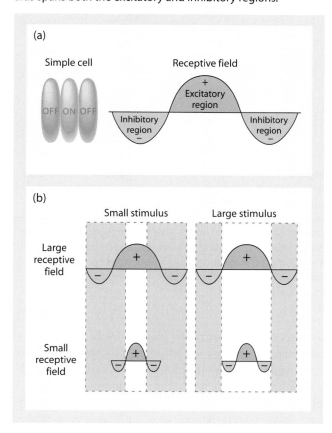

the unit corresponds to 1 degree of visual angle (about the size subtended by a thumb held at arm's length).

A cell with narrow center-surround regions operates like a high-frequency spatial filter that responds vigorously to high-frequency patterns. A cell with a wide center-surround region operates as a low-frequency filter and responds only to low-frequency patterns.

An appealing feature of the **spatial frequency hypothesis** is that any complex pattern can be described by its component spatial frequencies (Figure 10.26). Sergent (1982) conjectured that the spatial frequency hypothesis is a computational basis for hemispheric asymmetries in visual perception. She proposed that the left hemisphere is more adept at representing high-spatial-frequency information and that the right hemisphere is better with low-frequency information. Sergent assumed that the local structure of a hierarchical pattern would be carried in higher frequencies than would the global structure. In Figure 10.27, computer algorithms were used to recreate a hierarchical stimulus after either the high- or the

Figure 10.26 Fourier's theorem states that any complex pattern can be described as the composite of sinusoids that vary in power and phase (starting position). The grating pattern on the left is created by combining the three sinusoids on the right. For each position along the horizontal axis, the brightness in the compound grating is determined by averaging the brightness at the corresponding points in the sinusoids. In an identical manner, any complex pattern could be decomposed into a set of sinusoids. The set would have to include a large range of component frequencies that are oriented in all possible directions.

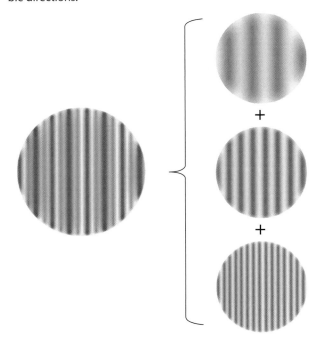

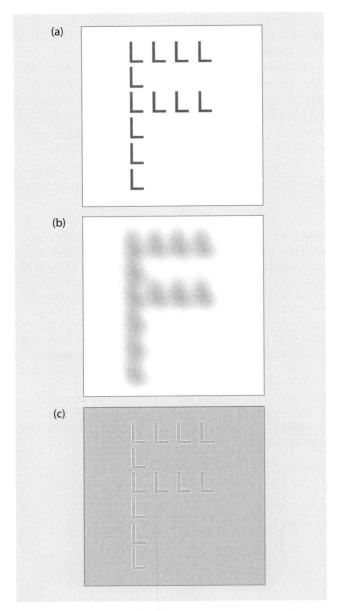

Figure 10.27 A hierarchical letter stimulus, before and after low- and high-pass frequency filtering. **(a)** The unfiltered stimulus. **(b)** The same stimulus after all of the high frequencies have been removed (low pass). Neurons sensitive only to these low frequencies would not be able to represent the local elements. **(c)** The stimulus after all of the low frequencies have been removed (high pass). High frequencies can still support local and global identification. However, the contrast of the stimulus is reduced.

low-frequency information was removed. In a low-pass stimulus, the identity of local elements cannot be distinguished even though the global shape persists. In a high-pass stimulus, both levels can be identified, but much of the contrast is absent. The advantage of identifying global targets is abolished with high-pass filtered patterns, presumably because the same units—such as receptive fields with narrow center-surround regions—are required for both kinds of stimuli.

Fred Kitterle and his colleagues (1990) at the University of Toledo provided more direct support for the spatial frequency hypothesis. They used sinusoidal gratings rather than hierarchical patterns. An advantage with sinusoids is their simplicity: Any complex pattern will have many component frequencies, but sinusoids have only one. Sinusoids of 1 cycle/degree or 3 cycles/degree were presented in either the left or the right visual field. Subjects were asked to respond as quickly as possible, pressing one key if the stimulus had "wide" stripes (1 cycle/degree) and the other key for "narrow" stripes (3 cycles/degree). While subjects were always faster when responding to the wide pattern, the results also showed the predicted side-by-size interaction. The reaction times for wide-striped, low-frequency patterns were faster when the stimulus was presented to the left visual field; narrow-striped, high-frequency patterns were identified fastest when presented to the right visual field (Figure 10.28).

APPLYING THE FREQUENCY HYPOTHESIS

The frequency hypothesis offers a powerful account of hemispheric asymmetries in visual perception. Rather than emphasize dichotomies that may be hard to define a priori because of stimulus characteristics (e.g., local-global) or processing modes (e.g., analytic-holistic), the hypothesis centers on an explicit source of information in the stimulus. Moreover, the hypothesis is grounded in a well-developed computational theory of how the visual system operates. An extension to laterality research builds on the idea that the visual system is composed of spatial frequency filters used in an asymmetrical manner by the two hemispheres (Ivry and Robertson, 1998).

The frequency hypothesis has been applied to assess perceptual asymmetries across a range of tasks. For example, it can account for the fact that numerous manipulations lead to greater impairments in the perception

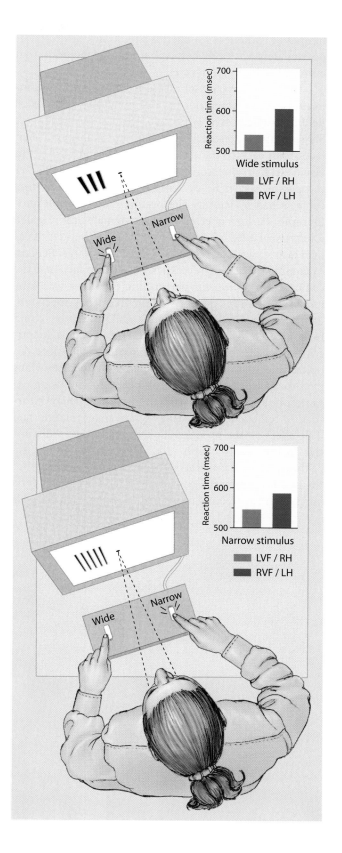

Figure 10.28 Asymmetries in how frequency information is represented in the two hemispheres may account for laterality effects in visual perception. A low-frequency (1 cycle/degree) or high-frequency (3 cycles/degree) stimulus was presented in either the left or the right visual field. Subjects judged whether the pattern was "Wide" or "Narrow." Reaction times to the low-frequency stimuli presented in the left visual field (right hemisphere) were much faster. This side also showed an advantage for the high-frequency stimulus, but the magnitude of the asymmetry was reduced. Adapted from Kitterle et al. (1990).

of stimuli presented to the right visual field in comparison to the left visual field. Blurring the stimulus, reducing its contrast, and restricting the exposure time reduce the fidelity of high-frequency content more than low-frequency content. As such, this should interfere disproportionately with left-hemisphere representations that amplify the high frequencies.

The frequency hypothesis offers a parsimonious interpretation of experiments that show reversals in visual field advantages even when stimuli remain unchanged. Sergent (1985) presented sixteen photographs of familiar faces, with each face shown in either the left or the right visual field. In one condition, subjects judged whether the person was male or female. In another condition, subjects had to identify the person shown in the photographs. Performance on the male-female task revealed a left-visual-field advantage and that on the identification task showed a right-visual-field advantage. Sergent maintained that this reversal reflected the fact that the critical information for performing each task was contained in different portions of the frequency spectrum. Determining whether a person is male or female could be done with low-spatial-frequency information. Identifying a specific person, however, requires the more detailed information in higher spatial frequencies. This hypothesis was tested in a second experiment, using the same photographs, but with the high-frequency information removed (Figure 10.29). This manipulation had little effect on the male-female task but was very disruptive to the identification task. A processing mode hypothesis would have to account for this by assuming that the two tasks required subjects to shift between analytic and holistic processing modes. The frequency hypothesis assumes that the processing remains unchanged. The laterality effects reverse simply because the critical information is different for the two tasks.

A similar asymmetry may hold in auditory perception, but here the distinction is based on sound frequencies rather than spatial frequencies (Ivry and Lebby, 1993). For instance, consider speech perception, which depends largely on high sound frequencies and for which the left hemisphere is clearly dominant. Patients with left-hemisphere lesions may be unable to discriminate simple consonant contrasts that depend on high-frequency information, such as "ba" and "da," whereas the problem is rare in patients with right-hemisphere lesions.

Nevertheless, the right hemisphere has been linked to one aspect of speech perception, prosody, which is carried primarily by fluctuations in the low-frequency portion of the speech signal. *Prosody* refers to the connotative aspects of oral language, the way we vary articulation to convey affect or intonation. The statement "We need to talk" can be perceived either as an encouraging solicitation or as a stern warning, depending on intonation. The importance of low-frequency information is evident from the way people can still make prosodic judgments even when high sound frequencies are eliminated. For example, consider an overheard conversation from the office next door. While we may fail to recognize the words because the walls effectively filter out the relatively high frequencies, the low-frequency information is sufficient to alert us when the voices have taken on an angry tone.

Lesions of the right hemisphere disrupt prosody much more than do lesions of the left hemisphere (Ley and Bryden, 1982; Blonders et al., 1991). A task-based ac-

Figure 10.29 Tasks vary in terms of which spectral cues are most informative. When shown pictures of faces that contain the full spectrum of frequency information **(left)**, subjects can state whether the person is male or female and, if familiar, the person's identity. Gender can still be determined when only the low frequencies are left intact **(right)**, but the person's identity has been obscured.

count of laterality must assume that the left hemisphere is activated for linguistic tasks and the right for paralinguistic tasks such as prosody. A frequency hypothesis assumes that the speech signal is projected to both hemispheres, but that the two hemispheres produce representations that amplify different frequency regions of the stimulus. Identifying words requires an analysis of the signal's high-frequency portion. Prosodic analysis can be performed on low-frequency portions of speech.

Asymmetries in Representing Spatial Relations

In the preceding section, we focused on hemispheric differences in how stimuli are identified. The central message is that different sources of information identify the attributes of a stimulus. In vision, high-frequency information is critical for identifying the local structure, or details; low-frequency information is more efficient for abstracting the global structure, or shape. In a similar fashion, if we extend the model to sound and speech, lexical analysis depends on the higher-sound-frequency components of the speech signal; prosodic information is carried in lower frequencies.

In Chapter 5 we noted an important distinction between cortical pathways involved in identification and those involved in localization (the what-where distinction). Studies of hemispheric asymmetries have not been limited to "what" problems; they also have been applied to "where" problems.

CATEGORICAL AND COORDINATE REPRESENTATIONS IN MEMORY

Stephen Kosslyn (1987) of Harvard University contended that a distinction can be made between two types of spatial representations. One type is essential for specifying the positions of objects or parts of objects. For example, when a guest begins to search through the kitchen drawers to locate the silverware, the host might instruct her to "look to the left" or "just under that drawer." Here the emphasis is on the relative spatial position between two objects—the drawer being searched and the drawer containing the silverware. Kosslyn termed this *categorical* because it assigns the relation to a broad equivalence class. With respect to some referent, an object is to the left or right, above or below, inside or outside. Categorical relations capture general properties in abstract terms.

Though categorical relations are usually sufficient, there are times when one needs precise spatial information. Consider the scene in Figure 10.30. We would know where to go if we were told to sit on the couch be-

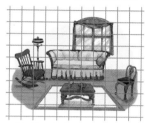

Categorical representation

Rocking chair LEFT of couch

Wine glass on TOP of table, candy INSIDE candy dish

Coordinate representation

Rocking chair is CLOSER than the dining chair to the couch

Figure 10.30 Spatial information can be represented in abstract, categorical terms, or in a precise coordinate manner. Categorical representations capture basic relational information such as the relative position of two objects from a particular point of view. Coordinate representations specify the exact positions of the objects and the distances between the objects.

hind the coffee table, but to accomplish this action we need to represent the exact locations of the couch, the coffee table, and any other obstacles that clutter up the living room. So we need coordinate, or metrical, spatial representations. Coordinate representations specify the exact location of objects, with respect either to one another or to the perceiver's position. A coordinate representation not only gives the relative position of the couch and coffee table but also specifies the couch's exact location.

Kosslyn maintained that **categorical representation** and **coordinate representation** have distinct purposes. A primary purpose of categorical representation is in using spatial information to classify objects. For example, the letters *b, d, q, p,* and *P* have the same component parts (or geons—see Chapter 5) but differ in terms of the relative position of the line and circle. Within a wide range of sizes and handwriting styles, we can make the appropriate classification once we learn the invariant categorical relations, for example, that the circle is to the left and at the base of the line for *d*. In this respect there is no strict segregation of "what" and "where." Spatial relations provide valuable cues to an object's identity.

Coordinate relations, by contrast, are essential for action. We need to know the exact location of a wine glass on the coffee table in order to pick it up. We do not just grope about, knowing the categorical relation that it is on top of the table. When navigating across a rocky beach, we have to estimate the exact distance between two rocks before we can decide if our step can span the distance.

Kosslyn hypothesized that the two hemispheres might differ in their contribution to categorical and coordinate representations. He proposed that the left hemisphere forms categorical spatial representations and the right hemisphere forms coordinate representations. This hypothesis is motivated by consideration of the similarity between categorical representations and language. The general equivalences given by categorical relations are captured by linguistic terms (e.g., *up, down, inside*), and categorical encoding is a prominent feature of language. For example, despite the many variations in how a word can be articulated, the cardinal feature of speech perception is that the surface attributes of an utterance are stripped away to identify the phonemes that form the word. In a similar sense the precise specification of an object's location might rely on, or underlie, the dominant role of the right hemisphere in perceiving spatial information.

To test this hypothesis, Kosslyn and his associates devised tasks that require judgments based on either categorical or coordinate spatial relations (Kosslyn et al., 1989). In one study, the stimulus consisted of a bar and a dot, presented to either the left or the right visual field. In the categorical condition, the subjects had to decide whether the dot was located above or below the line. In the coordinate condition, the judgment required metrical information: Here, the subject was required to judge the distance between the line and the dot, discriminating between distances that were near and those that were far. The clever twist in this experiment is that the same set of stimuli were used in both conditions. However, the assignment of stimuli to response categories differed as a function of the instruction (Figure 10.31).

Responses on the near-far task were faster when stimuli were presented to the left visual field, a result consistent with the hypothesis that the right hemisphere is more adept in making metrical judgments. The opposite trend was found for the above-below task. These categorical judgments were made more rapidly when the stimuli were presented in the right visual field.

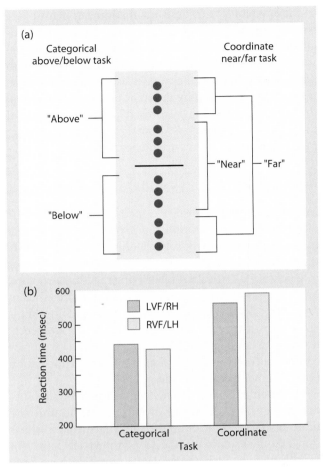

Figure 10.31 Hemispheric asymmetries on categorical and coordinate spatial judgments. **(a)** On each trial, the stimulus consisted of the bar and one of the twelve dots. For the categorical task, subjects judged whether the dot was "above" or "below" the bar. For the coordinate task, subjects judged whether the dot was "near" or "far" from the bar. **(b)** The reaction time data show a task by visual field interaction. Categorical judgments were faster when the stimuli were shown in the right visual field (left hemisphere). Coordinate judgments were faster when the stimuli were shown in the left visual field (right hemisphere). Adapted from Kosslyn et al. (1989).

These results call for a few caveats. First, as with many laterality studies, there have been problems with replication. Tasks such as the line-dot task appear to be especially problematic. Even in the original study, the effects disappeared after a few blocks of practice. Kosslyn suggested that subjects quickly form new categories that can be applied in the near-far task since the stimulus set is relatively small.

Second, it is important to ask whether the observed differences can be accounted for by alternative hypotheses such as the spatial frequency hypothesis. It turns out that there may be differences in the critical frequencies required for the two tasks. As shown in the low-pass filtered image in Figure 10.32, low-frequency information may be insufficient for determining whether dots at the closest locations are above or below the line. Thus, the above-below task may depend on the left hemisphere—not because of the categorical nature of the judgment but because of the need for examining high-frequency information.

These problems alert us to seek converging evidence. Bruno Laeng (1994) of the University of Michigan reported results of a study of neurologically impaired patients that support the categorical-coordinate dichotomy. At the beginning of each trial, the patients

Figure 10.32 Frequency-based interpretation of the categorical-coordinate dissociation. **(a)** Two of the stimuli. **(b)** Low-pass filtered representation of the stimuli. **(c)** High-pass filtered representation of the stimuli. The categorical, above-below task distinction is difficult to make from the low-frequency information alone. This suggests an alternative reason for the left-hemisphere advantage on this task. Adapted from Ivry and Robertson (1998).

(a) Actual stimulus

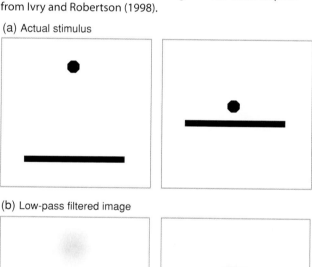

(b) Low-pass filtered image

(c) High-pass filtered image

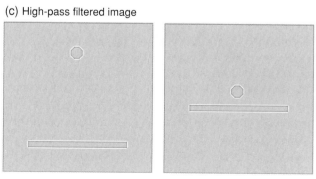

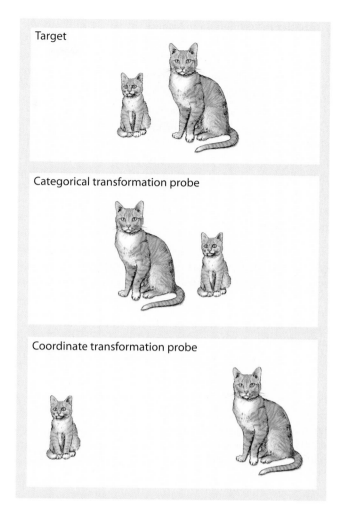

Figure 10.33 Memory test of spatial relations. Subjects were presented with a target stimulus consisting of a pair of objects such as the two cats. After a 5-second delay, a probe stimulus appeared and the subjects judged whether it was the same as the target. In these two examples, the correct answer would be no. The left-right reversal alters categorical relations; increasing the distance between the cats constitutes a coordinate transformation. Adapted from Laeng (1994).

were shown a drawing of a pair of items such as two cats. After a 5-second delay, a pair of drawings were shown, one that matched the original drawing and one that was a transformed version of the original. The transformations were of two types, as shown in Figure 10.33. In categorical transformations, the relation between the two items was changed; for example, the left-right position of the cats was reversed. In coordinate transformations, a metrical property was changed; for example, the distance between the two cats was increased. In both tasks participants were asked to choose the drawing identical to the original drawing.

Compared to control subjects, the patient groups made more errors with both types of transformations.

For patients with left-hemisphere impairment, however, most errors involved selecting a drawing that required a categorical transformation. The reverse was found for the patients with right-hemisphere lesions. For them, the errors involved coordinate transformations.

This experiment emphasized two advantages of patient studies in laterality research. First, because the stimuli were presented centrally, one does not need to make assumptions about the projection of visual information to the contralateral hemisphere. Second, effects that may be small and transient with normal subjects are frequently amplified when explored in people with neurological disorders.

GENERALIZING CATEGORICAL AND COORDINATE DISTINCTIONS TO OTHER PROCESSING DOMAINS

Laeng's task represents an interesting intersection of perception and memory. Subjects first encoded a target item and, after a short delay, were asked to match it to one of two probe items. The tendency for patients with left-

hemisphere lesions to select probes in which categorical relations were transformed may reflect an encoding problem—an inability to represent the information—or a memory problem—an inability to retain the information. And the difficulty patients with right-hemisphere lesions have with coordinate relations could be attributed to either an encoding or a memory problem.

While the parietal lobes are critical for abstracting spatial relations, the inferior part of the temporal lobe is the neural locus for memories of visual representations of shape. The format of the representations appears to differ for the two hemispheres: Whereas left-hemisphere memories appear to be abstract or categorical, right-hemisphere representations retain the specific metrics of the stimulus.

Think about what a dog looks like. Your image might be of a small animal with four legs, covered with shaggy fur, floppy ears, and a long, wagging tail. Of course, this picture does not describe all dogs. It might be apt for an Old English Sheepdog but not a Boxer, which has short fur and a short, stubby tail (Figure 10.34). The initial image is not necessarily of a particu-

Figure 10.34 Knowledge may be represented in an exemplar or prototype format. Exemplars would correspond to particular dogs such as the Old English Sheepdog or Chesapeake Bay retriever. The prototypical dog is a composite blend, the "average" dog.

Papillon

Smooth dachshund

Wire fox terrier

Soft coat

Long nose

Perky ears and high energy

Facial hair and big pink tongue

Prototypical dog

Coat color

Old English Sheepdog

General body proportions

Boxer

Chesapeake Bay retriever

lar species; rather, it reflects a *prototype,* a composite melded by combining features from lots of dogs. The dogs you are familiar with are *exemplars,* or specific examples of dogs. A long and contentious debate rages as to whether memory contains prototypes or exemplars. Results of recent laterality studies suggest that both views may be correct. The left hemisphere may rely on prototypes, or representations that categorize information into general equivalence classes. The right hemisphere may use an exemplar representational format.

This distinction is made clear in a study by Chad Marsolek of the University of Minnesota (1995). Eight sets of abstract line drawings were created by shifting the components of an initial set of eight prototypes (Figure 10.35). During a training phase, subjects viewed a subset of the drawings and were trained to categorize them. After training, the subjects were given a speeded classification task in which they categorized the drawings as quickly as possible. In this test phase, the stimuli were presented in either the left or the right visual field. When judging the studied subset, the subjects were faster when the stimuli were presented in the left visual field (right hemisphere). In contrast, when judging the previously unseen prototypes, the subjects were faster when the stimuli were presented in the right visual field (left hemisphere).

Kosslyn used the metaphor of a rolling snowball to account for the generalization of the categorical-coordinate distinction to multiple task domains. In his view, a fundamental asymmetry appears as a specialization to solve a

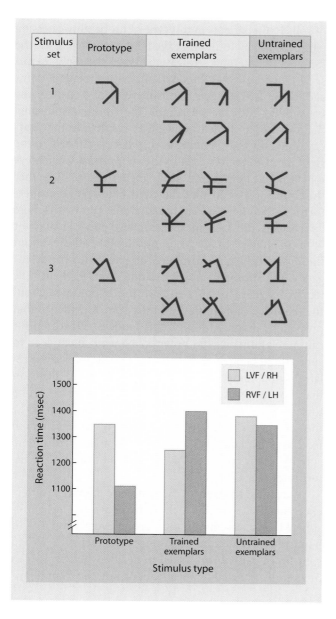

Figure 10.35 Hemispheric asymmetries in exemplar and prototype memory tasks. **(Top)** Stimulus sets were constructed by distorting a prototype pattern to create six exemplars. During the study phase, subjects were presented with a subset of the exemplars and learned to categorize them, using the arbitrary numbers assigned to each set. At the test, they performed the categorization task with the prototype, the trained exemplars, and the untrained exemplars. **(Bottom)** Classification of the prototype was faster when it was presented in the right visual field, supporting the idea that this form of representation is characteristic of left-hemisphere processing. The trained exemplars, in contrast, were more rapidly classified when presented in the left visual field. Adapted from Marsolek (1995).

problem. But this asymmetry then cascades into secondary specializations that build on the initial functional difference. While it is difficult to trace the evolutionary development of hemispheric specialization, Kosslyn conjectured that the key event was the establishment of a left-hemisphere specialization for language production. Unlike manual gestures that can be performed by one hand or the other, speech production requires the bilateral coordination of the muscles required for articulation. This created a pressure to localize control in a single focus since interhemispheric communication requires time and might limit how fast we can vocalize. A left-hemisphere specialization for language followed, since this hemisphere had primary access to the speech production system. As we noted, language is categorical in nature, with its emphasis on mapping percepts to broad semantic classes. The tendency for the left hemisphere to categorize is now found in many processing systems including language, perception, and memory.

Recent Theoretical Developments Concerning Hemispheric Specialization

After the human cerebral hemispheres are disconnected, the verbal IQ of a patient remains intact. And although there can be deficits in free recall capacity and in other performance measures, the problem-solving capacity of the left hemisphere, as seen in hypothesis formation tasks, appears unaffected also. In other words, isolating essentially half of the cortex from the dominant left hemisphere causes no major change in cognitive functions. The capacity of the left side remains unchanged from its preoperative level, yet the largely disconnected, same-size right hemisphere is seriously impoverished in cognitive tasks. While the right hemisphere remains superior to the isolated left hemisphere for some perceptual and attentional skills, and perhaps also for emotions, it is poor at problem solving and many other mental activities. A brain system (the right hemisphere) with roughly the same number of neurons as one that easily engages in cognitive functions (the left hemisphere) is incapable of higher-order thought processes, providing compelling evidence that cortical cell number by itself cannot fully explain human intelligence.

George Wolford and his colleagues at Dartmouth College (Wolford et al., 2000) recently demonstrated the difference between the two hemispheres in problem solving. In the task studied, subjects are presented with a simple task of trying to guess which of two events will happen next. Each event has a different probability of occurrence (e.g., a red stimulus might appear 75% of

the time and a green one 25% of the time), but the order of occurrence of the events is entirely random. There are two possible strategies for responding in this task: *matching* and *maximizing*. In the red-green example, frequency matching would involve guessing red 75% of the time and guessing green 25% of the time. Because the order of occurrence is entirely random, this strategy potentially could result in a great number of errors. The second strategy, maximizing, involves simply guessing red every time. That ensures an accuracy rate of 75% because red will appear 75% of the time. Animals such as rats and goldfish maximize. Humans match. The result is that nonhuman animals perform significantly better than humans in this task. The human's use of this suboptimal strategy has been attributed to a propensity to try to find patterns in sequences of events, even when told the sequences are random.

Wolford and colleagues tested the two hemispheres of split-brain patients in this type of probability-guessing paradigm. They found that the left hemisphere used the frequency-matching strategy, whereas the right hemisphere maximized. When patients with unilateral damage to the left or right hemisphere were tested on the probability-guessing paradigm, the findings indicated that damage to the left hemisphere resulted in the strategy of maximization being used, whereas damage to the right hemisphere resulted in the suboptimal frequency-matching strategy being used (Figure 10.36). Together the findings suggest that the right hemisphere outperforms the left hemisphere because the right hemisphere approaches the task in the simplest possible manner with no attempt to form complicated hypotheses about the task. The left hemisphere, on the other hand, engages in the human tendency to find order in chaos. The left hemisphere persists in forming hypotheses about the sequence of events even in the face of evidence that no pattern exists. Although this tendency to search for causal relationships has potential benefits, it can lead to suboptimal behavior when there is no simple causal relationship. Some of the common errors in decision making are consistent with the notion that we are prone to search for and posit causal relationships, even when the evidence is insufficient or even random. This search for causal explanations appears to be a left-hemisphere activity and is consistent with previous work on the **interpreter** (see Chapter 16).

A reconsideration of hemispheric memory differences suggests why this dichotomy between a left hemisphere that seems driven to interpret events, and a right hemisphere that does not, might be adaptive. When split-brain patients are asked to decide whether a series of stimuli appeared in a study set or not, the right hemi-

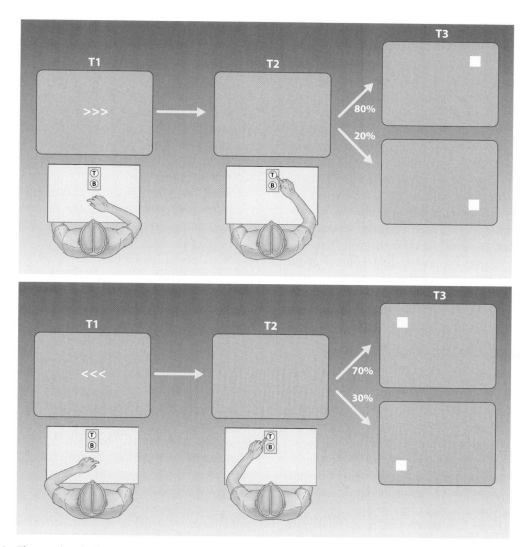

Figure 10.36 The two hemispheres respond differently when trying to predict where a stimulus will appear. At time 1 (T1), arrows indicate the side of the forthcoming stimulus. After a blank interval (T2), a light appears on the cued side. The light is presented either above or below the horizontal meridian. The position of the light is determined randomly on each trial, subject to the constraint that it will appear in the upper quadrant on 80% of the trials when presented in the right visual field and 70% of the trials when presented in the left visual field. Over time, each hemisphere learns that the light is more likely to appear in the upper quadrant. However, different decision-making strategies are adopted. The left hemisphere responds by matching the probabilities, predicting the upper quadrant on about 80% of the trials for right-visual-field trials. The right hemisphere responds by maximizing, predicting the upper quadrant on almost every trial for the left visual field. Interestingly, the maximizing strategy is more optimal. From Gazzaniga (2000).

sphere is able to correctly identify previously seen items and reject new items. The left-hemisphere, however, tends to falsely recognize new items as having occurred previously when they are merely similar to previously presented items, presumably because they fit into the schema it has constructed (Phelps and Gazzaniga, 1992; Metcalfe et al., 1995). This finding is consistent with the hypothesis that the left-hemisphere interpreter constructs theories to assimilate perceived information into a comprehensible whole. By going beyond simply observing events to asking why they happened, a brain can cope with such events more effectively should they hap-

pen again. In doing so, however, the process of elaborating (story making) has a deleterious effect on the accuracy of perceptual recognition, as it does with verbal and visual material. Accuracy remains high in the right hemisphere, however, because it does not engage in these interpretive processes. The advantage of having such a dual system is obvious. The right hemisphere maintains a veridical record of events, leaving the left hemisphere free to elaborate and make inferences about the presented material. In an intact brain, the two systems complement each other, allowing for elaborative processing without sacrificing veracity.

VARIATIONS IN HEMISPHERIC SPECIALIZATION

In this chapter we have reviewed general principles of hemispheric asymmetry in humans. The research showing asymmetries in how perceptual information is processed or in how spatial relations are represented has focused on differences that characterize the typical person. Most of us are right-handed and rely on the left hemisphere for speech production, but a significant percentage of the population is left-handed and may have a different organization of hemispheric specialization. Even within the right-handed population, there are significant individual differences in lateralization profiles. A complete understanding of hemispheric specialization must account for these differences as well as the prototypical organization.

The Relation Between Handedness and Left-Hemisphere Language Dominance

An ongoing debate in the laterality literature centers on whether there is a causal relationship between the predominance of right-handedness and left-hemisphere specialization for language. Some theorists point to the need for a single motor center as the critical factor. Though there may be benefits to perceiving information in parallel, our response to these stimuli must be unified. We cannot allow the left hemisphere to choose one course of action while the right hemisphere opts for another. Our brains may have two halves, but we have only one body. By localizing action planning in one hemisphere, the brain can achieve unification.

One hypothesis is that the left hemisphere is specialized for the production of sequential movements. Speech is obviously dependent on such movements. Our ability to produce speech is the result of many evolutionary changes in the shape of the vocal tract and articulatory apparatus. These adaptations make it possible for us to communicate at phenomenally high rates—the official record is 637 words/min, set on the British television show *Motor Mouth*. Such competence requires an exquisite control of the sequential gestures of the vocal cords, jaw, tongue, and other articulators.

Similarly, the left hemisphere has been linked to sequential movements in nonlinguistic domains. Left-hemisphere lesions are more likely to produce apraxia, a disorder in producing coherent actions that cannot be attributed to weakness or a problem in controlling the muscles (see Chapter 11). This deficit can manifest in movements with either hand. Oral movements also have a left-hemisphere dominance, regardless of whether the

movements create speech sounds or nonverbal facial gestures. In both domains, the gestures are more pronounced on the face's right side, and activation of the right facial muscles is quicker than that of the corresponding muscles on the left. Hence, the left hemisphere may have a specialized role in the control of sequential actions and this may underlie hemispheric asymmetries in both language and motor functions.

In his book *The Lopsided Ape: Evolution of the Generative Mind* (1991), Michael Corballis of the University of Auckland eloquently articulates this hypothesis. He takes an evolutionary perspective to account for the emergence of language and handedness. To Corballis, the main characteristic of human cognition is that we act and think in a generative manner. Thousands of human languages are based on a limited set of phonemes, or basic articulatory gestures. From these sixteen to forty-four elements (the number of basic speech sounds varies across languages), we have created vocabularies of over a million words and can combine these words to generate an infinity of sentences expressing perceptions, intentions, and desires. Corballis believes that the left hemisphere is uniquely equipped with a **generative assembling device** (GAD), a device that generates complex representations from a small vocabulary of primitive units (Figure 10.37).

Corballis assumes that the processing capabilities of the GAD are available to other left-hemisphere functions, including control of the right hand. With bipedalism, the hands are free to operate independently. In quadrupeds, the forelimbs and hind limbs are primarily used for locomotion. Symmetry is favored here: The animal moves along a linear trajectory. If the limbs on one side of the body were longer than the other, an animal would circle about. As our ancestors adopted an upright posture, however, the hands were no longer required to move symmetrically. One hand could hold a carcass while the other reached inside to extract morsels of meat.

The generative aspects of an emerging communication system could have been applied to the way hands manipulated objects, and the lateralization of GAD would have favored the right hand. This would be most evident in tool use. While nonhuman primates and birds can fashion primitive tools to gain access to foods that are out of reach or encased in hard shells, humans manufacture tools generatively. We not only design tools to solve an immediate problem but also can recombine the parts to create new tools. The wheel, an efficient component of devices for transportation, can be used to

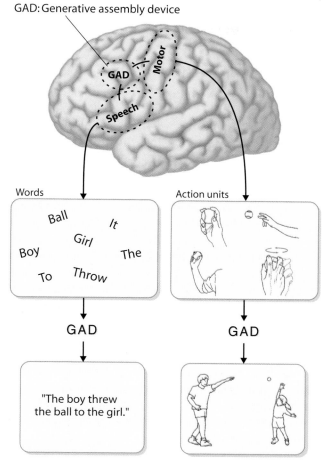

GAD: Generative assembly device

Words

Ball
It
Girl
Boy
The
To
Throw

GAD

"The boy threw the ball to the girl."

Action units

GAD

Figure 10.37 Corballis hypothesized that the left hemisphere has evolved a generative assembly device (GAD). This mechanism underlies the production of an infinite set of sequences from a relatively small set of units such as words or gestures.

extract energy from a flowing river or record information in a compact, easily accessible format. Handedness, then, is most apparent in our use of tools. As an example, right-handers differ only slightly in their ability to use both hands to block balls thrown at them. But when they are asked to catch or throw the balls, the dominant hand has a clear advantage.

In Corballis's scheme, specialization in language became accessible to nonlinguistic functions: As the left-hemisphere GAD allowed language to take on its generative quality, it also imbued the right hand with a functional advantage for skilled actions. Alternatively, the left hemisphere's dominance in language may be a consequence of a specialization in motor control. The asymmetrical use of hands to perform complex actions, including those associated with tool use, may have promoted the development of language. For example, as

our ancestors' use of tools became more advanced, a communication system to disseminate that knowledge was needed. From comparative studies of language we believe that most sentence forms convey actions. Not only will an infant first order commands such as "come" or "eat" before using adjectives ("hungry"), but this order is preserved in phylogeny. The earliest forms of communication centered on action commands such as "look" or "throw." If the right hand was being used for many of these actions, there may have been a selective pressure for the left hemisphere to be more proficient in establishing these symbolic representations.

It is also possible that the mechanisms producing hemispheric specialization in language and motor performance are unrelated. There is not a perfect correlation between these two cardinal signs of hemispheric asymmetry. Not only do a small percentage of right-handers exhibit either left-hemisphere language or bilateral language, but also in at least half of the left-handed population the left hemisphere is dominant for language.

These differences may reflect the fact that handedness is affected at least partly by environmental factors. Children may be encouraged to use one hand over the other, perhaps owing to cultural biases or to parental pressure. Yet it is possible that handedness and language dominance may reflect different factors. Fred Previc (1991), a researcher with the U.S. Air Force, proposed an intriguing hypothesis along these lines. According to Previc, the left-hemisphere dominance for language is primarily related to a subtle asymmetry in the skull's structure. In most individuals, the orofacial bones on the left side of the face are slightly larger, an enlargement that encroaches on middle-ear function and could limit the sensitivity to certain sound frequencies. Previc maintained that this enlargement has a deleterious effect on the projection of auditory information to the right hemisphere, especially in the frequency region that carries most of the critical information for speech. As such, the left hemisphere is favored for phonemic analysis and develops a specialization for language.

In contrast to this anatomical asymmetry, Previc argued that handedness is determined by the position of the fetus during gestation (Figure 10.38). Most fetuses are oriented with the head downward and the right ear facing toward the mother's front. This orientation leads to greater in vitro stimulation of the left utricle, part of the vestibular apparatus in the inner ear that is critical for balance. This asymmetrical stimulation will lead to a more developed vestibular system in the right side of the brain, causing babies to be born with a bias to use the left side of the body for balance and the maintenance of

To Approach or Withdraw: The Cerebral Tug-of-War

It is Friday night and you are heading to a party at the apartment of a friend of a friend's. You arrive and look around: Loud music and swirling bodies move about the living room, and a throng has gathered in the kitchen, around the counter laid out with chips and dips. Unfortunately, your friend is nowhere to be seen and you have yet to recognize a single person among the crowd.

Your reaction will depend on a number of factors: how comfortable you feel mingling with strangers, how lively you are feeling tonight, whether a host approaches and introduces you to a few of the guests. Unless you have a flair for flamboyance, it is unlikely you will head to the dance floor. A more likely response is to head for the kitchen and find yourself a soda or beer.

Richard Davidson (1995) of the University of Wisconsin proposed that the fundamental tension for any mobile organism is between approach and withdrawal. Is a stimulus a potential food source to be approached and gobbled up? Or a potential predator that must be avoided? Even the most primitive organisms display at least a rudimentary distinction between approach and withdrawal behaviors. The evolution of more complex nervous systems has provided mechanisms to modulate the tension between these two behavioral poles—we might overcome our initial reaction to flee from the party, knowing that if we stay we are likely to make a few new friends and have a few good laughs.

According to Davidson, this tension involves a delicate interplay between processing within the medial regions of the prefrontal cortex in the right and left cerebral hemispheres. The prefrontal cortex is a major point of convergence in the central nervous system, for information not only from other cortical regions but also from subcortical regions, especially those involved in emotional processing (see Chapters 12 and 14). In Davidson's theory, these inputs are processed asymmetrically. Left-hemisphere processing is biased to promote approach behaviors; in contrast, right-hemisphere processing is biased to promote withdrawal behaviors.

This theory has provided an organizing principle to evaluate the changes in behavior that follow neurological damage. For example, damage to the left frontal lobe can result in severe depression, a state in which the primary symptom is withdrawal and inactivity. While we might expect depression to be a normal response to brain injury, the opposite profile has been reported in patients with right frontal damage. These patients may appear manic. Damage to the right-hemisphere "withdrawal" system biases the patient to be socially engaging, even when such behaviors are no longer appropriate.

More compelling evidence comes from physiological studies that have looked at the brain's response to

posture. This frees the right side of the body for more exploratory movement, resulting in right-handedness.

According to Previc's theory, then, different factors determine language asymmetries and handedness. At present, the data are too scant for evaluating either of the mechanisms. Previc's own research concerns the demographics of the entire population, not whether language dominance or handedness can be predicted in any one person by orofacial asymmetries or fetal position. Nonetheless, it does raise the interesting possibility that numerous and unrelated factors might determine patterns of hemispheric specialization.

Hemispheric Specialization in Nonhumans

Because of the central role of language in hemispheric specialization, laterality research has focused on humans. It is possible that asymmetries between the left and right cerebral hemispheres arose only after the evolutionary divergence of the line leading to *Homo sapiens*. But the evolutionary pressures that underlie hemispheric specialization—the need for unified action, rapid communication, reduced costs associated with interhemispheric processing—can be applied to other species. It is now clear that hemispheric specialization is not a unique human feature (Bradshaw and Rogers, 1993).

Anatomical considerations might lead us to expect to find even greater specialization in some nonhuman species such as birds. In birds, almost all of the optic fibers cross at the optic chiasm, ensuring that all of the visual input from each eye projects to the contralateral hemisphere. The lack of crossed and uncrossed fibers probably reflects the fact that there is little overlap in the visual fields of birds, owing to the lateral placement of

affective, or emotional, stimuli (Gur et al., 1994). By their very nature, positive stimuli are likely to elicit approach while negative stimuli will elicit withdrawal or avoidance. Thus, depending on its valence, an affective stimulus is likely to differentially engage the two hemispheres. Davidson (1995) tested this idea by taking electroencephalographic (EEG) measurements while subjects viewed short video clips, chosen to evoke either positive (e.g., a puppy playing with flowers) or negative (e.g., a leg being amputated) emotional reactions. The EEG activity during these stimuli was compared to that during a baseline condition in which the subjects watched a neutral video segment. As predicted, more neural activity was observed over the left frontal lobe when the subjects watched the positive videos in comparison to the negative videos. In contrast, there was a huge increase in activity over the right frontal lobe while subjects viewed the disturbing video.

Similar results have been found in a variety of related experiments. For example, in one study, subjects were required to make rapid responses, either to acquire financial rewards or to avoid financial penalties. Left-hemisphere activation was greater during the reward trials and right-hemisphere activation was greater during the penalty trials, even when the analysis was restricted to the time period prior to the onset of the stimulus. At this point, the subjects had not experienced the reward or punishment; rather, they anticipated its consequences (see Chapter 12).

There are, of course, individual differences in this cerebral tug-of-war between approach and withdrawal. Depression has been linked to an abnormal imbalance favoring neural activity in the right hemisphere. Whether the imbalance preceded or followed the depression remains unclear. More provocative, EEG asymmetries in 3-year-old children are correlated with how well the kids tolerate being separated from their mothers. Children showing higher basal EEG activity in the right hemisphere are more inhibited, staying next to their mother even when surrounded by an array of new toys. Children with higher basal EEG activity in the left hemisphere are quite content to leave their mother to play with the toys.

The study of hemispheric asymmetries in emotion is in its infancy. Prior to the past decade, physiological studies of emotion generally focused on interactions between subcortical limbic systems and the cortex. In developing his account of cortical differences, Davidson started from a consideration of a marked behavioral dichotomy. What remains to be explored are the computations that might lead to one type of behavior over another, and whether these computations are related to those uncovered in the study of hemispheric specialization in other cognititve domains.

But in the interim, we might cull from this work one strategy to test the next time we find ourselves alone at a party: Start talking to someone, if just to get the left hemisphere active! Perhaps there is a reason why the left hemisphere appears specialized to promote approach behavior and is dominant in that most social of all behaviors, language.

the eyes (Figure 10.39). Moreover, birds do not have a corpus callosum. This limits the communication between the visual systems within each hemisphere and might lead to functional asymmetries.

Several asymmetries are known. Chickens and pigeons are better at categorizing stimuli viewed in their right eye. For example, they are more proficient in discriminating food from nonfood items when stimuli are presented to the right eye. In contrast, the left eye appears more adept when chickens are trained to respond to unique properties like color, size, and shape, or when the task requires that they learn the exact location of a food source. Thus, the difference between prototype- and exemplar-based representations that account for human memory asymmetries might apply to birds.

According to Kosslyn's snowball hypothesis (see above), the seed asymmetry for the categorical-coordinate distinction in memory was the evolution of a language system that exploited categorical representations. That is, the primary impetus for hemispheric specialization came from language and later was exploited by other processing domains such as perception and memory. If we assume that the categorical nature of human language is unique to our species, the same evolutionary story could not be applied to birds. The similar memory asymmetries in humans and birds may be fortuitous. This may well be plausible. If asymmetries such as the exemplar-prototype distinction independently arise in two evolutionary lines, half the time one would expect, by chance, a similar pattern of hemispheric organization.

Yet almost all birds have a communication system: They vocalize to scare away enemies, mark territory, and lure mates. In many species, the mechanisms of song

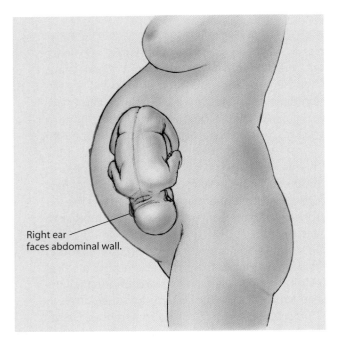

Figure 10.38 Functional asymmetries in manual coordination have been attributed to the prenatal environment of the fetus. The position of the fetus in the uterus influences prenatal vestibular experience. Most fetuses are oriented with the right ear facing outward, which results in a larger vestibular signal in the right hemisphere. At birth, the left side of the body is more stable, freeing the right hand for exploration. Adapted from Previc (1991).

Right ear faces abdominal wall.

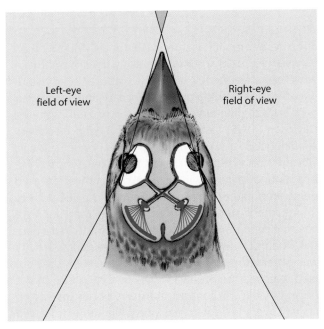

Left-eye field of view

Right-eye field of view

Figure 10.39 Visual pathways in birds are completely crossed. This organization reflects the fact that there is little overlap in the regions of space seen by each eye, and thus, the visual input to the left hemisphere is independent of the visual input to the right hemisphere. This anatomical segregation would be expected to favor the emergence of hemispheric asymmetries.

production depend on structures in the left hemisphere. The seminal finding here was by Fernando Nottebohm of Rockefeller University in the early 1970s (Nottebohm, 1980). He discovered that sectioning the canary's hypoglossal nerve in its left hemisphere severely disrupted song production. Right-hemisphere lesions had little effect. This happens in other species, although in some the lesions to either hemisphere can interfere with song production.

Nonhuman primates also have differences in hemispheric structure or function. In anatomical studies of Old World monkeys and apes, we find asymmetries similar to those of humans. For example, the sylvian fissure shows a greater upward slope in the right hemisphere and there is a similar forward skewedness of the right hemisphere.

Whether these anatomical asymmetries are associated with behavioral specializations remains unclear; the case for hemispheric specialization in nonhuman primates is not compelling. Unlike humans, nonhuman primates do not show a predominance of right-handedness. Individual animals may show a preference for one hand or the other, but there is no consistent trend for the right hand to be favored over the left hand, either when making manual gestures or when using tools.

Perceptual studies provide more provocative indications of parallel asymmetrical functions in humans and nonhuman primates. Rhesus monkeys are superior in making tactile discriminations of shape when using the left hand, as is true of humans. What is more impressive, split-brain monkeys show hemispheric interactions comparable to what is seen with humans on visual perception tasks. In a face recognition task, the monkeys have a right-hemisphere advantage; in a line orientation task, the monkeys have a left-hemisphere advantage. Also, left-hemisphere lesions in the Japanese macaque can impair the animal's ability to comprehend the vocalizations of conspecifics. But unlike the effects on some aphasic patients, this deficit is relatively mild and transient.

In summary, nonhuman species exhibit differences in the function of the two hemispheres. Hemispheric specialization is surely not a unique human characteristic. How should we interpret these findings? Does the left hemisphere, which specializes in bird song and human language, reflect a common evolutionary antecedent? If so, this adaptation has an ancient history because humans and birds have not shared a common ancestor since before the dinosaurs. But the fact that hemispheric specialization occurs in many species may reflect a general design principle for brain function. Specialized function between hemispheres may reflect similar advantages gained by specialization in each hemisphere. The auditory and visual cortices are each uniquely designed to solve problems. Hemispheric asymmetries may provide a similar benefit.

SUMMARY

Research on laterality has provided extensive insights into the organization of the human brain. The surgical disconnection of the cerebral hemispheres has produced an extraordinary opportunity to study which perceptual and cognitive processes are cortical in nature and which are subcortical. We have seen how visual-perceptual information, for example, remains strictly lateralized to one hemisphere following callosal section. Tactile-patterned information remains lateralized. Attentional mechanisms, however, can involve subcortical systems. Taken together, cortical disconnection produces two independent sensory information-processing systems that call upon a common attentional resource system in the carrying out of perceptual tasks.

Split-brain studies also have revealed the complex mosaic of mental processes that goes into human cognition. The two hemispheres do not represent information in an identical manner, as evidenced by the fact that each hemisphere has developed its own set of specialized capacities. In the vast majority of individuals, the left hemisphere is clearly dominant for language and speech and seems to possess a uniquely human capacity to interpret behavior and to construct theories about the relationship between perceived events and feelings. Right-hemisphere superiority, on the other hand, can be seen in tasks such as facial recognition and attentional monitoring. Both hemispheres are likely to be involved in the performance of any complex task, but with each contributing in its specialized manner.

Complementary studies on patients with focal brain lesions and on normal subjects tested with lateralized stimuli, and even comparative approaches, have underlined not only the presence but also the importance of lateralized processes active in cognition and perception. Recent work, such as the spatial frequency hypothesis, has moved laterality research toward a more computational account of hemispheric specialization, seeking to explicate the mechanisms underlying many lateralized perceptual phenomena. These theoretical advances are taking the field away from the popular interpretations of cognitive style and providing a scientific basis for these robust processes.

KEY TERMS

amobarbital

categorical representation

cerebral specialization

coordinate representation

corpus callosum

dichotic listening task

functional asymmetries

generative assembling device

handedness

hierarchical structure

homotopic areas

interpreter

lateral fissure

planum temporale

spatial frequency hypothesis

splenium

split-brain research

transcortical

THOUGHT QUESTIONS

1. What have we learned from over 40 years of split-brain research? What are some of the outstanding questions that remain to be answered?

2. What are the strengths of testing patients who have suffered brain lesions? Are there any shortcomings to this research approach? If so, what are they? What are some of the ethical considerations?

3. Why are double dissociations diagnostic of cerebral specializations? What pitfalls exist if a conclusion is based on a single dissociation?

4. Why do you think the human brain evolved cognitive systems that are represented asymmetrically between the cerebral hemispheres? What are the advantages of asymmetrical processing? What are some possible disadvantages?

5. What commonalities are shared between the spatial frequency hypothesis and the categorical-coordinate proposal of Kosslyn?

SUGGESTED READINGS

BROWN, H., and KOSSLYN, S. (1993). Cerebral lateralization. *Curr. Opin. Neurobiol.* 3:183–186.

CORBALLIS, M.C. (1991). *The Lopsided Ape: Evolution of the Generative Mind.* New York: Oxford University Press.

GAZZANIGA, M.S. (2000). Cerebral specialization and interhemispheric communication: Does the corpus callosum enable the human condition? *Brain* 123:1293–1326.

HELLIGE, J.B. (1993). *Hemispheric Asymmetry: What's Right and What's Left.* Cambridge, MA: Harvard University Press.

IVRY, R.B., and ROBERTSON, L.C. (1998). *The Two Sides of Perception.* Cambridge, MA: MIT Press.

11

The Control of Action

In July 1982, the physicians in the emergency room at hospitals in the San Jose, California, area were puzzled. Working in emergency rooms requires physicians to have uncommon skills. Not only do they encounter the carnage from cars and guns, but also they are generally the first point of contact for people who have suffered a neurological disturbance such as a stroke or severe discomfort, perhaps from exposure to a toxic substance. Life and death decisions hinge on these doctors' ability to make an accurate diagnosis, frequently on the basis of scant information—especially when the patient is too injured or sick to provide a sufficient history. Years of medical training and experience usually provide the best guidance. But this time, none of the attending physicians could recall any precedent for what they had seen over the past few days. Four adults, ranging in age from 26 to 42 years, showed symptoms that did not resemble any known disease. While still conscious, the patients were essentially immobile, were unable to speak, had frozen facial expressions, and showed extreme rigidity in their arms. It was as if they had encountered an evil fairy who had cast a spell on them, rendering them stone statues. The physicians knew they had to act fast, but they could not figure out what to do.

Interviews with the patients' friends and family uncovered a few clues. Even more puzzling was the fact that while two of the patients were brothers, they did not know the other two. This pointed toward exposure to a toxic agent. More questions revealed an important commonality among the cases: All were heroin users.

The symptoms, however, bore little resemblance to those of a heroin overdose. While this narcotic is a powerful central nervous system (CNS) depressant, these patients did not exhibit the typical flaccidity of a person under the influence of heroin. On the contrary, their rigidity was the opposite of what one expects from a powerful dose. No one could recall seeing a case of heroin overdose that produced these effects, nor did the symptoms resemble those of other street narcotics. A new substance was at work here. A few friends who had taken smaller doses confirmed this suspicion. When injected,

this heroin had unexpectedly produced a burning sensation at the site of injection, rapidly followed by a blurring of vision, a metallic taste in the mouth, and, most troubling, an almost immediate jerking of the limbs.

It was a few days before a breakthrough in this case occurred. Computed tomography (CT) and magnetic resonance imaging (MRI) revealed no structural abnormalities either in the brains of the patients, who were still rigid, or in the brains of those who had been lucky enough to have been injected with only a small dose. A neurologist at Stanford University, William Langston (1984), provided the first insight. When examining the patients, he was struck by how similar their symptoms were to those of a patient with advanced Parkinson's disease. This disorder, a common malady affecting approximately 0.5% of the population, is marked by muscular rigidity and disorders of posture and volitional

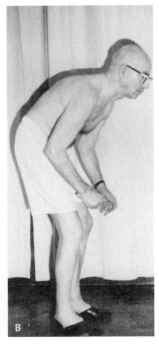

Figure 11.1 Parkinson's disease not only disrupts the production and flexibility of voluntary movement but also can distort posture. Facial expression, including blinking, is frequently absent, giving the person the appearance of seeming to be frozen. The person to the left has had Parkinson's disease for many years and is no longer able to maintain an upright posture. The people below developed symptoms of Parkinson's disease in their 20s and 30s after ingesting the drug MPTP.

movement, or akinesia, the inability to produce volitional movement (Figure 11.1). Everything about the patients' conditions matched this disorder. The biggest puzzles, though, were the age and history of the patients. The onset of Parkinson's disease is gradual and rarely becomes clinically evident until a person is over the age of 45. The heroin users had developed full-blown symptoms of advanced Parkinson's disease within days. Langston suspected that the drug users had come to inject a new synthetic drug being sold as

heroin, and that this drug had triggered the acute onset of Parkinson's disease.

This diagnosis proved to be correct. Parkinson's disease results from cell death in the **substantia nigra,** a brainstem nucleus that is part of the basal ganglia. These cells are the primary source of the neurotransmitter dopamine. Although Langston could not see any structural damage on CT and MRI scans, subsequent positron emission tomography (PET) studies confirmed hypometabolism of dopamine in the people who injected the toxic heroin, regardless of whether they developed Parkinson's disease. Of more immediate concern was how to treat the drug users. For this, Langston adopted the universal treatment applied in Parkinson's disease: He prescribed high doses of L-dopa, a synthetic cousin that is highly effective in compensating for the attrition of the endogenous source of dopamine. When Langston administered this medication to the drug abusers, they immediately showed a positive response. Their muscles relaxed and they could move, albeit in a limited form.

While this episode was tragic for the patients, the incident signified a breakthrough in research on Parkinson's disease. Researchers tracked down the tainted drug and performed a chemical analysis; it turned out to be a previously unknown synthetic substance, bearing little resemblance to heroin but similar in structure to meperidine, a synthetic opioid that creates the sensations of heroin. Based on its chemical structure, it was given the name *MPTP.* Laboratory tests demonstrating that MPTP is selectively toxic for dopaminergic cells led to great leaps forward in medical research on the basal ganglia and on treatments for Parkinson's disease. Prior to the drug's discovery, it had been difficult to induce parkinsonism in nonhuman species. Primates do not develop Parkinson's disease naturally, perhaps because their life expectancy is short. Moreover, it is difficult to access the substantia nigra with traditional lesion methods because of its proximity to vital brainstem nuclei. By administering MPTP, researchers can now destroy the substantia nigra and create a parkinsonian animal.

Two treatment methods directly tie into work with MPTP. In one, parkinsonian animals are given brain grafts with fetal tissue in the hope of regenerating dopaminergic cells (see Chapter 15). This method has met with some success, not only in MPTP-treated animals (Gash et al., 1996) but also in experimental clinical trials with humans who have true Parkinson's disease. Fetal grafts have been shown to survive for extended periods of time, and PET studies have shown that the dopamine activation is enhanced in the patients (Kordower et al., 1995). A second radical treatment method is based on the complex inhibitory and excitatory connections of the basal ganglia, a topic we will return to later in this chapter. As a result of dopamine depletion, the output from the basal ganglia ends up producing excessive inhibition on the motor structures in the cerebral cortex, thus inhibiting movement. Researchers have developed a variety of surgical methods to reduce this inhibition, either by lesioning output structures of the basal ganglia or by implanting brain stimulators within the basal ganglia. These procedures are widely practiced at many medical sites around the world, with stunning success (Samii et al., 1999; Krack et al., 1998).

The MPTP story exemplifies how neurological aberrations can elucidate the complicated patterns of connectivity in the motor structures of the CNS. Indeed, we might say that the complexity of our nervous system is designed for one purpose: to improve the efficiency of our actions. We are not passive processing machines but organisms built for interacting in the world. For organisms with simple nervous systems, we can readily see how sensory information is designed to promote efficient action. In the well-studied sea snail, the aplysia, the sensory neurons regulate the function of motor neurons with little intermediary processing. As we examine more complex nervous systems, the connections between sensation and action become more distant, and we may be fooled into thinking that we can study perception, attention, and memory in isolation. But we would be missing the forest for the trees. Elaborate sensory and memory capabilities are of use to an organism only in improving how it interacts with the world. No purpose is served by the capability of perceiving color unless this information imparts an advantage in how the organism responds to this stimulus. In a similar sense, learning and memory functions are of use only in enabling an animal to draw on its history to modify its future actions. Effective action is the ultimate goal for all internal processing. In this chapter, we will review the organization of the motor system, describing how the brain produces coordinated movement and, at a higher level, selects actions to achieve our goals.

MOTOR STRUCTURES

With these considerations, it is not surprising that so much of the CNS is implicated in controlling action. So let's take a look at the main structures associated with this control. We initially focus on anatomy: the essential components of the motor system and how they are connected. Later we develop a more detailed picture from a cognitive neuroscience perspective, when we consider what mental operations are performed by our action systems.

Muscles, Motor Neurons, and the Spinal Cord

A part of the body that can move is referred to as an **effector.** For most actions, we think of distal effectors, those far from the body center, such as the arms, hands, and legs. But we can also produce movements with more proximal or centrally located effectors, such as the waist, neck, and head. The jaw, tongue, and vocal tract are essential effectors for producing speech, and the eyes are effectors for vision.

All forms of movement result from changes in the state of muscles that control an effector or group of effectors. Muscles are composed of elastic fibers, tissue that can change length and tension. As shown in Figure 11.2, these fibers are attached to the skeleton at joints and are usually arranged in antagonist pairs, which enable the effector either to flex or to extend. For example, the biceps and triceps form an antagonist pair that regulates the position of the forearm. Contracting or shortening the biceps muscle causes flexion about the elbow. If the biceps muscle is relaxed, or if the triceps muscle is contracted, the forearm becomes extended.

The primary interaction of muscles and the nervous system comes about through **alpha motor neurons,** so called because of their large size. Alpha motor neurons originate in the spinal cord, exit through the ventral root, and terminate in the muscle fibers. As with other neu-

rons, an action potential in an alpha motor neuron releases a neurotransmitter; here, the transmitter is acetylcholine. The release of transmitter does not modify downstream neurons, however. Instead, it makes the muscle fibers contract. Thus, alpha motor neurons provide a physical basis for translating nerve signals into mechanical actions. Movement happens when the alpha motor neurons change the length and tension of muscles.

Input to the alpha motor neurons has a variety of sources. At the lowest level, alpha motor neurons receive input from sensory fibers located in the muscles themselves. When a muscle is unexpectedly stretched, a sensory signal is generated in the joints and muscle fibers and enters the dorsal roots of the spinal cord before synapsing directly on corresponding alpha motor neurons. This signal rapidly increases the output of the alpha motor neurons, a process that returns the muscle to its original length. Figure 11.3 shows the chain of reflexes that occur when the family physician raps just below the knee cap. This test assesses the integrity of rapid spinal reflexes.

Alpha motor neurons also receive input from the descending fibers of the spinal cord and interneurons

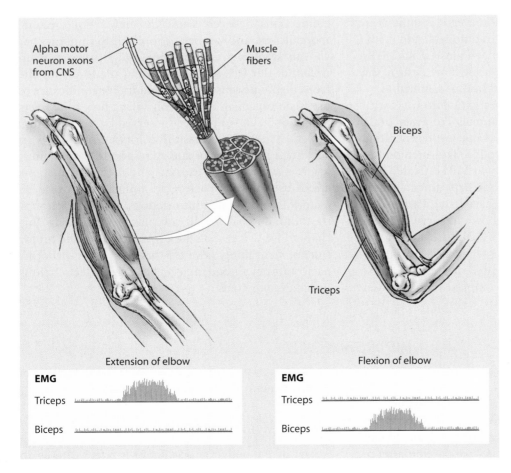

Alpha motor neuron axons from CNS

Muscle fibers

Biceps

Triceps

Extension of elbow

Flexion of elbow

EMG

Triceps

Biceps

EMG

Triceps

Biceps

Figure 11.2 Muscles are activated by the alpha motor neurons. Electrodes placed on the skin over the muscle can measure this electrical activity, producing an electromyogram (EMG). The input from the alpha motor neurons causes the muscle fibers to contract. Antagonist pairs of muscles span many of our joints. Activation of the triceps produces extension of the elbow; activation of the biceps produces flexion of the elbow.

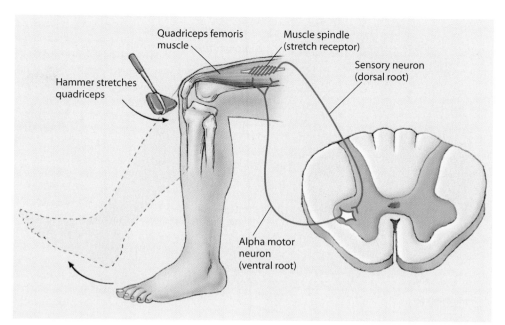

Figure 11.3 The stretch reflex. When the doctor raps your knee, the quadriceps is extended. This stretch triggers receptors in the muscle spindle to fire. The sensory signal is transmitted through the dorsal root of the spinal cord and directly activates an alpha motor neuron to contract the quadriceps. In this manner, the stretch reflex helps to maintain the stability of the limbs following unexpected pertubations.

within the spinal segment. The descending fibers originate in several subcortical and cortical structures. These signals can be either excitatory or inhibitory and are the basis for voluntary movements. For example, a central command for elbow flexion will require excitation of the biceps. This event by itself results in a passive stretch of the triceps. If unchecked, it would trigger, via a stretch reflex, excitation of the triceps and the limb would return to its original position. Excitatory signals to one muscle, the agonist, are accompanied by inhibitory signals to the antagonist muscle via interneurons. In this way, the stretch reflex that efficiently stabilizes unexpected perturbations can be overcome to permit volitional movement.

Subcortical Motor Structures

Many neural structures of the motor system are in the brainstem, which contains the twelve cranial nerves essential for critical reflexes involving breathing, eating, eye movements, and facial expressions. In addition, many nuclei within the brainstem send direct projections down the spinal cord. These tracts are referred to collectively as the **extrapyramidal tracts,** indicating that they do not originate in the pyramidal neurons of the motor cortex (Figure 11.4). Extrapyramidal tracts are a primary source of control over spinal activity; they receive input from subcortical and cortical structures.

Figure 11.5 shows the location of two prominent subcortical structures that play a key role in motor control: the cerebellum and basal ganglia. The **cerebellum**

is a massive structure that receives extensive sensory inputs, including information from somatosensory, vestibular, visual, and auditory channels. It also receives input from many association areas of the cortex. This input is projected primarily to the cerebellar cortex, perhaps the most densely packed neural region in the brain; indeed, more cells are in the cerebellum than in all the rest of the nervous system. The cerebellar cortex, though, does not send direct output to the brain. Rather, the cortex relays information to nuclei buried within the midst of the cerebellar enfolding; all cerebellar output comes from these nuclei. Within the cerebellum, there is a partitioning with respect to whether the output provides descending or ascending neural signals. Axons from more medial nuclei can synapse directly on spinal interneurons, but the majority terminate on the nuclei of the extrapyramidal tracts. In contrast, output from lateral regions influences motor and frontal cortical regions by way of relays in the thalamus.

The **basal ganglia** are a collection of five nuclei. Taken as a whole, the organization of the basal ganglia bears a similarity to that of the cerebellum. Input is mainly restricted to the two nuclei forming the striatum—the caudate and the putamen. Output is almost exclusively by way of the internal segment of the globus pallidus and part of the substantia nigra. The remaining components (the rest of the substantia nigra, subthalamic nucleus, and external segment of the globus pallidus) are in a position to modulate the output of the globus pallidus. As with the lateral cerebellum, basal ganglia output is primarily ascending. Axons of the globus pallidus terminate in the

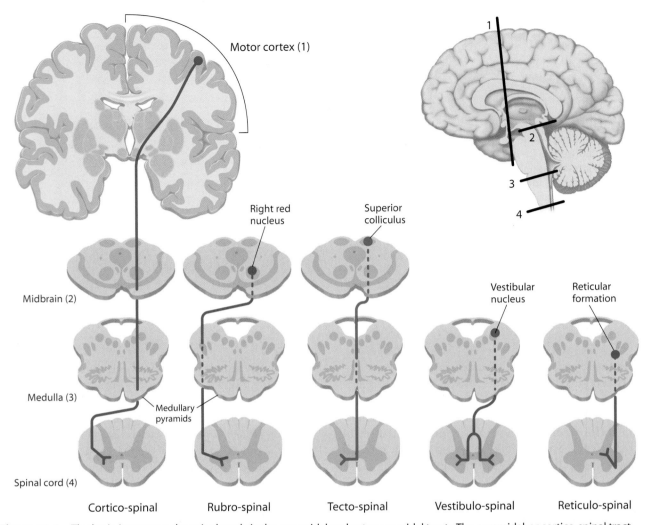

Motor cortex (1)

Right red nucleus

Superior colliculus

Vestibular nucleus

Reticular formation

Midbrain (2)

Medulla (3)

Medullary pyramids

Spinal cord (4)

Cortico-spinal

Rubro-spinal

Tecto-spinal

Vestibulo-spinal

Reticulo-spinal

Figure 11.4 The brain innervates the spinal cord via the pyramidal and extrapyramidal tracts. The pyramidal, or cortico-spinal tract originates in the cortex, and almost all of the fibers cross over to the contralateral side at the pyramids. Extrapyramidal tracts originate in various subcortical nuclei and terminate in both contralateral and ipsilateral regions of the spinal cord. Adapted from Bear et al. (1996).

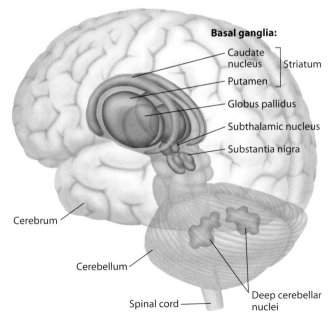

Basal ganglia:

Caudate nucleus ⎤ Striatum
Putamen ⎦

Globus pallidus

Subthalamic nucleus

Substantia nigra

Cerebrum

Cerebellum

Spinal cord

Deep cerebellar nuclei

Figure 11.5 The basal ganglia and cerebellum are two prominent subcortical components of the motor pathways. The basal ganglia proper includes the caudate, putamen, and globus pallidus, three nuclei that surround the thalamus. Functionally, the subthalamic nuclei and substantia nigra also are considered part of the basal ganglia. The cerebellum sits below the posterior portion of the cerebral cortex. All cerebellar output originates in the deep cerebellar nuclei.

thalamus, which in turn projects to motor and frontal regions of the cerebral cortex. Basal ganglia provide scant input to the brainstem nuclei forming the extrapyramidal tracts. Instead, motor control is cortically mediated by thalamic projections.

Cortical Regions Involved in Motor Control

The cerebral cortex can regulate the activity of spinal neurons in direct and indirect ways. Direct connections are provided by the **cortico-spinal tract.** As its name implies, this tract is composed of neurons that originate in the cortex and terminate directly or monosynaptically on alpha motor neurons or spinal interneurons. Thus, single axons of the cortico-spinal tract can stretch for more than 1 m. The cortico-spinal tract is frequently referred to as the *pyramidal tract* because it was once believed that the sole origin of these fibers was the giant pyramidal cells of the cortex. Now we know that cortico-spinal axons can arise from cells of different sizes in Layer V of the cortex.

Cortico-spinal fibers originate from many parts of the cerebral cortex (Figure 11.6). The most prominent of these is the **primary motor cortex,** or area 4. This area is anterior to the central sulcus, the anatomical division separating the frontal and parietal cortex. Cortico-spinal fibers also are found in portions of somatosensory cortex and area 6, divided functionally into a lateral region, **premotor cortex,** and a medial region, the **supplementary motor area.** Because the somatosensory cortex contains many direct projections to the spinal cord, it is clear that the brain is not divided into sensory, motor, and association areas; the processes must work in concert to produce coherent action.

While the cerebral cortex has direct access to spinal mechanisms via the cortico-spinal tract, it also can influence movement in four other prominent ways. First, the motor cortex and premotor areas receive input from most regions of the cortex by way of cortico-cortical connections. Second, many cortical axons terminate on brainstem nuclei, thus cortically influencing the extrapyramidal tracts. Third, the cortex sends massive projections to the basal ganglia and cerebellum. Fourth, the cortico-bulbar tract is composed of cortical fibers that terminate on the cranial nerves.

The Organization of Motor Areas

The anatomical organization of the motor areas follows two principles. First, within each motor structure is an organized, somatotopic representation of the body, a concept introduced in Chapter 3. In the motor cortex, this somatotopy is particularly clear. For example, an electrical

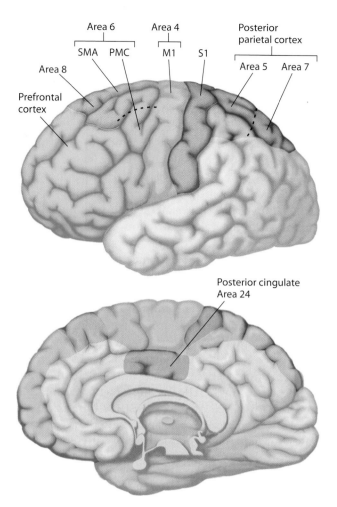

Figure 11.6 The motor areas of the cerebral cortex. Area 4 is the primary motor cortex. Area 6 on the medial surface is referred to as the *supplementary motor area* (SMA) and on the lateral surface as the *premotor cortex* (PMC). Area 8 includes the frontal eye fields. Lesions in many posterior areas also can produce severe problems in coordination.

stimulus applied to the medial wall of the precentral gyrus will create movement in the foot; the same stimulus applied at a ventral lateral site will elicit tongue movement. Representation of the effectors does not correspond to their actual size. Rather, the cortical area devoted to a certain effector reflects the importance of that effector for movement and the level of control required for manipulating the effector. For this reason, the fingers span a large portion of the human motor cortex because of the significance of manual dexterity. This somatotopy can even be demonstrated noninvasively with transcranial magnetic stimulation (TMS) by placing the coil over the motor cortex. As the center of the coil is moved downward, the effects of stimulation shift from movement of the upper arm to the wrist and then to individual fingers.

The somatotopic representation, particularly for distal effectors, is restricted to one side of the body. As with sensory systems, each cerebral hemisphere is devoted primarily to controlling behavior on the opposite side of the body. For cortico-spinal tracts, this contralateral organization occurs because almost all of these fibers cross, or decussate, at the junction of the medulla and spinal cord. Most extrapyramidal fibers also decussate so each side controls movements on the body's opposite side. The one exception to this crossed arrangement is the cerebellum. Fibers from the cerebellum cross as soon as they exit this structure; they project to either the contralateral cerebral hemisphere via the thalamus or the contralateral brainstem nuclei. The result is an ipsilateral organization of the cerebellum; the right side is associated with movements on the right side of the body and the left side with movements on the left side of the body.

The second general principle refers to the relation between motor areas. The components form a hierarchy with multiple levels of control. As can be seen in Figure 11.7, the lowest level of the hierarchy is the spinal cord. Not only do spinal mechanisms provide a point of contact between the nervous system and muscles, but also simple reflexive movements can be controlled at this level. At the highest level are premotor and association areas. Processing within these regions is critical for planning an action based on present perceptual information, past experience, and future goals. The motor cortex and brainstem structures, with the assistance of the cerebellum and basal ganglia, translate this action goal into a movement.

Viewing the motor system as a hierarchy enables us to recognize that motor control is a distributed process. The highest levels might not be concerned with the details of a movement, but they may allow lower-level mechanisms to translate motor commands into movements. Hierarchical organization also can be viewed from a phylogenetic perspective. Movement in organisms with primitive motor structures is based primarily on simple reflexive actions. An unexpected blast of water against the abdominal cavity of the sea slug automatically elicits a withdrawal response. Additional layers of control enable these reflexes to be modified, perhaps being expressed only in certain conditions. For example, a brainstem structure can inhibit spinal neurons so a

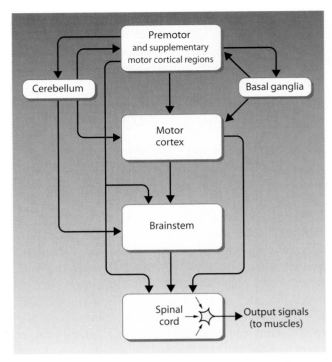

Figure 11.7 The motor hierarchy. All connections to the arms and legs originate in the spinal cord. The spinal signals are influenced by inputs from the brainstem and various cortical regions, whose activity in turn is modulated by the cerebellum and basal ganglia. Thus, control is distributed across various levels of the hierarchy.

change in a muscle does not automatically trigger a stretch reflex.

In an analogous fashion, the cortex can provide additional means for regulating the actions of the lower levels of the motor hierarchy. This ability offers an organism greater flexibility in its actions. We can generate any number of movements in response to a sensory signal. A tennis player can choose to hit a cross-court forehand, go for a drop shot, or pop a defensive lob. Cortical mechanisms also enable us to generate actions that are minimally dependent on external cues. We can sing aloud, wave our hands, or pantomime a gesture. Reflecting this greater flexibility, the cortico-spinal tract is one of the latest evolutionary adaptations, appearing only in mammals. It affords a new pathway over which the cerebral hemispheres can activate ancient motor structures.

COMPUTATIONAL ISSUES IN MOTOR CONTROL

We have seen the panoramic view of the motor system: how muscles are activated and how this activity can be influenced by spinal, subcortical, and cortical signals. Though we identified the major anatomical components, we have yet to consider the structures' functional roles. But it is useful first to examine certain behavioral phenomena to understand the information processing that must take place in the motor systems. This section elucidates computational issues in motor control and sets the stage for the final sections in which physiological and neurological evidence is assessed in developing a cognitive neuroscience of action.

Peripheral Control of Movement and the Role of Feedback

The notion of hierarchical control implies that higher-level systems can modulate the activity of lower-level mechanisms. One implication of this view is that lower levels can produce movement. The stretch reflex is one such example. Some spinal mechanisms can maintain postural stability even in the absence of higher-level processing.

Are spinal mechanisms a simple means for generating more complicated movements? At the end of the nineteenth century, Charles Sherrington, a Nobel laureate and British neurophysiologist, developed a procedure in which he severed the spinal cord in cats and dogs to disconnect peripheral motor structures such as alpha motor neurons and spinal interneurons from the cortex and subcortex (Sherrington, 1947). This procedure allowed Sherrington to observe whether the animals could produce any movement in the absence of higher-level commands. As expected, stretch reflexes remained intact; in fact, these reflexes were exaggerated due to the removal of inhibitory influences from the brain. More surprisingly, Sherrington observed that these animals could alternate the movements of their hind limbs. With the appropriate stimulus, one leg flexed while the other extended, and then the first leg extended while the other flexed. In other words, without any signals from the brain, the animal displayed movements that resembled walking. Sherrington subsequently observed that these movements depended on intact sensory processes. When the sensory signal to a limb was eliminated by cutting the dorsal root of the spinal cord, the animal ceased using that limb.

Sherrington concluded that the spinal reflex arc provided the essential components for movement. The alpha motor neurons could trigger movement, but the normal maintenance of descending, efferent motor commands required a continual sensory signal from the periphery concerning the consequences of these commands. More complicated movements might utilize a similar process in which sensory signals and motor outputs were joined in a stream of gestures. Each movement would change the sensory signal, and this new sensory signal could trigger movement; hence, complex movements could be seen as the successive chaining of stimuli and responses via reflex processes.

Sherrington was only partly correct. While movement does not require signals from the brain, it also does not need sensory signals, a phenomenon that T.G. Brown first demonstrated in 1911. When Brown sectioned the spinal cord and dorsal roots, he discovered that the animal still generated rhythmic walking movements (Figure 11.8). Hence, neurons in the spinal cord can generate an entire sequence of actions without any external feedback signal.

These neurons have come to be called *central pattern generators,* and they underscore two major features of motor control. First, they offer a powerful mechanism for the hierarchical control of movement. Consider, for instance, how the nervous system might initiate walking. Brain structures would not have to specify patterns of muscle activity. Rather, they would simply activate the appropriate pattern generators in the spinal cord, which in turn trigger muscle commands. The system is truly hierarchical in that the highest levels need be concerned only with issuing commands to achieve an action, while lower-level mechanisms translate the commands into a movement.

We cannot expect there to be central pattern generators for a wide range of movements. Locomotion probably represents a special situation, an action for which specialized mechanisms have evolved. But more arbitrary movements could exploit central pattern generators. When we reach to pick up an object, for example, low-level mechanisms could make the necessary postural adjustments to keep our body from tipping over as our center of gravity shifts.

The second important insight yielded by Brown's work is the observation that movements are not entirely dependent on peripheral feedback. As such, they imply internal representations of movement patterns. This hypothesis has been supported by a myriad of evidence over the past 50 years. One influential study demonstrating the central control of movement comes from an experiment on a colony of monkeys that lived in a large, enclosed environment (Taub and Berman, 1968). Their dorsal roots were cut; in other words, the animals were **deafferented,** deprived of all sensory or afferent feedback from the affected limb. In this study, the deafferentation procedure was conducted in two stages. In the first stage, the operation was restricted to a single arm. Following this surgery, the animals showed little evidence that they could still use the limb. They hung the limb by their side and never engaged in activities like climbing, which requires coordinated actions of the two forelimbs. In the second stage, the researchers severed the dorsal root in the remaining intact limb. Paradoxically, the animals now began to use

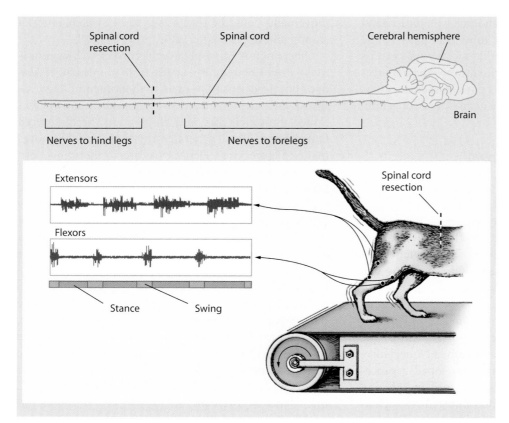

Figure 11.8 Movement is still possible following a resection of the spinal cord. In Brown's classic experiment, the spinal cord was severed so that the nerves to the hind legs were isolated from the brain. The cats were still able to produce stereotypic, rhythmic movements with the hind legs when walking on a moving treadmill. Since all inputs from the brain were eliminated, the motor commands must have originated in the lower portion of the spinal cord. Adapted from Kandel et al. (1991).

both limbs (Figure 11.9). Indeed, an outside observer would be hard-pressed to notice any abnormalities in the behavior of these animals. They would move about their cages in a quadrupedal gait and could be coaxed into climbing the sides of their cages to fetch a reward. These results demonstrate that sensory signals are not essential for movement, although, as we might expect, an animal prefers to use a limb that has intact sensory signals.

This same sort of phenomenon has been observed in humans with severe sensory deficits, or *neuropathies*. In some patients, the peripheral sensory fibers are destroyed or become nonfunctional for unknown reasons;

Figure 11.9 Movement without feedback. If the dorsal root on one side of the body is sectioned, the animal does not use the limb; it hangs limply by the animal's side. In a second operation, the dorsal root on the other side is sectioned. Now the brain does not receive sensory information from either upper limb. Paradoxically, the monkeys now begin to use the limbs, moving about on all fours, climbing fences, and using their hands to pick up objects.

however, the motor fibers are relatively spared, perhaps due to their smaller size and reduced metabolic demands. Careful clinical examinations are required to document that there is no residual perception of joint position, touch, or vibration in the distal extremities. One patient was reported to be oblivious to pinprick or even electrical stimulation of moderate intensity (Rothwell et al., 1982). Nonetheless, despite their severe sensory problems, these patients can still make complicated movements. Figure 11.10 shows the tracings of one patient who was asked to draw geometric shapes in midair. In each case, the shape is drawn correctly and consistently, even though the gestures require the coordinated interplay of many muscles.

The fact that they can make the movements emphasizes that movements can be centrally generated without feedback. This is not to say that feedback is unimportant, however. Errors accumulate quickly in its absence, and even in the initial tracings, the movements are not as precise as those of subjects who have intact sensation. These problems are especially pronounced in multijoint movements. The tracings shown in Figure 11.10 can be made by simple rotation about the shoulder. When the activity about two joints must be coordinated, accuracy becomes quite poor (Gordon et al., 1995). Moreover, feedback is essential for learning and keeping the motor system calibrated. The dart thrower is constantly using the results of his last toss to make subtle adjustments.

Figure 11.10 Movements produced by a patient with a severe sensory neuropathy. Time-lapse photography was used to record the patient's tracings of a circle, a figure eight, and a square. All of the movements were made in the dark. Despite the absence of proprioceptive, kinesthetic, and visual feedback, the shapes are readily identified.

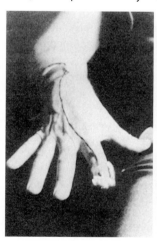
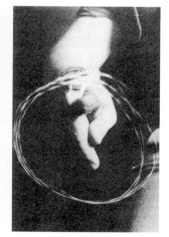

The Representation of Movement Plans

Research on central pattern generators and movement without feedback highlights the importance of representations in controlling movement. Patients who are functionally deafferented must have an internal representation of the expected consequences of a motor command. For example, to trace a circle or figure eight, they must have learned the required series of motor commands, commands that can be accessed without requiring any sort of feedback signal. A central issue is how to characterize properties of these representations and how they translate into actions (Keele, 1986).

THE CONTENT OF MOTOR PLANS

Consider a simple motor task. A light is presented at a certain location, and after it is turned off the subject must move her arm to the position of the light. With practice, she can improve on the task, even if visual feedback from the moving limb is precluded. For example, she might complete the movement more rapidly and with less error. This improvement implies that the representation of the movement has changed. But how? Two possibilities can be considered. Either she developed a distance-based representation of the desired movement—she might have learned that a certain pattern of muscular activity will displace the limb a desired distance—or she developed a location-based representation of the final position. By this latter hypothesis, practice might improve the representation of an endpoint configuration for the limb that matches the target location.

Emilio Bizzi and his colleagues at MIT (1984) provided evidence in favor of the hypothesis that the central representation is based primarily on a location code. In their experiments, deafferented monkeys were trained in a simple pointing experiment. On each trial, a light appeared at one of several locations. After the light was turned off, the animal was required to rotate its elbow to bring its arm to the target location.

The critical manipulation included trials in which an opposing torque force was applied just when movement started. These forces were designed to keep the limb at the starting position for a short time. Since the room was dark and the animals were deafferented, they were unaware that their movements were counteracted by an opposing force. The critical question centered on where the movement ended once the torque force was removed. If the animal has learned that a muscular burst will transport its limb a certain distance, applying an opposing force should result in a movement that falls short of the target. But if the animal generates a motor command specifying the desired position, it should achieve this goal once the opposing force is removed. As shown in Figure 11.11, the results clearly favor the location hypothesis. When the torque motor was on, the limb stayed at the starting location. As soon as it was turned off, the limb rapidly moved to the correct location.

An even more dramatic demonstration of **endpoint control** occurred in a follow-up experiment. Here the torque motor was used to transport the limb from the starting position to the final position just as the target location was illuminated. Because of the surgery, the animals were unaware of this passive displacement. Thus, the question centered on what would happen at the beginning of the movement. If the animal is programming muscular events that will result in movement of a desired distance, then the arm should be propelled well past the target. Yet if the animal is programming a desired final location, there should be no movement—the arm has already been moved to the target location. Neither of these predictions was confirmed. When the torque was removed, the limb moved toward the original location, followed by a reversal after a few hundred milliseconds toward the final location. Indeed, a similar return toward the starting position is also seen when this manipulation is done in the monkey with intact feedback mechanisms. These results provide even more compelling support for the location hypothesis and reveal insights into the dynamics of location-based representations. At the time the target was illuminated, the monkey was still generating a motor command corresponding to the initial location (Figure 11.12). Thus, the arm began to drift toward this position once the torque force was removed. When the target's location is identified, the representation of action commands corresponding to this location are instantiated and the arm moves toward the target location. The experiment demonstrated how movement can be viewed as a shift from one postural state to another.

Location planning is computationally appealing when considered from a hierarchical perspective. By this perspective, the highest level of the planning system is concerned with achieving a final configuration, a

Figure 11.11 Endpoint control. **(a)** Deafferented monkeys were trained to point in the dark to a target indicated by the brief illumination of a light. The top trace shows the position of the arm as it goes from an initial position to the target location. The bottom trace shows the EMG activity in the biceps. In the control condition, the animals were able to accurately make the pointing movements despite the absence of all sources of feedback. **(b)** An opposing force is applied at the onset of the movement, preventing the arm from moving (bar under the arm position trace). Once this force is removed, the limb rapidly moves to the correct target location. Since the animal could not sense the opposing force, it must have generated a motor command corresponding to the target location. Adapted from Bizzi et al. (1984).

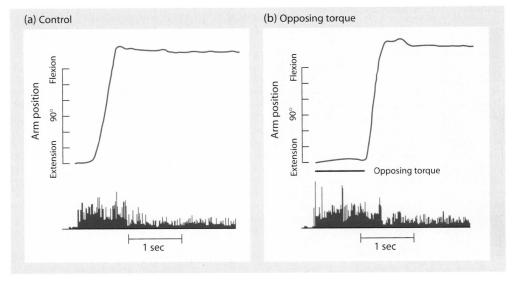

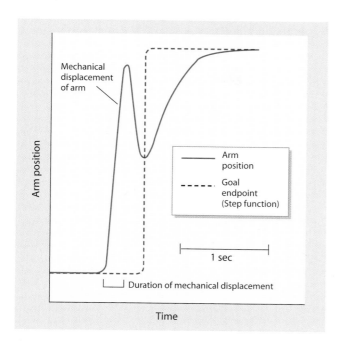

Figure 11.12 Pointing movements from a displaced starting location. Prior to the onset of EMG activity, the monkey's limb (solid line) was displaced mechanically to the target location. When this external force was removed, the limb began to move back to the initial location and then reversed direction to move to the target location. The central commands (dotted line) can be viewed as a step function from one posture to another, initially specifying the starting location and then abruptly shifting to the target location. Adapted from Bizzi et al. (1984).

movement that reaches a goal. The way the plan is executed can be assigned to lower levels of the motor hierarchy. Consider how this might apply to a simple reaching task like moving a hand from a computer keyboard to grasp a coffee cup at one side. The action can be achieved in many ways—we can glance at the cup to guide our hand movements, grope about as we remain engrossed in our text, or rotate about the waist so the cup is directly in front of us. This redundancy arises because we have multiple joints (elbow, shoulder, waist) that provide great flexibility; yet this redundancy can increase the complexity for a control system. According to the location hypothesis, the highest level of the hierarchy need represent only the ultimate goal—the elbow and hand assume a position where the cup can be grasped with minimal effort. How this goal is met does not have to be included in this representation. Lower levels of the motor hierarchy are concerned with translating a final goal into a certain trajectory.

While Bizzi and his colleagues provided impressive evidence in favor of location planning, it is also clear that we can represent other parameters of movement such as distance. For example, we can reproduce a movement to a fixed distance, even when the repro-

duced distance is to a novel location. Moreover, when movements are made to successive locations, errors that arise in the initial movements are manifest in the final movements. If the sequence is represented as independent locations, we should not expect to observe this persistence.

We also can control the form with which a movement is executed. For example, in reaching for the coffee cup, we can choose to extend only the arm or can bend forward by rotating about the waist. We also can vary how fast we execute the movement. Indeed, in many tasks, the trajectory is as important as the final goal. When a skilled figure skater performs an aerial maneuver, it is essential that she land properly to maintain her balance. But it is just as important that the jump follow a certain trajectory to achieve the desired number of rotations in the most graceful manner. Viewed in this way, endpoint control reveals the fundamental capabilities of the motor control system; the distance and trajectory planning demonstrate additional flexibility in the control processes.

Consider one other possibility. To this point, we have emphasized the hierarchical nature of the motor system. For example, at the highest level, the representation is in terms of locations and goals, with lower levels transforming this plan into a movement. An alternative conceptualization is that location and distance planning represent two independent forms of representation (Figure 11.13). This hypothesis provides an interpretation for why pointing movements generally have two phases: a rapid, ballistic phase when the arm is transported to the vicinity of the target, followed by a

Figure 11.13 Location and trajectory planning of motor commands. **(a)** In the hierarchical model, goals are specified initially as target locations. A translation process is then required to determine the trajectory and corresponding muscle activity required to move a limb to the goal. **(b)** An alternative hypothesis is that location and trajectory planning provide two different representations for determining muscle activity.

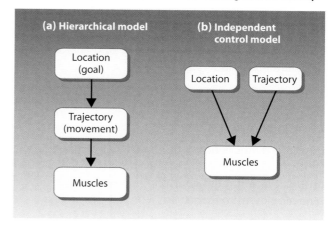

secondary phase when the target location is achieved. Although the second phase has been interpreted as showing the activation of feedback processes, it could reflect a transition from a representation based on distance to one based on location.

Some suggestive results favor the hypothesis of separate forms of representation. Richard Abrams and his colleagues (1994) at Washington University in St. Louis ran an experiment in which a dot was displaced across the computer screen, either by 4 or 7 degrees of visual angle. In separate blocks of trials, subjects were asked either to reproduce the distance that the dot had moved or to move to the final position of the dot. The critical manipulation was that on half the trials, background elements on the computer display moved in the opposite direction. This opposing motion creates a powerful illusion, amplifying the perceived motion of the dot (Figure 11.14). The effect of this illusion was obvious in the subjects' performance in the distance condition. They overestimated the distance the dot had moved relative to conditions where the background moved in the same direction as the target dot. But the background motion had minimal effect on their performance in the location condition. If responses in the two conditions were based on a single representational system, one would expect the illusion to affect both conditions similarly. The fact that the illusion produced differential effects in the two conditions indicates that the movements were guided by different representations.

In the Abrams study, only the distance-based representation was susceptible to the effects of the illusion. The location-based code appeared to be immune to these perceptual distortions. Results such as these suggest the provocative hypothesis that the human brain might have evolved two separate modes for planning movements. Perhaps location planning represents a more ancient and primitive system, one in which the representation is simply to specify the goal, or target location of an action. This system could produce movements that reach a final target location, but without much flexibility. In contrast, a second system might be capable of distance planning, of specifying the exact form with which an action will be achieved. This system would provide flexibility but with the additional costs of planning.

HIERARCHICAL REPRESENTATION OF ACTION SEQUENCES

In the preceding actions, the movements entailed simple gestures such as pointing to a location in space. Most of our actions are more complex, however, and involve a sequence of movements. In serving a tennis ball, we

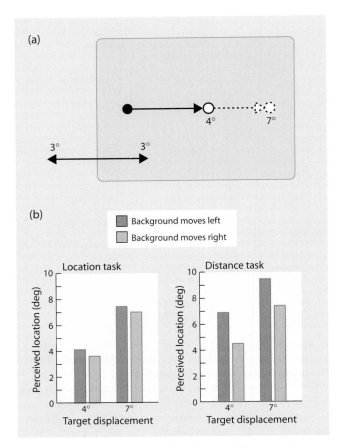

Figure 11.14 Dissociation of location and distance planning. **(a)** A dot was moved to the right, traversing a distance of either 4 or 7 degrees. In half of the trials, the frame moved to the right; in the other half, the frame moved to the left. Subjects were asked to move a lever, indicating in separate conditions the distance the target dot moved or the final location of the dot. **(b)** Distance judgments were more susceptible to the motion of the frame than were the location judgments. Adapted from Abrams and Landgraf (1990).

have to toss the ball with our nondominant hand and swing the racquet so it strikes the ball just after the apex of rotation. In playing the piano, we must strike a sequence of keys with appropriate timing and force. Are these actions simply constructed by linking independent movements, or are they guided by hierarchical representational structures that govern the entire sequence? The answer is the latter. These actions are guided by hierarchical representations, structures that organize the movement elements into integrated chunks. The idea of chunking, which was introduced in Chapter 8 in the discussion on capacity limits in memory, is also relevant to the representation of action. In piano playing such structure is obvious: The musical score defines measures and phrases. Similar principles apply to all skilled actions, albeit at a more implicit level.

Where Is It? Assessing Location Through Perception and Action

To demonstrate that spatial information can be represented differently in systems involved in conscious perception and those associated with guiding action, try the experiment outlined in the figure. While standing in an open area, have a friend place an object between 6 and 12 m from you. Then, have your friend move along the perpendicular direction and stop him or her when you perceive that you are both equidistant from the object. Measure your accuracy. Now, have your friend place the object in a new location, again between 6 and 12 m away. When ready, close your eyes and walk forward, attempting to stop right over the object. Measure your accuracy.

Assuming that your performance matches that of the average person, you will notice a striking dissociation (Loomis et al., 1992). You would have been quite in-accurate on the first task, underestimating the distance from you to the object. Yet on the second task you should be very accurate. This dissociation between two forms of judgment, one perceptual and the other motoric, is similar to that found in the Abrams experiment. In both situations, the results suggest that separate representational systems underlie judgments of location and distance. While location judgments are veridical, the representation of distance is subject to perceptual distortions. Our perception of distance is highly compressed—things almost always are farther away than they appear (perhaps a "safety" mechanism to ensure that we ready ourselves for an approaching predator?). But as this experiment demonstrates, our action systems are not similarly fooled. There is little, if any, compression of distance when we move to a target location.

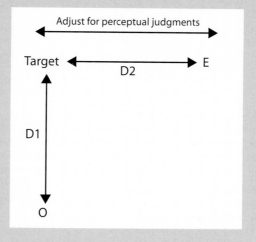

Perceptual judgment of distance. Two people are needed for this demonstration. The observer, O, should stand at a fixed location in an open area. The experimenter, E, places a target at some point in the area. E walks along the perpendicular direction and stops when O judges each person to be equidistant to the target (D1 = D2). The results will be quite striking.

Figure 11.15 presents a model developed by Donald MacKay (1987) of the University of California to show how the ideas of hierarchical representation can be applied to motor control. The top of the hierarchy, the conceptual level, corresponds to a representation of the goal of the action. In this example, the man intends to accept the woman's invitation to dance. At the next level, this goal has to be translated into a system. He could make a physical gesture such as offer his hand or start tapping his foot. Or he could verbally respond to the invitation, selecting one sentence from a large repertoire of appropriate responses: "Yes, that would be nice," "Yes, I've been dying to get out there and shake my booty," or "Yes, the music is really hot." MacKay referred to this as the lexical level to indicate that a common concept can be conveyed through a distinct set of actions. Lower levels of the hierarchy will then translate these units into patterns of muscular activation. For example, a verbal response will entail a pattern of activity across the speech articulators or the extension of the hand will require movements of the arm and fingers.

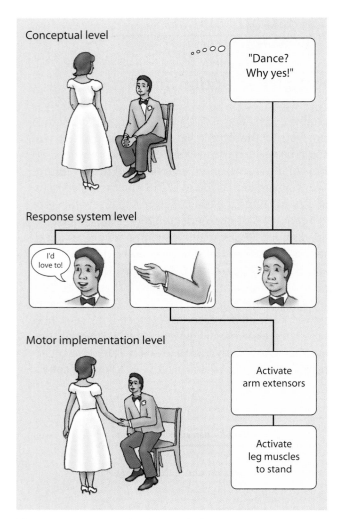

Figure 11.15 Hierarchical control of action. Motor planning and learning can occur at multiple levels. At the lowest level are the actual commands to implement a particular action. At the highest level are abstract representations of the goal for the action. Usually multiple actions can achieve the same goal. Learning occurs at all levels. If the motor commands (lowest level) are well established, then motor learning will be limited to strengthening the more abstract representations rather than involving the muscles themselves.

The hierarchical properties of this model are explicit. Each level corresponds to a different form for representing the action. Actions can be described in relation to the goals to be achieved, and this level need not be tied to a specific form of implementation. Whether our gentleman responds physically or vocally, the two forms of responding share a common level of representation. In a similar fashion, when we convey a linguistic message by speaking or by writing, a common level of representation is on both the conceptual and the lexical level. Higher levels in the hierarchy need not represent all of the information.

MacKay's model also has proved useful in explaining the learning of new motor patterns. People frequently attribute motor learning to low levels of the hierarchy. We speak of our muscles having learned how to respond: how to maintain our balance on a bike, or how to throw a dart with fine precision. The fact that we have great difficulty verbalizing how to perform these skills reinforces the notion that the learning is noncognitive. The Olympic gymnast Peter Vidman expressed this sentiment when he said, "As I approach the apparatus . . . the only thing I am thinking about is . . . the first trick . . . . Then, my body takes over and hopefully everything becomes automatic. . . ." (Schmidt, 1987).

But, on closer study, we find that some aspects of motor learning are independent of the muscular system used to perform the actions. Demonstrate this to yourself by taking a piece of paper and signing your name. Having done this, repeat the action but use your nondominant hand. Now do it again, holding the pen between your teeth. If you feel especially adventurous, you can take off your shoes and socks and hold the pen between your toes.

While the atypical productions are not as smooth as your standard signature, the more dramatic result of this demonstration is the high degree of similarity across all of the productions. An example from one such demonstration is given in Figure 11.16. This high-level representation of the action is independent of any particular muscle group. The differences in the final product reflect the fact that some muscle groups have more experience in translating an abstract representation into a concrete action.

When we view it this way, we see that motor learning can take place at multiple levels. When a learned action is produced by a new set of effectors, learning is greatest at the lower level. In contrast, if we are acquiring a new action, the effects of learning are likely to be at more conceptual levels of the action hierarchy. To show this, MacKay recruited a group of subjects who were fluent in English and German. These subjects were asked to produce novel sentences such as "This morning, I got out of bed eleven times." The subjects would repeat the sentence in English several times, and as expected, they became faster with practice. Then he would have them repeat the same sentence, but for this transfer test, they were required to speak the sentence in German. Even though they had never said the sentence in this language, the subjects showed perfect transfer on this task:

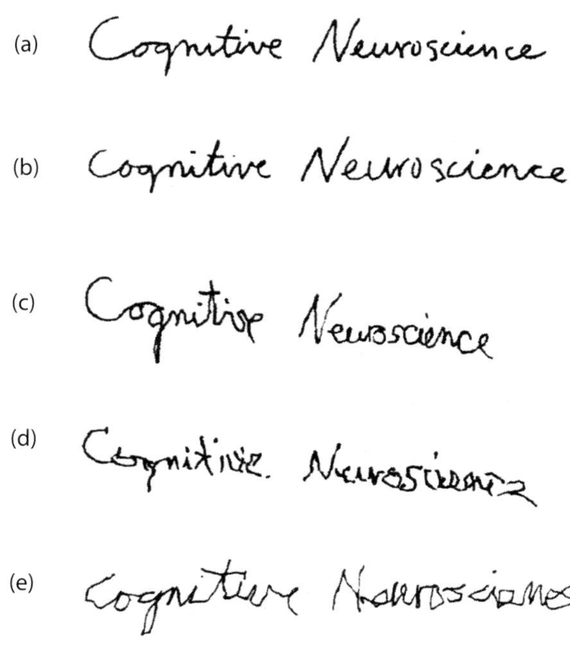

(a) *Cognitive Neuroscience*

(b) *Cognitive Neuroscience*

(c) *Cognitive Neuroscience*

(d) *Cognitive Neuroscience*

(e) *Cognitive Neuroscience*

Figure 11.16 Motor representations are not linked to particular effector systems. These five productions of "cognitive neuroscience" were produced by moving a pen with **(a)** the right hand, **(b)** the right wrist, **(c)** the left hand, **(d)** the mouth, and **(e)** the right foot. There is a degree of similarity in the productions despite the vast differences in practice writing with these five body parts.

They were just as fast as when they practiced the sentence in English.

MacKay argued that the effects of practice in this situation are restricted to the highest abstract level. Because the subjects are bilingual, the lower levels of the hierarchy are well established. They have a strong representation of the lexical items in both languages and of the translation of these items into articulatory patterns. What is learned, then, are the new concepts conveyed by these novel sentences. Since this level is abstract, the learning benefits can be seen in either the practiced or the unpracticed language.

Sports psychologists have long recognized the significance of hierarchical control of action. As part of their training, athletes are frequently taught to mentally rehearse movements. As with MacKay's transfer experiment, mental practice would be expected to benefit performance, so long as the practice can strengthen higher-level representations of the action. The skilled gymnast can improve by mentally simulating her routine, and the weekend golfer can benefit from watching a professional's videotape. Long hours of practice help to strengthen the lower levels of the hierarchy.

PHYSIOLOGICAL ANALYSIS OF MOTOR PATHWAYS

So far in this chapter we have stressed two critical points on movement: First, as with all complex domains, motor control depends on several distributed anatomical structures. Second, these distributed structures operate in a hierarchical fashion. We have seen that the concept of hierarchical organization also applies at the behavioral level of analysis. The highest levels of planning are best described by how an action achieves an objective; the lower levels of the motor hierarchy are dedicated to translating a goal into a movement. We now turn to the problem of relating structure to behavior: How can we best characterize the functional role of the different components of the motor system? In this section we will take a closer look at the neurophysiology of motor control. In the following section, we will address this same issue from the perspective of neurology.

The Neural Representation of Movement

Neurophysiologists have long puzzled over how best to describe cellular activity in the motor structures of the CNS.

Direct stimulation in primary motor cortex can produce discrete movements about single joints and has been useful for mapping the topography of the motor cortex. But this method does not provide insight to the activity of single neurons, nor can it be used to study how and when cells become active during volitional movement. To address these issues, we have to record the activity of single cells and ask what parameters of movement are coded by that activity. For example, is cellular activity correlated with movement direction or movement location?

DIRECTIONAL TUNING OF MOTOR CELLS

Apostolos Georgopoulos (1995) of the Veteran's Administration Medical Center in Minneapolis studied this question by recording from cells in the motor cortex of rhesus monkeys. The monkeys were trained with the apparatus shown in Figure 11.17, moving a lever to one of several targets. The animal initiates the trial by moving the lever to a designated starting position. After a brief hold period, a light comes on at a target position

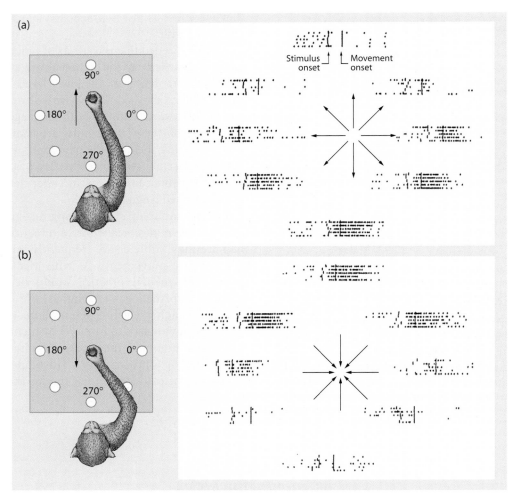

Figure 11.17 Motor cortex activity is correlated with movement direction. **(a)** The animal was trained to move a lever from the center location to one of eight locations. The activity of a motor cortex neuron is plotted next to each target location. Each row represents a single movement and the dots correspond to action potentials. **(b)** Movements originated at the eight peripheral locations and always terminated at the center location. The activity for the neuron is now plotted next to the starting locations. The neuron is most active for movements in the downward direction, regardless of starting and final location. Adapted from Georgopoulos (1990).

and the animal moves the lever to this position to obtain a food reward. This movement is similar to a reaching action and usually involves rotating two joints, the shoulder and the elbow.

The results of these studies convincingly demonstrate that movement direction provides a far superior correlate of motor cortex activity than does final location. Figure 11.17a shows a neuron's activity when movements were initiated from a center location to eight radial locations. This cell was most strongly activated when the movement was toward the animal. Figure 11.17b shows results from the same cell when movements were initiated at radial locations and always ended at the center position. In this condition, the cell was most active for movements initiated from the most distant position; movement was again toward the animal.

From such data, one could argue that motor cortex cells code movement direction and that movement in a certain direction requires activating appropriate cells. Indeed, cells within a cortical column have a common directional tuning, with that direction changing in a systematic manner across the cortex. As indicated in Figure 11.17, though, the cells' directional tuning is broad—there is significantly more activity in four of the eight directions.

Since the tuning is so broad, we should consider an alternative coding scheme. A command to move in a direction is distributed across all cells devoted to a certain limb. The response of each cell is a function of how closely the target direction corresponds to its preferred direction. If the target and preferred direction are identical, the cell will have a maximal increase in responsiveness. If the target and preferred direction are in opposite directions, the cell

Population Vectors and Prosthetic Limbs

The population vector reveals a tight correlation between the summed activity of a sample of motor cortex neurons and the movement of the corresponding limb. We can predict with high precision the direction of a movement by looking at the firing patterns of the neurons. Could these neural signals be used to drive a movement directly, bypassing the intermediate stage of muscles?

At first glance, this question seems nonsensical. What would it mean to produce movement without muscles? But this is exactly the kind of problem that scientists designing robots have faced for many years. All sorts of engineering solutions have been devised for robot control. Most of these have taken little inspiration from the study of human motor control. However, a group of scientists led by John Chapin at the Hahnemann School of Medicine in Philadelphia recently showed that a robot arm can be controlled on-line by the population vector (Chapin et al., 1999).

Chapin trained rats to use a lever to obtain drops of water. A computer measured the pressure on the lever and used this signal to adjust the position of a robot arm. By varying the pressure, the animals could replenish the water and then bring it to their mouth for a drink. Once the animals were trained, up to forty-six neurons, mostly in the primary motor cortex, were recorded from simultaneously. Population vectors during the lever presses were calculated and, similar to the monkey studies, showed a strong correlation between the actual movement and the summed activity of the task-related neurons.

It was at this point that the robot control was initiated. The connection between the lever and robot arm was disengaged. In its place, the electrical signals driving the robot arm were controlled directly in real time by the population vector. If the vector indicated a movement increasing the pressure on the lever, then the robot arm moved accordingly. Amazingly, population vectors generated from as few as twenty-five neurons proved sufficient for the rats to successfully control the robot arm to obtain water.

But what about the rat's paw? If the motor cortex is active, doesn't the paw continue to press? During the first phase of robotic control, the animals indeed did continue to generate pressing-like movements. Over time, however, these movements became sporadic, occurring only after trials in which the population vector failed to result in the water reward being delivered. The animals had learned that simply thinking about the correct movement was sufficient to move the robot arm.

These results promise new directions in the development of prosthetic devices such as artificial arms. While it might seem like an idea out of a Frankenstein movie, one can imagine control schemes in which amputees control their artificial limbs by simply thinking about the desired movement. The neural record of this process could be used to guide the limb. Of course, there are potential pitfalls in going directly from thought to action. When watching a diner at the next table eat an ice cream sundae, you might wish you could have a taste. And you might get what you wished for—at the cost of great social embarrassment!

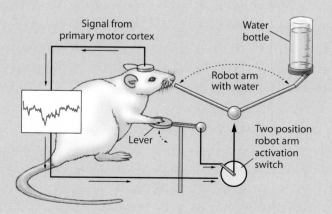

will have a maximal decrease. When the two axes are not parallel, the responsiveness is a function of their difference.

This coding scheme—the summed activity over all of the cells—turns out to be the best predictor of movement direction. Georgopoulos and his colleagues called it the **population vector** because it is a way to see how a global event, a movement in a certain direction, can result from the summed activity of many small elements, each con-

tributing its own vote (Figure 11.18). The population vector is powerful for analyzing motor neurophysiology. For example, once an experimenter has plotted the tuning curves for cells, the population vector for any movement closely agrees with the trajectory adopted by the animal—an analysis that holds for movement in three-dimensional space.

It is essential to factor time into the directional representation of movement. In the motor cortex, the popula-

Figure 11.18 The population vector provides a cortical representation of movement. The activity of a single neuron in the motor cortex is measured for each of the eight movements **(a)** and plotted as a tuning profile **(b)**. The preferred direction for this neuron is 180 degrees, the leftward movement. **(c)** Each neuron's contribution to a particular movement can be plotted as a vector. The direction of the vector is always plotted as the neuron's preferred direction, and the length corresponds to its firing rate for the target direction. The population vector (dotted line) is the sum of the individual vectors. **(d)** For each direction, the solid lines are the individual vectors for each of 241 motor cortex neurons; the dotted line is the population vector calculated over the entire set of neurons. While many neurons are active during each movement, the summed activity closely corresponds to the actual movements. Adapted from Georgopoulos (1990).

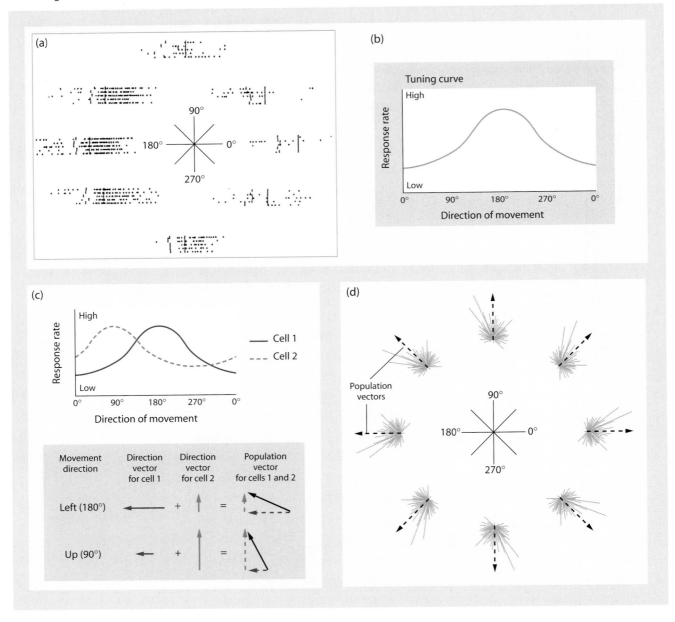

tion vector shifts in the direction of the upcoming movement before the movement is produced. This indicates that the cells help to plan the movement rather than simply reflect changes during a movement. To dissociate movement planning and movement execution, Georgopolous provided two signals on each trial: The first indicated the movement's target direction; the second served as a "go" signal indicating that the movement could now be initiated. Thus, the animals were trained to prepare the movement in advance so that they could initiate their responses as fast as possible. Figure 11.19 shows that, expressed as a population vector, cells in the motor cortex coded the intended direction well in advance of the "go" signal. The movement's direction could be precisely predicted when the population vector was recorded more than 300 msec prior to the movement. Such results clearly indicate that the motor cortex plans movement, and they also demonstrate that cellular activity in this area does not automatically lead to movement. Some downstream process must regulate when the activity translates into motor commands.

We have viewed the population vector in the context of how it corresponds to the direction of movement. Bear in mind that an alternative way of considering these representations is in relation to specific muscles; that is, move-

Figure 11.19 The direction of the population vector predicts the direction of a forthcoming movement. At the cue, one of the eight targets is illuminated, indicating the direction for a subsequent movement. The animal must refrain from moving until the "go" signal. The population vector was calculated every 20 msec. The population vector is oriented in the direction for the planned movement even though EMG activity is silent in the muscles. Adapted from Georgopoulos (1990).

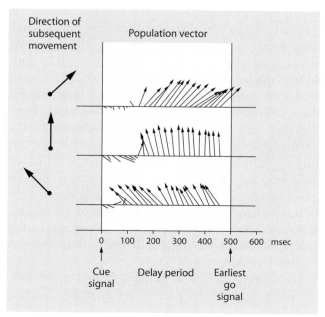

ments to the right from the center position occur because the monkey is extending the elbow and rotating the shoulder. The population vector could represent a pattern of activation for the muscles participating in these joint actions. Consider the situation in which a wrist rotation is required to move the hand from a starting position to a location closer to the body. If the arm is in its natural position with the hand facing inward, the movement requires wrist flexion. When the wrist is rotated, the same direction of movement requires wrist extension. By varying the starting position, muscle involvement and movement direction can be dissociated. When single-cell recordings are made in such experiments, both types of codes are observed in primary motor cortex (Kakei et al., 1999). The activity of some cells was tied to specific patterns of muscle activity, independent of direction. The activity of other cells was correlated with movement direction, independent of the muscle activity.

GOAL-BASED REPRESENTATION IN MOTOR STRUCTURES

Population analyses have been performed in other neural regions including the basal ganglia, cerebellum, and premotor cortex (Alexander and Crutcher, 1990). As with the motor cortex, the population vector in these other areas correlates with movement direction, as might be expected since all these areas contain topographic maps of the body surface and effectors. Yet the fact that a common basis of representation is coded across large areas of the cortex and subcortex is puzzling. From a hierarchical viewpoint, we might expect the areas to contribute to movement in distinct ways. For example, there might be an initial representation of the final target location; this location code would be transformed into a directional code.

One neurophysiological study showed how important the final location is in the cellular activity of single cells (Ashe et al., 1993). In this experiment, monkeys were trained to make a two-step movement by pushing a lever forward and then to the left (Figure 11.20). When the population vector was analyzed prior to the onset of movement, the initial shift was, surprisingly, to the second component. Only afterward did it point in the direction of the first component. It appears, then, that the motor plan unfolded in the reverse order of the actual movements—the initial representation was based on the final location. Thus, even with a direction-based representation, the time course of the population vector reveals the influence of the movement goal.

Goal-based representations also have proved useful in analyzing single-unit activity in the premotor cortex. Although cells in this area may still code the movement's direction, they are not activated uniformly in movements

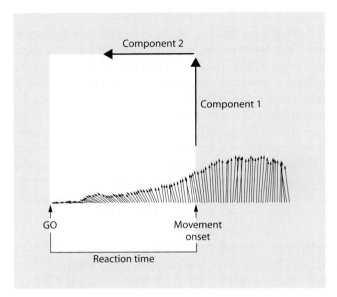

Figure 11.20 The population vector for a two-step movement. The forthcoming movement is shown by the two arrows. The initial direction of the population vector is toward the final component of the movement. It gradually changes so that by the start of movement, it corresponds to the direction of the first component. Adapted from Ashe et al. (1993).

requiring similar muscular events. Rather, the cells prefer certain motor acts. Giacomo Rizzolatti of the University of Parma, Italy, proposed that these neurons form a basic vocabulary of motor acts (Rizzolatti et al., 1988). Some cells are preferentially activated when the animal reaches for an object with its hand; others become active when the animal makes the same gesture to hold the object; and still others, when the animal attempts to tear the object, a behavior that might find its roots in the wild where monkeys break off tree leaves. Therefore, cellular activity in this area reflects not only the trajectories of a movement but also the context in which the movement occurs.

These contextual considerations remind us that motor gestures are not made in an abstract space. Most actions are designed to manipulate an object in space; that is, motor systems are powerful for interacting with the environment, for avoiding obstacles, and for obtaining rewarding stimuli. Such actions require that motor systems be constantly informed about the environment's layout to anticipate an action's consequences. Goal-based action depends on synthesizing sensory and motor information. Our skill in reaching for a desired object requires that we integrate perceptual information with what we know about the current state of the motor system.

Many cells in motor areas represent sensory and motor information. This is true not only for neurons in the parietal cortex but also for neurons in motor areas of the frontal cortex and in subcortical structures such as the basal ganglia, cerebellum, and superior colliculus. For example, cells in area 6 of the premotor cortex respond to visual stimuli. As with cells in visual areas of the cortex, these neurons have a receptive field: They respond to a visual stimulus only when it is presented in a certain region of space.

These receptive fields are dynamic. They depend on how the animal has positioned its arm in space and how the limb might interact with a seen object (Graziano and Gross, 1994). Consider the cell's behavior depicted in Figure 11.21. The receptive field shifts with the position of the arm. When the limb is placed in front of the animal, this cell will become active when a visual stimulus is presented in the vicinity of the hand. If the limb is moved to the side, the receptive field shifts to this new location. A visual stimulus presented at the initial location no longer produces a change in the activity of the cell. If the limb is positioned out of the animal's sight, the cell will not respond to any visual stimulus. This enables the animal to use visual information to coordinate actions. The cell can represent the position of the visual stimulus with respect to the limb's current position. Perceptual knowledge has been transformed from a representation based on external space to one based on internal space. The animal represents where an object is located in the three-dimensional world and the movement that would be required to reach this object.

When we think of the motor areas of the cortex, the focus is usually on areas in front of the central sulcus, the motor cortex, and secondary motor areas such as premotor cortex and the supplementary motor area. However, it is also clear that parts of the parietal cortex are also essential for coordinated movement. In Chapter 6 we saw that the dorsal visual stream plays an important role in visually controlled movements. Movement-related cells can be found in posterior parietal cortex. Interestingly, these cells provide a location-based code. For example, such cells will respond prior to movements to a particular location independent of the starting position of the hand. Similar to the cell described in the preceding paragraph, this location representation is based on an internal reference frame. The internal coordinates here are defined by the position of the animal's eyes (Batista et al., 1999). In this way, the action goal is defined with respect to where the animal is currently looking.

Behavioral studies in humans further underscore the importance of the parietal cortex in representing

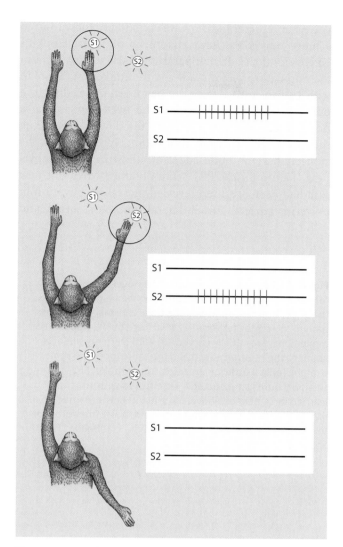

Figure 11.21 The activity of this neuron in the basal ganglia is dependent on the position of the hand. This neuron will respond to a visual stimulus falling within the circled region. When the hand is out of sight, the neuron does not respond to visual stimuli. Activity in this cell dynamically shifts as a function of hand position, and would be useful for coordinating visually guided movements. Adapted from Graziano and Gross (1994).

the spatial goal for movement. Consider a particularly diabolic experiment in which the target changes position just as the person begins to reach for it. People are quite accurate under such conditions, adjusting the trajectory of an ongoing movement within 100 msec or so in order to smoothly end up near the target. These corrections can occur outside awareness, likely utilizing the same processes involved in normal reaching where, as discussed previously, an initial trajectory plan is combined with a final location code to ensure that the target is reached in an efficient manner. When TMS is applied over the posterior parietal cortex, a selective disruption of the location code is observed. The

trajectory of the reach is unaffected, but the person fails to make the required adjustment after the target changes location (Figure 11.22). Interestingly, patients with posterior parietal lesions can move to the new position following a target displacement, but their trajectory adjustments require at least 450 msec (Pisella et al., 2000). It is as if they no longer can use an automatic location code to ensure the movement goal is achieved. Instead, they use a slow, conscious process to plan a change in the direction code.

In thinking about how goals influence actions, it becomes clear that most movements are not simply gestures

Figure 11.22 Transcranial magnetic stimulation (TMS) over the posterior parietal cortex (PPC) disrupts rapid corrections at the end of reaching movements. **(a):** Subjects reach toward an illuminated target presented at either the left, the center, or the right box. On some trials, the target jumps from the center location to one of the sides. The subjects are unaware of the target displacement because it occurs when they move their eyes just prior to the onset of the reach. **(b)** Movement paths for one subject when reaching with the right hand. On the left are trials without TMS. The solid traces show no displacement trials to the three locations. The dotted lines show the movement paths when the target jumped to the left or right. On the right are trials in which TMS was applied over the left hemisphere PPC. Note that the reach terminates at the center position, the initial target location on all displacement trials. From Desmurget et al. (1999).

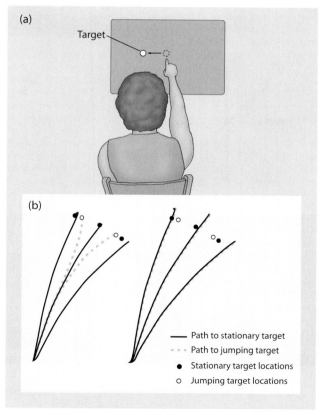

in space. They are designed to interact with the environment, and these interactions usually involve an object or tool. We reach to pick up our coffee cup or place our hand on the mouse by the keyboard. Once the object or tool is grasped, skillful manipulation is required. With such movements, we need to learn the relationship between a movement and the consequences of that movement. For example, it takes a few minutes to adjust to the clutch of a car we are not used to driving. Or the mouse on a new computer may be especially sensitive. Researchers have devised nonstandard input-output relationships as a way to identify neural structures in-

volved in representing how we learn to use these tools to achieve our goals. Reza Shadmehr of Johns Hopkins University devised an apparatus similar to that used in the population vector studies with monkeys, a device in which a two-jointed lever can be moved in any direction on a two-dimensional plane. The twist here is that a novel force field was applied to the device, perturbing the movements in a radial direction. With practice, the subjects became quite adept at adjusting to the novel field (Figure 11.23). PET scans obtained when this new skill had been acquired revealed increased activation in posterior parietal cortex, lateral premotor cortex, and

(a)

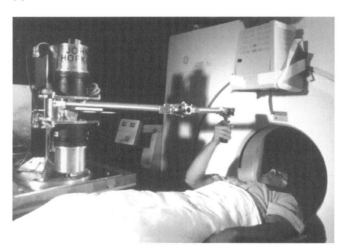

Figure 11.23 Learning to move in a novel environment. **(a)** Subjects moved the robot arm to one of eight locations. **(b)** Under normal conditions, the trajectory takes a straight path to the target. **(c)** When a radial force field is applied, the movements are initially curved. **(d)** With learning, subjects adapt to the new force field and the trajectories are again straight. **(e)** In a positron emission tomography (PET) scanning session 5 hours later, subjects were tested with either the learned force field or a novel force field. For the learned environment, activity was greater in three regions: premotor cortex, posterior parietal cortex, and anterior cerebellum. From Shadmehr and Holcomb (1997).

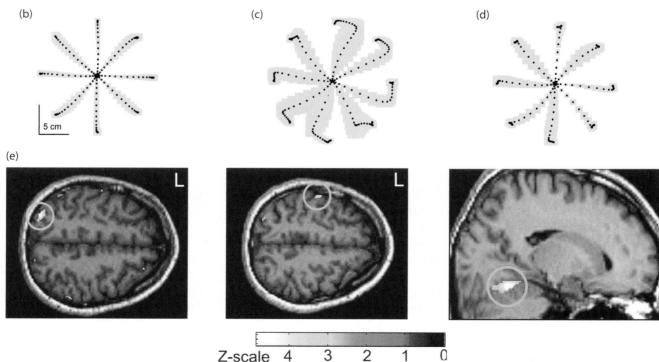

cerebellum (Shadmehr and Holcomb, 1997). Later in this chapter we will see that these three areas form a network involved in visually guided movements.

As these last paragraphs make clear, defining where perception ends in the brain and action starts may be an impossible task. Perceptual systems have evolved to support action; likewise, actions are produced in anticipation of sensory consequences. For a monkey in the wild, seeing a ripe banana on a tree engages the action systems to retrieve the food, actions designed to initiate and carry out the skillful climb among the branches and result in the satisfying taste of the fruit. The links between perception and action extend beyond this continuous cycle. The very act of perception frequently entails the activation of motor systems. This has been shown with a variety of techniques. Neurophysiologists have identified "mirror cells" in the premotor cortex of mon-

keys, cells that respond not only when the animal produces a particular movement but also when it perceives another animal making the same gesture (Rizzolatti et al., 2000). Brain stimulation in Broca's area disrupts both the production and the perception of speech sounds (Ojemann, 1983). Imaging studies consistently find similar patterns of activity in premotor, prefrontal, and parietal areas when people produce actions, imagine producing the actions, or observe others making the actions. Findings such as these have led to various "motor theories of perception," the idea that comprehension is achieved by activating the neural systems that would have produced the percept. Extending what was discussed in Chapter 6 concerning the relationship of perception and knowledge, we now see that the representations of actions are similarly interwoven into an overlapping network.

COMPARISON OF MOTOR PLANNING AND EXECUTION

Per Roland (1993) and his colleagues at the Karolinska Hospital in Stockholm, Sweden, used the single-photon emission computed tomography (SPECT) methodology to explore the functional organization of the motor pathways. SPECT is sometimes labeled the "poor man's PET" in that detection of radioactive decay is limited to a single two-dimensional field. Construction of the three-dimensional pattern of regional cerebral blood flow (rCBF) requires multiple scans, usually acquired at the expense of spatial resolution. Nonetheless, this method is useful when the areas are large.

Figure 11.24 summarizes areas of activation during a variety of motor tasks. When subjects repeatedly flexed their index fingers, significant increases in rCBF were restricted to the primary motor and somatosensory cortices in the contralateral hemisphere. These foci of activation were similarly present when the task required a complex sequence of finger movements. In addition, the sequencing task led to greater blood flow in the supplementary motor area and prefrontal cortex, as well as in the basal ganglia and ipsilateral cerebellum. Somewhat surprisingly, metabolic increases in the supplementary motor and prefrontal cortices were not restricted to the contralateral hemisphere, which projects to the moving fingers; these areas also were activated in the ipsilateral hemisphere. In the final condition, subjects imagined the sequencing task but did not produce any real movement (or muscle activation). For this task, activation increases were restricted to the supplemen-

tary motor area, and again this activation was bilateral. There was no reliable change in rCBF over the motor cortex.

Taken together, these results are in accord with a hierarchical control system. Simple movements require minimal processing, and as such, changes in rCBF are limited to primary motor and sensory areas. With greater complexity, cortical areas anterior to the primary motor area become activated. In addition, activation in these anterior areas occurs over both hemispheres. This bilateral activation can be interpreted in several ways. It might correspond to the activation of an abstract motor plan—one not tied to a specific effector. Or the bilateral distribution could reflect many potential motor plans, each viable candidates for achieving a common goal. For example, we could reach for a water glass with either the left or the right hand. With either of these hypotheses, the unilateral activation in the basal ganglia and motor cortex can be interpreted as the transformation of higher-level plans into a movement linked to specific effectors.

Because PET requires averaging metabolic activity over a 40-second period, it is not possible to observe the time course of activation through the motor pathways. However, with event-related functional magnetic resonance imaging (fMRI), we can examine when different neural regions become engaged as a person prepares and executes a response. In one such study (Chee et al., 1999), the subjects first were presented with a visual cue

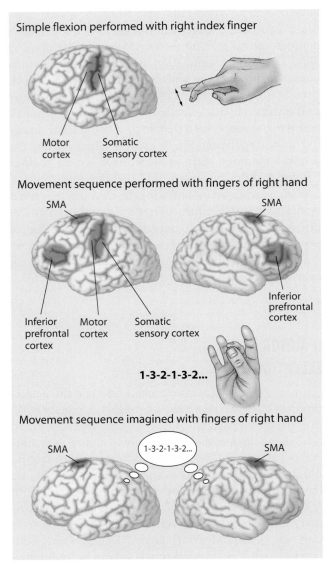

Simple flexion performed with right index finger

Motor cortex   Somatic sensory cortex

Movement sequence performed with fingers of right hand

SMA   SMA

Inferior prefrontal cortex

Inferior prefrontal cortex   Motor cortex   Somatic sensory cortex

1-3-2-1-3-2...

Movement sequence imagined with fingers of right hand

SMA   1-3-2-1-3-2...   SMA

Figure 11.24 Areas of metabolic activity associated with a variety of motor tasks. Blood flow increases were restricted to primary motor and somatic sensory cortical regions in the contralateral hemisphere during simple flexion and extension of the index finger on the right hand. When the subjects were asked to perform a complicated series of sequential finger movements with the right hand, blood flow increases also were observed bilaterally in the supplementary motor area (SMA) and the prefrontal regions. The supplementary motor area was also active, bilaterally, when the sequence was mentally rehearsed. During this imagery condition, no increases were present in the primary motor cortex. Adapted from Roland (1993).

indicating whether a forthcoming movement involving tapping the thumb and finger should be made with the right hand or left hand. Six seconds later, the word "GO" appeared and the subjects were to execute the response as quickly as possible. As shown in Figure 11.25, this task

produced an increase in the blood oxygen level dependent (BOLD) signal in many cortical areas, with the changes again occurring in premotor regions in both hemispheres. Most notably, the increase in the supplementary motor area and lateral premotor cortex is found after initial presentation of the cue, when planning begins. For primary motor cortex, the increase is locked to the time of movement execution. Even within the supplementary motor area, activity sweeps from the anterior to posterior direction as planning gives way to execution. Similar results have been reported in single-unit recordings within the supplementary motor area, leading researchers to propose that this area is actually composed of a number of functionally distinct motor areas.

A dramatic demonstration of the differential role of the primary motor cortex and supplementary motor area in motor planning and execution was shown with TMS (Gerloff et al., 1997). Subjects were trained to produce a complex sequence of finger movements in a continuous fashion. When TMS was applied over the area of the motor cortex that represents the hand, the next response in the sequence was disturbed frequently, either because the movement was halted in midstream or because the wrong key was pressed. Moreover, the subjects perceived the problem as a temporary loss of coordination, commenting that the finger suddenly seemed to jerk in the wrong direction. In contrast, when the coil was placed to target the supplementary motor area, the effects were delayed, occurring about three key presses after the TMS pulse. Here, the subjects reported that they lost track of their place in the sequence, that they temporarily forgot the order of key presses. TMS over the supplementary motor area appears to disrupt the goal of the movement. Without this representation of the action plan, one can know that something is not quite right but cannot pinpoint the problem. With an error of execution such as those seen following motor cortex stimulation, the exact form of the error can be articulated since the actual response can be compared to the intended response.

Internal Versus External Guidance of Movement

We have stressed that as a movement becomes more complex, structures beyond the primary motor cortex are activated. But how can we characterize the contribution of these areas? Is it just a matter of complexity, or are these additional contributions tied to certain tasks?

As we noted earlier, the supplementary motor area is critical in controlling sequences of finger taps. Perhaps other types of sequential actions require different types of processing. For example, to return a tennis shot, a

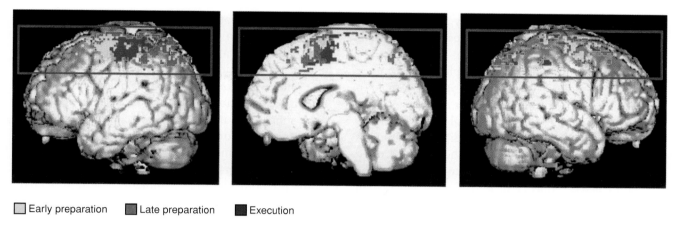

☐ Early preparation   ◼ Late preparation   ◼ Execution

Figure 11.25 Recruitment of brain areas during movement preparation and execution identified with event-related functional magnetic resonance imaging (fMRI). A cue indicating which hand to use was presented 6 seconds prior to the visual "GO" signal. Activation time locked to the cue in the frontal lobe is initially more anterior. Activation time locked to the "GO" signal is centered around primary motor cortex. Results are only shown for right-hand movements. From Lee et al. (1999).

player adjusts her posture and swings according to the velocity and angle of the ball and her opponent's position. Here, the action sequence for a successful return requires a rapid integration of external cues.

Some neuroscientists propose that the control of movement sequences depends on area 6, but that it varies as a function of whether the action is internally or externally guided. The medial portion, the supplementary motor area, will dominate when the task is internally guided, as in Roland's finger sequencing task. In contrast, the lateral portion, the premotor area, becomes more relevant when the task depends on external cues. This area is activated when people are asked to perform movements under the guidance of visual, auditory, or somatosensory feedback. It also is activated when extrapersonal sources of feedback are removed, but the action is performed with respect to an external frame of reference such as when tracing a path through a maze with the eyes closed. Even though the eyes are closed during the scanning sessions, visual feedback comes into play in the training session, and it is likely that the subjects use imagery.

Figure 11.26 provides an overview of the anatomical basis for the internal-external control hypothesis (Goldberg, 1985). The premotor area is innervated extensively by the parietal cortex and the cerebellum, two areas linked not only to visuomotor function but also with rich representations extracted from other sensory channels like audition and touch. The supplementary motor area, by contrast, receives extensive projections from the prefrontal cortex and basal ganglia. Due to their anatomical connections, these areas are in a position to allow limbic structures to convey information related to the animal's current motivational state and internal goals.

Physiological data in primate research support the hypothesized link of externally and internally directed actions to premotor and supplementary motor areas, respectively. Lesions of the supplementary motor area severely disrupt monkeys' ability to perform learned gestures without visual guidance. Lesions of the premotor cortex have no effect on performance. In contrast, the reverse pattern is observed when the monkeys are trained

Figure 11.26 Movements may vary in terms of the contribution of internal and external sources of information. The external loop, including the cerebellum, parietal lobe, and lateral premotor cortex (PMC), dominates during visually guided movements. The internal loop, including the basal ganglia and supplementary motor area (SMA), dominate during self-guided, well-learned movements.

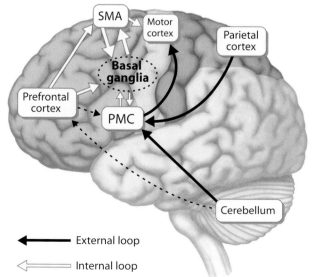

to move in the direction signaled by an external cue. Here, premotor lesions produce an impairment while ablation of the supplementary motor area has no effect.

To compare the roles of the supplementary motor area and the premotor cortex, monkeys were trained to generate a sequence of button presses according to two sets of instructions (Mushiake et al., 1991) (Figure 11.27). In the external-guidance condition, the buttons were illuminated in succession and the animal pressed each one in turn. In the internal-guidance condition, the animal learned a sequence of button presses and then reproduced the sequence after hearing an auditory "go" signal. Physiological recordings were made in the supplementary motor area and the premotor cortex.

This study produced two critical results. First, in both areas, some neural activity could not be linked to specific movements but depended on the entire sequence of movements; hence, cellular activity correlated with the direction of movement but only when that gesture was embedded in a certain sequence. When the same gesture was embedded in a different sequence, the activity of these cells showed only a slight increase compared to their baseline firing rate. This sequence specificity supports the hypothesis of a higher form of control for area 6. The other critical result is that supplementary motor area neurons were disproportionately active during internally guided movements, whereas premotor neurons was more active during externally guided sequences. Subsequent work showed a similar dissociation

in subcortical motor structures (Middleton and Strick, 2000). Activity in the basal ganglia and cerebellar neurons was linked more closely with internally generated and externally guided movements, respectively.

The distinction between internal and external control, however, is not absolute. Many cells in both cortical and subcortical motor areas were active under both types of conditions. And some supplementary motor area cells were more active in the external condition than in the internal condition, while the reverse held for some premotor neurons. Although these results might raise doubts about a strict dichotomy, it is also likely that movements in both conditions depend on external and internal cues. Consider the tennis player again. The decision to return a shot with a deep lob or a short dropshot depends not only on the location and velocity of the arriving ball but also on knowing the opponent's skills. If the opponent moves slowly, the dropshot might be preferred. Coordinated actions are guided by multiple sources of information.

Shift in Cortical Control with Learning

Learning plays a critical role in producing purposeful actions. We can describe the processes required to serve a tennis ball, but this description is likely to be different for the novice and skilled performer. The novice may verbalize each movement as a monk chants his mantra: "Toss the ball straight up, swing the racquet in an arc,

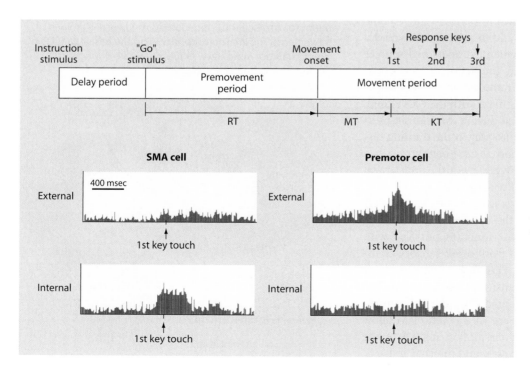

Figure 11.27 Monkeys were trained to make sequential button presses, either to a series of illuminated targets (external condition) or from memory (internal condition). Despite the fact that the movements were identical, the supplementary motor area (SMA) neuron was most active during the internal condition and the premotor neuron was most active during the external condition. Adapted from Mushiake et al. (1991).

make contact as the ball begins to descend, follow through." Through years of practice, the production of this sequence appears effortless to the expert. She ponders the finer aspects of the serve. Is her top-spin serve working today? Should she serve to her opponent's forehand or backhand?

Feedback after the service is treated differently by the novice and expert. After impact, the novice may have little idea of where the ball is heading—in the service box, he hopes, but more likely into the net. Feedback from observing his shot helps in modifying the next service. While the expert is also concerned with the consequences of her actions, she uses feedback obtained during the course of the service to make fine adjustments. If the arm is moving too fast, the wrist can be flexed earlier to ensure an appropriate trajectory. Whereas a novice essentially uses feedback to make changes in subsequent actions, an expert uses feedback to make on-line changes in performance.

Motor systems invoked in the action of serving a tennis ball clearly differ for the novice and the expert. Indeed, learning considerations have led to an alternative conceptualization of the functional difference between the lateral and the medial premotor areas. Rather than thinking these differences reflect whether a task is guided by external or internal cues, we could consider that learning involves a transition toward greater reliance on more medial motor structures. During the early stages of learning, the lateral premotor area, with its extensive connections to the parietal cortex, would be optimal for integrating external information with a motor plan—an essential job for developing any new skill. If we learn the mapping between the movements we produce and the effects they produce, we can anticipate the consequences of our actions, exploiting the functions of the more medial motor system.

A PET study explored metabolic changes associated with skill learning through trial and error (Jenkins et al., 1994). After hearing a tone, the subjects pressed one of four buttons. Correct responses were signaled by a high-pitched tone; incorrect ones were followed by silence. The subjects' task was to learn an eight-element pattern that was repeated cyclically. In the initial stages of training, the subjects simply had to guess the sequence. After six cycles, the subjects were averaging only one error per cycle. After one sequence was learned, the process was repeated with a new sequence. In this way, comparisons could be made between performance on new and performance on learned sequences without confounding the order of the scans with skill level. If only one sequence had been used, the earlier scan would always reflect initial learning.

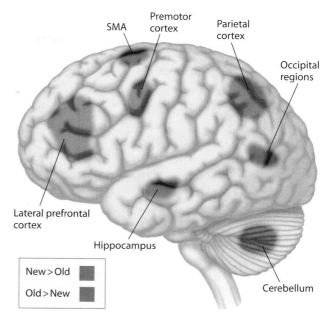

Figure 11.28 Shifts in metabolic activity during motor learning. PET scans were obtained under two conditions: while subjects performed a well-learned movement sequence (Old) and during the course of learning of a movement sequence (New). Learning was associated with blood flow increases in lateral premotor and prefrontal areas; in contrast, performance of previously learned sequences was correlated with blood flow increases in the supplementary motor area (SMA) and hippocampus. Adapted from Jenkins et al. (1994).

The metabolic results, shown in Figure 11.28, indicate that as learning progressed, activation shifted within area 6. The lateral premotor area was more activated during the acquisition of a new sequence in comparison to when an old sequence was being performed skillfully. In contrast, the supplementary motor area was more active during the skilled phase. Indeed, when the subjects began to learn a new sequence, activity declined in the supplementary motor area.

This study further revealed that the lateral and medial regions are embedded in distinct circuits related to skill acquisition. Higher metabolism in the lateral premotor cortex was correlated with increases in the superior parietal lobe and cerebellum. Changes in supplementary motor area activity were associated with changes in the temporal lobe and limbic structures.

Remember that there is usually a high degree of correspondence between the level of learning and the extent to which a behavior depends on external cues. If prompted to serve when blindfolded, the expert tennis player most likely will succeed on a high percentage of serves. The novice would be lucky to make contact with the ball. In a similar sense, we can see a shift in the reliance on external

An Interview with Steven Keele, Ph.D.

Dr. Keele is associated with the Psychology Department at the University of Oregon. His research helped bring the study of motor control into the mainstream of cognitive psychology.

Authors: The topic of motor control doesn't even get discussed in many textbooks on cognitive psychology. Why do you think this is so? Do the problems of motor control only become interesting when we consider the brain?

SK: It is not entirely clear why motor control has been so neglected, not only in textbooks of psychology, but even by psychologists themselves. The omission was noted by the famous learning theorist E.R. Guthrie in 1935, when he commented that psychology leaves an organism stranded in thought, unable to engage in action.

The basic neglect may derive from implicit but faulty assumptions that psychologists have had about motor control. They often think that motor control is an add-on, something beyond cognition itself. They often think as though the problems of motor control are simply problems of stringing together a series of motor actions, actions being defined in terms of movements and the muscles that power them. This prejudice reveals itself in many ways. Some scientists view the learning or control of a series of actions as being relatively primitive compared to such notable human achievements as declarative memory, the ability to reason, or language. This has led some to look for the engrams of motor learning in brain regions purely outside the cortex—in the basal ganglia or in the cerebellum—or within motor cortex. Indeed, this tendency has been reinforced by a belief that those brain systems are for motor control and motor control only, although the belief is beginning to come into serious question.

Authors: What do you see as a fundamental question for researchers in motor control in terms of cognitive function? For example, how does the study of coordination lend insights into other cognitive domains?

SK: Almost 50 years ago, the great neuropsychologist Karl Lashley pointed out a remarkable fact about human skill that should disabuse one of the notion that motor control and motor skill are less "human" than other remarkable capacities of our species. Humans speak, they type, they sign, they write—each an intricate motor skill. In the domain of music, people play the fiddle, they dance to it, and they may sing or hum along. People build cabinets, knit, and blow fine glassware. These diverse motor activities are beyond the realm of other animals, and suggest that motor capabilities are related to other intellectual capabilities.

Indeed, some psychologists, such as Jerome Bruner, have suggested that even human language capability is an outgrowth of capacities evolved to create new motor sequences. Consider this thought exercise. Imagine a person born deaf and dumb and totally without language. Could such a person learn to select boards, measure and saw each to size, nail them together to build a rafter, and attach the rafter to a house? Given this, might it be possible for the same person, given a capacity to hear and to speak, to attach symbols—that is, words—to each component of the action and substrate—a board, a saw, sawing, a hammer, a nail, nailing. Might it be possible for the words to be assembled in some manner approximating the action order itself? From this perspective, it might well be imagined that expressions

cues and feedback in just about any skill domain. Children have to assess carefully the outcome of each stroke as they learn to write their signature. With skill, handwriting suffers little when visual feedback is removed.

In summary, results from physiological studies using both single-cell recordings with primates and brain imaging techniques suggest that parallel circuits may be involved in motor planning. One circuit, including the parietal lobe, lateral premotor, and cerebellar pathways, is essential for producing spatially directed or guided movements. These movements are likely to dominate during the early stages of skill acquisition. A second circuit, associated with the supplementary motor area, basal ganglia, and perhaps the temporal lobe, becomes more dominant as the skill is well learned and driven by an internal representation of the desired action.

Both circuits converge on the motor cortex, the primary link between the cortex and limbs for voluntary

of language are themselves parasitic on more fundamental achievements in humans, the ability to creatively arrange actions into new skills.

Authors: Are you saying that there is a general-purpose sequencing module that is used in both language and action? How does this mesh with the belief that language involves a highly modular system?

SK: Yes, the general idea is that brain mechanisms responsible for action sequencing also participate in language—in a sense, a general-purpose module. This kind of argument has been forcefully stated by Patricia Greenfield (1991), who argues that brain areas around Broca's region are not only responsible for aspects of sequence in speech and language but in sequences of action in general. She points at the parallel development of hierarchical organization during infancy both in language and in action and at instances of brain injury in which lesions produce a loss in both domains. She points out that apes show the same progression of sophistication in language and in action, both peaking substantially below the capability of humans.

Greenfield's hypothesis contrasts with a more classic view of modularity in which a complex system, such as the speech perception system, makes use of a set of modules, with each module supplying a particular function. In this classic view, the modules operate in a very limited domain. In contrast, the view I'm discussing suggests that many functions are shareable among a variety of tasks. Paul Rozin developed this idea over 25 years ago, suggesting that the ability of new tasks to draw on computations that had evolved for other task domains was at the heart of human intelligence.

Authors: What about the unique aspects of high-level skills such as those exhibited by Michael Jordan on the basketball court or Issac Stern in the concert hall? Do you see the unique gifts shown by these individuals as reflecting general cognitive principles or are these people born with a special gift? For example, if we want to understand what makes Michael Jordan so special, should we study his muscles and physiology or does his expertise in basketball have something to tell us about learning at the cognitive level?

SK: Especially for sporting skills, it is unlikely that the highest levels of performance can be achieved without having special physical gifts that allow extraordinary bursts of speed or jumping ability. But at the same time, certain "cognitive" capabilities also are likely to underlie the development of extraordinary skill. Extensive evidence suggests that knowledge acquired as a result of extensive practice—thousands of hours of highly dedicated practice—is the key factor separating the most successful people in various motor and non-motor skill domains from the rest of us. This perspective grew initially out of analyses of chess expertise, but also has been found to apply to musical performance and basketball.

I'm not familiar with Michael Jordan's history of practice and knowledge, but some years ago I read a marvelous biography of Bill Russell, one of the most gifted basketball centers of all time and a defensive wizard. I recall three stories he told that helped me realize that the most talented of skilled performers often are so because of acquired knowledge and "cognitive" analysis. First, he and his college teammate, K.C. Jones, learned to steal the ball by working as a tandem, one distracting an opponent while the other would come in from the blind side to steal the ball. Second, he analyzed the geometry of ball flight and his own jumping and reaching to determine how best to block shots. Third, by imaging the moves of an opponent, he could learn to anticipate these moves during a game.

The surprising idea that stands out in the expertise literature is that the extraordinary motor capabilities of humans are best understood as an extension of their extraordinary cognitive capabilities.

movements. A subject of intense study in the motor learning literature centers on the question of whether representations within the motor cortex change as a function of practice. At one extreme is the idea that the motor cortex provides links to the muscles but plays a minimal role in the actual representation of actions. This hypothesis is consistent with the notion that learning occurs at an abstract level corresponding to action goals rather than movements per se. At the other extreme is the idea that motor learning is linked intimately to the muscles—perhaps captured best by the old Bell Telephone jingle, "Let your fingers do the walking."

Avi Karni and his colleagues (1995) at the National Institute of Mental Health examined this question with fMRI by training subjects on a simple motor sequencing task in which the thumb successively touched each of the four fingers. Two sequences were used, and they differed in the order the fingers were touched. The subjects were instructed to practice one sequence for 15 minutes each day over several weeks; the other was performed only during the scanning sessions. By the end of training, the subjects had doubled the speed with which they could complete the trained sequence. This learning was sequence specific; improvement on the other sequence was minimal. During the scanning sessions, the subjects were required to produce one movement every half second when performing either sequence, in order to equate movement speed. Despite this control, the metabolic activity was greater in the primary motor cortex when the subjects were producing the learned sequence (Figure 11.29).

These results suggest that learning-related changes can be seen in primary motor cortex. However, it is important to keep in mind the limitations of neuroimaging. The analysis in this study was restricted to the motor cortex. It is reasonable to assume that changes also were occurring in other motor areas. Thus, the increased activity in the primary motor cortex might not mean that the representations within this area have changed. It is possible that upstream areas such as the supplementary motor cortex had learned the sequential representation and were priming the motor cortex. Indeed, a priming effect might be especially pronounced here given that the subjects were forced to slow down during the imaging runs. As with all imaging studies, it is difficult to infer function from hemodynamic changes in a particular area. One result favoring the hypothesis that learning produced

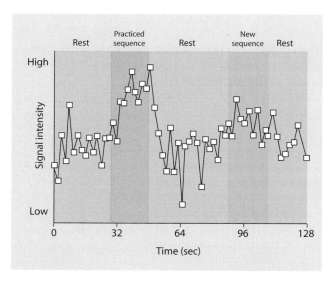

Figure 11.29 Increased activation in the primary motor cortex following extensive practice of a finger movement sequence. Each box indicates the signal strength from a single pixel in motor cortex, either during rest or during the performance of a well-learned (Old) or novel (New) finger sequence. Adapted from Karni et al. (1995).

changes directly in the motor cortex is that the extent of activation increased with learning. It appears that learning led to reorganization within the motor cortex such that a larger cortical area was recruited for the learned sequence. In Chapter 15 we will see how single-cell studies can provide a clearer picture on how experience modifies cortical representations.

FUNCTIONAL ANALYSIS OF THE MOTOR SYSTEM AND MOVEMENT DISORDERS

So far we have emphasized how motor areas might contribute to motor control according to the demands of the task. If an action requires the coordination of sensory and motor codes, the lateral, premotor loop will dominate. If sensory requirements are reduced, either by varying instructions or by practice, control shifts toward the medial, supplementary motor area loop.

What is lacking in this task-based analysis is an explanation of the computations performed by the different motor areas. For example, the neurophysiological work on population coding is an elegant tool for describing how neurons represent forthcoming actions. But several cortical and subcortical areas share this form of representation. One might consider this homogeneity and conclude that

motor systems should be viewed as distributed networks having little specialization.

Neurologists would dismiss such a hypothesis as utter foolishness. An hour in a movement disorders clinic would make it clear that the motor system is extensively specialized. Breakdowns in coordination can happen in many ways; such patterns of deficits afford clues to the contributions of the components of the motor pathways. Lesions' characteristic signatures enable neurologists to pinpoint the site and cause of pathology even in the absence of high-resolution MRI.

Before we turn to motor disorders, let's consider the computations required to perform a skilled action such as playing a difficult piano passage from Beethoven's *Ninth Symphony* (Figure 11.30). For one

Possible mental operations/computations	Example: playing piano
1. Select	Match fingers to keys, notes.
2. Sequence	Group notes into a phrase.
3. Force	Strike accented notes with greater force.
4. Timing	Establish rhythm.

Figure 11.30 Hypothesized mental operations that must be completed in order to perform a complicated action such as playing the piano.

thing, the pianist must figure out the mapping between notes and fingers. A single finger might suffice for the simplest melody line, but a skillful production requires all ten fingers. And as the context changes, the same note can be played by different fingers. Moreover, the notes must be linked in proper sequence. The popular nursery tunes "Go Tell Aunt Rhody" and "Lightly Row" share identical notes, but the sequencing of notes creates distinct tunes. Finally, for a piece to sound musical, the notes must be played with the proper timing and intensity.

Hypothetical components such as selection, sequencing, timing, and force can serve as useful heuris-

tics for analyzing disorders of the motor system. For example, if motor timing and motor selection are independent operations, then we might expect to find dissociations in the clinical problems of patients with motor disorders. A patient might select the appropriate fingers to play the notes of "Go Tell Aunt Rhody" but fail to provide the appropriate timing. Or the timing might be correct, but the notes out of order. In either case, we would fail to recognize this familiar tune. But the loss of coordination would reflect deficits in different component operations. In this way, we can develop a functional analysis of the cognitive neuroscience of motor control.

Cortical Areas

Some brain theorists would consider the entire cortex anterior to the central sulcus to be part of the brain's motor centers. And yet, many animals without a cerebral cortex are capable of complex actions. The fly can land with near-perfect precision or the lizard can flick its tongue at the precise moment required to snare its evening meal. Has evolution rendered a massive reorganization of the motor pathways? Or, should we consider the cortex as an additional piece of neural machinery superimposed on a more primitive apparatus?

HEMIPLEGIA

The primary motor cortex provides the most important signal for the production of skilled movement. This area receives input from almost all cortical areas implicated in motor control, including the parietal, premotor, and frontal cortices, and from subcortical structures such as the basal ganglia and cerebellum. In turn, the motor cortex's output constitutes the largest signal in the corticospinal tract—not to mention its indirect influence on spinal mechanisms by way of projections to the extrapyramidal pathway.

The preeminent status of the motor cortex is underscored by the fact that lesions to this area usually result in **hemiplegia,** the loss of voluntary movements on the contralateral side of the body. Hemiplegia most frequently results from a hemorrhage in the middle cerebral artery. It is perhaps the most telling symptom that a person has experienced a stroke. A person might wake up with a severe headache or experience a sudden loss of consciousness, and notice a complete loss of control in one limb. It is not a matter of will or awareness. The hemiplegic patient may exert great effort, but his limbs will not move. The loss of movement is most evident in the distal effectors.

Reflexes are absent immediately after a stroke that produces hemiplegia; within a couple of weeks the reflexes return and even become hyperactive. The hemiplegic patient presents a greater-than-normal reflexive response when a muscle is passively stretched. The muscles may even become spastic, reflecting the increased tone, or responsiveness, of the muscles. The spasticity is most pronounced in muscles that counteract the forces of gravity when we stand up or sit upright, leaving the patient with a contorted posture where the arm is maintained in a flexed position at the elbow with the fingers tightly curled into the palm. The legs become hyperextended, like pillars holding up the trunk.

Hyperactive reflexes and spasticity can be understood within a hierarchical framework. Muscular activity offsetting gravitational forces and reflexes comes from a simple form of motor control, a mechanism for ensuring postural stability. The cortex provides a way to inhibit these mechanisms. By inhibiting a tendency to maintain stability, the cortex allows us to willfully impose an action. We can move our arm without having the stretch reflex counteract the gesture. When the cortical influence is removed, primitive reflexive mechanisms become manifest again.

Recovery from hemiplegia is minimal. When the motor cortex has been damaged, the patient does not recover the use of the limbs on the contralateral side. Movement is possible, but only when carried out by gross movements that do not require the independent control and coordination of different joints. If a leg is affected, the patient may be able to walk again, but in a far-from-normal manner, perhaps by rotating the entire leg about the pelvis. The patient cannot coordinate flexion and extension about the knee and ankle. Arm movements also would be quite limited—any persisting control would be left to proximal joints like the shoulder and perhaps the elbow. It is unlikely that extensive rehabilitation training would help the patient to recover use of the fingers. At best, rehabilitation could reduce spasticity and thereby maximize the effectiveness of whatever volitional control the patient has retained.

Lesions of the motor cortex affect the corticospinal and extrapyramidal tracts. The loss of control over individual joints and distal extremities is attributed to damage in the cortico-spinal tract. Spasticity and hyperreflexivity are attributed to changes in the extrapyramidal system. When the cortical influence is removed, these more ancient systems become dominant, exerting their prominent role in maintaining postural stability. The hemiplegic patient no longer has the flexibility to generate an action based on internal goals and desires.

APRAXIA

Many cortical lesions result in coordination deficits that cannot be attributed to hemiplegia, motoric problems of weakness, sensory loss, or motivation. This syndrome is called **apraxia.** While the term *apraxia* technically means "no action," it is used in a more general sense to describe a loss of motor skill. For example, a patient with bilateral lesions of the parietal lobes was unable to continue her work as a fish filleter. When attempting to perform a routine that she had completed thousands of times, she correctly inserted the knife point into the head of the fish, began the first stroke, but then stopped. She claimed to know how the action should be completed but could not execute it. At home, she found herself putting the sugar bowl in the refrigerator or the coffee pot in the oven. In each case, she retained the ability to move her muscles but could no longer link gestures to a coherent act or recognize the appropriate use of an object.

The diagnosis of apraxia is based mainly on exclusionary grounds: A patient is said to have apraxia if he or she has a coordination problem that cannot be linked to a deficit in controlling the muscles themselves. Moreover, it is not always clear that the problems are specific to the motor system. They may reflect a general problem in comprehension. Apraxia is observed much more frequently in patients with lesions to the left hemisphere, and almost all apraxia patients have some aphasia (Keretsz and Hooper, 1982). Perhaps the motor problems represent a failure to understand instructions or a general loss of semantic knowledge. If we forget that sugar is best stored at room temperature, it is not so surprising that when cleaning up after a meal, a person might put the sugar bowl in the refrigerator. However, it is also true that many aphasia patients do not have apraxia.

Several standard tests for apraxia have been developed. These tests generally involve asking patients to produce goal-directed gestures. These gestures may consist of arbitrary behaviors such as saluting or following a sequence of arm gestures. Or they may be actions where the patient manipulates an external object such as lighting a pipe or slicing a loaf of bread. Apraxic patients are most impaired when trying to pantomime an action (Figure 11.31). For example, when requested to slice a loaf of bread, the patient may ball his hand in a fist and pound the table. If he is given a knife and a loaf of bread, his performance generally improves. Now the tool can be held appropriately, even if slicing is abnormal.

Neurologists distinguish between two general subtypes of apraxia, ideomotor and ideational. In

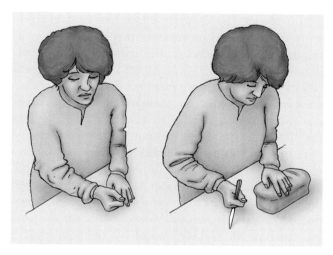

Figure 11.31 Apraxic patients are unable to produce coordinated actions, even though their strength is intact and they can move their limbs. The deficits are most pronounced when they are asked to pantomime actions. If given the object, they may succeed in producing the action, although with some clumsiness or inappropriate gestures.

ideomotor apraxia, the patient appears to have a rough sense of what the desired action is, but has problems in executing the action properly. If asked to pantomime how to comb his hair, the patient might knock his fist against his head repeatedly. **Ideational apraxia** is much more severe. Here, the patient's knowledge about the intent of an action is disrupted. She may no longer comprehend the appropriate use for a tool. For example, one patient used a comb to brush his teeth, demonstrating by the action that he could make the proper gesture but with the wrong object.

At the beginning of the twentieth century, one of Wernicke's students, Hugo Liepmann, made two key observations regarding apraxia: The disorder was generally associated with lesions of the left hemisphere and was most frequently observed when the lesions included the parietal cortex (see Heilman et al., 1982). Liepmann proposed that this area is the critical region for the control of complex movement (Figure 11.32). In his view, the representation of a desired action would originate in area 40 of the left parietal cortex. Output from this area would be projected to the frontal cortex and, through various cortico-cortical connections, would activate area 4. If the desired movement were executed with the contralateral right hand, the processing would remain within the left hemisphere. And if the desired movement were executed with the ipsilateral left hand, that information would transfer to the right hemisphere through frontal connections over the corpus callosum. In this manner,

unilateral lesions of the left parietal region would be associated with bilateral deficits.

In support of this model, Kenneth Heilman and his colleagues (1982) at the University of Florida reasoned that if the left parietal cortex contains the memory of actions, then damage to this area would impair performance on tests of either the production or the perception of gestures. That is, if representations of learned actions are destroyed, patients would find it difficult to produce movements and to recognize these movements when made by healthy individuals. To test this, two groups of apraxic patients were identified: those with anterior lesions and those with posterior lesions. On standard tests of apraxia, the two groups were indistinguishable. Patients in both groups made frequent errors when asked to mime gestures such as opening a door with a key or flipping a coin.

Figure 11.32 Liepmann's model of the neural regions associated with the production of skilled actions. The premotor areas of the contralateral hemisphere are essential for skilled movements of limbs. These areas receive input from the parietal lobe of the left hemisphere, an area assumed to store the representations of the actions. Thus, a lesion in this posterior region will lead to apraxic movements with both contralesional and ipsilesional limbs.

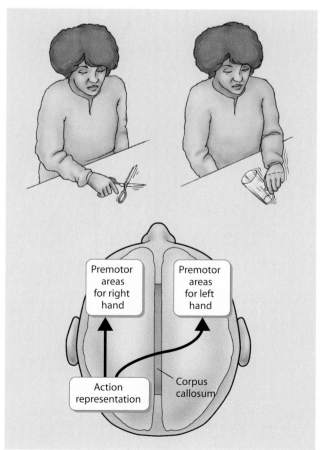

For the perception tests, the patients were shown short film segments in which an actor mimed these gestures. In one test, the actor performed three actions and the patient was asked to identify a target gesture; an example is given in Figure 11.33. For a second test, the actor performed the same task three times, but only one of the productions was well executed. In both tests, the patients with posterior lesions made significantly more errors than the patients with anterior lesions. The perceptual deficit for the posterior-lesion group could not be attributed to a general comprehension deficit. Patients with posterior lesions producing comparable aphasic problems but without apraxia did not have a problem on gesture comprehension tests.

To sum up, experimental studies of apraxia have provided support for the general picture outlined in Liep-mann's model. The left parietal lobe may contain representations of skilled movements, perhaps reflecting the fact that skilled movements involve dynamic transformations in space (Haaland et al., 1999). When the representations are damaged, the patient has difficulty generating these actions or recognizing them when produced by others. The implementation of an action, however, requires that the representations be transformed into specific motor plans. This involves additional processing by premotor and prefrontal areas involved in motor control. Damage to these areas also can produce apraxia but will not disrupt the ability to perceive such actions when produced by another individual. Viewed in this way, the parietal cortex is the highest level of the motor hierarchy; frontal areas are a means for translating an action memory into a pattern of muscular commands.

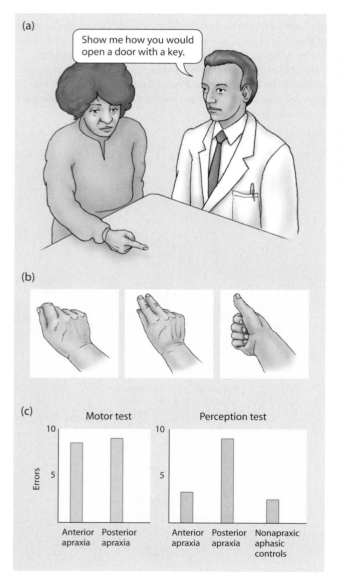

Figure 11.33 Patients with apraxia may not be able to recognize skilled movements. **(a)** In the motor test, the patient is asked to pantomime a gesture such as using a key to open a door. **(b)** In the perception test, the patient views an actor pantomiming an action in three different ways, only one of which is appropriate. The patient must choose which action is correct. **(c)** Patients with either anterior or posterior lesions who produced apraxic gestures on the motor task were selected. Only the patients with posterior lesions showed impairment on the perception test. The apraxic patients with anterior lesions performed as well as nonapraxic, aphasic control subjects on the perception test. Adapted from Heilman et al. (1982).

THE CORTEX AND THE SELECTION OF MOVEMENT

We can hypothesize that cortical processing for motor control is concerned primarily with selecting appropriate effectors for performing a movement. In planning an action, the initial representation is abstract—a goal of what the person hopes to achieve. For example, your goal might be to drink some coffee or ask a friend to join you at the movies—goals that can be achieved in many ways. Your right or left hand could be used to reach for the coffee cup or a pencil on which to write your friend a memo. In picking up these objects, you might grasp them with all your fingers or use a prehensile pinch of the thumb and index finger.

Translating a goal into movement might be best envisioned as a competitive process, an idea captured in Figure 11.34. The goal activates all movements that could produce the desired action. These candidates then compete over which movement will be selected (Rosenbaum et al., 1991). The competition can be driven by many sources, some of which reflect our internal states and past experience. For example, activating movements with the dominant hand may be achieved more efficiently than activating movements with the nondominant hand. Other sources can be external ones. Preference may be given to the hand closer to the cup or pencil because this movement requires less effort. Or, in an experiment, the task instructions might ask the subject to use one hand or the other. The motor cortex can be viewed as the final tallying point of the competitive process, given that it provides the primary motor signal from the cortex. As such, it represents a movement chosen to achieve the desired goal. If the motor cortex is lesioned, the selected movement cannot be implemented to activate the effectors.

The cortical selection hypothesis agrees with evidence that motor planning is a distributed process involving many neural regions. All these areas promote movement, but with their relative contribution varying as a function of the task demands. The supplementary motor area reflects the contribution of internal sources of activation—goals and motivational states—whereas the lateral premotor area is driven more strongly by external sources such as the position of the effectors and objects that might be manipulated. A task is likely to encompass both sources—we should not expect only one area to be activated. The differences will be quantitative rather than qualitative.

Neurological lesions can disturb the balance between different inputs. In the acute stage after a stroke, patients with lesions of the supplementary motor area

Figure 11.34 Response selection involves a competitive process between potential responses. The writer wishes to take a swig from his coffee cup. This goal can be achieved by reaching with the left or right hand, either from the current position or by rotating the trunk. Various sources of information—the location of the cup, what the hands are currently doing, experience—help resolve the competition.

reach out and grasp objects with the affected arm, even when they have not been instructed to do so or have been told to refrain from moving. This *alien hand syndrome* reflects a dominance of externally guided lateral premotor pathways. The sight of an object within reaching distance evokes a motor plan to grasp the object. We usually can inhibit movement if we are instructed to do so or if the movement is inappropriate. But when internal control sources are removed, the movement can be triggered by the appropriate external stimulus. The normal balance underlying movement selection is distorted.

If we view the process as competitive, we can understand why, under certain conditions, competing motor plans get activated. Consider an experiment by Steven Wise of the National Institute of Mental Health. Monkeys were trained to move a handle within one of two conditions: a compatible one and an incompatible one (Wise et al., 1996). In both conditions, a stimulus would appear at one of eight positions on a circle. For the compatible condition, the monkeys were trained to move the handle to the location of the stimulus; for the incompatible condition, the required movement was in the opposite direction of the stimulus. Population vectors were recorded in the lateral premotor cortex. In the compatible condition, the vector immediately pointed toward the forthcoming movement; in the incompatible one, though, the vector pointed first toward the stimulus—that is, the direction away from the forthcoming movement—and with more processing time, the vector

Patting Your Head While Rubbing Your Stomach

Recall the childhood challenge to pat your head while rubbing your stomach. This seemingly innocent task proved extremely frustrating. It is nearly impossible to generate the conflicting spatial trajectories, having one hand move in an up-down fashion while the other produces a circular movement. The two movements compete. We fail to map one direction for one hand and the other direction for the opposite hand. Eventually, one of the movements dominates and we end up rubbing the head or patting the stomach. Within the selection hypothesis outlined in this chapter, we can think of this bimanual conflict as competition between two movement goals. Each task activates both hemispheres and we cannot keep the cross talk from these activation patterns from interfering with one another.

If this hypothesis is correct, spatial interference should be eliminated when each movement goal is restricted to a single hemisphere and the pathways connecting the two hemispheres are severed. To test this idea, Elizabeth Franz and her colleagues (1996) at the University of California, Berkeley tested a patient who had undergone resection of the corpus callosum. The stimuli for their bimanual movement study were a pair of three-sided figures, with the sides following either a common axis or perpendicular axes. The stimuli were projected briefly, with one stimulus presented in the left visual field and the other in the right visual field. After viewing the stimuli, the subjects were instructed to produce the two patterns simultaneously, using the left hand for the pattern projected in the left visual field and the right hand for the pattern in the right visual field. The brief presentation was used to ensure that each stimulus was isolated to a single hemisphere in the split-brain patient. In control subjects, rapid transfer of information should occur via the corpus callosum.

Control subjects had little difficulty producing bilateral movements when the segments of the squares followed a common axis of movement. Yet when the segments required movements along perpendicular axes, their performance dramatically deteriorated, with long pauses before each segment and trajectories frequently deviating from the target—something you can demonstrate to yourself by trying this task.

In contrast, the split-brain patient did not significantly differ between the two movements. He initiated and completed movements in the two conditions with comparable speed, and the movements were accurate in both. Indeed, in a second experiment, this patient simultaneously drew a square with the left hand and a circle with the right hand. Each hemisphere produced the pattern without any signs of interference from demands presented to the opposite hemisphere.

These results indicate that the callosotomy procedure yields a spatial uncoupling in bimanual movements. As striking, it was also apparent that even for the split-brain patient, the actions of the two hands were not independent of one another. As with the control subjects, the two hands moved in synchrony. The segments of the squares were initiated and terminated at approximately the same time. This temporal coupling was seen more clearly when subjects were asked to produce oscillatory movements, with each hand moving along a single axis. Regardless of whether the two hands followed a common axis (e.g., both horizontal or both vertical) or perpendicular axes (e.g., one horizontal and the other vertical), the two hands reversed direction at the same time.

This study provided valuable insights to the neural structures underlying bimanual coordination. First, the spatial goals for bimanual movements are coordinated via processing across the corpus callosum. When a task requires conflicting directions of movement, interference is extensive as long as the callosal connections are intact. Second, these connections are not necessary for temporal coupling of movement. It may be that a single hemisphere regulates when movements are initiated or that the initiation is controlled by subcortical mechanisms. Third, the dissociation of spatial and temporal coupling emphasizes a distributed view of how the motor system's neural structures contribute to coordination. The neural structures that represent the spatial goals are separate from those involved in initiating the movements selected to meet these goals.

Bimanual movements following resection of the corpus callosum. While looking at a central fixation point, subjects were briefly shown the two patterns. They were instructed to simultaneously draw the pattern on the left with the left hand and the one on the right with the right hand. Normal subjects **(left)** were able to draw the mirror symmetrical patterns but had severe difficulty when the orientation of the two figures differed by 90 degrees. The split-brain patient **(right)** performed equally well in both conditions. Adapted from Franz et al. (1996).

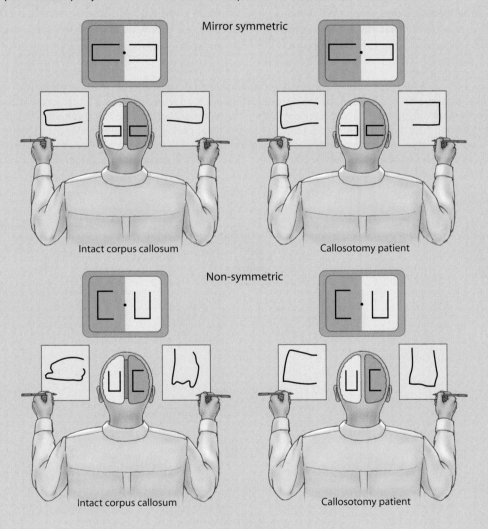

shifted toward the movement (Figure 11.35). This shift comes from two competing movement plans. Because we usually move to a stimulus's location, a target's onset automatically activates a motor plan toward that location. Because of their training, though, the monkeys learned that they must move in the opposite direction. As input from this internal set became dominant, the population vector's direction shifted.

The hypothesis of widespread competition among motor plans fits in with the activation patterns observed in premotor and motor cortices. As noted, imaging studies show that premotor activation is generally bilateral. This is surprising given that in most studies all of the movements were performed with a single hand. While it is possible that these areas represent both sides of the body, the bilateral activation may reflect the planning of all possible movements that could achieve an abstract goal. Activation eventually becomes lateralized over the contralateral motor cortex—the presumed outcome of a competition for one hand over the other. The transition of bilateral activation over premotor areas to unilateral activation over the motor cortex also is seen in measurements of evoked potentials (Figure 11.36).

The supplementary motor area is important in the selection process. While the primary source of output for the supplementary motor area is the ipsilateral motor cortex, this area also has callosal projections to the contralateral supplementary motor area and motor cortex. A good part of these crossed projections are inhibitory. In this way, activation of the supplementary motor area in one hemisphere promotes the selection

Figure 11.35 Physiological correlates of response competition. **(a)** A light comes on at a target position along the circumference of the circle. After a delay period, a "GO" signal indicates that the monkey should move a lever to the target location. The population vectors indicate the movement direction, becoming manifest during the delay period. **(b)** In the incompatible condition, the animal must move to the location opposite that signaled by the light. During the delay period, the population vector corresponds to a movement planned toward the light. The magnitude of this vector is less than that in the compatible condition, reflecting the simultaneous activation of the opposite movement. After the onset of the "GO" signal, the population vector orients in the direction of the (correct) movement. Adapted from Wise et al. (1996).

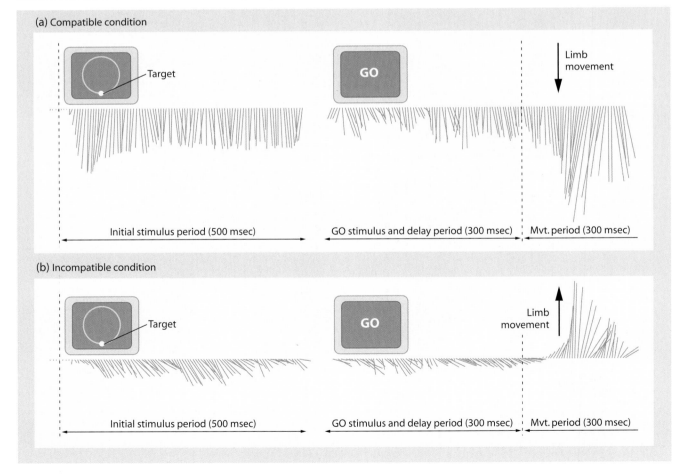

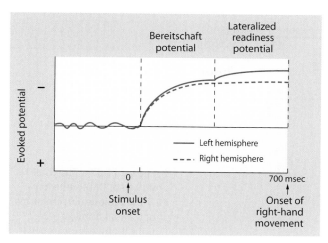

Figure 11.36 Evoked potentials recorded over the motor cortex before the right index finger presses a button. Following stimulus onset, the potential shows increased negativity, the Bereitschaft potential. This potential is observed bilaterally. During the last 300 msec of the reaction time period, the potential becomes larger over the contralateral hemisphere.

of a movement on the contralateral side and inhibits a similar movement on the ipsilateral side. This organization helps explain another unusual symptom associated with lesions of the supplementary motor area, the tendency to produce mirror movements. When a subject is asked to reach for an object with the hand contralateral to the lesion, the ipsilateral hand will make a similar gesture.

In most skilled behavior, we do not use just one hand or the other. Most motor tasks involve the coordinated use of both hands. The two hands may work in a similar fashion, as when we push a heavy object or row a boat. In other tasks, the two hands take on different, complementary roles, as when we open a jar or tie our shoes. The supplementary motor area also plays a central role in bimanual coordination. Damage to this area in both monkeys and humans can lead to impaired performance on tasks that require integrated use of the two hands, even though the individual gestures performed by either hand alone are unaffected (Wiesendanger et al., 1996). If a person is asked to pantomime opening a drawer with one hand to retrieve an object with the other, both hands may mime the opening gesture. Again, this deficit fits with the idea of a competitive process in which an abstract goal—retrieve an object from the drawer—is activated and a competition ensues to determine how the required movements are assigned to each hand. When the supplementary motor area is damaged, the assignment process is disrupted and execution fails, even though the person is still able to express the general goal.

Subcortical Areas: The Cerebellum and Basal Ganglia

In the preceding section we focused on how cortical motor areas select movement plans. The cortex is a vast processing network where input converges and helps an animal choose the action that achieves a goal. As such, the cortex is evolution's way of creating a flexible system. The animal need not slavishly respond to a stimulus but can modify its actions to meet current goals.

From this perspective, we can expect subcortical areas to participate less in selecting motor plans. Their role may be limited to ensuring that selected movements are executed efficiently. When the areas are lesioned, we would not expect representational disorders such as the ones observed in apraxia, a disorder that generates a wrong response. Instead, we would expect the response to be poorly executed.

This hypothesis accurately describes the motor disorders observed with subcortical lesions. In this section, then, we review the motor deficits of the primary subcortical structures associated with motor control: the cerebellum and basal ganglia. In examining these disorders, we develop functional hypotheses about how the structures contribute to the coordination of skilled action.

THE CEREBELLUM

Lesions of the cerebellum disrupt coordination in a variety of ways—a heterogeneity reflecting the distinct regions of the cerebellum. Figure 11.37 shows the three parts of the cerebellum: the vestibulocerebellum, the spinocerebellum, and the neocerebellum. This tripartite organization is reflected in their unique anatomical projections as well as by the fact that lesions within each area manifest as distinct clinical symptoms. Moreover, the three regions appear to have followed different courses in phylogeny. As the name implies, the neocerebellum has emerged most recently, coming after the development of the spinocerebellum and the more ancient vestibulocerebellum.

Anatomical and Functional Divisions of the Cerebellum
The vestibulocerebellum is innervated by the brainstem vestibular nuclei, and output is projected back to the same region. The vestibular system is essential for controlling balance and coordinating eye movements with body movements. Lesions of this area can affect the reflexes essential for balance and maintaining stability. For example, the vestibulo-ocular reflex (VOR) ensures that the eyes remain fixed on an object despite movements of the head or body. This is essential for an organism that

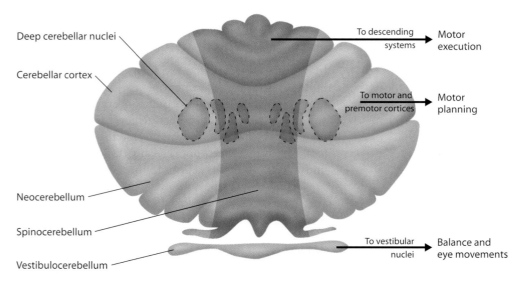

Figure 11.37 The three divisions of the cerebellum. These regions differ in terms of their anatomical projections to the deep cerebellar nuclei and extracerebellar target regions.

interacts with the environment: If the eyes were displaced with each movement, it would be difficult to monitor another organism or keep track of the location of a stimulus.

The main part of the cerebellum includes the spinocerebellum and the neocerebellum. As the name implies, the former receives extensive sensory information from the periphery via the spinal cord. In addition, cells in this area respond to auditory and visual stimuli, providing a basis for polysensory integration. This region developed later in vertebrate phylogeny than did the vestibulocerebellum. The emergence of the spinocerebellum was thought to correspond to the need for more precise and flexible control of the limbs used for locomotion. The output from the spinocerebellum originates in the medial cerebellar nuclei and innervates the spinal cord and nuclei of the extrapyramidal system. In addition, some of the fibers project to the motor cortex via the thalamus.

Lesions of the spinocerebellum create problems for the smooth control of movement. The medial part of the spinocerebellum, the vermis, is especially important for coordination involving axial muscles, those that control the body's trunk. Cells in this region are especially sensitive to the effects of alcohol. With chronic alcohol abuse, the vermis will atrophy, resulting in problems with balance and walking. Even with acute alcohol use, cerebellar signs can be observed: Tests used by police on suspected drunk drivers essentially record cerebellar function.

Damage to the spinocerebellum also will disrupt arm movements. These patients can select the appropriate action and initiate it in a normal manner, but their gestures will be clumsy, irregular, and erratic. Rapid pointing movements frequently extend beyond the target, a symptom referred to as **hypermetria.** Moreover,

the patient cannot smoothly terminate the movement. Rather, as the arm approaches the target, it goes into a series of oscillations (Hore et al., 1991).

The lateral zones of the cerebellar hemispheres constitute the neocerebellum. This area does not receive input from the spinal cord but is heavily innervated by descending fibers from the cerebral cortex originating from many regions in the parietal and frontal lobes. The output from the neocerebellum originates in the lateral deep nucleus, the dentate. While some afferents terminate on nuclei of the extrapyramidal system, much of the information ascends to the cortex via the thalamus, with the thalamic projections terminating in the primary motor, lateral premotor, and prefrontal cortices. As such, the neocerebellum contributes to the control of voluntary movements. This area has undergone tremendous expansion in primate evolution, reflecting the flexible and precise manner in which primates use their limbs. Indeed, in humans, the dentate nucleus contains over 90% of all neurons in the cerebellar nuclei.

Lesions of the neocerebellum produce symptoms similar to the ones in the intermediate cerebellum. Movements are clumsy (ataxic) and can be hypermetric, especially when an action consists of a sequence of gestures. Especially taxing can be movements that involve coordination across multiple joints such as in throwing a baseball. In addition, the initiation of movement is prolonged. The neocerebellum is involved in helping to plan movement while the intermediate cerebellum is essential in regulating the actual performance.

Interpreting Cerebellar Function Exactly how the cerebellum does these things remains unclear. One hypothesis is that the cerebellum has a special role in timing movement. While cortical areas primarily select the effectors needed to perform a task, the cerebellum provides the precise timing needed for activating these effectors. As such, cerebellar neurons also would code movement direction because the timing is part of planning a specific trajectory.

Consider how the timing hypothesis can account for the cerebellar symptom of hypermetria. As shown in Figure 11.38, rapid movements require both agonist and antagonist activity. While agonist activity is clearly needed to displace the limb, antagonist activity is, at least superficially, counterproductive. Why produce a force that counteracts a desired action? One argument

Figure 11.38 Arm movements and associated EMG activity during rapid elbow movements. **(a)** During rapid elbow flexion in a normal subject, the agonist, biceps muscle produces a burst to propel the limb. This movement is braked by the activation of the antagonist triceps. **(b)** The fine timing between the agonist-antagonist activity is disrupted in patients with cerebellar lesions. When the onset of the antagonist is delayed, the movement ends up overshooting the target because of the loss of the braking force. The oscillations seen at the end of pointing movements result from the alternation of agonist and antagonist activity.

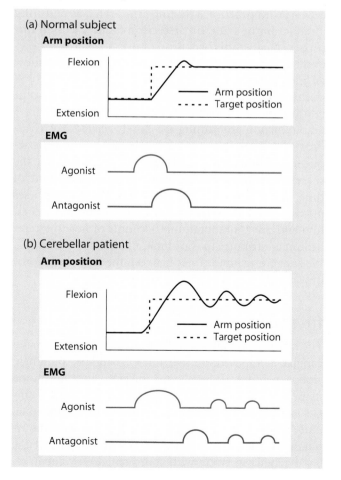

for the biphasic pattern is that it enables a person to move at fast speeds that are not reliant on sensory feedback. An initial agonist burst propels the effector rapidly in the correct direction with little opposing resistance. Then the antagonist brakes the movement. This anticipatory mode of control has a cost: The timing of the agonist and antagonist must be precise. In this view, the cerebellum's contribution to rapid movements establishes a temporal pattern across the muscles. The cerebellum can predict when the antagonist should be activated to terminate movement at the right location. If the cerebellum is lesioned, the agonist burst is sufficient to initiate the movement, but the animal cannot anticipate when the brake is needed. Other signals, such as feedback from the moving limb, may now be required to trigger the antagonist. Given delays in neural transmission, it is likely that the antagonist will be delayed and the animal will overshoot the target.

The timing hypothesis offers a novel slant on the role of the cerebellum in motor learning. Cerebellar lesions are most disruptive to highly practiced movements, which present the greatest need for precise timing. The novice tennis player may be pleased if he can simply get the ball over the net, and not care if his racquet strikes the ball a little early or a little late. But for the expert, it is essential that all gestures be exquisitely timed so the ball lands in a certain place on the court.

This point is highlighted in an experiment involving a simple model of motor learning, eye blink conditioning (see Chapter 4). When a puff of air is directed at the eye, a reflexive blink is produced, a response designed to minimize potential eye damage. If a neutral stimulus such as a tone is presented in advance of the air puff on a consistent basis, the animal comes to blink in response to the tone (Figure 11.39). The acquired response is highly adaptive: By blinking to the tone, the animal's eye closes before the air puff. This conditioned response is timed to reach maximal amplitude at the onset of the air puff.

Unlike lesions of the cerebral cortex or hippocampus, cerebellar lesions disrupt the acquisition and retention of a conditioned blinking response. The problem is not one of an impaired motor system, as eye blinks to the air puffs themselves are minimally affected by cerebellar lesions. As such, the selective disruption of the conditioned response is a sensorimotor learning problem, an idea in accord with the view that the cerebellum is an important part of the motor learning system. At a computational level, the timing hypothesis helps specify how the cerebellum contributes to motor learning. It is not only important for the animal to learn that the tone and air puff co-occur; the response is only adaptive if the animal learns that the tone predicts exactly when the

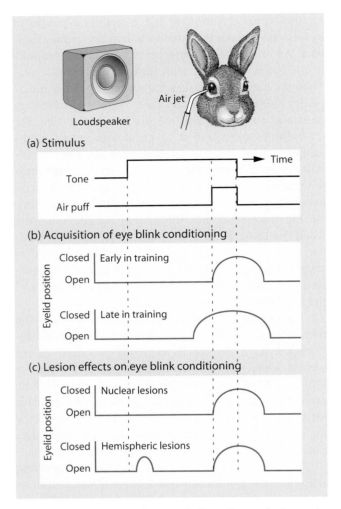

(a) Stimulus

(b) Acquisition of eye blink conditioning

(c) Lesion effects on eye blink conditioning

Figure 11.39 Lesions of the cerebellum disrupt the learned response in eye blink conditioning. **(a)** A neutral tone precedes and coterminates with an aversive air puff to the eye. **(b)** Early in training, the air puff causes the animal to blink. Late in training, the animal blinks in response to the tone, thus reducing the impact of the air puff. **(c)** Lesions of the deep cerebellar nuclei abolish the learned response. The fact that the animal continues to blink reflexively in response to the air puff indicates that the lesion has produced a learning deficit and not a motor deficit. The anticipatory, learned responses are still present following lesions of the cerebellar cortex. However, they are timed inappropriately and, thus, no longer adaptive.

air puff will occur. The animal must be able to represent the temporal relationship between the two stimuli. As shown in Figure 11.39c, rabbits with lesions in the cerebellar hemispheres continue to blink to the tone, but the response is no longer adaptive because it is not appropriately timed (Perrett et al., 1993). The eye is exposed at the time of the air puff.

The neocerebellum has undergone tremendous expansion in primate evolution, perhaps in parallel with the expansion of the prefrontal cortex. This has led researchers to consider possible nonmotor functions of this structure. One hypothesis is that the cerebellum's timing capabilities have come to be utilized by perceptual and cognitive systems (Ivry, 1993). For example, humans with cerebellar lesions are impaired on perceptual tasks that involve precise timing such as judging the duration of a sound. By this account, then, the cerebellum continues to perform a restricted computation—representing temporal information—but the computation is now available to other systems that require this information.

A second hypothesis, emphasizing the intimate links between perception and action, centers on the idea that the cerebellum generates predictions of expected sensory experience (Bower, 1997). Such expectancies would be important for motor control; consider the experience we all have when walking down a staircase and are surprised to find that one of the steps is missing. The correct anticipation of sensory events is also an important part of motor learning. The skilled tennis player can readily anticipate when a shot will be errant by the feel of the racquet striking the ball. Imaging studies of motor learning generally show that activation in the cerebellum decreases with practice, a finding interpreted as reflecting a reduction in error, or prediction failures as skill improves. Against this backdrop of overall decrease, however, are small areas within the cerebellum that show an increase in activity as a new skill develops. It is hypothesized that the focal increases result from the generation of new predictions (Imamizu et al., 2000).

Prediction and timing are closely related. For coordinated actions, it is not only important to know the sequence of limb movements but also essential that the successive movements be made at the right point of time in order to avoid errors. Thus, these hypotheses provide insight into nonmotor functions of the cerebellum and offer computational accounts of how the cerebellum contributes to coordination and motor learning. We see here again the links between the representational processes involved in perception and action.

THE BASAL GANGLIA

The other major subcortical motor structure is the basal ganglia, composed of five nuclei: caudate, putamen, globus pallidus, subthalamic nucleus, and substantia nigra (see Figure 11.5). These nuclei do not form a single anatomical entity; rather, they form a functional unity whose interconnected network of inputs and outputs is restricted. Moreover, lesions in any part of the basal ganglia interfere with coordinated movement, although the form of the problems will vary considerably with lesion location.

Before analyzing these disorders, we need to appreciate the complicated neuroanatomy of the basal ganglia, shown in Figure 11.40. All afferent fibers to the basal ganglia terminate in the caudate and putamen. Together, these nuclei are referred to as the *striatum*. The dominant input to the striatum is via the cortico-striatal projection, fibers that originate from the entire cerebral cortex including sensory, motor, and association cor-

tices. Efferent fibers from the basal ganglia are also restricted. Output from the basal ganglia originates in the internal segment of the globus pallidus (GPi) and the pars reticulata of the substantia nigra (SNr). SNr axons primarily terminate in the superior colliculus and provide a crucial signal for the initiation of eye movements. GPi axons, on the other hand, terminate in thalamic nuclei, which in turn project to the motor cortex, supplementary motor area, and prefrontal cortex.

In terms of physiology, processing within the basal ganglia is complex. Mahlon DeLong (1990) of Emory University distinguished between direct and indirect pathways: The direct pathway consists of direct inhibitory projections from the striatum to the GPi and SNr; the indirect pathway also connects the striatum to these output nuclei, but only through intervening processing stages involving the external segment of the globus pallidus and the subthalamic nucleus. All of the output signals from the basal ganglia are inhibitory. Neurons in the GPi and SNr have high baseline firing rates that inhibit target neurons in the superior colliculus and thalamus. The final pathway of note is the projection from the pars compacta of the substantia nigra to the striatum. Interestingly, this pathway has opposite effects on the direct and indirect pathways despite a common transmitter, dopamine. The nigra excites the direct pathway by acting on one type of dopamine receptor (D_1) and inhibits the indirect pathway by acting on a different type of dopamine receptor (D_2).

Figure 11.40 Wiring of the basal ganglia. Inputs from the cortex primarily project to the striatum. From here, processing flows along two pathways. The direct pathway goes to the output nuclei, the internal segment of the globus pallidus (GPi) and pars reticulata of the substantia nigra (SNr). The indirect pathway includes a circuit through the external segment of the globus pallidus (GPe) and the subthalamic nucleus (STN), and then to the output nuclei. The output projections to the thalamus are relayed to the cortex, frequently terminating close to the initial source of input. The dopaminergic projections of the pars compacta of the substantia nigra (SNc) modulate striatal activity by facilitating the direct pathway via the D_1 receptors and inhibiting the indirect pathway via the D_2 receptors. Adapted from Wichmann and DeLong (1996).

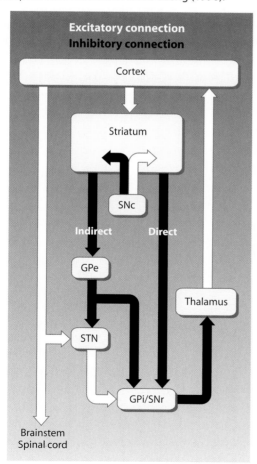

Functional Analysis of Basal Ganglia Function The functional consequences of the organization of the basal ganglia can be understood by tracing what happens when cortical fibers activate the striatum. Via the direct pathway, target neurons in the output nuclei of the basal ganglia are inhibited, leading to excitation of the thalamus and cortical motor areas. On the other hand, striatal activation along the indirect pathway results in increased excitation of the output nuclei, leading to increased inhibition of the cortex. It appears, then, that the direct and indirect pathways are at odds with one another; if processing along the indirect pathway is slower, however, the basal ganglia can act as a gatekeeper of cortical activity—less inhibition from the direct pathway is followed by more inhibition from the indirect pathway. The nigrostriatal fibers enhance the direct pathway while they reduce the effects of the indirect pathway. As such, this pathway is essential for promoting movement.

When seen in this way, the basal ganglia can be hypothesized to play a critical role in the initiation of actions (Figure 11.41). As we argued earlier in this chapter, processing in the cortical motor areas can be

Figure 11.41 The basal ganglia play a critical role in movement initiation. Potential responses are held in check until the basal ganglia provides a triggering response.

viewed as a competitive process in which candidate actions compete for control of the motor apparatus. The basal ganglia are positioned to help resolve the competition. The strong inhibitory baseline activity keeps the motor system in check, allowing cortical representations of possible movements to become activated without engaging the muscles. As a specific motor plan gains strength, the inhibitory signal is decreased for selected neurons.

Interestingly, computational analyses demonstrate that the physiology of the direct pathway in the basal ganglia is ideally designed to function as a winner-take-all system, a method for committing to one action from among the various alternatives. Greg Berns and Terry Sejnowski (1996) of the Salk Institute in La Jolla, California, evaluated the functional consequences of all pairwise connections of two synapses, either of which could be excitatory or inhibitory. As can be seen in Figure 11.42, a series of two successive inhibitory links is the most efficient way to make a selected pattern stand out from the background. With this linkage, the disinhibited signal stands out from a quiet background. In contrast, with a pair of excitatory connections, the selected pattern has to raise its signal above a loud background. Similarly, a combination of inhibitory and excitatory synapses in either order is not efficient in making the selected pattern distinct from the background. Berns and Sejnowski noted that the double inhibition of the direct pathway is relatively unique to the basal ganglia. This arrangement is useful for selecting a response in a competitive system.

Figure 11.42 The direct pathway in the basal ganglia consists of two, successive inhibitory links. The globus pallidus provides a strong, tonic inhibition on the thalamus, and in turn the cortex. The striatum sends a selective inhibitory signal onto this tonic inhibition, thus disinhibiting the selected response. The other pairwise combinations are less efficient in promoting a specific response. Adapted From Berns and Sejnowski (1996).

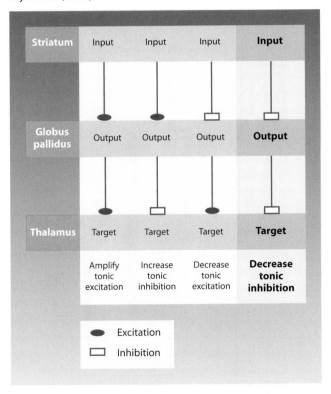

Striatum	Input	Input	Input	**Input**
Globus pallidus	Output	Output	Output	**Output**
Thalamus	Target	Target	Target	**Target**
	Amplify tonic excitation	Increase tonic inhibition	Decrease tonic excitation	**Decrease tonic inhibition**

● Excitation
□ Inhibition

Dopamine has long been known to be a critical transmitter in reward pathways. One can hypothesize that within the basal ganglia, dopamine serves to bias the system to produce certain responses over others. Dopamine release in the striatum follows successful actions—for example, a winning cross-court forehand in tennis. This transmitter might modify the input-output channels in the basal ganglia, making it more likely that the same response will be initiated when the rewarded input pattern is reactivated in the future.

Disorders of the Basal Ganglia Several neurological disorders affect the basal ganglia; in these we see a breakdown in the delicate balance between postural stability and movement. We will concentrate on two, Huntington's disease and Parkinson's disease, since they best demonstrate how movement is disrupted by basal ganglia dysfunction. While psychiatric conditions such as Tourette's syndrome and obsessive compulsive disorder have been linked to the basal ganglia, the neuroanatomical picture is murkier.

Huntington's disease is a progressive degenerative disorder that appears in the fourth or fifth decade of life. The onset is subtle, usually a gradual change in mental attitude where the patient is irritable, is absentminded, and loses interest in normal activities. Within a year movement abnormalities are noticed: clumsiness, balance problems, and a general restlessness. These involuntary movements, or *chorea,* gradually dominate normal motor function. The patient may adopt contorted postures, with the arms, legs, trunk, and head in constant motion. Indeed, during the seventeenth century, on at least two continents, Huntington's disease patients were executed for witchcraft because their chorea made them appear possessed by an evil spirit.

Neurological deficits in Huntington's disease are not restricted to motor function. As motor problems worsen, patients develop dementia of a subcortical type, unaccompanied by apraxia, aphasia, or agnosia—common signs of a cortical dementia such as Alzheimer's disease. Even so, the patients have impaired memory, especially in acquiring novel motor skills, and become easily confused on problem-solving tasks. The disorder is also accompanied by emotional and personality changes, although it is hard to know whether these are due to the disease itself or are a reaction to the onset of such a debilitating disease.

The genetic origin of Huntington's disease is reviewed in Chapter 3. Despite such advances on its cause, there is no known cure; patients usually die within 12 years of its onset. At autopsy, the brain of a Huntington's disease patient typically reveals widespread pathology in cortical and subcortical areas. Atrophy is most prominent in the basal ganglia, and the cell death rate is as high as 90% in the striatum. These changes are also evident from imaging studies performed as the disease unfolds.

It has been difficult to make inferences about normal motor function from studying patients with Huntington's disease. They frequently cannot complete experimental tasks, or do so with only the most labored movements. Even these are masked by involuntary movements. As such, selective deficits are hard to observe: The disease affects all measures of coordination. The excessive movements, or hyperkinesia, seen with Huntington's disease can be understood by considering how the pathology affects information flow through the basal ganglia. In the early stages, striatal changes are primarily in inhibitory neurons forming the indirect pathway. As shown in the left panel of Figure 11.43, these changes lead to a reduced output from the basal ganglia, and thus greater excitation of thalamic neurons. A second hyperkinetic disorder, *hemiballism,* is also associated with lesions of the indirect pathway, but here the lesion is centered in the subthalamic nucleus. Patients with these lesions produce violent and uncontrollable movements. These problems may persist for many years, requiring special precautions to ensure that the patients do not hurt themselves or those who come within reach.

Parkinson's disease is the most common and well-known disorder affecting the basal ganglia. The disorder is characterized by positive and negative symptoms, that is, motor disorders that heighten muscular activity and disorders that diminish it. Positive symptoms include resting tremor and rigidity. The former, which refers to the rapid shaking evident in distal effectors, is a tremor that becomes quieter, if not entirely quiescent, once the patient initiates a volitional movement. Rigidity results when agonist and antagonist muscles are activated simultaneously. When the limb is displaced passively, the neurologist can feel alternations between resistance and relaxation, a phenomenon known as *cogwheeling.*

Negative symptoms of Parkinson's disease include disorders of posture and locomotion, hypokinesia, and **bradykinesia.** Parkinsonian patients often lose their equilibrium. When the patient is sitting, the head may droop forward, or when standing, the forces of gravity will gradually pull the person forward until balance is lost. *Hypokinesia* refers to an absence or reduction in voluntary movement. Parkinson's disease patients act as if they are stuck in a posture and cannot change it. This problem is especially evident when patients try to initiate a new movement. Many patients develop small tricks

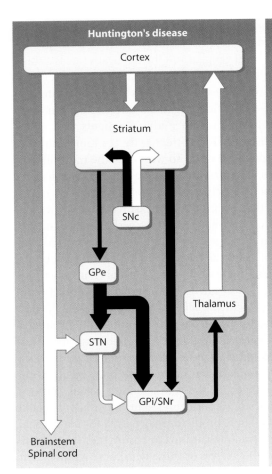 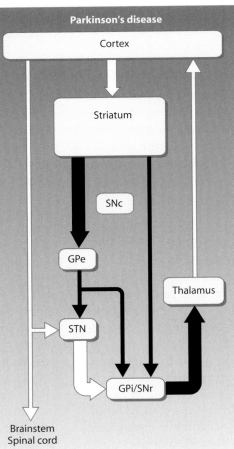

Figure 11.43 Differential neurochemical alterations in Huntington's and Parkinson's diseases. In Huntington's disease, the inhibitory projection along the indirect pathway from the striatum to the external segment of the globus pallidus (GPe) is reduced. The net consequence of this is reduced inhibitory output from the globus pallidus and thus an increase in cortical excitation and movement. Parkinson's disease primarily reduces the inhibitory activity along the direct pathway. This produces increased inhibition from the globus pallidus to the thalamus and thus a reduction in cortical activity and movement. Adapted from Wichmann and DeLong (1996).

to help them overcome the hypokinesia. For example, one patient walked with a cane, not because he needed help in maintaining his balance but because it provided him a visual target to give him a jump start. When he wanted to walk, he placed the cane in front of his right foot and kicked it—which caused him to overcome inertia and commence his walking. Once started, the movements may appear normal, although they are frequently slow, or bradykinetic. The parkinsonian patient can use his hand to reach for an object, though the entire sequence is in slow motion.

As noted at the beginning of this chapter, Parkinson's disease can develop following encephalitis or drug abuse. In most patients, however, the disease is *idiopathic,* a catch-all term that neurologists use when the cause is unknown. There is some evidence for a genetic origin, although the recent rise in early-onset Parkinson's disease has led researchers to suspect a rise in an environmental toxin. In all patients, the prominent pathology is a loss of dopaminergic fibers originating in the substantia nigra (SNc) and projecting to the striatum (Figure 11.43, right panel). As with most brain tissue, these neurons gradually atrophy with age. If the

percentage of lost neurons becomes too great, resulting in reduced dopamine levels of about 90%, Parkinson's symptoms will become obvious.

The role of dopamine in Parkinson's disease is incontrovertible (Figure 11.44). The first evidence came in the late 1950s when postmortem examinations correlated the disease with low dopamine levels. (The observations were the first time a brain disease was linked to a neurotransmitter deficiency.) Treatment programs were quickly developed in which patients were provided with synthetic precursors of dopamine, L-dopa, a therapy that is of great benefit to most Parkinson's disease patients. Many can function with little evidence of the disease, and Parkinson's disease is no longer considered a life-threatening condition. The drug's efficacy is far from perfect, though. Over the course of a day some patients experience rapid fluctuations in the severity of their symptoms. Over time, many patients develop hyperkinesias.

We noted earlier that input from the substantia nigra to the striatum can promote movement, both by its excitation of the direct pathway and by its inhibition of the indirect pathway. When the dopaminergic neu-

Figure 11.44 New neuroimaging techniques are able to label the distribution of specific transmitter systems. This procedure provides a new opportunity to visualize reduced dopaminergic activity in patients with Parkinson's disease. Healthy subjects and Parkinson's disease patients were injected with a radioactive tracer, fluorodopa. This agent is visible in the striatum, reflecting the dopaminergic projections to this structure from the substantia nigra. Compare the reduced uptake in a patient's scan (**right**) to the uptake in the scan from a healthy subject (**left**).

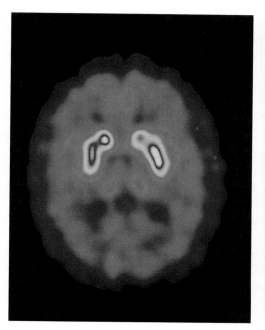
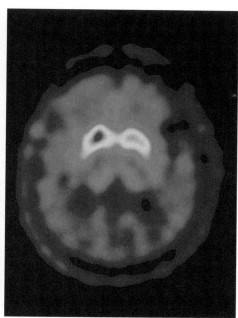

rons are depleted, the parkinsonian patient has great difficulty in initiating a movement. The patient looks frozen in place. The cortex may continue to select a movement plan, but the basal ganglia are needed to link that plan with commands to the motor effectors.

Parkinson's disease patients also show reduced flexibility in their volitional movements. For example, they may not be able to vary the force used to produce a movement. Normal subjects typically produce movements of different amplitude by scaling the magnitude of the initial agonist burst. A strong burst creates a larger force and enables one to move farther with minimal increase in the time required to complete the movement. In contrast, Parkinson's disease patients have to produce a series of small bursts to move a longer distance (Figure 11.45). Thus, the basal ganglia are significant not only in

Figure 11.45 EMG amplitude is not scaled appropriately in Parkinson's disease. **(a)** In healthy people, the magnitude of EMG activity increases to produce larger movements. **(b)** The EMG profiles of Parkinson's disease patients tend to be more invariant. To move the limb over a large distance, a series of EMG bursts have to be generated.

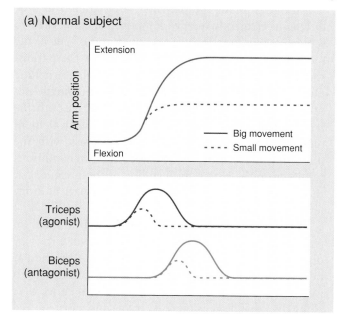
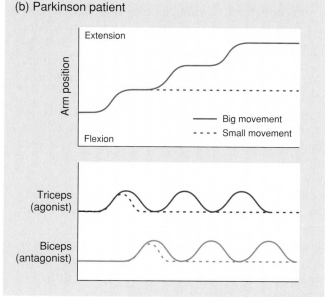

initiating movement but also in adjusting an action's force. Both problems relate to properly energizing a movement plan. For the initiation deficit, the releasing signal is absent, preventing the muscles from becoming active; for the force deficit, the plan is implemented but in a fixed manner.

Mike Samuel and his colleagues (1997a, 1997b) at Hammersmith Hospital in London have used PET to show how the balance between medial and lateral cortical motor pathways is disturbed in Parkinson's disease. Compared to normal individuals, Parkinson's disease patients exhibit greater activity in the lateral premotor and parietal areas when producing well-learned sequential movements. However, in patients who have undergone a pallidotomy, a procedure in which lesions are made in the globus pallidus, activity increases in the supplementary motor area and lateral prefrontal cortex.

Given these findings, we would expect that the movement deficits in Parkinson's disease patients would be most evident when actions are guided by internal cues. Indeed, the patients' problems can be virtually eliminated when their movements are guided by external cues. For example, parkinsonian patients are much poorer than control subjects in tracking a moving target when the target moves in a predictable, oscillatory pattern. If the target's motion is random, the patients perform as well as the control subjects. In the latter condition, movements must be guided visually since the target's future location is unknown. When the target oscillates, subjects anticipate its future location. Parkinson's disease patients, however, cannot take advantage of this information; in both conditions, their movements rely on external cues. They have difficulty selecting a movement pattern according to their own expectations. The patient who used a cane as a cue to initiate walking demonstrated his reliance on external cues. While he knew that he wanted to move from one location to another, he could not will it; he needed an external visual cue to activate his legs.

Whether or not the basal ganglia are central to generating movement sequences is a debatable issue. The slow movements in Parkinson's disease patients are most pronounced for the final elements of a movement pattern, and errors increase when the sequence is composed of heterogeneous gestures. Animal studies also point to the basal ganglia's role in generating movement sequences. In rats, lesions of the striatum disrupt their stereotyped grooming behavior (Berridge and Whishaw, 1992). It is not clear whether this is a specific problem in sequencing. The difficulty in producing learned or innate movement sequences

also might reflect an inability to energize movement plans or use internal cues to generate actions. For example, rats with striatal lesions do not produce random grooming patterns but fail to complete the sequence from start to finish.

Basal Ganglia Contributions to Learning and Cognition

As with the cerebellum, considerable controversy surrounds the question of whether and how the basal ganglia contribute to cognition. Patients who have had Parkinson's disease for a number of years perform below normal on various tests of neuropsychological function. These cognitive problems may, however, be secondary to effects of chronic L-dopa therapy. Or, the deficits may be the result of reduced dopaminergic input to the cerebral cortex.

Nonetheless, given the anatomical links between the basal ganglia and prefrontal cortex, researchers have explored how the motor problems seen in parkinsonian patients might manifest on cognitive tasks. One idea is that the basal ganglia perform an operation critical for shifting mental set. In the motor domain, a problem in initiating movements can be viewed as a deficit in set shifting. The parkinsonian patient gets stuck in one position or posture and cannot shift to a new one. Perhaps these patients' cognitive problems can be linked to shifting from one mental set to another.

To test this idea, Steven Keele and his colleagues at the University of Oregon (Hayes et al., 1998) developed two tasks that required a shifting operation (Figure 11.46). In the motor task, the patients were taught two short movement sequences of three elements each. After this training phase, the patients were required to produce a six-element sequence composed of either the two sequences in succession or two repetitions of one of the sequences. As predicted, the responses for the parkinsonian patients were especially slow at the switching point, the transition from the third to the fourth element, in the shifting condition. Note that in both the repetition and the shifting condition, the fourth element requires the same response—a finger press with the index finger. The difference between the two conditions results from the hierarchical coding of these sequences. In the shifting condition, this response is part of a different subsequence.

For the cognitive task, the patients were trained on reaction time tasks with two responses, involving color discrimination and shape discrimination. After the patients were trained on each dimension, pairs of trials were introduced in which the two responses either were along the same dimension (e.g., color-color) or required

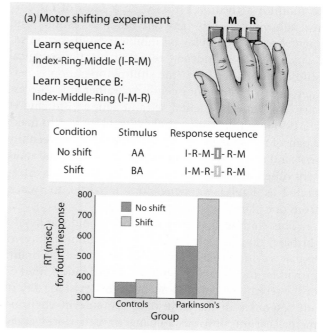

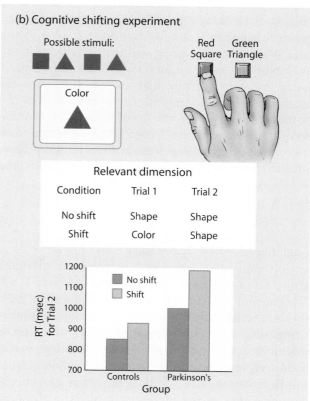

Figure 11.46 Motor and cognitive tests of set shifting. **(a)** In the motor task, subjects performed two successive sequences that were either identical or different. Although the movement at the transition point was the same in both the no-shift and shift conditions, Parkinson's disease patients were much slower in the latter condition. **(b)** In the cognitive task, subjects had to respond to either the color or the shape of a stimulus. Trials were paired such that the second response was either the same dimension (no shift) or the other dimension (shift). As in the motor task, Parkinson's disease patients were especially slow when they had to shift.

a shift from one dimension to the other (e.g., shape-color). As in the motor task, the parkinsonian patients were significantly slower when they had to shift dimensional set. This problem cannot be attributed to a motor problem as the responses on the second trial were iden-

tical in all conditions (e.g., decide whether the stimulus color was red or blue).

The shifting hypothesis offers a unified framework for understanding basal ganglia function in both action and cognition. The basal ganglia are in a position to monitor activation across wide regions of the cortex, allowing a shift between different actions and mental sets by removing an inhibitory influence in selected neurons. Indeed, recent neurophysiological studies emphasized that the shifting operation is neither strictly motoric nor cognitive; rather, it is a necessary link between mental set and action (Brotchie et al., 1991). In this work, monkeys were trained to make a pair of movements, with a variable delay between the first and second movements. A burst in pallidal neurons observed at the end of this delay period was interpreted as a releasing signal for the cortex to switch from one plan to another.

This shifting hypothesis also may hold the key to the basal ganglia's role in learning. Dopamine is known to play a critical role in the reward systems of the brain, providing the organism with a neurochemical marker of the reinforcement contingencies that exist for different responses in the context of the current environment. Learning involves a change in behavior—either acquiring the appropriate response in an unfamiliar context or breaking a habitual response when contingencies change in a familiar context. For a rat in the wild, this might mean being sensitive to a change in the availability of food at a foraging site. For a human, it might mean recognizing that a demanding problem cannot be solved by conventional means. One can hypothesize that within the basal ganglia, dopamine serves to bias the system to produce certain responses over others. Dopamine release in the striatum follows successful actions. This transmitter may modify the input-output channels in the basal ganglia, making it more likely that the animal will shift to the previously rewarded action when the associated input pattern is reactivated in the future.

Viewed this way, the ability to shift is required for producing novel behavior or for combining patterns of behavior into novel sequences. We can now see a link between basal ganglia dysfunction and psychiatric disorders such as Tourette's syndrome and obsessive compulsive disorder. A cardinal feature in each syndrome is the repetitive production of stereotyped movement patterns. For the patient with Tourette's syndrome, this might be a simple tic, a flick of the shoulder, or a hand brushing across the face. For someone with obsessive compulsive disorder, an entire behavioral sequence such as hand washing can be performed over and over again. A failure to shift may result in an absence of movement, the problem of the Parkinson's disease patient, or in the repeated production of a single pattern. In either case, basal ganglia dysfunction makes it difficult to select new actions that arise when sensory input or internal goals change.

Whether or not the shifting hypothesis is correct, it reveals a major feature of cognitive neuroscience's approach to brain function. Traditional neuroscience and neurology generally rely on task-based taxonomies such as motor or sensory systems. For the cognitive neuroscientist, the prime goal is to understand a neural structure's computational role, which can begin in a narrow domain but may also generalize to tasks that utilize a particular mental operation.

SUMMARY

Cognitive neuroscience has had a major impact on our conceptualization of how the brain produces skilled action. Consider the two halves of Figure 11.47. The diagram in Figure 11.47a, first introduced in 1974, shows the critical circuits of the motor pathway, emphasizing patterns of anatomical connectivity with a crude partitioning of function into motor planning, movement preparation, and movement execution. Figure 11.47b retains the basic circuitry but offers a functional decomposition of the processes involved in planning and programming.

As can be seen in this figure, the control of action involves a number of distributed systems. Nonetheless, we need not conclude that this distributed pattern suggests that all of the systems operate in a similar manner. As with other processing domains such as attention and memory, the different motor structures have their unique specializations. The cortical pathways for movement selection are biased to provide particular sources of information. The subcortical loops through the basal ganglia and cerebellum are essential for movement preparation, but in quite different ways.

By specifying a functional role for these structures, we can appreciate the limitations of brain theories that focus on the task rather than the internal computations. For example, the loss of fine coordination and erratic movements seen in patients with cerebellar lesions has led to a view that this structure is essential for skilled movement, that somehow the representation of an action shifts from one neural locus to another with practice. But when the cerebellum is viewed as a structure that is specialized to represent the temporal properties of a movement, we

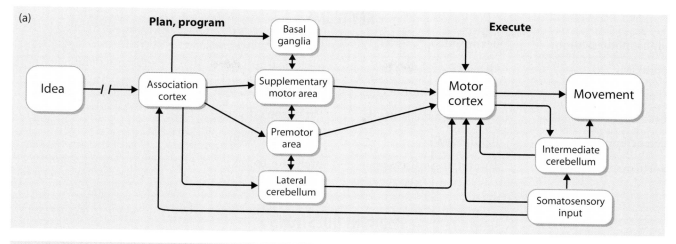

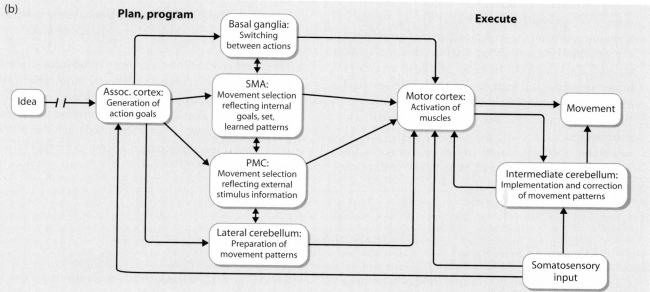

Figure 11.47 Summary of the functional architecture of the motor system. **(a)** Major neural structures, partitioned into areas associated with the planning and execution of movement. **(b)** Functional hypotheses regarding how these different structures contribute to actions.

can see that the loss of skilled movements is a consequence of a breakdown in the fine timing. This computation would not be as essential during the early phases of skill acquisition when the person builds the representations that underlie the skill. Moreover, this functional analysis has made clear that the boundaries between perception and action are murky. In the same way that the parietal lobe is essential for both perceiving and acting in space, the timing functions of the cerebellum or shifting functions of the basal ganglia are not restricted to motor control. As noted at the beginning of this chapter, perceptual information is only useful to the extent that it facilitates behavior.

We are not simply robots that respond in a fixed manner to the information being delivered by our perceptual apparatus. Flexibility provides one metric for comparing the sophistication of the cognitive capabilities across species. Is the animal's behavior completely dictated by the environment? Or can it modify its behavior in order to attain goals that require more complex planning? In the next chapter, we will continue with our emphasis on the goal-oriented aspects of behavior. This will take us to the highest level of abstraction in the hierarchical representations of actions, the level at which we form the plans that guide our most complex behaviors. To do so, we must consider the functions of the prefrontal cortex.

KEY TERMS

alpha motor neurons	deafferented	hypermetria	primary motor cortex
apraxia	effector	ideational apraxia	substantia nigra
basal ganglia	endpoint control	ideomotor apraxia	supplementary motor area
bradykinesia	extrapyramidal tracts	Parkinson's disease	
cerebellum	hemiplegia	population vector	
cortico-spinal tract	Huntington's disease	premotor cortex	

THOUGHT QUESTIONS

1. Motor control, when viewed both from a functional perspective and from a neuroanatomical/neurophysiological perspective, involves a hierarchical organization. Outline this hierarchy, starting with the most basic or primitive aspects of motor behavior and progressing to the highest level or most sophisticated aspects of motor behavior.

2. What is the difference between the pyramidal and extrapyramidal motor pathways? What type of movement disorder would you expect to see if the pyramidal tract is damaged? How would extrapyramidal damage differ?

3. What types of movements are possible without feedback? How might feedback be used by the motor system, both for on-line control and for learning?

4. Explain the concept of the population vector. How could it be used to control a prosthetic (artificial) limb?

5. Why do people with Parkinson's disease have difficulty moving? Provide an explanation based on the physiological properties of the basal ganglia. Why does dopamine replacement therapy improve their condition?

SUGGESTED READINGS

GEORGOPOULOS, A.P. (1995). Motor Cortex and Cognitive Processing. In M.S. Gazzaniga (Ed.), *The Cognitive Neurosciences* (pp. 507–517). Cambridge, MA: MIT Press.

JEANNEROD, M. (1997). *The Cognitive Neuroscience of Action*. Cambridge, MA: Blackwell Science.

PASSINGHAM, R. (1993). *The Frontal Lobes and Voluntary Action*. New York: Oxford University Press.

RIZZOLATTI, G., FOGASSI, L., and GALLESE, V. (2000). Cortical Mechanisms Subserving Object Grasping and Action Recognition: A New View on the Cortical Motor Functions. In M. Gazzaniga (Ed.), *The New Cognitive Neurosciences*, 2nd edition (pp. 539–552). Cambridge, MA: MIT Press.

ROSENBAUM, D.A. (1991). *Human Motor Control*. San Diego: Academic Press.

12

Executive Functions and Frontal Lobes

The attending physician at a neurology clinic had heard many bizarre reports over the 20 years of his practice. Most of the people he had seen in his outpatient clinic had symptoms of backaches or headaches. And he had seen his share of stroke and head trauma patients, people who had suffered an acute neurological disorder that had immediately wrought dramatic changes in mental function. But this morning a new patient, W.R., had come to his office accompanied by his brother. The patient recounted his life history and hit upon what he saw as his novel ailment. It was one that was unlikely to be prominent among the medical textbooks that lined the neurologist's shelves. The patient's chief complaint was that "he had lost his ego" (Knight and Grabowecky, 1995).

The neurologist observed the well-dressed man describe the past 10 years of his life. In many ways it was typical of a young professional. Early on, W.R. had decided to become a lawyer. He described a college tenure that was focused on this goal, with intense study in a pre-law program, and a time of social indulgences—hours on the tennis court, parties, and numerous girlfriends. He had been admitted to the law school of his choice and had completed the program with a solid, if not stellar academic record. But when he earned his degree, his life suddenly seemed to change course. He no longer found himself driven to secure a job with a top law firm. Indeed, 4 years had passed and he had still not taken the bar exam or even looked for a job in the legal profession. Rather he had been hired as an instructor at a tennis club.

The family had been extremely disturbed to see the changes in W.R.'s fortunes. After law school, they thought he was experiencing an early midlife crisis that was not atypical of the times. Many of W.R.'s peers were having second thoughts about entering the mainstream of American life, and they only hoped that he either would find satisfaction in his passion for tennis or would eventually resume his pursuit of a career in law.

But as time passed, it became more and more difficult to tolerate the changes. W.R. seemed to drift along. He was unable to support himself, hitting his brother up with increasing frequency for another "temporary" loan.

It was clear to the neurologist that W.R. was a highly intelligent man. Not only could W.R. clearly recount the many details of his life history, but also he was cognizant that something was amiss. He realized that he had become a burden to his family, even expressing repeatedly that he wished he could pull things together. He also knew that he really did not care, that when he was off on his own, he just could not find the motivation to take the necessary steps to find a job or get a place to live. W.R. had even given up playing tennis. His opponents would quickly become frustrated because shortly after commencing a match, W.R. would project an aura of nonchalance, forgetting to keep track of the score or even whose turn it was for service.

His brother noted another radical change in W.R.'s interests. He pointed out that his brother had not been on a date for a number of years and seemed to have lost all interest in romantic pursuits. W.R. sheepishly agreed. He recalled how he had been sexually active throughout his college years and had even lived with a woman for a number of years. But now that his brother mentioned it, why yes, it was true, he had not dated anyone in years and had not even given it a second thought. It was another sign of the lost ego. W.R. had little regard for his own future, for his successes, even for his own happiness. While aware that his life had drifted off course, he just was not able to make the plans to effect any changes.

If this was the whole story, the neurologist might have thought that a lost ego was not something he was equipped to treat. A psychiatrist might be a better option. But W.R. had suffered a seizure during his last year in law school, after staying up all night and drinking large amounts of coffee so he could study for a midterm exam. An extensive neurological examination at that time, including positron emission tomography (PET) and computed tomography (CT), had failed to identify the cause of the seizure. But given the claims of a lost ego and the patient's obvious distractability, the neurologist was suspicious. A CT scan that day confirmed his worse fears. W.R. had an astrocytoma. Not only was the tumor extremely large, but also it had followed an unusual course, traversing along the fibers of the corpus callosum so that it now extensively invaded the lateral prefrontal cortex in the left hemisphere and a considerable extent of the right frontal lobe. This tumor was very likely the cause of the initial seizure, even though it was not detected at the time. Over the past 4 years, it had spread slowly.

The next day, the neurologist informed the two brothers of the diagnosis. Containment of the tumor was all but impossible. They could try radiation, but the prognosis was poor: W.R. was unlikely to live for more than a year. His brother was devastated. Not only did he face the loss of W.R., but also he felt guilty over the frustration he had felt and expressed for the past 10 years. More perplexing, though, was the response of W.R. While his brother shed tears upon hearing the news, W.R. remained relatively passive and detached. He understood that the tumor was the culprit behind the dramatic life changes he had experienced. But there was no rage, minimal anguish, and indeed, a seeming absence of great concern. He understood the seriousness of his condition, but the news, as with so many of his recent life events, failed to evoke a clear response, a resolve to take some action. W.R.'s self-diagnosis seemed to be right on target: He had lost his ego and with it, the ability to take command of his own life.

The study of the ego is something that would make most neuroscientists squeamish. Compared to color perception or procedural learning, it is difficult to define this concept, let alone figure out how to investigate the neural basis. Animal models seem unlikely. And yet, from W.R.'s actions, or rather nonactions, we can distill a sense of an essential aspect of having an ego. W.R. had lost the ability to engage in goal-oriented behavior. He could handle the daily chores required to groom and feed himself, although even here he had needed the financial assistance of his family. But these actions were performed without the context of an overriding goal. He was not getting up in the morning, brushing his teeth, and eating breakfast so that he would be full of energy, ready to tussle in the legal system. He had few plans beyond satisfying his immediate needs, and even these seemed minimal. He could step back and see that things were not going as well for him as others hoped, but on a day-to-day basis, the signals that he was not making progress just seemed to pass him by.

In this chapter, our focus turns to the more complex aspects of cognition. We will consider the processes that are essential for goal-oriented behavior, the behaviors that allow us to interact in the world in a purposive manner. As we will see, there are many challenges to the successful completion of **goal-oriented behavior.** We must formulate a plan of action, one that draws on our past experiences yet is tailored to the current environment. Such actions must be flexible and adaptive. The best route from the dorm room to the eatery might be blocked by a construction project, but a highly evolved system like a hungry college student will still succeed in getting to dinner. To switch from one plan to another requires a means for monitoring the success of our ongoing actions. These operations are commonly referred to as **executive functions,** based on the idea that their operation serves to control and regulate information processing across the brain. The study of executive functions brings us to the part of the cerebral cortex that has received little attention in the preceding chapters, the prefrontal cortex.

SUBDIVISIONS OF THE
FRONTAL LOBES

The frontal lobes comprise about a third of the cerebral cortex in humans (Figure 12.1). The posterior border with the parietal lobe is marked by the central sulcus, the only sulcus to extend the entire length of the brain's lateral surface and down along the medial surface. The frontal and temporal

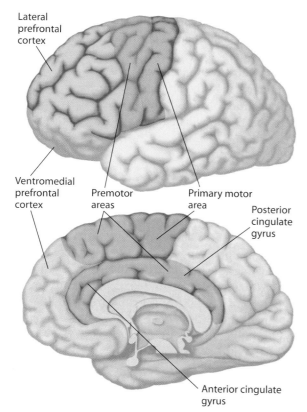

Lateral prefrontal cortex

Ventromedial prefrontal cortex

Premotor areas

Primary motor area

Posterior cingulate gyrus

Anterior cingulate gyrus

Figure 12.1 The areas of the frontal lobe. The prefrontal cortex includes all of the areas in front of the primary and secondary motor regions. The three major subdivisions of prefrontal cortex are the lateral prefrontal, ventromedial prefrontal, and the anterior cingulate cortex.

lobes are separated by the lateral fissure. Though the frontal cortex is present in all mammalian species, it has undergone tremendous expansion in human evolution, especially in the most anterior aspects. Since the development of functional capabilities parallels phylogenetic trends, the frontal lobe's expansion is related to the emergence of capabilities associated with cognition.

The most posterior part of the frontal lobe is the primary motor cortex (area 4), encompassing the gyrus in front of the central sulcus and extending into the central sulcus itself. Anterior and ventral to this are the secondary motor areas, including the lateral premotor cortex and supplementary motor area (area 6), the frontal eye field (area 8), Broca's area (area 44, perhaps area 45), and the posterior portion of the cingulate cortex. The remainder of the frontal lobe is termed the *prefrontal cortex*. The extent of prefrontal cortex includes half of the entire frontal lobe in humans. The ratio is considerably smaller for subhuman species (Figure 12.2).

Prefrontal cortex constitutes a massive network that links the brain's motor, perceptual, and limbic regions (Goldman-Rakic, 1995; Passingham, 1993). There are extensive projections to the prefrontal cortex from al-

most all regions of the parietal and temporal cortices, and even some projections from prestriate regions of the occipital cortex. Subcortical structures including the basal ganglia, cerebellum, and various brainstem nuclei project indirectly to the prefrontal cortex via thalamic connections. Indeed, almost all cortical and subcortical areas influence the prefrontal cortex either directly or within a few synapses. The prefrontal cortex also sends reciprocal connections to most areas that project to it, and to premotor and motor areas. The prefrontal cortex has many projections to the contralateral hemisphere—not only projections to homologous prefrontal areas via the corpus callosum but also bilateral projections to premotor and subcortical regions. From these neuroanatomical considerations we can assume that the prefrontal cortex is in an excellent position to coordinate processing across wide regions of the central nervous system (CNS).

In this chapter, we will focus on two regions of prefrontal cortex. The first region, the **lateral prefrontal cortex,** includes the lateral aspects of areas 9 to 12, all of areas 45 and 46, and the superior portions of area 47. The second region, the **anterior cingulate,** includes

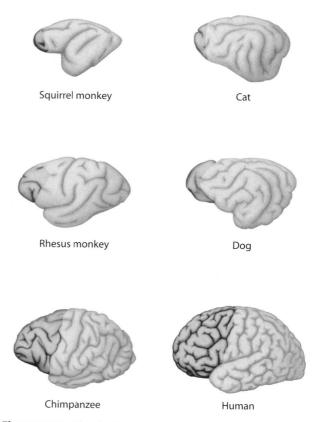

Squirrel monkey   Cat

Rhesus monkey   Dog

Chimpanzee   Human

Figure 12.2 The shaded areas show the extent of prefrontal cortex in six species. Note how small this region is in the cat, dog, and squirrel monkey. It is greatly enlarged in humans. The brains are not drawn to scale. Adapted from Fuster (1989).

areas 24, 25, and 32. Note that the cingulate is not always included in considerations of prefrontal function, perhaps because of its older, phylogenetic history and because it lacks the granular, six-layer structure of other cortical areas. Yet the cingulate plays a major role in executive functions. Moreover, granularity appears to be a questionable criterion since only primates have a granu-

lar prefrontal cortex. For all other mammals, prefrontal and cingulate regions have an agranular structure, an architecture quite distinct from more posterior cortical zones. The third region, the *ventromedial zone*, or what is sometimes called the *orbitofrontal cortex*, involves the inferior portions of area 47 and the medial parts of areas 9 to 12. This region will be discussed in Chapter 13.

THE LATERAL PREFRONTAL CORTEX AND WORKING MEMORY

Patients with frontal lobe lesions present a paradox. From their everyday behavior, it is frequently difficult to detect a neurological disorder. They do not display obvious disorders in any of their perceptual abilities, and their speech is fluent and coherent. On conventional neuropsychological tests of intelligence, the patients may perform normally. For example, when matched for education and age, patients with frontal lobe lesions generally score within the normal range on IQ tests such as the Wechsler Adult Intelligence Scale and most subtests of the Wechsler Memory Scale. Results such as these may contribute to the perpetuation of myths claiming that we fail to use vast portions of our cerebral capabilities. How else can we claim that lesions which destroy millions of cells fail to produce behavioral problems? With more sensitive and specific tests, though, it becomes clear that frontal lesions disrupt normal cognition and produce a host of problems, as we saw in the case of patient W.R.

Lesion studies in animals have revealed a similar paradox. Unilateral lesions of prefrontal cortex also tend to produce relatively mild deficits. However, when the lesions are bilateral, dramatic changes can be observed. Consider the observations of Leonardo Bianchi (1922), an Italian psychiatrist of the early twentieth century: "The monkey which used to jump on to the window-ledge, to call out to his companions, after the operation jumps to the ledge again, but does not call out. The sight of the window determines the reflex of the jump, but the purpose is now lacking, for it is no longer represented in the focal point of consciousness. . . . Another monkey sees the handle of the door and grasps it, but the mental process stops at the sight of the bright colour of the handle. The animal does not attempt to turn it so as to open the door. . . . Evidently there are lacking all those other images that are necessary for the determination of a series of movements coordinated towards one end." As with W.R., there is a loss of goal-oriented behavior.

The behavior of the monkeys underscores an important aspect of goal-oriented behavior. Following the lesions, the behavior is **stimulus driven.** The animal sees the ledge and jumps up; another grasps the door but fails to open it. What is lacking is the knowledge of how these actions have served the animal in the past. The animal acts as if it does not remember the purpose for these actions. The sight of the door is no longer a sufficient cue to remind the animal of the food and other animals that can be found in the other room. Is this deficit a motivational problem? Perhaps. The animals continue to be active, but their actions lack purpose. Alternatively, the problem could be a type of memory deficit. The animals may fail to remember the rewards that can be found in the other room.

Distinguishing Between Stored Knowledge and Activated Information

We need to distinguish between memory functions associated with the long-term storage of information and its activation when it is relevant for on-line processing. The initial impetus for this work grew out of cognitive psychology when researchers emphasized a distinction between long-term and short-term memory. Here, too, we face a paradox. There are many dramatic demonstrations of the unbounded capacity of humans and animals to remember even the most minute details of past experiences. We can recall our first-grade classroom or recognize the face of a person only briefly encountered at a party. In contrast, sometimes when we are introduced to a person or given a new phone number, within seconds this information slips away—in one ear and out the other. (As mentioned in Chapter 8, studies of implicit learning convincingly demonstrate that the information often is not lost but is buried in the cortical abyss.) Short-term memory studies have shown that our ability to keep information active is severely limited—

An Interview with Patricia Goldman-Rakic Ph.D.

Dr. Goldman-Rakic is affiliated with the Department of Neurobiology at Yale University. She has studied prefrontal cortex at many levels, including the precise anatomy of its projections to the rest of the brain as well as its function, through neuroimaging research.

Authors: You have studied the frontal lobes for many years. At a time when most neurobiologists were examining the visual system in detail, you dove into the most complex cortical zone in the brain. How did your decision to study complex processes come about?

PG-R: I began my studies on the frontal cortex when Hubel and Wiesel were conducting their pioneering and celebrated work on the cat and monkey visual cortex. As interesting and influential as the visual system research was, I was never attracted to it. I was captivated by the opportunity to study the unexplored frontal lobe and was especially intrigued by the challenge of understanding its relation to higher cortical functions. It never occurred to me to be concerned with the complexity of the frontal lobes—I never thought of it as an obstacle; rather, it's an attractive feature of the subject.

Authors: When approaching a cognitive capacity in the setting of brain research, it is important to have a good test of that capacity. That is sometimes hard to do. How do you go about zeroing in on what tests you will use when studying how the frontal lobes enable this or that cognitive function?

PG-R: When I started in primate research, I used delayed-response tasks to assay functional involvement without fully appreciating their significance. At the time, the tests had a host of interpretations: They were considered tests of spatial memory, spatial discrimination, tests of attention, of motor preparation or kinesthetic discrimination, response perseveration, and tests of immediate memory. Bolstered in part by comparative research in humans and monkeys, I realized that delayed-response tests tap into a component of the working memory. Once I understood this connection, it reinforced the functional significance of delayed-response tests vis-à-vis human cognition and convinced me that the attempt to weave cellular and behavioral levels together would elucidate human brain and behavior. Once you take a process view of cognition, test design becomes theoretically driven rather than strictly empirical and descriptive.

Anyway, delayed response is an example of an "oldie but goodie"; this old test, devised by Walter Hunter in 1913, has achieved growing significance. These paradigms are used widely in studies of rodent learning and memory, in lesion and single-cell analysis of primate cortical function, and in human neuroimaging research. A common set of paradigms is what allows information from experiments in different laboratories to be cross-validated and built upon. But it's also productive to design new tasks when guided by a process theory of cognition. The most useful tests of brain and cognition research are those that comport with anatomical, physiological, and behavioral-clinical data bases and have cross-species validity.

Authors: Since the prefrontal cortex is so important for human cognition, how can we meaningfully study it in monkeys, where it represents such a small fraction of total cortical processes?

PG-R: Yes, the human cortex is larger than the monkey cortex. So the issue is whether one or more areas of a monkey's prefrontal cortex can be used for understanding the remaining areas of the macaque's prefrontal cortex and for understanding functional areas in the human prefrontal cortex not shared with the monkey. I believe that they can, and recent studies support that view. For example, the rhesus monkey has one area dedicated to visuospatial processing and a distinctly different area dedicated to processing the features of objects. PET and functional MRI studies in humans are showing this general compartmentalization of information processing to also be relevant when considering the functional role of different areas of prefrontal cortex in humans. The human obviously has additional modules for semantic processing that the monkey does not share. Yet, understanding how Baddeley's visuospatial sketchpad works (in monkeys) is likely to shed light on the workings of the phonological loop. We may also be able to approach the nature and organization of the central executive given the assumption that the functional architecture of the prefrontal cortex is conserved across species and that what is added in evolution are new or refined information-processing systems but not necessarily new principles of their organization and function.

we can keep track of only a few pieces of information at a time.

Short-term memory and long-term memory depend on dissociable systems. Cognitive studies have indicated that the way we represent information may shift over time. For example, when people are asked to immediately recall a visually presented list of words, memory failures are much more frequent when the list contains phonologically similar words (e.g., *man, mat, cap*) than when the list contains semantically similar words. But if the recall is delayed, the pattern of errors reverses. Now, performance is poorer for lists whose words are similar in meaning. Hence, we translate verbal information into an acoustic code immediately after its presentation, although long-term storage may be in a network that represents semantic properties.

Patient research has supplemented the cognitive evidence for the neural system's dissociation in short- and long-term memory. Many amnesic patients have severe deficits in their ability to develop new long-term memories while showing little impairment on tests of short-term memory. Although rarer, the reverse situation occurs. For example, a patient who is unable to remember lists of more than two digits may have normal long-term learning.

Early work on memory dichotomies emphasized that the critical function of short-term memory was to assist in the translation of newly acquired information into a format for longer-term storage. More recent work has reconceptualized the functional role of a different form of short-term memory, what is called **working memory.** Working memory refers to transient representations of task-relevant information. These representations may be from the distant past: When returning to a high school reunion, we can anticipate what the cafeteria will look like or where our old homeroom was as we enter the front door. This knowledge had been buried deep within the recesses of our long-term memory. But as we approach the school, it is reactivated into working memory, with the representations now available to guide our actions. Representations in working memory also might be closely related to something that is currently in the environment or has been experienced recently. If we are introduced to the spouse of an old classmate, our ability to recall her name 20 minutes later will depend on bringing that information back into working memory.

The emphasis here is on how past knowledge influences and constrains current behavior. We do not process information simply so we can put it to use at a later time, a view that would be oriented only toward future behavior. Rather, representations are activated to shape behavior in the present. Reflecting this, the concept of a working memory system has surfaced as an apt characterization of processes that represent the current contents of cognition. Patricia Goldman-Rakic has called working memory the "blackboard of the mind."

Working memory is critical for animals that are not stimulus driven in their behavior. What is immediately in front of us surely will influence our behavior, but we are not automata. We can hold off eating until all the guests are sitting about the table. We can resist jumping onto the city bus that pulls over to the curb as we walk down the street. This capacity reflects the fact that in addition to reacting to stimuli that currently dominate our perceptual pathways, we also can represent information that is not immediately evident. We can mind our dinner manners or choose to respond to some stimuli while ignoring other stimuli. This process requires integrating current perceptual information with stored knowledge.

Working Memory Versus Associative Memory

The lateral prefrontal cortex is the primary repository for the interaction between current perceptual information and stored knowledge, and thus constitutes a major component of the working memory system. A classic demonstration comes from delayed-response tasks. In the simplest version, sketched in Figure 12.3, a monkey is situated within reach of two food wells. At the start of a trial, the monkey observes the experimenter placing a food morsel in one of the two wells. Then the two wells are covered and a curtain is lowered to prevent the monkey from reaching toward either well. After a delay period, the curtain is raised and the monkey is allowed to choose one of the two wells and recover the food. While this appears to be a simple task, it demands one critical cognitive capability: The animal must continue to represent the location of the unseen food during the delay period. Monkeys with lesions of the lateral prefrontal cortex do poorly on the task.

This inability does not imply a general deficit in forming associations. In an experiment to test associative memory, the food wells are covered with distinctive visual cues: One cue is associated with the food location and the other with the position of the empty well. In this condition, the food morsel's location may be shifted during the delay period, but the visual cue also will be relocated so it continues to cover the food. Prefrontal lesions do not disrupt performance in this task.

These two tasks clarify the concept of working memory (Goldman-Rakic, 1992). In the **delayed-response task,** no explicit cue remains constant from one trial to the next. In half of the trials, the food is placed in the left

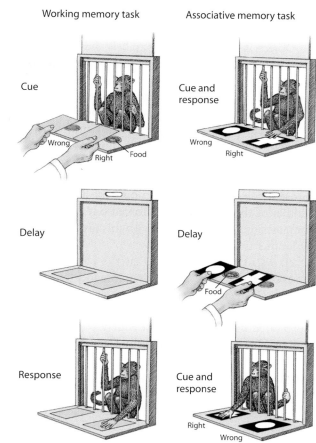

Working memory task   Associative memory task

Cue

Cue and response

Wrong Right Food

Wrong Right

Delay

Delay

Food

Response

Cue and response

Right

Wrong

Figure 12.3 Monkeys with prefrontal lesions demonstrate selective impairment on the working memory delayed-response task. **(Left)** In the working memory task, the monkey sees one well baited with food. After a delay period, the animal retrieves the food. The location of the food is determined randomly. **(Right)** In the associative memory task, the food reward is always associated with one of the two visual cues. The location of the cues (and food) is determined randomly. Working memory is required in the first task because, at the time the animal responds, there are no external cues indicating the location of the food. Long-term memory is required in the second task since the animal must remember which visual cue is associated with the reward. Adapted from Goldman-Rakic (1992).

well; in the other half, the right well is baited. Thus, there is no bias for either side to become associated with the reward over the long term. Rather, the associations fluctuate from trial to trial and the animal must remember the currently baited location during the delay period. In contrast, in the associative learning condition, it is necessary only that the visual cue reactivate a long-term association of which cue is associated with the reward. The reappearance of the two visual cues can trigger recall and guide the animal's performance.

Prefrontal lesions do not disrupt recognition memory. In another experiment, monkeys are shown three objects and allowed to select one. A curtain is lowered, and the experimenter quickly rearranges the display. In the working memory condition, two objects are presented, one of which had just been selected. The monkey is rewarded for choosing the other object. In the recognition memory condition, the previously selected object is again presented, but now it is paired with a novel object. The animal is rewarded for choosing the novel object. In both conditions, then, the animal is rewarded for picking a previously unselected object.

Despite this similarity, animals with lesions that encompassed areas 46 and 9 of the dorsolateral prefrontal

cortex had selectively impaired working memories. As a group, their performance was barely above the level of chance. In the recognition memory task, this group performed as well as control subjects. Here, it was sufficient to recognize that one object was novel—or that the other was familiar. In the working memory condition, both objects were familiar and the animal had to keep track of which had been previously selected. An important control in this study was the inclusion of animals who received frontal lesions in the vicinity of areas 6 and 8. These animals performed as well as the control animals on both versions of the task.

HUMAN STUDIES OF WORKING MEMORY

Humans also display frontal involvement in working memory. Adele Diamond of the University of Pennsylvania (1990) pointed out that a common marker of conceptual intelligence, Piaget's Object Permanence Test, is logically similar to the delayed-response task. In this task, a child observes the experimenter hiding a reward in one of two locations. After a delay of a few seconds, the child is encouraged to find the reward. Children younger than 1 year are unable to accomplish this task.

At this age, the frontal lobes are still maturing. Diamond maintained that the ability to succeed in tasks such as the Object Permanence Test parallels the development of the frontal lobes. Prior to this development, the child acts as though "out of sight, out of mind." As the frontal lobes mature, the child can be guided by representations of objects and no longer depends on their presence.

Tasks similar to the ones used with animals do not sufficiently elicit deficits in adult human subjects. To challenge adults, a variant of the delayed-response task, the delayed-alternation task, is used. Here, a correct response requires the subject to choose the location opposite that of the previously reinforced one. Patients with frontal lesions have difficulty with this task, especially if the lesions are bilateral.

Errors in this task imply that the patients tend to perseverate, that is, return to the same location on successive trials. **Perseveration** is one of the most common symptoms of the frontal lobe syndrome. This tendency is exploited in one of the most widely accepted tests of frontal disorders, the **Wisconsin Card Sorting Task,** which involves cards containing objects that vary along three dimensions: shape, color, and numerosity (Figure 12.4). The cards are presented one at a time and the subject is instructed to sort the cards according to an experimenter-defined sorting rule. The task has two catches. First, the subject is not informed of the sorting rule but must discover it through trial and error; the experimenter simply says correct or incorrect after each card is played. Second, and especially devilish, once the subject has learned to sort by one dimension, the experimenter changes the rule without informing the subject. So the subject not only must seek the appropriate sorting rule but also must be flexible enough to discard a previously reinforced hypothesis and begin the discovery anew. Patients with frontal lobe lesions perseverate, applying the initial rule over and over again after the switch, despite the experimenter's continued admonishment that the responses are incorrect.

There are obvious differences between the demands imposed by delayed-response tasks and the Wisconsin Card Sorting Task, but one essential feature is common: Recognition of the stimulus by itself is insufficient to guide subjects to the correct response but must be integrated with how that information was relevant on previous trials. The subject's working memory must retain knowledge about the relevance of certain features in their previous responses.

During the last 10 years there has been a barrage of neuroimaging studies designed to examine the neural correlates of working memory. In a functional magnetic resonance imaging (fMRI) study, subjects viewed four-

Figure 12.4 Patients with damage in the lateral prefrontal cortex have difficulty on the Wisconsin Card Sorting Task. On each trial, the subjects place the top card of the deck under one of the four target cards. The experimenter indicates whether the response is correct or incorrect, allowing the subject to learn the sorting rule by trial and error. The sorting rule changes whenever the subject makes ten consecutive correct responses.

teen or fifteen abstract shapes that were either white or red (McCarthy et al., 1994). The shapes could appear in one of twenty locations and were presented one at a time at a rate of one every 1.5 seconds. In the working memory task, subjects were instructed to raise an index finger whenever a shape appeared in a location that had been occupied by another stimulus; thus, the subjects had to continually update a record of all the stimulus locations (Figure 12.5). For the control task, the same stimuli were used, but the memory demands were eliminated: The subjects responded whenever a red shape appeared. These two tasks were compared to a baseline condition in which the subjects were instructed to relax with their eyes open. In this way, the stimuli were identical for the two experimental tasks, and assuming that the subjects correctly responded in all the trials, the total number of responses was equated. Yet the blood oxygen level dependent (BOLD) signal increased in the prefrontal cortex during the memory task, and the center of activation was in area 46. While this activation was

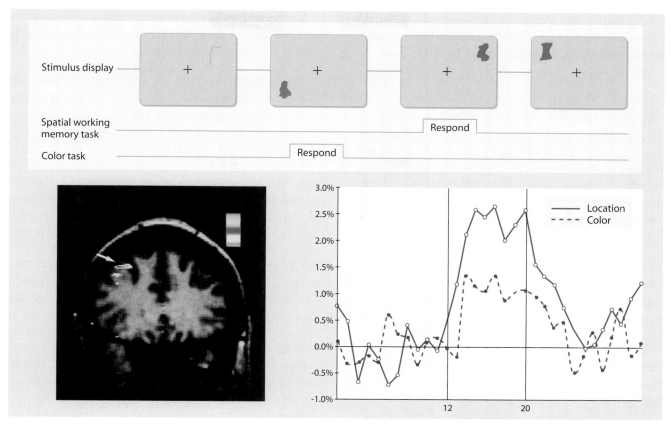

Figure 12.5 Lateral prefrontal activation revealed by functional magnetic resonance imaging (fMRI) in humans during a working memory task. **(Top)** Subjects viewed a series of colored, abstract shapes, appearing one at a time at various locations on the screen. In the spatial working memory condition, responses were required whenever a stimulus appeared at a location that had been used previously. In the control, color task, responses were required to all of the red objects. **(Bottom left)** During the spatial working memory task, there was a pronounced increase in activity in the lateral prefrontal cortex. This scan, obtained from a single subject, shows a prominent focus in the right prefrontal cortex (right hemisphere is on left). **(Bottom right)** The fMRI signal increased in the right prefrontal cortex during the 8-second stimulus period for both tasks. Most notable, the percentage increase in this area relative to the baseline was more pronounced during the spatial working memory task. Adapted from McCarthy et al. (1994).

bilateral, the effect was greatest in the right hemisphere. Similar asymmetries in other memory tasks have led to the hypothesis that the right prefrontal cortex may have a special role in memory retrieval. Alternatively, as we have seen in Chapters 7 and 10, the right hemisphere is implicated in spatial processing. The laterality effects may reflect the requirements in this task that the subjects remember the location of the stimuli.

It would be misleading to assume that working memory tasks only engage prefrontal cortex. In most imaging studies, task-related activation is observed across many cortical areas. What does stand out in these studies, though, is that the sustained activation during the memory delay period is especially pronounced in prefrontal regions. A working memory study involving face perception (Courtney et al., 1997) demonstrates this nicely. In the working memory task, subjects were shown a face and had to keep this image in mind for an 8-second delay

period, deciding if a subsequent face was the same or different. In a control condition, the face stimuli were scrambled and no memory task was required. Task-related activation foci were identified in the occipital and frontal lobes. To analyze the time course of the BOLD signals, three analyses were performed. One analysis identified areas that responded similarly to the faces and the scrambled faces. Presumably these areas are involved in the initial visual analysis of the displays. A second analysis identified regions that responded more strongly to faces. A third identified areas that showed sustained activation over the memory delay period. As predicted, the most posterior occipital region responded equally to both types of visual stimuli. This response was transient, only apparent immediately following the stimulus presentation. A similar transient response also was found in the inferior occipital lobe near the fusiform face area, and this response was selective for faces. In contrast, the prefrontal areas not only responded selectively to the faces but also remained active across the delay period. Interestingly, a similar pattern, although less strong, was also found in a prestriate region, posterior to the fusiform cortex.

CELLULAR MECHANISMS OF WORKING MEMORY

How is information activated and maintained in working memory? In Chapter 8 we described how the hippocampus plays a critical role in forming long-term memories. The idea was that the hippocampus operated as a consolidation device, strengthening connections between conceptual and perceptual processing centers so further experiences with similar stimuli would produce more efficient activation and thus facilitate recognition. In the working memory tasks just described, it is not enough for a stimulus to be recognized; subjects need to retain a record of its relevance, regardless of whether it pertained to the task at hand.

Two conditions are sufficient for a working memory system: First, it should have a mechanism to access stored information; second, there should be a way to keep the information active. The prefrontal cortex can perform both operations. Clues to the neural basis for working memory have been provided by single-cell recordings in the lateral prefrontal cortex. Recall that in the delayed-response task, the animal first is shown a cue, the placing of a food morsel in one of two food wells. This is followed by a delay when the animal is prevented from viewing the food wells. The occluding blind is then removed and the animal tries to retrieve the morsel. Cells in the prefrontal cortex such as the ones in Figure 12.6 become active during this task and have sustained activity throughout the delay period (Fuster, 1989). Indeed, for some cells activation does not commence until after the delay begins and can be maintained up to 1 minute. These cells, then, provide a neural correlate for keeping active a representation after the triggering stimulus is no longer visible. The cells provide a continuous record of the response required for the animal to obtain the reward.

Earl Miller and his colleagues at MIT have conducted a series of experiments to determine the specificity of prefrontal neurons (Rao et al., 1997). Do prefrontal neurons simply provide a generic signal that supports representations in other cortical areas, or do the neurons code specific stimulus features? To look at this question, monkeys were trained on a working memory task that required successive coding of two stimulus attributes: identity and location. The sequence of events in each trial is depicted in Figure 12.7. A sample stimulus is presented, and the animal must remember the identity of this object for a 1-second delay period in which the screen is blank. Following this, two objects are shown, one of which matches the sample. The position of the matching stimulus indicates the target location for a forthcoming response. However, this response must be withheld until the end of a second delay. Within the lateral prefrontal cortex, cells characterized as what, where, and what-where were observed. For example, "what" cells responded to specific objects, and this response was sustained over the delay period. "Where" cells showed selectively to certain locations. In addition, about half of the cells responded to specific combinations of what and where information. A cell of this type exhibited an increase in firing rate during the first delay period when the target was their preferred stimulus. Moreover, the same cell continued to fire during the second delay period if the response was directed to a specific location.

These results indicate that cells in the prefrontal cortex exhibit selectivity in terms of stimulus attributes. Such selectivity is task specific. In these studies, the cell activity is dependent on the monkey using that information to

Figure 12.6 Prefrontal neurons can show sustained activity during delayed-response tasks. Each line represents a single trial. The cue indicated the location for a forthcoming response. The monkey was trained to withhold the response until a "Go" signal (arrows) appeared. Each vertical tick represents an action potential. This cell did not respond during the cue interval. Rather, its activity increased when the cue was turned off, and persisted until the response. Adapted from Fuster (1989).

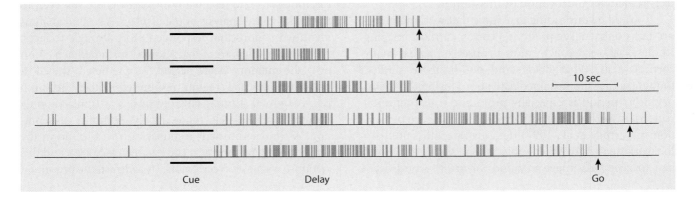

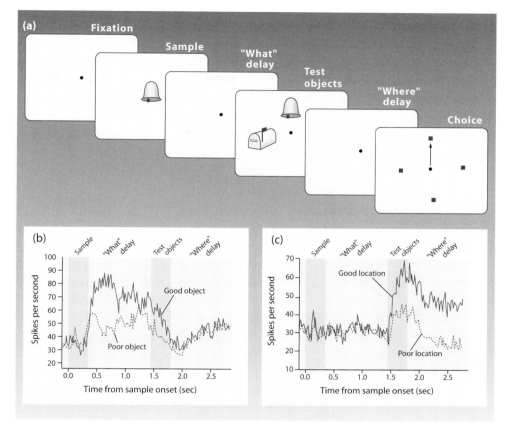

Figure 12.7 Coding of "what" and "where" information in single neurons of the prefrontal cortex in the macaque. **(a)** Sequence of events in a single trial. See text for details. **(b)** Firing profile of a neuron that shows a preference for one object over another during the "what" delay. The neural activity is low once the response location is cued. **(c)** Firing profile of a neuron that shows a preference for one location. This neuron was not activated during the "what" delay. Adapted from Rao et al. (1997).

obtain a response. If the animal is simply required to passively view the stimuli, the response of these cells is minimal right after the stimulus is presented and entirely absent during the delay period. Moreover, the response of these cells is malleable. If the task conditions change, the same cells become responsive to a new set of stimuli (Freedman et al., 2001). Thus, these cells are distinct from what is seen in the ventral visual pathway. Cells in the inferior temporal lobe respond much more strongly when the stimulus is present, and these responses can be elicited under passive viewing conditions (Desimone, 1996).

One could argue that this activity should not be interpreted as a mnemonic device but rather as a mechanism for maintaining a motor response. It is difficult to distinguish between remembering an event and anticipating a response triggered by that event. Having a record of recent processing can facilitate an upcoming action. Yet these cells are more closely linked to memory than to motor processes because their activity is modified by past experience. Fewer sustained responses are observed in an untrained animal despite the fact that the animal will respond on every trial. As the animal becomes skilled in using the cue to guide performance, the activity during the delay period becomes more pronounced. Sustained activity cannot be attributed to arousal. Some cells are activated after one cue; others fire more after the other cue appears.

Cellular responses like these could indicate that long-term representations are stored in the prefrontal cortex and that the cues activate them. By this hypothesis, the prefrontal cortex not only plays a role in working memory but also is implicated in the long-term storage of knowledge. While this idea is in accord with the single-cell results, it cannot account for the fact that prefrontal lesions have little effect on long-term memory.

Figure 12.8 presents an alternative view: Prefrontal areas are a temporary repository for representations accessed from other neural sites. Information is not permanently stored in the prefrontal cortex but is maintained there while it is relevant for performing a task. This hypothesis jibes nicely with the fact that the prefrontal cortex is connected intimately with postsensory regions of the temporal and parietal cortices. When a stimulus is perceived, a temporary representation is instantiated in the prefrontal cortex via these connections to posterior brain regions. The sustained activation of prefrontal cells probably requires continuous reverberation across this network.

This hypothesis is supported by a metabolic imaging method with excellent spatial resolution (Friedman and Goldman-Rakic, 1994). Animals were trained to perform either working memory tasks or control tasks that relied on associative memory. Prior to performing the task in the experimental session, the animals were injected with a slow

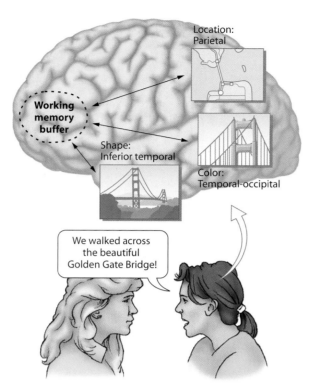

Figure 12.8 Lateral prefrontal cortex may provide a transient buffer for sustaining information stored in other cortical regions. In this example, the person is telling a friend about her walk across the Golden Gate Bridge during a visit to San Francisco. Long-term knowledge is reactivated and temporarily maintained through the reciprocal connections between the prefrontal cortex and the more posterior regions of the cortex. Note that the long-term memories of the Golden Gate Bridge are stored in dimensions-specific cortical regions.

radioactive tracer (^{14}C-2-deoxyglucose). A unique feature of this tracer is that it gets trapped in metabolizing brain tissue and thus can be measured with a photographic technique that highlights the radioactive agent. Glucose utilization was higher in prefrontal area 46 and in parietal area 7 in the working memory group compared to the control groups (Figure 12.9). Glucose utilization in temporal cortical regions related to auditory perception was comparable for the two groups, which demonstrated that the memory groups did not simply have more active brains.

Two other results are noteworthy. First, glucose utilization in the parietal cortex was correlated with task performance. Animals that performed well had higher rates of metabolism in the parietal cortex. In contrast, the degree of prefrontal metabolism depended on task difficulty, regardless of how well it was performed. This highlights the fact that memory tasks require that prefrontal areas devoted to memory functions interact with posterior areas that support long-term representations. More parietal activation reflects improved fidelity of the representations in working memory and thus more accurate performance. Prefrontal areas engage more when the task's memory demands increase.

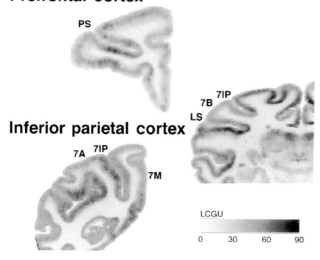

Figure 12.9 A radioactive tracer can reveal correlated activity in the prefrontal and inferior parietal cortex during a spatial working memory task. After being injected with the tracer, the animal performed the memory task. Upon completion, the animal was killed. Histological analysis revealed how the slow-decaying tracer was trapped in different brain regions. The results are coded on a gray scale in units of local cerebral glucose utilization (LCGU). PS in the top panel refers to the principal sulcus of the prefrontal cortex. The abbreviations with numeral 7 refer to area 7 regions of the parietal lobe. LS is the lateral sulcus, the division of parietal and temporal lobes. Embedded within this sulcus is the auditory cortex area measured for control purposes.

THE PREFRONTAL CORTEX PARTICIPATES IN OTHER MEMORY DOMAINS

To this point, the studies of working memory have focused on variants of the delayed-response task. However, a number of other tasks have been used to study the contribution of prefrontal cortex to memory. In this section, we will review this literature. We then will consider whether these problems reflect a common processing deficit, or whether they suggest a heterogeneity of function within the prefrontal cortex.

The Frontal Lobes and the Temporal Organization of Memory

Memory fades with time. It is impossible to remember all the details of our lives. We have a few memories of our childhood, perhaps a special birthday party, or a cross-country trip, or our first-grade classroom. But the details of most experiences become difficult to reconstruct as time passes. This information is not necessarily lost—it is absorbed into new experiences. We remember how to play Monopoly or how to bake a favorite cake, not by recalling our initial attempts but by repeating the activities. Practice makes perfect. By measuring and sifting flour, salt, and baking powder, by blending egg yolks and sugar, and by folding this mixture into the dry ingredients, we remember how to prepare the cake. The sequence for these actions must be executed in proper order for the cake to come out as planned. Something does not taste right if the sugar is added to the flour mix without first having been stirred with the eggs.

Even when it is unnecessary to remember a sequence of actions, our memories include temporal tags. Think about what you did yesterday. Perhaps the events have not yet faded from your memory. Most likely, you can recall waking up and what you ate for breakfast, which classes you attended, where you ate lunch, and how you spent the evening, perhaps pouring over books in preparation for an exam or relaxing with friends. While no one required that the events be recalled in order, you probably used temporal cues to organize the information. Time continually moves forward, and our reconstruction of the past obeys this principle.

People with frontal lobe lesions may be impaired in their ability to organize and segregate events in memory, an ability referred to as **recency memory** (Milner, 1995). A recency discrimination task has been used to study temporal memory. In this task, the subjects are presented with study cards, each having two stimuli such as a pair of pictures. The subjects are instructed to study the pictures. Every so often, a probe card is presented with a question mark in addition to two pictures (Figure 12.10). The subject must decide which of the two pictures was seen most recently. For example, one of the pictures might have been on a study card presented four trials previously, whereas the other picture was on a study card shown thirty-two trials ago. For a control task, the procedure is modified: The probe card contains two pictures and the question mark, but only one of the two pictures was presented previously. Following the same instructions, the subject should choose that picture since, by definition, it is the one seen most recently. Note, though, that the task is really one of recognition memory. There is no need to evaluate the temporal position of the two choices.

Frontal lobe lesions were associated with a selective deficit in recency judgments. Patients performed as well as control subjects on the recognition memory task, with both groups scoring above 90% correct. The recency task proved considerably more difficult, and most importantly, this effect was even more marked in patients with frontal lobe lesions, which provided a single dissociation. Temporal lobe lesions did not affect performance on either task.

All the patients had undergone a unilateral brain operation for the treatment of severe, focal epilepsy. With this procedure, the extent of the surgically induced lesion can be limited and is precisely known. Hence, one can distinguish between patients in whom the lesion encompassed dorsolateral prefrontal regions (area 46) from those in whom this area was spared. This analysis revealed that deficits in the recency task were restricted to patients in the dorsolateral group. Thus, the same area implicated in working memory tasks, such as delayed response, is associated with the memory for temporal order. Moreover, a laterality effect was a function of stimuli. With word stimuli, the deficit was most evident in patients with left-hemisphere lesions, whereas right-hemisphere lobectomies were associated with impaired performance when drawings of common objects or abstract paintings were substituted for words.

A variant of the recency discrimination task is the self-ordered pointing task. Here, a series of n cards is presented, each with the same set of n objects (with n ranging from 6 to 12). The positions of the objects are shuffled about from card to card. The subject must point to a new object on each card. Thus, successful performance requires that the subject keep track of which items had been responded to on previous cards. As in

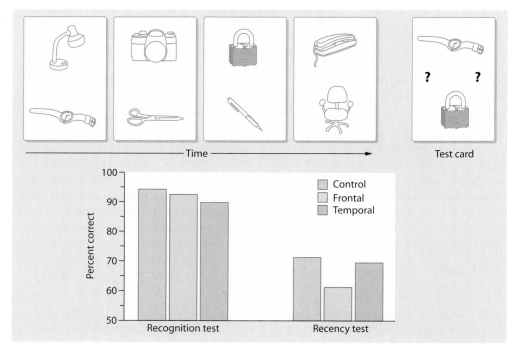

Figure 12.10 Recency memory is impaired in patients with prefrontal lesions. **(Top)** Subjects are presented with a series of cards, each one showing a pair of objects. On test cards, the objects are flanked by question marks, and the subject must indicate which object was seen most recently. In the recency test, both objects on the test cards had been seen previously. In the item recognition test, only one object had appeared previously. **(Bottom)** The results revealed a single dissociation. Patients who had had a frontal lobectomy performed more poorly on the recency task compared to both control subjects and patients who had had a temporal lobectomy. The frontal group was not impaired on the item recognition task. Adapted from Milner et al. (1991).

the recency discrimination tasks, patients with frontal lesions make more errors than do control subjects, and the deficit becomes more pronounced as the number of items per card increases.

The breakdown in the temporal structure of memory may account for more bizarre aspects of frontal lobe syndrome. For example, in a classic report of frontal lobe dysfunction, Wilder Penfield described a patient who was troubled by her inability to prepare her family's evening meal. This patient could remember the ingredients for dishes, but she could not organize her actions into a proper temporal sequence. Her behavior was haphazard. She might assemble all of the ingredients but become flustered and switch her preparation from one dish to the other, or mix up which items belonged together. No longer could she generate a temporal plan to achieve a coherent goal (Jasper, 1995).

Source Memory

Remembering when we learned a fact is a part of *episodic memory*. Our knowledge is not limited solely to content but includes the context in which learning took place. In essence, to remember a learning episode is to remember

details about time and place, and the episode itself: who was present, what they wore, what kind of day it was. Sometimes source information is essential. The alert private detective has to recognize a face and recall the context in which that person was encountered. At other times these details may be irrelevant. It is not obvious what benefit there is in remembering that the professor who lectured on Freud and his theory of dream interpretation was wearing a bright-green suit. Are these details simply information cluttering up our cranial warehouse? Or can they facilitate memory? These obscure details are actually useful retrieval cues (Figure 12.11).

Source memory depends on the integrity of the frontal lobes. To show this, Larry Squire and his colleagues (Janowsky et al., 1989) at the Veteran's Administration Medical Center in San Diego had control subjects and frontal lobe patients learn facts such as "The name of the dog on the Cracker Jacks box is Bingo," or "The body of water between Russia and Iran is the Caspian Sea." After a 6- to 8-day retention interval, the subjects were tested on these statements plus new ones with equally obscure facts or easy questions. Whenever subjects correctly answered a question, they were asked to recollect how they had learned the information. Subjects could

Figure 12.11 *Source memory* refers to knowledge concerning the source of information or the context in which the information was learned. This student recalls the specific episode in which she learned about Freud's theory of the unconscious.

answer that they had learned it during the previous testing session, or they could attribute it to something they learned in school or read in the paper. There was a dissociation between recall and source memory tasks. Patients performed as well as control subjects on the recall task but made more errors on the source task.

Healthy subjects also provided novel evidence linking the frontal lobes to source memory (Glisky et al., 1995). Although the subjects never had a neurological disorder, they were given neuropsychological tests. According to their performance on tests designed to assess frontal lobe function, they were divided into groups based on "high" and "low" frontal function (the "low" group still performed higher than patients who had incurred a neurological insult in the frontal lobes). A similar division was based on their performance on tests designed to assess temporal lobe function. The two tests had minimal correlation. Subjects who scored well on the frontal test did not necessarily score well on the temporal lobe test.

In the primary experiment, subjects were tested on item and source memory. During the study phase, the participants heard a series of sentences expressing common events (e.g., "The boy went to the store to buy some apples and oranges"). Half of the sentences were read by a woman and the other half by a man. Item memory was tested by pairing a study sentence with a novel sentence and asking subjects to choose the sen-

tence they heard in the study phase. Source memory was tested by having the subjects hear the same sentence read by both speakers and then judge which voice matched the original presentation.

The results, presented in Figure 12.12, reveal a double dissociation. High temporal lobe function was associated with good performance on the item memory test, whereas the two frontal lobe function groups did not differ significantly. In contrast, only the frontal lobe function groups differed on the source memory task. Participants rated as having low frontal function performed barely above chance, whereas those whose function was rated high did considerably better. Thus, healthy people dissociate item and source memory, and the two tasks correlate with differences in temporal and frontal lobe function.

Figure 12.12 Double dissocation on tests of item and source memory in healthy elderly adults who were rated as having "high" or "low" function on tests of frontal and temporal lobe function. Low frontal lobe function was associated with poor performance on the source memory test. Low temporal lobe function was associated with relatively poor performance on the item memory test. Adapted from Glisky et al. (1995).

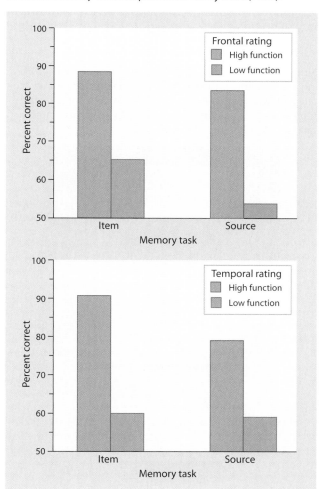

COMPONENT ANALYSIS OF PREFRONTAL CORTEX

Working memory is conceptualized as a temporary network to sustain the current contents of processing. As reviewed earlier, the evidence overwhelmingly demonstrates that the lateral prefrontal cortex is a key component of working memory. But as can be seen in Figure 12.1, the lateral prefrontal cortex encompasses a large area and includes numerous cytoarchitectonic areas. A question of intense debate has been how to best characterize the different contributions of the prefrontal cortex.

Content-Based Accounts of Functional Specialization Within Lateral Prefrontal Function

One approach focuses on the content of the represented information. **Content-based hypotheses** have been motivated by psychological models of working memory. Alan Baddeley (1995) of the Applied Psychology Unit in Cambridge, England, promoted one influential view. His model is summarized in Figure 12.13: It consists of two subsystems that compete for access to a central

Figure 12.13 Baddeley's model of working memory. Working memory entails three critical components: a central executive and two "slave" systems, one for sustaining visuospatial representations and the other for sustaining verbal representations in a phonological format.

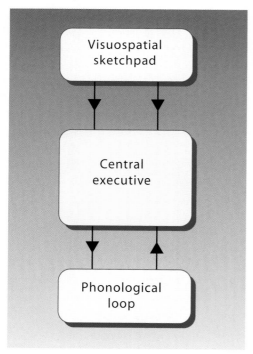

executive. One system, labeled the visuospatial sketchpad, activates representations of objects and their spatial positions. The other system is linguistic in nature and is referred to as the *phonological loop,* reflecting Baddeley's belief that we maintain linguistic representations via covert articulatory rehearsal (i.e., subvocal speech). A key motivation for postulating two working memory components comes from studies of subjects who were instructed to remember a list of words by using either a verbal strategy such as rote rehearsal or a visuospatial strategy based on an imagery mnemonic. Under control conditions in which the memory rehearsal was the only task, subjects were better on the memory test when they used the visuospatial strategy. However, the verbal strategy proved better when the subjects were required to track concurrently a moving stimulus by operating a stylus during the retention interval. This reversal cannot be explained by assuming a unitary memory system.

One expectation is that the phonological loop and visuospatial sketchpad correspond to working memory functions of the left and right hemispheres, respectively, an idea consistent with the general picture of hemispheric specialization. Indeed, across a large number of neuroimaging studies, activation in the lateral prefrontal cortex, especially the more ventral regions anterior to Broca's area, is found in working memory tasks with verbal stimuli (Gabrieli et al., 1998). For example, when subjects are asked to generate a verb associated with a target noun, working memory is required to sort the possibilities (Petersen et al., 1988). Similarly, right prefrontal activation is observed consistently in spatial working memory tasks (see Figure 12.5).

A different content-based hypothesis builds on the idea of two cortical visual systems. As described in Chapter 6, the dorsal visual pathway is prominent in the analysis of spatial information; the ventral pathway is essential in the analysis of object information. It has been hypothesized that this dorsal-ventral distinction may extend into the prefrontal cortex (Wilson et al., 1993). However, both single-cell and imaging studies have failed to consistently demonstrate this dissociation in prefrontal cortex. A more consistent account of the neural correlates of spatial and nonspatial working memory is found by considering laterality differences. Figure 12.14a shows the prefrontal areas of activation from twenty different PET and fMRI studies of working memory. There are more than twenty foci in the figure, owing to the fact that most of the studies demonstrated multiple areas of

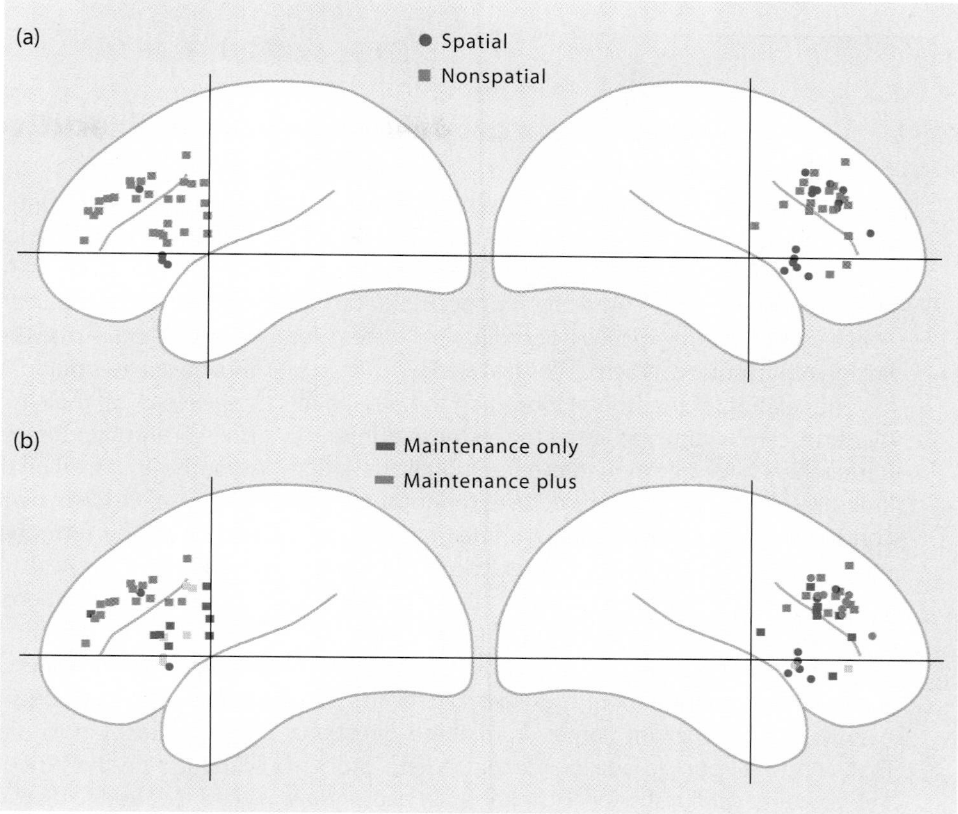

Figure 12.14 Functional organization within the lateral prefrontal cortex as revealed by meta-analysis of imaging studies. **(a)** Activation foci for tasks involving either spatial or nonspatial working memory. **(b)** The same data, coded to discriminate between tasks that involved maintaining information and those that also involved manipulating the information. The maintenance-manipulation dichotomy provides a more parsimonious account of the results than the spatial-nonspatial one. Gray symbols indicate foci activated during maintenance plus tasks that also led to activation in a more dorsal prefrontal region. From D'Esposito et al. (1998).

activation. The meta-analysis supports the hypothesis that spatial tasks are associated with activation in the right prefrontal cortex. In contrast, nonspatial tasks generally activate the left hemisphere, although a number of the studies reported bilateral activation.

The nonspatial tasks involved both linguistic and nonlinguistic stimuli. As an example of the latter, at the start of the trial the subjects view three objects such as a bell, hammer, and chair. After a memory delay, a single probe stimulus is presented and the task is to judge whether or not the probe is one of the three items. For this task, it is quite likely that the subjects use a verbal coding strategy when memorizing the stimuli. To minimize this possibility, researchers have used nonsense figures, but even here, the subjects may still label the items with words (e.g., "the squiggles remind me of my sister's hair"). Thus, at present it is difficult to conclude that prefrontal cortex in the left hemisphere plays a key role in nonspatial working memory tasks in general, or just those in which the task is performed through verbal recoding. Nonetheless, the overall picture does support the evidence of a content-based distinction between the two hemispheres in working memory, a distinction that parallels the laterality effects seen in perception (see Chapter 10).

Process-Based Accounts of Functional Specialization Within Lateral Prefrontal Function

A second approach for partitioning prefrontal cortex centers is process-based: Are different regions recruited as a function of the type of processing required for a particular task? Consider two types of working memory tasks. In a standard delayed matching to sample task, a stimulus is presented and the subjects must internally maintain a representation of that stimulus until the probe is presented. As we have seen, such tasks involve the lateral prefrontal cortex. These tasks can be made more challenging by requiring the subjects to internally manipulate these transient representations. One favorite variant is the "*n*-back" task (Figure 12.15). Here the display consists of a continuous stream of stimuli. Responses are required only when the current stimulus matches a stimulus that had been presented *n* items previously. In the simplest version, *n* equals 1, responses are made when the same stimulus is presented on two successive trials. In more complicated versions, *n* can equal 2 or more. With *n*-back tasks, it is not sufficient to simply maintain a representation of recently presented

Cortical-Subcortical Interactions in Executive Functions

Executive functions are not restricted to the frontal lobes. For example, the parietal lobe participates in spatial attention, ensuring that we focus on the most important stimuli. And the hippocampus can be conceived of as an executive coordinating system for linking representations across cortical areas.

The relation of the frontal lobes and the basal ganglia and cerebellum has attracted extensive interest. Researchers have questioned the traditional views that limit these two subcortical structures to motor control, contributing to the preparation, implementation, and monitoring of movements. Rather, the basal ganglia and cerebellum may form an integrated network with the prefrontal cortex subserving the executive functions of higher cognition.

The evidence for a more cognitive role for the basal ganglia and cerebellum comes from three directions. First, using a retrograde labeling technique, Peter Strick of the Veteran's Administration Hospital in Syracuse, New York, explored subcortical projections to the monkey's dorsolateral prefrontal cortex (Middleton and Strick, 1994). The method works transynaptically; thus, we can confirm that areas within the basal ganglia and cerebellum innervate the dorsolateral cortex via their projections to the thalamus. Second, PET studies consistently show activation in the basal ganglia, and especially the cerebellum, even when the experimental and control tasks require the same amount of overt movement. Activation is greater in the cerebellum when subjects generate a semantic associate to a word (e.g., *eat* to *apple*) in comparison to a control condition where subjects simply repeat the target word (e.g., *apple*). Third, patients with Parkinson's disease and cerebellar disorders perform poorly on neuropsychological tests of frontal lobe function.

The contributions of the subcortical components of cortico–basal ganglia and cortico-cerebellar networks remain unclear. One possibility is that the frontal-like problems in patients with basal ganglia or cerebellar pathology are indirect. Hypometabolism has been observed in the frontal lobe following such pathology. Thus, the deficits may not reflect abnormal processing within the basal ganglia or cerebellum; rather, they may come from secondary alterations in frontal lobe activity.

A second possibility is that the basal ganglia and cerebellum contribute in a direct way to the executive functions of the frontal lobe. Here the analytic tools of cognitive neuroscience are promising. Neuropsychological tests such as the Wisconsin Card Sorting Task only provide crude comparisons of functional deficits. The problem is that the task is quite complex. Subjects must evaluate multidimensional stimuli, keep an internal record of their most recent responses, generate rule-sorting hypotheses, and be flexible about altering hypotheses based on the feedback provided after each response. Problems with this task could arise from an inability to perform the operations. More sensitive, theoretically motivated tasks are required to isolate the component operations.

Adrian Owen and his colleagues (1993) at Cambridge University identified one intriguing dissociation between frontal lobe and parkinsonian patients by using the two tasks in Figure A. Both tasks involved a slight variation on the Wisconsin Card Sorting Task: Subjects were required to discover that the sorting rule changed from one dimension to another. The key modification in this experiment was that stimuli did not simultaneously vary in all dimensions. For instance, during the early trials, the target dimension might be shape; the irrelevant (distractor) dimension, color; and the constant (invariant dimension) might be the size. The conditions differed in how the dimensions were manipulated when the sorting rule changed. In the perseveration condition, size would become the target dimension and shape the irrelevant dimension. Thus, the subjects could continue to respond erroneously on the basis of shape. In the learned irrelevance condition, the target would become color and the irrelevant dimension would be the size. With this condition, it would not be possible to perseverate since shape is now constant. Rather, errors could be due to either an inability to respond to a previously irrelevant dimension or a bias to respond on the basis of the novel dimension.

In comparison to control subjects, patients with frontal lesions were selectively impaired on the perseveration condition. After the shift, they continued to respond on the basis of the dimension that had been the target, perhaps reflecting an inability to inhibit recently activated and task-relevant information. These patients were not impaired on the learned irrelevance condition. In contrast, parkinsonian patients receiving L-dopa

	Training	Transfer conditions	
		Perseveration	Learned irrelevance
Target dimension	Shape	Size	Color
Distractor dimension	Color	Shape	Size
Invariant dimension	Size	Color	Shape

Figure A Dissociation of the contributions of the frontal lobe and basal ganglia to task shifting. Both tasks are variants of the Wisconsin Card Sorting Task. Subjects must learn the correct sorting rule through trial and error. On each trial, they are shown a card with two stimuli that vary on two of the three dimensions; shape, color, and size. The value on the third dimension is fixed (invariant) for a given condition. After they complete training, they are tested on two transfer conditions. In the perseveration condition, the target dimension becomes the distractor dimension after the shift. The new target dimension was not varied prior to the shift. In the learned irrelevance condition, the target dimension had been the distractor dimension prior to the shift. The new distractor dimension had not been varied prior to the shift. Frontal lobe patients show selective impairment on the perseveration condition. Patients with Parkinson's disease show selective impairment on the learned irrelevance condition. Adapted from Owen et al. (1993).

medication exhibited the opposite profile. They made few perseverative errors but had difficulty in responding to a previously irrelevant dimension. Whether this reflects excessive inhibition of this dimension or a bias to respond to the novel dimension requires further study. Nonetheless, this study elegantly demonstrates how superficial similarities on standard neuropsychological tests might arise from deficits in dissociable component operations.

A similar strategy is being pursued to tease apart the relation between the prefrontal cortex and the cerebellum (Fiez et al., 1996). Julie Fiez of Washington University in St. Louis noted that, based on several PET studies, tasks that require verbal rehearsal are associated with higher blood flow in inferior regions of the lateral prefrontal cortex (areas 44 and 45) and the right cerebellum. Fiez proposed that this network might constitute part of the phonological loop of Baddeley's working memory model. A processing model of the phonological loop sketched in Figure B consists of three operations: a phonological store that contains phonological representations of words, an articulatory

process that refreshes these representations through internal rehearsal, and for visual stimuli, a recoding process that translates written letter strings into phonological representations. So, with a visual list of words to remember, we recode stimuli into phonological representations and silently articulate the list as a means of keeping the words active in working memory. If stimuli are presented auditorily, there is no need for recoding; the stimuli activate representations sustained through internal rehearsal.

We can now ask whether neural structures are associated with these component operations of verbal working memory. Previous work had suggested a parietal locus for the phonological store. Fiez explored functional dissociations between the prefrontal cortex and cerebellum. In particular, she proposed that areas 44 and 45 in the left hemisphere are essential for phonological recoding. In support of this, these areas are active, not only during working memory tasks that may depend on internal rehearsal but also during tasks that require a phonological judgment (e.g., is the e pronounced as a long or short vowel in the word *held*). In

continued on the following page

continued from the previous page

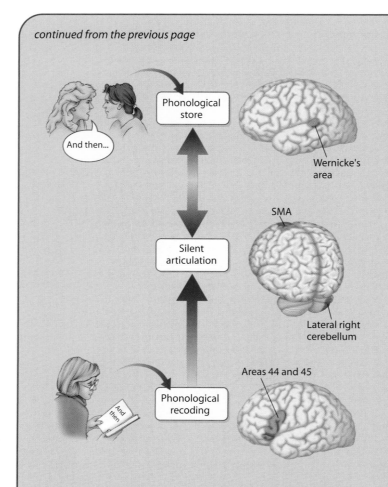

Figure B Component analysis of the phonological loop system of working memory. For the task of remembering a list of read words, the phonological loop is hypothesized to be composed of (at least) three components associated with the premotor/prefrontal cortex, cerebellum, and posterior region of the left hemisphere. The recoding process would not be required if the stimuli were presented auditorally.

contrast, the cerebellum's role may be restricted to rehearsal. Numerous PET studies revealed that similar structures are active during overt and covert (or imagined) movement. Thus, the cerebellum may be activated during silent articulation in a manner similar to its contribution to overt speech.

This analysis can account for the cerebellum's being activated in many PET studies of working memory, even when the amount of overt movement is equated in experimental and control tasks. What remains to be seen is whether the cerebellum's role in this aspect of working memory is essential, or simply part of a net-

work for internal speech. If the latter, the cerebellum's role would still remain closely linked to its motor functions. Yet it may be that the capability to sustain phonological representations depends on internal rehearsal. As such, we might expect that patients with cerebellar lesions are impaired on phonological analyses.

These questions await further study. For now, we can appreciate how cognitive neuroscience has opened new frontiers for exploring how the brain supports complex behaviors. Executive functions do not reside in a single structure but result from the interplay of diverse cortical and subcortical neural systems.

items; the working memory buffer must be updated continually to keep track of what the current stimulus must be compared to. Tasks such as *n*-back tasks require both the maintenance and the manipulation of information in working memory.

Figure 12.14b depicts the same set of activations from the meta-analysis of imaging studies, but now the data are recoded. As shown, the difference between maintenance-only and maintenance plus manipulation tasks map on to a dorsal-ventral distinction with the prefrontal cortex.

The *n*-back tasks capture an essential aspect of prefrontal function, putting the emphasis on the active part of working memory. Think about what happens when you reach for your wallet to pay a bill after dinner at a restaurant. Not only must you remember the price of the

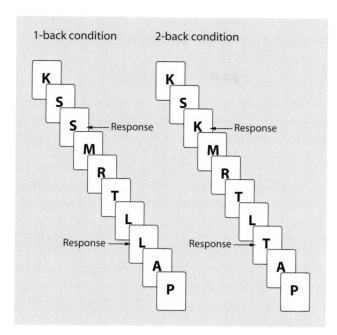

Figure 12.15 In *n*-back tasks, responses are required only when a stimulus matches one shown *n* trials before. The contents of working memory must be manipulated constantly as the target is updated on each trial.

dinner, but also you have to calculate the tip. Working memory is not only about keeping task-relevant information active; it also is about manipulating that information to accomplish behavioral goals (Petrides, 2000). The recency and self-oriented pointing tasks also reflect the updating aspect of working memory. Both require that the subjects remember whether something is familiar and the context in which the stimulus was encountered.

The link between source memory and working memory is not as obvious. On the one hand, source memory is a form of contextual memory. By definition, it requires remembering the context in which something was experienced. But source memory is a form of a long-term memory, and in general, patients with prefrontal lesions do not show impairment on long-term memory tasks. However, the deficit on source memory tasks is primarily in terms of learning source information. It is not clear that the patients would have similar problems if they are asked about the source of information they had learned prior to the onset of their neurological condition. Nonetheless, if we think of working memory as a repository of transient representations and the manipulation of this information, why should source memory be especially disrupted? One possibility is that source memory requires integrating distinct sources of information. The prefrontal cortex is important in this integration process, perhaps because the representations

must be sustained to allow these links to occur (Prabhakaran et al., 2000).

The Selection of Task-Relevant Information

As described earlier, goal-oriented behavior requires the **selection of task-relevant information.** Indeed, this selection process is a cardinal feature of tasks associated with the lateral prefrontal cortex. Within this framework, the emphasis on prefrontal function shifts from a role in memory to a role involved in the allocation of attentional resources.

Suppose you are telling a friend about a recent trip to San Francisco (Figure 12.16). One highlight was your walk across the Golden Gate Bridge. In describing the beauty of this structure, how its two towers span the mouth of the bay linking the city with the Marin Headlands, you will have activated a lot of semantic information from long-term memory, information about the location, shape, and color of the bridge, as well as episodic information related to your visit to the bridge. In the working memory model, a transient file is established to

Figure 12.16 Prefrontal cortex not only provides a working memory buffer but also may use an inhibitory mechanism to highlight the information that is most relevant to the current task demands. When the subject is asked about the color of the Golden Gate Bridge, information regarding the location and shape of the bridge is inhibited.

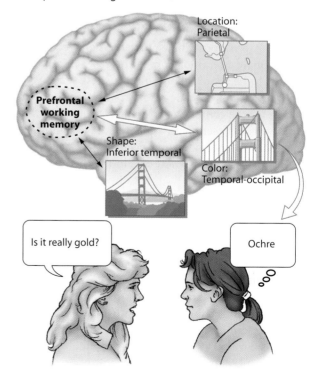

hold this information. Suppose your friend asks you to focus on a particular feature such as the color of the bridge. This query requires that one piece of information be made more salient than others. To answer the question, you must be able to select this information from the contents of working memory.

Art Shimamura (2000) of the University of California has argued that the prefrontal cortex can be conceptualized as a **dynamic filtering** mechanism. The frontal cortex is a repository of representations and selects information most relevant for meeting the task's demands. Working memory is more than the passive sustaining of representations; it requires an attentional component in which the subject's goals modify the salience of sources of information. In a spatial delayed-response task, the critical information is the object's location. In an object version of this task, the critical information is the stimulus's shape.

Can a deficit in dynamic filtering account for the range of deficits exhibited by patients with lesions of the lateral prefrontal cortex? Could it provide an alternative account for the difficulty these patients have on source memory tasks? Perhaps, if we consider that memory for context is generally secondary to memory for content. We rarely need to recall where we learned something; it is usually sufficient to remember the information itself. We can successfully bake a cake by remembering how to combine the ingredients without recalling that we first learned to make this recipe from *Joy of Cooking.* (But when we forget a step, it is useful to know where the information can be obtained.) With the asymmetry between memory for content and memory for context, it is reasonable to suppose that source memory requires disproportionate attention. Content representations can be expected to be more salient and thus less sensitive to a loss of attention. Greater resources are required when we must ignore content and attempt to retrieve context. When we read a sentence, we derive its meaning while paying little attention to superficial aspects such as what font it is printed in or where it is in the text. We can recall this surface information, but only when we make an effort to encode it.

The hypothesis that the frontal lobes play a critical role in selecting task-relevant information can account for frontal lobe patients' problems with the Wisconsin Card Sorting Task, in which subjects sort multidimensional stimuli (see Figure 12.4). They learn by trial and error; they hypothesize which information is relevant and which is not. Subjects, then, must learn to attend to the correct dimension while filtering the other two dimensions. Frontal lobe patients have difficulty with this. They are especially prone to perseverative errors—con-

tinuing to sort by an old rule even when told it is no longer appropriate. This tendency can be viewed as a failure in selection. After learning to attend to one dimension, a person with frontal lobe damage continues to focus on it.

The filtering hypothesis also offers a way to appreciate the role of the frontal lobe in tasks where memory demands are minimal. Frontal lobe patients display heightened interference on the Stroop task. For the Stroop task, subjects are shown a list of colored words; the words spell color names such as *red, green,* or *blue*. In the congruent condition, the colors of the words correspond to their names; in the incongruent one, the word names and colors do not correspond (see Figure 4.5). With years of reading experience, we have a strong urge to read words even when the task requires us to ignore them in favor of color; thus, everyone is slower in responding to incongruent stimuli in comparison with congruent stimuli. This difference is even greater in patients with frontal lobe lesions.

In an elegant series of experiments, Sharon Thompson-Schill, Mark D'Esposito, and their colleagues (1997, 1998) tested the dynamic filtering hypothesis. The studies involved the semantic generation task. On each trial, a noun is presented and the subjects are required to generate a word that is semantically associated. As described in Chapter 9, early PET studies on language had found that in comparison to a task in which subjects simply read the nouns, the generation task produced increased activation in area 44, the inferior frontal gyrus of the left hemisphere. This region is just anterior to Broca's area. One interpretation of this finding is that this region is a critical component of semantic memory. Alternatively, the prefrontal activation might reflect a retrieval into working memory of the semantic associates to the target item, information that is stored in posterior regions of the cortex.

To discriminate between these two hypotheses, the researchers compared brain activation during two types of verb generation tasks in which the selection demands were varied (Figure 12.17). In the low-selection condition, each noun was associated with a single verb. For example, when asked to name the action that goes with "scissors," almost everyone will respond, "cut." In the high-selection condition, each noun had numerous associates. For example, for the word *rope*, multiple answers are reasonable: *tie, lasso, twirl*. In both conditions, the demands on semantic memory are similar. The subject must comprehend the target noun and retrieve semantic information associated with that noun. Thus, one would expect similar activation in the inferior prefrontal cortex, and this activation would be higher than

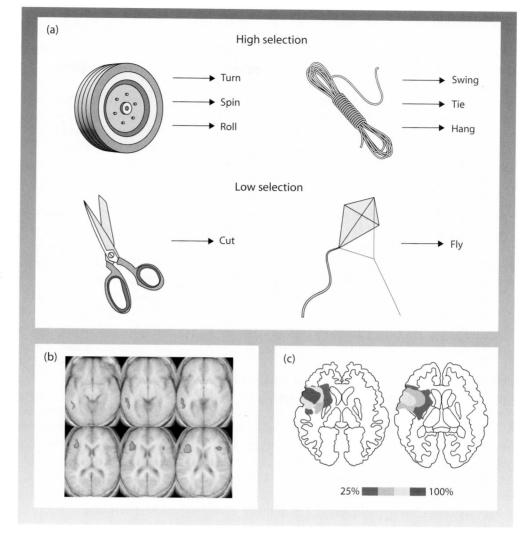

Figure 12.17 Involvement of inferior frontal cortex in response selection. **(a)** The verb generation task can be performed with nouns that are associated with many actions (high selection) or few actions (low selection). **(b)** Areas showing higher activity in the high-selection condition are shown in yellow. **(c)** Overlap in lesion location for patients who had difficulty in the high-selection condition. (b) From Thompson-Schill et al. (1997). (c) From Thompson-Schill et al. (1998).

in a baseline condition requiring the subjects to read the nouns. However, if this region is involved in the selection operation, then activation should be greater in the high-selection condition. The results clearly favored the selection hypothesis.

Converging evidence was obtained in a neuropsychological study (Thompson-Schill et al., 1998). Patients with lesions that spanned the inferior frontal cortex in the left hemisphere failed on about 15% of the trials in the high-selection condition. Interestingly, these errors did not result from the generation of an erroneous verb. Rather, the patients either failed to come up with a single verb within 30 seconds or simply read the nouns. Such errors did not occur in the low-selection condition; in fact, the patients made essentially no errors in this condition. The inability of the patients to come up with a verb points to a selection problem. These same patients would have no trouble judging that the words *rope* and *twirl* go together. But when they

have to select one response from among a set of viable candidates, they get stuck.

It is important to note that the generation task also produced increased activation in the left temporal lobe in comparison to a baseline repeat task. This area is a reasonable candidate for the long-term storage of semantic information. Indeed, the results of a follow-up study support this hypothesis (Thompson-Schill et al., 1999). Subjects were trained to make two types of generation responses, one based on naming an action associated with the noun and another based on naming the color associated with the noun. The initial scanning run revealed a replication of the prefrontal and temporal cortical engagement during the generation tasks, demonstrating that the same inferior frontal region was recruited for both types of semantic associations. Of interest here is what happened in subsequent scanning runs. The list of nouns was repeated. In one condition, the subjects performed the same generation task as for the first run; in the other, they were required to perform the alternative generation task. This manipulation led to an interesting dissociation between the BOLD response in the prefrontal and temporal cortices. Prefrontal activation increased in scanning runs in which the generation requirements changed. Selection and filtering would likely be high under such conditions. A different pattern was seen in the temporal lobe. Here, the activation decreased on the second run for both the same and the different generation conditions. Such decreases with repetition have been seen in many imaging studies of priming (see Chapter 8). The fact that the decrease was observed even when the generation requirements changed is consistent with the idea that semantic attributes, be they relevant or irrelevant to the task at hand, were automatically activated upon presentation of the nouns. The prefrontal cortex applies a dynamic filter to select the information that is relevant to the current task requirements.

FILTERING AS AN INHIBITORY PROCESS

Dynamic filtering can influence the contents of information processing in at least two distinct ways. One is to accentuate the attended information. For example, when we attend to a location, our sensitivity to detect a stimulus at that location is enhanced. Or, we can selectively attend by excluding irrelevant information. Stroop interference can be eliminated by squinting one's eyes so the words are no longer legible. As seen in times of budgetary crises, the hypotheses are not mutually exclusive. If we have fixed resources, allocating resources to one thing places a limit on what is available for others.

In behavioral tasks, it is often difficult to distinguish between facilitatory and inhibitory modes of control. But electrophysiological studies have indicated that losing **inhibitory control** may be a more appropriate descriptor of frontal lobe dysfunction. Robert Knight of the University of California recorded the evoked potentials in groups of patients with localized neurological disorders (Knight and Grabowecky, 1995). In the simplest experiment, subjects were presented with tones, and no response was required. As might be expected, the evoked responses were attenuated in patients with lesions in the temporoparietal cortex in comparison to control subjects. This difference was apparent about 30 msec after stimulus onset, the time when stimuli would be expected to reach the primary auditory cortex. The attenuation presumably reflects tissue loss in the region that generates the evoked signal. A more curious aspect is shown in Figure 12.18: Patients with frontal lobe lesions have en-

hanced evoked responses. This enhancement was not seen in the evoked responses at subcortical levels; the effect did not reflect a generalized increase in sensory responsivity but was limited to the cortex.

The failure to inhibit irrelevant information was more apparent when the subjects were instructed to attend to auditory signals in one ear and ignore similar sounds in the opposite ear, with the attended ear varied between blocks. In this way, one can assess the evoked response to identical stimuli under different attentional sets (e.g., response to left-ear sounds when they are attended to or ignored). With healthy subjects, these responses diverge at about 100 msec; the evoked response to the attended signal becomes greater. This difference is absent in patients with prefrontal lesions, especially for stimuli presented to the ear contralateral to the lesion. What happens is that the unattended stimulus receives a heightened response, which accords with the notion that the frontal lobes modulate the salience of perceptual signals by inhibiting unattended information. We are bombarded with stimuli: Our ability to respond appropriately requires that we select information relevant to the task.

In the study just described, we can see inhibition operating to minimize the impact of irrelevant perceptual information. This same mechanism can be applied to memory tasks for which information must be internally maintained. Again, consider the monkey attempting to perform the delayed-response task. The monkey views the target being placed in one of the food wells, and then the blind is closed during the delay period. The monkey's mind does not just shut down; it sees and hears the blind being drawn, looks about the room during the delay interval, and perhaps contemplates its hunger. All such intervening events can distract the animal and cause it to lose track of which location is baited. To succeed, it must ignore the distractions and sustain the representation of its forthcoming response. We have all experienced failures in similar situations. A friend gives us her telephone number, but we forget it. The problem is not a failure to encode the number. Something else captures our attention. We fail to block out the distraction. This point is underscored by the fact that primates with prefrontal lesions perform better on delayed-response tasks when the room is darkened during the delay (Malmo, 1942) or when given drugs that decrease distractibility.

Failure of inhibition also can account for the dissociation between recognition and recency memory. Subjects must remember all stimuli since they do not know which will be tested in probe trials. As we encode each stimulus, we pay attention to it, with one cost being the

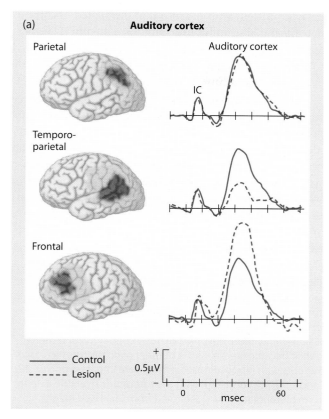

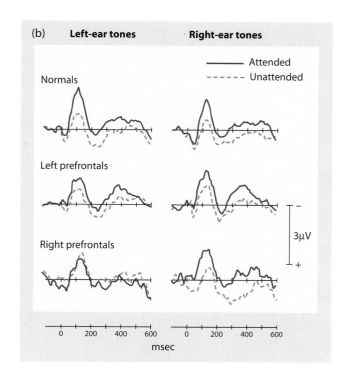

Figure 12.18 Evoked potentials reveal filtering deficits in patients with lesions in the lateral prefrontal cortex. **(a)** Evoked responses to auditory clicks in three groups of neurological patients. The subjects were not required to respond to the clicks. The first positive peak occurs at about 8 msec and reflects neural activity in the inferior colliculus. The second positive peak occurs around 30 msec, the P30, reflecting neural responses in the primary auditory cortex. Both responses are normal in patients with parietal damage. The second peak is reduced in patients with temporoparietal damage, reflecting the loss of neurons in the primary auditory cortex. The auditory cortex response is amplified in patients with frontal damage, suggesting a loss of inhibition from frontal lobe to temporal lobe. Note that the evoked response for control subjects is repeated in each panel. **(b)** Difference waves for attended and unattended auditory signals. Subjects were instructed to monitor tones in either the left or the right ear. The evoked response to the unattended tones is subtracted from the evoked response to the attended tones. In healthy individuals, the effects of attention are seen at approximately 100 msec, marked by a larger negativity (N100). Patients with right prefrontal lesions show no attention effect for contralesional tones presented in the left ear but show a normal effect for ipsilesional tones. Patients with left prefrontal lesions show reduced attention effects for both contralateral and ipsilateral tones. Adapted from Knight and Grabowecky (1995).

inhibition of previous stimuli. The degree of activation of a representation is inversely related to how long ago the stimulus was presented. As shown in Figure 12.19, recency judgments can be made by comparing the magnitude of the two target items. Whichever item has the strongest residual activation would be judged most recent. For healthy subjects, inhibition could quickly lower the activation of distinct items. But for frontal lobe patients, the loss of inhibition would result in lingering activations and render such judgments difficult. A paradox is that they would not have a problem with recognition because their memory is so high.

The dynamic filtering hypothesis demonstrates how lateral prefrontal cortex contributes in a similar way to

the control of on-line behavior and memory. In Chapter 8 we saw that both the prefrontal cortex and the hippocampus become activated during neuroimaging studies of memory encoding. A patient study offers a way to see dissociations in the functional contributions of these areas (Chao and Knight, 1995). Patients with either lateral prefrontal or hippocampal lesions were tested on an auditory version of a delayed matching to sample task. The stimuli were tape recordings of common environmental sounds such as a dog barking or a running faucet. The delay between the sample and test stimulus varied from 4 to 12 seconds. When no distractor was presented during this interval, performance was near perfect for both patient groups and indistinguishable from that of healthy

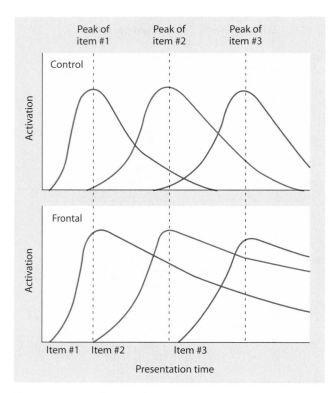

Figure 12.19 When a subject is presented with a series of items, activation for each item decays. A loss of inhibitory mechanisms following frontal lobe damage will lead to a slower decay process. Judgments on a recency memory task may be based on a comparison of the residual activation of the series of stimuli. In healthy people, the rapid decay of activation allows the temporal tag for each item to be distinct. In frontal lobe patients, the sustained activation leads to errors due to the similar activation states associated with successive items.

control subjects. However, the results for the two groups were quite different when a series of irrelevant distractor tones was presented during the delay period (Figure 12.20). The patients with hippocampal lesions performed similar to the control subjects after the shorter delays. Their performance abruptly deteriorated after the longest

delay, consistent with the idea that they failed to consolidate the initial stimulus into long-term memory. The prefrontal group was affected by the distractors. They failed to filter out the irrelevant tones, and this problem became exacerbated as the number of tones increased.

The loss of dynamic filtering captures an essential feature of prefrontal damage. Basic cognitive capabilities are generally spared. The patients' intelligence shows little evidence of change, and they can perform normally on many tests of psychological function. But they are in a particularly vulnerable condition. In an environment in which multiple sources of information compete for attention, the patients have difficulty maintaining their focus. With this in mind, we are ready to return to the links between working memory and goal-oriented behavior.

Figure 12.20 Susceptibility to distraction in patients with lateral prefrontal lesions. Subjects performed a delayed auditory matching to sample task. Unrelated distractor tones were presented during the delay period. The group with prefrontal lesions made more errors for all delay conditions, and the deficit became greater as the number of distractors increased. Patients with hippocampal damage were impaired only at the longest delay, consistent with the role of this structure in long-term memory formation. Adapted from Chao and Knight (1995).

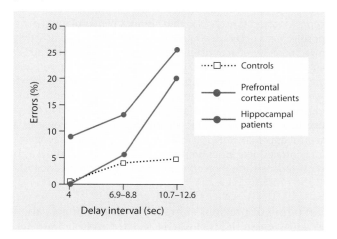

GOAL-ORIENTED BEHAVIOR

Our actions are not aimless, nor are they entirely dictated by events and stimuli immediately at hand. We choose to act because we want to accomplish goals, to gratify personal needs. Goal-driven behavior can be as mundane as turning on a computer to play a game, or as complex as attending lectures, reviewing notes, and reading to learn. Goals dwell within a hierarchy. We can describe the immediate goal of an action—to learn—but also recognize that this is really a subgoal for a larger plan. Learning subjects is necessary for admission to graduate or professional school, and these institutions can be a ticket to fame, fortune, and personal satisfaction.

The ability to form a coherent plan of action is compromised after damage to the prefrontal lobe. Patients like W.R., the wayward lawyer, are plagued by this problem. They are unable to resurrect a normal life even after the acute symptoms of their illness have passed. It is difficult to attribute the problem to a lack of knowledge. They are aware of their deteriorating social situation and have the intellectual capabilities to generate ideas that may alleviate their condition. But their efforts to overcome their inertia are haphazard at best. They are unable to sustain a plan of action and meet their goals.

Tim Shallice at the National Hospital in London (Shallice and Burgess, 1991) documented this problem in three patients who suffered frontal lesions from head traumas. On many neuropsychological assessment procedures, the patients demonstrated intact, perhaps even superior, cognitive abilities. Not only did they all score at least one standard deviation above average on an IQ test, but also they had few problems on standard tests of frontal lobe function, including the Wisconsin Card Sorting Task. To examine their ability to engage in goal-oriented behavior, the experimenters designed tasks that mimicked the errands a person might have to run on a Saturday morning. Patients were asked to go to a shopping center and purchase items (a loaf of bread, a packet of throat lozenges), keep an appointment at a certain time, and collect four pieces of information such as the price of a pound of tomatoes or the exchange rate of the rupee. The task was not designed to tap memory. The patients were given a list of the errands and instructions to follow, such as spend as little money as possible.

Despite their superior intellectual abilities, all three patients had difficulty executing this assignment. One patient failed to purchase soap because the store she visited did not carry her favorite brand; another wandered outside the designated shopping center in pursuit of an item that could be found within the designated region. All became embroiled in social complications. One succeeded in obtaining the newspaper but was pursued by the merchant for failing to pay! In a related experiment, patients were asked to work on three tasks for 15 minutes. Whereas control subjects successfully juggled their schedule to ensure that they made enough progress on each task, the patients got bogged down on one or two tasks.

Planning and Selecting an Action

In preparing for an exam, a good student develops an action plan such as the one in Figure 12.21. This plan can be represented as a hierarchy of subgoals, each requiring actions to achieve the goal. At the top is the goal of doing well on the exam. To do so, subgoals are designed: Reading must be completed, lecture notes reviewed, and material integrated to identify themes and facts. Perhaps the student will generate essay questions and practice writing answers. A timeline might include

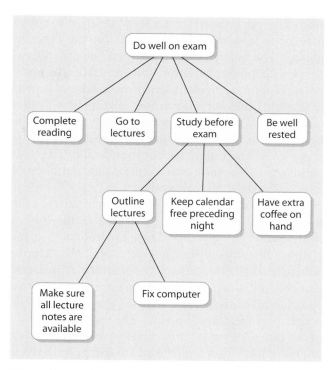

Figure 12.21 An action hierarchy. Successfully achieving a complex goal such as doing well on an exam requires planning and organization at multiple levels of behavior.

when the subgoals will be completed and how to devote the day before the exam to study.

Three components are essential for successfully executing an action plan (Duncan, 1995). First, one must identify the goal and develop subgoals. Perhaps the student has noticed that the professor often includes readings on exams, and so a subgoal might be to maximize the effort devoted to reading. Second, in choosing among goals, consequences must be anticipated. Will the information be remembered better if the student sets aside 1 hour a day for study during the week preceding the exam, or is it better to cram intensively the night before? Third, the student must determine what is required to achieve the subgoals. A place must be identified for study. The coffee supply must be adequately stocked. A pen must be available for marking critical passages. It is easy to see that these components are not entirely separate. Purchasing coffee can be an action and a goal.

THE SELECTION OF APPROPRIATE SUBGOALS

When viewed as a hierarchy, it is easy to see that failure to achieve a goal can happen in many ways. If reading is not completed, the student may lack knowledge essential for an exam. If a friend arrives unannounced to celebrate his birthday the weekend before the exam, critical study time can be lost. If coffee is not handy, the student may lack the stamina to stay awake for the final all-nighter. The failures of goal-oriented behavior in patients with prefrontal lesions can be traced to many potential sources. Problems can arise because of deficits in filtering irrelevant information, in keeping one's eyes on the prize. Or the challenge may come in selecting the best way to achieve a particular goal. Developing an action plan means a simultaneous consideration of possible subgoals. We must evaluate these different plans of attack to establish sensible goals.

A clever demonstration of the importance of the frontal lobes in this evaluation process comes from a study carried out in Jordan Grafman's laboratory at the National Institutes of Health (Goel et al., 1997). The researchers sought to capture the real-world problems that patients with penetrating head injuries face. To this end, they asked the patients to help an upwardly mobile couple plan their family budget. The patients were told of the family's long-term financial goals: to purchase a home within 2 years, send the kids to college in 15 years, and retire in 35 years. But similar to many young couples, this family was having trouble living within their means. They were actually falling further into debt rather than building up a nest egg. When shown the list of family expenditures, control participants focused on a few categories. The expenses for clothing, $175 a month, seemed like a lot for a family of four with two young children. Perhaps substantial savings could be found by dressing the younger child in hand-me-downs from the oldest.

The responses of the frontal patients were revealing. They understood the overriding need to identify places where savings could be achieved. Appreciating the need to budget was something they could relate to. But their solutions did not always seem reasonable. One patient focused on the family's rent. Noting that the $10,800 yearly expense for rent was by far the biggest expense in the family budget, he proposed that it be eliminated. When the experimenter pointed out that the family would need a place to live, the patient was quick with an answer: "Yes. Course I know a place that sells tents cheap. You can buy one of those." We see in this naturalistic domain another manifestation of the selection problems reviewed earlier.

EXECUTIVE CONTROL OF GOAL-ORIENTED BEHAVIOR

By focusing on the housing costs, the patient is demonstrating a certain inflexibility in his decision making. The large price tag assigned to rent was a particularly

salient piece of information, and the patient's budgeting efforts were captured by the potential savings to be found here. From a strictly monetary perspective, this decision makes sense. But at a practical level, we realize the inappropriateness of this choice. Making wise decisions with complex matters such as one's long-term financial goals requires keeping an eye on the overall picture. We must not lose track of the forest for the trees. It is essential that we monitor and evaluate the different subgoals. To do so requires an ability to shift our focus from one subgoal to another. In more general terms, this ability is referred to as *task control.*

We are all familiar with the importance of task control. It is common to hear someone describe a particularly active friend as someone who is good at "multitasking." Globally, this description might apply to someone who can keep up with all of her studies, manage a part-time job, and still find time to exercise 2 hours each afternoon. Locally, this phrase is used to describe someone who, while checking e-mail and transferring files on the computer, is also able to scan her phone messages. Task control represents the interface through which goals influence behavior. Complex actions require that we shift from one subgoal to another in a coordinated manner.

Task-switching experiments have been designed to examine this aspect of executive control. We already have introduced one such example, the Wisconsin Card Sorting Task. This task tests how flexible a person can be in sorting multidimensional stimuli. However, as with many neuropsychological tests, the Wisconsin Card Sorting Task is limited in terms of how well it can isolate particular mental operations. It was developed as a diagnostic tool: Poor performance on the test correlates with frontal lobe damage. Thus, it is an important tool for clinical neuropsychologists who wish to assess "frontal" function in patients with head trauma or degenerative disorders, but the source of the patients' failure remains unclear. It may reflect the strategic requirements involved in trial-and-error learning, a tendency to focus on one stimulus dimension at the expense of others, or an inability to ignore a dimension that was reinforced previously.

To get around this limitation, simplified **task-switching** experiments are used. In Chapter 11 we described a task-switching study with Parkinson's disease patients, showing that they had a deficit in their ability to change mental set that is similar to their problems in changing motor set. Another example of a task-switching experiment is shown in Figure 12.22. On each trial, a letter-digit pair is presented. The task goal is switched every two trials, alternating between trials in which the subject is required to name the digit and trials in which

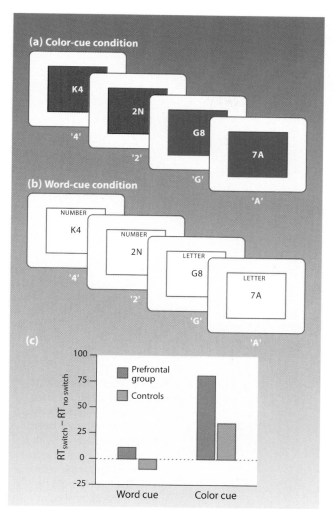

Figure 12.22 Task-switching experiment, with the task cued by either **(a)** a color or **(b)** a word. **(c)** Switching cost, the time required to switch from one task to the other (e.g., from naming the digit to naming the letter), is measured as the difference in reaction time on switch trials and no-switch trials. Patients with prefrontal lesions showed impairment only on the color cue condition. (a), (b) From Rogers et al. (1998).

the subject is required to name the letter. In this manner, the first trial for each pair represents a trial in which the task goal switches (e.g., changes from naming letter to naming digit) and the second represents a trial in which the task goal remains the same. The time required to change from one goal to the other, the *switching cost,* is measured by the difference in reaction time on these two types of trials.

A second important variable in this experiment is the manner in which the task goal is specified. Given that the trials alternate in a consistent manner, it is possible for subjects to keep track of their place. But this would increase the processing requirements for the subjects. To avoid this, the task goal is cued by an external cue. In one

condition, this cue is indicated by the background color. In the other condition, a visual word cue is used.

The cueing variable turned out to be critical. When a visual word cue was used to specify the task goal, the patients with lateral prefrontal lesions performed similarly to matched control subjects. However, when a color cue was used, the patients were slow on the switch trials. This dissociation reinforces the idea that the prefrontal cortex is important for coordinating goal-oriented behavior. Moreover, this form of control is needed especially when the goal must be retrieved from working memory. With the color cue, the patients must remember the associations between the colors and the tasks (e.g., blue with digit naming). The word cues do not require this referencing back to working memory.

One result of this experiment is noteworthy. Parkinson's disease patients who were tested on the same task had no difficulty in either the color or word cue conditions. This finding would seem to be at odds with that shown in Figure 11.46. One difference between the two studies is the use of a fixed sequence in which the two tasks alternated every other trial. Perhaps the Parkinson's disease patients were able to anticipate the switch from one goal to the next. Frontal patients failed to exhibit the kind of insight needed to come up with such strategies (see Cortical-Subcortical Interactions in Executive Functions).

Seiki Konishi of the University of Tokyo School of Medicine and colleagues (1998) used fMRI and a computerized version of the Wisconsin Card Sorting Task to specify regions within the prefrontal cortex that are involved in task switching. Recognizing the complexity of the mental operations involved in this task, they focused their analysis on neural responses following a feedback message indicating that a dimensional shift was required (Figure 12.23). The most prominent foci were in the inferior frontal sulci in both hemispheres. The magnitude of the hemodynamic response during task switching was a function of the number of possible dimensions. Larger responses were observed when there were three potential new dimensions rather than just one or two new dimensions. The function of this area is not simply related to inhibiting the dimension that is no longer relevant. Rather it appears related to the demands of dynamically filtering between the various new courses of action. Presumably, when cued to shift, the subject retrieves into working memory the other potential categorization rules. Selection is now required to determine which goal should be used to guide the subject in processing subsequent stimulus information.

Tim Shallice, together with Donald Norman of the University of California (Norman and Shallice, 1986),

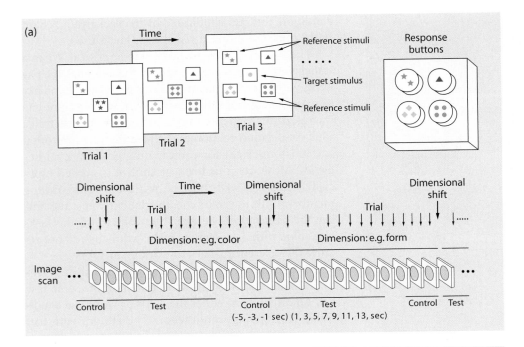

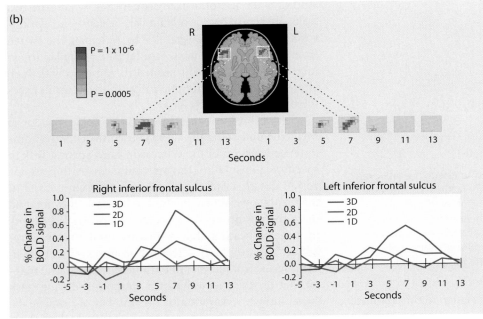

Figure 12.23 Modified Wisconsin Card Sorting Task for event-related fMRI. **(a)** Stimulus displays and response board. Subjects matched the center object to one of four objects in the corners, using the response board. The match could be made on the basis of color, form, or number. After ten correct responses, the matching rule would change, indicating a dimensional shift. **(b)** Increased activation was observed bilaterally in the inferior frontal cortex following the signal to shift dimensions. Note that the hemodynamic response peaks about 7 seconds after the shift. From Konishi et al. (1998).

developed the model in Figure 12.24 to account for goal-oriented behavior. This model conceptualizes the selection of an action as a competitive process, similar to concepts developed in Chapter 11. At the heart of the model is the notion of *schema control units* or *representations of responses* (a term used in a generic sense here). These schemas can correspond to explicit movements or to the activation of long-term representations that lead to purposeful behaviors. For example, when we see a word printed on paper, an action schema could be the articulatory gestures required to pronounce the word.

Or, in reading the word, its semantic meaning and associated representations may be activated.

Schema control units receive input from many sources. Norman and Shallice emphasized perceptual inputs and their link to these control units. The strength of the connections, however, reflects the effects of learning. If we have had experience in restaurant dining, walking into a restaurant will activate behaviors associated with waiting for the hostess or looking at the menu. Moreover, walking into the restaurant can elicit varied affective responses—which is how somatic markers in-

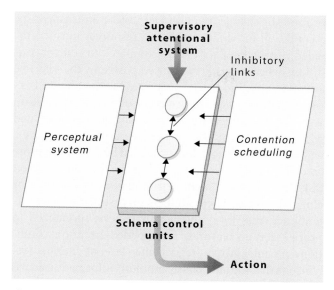

Figure 12.24 Norman and Shallice's model of response selection. Actions are linked to schema control units. The perceptual system produces input to these control units. However, selection of these units can be biased by the contention scheduling units and the supervisory attentional system (SAS). The SAS provides flexibility in the response selection system. Adapted from Shallice et al. (1989).

fluence schema control units. A restaurant setting is often associated with pleasant memories of fine meals and good company. For some, though, it may reawaken painful thoughts of washing dishes or waiting for grown-ups to end dull conversations.

External inputs can be sufficient to trigger schema control units. For example, it is hard not to track a moving object by moving one's eyes. But in most situations our actions are not dictated solely by the input; many schema control units can be activated simultaneously, and a control process is needed to ensure that the appropriate control units are selected. Norman and Shallice postulated two types of selection. One type, which is rather passive, is what they call *contention scheduling*. Schemas not only are driven by perceptual inputs but also compete with one another, especially when two control units are mutually exclusive. We cannot look at two places at the same time or move the same hand to simultaneously pick up a glass and a fork. By having inhibitory connections between schemas, the model accounts for why we act coherently. Only one schema (or nonoverlapping schemas) can win the competition. If competition does not resolve the conflict, the result is no action. None of the schemas is activated enough to trigger a response.

The second means for selection comes by way of the **supervisory attentional system** (SAS). The SAS is es-

sential for ensuring that behavior is flexible. It is a mechanism for favoring certain schema control units, perhaps to reflect the demands of the situation or to emphasize some goals over others. We can postulate types of situations where selection would benefit from an SAS:

1. When the situation requires planning or decision making
2. When links between the input and schema control units are novel or not well learned
3. When the situation requires a response that competes with a strong, habitual response
4. When the situation requires error correction or troubleshooting
5. When the situation is difficult or dangerous

The SAS is a psychological model of executive control. It specifies some of the key situations in which control operations would be useful. Although the SAS in this model is sketched as a single entity, it is unlikely that a single neural structure would be involved in all of these operations. It would be more plausible to expect that the functions embodied in the SAS are part of a distributed network, a set of neural regions that as a group come into play in the situations described above. Much of the discussion in this chapter has emphasized the first type of situation. We have seen various ways in which prefrontal cortex is important for goal-oriented behavior: It provides a way to represent the goals that guide our actions and thus, select, maintain, and manipulate the information that is essential for accomplishing these goals. Complex behaviors require an ability to switch from one subgoal to another, or in multitask situations, to switch between the subtasks. Again, the evidence suggests a key role for prefrontal cortex in this function.

The last four situations listed share one aspect of executive control that has not been discussed in detail to this point. For a person engaged in goal-oriented behavior, especially a behavior that involves subgoals, it is important that there be a way to monitor progress. If this is a well-learned process, then there should be a means for signaling deviations from the expected course of events. The cook must recognize that the stove thermostat is off when his favorite soufflé has failed to puff up. If the behavior is novel, or being performed in an unusual context, then different actions must be evaluated to determine if progress is being made toward achieving the goal. The cook may experiment with a few new spices in the soufflé for a special dinner. But if the combination turns out all wrong, he will want to whip together a last-minute concoction.

The Anterior Cingulate as a Monitoring System

In the last decade and a half we have witnessed burgeoning interest in possible executive functions of the anterior cingulate cortex. Buried in the depths of the frontal lobes and characterized by a primitive cytoarchitecture, this structure was assumed to be a component of the limbic system, helping to modulate autonomic responses during painful or threatening situations. Whereas functional roles for most cortical regions have been inspired by behavioral problems associated with neurological disorders, interest in the anterior cingulate has been inspired by serendipitous activations found in this region during PET studies. For example, metabolic activity increases in the anterior cingulate during semantic generation tasks, similar to what is found in the inferior prefrontal cortex. The cingulate activation was especially surprising in these initial studies given that the lesions of this region were not associated with language disorders.

These findings have led to a reconceptualization of this area as part of an attentional hierarchy. In this view, the anterior cingulate occupies an upper rung on the hierarchy, playing a critical role in coordinating activity across attentional systems (Figure 12.25). Consider a PET study of visual attention in which subjects must selectively attend to a visual feature (color, motion, shape)

Figure 12.25 The anterior cingulate has been hypothesized to operate as an executive attention system. This system serves to ensure that processing in other brain regions is most efficient given the current task demands. Interactions with the prefrontal cortex may select working memory buffers; interactions with the posterior cortex can serve to amplify activity in one perceptual module over others. The interactions with the posterior cortex may be direct or they may be mediated by connections with the prefrontal cortex. Adapted from Posner and Raichle (1994).

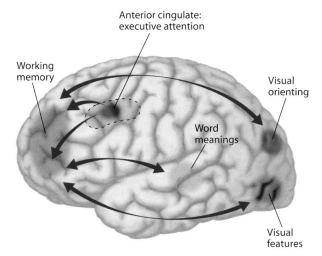

or monitor changes in all three features simultaneously, a condition in which attentional resources must be divided (Corbetta et al., 1991). Compared to control conditions where stimuli are viewed passively, the selective attention conditions were associated with enhanced activity in feature-specific regions of visual association areas. For example, attending to motion was correlated with greater blood flow in the lateral prestriate cortex, whereas attending to color stimulated blood flow in more medial regions. In contrast, during the divided-attention task the most prominent activation was in the anterior cingulate cortex. These findings suggest that selective attention causes local changes in regions specialized to process certain features. The divided-attention condition, in contrast, requires a higher-level attentional system, one that simultaneously monitors information across these specialized modules.

An association between the anterior cingulate and attention is further shown by how activation in this region changes as attentional demands decrease. If the verb generation task is repeated over successive blocks, the primary activation shifts from the cingulate and prefrontal regions to the insular cortex of the temporal lobe (Raichle et al., 1994). This shift reflects the fact that the task has changed. In the initial trial, the subjects have to choose between alternative semantic associates. If the target noun is *apple*, then possible responses are "peel," "eat," "throw," or "type," and the subject must select between these alternatives. On subsequent trials, however, the task demands change from semantic generation to memory retrieval. The same semantic associate is almost always reported. Thus, if the subject reports "peel" on the first trial, he invariably will make this same choice on subsequent trials.

The anterior cingulate activation during the first trial can be related to two of the functions of an SAS: responding under novel conditions and with more difficult tasks. The generation condition is more difficult than the repeat condition since the response is not constrained. But over subsequent trials, the generation condition becomes easier (as evidenced by markedly reduced response times) and the items are no longer novel. Thus, the higher blood flow in the cingulate disappears, which reflects a reduced need for the SAS. That this shift indicates the loss of novelty rather than a general decrease in cingulate activity with practice is shown by the fact that when a new list of nouns is used, the cingulate activation returns.

The poor temporal resolution of PET makes it difficult to distinguish between functions associated with activations in the cingulate, lateral prefrontal cortex, and posterior foci. An event-related potential (ERP) study sheds some light here (Snyder et al., 1995). The

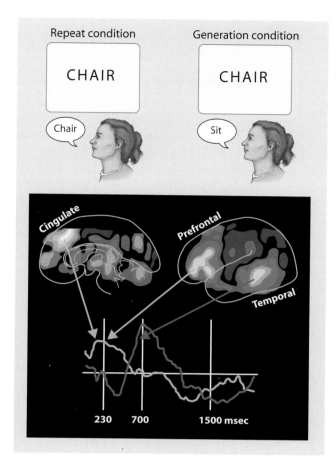

Figure 12.26 Neural generators associated with each peak in the difference waveform between generation and repeat tasks. **(Top)** Subjects hear a noun. In the repeat condition, they simply repeat the word; in the generation condition, they name a word that is a verb associate. To avoid including motor activity in the evoked potentials, subjects were instructed to withhold their responses until a "go" signal appeared, about 1500 msec after the stimulus. **(Bottom)** The difference waveform is obtained by subtracting the evoked potential in the repeat condition from the evoked potential in the generation condition. Dipole modeling techniques were used to identify the neural regions associated with each peak. Adapted from Snyder et al. (1995).

same subtractive logic was applied, but the difference was between ERP waveforms obtained during the generation and repeat conditions (Figure 12.26). Using ERP localization methods, the researchers sought to identify the source of the differences and their time course. The PET foci constrained how the generators were modeled. The first difference was observed about 180 msec after the onset of the target noun and was attributed to a single generator in the anterior cingulate. About 30 msec later, a second generator was required to model the data. This generator was localized to the lateral prefrontal cortex in the left hemisphere. Finally, around 620 msec

after stimulus onset, a third generator was linked to the posterior cortex in the left hemisphere.

The time course fits well with the general picture developed in this chapter. We can hypothesize that the initial cingulate activity reflects the allocation of attentional resources to these novel stimuli. The cingulate may establish a node in the working memory system of the lateral prefrontal cortex to hold representations retrieved from the longer-term semantic representations of word meanings in the posterior cortex. As processing spreads among the semantic network in the posterior cortex, the working memory system will inhibit representations of irrelevant associates (e.g., "red" or "round") and allow a task-relevant associate to become sufficiently activated (e.g., "peel") over alternative choices (e.g., "eat" or "throw"). In this way, we see how a monitoring system can help ensure the successful interactions between working and long-term memory in goal-oriented behavior. This study elegantly demonstrated the analytic power of combining the spatial resolution of PET and the temporal resolution of evoked potentials. PET had identified three foci of activity during the generation task. The ERP study revealed the time course of processing across these three areas.

Figure 12.27 summarizes evidence linking the anterior cingulate to some of the hypothesized functions of the SAS (Posner, 1994). The divided attention and generation tasks show how the cingulate is engaged under

Figure 12.27 Four ways in which a supervisory attentional system (SAS) would be involved in monitoring functions. Evidence for a role of the cingulate in each function is listed on the right. (See text for details.)

Required function of the SAS	Evidence indicating function related to anterior cingulate
Difficult situations	Blood flow increases during divided attention studies in comparison to focused attention studies
Novel situations	Blood flow increases during word generation task in comparison to word repeat task
Error correction	Evoked potential studies
Overcoming habitual responses	Blood flow increases during incongruent Stroop trial in comparison to congruent Stroop trials

The Brain and the Computer: Is the Marriage Working?

Authors: We asked Jonathan Cohen, professor of psychology at Princeton University and director of the Center for the Study of Mind, Brain, and Behavior, to address the question posed by the title of this box. Perhaps more than anyone else in the field, Cohen's research involves a combination of imaging and modeling work. We asked him to discuss the relationship between these two approaches. Do the results from one approach guide the work with the other? Or is there an asymmetry, with the results from one approach driving the questions one tests with the other? Here is his answer:

JC: This is a great question, and it helps call attention to the promise of the two big breakthroughs that have occurred over the past two decades: the development of computational (and mathematical) methods for formalizing theories about the relationship between brain mechanisms and psychological function, and the ability to test these in intact human subjects. The promise of both of these approaches, used alone and especially in combination, is as bright as ever, and major progress has been made in the application of each. However, perhaps because both are so new, the combined or interactive use of the two together is still rare. One specific obstacle is that computational modeling—especially at the neural network level and below—often addresses a level of detail that in many cases may be beyond the spatial and temporal resolution of the imaging techniques currently in widest use. For example, a fundamental insight that has been provided by connectionist models is the power, and neural relevance, of distributed representations. One characteristic of such representations is that differences in the representation between different items, or even between different categories of items, may be very subtle, and thus hard to identify with functional magnetic resonance imaging (fMRI). Similarly, many psychological

processes transpire over relatively brief time scales (tens to hundreds of milliseconds). Computational modeling has been extremely useful in exploring the dynamics of such processes, in some cases making detailed predictions about them. However, the temporal resolution of fMRI is too crude to be useful in testing many such predictions. Event-related potentials (ERPs) may offer some hope here, but our understanding of the mapping from neural activity to scalp electrical activity is still limited, thus making this difficult as well.

That said, there are more general ways in which modeling and imaging are beginning to impact one another, and in some cases we have begun to see direct contact. Let me provide an example of each.

As I noted, connectionist modeling has provided a powerful tool for understanding how the structure of representation can affect processing, especially when we consider how processing occurs in the brain. I think Jim Haxby's recent work on visual object recognition provides a beautiful example of how a conceptual framework borne out of connectionist modeling—in this case, the idea of distributed representations—can motivate a novel approach to imaging. For example, Haxby has argued that face recognition may rely on multiple brain areas within the visual system and is not localized to the specific regions associated with maxima of activity (i.e., loci showing the greatest difference in activity to face versus nonface stimuli). To demonstrate this point, he has identified the maxima of brain activity, as well as the patterns of activity over the entire visual system, associated with different categories of visual stimuli (e.g., faces, houses, etc.). He has found that when he eliminated the regions of maximal activity, the rest of the pattern associated with a given category does as good a job (and in the case of face stimuli, a slightly better one) of predicting what the subject is observing as did the area of maximum activity. This is a

demanding and difficult conditions. Cingulate activation is almost always found during imaging conditions in which the attentional demands are high. Such conditions require close **monitoring.**

When we fail to pay close attention during demanding tasks, errors occur. We miss a target in the divided attention study or generate an adjective when asked to come up with verb associates. Evoked potential studies have shown that the anterior cingulate provides an electrophysiological signal correlated with the occurrence of errors. Figure 12.28 shows that when people make an incorrect response, a large evoked response sweeps over the

striking finding, suggesting that information about stimuli is distributed over a wide area of cortex, and that conspiracies of regions with lower levels of activity may be as (or more) important to perception and representation as are the areas of highest activity. This idea was inspired by and is a fundamental characteristic of distributed representations in connectionist models.

This work suggests a somewhat radical reconceptualization of how we interpret the so-called fusiform face area. One might suppose that this area responds to what all faces have in common. On this view, that area might be good for distinguishing faces from other types of stimuli but might be bad at discriminating between different faces. This idea contrasts with the view that this area represents a dedicated face processing module, which one would then expect to be the area primarily responsible for the identification of individual faces. On the distributed view, all of the other activated areas are the ones that "shape" and "color" the interpretation of a given face and distinguish it from others.

Returning to the general topic at hand, connectionist models have shown us that we should be especially cautious in looking to the brain for a reification of the constructs that we develop at the psychological level (e.g., a "face processor"). Rather, we must consider how these constructs might arise as emergent properties of the coordinated activity of various parts of a complex system working together.

An example of more direct and interactive contact between modeling and imaging can be found in our recent work on the anterior cingulate cortex (ACC). We have hypothesized that one function of this structure is to monitor conflict in processing (e.g., the coactivation of competing and incompatible response alternatives in a simple forced-choice task). First, I should note that this hypothesis arose directly from an examination of the neuroimaging literature. Previous work—primarily using ERP—had suggested that the ACC was responsive to errors in performance. This idea was consistent with the accruing evidence from imaging studies that showed that the ACC consistently activates under conditions of increased task difficulty (i.e., when errors are most likely).

However, we were puzzled by the fact that the ACC also appeared to activate when a task was difficult but subjects made few if any errors (e.g., in the Stroop task, which, as it happens, was the first task used to elicit ACC activation in a neuroimaging experiment). Viewed from a computational perspective, it occurred to us that processing conflict (i.e., competition between mutually inhibitory units) could be a precursor, or a close correlate of errors, and that in fact the ACC may be responding to the conflict associated with errors rather than the errors as such. In an effort to explore this hypothesis, we have conducted a number of simulations, both to better understand the circumstances under which ACC activity has been reported and to generate predictions about new conditions under which it should be observed.

One thing that has become evident in this work is that the pattern of simulated ACC activity in our models can become very complex and is influenced by a number of interacting variables (e.g., the particular sequence of previous stimuli, priming effects, the subject's explicit expectations, motivational factors, etc.). These variables, in principle, can be captured in the model. Doing so allows us to generate detailed patterns of predicted ACC activity—unique to each particular run of an experiment—that can be tested in an fMRI experiment. One fortuitous feature of this model is that it can make predictions about changes in the ACC that are within the temporal resolution of fMRI (e.g., trial-to-trial differences). We are presently conducting studies of this sort (see Figure 12.29 for an example). Whether or not our hypothesis is correct, I think it provides a nice example of how a model can be used to simulate the complex interactions among factors considered to be relevant to the functioning of a brain area, and to generate predictions that are significantly more complex than those arrived at by intuition or verbal description.

prefrontal cortex just after the movement is initiated. This signal, referred to as the **error-related negativity,** or ERN response, has been localized to the anterior cingulate (Dehaene et al., 1994). The ERN, interestingly enough, is absent when the subject is unaware that the response is erroneous, and its magnitude is correlated with the intensity of the incorrect response (Gehring et al., 1993). One might suppose that a monitoring system detects an incorrect schema control unit being fired and attempts to negate the response. Usually it is too late to inhibit it. Skilled typists experience this phenomenon with troubling frequency. In completing the word *with,*

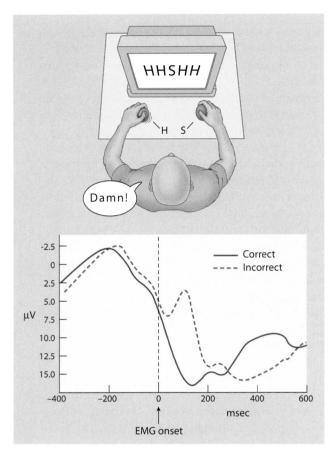

Figure 12.28 Subjects were tested on a two-choice letter discrimination task in which they made speeded responses with either the right or the left hand. Errors were obtained by emphasizing speed and by flanking the targets with irrelevant distractors. Evoked potentials for incorrect responses deviated from those obtained on trials with correct responses just after the onset of peripheral motor activity. This error detection signal is maximal over a central electrode positioned over the prefrontal cortex, and has been hypothesized to originate in the anterior cingulate. The zero position on the x axis indicates the onset of electromyographic (EMG) activity. Actual movement would be observed about 50 to 100 msec later. Adapted from Gehring et al. (1993).

the middle finger on the left hand may move toward the letter *e*, reflecting the thousands of times that the word *the* has been typed. The schema for *the* may have been falsely activated by the final keystrokes *th,* and the error detection system is forced to play catch-up.

The ERN is an especially salient signal of a monitoring system. But as shown in Figure 12.27, the functions of the anterior cingulate are not limited to the detection of errors. Cingulate activation is prominent in many tasks in which errors rarely occur. One such example is the Stroop task (see Figure 4.5). A subject is given a list of words that spell out the names of colors that are either congruent or

incongruent with the ink color. In either case, the subject must name the color of the words, inhibiting the natural tendency to read the word names themselves. In the congruent condition (e.g., saying "blue" when seeing the word *blue* written in blue ink), there is no conflict; while the subject may focus on the color, the schema control unit activated by the words also promotes the same response. In the incongruent condition (e.g., saying "blue" when seeing the word *red* written in blue ink), the two schema control units are in conflict. Here, the subject must overcome the inclination to read the words and focus on the ink color. As would be expected by Norman and Shallice's model, the anterior cingulate is more active in this condition (Bush et al., 2000). This activity occurs despite the fact that people make few errors on the task. The Stroop effect is manifest in the reaction time data and only minimally, if at all, in measures of accuracy.

Jonathan Cohen and his colleagues (2000) at Princeton University have hypothesized that a key function of the anterior cingulate is to evaluate **response conflict**. This hypothesis is intended to provide an umbrella account of the monitoring role of the ante-

Figure 12.29 A flanker task in which the response is based on the direction of the central arrow. **(a)** Conflict is introduced on incompatible trials because the flanking items point in the other direction. **(b)** Anterior cingulate activity is high on incompatible trials even when the response is correct. (b) From Botvinick et al. (1999).

(a)

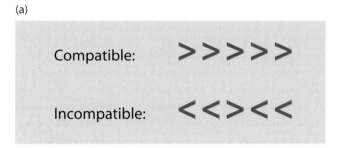

(b)

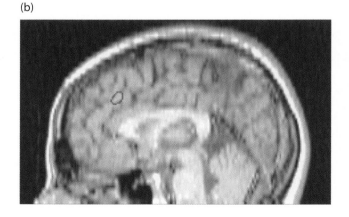

rior cingulate. Difficult and novel situations are ones that should engender high response conflict. In the verb generation task, there is a conflict between acceptable alternative responses. Errors, by definition, are situations in which conflict exists. Similarly, tasks such as the Stroop task entail conflict in that the required response is in conflict with a more habitual response. In Cohen's view, conflict monitoring is a computationally appealing way to allocate attentional resources. When the monitoring system detects that conflict is high, there is a need to increase attentional vigilance. Increases in anterior cingulate activity can be used to modulate activity in other cortical areas.

Event-related fMRI was used to assess the role of the anterior cingulate in monitoring response conflict (Botvinick et al., 1999). This study used the flanker task, similar to that shown in Figure 12.28. On each trial, a row of five arrows was presented (Figure 12.29). Subjects responded to the direction of the central arrow, pressing a button on the right side if this arrow pointed to the right and pressing a button on the left side if this arrow pointed to the left. On compatible trials, the flanking arrows pointed in the same direction; on incompatible trials, the flanking arrows pointed in the opposite direction. Neural activity in the anterior cingulate was higher on the incompatible trials compared to the compatible trials. Importantly, this increase was observed when subjects responded correctly. It also likely

occurred when subjects responded incorrectly, but this did not happen often enough to analyze the imaging results. However, an earlier imaging study showed that a similar region was activated on trials with incorrect responses (Carter et al., 1998). These results strongly suggest it is the monitoring demands that engage the anterior cingulate rather than the occurrence of an error. An additional comparison bolsters this argument. Activity on incompatible trials was lower when the preceding trial also was incompatible. According to Cohen's theory, the anterior cingulate has just signaled the presence of a conflict and recruited the necessary attentional resources required to handle this conflict. The need to re-engage this network is reduced.

The monitoring hypothesis emphasizes how the anterior cingulate is part of an executive control network. The cingulate does not operate in isolation. Rather, it works in tandem with prefrontal cortex to ensure coordinated goal-oriented behavior. In the flanker task, activity in the cingulate signals the presence of conflict. In turn, activation in prefrontal regions can be enhanced to ensure that selection processes operate carefully. For example, activity also increases in the inferior prefrontal cortex during incompatible trials (Hazeltine et al., 2000). We can hypothesize that the effects of cingulate activity modulate the selection operation of the prefrontal cortex, ensuring that the irrelevant flanking stimuli are ignored. A dynamic filtering system is in operation.

SUMMARY

In this chapter, we described the crucial role the prefrontal cortex plays in complex behavior and identified some of the critical requirements for goal-oriented behavior. Working memory is essential for representing information that is not immediately present in the environment. It allows for the interaction of current goals with perceptual information and knowledge accumulated from past experience. Not only must we be able to represent our goals, but also it is essential that these representations persist for an extended period of time. Working memory must be dynamic. It requires the selection and amplification of representations that are useful for the task at hand, as well as the ability to ignore potential distractions. Yet, we also must be flexible. If our goals change, or if the context demands an alternative course of action, we must be able to switch from one plan to another. These operations require a system that can monitor ongoing behavior, signaling when we fail or when there are potential sources of conflict.

Two functional systems have been emphasized. The

lateral prefrontal cortex is conceptualized as a working memory system devoted to sustaining representations of information stored in the cortex's more posterior regions through selection mechanisms. The anterior cingulate is hypothesized to work in tandem with prefrontal cortex, monitoring the operation of this system. As we emphasized in this chapter and the preceding one, the control of action has a hierarchical nature. Just as control in the motor system is delegated across many functional systems, an analogous organization characterizes prefrontal function. With control distributed in this manner, the need for an all-powerful executive, a homunculus, is minimized. The prefrontal cortex can be a reservoir of the current contents of processing by linking up to stored representations in the cortex's more posterior regions—representations that help to select actions.

But these representations, the content of ongoing processing, are embedded in a context that reflects the history and current goals of the actor. Our focus to this point has been on relatively impersonal goals—naming words, attending to colors, remembering locations. But most of our actions are socially oriented. They reflect our personal desires, both as individuals and as members of social groups. To gain a more complete appreciation of goal-oriented behavior, we must turn to the study of affect, the emotional reactions that guide our own behavior and arise in response to our interactions with others. In Chapter 13, we will address the topic of affect and bring the spotlight on to the third division of the prefrontal cortex, the ventromedial cortex. By recognizing the intimate connections between the regions of prefrontal cortex, we can start to appreciate how a mind emerges from the architecture of the human brain.

KEY TERMS

anterior cingulate

content-based hypotheses

delayed-response task

dynamic filtering

error-related negativity

executive functions

goal-oriented behavior

inhibitory control

lateral prefrontal cortex

monitoring

perseveration

recency memory

response conflict

selection of task-relevant information

source memory

stimulus driven

supervisory attentional system

task switching

Wisconsin Card Sorting Task

working memory

THOUGHT QUESTIONS

1. What memory functions are associated with prefrontal cortex? How do these mnemonic functions differ from other types of memory?

2. Compared to the visual cortex, it has been difficult to identify subregions of the prefrontal cortex. What are some of the current hypotheses concerning functional specialization within the prefrontal cortex? Does the lack of specificity reflect the current lack of knowledge or some difference between anterior and posterior cortex? Explain your answer.

3. One of the cardinal features of human cognition is that we exhibit great flexibility in our behavior. Flexibility implies choice, and choice entails competition between alternatives. Describe some of the neural systems involved in response selection.

4. Review and contrast some of the ways in which the prefrontal cortex and the anterior cingulate are involved in monitoring and controlling processing.

5. The notion of a supervisory attentional system does not sit well with some researchers because it seems like a homuncular concept. Is such a system a necessary part of an executive system? Explain your answer.

SUGGESTED READINGS

BADDELEY, A. (1995). Working Memory. In M.S. Gazzaniga (Ed.), *The Cognitive Neurosciences* (pp. 755–764). Cambridge, MA: MIT Press.

FUSTER, J.M. (1989). *The Prefrontal Cortex: Anatomy, Physiology, and Neuropsychology of the Frontal Lobe,* 2nd edition. New York: Raven Press.

MACDONALD, A.W., COHEN, J.D., STERGER, V.A., and CARTER, C.S. (2000). Dissociating the role of the dorsolateral prefrontal and anterior cingulate cortex in cognitive control. *Science* 288:1835–1838.

MONSELL, S., and DRIVER, J. (2000). *Control of Cognitive Processes. Attention and Performance XVIII.* Cambridge, MA: MIT Press.

SHIMAMURA, A.P. (1995). Memory and Frontal Lobe Function. In M.S. Gazzaniga (Ed.), *The Cognitive Neurosciences* (pp. 803–813). Cambridge, MA: MIT Press.

SMITH, E.E., and JONIDES, J. (1999). Storage and executive processes in the frontal lobes. *Science* 283:1657–1661.

13

Emotion

On a late-summer day in 1848, Phineas Gage woke in his campsite near the Vermont hamlet of Cavendish, ready for another demanding day of work building the extension of the Rutland & Burlington Railroad. Gage was proud to be participating in the expansion of America's vast transportation system through uneven and rocky terrain, the remnants of unrelenting glaciers that had shaped northern New England. Laying track required blasting the rock so graders could smooth the surface for the rail lines. Gage was foreman of the construction crew, a position of responsibility that reflected his years of experience and his expertise with a risky task: setting dynamite charges.

Though the job was straightforward, great care had to be taken to ensure that each step was done properly. A hole was drilled in the rock and filled with explosive powder, and then a fuse was placed on top of the powder. To ensure that the explosion impacted the rock itself, the powder had to be covered with a dense layer of sand and tamped down with an iron rod. Over the years, Gage had given much thought to how these blasts could be made most efficient; he even had a tamping iron manufactured to his specifications.

Many blasts had been set that late-summer day, and the men were encouraged by their progress through the rugged landscape. As the afternoon heat waned, their thoughts turned to dinner and perhaps a cooling swim in the creek. Around 4:30, Gage set another charge. As he waited for his assistant to pour in the sand, however, he was distracted and failed to check whether the powder was covered. When he thrust down his tamping iron, the consequences were disastrous. The iron set off a spark and the explosion ripped through the worksite.

The work crew turned to find Gage sprawled on the ground. They gasped in horror, shocked by the blood seeping from two large holes, one where his left cheek had been and the other from an opening in the top of his head. The iron rod lay nearby, having careened through the tis-

sue and bone of Gage's brain and skull. Equally shocking was the fact that their boss had not been killed. Though stunned, Gage was conscious! At first his body twitched convulsively, but within a few minutes he sat up and spoke to the men as they fetched a cart to take him to Cavendish for medical assistance. In town, he greeted the doctor in a most understated fashion: "Doctor, here is business enough for you." The physician, Dr. John Harlow, stopped the bleeding and sutured the wounds, liberally administering disinfectants to diminish the chances of infection. Within 2 months Gage was declared to be cured. As Harlow later wrote, "I dressed him, God healed him."

Unfortunately, the cure was only superficial. As soon as the injury's acute effects had subsided, it became apparent to all who knew Gage that his personality had undergone a radical transformation. Before the accident, Gage had been an exemplary citizen, hard working and energetic, a clear thinker who was a shrewd manager of his personal and financial affairs. Afterward, he was impatient and rude, given to outbursts of anger and rage. He brushed off the well-intended advice of friends and medical advisors with shocking profanity. He could not follow a coherent plan of action; instead, he reeled off a constant stream of ideas that were discarded almost as soon as they were vocalized. His employers with the railroad soon discharged him.

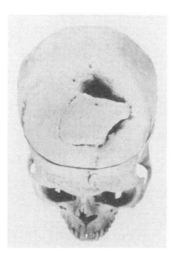

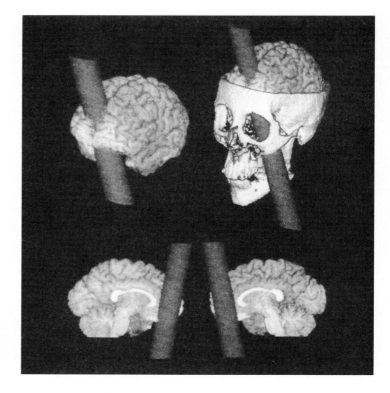

Figure 13.1 Phineas Gage's skull, and computer reconstructions showing how the tamping iron passed through his brain. The iron entered just below the left eye and exited from the top. It destroyed much of the medial region of the prefrontal cortex.

The new Gage embarked on a picaresque journey including stints with Barnum's freak show in New York, roping horses in South America, and laboring in California towns at the height of the Gold Rush. Fortune would never come his way because he was content with a transient lifestyle. The end came in 1861, when Gage died at age 38, a victim of violent epileptic seizures that resurfaced as remnants of the neurological trauma he had suffered 13 years previously. Gage was buried with the tamping iron placed alongside his body.

Five years after Gage's death, Dr. Harlow had the body exhumed from its burial site in California and brought the skull and iron back East, where it made its way to the museum of the Harvard Medical School. Hanna Damasio and her colleagues (1994) at the University of Iowa recently used modern brain imaging techniques to reconstruct Gage's lesion. They carefully measured the skull and used computer simulations to create a brain that best fit the skull. Then, as shown in Figure 13.1, they observed which parts of the brain would have been impacted by the iron's blow. While the rod had passed through the frontal cortex, the motor and premotor cortices were spared, accounting for reports that Gage had no obvious motor problems. Instead, the iron had destroyed the ventromedial aspects of the most anterior portions of the frontal cortex in the left and right hemispheres.

The sad tale of Phineas Gage forces us to think about the essence of what makes a person act the way he or she does. Every action we take is colored by how we value that action and the emotional response that action engenders in ourselves and others. We each have a unique personality with a consistent manner of thinking and acting in a social world. This personality reflects our innate characteristics and how these dispositions are shaped by our experiences and the social forces we encounter. We all have aspirations—the desire to be loved, to succeed, to procreate—and recognize that our actions as individuals are generated not only with regard to their impact on us but also with regard to their effects on those around us. While we recognize that our emotions, aspirations, and choices reflect the brain's operation, it is difficult to attribute our uniqueness to a physical substance. It is this dilemma that led Descartes to postulate a duality of mind and soul.

Experimenters in psychology and the neurosciences shy away from such issues, content to leave them to the domain of philosophy. But cases such as Phineas Gage bring these problems to the forefront. It is not sufficient to establish a cognitive neuroscience of how we perceive information, how this information is stored, and how we move. We also need to understand how emotion and motivation influence our ability to process information and choose actions. The case of Phineas Gage represents a situation in which a neurological event produced a dramatic change in personality. Although he showed no obvious impairment in his intelligence and perceptual and motor abilities, he was no longer able to evaluate appropriately the significance of events and regulate his emotional responses. As a result one of his peers commented, "Gage was no longer Gage." The case of Phineas Gage tragically demonstrates that when we turn to the question of what constitutes the self, we must consider the neuroscientific basis of emotion.

Cognitive neuroscience research traditionally has addressed issues such as how we perceive, attend, remember, and use language. However, if we hope to speculate on how this research extends to everyday life, we also need to understand how these behaviors interact with one of the most fundamental human characteristics: the ability to feel. We are not information-processing machines but rather motivated, emotional, and social animals. It is not possible to understand who we are or how we interact with the world without considering our emotional life. The cognitive neuroscience of **emotion** has been slow to emerge because emotion is a behavior that is difficult to study systematically. Nevertheless, researchers have been taking on the challenge, and emotion is emerging as a critical research topic. Here we will review what we know about the cognitive neuroscience of emotion.

ISSUES IN THE COGNITIVE NEUROSCIENCE OF EMOTION

Although Harlow's description of Phineas Gage and his emotional and social deficits was published over a century ago, the systematic study of emotion and motivation in cognitive neuroscience is relatively new. One primary reason for this is that emotion, relative to other behaviors, seems especially difficult to pinpoint. In order to develop a cognitive neuroscience of emotion we need to address some basic issues: How do we define *emotion?* How do we manipulate and measure emotion in scientific studies? What is the relation between cognition and emotion? Unfortunately, we do not have easy answers to any of these questions. Nevertheless, investigations into the cognitive neuroscience of emotion are thriving, with researchers doing the best they can to deal with these uncertainties. Before describing the fruits of this research, we will briefly consider each of these issues.

Defining Emotion

Happy, sad, fearful, anxious, elated, smitten, disappointed, angry, pleased, disgusted, excited, shameful, guilty, and *infatuated* are just some of the terms we use to describe our emotional life. Unfortunately, our rich language of emotion is difficult to translate into discrete states and variables that can be studied in the laboratory. In an effort to create some order and uniformity to our definition of emotion, researchers have adopted two primary approaches.

BASIC EMOTIONS

Although we may use words like *elation, joy,* and *glee* to describe how we feel, most people would agree that these terms represent a variation of feeling happy. Since Darwin's work on the evolutionary basis of human behavior, investigators have been proposing that we define a finite set of universal, basic emotions. Darwin approached this problem by questioning people who were familiar with a variety of different cultures around the world about the emotional life of these varied cultures. From this, Darwin determined that humans have evolved to have a finite set of basic emotional states, each of which is unique in its adaptive significance and physiological expression.

One of the more recent attempts to characterize basic emotions examined the universality of **facial expression** (Ekman and Friesen, 1971). By studying different cultures around the world, Paul Ekman of the University of California at San Francisco discovered that the means by which emotion is conveyed through facial expression does not vary. Whether we are from the Bronx, Beijing, or Papua New Guinea, how we show with our facial expression that we are happy, sad, fearful, disgusted, angry, or surprised is pretty much the same (Figure 13.2). From this work Ekman and others suggested that anger, fear, disgust, happiness, sadness, and surprise are the six basic human facial expressions representing emotional states.

Although there is still considerable debate as to whether any single list is adequate to capture emotional experience, most people accept the idea that there are basic, universal human emotions. Scientists have adopted the premise of basic emotions to investigate the different neural systems underlying reported emotional states or moods, as well as the neural and developmental basis of facial expression and evaluation.

DIMENSIONS OF EMOTION

Another way of approaching the categorization of emotions is not to describe them as discrete states but rather as reactions to events in the world that vary

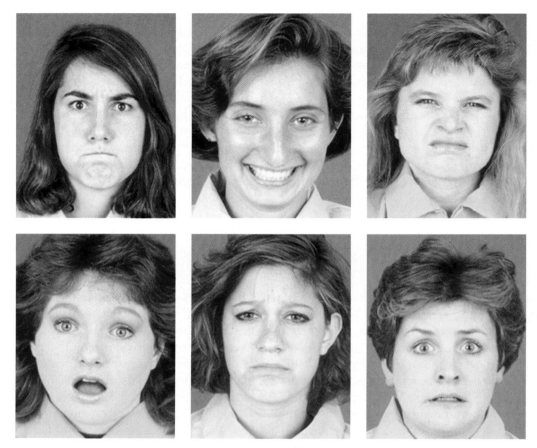

Figure 13.2 The six emotional facial expressions Ekman and colleagues found to be universal across cultures. See how well you can pick out the faces showing anger, happiness, disgust, surprise, sadness, and fear. Adapted from Ekman (1973).

along a continuum. For instance, most would agree that being happy is a pleasant feeling and being angry is an unpleasant feeling. However, we are happy when we find a penny on the sidewalk. We also are happy when we win $10 million in a lottery. Although in both situations we experience something that is pleasant, the intensity of that feeling is slightly different. A dimensional approach that has been used proposes that emotional reactions to stimuli and events can be characterized by two factors: valence (pleasant–unpleasant or good–bad) and arousal (how intense is the internal emotional response, high–low) (Osgood et al., 1957; Russell, 1979). We might be pleasantly surprised at finding a penny on the street, but most people would not describe their reaction to this event as intense or arousing. Winning the lottery, however, likely would lead to a very intense, arousing, and elated state. By using this dimensional approach, researchers can achieve a more concrete assessment of the emotional reactions elicited by stimuli.

A second dimensional approach that has been used in cognitive neuroscience is to characterize emotions by

the actions and goals they motivate. Davidson and colleagues (1990) suggested that different emotional reactions or states can motivate us either to approach or to withdraw. For example, happiness and surprise may excite a tendency to *approach* or engage in the eliciting situations, whereas fear and disgust may motivate us to *withdraw* from the eliciting situations.

Clearly the basic and dimensional approaches to defining emotions are not adequate to capture all of our emotional experience. However, they provide a beginning framework we can use in our scientific investigations of emotion. No single approach is correct. Depending on the question or issue being addressed, one of the approaches described here, or another approach, may be preferable. What is essential is that we are clear about how we define what we call "emotion" in our scientific investigations.

Manipulating and Measuring Emotion

Throughout the rest of this book, there are descriptions of how we can study behaviors such as memory, attention, and vision in the laboratory. In comparison, studying emotion in the laboratory seems especially challenging. Emotional reactions vary significantly between individuals and often can be unpredictable. However, in everyday life we try to manipulate and measure emotion all the time. We motivate people with money or prestige. Advertisers try to convince us that a certain product is better than another by linking it to things we value. On Halloween, we playfully try to frighten our friends. And, if someone we love is sad, we try to cheer him up. These are all situations in which we attempt to elicit emotional reactions and measure our success. Measuring and manipulating emotion is not always an easy task, and doing so in the laboratory adds some difficult ethical and practical constraints. In spite of these constraints, there have emerged a few common techniques by which emotion can be elicited and measured in laboratory studies.

TECHNIQUES TO ELICIT EMOTION

Mood Induction   To measure the effect of a particular emotional state on a subject's cognitive and neural responses, investigators induce moods. They get the subject into a particular mood or state, usually by encouraging the subject to use her will to achieve the emotional state (e.g., "Think of things that make you sad") and by presenting stimuli (e.g., movies or music) to aid in eliciting the desired emotional state.

Reward and Punishment   One of the most common means of eliciting emotional reactions in nonhuman animals is reward or punishment. For instance, when train-

ing a dog, we might reward him with doggie treats. Food is a primary reinforcer in that it has an intrinsic value and its rewarding properties are innate. In human research, we more often use a secondary reinforcer, money. Money is a secondary reinforcer because, by itself, it is meaningless. Only by being linked to things we value (e.g., food, shelter) has money become a stimulus that can be both punishing and rewarding.

Presentation of Emotionally Evocative Stimuli   One of the more common means used in the laboratory to evoke an emotional response in subjects is to present emotional stimuli. Typical emotional stimuli include pictures of emotional scenes, words that represent emotional concepts, loud noise, and mild electrical shock. One set of stimuli that has been widely used is the International Affective Picture System (IAPS). Peter Lang of the University of Florida and his colleagues (1995) have gathered over a thousand pictures of scenes from published materials (Figure 13.3). These pictures evoke a range of emotional responses and are of accident scenes, scenes of babies and families, sexual scenes, scenes of violence, and neutral scenes. Hundreds of subjects have rated these pictures for valence and arousal, and investigators have used these ratings to select scenes that will evoke a desired emotional response in their studies.

TECHNIQUES TO MEASURE EMOTION

Direct Assessment   The most obvious means to assess an emotional reaction or state is to ask the subject. We use this technique most often in everyday life. If we want

Figure 13.3   Examples of positive and negative emotional scenes from the International Affective Picture System developed by Peter J. Lang and colleagues (1995).

An Interview with Joseph Le Doux, Ph.D. Dr. Le Doux is a professor at New York University. He is the leading expert on the emotional brain.

Authors: You started out your career working on issues in brain laterality and then switched and are responsible for energizing the field of the neurobiology of emotion. How did that come about?

JL: A chance observation, a bit of luck, and a lot of work. It started with a study of split-brain patient P.S. When we put emotional stimuli into the right hemisphere, the left had no idea what the right saw, but could tell us how it felt about the stimulus, whether it represented something good or bad. This suggested that the cognitive and emotional representations of a stimulus are handled by different pathways in the brain. And since this patient had a corpus callosum section but his anterior commissure was intact, we speculated that the emotional information might be traveling from one amygdala to the other through this set of connections. The idea that cognition and emotion have somewhat different neural systems intrigued me, and I've been trying to figure out how it works ever since. That was the observation. The lucky part was that I was able to redirect my career in a way that would let me pursue the questions through animal research, since that was and still is the only way to get down to the nitty gritty of how the brain does something. Over the past 15 or so years, I have been working on the rat. It is interesting how hostile the neuroscience community was to this kind of thing 15 years ago. When I wrote my first grant to study emotion in the brain, I proposed using classical fear conditioning as the behavioral model. The grant was disapproved, the lowest possible rating. The reviewers said that I was investigating con-

ditioning, not emotion, and that I should rewrite the grant and focus on learning and memory. I did and it was funded, which had two implications. One was that I was able to use learning and memory as the front for my emotion work, and the other was that I got interested in learning and memory and now regard myself as involved in the neurobiology of memory and emotion. I see the two as sort of the same. Most emotions involve memory, and many memories involve emotion. So that's why I do what I do today.

Authors: Animal research is essential. At the same time how does one study the subtle human states of emotion, like sorrow, in a rat?

JL: I think most scientists work somewhere in between what they want to do and what they can do. In my case, I realized pretty early on that if I want to get at the nuts and bolts of emotion, I would have to pick an approach that would let me study emotion in the animal brain. It took me a little longer to realize that my approach, which is to use conditioned fear in the rat, might not tell us about how emotion works so much as about how fear works. Many emotions—enough to make this a meaningful statement—involve phylogenetically old brain systems that evolved to control the body behaviorally and physiologically in response to environmental challenges. These systems take care of things like defense against danger, sexual behavior, maternal behavior, eating, and other things like this. These are the kinds of emotional systems we can study in the animal brain. There are two points to make about this. One is that these emotions need to be stud-

to know how someone is feeling, or whether he likes something, we ask him. This self-report method is also the most common technique for emotional assessment in the laboratory.

Indirect Assessment Even though self-report may be the most direct way to assess emotion, it is not always the most informative or accurate. For instance, although subjects might believe that a stimulus (like a dirty word) is arousing and report that it is, they may not experience

physical arousal when encountering this stimulus. In addition, subjects may not always be aware of emotional manipulation, such as when a stimulus is presented subliminally, which would make the self-report of an emotional response particularly suspect. Finally, self-report is not a viable means to assess emotional reactions in nonhuman animals and in infants. To overcome some of the difficulties of the self-report method, researchers have used a range of indirect means to assess an emotional reaction or state.

ied one at a time because they evolved for different reasons to do different things and have different brain systems controlling them. The other is that these systems evolved before consciousness; the conscious feelings we know our emotions by are not the reason they evolved. Brains were nonconscious and nonverbal long before they were conscious and verbal, but we use the human state (conscious and verbal), especially the negation of the human state (nonconscious and nonverbal). When we consider emotion in animals, we are thinking of human negations that may not exist, and even if they do, they're going to be difficult to pin down. It's important to take the similarity of emotional behavior in animals and humans at face value. It tells us that when, for example, rats and people are in danger, they freeze and their blood pressure goes up and they release stress hormones. These similarities don't tell us that rats and humans experience fear. For that you need to be aware of the state you are in, and it's not clear that rats are. Yet there is an awful lot we can learn about emotion systems from studying animal brains, and this information can be useful for understanding conditions that humans experience as well. But we've got to be careful and not generalize too freely from one emotion to another. There is no unitary emotion system; lots of systems take care of things we call emotions. Some subtle human emotions may not be easily studied through animal experiments, but there are still plenty that can be.

Authors: Emotions like fear cue one to the nature of a situation. The brain decides that a situation is dangerous. Why couldn't the brain have evolved in a way that allows for such decisions but in the absence of the emotion of the decision?

JL: The brain did evolve in a way that allows for decisions about danger and perhaps other challenging situations to be dealt with in the absence of emotion. A fruit fly and a snail, to use two common examples from neurobiology, defend themselves from danger by using cues in their environment. And both can learn that novel cues are signals for danger. They use the same conditioning processes we do: The time overlap of a neutral stimulus and an arousing one can modify how the brain copes with the previously neutral one. When the stimulus occurs, the organism (snail, rat, person) reacts the way its species normally responds to danger. No conscious awareness is needed in the snail or the human. These are just responses. And what evolved to evaluate stimuli and produce responses is the system that we have to understand, to see where our emotional reactions originate. The difference between a rat and a person is not so much the system that produces the responses but instead is the cognitive layering above the more basic systems. Along with understanding emotional reactions, we also need to understand emotional actions. That's where cognition and consciousness come into play and make a difference. These systems provide more flexibility to deal with reactions. We react to emotional situations unconsciously, but we then figure out what to do, make plans, use strategies, and so on. From the point of view of brain research, we are much farther along in understanding emotional reaction than emotional action, but now that we have a sense of how reactions occur, we can begin to look at action.

So I think that the brain did evolve to deal with danger and other so-called emotional situations without cognition and consciousness. After all, when evolution was putting defense response networks together in invertebrates, she didn't know that she would later make vertebrates and consciousness. I'm taking a chance and assuming that invertebrates are not consciously aware of their emotional reactions, but I'm willing to take that chance.

CHOOSING AMONG POSSIBLE ACTIONS. One means of indirectly assessing an emotional response or evaluation is to give a subject a few options and see what she chooses. If we wanted to know which kind of toy an infant prefers to play with, a doll or a truck, we could place both toys in front of the infant and see what he reaches for and how long he chooses to play with it before losing interest. Examining the choices people and animals make reveals the value they assign to those choices.

FACILITATION OR INHIBITION OF A RESPONSE. Emotion can facilitate or inhibit a response, depending on the task. For instance, in the classic Stroop task (see Chapter 4), subjects are asked to report the color of the ink in which individual words are printed. Subjects take longer to report the color of the ink if the word is the name of another color (e.g., the word *red* printed in green ink). Once we learn to read, the reading of the word is automatic, even if that is not the task we are asked to perform. When the meaning of the word we read conflicts with the color of the ink, the correct response is inhibited and takes longer. In an emotional variant of the Stroop task, subjects are asked to perform the same task (naming the color of the ink), but the words are either emotional or neutral. The emotional content of the words also inhibits the latency of the response. It has been suggested that the emotional content of stimuli is processed more automatically and sometimes can capture our attentional resources, thus inhibiting our ability to respond to other stimulus properties.

PSYCHOPHYSIOLOGICAL VARIABLES. One factor that differentiates emotion from other behaviors discussed in this book is that emotion alters not only our mental and neural state but also our bodily state. Emotional responses can cause a number of bodily reactions. For instance, when we are scared, our heart starts beating faster and we start sweating. This reaction is the result of arousal of the **autonomic nervous system.** When we are scared, we also may startle more easily. The startle reflex is enhanced or potentiated by a negative emotional state. If we are walking down the street in the middle of the day and hear a sudden loud noise, we might be startled. However, if we hear the same loud noise late at night on a dark and empty street when we are feeling a little anxious, we probably will be startled even more. This is the emotional potentiation of our startle reflex.

In the laboratory, we can measure these variables as an indication of emotional state or reaction. A typical means of assessing an arousal response to a stimulus is the *skin conductance response,* or SCR (also called the *galvanic skin response,* or GSR). The SCR can be measured by placing electrodes on the subject's fingers and sending a mild electrical current through the skin. The SCR reflects a change in the electrical conductivity of the skin as a result of the activity of the sweat glands. It can be measured even if there is a subtle, transient arousal response and the subject is not noticeably sweating.

This idea that emotion alters our physiological state and that we can determine an emotional reaction by measuring our body's response is the principle behind the classic lie detector test. The assumption underlying this test is that lying creates an uncomfortable emotional state that is arousing. Therefore, if someone is lying, he will show signs of autonomic nervous system arousal that can be measured by the SCR and other techniques. This assumption may be fairly accurate for most people, most of the time. However, criminal psychopaths or patients with specific brain lesions may not show normal emotional and physiological reactions under certain circumstances. In addition, normal subjects may not always

show the expected emotional reaction in all situations. These are some of the reasons why the results of lie detector tests can be difficult to interpret.

Emotion and Cognition

"I was so *mad* I couldn't think straight!" "I would never forget something like *that*!" Both of these phrases refer to situations where an emotional response interacts with cognitive processes. An angry state may affect our ability to reason. An emotionally arousing or significant event may lead to enhanced memory. As mentioned earlier, emotion clearly differs from other, "cognitive" behaviors discussed in this book in that it can involve changes in the bodily state. But is there a clear dividing line between emotion and cognition? Philosophers and scientists have widely debated the nature of the relation between cognition and emotion since the days of Aristotle, who suggested that emotion (the sensitive soul) and cognition (the rational soul) are separate grades or components of the soul, with only the rational soul being unique to humans.

One of the more recent debates about the relation between emotion and cognition was between Robert Zajonc of Stanford University and Richard Lazarus of the University of California at Berkeley. Zajonc conducted research showing dissociations between evaluation and awareness. For example, emotional stimuli presented subliminally, so quickly that subjects did not report seeing them, influenced how the subjects evaluated neutral stimuli that followed. Based on his work, Zajonc (1980, 1984) argued that **affective** judgments occur before, and independently of, cognition. Lazarus (1981, 1984) argued that emotion could not occur without cognitive appraisal. If we experience signs of autonomic nervous system arousal, such as sweating and an increased heart rate, our emotional response will depend on whether we are exercising, are in the presence of a person we find sexually attractive, or are looking down from the roof of a tall building. In other words, the emotional response depends on the reason we believe we are experiencing the arousal (see Schacter and Singer, 1962). Lazarus believed that emotion was a subset of cognitive processes. The primary issues in this debate hinged largely on how one defined *cognition.* Zajonc (1984) defined *cognition* as a slower mental transformation of sensory input or information processing, whereas Lazarus defined *cognition* as including early evaluative perception as well as later stages of information processing.

One benefit of the recent work on the cognitive neuroscience of emotion is that by identifying some of the brain systems of emotion, we have shifted the debate about the relation between cognition and emotion to neural terms. For instance, we know that some brain structures, such as the **amygdala,** are specialized for processing emotional stimuli and can respond very quickly and early in the processing of these stimuli (Le Doux, 1996). These findings are consistent with Zajonc's position that we have separate systems for the processing of emotion. At the same time, the neural structures that are specialized for emotion can interact with and be influenced by neural systems known to be specialized for other, cognitive behaviors.

These results suggest that emotion and cognition are interdependent, consistent with Lazarus's proposal. It no longer appears useful to study emotion without cognition and vice versa. Emotion, like other mental behaviors, has unique and defining characteristics. However, our understanding of the neuroscience of emotion cannot be separated from other behaviors considered more "cognitive." Research in the cognitive neuroscience of emotion has essentially answered the Zajonc-Lazarus debate by saying they both are right. The neural systems of emotion and cognition are independent and interdependent.

NEURAL SYSTEMS IN EMOTIONAL PROCESSING

Early Concepts: The Limbic System

One goal of the research on the cognitive neuroscience of emotion is to identify and understand the neural systems underlying different emotional states and processes (e.g., see Damasio et al., 2000). However, the notion of a network of brain structures underlying emotional behavior is not new. In 1937, James Papez proposed a circuit theory of the brain and emotion, suggesting that emotional responses involve a network of brain regions including the hypothalamus, anterior thalamus, cingulate gyrus, and hippocampus. Paul MacLean (1949, 1952) later

named these structures the "Papez circuit." He then extended this emotional network to include the amygdala, **orbitofrontal cortex,** and portions of the basal ganglia. He called this extended neural circuit of emotion the **limbic system** (Figure 13.4).

MacLean's early work identifying the limbic system as the "emotional" brain was very influential. To this day, it is not uncommon to see references to the "limbic system" or "limbic" structures in studies on the neural basis of emotion. The continued popularity of the term *limbic system* in more recent work is primarily due to the inclusion of the orbitofrontal cortex and amygdala (Damasio, 1994;

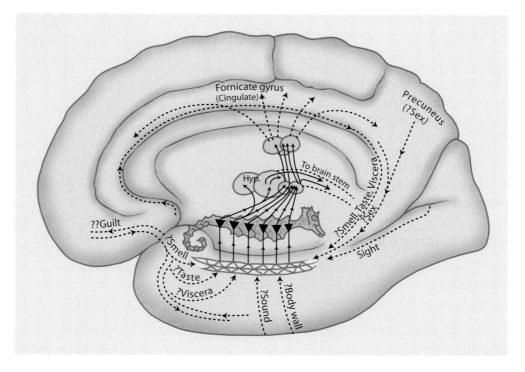

Figure 13.4 MacLean's limbic system. The hippocampus was the centerpiece of the limbic system, represented here by the seahorse, with the black triangles representing the pyramidal cells. It was believed to receive inputs from external sensations as well as the internal and visceral environment. The integration of these internal and external inputs was believed to be the basis for emotional experience. From MacLean (1949).

Le Doux, 1992). However, the limbic system concept as outlined by MacLean has not been supported over the years (Brodal, 1982; Swanson, 1983; Le Doux, 1991; Kotter and Meyer, 1992). Although several of the limbic structures are known to play a role in emotion, it has been impossible to determine criteria for defining which structures and pathways should be included in the limbic system. At the same time, classic limbic areas such as the hippocampus have been shown to be more important for other, nonemotional processes, such as memory (see Chapter 8). Without any clear understanding as to why some brain regions, and not others, are part of the limbic system, MacLean's concept has proved to be more descriptive and historical than functional in our current understanding of the neural basis of emotion.

These early attempts to identify neural circuits of emotion tended to view emotion as a unitary concept that could be localized to one specific circuit, such as the limbic system, thus separating this "emotional brain" from the rest of the brain. Over the years our investigations of emotion have become more detailed and complex. We acknowledge that emotion is a multifaceted behavior that may not be captured by a single definition or instantiated in a single neural circuit or brain system. Recent research has focused on specific types of emotional tasks and identifying the neural systems underlying specific emotional behaviors. We no longer think there is only one neural circuit of emotion. Rather, depending on the emotional task or situation, we can expect different neural systems to be involved. These systems might involve some brain regions that are more or less specialized for emotional processing, along with others that serve many functions. Studies of the cognitive neuroscience of emotion invoke a number of brain regions that play a role in different emotional tasks, some of those mentioned earlier (e.g., anterior cingulate gyrus, hypothalamus, basal ganglia), as well as others (e.g., insular cortex, somatosensory cortex). However, the orbitofrontal cortex and the amygdala have emerged as two brain regions whose primary functions are related to the processing of emotion (Figure 13.5). Understanding these two regions is crucial to our investigation into the cognitive neuroscience of emotion.

Orbitofrontal Cortex

The orbitofrontal cortex is the part of the prefrontal cortex that forms the base of the frontal lobe and leans on the upper wall of the orbit above the eyes. This region is sometimes broken down into two areas. The more central part of the orbitofrontal cortex is referred to as the *ventromedial prefrontal cortex*. This was the re-

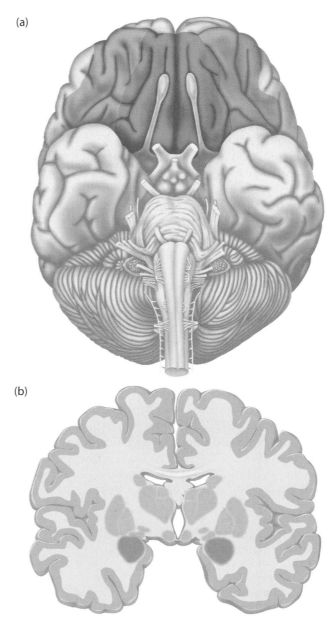

(a)

(b)

Figure 13.5 **(a)** The human orbitofrontal cortex, which is often divided into the ventromedial prefrontal cortex (red) and the lateral orbitofrontal cortex (green). **(b)** The human amygdala is highlighted in orange. From Davidson et al. (2000).

gion damaged in Phineas Gage, as described at the beginning of this chapter. The remaining, more lateral portion sometimes is referred to as the *lateral-orbital prefrontal cortex*.

The functional role of the orbitofrontal cortex is hard to define precisely, because the behaviors depending on this region cannot be categorized easily. As the case study of Phineas Gage points out, those who knew him did not refer to his deficit as a specific problem, but rather simply stated, "Gage is no longer Gage." Harlow, the physician who treated Gage, could list a series of uncharacteristic behaviors but could not sum them up by identifying a single deficit. The behavioral function of the orbitofrontal cortex is elusive because it fails to fit into a single, identifiable category. Yet at the same time, all of the behaviors known to rely on the orbitofrontal cortex seem somehow related. The challenge facing cognitive neuroscientists is to find a precise way to characterize the functional role of the orbitofrontal cortex so that it encompasses the wide range of behaviors that depend on this critical brain region. Presently, research on the role of the orbitofrontal cortex falls into separate but related domains. Each of these domains is related to the role of the orbitofrontal cortex in regulating our abilities to inhibit, evaluate, and act on social and emotional information. These abilities manifest themselves in a number of ways, which, for the sake of clarity, we will classify as social and emotional **decision making.** Before we discuss these, let's briefly consider how complicated even everyday decisions can be.

DECISION MAKING

Choosing how to act does not simply require discriminating between incoming stimuli. When choosing how to act, we must integrate incoming stimuli with our values, current goals, emotional state, and social situation. Consider the daily commuter as she chugs along the highway during the evening rush hour. Familiar visual signs along the route will dictate certain actions. The golden arches of a local McDonald's, the "Main Street" exit sign, and the arrows signaling this exit, which is near her home, indicate that she must move to the right lane and begin to decelerate. These stimuli are familiar and her actions are habitual. She may not even be aware of her preparation for taking the exit as she tunes in to the latest news report on the radio. Habitual actions are carried out in conjunction with the appearance of each familiar landmark.

Even in this routine it is essential to maintain flexibility for alternative actions. Sometimes an external cue provides the critical trigger. Our commuter scans ahead and notices that the traffic is at a standstill and emergency vehicles are driving along the shoulder lane. An accident must have happened—this is sure to add at least 40 minutes to the normal commute. Figuring that the surface road traffic will be moving, she quickly moves to the right to take the "Central Avenue" exit. Or, flexibility may be necessary because of a change in internal goals. It is Wednesday evening, her kids' time for soccer practice. She must drive past her usual exit to reach the playing fields. If there is a particularly interesting news report, she may suddenly find herself turning off the highway at her usual exit, not remembering the atypical goal required this evening. Internal goals reflect personal desires and aspirations and can also be socially mediated. Our commuter is eager to get home and is frustrated that she is late. On an evening when her last meeting ran an hour too long, she considers using the lane reserved for carpools (Figure 13.6). But after mulling it over, she stays in the slow lanes filled with single drivers. She dreads the $250 ticket that she might get if she is caught by a cop, not to mention the scorn from fellow drivers for this socially inappropriate action.

This example illustrates the wide range of factors that go into everyday decision making. There are habitual actions cued by perceptual information, new information that requires flexibility in planning, internal goals, emotional information in the environment, internal emotional cues, and social cues. These factors combine when decisions are made. Depending on the decision task, some of these factors may be more important than others. If our ability to process any of these types of information is impaired, our decision making will be altered. The orbitofrontal cortex seems to be especially important for processing, evaluating, and filtering social and emotional information. The result is that damage to this region impairs the ability to make decisions that require feedback from social or emotional cues.

Figure 13.6 Our behavior reflects the combined influences of our personal desires and social constraints. The driver is tempted to take the carpool lane to get home on time but is restrained by her fear of the potential fine as well as the scorn of other drivers.

SOCIAL DECISION MAKING

The frontal lobes play a critical role in how we select among a multitude of information. In Chapter 8, we outlined how some regions of the frontal lobe, such as the dorsolateral prefrontal cortex, are important for selecting information in working memory tasks. One potential criterion for the selection of an action is based on evaluating cues within a social context. The orbitofrontal cortex appears to be specialized for this type of information processing, so that patients with damage to this region show a number of deficits in social decision making. In some situations, their responses seem overly dependent on perceptual information, ignoring social cues. In other situations, they appear to be insensitive to social norms or goals, resulting in an inability to follow through with social expectations. At times, patients with damage to this region have difficulty inhibiting inappropriate social responses, such as aggressive impulses.

Many anecdotes demonstrate how the behavior of frontal lobe patients can be dominated by perceptual information. F. Lhermitte of the Hôpital de la Salpêtrière in Paris demonstrated a black humor for exploiting this tendency (Lhermitte, 1983; Lhermitte et al., 1986). He would invite a patient to a meeting and place a hammer, nail, and picture on a table in the entryway. When encountered by this array of objects, a frontal lobe patient might pick up the hammer and nail and hang the picture on the wall. In another instance, Lhermitte put a hypodermic needle on his desk, dropped his trousers, and turned his back to the patient. While most would consider filing ethical charges in this situation, the patient was unfazed. He simply picked up the needle and gave his doctor a healthy jab in the buttocks!

Lhermitte coined the term *utilization behavior* to describe the fact that the patients demonstrated an exaggerated dependency on environmental cues in guiding their behavior. Their gestures indicate intact knowledge of the uses of the objects such as the hammer or needle. What is lacking is an ability to evaluate the social context and to determine if the action is appropriate. Further evidence of this reliance on external cues is that some orbitofrontal patients are prone to imitative behaviors. In one study, Lhermitte sat opposite a patient and, without explanation, produced gestures. Some were innocuous: folding paper, combing his hair, or tapping his leg. Others were socially inappropriate: thumbing his nose, chewing paper, or kneeling in prayer. The patients with orbitofrontal lesions mimicked these actions (Figure 13.7). While we might suppose that this reflects the revered status of doctors, control subjects and patients with lesions outside the prefrontal cortex did not engage in such imitative actions. Indeed, we can imagine that the control subjects were contemplating engaging the services of a new physician. The patients, in comparison, did not question or seem puzzled by Lhermitte's violation of social conventions. Although not all patients with orbitofrontal cortex damage show this type of imitative behavior, it is not uncommon for these patients to show an overreliance on perceptual cues and to be unaware of violations of social customs.

Patients with orbitofrontal lesions also demonstrate a range of other social impairments. They often show a change in personality, irresponsibility, and lack of concern for the present or future. They have been reported to show diminished social awareness and empathy, and a lack of concern for social rules (Hecaen and Albert, 1978; Stuss and Benson, 1986). A modern Phineas Gage, named Elliot, lost his ability to make decisions within a social context (Damasio, 1994). Elliot had a brain tumor that bilaterally invaded his brain's orbital surface, the cortical region just above the orbits of the eyes. After surgery, he demonstrated superior intellectual abilities, performing above normal on long-term memory and working memory tasks. Nonetheless, Elliot could not perform the routines necessary to survive in today's world. He needed prompting to get up and go to work. He lost all sense of a schedule, and his fellow workers found him immersed in a mundane task for hours on end. After losing his job, he initiated risky ventures, against the advice of friends and family; he eventually went bankrupt. As striking as Elliot's lost sense of social norms was his personal detachment from his problems. He had no difficulty recounting the minutest details of his many failings, but spoke of them as if he were a dispassionate observer—a striking dissociation between decision making in the abstract and decision making about personal and social involvement.

A similar loss of social guidance of behavior can be observed in primates after they receive lesions to their prefrontal cortex. While social stratification is implicit in human society, animals like the rhesus monkey live in a highly structured social setting. Each animal has an assigned place in the social hierarchy, a position mainly inherited at birth. Social position determines many behaviors such as access to food and sexual partners—critical factors considering that morbidity rates are high in natural settings. The social status of animals that receive lesions of the orbitofrontal cortex plummets immediately (Myers et al., 1973). They are treated as outcasts by the group, incurring **aggression** and being forced to withdraw to a solitary existence. It is not clear what cues the healthy animals use to reject the lesioned animals so quickly. Aberrant behaviors can be seen after surgery. The animals demonstrate

(a)

(e)

(b)

(f)

(c)

(g)

(d)

Figure 13.7 Imitative and utilization behaviors are two signs of prefrontal damage, usually associated with lesions in the ventromedial region. **(a–d)** Imitative behaviors. The patient mimics the physician making a threatening gesture **(a)**, putting on spectacles **(b)**, smelling a flower **(c)**, and praying **(d)**. **(e–g)** Utilization behaviors. When objects are placed in front of him, the patient puts on three pairs of glasses **(e)** or proceeds to use the makeshift urinal **(f, g)**.

restlessness, aimless pacing, occasional displays of aggression, and a failure to groom themselves. It is likely that following frontal surgery, the lesioned animals cannot act in a manner appropriate for their position and therefore pose a threat to the social order.

Although most patients with orbitofrontal cortex damage exhibit inappropriate social responses and may occasionally show inappropriate aggressive reactions, their actions are more often hurtful to themselves than others. However, the role of the orbitofrontal cortex in evaluating social cues also has been linked to antisocial behavior disorders and difficulties controlling violent or aggressive impulses. For instance, positron emission tomography (PET) studies have shown reduced glucose metabolism in the orbitofrontal cortex in individuals with violent and antisocial histories (Raine et al., 1998). Some patients with orbitofrontal lesions are more prone to exhibit antisocial behaviors, such as stealing or violent outbursts. In these cases, it has been suggested that these patients suffer from "acquired sociopathy" (Damasio, 1994), because their tendency toward violence and lack of concern for social consequences somewhat mimics that of individuals with antisocial personality disorders.

An example of this is patient J.S. (Blair and Cipolotti, 2000). J.S. suffered a head trauma that resulted in bilateral damage to the orbitofrontal cortex along with some damage to the left temporal lobe, including the amygdala. His intellectual abilities were normal following his accident. Although he was described as relatively quiet and withdrawn prior to his injury, during his hospital stay he assaulted and wounded a member of the staff, threw objects and furniture at people, and was aggressive toward other patients. When tested, he demonstrated severe impairment at processing social stimuli, such as facial expressions. He also showed impairments in interpreting emotional reactions in other individuals. Although J.S. fit the standard criteria for antisocial personality disorder, his behavioral pattern differed somewhat from that of individuals with this disorder. When such individuals are violent or aggressive, there tends to be a reason or plan for this behavior. In contrast, J.S.'s aggressive or violent acts seemed to be impulsive reactions with no particular goal.

As mentioned earlier, most patients with orbitofrontal lesions display deficits in social response but do not exhibit this type of aggressive behavior. It is unclear what differentiates patients like J.S. from these other patients. It has been suggested that damage to the orbitofrontal cortex early in life is likely to lead to more severe antisocial behavior (Anderson et al., 2000). It has also been suggested that the additional medial temporal lobe damage in J.S. contributes to his violent tendencies.

Further research is needed to understand the precise role of the orbitofrontal cortex in violent or aggressive behaviors. However, it is clear from functional neuroimaging studies in individuals with antisocial personality disorders, and the behavior of patients like J.S., that the role of the orbitofrontal cortex in social decision making is linked to the ability to monitor and control aggressive impulses.

EMOTIONAL DECISION MAKING

Deciding and acting in a social context cannot be separated easily from our ability to evaluate and process emotional information, since social cues often provide emotional feedback. But how might the orbitofrontal cortex be specialized for processing emotional feedback? We are just beginning to understand the mechanisms by which the orbitofrontal cortex uses emotional information to aid in decision making.

Edmond Rolls of the University of Oxford (1999) proposed one theory. He suggested that the orbitofrontal cortex is necessary for the on-line, rapid evaluation of stimulus-reinforcement associations. He argued that the orbitofrontal cortex is involved in learning to link a stimulus and action with its reinforcing properties. An important aspect of this theory is that as we interact with the environment, the reinforcing properties of certain stimuli or potential actions may change. For example, it may be appropriate to laugh loudly at a friend's joke at a party. However, it would not be appropriate to laugh loudly if the friend whispers the joke during a lecture. At times, deciding on an action requires that we correct stimulus-reinforcement associations when they become inappropriate based on new information. The appropriate and rewarding response to hearing a joke changes depending on the social context. Rolls argued that this on-line, rapid ability to evaluate the **reinforcement** properties of a stimulus and a subsequent action requires the orbitofrontal cortex.

An example is illustrated in a task called *reversal learning of stimulus-reinforcement associations*. In this task subjects are told that they can earn points by touching one stimulus when it appears on a video monitor. At the same time, they have to withhold a response when a different stimulus appears, or they lose a point. After subjects learn to make this discrimination, the stimulus-reinforcement contingencies are unexpectedly reversed. Now in order to avoid losing a point, the subjects have to withhold a response to the previously rewarding stimulus, and to gain points, they have to touch the stimulus that they previously had to avoid. Patients with orbitofrontal cortex lesions were able to learn the initial task, but they showed difficulty in correcting their responses when the stimulus-reinforcement properties were reversed. The extent to which they had difficulty adapting to these reversals was correlated with the extent to which they showed socially inappropriate or disinhibited behavior (Rolls et al., 1994).

The ability of the orbitofrontal cortex to interpret the emotional properties of stimuli is also at the heart of another popular theory of decision making. Antonio Damasio (1994) of the University of Iowa articulated a theory of how emotion can influence even everyday, rational decision making. He argued that our everyday actions do not occur in an abstract, impersonal state. We know that they have personal and social consequences. Damasio dismissed the belief that reasoning and emotion are separate cognitive domains. He maintained that reasoning is guided by the emotional evaluation of an action's consequences, an idea captured by the main title to his book, *Descartes' Error*. Descartes argued for a duality of the mind and the body: The mind is a conscious entity of pure reasoning and thinking, and the body is limited by its striving to satisfy physical needs. For Damasio, such a segregation is a myth. The mind is an adaptation designed to better our chances of satisfying physical and psychological needs. To do so, it must be informed by neural structures that process affective responses to stimuli and memories.

Decisions on how to act require an analysis of the costs and benefits of the options, which are played out in working memory. While we have to consider the future, we also have to base our decision on knowledge accumulated from the past. In the past we may have encountered the same situation or stimulus and our action may have resulted in positive or negative consequences. For instance, imagine you are faced with the possibility of going to a championship game for your favorite sports team, but the ticket is very expensive. You can anticipate the excitement at the coliseum because you have been to thrilling sporting events in the past. This year your team is favored to win and if you are there, it will likely be an experience to talk about for years to come. However, you had a similar option a few years earlier and decided to spend more than you could afford to attend a championship game. At that game, your team lost horribly. It was a miserable and very expensive experience. You were angry that you had to make financial sacrifices to pay for the ticket to this disappointing event. A rational economist might perform a cost-benefit analysis of the possible scenarios. You have to compare the thrill of being at the game with the potential disappointment and cost. You reason through the choices and decide. Damasio argued that the ideal of a rational decision maker is not appropriate for an organism that con-

tinually faces choices. Our choices may not always be as emotional or costly as in this example, but we constantly decide between courses of action. And these decisions frequently must be made quickly. We need a mechanism that will help sort through the options, a mechanism that provides a common metric for evaluating options with respect to their potential benefit.

Damasio referred to this mechanism as the **somatic marker.** Somatic events are bodily sensations; thus, a somatic marker implies a link to a physiological experience. For Damasio, the phrase "gut feeling" is almost literal. When we watch a horror movie, our reaction is not detached and purely intellectual; it invokes a physical reaction. Our hands may become sweaty or we might experience a tightening in the muscles of our face and stomach. In a similar way, our memories of these events reactivate these physical reactions, or at least our memories of these reactions. An upcoming decision calls for activating representations of similar events experienced in the past. But these memories are not generated as abstract entities. Rather, they are imbued with emotional associations. How did similar events make us feel in the past, and what were the affective consequences of our prior experiences? Whereas common wisdom tells us not to let our emotions get "in the way," Damasio argued that affective memories are essential for decisions. They allow us to sift through options, alert us to plans linked to negative feelings, and bias us toward ones connected with positive feelings. Somatic markers rapidly narrow the options by automatically anticipating the affective consequences of each action. They may not enable us to make an unambiguous choice, but they constrain the playing field. Working memory is a limited resource. We cannot consciously mull over the multitude of options that any situation offers. Somatic markers focus on restricted possibilities.

The somatic marker hypothesis is a fresh slant on the paradoxical behavior of patients with lesions of the orbitofrontal cortex. When this region is damaged, the representations required to guide and produce an action are brought into working memory, but they are stripped of emotional content. A patient may still mull over problems, albeit in an impersonal manner. He may be aware of the death of a close relative and understand the finality of it, but he is divested of the emotional pain that accompanies the loss.

For the somatic marker hypothesis to be viable, Damasio and his colleagues needed to demonstrate that orbitofrontal lesions can disrupt the emotional processing of affective memories. Their work initially focused on an autonomic nervous system response, the skin conductance response (SCR). The mechanism generating this response is intact in frontal lobe patients in that they show an autonomic response to an innately negative stimulus, such as a sudden loud sound. Where these patients differ from control subjects is when SCR changes are evoked by stimuli with emotional qualities that are not innate, and can only be determined from past experience. In one experiment, the stimuli were slides of mostly neutral images: scenery from the Iowa countryside or abstract paintings. Other slides were of disturbing or socially taboo images such as disaster scenes, mutilations, or nudity. When subjects viewed the latter images, the SCR for control subjects had a consistent spike in comparison to the response to neutral images. As can be seen in Figure 13.8, the SCR was flat for the frontal lobe patients. Again, patients showed a dissociation between their explicit description and physiological, emotional reaction. They could describe in detail the disturbing images and use words that convey emotional experience (e.g., "It was a disgusting image of a mutilated body"), but their lack of an affective response meant that their words reflected semantic associations, not emotional ones.

Figure 13.8 Patients with ventromedial cortical damage fail to show autonomic, emotional responses to arousing stimuli. Subjects were shown a series of stimuli while measurements were made of their skin conductance response (SCR), a measure of emotional responsivity. Some of the stimuli were affectively neutral (N) such as photographs of the Iowa countryside. Others were expected to evoke strong emotional responses (E). The control subjects showed a large SCR to the emotional stimuli, whereas the prefrontal lesion patients had a "flat" SCR. Bottom panel is adapted from Damasio (1994).

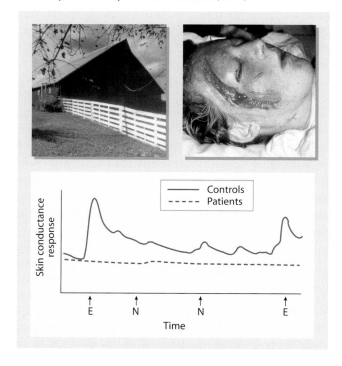

The lack of emotionality is obvious when patients view affect-laden stimuli. But an important question remains: Do physiological, emotional responses moderate decision making, as Damasio proposed? To explore this, a risk-taking task was devised where (pretend) monetary rewards and penalties were associated with stimuli. Subjects were free to select from piles of cards and learned through trial and error the payoffs connected with each deck. You can try this yourself by covering the two columns in Figure 13.9. On each trial, choose one pile and unveil the top item on that pile. The goal is to maximize your total amount of money.

The cards in two decks usually provide large payoffs ($100) but can demand hefty payments (up to $1250). For other decks, rewards and penalties are milder (win $50 and lose up to $100). Control subjects gradually come to choose from the latter decks. Patients with orbitofrontal lesions favor riskier decks, perhaps because they are attracted to the frequent $100 payments even though they eventually would be offset by a severe penalty.

Most intriguing were the SCR responses for the two groups. On turning over a card, both groups displayed transient increases in SCR, hence, an autonomic response to the rewards and penalties. Over time, though, these changes became anticipatory for the control subjects; that is, when the subjects were contemplating choosing a card from the risky decks, their SCRs would skyrocket. For patients with orbitofrontal lesions, the SCRs remained reactive and failed to show anticipatory changes. Thus, no physiological evidence proved that their decision was mediated by emotion.

The idea that memories have emotional associations that can aid in guiding our actions is not controversial. What is less clear is whether evoking the associated affect depends on reactivating physiological responses, as Damasio suggested. What we face is a problem of inferring causation from correlation. Are emotional responses in anticipation of an action absent because patients with orbitofrontal lesions are impaired at associating the reinforcement properties of the stimulus to the potential action, as Rolls might suggest? Or do they make inappropriate decisions because they fail to generate somatic markers? Although this question remains, Damasio's hypothesis raises the provocative idea that emotion plays a central role in guiding rational decision making.

SUMMARY OF THE ORBITOFRONTAL CORTEX

The orbitofrontal cortex plays a significant role in our ability to respond and act in a social and emotional world. Although we have a fairly good understanding of some of the deficits that result from damage to the or-

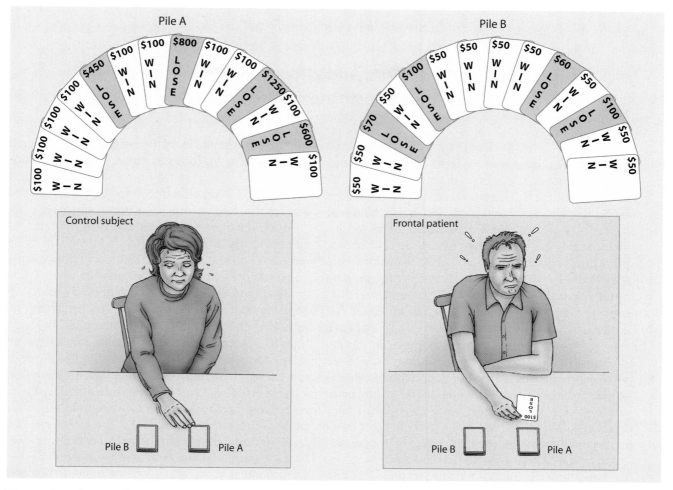

Figure 13.9 Emotional responses occur in reaction to stimuli but also are useful in guiding our decision processes. Subjects were required to choose cards from one pile or the other, with each card specifying an amount won or lost. Through trial and error, the subjects could learn that pile A was riskier than pile B. Control subjects not only tended to avoid the high-risk pile but also showed a large SCR when considering choosing a card from this pile. The patients with prefrontal lesions failed to show these anticipatory SCRs. Interestingly, they did show a large SCR upon turning over a card and discovering they had lost $1000 (of play money).

bitofrontal cortex, our understanding of the precise role of this region in normal decision making is still largely theoretical. One aspect that is clear, however, is that the orbitofrontal cortex must rely on learned information about the emotional qualities of stimuli in order to assess the utility of our actions. Emotional learning and memory are thought to rely on other, interconnected brain structures. One of the primary neural structures thought to interact with the orbitofrontal cortex in this role is the amygdala.

Amygdala

The amygdala is a small, almond-shaped structure in the medial temporal lobe adjacent to the anterior portion of the hippocampus. The medial temporal lobe

structures were first proposed to be important for emotion in the early part of the last century when Klüver and Bucy (1939) documented unusual emotional responses in monkeys following damage to this region. The observed deficit was called "psychic blindness," and one of the prominent characteristics was a tendency to approach objects that would normally elicit a fear response. It was not until the 1950s that the amygdala was identified as a primary medial temporal lobe structure underlying the deficits observed with what became known as "Klüver-Bucy syndrome" (Weiskrantz, 1956). Since that time, the amygdala has been a focus of research on emotional processing in the brain. Although humans with amygdala damage do not show the classic signs of Klüver-Bucy syndrome, they do show several, more subtle deficits in emotional processing.

Approaches to Drug Addiction: Steven Grant, Ph.D., National Institute on Drug Abuse

Alcohol or other drugs are attractive because they can alter how we feel—they can make us feel good. In fact, using drugs is one of the oldest human pursuits. It is considered normal in our society to have the occasional glass of wine with dinner or a beer at a football game. However, most of us also know people, maybe ourselves, whose use of alcohol or other drugs has gotten out of hand to the point where they risk losing everything they value. Although there is not a clear line in the transition between casual drug use and drug abuse, there is a general consensus that individuals have a drug addiction if they take more of a drug than intended, if they focus their life around drugs, and if they persist in using a drug despite adverse consequences. However, categorizing individuals as addicts does not explain the cause of substance abuse, or the strong tendency to start using drugs again even after they have quit. One potential benefit of cognitive neuroscience research is to help us understand how neural systems and psychological functioning interact during emotional states. This understanding eventually may aid the development of more effective treatments for those suffering from the pervasive and costly problem of addiction.

The emotions individuals experience over the course of using drugs can change dramatically. A familiar view of the course of addiction as portrayed in the media and popular press is the experience of heroin addicts. When first taken, heroin and other opiates (morphine and codeine) produce a "high" consisting of euphoria and a sense of well-being and contentment. With continued use, the high or euphoria becomes less and less pronounced, and the addicts must take a larger dose to attain the desired state. When the addicts stop taking opiates, they not only undergo severe physical discomfort resembling a bad case of the flu but also have strong negative emotions, including depression and anxiety. The addicts' continued use of drugs therefore was thought to be motivated by the avoidance of the onset or escape from the presence of this adverse emotional and physical condition. The stereotype of drug addicts engaging in theft or violence to feed their drug habit results from the presumption that drug abusers would go to any length to avoid withdrawal. However, even after addicts no longer suffer from withdrawal symptoms, they often start using drugs again. Clearly, the adverse emotional and physical state associated with withdrawal could no longer be considered the sole cause of substance abuse.

When psychostimulant drugs such as cocaine (or "crack"), amphetamine, and methamphetamine became popular in the 1980s, it came to be understood that the pursuit of the positive emotional effects of drugs rather than the withdrawal syndrome is the basis for addiction. When a person stops taking these drugs after long-term use, a severe withdrawal syndrome does not occur, although there may be a period of depression and sleep disturbances. But the social and personal costs of widespread cocaine use, especially the "crack" epidemic of the last two decades, led to the widespread acceptance of the addictive nature of these drugs. This new view of addiction was rooted in animal experiments where animals

The amygdala has been described primarily as being important for emotional learning and memory. This is because damage to the amygdala does not usually impair emotional responses to innately aversive or rewarding stimuli, but rather a subset of learned emotional responses. This role for the amygdala in emotional learning affects a number of emotional behaviors related to implicit learning, explicit memory, social responses, and vigilance. We will discuss each of these.

IMPLICIT EMOTIONAL LEARNING

Imagine a young man is taking the train to work one day and starts a conversation with another commuter whom he has not met before. After he has been talking with this fellow commuter for a few minutes, a terrible accident occurs. The train hits a car. Everyone on the train is thrown, several passengers on the train are injured, and the driver of the car is killed. Immediately after the accident, the frightened young man, who sustained only

learned to press a lever to receive a small intravenous injection of a drug. This experimental paradigm was identical to the operant learning procedures that B.F. Skinner established.

These studies established that drugs could serve as reinforcers of behavior, just like natural rewards such as food, water, and sex. In fact, drugs were seen as more powerful than natural rewards because the animals would press levers as much as possible to obtain drugs. These studies eventually established that self-administration of drugs was critically dependent on a forebrain circuit called the *meso-limbic dopamine system*. The reinforcing properties of psychostimulant drugs were found to be proportional to their ability to increase the action of the neurotransmitter dopamine in the structure known as the *nucleus accumbens.*

Despite the long-standing controversy as to whether behavioral reinforcement is equivalent to the experience of pleasure, these studies were interpreted as showing that people use drugs to produce the euphoria that comes from an increased release of dopamine. In this view, treatment of substance abuse would be just a matter of restoring normal dopamine function. Although appealing in its simplicity and neurobiological basis, this approach has yet to yield any effective treatments for substance abuse. Nor does this view account for the fact that many people who take drugs do not become addicted. Contemporary theories suggest that the presence of other emotional disorders (anxiety, depression, posttraumatic stress disorder, partner abuse) may be as important in substance abuse as the specific pharmacological action of drugs.

Neuroimaging studies of addiction have begun to show complex ways that drugs work in the human brain. In turn, this research can lead to explanations of how drugs alter how we feel and even how the effects of drugs can be altered by how we expect to feel. These studies have indicated that the cerebral cortex in humans, and by inference cognitive processes, may be as important in addiction as the subcortical systems emphasized in animal studies. One major advance has been the discovery that drug addiction involves the same brain networks that are associated with normal emotional processes. When drug addicts are shown items related to drug use, such as pipes and syringes, or movies of people using drugs, they report an increased craving or urge for drug use. Imaging studies have revealed that both cortical and subcortical brain regions associated with emotional memory are activated during exposure to drug-related cues. These regions include the amygdala, anterior cingulate, orbitofrontal cortex, and dorsolateral prefrontal cortex. This set of regions is similar to the brain areas activated during normal emotional processes, but drug-related cues produced more intense activation of these areas in drug abusers than did other emotional processes (e.g., sexual stimuli).

Apart from producing a craving for drugs, exposure to drug-related stimuli is problematic because these stimuli are intrusive and distracting. When asked to perform a simple attention task in the presence of drug-related stimuli, drug addicts' ability to perform the task is decreased in comparison to the ability of non-drug-using subjects. These results suggest that top-down attentional control mechanisms may be disrupted in drug abusers. Other studies have suggested that drug abusers have problems with other cognitive processes such as memory and decision making.

Substance abuse in humans is more complex than simply avoiding discomfort or pursuing pleasure. It is now clear that in addition to understanding the pharmacological mechanism of the action of drugs of abuse, effective treatment of substance abuse will require understanding the interaction of emotional systems in the brain with cognitive processes such as memory, expectancy, and attention.

minor cuts and bruises, leaves the train. He decides to go home to try to calm down. A few months later, the young man is invited to a cocktail party. At the party is a guest he does not immediately recognize but who looks familiar. This guest starts talking to him, and for some reason the young man becomes immediately uneasy and nervous and excuses himself. He later asks the host who this person is, and he realizes it is the fellow commuter from the train on the day of the horrible accident. Although the young man at first could not consciously identify the party guest as being the fellow commuter, when the party guest started talking to him, his emotional response indicated that he had some memory of this person. He showed signs of physiological arousal that left him feeling uneasy and nervous. His body's response indicated that the image of this fellow commuter/party guest was linked to the train on that fateful day and the aversive consequences of the accident.

This type of learning, in which a neutral stimulus acquires aversive properties by virtue of being paired with an aversive event, is an example of fear conditioning, and it is the primary paradigm used to investigate the amygdala's role in emotional learning. **Fear conditioning** is a form of classical conditioning in which the unconditioned stimulus is aversive. One advantage of using the fear conditioning paradigm to investigate emotional learning is that it works essentially the same across a range of species, from fruit flies to humans.

An example of a laboratory version of fear conditioning is illustrated in Figure 13.10. A rat is put in a cage, and every so often a light is turned on for about 4 seconds and then it is turned off. The light is the *conditioned stimulus,* or CS. At first, the light turning on and off might surprise the rat a little, but this is a neutral stimulus and eventually the rat becomes used to it. This stage is called *habituation.* After the rat is habituated to the light (the CS), the presentations of the light are paired with a shock that is delivered immediately before the light is turned off (Figure 13.10a). The shock is the *unconditioned stimulus,* or US. The rat has a natural fear response to the shock, called the *unconditioned response,* or UR. This stage is referred to as *acquisition.* After a few pairings of the light (the CS) and the shock (the US), the rat learns that the light predicts the shock and eventually the rat exhibits a fear response to the light alone. This fear response is the *conditioned response,* or CR. The conditioned response illustrated in Figure 13.10 is potentiated startle. As mentioned earlier, the startle reflex is a natural response to surprising stimuli, such as the presentation of a loud noise. The startle reflex is enhanced or potentiated in the presence of a fearful stimulus or an anxious state. Figure 13.10b illustrates a startle reflex in a rat when it hears a loud noise. Figure 13.10c shows how this startle response is potentiated in the presence of the light (the CS), indicating the conditioned response. Following habituation and acquisition, a third stage in the standard fear conditioning paradigm is *extinction.* During this stage, the light (the CS) is presented alone without the shock (the US). After several presentations of the light alone, the rat no longer exhibits startle potentiation (the CR) in the presence of the light. At this point the conditioned response is considered extinguished.

A number of responses can be assessed as the conditioned response in this type of fear learning paradigm. In addition to potentiated startle, signs of autonomic nervous system arousal, such as a change in heart rate or SCR, can indicate a conditioned response. In the earlier example of the young man on the train, the conditioned stimulus is the fellow commuter, the unconditioned stimulus is the accident, the unconditioned response is being frightened and hurt during the accident, and the conditioned response is the uneasy and nervous feeling the young man experienced in the presence of the party guest/commuter.

Regardless of the stimulus (CS, US) or response (UR, CR) used, one consistent finding has emerged across

Figure 13.10 An illustration of fear conditioning. Adapted from Davis (1992).

(a) Training: Light and shock paired

(b) Testing: Noise-alone trials

(c) Testing: Light-noise trials

species: Damage to the amygdala impairs conditioned fear responses. Lesions to the amygdala do not usually block the unconditioned response to the aversive event, indicating that the amygdala is not necessary to exhibit a fear response. However, amygdala lesions block the ability to acquire and express a conditioned response to the neutral conditioned stimulus that is paired with the aversive unconditioned stimulus.

By using the fear conditioning paradigm, researchers such as Joe Le Doux (1996) of New York University, Mike Davis (1992) of Emory University, and Bruce Kapp (1984) of the University of Vermont and his colleagues have been able to map the neural circuits of fear learning from stimulus perception to emotional response. From these studies it has become clear that the amygdala is a complex structure consisting of several subnuclei. As seen in Figure 13.11, the lateral nucleus of the amygdala seems to be a convergence area for information from a number of brain regions, allowing for the formation of associations underlying aversive conditioning. The lateral nucleus then projects to the central nucleus. It is projections from the central nucleus that initiate an emotional response if a stimulus, after being analyzed and placed in the appropriate context, is determined to represent something threatening or potentially dangerous.

An important aspect of this circuitry of fear conditioning is that information about a stimulus or a conditioned stimulus can reach the amygdala through two separate, and simultaneous pathways (Le Doux, 1996). One is sometimes called the "low road"; it is quick but dirty. This is a subcortical pathway in which sensory information about a stimulus projects to the thalamus, which in turn sends a signal directly to the amygdala. The thalamus does not produce a sophisticated analysis of this sensory information, but it can send a crude signal to the amygdala indicating whether this stimulus roughly resembles the conditioned stimulus. At the same time, sensory information about the stimulus is being projected to the amygdala via another, cortical pathway, sometimes referred to as the "high road." The "high road" is somewhat slower, but the analysis of the stimulus is more thorough and complete. The sensory information projects to the thalamus. The thalamus then sends this information to the sensory cortex for a finer analysis. The sensory cortex projects the results of this analysis to the amygdala. The "low road" allows for the amygdala to receive information quickly in order to prime, or ready, the amygdala so that it can respond right away if the information from the "high road" confirms that the sensory stimulus is the conditioned

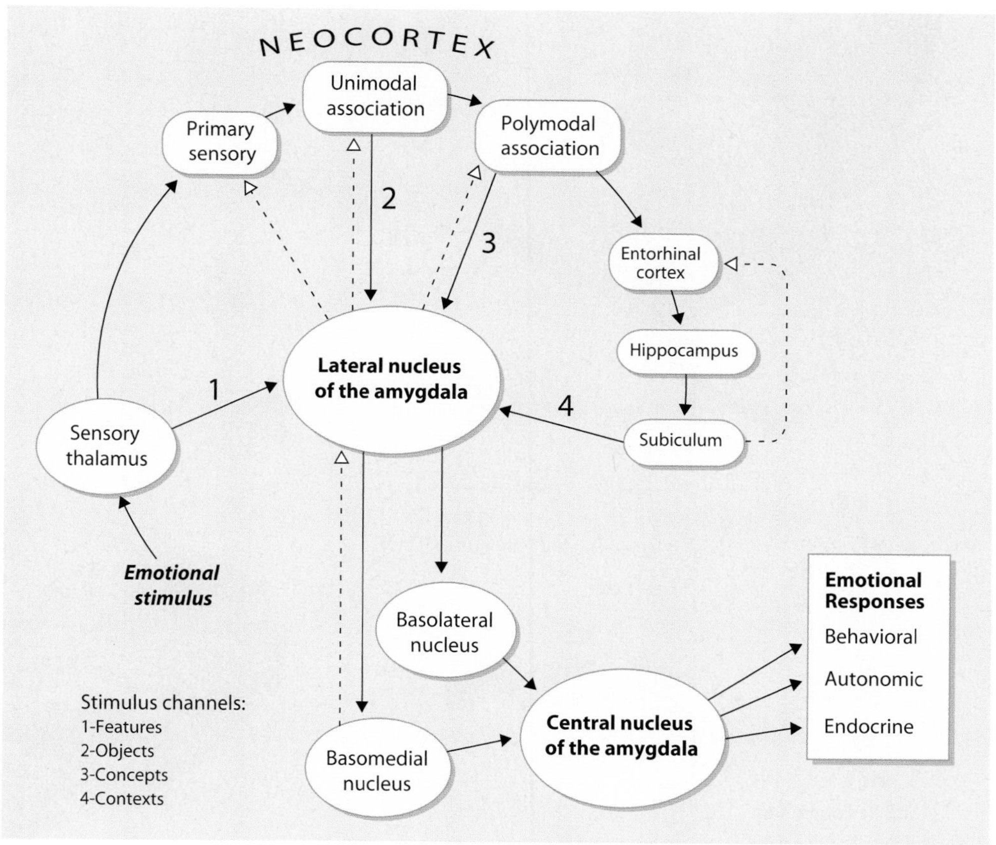

Figure 13.11 Amygdala pathways and fear conditioning. Adapted from Le Doux (1995).

stimulus. Although it may seem redundant to have two pathways to send information to the amygdala, when it comes to responding to a threatening stimulus, it is adaptive to be both fast and sure. An example of the neural circuits involved in the amygdala's response to a threatening stimulus is illustrated in Figure 13.12.

The role of the amygdala in learning to respond to stimuli that have come to represent aversive events through fear conditioning is said to be implicit. This term is used because the learning is expressed indirectly through a behavioral or physiological response, such as autonomic nervous system arousal or potentiated startle. When studying nonhuman animals, we can assess the conditioned response only through indirect or implicit means of expression. In humans, however, we also can assess the response directly, by asking the subjects to report if they know that the conditioned stimulus represents a potential aversive consequence (the US). Although patients with amygdala damage fail to demonstrate an indirect conditioned response, when they are asked to ex-

Figure 13.12 Emotional processing in the amygdala projects to the ventromedial prefrontal cortex and the anterior cingulate. When the hiker chances upon the rattlesnake, the visual information activates affective memories through the projections to the amygdala. These memories not only produce autonomic changes such as an increase in heart rate and blood pressure, but also can influence subsequent actions through the projections to the prefrontal cortex. The hiker will use this emotion-laden information in choosing his next action: Turn and run? Or slowly move around the snake? Adapted from Le Doux (1994).

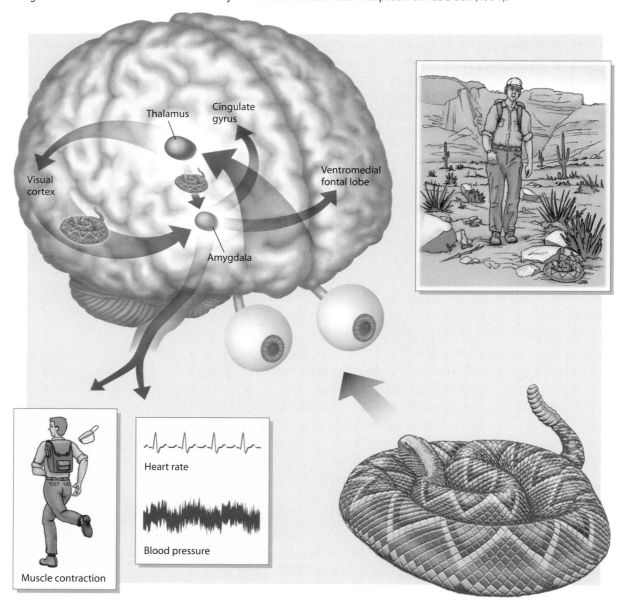

plicitly or consciously report the parameters of fear conditioning, they appear to be completely intact.

This concept is illustrated in Figure 13.13, which presents data from patient S.P., a woman with bilateral amygdala damage (Phelps et al., 1998). S.P. participated in a fear conditioning study in which a picture of a blue square on a computer screen (the CS) was periodically presented for 10 seconds. During the acquisition phase, S.P. was given a mild electrical shock to the wrist (the US) at the end of the 10-second presentation of the blue square (the CS). As expected, S.P. showed a normal fear response to the shock (the UR), which was assessed by measuring her SCR. Also as expected, she failed to show any change in her SCR when the blue square (the CS) was presented, even after several acquisition trials. This lack of change in the SCR demonstrates that she failed to acquire a conditioned response.

Figure 13.13 Example of the SCR in a normal control subject and patient S.P., who has amygdala damage. The patient showed no response to the blue square (the conditioned stimulus, or CS), indicating a lack of a conditioned response. However, S.P. did respond to the shock (the unconditioned stimulus, or US), indicating a normal unconditioned response. The control subject showed a response to both the conditioned and the unconditioned stimulus. Even though S.P. failed to show an indirect conditioned response as measured by the SCR, she demonstrated intact declarative knowledge of the fear conditioning procedure. When shown the data and asked to comment on her performance, S.P. replied, "I knew that there was an anticipation that the blue square, at some particular point in time, would bring on one of the volt shocks. But even though I knew that, and I knew that from the very beginning, except for the very first one where I was surprised. That was my response—I knew it was going to happen. I expected that it was going to happen. So I learned from the very beginning that it was going to happen: blue and shock. And it happened. I turned out to be right, it happened!"

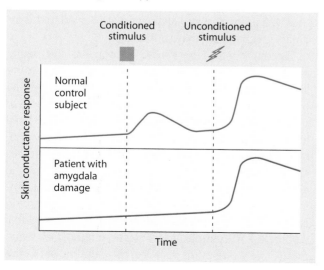

Figure 13.13 shows the SCRs for S.P. and a control subject. S.P. was shown her data and the control subject's data and was asked what she thought about it. She was somewhat surprised that she did not show a change in SCR (the CR) to the blue square (the CS). She reported that she knew after the very first acquisition trial that she was going to get a shock to the wrist when the blue square was presented. She claimed to have figured this out early on and expected the shock whenever she saw the blue square. She was not sure what to make of the fact that her SCR did not reflect what she consciously knew to be true.

This dissociation between intact explicit knowledge of the events that occurred during fear conditioning and impaired conditioned responses has been shown in a number of patients with amygdala damage (Bechara et al., 1995; LaBar et al., 1995). As discussed in Chapter 8, explicit or declarative memory for events depends on another medial temporal lobe structure, the hippocampus. Damage to the hippocampus impairs the ability to explicitly report memory for an event. When the conditioning paradigm described with S.P. was conducted with patients who had bilateral damage to the hippocampus but an intact amygdala, the opposite pattern of performance emerged. These patients showed a normal SCR to the blue square (the CS), indicating the acquisition of the conditioned response. However, when asked what occurred during conditioning, they were unable to report that the presentations of the blue square were paired with the shock, or even that a blue square was presented at all (LaBar et al., in press). This double dissociation between patients who have amygdala lesions and patients with hippocampal lesions highlights the fact that the amygdala is necessary for the implicit expression of emotional learning, but not all forms of emotional learning and memory. The hippocampus is necessary to acquire explicit or declarative knowledge of the emotional properties of a stimulus, whereas the amygdala is critical for the acquisition and expression of a conditioned fear response.

EXPLICIT EMOTIONAL LEARNING AND MEMORY

The amygdala is necessary for emotional learning as measured by fear conditioning. However, its role in learning and memory does not stop there. The amygdala interacts with other memory systems, particularly the hippocampal memory system, when there are emotional events or information. There are two primary ways the amygdala interacts with hippocampal-dependent declarative memories. First, the amygdala is necessary for normal indirect emotional responses to stimuli whose emotional properties are

Social Neuroscience: Example of an Emerging Field

Danny Glover, the stage and movie actor from the *Lethal Weapon* movie series, recently filed a complaint with the New York City Taxi and Limousine Commission (TLC) charging that local cab drivers repeatedly refused to pick him up because of his race. He claims that although he is a movie star in Hollywood, on the streets of New York City he is, first and foremost, a black man. This social group membership, he contends, is his primary identification in society, and because of negative attitudes toward his social group, he is treated unfairly.

In this new millennium it is somewhat shocking, yet at the same time not surprising, that incidents like those alleged by Mr. Glover still occur. Although at one time social norms and laws dictated the attitudes toward racial groups, since the civil rights movement of the 1960s these social norms have changed. One result of these changes is that the endorsement of prejudicial attitudes has dramatically declined over the last several decades. However, as Danny Glover's experiences illustrate, this dramatic decline in stated racial attitudes has not been accompanied by a similar decline in societal racial problems.

Social psychologists have suggested that this discrepancy between reported attitudes and behavior may be due to the fact that the influence of attitudes is much more subtle and pervasive than is suggested by measures of self-report. There are two reasons why these self-report measures may not mirror behavioral responses. First, individuals may be reluctant to admit that they endorse certain prejudicial attitudes. It is likely that the cab drivers in New York City would not express negative attitudes toward blacks (especially to the TLC), yet they still intentionally avoided Mr. Glover. Second, a more recent suggestion is that we may not always be consciously aware of our biased behavioral responses. This type of unintentional social group bias is reflected in the recent MIT report on gender discrimination, which found a "pervasive and unintentional" pattern of discrimination against female faculty. In his commentary, MIT Dean Robert Birgeneau stated, "I believe that in no case was this discrimination conscious or deliberate. Indeed it was totally unconscious and unknowing."

Using indirect behavioral measures, social psychologists are beginning to differentiate between social responses and attitudes that are purposeful and direct (explicit) and those that are unintentional and indirect (implicit). Explicit attitudes are those we consciously express and often believe to be true. Implicit attitudes are expressed through our behavior, without our conscious awareness or control. While our explicit attitudes may reflect the ideals that we endorse, our implicit attitudes may reflect our underlying knowledge of cultural norms, our own social group membership, and our personal experience. These implicit attitudes may or may not be consistent with our explicit attitudes. For instance, in contrast to studies showing relatively unbiased explicit evaluations of black Americans by white Americans, there is robust evidence for negative evaluations using implicit or indirect measures.

This research provides scientific confirmation of the sentiments long expressed by members of underrepresented social groups. Even though we now have laws against prejudicial acts, and even though many individuals may be well intentioned, cultural stereotypes of social groups exert a subtle but real influence on behavior. Social psychologists have approached this problem by identifying personal and social group factors related to the direct and indirect expression of social attitudes. Researchers in the emerging field of social neuroscience have another approach. They believe that it is equally important to understand the specific mechanisms of learning and expression that underlie these behaviors. Are we endowed with special cognitive and neural mechanisms to learn and express our social attitudes? Or do we learn and express social attitudes using the same systems we use to evaluate nonsocial, emotional stimuli? In other words, by combining research in social psychology with the cognitive neuroscience of emotion, can we show how social learning and evaluation are rooted in the ordinary mechanics of mind and brain?

Recent research on the basic cognitive and neural mechanisms of emotional learning has suggested some links to studies of social attitudes. Much in the same way that social psychologists differentiate direct and indirect means of social evaluation, cognitive neuroscientists have identified brain mechanisms related to the direct or indirect expression of learned emotional responses. Studies conducted across a range of species have shown that the amygdala is a critical brain structure in emotional learning. In humans, the amygdala's role in emotional learning is often limited to the indirect expression of the emotional response. For example, it is a classic finding that when people are startled (for instance, by a loud noise) in the presence of negative stimuli (e.g., a dark, empty street), this startle response will be enhanced or potentiated. This startle potentiation indirectly indicates the emotional evaluation of the stimulus. Patients with amygdala damage, in contrast to normal control subjects, do not exhibit this startle potentiation in the presence of negative, nonsocial stimuli. At the same time, these patients rate these stimuli as equally arousing and negative as control subjects. In other words, they are able to explicitly report the emotional evaluation of the stimuli, even though they do not show an implicit or indirect emotional response. The performance of these patients is in contrast to the performance of patients with damage to a neighboring brain structure, the hippocampus. Hippocampal damage results in the opposite pattern; patients are able to indirectly express an emotional response to a stimulus through behavior, even though they cannot explicitly report the emotional significance of the stimulus. Thus, different cognitive and neural systems underlie the direct and indirect expression of affective evaluation.

Investigations into both social attitudes and the cognitive neuroscience of emotional learning have found dissociations between indirect and direct affective evaluation. These common results led Elizabeth Phelps, a cognitive neuroscientist from New York University, and Mahzarin Banaji, a social psychologist from Yale University, to wonder if similar mechanisms underlie these two types of findings (Phelps et al., 2000). Could amygdala activity be related to the indirect evaluation of social groups? To test this hypothesis, they used functional magnetic resonance imaging (fMRI) to examine activity in the amygdala in white American subjects while they viewed pictures of unfamiliar white and black male faces. Subjects were given three measures of social evaluation. One was an explicit, self-report of racial attitudes. The other two were indirect, behavioral assessments of racial evaluation.

Although the majority of white subjects showed greater amygdala activity in response to the black compared to the white faces, there was significant variability among the subjects. When the researchers examined the relationship between amygdala response and measures of social evaluation, they found that the white subjects who showed greater amygdala activity in response to the black faces also showed stronger biases in social responses on the indirect measures. This same relationship was not found with the direct, explicit measure of social attitudes. In other words, the amount of amygdala activity in response to the black versus the white faces was related specifically to indirect social responses. These results suggest that the same neural sys-

tems that underlie the indirect expression of emotional learning to nonsocial stimuli also are related to the indirect expression of social responses.

In an effort to understand some of the variables related to these results, Phelps and Banaji performed a second study identical to the first, with one exception. The black and white faces belonged to well-known and positively regarded individuals (e.g., Martin Luther King, John F. Kennedy). In this study, there was no consistent pattern of amygdala activity in response to black versus white faces in the white subjects and no relation between amygdala activity and any of the behavioral measures of evaluation. These results suggest that the amygdala's response to black faces in white subjects is diminished with specific experience, familiarity, and positive evaluation.

In a related study by Alan Hart of Amherst College, Scott Rauch of Massachusetts General Hospital, and their colleagues, the amygdala in some black American subjects showed a greater response to unfamiliar white faces than to black faces (Hart et al., 2000).

This research represents the emerging field of social neuroscience. These studies are the first to link the evaluation of social groups to brain activity. By creating this link, we can add what we know about the cognitive function of these neural systems for emotional learning to our understanding of social evaluation and attitudes. The preliminary findings not only start to delineate the mechanisms involved in social evaluation but also raise several crucial questions. For instance, what factors are related to the variability observed among the subjects? What kind of responses would we expect from members of other racial groups, such as Asian Americans and Hispanic Americans or those with mixed racial backgrounds? What other brain regions are related to social evaluation and what are their behavioral roles? Precisely how does familiarity, positive information, or specific experience with members of different social groups alter the neural systems of social evaluation? By identifying the cognitive and neural mechanisms of social responses, we can initiate discovery of the means by which they are learned and modulated.

In his complaint to the TLC, Mr. Glover requested that New York City cab drivers receive diversity training. How can we be sure this diversity training changes the explicit and implicit mechanisms of social evaluation? Given what we know about learned emotional responses in general, what types of training might be most effective at altering both direct and indirect social evaluations? Researchers in social neuroscience believe that understanding the complex mechanisms underlying social responses will help us answer these questions.

learned explicitly, by means other than fear conditioning. Second, the amygdala can act to enhance the strength of explicit or declarative memories for emotional events by modulating the storage of these memories.

Imagine that a young woman is walking down the street in her neighborhood and sees a neighborhood dog on the sidewalk. Even though this woman is a dog owner, she is afraid of this particular dog. When she encounters it, she gets nervous and fearful and decides to walk on the other side of the street. Why might this woman, who likes most dogs, be afraid of this particular dog? There are a few possible reasons. It could be that this dog once bit her. In this case, her fear response to the dog was acquired through fear conditioning. The dog (the CS) was paired with the dog bite (the US), resulting in pain and fear (the UR) and an acquired fear response to the dog (the CR). However, the woman may fear this dog for another reason. She may fear the dog because she heard from her neighbor that this was a mean dog that *might*

bite her. In this case, she has no aversive experience linked to this particular dog. Instead, she learned about the aversive properties of the dog explicitly. Her ability to learn and remember this type of information is dependent on her hippocampal memory system. It is unlikely she experienced a fear response when she learned this information during a conversation with her neighbor. She did not experience a fear response until she actually encountered the dog. Her fear response is not based on actual experience with the dog, but rather is imagined and anticipated based on her explicit knowledge of the potential aversive properties of this dog. This type of learning, in which we learn to fear or avoid a stimulus because of what we are told, as opposed to our actual experience, is a common means of emotional learning in humans.

This explicit learning of the emotional properties of a stimulus, in the absence of aversive experience, is modeled in the instructed fear paradigm illustrated in Figure 13.14. There are two means by which a person can learn

Figure 13.14 Two methods by which humans can learn about the aversive properties of an event. **(a)** Fear conditioning, in which the aversive properties of the blue square are learned by the pairing of the square and shock. **(b)** Instructed fear, in which the blue square is linked to the shock by verbal instruction. In both cases, the amygdala plays a role in the expression of the fear response.

that the presentation of a blue square predicts a shock to the wrist. The upper figure shows an example of fear conditioning, in which a person fears the presentation of a blue square because it was previously paired with a shock. The lower figure shows an example of instructed fear, in which a person fears a blue square because he was told it might be paired with a shock. As mentioned earlier, we know the amygdala is critical to show a conditioned fear response during fear conditioning. The question is, Does the amygdala play a role in the indirect expression of the fear response in instructed fear?

Elizabeth Phelps of New York University and her colleagues (Phelps et al., 2001; Funayama et al., in press) addressed this question. They found that even though explicitly learning the emotional properties of the blue square is dependent on the hippocampal memory system, the amygdala is critical for the expression of some fear responses to the blue square. During the instructed fear paradigm, patients with amygdala damage were able to learn and explicitly report that some presentations of the blue square might be paired with a shock to the wrist, even though none of the subjects ever received a shock. However, unlike normal control subjects, they did not show a potentiated startle response when the blue square was presented. In addition, normal control subjects showed an increase in SCR to the blue square that was correlated with amygdala activity. These results suggest that in humans the amygdala is sometimes critical for the indirect expression of a fear response in situations where the emotional learning occurs explicitly, by means other than fear conditioning. Similar deficits have been observed when patients with amygdala lesions respond to emotional scenes (Angrilli et al., 1996; Funayama et al., in press).

Although animal models of emotional learning highlight the role of the amygdala in fear conditioning and the indirect expression of the conditioned fear response, human emotional learning can be much more complex. We can learn that stimuli in the world are linked to potentially aversive consequences in a variety of ways, including instruction, observation, and experience. Whatever the manner by which we learn the aversive or threatening nature of stimuli, whether it is explicit and declarative, implicit, or both, the amygdala may play a role in the indirect expression of the fear response to that stimuli.

The instructed fear studies indicate that a hippocampal-dependent, declarative representation about the emotional properties of stimuli can influence amygdala activity, which then modulates some indirect emotional responses. James McGaugh (1992) of the University of California at Irvine and his colleagues (Ferry et al., 2000) demonstrated another means by which these two brain systems interact. They found that the amygdala modulates the strength of declarative or explicit memories for emotional events.

How do we use declarative memory in everyday life? The types of things we usually recollect are things like where we left the keys, what we said to a friend the night before, or whether we turned the iron off before leaving the house. However, when we look back on our lives, we do not remember these types of events. We remember our first kiss, being teased by a friend in school, the pride of our family at graduation, or hearing about a horrible accident. The memories for events that last over time are those of emotional or important events. These memories seem to have a persistence and vividness that other memories lack. One reason for this persistence may be related to the action of the amygdala.

In Winston Churchill's autobiography *My Early Life: A Roving Commission* (1930), he described in great detail an exam he took to gain admittance to secondary school. "I wrote my name at the top of the page. I wrote down the number of the questions 'I'. After much reflection I put a bracket around it thus '(I)'. But thereafter I could not think of anything connected with it that was relevant or true. Incidentally there arrived from nowhere in particular a blot and several smudges. I gazed for two whole hours at this sad spectacle: and then merciful ushers collected my piece of foolscap with all the others and carried it up to the Headmaster's table." Although this event occurred decades before Churchill wrote about it, he still remembered very specific details. What was it about this particular event that made it stand out in Churchill's memory? It was not the first exam he had taken, or evidently, the first exam he failed. However, given the circumstances (an admittance exam for a prestigious school), it was a particularly important and embarrassing event. He was nervous about taking the exam and especially aroused when he discovered that he was unable to answer the questions.

An arousal response can influence the ability to store declarative or explicit memories, which may partially explain Churchill's detailed recollection of the events that occurred during the exam. This effect has been demonstrated in studies across species. For example, it is known that creating an arousal response will enhance performance on declarative, hippocampal-dependent memory tasks in rats. In a series of studies, McGaugh and colleagues (1999) demonstrated that this memory enhancement effect of arousal is blocked by lesions to the amygdala.

There are two important aspects of this work that help us understand the mechanism underlying the role of the amygdala in the enhancement of declarative memory that has been observed with arousal. The first is that the amygdala's role is modulatory. The tasks used in these studies are dependent on the hippocampus for acquisition. For example, one task used is the Morris water maze described in Chapter 7. A lesion to the amygdala does not impair the ability of rats to learn this task under ordinary circumstances. However, if the rat is aroused immediately after training, by either a physical stressor or the administration of drugs that mimic an arousal response, then the rat will show improved retention of this task. The memory is *enhanced* by arousal. It is this arousal-induced enhancement of memory that is blocked by lesions to the amygdala. In other words, the amygdala is not necessary for learning this hippocampal-dependent task, but it is necessary for the modulation of memory for this task due to arousal.

The second important facet of this work is that this effect of modulation with arousal occurs *after* initial encoding of the task, during the retention interval. The administration of drugs that mimic an arousal response immediately after learning leads to enhanced memory for the task. The same is true if the rat is physically stressed after learning. Similarly, using pharmacological lesions to temporarily disable the amygdala immediately after learning eliminates any arousal-enhanced memory effect (Teather et al., 1998). All of these studies point to the conclusion that the amygdala's modulation of hippocampal, declarative memory occurs by enhancing retention, not by altering the initial encoding of the stimulus. Because this effect occurs during retention, it is thought that the amygdala enhances hippocampal consolidation. As described in Chapter 8, *consolidation* is a process that occurs over time, after initial encoding, and leads to memories becoming more or less stable. When there is an arousal response, the amygdala acts to alter hippocampal processing by strengthening the consolidation of memories. McGaugh and colleagues showed that the basolateral nucleus of the amygdala is important for this effect.

This role for the amygdala in enhancing emotional, declarative memory also has been demonstrated in humans. A number of studies over the years have indicated that a mild arousal response can enhance declarative memory for emotional events (see, e.g., Christianson, 1992). This effect of arousal on declarative memory is blocked in patients with amygdala damage (Cahill et al. 1995). In addition, functional neuroimaging studies have shown that activity observed in the human amygdala during the presentation of emotional stimuli is correlated with the arousal-enhanced recollection for these stimuli (Cahill et al., 1996; Hamann et al., 1999). These studies indicate that normal amygdala function plays a role in the enhanced declarative memory observed with arousal in humans.

The mechanism for this effect of arousal appears to be related to the amygdala's role in modifying the rate of forgetting for arousing stimuli, consistent with the notion of a postencoding effect on memory, such as enhancing hippocampal storage or consolidation. One effect of arousal on declarative memory is to alter how quickly we forget. Although the ability to recollect arousing and nonarousing events may be similar immediately after they occur, arousing events are not forgotten as quickly as nonarousing events (Kleinsmith and Kaplan, 1963). Unlike normal control subjects who show less forgetting over time for arousing compared to nonarousing stimuli, patients with amygdala lesions forget arousing and nonarousing stimuli at the same rate (LaBar and Phelps, 1998).

These results suggest that the animal models outlined by McGaugh and colleagues, demonstrating that the amygdala may act to modulate hippocampal consolidation for arousing events, are also applicable to humans. However, this mechanism does not underlie all of the effects of emotion on human declarative memory. Emotional events are more distinctive and unusual than everyday life events. They also form a specific class of events. These and other factors may enhance declarative or explicit memory for emotional events in ways that are not dependent on the amygdala (Phelps et al., 1998). In addition, Robert Sapolsky of Stanford University (1992) and his colleagues demonstrated that extreme arousal or chronic stress may actually *impair* performance of the hippocampal memory system. This memory impairment is due to the effect of excessive stress hormones, such as glucocorticoids, on the hippocampus. The precise role of the amygdala in this impairment of hippocampal memory during chronic or excessive stress is not fully understood.

The amygdala's interactions with the hippocampal memory system and explicit or declarative memory are specific and complex. The amygdala acts to modulate the storage of arousing events, thus ensuring they will not be forgotten over time. In addition, we can learn explicitly that stimuli in the environment are linked to potential aversive consequences, without having to experience these consequences. This explicit, hippocampal-dependent representation of the emotional properties of events can affect amygdala activity and some indirect fear responses. The interactions of the amygdala and hippocampus help to ensure that we remember things that are important and emotional for a long time, and that our bodily response to threatening events is appropriate and adaptive.

SOCIAL RESPONSES

The human amygdala does not appear to be important for most forms of explicit evaluation. That is, the ability to consciously label a stimulus as good, bad, arousing, or neutral does not depend on the amygdala. However, there is one exception. The human amygdala is important for normal responses to a subset of social stimuli: facial expressions. Ralph Adolphs of the University of Iowa and his colleagues (1994, 1999) demonstrated in a series of studies that patients with amygdala damage are particularly impaired at evaluating fearful facial expressions. As illustrated in Figure 13.15, if a normal control subject is shown a picture of a fearful face and is asked to rate on a scale of 1 (not at all) to 6 (very much) "How fearful does this person look?," she will say 5 or 6. If a patient with amygdala damage is given the same question and stimulus, she will say 2 or 3. This impairment is most pronounced for fear. If the control subject and patient are instead shown a picture of a happy face and are asked to rate on a scale of 1 (not at all) to 6 (very much) "How happy does this person look?," both will respond with a 5 or 6.

This sensitivity of the human amygdala to fearful facial expressions also has been demonstrated with functional neuroimaging. Activity in the amygdala increases in response to fearful expressions (Breiter et al., 1996). Although the amygdala will show an activation response to other emotional expressions, such as happy or angry, the activation response to fear is significantly greater. One interesting aspect of the amygdala's response to fearful facial expressions is that the subject does not have to be aware of seeing the fearful face for the amygdala to respond. Fearful facial expressions that are presented subliminally, so quickly that the subject is unaware of their presentation, will result in activation of the amygdala. This activation to subliminal fearful faces is as strong as the response to fearful faces the subject is aware of seeing (Whalen et al., 1998).

This critical role for the amygdala in explicitly evaluating fearful faces also extends to other social judgments about faces, such as indicating from a picture of a face whether or not the person appears to be trustworthy or approachable (Adolphs et al., 2000). However, the amygdala does not appear to be critical for all types of social communication. Unlike patients with damage to the orbitofrontal cortex, patients with amygdala lesions do not show gross impairment in their ability to respond to social

Figure 13.15 Example of the deficit in identifying and responding to fearful facial expressions by a patient with amygdala damage (blue person) compared to a control subject (green person).

Psychiatric Disorders and the Frontal Lobes

Psychiatric disorders such as schizophrenia and depression represent a widespread breakdown in mental function. Problems faced by patients affect almost all aspects of their behavior. It is unlikely that their problems are linked to a simple physiological mechanism. Rather, the disorders arise from a delicate interplay of physiological mechanisms that reflect endogenous dispositions and a person's idiosyncratic experiences.

One of the most promising aspects of cognitive neuroscience is that it may offer new insights concerning the functional deficits associated with severe psychiatric disorders. Simple neuropsychological descriptions do not adequately account for these disorders. Schizophrenia cannot be thought of as a temporal lobe or frontal lobe problem; it arises as a disturbance in cognitive systems that span cortical and subcortical systems. Though schizophrenia has been linked to abnormal dopamine levels, we still need to know how this neurotransmitter affects cognition if we want to understand the functional consequences of this debilitating disorder.

As integrators of all cortical regions, the frontal lobes are, unsurprisingly, abnormal in psychiatric patients. One source of evidence is the comparative blood flow patterns in psychiatric patients and in control subjects (Drevets and Raichle, 1995). These results are, at best, weak predictors of dysfunction, perhaps because psychiatric labels encompass heterogeneous disorders and perhaps because such patients are almost always under a cornucopia of medications. Yet we can tease out intriguing dissociations in metabolic profiles of schizophrenics and depressives: Schizophrenia is often tied to hypometabolism of the prefrontal cortex, and depression is linked to hypermetabolism in the same region (Figure A).

The first reports of hypofrontality, or reduced blood flow in the frontal cortex, in schizophrenia came when patients were scanned while at rest. With the distribution of blood flow in normal subjects used as a baseline, patients with schizophrenia had less prefrontal blood flow in comparison to posterior blood flow. Also, the degree of hypofrontality was correlated with the severity of the patients' symptoms. Studies now focus on blood flow changes during activities associated with frontal lobe function. For example, Daniel Weinberger (1988) at the National Institute of Mental Health scanned patients and control subjects during rest and in two behavioral conditions: first, when patients performed the Wisconsin Card Sorting Test; second, when the task had a number-matching rule that excluded working memory but controlled visual processing and motor response. Unlike in earlier findings, blood flow in the two groups did not differ during the rest condition or the nonfrontal behavioral condition. During the Wisconsin Card Sorting Test, though, blood flow in the groups sharply diverged. Frontal patients had less blood flow in the dorsolateral prefrontal cortex than did the control subjects.

In a follow-up study the Wisconsin Card Sorting Test was performed twice, once after the patients had been given a placebo and once after administration of apomorphine, a morphine derivative with antipsychotic properties. For each patient, blood flow in the dorsolateral prefrontal cortex increased when they received drugs; hence, reducing dopamine levels heightens activity in the prefrontal cortex.

The chemical basis of depression is even more mysterious. Even so, patients have hypermetabolic prefrontal cortices during depressions in contrast to when they are asymptomatic. Hypermetabolism is widespread, encompassing much of the prefrontal cortex and the anterior cingulate and amygdala. Normal subjects also evince hypermetabolism in prefrontal regions when they are asked to think sad thoughts while being scanned.

It would be premature to definitively interpret these divergent metabolic profiles. Even so, we can make intriguing connections to themes in this chapter. Schizophrenics have an underactive frontal cortex, especially in lateral regions. Losing their working memory and inhibitory capabilities renders them more reliant on activity in the posterior cortex. They may be easier to distract, and hence fail to inhibit irrelevant representations such as the ones related to persistent hallucinations.

Depressed patients exhibit a profile of overactivity in prefrontal regions associated with working memory and in areas linked to the generation of affective memories. With these people, representations persist for a long time and are imbued with heightened affect. A situation that a normal person might find neutral, or at most mildly aggravating, becomes amplified and takes on onerous over

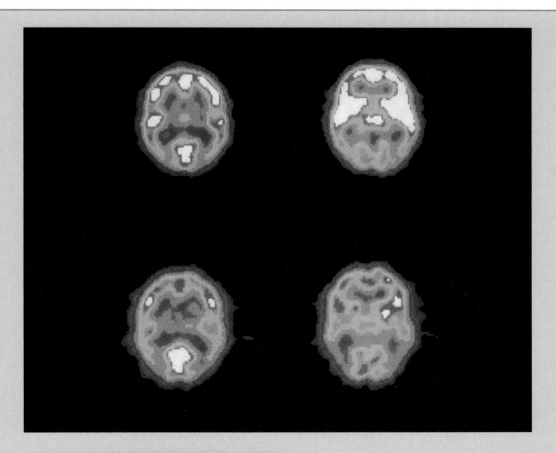

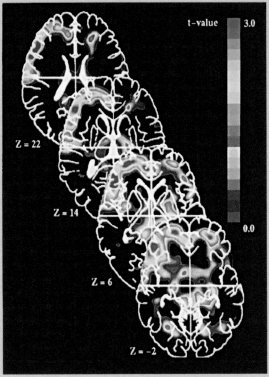

Figure A Positron emission tomography (PET) reveals abnormal patterns of blood flow in patients with psychiatric disorders. **(Top)** Schizophrenic patients show hypometabolism in the prefrontal cortex. This abnormality is especially marked during tasks that produce increased blood flow in this area in healthy subjects. In this study, subjects were involved in a continuous auditory discrimination task. Compared to the control subjects (top slices), uptake of the tracer is much lower in schizophrenic patients (lower slices). **(Left)** Blood flow at rest was measured in control subjects and patients with depression. Colored areas indicate regions of increased blood flow in the depressed patients, and are centered in the lateral prefrontal cortex in the left hemisphere.

continued on the following page

continued from the previous page

tones. The depressed patient cannot let a situation go; the representation of a thought or obsession persists, sustained by input from inappropriate somatic markers.

With a cognitive neuroscience perspective, we can make sense of the outcome of one of the great debacles of neurosurgery: frontal lobotomies for treating psychiatric disorders (Valenstein, 1986). Prior to the implementation of drug therapies in the 1950s and 1960s, mental institutions were overflowing with desperate patients and doctors, eager to try any procedure that promised relief. In the 1930s, Egas Moniz, a renowned Portuguese neurologist who had developed cerebral angiography in 1927, introduced a psychosurgical procedure for treating patients with severe schizophrenia and obsessive-compulsive disorder.

Moniz's inspiration came from an international scientific conference where two American researchers reported the effects of frontal lobectomy in chimpanzees. One animal appeared to have undergone a personality change. Before the operation, the chimp was uncooperative and would throw temper tantrums. After removal of most of her frontal lobes, the animal was cheerful and participated in experimental tests without hesitation. Moniz reasoned that the procedure might bring relief to severely agitated patients, a well-intended thought given the paucity of alternatives.

Removing large amounts of tissue from the frontal lobes seemed excessive. Instead, Moniz decided to isolate the prefrontal cortex from the rest of the brain by severing the white matter's connecting fibers. In his early efforts he applied toxic levels of alcohol through holes in the skull's lateral surface. He later switched to a procedure in which a leukotome, a plunger with an extractable blade, could be lowered into the brain to sever fibers in targeted regions.

Walter Freeman at Georgetown University refined this procedure. He developed a simple technique that did not require a surgeon. The patient was first given an anes-thetic consisting of a severe electrical shock. While the patient was unconscious for 15 minutes, the lobotomy was performed by jabbing an ice pick through the bone above each eye and wiggling it back and forth. Freeman assembled a portable kit containing his electroshock apparatus, ice picks, and a small hammer, and set off on a barnstorming trip to promote the benefits of this miracle cure (Figure B). The public and scientific community were welcoming. Thousands of procedures were performed over the next few decades, and Moniz received the Nobel Prize in physiology and medicine in 1949.

With the advantage of hindsight, we now recognize the abject failings of the lobotomy craze. There were few outcome studies, and those few revealed that the discharge rate from mental institutions was no greater for lobotomy patients than it was for control subjects. Scant concern was given to the patients selected; the procedure had minimal effect on schizophrenics but drastically altered patients with affective disorders like depression or severe neurosis, who felt much less anxious, impulsive, and depressed. But these feelings brought new problems that rendered them incapable of functioning outside the institutional setting. They were now withdrawn and underactive, lacking in affect or responsiveness. The benefits, if any, were experienced by attendants who rejoiced that the patients were docile and easy to manage. As with Phineas Gage, the patients' souls had been transformed.

These differential outcomes make sense in light of metabolic studies. Lobotomies targeted the prefrontal cortex, a region already underactive in schizophrenia. As such, we might expect little effect on schizophrenics, or maybe new problems for those with an excessive dominance of posterior brain function. For affective disorders, though, lobotomies isolated an overactive region. Moreover, the primary foci were on medial regions, so the procedure may have eliminated behaviors associated with exaggerated emotionality but turned patients into affectless zombies.

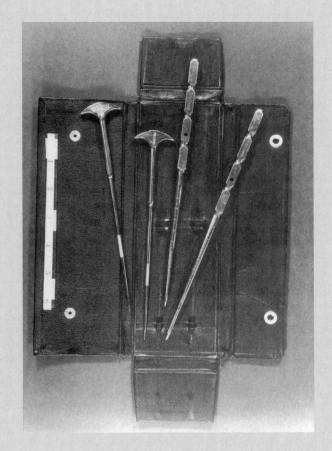

Figure B   (Left) The portable tool kit for performing prefrontal lobotomies. The kit included various tools for penetrating the skull and cutting through brain tissue. **(Below)** Freeman's sketch of two lobotomy procedures. In one, the blade is rotated back and forth, forming a radial cut. In the other, the blade is inserted at various angles. From the book *Great and Desperate Cures* by Elliot S. Valenstein (1986). By permission of the author.

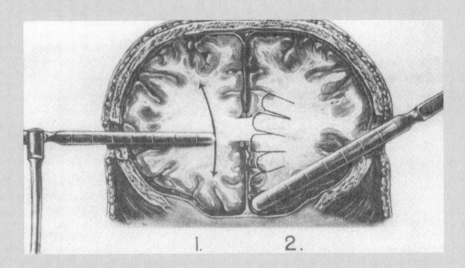

stimuli. They can interpret descriptions of emotional situations correctly, and they can give normal ratings to emotional prosody, or the speech sounds that indicate emotion, such as when a person is speaking in an angry or fearful tone of voice (Anderson and Phelps, 1998; Scott et al., 1997). They do not show other, nonemotional face recognition deficits and are even able to recognize the perceptual similarity between facial expressions whose emotional content they label incorrectly. In addition, their deficit appears to be restricted to the perception of facial expressions. They are able to generate and communicate a full range of facial expressions (Anderson and Phelps, 2000). Although functional imaging studies demonstrated that the amygdala can respond to a wide range of social or emotional signals, it appears to be critical only for the explicit evaluation of a small range of facial stimuli.

In humans with amygdala lesions the deficit in responding to social stimuli appears to be fairly subtle, but in nonhuman animals the deficit is more severe. Monkeys with amygdala lesions have more significant social problems, such as difficulties with aggression and social rank (Kling and Brothers, 1992). These deficits impair the ability of monkeys with amygdala lesions to survive in a normal monkey social hierarchy. It is somewhat puzzling that the social response deficits observed in humans with amygdala lesions would be much less devastating than those observed in other species. One possible explanation for this discrepancy comes from functional neuroimaging studies. The results of these studies indicate that although damage to the human amygdala does not impair normal explicit evaluation of varied stimuli, in normal subjects the amygdala shows an increase in activity when they are presented with a wide range of emotional and social stimuli. These results suggest that the amygdala is sensitive to these stimuli. Just because a patient with a lesion to the amygdala can demonstrate a normal response to an emotional stimulus under some circumstances does not mean that this stimulus was processed normally. It may be that humans have developed different forms of responding, and when one ability is lost, humans may be able to compensate with other mechanisms more efficiently than nonhuman animals. The functional imaging studies would suggest that if we look more carefully, there may be more extensive, subtle deficits in social responding following amygdala lesions in humans. Future research will determine if this is the case. By combining the results of studies on brain-injured patients with those from functional neuroimaging studies in normal subjects, we can begin to understand more completely the complex role the amygdala plays in social responding.

VIGILANCE

Functional neuroimaging studies have shown that activity in the amygdala increases in response to a wide range of social and emotional stimuli. But does the amygdala cause an emotional response to a stimulus? It appears that the amygdala is not necessary for the conscious experience of emotional states. When patients with amygdala damage are asked to rate the intensity and frequency with which they experience different emotional states (including fear/anxiety states), their ratings appear to be normal (Anderson and Phelps, in press). In some tasks, the amygdala's activation response to an emotional stimulus is correlated with an indication of an emotional response, such as the SCR. Lesions to the amygdala also might block some indirect emotional responses. However, there are other brain regions whose activity appears to be linked more closely to these types of arousal responses (Critchley et al., 2000). In other tasks, the amygdala demonstrates strong activation in response to a stimulus that does not evoke any significant change in the subjects' emotional and physiological state. For instance, activity in the amygdala increases in response to pictures of individuals with posed facial expressions, such as fear. Just looking at these pictures does not yield much of an emotional response in subjects, even though the amygdala demonstrates robust activation. Subjects also show an activation in response to these pictures when they are presented subliminally. In these studies, the subjects are not aware of emotional stimuli or any alteration of an emotional state. It is clear from these studies and others that amygdala activity by itself does not cause a wide range of emotional responses, but somehow it must be involved in emotional responding.

It has long been thought that humans have a greater sensitivity to perceiving or processing emotional compared to nonemotional information in the environment (Zajonc, 1980). This advantage makes evolutionary sense, since it is likely that emotionally charged stimuli are important for survival, especially if these stimuli are fearful or threatening. It has been proposed that one role of the amygdala is to increase the vigilance or readiness of cortical response systems when emotional stimuli are present (Whalen, 1998). In a series of studies with rabbits, Bruce Kapp and colleagues (1994) showed that the activity of amygdala neurons that are sensitive to threatening stimuli is correlated with a spontaneous increase in the excitability of cortical neurons. The amygdala might influence the processing of sensory cortical systems through connections with the basal forebrain. Basal forebrain neurons release the neurotransmitter acetylcholine to sensory cortical neurons; acetylcholine is known to facilitate the response of these neurons.

Studies with humans also have implicated the amygdala in enhancing the responsiveness of sensory cortical regions. Functional neuroimaging studies have shown an increase in activity to emotional stimuli in cortical regions known to be involved in the sensory processing of these stimuli (Kosslyn et al., 1996). For example, activity in the extrastriate cortex, which is involved in the processing of visual stimuli, increases in response to faces with fearful expressions compared to faces with neutral expressions (Morris et al., 1998). The strength of this response is correlated with the amygdala's response to these same stimuli.

Although studies have shown correlations between activity in the sensory cortices and activity in the amygdala, they have not shown that the amygdala is critically involved in an enhanced sensitivity to emotional stimuli. This was demonstrated in the study illustrated in Figure 13.16. This study used a paradigm called *rapid serial visual presentation*. In this paradigm, a series of approximately fifteen stimuli, words in this case, are presented very quickly, at the rate of approximately one word every 100 msec. To the subject, the words just seem to fly by so quickly, that they cannot all be noticed or remembered. However, the subject is instructed to ignore most of the words and concentrate on reporting only two words, the two words that are presented in green ink (the rest are in black ink). Most of the time, a control subject has no problem noticing and identifying these two words and reporting them to the experimenter. However, if the second

target word (target 2) comes soon after the first target word (target 1), a normal subject has a hard time identifying and reporting this second target word. It is thought that noticing and processing the first target word leads to a short refractory period during which time there are fewer resources available to notice and process the second one. This effect has been called the *attentional blink* because it is as if the ability to attend and notice has blinked. Adam Anderson and Elizabeth Phelps (2000) used this paradigm and varied the emotional salience of target 2. They found that when target 2 was a negative word, the ability of control subjects to notice and report this word was significantly improved relative to neutral words. This improvement was most pronounced when target 2 was most difficult to report, that is, when it was presented soon after target 1. However, patients with amygdala lesions did not show any difference in their ability to report emotional and nonemotional target 2 words. Their performance for both negative and neutral target 2 words was similar to the control subjects' performance for the neutral words. These results indicate that the amygdala may act to enhance the perceptual encoding of emotional stimuli, and are consistent with the hypothesis that perceptual processing systems are more vigilant with emotion.

Although there may be circumstances where the activity of the amygdala can be *linked* to an emotional response, amygdala activity does not necessarily *lead* to an emotional response. The amygdala's primary role in

Figure 13.16 **(a)** A representative trial of the rapid serial visual presentation procedure. Subjects are told to report the green words after all fifteen words are presented. **(b)** Percentage of negative and neutral target 2 words correctly reported for normal control subjects and a patient with damage to the amygdala. Control subjects show an increased sensitivity to negative words in comparison to neutral words, which is most pronounced at early lags when the target 2 words are the most difficult to identify. The patient with amygdala damage does not demonstrate any increased sensitivity or vigilance for the negative words in comparison to the neutral words. Adapted from Anderson and Phelps (2001).

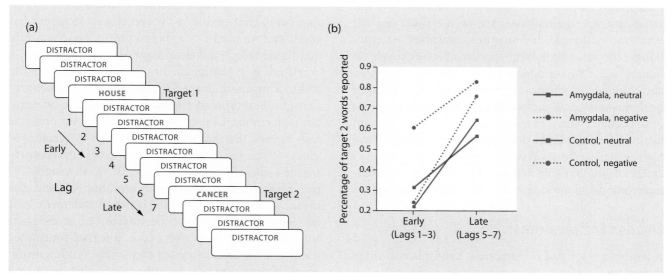

emotional responding may be related to processing emotional information and sending signals to other response systems that should be sensitive to this information. By increasing the vigilance of other response systems, the amygdala is increasing the likelihood of an emotional response, if it is appropriate given the circumstances.

SUMMARY OF THE AMYGDALA

The amygdala is involved in a variety of emotional tasks, ranging from fear conditioning to social responses. We have focused on its role in responding to fear or threatening emotional events. However, it would be misleading to suggest that the amygdala only responds to negative stimuli. The amygdala also plays a role in responding to positive stimuli in somewhat more limited circumstances. For instance, the effects of the amygdala modulating hippocampal consolidation appear to be mediated by arousal, a response to positive or negative events (Hamann et al., 1999). In addition, the amygdala plays a role in some learning tasks in which a rewarding stimulus is associated with a neutral stimulus (Gaffan and Harrison, 1987; Johnsrude et al., 2000; Gallagher and Holland, 1992). It would be inaccurate to say the amygdala is only involved in responding to negative or fear stimuli, although it does appear to be particularly sensitive to these types of stimuli.

Our discussion of the amygdala and the orbitofrontal cortex has focused on how these two brain structures operate independently. However, some of the more exciting current research in the cognitive neuroscience of emotion is starting to outline how these regions work together to produce normal emotional responses. For example, while acquisition of fear conditioning requires the amygdala, normal extinction of a conditioned response involves interactions of the amygdala and the prefrontal cortex (Morgan and Le Doux, 1999). Studies examining the ability to associate a reward with a stimulus found that the amygdala and orbitofrontal cortex may act together on this type of task (Baxter et al., 2000). A neuroanatomical model of depression suggests that a circuit of structures (amygdala, orbitofrontal cortex, and thalamus) are overactive in depression, and that these structures working in concert lead to some of the symptoms of depression (Drevets, 1998). Finally, Damasio's somatic marker hypothesis proposes that the amygdala and orbitofrontal cortex interact and make unique contributions to emotional decision making. These are just some examples of the direction of future research in the cognitive neuroscience of emotion, where the emphasis is switching from understanding how individual neural structures work, to understanding how they act together to produce normal and adaptive emotional responses.

LATERALITY

One approach to studying the cognitive neuroscience of emotion is to examine the specific contributions of neural structures like the orbitofrontal cortex and amygdala, and then to determine how these structures work together or with other neural systems. Another approach does not focus on specific neural structures but rather attempts to understand how the right and left hemispheres interact and uniquely contribute to emotional experience. This latter approach focuses largely on human subjects, examining the perception, expression, and conscious experience of emotion. What has emerged from this work is an understanding that our descriptions of the neural systems of human emotion have to account for the lateralized input that these systems receive. There have been two primary types of investigations into the laterality of emotion. The first has examined **emotion communication** and the second has explored **affective style.**

Emotion Communication

It has been suggested that the right hemisphere is more important for emotion communication than the left

hemisphere (Bowers et al., 1993). This right-hemisphere dominance hypothesis has been supported primarily by studies of neuropsychological patients with damage to one or the other hemisphere. To communicate emotion, we need to have two basic abilities. We need to be able to comprehend the emotional information being conveyed to us by speech and facial expressions, and we need to be able to produce and generate emotional speech and facial expressions.

There is a significant body of evidence that the right hemisphere is specialized for emotion comprehension. Two types of emotional stimuli have been examined: emotional prosody and facial expressions. It is well known that patients with damage to certain regions of the left hemisphere have language comprehension difficulties (see Chapter 10). Speech, however, can communicate emotion information beyond the meaning and structures of the words. A statement such as "John, come here" can be interpreted in different ways if it is said in an angry tone, a fearful tone, a seductive tone, or a surprised tone. This nonlinguistic, emotional component of speech is called *emotional prosody.* There are reports of patients with left-hemisphere damage who have difficulty comprehending words but show little deficit in interpreting the meaning of emotional prosody (Barrett et al., 1997). At the same time, there are several examples of patients with damage to the temporal-parietal lobe in the right hemisphere who comprehend the meaning of language perfectly but have difficulty interpreting phrases when emotional prosody plays a role (Heilman et al., 1975). This double dissociation between comprehending meaning based on language and emotional prosody suggests that the right hemisphere is specialized for comprehending emotional expressions of speech.

Studies of patients with right- and left-hemisphere damage have found that the right side also is specialized for comprehending facial expressions. The deficit in interpreting fearful facial expressions that we described earlier in patients with amygdala lesions is limited to those patients whose damage includes the right amygdala (Anderson et al., 2000). Several studies have indicated that more extensive damage to the right hemisphere results in a more comprehensive deficit that includes all facial expressions of emotion (Borod et al., 1986).

The right-hemisphere dominance hypothesis for emotion communication is consistent with the data on emotion comprehension. However, the ability to produce emotional expressions is not completely dependent on the right hemisphere. There is little research examining the production of emotional prosody, although what evidence there is supports the idea that the right hemisphere is more important for prosody production (Ross, 1993). However, the production of facial expression can be dependent on the left or the right hemisphere, or both, depending on whether the expression is willful and voluntary, or spontaneous.

There appear to be two neural systems for controlling facial expressions (Figure 13.17). The system that controls voluntary expression is managed by the left hemisphere (Gazzaniga and Smylie, 1990). It sends its messages to the contralateral facial nucleus (cranial nerve VII), which in turn innervates the right facial muscles. At the same time, the left hemisphere sends a command over the corpus callosum to the right half of the brain. The right hemisphere sends the message down to the left facial nucleus, which in turn innervates the left half of the face. The result is that one can make a symmetrical facial response such as a smile or frown.

Spontaneous facial expression is managed by a different neural pathway. First, unlike voluntary expressions, which only the left hemisphere can trigger, spontaneous expressions can be managed by either half of the brain. When either half triggers a spontaneous response, the

pathways that activate the brainstem nuclei are signaled through another pathway, one that does not course through the cortex. Each hemisphere sends signals straight down through the midbrain and out to the brainstem nuclei. Clinical neurologists know of the distinction between these two ways of controlling facial muscles. For example, a patient with a lesion in the part of the right hemisphere that participates in voluntary expressions is unable to move the left half of the face when told to smile. At the same time, the very same patient can easily move the left half of the face when spontaneously

Figure 13.17 The neural pathways that control voluntary and spontaneous facial expression are different. **(a)** Voluntary expressions that can signal intention have their own cortical networks in humans. **(b)** The neural networks for spontaneous expressions involve older brain circuits and appear to be the same as those in chimpanzees.

(a)

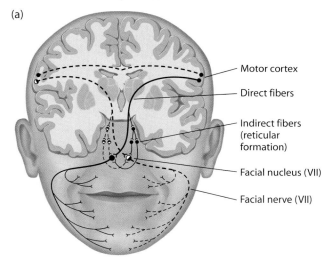

(b)

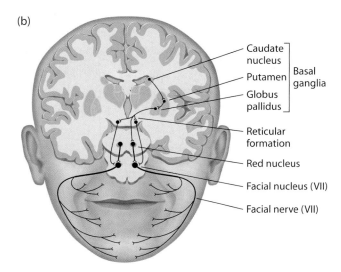

smiling because those pathways are unaffected by right-hemisphere damage. Also with Parkinson's disease, the pathways supporting spontaneous facial expressions do not work, whereas the pathways that support voluntary expressions do work. Such patients can lose their masked-face appearance when told to smile (Figure 13.18).

The right-hemisphere dominance hypothesis for emotion communication is supported by most but not all of the neuropsychological data. It appears that emotional facial expression in particular is more complex and can be generated in either the right or the left half of the brain, depending on the eliciting context. In addition, neuropsychological research has relied primarily on one technique to get at the laterality of emotion communication. Research with other techniques, such as functional neuroimaging, suggests that the neural circuits of emotion communication in normal subjects may not be as lateral-ized as the neuropsychological data suggest (Breiter et al., 1996). Although the right-hemisphere dominance hypothesis for emotion communication is not unequivocally supported by all of the data, it is clear that the right hemisphere plays a more comprehensive role in emotion communication than the left hemisphere.

Affective Style

In our discussion of the cognitive neuroscience of emotion thus far, we have focused on describing how people, in general, function in emotional circumstances. However, one of the more striking characteristics of emotion and emotional reactions is how variable individuals can be. Although everyone may react negatively when hearing bad news, some people may bounce back immediately and look on the bright side, whereas others might mull over the bad news for days and sink into a depression. We all have distinct personalities that are described in part by our disposition and emotional temperament. Richard Davidson (2000) of the University of Wisconsin at Madison has called these differences in emotional tendencies *affective style*. He is one of the few investigators attempting to understand the neural systems related to the variability in affect across individuals. His work has focused on the relative contributions of the right and left hemispheres to affective style.

Early work on the emotional disposition of brain-injured patients demonstrated a marked difference in reaction between those suffering from right-hemisphere and those suffering from left-hemisphere damage. Patients with right-hemisphere damage were often described as being not sufficiently upset or concerned about their injury, whereas patients with left-hemisphere damage were sometimes described as being overly catastrophic and weepy in reaction to their injury. This laterality difference in the emotional response to sustaining a brain injury suggested that the right and left hemispheres might have different contributions to our style of emotional reaction.

Davidson and colleagues (1990, 2000) examined the effect of lateralized brain function on affective style in normal subjects by measuring electrical activity from the scalp. They focused on measuring electroencephalographic (EEG) responses from the anterior portions of the scalp overlaying the frontal lobes. Although all individuals have signs of electrical activity in the anterior right and left hemispheres, it appears that some individuals show relatively more baseline activity in the anterior right hemisphere, whereas others show relatively more baseline activity in the anterior left hemisphere. Davidson and colleagues found that these relative differences in the laterality of the EEG response were related to affective style.

Figure 13.18 Facial expressions of two kinds of patients. **(Top)** The patient suffered brain damage to the right hemisphere; the lesion interfered with voluntary facial expression. **(Bottom)** A Parkinson's disease patient with a typical masked face. Since Parkinson's disease involves the part of the brain that controls spontaneous facial expression, the faces of these patients, when told to smile, light up, since the other pathway is still intact.

An example of this effect was demonstrated in a study in which subjects were given the Positive and Negative Affect Scale (PANAS). This scale asks them to rate the extent to which a series of positive (e.g., happy) and negative (e.g., anxious) traits are descriptive of their personality. As can be seen in Figure 13.19, subjects who showed relatively greater EEG responses in the left midfrontal region rated themselves higher on positive affective traits than did subjects with relatively greater EEG responses in the right midfrontal region. The self-ratings for negative personality traits showed the opposite pattern of results. Taken together, these results demonstrate that subjects who described themselves as having a more sunny disposition showed more left-hemisphere activity, whereas subjects who rated themselves as having a more negative disposition showed more right-hemisphere activity. This difference in the relative EEG activity of the right and left hemispheres also has been linked to how we respond to emotional events. In one of the more interesting studies of this type, Davidson and colleagues investigated EEG responses from the scalps of infants. They found that infants with more dominant right-hemisphere activity were more likely to cry and be upset when separated from their mothers, in comparison to infants with more dominant left-hemisphere activity. These results suggest that hemispheric asymmetry in brain activity may be linked to affective style from a very early age.

Davidson (2000) characterizes the relative contributions of the right and left hemispheres using a dimensional distinction of emotion described at the beginning of this chapter. He believes there may be separate neural circuits involved in emotional reactions that are more positive and elicit goals to *approach* or engage in a situation, and emotional reactions that are more negative and elicit goals to *withdraw*. These approach-withdrawal circuits are lateralized so that approach behaviors are

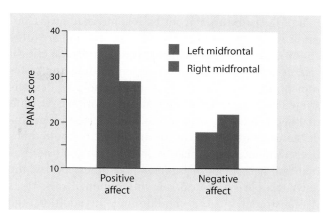

Figure 13.19 Ratings on the Positive and Negative Affect Scale (PANAS) by subjects with relatively more baseline electroencephalographic (EEG) activity in the left midfrontal region and those with more baseline EEG activity in the right midfrontal region. The subjects with more left midfrontal activity rated themselves higher on positive personality traits and lower on negative personality traits, in comparison to the subjects with more right midfrontal EEG activity. Adapted from Tomarken et al. (1992).

more dependent on the left hemisphere, and withdrawal behaviors are more dependent on the right hemisphere. Although all of us are capable of approach and withdrawal emotional reactions, the magnitude and frequency of these different types of reactions in any given individual may be related to the relative baseline asymmetry in that individual's right and left frontal activity.

Davidson's work is some of the first to systematically explore differences in emotional traits in normal individuals. Although we have identified some general neural circuits of emotion throughout this chapter, it is likely that as our understanding of the cognitive neuroscience of emotion becomes more sophisticated, the role of individual differences will receive greater emphasis.

SUMMARY

Since the description of Phineas Gage over a century ago, scientists have recognized that emotion can be linked to brain function. However, progress in understanding the cognitive neuroscience of emotion has been relatively slow, partially due to the fact that emotion is a behavior that, on the surface, seems difficult to manipulate and study in a scientific manner. Another challenge to emotion researchers has been finding the proper place for studies of emotion in cognitive neuroscience. Earlier research and theories tended to view emotion as separate from cognition, implying that they could be studied and understood separately. However, as

research in the neuroscience of emotion proceeded, it became clear that emotion cannot be considered independently from other, more "cognitive" behaviors and vice versa. The neural systems of emotion and other mental behaviors are interdependent. Although emotion, like all behaviors, has its unique and defining characteristics, current research into the cognitive neuroscience of emotion suggests that there is not a clear emotion-cognition dichotomy.

Studies in the cognitive neuroscience of emotion have tended to emphasize the importance of two brain regions, the orbitofrontal cortex and the amygdala. The role of the orbitofrontal cortex is somewhat elusive because the behaviors linked to this brain region are difficult to classify. It is clear that this region plays a critical role in aspects of the evaluation, inhibition, and selection of social and emotional information, as demonstrated by impairments in social and emotional decision making in humans with orbitofrontal lesions. Our understanding of the role of the amygdala in emotion has been influenced significantly by research with nonhuman animals. In both humans and other species, the amygdala plays a critical role in emotional learning as demonstrated by fear conditioning. In addition, the amygdala is involved in explicit or declarative emotional learning and memory, through interactions with the hippocampus; it is also involved in social responses, and plays a role in vigilance and emotion.

There is an emerging shift in our approach to studying the cognitive neuroscience of emotion that is moving the emphasis from the study of neural structures to the investigation of neural systems. As we achieve a greater understanding of the relative roles of the amygdala and orbitofrontal cortex in emotional processing, it has become more apparent that we need to understand how these neural systems, and others, interact to produce normal and adaptive emotional responses.

And finally, we've learned that there are unique contributions of the right and left hemispheres to emotional communication and affective style. Research on the laterality of affective style highlights another important trend in the future of cognitive neuroscience, which is investigating the neural systems of individual differences in emotion.

KEY TERMS

affective	autonomic nervous system	emotion communication	orbitofrontal cortex
affective style		facial expression	reinforcement
aggression	decision making	fear conditioning	somatic marker
amygdala	emotion	limbic system	

THOUGHT QUESTIONS

1. Briefly describe the limbic system hypothesis and its historical role in the cognitive neuroscience of emotion.
2. What are three possible impairments in social decision making that result from damage to the orbitofrontal cortex?
3. Explain the amygdala's role in fear conditioning. Be sure to include what is known about the neural pathways for emotional learning based on nonhuman animal models and also why the amygdala's role in emotional learning is said to be implicit.
4. In what two ways do the amygdala and hippocampus interact in emotional learning and memory?
5. What is the relation between affective style and laterality?

SUGGESTED READINGS

DAMASIO, A.R. (1994). *Descartes' Error: Emotion, Reason, and the Human Brain.* New York: Putnam.

DAVIDSON, R.J., JACKSON, D.C., and KALIN, N.H. (2000). Emotion, plasticity, context, and regulation: Perspectives from affective neuroscience. *Psychol. Bull.* 126:890–909.

LE DOUX, J.E. (1996). *The Emotional Brain: The Mysterious Underpinnings of Emotional Life.* New York: Simon and Schuster.

ROLLS, E.T. (1999). *The Brain and Emotion.* Oxford, UK: Oxford University Press.

SAPOLSKY, R.M. (1992). *Stress, the Aging Brain, and the Mechanisms of Neuron Death.* Cambridge, MA: MIT Press.

WHALEN, P.J. (1998). Fear, vigilance, and ambiguity: Initial neuroimaging studies of the human amygdala. *Curr. Direct. Psychol. Sci.* 7:177–188.

14

Evolutionary Perspectives

In the early 1990s, Giacomo Rizzolatti and his colleagues at the University of Parma studied how single cells in the prefrontal cortex of a monkey respond when the animal moves its hand to grasp an object, in fact, a nice juicy grape. Research has shown that the area they were testing may be homologous to Broca's area in humans, which you may recall from Chapter 9 is identified with language. The researchers found that these cells fire when the monkey grasps the grape with its hand. They do not fire when the monkey sees the grape or is *about* to grasp the grape, but only when it grasps it. Now comes the ingenious part. Rizzolatti showed that the neurons also fire when the monkey observes a human experimenter reaching out to grasp the grape! These "mirror neurons" respond to both the monkey's actions and the actions of others. Rizzolatti suggested that these neurons play a role in "understanding motor events." His view is that the mirror neurons may help the monkey understand another individual's motor actions by mapping the same meaning it associates with its own action onto its observations of the other individual's action.

Is this the ancestral beginnings of what has been called the "theory of mind" module (see Chapter 16)? This system that humans possess allows them to have a theory about another organism's intentions. These mirror neurons are recognizing a movement in another and mapping it to the monkey's own movement. So the meaning of the observed action is matched with that of a self-action. Although the monkey's neurons operate in a system without language, the fact that the mirror neurons are in an area that is considered homologous to Broca's speech area in humans could indicate that these neurons may be the ones that have evolved in humans and allow us to make an assumption about another's intentions. In short, mirror neurons, or what is called a *preadaptation,* may be the beginnings of a neural system that sustains a more complex function in a higher animal.

The fascinating experiments of Rizzolatti raise the questions, How did we get here? Why are our brains the way they are? What do we have in common with other animals? What makes us unique? What can we learn from existing animals about our origins? What does evolutionary theory have to say about the nature of human cognition?

Neuroscience and cognitive science are beginning to incorporate the facts and theories of comparative anatomists and evolutionary biologists to address these questions. We will start this chapter by reviewing the history of this approach. We then will learn the basic terminology and concepts used by comparative neuroanatomists to illuminate how examining the brains of existing animals can teach us about our evolutionary roots. This enterprise is a tricky one because the structure of animal bodies and brains can vary dramatically, and trying to deduce similar features of brain organization and unique features is difficult. Therefore, we must learn the strategy that comparative neuroscientists use to avoid the pitfalls of applying conclusions from the brain of one species to that of another.

In the second half of the chapter we will show how evolutionary theory is important for understanding learning and cognition. The mind is a collection of old adaptations designed to solve problems that our prehistoric ancestors faced, and newly developed features that take shape via activity-dependent mechanisms. Appreciation of the roots of our mental behavior is essential in exploring current cognitive capabilities and limitations.

EVOLUTION OF
THE BRAIN

The brain is a compromise, and therefore the functions that it generates are absolutely imperfect, although relatively optimal. This statement may be bothersome to some of you, possibly because it implies that we humans must be imperfect. This notion, and our long history of struggles to prove just the opposite, has a dramatic impact on our current thinking about the mind, the brain, and the complex behaviors it generates. In addition, the notion goes against what can only be considered a history of dogma regarding two major tenets of human design that shaped the direction of science in earlier times and often form the basis of modern theory. The first is that humans are the pinnacle of life or the center of the cosmos, and the second is that the mind or soul is somehow distinct from the body. The former idea affects how we view humans, and the human brain compared to the brains of other animals. We tend to view evolution as linear, with humans at the top and other animals as degraded forms of ourselves. Therefore, we view brain evolution as additive rather than in the nonlinear fashion in which it should be regarded. For example, complex brains like those of humans are not simply monkey brains or chimpanzee brains with a few new parts added, such as Broca's area and prefrontal cortex. Rather, the human brain is a unique amalgamation of evolutionary old areas and new areas that have been modified in predictable ways through expansion or reduction of existing parts, formation of new connections, and adaptations for new environmental demands. The latter notion, that there is some unique aspect of the human condition that is different from the condition of other animals, affects how we compare animal brains with human brains, and the conclusions we reach regarding cortical evolution in mammals.

Before we can actually talk about the nuts and bolts of evolution and its impact on our thinking about the brain and cognition, it is necessary to discuss the history of evolutionary thought. History shapes our current understanding of the brain and our beliefs about how things work. Indeed, as already noted, two very old ideas are embedded in our culture and form the basis of our current thinking on issues regarding cognition and the evolution of the nervous system.

The Historical Underpinning of Contemporary Evolutionary Neurobiology

Probably the most significant event in the biological sciences during the nineteenth century was the discovery by Charles Darwin (1859) and Alfred Russell Wallace that **natural selection** was the vehicle of evolution. The basic observations were that individuals within a population vary, and that changes in individuals arise via natural selection of these variants. The trait or characteristic that is selected for has no absolute value, but its value is measured by how well it contributes to the individual's fitness within a particular environment. The details of this theory deserve a more in-depth discussion and are dealt with later. Here, we only wish to underscore the historical importance of this discovery, and its impact on modern evolutionary biology.

A number of twentieth-century scientists contributed to our understanding of the brain and its evolution. Investigators early in this century, such as Korbinian Brodmann (1909), utilized the technique of Franz Nissl to distinguish the architecture of the cerebral cortex in humans from that in other mammals. Most of us know Brodmann for his cytoarchitectonic descriptions of the human brain. Although he made a number of errors regarding the divisions of the cerebral cortex, an illustration of the divisions of the human brain likely exists in every textbook on the brain that has ever been printed. Perhaps his most important contribution was his cross-species comparisons of the cytoarchitecture of the cerebral cortex (Figure 14.1). He observed that some cytoarchitectonic areas of the brain, such as area 17, could be found in all species examined. The ubiquity of these fields across phyla suggested that they must be inherited from a common ancestor. He proposed that these cytoarchitectonic areas had different functions, although there was no direct evidence for this hypothesis. We now know that area 17 corresponds to the primary visual area

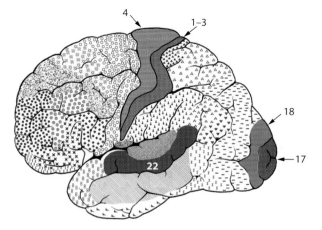

Homo sapiens (human)

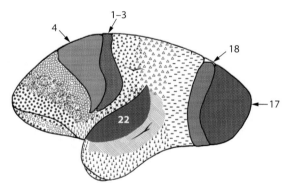

Callithrix jacchus (common marmoset)

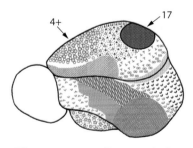

Erinaceus europaeus (European hedgehog)

Figure 14.1 Cytoarchitectonic divisions of the human, marmoset, and hedgehog neocortex as described by Brodmann (1909). The different patterns of black and white stipple denote separate regions of the neocortex that have different architectonic appearances. Some cortical areas, such as areas 17 and 4, were present in all animals investigated, suggesting that these regions are homologous. Other regions, such as area 22, were present only in primates.

in all mammals. Contemporaries of Brodmann such as Joseph Shaw Bolton, George Elliot Smith, and Constantin von Economo devoted much time to parceling the human cerebral cortex into separate divisions based on cytoarchitecture. These early investigators established a

common view of the functional organization of the human brain that still persists today.

This view, outlined by Jon Kaas (Merzenich and Kaas, 1980), is that sensory input is relayed from subcortical structures to primary receiving areas of the neocortex (such as V1). These primary areas send this sensory information to secondary sensory or "psychic" cortex in which more complex perceptual functions are performed. This information is finally relayed to multimodal "association areas" for higher-order processing. This traditional view, with the caveat that humans possess more association cortex than any other mammal, still dominates neuroscience. However, as Kaas (1999) pointed out, a dramatic revision of this view is necessary as most of the neocortex is sensory or motor rather than associative in nature.

Major events occurring in other disciplines in the middle of the twentieth century were to have a tremendous impact on neuroscience in general and evolutionary neurobiology in particular. James Watson, Francis Crick, and Maurice Wilkins discovered the molecular structure of deoxyribonucleic acid (DNA), the "double helix," and its significance for information transfer in living organisms. Their discovery that life could be reduced to this very basic unit of organization, which continually replicates itself, rocked the world. On the heels of this discovery, a quiet revolution was brewing. From the 1950s through the 1970s Konrad Lorenz, Niko Tinbergen, and Karl Von Frisch, all of whom shared the Nobel Prize in medicine (physiology) in 1973, led the fields of **ethology,** the study of animal behavior, and its counterpart, **neuroethology,** the study of the neurobiology of animal behavior. Lorenz and Tinbergen contributed to the fields in two fundamental ways. They focused attention on the innate aspects of behavior, and they provided empirical data from animal studies that allowed behavior to be understood from a more biological perspective.

Probably the most famous study of Tinbergen and Lorenz (see Tinbergen, 1957) was that on the egg-rolling response in greylag geese. When the brooding goose notices that an egg has rolled outside of the nest, a very stereotypic behavior ensues. First, the goose fixates on the misplaced egg and then rises and extends her neck to touch the egg. She then places her bill over the egg and carefully rolls it back into the nest. Tinbergen and Lorenz at first believed that the goose was thoughtfully performing a behavior to return a lost egg to the nest. However, when these scientists removed the egg once the goose had initiated the behavior, the goose continued with the behavior despite the absence of the egg. Thus, this behavior has an initial releaser (the sight of the egg out of the

An Interview with Steven Pinker, Ph.D.
Steven Pinker is a professor of brain and cognitive science at MIT. He has worked on problems in vision, attention, development of language, and evolution.

Authors: Why is it important for students of the mind to understand the principles of evolution?

SP: The brain is a highly organized, nonrandom system, and it can't be understood without knowing the forces that gave rise to that organization. We know the brain did not fall out of the sky; like other parts of the body, its functional complexity—the fact that it can do interesting things like see, think, and act—is a product of evolutionary forces, particularly natural selection.

Authors: OK, let's get down and dirty. Suppose a student is interested in the problem of memory or attention or even morality. These issues can be studied without mentioning natural selection. How does natural selection inform them about their chosen topic?

SP: These issues cannot be understood without natural selection. Natural selection is the rationale for reverse-engineering the brain—figuring out what it was designed to accomplish. Why do we remember recent and frequent items best? Is it some inherent property of the stickiness and softness of neural tissue? Or could evolution have built a brain that remembers everything equally well, but steps in that direction were selected against? When you compare the computer information retrieval system at the library, which spills hundreds of useless titles in your lap, to a human expert, who homes in on the five or six most appropriate ones, you appreciate that human memory might be close to optimal in trading off the likelihood of finding needed information against the costs in time of considering unneeded information. Since organisms operate in real time, this is not a trivial trade-off. John Anderson has shown that retrieving frequent and recent items is the optimal strategy for any information access system, so the explanation for the human case is quite likely to be that the brain is specially organized to be frequency- and recency-sensitive because of the selective advantages it brought, not that calcium channels or whatever make it inevitable. As for morality, the necessity of evolutionary thinking is even stronger. Evolutionary game theory has made very strong predictions of what kinds of algorithms have to be in the mind of an organism that can engage in moral behavior. Many commonsense notions prevalent among academics (e.g., that morality evolved for group cohesion, or that there is an instinct for aggression) are literally unevolvable.

Authors: Unevolvable? Come at that point one more time.

---

nest) that triggers it, and then an innate, highly stereotypic pattern that they termed a *fixed action pattern*. A fixed action pattern, once initiated, continues through to completion, independent of feedback.

Ethologists and neuroethologists stressed the importance of invoking evolutionary theory to understand animal behavior. Unfortunately, while evolutionary theory was used to understand animal behavior, it was still excluded from explanations of human behavior. Indeed, at this point, sociology and psychology were all but devoid of any theory that incorporated evolutionary principles, as Hodos and Campbell (1969) pointed out.

In the early 1970s, E.O. Wilson sought to "reformulate the foundations of the social sciences" and "biologicize" them, and in essence contracted a marriage between zoology and population biology. He invoked evolutionary theory to explain social phenomena, and in 1975 promoted the new field of **sociobiology,** which he defined as the systematic study of the biological basis of all social behavior. Sociobiology allowed for a logical explanation of animal behavior that at first glance seemed counter to natural selection. For instance, highly social insects, such as ants, have cooperative care for the young and a reproductive division of labor, with a number of sterile individuals working on the behalf of fecund nest mates. A lack of reproduction by all individuals in the colony seemed at odds with what was known about evolution and natural selection. However, others described the reproductive benefits of colonial life by examining the relatedness of individuals within the colony, and the high net contribution to their reproductive fitness that such lifestyles allow (i.e., the propagation of their genes into future generations).

Wilson (1975) observed that ". . .in evolutionary time the individual organism counts for almost nothing. In a Darwinist sense the organism does not live for itself. Its

SP: One might think a group of indiscriminate altruists, all helping each other out, would do better than a group of selfish creatures who refuse to sacrifice for the benefit of all. But the problem is getting the altruistic group to begin with. A mutant with a tendency toward selfishness would enjoy all the benefits of his altruistic buddies without paying the costs. Nothing could stop it from proliferating through the group, given that individuals reproduce faster than whole groups. As for aggression, again the text of other organisms has to be taken into account. A bully mutant would do fine at first, but after a bunch of generations everyone will be a bully, and the advantage is gone. It's not that altruism and aggression can't evolve; it's just that they can evolve only in conjunction with information-processing mechanisms that strategically assess how and when to deploy them.

Authors: Finally, with the new awareness of the importance of evolutionary thinking for understanding mental processes, how might experiments be executed in the future? If, for example, one hypothesized that human memory systems were built to aid in finding food sources scattered about a home base, might one reject using word-pair associates as a way of understanding human memory? Instead wouldn't one want to study the efficiency of memory with and without a lot of spatial cues?

SP: Certainly, it would do everyone good to pay more attention to the ecological validity of the stimulus materials and the task used in experiments. It also is important to think of the brain as a family of systems engineered to solve the kinds of problems the organism faced in its evolutionary history (e.g., foraging, mating, language or other forms of communication, etc.) rather than hoping to explain intelligence exclusively with very crude general mechanisms like forming associative bonds. And attention to phylogeny and speciation would correct the lamentable tendency to treat all animals as half-baked humans we can cut up, rather than as cohesively functioning species that are well adapted to their own niches.

But ultimately, evolutionary thinking isn't a specific theory that one goes out and tests like a hypothesis about shape recognition. A cognitive neuroscientist should understand evolution for the same reason a biologist should understand chemistry, or a chemist should understand physics. The chemist doesn't ask, "How will knowing physics help me to design my next experiment?" He or she had better know physics because everything done in chemistry ultimately has to make sense in the light of physics. Similarly, cognitive science and neuroscience are studying the products of specific causal processes (natural selection and other evolutionary forces) and ultimately nothing in those fields makes sense—no explanation, no experiment, no choice of an organism to study—until it is made consistent with what we know about those processes.

---

primary function is not even to reproduce other organisms; it reproduces genes, and it serves as their temporary carrier." This idea represented a large shift in paradigm from traditional psychology, sociology, and ethology, and constituted a virtual revolution in perspective. The contributions of Maynard Smith, William Hamilton, Robert Trivers, and David Barrash, along with many others scientists, helped shape the field of sociobiology in its infancy in the late 1970s and 1980s. In 1976, the highly popular book *The Selfish Gene*, by Richard Dawkins, drew attention to the field and put the gene at the center of importance. Dawkins's contention was that life is simply about the replication of genes and the propagation of "good" genes into the future.

Since its introduction, sociobiology has undergone a rebirth and has been strongly embraced by psychology and sociology. Modern proponents of the newly evolved field of **evolutionary psychology** are Steven Pinker (see An Interview with Steven Pinker, Ph.D.), Leda Cosmides, and John Tooby, who use an evolutionary framework to explain cognitive behavior. Evolutionary psychologists have painted a picture different from that portrayed by sociobiologists. They do not believe all behaviors are driven by genetic mechanisms. Rather they believe the brain has built into it adaptations that are of a more general nature. These adaptations are a set of rules that govern behavior. However, since there are an infinite number of environments, the rules can be applied differently, resulting in an infinite number of behaviors. This view is quite different from that of traditional sociobiologists; it allows for a more objective, biologically compatible view of human behavior compared with traditional psychology-based interpretations.

Cosmides and Tooby (1995) stated their case in terms of answering the question, What are our brains built for?

Understanding the neural organization of the brain depends on understanding the functional organization of its cognitive devices. The brain originally came into existence, and accumulated its particular set of design features only because these features functionally contributed to the organism's propagation. This contribution, that is, the evolutionary function of the brain, is obviously the adaptive regulation of behavior and physiology on the basis of information derived from the body and from the environment. The brain performs no significant mechanical, metabolic, or chemical service for the organism; its function is purely informational, computational, and regulatory in nature. Because the function of the brain is informational in nature, its precise functional organization can be described accurately only in a language that is capable of expressing its informational functions, that is, in cognitive terms, rather than in cellular, anatomical, or chemical terms. Cognitive investigations are not some soft, optional activity that goes on only until the real neural analysis can be performed. Instead, the mapping of the computational adaptations of the brain is an unavoidable and indispensable step in the neuroscience research enterprise....

We shall return to these ideas after considering how current neurobiologists approach studying the brain from a comparative perspective.

Modern Evolutionary Neurobiology: Assumptions and Aims

Modern evolutionary neurobiology or **comparative neuroscience** (Bullock, 1984a), while not incompatible with sociobiology, evolutionary psychology, or neuroethology, differs from these disciplines in that it is more brain centered and systems oriented. Further, while it has an appreciation of the fundamental role of genes in evolution, it also has an appreciation of the complex interaction between genes and the environment in the construction of a nervous system. Comparative neuroscience probably began with Brodmann (1909) and his demonstration that mammals have neocortical areas in common, despite very distant phylogenetic relationships. Unfortunately, evolutionary neurobiology was put on the back burner for some 40 years.

Around the middle of the twentieth century, Clinton Woolsey (1952, 1958) and colleagues used electrophysiological recording techniques to subdivide the neocortex into functional subdivisions in a variety of mammals (Figure 14.2). His work empirically demonstrated what Brodmann only proposed: that the separate architectonic areas of the neocortex were indeed related to different functions.

Figure 14.2 Functional subdivisions of the neocortex as described by Woolsey and colleagues (1952, 1958) in the middle of the twentieth century. Evoked potentials in response to mechanical stimulation of the sensory surface were used to determine the number and topographic organization of different sensory neocortical areas in a variety of mammals. Some areas, such as the primary visual area (V1), somatosensory area (S1), and auditory area (A1), were in all the mammals they investigated, suggesting descent from a common ancestor. Additional sensory areas, such as S2 and V2, also were present in most of the mammals investigated.

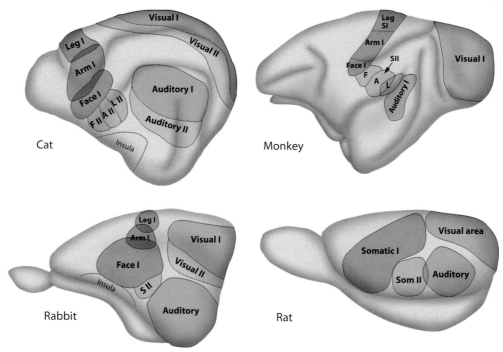

Charles Judson Herrick, a contemporary of Woolsey, also advanced the field of comparative neuroanatomy. As a result of his numerous cross-species comparisons of neural structures, he reached profound conclusions regarding how complexity is achieved in the nervous system. After examining the connections of the brain in a variety of species, he proposed that a slight increase in the connectivity of homologous structures could effectively increase the processing capacity of the brain exponentially. Herrick (1963) proposed that "during a few minutes of intense cortical activity the number of interneuronic connections actually made (counting also those that are activated more than once in different associational patterns) may well be as great as the total number of atoms in the solar system. Certainly not all anatomically present connections of nervous elements are ever used, but the *potentialities* of diversity of cortical associational combinations are practically unlimited and the personal experience of the individual is probably an important factor in determining which of these possibilities will be actually realized."

From the 1960s through the 1980s, Irving Diamond, influenced by George Bishop and Le Gros Clark in the 1950s, focused his research on the evolution of the mammalian brain and the use of the comparative approach to understand the brain, particularly the evolution of thalamocortical relationships. He was a pioneer in the use of modern neuroanatomical techniques and trained a number of contemporary comparative neuroscientists, including Jon Kaas.

Kaas, a graduate student of Diamond's and a postdoctoral student of Woolsey's, elegantly applied the comparative approach to understanding brain organization and evolution. He has combined the techniques used in each of his mentors' laboratories to examine the functional and anatomical organization of the neocortex in mammals. Using multiunit electrophysiological recording techniques, along with examinations of cortical architecture and connections, Kaas has studied species ranging from tenrecs (small insectivores from Madagascar) to lorises (small primates found in Southeast Asia) to humans, and has become the leading figure in the evolutionary neurobiology of the mammalian brain.

Kaas has advanced our understanding by challenging the traditional view that primate neocortex, particularly human neocortex, is composed predominately of "association areas." Association cortex was defined by default as cortex that was not sensory, and presumably was involved in high-level perceptual and cognitive processing. His work in primates over the past several decades has demonstrated that almost all of the neocortex is sensory and motor in nature, and that complex brains evolve not by simply expanding association cortex, but by increasing the number of

sensory and motor areas and the interconnections between them. As Kaas (1997) himself put it, "The validity of a comparative approach depends on the basic premise that theories of brain organization applied to any given species, say humans, should be compatible with evidence and theories of brain organization for other species, with the evidence and theories for closely related species being most relevant."

Scientists who compare brains of animals to deduce how brains are constructed in evolution use a variety of names to identify themselves, including *comparative neuroanatomists, physiological psychologists, evolutionary neurobiologists,* and *comparative neuroscientists.* However, all modern comparative studies are undertaken to address several aims based on a few underlying assumptions. The first assumption is that *all behavior in all animals is generated by the nervous system.* This definition of behavior is not limited to sensory, perceptual, and motor behaviors but includes cognitive behaviors and even those that are difficult to define and measure, such as consciousness. The second assumption is that *brains evolve, and therefore behaviors evolve.* Thus, behavior cannot be completely understood without an evolutionary perspective. The final assumption is that *we can understand the process of the evolution of the brain and the behavior it generates by examining the products of the process, the products of the process being extant (existing) animals.*

Theodore Bullock (1984a) has led the field of comparative neurobiology for nearly half a century. As he put it, "Long before the human species appeared, the pinnacle and greatest achievement of evolution was already the brain—as it had been before mammals appeared, before land vertebrates, before vertebrates. From this point of view, everything else in the multicellular animal world was evolved to maintain and reproduce nervous systems—that is, to mediate behavior, to cause animals to *do* things. Animals with simple and primitive or no nervous system have been champions at surviving, reproducing, and distributing themselves, but they have limited behavioral repertoires. The essence of evolution is the production and replication of diversity—and more than anything else, diversity in behavior."

Bullock describes well the aims of comparative neuroscience, and lists them under three major headings: roots, rules, and relevance. By phylogenetic *roots,* he means the evolutionary history of the brain and behavior. How are brains similar and different? What has evolution produced? He describes the *rules* of change as the mechanisms that give rise to changes in the nervous systems in the course of evolution. Are there constraints under which evolving nervous systems develop? Finally, the *relevance* of our observations refers to the general principles of organization and functions that can be extrapolated from a particular animal studied to all animals, including humans.

The choice of animal that researchers use for their experiments depends on their aim. For instance, if they are interested in how sodium channels work in neurons, the squid is an excellent choice because it contains axons that are large and easily accessible. Further, because sodium channels evolved very early in animal evolution, and their basic contribution to the action potential has not changed, findings in the squid are relevant to human neurons.

If researchers are interested in when the middle temporal (MT) visual area, an area linked to motion perception in primates, first appeared (i.e., the roots of MT),

then their choice of animals should include representatives of early primate lineages such as lemurs, galagos, and lorises, as well as other archontans such as tree shrews (Figure 14.3). Tree shrews (squirrel-like mammals) belong to a family of mammals termed Tupaiidae, whose ancestors are believed by many to form an early branch of the primate lineage. Lemurs form a unique family of primates and exist only on the island of Madagascar off the southeastern coast of Africa. Galagos and lorises are members of another primate family (Lorisidae) and are found in Africa and Southeast Asia. The ancestor of these

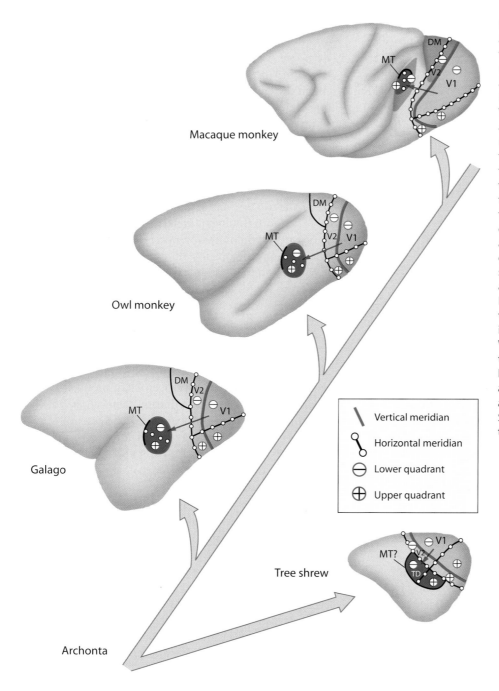

Figure 14.3 A comparative analysis can be used to determine when area MT (blue) evolved in primates. The visuotopic organization and relative location of MT are similar in prosimians (galagos), New World monkeys (owl monkeys), and Old World monkeys (macaque monkeys). In addition, the architectonic appearance of area MT and the presence of direct connections from the primary visual area add further support to the hypothesis that area MT is homologous in all of these primates. An out-group analysis of a more distantly related group (tree shrews) indicates that some features of an MT-like area are present in the area labeled TD. However, further evidence needs to be gathered to determine if TD is homologous to MT in primates. Without this evidence, it appears that area MT evolved some time before the emergence of prosimians, but after the radiation of early archontans such as tree shrews. Shaded gray areas indicate cortex that resides in fissures.

primates diverged very early in primate evolution, and extant species are believed to represent, at least to some extent, the primate ancestor.

Kaas has addressed the question of when area MT arose in primate evolution. He has examined primates from different infraorders, including Lorisiformes (primates that include galagos and lorises), New World monkeys (South American monkeys such as squirrel monkeys and marmosets), and Old World monkeys (African and Asian monkeys such as macaque monkeys and talapoin monkeys), and has determined that all of these primates have an area MT. Although there have been no studies on lemurs, his examination of tree shrews demonstrated a V1 projection zone that shares some features of MT. However, a distinct MT, like that in primates, does not appear to be present. Thus, he concluded that MT may have had some primitive underpinning early in primate evolu-

tion, but probably arose after the Tupaiidae radiation and some time before the emergence of Lorisiformes.

If researchers are interested in the rules of change for specialized peripheral structures and the behaviors associated with these structures, they might use a variety of highly specialized mammals such as the duck-billed platypus (Krubitzer et al., 1995; Krubitzer, 1998) or the star-nosed mole (Catania et al., 1993; Catania and Kaas, 1995) to determine how the somatosensory and motor cortex is modified in relation to specialization, and if the changes take a similar form. For instance, the duck-billed platypus has a specialized bill, and approximately 80% of its nervous system is devoted to processing inputs from the bill. Likewise, the star-nosed mole has an extraordinary specialization of the nose, and like the platypus, a large portion of its nervous system is devoted to processing inputs from the nose (Figure 14.4). This neural magnification of

Figure 14.4 Some mammals are highly derived (have extreme specializations of some body part, behavior, or piece of neural tissue). **(a)** The bill of the platypus contains both mechanosensory (gray) and electrosensory (white) receptors arranged in stripes. In the primary somatosensory area (S1), neurons are aggregated in groups of mechanosensory (gray) and electrosensory (white) inputs. The magnification of the representation of the bill in the neocortex is extreme, assuming about 75% of the entire neocortex. **(b)** Specializations of the face, including the nose of the star-nosed mole, have evolved independently in a number of lineages. The nose of this mole has evolved a number of movable appendages used for exploration, prey capture, and feeding. An independent magnification of the representation of this highly derived nose is found in the primary (S1) and secondary (S2) somatosensory areas in the neocortex. These examples illustrate that the types of modifications to the neocortex are highly constrained. (a) Adapted from Krubitzer (2000). (b) Adapted from Catania et al. (1993) and Catania and Kaas (1995).

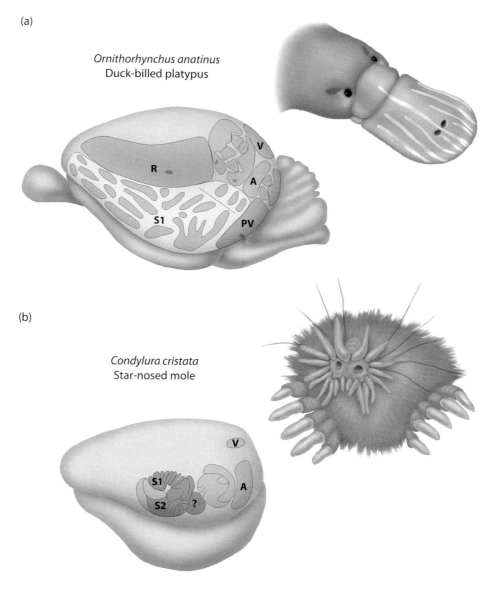

(a)

Ornithorhynchus anatinus
Duck-billed platypus

(b)

Condylura cristata
Star-nosed mole

specialized body parts arose independently in these mammals and is a feature of modifications shared by a number of mammals. Because the neural specialization looks remarkably similar, it seems likely that its nervous system implements certain rules when it is generating representations of specialized body parts. Thus, studying the neural organization and connections in these mammals may provide some insight into what these rules are.

Now that we have an overview of the theoretical basis of an evolutionary perspective, we need to delve more deeply into evolution via natural selection, the terms associated with the field of evolutionary neurobiology, and the use of the comparative approach to formulate hypotheses and make valid inferences regarding the human condition.

FIRST PRINCIPLES

Charles Darwin (1859) is a powerful and pervasive figure in modern biology. His theory of evolution via natural selection is one that everyone believes they understand. After all, isn't the idea of natural selection that only the fittest survive? Doesn't it have something to do with our evolution from lower animals? It turns out that when most people are asked about the significance of evolution, a hodgepodge of ideas bubble up, unstructured and diffuse. If we are to thoroughly understand its importance, and how we can apply this evolutionary perspective to our theories of cognition, we must learn a few basic principles of the evolutionary theory (Williams, 1966).

The concept of evolution was appreciated well before Darwin's time. Early scientists noted that animals had a number of parts in common and therefore all animals must be related. In other words, there must be a common ancestor. Early scientists further reasoned that if God had indeed created all animals simultaneously, and if all animals are complete, why do some animals have features or parts that they apparently do not use? However, while speculation regarding evolution was rife, the vehicle by which it proceeded was not appreciated. In the mid-1800s, both Charles Darwin and Alfred Russell Wallace hit upon this vehicle. Three observations led to Darwin's theory of evolution by natural selection. The first was that individuals within a population of animals vary, the second was that some of this variation is **heritable,** and the third was that not all individuals within a population survive. How this variability was generated was not appreciated until Gregor Johann Mendel's (1822–1884) principles of heredity were rediscovered in the early twentieth century. Evolution proceeds by differential reproductive success, by the natural selection of some **traits** or characteristics over others. The traits or characteristics that are selected for endow some benefit. It should be noted that there is no absolute value given to any trait selected for (i.e., big brains with many parts are not absolutely better).

Natural selection acts on the variations of the phenotype within the population. The **phenotype** is an observable trait, or set of characteristics of an organism. A phenotype can refer to a morphological structure, like the hand, foot, leg, heart, or kidney, as well as a neural structure, such as the lateral geniculate nucleus, the superior colliculus, the neocortex, or a cortical field within the neocortex. We also can describe neural properties as having a particular phenotype, such as the on/off responses of ganglion cells in the retina, and behaviors as being phenotypic, such as the caching of food by squirrels. A **genotype** refers to the genetic composition of an organism. Any particular phenotype is the result of the genotype (which is not directly observed) as well as environmental or activity-dependent mechanisms (see Chapter 15).

The measure of evolutionary success in terms of some gene, trait, or behavior being represented in future generations is a measure of its **fitness.** An **adaptation** is a characteristic of a living thing that contributes to its fitness. Thus, it is a process by which a species adjusts to environmental change, or a feature of an organism that is suited to the environment. For example, the hand in primates is an adaptation of the distal forelimb used for tactile exploration and active reaching. This same (i.e., homologous—see below) structure has been adapted in bats for flying, in dolphins for swimming, and in moles for digging. The acquisition of adaptations through natural selection is the heart of the evolutionary process. This applies to morphological structures, behaviors, and the underlying nervous system that generates them.

A related term is exaptation, first coined by Stephen Jay Gould of Harvard University. **Exaptation** refers to a structure that serves a particular function but then is co-opted for some other, very different function. For instance, the lateral line in aquatic anamniotes (e.g., jawless vertebrates, cartilaginous fish, amphibians) contains hair cell–like receptors used for electroreception (the ability to detect changes in electrical current) or mechanoreception (the ability to detect mechanical

Darwin's Big Idea

Charles Darwin was a genius by any measure. He provided humankind with the key idea about our origins, the idea of natural selection. He was the first one who brought together extensive data that the idea of natural selection had sustaining value in understanding human origins. He summarized his idea in two long sentences, as recently recounted by Daniel Dennett in his fascinating book, *Darwin's Dangerous Idea* (1995). Here is that idea:

> If during the long course of ages and under varying conditions of life, organic beings vary at all in the several parts of their organization, and I think this cannot be disputed; if thereby, owing to the high geometric powers of increase of each species, at some age, season, or year, a severe struggle for life, and this certainly cannot be disputed; then considering the infinite complexity of the relations of all organic beings to each other and to their conditions of existence, causing an infinite diversity in structure, constitution, and habits, to be advantageous to them, I think it would be a most extraordinary fact if, no variation ever had occurred useful to each being's own welfare, in the same way as so many variations have occurred useful to man. But if variations useful to any organic being do occur, assuredly individuals thus characterized will have the best chance of being preserved in the struggle for life; and from the strong principle of inheritance they will tend to produce offspring similarly characterized. This principle of preservation, I have called, for the sake of brevity, Natural Selection.

Now, while Darwin could have used an editor, he did condense the most powerful idea in biology into two sentences. And yet, as Dennett pointed out in his book, his brilliant idea was not born from whole cloth. In his *Dialogues*, the great philosopher David Hume had three characters carry out a fictional debate about whether the world exists as a result of a design; that is, any complex entity must have a designer and in this case it is God. Cleanthes, the Greek philosopher, had previously defended the Argument for Design:

> Look round the world: Contemplate the whole and every part of it: You will find it to be nothing but one great machine, subdivided into an infinite number of lesser machines, which again admit of subdivisions to a degree beyond what human senses and faculties can trace and explain. All these various machines, and even their most minute parts, are adjusted to each other with an accuracy which ravished into admiration all men who have ever contemplated them. The curious adapting of means to ends, throughout all nature, resembles, exactly, though it much exceeds, the productions of human contrivance, of human design, thought, wisdom, and intelligence. Since therefore the effects resemble each other, we are led to infer, by all the rules of analogy, that the causes also resemble, and that the Author of Nature is somewhat similar to the mind of man, though possessed of much larger faculties, proportioned to the grandeur of the work which he has executed. By this argument a posteriori, and by this argument alone, do we prove at once the existence of a Deity and his similarity to human mind and intelligence.

Translate "Deity" to "Natural Selection" and a tangible mechanism has been articulated, which was Darwin's genius. And yet, the idea was in the air 80 years before he wrote about it.

Charles Darwin

displacement of some structure, e.g., hair follicle, skin, and cilia of hair cells). In mammals, the inner ear is an exaptation of the lateral line. The hair cells in the inner ear of mammals are involved in transducing air pressure changes (sound) into neural activity via hydraulic and mechanosensory mechansims.

The heritable part of the natural selection equation is the gene. **Genes** are composed of a complex organic molecule termed *deoxyribonucleic acid* or DNA, and have alternative forms called **alleles.** For instance, if two alleles for a gene responsible for height exist in the population (A for tall and a for short), the phenotype will include one of three allelic combinations (AA, aa, or Aa). *Homozygous* refers to the occurrence of identical alleles at one or more genetic loci (e.g., AA or aa), while *heterozygous* refers to the occurrence of different alleles at one or more genetic loci (Aa). From an evolutionary perspective, over time different alleles compete with each other for space on the chromosome. A **chromosome** is a structure composed of nucleic acids and proteins within the nucleus of a cell, which contains the genetic material (DNA) of an organism. Selection is the ultimate process of the differential reproductive success of alternative alleles.

At this point it is useful to introduce two terms regarding genes, and ultimately the characteristics or functions that they generate. The first is genetic specificity. **Genetic specificity** means that one gene is responsible for a single function or behavior. For instance, a single gene encodes the egg-laying peptide prohormone of molluscs such as aplysia, the sea slug (Geraerts et al., 1994). Thus, a very specific behavior, egg-laying, appears to be encoded by a single gene. The general idea in sociobiology that complex behaviors such as altruism, aggression, and even some aspects of cognition are controlled by a gene, or a set of genes directly responsible for that behavior, at least in part, subscribes to genetic specificity. While this notion is attractive and has gained much notoriety because of its simplicity, most consider genetic specificity to be the exception to the rule, rather than a common occurrence. The second concept is **genetic pleiotropy,** or one gene that has many functions. From a developmental perspective, one gene can participate in the generation of a number of different events and can be expressed differentially at various times for different purposes.

For example, one might ask why diseases like Alzheimer's or even schizophrenia survive in the population. Alzheimer's disease appears late in life, and accumulating evidence indicates that this disease is mediated genetically. Why would such genes be selected for? The likely answer to this question is that their selection is due to genetic pleiotropy. The genes responsible for unleashing the neuropathological state of this disease are selected for because they play some positive role early in life. That the same gene has detrimental effects later in life is irrelevant in evolutionary terms, because its negative effects occur well after the age of reproduction. Even the genes for schizophrenia, which has a wide distribution throughout the species, most likely have a positive role as well as the relatively negative effect in the environment in which they currently are expressed.

Because of genetic pleiotropy, structures, or components of structures, functions, and behaviors get carried along in evolution as a package of positive, negative, and neutral events. The assignment of a trait as negative or positive depends on the environment in which it is expressed. This is why the brain is a compromise. It is a well-adapted structure, yet it consists of a combination of positive, negative, and neutral characteristics, with the net effect based on the particular environment in which it finds itself. Should the environment change, the value or fitness of the package (although genetically static) can change dramatically and lead to eventual genetic changes, speciation, or extinction.

The problem for neuroscientists is to decipher whether a structure, receptor type, connection, neural property, or behavior under investigation is functionally significant to the organism in the environment to which it is adapted, or whether it is an epiphenomenon of evolution. *Epiphenomenon* refers to a secondary symptom or effect, occurring with but not necessarily the cause or result of a phenomenon or event. Thus, the characteristic or feature we observe, or are studying, may be a by-product of selection for something quite different. For example, ocular dominance columns (ODCs) are the result of correlated activity within each eye and discorrelated activity between the two eyes during development (see Chapter 5). Thus, ODCs are an epiphenomenon of development. Indeed, this system in cats and monkeys has been used extensively as a model to study the role of correlated activity in wiring the nervous system (i.e., rules). While the current organization in primate brains, which include ODCs, is functionally optimal, such an arrangement is not a requisite for stereopsis, for instance.

Another example is consciousness. Has the actual behavior that we term *consciousness* been selected for in evolution, or is it an epiphenomenon of the neural circuitry of a brain that evolved to solve complex sensory problems? Although consciousness indeed may be a by-product of complexly organized brains, it does not mean that, as a characteristic itself, it cannot ultimately be selected for. Thus, what was initially an epiphenomenon can become highly adaptive in a particular environment. This is where the story becomes extremely convoluted, and why evolution is likened to a tinkerer (see Jacob, below). If the above example is true, then in order to select for something like consciousness, it is necessary to select for all the bits and pieces of the original system that generated this characteristic. While there may be more than one method to neurally achieve the conscious experience, in terms of evolution it is tied to the expansion of our sensory systems and all the individual components selected for over generations.

From an evolutionary perspective complex behaviors are very difficult to study. The genetic contribution to some complex phenomenon like consciousness is rarely straightforward. As François Jacob (1977) noted, "Natural selection has no analogy with any aspect of human behavior. However, if one wanted to play with a comparison, one would have to say that natural selection does not work as an engineer works. It works like a tinkerer—a tinkerer who does not know exactly what he is going to produce but uses whatever he finds around him whether it be pieces of string, fragments of wood, or old cardboards; in short it works like a tinkerer who uses everything at his disposal to produce some kind of workable object . . . The tinkerer gives his material unexpected functions to produce a new object . . . Tinkerers who tackle the same problem are likely to end up with different solutions."

As noted earlier, because of genetic pleiotropy, what we observe, while functionally optimal, may not be the best or only method of accomplishing a particular task (such as ODCs). For example, the method of cell-cell communication that has evolved in animal nervous systems, the action potential, is only one of a number of possible ways in which cells can communicate. Indeed, synaptic transmission, with its highly intricate molecular mechanisms, is a superb example of evolutionary tinkering.

Our nervous system then, and that of any animal that we study, is a snapshot in evolutionary time of a very fluid process, and is composed of a number of adaptations for previous environments, exaptations, and epiphenomenal characteristics that may or may not be functionally important and ultimately selected for themselves.

Evolutionary Mechanisms

How is phenotypic variability achieved within a population? A number of epigenetic, or activity-dependent factors contribute to the phenotype, but we will deal with these in Chapter 15 when we examine development and plasticity. We know that genes are ultimately responsible for evolutionary changes in the phenotype, and that phenotypic variability is achieved by genetic mechanisms. The first mechanism is mutation. **Mutation** refers to a sudden change in the genetic structure of an organism that affects the development of the organism. Mutations are rare and can have a positive, negative, or neutral effect on the organism. Recent work on mutant and transgenic mice demonstrates well the mixed effects that a mutation can have on the phenotype. For instance, molecular neurobiologists have begun to use mutant mice to "genetically lesion" a molecule, brain area, or connection pathway, in an effort to understand the genetic contribution to the generation and maintenance of the phenotype.

Unfortunately, there are several problems with this marvelous new tool. The first is that a large percentage of mutants die very early in development. The second, related problem is that the "genetic lesion" of the characteristic under study is never really specific. Thus, the mutants, if they are viable, have a plethora of other phenotypic changes (both peripheral and central) associated with the lesion. Both problems are due to genetic pleiotropy. In the first instance of nonviable offspring, the mice generally do not die from the targeted characteristic that was genetically lesioned (e.g., loss of thalamo-cortical connections), but from other effects such as kidney failure, respiratory abnormalities, or cardiac dysfunction. The point is that the single gene controls not only the characteristic in question but also a number of other functions. This is why mutations are relatively rare, and genetic change across species is small. Indeed, modern DNA sequencing techniques indicate that humans are very closely related to chimpanzees, and our genetic relationship to something as simply organized as a sea anemone is relatively close considering the rather large phenotypic differences in the two species.

Despite the very close genetic relationship we share with other primates, and even other animals, there is a remarkable degree of phenotypic variability in both external morphology and organization of the brain and behaviors. This indicates that (1) phenotypic change via genetic change is difficult to accomplish, (2) a slight change to the genotype can account for very large changes in the phenotype, and (3) there must be a relatively large epigenetic component that contributes to phenotypic variability.

Another genetic mechanism that increases phenotypic variability is recombination. Genetic **recombination** refers to a change in the array of alleles of existing genes. Here, the combination of existing alleles on a chromosome, rather than the frequency of any given allele in a population, is changed. Factors that affect the frequency of any given allele within a population include genetic flow and genetic isolation. *Genetic flow* refers to individuals within the same species migrating into particular populations and interbreeding with its members. *Genetic isolation* is usually the result of geographic discontinuity. A population of individuals of the same species can become isolated due to shifts in landmasses or water, and ultimately are unable to mate with others outside of the geographically isolated population. Genetic isolation may be an important factor in speciation.

THE COMPARATIVE APPROACH

Neuroscientists who work on animals other than humans in order to discover the basic principles of neural organization that can be applied to humans are comparative neurobiologists. While you may actually be more interested in studying the human brain directly, the tools currently available are limited. For example, functional magnetic resonance imaging (fMRI) and positron emission tomography (PET) are only indirect measures of neural activity. While magnetoencephalography (MEG) and electroencephalography (EEG) allow us to appreciate the latency of processing of some stimulus with varying degrees of localization specificity, we can only infer connectivity and the hierarchy of networks involved in some given behavior. Unfortunately, directly examining neural activity, connections, neurotransmitter distribution, or the molecular mechanisms that mitigate information transfer sometimes cannot be done adequately or at all in humans. Therefore, we must rely on studies of other animals to provide insight into a number of aspects of organization and function of the human brain. For these reasons, understanding the theory behind the comparative approach is critical.

When one truly embraces an evolutionary perspective for solving problems in neuroscience, the strength of the comparative approach is unquestionable. What we cannot test directly, we can infer with varying degrees of accuracy, our extrapolations can be more precise, and our hypotheses can be biologically compatible.

Perhaps the best place to start is with the most ill-used word in the field of neuroscience, homology. **Homology** simply refers to a structure, behavior, or gene that has been retained from a common ancestor. Homology is a relative term. That is, some structure in one species is homologous to some structure in another species. The hand of a monkey and the hand of a human are homologous. The wing of a bat and the hand of a human are also homologous because both structures have the same evolutionary descent. *Homology* is also used to refer to structures in the brain. For instance, area MT in the macaque monkey is homologous to area MT in the owl monkey and galago (see Figure 14.3).

It is important to keep in mind that homologous structures do not necessarily have the same function (i.e., they are not always analogous). For example, V1 in humans is homologous to V1 in monkeys, and because of similarities in organization it is likely to be analogous in humans and monkeys. V1 in humans and other primates and V1 in the duck-billed platypus are also homologous. However, the internal organization and connections of V1 in each species vary dramatically. Thus, while V1 is homologous in both groups, it is unlikely to be analogous. A homologous structure may or may not look the same (homoplaseous). For example, although the wing of a bat and the hand of a monkey are homologous, the wing of a bat is considered to be a derived feature that has been co-opted for different functions (an exaptation). Thus, the two appendages look quite different and serve very different functions. A *derived trait* is one that is specialized and limited to a particular group or species. Examples of derived features include the bill of a duck-billed platypus, the flipper of a dolphin, and the nose of the star-nosed mole. We also can examine derived features of the nervous system. For example, the primary somatosensory cortex of murine rodents (mice and rats) contains a highly derived feature called *barrel fields,* which are related to whiskers. A **plesiomorphic** trait is one that is a general feature of some group or lineage. For instance, area MT is a general feature of primates.

Homoplasy refers to structures that look the same but do not necessarily have a common ancestry. For instance, the wing of a bat and the wing of a fly are homoplaseous and analogous but not homologous. This also holds true for features of the brain. A barrel field has been identified in the primary somatosensory area of murine rodents (Woolsey et al., 1975) and in some marsupials such as possums (Weller, 1993). The barrel fields in each group are considered to be homoplaseous but not homologous (Figure 14.5). The remarkable similarity in appearance of the barrel fields in these very distantly related mammals (approximately 125 million years of independent evolution) indicates that the rules governing neocortex development and evolution are

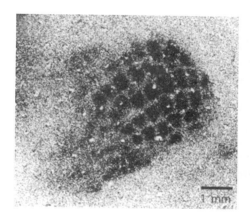
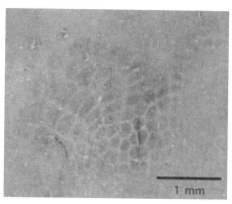

Figure 14.5 The barrel cortex in the primary somatosenory area (S1) in the brush-tailed possum **(left)** and the mouse **(right)** as demonstrated with Nissl stains and cytochrome oxidase stains, respectively. The barrel cortex represents the mystatial vibrissae in these mammals. Despite over 100 million years of independent evolution, the neural organization of this peripheral specialization looks remarkably similar. (a) Adapted from Weller (1993).

very highly constrained. Likewise, visual motion areas have been identified in other mammals as homoplaseous and analogous to MT; an example is the posteromedial lateral suprasylvian (PMLS) area in cats (Figure 14.6). The question of whether these fields in cats and monkeys are homologous is still contentious. The implication that similarities can arise independently in different lineages (**convergent evolution**) is a sign of the limited and rigid rules by which brains evolve.

To determine whether features of the brain are homologous, homoplaseous, or analogous, an **out-group comparison** is done. An out-group comparison defines phylogenetic relationships among animals (Figure 14.7). Basically, close sister groups of a species are examined to determine if they possess the structure in question. The probability that close sister groups will have the characteristic in question is higher than for a distantly related species. For instance, the middle temporal visual area (area MT) was first identified in owl monkeys. The question that arose after its discovery was: Is this a derived feature of owl monkey brains, or is this a general feature of primate brains, including humans? Decades of research on a number of sister groups (such as New World monkeys, Old World monkeys, and prosimians) indicate that MT is indeed a general feature of most, if not all, of the primate brains studied, and is likely to be a feature of human brains as well. This hypothesis recently was tested using fMRI in humans and was proved correct. MT is also homoplaseous in all primates because it looks the same across species. The question of whether MT is analogous across primates is not certain. However, because the connections of MT appear similar across primates, and neurons in area MT have similar response properties, the most parsimonious conclusion is that MT has the same function in all primates.

When researchers use the comparative approach, the **principle of parsimony** is applied to arrive at the most accurate inference regarding the unknown condition. This principle asserts that after researchers have done an out-group comparison, the best hypothesis is the one that requires the fewest number of transformations to explain a phenomenon. For example, let's refer

Figure 14.6 A motion area has been identified in both primates and cats. The area in primates is termed MT, while the area in cats is termed the posteromedial lateral suprasylvian (PMLS) area . While neurons in both of these regions respond to moving stimuli, and these areas receive direct inputs from the primary visual area (V1), it is unlikely that they are homologous because of the very distant phylogenetic relationship between cats and primates, and the lack of such an area in a number of intervening groups. This area is homoplaseous, and likely emerged owing to similar selective pressures for detecting moving objects in the environment.

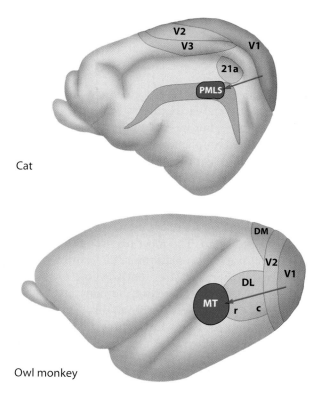

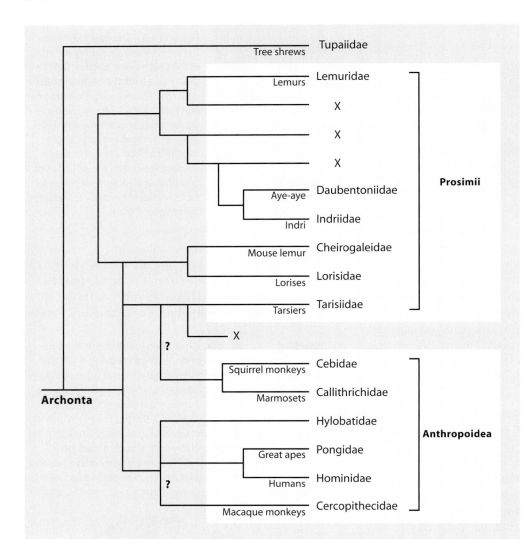

Figure 14.7 A simplified diagram of the phylogenetic relationships between extant primates. Archonta is the grand order to which the order of primates belong. Primates are divided into two suborders, Prosimii and Anthropoidea. Each suborder is divided into several families (e.g., Lorisidae and Tarisiidae for Prosimii, and Hominidae and Cercopithecidae for Anthropoidea). The common names of some representatives, or groups are found just to the left of the family names. Although no dates are given, and some of the details of the relationships are still in dispute, this diagram gives a general overview of relationships and is used in comparative analysis. Tupaiidae diverged early in the archontan line and is not considered to be a primate, but its neocortex does possess some primate-like features. Adapted from Eisenberg (1984), Walker (1964), and Grizimek (1990).

to a hypothetical brain trait or cognitive behavior as trait A (Figure 14.8). We are interested in whether trait A exists in humans, but we cannot directly test for it in humans. An out-group analysis of three great apes indicates that trait A is present in two of these animals but not the other. A further analysis of Old World monkeys indicates that trait A is present in four of the five species examined. Finally, an examination of six species of New World monkeys indicates that trait A is present in all of them. The most parsimonious explanation is that trait A is a plesiomorphic feature of primates (general characteristic) that was lost in two of the species examined. Therefore, humans are highly likely to have trait A. A less parsimonious explanation, because the number of transformations is considerably greater, would be that trait A is not a plesiomorphic feature of primates but has arisen independently in six New World monkeys, four Old World monkeys, and

two great apes. Based on this explanation, trait A is not likely to be present in humans.

Let's apply the principles of parsimony to the previous example of area MT. We have concluded that MT is a general feature of all primates that looks the same and is likely to have a similar function. Is the motion area identified in cats (PMLS) homologous to MT in primates? This question has been vigorously contested. An out-group comparison, which includes representatives of all of the major mammalian radiations, indicates that MT is not present in rodents, ungulates, or other carnivores. The most parsimonious explanation is that a motion area evolved independently in primates and cats. MT and PMLS are homoplaseous and may be analogous but are not homologous. Does this mean that it is not worthwhile to study the motion area in cats? Absolutely not. It is not a simple coincidence that an area that has a number of similar features of organization arose inde-

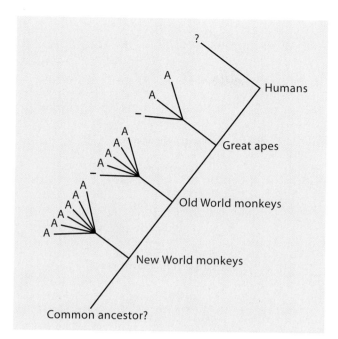

Figure 14.8 An example of how an out-group analysis can be used to make inferences about the unknown condition (e.g., the common ancestor, or an extant group that cannot be studied, such as humans). Great apes are relatively closely related to humans, while Old World and New World monkeys are less closely related. If we are interested in whether some brain characteristic (A) exists in humans, we can do an out-group comparison to make a fairly accurate inference about humans, even though we cannot study them directly. For instance, if two of three great apes, four of five Old World monkeys, and six of six New World monkeys examined possess feature A, then the most parsimonious conclusion is that the common ancestor of these primates possessed this feature, and that it was lost in one group of Old World monkeys and one group of great apes.

pendently in two distinct lineages. The implication here is that while the roots of the motion area are different, the rules of how the brain constructs a motion area in evolution must be highly constrained. Homoplasy allows us to examine the rules of how brains can change over time, and the limitations inherent in constructing nervous systems.

The Scale of Nature Revisited

In 1969, Hodos and Campbell wrote the following: "The concept that all living animals can be arranged along a continuous 'phylogenetic scale' with man at the top is inconsistent with contemporary views of animal evolution. Nevertheless, this arbitrary hierarchy continues to influence researchers in the field of animal behavior [neurophysiology, neuroanatomy, neuroscience and cognitive neuroscience] who seek to make inferences about the evolutionary development of a particular type of behavior.... The widespread failure of comparative psychologists [and most neuroscientists] to take into account the zoological model of animal evolution when selecting animals for study and when interpreting behavioral similarities and differences has greatly hampered the development of generalizations with any predictive value."

Despite the advances that we have made in understanding the brain, we still find it hard to move away from a human-centered or anthropocentric view of life, the brain, and the behavior it generates. We all have some internal ranking of animals on a continuous dimension of low, intermediate, and higher levels. Generally this ranking puts insects as the least important and progresses in importance or value from fish, to reptiles, birds, small rodent-like mammals, cats, dogs, monkeys, apes, and finally humans. This general thinking is reflected in our everyday life, our value system or moral codes, and our political system. For instance, we would feel much worse if we ran over a dog with our car than if we stepped on a bug. We eat tuna that has been caught without endangering dolphins. Animal welfare activists generally do not target scientists who work on sea slugs, but do target those who work on cats and monkeys. We feel it critical to save whales from extinction, but include few, if any insects on the list of species to save from extinction. In short, our very socialization does not allow us to be objective about the biology of our behavior, our own nervous system, or that of other animals. It affects the choice of animals we study; promotes the use of inappropriate comparisons between rats, cats, and monkeys; and propels us to make erroneous extrapolations from monkeys to humans without considering what is derived.

Lineages have changed variably over time. Thus, some animals reflect more primitive states (less changed), and others more advanced states (more changed). The latter condition is not due to a linear process across extant groups but has been achieved, often independently, in a number of different lineages (e.g., dolphins, apes, and humans). Unfortunately, most scientists still view human brains as being "most evolved" and other animal brains as some lesser or degraded version of the human brain. Not only does this view incorporate elements of the "ontogeny recapitulates phylogeny" (developmental stages reflect evolutionary stages) theory, but it suggests that evolution is linear or simply additive, and promotes the idea that "higher-order" brain areas that subserve behaviors such as language, cognition, and even consciousness should be distinguishable as some part that was added on to "lower animal" brains.

The following two examples of brain areas that are considered higher-order areas in humans are reinterpreted from an evolutionary perspective. We can use a comparative analysis of other mammal brains to make inferences about the roots and rules of "higher-order" areas. The first example is prefrontal cortex. We view this region as the hallmark of human brain evolution and consider it to have been added on to human brains during the course of human evolution. A comparative or evolutionary perspective suggests that the prefrontal cortex actually arose from a very old system, the olfactory system, and has been modified in the human lineage with the expansion of sensory cortex, to which it is interconnected via multisynaptic pathways. All mammals have some retained (homologous) orbitofrontal pathways that are similar to those described in nonhuman primates and presumably humans as well. For instance, all mammals that have been investigated have connections between the pyriform cortex and orbitofrontal cortex, and both the pyriform cortex and orbitofrontal cortex are densely interconnected with the amygdala. In addition, orbitofrontal cortex receives indirect inputs from higher-order sensory cortex in primates (Kupfermann, 1991). In humans, orbitofrontal cortex may have expanded to become prefrontal cortex (which consists of several divisions), and this expansion likely was tied to the expansion of sensory neocortex. Although there may be some overlapping functions of orbitofrontal cortex mediated by homologous neural pathways among mammals, this region of cortex has been co-opted for more varied functions in the human brain.

Interestingly, other mammals, such as echidnas, have independently "added on" a large prefrontal cortex (Figure 14.9). However, in this species we do not automatically liken this expansion to increased intelligence or a "more evolved" state. Echidnas are monotremes, mammals whose ancestor diverged very early in mammalian evolution (some 200 million years ago). These egg-laying mammals are thought to represent a primitive form of mammals (not to be confused with being a primitive form of the common ancestor), and they possess a fairly limited behavioral repertoire compared to that of humans. It is only our anthropocentric view that compels us to equate this expansion with a more highly evolved intellect. Although there is clearly a tie between evolution of the prefrontal cortex and complex human abilities, its presence is not the hallmark of advanced neocortical evolution.

Another example is Broca's area. Broca's area generally is considered to be an added-on language area of the human cortex. An evolutionary perspective provides an alternative view of Broca's area. We know from a number

Homo sapiens (human)

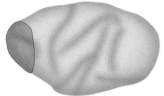

Macaca mulatta (rhesus monkey)

Tachyglossus aculeatus (echidna)

Figure 14.9 The location and extent of "prefrontal" cortex (gray) in three different species. The brains have been scaled to approximately the same size. The proportion of the entire neocortex that the prefrontal cortex assumes in humans is larger than the proportion in other primates such as macaque monkeys. In other species, such as the spiny anteater or the echidna, an expanded prefrontal cortex (relatively larger than that in humans) has evolved independently. The pressure that drove this expansion is likely to be quite different in humans and echidnas, and therefore the prefrontal cortex in each group is not analogous.

of experiments on sensory and motor cortex in a variety of mammals that morphological and behavioral specializations are associated with cortical specialization. Such specializations usually take the form of enlargement or modularization of the cortical area representing the specialized portion of the sensory epithelium or muscle group (Figure 14.10).

For instance, the duck-billed platypus has evolved a bill with a unique distribution of mechanosensory (touch) receptors interfaced with a novel type of sensory receptor called *electroreceptors*. Specialized motor programs have evolved in the platypus in which oscillatory movements of the bill are used to help localize and identify prey. The representation of the bill along the entire neuroaxis has been modified, particularly in the neocortex. This modification takes the form of an enlargement in the amount of neural

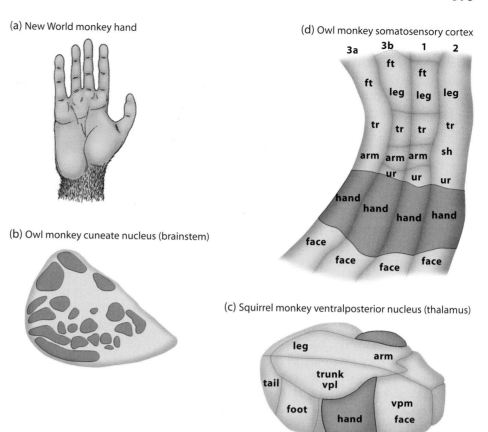

(a) New World monkey hand

(d) Owl monkey somatosensory cortex

(b) Owl monkey cuneate nucleus (brainstem)

(c) Squirrel monkey ventralposterior nucleus (thalamus)

Figure 14.10 The organization of the hand representation in New World primates at all levels of the nervous system. **(a)** The glabrous hand is a specialized structure in primates adapted for tactile exploration, recognition, and goal-directed grasping and reaching. A greatly magnified representation of this structure appears in the cuneate nucleus of the brainstem **(b)**, the ventral posterior nucleus of the thalamus **(c)**, and the anterior parietal somatosensory areas of the neocortex **(d)**. The red indicates the amount of each of these nuclei and fields associated with the specialized representation. (b) Adapted from Florence and Kaas (1995). (c) Adapted from Kaas et al. (1984). (d) Adapted from Merzenich et al. (1978).

space devoted to the specialized structure and its receptors, as well as specialized motor programs that generate foraging behaviors associated with the bill.

Other examples of neural specializations that reflect morphological specializations abound. For instance, the glabrous hand of most primates and the raccoon shows a similar magnification of representation along the entire neuroaxis, as does the nose of the star-nosed mole. All of these neural specializations have arisen independently, although all reside in the primary somatosensory area and appear to follow the same rules of specialization. Thus, while all mammals have a primary somatosensory area (S1), only in the platypus cortex has the bill representation been modified to such a large extent. In the star-nosed mole, the nose representation has been modified in a similar fashion, and in the raccoon and some primates the hand representation has been modified similarly.

In humans, if we view the changes in the density and distribution of receptors for the tongue, lips, and teeth, and the physical changes to the structure of the larynx and the motor programs associated with it, in the manner in which we view other mammal specializations, then we would view Broca's area quite differently. We would not view this region of the brain as an area that

has been added on in humans, but as a region that is present in most mammals, and certainly in all primates, but is highly derived in humans. The derived state in humans is a magnification of the cortical representation of the face and oral structures in the premotor cortex (PM), as well as M1 and S1, as a consequence of, or in conjunction with, changes in the periphery. Like in other species with sensory and motor specializations, changes in the connections and the motor programs direct the use of these structures (e.g., nose in star-nosed mole is used for exploration; platypus bill for oscillation; hand in primate, for reaching and grasping). In the case of Broca's area, interconnections with specialized auditory regions of the brain (such as Wernicke's area) have formed. Thus, while all mammals have an M1, and certainly all primates have a PM, only in humans have the motor and sensory representations of oral and facial structures been modified in such a manner.

A comparative analysis suggests that the most parsimonious and biologically compatible explanation is that the modification to the human brain known as Broca's area has homologous counterparts in other primates, and certainly follows highly restricted rules of cortical field modification. However, the formation of new patterns of connections, and the clear behavioral differences this region of cortex generates, indicate that it is not analogous. Interestingly, Broca himself appreciated that there were two alternative explanations for areas in the brain involved in articulate language (see Greenblatt, 1984). The first (which he favored) was that they are part of the human intellectual function. The second explanation was that they are part of cerebral functions that have to do with motion or the motor system.

If we throw out the "scale of nature" and invoke the history of change (roots) and the proposed mechanisms (rules) that generate these changes over time in other mammals, we can allow for richer and more varied hypotheses regarding particular cognitive behaviors. These hypotheses will be more objective, less anthropocentric, and more compatible with evolutionary neurobiology.

ADAPTATION AND THE BRAIN

Evolutionists interested in human behavior think that the modern human brain was adapted to deal with the world as it was in Pleistocene hunter-gatherer societies 100,000 years ago. This period was picked because of the slowness of new adaptations. Our auditory system, for instance, is not adapted to sensory events such as loud rock and roll music. This is why hard-rock musicians and their audiences wear earplugs, raising the question as to why they play their music that loud to begin with. Or consider social structures. In hunter-gatherer times, groups were rarely larger than fifty. With today's vast cities and social structures like government bureaucracies, our brains are severely challenged to cope. So, when we think about the functions of the modern brain and what it does and does not do well, we should take into account what the early hunter-gatherers had to solve.

This proposal takes a novel tack in the study of decision making. W.T. Wang (1996) argues that mechanisms for decision making evolved in a particular social context. In particular, he focuses on the idea that for most of our evolutionary history, humans lived in small groups. As such, he proposes that our decision-making capacities are tailored for such contexts and that there may well be differences in how we make decisions in different social contexts. This idea is intriguing. There has been a movement toward bringing evolutionary thinking into the mainstream of cognitive psychology. Wang's experiments to date have focused on the framing effect. In a framing experiment, people are presented with two choices regarding some risky situation, such as which of two treatments should be applied to combat a potential medical disaster. In the classic demonstration by Tversky and Kahneman (1988) of the framing effect, the treatments are posed in either a positive or a negative light (e.g., treatments to save lives or treatments to minimize death). The basic finding here is that risk-taking behavior differs in these two situations. When the treatment is framed in a positive manner, people tend to choose sure bets; when it is framed in a negative manner, people tend to choose more probabilistic options. The framing effect is one demon-

stration of human irrationality, or more specifically, that a purely cost-benefit analysis will not be sufficient to account for human decision making.

This example illustrates well that in a number of respects our behaviors are adapted for a simpler life, the life of the Stone Age human. Back then we had to be ready to defend ourselves, detect cheaters, read other people's facial expressions, forage, avoid incest, recognize kin, and read other people's minds and their intentions. Our mind does not instinctively share common goals with others. Have you ever been to a meeting that sets policies? It takes luck, persuasion, and brute force to get a group of independently minded people to agree on anything.

According to evolutionary thinking, these special capacities grow from separate and individual adaptations. The cognitive system that evolved is not a unified system that can work by applying special solutions to individual problems. This fundamental point is at the heart of the evolutionary perspective, and concurs with a vast amount of neuropsychological research. Localized brain lesions can lead to a loss of some capacity, say, facial recognition, but local brain lesions also can have a maddening, mild effect on specific functions. This latter truth most likely reflects the observer's inability to present the right challenge to the patient. That is, the patient probably does possess deficits, but the examiner's tests are either incorrect or not sensitive enough to detect the disorder. Alternatively, the failure to find a deficit may mean that devices built into the brain for other functions can solve other challenges. Just as a screwdriver can loosen screws, it can also open paint cans.

The adaptations built into our brains are the physical, or neural, structural devices we should try to understand when trying to figure out how the brain works. As already pointed out, an important rule to remember, one commonly overlooked, is to focus on the adaptation, not on the ancillary events associated with the adaptation. Bones are an adaptation, but their whiteness is a by-product, or epiphenomenon, of the calcium that gives bones their strength. Calcium was available in the environment and was used to build the bones' rigid structure. If we want to study bones and how they came to be, it would be a mistake to delve into the fact that they are white. We should simply distinguish between what are proximate factors versus ultimate ones when we consider why something evolved. *Proximate factors* are those at hand, and part of a structure. But their presence may not be why the structure evolved.

Adaptations at Multiple Brain Levels

It is easy to see how adaptations gradually developed in primates; they occurred with primary sensorial systems like our visual and auditory systems. Furthermore, we have adaptations for more complex behaviors like our capacity to have a theory of mind about someone else. We look at another human and quickly theorize about their intentions toward us and how we should respond. At these levels evolutionary processes operate—and deserve comment.

At the simpler level of understanding vision, today's researcher may have insights that are consonant with the realities of the Pleistocene landscape. After all, the physical world of light and object has not changed all that much. Our visual system is built to take a two-dimensional retinal image and turn it into a real-world representation of the visual scene. David Marr of MIT first articulated the true problems associated with understanding how vision must work. As Marr (1982) put it, "Trying to understand perception by studying only neurons is like trying to understand bird flight by studying only feathers: It just cannot be done. To understand bird flight, we have to understand aerodynamics; only then do the structure of feathers and the different shapes of birds' wings make sense."

Marr is widely recognized as a genius, and the field was devastated when he died a young man. He was pursuing a computational analysis of what he was interested in understanding. In general terms, he liked three levels of description for any information-processing device, whether it be a cash register or a brain. In brief, he based his ideas on the following logic: First, information-processing devices are designed to solve problems. Second, they solve problems by virtue of their structure. And, third, they explain a device's structure when one knows what problem it was designed to solve and why it was designed for that problem and not another one.

These issues were partly illuminated when researchers in artificial intelligence tried to build a device that could analyze a visual scene just as a human does. Marr first realized that the evolutionary function of vision is scene analysis: The brain must reconstruct a model of real-world conditions from a two-dimensional visual array—the information on the retina.

The first discovery was that scene analysis is far more complicated than had been imagined. A simple object sitting on the horizon in the morning light looks completely different when the sun reaches high noon and then falls over the opposite horizon. An intricately specialized system must have been built into the primary visual system to allow for this natural progression and enable us to see the object as constant. Second, investigators discovered that our visual system apprehends far more information about a scene than can any artifact. It has many adaptations built in for this, all specific to vision. Finally—and this is the tricky part—our evolved visual system must have a cognitive, built-in component that has deduced that objects in the world have regularities which allow for the proper computations. This built-in component is what developmentalists mean by biologically prepared implicit knowledge in infants and children.

As we move up the scale to more complex adaptations like those associated with problem solving, social exchange, and the like, the insights we might gain by looking around us in our present world are probably not as helpful. Most studies do not take into account the sorts of ways that Stone Age humans dealt with social problems. In many of today's attempts to understand human rational processes, subjects are presented with artificial thought problems developed in the laboratory. Yet our brains are adapted only for real-world problems. For instance, many logic tests on issues of social exchange stump college sophomores because they are posed in the abstract. Does their failure mean that sophomores are illogical? No, because when the same logic problem is based on a real-life story about obtaining beer or food, their logic systems work just fine.

Leda Cosmides (1984) worked out a telling example of this fact. She built on a test first developed by Peter Wason, who showed how poorly-educated people can perform a simple logical task. Try it yourself: Each card has a number on one side and a letter on the other. Examine the following four cards:

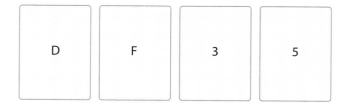

The task is simple. You are to determine if the following rule has any exceptions: If a card has a D on one side, it has a 3 on the other side. Which cards do you

HOW THE BRAIN WORKS

Lessons from a Frog

Wonderful examples illustrate how evolutionary theory has led to great insights into brain function. Perhaps the most famous example is the work of Jerome Lettvin, one of the fathers of modern neuroscience and an extremely lively intellect. In 1959 he wrote a famous paper with colleagues, titled "What the Frog's Eye Tells the Frog's Brain." Lettvin broke with his predecessors and asked questions about the visual system based on his understanding of how a frog views the world.

Prior to Lettvin's work, the eye was regarded as an organ that translated an image into electrical impulses, and then the brain sorted out and interpreted this retinal information. The great American physiologist, H.K. Hartline, gave support to this idea in 1938 by studying the retina with simple points of light and dark. Using these abstract stimuli, Hartline and his colleagues concluded that the frog's retina passed on to the brain information about the tone of various objects. They concluded that the retina's ganglion cells had but one function, and that was it.

Lettvin changed this interpretation. While recording from the frog's ganglion cells and stimulating the frog's visual system with bugs, twigs, and other ecologically relevant material, he and his colleagues discovered several types of ganglion cells—five, to be exact. In simple terms, Lettvin showed how much information is weeded out by biological systems like the retina. Evolution has seen to it that the frog's brain does not detect things about the visual environment that it does not need in order to function and survive. Hence, the male frog's brain does not spend energy on noticing the actor Melanie Griffith, but does detect the movement of female frogs.

Table 14.1 Types of Detectors

1. *Sustained-edge detectors* (SEDs): These showed the greatest response when a small, moving edge entered and remained in their receptive field. Immobile or long edges did not evoke a response.
2. *Convex-edge detectors* (CEDs): These were stimulated mainly by small, dark objects with a convex outline like beetles and other bugs.
3. *Moving-edge detectors* (MEDs): These were most responsive to edges moving in and out of their receptive fields.
4. *Dimming detectors* (DDs): These responded most to decreases in light intensity, such as a shadow cast suddenly over the frog.
5. *Light-intensity detectors* (LIDs): These cells' responses were inversely proportional to light intensity, being most responsive in dim light.

---

need to turn over to discover if this is true? Feel your mind rattling? But now consider the following problem: You are a bouncer at a bar and your job is to make sure no one under 21 drinks beer at the bar. The cards below have information about age on one side and what the patron is drinking on the other. Again, which cards do you need to turn over?

| Beer | Soda | 25 | 17 |

The mind springs to action on this task. It is obviously the first and last card, just like in the more abstract example earlier. Why? Cosmides and Tooby (1992) maintained that we have a built-in cheater detector system that has been a necessary part of our brains ever since we as a species began to exist in social groups. The moment survival becomes conditional on what a group does, as opposed to what an individual does, there must be a way to make sure the collective idea works as it is supposed to. In other words, we have the beginnings of a social contract.

The argument here is that social exchange can be expressed by a conditional statement such as, If you take the benefit, then you must pay the cost. For example, if you play on the hotel's golf course, then you must be a guest at the hotel. A cheater is someone who takes a ben-

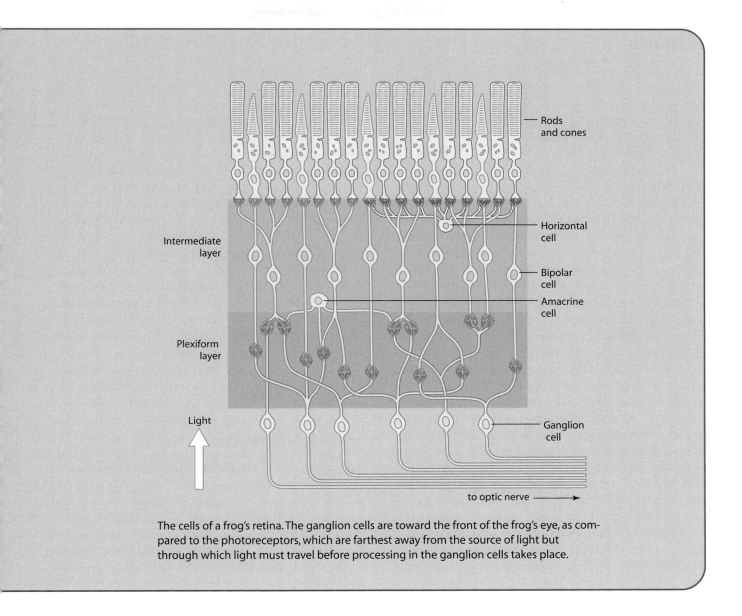

The cells of a frog's retina. The ganglion cells are toward the front of the frog's eye, as compared to the photoreceptors, which are farthest away from the source of light but through which light must travel before processing in the ganglion cells takes place.

efit without paying the cost. People are quite good at detecting potential violations of these kinds of conditional rules (i.e., catching potential cheaters), whereas they cannot detect potential violations of purely descriptive rules. An example of the latter being, If a man wears a tuxedo, then he must wear a bow tie. In studies of undergraduates in the United States and Germany, 75% to 90% of people reason correctly about social exchange, compared with only 4% to 25% of people reasoning correctly about abstract or descriptive rules. Detecting potential cheaters is an important evolutionary problem—those who could not detect cheaters got ripped off more often, and did not end up as our ancestors.

This intriguing new work has built on the insight of Robert Trivers (1971). Many years ago Trivers, now at Rutgers University, showed that the elements of social exchange are based on the evolution of reciprocal altruism. Reciprocity amounts to roughly equal amounts of give and take in social relationships. The associated cognitive, psychological, and emotional systems allow us to develop and maintain friendships with nonkin. Kids implicitly recognize the reciprocal nature of friendships, and socially reject kids who do not reciprocate.

Social exchanges and titrating of reciprocity give rise to many emotions. Indeed, Trivers argued that many emotional reactions regulate social exchange. If we give more than we take in a social relationship, we get angry; if we take more than we give, we feel guilty. Guilt motivates returning the favor, while anger motivates us to break off relationships with people who cheat, who do not reciprocate.

Sexual Selection and Evolutionary Pressures on Behavior

Sex has realities that are at once clear and puzzling. Darwin knew natural selection was at work, but he was bothered by why males and females differ so much, given their mutual goal of trying to survive in the same niche. It is easy to understand why genitalia are different, but why such big differences in behavior and bodily structure?

Some dispute whether sexual selection should be distinguished from natural selection, but leading researchers such as Steven Gaulin at the University of Pittsburgh (1995) argued that it should be. Gaulin, who has puzzled over the problem of sexual selection for years, maintained that a distinction between sexual selection and natural selection explains how ecologically useless characteristics such as antlers on a deer evolved and have survived. While not necessary for functions such as food gathering, these characteristics are important in enhancing sexual contact. An analogy would be that one can get around town in a Honda, but a red Corvette conveys a different message.

The sexual life of the deer readily illustrates how selection starts the two sexes down diverging paths. Once a female deer has mated and conceived her maximum litter, she has no need for further sexual contact. The male, though, can maximize his reproductive success by continuing to impregnate as many females as he can. Thus, any somatic event that would enhance a male's reproductive fitness would be selected for, while a similar change for the female would not. The female would not gain, because her reproductive limit has been met (Figure 14.11).

This pattern for mammals has exceptions, and it is the exceptions that bring strength to the idea that sexual selection is distinct from natural selection. First, male

Figure 14.11 The typical morphology of **(a)** a female doe that invests her energy into raising her offspring and that of **(b)** a polygamous male elk with its large investment of energy into the massive set of antlers, which aid his ability to attract mates and fight off competitors.

mammals have small and many sex cells, whereas females have large cells. In other animals, the quickly reproducing sex is the female, and the male stays home with the young. In some shorebirds, the male stays in the nest to incubate the egg. The female, right after laying the egg, leaves the nest and seeks other males, so she can lay yet more eggs for them to incubate. In this turnabout of roles, the fast-sexed females are larger, more brightly colored, and aggressive. It is clear, then, that the flamboyant characteristics of most males and of the less common fast-sex females enhance their capacity for a higher reproductive rate—which is not to say that there are no monogamous species. Over 90% of foxes and birds are deeply committed to each other because both must participate in child rearing if their species is to survive.

Another significant difference between mammals and birds is that mammals have an internal gestation period. This means that female mammals are making a larger commitment to offspring, which may underscore the known difference in how much the male and female invest in their offspring as compared with spending time on mating. For birds and many fish species, external gestation creates the potential for females to force parental care onto males, a more difficult trick with mammals.

Sexual Abilities and Spatial Abilities

Too many jokes to recount here are made about the spatial skills of women versus men. While once good-humored, they now take on a social significance that becomes lost in current social values. Still, some facts are intriguing. Spatial abilities do vary according to sex—and they do so because of selective pressures. These spatial skills differ in humans and animals of all kinds, which suggests that brains manage spatial skills differently in males and females.

Natural selection sees to it that males and females have basic navigational skills. Both meet the same challenges for reward and risk in food gathering and other life-perpetuating activities. Where sexual selection might start to mold differences is when males of a species are polygamous. Here the males might need better spatial skills to find available females for sex and yet return home. This phenomenon pertains to polygamous rather than monogamous mammals (Figure 14.12).

In trying to test the hypothesis that polygamous males would have greater spatial skills as compared to females, we run into a problem: Over 95% of mammalian species are polygamous. In fact, verifiable sex differences in spatial skills have been discovered in rats, mice, and people. To truly test the hypothesis, we would

Typical polygamous mammal

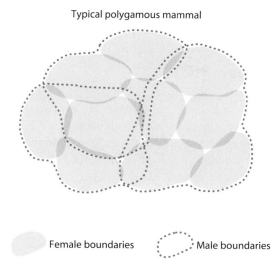

Female boundaries   Male boundaries

Typical monogamous mammal

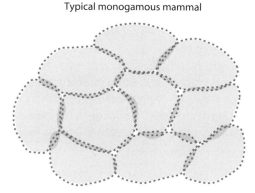

Figure 14.12 Typical home-range patterns for polygamous versus monogamous mammals. For many polygamous species, male home ranges overlap with numerous, smaller female ranges. Females and males in monogamous species where parenting by both sexes is necessary for the raising of young tend to have smaller, isomorphic home ranges.

need a species with monogamous and polygamous strains, to evaluate each group's spatial skills.

Studies of wild voles have elucidated these issues. These rodents were chosen because we have a myriad of well-developed spatial tests for them. Yet the sizes of the ranges of the two monogamous pine and prairie voles did not differ (Figure 14.13). At the same time, the free-wheeling, polygamous meadow vole demonstrated huge differences in range sizes between females and males. Gaulin and his colleagues confirmed these spatial skills by testing the same species in a laboratory. Males with larger range sizes had better spatial skills (Figure 14.14).

With the spatial abilities of the two sexes clearly demonstrated, researchers became eager to establish whether brain structures for spatial abilities varied. Over

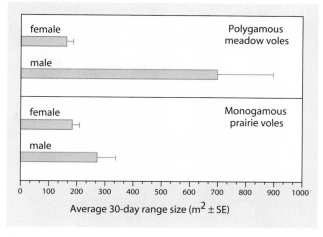

Figure 14.13 Thirty-day average range area of monogamous prairie voles and polygamous meadow voles. Adapted from Gaulin and Fitzgerald (1989).

Figure 14.14 Symmetrical maze performance by monogamous prairie voles and polygamous meadow voles. Adapted from Gaulin and Fitzgerald (1989).

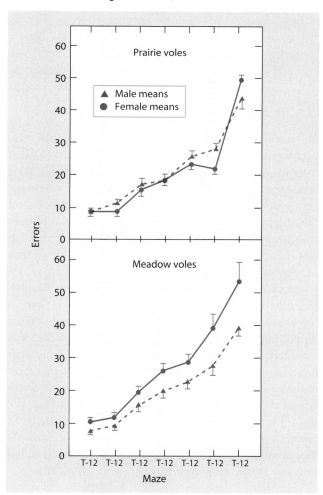

Sexual Selection and Mathematics

The perennial topic of sex-related differences in mathematical abilities has been exhaustively studied by David Geary at Columbia University (1995). He observed that there are no sex-related differences in biological primary mathematical skills, even in nonhuman primates. This is true for all cultures. Yet there are sex-related differences in secondary mathematical skills, the kinds of math taught in schools in the industrialized world. Males consistently outperform females in word and geometry problems. Geary suggested that this capacity builds on the sort of male superiority in spatial skills seen in many species, such as the ones described in this chapter. In short, it is a secondary benefit to males, arising out of the sexual selection process we have described.

Geary went on to show that sex hormones, which are the proximate mechanisms associated with sexual selection, appear to indirectly influence mathematical ability. He argued that sexual selection resulted in greater elaboration of the neurocognitive systems that support navigation in three-dimensional space in males than in females. These navigational systems have evolved in the three-dimensional physical world, so some information about the structure of the physical world is built into these systems. It appears that features of Euclidean geometry are a mathematical representation of the organization of the physical world; thus, an implicit understanding of aspects of geometry is built into the spatial system. Males do better in geometry than do females because evolution has provided males with more built-in knowledge of geometry. These same spatial skills can solve other types of math problems, such as word problems, because solving word problems is much easier if important information in the problems is diagrammed or spatially represented.

This same study also observed that many sex differences are found in social styles and interests; these play into the superior math skills. Because of these differences, males are more likely to engage in mathematical problems, which further enhances their ability.

the years, many investigators discovered that the hippocampus is crucial to spatial memory tasks. Birds such as titmouses, nuthatches, and loud-mouth jays all cache their food over a large spatial area. Because they must have superb spatial skills to retrieve their food, one might expect their hippocampus to be larger, corrected for whole-brain volume, when compared to birds that do not store food in this manner. This is exactly what researchers found in a number of species such as the vole (Figure 14.15).

Lucy Jacobs and colleagues at the University of California at Berkeley (1990) asked similar questions about kangaroo rats. In two species of kangaroo rats, she found that the hippocampus is larger in males that, during their breeding season, range more widely than do females. Just as fascinating is the cowbird, a species whose females sneak their eggs into the nests of other species. The cowbird's timing has to be just right, and she has to search a wide area to find the unsuspecting host nest. Sure enough, a female cowbird's hippocampus is much larger compared to a male cowbird's hippocampus. Males do none of these wide-ranging activities; they are the avian equivalent of couch potatoes.

Hormones play a fascinating role in the development of spatial abilities (Dawson et al., 1973). Investigators showed that sex-typical patterns of maze performance can be reversed by the early administration of appropriate hormones. How, we might ask, can hormones affect the genetic blueprint for sex differences such as spatial skills?

The expression of sexual differences involves ontogenetic forces, which are the forces at work during an organism's development—in this case, in the fetal stage. The true genetic differences between males and females are slight, and they are expressed by factors such as the local hormone environment. For example, in humans who have Turner's syndrome, a neuroendocrine genetic disorder, the gonads remain undifferentiated. As a result, a phenotypic female who was deprived of androgen and estrogen during development has spatial abilities significantly below her verbal skills. The same is true for males who experience low androgen levels during development, such as in Klinefelter's syndrome, another genetically based neuroendocrine disorder. They too exhibit depressed spatial skills.

Evolution and Physiology

By now, it should be clear that natural selection is crucial in shaping each species's brain (Gazzaniga, 1992). Special

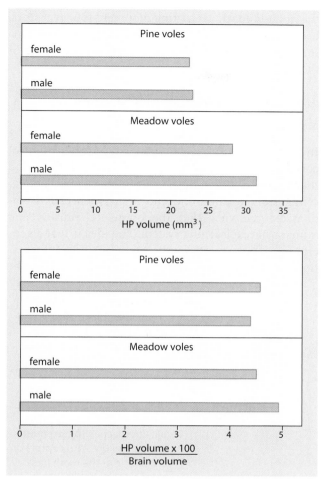

Figure 14.15 Comparison of hippocampal volume for monogamous pine voles and polygamous meadow voles. Adapted from Jacobs et al. (1990).

devices built into each brain enhance the species's capacity to reproduce and to promulgate itself; the devices vary tremendously, as each species has different niches and predators. Since no two species utilize the exact same kind of resources, and since no two species have the exact same predators, each species also has special isolating reproduction mechanisms. This is important to understand, as each species has unique traits to pass on, and thus sensory receptors and brain decoders in each species are rather idiosyncratic. This recurring theme is a fundamental one for students of the nervous system. Even though nerve cells may share similarities, and nervous systems may be universally composed of certain nerve cells, each nervous system varies from one species to another in ecologically appropriate ways (Bullock, 1993).

There is a wonderful example of how biological structures coevolve, each trying to adapt to meet its own needs. In a predator-prey relation, as the predator develops an edge by evolving toward better fitness, the prey responds by evolving a mutated member that counters the new predator's skill and subsequently enhances reproduction—much as the bat and the moth it eats.

As contrasted with the visual system we predominantly use to navigate, bats maneuver through their environments by emitting high-frequency sound waves and detecting their weak echoes off surfaces. The bat's nervous system has evolved sophisticated adaptations that enable the bat to be sensitive to these weak echoes. As a bat approaches an object like a moth, it emits more sound waves to gain exact information on the object's direction and distance. Because of their mode of transportation, flying insects give more clues to the bat. As an insect moves its wings for flight, rhythmic reflections of the bat's sound waves striking the upper and lower surfaces of the insect's wings return to the bat. The bat uses these to distinguish between moths and tree leaves. With such an accurate detection system, it is hard to believe bats' prey could stand a chance of survival.

But prey such as the noctuid moth have evolved antipredator adaptations to prevent them from becoming the next meal of a strong hunter like the bat. The noctuid moth has two ears, each with only two receptors, the A1 and A2 receptors, with which to perceive an approaching bat (Figure 14.16). The A1

Figure 14.16 The noctuid moth's (*Agrotis ypsilaon*) ear, and location of receptors. When sounds are of a sufficient frequency, the auditory receptors stimulate the tympanic membrane, which in turn induces the auditory neurons to fire.

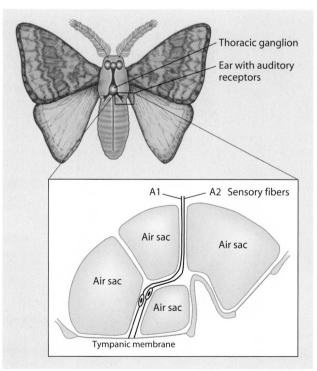

receptor responds to low-intensity sounds such as the ones from a bat 3 to 30 m (10 to 100 feet) away. The moth's ears work as ours do to locate a sound source; the moth relies on its knowledge that whichever receptor is closer to the sound will be activated slightly before the receptor that is farther from it. In this way, the moth can tell if the bat is to its left or right. Because the beating of the moth's wings makes small interruptions in the reception of the bat's signal, the moth can determine whether the bat is above or below it (Figure 14.17). If the bat approaches within 3 m (10 feet), the moth's A2 receptors, which are sensitive to high-intensity sounds, start firing, and the moth responds by beating its wings irregularly, thereby throwing off the bat's detection strategy and prompting the moth to dive for safety (Figure 14.18). This simplistic system of

the noctuid moth, consisting of a mere four receptors, has coevolved with the bat's more sophisticated sensory system. The result is a balanced coexistence of these two species.

Adaptive Specializations and Learning Mechanisms

In cognitive neuroscience there is no greater point of contact between evolutionary theory and the mind than on the question of the nature of learning. The recent history of the mind and brain sciences has emphasized the view that the brains of animals and humans have a learning system. The idea is that a general-purpose capacity in the brain acquires information during any and all kinds of learning tasks. The

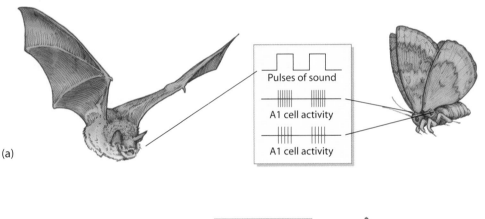

(a)

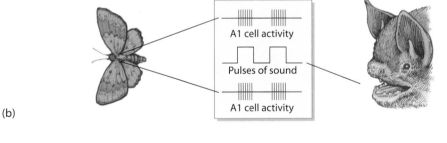

(b)

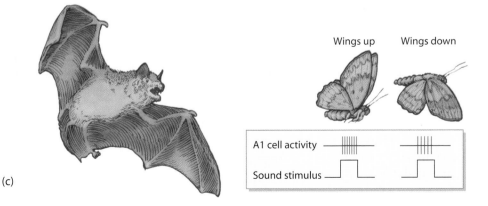

(c)

Figure 14.17 **(a)** Differential firing of the bilaterally located A1 receptors indicates from which direction, left or right, a bat is approaching. The receptor that fires with a higher frequency is closer to the source of the sound and thus indicates the direction from which the bat approaches. **(b)** Symmetrical firing of the A1 receptors on either side of the moth indicates that the bat is directly behind the moth. **(c)** The interruption of the bat's sounds due to the location of the moth's beating wings indicates whether the bat approaches from above or below.

Figure 14.18 **(a)** The simplistic A1 and A2 receptors work together to process information about sounds of different intensities and frequencies. When the bat is farther away, sounds of lower intensity affect the A1 receptor but not the A2 receptor. As the intensity increases, the A2 receptor starts firing to give the moth more detailed information about the location of the sound source. **(b)** The A1 receptor reacts strongly to high-frequency sounds that are detected in pulses. If the stimulus is a steady sound of the same frequency and intensity, the A1 receptor will cease firing after a short while. This prevents the moth from being overly sensitive to persistent, irrelevant sounds in its environment, and reserves the functioning of these receptors for ecologically important situations such as detecting a hunting bat.

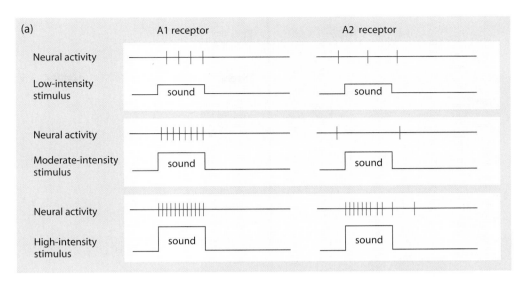

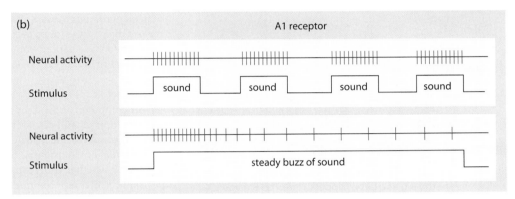

idea of simple associationism was prevalent and, in early American experimental hands, became known as *stimulus response* or *SR psychology*. With the right reinforcement contingencies and the right brain state, learning could proceed with ease.

This dominant view has changed over the past few years. Leading the charge has been Randy Gallistel (1995) at the University of California, Los Angeles. He argued for the idea that there are many learning mechanisms, each computationally specialized for solving problems. He took the strong view that the association formation mechanisms so ubiquitously touted by psychologists are not even responsible for classic and instrumental conditioning. To support his ideas, Gallistel recounted how migratory thrushes must learn the center of rotation of the night sky when they are mere nestlings. This knowledge, gained as a young bird, is called upon only when they grow up and use their knowledge of the night sky's celestial pole to maintain their southerly route during their first migratory flight (Figure 14.19). A simple associationism could never explain this behavioral capacity. The knowledge gained as

a young nestling is not used at this stage of its life. There are no contingencies; the knowledge is called on only at a later time.

We have no shortage of examples of what is called *nonassociative learning*. The field of *ethology*—the study of animal behavior in the real world as opposed to the world of the laboratory rat maze—has many rich examples. An often-cited example has to do with the capacity of insects to learn *dead reckoning*, which is the capacity of all kinds of animals, including the lowly insect, to find their way home after they have been out foraging. When the ant leaves its home base, it computes and stores information on how to get back home. As Gallistel (1995) put it, "Like all learning mechanisms, it computes and stores the value(s) of variables. In this case, the mechanisms compute the values of the variables that represent the animal's position relative to its nest or home base. The computation is equivalent to integrating velocity with respect to time" (Figure 14.20).

These sorts of examples have led several ethologists to the view that an animal has a constellation of

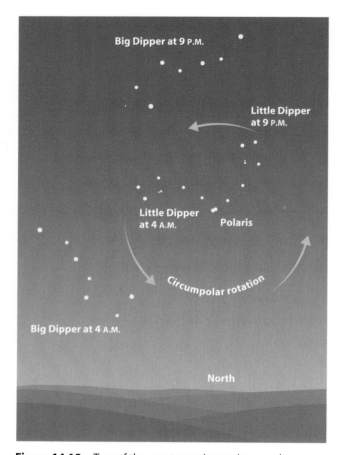

Figure 14.19 Two of the most prominent circumpolar constellations are the Big and Little Dippers. Here, the Big and Little Dippers are seen as they would appear at 9:00 P.M. and 4:00 A.M. from temperate latitudes in the northern hemisphere in the spring. Migratory birds learn the directions North and South by observing the rotation of the circumpolar stars around the celestial pole (near Polaris now). Adapted from Gallistel (1995).

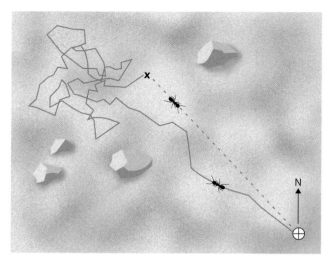

Figure 14.20 A foraging ant's path. The solid line represents the outward, searching journey until food was found at X. After making numerous turns in all directions, the ant is able to make a direct route (dead reckoning) home, as designated by the dashed line. Adapted from Harkness and Maroudas (1985) and Gallistel (1995).

specialized learning mechanisms. The learning capacity exhibited in one situation works only in that situation and no other. Peter Marler (1991) at University of California at Davis and others called these examples of *problem-specific learning mechanisms*. The importance of this to issues in cognitive neuroscience is that when we are trying to understand how the brain enables learning, we must realize that there may be several mechanisms, not just one. We saw in Chapter 11 another example of this: the idea that the cerebellum is essential for representing the temporal relation between a tone and a forthcoming air puff in order to respond at the right time.

Gallistel argued that although there are many different learning mechanisms, there may still be some commonalities. The basic computational operations may be the same. As he put it:

In computer terms, they may all use the same basic instruction set. (Of course, they may not. At a time when we cannot specify the instruction set underlying any computation of any substantial complexity, we are in no position to say whether different complex computations use the same elementary computational operations.)

The other thing that all learning mechanisms must do is store and retrieve the values of variables. The bird has to store values that represent the center of rotation of the night sky and retrieve them when it determines what orientation to adopt with respect to whatever constellation of circumpolar stars it can see at the moment. The foraging bee must store the distance and direction of the food source and retrieve that information when it gives the dance that transmits the values to other foragers. The dead-reckoning mechanism must store the values that represent the animal's current position and then add to them whenever the animal moves, because dead reckoning amounts to keeping a running sum of displacements.

As we ascend into the human brain, we can see from an evolutionary perspective how humans must possess special devices for learning. William James stated that the human has more instincts than animals, not fewer. It is in this setting that Noam Chomsky and Steven Pinker argued for the view that there is a special learning module for human language. It has specific features and capacities, and most likely a definite neural organization. In short, the work on animals, where specialized systems are easily identified, raises provocative notions about the human brain's organization and cognitive powers.

EVOLUTIONARY INSIGHTS INTO HUMAN BRAIN ORGANIZATION

A major assumption in neuroscience is being challenged: the idea that a larger brain with more cells is responsible for the greater computational capacity of the human being. Consider Passingham's (1982) main conclusion to his fascinating book, *The Human Primate:*

> Simple changes in the genetic control of growth can have far-reaching effects on form. The human brain differs from the chimpanzee brain in its extreme development of cortical structures, in particular the cerebellar cortex and the association areas of the neocortex. The proportions of these areas are predictable from rules governing the construction of primate brains of differing size. Furthermore, there appears to be a uniformity in the number and type of cells used in building neocortical areas; the human brain follows the general pattern for mammals. Even with two speech areas we believe we can detect regions in the monkey brain that are alike in cellular organization. The evolution of the human brain is characterized more by expansion of areas than by radical reconstructions.

The uniqueness of the human brain, it is commonly believed, can be traced to its large size. It has more neurons and more cortical columns, and in that truth lies (somewhere) the secret to human experience. This is entirely consistent with many other observations of humans and animals. The disproportionately large representation of some sensory and motor regions of the cortex in animals and humans is well established. The recognized correlation between the large inferior colliculus for echo-locating bats and dolphins, and the enlarged optic lobes for visual fish is well known. In short, the idea that a larger brain structure reflects an increase in function is ubiquitous.

Even Charles Darwin promoted the idea that big brains explained the uniqueness of the human condition. In *The Descent of Man and Selection in Relation to Sex* (1871), he wrote that there is no fundamental difference in the mental faculties of humans and the higher mammals. He went on to add that the difference in mind between humans and the higher animals, great as it is, is certainly one of degree and not of kind. He did not want to be part of any thinking that there may be critical qualitative differences between the subhuman primate and humans. Darwin left the actual anatomy to his colleague Thomas Henry Huxley. At that time, Richard Owen, another anatomist, maintained that there was a special structure in the human brain, the hippocampus minor. Yet Huxley proved that this structure was also found in other primates, thereby undercutting the idea that the human brain was qualitatively different from the primate brain. So here we had Darwin, the genius who articulated natural selection and diversity, arguing for a straight-line evolution between primates and humans. Organisms, the product of selection pressures, displayed rich diversity in the evolution of species. But when it came to brain and mind, Darwin thought the human brain to be a blown-up monkey brain, a nervous system that had a monotonic relation to its closest ancestor.

Nonetheless, a lot of evidence shows that the human brain's unique capacities do not rely on cell number so much as the appearance of specialized circuits. That the human brain has more cells does not explain greater capacities. Evolutionary perspectives would find the human brain adapted to its own biological niche, and one would predict differences in brain organization from other animals. After millions of years of natural selection, we have accumulated circuits that enable us to carry out specific aspects of human cognition. In short, just as comparative neurobiologists have demonstrated the presence of specialized circuitry in lower animals that reflect adaptations to niches, similar demonstrations can be made in humans.

Let's look at more evidence for specialized circuits. Consider the human brain's two halves, left and right. We know the left cortex is specialized for language and speech, and the right has specializations of its own (see Chapter 10). Each half is the same size and has roughly the same number of nerve cells. The cortices are connected by the corpus callosum. The total, linked cortical mass contributes to our unique human intelligence. What would happen to intelligence if the halves were disconnected, leaving the left operating independently of the right and vice versa? Would split-brain patients lose half of their cognitive capacity because the left, talking hemisphere would now operate with only half of the total brain cortex?

A cardinal feature of split-brain research is that after cerebral hemispheres are disconnected, the patient's verbal IQ remains intact, and the problem-solving capacity of the left hemisphere, such as hypothesis formation, remains unchanged. While there can be deficits in recall capacity and in other performance measures, the total capacity for problem solving remains unaffected. Isolating essentially half of the cortex from the dominant left hemisphere, then, causes no major change in cognitive functions. Following surgery, the integrated 1200- to 1300-gm brain becomes two isolated 600- to 650-gm brains, each about the size of a chimpanzee's brain. The capacity of the left half remains unchanged from its preoperative level, while the largely disconnected, equally-sized right hemisphere is seriously hampered from performing tasks. Although the largely isolated right hemisphere remains superior to the isolated left hemisphere for things like recognizing upright faces, attentional skills, and perhaps emotions, it is poor at problem solving and many other mental activities. A brain system (the right hemisphere) with roughly the same number of neurons as one that easily cognates (the left hemisphere) is incapable of higher-order cognition—strong evidence that cortical cell numbers do not fully account for human intelligence.

Perhaps the most influential and dominant idea that more cortical area means higher-level function came from Norman Geschwind and Walter Levitsky (1968). Over the past 30 years, their report that the left hemisphere has a larger planum temporale solidified the belief that more brain area meant higher-level function. They concluded their classic paper by stating, "Our data show that this area is significantly larger on the left side, and the differences observed are easily of sufficient magnitude to be compatible with the known functional asymmetries." In other words, the belief was that the greater brain area in this language zone was responsible for language.

Because this classic finding makes a strong case for a relation between cortical area and function, the issue of whether the left planum temporale is larger than the right planum has been re-examined. With three-dimensional reconstructions of normal brains provided by magnetic resonance imaging, the posterior temporal region was carefully measured using the same methods Geschwind used; approximately the same percentage of brains had apparent asymmetry, with the left side being larger. But this measurement is not a true three-dimensional reconstruction since it does not take into account the natural curvature of the cortical surface from one coronal slice to another (see Figure 10.4). When a true three-dimensional reconstruction algorithm is applied to this region, its cortical surface area is not reliably asymmetrical. In a sample of ten brains, as many had a larger cortical surface area in the right as in the left hemisphere.

Many lines of anatomical and physiological research suggest that cortical areas within a species contain variable proportions of morphologically and neurochemically defined cell types. For example, primary and secondary visual, somatosensory, and auditory cortices express varying distributions of specific nerve fibers, and the density of certain nerve cells, called *chandelier cells,* differs between prefrontal and visual cortical regions (Lewis and Lund, 1990).

Cortical connectivity varies among species, which may reflect the organism's niche. The squirrel monkey and bush baby have differing connections of the interblob region in their visual cortices. In the bush baby, layer IIIB nonblobs receive input from lamina IV alpha, while in the squirrel monkey this layer receives input from lamina IV beta. The effect is altered inputs to lamina IIIB from magnocellular pathways in the bush baby, and altered inputs to parvocellular pathways in the squirrel monkey (Lachica et al., 1993).

Blob regions in these two species are connected almost identically. The significance of the species' difference in the nonblob regions of visual cortex is likely related to their activity patterns (Livingstone and Hubel, 1984): Bush babies are nocturnal and squirrel monkeys are diurnal. Bush baby layer IIIB receives input from layer IV alpha (the magnocellular stream), while squirrel monkey layer IIIB receives input from layer IV beta (the parvocellular stream).

Fascinating clues have emerged from work on human brain tissue. For example, the physiological properties of dendritic spines in the human might differ from those in other animals. Gordon Shepherd and his colleagues (1989) at Yale University studied presumed normal cortical tissue removed from epileptic patients. Comparing the membrane and synaptic properties of human and rodent dentate granule cells, they noticed important variations. First, humans had less spike-frequency adaptation in comparison with rodents; second, feedback was inhibited in human tissue, while rodent tissue showed feedforward and feedback inhibition—consistent with Shepherd's neuronal modeling. This work suggests that by simply adding a few calcium channels to the dendritic spine, vastly complex computational capacities can result in the spines and lead to more information-processing capability. These suggestive study results are exciting, and may point to new ways of thinking about variations in neuronal physiology among species (Williamson et al., 1993).

Nonhuman primate and human visual systems also have different organizational properties. When comparing, for example, the anterior commissure between humans and other primates, it is easy to see how the species differ in neural organization. The anterior commissure is one of the neural connections between the two halves of the brain. It is the smaller of the two cortical connections, second in size to the huge corpus callosum. When this structure is left intact but the corpus callosum is sectioned, visual information easily transfers in monkeys but not in humans. Thus, species vary considerably in how they transfer visual information between hemispheres.

What is more, lesions to the human primary visual cortex render patients blind, whereas monkeys with similar lesions are capable of residual vision. When residual vision is discovered in a human, as with blindsight, it likely reflects incomplete damage to the primary visual cortex. When a monkey has residual vision, it reflects capacities of other secondary visual system processes.

Examples abound of system-level variations between primates and other lower animals, but less attention is paid to those between nonhuman primates and humans. Yet the preceding observations elucidate major differences in anatomical organization, even though the monkey's visual system and the human's visual system have virtually identical sensory capacities. Careful psychophysical measurement of acuity, color, and other parameters reveals identical sensitivities. In addition, at the level of anatomical processes, both have approximately 1.2 million retinal ganglion cells. Even though the gray matter volume of the primary visual cortex in humans is three times larger than it is in the *Macaca mulatta* and five times larger than that in owl monkeys, this cortical area has the same number of cells in rhesus monkeys and humans.

In grasping the differences between monkey and human behavior, one has to consider the variations between the neuronal organization of each visual system. Can these differences be understood in relation to the connectivity of major processing areas or to the level of synaptic function? We do not know.

Arguing about similarities between species has been criticized by many. Take the problem of intelligence. It is naive anthropomorphism to apply the concept of human intelligence to the behavior of animals. It is simply a fact that each species has developed behavioral capabilities that are advantageous to its own survival, and that each member of that species possesses these capacities. There have been many attempts to raise the intelligence of a rat by selective breeding. All have failed. A rat that might run a maze better turns out to be lousy at discrimination learning. Our human brains are larger because they have more devices for solving problems, and the devices are shared by all members of the species. It is not likely that the variations seen in our own species's capacity to solve problems will vary with brain size; recent direct measures have shown there is no correlation.

Even though brain size cannot explain the unique capacities of the human, Noam Chomsky (1957) favored the view that, although language is deeply biological in nature, it is not a product of natural selection. Chomsky left open the possibility that language is the result, the concomitant, of massive interactions of millions of neurons. So, in the heart of the great Chomsky, the one who argued deeply for the biological basis of language, there lingers the idea that bigger is better.

Steven Pinker (1997a) challenged this bit of backsliding by Chomsky. Cranking up his unusually insightful and lively style, Pinker chided his colleague: "If Chomsky maintains that grammar shows signs of complex design, but is skeptical that natural selection manufactured it, what alternative does he have in mind? What he repeatedly mentions is physical law. Just as the flying fish is compelled to return to the water and calcium-filled bones are compelled to be white, human brains might, for all we know, be compelled to contain circuits for Universal Grammar."

Chomsky wrote (see Pinker, 1994):

These skills [e.g., learning a grammar] may well have arisen as a concomitant of structural properties of the brain that developed for other reasons. Suppose that there was selection for bigger brains, more cortical surface, hemispheric specialization for analytic processing, or many other structural properties that can be imagined. The brain that evolved might well have all sorts of special properties that are not individually selected; there would be no miracle in this—only the normal workings of evolution. We have no idea, at present, how physical laws apply when 1010 neurons are placed in an object the size of a basketball, under the special conditions that arose during human evolution. We may have no idea—just as we do not know how physical laws apply under the special conditions of hurricanes sweeping through junkyards—but it seems unlikely that an undiscovered corollary of the laws of physics causes human-size and -shaped brains to develop the circuitry for Universal Grammar.

Neuroscientists have had a hard time accepting the view that big brains may be a by-product of other processes for establishing the uniqueness of each species's nervous system. Yet biologists have known for years how specialized circuits define differences between fish and reptile, reptile and mammal, snail and octopus, worm and jellyfish. It is only logical that this information would help to define the neural processes supporting unique human capacities, especially language, and that big brains (corrected for body size) may get bigger because they collect more specialized circuits.

SUMMARY

The lesson of this chapter is simple: Complex capacities like language and social behavior are not constructs that arise out of our brain simply because it is bigger than a chimpanzee's brain. No, these capacities reflect specialized devices that natural selection built into our brains through blind trial and error. Mutations create variations in capacities. If the variations produce a slightly unique state of affairs that helps our brains make better decisions about enhancing reproductive success, the new capacities will survive. Variations that further enhance the capacity in question will also survive. An eye was not built in a day. It started as something that worked a little bit, which was better than something that did not work at all. As it evolved, the visual system became the finely tuned device it is now. So, too, with language and other mental abilities. The positive-feedback mechanism of natural selection, not experience, builds complexity into organisms.

The structure of animal bodies and brains can vary dramatically, and trying to deduce similar features of brain organization and unique features is difficult. Therefore, the strategy comparative neuroscientists use is crucial. Bullock proposed three main aims for comparative neuroscience: roots, rules, and relevance. Roots refer to the evolutionary history of the brain and behavior. How are brains similar and different? What has evolution produced? Rules of change are the mechanisms that give rise to changes in the nervous systems in the course of evolution. Are there constraints under which evolving nervous systems develop? Relevance refers to the general principles of organization and functions that can be extrapolated from a particular animal studied to all animals, including humans.

Finally, a multitude of commonalities connect all species and lend strength to much of biological research. At the same time, species exhibit crucial differences, such as those reviewed here, and human brain research has uncovered unique aspects of human behavior that may be supported by specialized neural circuitry.

KEY TERMS

adaptation	exaptation	homology	plesiomorphic
alleles	fitness	homoplasy	principle of parsimony
chromosome	genes	mutation	recombination
comparative neuroscience	genetic pleiotropy	natural selection	sociobiology
convergent evolution	genetic specificity	neuroethology	traits
ethology	genotype	out-group comparison	
evolutionary psychology	heritable	phenotype	

THOUGHT QUESTIONS

1. A hypothetical cortical area, DF, has been newly discovered in macaque monkeys. The investigators who first described it hypothesize that it is involved in tactile object discrimination and propose that it might be present in humans as well. How can they test this hypothesis without performing experiments using fMRI or PET?

2. What are the three major aims of comparative neuroscience? Briefly explain each.

3. Why is the study of convergent evolution of brain structures or fields important?

4. An area MT has been described in humans. This region does not reside at the tip of the superior temporal sulcus as it does in all other primates that have been investigated. Further, the cortical architecture (how it looks in neural tissue that has been histologically processed) is quite different from that in other primates. Although it becomes active in response to moving stimuli, so do a number of other areas of human neocortex. Is this field homologous to the area MT described in other primates? Why or why not?

5. Should one look for a genetic explanation for all behaviors? Explain your answer.

SUGGESTED READINGS

BULLOCK, T.H. (1984). Comparative neuroscience holds promise for a quiet revolution. *Science* 225:473–478.

JACOB, F. (1977). Evolution and tinkering. *Science* 196:1161–1166.

MERZENICH, M.M., and KAAS, J.H. (1980). *Principles of Organization of Sensory-Perceptual Systems of Mammals.* New York: Academic Press.

PINKER, S. (1997). *How the Mind Works.* New York: W.W. Norton.

WILSON, E.O. (1975). *Sociobiology, The New Synthesis.* Cambridge, MA: Belknap Press of Harvard University Press.

15
Development and Plasticity

In a set of seemingly mundane experiments, Professor Vilayanur Ramachandran (1993) at the University of California at San Diego (Figure 15.1) and his colleagues studied the sensory abilities of a healthy young man. While the man had his eyes closed, Ramachandran stroked parts of his body with a Q-tip and asked the man to say where he was being touched. Ramachandran touched him and the young man said, "Left finger." He touched him in another spot and the reply was, "Left thumb." In response to additional touches in various locations, the young man reported sensations in different areas of the left hand and arm. What is so interesting about these reported sensations is that despite the claims by the man that he felt the good doctor touching his left arm, the man in fact *had no left arm*—it was lost in an accident some time before. What is even more amazing is that Ramachandran had been stroking parts of the young amputee's cheek! When his face was stroked by the Q-tip, the young man responded that he felt sensation in parts of his amputated left arm and hand (Figure 15.2). With patients of this type, Ramachandran uncovered a remarkable example of plasticity in the human somatosensory system. The precise mechanisms and the details of this story are presented later in this chapter. For now, we will provide a hint for solving Ramachandran's mysterious findings. Consider the mapping of body regions onto the somatosensory cortex in the human shown in Figure 15.3—the answer lies here.

THE SHAPING OF THE BRAIN

The fertilization of a single egg leads to the creation of an entire human being. This developmental process remains largely mysterious, but one characteristic that dominates is change. Between conception and birth, a complete human infant is constructed by the myriad of biological mechanisms specified by the genetic code, and yet it is far from being a finished product, as any parent can tell you. During growth and maturation the young human changes significantly, and the most impressive changes may well be those related to cognition. A host of new abilities mature at an astounding rate over the first years of life. The development of these abilities can be affected by experience during **critical periods,** specific time ranges early in life when experience can maximally influence the organization and function of the nervous system. Critical periods for behavioral development likely correspond to specific stages of neural development such as the formation of synaptic connections or the presence and activity of molecular cues for specifying neuronal connectivity. But even after critical periods in development pass, some forms of neural plasticity remain. **Plasticity** refers to the ability of the nervous system to change in subtle, and sometimes not so subtle, ways. One tantalizing example

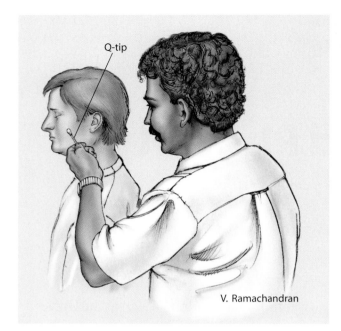

Figure 15.1 An example of mapping sensation. Professor Ramachandran of the University of California at San Diego has tested the ability of persons to discriminate sensory stimuli. He performs these tests using a simple stroke of a cotton swab and asks the person to report where he feels the touch. During strokes to the cheek of one man who had his eyes closed, the man reported the sensation of being touched on the arm.

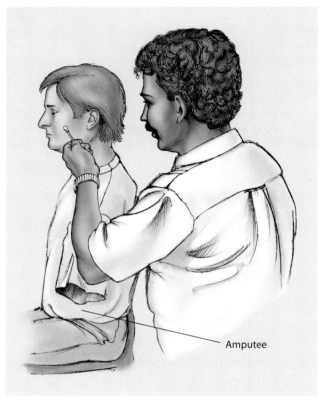

Figure 15.2 This is the same scenario as in Figure 15.1, but with a surprising twist. Stroking his cheek led the man to report sensation in his arm. However, the man lost his arm in an accident some time earlier. This amazing demonstration is one that involves phantom limb sensation and yet something else as well. Why would touching the man's cheek lead him to feel touch to his missing arm?

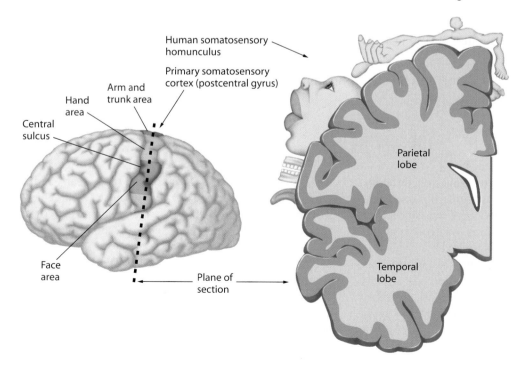

Figure 15.3 A human somatosensory homunculus. The representation of the body surface is mapped in a somatotopic fashion onto the somatosensory cortex of the parietal lobe. At the right is a coronal view of the left hemisphere showing the approximate location and relative cortical area (as indicated by distorted sizes of various body parts) dedicated to body parts. The anterior-posterior plane of the coronal section is indicated at the left. Of special note for the story told in Figures 15.1 and 15.2 is that the face representation and the hand representation are located in adjacent cortical areas of somatosensory cortex. Adapted from Kandel et al. (1991).

was presented in the story of Professor Ramachandran and the amputee, but less dramatic examples can be found in everyday life. Consider our continued ability to learn new things and remember them, which tells us that even the adult brain is plastic to some extent. One of the greatest challenges facing cognitive neuroscience is to understand neural plasticity and its effect on perceptual and cognitive processes in the developing, adult, and diseased nervous system.

In this chapter we will look at plasticity from different perspectives. We will begin by looking at the plasticity that accompanies normal development, especially with respect to the theories of perceptual and cognitive development that have emerged over the past 50 years. In so doing, we will examine how changes in the developing brain might enable cognition. Then we will take a tour through the biological processes involved in the development of the nervous system at the cellular level, finally coming back to a discussion of plasticity in the adult nervous system, and perhaps an answer to the mystery presented by Ramachandran's work.

PERCEPTUAL AND COGNITIVE DEVELOPMENT

Knowledge about the development of mental abilities is essential for understanding the organization and function of the adult mind. This investigation can be undertaken in humans and animals. In humans, one approach is to correlate the maturation of a specific cognitive function with a particular stage of neural development. If a behavior like, for example, numerical ability appears at a certain postnatal time, brain structures that become functional (e.g., by way of myelination of input and output axons) at that same time might be implicated in that behavior. In this way, the analysis of development enlightens us about relations between specific brain circuits, structures, and systems and the cognitive processes they support. But it is also possible to go beyond simple correlational inferences. Research on the origins and development of cognitive phenomena can inform us about adult cognition by helping us to choose between competing models. For example, knowing that reflexive orienting of attention develops prior to voluntary attentional orienting informs us about the possible mechanistic relationships between these two related but also different forms of attentive behavior. From a different perspective, developmental science can elucidate the role of experience in shaping the mind and brain. For example, balanced input (or lack thereof) from both eyes is necessary for the normal development of ocular dominance columns in primary visual cortex (see Chapter 5). If input from one eye is disrupted during the critical period for visual system development, column formation is also disrupted. However, column formation is not disrupted if input to both eyes is similarly disrupted. These findings tell us that there is much prespecification of the wiring of ocular dominance columns (see Experience and Activity in Development of Visual Cortex). The different developmental approaches are being applied to investigate perception, attention, memory, language, motor skills, and higher mental processes like numerical cognition.

The difference between the capabilities of newborns and those of adults are apparent to anyone taking the time to look. Newborns do not walk, hold objects, speak, or understand us when we speak to them. These differences could be explained in at least two ways. Newborns might have all the capabilities of adults but have not yet attained—via experience—the abilities of adults. In contrast, newborns may differ radically from adults in neural or cognitive capabilities, or both. The former view amounts to newborns possessing full-formed neural circuitry that simply awaits input in order for development to occur. The latter suggests that newborns do not yet possess the neural and cognitive structures necessary to perform as adults do and that development will involve radical, qualitative change. This view has dominated theories of development based on both neural and psychological evidence. One classic view that newborns differ significantly from adults comes from the Swiss scientist Jean Piaget (Figure 15.4). Let's begin by reviewing Piaget's theory, and then continue by contrasting his views with emerging evidence from a variety of development studies that have significantly modified **Piagetian theory**, and hence, how we think about the development of mind.

A Classic Theory of Cognitive Development

Jean Piaget is considered by many to be the father of modern developmental psychology. Piaget believed, and developed ingenious tasks to test, that human infants and children perceive and comprehend the world

Experience and Activity in the Development of the Visual Cortex

The development of the nervous system is under the control of genetic instructions that wire the brain in intricate splendor. In addition, experience can modify the functional organization of the nervous system. In a series of now classic studies, David Hubel and Torsten Wiesel (1977) and their colleagues demonstrated that the organization of the ocular dominance columns in the primary visual cortex of cats and monkeys can be altered by preventing vision through one eye; that is, by sewing the eyelid shut after birth. As described in Chapter 5, ocular dominance columns represent alternating groups of cells in the primary visual cortex (striate cortex or V1) that respond preferentially to stimulation of one or the other eye. When vision from one eye is prevented, systematic changes in ocular dominance columns are observed in the visual cortex. These changes are noted as an abnormal bias in the responsiveness of neurons for stimuli delivered to the open eye, as determined by single-cell recording methods (and backed up later by neuroanatomical staining methods). Stimuli presented to the previously closed eye do not elicit much activity in the primary visual cortex. This manipulation of ocular dominance in the primary visual cortex only occurs when the eyelid is closed within the critical period for visual system development, which in the cat is between the ages of 3 weeks and about 3 months and in the primate is from birth until about the age of 6 months. These findings demonstrate that the nervous system is highly plastic early in its development, during the critical periods (restricted windows of time early in development).

This raises the important question of how experience alters the functional organization in the nervous system.

Current theories hold that neuronal activity drives the fine-tuning of nervous system organization. For example, the neural projections from the two eyes through the lateral geniculate nucleus of the thalamus and on to the primary visual cortex are thought to overlap much more at birth than later when the system is fully developed, and activity in the two eyes may help to refine the projections into the ocular dominance columns. Even when no visual stimuli reach the eyes, however, ocular dominance columns form, at first leading to the suggestion that activity is not essential. Later studies, though, showed that waves of intrinsic neuronal activity (not driven from the outside by stimulation) might provide a mechanism for activity-dependent control over the formation of ocular dominance columns. These *retinal waves* send periodic bursts of action potentials to the thalamus during development. Recently, Justin Crowley and Larry Katz (1999) and their colleagues at Duke University challenged this view. They used anatomical tracing methods to show that ocular dominance columns can develop in the ferret even when the eyes are removed at birth. Since ocular dominance columns form after birth in the ferret, these findings suggest that retinal activity is not required to form ocular dominance columns. An alternative mechanism would be that genetically prespecified molecular cues are responsible for guiding thalamic axons to project in alternating ocular dominance patterns to layer IV of the primary visual cortex.

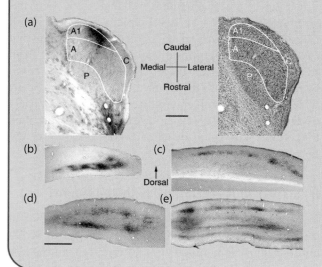

Ocular dominance columns in normal ferrets and ferrets whose eyes were removed. **(a)** Horizontal section through the lateral geniculate nucleus (LGN) following local injection (left side, dark region) of biotinylated dextran amine (BDA), an anterograde label that is taken up by LGN neurons and transported to visual cortex. A1 is the layer in LGN that recieves ipsilateral eye inputs. P stands for perigeniculate regions, and C for layer C of LGN. The right side of (a) shows the corresponding Nissl stain section of the LGN. **(b–d)** Coronal sections through the primary visual cortex of ferrets. Normal ocular dominance columns (alternating dark patches) are shown in (b). The pattern for an animal whose eyes were enucleated on postnatal day 18 (P18) is shown in (c), and cortex from two ferrets whose eyes were enucleated on postnatal day 0 (P0) is shown in (d) and **(e)**. All show ocular dominance columns. From Crowley and Katz (1999).

Figure 15.4 Swiss scientist Jean Piaget.

Table 15.1	The Divisions of Cognitive Development in Humans as Formulated by Piaget	
Stage	**Age**	**Characteristics**
Sensory-motor intelligence	0–2 years	Unconnected sensations, representational thought
Preoperational period	2–7 years	Conservation of quantity and number, ego centrism
Concrete operations	7–11 years	Concrete concepts/ no abstract thinking
Formal operations	11 years and older	Development of abstract thought

differently than adults do. He was a keen observer of infant and child behavior and took wonderfully detailed notes of his observations. In addition to being a skilled observer, he performed well-controlled experiments to test new hypotheses derived from those observations. His ideas were based on a biological view of development and thus differed from prior behaviorally based theories. On the basis of his observations he characterized the cognitive development of humans in a four-stage model, that as we will learn, has been and is undergoing rapid modification (Table 15.1).

Piaget termed the period from birth until the age of about 2 years the *sensory-motor intelligence stage.* He conceived of the newborn as a work in progress and held that the newborn experienced a disconnected welter of ill-formed sensory percepts and generated only random motor acts. Thus, according to Piaget, in this first stage of postnatal life, the developing nervous system aims to achieve sensory-motor integration and an integration across different sensory modalities (e.g., sound, touch, vision). Piaget also believed that the newborn could not form a concept of self that could distinguish between it and the outside world, and thus the development of self-identity had to begin during this period.

Piaget proposed that to achieve sensory-motor and cross-modality integration, infants develop sensory-motor schemas during the sensory-motor intelligence period. That is, they learn to relate sensory inputs with motor acts and thus can purposively grasp objects they are looking at. During this stage, Piaget argued, infants have poor concepts of objects in the world. Even when they are old enough to interact with objects, they do not exhibit abilities such as object permanence, according to Piaget's model. *Object permanence* is the ability to know that an object does not cease to exist merely because it is out of view. Obscuring an object from an infant during this period will at first lead the infant to ignore it. Later the infant may learn to look for it but fail to integrate new information about where it might be. For example, in repeated trials, if an investigator hides a toy from a child in her plain view, she will explore the hiding place to retrieve the toy. If the same hiding place is used over consecutive trials but then a new hiding place is used, she will continue to search the original, well-practiced location, even though she watched the toy being hidden in the new hiding place (Figure 15.5). As the child ages, this perseverative behavior diminishes. Piaget proposed that success in tasks such as this marks the end of the sensory-motor intelligence stage and is the result of a newly developed ability to represent objects and events internally; that is, infants can think about objects and acts that are no longer within sight. Thus, infants are said to exhibit object permanence when they no longer have difficulty conceptualizing the presence of an unseen object.

Many investigators have challenged Piaget's concept of the limited nature of a newborn's capabilities in the realms of sensory-motor integration, cross-modal integration, and object perception. The nature

Development and Evolution of Numerical Ability

Piaget argued that children do not possess quantitative skills until the concrete operational stage, beginning around the age of 7 years. But more recent research has shown that very young infants may be able to distinguish between sets of items based on their number. For example, when infants are shown a series of diverse pictures that have the same number of elements (e.g., 2 fish, 2 hearts, 2 balls, 2 shoes), they rapidly get bored. Their boredom is reflected in a reduction in the time they spend looking at the pictures (habituation). If the infants then are shown a new picture that has the same number of elements (e.g., 2 cats) or a new number of elements (e.g., 3 cats), they typically look longer at the one with the new number of elements (dishabituation). Using this method, researchers have shown that infants can tell the difference between pictures based solely on the number of items present. But what is the relationship between these numerical abilities and adult numerical knowledge?

An important component of a numerical concept is ordinality. *Ordinality* is the ability to recognize the greater-than and less-than relationships between numerical values. Recent research by Elizabeth Brannon at Duke University and Gretchen van de Walle at Rutgers University (in press) showed that toddlers as young as 2 years recognize the ordinal relationships between numerical values as large as 4 and 5. In their study, children were shown two trays, each containing a number of boxes. Initially the children were shown trays with 1 or 2 boxes and were shown (by demonstration) that the tray with 2 boxes was always the "winning" tray because there was a sticker (with a picture on it) beneath each box on that tray (Figure A). The stickers served as a reward because the children like them and want to find them. The children were then presented with two trays that contained larger numbers of boxes (e.g., 3 versus 4, and 4 versus 5) and were asked to point to the winning tray. The children reliably chose the tray with the larger number of boxes. They chose the larger number even when the boxes from the larger set were smaller in physical size than the boxes from the smaller set! At this age, however, the same children cannot verbally label sets of 3, 4, and 5 objects or reliably count out that number of toys. Brannon and Van de Walle's research shows that children understand the ordinal relations between numerical values that they have not yet learned to verbally label. This finding suggests that fully functional language ability is not necessary to support numerical thinking, raising the possibility that animals might possess numerical abilities too.

In related work, Brannon, with Herb Terrace at Columbia University (1998), investigated this issue in a series of studies in rhesus monkeys. The monkeys were trained to point to pictures of items representing the numerical values 1, 2, 3, and 4 in that order (Figure B). Once the monkeys learned to order the pictures according to their numerical values, 1 through 4, the researchers then tested the monkeys' ability to order new pictures representing different numerical values (from the range 5 through 9). The monkeys were up to the challenge! They reliably pointed to the set of elements representing the smaller numerical value first, even when the elements representing the smaller number were physically larger than the elements representing the larger number. These and related findings tell us that monkeys know the relationships between numerical values, even when they have not been trained with the specific values previously. By demonstrating these skills in nonverbal animals, these researchers showed that language is not necessary for the development of ordinal numerical abilities.

of the challenge has to do with how quickly after birth an infant displays a particular ability. Piaget's critics have argued that the newborn has some form of integration of sensory experiences across the modalities of sight, sound, and touch. For example, newborn infants, when provided with adequate head support, can

Figure A   Toddler pointing to the tray containing 2 boxes, in order to retrieve two reward stickers under the boxes. Children as young as 2 years can distinguish sets of more boxes (larger numbers) from sets of fewer boxes even when the boxes in the larger set are actually smaller than those in the smaller set.
Figure B   Monkey looking at pictures on a screen and pointing to sets of 1, 2, 3, and 4 items in ascending value.

Figure 15.5   Testing a child's knowledge of the world—object permanence. **(a)** A baby 2 years old or less is playing ball with an investigator. Psychologist Adele Diamond (1991) showed that if the investigator takes the ball and places it behind a screen in full view of the baby, the baby might ignore it. As the infant ages, however, he will eventually come to understand that the ball is still there but is merely occluded, and may reach for it at the hidden location. This is the concept of object permanence. **(b)** However, if the baby is shown the ball being hidden in the same location over repeated trials, but is then shown the ball being hidden at a new location during this period, the baby may look at the new location but continue to reach to the old hiding place. This perseverative reaching behavior changes as the infant ages, and is replaced by normal looking and reaching to the new hiding place.

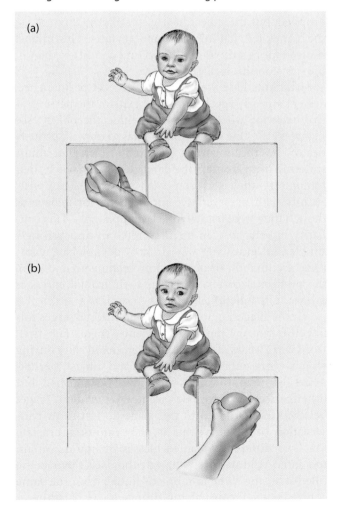

visually track sounds moving across space. This suggests a well-developed skill at cross-modal visual and auditory integration, and the ability to link motor actions with cross-modal perceptions. In line with this idea, studies have shown that infants only a few months old prefer to look at films of people speaking with the soundtrack in proper synchronization with mouth movements than at identical films with the speakers' voices out of synchronization with mouth movements.

Young infants, only a few months old, also evince knowledge of real-world objects. They perceive partially occluded objects normally and also can represent the relation between objects in three-dimensional space. René Baillargeon (1991) demonstrated this in an object occlusion task (see also the later section on perceptual development). She showed infants an object and then placed it behind a vertical panel that occluded their view. The panel was then dropped under two conditions. In one condition the panel was dropped and hit the object placed behind it, as would be expected. In the other condition the panel was dropped, but the object had been secretly removed, so the panel fell flat to the table surface. The infants showed more surprise in the second condition than in the first condition (Figure 15.6).

If infants have well-developed object permanence, even at an early age, how can we explain the perseverative behavior when investigators hide the object (see Figure 15.5)? One interpretation has to do with properties of the frontal cortex. It is well known that adults suffering from frontal lobe damage cannot switch their motor set—they persevere with a previous response. Infants with perseverative motor behavior behave as though they were frontal lobe lesion patients. This similarity in behavior can be interpreted in a surprisingly simple and gratifying way: Infants do not have complete myelination of neuronal projections to and from the prefrontal cortex, and thus their frontal cortex is not yet fully functional.

In the Piagetian model, three additional stages follow the sensory-motor intelligence stage. The first, from 2 to 7 years old, is the *preoperational stage* during which representational thought and object permanence are hypothesized to be well established but other conceptual processes are not yet evident. Piaget believed that children in this stage do not show conservation of quantity; that is, they cannot appreciate that two differently shaped glasses of liquid contain the same volume, even though they see them being filled with the same amount of liquid from the same source. Thus, the visual appearance of a taller thinner glass versus a shorter fatter glass dominates the chil-

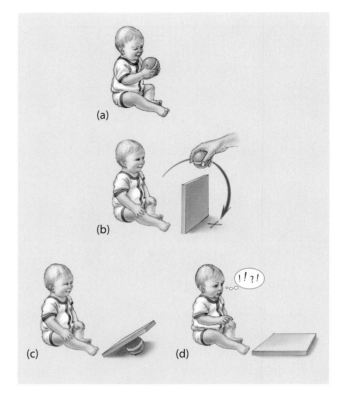

Figure 15.6 Physical reasoning in infants. Psychologist René Baillargeon (1991) tested infants' knowledge about the relationships between physical objects in the world. Babies 4½ to 6½ months old were tested using the "looking time" measure—babies look longer at surprising events. **(a–b)** To test babies' knowledge of physical objects, they are shown an object being placed behind an occluding screen. **(c)** Then the screen is allowed to fall onto where the object is placed. In one condition, the screen strikes the hidden object. **(d)** In a second condition the object is secretly removed, and the screen falls to the floor. Babies look longer at and are perplexed by the condition that violates the physical laws they believe to be correct. That is, the screen should have hit the object and stopped. This shows that babies have object permanence and the ability to compute the size and relative locations of objects in the world, even when some are out of sight.

dren's decisions about quantity; they believe that the taller glass holds more liquid than the shorter fatter glass, even though they actually have the same volume. Piaget proposed that a similar effect happens with numbers of objects. It is not until near the end of this stage, at about 7 years old, according to Piagetian theory, that children learn these abstract concepts and are rarely fooled if given all the information needed to make the correct decision.

From 7 to 11 years old, Piaget held that children become capable of some forms of quantitative conceptual thinking. He argued, however, that during this period they initially can perform quantitative operations only on concrete items or events; that is, they fail to make abstract inferences. Piaget called this period the stage of *concrete operations.* Then from 11 years onward, during the stage of *formal operations,* children learn to make abstract representations of relationships, according to Piaget. Children at this age can generalize mathematical relationships and manifest hypothetical-deductive thought—the ability to generate and test hypotheses about the world.

New research challenges Piaget's conceptualization about these last three stages. Infants show evidence of a rudimentary sense of number, or amount, very early in life. They can detect the difference between two of something and three of something. Thus, infants appear to be sensitive to the concepts of "more" and "less" (see Development and Evolution of Numerical Ability).

Piaget's contribution was to chart the timeline of cognitive development and to attempt to show when infants and children are able to perform complex perceptual, motor, and cognitive tasks. The fact that the precise age when a particular process is manifest may be earlier than what Piaget proposed, or that the discrete stages of Piaget may be more gradual than what is suggested in Table 15.1, does not detract significantly from his concept of cognitive development. Indeed, it is the job of science to modify theories with new concepts based on fresh data. In perceptual and cognitive development, recent findings suggest greater abilities in newborns and young infants than Piaget had proposed. Still, a timeline of cognitive maturation is, with proper modification, useful because one goal of cognitive neuroscience is to relate the timeline of cognitive development to neural development, to elucidate the biological bases of cognition. In the following sections, we will first look at the development of higher-level perceptual abilities, especially object perception and representation. Then we will consider how attentional mechanisms come on line as the infant develops. Finally, in a section on language acquisition we will ask how much experience is necessary to acquire a skill such as human language.

Development of Visual Cognition—Object Recognition

Like Piaget, William James (see Chapter 1), one of the founding fathers of the field of psychology, speculated that newborn infants' perceptual experiences amounted to one great "blooming, buzzing confusion." Over the past 25 years, researchers have devised methods to investigate whether infants' perceptual experience is indeed as confused as James and Piaget believed or whether it is more organized. These methods involve capitalizing on and quantifying overt behaviors in which babies engage naturally.

You might think that young infants do not do anything that is systematic enough for scientific study, but if you think that, you are wrong! Extensive research has made use of babies' **looking time**—that is, how long they look at a stimulus. It turns out that there are very predictable aspects of infants' looking behavior that render measures of looking time as powerful tools in developmental science. For example, infants' looking time (like many aspects of human and nonhuman animal behavior) habituates; it decreases over time when the same stimulus is presented repeatedly, and then shows a notable increase (dishabituates) when infants notice a change in the stimulus. Moreover, infants tend to look longer at events that are unexpected than at expected events.

These characteristics of looking behavior have been used to explore various aspects of infants' perceptual and cognitive functioning. For example, researchers have been interested in how early after birth infants organize visual information into objects. Philip Kellman and Elizabeth Spelke (1983), then at the University of Pennsylvania, systematically studied this question. They observed that by the age of 4 months, infants used the common motions of objects to group them together into the same unit, and the spatial separations between objects to group them into separate units. Further research by Alan Slater and his colleagues (1990) in Exeter, England, showed that although 4-month-old infants do use common motion to help them unify two portions of an object, newborn infants do not. Thus, the ability to use common motion to organize a display into objects must develop between birth and the age of 4 months, and most researchers believe this ability may begin to develop around the age of 2 months.

More recent research has investigated other kinds of information that infants might use to determine the objects in a display. Amy Needham (1999) at Duke University studied infants and adjacent objects, objects that are touching each other but are fully visible to the observer. Her work showed that infants as young as 4 ½ months can use differences in object shape to determine that two adjacent objects are actually separate pieces that would move separately if pulled (Figure 15.7). These findings show that infants can make sense of complicated scenes of objects even when there are no spaces between the objects and even

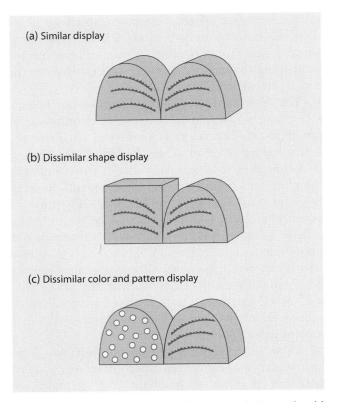

(a) Similar display

(b) Dissimilar shape display

(c) Dissimilar color and pattern display

Figure 15.7 Test showing that infants around 4½ months old can use object shape to segregate objects. **(a)** When objects are touching and are of similar shapes, infants at this age do not recognize them as distinguishable. **(b)** They do recognize them as distinguishable if the objects are of dissimilar shapes. **(c)** At 4½ months, infants cannot use color to distinguish objects; this becomes possible for the infant between 5 and 9 months old. Courtesy of Amy Needham.

when the objects are not moving. Needham's work also showed that other factors, such as the prior experience an infant has had with an object or collection of objects, and the physical relations between the objects in a display, play an important role in an infant's segmentation of a display into its component parts. This research suggests that infants' visual world is richer and much better organized than any of us (including Jean Piaget or William James) may have suspected.

Development of the Human Attention System

The visual perception system shows dramatic developmental changes early in life, as demonstrated by the acquisition of the ability to recognize objects in the world. What changes in the developing brain support the acquisition of such abilities? Primate visual systems have been investigated extensively over the past 40

years. We now know more about the visual system than perhaps any other part of the brain. Yet, we cannot answer the question of what changes in the brain enable the development of object perception at present. We do, however, have a relatively clear picture of the developmental time course of some visuomotor processes. The structures and systems of relevance to the oculomotor system are well mapped, and as a result, observations of how oculomotor behavior develops have enlightened us about the neural substrates of a key cognitive mechanism—attentional orienting.

The idea is simple, as Mark Johnson (1993) of University College, London, clearly articulated. The neural circuits supporting visuomotor behaviors become functional at varying postnatal times. An overt oculomotor behavior that develops at the same time as a specific circuit must be subserved by that circuit. Let's first review the time course of visual system development.

MATURATION OF SUBCORTICAL VISUAL CIRCUITS

The foveal region in the human retina is immature at birth, whereas the peripheral retina is more developed. Hence, newborn vision is driven predominantly by peripheral inputs. In a similar way, the optic nerve is not myelinated completely in the newborn; however, myelination in the optic nerve occurs rapidly during the first 4 months of life and reaches adult patterns at about the age of 2 years. The lateral geniculate nucleus of the thalamus—the main relay from the retina to the cortex—also experiences rapid growth in the first 6 months after birth, almost doubling in volume. The time of maturation of the primary visual cortex varies with respect to different cortical layers. The deep layers of cortex, which project to the superior colliculus, develop earlier than the superficial layers. The superior colliculus is the primary subcortical target of retinal ganglion cells and is the structure involved in saccadic eye movements and involuntary oculomotor movements toward salient events in the environment. The superior colliculus has a virtually normal pattern of neuronal lamination even before birth, and the fibers of the retino-collicular projection are partially myelinated prenatally and completely myelinated by 3 months postnatally. Thus, the first neural circuit supporting visual behavior to develop and to become myelinated is the subcortical projection of the retina to the superior colliculus oculomotor system (Figure 15.8a).

The visuomotor behaviors of newborns have been well characterized and have the following patterns: Newborns track moving objects but do not yet display smooth pursuit eye movements (smooth eye move-

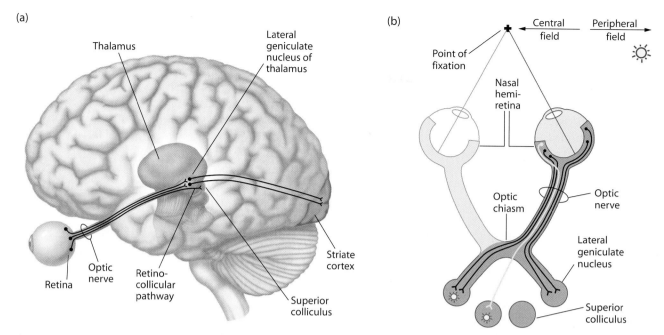

Figure 15.8 **(a)** Diagram of the visual system showing the separate retino-geniculostriate pathway that carries visual information to cortical visual processing areas, and the retino-collicular pathway that projects from the retina to the midbrain superior colliculus, part of the visuomotor system. **(b)** Infants show a greater tendency to orient to stimuli in the temporal (peripheral) visual field of each eye. This may be the result of a preferential projection of retinal ganglion cells from the nasal hemiretina to the superior colliculus. Ganglion cells in the nasal hemiretina are stimulated by stimuli presented in the temporal hemifield.

ment when tracking a moving object); rather, they use a saccadic pattern of eye movements (small jumps in fixation from point to point) to follow a slowly moving object. Stimuli presented to their temporal visual field (stimuli on the right for the right eye, and stimuli on the left for the left eye) are more likely to elicit overt orienting of the head and eyes. This occurs because stimuli in the temporal hemifields impinge on portions of the retina that project more strongly to the superior colliculus and thus are more likely to induce oculomotor orienting (Figure 15.8b). Newborns also tend to ignore the internal features of complex stimuli, owing to their poor acuity, partly because the fovea lags behind in development but also because processing at higher stages of the visual system is limited. One could say, then, that newborn visual behavior is driven primarily by the subcortical visuomotor system. How does this biological fact influence the development of attentive behavior?

When infants reach the age of 1 month, there is a change in their overt visual orienting to stimuli. They often manifest obligatory overt attention, during which they fixate their eyes on objects for long periods. Coincident with this is the development of projections from striate cortex to subcortical structures that inhibit activity in the superior colliculus. The state where periph-

eral stimuli could trigger automatic overt orienting is replaced by one in which the superior colliculus is less able to effect saccadic eye movements, and hence 1-month-old infants fixate and become locked onto a stimulus event.

By 2 months, infants develop smooth pursuit tracking of moving objects with their eyes. They also start to show normal orienting to new stimuli presented in the visual field. Moreover, at this age infants also begin to attend to the internal features of complex stimuli. The pattern in smooth pursuit may result from coincident development and maturation of the striate cortex projection to the middle temporal (MT) motion areas of the extrastriate cortex. The MT pathway is essential for smooth pursuit. As well, paying more attention to the internal features of stimuli likely results from improving acuity within the visual field's central regions.

Between the ages of 3 and 6 months, infants begin to make anticipatory eye movements. This development is likely the result of maturation of the projection from upper layers of striate cortex to the frontal eye fields, which are the last to develop. The frontal eye fields, regions in dorsal-lateral frontal cortex, participate in voluntary eye movements; therefore, frontal eye field maturation fits closely in time with the onset of

HOW THE BRAIN WORKS

Development of Face Recognition

Face processing, like language, is an exquisitely developed skill in humans that has its origins in the first days of life. Newborn babies seem to like looking at faces or face-like stimuli more than they like looking at many other interesting stimuli such as bull's-eyes and checkerboards. Infants just a few weeks old can distinguish their mother's face from other women's faces, but when doing so, they rely principally on global aspects of the face such as the shape of the hairline. They rely on global aspects primarily because their visual acuity is poor at birth and does not approach adult capability until the age of 3 to 4 months. Once normal acuity develops, babies begin to recognize and distinguish faces on the basis of their features (eyes, nose, mouth, etc.). A hallmark of mature face processing is the inversion effect, whereby recognition of faces presented upside down is markedly slow and difficult in comparison to faces presented right side up. This inversion effect does not occur until the early school years, suggesting that increasing skill and experience in recognizing faces create the effect. Event-related potentials recorded in response to faces suggest that face processing does not fully mature until puberty. Thus, although precocial skills are seen in newborns, face processing requires a great deal of time and experience before it is fully developed.

Mark Johnson (1997) proposed a neural theory to explain the development of face processing in two "stages." In the early stage, subcortical visual pathways are fully functional in the neonate, and these pathways are predisposed toward processing face-like stimuli. It is because of this predisposition that babies spend a lot of time fixating on faces. The time spent fixating on faces produces a great deal of activity in the subcortical pathways, which in turn helps to shape the development of the cortical structures to which they send neural projections. Over time, this looking behavior and correlated neural activity set up the cortical regions that are known to be involved in face processing in adults. Thus, Johnson proposed that the development of the face processing system involves simple but profound interactions between elementary predispositions in subcortical visual pathways, the behavioral result of these predispositions, and the consequences of the behavior for cortical development. Based on available developmental data, the early precortical stage is relatively short-lived in comparison to the later cortical stage, which appears to be quite protracted. The question of when cortical areas such as the fusiform gyrus, in the posterior-ventral cortex, display face-specific activity is still unanswered!

A toddler gazing intently at a photograph containing a face. Evidence suggests that face processing develops early and in two stages, each correlated with the development and maturation of specific circuits in the visual system.

more controlled oculomotor programs. Also during this period infants develop fully normal acuity as well as binocular vision, in part because of maturation of striate cortex and experience-dependent tuning of ocular dominance columns. The acquisition of normal acuity has behavioral consequences, particularly for face processing (see Development of Face Recognition). In summarizing his data, Mark Johnson made three proposals about visuomotor behaviors:

1. The behavioral and neural sequence described holds for a normal infant.
2. Specific neural events coincide with or just precede observed behavioral changes.
3. The main limitation to the development of visuomotor behaviors is the delay in development of the primary visual cortex in comparison to the subcortical visuomotor system.

What is the relation between overt signs of visuomotor behavior and covert attention? Michael Posner and Mary Rothbart (1980) proposed that covert attention should parallel the development of overt attention; they hypothesized that covert attention is involved in overt eye movements to locations in the visual field (see Chapter 7). According to their model, covert attention is thought to precede eye movements and may help to program the oculomotor system to make the correct eye movement.

The effort to correlate the timeline of neural development to cognitive development is sometimes highly successful, as we have just learned from studies of the development of visual attention. But in many cases, it is far more challenging. Take the case of language, where the neural substrates are less well defined in the adult in comparison to our knowledge of the visual system. As a result, the developmental changes in the human brain that parallel the development of language abilities are not very well understood, although progress is being made using modern cognitive neuroscience tools; some of this work will be described in the next section.

Another way that investigations of cognitive development can enlighten us about biological mechanisms is to study the issue of whether a cognitive ability is acquired by learning or is biologically predetermined (**innate**). Or perhaps more generally we can ask, How much experience is necessary for a cognitive faculty to develop? Addressing this type of question will probe the biology of cognitive processes but will not necessarily directly elucidate the neural circuitry involved.

Nonetheless, whether an ability such as language is learned by a general-purpose cognitive-brain system or is instead part of a neural circuitry predestined for communication is a fundamental issue that can be addressed by consideration of cognitive development. In the next section we will investigate the cognitive development of language from the perspective of whether language, as MIT psycholinguist Steven Pinker suggested, is an instinct.

Language Acquisition During Development

Language is unique to humans, as most of us would agree. Whether it is an innate or acquired ability is a question that scientists have struggled with for decades. Humans are not born with the ability to speak and understand a language; they must learn it through exposure—which means that language is a learned talent. But this does not provide the complete answer; we must still ask whether there is a learning mechanism for language, which is itself innate. We cannot say until we establish criteria for what qualifies as an innate ability.

Innate abilities should be present in all normal individuals in a population. The development of abilities should follow a common course among different individuals, and they should develop without overt training of one individual by another. By definition, there should be a genetic component to innate behaviors; thus, language-related diseases might be inherited. We might add other criteria, but the foregoing require the fewest assumptions. If the brain has a modular organization for cognition and sensation, and if this gross organization is similar across individuals, then an innate ability might be expressed in specific brain areas that are the same across individuals. But this characteristic is less crucial to the definition of *innateness* than the others. Overall, these defining characteristics of innate behavior are useful in evaluating the role of experience in the acquisition of cognitive abilities.

Is swimming, for instance, innate or learned? It appears to be innate for dogs (they swim the first time you place them in water) and other animals. Humans do not do a good job when they are thrown into water without prior experience with swimming, which usually requires some instruction. What about walking? This is slightly more difficult to intuit. Didn't our parents teach us to walk? They probably taught us a bit, but even without any instruction, humans eventually stand and walk—a normal characteristic that

Gestural Communication in Infants Before Speech

Babies do not talk until between the ages of 1½ and 2 years, and so they cannot communicate, right? Wrong! You do not need to talk to communicate, and neither do babies. Researchers have known for some time that gesturing is an important form of communication between the baby and the parents. In a comprehensive study, Linda Acredolo of the University of California at Davis and Susan Goodwyn of California State University at Stanislaus (1996) investigated preverbal communication in babies in a sample of 140 families. The babies were studied for 3 years beginning at the age of 11 months. One-third of the families were encouraged to try to teach their babies baby signs, and the rest were not.

From tests on these children, the researchers discovered that babies who were encouraged to sign outperformed the other babies in many comparisons of communication skills, including the verbal ones that all babies come to develop naturally. Much of the reason for this can be attributed to the positive effects that preverbal communication has on the parent-child relationship. Babies and parents who sign, communicate more and have more interactions. What are baby signs?

The following is a story from the research files of the baby sign team:

Even though she is too young to say more than a few words, 13-month-old Jennifer loves books. As her dad, Mark, settles on the couch after dinner, she toddles over. Holding her palms together face up, she opens and closes them. Mark's immediate, "Oh, OK. Let's get a book to read," satisfies her and she soon returns with her favorite animal book, cuddles up close, and begins turning the pages. With delight she looks at a picture, scrapes her fingers across her chest, and looks up with a broad smile at Mark. "You're right. That's a zebra!" (Acredolo and Goodwyn, 1996).

The scene continues with Jennifer identifying animals and what the animals are doing without uttering a word. What has occurred? According to Acredolo and Goodwyn, Jennifer has just told her father what was in the book and he understood. Jennifer was using simple nonverbal gestures, or baby signs. What these data show is that the need and desire to communicate arises in babies before they have sufficient control of their vocal apparatus to speak for speech. Their relatively well-developed manual skills come in handy, indeed, as their communicative needs arise earlier in life.

Infant children signing before they can verbally communicate their needs. Courtesy of Linda Acredolo.

occurs at a certain stage of development. And walking has dedicated neural structures within all individuals of a population. Walking is an innate ability in humans and swimming is probably not. So what about language?

COMMONALITY OF LANGUAGE IN HUMANS

How common is language as a behavior on this planet? Wide varieties of distinct languages are recognized. If each separate language represented different acquired knowledge and skills, then we would not want to argue that language is innate. It could be that we have innate predispositions to learning the language of our cultural heritage. If this were so, we would have to modify the characteristics that define an innate ability such that the word population did not refer to the entire species but merely the local genetic line in a region that spoke a certain language. It is not necessary to resort to such extreme positions because the data are unambiguous: Humans can acquire any language they are confronted with during childhood. Children of Chinese parents born in Beijing and raised speaking a Chinese language can acquire English without difficulty if they are raised in the United States speaking English. Indeed, if they are raised by native English speakers and only English is spoken, their language skills will be indistinguishable from those of Americans of English descent raised learning English. There is no mystery here; language is a system for communication and representation of events, items, and ideas that have universal qualities. Therefore, language satisfies one defining characteristic of innate behavior in that it is shared by all members of the population.

This fact has been long understood by linguists who search for the universal features of language. Some, like Noam Chomsky, argued that all human languages are similar to one another. One such view is that a **universal grammar** is common to all languages and built into each person's language system. What differs from one language to the next are merely local characteristics of the grammar—how it is implemented in a specific language. This general truth is consonant with some knowledge about the neural substrates for language. Some language structures in the human brain appear to be common across languages. Semantics and grammar appear to be subserved by nonidentical neural systems, regardless of the speaker's native tongue. And speech is lateralized to the left hemisphere for everyone (except for a subset of left-handers, of course).

LANGUAGE ACQUISITION IS SIMILAR IN EVERYONE

The innateness of language ability also can be assessed by examining whether all individuals of the population express or acquire the ability in a similar fashion. Do children acquire language over a similar time period and by similar stages, or is there no norm? The answer is easy: All normal individuals acquire language in a similar way.

By their first birthday, children usually speak their first real words. Then over the next half year or so, they slowly gain about fifty more words. These first words usually refer to single objects in the world such as parents, food items, toys, pets, and so on. They also acquire verbs such as *eat* and words for interactions with others such as *hello*. Over the next 6 to 12 months, their acquisition of new words speeds up tremendously, the normal range being seven to nine new words a day, and each new word is learned immediately and correctly. This period of rapid change in word learning is called the *naming explosion*. What is happening to the toddler's brain during this period of prolific word learning? Event-related potentials (ERPs) recorded in response to words that a toddler knows become lateralized to the left hemisphere, just as they are in adults, but slightly earlier during language acquisition in the same toddler these same words would have elicited bilateral brain responses. Moreover, among a group of normal toddlers who are 20 months old, those who have large vocabularies tend to produce left-lateralized ERPs in response to words they hear, while those who have relatively small vocabularies tend to produce bilateral ERPs. The development of lateralization is correlated with the size of their vocabulary, not their chronological age (see An Interview with Helen Neville).

The use of words in combination soon follows the acquisition of single words. First come two-word combinations, then three words and more, until sometime between the ages of 3 and 4 years, children start speaking complete sentences. If you studied a foreign language in high school or college, you know that it is difficult or impossible to achieve that rate of acquisition of a new language with such ease. A unique biological miracle happens to all of us in early life: We acquire language.

Language is more than strings of words, and children not only must acquire tags for items and actions but also must learn the rules for putting the words together into grammatically correct sentences—not, as Pinker (1994) put it, "the prescriptive grammar of school marms and style manuals, which list differences between standard and nonstandard dialects of

An Interview with Helen J. Neville, Ph.D.

Dr. Neville is a professor of psychology and director of the Brain Development Laboratory at the University of Oregon and a leading authority in developmental cognitive neurosciences.

Authors: You have studied deaf people by using event-related potentials (ERPs) and have discovered that their brains have a different organization than that of normal subjects. Could you tell us what this amazing finding entails?

HJN: Yes, we've been studying the role of experience on the development of neural systems important to aspects of cognition, and the people in our research are individuals who were born bilaterally deaf due to a genetic failure in the development of the cochlea. We have tested the idea that the lack of auditory input might lead to a reorganization of cortical areas that normally process auditory information. It appeared plausible that visual processing might spread into the "abandoned" auditory cortex, and we've studied how much this happens, the time that it might occur, and the mechanisms that might mediate such a change. Our deaf subjects also had a very different language experience: None learned an aural-oral language, but all learned a visuomanual language, American Sign Language (ASL). We examined the effects of this language on the development of the brain's language systems. One way we separately assess the effects of auditory deprivation from the effects of acquiring a visuospatial language is to study hearing people who were born to deaf parents. These people haven't had any auditory deprivation, but all learned ASL as a first language, just as our deaf subjects. Any effects from auditory deprivation shouldn't be evident in the hearing person born of deaf parents, but effects due to the acquisition of ASL should appear in this group as well as the deaf subjects. We have used ERPs extensively to study these issues, and also functional magnetic resonance imaging (fMRI) techniques. The former provide excellent temporal resolution, and the latter great spatial resolution. So they are highly complementary techniques.

Authors: Does the brain reorganize itself as a result of the loss of auditory input?

HJN: Our studies of visual processing in the deaf have indicated that there are indeed alterations in visual processing following auditory deprivation. Visual ERPs are several times larger in deaf than in hearing subjects. These effects are absent in the hearing subjects born of deaf parents. And the changes were specific: They were found in response to peripheral visual stimuli but not stimuli presented foveally. Such changes happen in visual sensory paradigms and visual attention tasks.

English, and lay down conventions of written prose . . . ," but the mental grammar that describes how "words can be combined into bigger words, phrases, and sentences by rules that give a precise meaning to every combination." Children must learn morphology (combining words and fragments of words into larger words), syntax (combining words and phrases into sentences), and phonology (combining sounds into legitimate patterns appropriate for the language). English speakers have to differentiate between "man bites dog" and "dog bites man," and other, more complex formulations.

Children begin to acquire the mental grammar of a language early, and they learn these rules at similar ages. For example, children learning English have a highly similar order of acquiring grammatical morphemes, auxiliary verbs, and complex constructions such as negations or passive constructions. Their errors are also strikingly similar because they overgeneralize rules or apply them when exceptions exist, as with plurals like *mouses* instead of *mice*.

Acquiring categories for words begins early, when children learn to categorize words as verbs or nouns, and they probably learn this by experience. By combining cues, infants learn which category a word belongs to; for example, semantic cues and correlational cues tip infants to a word's syntactic category. Pinker proposed that semantic bootstrapping permits children to

Authors: Yes, but do these physiological changes reflect any behavioral change?

HJN: You bet! The physiological effects have functional consequences: Deaf subjects are faster and more accurate at detecting moving targets in the peripheral visual fields. Anatomical investigations suggest that the periphery of visual space is largely represented along the dorsal visual pathway that projects from the striate cortex to the parietal cortex that includes motion-sensitive areas. By contrast, central visual space is represented mainly along the ventral pathway that projects from V1 to anterior temporal cortex. It's important for high-resolution form discrimination and color discrimination. We tested the ideas that big alterations in visual processing secondary to auditory deprivation happen along the dorsal pathway, and more generally that it displays more developmental plasticity than does the ventral visual pathway. We presented stimuli designed to stimulate the magnocellular and parvocellular pathways that project to the dorsal and ventral pathways. ERPs from deaf and hearing subjects are similar in response to the parvocellular stimuli (colored, high spatial frequency gratings, flickering at low rates), but responses from deaf subjects are several times larger than those from hearing subjects in response to the magnocellular stimuli (moving, black and white, low spatial frequency gratings, flickering at high rates).

Authors: How do the ERPs become larger in the deaf? Are there more neurons involved or the same ones yielding a more intense response?

HJN: We are now examining these issues by using fMRI to determine where within the temporal lobe these changes occur. What we are finding is that deaf persons produce more activated voxels in MT during peripheral attention to a field of moving dots. Knowing the location of changes raises hypotheses about how visual processing might be enhanced after auditory deprivation. A likely area for change is the superior temporal polysensory area that normally receives auditory and visual input. In the absence of auditory input, visual afferents may take over what would normally be auditory cells and enhance visual responsiveness. Animal research shows changes within multimodal areas after congenital visual deprivation.

Authors: But couldn't the change reflect an unmasking of preexisting circuits?

HJN: Yes. There may be, in the immature human brain as in other animals, a transient redundancy of connections between sensory areas that is normally competitively displaced (anatomically or functionally) with normal sensory experience during development. In deaf subjects it may be that visual afferents projecting to auditory areas become stabilized in the absence of competition from normal auditory inputs. We are investigating this by studying deaf and hearing infants in the first few years of life to see whether auditory and visual systems differentiate more along a timeline that parallels the sensitive period when auditory deprivation leads to these changes in the visual system.

---

learn the differences between nouns and verbs. The idea is that the brain comes equipped to know that objects, actions, and attributes are different; hence we need only observe and infer the syntactic category of words. Some of the cues young children use are the perceptual features of objects, such as their shape or color. Research has shown that if a novel object is labeled as a noun ("that's a dax"), children generalize that label to other objects of similar shape. On the other hand, if an object is labeled with an adjective ("that's daxy"), children tend to generalize that category label to similarly colored objects. These results suggest that word categories can bias children's attention to different features of objects and events. Inter-

estingly, these kinds of biases appear after the naming explosion, once ERPs in response to words have become lateralized. But all of this presents a paradox: How can we learn syntax if we have to know it before being exposed to it? Put another way, recognizing and becoming facile with knowing a noun from a verb requires that we are already born with that knowledge! How can we test this?

One way to test the idea that linguistic knowledge is innate rather than acquired is to examine situations where language errors predict whether a general rule is being applied. Take the relation between lexical and auxiliary verbs. A *lexical verb* is the prototypical verb (e.g., *eat, run, jump*), whereas an *auxiliary* or *helping verb* is one that varies tense, voice, or mood or modifies aspects of another verb (e.g., *may* in "I may leave tonight"). Lexical and auxiliary verbs are quite similar in meaning and syntax and in their lexical form; a person learning these might confuse them and make predictable errors. Consider the lexical and auxiliary forms of the verb *to have:* "He *has* courses in psychology" (lexical) versus "He *has* taken courses in psychology" (auxiliary). These two forms must be used properly or many types of incorrect sentences will be constructed. Linguists predict that children's speech should include errors like an incorrect order of lexical and auxiliary verbs ("She have should eaten"). However, this does not occur.

Children learn auxiliary verbs rapidly and accurately. One might argue that learning the correct use of auxiliary verbs results from parental feedback, correction, and instruction. But interactions between children and parents reveal that parents primarily correct their children's speech when errors are made in meaning, not grammar. If you have had experience with infants and young children, you might have noticed that you try to adopt their manner of speaking rather than correct their speech. Humans therefore have an innate, linguistic knowledge that enables them to employ syntactic structures of language, and thus develop complex forms of linguistic representation such as the use of auxiliary verbs. Humans do not have an innate ability to distinguish between lexical and auxiliary verbs per se; instead, the brain apparently can distinguish between lexical and functional categories. So we have at least one form of innate linguistic knowledge, which is part of the human brain's specialization for language. In summary, language has many characteristics of an innate property: as Pinker insisted, "Language is an instinct." The brain is predisposed to manifest the complex cognitive processes that comprise language. This proposal is supported by findings like the discovery that lateralization of speech to the left hemisphere of the brain is universal. Ongoing research will provide increasing detail about which brain structures and systems supporting language are common to all normal language users, which structures and systems are modified by experience and therefore may differ across cultures with different linguistic structures, and which may be nonspecific to language per se but are co-opted by experience to support human language (see also Chapter 10).

Summary of Cognitive Development

Cognitive development takes place over many years from birth to adulthood, but some cognitive functions mature earlier than others. This is not to say that the human infant is not capable of significant processing at a very early age. Indeed, the classic view of Piaget has been challenged and continues to be modified as researchers demonstrate that even young infants may have knowledge about objects around them, and that at an early age they develop sophisticated representations for the things they encounter and for mental abilities such as numerical knowledge. The evidence argues for a nervous system that acquires knowledge quickly by selecting from preconstructed neural circuits as soon as they physically come on-line. The result is that perception, action, and reasoning develop early and in parallel, and not in a simple progression from sensation to higher cognition. Some of the biological correlates of these fantastic developments are known, such as in the case of the visuomotor system and attentional orienting. In the next section we will review the developmental biology of the nervous system in more detail and learn how the nervous system, especially the cerebral cortex, comes to be.

DEVELOPMENT OF THE NERVOUS SYSTEM

So far in this chapter we described some key aspects of perceptual and cognitive development and how postnatal maturation can be correlated with aspects of behavior. By the time of birth, though, the fetal brain is well developed and shows cortical layers, neuronal connectivity, and myelination; in short, it is already extremely complex. To find out how this complex brain develops prenatally, let's examine the development of the human nervous system, with special reference to the neocortex, to learn about the rules governing development.

One rule that has emerged from the vast data on brain and behavior is that the brain is not an equipotential mass of randomly interconnected neurons but instead is an intricately detailed and precise set of neuronal structures, circuits, and systems. Indeed, this knowledge began to emerge from the work of the great anatomists Cajal and Golgi and their colleagues long ago. Today we know much more about the intricacies of the nervous system's neuronal circuits. These precise and complex circuits arise from a careful developmental plan that, if disrupted, may lead to disastrous consequences for the organism. The bulk of the brain consists of cortex, and without minimizing the role of subcortical circuits, we can state that most cognitive activity at least involves cortical circuits and systems.

Animal models, such as those making use of the rhesus monkey, have permitted a close look at how neurons in the cortex achieve their final connectivity with each other and other brain systems. The monkey has a large cortex and well-developed perceptual and motor skills. In these animals one can use experimental methods to determine the mechanism of cortical development that would not be possible by observing only humans. In recent decades, a wealth of information has been derived about the cellular, biochemical, and hormonal influences on the genesis of the cerebral cortex. New and sophisticated methods permit us to do more than merely observe the changes in the developing brain. They provide us with the chance to manipulate the course of development in ways that inform us about the underlying processes. For example, in addition to anatomical studies that observe the microanatomical organization of brain tissue at various developmental stages, investigators now also use advanced genetic techniques to learn about the formation of brains.

Overview of Gross Development

Before we discuss development at the cellular level, recall that from a single fertilized egg, an organism of billions of cells with specialized functions will arise. The peak of this complexity is clearly in the nervous system. Figure 15.9 depicts a few stages of human development between fertilization and birth. Recall that following fertilization, events lead to the multicellular *blastula* that has already begun to specialize. The blastula contains three main types of cell lines: the **ectoderm**, from which neural ectoderm will form; the *mesoderm*; and the *endoderm*. In a broad sense, these respectively form (1) the nervous system and the outer skin, lens of the eye, inner ear, and hair; (2) the skeletal system and voluntary muscle; and (3) the gut and digestive organs. The blastula undergoes further development during gastrulation, when invagination and cell migration prompt the ectoderm to surround the entire developing embryo. The embryo now has mesoderm and endoderm layers segregated dorsally and ventrally, and thence undergoes *neurulation*, as shown in Figure 15.10. During this stage, the ectodermal cells on the dorsal surface form what is called the **neural plate.**

The development of the nervous system continues as the neural plate invaginates via neural folds being

Figure 15.9 Photographic sequence of a developing human fetus, at 7 weeks **(a)**, 9 weeks **(b)**, and 7 months **(c)**.

(a)

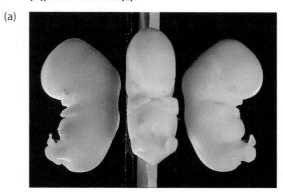

(b)

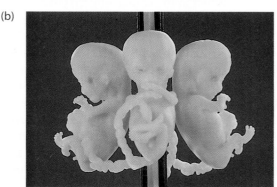

(c)

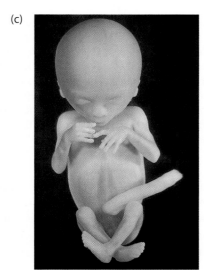

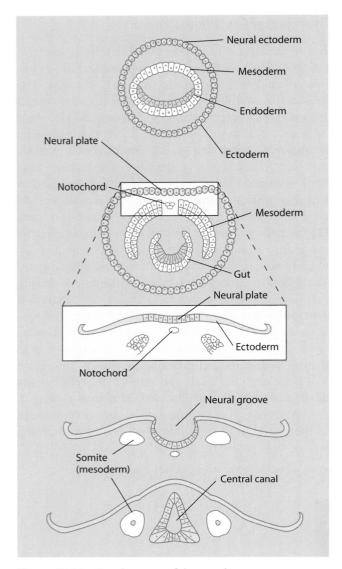

Figure 15.10 Development of the vertebrate nervous system. Cross sections through the blastula and embryo at various developmental stages during the first 21 days of life. Early in embryogenesis, the multicellular blastula **(top)** contains cells destined to form various body tissues. Migration and specialization of different cell lines leads to formation of the primitive nervous system around the neural groove and neural tube on the dorsal surface of the embryo. The brain is located at the anterior end of the embryo and is not shown in these more posterior sections, which are taken at the level of the spinal cord. Adapted from Carpenter (1976), Goldsby (1976), and Kandel et al. (1991).

pushed up at its border. At this point there is already an axis of symmetry where the neural folds form a groove, the *neural groove*. As this groove deepens, the cells of the neural fold region eventually meet and fuse, forming the *neural tube* that runs anteriorly and posteriorly along the embryo. The adjacent nonneural ectoderm then reunites to seal the neural tube within an ectodermal covering that surrounds the embryo. At both ends of the neural tube are openings (the anterior and the posterior neuropore) that eventually close. When the anterior neuropore is sealed, this cavity forms the primitive brain consisting of three spaces or ventricles (Figure 15.11a, top). From this stage on, the brain's gross features are formed by growth and flexion (bending) of the neural tube's anterior portions (Figure 15.11a, bottom). The result is a cerebral cortex that envelops the subcortical and brainstem structures that started out in line with the cortex along the neural tube. The final three-dimensional relations of the brain's structures are the product of continued cortical enlargement and folding. One interesting feature of the flexions during fetal development is that at these early stages there is a striking similarity between fetuses of humans and those of other mammals (Figure 15.11b).

Genesis of the Cerebral Cortex

In the preceding section we briefly reviewed the development of the nervous system's gross neuroanatomical features. Many details were omitted, but the general scheme laid out here is nonetheless informative, if not actually fascinating! We will now consider mechanisms underlying the brain's growth and connectional specificity: **neuronal proliferation, neuronal migration, neuronal determination and differentiation, synaptogenesis,** and **synapse elimination.** Decades of careful studies in animals and humans have provided a relatively detailed understanding of the sequence of these cellular events as they take place on the path to building a mammalian brain.

NEURONAL PROLIFERATION

The first question about the brain's development is, When in the course of prenatal and postnatal development are neurons in the brain "born"? Examination of the brains of newborn monkeys or humans reveals that virtually the entire adult pattern of gross and cellular anatomical features is present at birth. With the exception of complete myelination of axons in the brain, the newborn has a well-developed cortex that includes the cortical layers and areas characterized in adults. Area 17 (the primary visual cortex) can be distinguished from the motor cortex by cytoarchitectonic analysis of its neuronal makeup. Indeed, in primates there is little

(a)

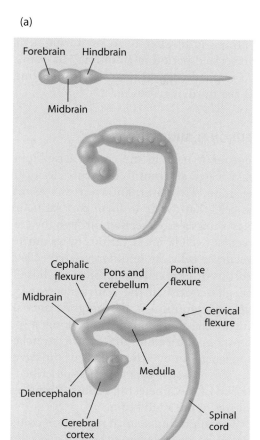

(b)

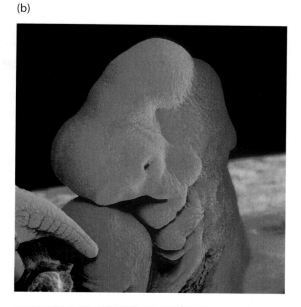

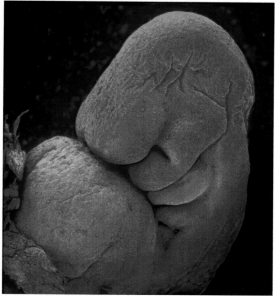

Figure 15.11 **(a)** Diagram of a developing embryo. The developing embryo goes through a series of folds, or flexures during development. These alterations in the gross structure of the nervous system give rise to the compact organization of the adult brain and brainstem in which the cerebral cortex overlays the diencephalon and midbrain within the human skull. **(b)** There is significant similarity between the gross features of the developing fetuses of mammals, as shown by this comparison of human **(top)** and pig fetuses. (a) Adapted from Carpenter (1976).

if any additional generation of neurons after birth, or apparently at any time during their decades of life. Virtually all neurons are generated prenatally, although some recent evidence suggests a modification of this story is required (see later portions of this section). Again, the question remains, When are specific groups of neurons generated?

The timeline of neuronal development in nonhuman primates has been tracked by clever cell-labeling methods. The classic anatomical methods are unable to provide this information; by merely staining sections of cortex and observing the neurons present during embryogenesis, we cannot accurately determine which neurons arose first. The method developed, known as ^3H-thymidine labeling, involves injecting radioactively labeled thymidine into an embryo early in development. The thymidine is taken up by neurons and used to form DNA in cells undergoing cell division. Because the thymidine is labeled with a radioactive tag, the DNA of only the cells undergoing division at the time of injection will contain the radioactive label; hence, the label's distribution in the brain tissue can permit the localization of the final fate of the neurons born at the time of injection (Figure 15.12). Determining the distribution is done by the autoradiographic method. Sections of brain tissue are placed against photosensitive film and the radiation exposes the film, creating an image that can be developed and viewed as a picture. The resultant picture shows the distribution within the brain section that contains the radiolabel and therefore the cells that were born at injection time.

Based on these cell-labeling studies on embryos during gestation, a clear view of the timing of **corticogenesis** is formed. The genesis of the cerebral cortex in primates begins during the first quarter of gestation. All neurons in the primate cortex are derived within 1 to 2 months after the process begins, depending on which cortical region is under investigation. For example, production of the neurons of area 17 (striate cortex) is not finished until long after the neurons in the other brain areas have been born. Nevertheless, in primates the middle third of the gestational period accounts for the production of all cortical neurons present at birth. This is untrue for other mammalian species, whose cortical neuronal genesis may continue until after birth or during a different circumscribed period in gestation. For example, in mice and rats, all cortical neurons arise during 1 week of the last third of gestation.

NEURONAL MIGRATION

Neurons that form the cortex arise from a layer of cells located adjacent to the ventricles of the developing brain. This layer, known as the *ventricular zone,* has cells that divide to form cortical neurons. Figure 15.13 shows a cross section through the cortex and the precursor cell layers at various times during gestation. **Precursor cells** are undifferentiated cells from which neurons or glial cells are produced; those for neurons and glia coexist in the ventricular zone. After these cells undergo mitosis, they migrate outward from the ventricular zone by moving along a peculiar cell known as the **radial glia,** which stretches from the ventricular zone to the surface of the developing cortex. This unusual glial cell, first described by Cajal at the end of the nineteenth century, transforms into astrocytes in the adult brain.

The migrating neuron remains in contact with the framework provided by the radial glia cell via interactions of cell-surface molecules that keep the two cells intimately associated. The movement of the migrating cells is believed to result from contractions of skeletal-like intracellular molecules initiated by signals conducted across the membrane via ion channels. These intracellular molecules include microtubules and actin-like contractile proteins that contract to move the nucleus and intracellular fluids toward a so-called leading process, which is the portion of the migrating neuron that moves out along the radial glia, contacting it and adhering to it via molecular "glue" proteins. Both voltage-gated channels and NMDA-type receptors (see Chapters 2 and 8) on the migrating neuron mediate the cell signaling that leads to migration via movements produced by the contractile molecules.

Once the migrating neurons approach the surface of the developing cortex—a point known as the *cortical plate*—they stop short of the surface. Then neurons that migrate later pass the earlier neurons and end up in more superficial positions, positions nearer to the outer cortical surface. Thus, it is said that the cortex is built from the inside out because the first neurons to migrate lie in the deepest cortical layers, whereas the last to migrate move farthest out toward the cortical surface. This timeline of cortical cell genesis has been

Figure 15.12 Diagram of the ³H-thymidine method used to determine the fate of neurons that arise at a particular time in prenatal development. ³H-thymidine is incorporated into dividing cells and remains within the cell body. The radioactive label can be viewed in the mature neurons and because the investigator knows when the injection was made, can be used to trace when the neurons arose during development.

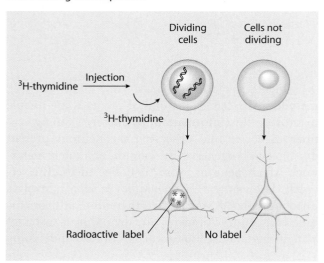

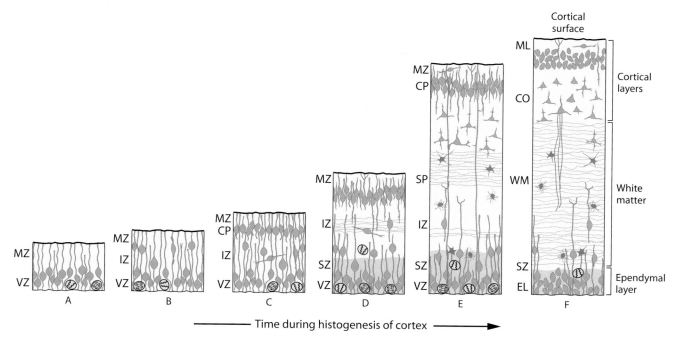

Figure 15.13 Histogenesis of the cerebral cortex. Cross-sectional views of developing cerebral cortex at early **(left)** and late **(right)** times during histogenesis. The cortex of mammals develops from the inside out as cells in the ventricular zone (VZ) divide, and some of the cells migrate to the appropriate layer in the cortex. Radial glial cells form a superhighway along which the migrating cells travel en route to the cortex. Adapted from Rakic (1995a).

demonstrated with the thymidine-labeling method, as in Figure 15.14. Early in corticogenesis, injection of radioactive thymidine leads to labeling of neurons in the cortex's deepest layers, V and VI, and the underlying white matter, whereas at later stages of corticogenesis, injections of radioactive thymidine label neurons in more superficial cortical layers, II and I. As noted earlier, the timeline of cortical neural genesis differs across cortical cytoarchitectonic areas (e.g., area 17 versus area 24), but the inside-out pattern is the same for all cortical areas. Because the timeline of cortical neurogenesis determines the ultimate pattern of cortical lamination, anything that affects the genesis of cortical neurons will lead to an ill-constructed cortex. A good example of how neuronal migration can be disrupted in humans is fetal alcohol syndrome. In cases of chronic maternal alcohol abuse, neuronal migration is severely disrupted and results in a disordered cortex, and apparently a consequent plethora of cognitive, emotional, and physical disabilities.

DETERMINATION AND DIFFERENTIATION OF NEURONAL TYPES IN THE CORTEX

So far we have regarded the cells in the ventricular zone as a single population. But how does this population of virtually identical precursor cells give rise to the variety of neurons in the adult cortex? Moreover, where do the glial cells come from? The answer is known: All cortical cells, including neuronal subtypes and glial cells, arise from the precursor cells of the ventricular zone through cell division and differentiation.

For the first 5 to 6 weeks of gestation, the cells in the ventricular zone divide in a symmetrical fashion. The result is an exponential growth in the number of precursor cells. After this time, though, asymmetrical division begins and one of the two cells present after division becomes a migratory cell. The other remains in the ventricular zone and continues to undergo cell division. This subsequent division is also asymmetrical, yielding one remaining and one migratory cell. This process contributes the cells migrating to cortical layers. In later gestational periods, the proportion of cells that migrate increases until eventually the final state yields a laminar cortex, with an epithelial layer that becomes the cell lining of the ventricles known as the *ependymal cell layer*.

Which type of neuron (e.g., layer IV pyramidal cell or layer III stellate cell) a migrating neuron will eventually become is determined at the point of cell division. Once the cell has been fated to migrate, the type of cell it will become and in which cortical layer it will reside are determined. This determination correlates with the time

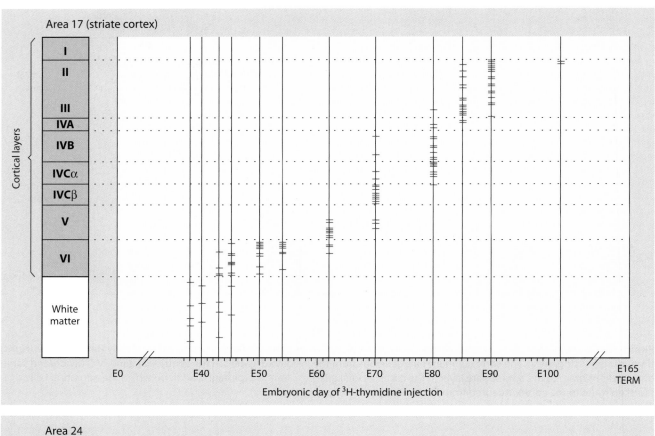

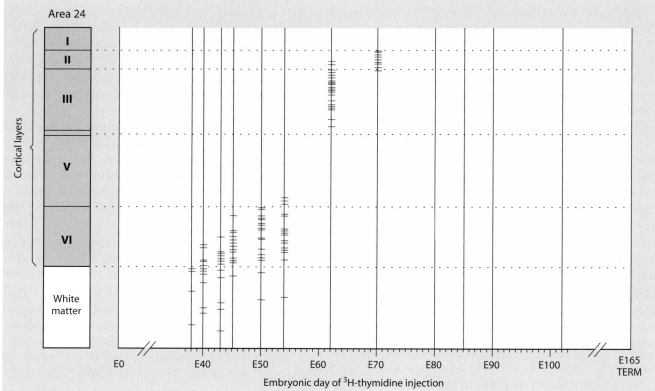

Figure 15.14 Birth ages of cortical neurons. Radiolabeled thymidine was used to label cells at different embryonic days in two cortical areas, Brodmann's area 17 and 24. The cortical layers present at birth are shown on the axis at the left, with the cortical surface at the top of each plot. Cells with birth dates later in gestation are found in more superficial cortical layers, but the time course of this development differs for different cortical regions. Adapted from Rakic (1995a).

of its creation during gestation. Neurons that were supposed to migrate but were prevented from doing so by experimental intervention, such as exposure to high-energy x-rays, eventually develop patterns of connectivity that would be expected from neurons arising at the same gestational stage. Even though these neurons might remain in the ventricular zone, they display interconnections with other neurons that would be normal had they migrated to the cortical layers.

Similar evidence comes from transplanting cells from one animal to another. If embryonic cells from the ventricular region are removed at a certain stage of gestation and transplanted to the cortex of newborn host animals, such as ferrets, these transplanted neurons migrate to the proper cortical layer expected for their gestational age, regardless of the gestational period the host tissue is in at the time it receives the graft (Figure 15.15). Even if the host cortex is past the stage when neurons migrate to layers V and VI—for example, when the neurons in the host are migrating past these layers to form layers II and III—a transplanted neuron whose gestational age dictated that it should migrate to layers V and VI will indeed move to these layers. What is more, these transplanted neurons take on the morphological form (i.e., stellate cell, pyramidal cell, etc.) and the connectional pattern predicted by their age; that is, they are predetermined to be a certain neuronal type. The alternative would have been to have the properties (morphology, connectivity, biochemistry, etc.) of each cortical neuron determined by the neuronal environment where they reside, as with cells that form the neuronal and glial cells of the peripheral sensory systems and the autonomic nervous system. The evidence is strong that, for mammalian cortical neurons, a prespecification of neuronal properties takes place long before the migrating neuron reaches its destination in the cortex.

THE RADIAL UNIT HYPOTHESIS

We now have a picture of how cortical neurons are born and how they migrate radially from the ventricular zone toward the surface of the developing cortex. The migration is along the radial glial cells that form a pathway for neurons. Because the radial glial highway is organized in a straight line from the ventricular zone to the cortical surface, there is a topographic relation between precursor and proliferating neurons in the ventricular area, and the cortical neurons they yield in the adult. Hence, cells born next to each other in the ventricular zone end up near each other (in the plane perpendicular to the surface of cortex) in the cortex. As well, cells derived from precursor cells distant from one another will ultimately be distant in the cortex.

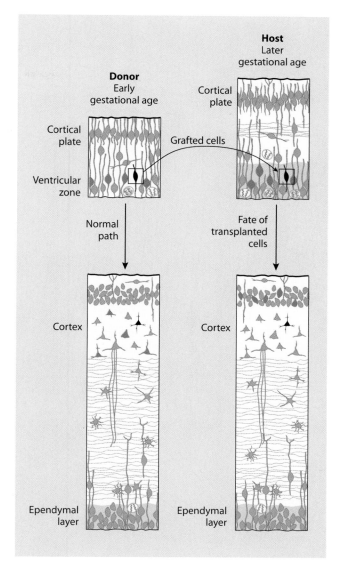

Figure 15.15 Determination of neuronal types in cortex. Transplantation of fetal cells from one animal to another has demonstrated that neurons migrate to the region of cortex that is specified by the developmental stage at which they arose. Thus, neurons born at the same age migrate to their prespecified cortical layer (see black pyramidal neuron in each half of figure) regardless of whether they remain in the donor animal **(bottom left)** or are transplanted to a host animal's brain that is actually older than the neurons being transplanted into it **(bottom right).**

This concept, termed the "radial unit hypothesis" by Yale neuroscientist Pasko Rakic (1995a), is the idea that the columnar organization in the adult cortex is derived during development from cells that divide in the ventricular region (Figure 15.16). The cortical column is thus a principal unit of organization that has functional consequences and a developmental history. The radial unit hypothesis also provides a method for the evolutionary expansion of cortical size; the idea is that

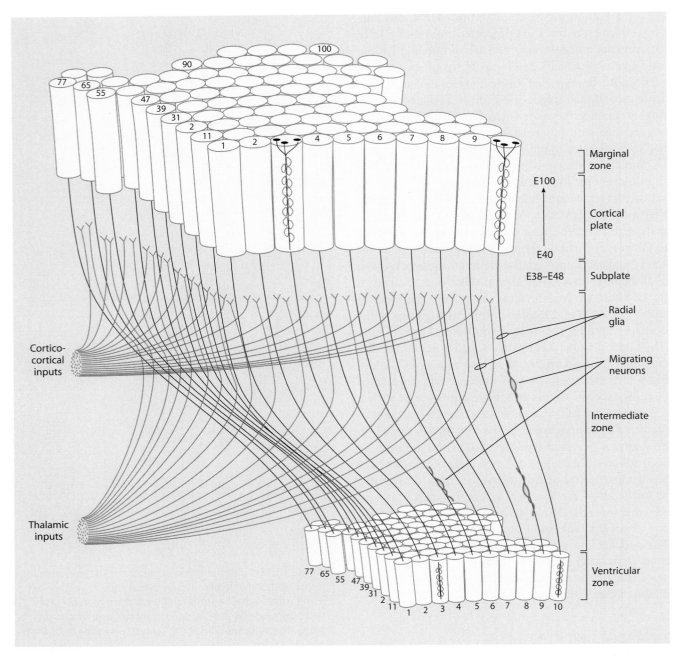

Figure 15.16 Diagram of radial unit hypothesis. Radial glial cells in the ventricular zone project their processes in an orderly map through the various cortical layers, thus maintaining the organizational structure specified in the ventricular layer. Adapted from Rakic (1995a).

rather than enlarging each unit, the number of units is increased. The radial unit and the cortical columns that arise from these groupings have functional-anatomical consequences in the adult. For example, the intracortical interconnectivity among local neurons appears to be well suited to the size of cortical columns, which vary in adults from about 100 microns to 1 mm on a side, depending on the species and cortical area.

CYTOARCHITECTONIC VARIATION ACROSS CORTICAL AREAS

We discovered in Chapter 3 that distinctive variations appear in the cellular organization of the cortical regions. These cytoarchitectonic differences segregate the cortex into distinct areas. Thus, Brodmann's area 17 (primary visual cortex) varies in cellular composition

Mutant Mice and Determination of Cortical Neuron Phenotypes

The reeler mouse has a genetic mutation that undermines the development of its cerebral cortex. This animal's cortex contains all classes of neurons, but they vary in position compared to normal mice, and the percentages of neuronal classes are altered—some are reduced from normal. The result is implied by the name "reeler"; these mice have severe problems with motor control: ataxic gait (clumsy walking), dystonic posture (improper muscle tone when standing), and tremors (shaking). These mice also may have a smaller cerebellum. All these problems are caused by an autosomal recessive gene.

The reeler mouse brain has another, more striking feature of its genetic failure. The radial glial cells in the reeler cortex are present during development but are oriented at unusual angles, rather than vertically as in the normal brain. Despite this strange alignment of the radial glia, the migrating neurons in the reeler still crawl along the glia, indicating that the cell-cell molecular interactions in normal mice are also present in the reeler mouse. The migratory process is relatively normal in the reeler; the neurons migrate as they should, outward along the radial glia toward the cortical surface. Yet the final positions of the neurons are not correct. In adults, the laminar organization of neurons is inverted in the reeler cortex. Large pyramidal cells are in deeper cortical layers, while smaller polymorphic neurons are located more superficially—the opposite of the normal mouse. The final positions of migratory neurons in the cortex are determined during a postmigratory stage. Once again, cellular morphology and patterns of connectivity are related to the time of cell birth at the ventricular zone, and not by the final position in the cortex, which in the reeler is incorrect.

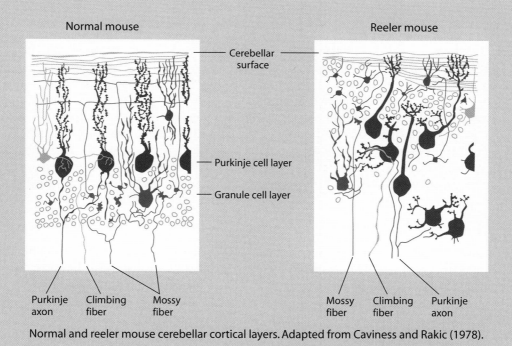

Normal and reeler mouse cerebellar cortical layers. Adapted from Caviness and Rakic (1978).

from area 4, even though each of these cortical areas shares a general laminar organization and the laminar positions of different neurons in each cytoarchitectonic area are similar. A question from the developmental perspective is how this variation occurs. What factors specify the differences between cortical areas?

One influential model, the "protomap hypothesis" of Rakic, posits that prespecified instructions inherent

Innovations in Cellular Labeling in the Study of Development

The developing cortex carries in its genetic makeup intrinsic instructions for controlling the differentiation of precursor neurons into mature cortical neurons of various types: astroglia, pyramidal cells, stellate cells, and so on. Investigating this differentiation calls for powerful biochemical and molecular techniques. These methods permit scientists to track the differentiation of neurons back to the first stages of development when different cell lines can be distinguished.

Cell surface molecules such as proteins can be targets (antigens) for antibodies, proteins produced by the body for protection against infection. Antibodies can be harvested by researchers and labeled to be used as probes for specific proteins on cell surfaces. One simply adds a label (tag), either fluorescent or radioactive, to antibodies so they can be seen through a microscope. Visualization of the reaction of a tagged antibody with cell surface proteins expressed by a subpopulation of cells is an ideal method for identifying those cells in the brain.

Using the antibody label specific to an adult cell, investigators found that cell surface proteins in adult cells of animals' brains are sometimes expressed quite early in development. Monkeys' radial glial cells contain a protein called *glial fibrillary acidic protein* (GFAP). Using an antibody to GFAP, scientists showed that only some neurons in the ventricular zone had this protein—key support for the idea that a separate, prespecified group of precursor cells in the ventricular zone gives rise to radial glia and astrocytes.

Another powerful tool for investigating the lineages of neurons and glia derives from the use of certain viruses (retroviruses) to label the genetic material of neurons. The retrovirus is modified to prevent it from replicating, and to yield a protein that can be visualized later by using chemical reactions in histological slices. The retrovirus is injected locally and allowed to infect cells in the developing brain. (Infection means that the retrovirus inserts itself into the cell's DNA, effectively becoming part of the cell's genetic structure.) Careful quantitative methods ensure that the virus infects only a single precursor cell. When the cell divides to form two daughter cells, they also contain the retrovirus label, as will all descendants of the infected cell. With this approach, one can search later in development for neurons that contain this retrovirus marker. We have discovered that a precursor cell gives rise to either glia or neurons, but not both. Further, neurons that descend from the single infected cell are all of the same morphological type. It is curious that the timeline of generating descendant cells is unimportant. A single precursor cell can give rise to neurons that later appear in various cell layers, which means that they were born at different gestational times. Although the time when a neuron is born determines where it will reside, this alone does not indicate that it was derived from a certain precursor cell, because a cell line may produce migrating neurons at various gestational times and contribute to different cortical layers, but all these will be of the same cell types such as the pyramidal neuron.

in developing neurons interact with signals derived from neuronal inputs from subcortical brain areas (Figure 15.17). According to this view genetic factors predetermine the organization. Neurons located in the ventricular zone will establish a protomap that attracts the thalamic afferent fibers appropriate to the function that the region is destined to perform. For example, the protomap in the region of the adult visual cortex will attract axons of neurons in the lateral geniculate nucleus of the thalamus, the thalamic relay that receives ascending projections from the retina. As well, the region of the protomap that becomes the primary auditory cortex attracts thalamo-cortical projections from the medial geniculate of the thalamus—the auditory relay nucleus. Recent work also showed, however, that neurons disperse laterally in addition to radially. Labeling studies using, for example, retroviruses have shown that cells derived from individual progenitor cells can disperse laterally, to regions of cortex that do not directly overlay the location of the original progenitor, although this does not appear to be the case for progenitor cells at very early stages of development (see Innovations in Cellular Labeling in the Study of Development). A general principle may be that neurons involved in specific cortical-subcortical connectivity (a radial pattern) tend to migrate radially, whereas other neurons, such as

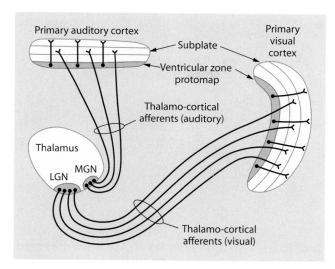

Figure 15.17 The protomap hypothesis of cytoarchitectonic diversity. The protomap model proposes that the neurons in the ventricular layer form a protomap that attracts the axons of neurons in the subcortical projection structure in a modality-specific manner. Thus, auditory fibers are attracted to auditory cortex, and visual fibers are attracted to visual cortex during embryogenesis. Adapted from Rakic (1995a).

GABAergic interneurons, may be able to disperse more broadly within the brain, coming to rest in distant locations where they become part of local circuitry. Interestingly, some knockout mice (genetically altered mice) do not display widespread dispersion of GABAergic neurons, suggesting a degree of genetic control over the dispersion of some neuronal types.

Demonstrating that normal cortical organization is possible without experience demonstrates the powerful prespecification that takes place in neural development. Several lines of experimental evidence converge to support the view that the visual cortex in mammals, for example, can develop normal cellular makeup, intracortical synaptic organization, and neurotransmitter expression in the absence of activity from the retina, as do animals born without eyes or whose eyes were removed prenatally (see Experience and Activity in the Development of the Visual Cortex). The loss of visual input does affect functional organization, however. For example, as noted earlier and described in Chapter 5, removal of one eye changes the pattern of ocular dominance columns in the cortex.

In cases of eye removal, the thalamic neurons still project axons to the visual cortex. But cortical organization only partly depends on the type of thalamo-cortical inputs. In mice, cortical regions can be transplanted prenatally by microsurgical techniques. For example, we can transplant cells from the somatosensory cortex to

another region of the cortex and then allow the fetus to develop. The transplanted cells survive and make connections with the thalamus. Genetic markers in the adult can tell us about the expression of genes in the transplanted tissue. With this procedure, we can see that although the transplanted cortex received nonsomatosensory afferents from the thalamus, the cortical neurons still express genes typical of the ones in the normal somatosensory cortex.

All features of the transplanted tissue are not expressed as they would be if the tissue had been left in the somatosensory cortex. Indeed, the final structure of the cortex is determined by the interactions of the intrinsic neuronal signals that are genetically specified, and the pattern of connections the neurons form with each other and thalamic afferents. Studies of transplantation of embryonic cortex from the mouse visual cortex to the somatosensory cortex, and vice versa, reveal the degree to which environmental factors alter the organization expressed by cortical neurons (Figure 15.18). These studies used the barrel fields of the somatosensory cortex—characteristic circular arrangements (barrels) that are groups of neurons which receive input from a rodent's single whisker. Regions outside the somatosensory cortex do not have these anatomical features. When

Figure 15.18 Transplanted tissue interacts with host neurons. The mouse whisker zones, barrels, in somatosensory cortex are characteristically identifiable anatomical structures. If portions of barrel cortex in fetal somatosensory cortex are transplanted to the visual cortex before the barrels arise, this tissue does not form the characteristic barrels. However, visual cortical neurons can be induced to form barrels if transplanted from visual to somatosensory cortex.

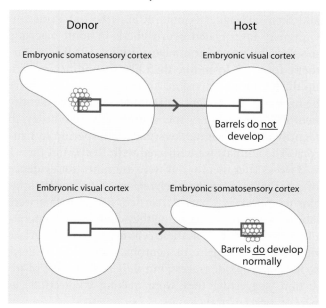

the somatosensory cortex is transplanted into the developing visual cortex, barrel fields do not develop. In contrast, visual cortex grafts placed in the somatosensory cortex do develop the barrel field organization typical in this cortical area. In both instances, the transplanted tissue is innervated by axons from the thalamus that are appropriate for the area where the transplant is placed, not by those appropriate for the region where the transplant was taken.

Together, these data tell us that the structure of the cortex is controlled by a complex interplay between intrinsic genetic factors that prespecify the fate of neurons and glia during gestation, and extrinsic factors that influence cortical connectivity. The result is that although the genetic and biochemical properties of neurons may well be prespecified very early in the developmental process, the extent to which these prespecifications reflect the final phenotype of the neurons can be influenced by the environment.

Birth of New Neurons Throughout Life

One strongly held belief about the human brain has been the idea that the adult brain produces no new neurons. This belief is held despite a variety of claims of new cells being observed in the brain in histological studies dating as far back as the time of Cajal, warranting cautious commentary by the great neuroanatomist (Cajal, 1913): "None of the methods used by these investigators are capable of distinguishing absolutely a multiplying neuroglia cell from a small mitotic neuron."

However, recent findings based on an array of modern neuroanatomical techniques are challenging this belief. In one line of work, Elizabeth Gould, Charlie Gross, and their colleagues (Gould et al., 1999) at Princeton University investigated neurogenesis in adult macaque monkeys by using bromodeoxyuridine (BrdU—a synthetic form of thymidine that, like thymidine, is taken up during mitotic division) to label neurons that were being born, fluorescent retrograde tracing to identify neuronal connections, and immunohistochemistry to identify cell surface markers that were specific to neurons. The animals were injected with BrdU, and then 1 to 3 weeks later their brains were examined for evidence of cells that took up the label; regions of prefrontal, inferior temporal, and parietal cortices had cells that were labeled with BrdU, suggesting they contained proliferating neurons. Moreover, these cells expressed cell surface markers characteristic of neurons, and some of these took up retrograde tracers, but only near the site of injection, suggesting they were making connections as part of local cortical circuits.

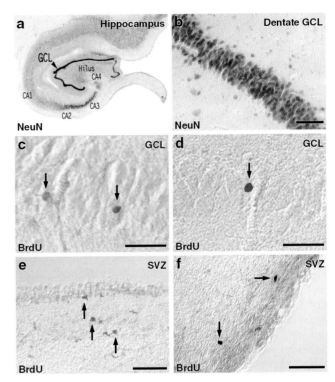

Figure 15.19 Newly born neurons in adult human. **(a)** The hippocampus of the adult human brain stained for a neuronal marker (NeuN). **(b)** The dentate gyrus granule cell layer (GCL) in a NeuN-stained section. **(c)** Bromodeoxyuridine (BrdU)-labeled nuclei (arrows) in the granule cell layer of the dentate gyrus. **(d)** BrdU-labeled cells (arrow) in the granule cell layer of the dentate gyrus. **(e)** BrdU-stained cells (arrows) adjacent to the ependymal lining in the subventricular zone (SVZ) of the human caudate nucleus. These neurons have elongated nuclei resembling the migrating cells that typically are found in the rat subventricular zone. **(f)** BrdU-stained cells (arrows) with round to elongated nuclei in the subventricular zone of the human caudate nucleus. The horizontal black bars are scale bars representing 50 microns. From Eriksson et al. (1998).

Despite these tantalizing initial findings, the story regarding neurogenesis in adult neocortex is still in the making. If these findings hold, then it remains to be demonstrated whether the connections formed by these developing neurons in the adult monkey are actually functional, leading the new neurons to process information in the service of sensory, motor, or cognitive processes. However, prior studies from a host of different investigators, in a variety of species from birds to primates, have provided strong evidence for the birth of new neurons in the hippocampus of the adult brain. These findings are particularly interesting because the number of new neurons correlates positively with learning or enriched experience (more social contact or challenges in the physical environment)

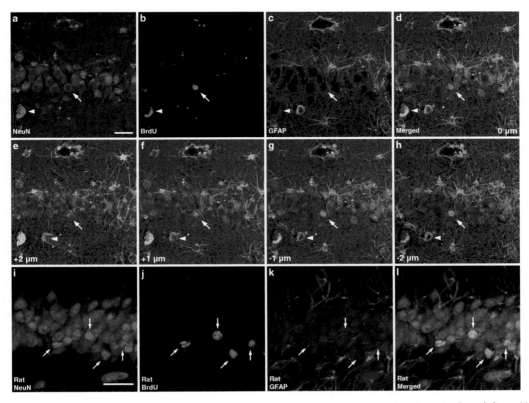

Figure 15.20 The birth of new neurons in the dentate gyrus of the adult human compared to those in the adult rat. New neurons show simultaneous labeling for different stains. **(a)** A neuron is labeled for NeuN, a neuronal marker. **(b)** The same cell is labeled with BrdU, indicating it is newly born (full arrow). (Note that the lone arrowheads in (a) through (d) are pointing to neurons that are fluorescing red or green, owing to nonspecific staining; i.e., these are not newly born neurons). **(c)** This same cell is not stained by glial fibrillary acidic protein (GFAP), indicating it is not an astrocyte. **(d)** The three stained sections are merged. The image shows that a BrdU-labeled cell could specifically co-express NeuN without expressing GFAP. Confocal microscopy permits examination of the co-expression of NeuN and BrdU in the neuron by focusing the image above **(e, f)** and below **(g, h)** the level of the section shown in panel (d). Note that red blood cells and endothelial cells, present in several small blood vessels, show nonspecific staining, as indicated by the asterisks in (e) through (h). Panels **(i)** through **(l)** show the similarity of the BrdU-labeled neurons in rat dentate gyrus. Note: The scale bar in (a) is 25 μ and the scale is the same for panels (a) though (h). The scale bar in panel (i) is also 25 μ and is the scale for (i) through (l), but the magnification for (i) through (l) is higher than for (a) through (h). From Eriksson et al. (1998).

and negatively with stress (e.g., living in an overcrowded environment).

What about the adult *human* brain? Does neurogenesis also occur? In a fascinating line of research, Fred Gage of the Salk Institute in San Diego, California, Peter Eriksson of Sahlgrenska University Hospital in Göteborg, Sweden, and their colleagues (Eriksson et al., 1998) used very similar methods in a group of cancer patients who were being treated for their disease but later died. As part of a diagnostic procedure related to their treatment, the patients were given BrdU, the same synthetic form of thymidine used as a label to identify neurogenesis in animal studies. The purpose was to assess the extent to which the tumors in the cancer patients were proliferating; tumor cells that were dividing would take up BrdU, and this could be used as to quantify the progress of the disease.

As in the animal studies described, neurons undergoing mitotic division during neurogenesis in these patients also took up the BrdU, which then could be observed in postmortem histological examinations of their brains. The postmortem tissue also could be stained using immunohistochemical methods to identify neuron-specific cell surface markers. Upon investigation of the brains of these patients, the scientists found cells labeled with BrdU in the subventricular zone of the caudate nucleus and in the granular cell layer of the dentate gyrus of the hippocampus (Figure 15.19). By staining the tissue to identify neuronal markers, the researchers showed that the BrdU-labeled cells were neurons (Figure 15.20). These findings demonstrate that new neurons are produced in the adult human brain, and that our brains renew themselves throughout life to an extent not previously thought possible.

Postnatal Brain Development

As we touched on in the section on perceptual and cognitive development, a host of behavioral changes take place during the first months and years of life. What neurobiological changes take place during this period to enable these developments? Even if we assume that neuronal proliferation continues, that does not discount the fact that at birth the human has a fairly full complement of neurons, and these are organized intricately to form a normal human nervous system, even if not complete in all the details. So what details are incomplete, and what is known about the time course of the maturation of the brain?

One important aspect of neural development is synaptogenesis. Synapses in the brain begin to form much before birth, prior to week 27 in humans, counting from conception, but do not reach peak density until after birth, during the first 15 months of life. Synaptogenesis is more pronounced early on in the deeper cortical layers and occurs later in more superficial layers, following the pattern of neurogenesis described earlier. At roughly the same time that synaptogenesis occurs, the neurons of the brain are increasing the size of their dendritic arborizations, extending their axons, and undergoing myelination (see also An Abundance of Neurons). Synaptogenesis is followed by synapse elimination (sometimes called *pruning*), which continues for more than a decade. Synapse elimination is a means by which the nervous system fine-tunes neural connectivity, presumably eliminating the interconnections between neurons that are redundant or do not remain functional. An example comes from ocular dominance columns in the visual cortex where there is initially more overlap between the projections of the two eyes onto neurons in the primary visual cortex than there is when synapse elimination is complete.

One of the central hypotheses about the process of human synaptogenesis and synapse elimination is that the time course of these events differs in different cortical regions. This is in contrast to brain development in other primates, in which synaptogenesis and pruning appear to occur at the same rates across different cortical regions. However, differences in methodology must be resolved before these interspecies variations will be wholly accepted. Nonetheless, there is compelling evidence for the idea that different regions of the human brain reach maturity at different times.

Peter Huttenlocher and his colleagues (e.g., Huttenlocher and Dabholkar, 1997) at the University of Chicago investigated the time course of synaptogenesis in auditory cortex and prefrontal cortex (middle frontal gyrus) in postmortem human brains ranging in age from 28 weeks from conception (one premature infant) to 59 years. They stained the tissue from the two cortical areas using the phosphotungstic acid (PTA) method, which stains proteins associated with synapses, and visualized the stained synapses using electron microscopy; the method involved counting the numbers of synapses per unit of cortical volume.

They found that synapses in the superior temporal region, in the auditory cortex, reached peak density earlier in postnatal development (around the age of 3 months) than did synapses in the association cortex of the frontal lobe, the density of which peaked around the age of 15 months. Synapse elimination, which occurs later in life, also appeared to end earlier in the auditory cortex than in the prefrontal cortex, although this finding remains less well established. The general pattern is shown in Figure 15.21 as the biphasic plot of the difference between the synaptic density values in the prefrontal cortex and those in the auditory cortex; initially density is greater in the auditory cortex and later it is greater in the prefrontal cortex. These data suggest that in humans, synaptogenesis and synapse elimination follow different time courses in sensory (and motor) cortex than in association cortex, like that in the prefrontal cortex.

Measurements of glucose metabolism also can be used in living persons to investigate development of the brain, and a time course similar to that in the foregoing has emerged from studies using positron emission tomography (PET) in infants, children, and adolescents. Harry Chugani (1998) at Wayne State University School of Medicine showed that glucose metabolism measured at rest rapidly increases with age in young infants and decreases during the teenage years. In newborns, glucose metabolism is highest in sensory and motor cortical areas, in the hippocampus, and in subcortical areas including the thalamus, brainstem, and cerebellar vermis. By the age of 2 to 3 months, more glucose starts to be used in parietal and temporal cortex, the primary visual cortex, as well as the basal ganglia and cerebellar hemispheres. Glucose utilization in the frontal cortex increases between the ages of 6 and 12 months. The overall level of glucose use in the developing brain increases until about the age of 4 years, plateaus through about age 10, and decreases gradually to adult levels between the ages of 16 and 18 years. These data argue for a similar developmental time course in the human brain as that described from the results of histological methods, such as measures of synaptic density, and therefore reinforce the idea that different regions of the human brain develop at different rates, with association areas lagging behind sensory and motor structures.

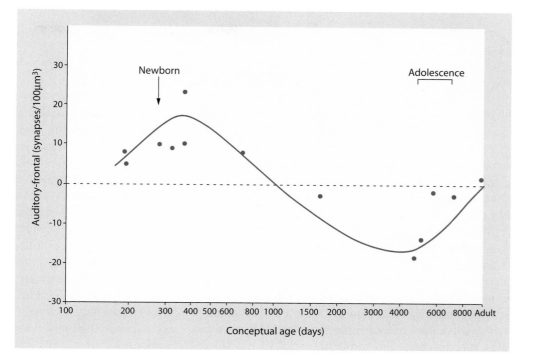

Figure 15.21 Plot of the difference score obtained by subtracting the value for synaptic density in the frontal cortex from that for synaptic density in the auditory cortex, as a function of the conceptual age of humans. Prenatal synaptic density peaks earlier for the auditory cortex, as reflected by the positive difference score, and later for the frontal cortex, as reflected by the negative difference score. From Huttenlocher and Dabholkar (1997).

Another trend in postnatal development is the significant increase in brain volume during the first 6 years. The increase appears to be the result of both myelination and the proliferation of glial cells. Recently, Jay Giedd, Judy Rappoport, and their colleagues at the National Institutes of Health, with Tomas Paus and Alan Evans and their colleagues at McGill University in Montreal, Canada, calculated growth curves for gray and white matter volume in the lobes of the developing human brain (Giedd et al., 1999). The research team used magnetic resonance images to quantify white and gray matter volumes in the same living volunteers as they aged from childhood through adolescence (ages 4 through 20 years); most of the group was scanned twice, but a few were studied three, four, or five times.

The researchers found that white matter volume increased linearly with age and that the time course was not different for different cortical regions. In contrast, gray matter volume increased nonlinearly, showing a preadolescent increase followed by a postadolescent decrease, a finding in accord with the results from PET studies and postmortem histological measures. In addition, the time courses of the gray matter increases and decreases were not the same across different cortical regions. In general, these data support the idea that postnatal developmental changes in the cerebral cortex may not occur with the same time course across all cortical regions.

What is the role of experience in the postnatal development of the brain? William Greenough and his colleagues (Kleim et al., 1998) at the University of Illinois showed that rats living in environments that require motor activity produce more glial cells than those that do not. Furthermore, rats living in environments that foster learning produce more dendritic branching than those that live in relatively impoverished environments. Similarly, sensory input and normal neural activity may be required for pruning to occur.

Summary of Cortical Development

From a single fertilized egg an entire human is constructed, and the intricacies of this process are controlled initially by powerful genetic prespecifications. Neurons are born, migrate, and then reside in their final locations in the brain to lay down the initial organization of the nervous system. Throughout the developmental process, however, the organization of the nervous system can be influenced by the environment. For the growing embryo and fetus, these influences are focused on the effects of cell-cell interaction. That is, the "environment" is the local neuronal environment, such that cells in the brain interact as they become connected to realize the final specialized organization and interconnectivity. Yet, even prenatally, external environmental factors can alter the normal organization of the brain, as, for example, when the fetus is exposed to repeated high doses of alcohol.

Postnatally, development is marked by several serial but temporally overlapping processes. There is a

HOW THE BRAIN WORKS

An Abundance of Neurons

The developmental events that lead to the adult brain include a fascinating wonder that is now a fundamental principle of corticogenesis: The developing brain produces more neurons and connections than will be used, and the numbers decline from this initial developmental explosion to the adult level. Between 30% and 60% more neurons are in the primate fetus than in the adult brain. More axons are present in the fetal and newborn primate than in the adult. Examinations of brain commissures (white matter tracts that connect the left and right hemispheres) reveal that the corpus callosum has 400% more axons in the prenatal monkey than in the 3-month-old infant or the adult. Peak numbers are found at birth. Axon loss occurs only after the cortex has reached its final connectional pattern, and thus axon loss does not reflect large-scale cortical reorganization. Rather, local circuits are primarily affected.

An exuberance of synaptic connectivity does not correlate with axon numbers in the developing brain.

The initial overproduction of synapses happens when axons are reducing in number. Unlike neuronal and axonal overproduction, synaptic overproduction occurs just prior to and after birth; the number of synapses declines through puberty (in macaque monkeys at about ages 3–4). Studies of the primate visual system showed that competition between inputs to a neuron leads to the strengthening of one contact and the weakening of others, which subsequently are eliminated. This competition is driven by activity in the afferent inputs, which appears to modify gene transcription in the neurons. The details of the mechanism are not fully understood, but we find a paradoxical pattern: The earlier in development, the more numerous the neurons, axons, and synapses in the brain. This coincides with periods of minimal motor, perceptual, and cognitive skills in the developing primate, and with a time of greatest plasticity in the brain, as in the ability to learn. Thus, as with most things, more is not necessarily better in the brain.

continuation of synaptogenesis, which begins prior to birth and in humans occurs at different rates in different brain regions, even in different cortical areas. Such heterogeneous developmental changes correlate with the developmental time courses of the behaviors we observed earlier in this chapter. The human brain changes until the late teens, when in most respects it resembles the adult brain. Throughout adulthood, the

brain changes very little if at all in terms of volume, myelination, and synaptic density, only showing reductions in old age. Yet, recent evidence has begun to provide increasing support for the idea that new neurons may well be produced throughout adult life, but future studies will have to determine whether these new neurons serve any functional roles in the adult nervous system.

PLASTICITY IN THE NERVOUS SYSTEM

It is obvious from the dramatic cellular events that occur during gestation that the nervous system is tremendously plastic during development. It can change its form, including the type and location of cells and how they are interconnected with one another. This developmental plasticity stands in stark contrast to the relative rigidity of the adult brain. Throughout development, however, some nonplastic-

ity is also a hallmark of the brain. For example, early in gestation undifferentiated precursor cells are fated to express the characteristics of the brain region where they migrate to and remain. Thus, both plasticity and nonplasticity are present during prenatal development.

During postnatal development, the brain changes. In some notable instances, such as the orientation

Plasticity in Fetal Tissues

Parkinson's disease affects approximately 500,000 Americans and is characterized by resting muscle tremors, rigidity of the limbs, difficulty in movement initiation, and slowness of movement execution. This neurodegenerative disease is most common in the elderly, but similar syndromes also have been induced by drugs in younger persons—street drugs are not pure, which can create catastrophic problems when impurities poison neurons in the brain.

Limb and body movements are controlled by the frontal cortex and the motor circuits of the basal ganglia. From the substantia nigra, dopaminergic neurons connect to the putamen, caudate, and globus pallidus of the basal ganglia (the nigrostriatal bundle). These areas also receive connections from the primary motor cortex and send neuronal connections back to the primary motor cortex and substantia nigra. Dopamine is synthesized in the nerve endings of the dopaminergic neurons whose cell bodies rest in the substantia nigra. Approximately 80% of the dopamine is found in the basal ganglia, which comprises less than 0.5% of the total brain weight.

With Parkinson's disease, the dopaminergic pathway from the substantia nigra to the basal ganglia is disrupted. Up to 90% of the dopaminergic neurons in the substantia nigra degenerate and there is a reduction in the synthesis of dopamine. L-Dopa (L-3, 4-hydroxyphenylalanine), the amino acid precursor of dopamine, can be injected into Parkinsonian patients to alleviate the disease's symptoms, but this leads to negative side effects, and its effectiveness diminishes with time.

Medical science has long searched for a way to protect or replace the damaged neurons of the Parkinsonian patient's substantia nigra, or at least to mimic their function by other means. One way is by fetal tissue transplantation. The object of fetal tissue transplantation is to introduce dopamine-producing cells, which have not fully differentiated, into brain areas where the cells can proliferate, form extensive neural connections, and release the deficient neurotransmitter.

Since the ban on fetal tissue transplantation was lifted in 1992, hundreds of operations worldwide have been performed. During transplantation, fetal mesencephalon cells rich in dopaminergic neurons are ectopically grafted to key areas in the motor circuit of the basal ganglia, namely, the caudate and putamen. Although success has varied, which makes generalizations difficult, the results look promising.

Widner and colleagues (1993) removed brain tissue from six to eight fetuses 6 to 8 weeks old, and placed it bilaterally into the caudate and putamen of two patients with MPTP (1-methyl-4-phenyl-1,2,3,6-tetrahydropyramidine) neurotoxin–induced Parkinson's disease. Positron emission tomography scans showed the uptake of radioactive L-dopa in the basal ganglia before transplantation (March 1988) and at two times after surgery (January and October 1990). The increased uptake of L-dopa indicated that the transplanted fetal cells successfully converted L-dopa to dopamine.

Parkinson's disease and its physical manifestations are not the only ailments treatable by transplanting fetal tissue. Cognitive abilities in rats with deficits induced by lesions and alcohol ingestion were restored almost to the levels of control animals, by grafting cholinergically rich fetal cells into the neocortex or hippocampus. The central nervous system's inability to regenerate after damage may be reversed by an increasing array of alternative therapies. We can predict a bright future for regeneration therapies.

---

columns of the visual cortex, factors concerned with normal activity via afferent inputs fine-tune the functional connectivity within the cortex. But postnatal plasticity is limited: Cells are not free to migrate to new areas or to make large changes in long-distance connectivity. In contrast, local cortical connectivity can be affected during **sensitive periods** when extrinsic influences can alter brain organization. Once these sensitive periods have passed, the central nervous system can be characterized by its marked lack of plasticity. Thus, damage to the central nervous system in the adult leads to irreversible damage; neurons do not regenerate damaged connections nor does the brain replace lost neurons, even though recent evidence suggests that new neurons are being born in the adult primate brain.

The inability of neurons to regenerate in the central nervous system may be an adaptive strategy to prevent the wiring of the nervous system from changing too much during life. This inability, in part, might be due to the interfering effects of glial cells, which form impenetrable scars after brain damage (Figure 15.22). In contrast to the central nervous system, the peripheral nervous system does regenerate severed axons. If no obstructions are present, a severed nerve in the arm will regenerate many millimeters or centimeters in order to reinnervate a denervated target, like a skeletal muscle. This growth can be as fast as 1 mm/day.

But, what if you could rewire a brain? What if you took visual input and rewired the connections so that

input to the eyes ended up in auditory cortex? Could the auditory cortex learn how to process visual inputs in order to "see"? Mriganka Sur and his colleagues at MIT (von Melchner et al., 2000) have asked these questions. They did so in experiments that addressed whether the perceptual modality of cortex can be shaped by the nature of the sensory-neural inputs it receives. In neonatal ferrets, they redirected retinal axons destined to the superior colliculus, to the auditory nucleus of the thalamus, the medial geniculate nucleus (MGN), in one hemisphere and left the other hemisphere untouched to serve as a control (Figure 15.23). The neurons in the MGN send their axons to the auditory cortex, even in the "rewired" hemisphere. After these animals were raised to adulthood, Sur's team found that the neurons in the auditory cortex of the

Figure 15.22 Inducing regeneration in the central nervous system (CNS). Neurons in the CNS **(top)** do not regenerate their connection with their target cells following damage **(middle)**, in part because glial cells interfere with regrowth. However, researchers have demonstrated that neurons in the CNS can regenerate their axons and can reinnervate their old targets if a proper pathway is provided. Grafts of the sciatic nerve from the leg of rats **(bottom)** can provide such a pathway, permitting the regrowth, in this example, of retinal ganglion cells innervating the superior colliculus (SC).

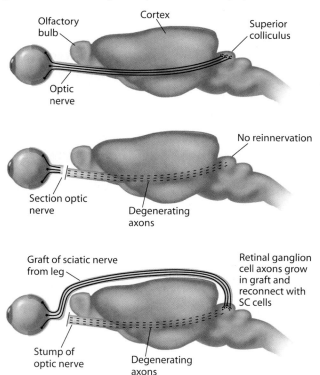

Figure 15.23 Surgical rewiring of visual inputs in the ferret brain. The diagram shows the rewired visual system of a ferret where right visual hemifield inputs from the retina destined for the superior colliculus (SC) were redirected to the medial geniculate nucleus (MGN) in the left hemisphere. In addition, the superior colliculus and the brachium (b) of the inferior colliculus (IC) in the left hemisphere were removed. Projections from the right hemifield to the lateral geniculate nucleus (LGN) were untouched, as were all connections from the left hemifield going to the right hemisphere, which served as a control. The rewired visual projections to the MGN in the left hemisphere innervated MGN neurons and influenced the properties of the auditory cortex, turning it into a visually responsive region of cortex. From von Melcher et al. (2000).

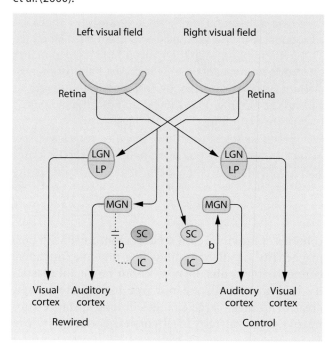

rewired hemisphere demonstrated many of the characteristics of visual neurons in their responses to visual stimuli, including showing orientation and direction selectivity. They were also organized in a retinotopic map of visual space. That is, these auditory cortex neurons behaved as visual neurons!

But what do the animals perceive? Does visual stimulation of these rewired auditory cortex neurons produce auditory or visual percepts in the animals? To find out, Sur's group performed a clever study. Using juice as a reward, they trained the ferrets to distinguish sound from visual stimuli. Importantly, they only trained the ferrets by presenting visual stimuli to the control hemisphere, so that the rewired hemisphere was not trained in the task. Once the animals were well trained in the task, the researchers presented the visual stimuli to the rewired hemisphere. They found that the animals acted as though the stimulus was visual, thereby demonstrating that the percept they experienced was indeed visual and hence driven by the rewired inputs to the cortex, not by the original pre-specification of the neurons as auditory. These results indicate that cortical specialization is determined, at least in part, by the kind of sensory input received by the cortex. Or to put it another way, sensory afferents can instruct cortex how to behave.

Plasticity in the Normal Adult Brain

There is clearly plasticity in the adult brain. After all, we learn, don't we? Learning appears to involve changes in the synaptic weights between neurons in the brain's circuitry, as, for example, in long-term potentiation (see Chapter 8). So there must be some plastic change in the adult brain given the fantastic behavioral plasticity that adults display. The questions are, How great is this change? To what extent might this reflect cortical reorganization in the adult brain? Fascinating experiments hold the answers.

CORTICAL MAPS AND EXPERIENCE

An amazing revolution in neuroscience has taken place since the mid-1980s. It has grown out of the work of Michael Merzenich at the University of California, San Francisco, and Jon Kaas (1995) at Vanderbilt University (Merzenich et al., 1988; Merzenich and Jenkins, 1995). They and their colleagues launched a research program on how sensory and motor maps in the cortex can be modified with experience. To understand the story, we need details about cortical organization.

The body's sensory surface and our external auditory and visual worlds are represented in cortical maps. Part of our cerebral cortex, for example, is where neurons respond to the stimulation of points on our body—the somatosensory cortex. When we analyze how all the neurons respond, we discover maps in the cortex that provide a point-for-point re-representation of the body's surface. There is a map for the hand, the face, the trunk, the legs, the genitals, and so on. It is also true that the higher the resolution a sensory surface has, the more neurons there are to represent that area. The neurons that respond to fingertips are greater in number and more densely packed than the neurons that respond to the back of the hand. Hence, fingertips are more sensitive than the back of the hand; this is known as the *cortical magnification factor.*

Body maps are also nicely organized such that the index finger is coded by neurons that are next to middle-finger neurons, which are next to ring-finger neurons, which in turn are next to the neurons coding inputs from the little finger. This is known as **somatotopy,** and these cortical areas are somatotopic maps. The maps are present in adult animals and humans and appear to be the basis for the ordering of our perceptions; they reflect the receptive field properties of the cortical neurons. It is not obvious why these maps exist the way they do. There is no inherent necessity for the organization. They are simply there—perhaps as a parsimonious organizational feature or perhaps as a means of coding relative relations as a place code—and because of this, fascinating phenomena have been discovered. Merzenich's and Kaas's seminal observation was that these maps change after peripheral receptors and nerves have been manipulated in adult animals. This manipulation can be achieved by cutting the nerves innervating a portion of a limb, by tampering with the normal relations between adjacent regions of the limbs (e.g., by sewing together the fingers of one hand), or by altering the demands on the somatosensory system by increased use. They found, for example, that when a finger of a monkey is denervated, the relevant part of the cortex no longer responds to the touch of that finger (Figure 15.24). This is perhaps no big surprise—after all, the cortex is no longer receiving input from that finger. But here comes the strange part: The area of the cortex that formerly represented the denervated finger soon becomes active again and responds to stimulation of the finger adjacent to the amputated finger. The surrounding cortical area fills in the silent area and takes it over. This functional plasticity suggests

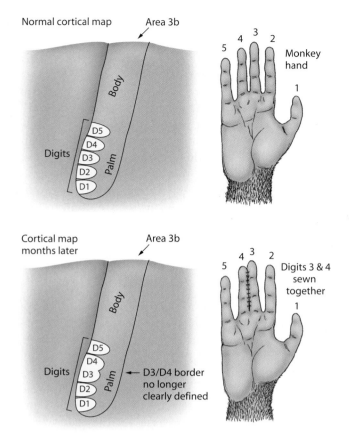

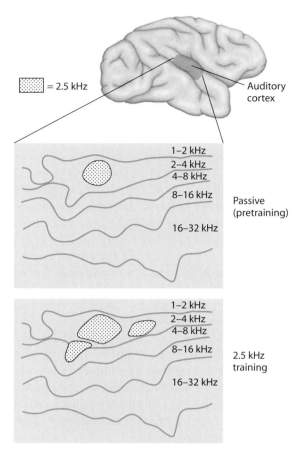

Figure 15.24 Reorganization of sensory maps in the primate cortex. At the top left is a mapping of the somatosensory hand area in a normal monkey cortex. The individual digit representations can be revealed using single unit recording. If the two fingers of one hand are sewn together, months later the cortical maps change such that the sharp border once present between the sewn fingers is now blurred. Adapted from Kandel et al. (1991).

Figure 15.25 Changes in auditory frequency mapping following training. Gregg Recanzone and colleagues (1993) showed that training animals to discriminate specific tone frequencies leads to an enlargement of the cortical regions mapping the trained frequency.

that the adult cortex is a dynamic place where changes can still happen. Such phenomena demonstrate a remarkable plasticity.

This cortical plasticity is not limited to the somatosensory system. It can also be observed in the auditory system. Gregg Recanzone, now at the University of California at Davis, and Michael Merzenich trained owl monkeys to perform a tone discrimination task for certain frequencies (Recanzone et al., 1993). They examined the tonotopic organization of the auditory cortex in the trained monkeys and compared it to control monkeys. The monkeys who had been performing the tone discrimination task had increased cortical representation for those relevant frequencies in the auditory cortex (Figure 15.25). Thus, as in the somatosensory cortex, the auditory cortex has functional plasticity.

The visual system also can have such changes, as Charles Gilbert and Torsten Wiesel (1990) demonstrated. The visual cortex, just like the somatosensory cortex, has a beautiful topographic map of the visual world, called *retinotopic organization*. Stimulation of one point on the retina leads to a certain group of neurons responding in the visual cortex. Stimulation at an adjacent point on the retina causes a response in an adjacent area in the cortex, and so on. Gilbert and Wiesel created lesions in each retina to see what happened to the visual maps based on new recordings from the visual cortex. It is important to understand that when one is recording from a single neuron in the visual cortex, that neuron is responding to a stimulus in the real world. Remember that early in the visual system a neuron responds only to a certain narrowly defined area in the visual world, a space called the *receptive field* of that neuron. Gilbert and Wiesel found that almost instantly the size of the receptive fields changed for neurons adjacent to the lesioned area (Figure 15.26). After a few months, the cortical area that became silent due to the retinal lesions began to respond when the unlesioned but adjacent retinal regions were stimulated. However, note one crucial point of anatomy: After creation of a retinal lesion, there is also a silent area created in the lateral geniculate nucleus, the thalamic relay to the striate cortex. Gilbert and Wiesel showed that this region of the lateral geniculate remained silent even after changes were seen in the cortex, which means all the changes at the higher-level were accomplished by plasticity in cortical neurons.

Reorganization in Human Cortex

Are phenomena of cortical functional plasticity limited to animals? How would one determine the answer to such a difficult question? Remember Professor Vilayanur Ramachandran who was introduced at the beginning of this chapter? His studies of amputees brought these dramatic animal results into the realm of human phenomena. Were you able to figure out the reason for the findings he obtained from his patients? Humans have maps of the body surface just like the experimental animals. When the whole map is considered, strange juxtapositions emerge. Consider the cortex's surface. Starting at one point in the somatosensory cortex, the pharynx is represented, then the face, with the sensitive lips having a huge area of representation, and next to the face are the fingers and hand, followed by the arm and leg (Figure 15.27).

Ramachandran reasoned that a cortical rearrangement ought to take place if an arm is amputated, just as Merzenich, Kaas, Recanzone, and their colleagues found. Such a rearrangement might be expected to create bizarre patterns of perception. Remember that the face area is next to the hand and arm area—an important point for understanding this story. Suppose there was no lower arm or hand to send sensory signals to the brain. According to animal research, the region coding for the arm that had been amputated might become functionally innervated by the surrounding cortex. That is exactly what happened in several spectacular cases like the one we described earlier. Here is the whole story.

Ramachandran studied a young man who recently had had his arm amputated just above the elbow. About 4 weeks after the amputation, he was tested.

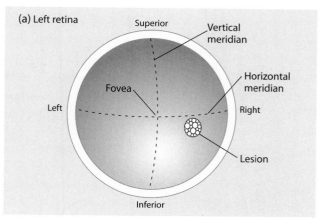

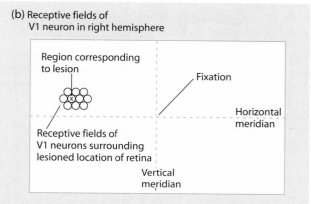

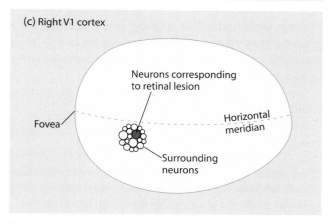

Figure 15.26 Reorganization of cortical visual receptive fields in response to retinal damage. **(a)** A representation of the retina shows the area of the lesion and the fovea. **(b)** The lesion in the lower right portion of the retina produces a scotoma (blindspot in the visual field); the lesioned region is indicated by an x. **(c)** The sizes of the receptive field of neurons in the visual cortex change when the fields are adjacent to those corresponding to the retinal lesions. This process is called *filling in*. It occurs over time and mitigates the negative impact of the lesions by allowing nearby neurons to assume some of the processing responsibilities of the neurons that lost their inputs owing to the retinal lesion.

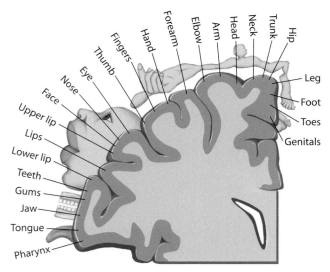

Figure 15.27 Map of a human homunculus in somatosensory cortex. This cross section (coronal section) shows only the dorsal half of one hemisphere. The medial surface of the hemisphere is at the right and the lateral surface, at the left. The cortex is indicated by the darker green color. The section is at the anterior-posterior level just posterior to the central sulcus, within the somatosensory cortex of the parietal lobe.

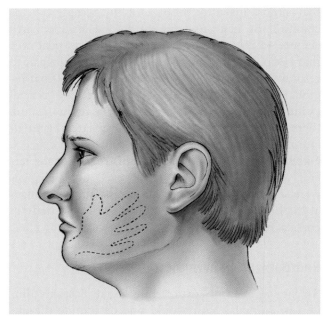

Figure 15.28 The drawing shows the hand representation sketched on the face of the amputee studied by Professor Ramachandran. The hand map on the face was obtained as described in Figures 15.1 and 15.2. Do you now have an idea as to why this might occur? See text for discussion.

When a Q-tip was brushed lightly against his face, he reported feeling his amputated hand being touched! Feelings of sensation in missing limbs are the well-known phenomenon of *phantom limb sensation*; however, this case is different because the sensation was introduced by stimulating the face.

Indeed, with careful examination, a map of his hand could be demonstrated on his face (Figure 15.28)! Ramachandran (1993) reported another case, which is too good not to quote verbatim:

> A neuroscience graduate student wrote to us that soon after her left lower leg was amputated she found that sensation in her phantom foot was enhanced in certain situations—especially during sexual intercourse and defecation. Similarly an engineer in Florida reported a heightening of sensation in his phantom (left) lower limb during orgasm and that his experience . . . actually spread all the way down into the [phantom] foot instead of remaining confined to the genitals: so that the orgasm was much bigger than it used to be.

These are dramatic examples of plasticity in humans following significant injury or altered experience (see also An Interview with Helen J. Neville). But is there evidence that changes in experience within the normal range, say, due to training, have any ability to influence the organization of the adult human brain? The evidence from the animal literature would tend to suggest that the answer is yes. Even though evidence for normal plasticity in humans might be more difficult to demonstrate, perhaps being smaller in scale than changes induced by events such as the loss of a limb, there is some early evidence to support the idea.

Thomas Elbert, Brigitte Rockstroh, and their colleagues (1995) at the University of Konstanz and University of Muenster in Germany, and the University of Alabama at Birmingham, used magnetoencephalography (MEG) to investigate the somatosensory representations of the digits on the left hand in string musicians and nonmusician control persons. They stimulated the thumbs and fingers of the volunteers with a precisely controlled device that touches the skin, and found that the responses were larger in the musicians (Figure 15.29). These findings suggest that a larger cortical area was dedicated to representing the sensations from the fingers of the musicians, owing to their altered but otherwise normal sensory experience. Interestingly, these musicians typically began training early in life, and the size of the effect (the enhancement in the response) correlated with the age at which they began their musical training. Would these types of training effects be evident in the adult?

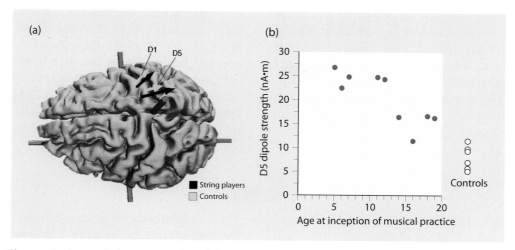

Figure 15.29 Changes in the cortical representation of the human fingers in musicians who play string instruments. **(a)** Top view of a three-dimensional reconstruction of a human brain from MRI; anterior is to the left. The arrows over the right hemisphere indicate the calculated locations of neural activity recorded using MEG for the thumb (D1) and fifth finger (D5). The black arrows represent the thumb and fifth finger for musicians, while the yellow arrows represent nonmusicians. The size of the arrows shows the strength of the responses (to stimulation of the thumb and finger). Musicians show larger responses and a larger area of cortical representation for the thumb and fifth finger (distance between arrows). **(b)** The size of the cortical response plotted as a function of the age at which the musicians begin training. Responses were larger for those who began training prior to the age of 12 years; controls are shown at the lower right of the graph. From Elbert et al. (1995).

Leslie Ungerleider, Avi Karni, and their colleagues (Karni et al., 1995) at the National Institute of Mental Health asked this question with regard to the motor system. They asked a group of volunteers to perform a simple motor task with one hand; the task required the subjects to touch their finger to thumb in a particular sequence. They only had to perform the task a few minutes each day, but their accuracy and speed of finger-thumb touching improved with this practice. Using functional magnetic resonance imaging, the scientists measured the size of activity-related changes in blood flow in the motor cortex for the practiced sequences and untrained sequences. Figure 15.30 shows the results; there were greater changes in blood flow in the corresponding motor cortex for trained than for untrained sequences after only a few weeks. These data argue that in the normal brain, training can induce relatively rapid changes in cortical organization that reflect the plastic ability of the nervous system to acquire and retain new information and skills.

Together, these studies indicate that a significant degree of functional plasticity exists in the human cerebral cortex. The term *functional plasticity* is used

Figure 15.30 Changes in the size of cortical activation of hand movements with training. **(a)** A sagittal section through the right hemisphere of a representative subject shows greater activation (larger area activated—yellow-orange color) for motor sequences that were repeated over many sessions in comparison to motor-evoked activity for untrained sequential movements **(b)**. These activations were in M1, the primary motor cortex. **(c)** The increased activation for trained movements in comparison to untrained movements **(d)** lasted up to 8 weeks, even when no training ensued on the task in the interim. Adapted from Karni et al. (1998).

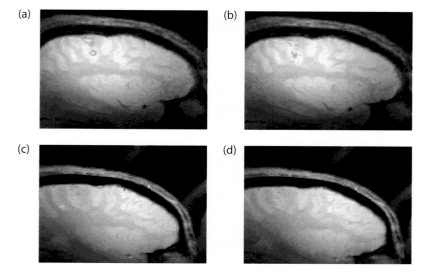

because the effects may not be caused by a physical reorganization in cortical neuronal circuitry. Rather, in the normal case, the receptive fields of neurons overlap more than what is apparent from recordings from single neurons. Most of the time this overlapping connectivity produces no sensory effects, but when the main input to a cortical region is removed, the secondary inputs from nearby parts of the sensory map appear to become functional. This effect leads to changes in the observed shapes of the cortical maps, in, for example, the case of amputation. For normal changes that occur with learning, such as in the case of motor skill learning, the precise mechanism has yet to be firmly established but may involve long-term experience-induced changes in the synaptic strengths among participating neurons.

SUMMARY

The development and plasticity story is complex, beginning with fantastic plastic changes as the brain and body form during embryogenesis and concluding with a much more rigid and less plastic nervous system in the adult. In between, during early development, important changes take place in correspondence with critical periods during maturation. As a result, permanent changes in human brain organization occur by the loss or alteration of sensory inputs during these sensitive periods of early life.

Even in the adult, however, plastic changes occur, as we are still capable of learning throughout life. Plastic changes can be observed in the sensory cortical areas of animals or humans with lost or denervated limbs, and after training, as in the case of motor learning. The story is not one of a completely plastic brain but of controlled development and maturation that lead to the marvelous complexity of the adult brain, which we now know shows various forms of plasticity. Throughout development and maturation, development and plasticity support the rise of cognition in a steady march toward language, literature, and reason—the province of the human brain.

KEY TERMS

corticogenesis

critical periods

ectoderm

innate

looking time

neural plate

neuronal determination
   and differentiation

neuronal migration

neuronal proliferation

Piagetian theory

plasticity

precursor cells

radial glia

sensitive periods

somatotopy

synapse elimination

synaptogenesis

universal grammar

THOUGHT QUESTIONS

1. How do developing neurons migrate from the locations where they are dividing and being "born" to where they must finally reside in cortex?

2. Can a visual neuron process auditory information if it is transplanted from visual to auditory cortex? If so, when must such transplantation occur and what does it tell us about developmental specification?

3. Infants born anywhere in the world to parents of any racial or ethnic group can learn their native language and become speakers of their native language. However, if they move to another country later in life, they have accents in their speech. But why is it that if they move to another country when young enough, they can learn to speak like a native speaker? Is this merely a practice effect because of the extra time spent in the new country? What happens in the latter case to their ability to learn their native language if they move back later in life?

4. Once the human brain is past the critical period in development and is fully developed, damage has less impact on its organization. But how would a person losing a limb in an accident in adulthood feel the missing limb, and why might this be triggered by stimulation of the remaining body parts?

5. Expert violinists may have larger regions of the brain dedicated to somatomotor processing in the hand and finger region of the brain. Are these individuals born with more cortex dedicated to somatomotor processing? What evidence in animals helps resolve the story of this form of plasticity?

SUGGESTED READINGS

BAILLARGEON, R. (1995). Physical Reasoning in Infancy. In M. Gazzaniga (Ed.), *The Cognitive Neurosciences* (pp. 181–204). Cambridge, MA: MIT Press.

GOLDMAN-RAKIC, P.S. (1987). Development of cortical circuitry and cognitive function. *Child Dev.* 58:601–622.

GROSS, C. (2000). Neurogenesis in the adult brain: Death of a dogma. *Nature Rev. Neurosci.* 1:67–73.

JOHNSON, M.H. (1997). *Developmental Cognitive Neuroscience.* Cambridge, MA: Blackwell Science.

NEVILLE, H., and BAVELIER, D. (2000). Specificity and Plasticity in Neurocognitive Development. In M. Gazzaniga (Ed.), *The New Cognitive Neurosciences*, 2nd edition (pp. 83–98). Cambridge, MA: MIT Press.

RECANZONE, G., SCHREINER, C.E., and MERZENICH, M. (1993). Plasticity in the frequency representation of primary auditory cortex following discrimination training in adult owl monkeys. *J. Neurosci.* 13:87–103.

16

The Problem of Consciousness

It was a boiling hot day in southern California in the early 1960s. One of us (M.S.G.) was playing horseshoes in the backyard of patient W.J.'s home. W.J. was the first human in recent times to have his brain split to control his otherwise intractable epilepsy.

In those years the race was on to study the effects of disconnecting the two halves of the brain of animals. Research on cats, monkeys, and even chimpanzees showed that when one half of the brain was trained on visual or tactile discrimination problems, the other half remained ignorant of the training. It was as if someone had trained you such that every time you picked up an apple (and not an orange) with your left hand, you got five dollars, but if you used your right hand, you would not know to pick up the apple. It seemed too much to believe. Maybe animals were somehow different. But humans? No way.

The surgical idea was that by cutting the fibers that connect the two halves of the brain, an epileptic seizure starting in one hemisphere would not spread to the other. In such an instance, while one half of the body might seize, the other would remain seizure free, so the patient would not lose consciousness. W.J. was tested prior to his split-brain surgery and showed no disconnection effects. When an apple was placed in his left hand, he could name it, he could find the same apple with his right hand, he could name visual stimuli presented to either half of the brain, and so on. In short, he was normal with respect to sensory, motor, perceptual, and cognitive processes.

After his split-brain surgery, W.J. named and described information presented to his left (speaking) hemisphere. What was surprising was his apparent lack of response to stimuli presented to his surgically isolated right hemisphere. It was as if he was blind to visual stim-

uli presented to the left of where his eyes were fixated. Yet it became obvious that while the left (talking) hemisphere could not report on stimuli presented to the right hemisphere, the right hemisphere could easily react to a simple visual stimulus via its ability to control the manual responses of the left hand.

An early conclusion about these phenomena suggested that each half of the brain behaved independently of the other. Information experienced by one side seemed unavailable to the other. Moreover, each half appeared to be specialized for certain mental activities. The left was superior for language, while the right was more adept at carrying out visuospatial tasks. The surgeon had separated structures with specific and complex functions.

The capacities demonstrated by the left hemisphere were no surprise. But when the first patients were able to read (but not speak) from the right hemisphere and were able to take that information and choose between test alternatives, the case for a dual, and now independent, conscious system appeared to be strong. After human cerebral hemispheres were separated, each half functioned outside the conscious realm of the other. Each could learn, remember, emote, and carry out planned activities.

Back to the story. As the game of horseshoes wore on, W.J. suddenly paused. He had been playing with his right hand, which meant his left (dominant) hemisphere—the hemisphere that talks and thinks and generally runs our mental life—had been in charge. W.J. had been losing the game and outwardly seemed nonchalant about it. He walked over to the side of his yard, picked up an ax with his left hand, gripped it firmly, and swung it around a couple of times, all the while smiling.

The left hand gains its major control from the silent, generally intellectually limited right hemisphere. Was it annoyed with its status in the game? Was it an independent conscious entity, different from the left mind? If it attacked, which half of the brain would be prosecuted and charged with the crime? In short, had surgical divisions of the cerebral connections between the two halves produced a person with two conscious systems?

Conscious experience is a wonderful thing. We all have it, we all talk about it, and we pity people who do not have it. Watching a trauma patient in a coma is grim. As his eyes stare off into space, even though the heart beats strongly and the muscles remain firm, we quickly realize that consciousness is the most precious jewel of our existence. When conscious experience begins to fade, as in Alzheimer's disease, the sight is stressful for everyone. The patient is awake, even alert, but out of touch with what is happening around her, even unsure about her own identity.

In normal activity, we speak of raising people's consciousness. That is what education is all about. It is one thing to be able to read a comic strip. It is another to understand the nature of gravity and to appreciate where planet Earth is located in the universe. Having such knowledge affects one's conscious awareness.

Yet no one has been able to define consciousness to anyone's satisfaction. Consciousness is the key concept of mind-brain research, and over the years it has stayed on the sidelines. Some say that the origin of consciousness is so complex that the human brain cannot grasp it. For humans to understand consciousness would be like a flatworm trying to fathom a monkey.

People certainly do try to understand what is meant by consciousness. In the last 15 years, more than 15,000 articles have been published on the topic. Yet headway is limited, at best. The last 15,000 articles in molecular biology produced great advances in our appreciation of the molecular nature of biological processes. In some instances, this made a real difference in understanding diseases or in gene identification. But on the subject of consciousness, the publications appear to be only words and statements about viewpoints on the topic.

Some of the leading philosophers of our time, such as the University of California's John Searle, maintained that science will never understand the nature of subjective experience. Searle (1992) claimed that **subjectivity** is beyond the descriptive resources of objective science as we now conceive it. It follows from this that we can never have an adequate theory of consciousness unless we treat irreducibly subjective concepts such as feelings as basic objects or explanatory constructs, just as atoms or ions or force fields are used in the physical sciences.

But most scientists are not swayed by this argument. The quickness with which our ignorance evaporates is exemplified by merely noting what a leading academician said a mere 145 years ago about the creation of the world. As Robert J. Wenke (1980) pointed out in his fascinating book on prehistory, "Indeed, in the 1850s the eminent Dr. Lightfoot of Cambridge University, on the basis of his study of the Book of Genesis, proclaimed that the world had been created on October 23, 4004 B.C., at the civilized hour of 9:00 A.M."

We have come a long way in a short time. Leading philosophers such as Patricia Churchland of the University of California, San Diego, have raised the hope that we are dealing with a tractable problem. During her Presidential Scholar address at the year 2000 annual meeting of the Cognitive Neuroscience Society, Churchland named the scientific pursuit of consciousness as the most important outstanding problem for neurobiology. She related the truth that most knowledge we now receive from our cultural history was once held out as impossible to know. The mysteries of evolution, genetics, cellular mechanisms, and vision were stupefying only a few-dozen years ago. Now explanations abound for all these processes. Why, she asked, can the same not follow for the problem of consciousness?

Naysayers may ask, how can we study a problem that does not yet have an agreed-on definition? On the surface, such complaints seem plausible. Yet, time and time again in science, people work on poorly defined problems, and as they stumble through the darkness, precision comes to their observations and to the definitions that motivate their initial reasons for doing experiments. The argument that we cannot learn about something because we do not know enough about it is an oxymoron. A much more productive posture is to study consciousness from every angle possible and come up with definitions and new methods for studying the mind.

Understanding what makes us uniquely human has been a hotly debated topic ever since humans have been debating. What is it that sets us apart from other animals? Certainly, it is not that we can think, because other animals think too. Maybe it is that we can think about thinking. Or that we can write about thinking about thinking. But why? The ability for self-reflection, the ability to articulate a theory of mind, is at the root of understanding consciousness.

Why is understanding **consciousness** important? It certainly seems like conscious awareness drives our behavior. But is this truly the case? How do we know how to respond? Do we even have to know how to respond? Often we simply react to survive. In the past, we might have dodged a tiger in a jungle. Now, we are dodging cars in the street. What is even more intriguing is that we are often dodging even before we know we are dodging! Certainly, we cannot be aware of everything that drives our behavior—behavior without conscious awareness is necessary for survival. If we waited to see the car clearly, we might not survive to dodge another car. Of course, we are also aware of lots of behavior before it even occurs. If not careful, we could become paralyzed by indecision. Thus, understanding how the intricate interplay of processing with and without awareness gives rise to our conscious percept is important for understanding how we manage to survive in such a complex, information-rich world.

In this chapter we will discuss several cognitive neuroscience approaches to understanding human consciousness. We will start by reviewing a major divide in the philosophical views of consciousness—dualism versus materialism. Next we will consider how studies on deficits in visual awareness help us understand vision with and without awareness, a topic we will explore in the intact, healthy brain. We also will compare and contrast theory of mind research in chimpanzees and human infants. Finally, we will tie it all together with speculation concerning the evolutionary component of consciousness.

PHILOSOPHICAL PERSPECTIVES

The problem of consciousness has been called the mind-brain problem or the ontological problem. It encompasses many questions: What is the real nature of mental states and processes? In what medium do they take place? How are they related to the physical world? Does consciousness survive death? Can a purely physical system construct conscious intelligence?

Two philosophies attempt to give direction to this problem—dualism and materialism. **Dualism** takes the stand that mind and brain are two distinct phenomena, whereas **materialism** asserts that both mind and body are physical mediums. Within each of these philosophies, views differ on the specifics. Paul Churchland, at the University of California, San Diego (1988), neatly outlined the distinctions.

All forms of dualism have a common premise: Conscious experience is nonphysical and beyond the scope of the physical sciences. The pure form of this view comes from Descartes, who believed the mind and the body are two completely separate entities. Descartes was not too clear on how these two entities interacted: When the mind decided to move a hand, the hand moved.

Another form of dualism is referred to as *popular dualism;* the idea is that people are "ghosts in the machine" (brain) and the spatial properties of the ghost interact with the spatial properties of the brain. This idea moves from the contention that matter is merely a manifestation of energy. The problem is that we have no evidence for a nonmaterial thinking substance that survives death.

A more interesting form of dualism is called *property dualism.* In this view there is no substance beyond the physical brain, but it has unique nonphysical properties possessed by no other physical object. Over the years this view has gradually changed. The idea was first referred to as *epiphenomenalism,* which meant that mental phenomena are not part of the physical phenomena in the brain but rather ride above the fray. Mental phenomena are caused by various brain activities but do not have any causal effects on the brain itself.

Epiphenomenalism has given away to what Churchland called *interactionist property dualism.* This view suggests that mental phenomena can affect the brain and thereby behavior. As such, mental properties are emergent ones that do not appear until ordinary physical matter has managed to organize itself through evolution into complex systems. Also, mental properties are irreducible. They are not just organizational features of physical matter; they are novel properties of the brain. Examples include the concept of something being painful or fragrant or colorful. These mental states emerge from the brain's physical processes and, once triggered, can turn back on the brain and guide lower information processing.

With advances in cognitive neuroscience, most philosophers and scientists do not champion the dualists' idea in any of its forms. More typically they are materialists. Science has proved how parts of the brain have specific roles in our mental life. Lesion one region, and a person's emotional state can change. Change another part, and the patient loses the ability to recognize faces. While these findings do not completely rule out dualism, they do suggest that the brain enables mind. And, just as with dualism, there are many forms of materialism.

Philosophical behaviorism is one form of materialism. Here the view is that one cannot talk about inner experience at all. Rather, one simply talks about a person's capabilities and dispositions because these can be measured. This simplistic approach to the problem of the nature of conscious experience has been abandoned in recent years. After all, we do have inner experience, mental imagery, thoughts that are never expressed, and so forth.

According to *reductive materialism,* on the other hand, mental states are physical states of the brain, and each type of mental state or process is numerically identical with some physical state or process in the brain. While most brain scientists believe this, it has been difficult to tie the myriad of mental processes to specific brain locations. Although we can catalogue the large-scale systems in mental activities such as language, we are not sure where mental states like ennui are managed.

Perhaps the most favored theory of mind and brain and the phenomena of consciousness is that of **functionalism.** This theory is widely adopted by psychologists, philosophers, and the artificial intelligence community. Its biggest proponent is Daniel C. Dennett (1981), one of the leading philosophers of the twentieth century. Functionalism differs from behaviorism in an important way. The behaviorist believes that defining environmental input and behavioral output will be sufficient for understanding mind; the functionalist does not. For the functionalist any one mental state makes reference to other mental states. In trying to understand a mental state with this view, there could be no possible explanation in solely behavioral terms.

Functionalists also believe that anything that looks like it feels pain or sees red or thinks appropriately is functionally equivalent to the human brain. Thus, an artifact, such as the one being built at MIT by Rodney Strong, can become an equivalent human. Cog, as the robot at MIT is dubbed, is being built in an attempt to construct an entity that does all things human conscious agents do. If the developers succeed, they believe the system will be equivalent to a human mind and should receive all due rights and honors.

But the functionalist approach is challenged by many because it ignores subjective experience, or **qualia** as it is sometimes called. When this issue is put to leading proponents of functionalism such as Dennett, they maintain that we are worried over nothing. When the cognitive neuroscience of intelligence, language, feelings, memory, attention, and perception is explained, qualia will come along free. It, too, will be understood.

As the battle of theories rages on, we are left with trying to explain how the brain enables mind. Dualism tends to ignore biological findings, and materialism overlooks the reality of subjective experience. As we move forward in our understanding of conscious process, we will be best served by collecting new data and observations. To this end, Searle (2000) has proposed completely revamping our approach to understanding consciousness. He suggests discarding the whole dualism versus materialism dichotomy in favor of what he calls "biological naturalism." He states that we should reject dualism outright because consciousness is obviously a biological phenomenon that is realized in the brain. However in rejecting dualism, Searle does not believe it is necessary to embrace materialism. He feels that materialism is also an untenable explanation for consciousness since it denies the existence of subjectivity. Instead, Searle proposes that we accept consciousness as a naturally occurring biological phenomenon that we all view with individualistic subjectivity. To explore biological naturalism, Searle says that we should ignore the notion that consciousness is made up of various component parts—what he calls the *building block approach*—and should adopt a unified field theory approach. The unified field theory approach would take a more basal approach to understanding consciousness, by first exploring consciousness as a unified concept at the sleep/wakefulness border. By determining the differences between unconscious sleep and conscious wakefulness, we can investigate consciousness as a whole and therefore understand it as a whole. Unfortunately, such basal exploration disallows investigations of individualistic subjectivity—the very feature used to define biological naturalism.

Taking a more succinct and cognitive neuroscience approach, Steven Pinker of MIT pulled together a framework for thinking about the problem of consciousness. In his book *How the Mind Works* (1997a), Pinker reviewed the work of the linguist Ray Jackendoff of Brandeis University and the philosopher Ned Block at New York University. These mind scientists observed how people who write on the topic are guilty of using the term *consciousness* in so many ways that it becomes impossible to ascertain what each is talking about. The proposal for ending this confusion consists of breaking the problem of consciousness into three issues. Pinker summarized and embellished the view as follows:

- **Sentience: Sentience** refers to subjective experience, phenomenal awareness, raw feelings, first-person tense, what it is like to be or do something. If you have to ask, you will never know.

An Interview with Daniel C. Dennett, Ph.D.   Professor Dennett is a leading philosopher of consciousness and the mind-brain problem. He is director for the Center for Cognitive Studies at Tufts University.

Authors: Many scientists like to use the term *qualia* when discussing consciousness. It seems to refer to phenomenal awareness, raw feelings, that sort of thing. Can we get out on the table what you mean by *qualia*?

DCD: I thought you'd never ask. *Qualia* are the souls of experiences. Now do you believe that each human experience has its own special and inviolable soul?

Authors: What are you getting at? What on earth does that even mean?

DCD: That's just my wake-up call for people who think they know what qualia are. It's frustrating to learn that in spite of my strenuous efforts, people keep using the term *qualia* as if it were innocent. Consider a parallel: According to Descartes (and many churches) the difference between us and animals is that animals have no souls. Now when Darwin showed that we are a species of hominid, did he show that there really aren't any people after all—just animals? If Darwin is saying we're just animals, he must be denying we have souls! So he must be saying that people aren't really people after all!

In spite of tradition, the very real and important differences between people and (other) animals are not well described in terms of the presence or absence of souls fastened to their brains. At least I would hope most of your readers would agree with me about that. Similarly, the differences between some mental processes and others are not well described in terms of the presence or absence of qualia—for what are they? Not only is there no agreed-upon definition among philosophers; controversies rage. Until they get settled, outsiders would be wise to avert their gaze, and use some other term or terms—some genuinely neutral terms—to talk about properties of subjective experience. In fact the term *qualia,* which is, after all, a term of philosophical jargon, not anything established in either common parlance or science, has al-

ways had a variety of extremely dubious connotations among philosophers. Denying there are qualia is more like denying there are souls than denying that people are much smarter than animals. If that makes qualia sound like a term one would be wise to avoid, good!

To put it bluntly, nobody outside of philosophy should take a stand on the reality of qualia under the assumption that they know what they're saying. You might as well express your conviction that trees are alive by saying they are infused with *elan vital.* So when Francis Crick, for instance, says that he believes in qualia, or when Gerald Edelman contrasts his view with mine because his view, unlike mine, allows for qualia, these pronouncements should be taken with more than a grain of salt. I'd be very surprised if either Crick or Edelman—to take two egregious examples—believes in what the philosophical fans of qualia believe in. If they do, they have a major task ahead of them: sorting out and justifying their claims against a mountain of objections they've never even considered. I would think they'd be wise to sidestep the mess.

I fear I'm losing the battle over the term *qualia,* however. It seems to be becoming the standard term, a presumably theory-neutral way of referring to whatever tastes and smells and subjective colors and pains are. If that's how it goes, I'll have to go along with the gang, but that will just make it harder to sort out the issues, since it means that all the controversies will have to be aired every time anybody wants to ensure that others know what is being asserted or denied. Too bad. Don't say I didn't warn you.

Authors: Well, OK. Qualia is doomed to mean the feeling about the specialized perceptual and cognitive capacities we humans enjoy. Put directly, should we not distinguish between the task of characterizing the cog-

---

• **Access to information:** This is the ability to report on the content of mental experience without the capacity to report on how the content was built up by the nervous system. Information processing in the nervous system falls into two pools: One pool, which includes the products of vision and the contents of short-term memory, can be accessed by the systems underlying verbal reports, rational thought, and deliberate decision making. The other, which includes autonomic (gut-level) responses, the internal operations of vision, language, and motor control, and re-

nitive operations of the human mind and the, here we go, qualia we have about them?

DCD: Certainly we should divide and conquer. So we should distinguish between the task of characterizing some of the cognitive operations of the human mind, and the rest (which we conveniently set aside till later); but if we call the latter *qualia* and think that they are somehow altogether different from the cognitive operations we are studying now, we prejudge a major question.

Take experience of color, every philosopher's favorite example of a qualia. Suppose what interests you as a cognitive scientist are the differences in people's responses to particular colors (Munsell color chips will do for standard stimuli, at least for this imaginary example). But instead of looking at such familiar measures of difference as size of JNDs [just noticeable differences], or latency of naming, or choice of color words (where does each subject's pure red lie on the spectrum, etc.), or galvanic skin response, or an ERP [evoked-response potential] difference, suppose you looked at variations in such hard-to-measure factors as differences in evoked memories, attitude, mood, cooperativity, boredom, appetite, willingness to engage in theological discussion—you name it. Until you've exhausted all these imponderable effects, you haven't covered all the cognitive or disposition-affecting factors in subjective color experience, so there will be features of color experience, features of what it is like for each individual, that you are leaving out of your investigation. Obviously. But if you then call these unexamined residues *qualia* and declare (or just assume) that these leftovers are somehow beyond the reach of cognitive science, not just now but forever, you are committing a sort of fallacy of subtraction. There need be nothing remarkable about the leftovers beyond their being leftovers (so far). When some qualia freak steps up and says, "Well, you've got a nifty account of the cognitive side of color vision, but you still have a mystery: the ineffable what-it-is-likeness of color QUALIA," you needn't concur; you are entitled to demand specifics.

To cut to the chase, I once got Tom Nagel in discussion to admit that given what he meant by qualia, there could be two identical twins whose scores on every test of color discrimination, color preference, color memory, effects of color on mood, etc., etc., came out the same, and there would still be a wide-open question of whether the twins had the same color qualia when they looked into a particular can of paint! (By Nagel's lights, neither twin would have any grounds at all for supposing that now he knew that he and his twin brother had the same color qualia.) Nagel's position is an available metaphysical position, I guess, but I hope it is obvious that it doesn't derive any plausibility from anything we have discovered about the nature of color experience, and hence no cognitive neuroscientist needs to be shackled by any such doctrine of qualia.

There are obviously large families of differences and similarities in experience that are best ignored at this stage of inquiry—no one can get a good scientific handle on them yet. One can admit that there is a lot more to color experience, or any other domain of subjectivity, than we have yet accounted for, without thereby endorsing the dubious doctrine that qualia are properties that elude objective science forever. But that doctrine is the standard destination of all the qualia arguments among philosophers.

Authors: So what is the task of future students of the problem of consciousness? What should be the content of their research? Is it to solve the brain mechanisms enabling, say, problem solving, and along with that will come some deeper understanding of the ol' ineffable qualia?

DCD: That's roughly right, in my opinion. Here is one place—not the only one, of course—where cognitive neuroscientists could take a hint from AI [artificial intelligence]. The people in AI have almost never worried about consciousness as such, since it seemed obvious to them that if and when you ever got a system—an embodied robot, in the triumphal case—that actually could do all the things a person can do (it can reflect on its reflections about its recollections of its anticipations of its decisions, and so forth), the residual questions about consciousness would have fairly obvious answers. I have always thought they were right.

pressed desires or memories (if there are any), cannot be accessed.

• **Self-knowledge:** Among the people and objects that an intelligent being can have accurate information about is the being itself (**self-knowledge**). As Pinker said, "I cannot only feel pain and see red, but think to myself, 'Hey, here I am, Steve Pinker, feeling pain and seeing red!'"

These three categories enable us to bring present cognitive neuroscience knowledge to bear on the topic of consciousness. Right from the start we can say that science has little to say about sentience. We are clueless on how the brain creates sentience. Dennett would not worry about this, but other scientists still believe it is an issue to be understood in scientific terms.

At the same time, cognitive neuroscience has much to say about access and self-knowledge. It does so because the cognitive neuroscience approach to the problem is not driven by a philosophical viewpoint. Cognitive neuroscience data and observations on the

topic are used with great enthusiasm by philosophers. But the day-to-day work studying patients with broken brains or studying normal brains with brain imaging technologies goes on in the simple spirit of learning more about human brain organization and how it enables mind. In what follows, we present many studies that have implications, direct and indirect, for the problem of access and self-knowledge. We examine observations that most assuredly deal with conscious experience.

CONSCIOUS VERSUS
UNCONSCIOUS PROCESSING

The insights gained from evolutionary theory are vast. It is essential to remember how many domain-specific specialized systems we have in our brains because they all interact in unique ways that produce our sensation of conscious experience. At the same time, the vast majority of mental processes that control and contribute to our conscious experience happen outside our conscious awareness. We really have little or no insight into what prepares us to throw a baseball, to see a colorful flower, or to speak a grammatical sentence. The staging of this behavior is not part of our conscious life. We do not have access to how the brain does those things.

A vast amount of research in cognitive science clearly shows that we are conscious only of the content of our mental life, not what generates the content. It is the products of mnemonic processing, of perceptual processing of imaging, that we are aware of—not what produced the products. Sometimes people report on what they think were the processes, but they are reporting after the fact on what they thought they did to produce the content of their consciousness, as opposed to reporting on the unconscious processes that presumably preceded and gave rise to the conscious processes. Thus, when considering conscious processes, it is also necessary to consider **unconscious processes** and how such processes interact. A statement about conscious processing involves *conjunction*—that is, putting together awareness of the stimulus with the identity, or the location, or the orientation, or some other feature of the stimulus. A statement about unconscious processing involves *disjunction*—that is, separating awareness of the stimulus from the features of the stimulus such that even when unaware of the stimulus, subjects can still respond to stimulus features at an above-chance level.

The cognitive neuroscience approach also has revealed some evidence on sentience. When Block originally drew distinctions between sentience and access, he suggested that the phenomenon of blindsight provided a possible paradigm. **Blindsight,** a term coined by Larry Weiskrantz at Oxford University (Weiskrantz et al., 1974; Weiskrantz, 1986), refers to the phenomenon that patients suffering a lesion in their visual cortex can respond to visual stimuli presented in the blind part of their visual field (Figure 16.1). Most interestingly, these activities happen outside the realm of consciousness. Patients will deny that they can do a task, yet their performance is clearly above that of chance. Such patients have access to information but do not experience it.

Weiskrantz believed that subcortical and parallel pathways and centers could now be studied in the human brain. A vast primate literature had already developed on the subject. Monkeys with occipital lesions not only can localize objects in space but also can make color, luminance, orientation, and pattern discriminations. It hardly seemed surprising that subjects could use visually presented information not accessible to consciousness. Subcortical networks with interhemispheric connections provided a plausible anatomy on which the behavioral results could rest.

Since blindsight demonstrates vision outside the realm of conscious awareness, this phenomenon has often been invoked as support for the view that perception happens in the absence of sensation, as sensations are presumed to be our experiences of impinging stimuli. Because the primary visual cortex processes sensory inputs, advocates of the secondary pathway view have found it useful to deny the involvement of the primary visual pathway in blindsight. Certainly, it would be difficult to argue against the concept that perceptual decisions or cognitive activities routinely result from processes outside of conscious awareness. But it would be even more difficult to argue that such processes do not involve primary sensory systems.

Congruent with such thinking, recent evidence further underscores this notion by demonstrating the involvement of the damaged primary pathway in blindsight. In an exemplary piece of modern cognitive brain science, Mark Wessinger teamed up with Robert Fendrich at Dartmouth College (Wessinger et al., 1997) to investigate blindsight. They investigated this fascinating phenomenon using a dual Purkinje image eye tracker that was augmented with an image stabilizer, allowing for the sustained presentation of information in discrete parts of the visual field (Figure 16.2). Armed with

this piece of equipment and with the cooperation of C.L.T., a robust 55-year-old outdoorsman who had suffered a right occipital stroke 6 years prior to his examination, they began to tease apart the various explanations for blindsight.

Standard perimetry indicated that C.L.T. had a left homonymous hemianopia with lower-quadrant macular sparing. Yet the eye tracker found small regions of residual vision (Figure 16.3). C.L.T.'s scotoma was explored carefully, using high-contrast, retinally stabilized stimuli and an interval, two-alternative, forced-choice procedure. This procedure requires that a stimulus be presented on every trial and that the subject respond on every trial. Such a design is more sensitive to subtle influences of the stimulus on the subject's responses. C.L.T. also indicated his confidence on every trial. The investigators found regions

Figure 16.1 Weiskrantz and colleagues reported the first case of blindsight in a patient with a lesion in the visual cortex. The hatched areas indicate preserved areas of vision for the left and right eyes for patient D.B. Adapted from Weiskrantz (1996).

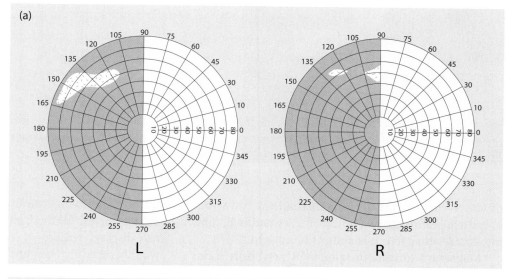

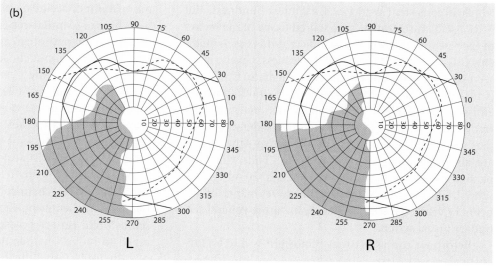

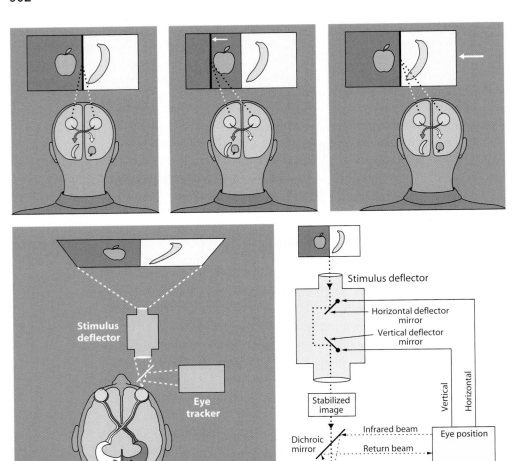

Figure 16.2 Schematic of the Purkinje image eye tracker, which compensates for a subject's eye movements by moving the image in the visual field in the same direction as the eyes, thus stabilizing the image on the retina.

of above-chance performance surrounded by regions of chance performance within C.L.T.'s blindfield. Simply stated—they found islands of blindsight.

Magnetic resonance imaging (MRI) reconstructions revealed a lesion that damaged the calcarine cortex, which is consistent with C.L.T.'s clinical blindness. But MRI also demonstrated some spared tissue in the region of the calcarine fissure. We assume that this tissue mediates C.L.T.'s central vision with awareness. Given this, it seems reasonable that similar tissue mediates C.L.T.'s islands of blindsight. More importantly, both positron emission tomography (PET) and functional magnetic resonance imaging (fMRI) conclusively demonstrated that these regions are metabolically active—these areas are alive and processing information! Thus, the most parsimonious explanation for C.L.T.'s blindsight is spared, albeit severely dysfunctional, remnants of the primary visual pathway, rather than a more general secondary visual system.

Before we can assert that blindsight is due to subcortical or extrastriate structures, we first must be ex-

tremely careful to rule out the possibility of spared striate cortex. With careful perimetric mapping, we can discover regions of vision within a scotoma that would certainly go undetected with conventional perimetry. Through such discoveries we can learn more about consciousness.

There are similar reports of vision without awareness in other neurological populations that can similarly inform us about consciousness. It is commonplace to design demanding perceptual tasks on which both neurological and nonneurological subjects routinely report low confidence values for tasks they perform at a level above chance. Yet it is unnecessary to propose secondary visual systems to account for such reports, since the primary visual system is intact and fully functional. For example, patients with unilateral neglect as a result of right-hemisphere damage are unable to name stimuli entering their left visual field. The conscious brain cannot access this information. But when asked to judge whether two lateralized visual stimuli, one in each visual field, are the same or different

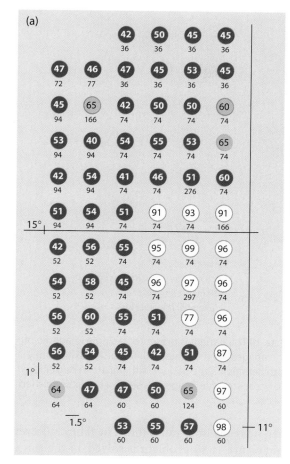

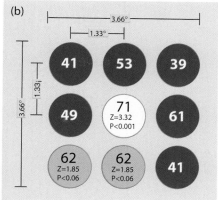

Figure 16.3 **(a)** Results of stabilized image perimetry in C.L.T.'s left visual hemifield. The large numbers in the circles represent the percentages for correct detection. The total number of trials for each location are indicated under the circles. The white circles indicate unimpaired detection, the green circles with black borders indicate impaired detection that was still above the level of chance with Bonferroni's correction, and the green circles without borders show detection that was better than chance without Bonferroni's correction. **(b)** Further detail of the retinal area containing C.L.T.'s island of preserved visual function. The large numbers in circles represent percentages for detection out of sixty-six trials. At the center position, detection was above the level of chance regardless of Bonferroni's correction for nine tests. The green areas indicate elevated but statistically insignificant detection rates. Adapted from Wessinger et al. (1997).

(Figure 16.4), these same patients can do so. When they are questioned on the nature of the stimuli after a trial, they easily name the stimulus in the right visual field but deny having seen the stimulus in the neglected left field. In short, patients with parietal lobe damage but spared visual cortex can make perceptual judgments outside of conscious awareness. Their failure to consciously access information for comparing the stimuli should not be attributed to processing within a secondary visual system, because their geniculo-striate pathway is still intact.

The Extent of Subconscious Processing

A variety of reports extended these initial observations that information presented in the extinguished visual field can be used for decision making. A central question is, how sophisticated can the processing outside of conscious awareness be? Recent work showed that quite complex information can be processed (Figure 16.5). In one study, a picture of a fruit or animal was quickly presented to the right visual field. Subsequently, a picture of the same item or of an item in the same category was presented to the left visual field. In another condition of the experiment, the pictures presented in

each field had nothing to do with each other (Volpe et al., 1979). All patients in the study denied that a stimulus had been presented in the left visual field. But when the two pictures were related, patients responded faster than they did when the pictures were different. The reaction time to the unrelated pictures did not increase. In short, high-level information was being exchanged between processing systems, outside the realm of conscious awareness.

The vast staging for our mental activities happens largely without our monitoring. This can be identified in many experimental venues. The study of blindsight and neglect provides important insights. First, it underlines a general feature of human cognition: Many perceptual and cognitive activities can and do go on outside of the realm of conscious awareness. We can access information that we are not sentient about. Further, this feature need not depend on subcortical or secondary processing systems; it is more than likely that unconscious processes related to cognitive, perceptual, and sensory-motor activities happen at the level of the cortex. To help understand how conscious and unconsciousness interact within the cortex, it is necessary to investigate both conscious and unconscious processes in the intact, healthy brain.

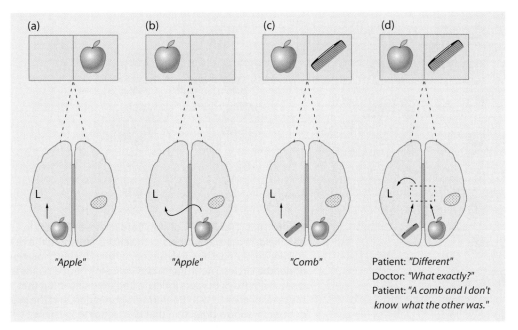

Figure 16.4 The same-different paradigm presented to patients with neglect. **(a, b)** The patient is presented a single image, first to one hemifield then to the other. The patient subsequently is asked to judge if the images are the same or different, a task that they are able to perform. **(c, d)** When the images are presented simultaneously to both hemifields, the patient with unilateral neglect is able to determine whether the images are the same or different but is unable to verbalize what image was seen in the extinguished hemifield that enabled them to make their correct comparison and decision.

Richard Nisbett and Lee Ross (1980) at the University of Michigan made this point most clearly. The work is all cleverly done with the tried-and-true technique of learning word pairs. They first exposed subjects to word associations like *ocean–moon*. The idea is that subjects might subsequently say "Tide" when asked to free associate the word *detergent*. That is exactly what they do, but they do not know why. When asked, they might say, "Oh, my mother always used Tide to do the laundry."

Now any student will commonly and quickly declare that he is fully aware of how he solves a problem even when he really does not know. The famous Tower of Hanoi (Figure 16.6) problem is solved all the time. And when researchers listen to the running discourse of students articulating what they are doing and why they are doing it, the result can be used to write a computer program to solve the problem. The subject calls on facts known from short- and long-term memory. These events are accessible to consciousness and can be used to build a theory for their action. Yet no one is aware of how the events became established in short- or long-term memory.

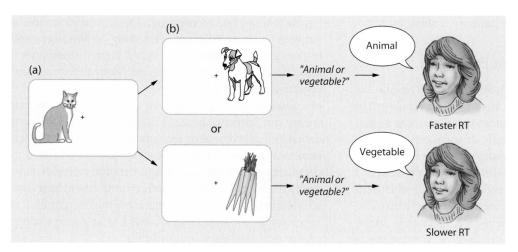

Figure 16.5 **(a)** A picture of an item such as a cat was flashed to the left visual field. **(b)** A picture of the same item or a related item such as a dog was presented to the right visual field, and the subject was to discriminate the category to which the second item belonged. If the items were related by category, the time needed for the categorization of the second word was facilitated.

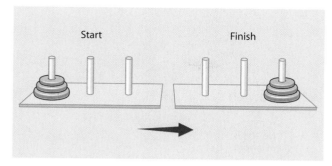

Figure 16.6 The Tower of Hanoi problem. The task is to rebuild the rings on another tower without ever putting a larger ring on top of a smaller ring. It can be done in seven steps, and after much practice, students learn the task. After they have solved it, however, their explanations for how they solved it can be quite bizarre.

Figure 16.7 One technique for testing subliminal perception. A subject is quickly presented a picture of a boy, as viewed in either picture **A1** or **A2**, in such a way that the subject does not have conscious awareness as to the content of the picture. The subject is then shown a neutral picture **(B)** and is asked to judge the character of the boy. These judgments of the boy's character have been found to be biased by the previous subthreshold presentation.

Cognitive psychologists also have examined the extent and kind of information that can be processed unconsciously. Freud staked out the most complex range, where the unconscious was hot and wet. Deep emotional conflicts are fought and their resolution slowly makes its way to conscious experience. Other psychologists placed more stringent constraints on what can be processed. Many researchers maintain that only low-level stimuli—like the lines forming the letter of a word, not the word itself—can be processed unconsciously. Over the last century these matters have been examined time and again; only recently has unconscious processing been examined in a cognitive neuroscience setting.

The classic approach was to use the technique of **subliminal perception.** Here a picture of a boy either throwing a cake at someone or simply presenting the cake in a friendly manner is flashed quickly. A neutral picture of the boy is presented subsequently, and the subject proves to be biased in judging the boy's personality as a function of the subliminal exposures he received (Figure 16.7). Hundreds of such demonstrations have been recounted, although they are not easy to replicate. Many psychologists maintain that elements of the picture are captured subconsciously and that this is sufficient to bias judgment.

Cognitive psychologists have sought to reaffirm the role of unconscious processing through various experimental paradigms. A leader in this effort has been Tony Marcel of Cambridge University (1983a, 1983b). Marcel used a masking paradigm in which the brief presentation of either a blank screen or a word was followed quickly by a masking stimulus of a cross patch of letters (Figure 16.8). One of two tasks followed presentation of the **masking stimulus.** In a detection task, subjects merely had to choose whether a word had been presented. On

this task, subjects responded at a level of chance. They simply could not tell whether or not a word had been presented. If the task became a lexical decision task, however, the subliminally presented stimulus had effects. Here, following presentation of the masking stimulus, a string of letters was presented and subjects had to specify whether the string formed a word. Marcel cleverly manipulated the **subthreshold** words in such a way that some were related to the word string and some were not. If there had been at least lexical processing of the subthreshold word, related words should elicit faster response times, and this is exactly what Marcel found.

Investigations of conscious and unconscious processing of pictures and words recently have been combined

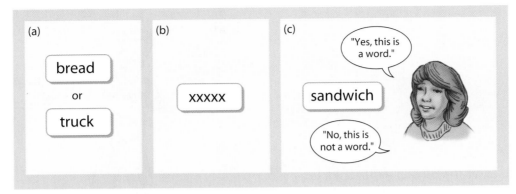

Figure 16.8 To test subliminal processing of words, Tony Marcel presented a word such as *bread* at a subthreshold level **(a)**. This word was quickly replaced by a cross patch of letters **(b)**. Then, in a lexical decision task, the subject was presented with a string of letters and asked if the letters formed a word or not **(c)**. Presentation of subthreshold words influenced the response times to related words but had no effect on the response to unrelated words. Adapted from Marcel (1983).

successfully into a single cross-form priming paradigm. The paradigm involves presenting pictures for study and word stems for the test (Figure 16.9). Using both extended and brief periods of presentation, the investigators also showed that such picture-to-word priming can occur with or without awareness. In addition to psychophysically setting the brief presentation time at identification threshold, a pattern mask was used to halt conscious processing. However, apparently not all processing was halted, since priming occurred equally well under both conditions. Given that subjects denied seeing the briefly presented stimuli, unconscious processing must have allowed them to complete the word stems (primes)—that is, they were extracting conceptual information from the pictures, even without consciously seeing them. How often does this happen in everyday life? Considering the complexity of the visual world, and the speed with which our eyes look around, briefly fixating from object to object (approximately 100–200 msec), this probably happens quite often! These data further underscore the need to consider both conscious and unconscious processes when developing a theory of consciousness.

Figure 16.9 Cross-form (picture-to-word) priming paradigm. **(a)** During study, either extended, unmasked **(top)** or brief, masked **(bottom)** presentations were used. **(b)** During the test, subjects were asked to complete word stems. Priming performance was identical between extended and brief presentations. **(c)** Afterward, subjects were asked if they remembered seeing the words as pictures. Here performance was not identical—subjects usually remembered seeing the extended presentations but regularly denied having seen the brief presentations.

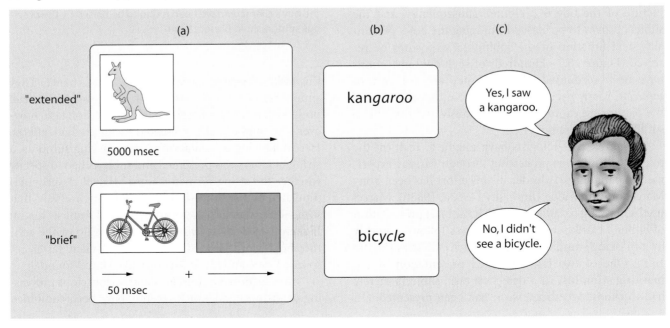

Gaining Access to Consciousness

As further work attempts to understand the links between conscious and unconscious processing, it becomes clear that these phenomena remain elusive. We now know that obtaining evidence of subliminal perception depends on whether subjective or objective criteria set the threshold. When the criteria are subjective (i.e., introspective reports from each subject), priming effects are evident. When criteria are set objectively by requiring a forced choice as to whether a subject saw any visual information, no priming effects are seen. Among other things, these studies point out the gray area between conscious and unconscious. Thresholds clearly vary with the criteria.

Pinker (1997) presented an enticing analysis on how evolutionary pressures gave rise to access-consciousness. The general insight has to do with the fact that information has costs and benefits. He argued that at least three dimensions must be considered: cost of space, cost of time, and cost of resources. Regarding space, while the human cortex has exploded in size, it is not contributing simply because of its size. The brain is built by accumulating special processors that solve problems only when an answer must be calculated. As Pinker pointed out, a simple chess game has between 30 and 35 moves for each turn, and each game has about 40 turns, which yields 10^{120} different moves. (Only 10^{70} particles are in the visible universe.) In short, the brain must be built so specialized processes do calculations as needed. As for the cost of time, information processing takes time, and if all decisions were the products of rational conscious consideration, there would not be enough time in one's life to carry out what we do freely and easily in an hour. And finally, thinking is expensive. It takes energy in the form of oxygen and glucose.

The point is that any complex organism that is an information processor working in real time must have, as Pinker said, restrictions on the information it accesses. Only information relevant to the problem at hand should be allowed, which seems to be how the brain is organized.

Access-consciousness has four obvious features that Pinker recounted. First is the rich field of sensation we all live in. Second is the capacity to move information into and out of our awareness, into and out of short-term memory but turning our attentional spotlight on it. Third, such information always comes with salience, some kind of emotional coloring. Finally is the "I" that calls the shots on what to do with the information as it comes into the field of awareness.

Jackendoff (1987a) argued that for perception, access is limited to the intermediate stages of information processing. We do not ponder the elements that go into a percept, only the output. Consider the patient described in Chapter 6 who could not see objects but could see faces,

indicating he was a face processor. When this patient was shown a picture that deployed pieces of vegetables arranged in such a way as to make them look like a face, the patient immediately said he saw the face but was totally unable to state that the eyes were tomatoes and the nose a banana. He only had access to output of the module.

Concerning attention and its role in access, the work of Anne Treisman (1991) at Princeton University reveals that unconscious parallel processing can go only so far. Treisman proposed a candidate for the border between conscious and unconscious processes. In her famous pop-out experiments, a subject picks a prespecified object from a field of others. The notion is that each point in the visual field is processed for color, shape, and motion, outside of conscious awareness. The attention system then picks up elements and puts them together with other elements to make the desired percept. Treisman showed, for example, that when we are attending to a point in space and processing the color and form of that location, elements at unattended points seem to be floating. We can tell the color and shape, but make mistakes about what color goes with what shape. The illusory conjunctions of stimulus features are prima facie evidence for how the attentional system combines elements into whole percepts.

The other senses of access outlined by Pinker, the notion of salience and executive controller, have been touched on elsewhere. His notion of self-knowledge and its role in conscious experience and our own personal narrative are addressed at the end of this chapter. However, before we turn to such musings, it is important to consider another, often overlooked, aspect of consciousness, namely, the ability to move from conscious, controlled processing to unconscious, automatic processing. Such conscious to unconscious "movement" is necessary when we are learning complex motor tasks such as riding a bike or driving a car, as well as for complex cognitive tasks such as verb generation and reading.

Marcus Raichle and Steven Petersen, two pioneers in the brain imaging field, at Washington University in St. Louis, proposed a "scaffolding to storage" framework to account for this movement (Petersen et al., 1998). Initially, according to their framework, we must use conscious processing during practice while developing such skills (or memories)—this can be considered the scaffolding process. During this time the memory is being consolidated, or the skill is being developed and honed. Once the task is learned, brain activity and brain involvement change. This change can be likened to the removal of the scaffolding, or the disinvolvement of support structures and the involvement of more permanent structures as the tasks are "stored" for use.

Petersen and Raichle demonstrated this scaffolding to storage movement in the awake behaving human brain.

Using PET techniques, with a verb generation task as well as a maze tracing task, they clearly demonstrated that early, unlearned, conscious processing uses a much different network of brain regions than does later, learned, unconscious processing (Figure 16.10). They hypothesized that during learning a scaffolding set of regions is used to handle novel task demands. Following learning, a different set of regions is involved, perhaps regions specific to the storage or representation of the particular skill or memory. Furthermore, once this movement from conscious to unconscious has occurred, that is, once the scaffolding is removed, it is sometimes difficult to reinitiate conscious processing. A classic example is learning to drive with a clutch. Early on, you have to consciously practice the steps of releasing the gas pedal, depressing the clutch, moving the shift-lever, releasing the clutch, and applying pressure to the gas pedal again—all without stalling the car. However, once you know the procedures well and the process has been stored, it is rather difficult to articulate the steps.

Similar processes occur in learning other complex skills. Chris Chabris, a young cognitive psychologist at Harvard University, has studied chess players as they progress from the novice to the master level (Chabris and Hamilton, 1992). During lightning chess, masters play many games simultaneously and very fast. Seemingly, they play by intuition as they make move after move after move, and in essence they *are* playing by intuition—"learned intuition," that is. They intuitively know, without really knowing how they know, what the next best move is. For novices, such lightning play is not possible. They have to painstakingly examine the pieces and moves one by one. But following many hours of practice and hard work (hard play?) as the novices develop into chess masters, they see and react to the chess board differently.

They now begin to view and play the board as a series of groups or clumps of pieces and moves, as opposed to separate pieces with serial moves. Chabris's research has shown that during early stages of learning, the talking, language-based, left brain is consciously controlling the game. But with experience, as the different moves and possible groupings are learned, the perceptual, feature-based, right brain takes over. When asked, the speaking left brain can assure us that it can explain how the moves are made, but miserably fails to do so—as often happens when you try to explain how to use a clutch.

Such controlled, conscious processing, transitioning to automatic, unconscious processing is analogous to the implementation of a computer program. Early stages require multiple interactions among many brain processes, including consciousness, as the program is written, tested, and prepared for compilation. Once the process is well underway, the program is compiled, tested, recompiled, retested, and so on. Eventually, as the program begins to run and unconscious processing begins to take over, the scaffolding is removed, and the executable file is uploaded for general use.

This theory seems to imply that once conscious processing has effectively allowed us to move a task to the realm of the unconscious, we no longer need conscious processing. If true, then why aren't we more unconscious than conscious? Perhaps we are but we just do not know it until after the necessary processing has occurred. One evolutionary goal of consciousness may be to provide for greater efficiency of unconsciousness. The ability to relegate learned tasks and memories to unconsciousness allows us to devote more consciousness to recognizing and adapting to changes and novel situations, ultimately allowing our continued survival.

Figure 16.10 Positron emission tomography (PET) images showing localized changes in brain activation as a function of practicing verb generation **(a)** and maze tracing **(b)** tasks. (a) The upper images show that left frontal activation subsides with practice. The lower images show increases in blood flow (decreases in deactivation) in the left insular region as a result of practice. (b) The upper images show that activation in the premotor and parietal areas subsides with practice. The lower images show increases in blood flow in the primary and supplementary motor areas as a result of practice. Adapted from Petersen et al. (1998).

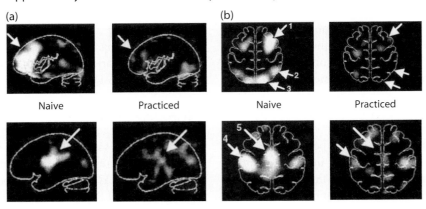

(a) (b)

Naive   Practiced   Naive   Practiced

NEURONS, NEURONAL GROUPS, AND CONSCIOUS EXPERIENCE

Neuroscientists interested in higher cognitive functions have been extraordinarily innovative in analyzing how the nervous system enables perceptual activities. Recording from single neurons in the visual system, they not only have tracked the flow of visual information and how it becomes encoded and decoded during a perceptual activity, but also have directly manipulated the information and influenced an animal's decision processes. One of the leaders in this approach to understanding the mind is William Newsome at Stanford University. He recorded information, stimulated small neuronal groups, and showed how this influences the decision an animal makes.

Newsome studied how neural events in area MT of the monkey cortex, which is quite involved in motion detection, correlate with the actual perceptual event. One of his first findings was striking. The animal's psychophysical performance capacity to make a motion discrimination was predictable by the neuronal response pattern of a single neuron (Figure 16.11). In other words, a single neuron in area MT was as sensitive to changes in the visual display as was the monkey!

This finding stirred the research community because it raised a fundamental question about how the brain does its job. Newsome's observation challenged the common view that the signal averaging that surely goes on in the nervous system eliminated the noise carried by individual neurons. Thus, the capacity of pooled neurons making a decision should be superior to the sensitivity of single neurons. And yet Newsome did not side with those who believe that a single neuron is the source for any one behavioral act. Since it is well known that killing a single neuron, or even hundreds of them, will not impair an animal's ability to perform a task, a single neuron's behavior is clearly redundant. Researchers are now building models of how many neurons would be needed to mimic an animal's performance.

An even more tantalizing finding of particular interest to the study of conscious experience is that careful **microstimulation** of these same neurons, altering their response rates, can tilt the animal toward making the right decision on a perceptual task. The maximum effects are seen during the interval the animal is thinking about the task. Newsome and his colleagues, in effect, inserted an artificial signal into the monkey's nervous system and influenced how it thinks.

This discovery immediately raises the issue of whether one should consider the site of the microstimulation as the place where the decision is made. Researchers are not convinced that this is the way to think about the problem. In-

stead, they believe they have tapped into part of a neural loop involved with this particular perceptual discrimination. They argue that stimulation at different sites in the loop creates different subjective experiences. For example, let's say that the stimulus was moving

Figure 16.11 **(a)** The experimental design of Newsome and Pare (1988). The stimuli consisted of random patterns of dots that were briefly illuminated and then replaced by a dot at another location on the screen. If the successive replacement dots are slightly offset from the previously illuminated dot in time and location (far right), then the dots appear to move coherently, much like consecutive frames in a movie result in apparent motion. If the replacement dots are repositioned randomly on the screen (far left), then no motion is seen. By adjusting the number of dots carrying this motion signal, and thus manipulating how much motion is apparent in the stimulus, researchers can determine experimentally the threshold for motion direction discrimination in humans, monkeys, or single cells in monkeys' brains. **(b)** The stimuli were presented, with varying levels of coherent motion, to rhesus monkeys trained in a motion-direction discrimination task. The monkey's decision regarding the direction of apparent motion, and the responses of single MT cells (which are selective for direction of motion) were recorded and compared to the stimulus coherence on each trial. Remarkably, on average, individual cells in MT were as sensitive as the entire monkey. Further, in subsequent work the firing rate of single cells actually predicted (albeit weakly) the choice of the monkey on a trial-by-trial basis. Adapted from Newsome and Pare (1988).

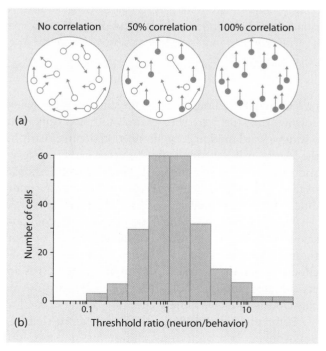

An Interview with William T. Newsome, Ph.D.

Dr. Newsome is a professor of neuroscience at Stanford University. His neurophysiological studies on the monkey visual system have led to new insights on how the brain processes information.

Authors: What are the limitations of the single-neuron analysis? One might think assessing the behavior of one neuron at a time would severely limit the kind of analysis you could do.

WTN: The most obvious limitation of the single-unit approach is that we cannot analyze effectively neural representations that involve patterns of activity existing simultaneously at multiple locations in the brain. This limitation is, of course, rather severe since even simple sensory stimuli or motor acts evoke complex patterns of neural activity. A second limitation, which receives somewhat less attention, is that the single-unit approach provides little information about the relative timing of neural events at different locations in the brain. Simultaneous measurement of the onset and offset of neural activity in different brain structures can provide critical information about cause-and-effect relationships between those structures as cognitive operations unfold. A final limitation is that single-unit recording in the central nervous system virtually precludes precise analysis of input-output transfer functions. For any given neuron, we have only the most general notion of what its inputs might be, such notions being based largely on population studies of anatomical connections and physiological properties. The strongest statements that single-unit studies permit concerning transfer functions are usually of the following form: This type of response selectivity has not been observed at prior levels of the pathway, and is probably synthesized from simpler inputs for [some specific computational or behavioral purpose].

Having criticized the single-unit approach, let me hasten to add that we have not yet begun to exhaust its usefulness. Single-unit analyses are still employed profitably in conjunction with anatomical techniques to identify basic processing modules in different brain structures—what Hubel and Wiesel termed *functional architecture.* I suspect that this enterprise will continue to be productive, especially if molecular techniques can provide increasingly precise anatomical markers for neural circuits. Even more exciting to me is the recent trend toward applying the single-unit approach in behaving animals to identify neural correlates of simple cognitive operations. A growing number of laboratories are employing clever behavioral paradigms (frequently adapted from the experimental traditions of psychophysics and behavioral psychology) to investigate neural substrates of perception, attention, learning, memory, and motor planning, to name but a few. A wealth of new insights is emerging from these efforts, and I believe we have only scratched the surface of what can be learned with this approach.

---

upward and the response was as if the stimulus was moving downward. If this were your brain, you might think you saw downward motion if the stimulation occurred early in the loop. But if the stimulation occurred late in the loop and merely found you choosing the downward response instead of the upward one, your sensation would be quite different. Why, you might ask yourself, did I do that?

This question raises the issue of the timing of consciousness. When do we become conscious of our thoughts, intentions, and actions? Do we consciously choose to act, then consciously initiate an act? Or is an act initiated unconsciously, then only afterward do we consciously think we initiated it?

Benjamin Libet (1996), an eminent neuroscientist-philosopher, has been researching this question for nearly 30 years. In a ground-breaking and often controversial extensive series of experiments, he investigated the neural time factors in conscious and unconscious processing. These experiments are the basis for his **backward referral hypothesis.** Libet and colleagues (Libet et al., 1979) have concluded that awareness of a neural event is delayed approximately 500 msec following the onset of the stimulating event, and more importantly, this awareness is referred back in time to the onset of the stimulating event. Thus, we are not aware of the event until after it occurs, though we often think we are aware of the event from the onset.

Rodney Cotterill (1997), a Danish biophysicist, has proposed a neural circuit that could account for this delay. He refers to this circuit as the "vital triangle." The triangle consists of sensory cortex, frontal cortex, and various thalamic nuclei (Figure 16.12). However, the vital part is not the regions at the apices. Rather, it is the various feed-forward and feedback connections within the system that enable consciousness. More specifically, Cotterill posits that comparisons within the brain of the original input signal with a copy of that same signal, modified by our interaction with the environment, allows us to react to the environment. Cotterill believes that this reaction to our interaction with the environment leads to our consciousness of action. The fact that comparison of the original signal to an environmentally modified copy of the signal takes time coincides nicely with Libet's backward-referral hypothesis.

Fortunately, backward referral of our consciousness is not so delayed that we act without thinking. The actual beginning of the act occurs sufficiently after the awareness of the intent to act, giving us time to override inappropriately triggered behavior. This ability to detect and correct errors is what Libet believes is the basis for free will.

Whether or not error detection and correction are indeed experimental manifestations of free will, such abilities have been linked to brain regions (Figure 16.13). Not all people can detect and correct errors adequately. In a model piece of brain science linking event-related potentials (ERPs) and patient studies, Robert Knight at the University of California, Berkeley, and William Gehring at the University of Michigan, Ann Arbor, characterized the role of the frontal lobe in checking and correcting errors (Gehring and Knight, 2000). By comparing and contrast-

ing the performance of patients and healthy volunteers on a letter discrimination task, they conclusively demonstrated that the lateral prefrontal cortex was essential for corrective behavior. The task was arranged such that responses to "targets" often were disrupted by flanking "distracters." Healthy volunteers showed the expected "corrective" neural activity in the anterior cingulate (see Chapter 12). Patients with lateral prefrontal damage also showed the corrective activity for errors. However, the patients also showed the same sort of "corrective" activity for nonerrors; that is, patients could not distinguish between errors and correct responses. It seems that such patients with lateral prefrontal damage no longer have the ability to monitor and integrate their behavior across time. Perhaps they have even lost the ability to learn from their mistakes. It is as if they are trapped in the moment, unable to go back yet unable to decide to go forward. They seem to have lost a wonderful and perhaps uniquely human benefit of consciousness—the ability to escape from the here and now of linear time or to "time shift" away.

Figure 16.13 Model of the conflict monitoring system suggested by Gehring and Knight (2000). Basic task-related components are in red, and control-related components are in blue. Patients with damage to control-related components, particularly the prefrontal cortex, have problems recognizing and correcting their mistakes. Adapted from Cohen et al. (2000).

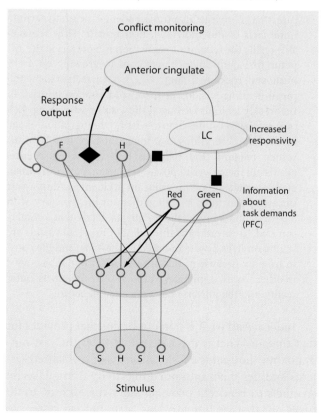

Figure 16.12 Schematic of the "vital triangle" as proposed by Cotterill. The apices are the thalamic region, the sensory cortical regions, and the frontal lobe. Connections among the apices that comprise the "triangle" are represented by the hatched arrows. Such connections are believed to enable consciousness by allowing reactions to interactions with sensory input and the environment. Adapted from Cotterill (1997).

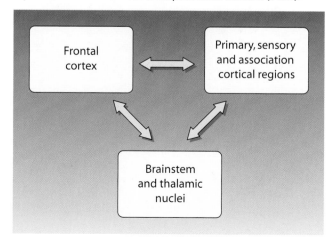

THE EMERGENCE OF
THE BRAIN INTERPRETER
IN THE HUMAN SPECIES

Even with the exciting advances in systems neuroscience and with the insight that many of our cognitive capacities, so heavily a part of our conscious experience, appear to be built-in domain-specific operations, we think we are a unified conscious agent with a past, a present, and a future. The domain-specific or modular systems are fully capable of producing behaviors, mood changes, and cognitive activity. With all of this apparent independent activity, what allows for the sense of conscious unity we possess?

A private narrative appears to take place inside us all the time, and it partly consists of the effort to tie together into a coherent whole the diverse activities of thousands of specialized systems we have inherited to handle challenges. The great American writer John Updike mused on the subject in his 1989 book *Self-consciousness: Memoirs:*

> "Consciousness is a disease," Unamuno says. Religion would relieve the symptoms. Religion construed broadly, not only in the form of the world's barbaric and even atrocious religious orthodoxies but in the form of any private system, be it adoration of Elvis Presley or hatred of nuclear weapons, be it fetishism of politics or popular culture, that submerges in a transcendent concern the grimly finite facts of our individual human case. How remarkably fertile the religious imagination is, how fervid the appetite for significance; it sets gods to growing on every bush and rock. Astrology, UFOs, resurrections, mental metal-bending, visions in space, and voodoo flourish in the weekly tabloids we buy at the cash register along with our groceries. Falling in love—its mythologization of the beloved and everything that touches her or him is an invented religion, and religious also is our persistence, against all the powerful post-Copernican, post-Darwinian evidence that we are insignificant accidents within a vast uncaused churning, in feeling that our life is a story, with a pattern and a moral and an inevitability—that as Emerson said, "a thread runs through all things: all worlds are strung on it, as beads: and men, and events, and life come to us, only because of that thread." That our subjectivity, in other words, dominates, through secret channels, outer reality, and the universe has a personal structure.

Indeed. And what is it in our brains that provides for that thread? What is the system that takes the vast output of our thousands upon thousands of specialized systems and ties them into our subjectivity through secret channels to render a personal story for each of us? It turns out that we humans have a specialized system to carry out this interpretive synthesis, and it is located in the brain's left hemisphere. The **interpreter** is a system that seeks explanations for internal and external events in order to produce appropriate behaviors in response. We know it to be only in the left hemisphere, and it appears to be tied to our capacity to see how contiguous events relate to one another. The interpreter, a built-in specialization in its own right, operates on other adaptations built into our brains. The adaptations are most likely cortically based, but they work largely outside of conscious awareness, as do the vast majority of our mental activities. It is the conse-

Figure 16.14 Method for presenting different cognitive tasks simultaneously to each hemisphere. In the split-brain patient, the right hemisphere processes the information in the left visual field, in this case the snow scene, and the left hemisphere processes the information in the right visual field, the chicken claw. Using each hemisphere, the patient is asked to associate the presented image with one of several choices. The patient verbalizes through the language-dominated left hemisphere that he sees a chicken claw, and thus creates reasons based on this knowledge as to why he picked the item that related to the image presented to the right hemisphere.

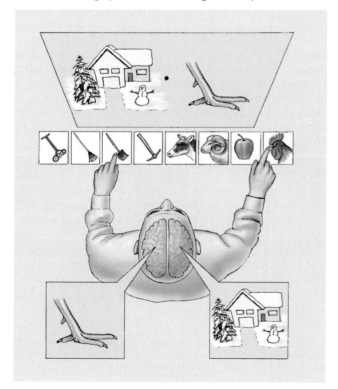

quences of their activity that are interpreted and that provide the thread for our personal story.

The interpreter was discovered by Michael Gazzaniga and Joseph Ledoux, using a simultaneous concept test (Figure 16.14). A split-brain patient was shown two pictures, one exclusively to the left hemisphere and one exclusively to the right, and was asked to choose from an array of pictures placed in full view in front of him the ones associated with the pictures lateralized to the left and right sides of the brain. In one example of this test, a picture of a chicken claw was flashed to the left hemisphere and a picture of a snow scene, to the right hemisphere. Of the array of pictures placed in front of the subject, the obviously correct association is a chicken for the chicken claw and a shovel for the snow scene. Split-brain subject P.S. responded by choosing the shovel with the left hand and the chicken with the right. When asked why he chose these items, he (his left hemisphere) replied, "Oh, that's simple. The chicken claw goes with the chicken, and you need a shovel to clean out the chicken shed." Here the left side of the brain, observing the left hand's response, interprets that response into a context consistent with its sphere of knowledge—one that does not include information about the left-hemifield snow scene.

Another example of this phenomenon of the left hemisphere interpreting actions produced by the dis-connected right one involves lateralizing a written command, such as "Laugh," to the right hemisphere by tachistoscopically presenting it to the left visual field (Figure 16.15). After the stimulus is presented, the patient laughs and, when asked why says, "You guys come up and test us every month. What a way to make a living!" In still another example, if the command "Walk" is flashed to the right hemisphere, the patient typically stands up from the chair and begins to take leave from the testing van. When asked where she is going, she (the left side of the brain) says, "I'm going into the house to get a Coke." However, this type of test is manipulated—it always yields the same result.

There are many ways to influence the left-hemisphere interpreter. As just mentioned, Gazzaniga and colleagues wanted to know whether the emotional response to stimuli presented to one half of the brain would affect the other half. In this study, they presented, by lateralized stimulus procedures, film vignettes that included either violent or calm sequences. The emotional valence of the stimulus clearly crossed over from the right to the left hemisphere. The left hemisphere remained unaware of the content that produced the emotional change, but it interpreted and experienced the emotion.

The brain's modular organization has now been well established. The functioning modules do have some kind of physical instantiation, but brain scientists

Figure 16.15 When the nonverbal right hemisphere is given a command, the person carries it out without specific knowledge as to why she is doing so. After performing the activity, the person is quickly able to generate a reason for this activity.

An Interview with Simon Baron-Cohen, Ph.D.   Professor Baron-Cohen has led the way in describing how brain pathologies disrupt the capacity to have a theory of mind.

Authors: David Premack introduced the concept of theory of mind when he asked the question about whether or not the chimp possessed such a thing. You have taken this same issue into the clinical setting by examining autistic children. You have also studied the emergence of such a module in children. Could you tell us a little about this fascinating work?

SB-C: I had just spent a fascinating year as a teacher of children with autism, in a small unit north of London, and had concluded that they were un-self-conscious. They did as they pleased, not in an antisocial way, but simply oblivious of how they might appear to others. They seemed totally unconcerned by what others might be thinking (about them, or about anything else, for that matter). The idea occurred to me that as well as being unaware of other people's thoughts, they might also be unaware of their own thoughts or mental lives.

Moving to the Cognitive Development Unit in London after that year, I began my Ph.D. on this topic. With Alan Leslie and Uta Frith, we sharpened the question to, Does the autistic child have a theory of mind? The wording of this question (the title of our first paper too) was intended to exactly mirror the wording from Premack and Woodruff (1978), about the chimpanzee. Even down to the scare quotes around the phrase *theory of mind*. This was because we realized the importance of situating the autism research in an evolutionary context. A theory of mind was so central to human social behavior, that in all

likelihood it had evolved to support social behavior—and its potential impairment in autism might be just the strand of neurological evidence to allow this evolutionary question to become tractable.

But my interest in this topic was never solely on this basic science. It was also to try to see how a new understanding of autism might lead to new ways of intervening. And the basic and applied sides to my research continue to intertwine.

Authors: But if an autistic child is the product of pathology, why does it make sense to place it in an evolutionary framework?

SB-C: Autism as a form of neuropathology seems to be highly specific: an impairment in the brain system's underlying theory of mind. The value of considering this in an evolutionary framework is that since theory of mind seems to be a universal human ability—as pervasive as language, for example—autism may provide a strong clue about where in the brain our theory of mind is located, or which brain systems we employ during this cognitive activity. This is of interest in terms of tracing the origins of such brain systems, both phylogenetically and ontogenetically.

If one doubts whether theory of mind is indeed universal, consider Fodor's thought experiment: Try imagining a culture in which people did not use belief-desire reasoning. It seems impossible. And so far, the anthropological evidence suggests no such human cultures exist.

cannot yet specify the nature of the neural networks. What is clear is that these networks operate mainly outside the realm of awareness and announce their computational products to executive systems that result in behavior or cognitive states. Catching up with all of this parallel and constant activity appears to be the responsibility of the left hemisphere's interpreter module. The interpreter is a system of primary importance to the human brain: It allows for the formation of beliefs, which in turn are mental constructs that free us from simply responding to stimulus-response aspects of everyday life. In many ways it is the system that provides the story line or narrative of our lives.

The problem of consciousness, like the problems of language, sexual selection, and visual motion, should always be considered from an evolutionary perspective. Then certain truths emerge that the core of human consciousness is a feeling—a felt sense about specialized capacities.

Looking at the past decades of split-brain research, we find one unalterable fact. Disconnecting the two cerebral hemispheres, an event that finds one half of the cortex no longer interacting in a direct way with the other half, does not typically disrupt the cognitive-verbal intelligence of these patients. The left dominant hemisphere remains the major force in our conscious experience and

Authors: Well, what has the neuroscientific literature taught us about the location of the theory of mind (TOM) module? Couldn't it be that an adaptive specialization like the TOM module was represented by a highly distributed system which had specific function?

SB-C: At the moment it's too early to say much about the neuroscientific literature on theory of mind, because there are too few studies. One SPECT [single-photon emission computed tomography] study implicates right orbitofrontal cortex, another PET [positron emission tomography] study (using a different task) implicates left medial frontal cortex, and our model assumes theory of mind is highly likely to be distributed in that it makes use of emotion processing (amygdala?) and aspects of face processing (superior temporal sulcus?). But this is a new model that needs a lot more empirical investigation. Evidence from acquired brain damage might also help elucidate its brain basis, though so far nothing has been published concerning such cases.

Authors: Doesn't that suggest the TOM module could come perilously close to being the whole brain? Put differently, what are the studies that suggest it is an added chip into the human brain as opposed to an emergent property of a big brain?

SB-C: TOM could be a high-level emergent property of the whole brain—this is not yet ruled out in any definitive way. But then a new puzzle would need solving: Why are children with autism able to solve some tasks that appear to involve high-level reasoning and even forms of meta-representation (such as understanding nonmentalistic forms of representation), and yet have such basic deficits in understanding mental representation? For example, they can understand that photos or drawings or maps or models can all represent the world (or misrepresent the world), and yet they fail to understand beliefs as representations (and misrepresentations) of the world. And in their reasoning, they can perform well on analogies and syllogisms, and yet fail on tests of psychological reasoning. Such neuropsychological dislocation strongly suggests TOM is not just an emergent property of the whole brain, but rather is subserved by highly specific neural systems, which are open to selective deficit.

Authors: Finally, is there any evidence for a TOM module in nonhuman primates?

SB-C: That's an interesting question. Daniel Povinelli has carried out the most work on nonhuman primates in relation to TOM, and so far concludes that they have only rudimentary aspects of TOM, if at all. For example, a range of animals show sensitivity to when they are being looked at, though few seem to show any awareness of the importance of perception in communication, or as a source of knowledge. As for evidence of awareness of epistemic states (beliefs), there are no persuasive grounds yet for concluding that this is within nonhuman primate ability. There is some evidence (from Premack) that chimpanzees might be sensitive to intentions (goal states), though. So the answer to the question could be summarized as follows: Some nonhuman animals may be aware of the volitional and perceptual mental states, but none appear aware of the informational or epistemic ones.

I would end by pointing out that some people think that if other animals had a TOM, we should change our moral stance towards them: Accord them human rights. This is an interesting example of the relation between morality and science. Personally I wouldn't eat a primate whether it had a TOM or not.

that force is sustained, it would appear, not by the whole cortex but by specialized circuits within the left hemisphere. In short, the inordinately large human brain does not render its unique contributions by simply being a bigger brain.

With the realization that the accumulation of specialized circuits accounts for the unique human experience, consciousness can be viewed from two perspectives, one that comes from realizing that the brain is a constellation of specialized circuits and the other from understanding that beyond early childhood our sense of being conscious never changes. When these views are taken together, it becomes evident that what we mean by consciousness is how we feel about our specialized capacities. We have feelings about objects we see, hear, and feel. We have feelings about our capacity to think, to use language, to apprehend faces. Hence, consciousness is not another system. It reflects the affective component of specialized systems that have evolved to enable human cognitive processes. With an inferential system in place, we have a system empowering all sorts of mental activity. And again, our consciousness of those mental activities is related to our capacity to assign feelings to them, which is what distinguishes us from the electronic artifacts surrounding us.

An Interview with David Premack, Ph.D.   Professor Premack, who now lives in Paris, is noted for his work in learning and motivation, and in chimpanzee language and social development.

Authors: Over the years you have devised clever tests to examine complex issues in the area of motivation and language. It is that cleverness that has unlocked truths about the nature of organisms. When it comes to the issue of consciousness, can clever experimental questions be asked? Take an aspect of conscious experience, morality. Can such abstract issues be studied in the laboratory?

DP: Abstract, morality, consciousness—too many issues. Let's push consciousness aside for the moment, not for entirely inconsequential reasons. Some factors important for morality may owe little to consciousness. Further, if the word *consciousness* leads you to contemplate subjective states—of a kind you might grant apes but would deny earthworms—I'm not sure we're in experimental territory; I don't know how to do tests on subjective states. But *abstract* and *morality* shouldn't scare off experimentalists.

Authors: Are you saying the experimentalist can begin to pursue the content of conscious experience, like moral reasoning, and look to its roots and mechanisms? As an example of the content of conscious experience, that seems rich but horridly complex. How do you break the problem down?

DP: Developmental psychology argues that infants divide the world into two basic domains, physical objects on the one hand and intentional or psychological objects on the other. Twenty years of work by Spelke, Carey, Leslie, Baillergeon, Gelman, and a host of other gifted people suggest that infants have definite expectations concerning physical objects. For example, they don't expect them to move unless caused to move by the action of another object. And they expect them to move continuously in

space and time, not to appear at one location, disappear, and then reappear at another location. And not to come apart. And if two of them touch, not to stick together like drops of water. How do you know that that's what they expect? Looking-time experiments. When infants are shown objects that deviate from these properties—objects that move discontinuously or that come apart, etc.—they show increased looking time. But let's put physical objects aside, and move to the infant's other module, the one for psychological or intentional objects. That's what my wife Ann and I have been looking at.

Authors: What have you found?

DP: We've mainly been working on a model, though we've made a few tests of it. The model assumes that infants deal with intentionality in terms of two basic concepts: goal directedness and value, positive or negative. Infants younger than about 2 years don't appear to have a theory of mind. They don't attribute mental states like perceive, want, believe. They understand psychological objects in a far simpler way. But let's make it official. Here are the four basic assumptions of the model.

1. A physical object moves only when caused to move by another object; however, a psychological object starts and stops its own motion; it appears to be self-propelled.
2. Psychological objects display goal-directed action.
3. An infant who perceives an object that is both self-propelled and goal directed interprets the object as intentional and assigns unique properties to it.
4. Infants attribute value (either positive or negative) to the appropriate interaction between intentional objects.

Is Consciousness a Uniquely Human Experience?

Given our capacity for consciousness and our biological similarity to other primates, it is not surprising that the question of nonhuman primate consciousness has been debated for years. If our conscious state has evolved as a product of our brain's biology, is it possible that our closest relatives might also possess this mental attribute or a developing state of our ability? David Premack coined the term **theory of mind,** which refers to the ability to represent and infer unobservable mental states such as desires, intentions, and beliefs from the self and others (Premack and Woodruff, 1978).

If theory of mind is a biologically based human cognition, it follows that there should be corresponding neural components. From this reasoning, perhaps the best way to

Incidentally, it's the motion of objects—and not their properties—that activates the modules. Motion is the key to activating the infant's modules in both the physical and psychological domains. That's our argument.

Authors: What do you mean by *goal*? What does an infant think a goal is?

DP: According to the model, they use four properties: trajectory, target, greater than default values, and satisfaction—but let's skip the middle two. Trajectory is the direction in which an object moves. Because a self-propelled object can move in many directions, its consistent movement in only one direction is significant. Consider the parallel between the trajectory of an object and that of both gazing and pointing. Infants follow the mother's gaze; rather than look at her, they look where she is looking. They react in the same manner to pointing; rather than look at the end of the finger (as many species do), they look where the finger is pointing. We argue that infants react in the same manner to the trajectories of gazing, pointing, or object movement, i.e., by anticipating the target, whether it is a target being looked at, pointed to, or moved toward. Two simple experiments would bear that out. Change the location of the target toward which the object has been heading; if the object doesn't change its trajectory to accommodate this change, the infant should be surprised—show increased looking time. Or show the infant a trajectory that, in effect, has no target; that too should surprise the infant. When shown a trajectory—be it that of a finger, eye, or self-propelled object—infants expect a target—that's the heart of the matter. Satisfaction refers to those conditions that bring goal-directed action to an end. The infant who understands "goal" as a satisfiable state expects certain conditions to terminate goal-directed action, and would be surprised if, for example, an object that succeeded in escaping reinstated its confinement.

Authors: Aside from movement and goals, what's the main difference infants see between physical and intentional objects?

DP: In a word, value positive or negative. Infants not only expect intentional objects to interact, they assign value positive or negative to the interaction. They do not assign value positive or negative to interactions between intentional and physical objects. The model claims that the infant uses two criteria in distinguishing between positive and negative value. The simpler criterion is based on intensity. The hard action of hitting is coded negative; the soft action of caressing, positive. The second criterion is the functional equivalent of helping and hindering. When one object is engaged in goal-directed action—for example, seeking to escape confinement—a second object can be seen as helping or hindering the first object achieve its goal. Helping is coded positive; hindering, negative.

Authors: How do you prove those claims?

DP: By using Ann's animations in a standard habituation-dishabituation design. For example, habituate infants on each of the four cases—caress, hit, help, hinder—then transfer all of them to the same, say, negative condition—a new case of hit. Infants shifted from negative to negative should show little dishabituation, whereas those shifted from positive to negative should show far more; that's what the model predicts and that's what we found.

Authors: What has all this to do with morality and consciousness? You remember that's what we were going to talk about—but we're out of time.

DP: We argue that the infant's positive and negative, the value positive or negative that it automatically assigns to the interactions between intentional objects, are the moral primitives, the concepts that cultures transform into good and bad. A good thing we've run out of time, for it's exceedingly hard to give even an approximate account of the cultural transformation of biological primitives—almost as hard as explaining the brain!

tackle the question of nonhuman primate consciousness would be to compare different species' brains to those of humans. Comparing the species on a neurological basis has proved to be difficult, however, except for the fact that the human prefrontal cortex is much larger in area than that of other primates. Another approach, one taken by comparative psychologists, is to study the theory of mind in nonhuman primates. Instead of comparing pure biological elements, their strategy is to focus on the behavioral manifestation of the brain. This approach parallels that of developmental psychologists who study the development of the theory of mind in children.

Studies that explore a primate theory of mind have drawn from the idea that children develop abilities that outwardly indicate conscious awareness of themselves and their environment. For example, once children understand the concept of a false belief or the mind's ability to incorrectly represent the world, they realize that beliefs are mental representations and hence can distinguish between the mind and the world. This comprehension often happens between the ages of 3 and 5 years.

This awareness has been tested in nonhuman primates and stems from the assumption that along with comprehending the difference between mind and world comes an understanding of the sources of knowledge and the representation of beliefs by others. Children learn that direct contact with a situation can give an individual a privileged state of knowledge, such as when a child sees a person looking into a box. A 4-year-old child can grasp that this person now knows what the box contains, whereas a younger child may not.

Daniel Povinelli, a rising cognitive neuroscience star (as evidenced by his being chosen as one of the select few to receive a James S. McDonnell Foundation Centennial award), and his colleagues at the New Iberia Research Center at the University of Louisiana at Lafayette have studied the theory of mind in chimpanzees (Povinelli and Eddy, 1996). They conducted an excellent series of experiments designed to explore the primate theory of mind by comparing and contrasting the **mentalistic** and **behaviorist** frameworks in terms of understanding the connection between seeing and knowing (Figure 16.16). The mentalist framework posits that chimpanzees have a subjective understanding of what they and others see and can act on this understanding; that is, they appreciate that seeing what

someone else sees conveys common information about that object. Conversely, the behaviorist framework posits that chimpanzees do not subjectively understand what others see; that is, they do not appreciate that seeing is about anything. Despite great effort expended on a wide range of tasks specifically designed to identify any mentalistic abilities, the data fail to support this framework. As always, one must be cautious when interpreting an absence of supporting data—it is possible that additional tests with additional chimpanzee groups, looking at a wider range of tasks, may produce data that support the mentalistic framework. Along these lines, the results of a recent study that expanded the definition of pointing, from only considering extension of the index finger to including hand gestures, suggest that chimpanzees can effectively communicate information about commonly seen objects (Krause and Fouts, 1997). However, such "communication" does not necessarily imply intention to act, a necessary condition for accepting that chimpanzees have a theory of mind.

Another test of chimpanzees' conscious abilities concerns the emergence of conceptual knowledge of the self and others, which in children occurs at 18 to 24 months. It has been shown that infants understand the notion that when another person looks or points at an object or event, that person becomes connected to it subjectively through the mental state of attention.

That chimpanzees are capable of self-recognition, as seen when a mirror is present, and can follow a person's gaze, might suggest that they have an understanding, as

Figure 16.16 **(a)** One experimenter leaves the room (the guesser). **(b)** One experimenter hides food under one cup (the knower). **(c)** Both experimenters then point to different cups. Chimpanzees initially randomly choose which cup the food is under. In time, their performance improves, suggesting that they solve the task by making associations with the roles of the experimenters as opposed to inferring the seeing and knowing connection. By comparison, 4-year-olds perform correctly from the onset, whereas 3-year-olds respond randomly.

(a)

(b)

(c)

(a)   (b)   (c)

Figure 16.17 **(a, b)** When initially presented with a mirror, chimpanzees react to it as if they are confronting another animal. After 5 to 30 minutes, however, chimpanzees will engage in self-exploratory behaviors, indicating that they know they are indeed viewing themselves. **(c)** Chimpanzees also are able to follow the gaze of an experimenter, which indicates that they might know that eyes direct attention.

children do, of how the eyes and internal states of attention are connected (Figure 16.17). Drawing from chimpanzees' awareness of eyes, Povinelli and Eddy (1996) studied chimpanzees' understanding of the eyes as portals through which the mental state of attention emanates (Figure 16.18).

Nonhuman primates demonstrate in countless ways their intelligence and ability to crudely communicate with humans; hence, it is not surprising that most people would like to believe that these relatives of ours also possess some form of consciousness like ours. But because the chimpanzees failed to fully appreciate knowledge communicated in the aforementioned looking and pointing tasks, we must conclude that the chimpanzee theory of mind is much less advanced than the human theory of

mind. In fact, chimpanzees do not even demonstrate conscious abilities as do growing children—at least in the paradigms set up to test assumptions about the expression of conscious awareness that emerges in children. That chimpanzees do not have conscious abilities makes perfect sense—humans, not chimpanzees, are the primates that continued to evolve. Perhaps this continued evolution was due to the advanced theory of mind with which humans appear to be uniquely endowed.

Left- and Right-Hemisphere Consciousness

When consciousness is viewed as a feeling about specialized abilities, one would expect the quality of consciousness emanating from each hemisphere to differ

(a)   (b)

Figure 16.18 **(a)** Chimpanzees can learn to request food from an experimenter by reaching through the left hole when an experimenter stands on the left, and vice versa. **(b)** When two experimenters are present, one wearing a blindfold, chimpanzees respond randomly as to which hole to gesture through for food, suggesting that they do not have a firm understanding of the eyes as a means of communicating visual attention, as children 2½ years old indicate in similar tests.

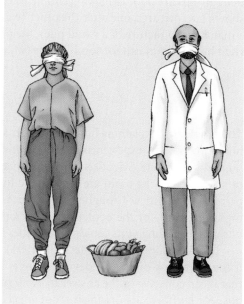

radically. While left-hemisphere consciousness would reflect what we mean by normal conscious experience, right-hemisphere consciousness would vary as a function of the specialized circuits that half brains possess. Mind Left, with its complex cognitive machinery, can distinguish between sorrow and pity and appreciate the feelings associated with each state. Mind Right does not have the cognitive apparatus for such distinctions and consequently has a narrower state of awareness. Consider the following examples of reduced capacity in the right hemisphere and the implications this has for consciousness.

Split-brain patients without right-hemisphere language have a limited capacity for responding to patterned stimuli, ranging from no capacity to the ability to make simple matching judgments above the level of chance. Patients with the capacity to make perceptual judgments not involving language were unable to make a simple same-difference judgment within the right brain when both the sample and the match were lateralized simultaneously. Thus, when a judgment of sameness was required for two simultaneously presented figures, the right hemisphere failed.

This minimal profile of capacity stands in marked contrast to patients with right-hemisphere language. One patient, J.W., who understood language and had a rich right-hemisphere lexicon as assessed by the Peabody Picture Vocabulary Tests and other special tests, until recently could not access speech from the right hemisphere. Patients V.P. and P.S. could understand language and speak from each half of the brain. Would this extra skill lend a greater capacity to the right hemisphere's ability to think, to interpret the events of the world?

It turns out that the right hemispheres of both patient groups are poor at making simple inferences. When shown two pictures, one after the other (e.g., a picture of a match and a picture of a wood pile), the patient (or the right hemisphere) cannot combine the two elements into a causal relation and choose the proper result (i.e., a picture of a burning woodpile as opposed to a picture of a woodpile and a set of matches). In other testing, simple words are presented serially to the right side of the brain. The task is to infer the causal relation between the two lexical elements and pick the answer from six possible answers in full view of the subject. A typical trial consists of words like *pin* and *finger* being flashed to the right hemisphere, with the correct answer being *bleed*. Even though the patient (right hemisphere) can always find a close lexical associate of the words used, he cannot make the inference that *pin* and *finger* should lead to *bleed*.

In this light, it is hard to imagine that the left and right hemispheres have similar conscious experiences. The right cannot make inferences; consequently, it is extremely limited in what it can have feelings about. It deals mainly with raw experience in an unembellished way. The left hemisphere, though, is constantly, almost reflexively, labeling experiences, making inferences as to cause, and carrying out a host of other cognitive activities. The left hemisphere is busy differentiating the world, whereas the right is simply monitoring the world.

Again as John Updike (1989) put it:

> Perhaps there are two kinds of people: those for whom nothingness is no problem, and those for whom it is an insuperable problem, an outrageous cancellation rendering every other concern, from mismatching socks to nuclear holocaust, negligible. Tenacious of this terror, this adamant essence as crucial to us as our sexuality we resist those kindly stoic consolers who assure us that we will outwear the fright, that we will grow numb and accepting and, as it were, religiously impotent. As Unamuno says, with the rhythms of a stubborn child, "I do not want to die—no: I neither want to die nor do I want to want to die; I want to live forever and ever and ever. I want this 'I' to live, this poor 'I' that I am and that I feel myself to be here and now."

SUMMARY

The problem of explaining how the brain enables human conscious experience remains a great mystery of human knowledge. Scientists are gaining vast knowledge of how parts of the brain are responsible for mental and perceptual activities. Though great advances are happening in the study of the content of conscious experience, our understanding of its subjective qualities is scant.

The philosopher Colin McGinn (1991) believed that if the mind is a biological device, there is no guarantee that it can conceive of the answer to every problem it can pose for itself. We are organisms, not angels, so some an-swers may not be thinkable by our limited brains. Just as a rat could never learn a maze in which it had to turn left at all the prime-number arms—a rat brain cannot entertain the notion "prime number"—people may never figure out why some kinds of information processing in the brain throw off sentient experience instead of just input and output. The rat might think that the food in the maze was placed by God, and might invent all kinds of cockamamie theories as to why it gets there—all because it does not have the kind of brain that thinks in terms of prime numbers. In a similar way, humans invent religion and philosophy, and go round and round pontificating about these questions for thousands of years, with no progress. Of course, McGinn cheerfully admitted that his view is not easily falsifiable, but then almost everything we say about sentience is nonfalsifiable at this point in human knowledge.

We believe there are a myriad of problems embedded in what we generally mean by the problem of consciousness. Issues of access, self-knowledge, attention, perception, salience, and history are addressable—and current research is illuminating these issues. The study of conscious experience is central to understanding the mind.

KEY TERMS

access-consciousness

backward referral hypothesis

behaviorist

blindsight

consciousness

dualism

functionalism

interpreter

masking stimulus

materialism

mentalistic

microstimulation

qualia

self-knowledge

sentience

subjectivity

subliminal perception

subthreshold

theory of mind

unconscious processes

THOUGHT QUESTIONS

1. Given that the key premise of dualistic theories of consciousness is that conscious experience is beyond the realm of physical sciences, how can you reconcile this view with scientific investigation? If consciousness is nonphysical, then presumably it cannot be measured. If it cannot be measured, how can it be studied?

2. What if Searle is right? Should we toss out both dualism and materialism, ignoring the notion that consciousness is made up of many hierarchical components, and start over? Discuss your answer.

3. We know that with practice we get better at performing tasks (such as driving a car, or reading upside-down text) and often eventually perform the tasks unconsciously. But we also know it takes more than just practice to improve performance. What else is happening?

4. Since blindsight subjects have deficits in visual awareness, they are often held up as archetypal cases for consciousness investigations. What is wrong with this approach? Can studying unconsciousness in the damaged brain really tell us anything about consciousness in the intact, healthy brain? Explain your answer.

5. Can Libet be right? Do we actually live 500 msec in the past? If so, do we really control our actions, or are we just reacting and then interpreting our behavior afterward? How does this fit in with Gazzaniga's views on the left-brain interpreter that has been demonstrated in split-brain patients (see Chapter 10)?

SUGGESTED READINGS

CHURCHLAND, P. (1988). *Matter and Consciousness.* Cambridge, MA: MIT Press.

DENNETT, D.C. (1991). *Consciousness Explained.* Boston: Little, Brown.

GAZZANIGA, M.S. (1995). Consciousness and the Cerebral Hemispheres. In M.S. Gazzaniga (Ed.), *The Cognitive Neurosciences* (pp. 1391–1399). Cambridge, MA: MIT Press.

LIBET, B. (1996). Neural Processes in the Production of Conscious Experience. In M. Velmans (Ed.), *The Science of Consciousness* (pp. 96–117). London: Routledge.

PREMACK, D., and WOODRUFF, G. (1978). Does the chimpanzee have a theory of mind? *Behav. Brain Sci.* 1:515–526.

SEARLE, J.R. (2000). Consciousness. *Annu. Rev. Neurosci.* 23:557–578.

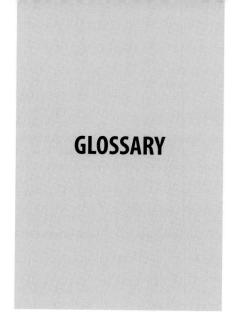

GLOSSARY

access-consciousness   The concept that the brain restricts the information that enters awareness to what is relevant for the current process.

achromatopsia   Selective disorder of color perception resulting from a lesion or lesions of the central nervous system, typically in the ventral pathway of the visual cortex. In achromatopsia, the deficit in color perception is disproportionately greater than that associated with form perception. Colors, if perceived at all, tend to be muted.

acquisition   To register inputs in sensory buffers and sensory analysis stages.

action potentials   The active or regenerative electrical signals long-distance communication requires.

adaptation   A characteristic of a living thing that contributes to its fitness.

affective   Having an emotional dimension, either positive or negative; not neutral.

affective style   An individual's unique tendencies in emotional expressions and reactions.

aggregate field theory   The belief that the whole brain participates in behavior.

aggression   A socially inappropriate expression of emotion, marked by an intent to dominate or control another's actions through the infliction of physical harm or verbal assault.

agnosia   Neurological syndrome in which disturbances of perceptual recognition cannot be attributed to impairments in basic sensory processes. It can be restricted to a single modality such as vision or audition.

agrammatic aphasia   Difficulty in producing and/or understanding the structure of sentences. Seen in brain-damaged patients who may speak using only content words, leaving out function words such as "the" and "a."

akinetopsia   Selective disorder of motion perception resulting from a lesion or lesions of the central nervous system. Patients with akinetopsia will fail to perceive stimulus movement, either created by a moving object or their own motion, in a smooth manner. In severe cases, motion may only be inferred by noting that the position of objects in the environment has changed over time, as if the patients are constructing dynamics through a series of successive static snapshots.

alexia   Neurological syndrome in which the ability to read is disrupted. It is frequently referred to as acquired alexia to indicate that it results from a neurological disturbance such as a stroke, usually including the occipital-parietal region of the left hemisphere. In contrast, developmental dyslexia refers to problems in reading that are apparent during childhood development.

allele   Alternate forms of genes composed of a complex organic molecule termed deoxyribonucleic acid (DNA).

alpha motor neurons   The neurons that terminate on muscle fibers, causing contractions that produce movements. They originate in the spinal cord and exit through the ventral root of the cord.

amobarbital   A barbiturate used to produce a rapid and brief anesthesia.

amygdala   Groups of neurons anterior to the hippocampus in the medial temporal lobe that are involved in emotional processing.

analytic processing   Perceptual analysis that emphasizes the component parts of an object. Reading is thought to be a prime example of analytic processing in that the recognition of words requires the analysis of at least some of the component letters.

anterior cingulate   Anterior portion of the cingulate cortex, located below the frontal lobe along the medial surface. This region is characterized by a primitive cytoarchitecture (three-layered cortex) and is part of the interface between the frontal lobe and limbic system.

Implicated in various executive functions such as response monitoring, error detection, and attention.

aphasia Language problems following brain damage or disease.

apperceptive agnosia Form of agnosia (loss of ability to interpret stimuli) associated with deficits in the operation of higher-level perceptual analyses. A patient with apperceptive agnosia may recognize an object when seen from a typical viewpoint. However, if the orientation is unusual, or the object is occluded by shadows, recognition deteriorates.

apraxia Neurological syndrome characterized by the loss of skilled or purposeful movement that cannot be attributed to weakness or an inability to innervate the muscles. It results from lesions of the cerebral cortex, usually in the left hemisphere.

apraxia of speech Difficulty in pronouncing words.

arcuate fasciculus A white matter tract that interconnects the posterior temporal region with frontal brain regions and is believed to transmit language-related information between the posterior and anterior brain regions.

area MT Region in the visual cortex containing cells that are highly responsive to motion. Area MT is part of the dorsal pathway, thought to play a role not only in motion perception but also in representing spatial information.

association cortex The volume of the neocortex that is not strictly sensory or motor, but receives inputs from multiple sensory-motor modalities.

associationism The theory that the aggregate of a person's experience determines the course of mental development.

associative agnosia Form of agnosia (loss of ability to interpret stimuli) in which the patient has difficulty in linking perceptual representations with long-term knowledge of the percepts. For example, the patient may be able to identify that two pictures are of the same object, yet fail to demonstrate an understanding of what the object is used for or where it is likely to be found.

auditory brainstem response (ABR) Electrical responses from ascending auditory pathways in the brainstem recorded, from electrodes on the human scalp using signal averaging to extract the small signals from the ongoing background EEG.

autonomic nervous system System that regulates heart rate, breathing, and glandular secretions and may become activated during emotional arousal, initiating a "fight or flight" behavioral response to a stimulus. It has two subdivisions, the sympathetic and parasympathetic branches.

axon The process extending away from a neuron down which action potentials travel, and the terminals of which contact other neurons at synapses.

backward referral hypothesis Hypothesis that states that our conscious awareness of a neural event is delayed (by approximately 500 msec), but we refer this awareness back in time so that we think we are aware of the event from its onset.

Balint's syndrome A disorder following bilateral occipital-parietal stroke, characterized by difficulty in perceiving visual objects. Patients with the disorder can correctly identify objects but have difficulty relating objects to one another. They tend to focus attention on one object to the exclusion of others when the objects are presented simultaneously.

basal ganglia Collection of five subcortical nuclei, the caudate, putamen, globus pallidus, subthalamic nucleus, and substantia nigra. The basal ganglia are involved in motor control and learning. Reciprocal neuronal loops project from cortical areas to the basal ganglia and back to the cortex. Two prominent basal ganglia disorders are Parkinson's disease and Huntington's disease.

behaviorism The theory that environment and learning are the primary factors in mental development, and that people should be studied by outside observation.

benefits Improvements in performance or physiological responses as a function of selective attention.

biased competition model A theoretical model that posits that attention acts at each stage of information processing in order to enable the relevant stimulus to influence the responses of neurons at that stage of processing.

blindsight Residual visual abilities within a field defect in the absence of acknowledged awareness.

blood-brain barrier (BBB) A physical barrier formed by the end feet of astrocytes between the blood vessels in the brain and the tissues of the brain. It limits which materials in the blood can gain access to neurons in the nervous system.

blood oxygenation level dependent (BOLD) effect Hemoglobin carries oxygen in the bloodstream, and when the oxygen is absorbed, the hemoglobin becomes deoxygenated. Deoxygenated hemoglobin is more sensitive, or paramagnetic than oxygenated hemoglobin. The magnetic detectors in a magnetic resonance imaging machine measure changes in the ratio of oxygenated to deoxygenated blood, or what is called the BOLD signal. When neural regions are active, this ratio increases as the blood supply to the active tissue increases.

bottleneck Stage of processing where not all of the inputs can gain access or pass through.

bradykinesia Symptom associated with Parkinson's disease, characterized by slowness in the initiation and execution of movements. A patient with bradykinesia appears to have difficulty in energizing movements.

brain lesions Structural damage to the white or gray matter of the brain. Lesions result from many causes, including tumor, stroke, and degenerative disorders such as Alzheimer's disease.

brainstem The region of the nervous system that contains groups of motor and sensory nuclei, nuclei of widespread modulatory neurotransmitter systems, and white matter tracts of ascending sensory information and descending motor signals.

Broca's aphasia The oldest and perhaps well-studied form of aphasia, characterized by speech difficulties in the absence of severe comprehension problems. However, Broca's aphasics may also suffer from problems in fully comprehending grammatically complex sentences.

categorical representation One manner of representing spatial information of objects. Categorical representations capture basic relational information such as the relative position of two objects from a particular viewpoint.

category-specific deficits Recognition impairments that are restricted to certain classes of objects. Although rare, there are reports of individuals who demonstrate a selective impairment in their ability to recognize living things, yet exhibit a near normal level of performance in recognizing nonliving things. Such deficits are useful in the development of models about how perceptual and semantic knowledge is organized in the brain.

central nervous system (CNS) The brain and spinal cord.

cerebellum Also known as "little cerebrum," a large, highly convoluted (infolded) structure located dorsal to the brainstem at the level of the pons. The cerebellum maintains (directly or indirectly) interconnectivity with widespread cortical, subcortical, brainstem, and spinal cord structures, and plays a role in various aspects of coordination ranging from locomotion to skilled, volitional movement.

cerebral cortex The layered sheet of neurons that overlays the forebrain. Also consists of neuronal subdivisions (areas) interconnected with other cortical areas, subcortical structures, and the cerebellum and spinal cortex.

cerebral specialization A particular brain region's activity may specifically subserve a given cognitive function or behavior.

chromosome A structure composed of nucleic acids and proteins within the nucleus of a cell, which contains the genetic material (DNA) of an organism.

cochlear nucleus   Midbrain nucleus that is one of the primary receiving regions of the output from the cochlea of the inner ear. Axons from the cochlear nucleus project to the inferior colliculus, and from there on to the medial geniculate nucleus of the thalamus and the auditory cortex.

cognitive psychology   Branch of psychology that studies how the mind internally represents the external world and performs the mental computations required for all aspects of thinking. Cognitive psychologists study the vast set of mental operations associated with such things as perception, attention, memory, language, and problem solving.

commissure   White matter tracts that cross from the left to the right side, or vice versa, of the CNS.

comparative neuroscience   Evolutionary neurobiology.

computed tomography (CT or CAT)   Method that provides noninvasive images of internal structures such as the brain. CT is an advanced version of the conventional x-ray. Whereas conventional x-rays compress three-dimensional objects into two dimensions, CT allows for the reconstruction of three-dimensional space from the compressed two-dimensional images through computer algorithms.

concurrent processing   Hypothesis that analysis within the perceptual systems of the brain involves a set of specialized systems. These subsystems may operate in parallel, extracting different properties of the stimulus, but also interact at various stages to converge on a coherent percept.

conduction aphasia   A form of aphasia that is considered a disconnection syndrome. May occur when the arcuate fasciculus, the pathway from Wernicke's to Broca's area, is damaged, thereby disconnecting the posterior and anterior language areas.

consciousness   The human ability to assign feelings to specialized mental capacities (e.g., thinking, feeling, and talking).

consolidation   The process by which memory representations become stronger over time. Believed to include changes in the brain system participating in the storage of information.

content-based hypothesis   Proposal that memory systems are defined in terms of their content. An influential model of working memory proposes that there are two distinct systems, one that supports verbal representations and a second that supports visuospatial representations.

convergent evolution   Similarities that arise independently in different lineages.

coordinate representation   One manner of representing spatial information of objects. Coordinate representations specify the exact positions of the objects and the distances between them.

corpus callosum   Fiber system composed of axons that connect the cortex of the two cerebral hemispheres.

cortical visual areas   Regions of visual cortex that are identified on the basis of their distinct retinotopic maps. The areas are specialized to represent certain types of stimulus information, and through their integrated activity, provide the neural basis for visually-based behavior.

corticogenesis   The creation of the cortex during development. It describes how neurons are born, are specified to be a certain neuronal type or class, and are placed in the correct anatomical position accordingly.

cortico-spinal tract   Bundle of axons that originate in the cortex and terminate monosynaptically on alpha motor neurons and spinal interneurons in the spinal cord. Many of these fibers originate in the primary motor cortex, although some come from secondary motor areas. The cortico-spinal tract (also referred to as the pyramidal tract) is important for the control of voluntary movements.

covert attention   The ability to direct attention without overt alterations or changes in sensory receptors. For example, attending to a conversation without turning the eyes and head toward the speakers.

critical period   Specific time ranges early in life when experience can maximally influence the organization and function of the nervous system. Critical periods for behavioral development likely correspond to specific stages of neural development such as the formation of synaptic connections or the presence and activity of molecular cues for specifying neuronal connectivity.

cytoarchitectonic map   A diagram that shows regions of the cerebral cortex containing neurons and neuronal organization of similar type. There are as many as 50 different cytoarchitectonic areas, defined by histological analysis of the cortex.

cytoarchitectonics   How cells differ between brain regions.

deafferentiation   Condition in which sensory, or afferent, signals from the limb are blocked from reaching the central nervous system. This condition is sometimes created in animals by severing the dorsal root of the spinal cord, the pathway for information coming from the limbs to the cord. Patients may also become functionally deafferented due to damage in the sensory nerves.

decision making   The process of selecting and evaluating social and emotional information relevant to a particular problem in order to generate possible solutions to that problem, followed by selection of the most appropriate solution.

declarative memory   Knowledge to which we have conscious access, including personal and world knowledge (events and facts). The term "declarative" signals the idea that declarations can be made about the knowledge, and that for the most part, we are aware that we possess the information.

delayed-response task   Task in which the correct response must be produced after a delay period of several seconds. Such tasks require the operation of working memory since the animal or person must maintain a record of the stimulus information during the delay period.

dendrite   Large treelike processes of neurons that receive inputs from other neurons at locations called synapses.

diaschisis   The idea that damage to one part of the brain can create problems for another.

dichotic listening   To listen to a different message in each ear at the same time.

dichotic listening task   An auditory task involving two competing messages presented simultaneously, one to each ear, while the subject tries to report both messages. The ipsilateral projections from each ear are presumably suppressed when a message comes over the contralateral pathway from the other ear.

distributed representation   The idea that information may be stored in large populations of neurons located in relatively widespread regions of the brain. Contrasts with the idea that the representation of some items in memory are stored in a discrete, highly localized set of neurons.

dorsal pathway   One of the two main processing streams for visual information in the cerebral cortex. The dorsal pathway extends from the occipital lobe into the superior temporal and parietal lobes. Functional hypotheses have centered on the idea that this pathway is important for the representation of spatial information and for guiding visually directed actions.

double dissociation   Method used to develop functional models of mental and/or neural processes. Evidence of a double dissociation requires a minimum of two groups and two tasks. In neuropsychological research, a double dissociation is present when one group is impaired on one task and the other group is impaired on the other task. In imaging research, a double dissociation is present when one experimental manipulation produces changes in activation in one neural region and a different manipulation produces changes in activation in a different neural region. Double dissociations provide a strong argument that the observed differences in performance reflect functional differences between the groups, rather than unequal sensitivity of the two tasks.

dualism   A major philosophical approach to describing consciousness, which holds that the mind and brain are two separate phenomena.

Variations include: popular dualism, property dualism, epiphenomenalism, and interactionist property dualism.

dura mater Dense layers of collagenous fibers that surround the brain and spinal cord.

dynamic filtering The hypothesis that a key component of working memory involves the selection of information that is most relevant given current task demands. This selection is thought to be accomplished through the filtering, or exclusion of, potentially interfering and irrelevant information.

early selection The theoretical model that posits that attention can select (partially or completely) incoming information prior to complete perceptual analysis and its encoding as categorical or semantic information. This can be contrasted with "late selection."

ectoderm Cell line in blastula of developing embryo from which neural ectoderm will form, leading to the development of the nervous system.

effector Any part of the body that can move, such as the arms, fingers, or legs.

electrical gradient A force that develops when a charge distribution across the neuronal membrane develops such that the charge inside is more positive or negative than the one outside. Results from asymmetrical distributions of ions across the membrane.

electroencephalography (EEG) Technique to measure the electrical activity of the brain. It usually involves taking surface recordings from electrodes placed on the scalp, although it can also be measured from the cortical surface directly. The EEG signal will include endogenous changes in electrical activity, as well as those triggered by specific events (e.g., stimuli or movements).

electrotonic conduction Passive current flow through neurons that accompanies activated electrical currents.

emotion An affective (positive or negative) mental response to a stimulus that also may be expressed physically (e.g., by change in heart rate, facial expression, and speech).

emotion communication Conveying one's own affective feelings to others using emotional speech and facial expressions, and comprehending others' affective feelings by accurately interpreting their emotional speech and facial expressions.

empiricism The idea that all knowledge comes from sensory experience.

encoding Processing incoming information to be stored. It includes the two stages of acquisition and consolidation.

endpoint control Hypothesis concerning how movements are planned in terms of the desired final location. Endpoint control models emphasize that the motor representation is based on the final position required of the limbs to achieve the movement goal.

equilibrium potential The membrane potential at which a given ion (e.g., K+) has no net flux across the membrane; that is, as many K+ ion move outward as inward across the membrane.

error-related negativity (ERN) Electrical signal that is derived from the EEG record following an erroneous response. The ERN is seen as a prominent negative deflection in the ERP, and is hypothesized to originate in the anterior cingulate.

ethology The study of animal behavior.

event-related potential (ERP) Change in electrical activity that is time-locked to specific events such as the presentation of a stimulus or the onset of a response. When the events are repeated many times, the relatively small changes in neural activity triggered by these events can be observed by averaging the EEG signals. In this manner, the background fluctuations in the EEG signal are removed, revealing the event-related signal with great temporal resolution.

evolutionary psychology Field of study that uses an evolutionary framework to explain cognitive behavior.

exaptation The process whereby a structure that serves a particular function is co-opted for some other, very different function.

executive functions Refers to the set of higher-level cognitive operations that are essential for the production of goal-oriented behavior. Executive functions involve the maintenance and manipulation of information that is essential for dealing with situations in which the appropriate response is not dictated by the current stimulus information. Includes processes such as working memory, goal representation and planning, response monitoring, and error detection.

exogenous orienting The control of attention by external stimuli and not by internal voluntary control.

extinction The failure to perceive or respond to a stimulus contralateral to the lesion (contralesional) when presented with a simultaneous stimulus ipsilateral to the lesion (ipsilesional).

extrapyramidal tracts Collection of motor tracts that originate in various subcortical structures including the vestibular nucleus and the red nucleus. These tracts are especially important for maintaining posture and balance.

extrastriate Visual areas that lie outside striate cortex (Brodmann's area 17, the primary visual cortex) and are considered secondary visual areas since they receive input either directly or indirectly from the primary visual cortex.

facial expression The nonverbal communication of emotion by manipulation of particular groups of facial muscles. Research findings suggest that there are six basic human facial expressions that represent the emotional states: anger, fear, disgust, happiness, sadness, and surprise.

fear conditioning Learning, in which a neutral stimulus acquires aversive properties by virtue of being paired with an aversive event.

fitness The measure of evolutionary success in terms of a gene, trait, or behavior that is represented in future generations.

flanker task Behavioral task that measures the effects of spatial selective attention to a stimulus by assessing the degree of interference, measured in slowed reaction time, produced by distracters flanking the to-be-discriminated target stimulus. Typically, the distracters require no response, being at a location that is to be ignored, but contain features that would require a response different from that required to the target.

frontal lobe The mass of cortex anterior to the central sulcus and dorsal to the Sylvian fissure. Contains two principal regions, the motor cortex and the prefrontal cortex, each of which can be further subdivided into specific areas both architectonically and functionally.

functional asymmetries Limited communication between the visual systems within each hemisphere.

functionalism A variation of materialism. The dominant feature is the belief that anything that responds, feels, or thinks appropriately is functionally equivalent to a human brain. This theory rejects the possibility of understanding mental states in terms of behavioral output, and proposes that subjective experience will be understood after all other underlying systems are explained.

fusiform gyrus Gyrus located along the ventral surface of the temporal lobe. This area has been shown in neuroimaging studies to be consistently activated when people view face stimuli. Neurological lesions that include the fusiform gyrus are associated with prosopagnosia, although the damage also extends to other regions of the cortex.

gate Conceptualization of the attentional mechanism that prevents ignored stimuli from gaining access to further processing.

gene A hereditary unit composed of a complex organic molecule termed deoxyribonucleic acid (DNA) and has alternate forms called alleles.

generative assembling device A device that generates complex representations from a small vocabulary of primitive units.

genetic pleiotropy The idea that a single gene has many functions.

genetic specificity The idea that one gene is responsible for a single function or behavior.

genotype The genetic composition of an organism.

glial cells Also called neuroglial cells or simply glia. The other cell type in the nervous system, glia, are more numerous than neurons, by perhaps ten times, and may account for more than half of the brain's volume. They typically do not conduct signals themselves; but without them, the functionality of neurons would be severely diminished.

gnostic unit Refers to the concept that the final product of perceptual recognition reflects activity in a neuron or small set of neurons that are tuned for a specific percept. Based on the idea that hierarchical models of perception imply that at higher levels in the system, neurons become much more selective in terms of what they respond to.

goal-oriented behavior Behavior that allows us to interact in the world in a purposeful manner. Goals reflect the intersection of our internal desires and drives, coupled with the current environmental context.

gray matter Regions of the nervous system containing primarily neuronal cell bodies; for example, the cerebral cortex, the basal ganglia, and the nuclei of the thalamus. So called because in preservative solution, these structures look gray in comparison to the "white matter" where myelinated axons are found (which look more white).

gyri The protruding rounded surfaces of the cerebral cortex that one can see upon gross anatomical viewing of the intact brain.

handedness Term that refers to whether the majority of one's manual actions are performed with the right or left hand.

Hebbian Learning Theory that if a weak and a strong input act on a cell at the same time, the weak synapse becomes stronger. Named for Donald Hebb, who postulated this mechanism as a means for the connectional strength between neurons to change in order to store information.

hemiplegia Neurological condition characterized by the loss of voluntary movements on one side of the body. Typically results from damage to the cortico-spinal tract, either from lesions to the motor cortex or to white-matter lesions that destroy the descending fibers.

heritable Capable of being inherited or passed on to successive generations.

hierarchical structure A configuration that may be described at multiple levels, from global features to local features; the finer components are embedded within the higher-level components.

hippocampus The "seahorse" of the brain, the hippocampus is a layered structure in the medial temporal lobe that receives inputs from wide regions of the cortex via inputs from the surrounding regions of the temporal lobe, and sends projections out to subcortical targets. Involved in learning and memory, particularly spatial memory in mammals and episodic memory in humans.

holism The belief that brain functions are carried out by the whole brain, rather than by distinct or "local" delegate regions.

holistic processing Perceptual analysis that emphasizes the overall shape of an object. Face perception has been hypothesized to be the best example of holistic processing, in that the recognition of an individual appears to reflect the composition of the person's facial features rather than being based on the recognition of the individual features themselves.

homology A structure, behavior, or gene that has been retained from a common ancestor.

homoplasy Structures that look the same, but not necessarily due to common ancestry.

homotopic areas Areas in corresponding locations in the two hemispheres.

Huntington's disease Genetic degenerative disorder in which the primary pathology involves atrophy of the striatum (caudate and putamen) of the basal ganglia. Prominent symptoms include clumsiness and the presence of involuntary movements of the head and trunk. Cognitive impairments are also seen and become pronounced over time.

hypermetria Motor deficit associated with cerebellar lesions, especially those involving the cerebellar hemispheres. Hypermetria refers to movements that extend beyond the target, and are usually seen when the patient attempts to move rapidly.

hypothalamus A small collection of nuclei that form the floor of the third ventricle. The hypothalamus is important for the autonomic nervous system and the endocrine system, and controls functions necessary for the maintenance of homeostasis.

ideational apraxia Severe form of apraxia in which the patient's knowledge about the intent of an action is impaired. For example, the patient may no longer comprehend the appropriate use for a tool even though she is still capable of producing the required movement.

ideomotor apraxia Form of apraxia in which the patient has difficulty in executing the desired action properly. Unlike ideational apraxia, patients with ideomotor apraxia appear to have a general idea about how the action should be performed and how tools are used, but are unable to coordinate the movements to produce the action in a coherent manner.

information processing system A conceptualization of sensory, perceptual, conceptual, and response processes in the brain that focuses on the flow of information from input through storage and analysis to output.

inhibition of return (IOR) Hypothesized process underlying the slowing in reaction time observed over time when attention is reflexively attracted to a location by a sensory event (i.e., reflexive cue). IOR is, as the name implies, conceptualized as inhibition of recently attended locations such that attention is inhibited in returning to that location (or object).

inhibitory control Hypothesis that one aspect of executive functions is the regulation of actions by actively inhibiting the tendency to produce habitual responses or environmentally-dictated actions. A loss of inhibitory control is assumed to underlie the tendency of some patients with prefrontal lesions to produce socially inappropriate behavior.

innate An ability or behavior that an organism is born capable of exhibiting, and that therefore is under genetic control, being prespecified by the genetic code (i.e., as opposed to learned).

integrative agnosia Form of agnosia associated with deficits in the recognition of objects due to the failure to group and integrate the component parts into a coherent whole. Patients with this deficit can faithfully reproduce drawings of objects; however, their percept is of isolated, unconnected parts or contours.

interaural time Difference in time between when a sound reaches the two ears. This information is represented at various stages in the auditory pathway, and provides an important cue for sound localization.

interpreter A left-brain system that seeks explanations for internal and external events in order to produce appropriate response behaviors.

ion channels Formed by transmembrane proteins that create pores, these are actual passageways through the membrane via which ions (charged atoms in solution) of sodium, potassium, and chloride (Na+, K+, and Cl-) might pass.

knock-out procedures Technique to create a genetically altered version of a species. In the knock-out species, specific genes are altered or eliminated. The method can be used to study behavioral changes that occur in animals that have developed without the targeted gene, or to observe how genes code the development of the nervous system.

late selection The theoretical model that posits that all inputs are equally processed perceptually, but attention acts to differentially filter these inputs at later stages of information processing. This can be contrasted with "early selection."

lateral fissure The large sulcus that defines the superior border of the temporal lobe.

lateral geniculate nucleus (LGN) Thalamic nucleus that is the main target of axons of the optic tract. Output from the LGN is primarily directed to the primary visual cortex (Brodmann's area 17).

lateral prefrontal cortex The region of the cerebral cortex that lies anterior to Brodmann's area 6, along the lateral surface. This region has been implicated in various executive functions, such as working memory and response selection.

laterality effect A research finding that indicates that a particular cognitive function or behavior is primarily dependent upon activity in only one of the two hemispheres (right or left).

learning The process of acquiring new information.

lexical access The process by which perceptual inputs activate word information in the mental lexicon, including semantic and syntactic information about the word.

lexical selection The selection of the activated word form representations that best matches the sensory input.

limbic system Several structures that form a border (limen in Latin) around the brainstem, named the "grand lobe limbique" (limbic lobe) by Paul Broca. It is the emotional network that includes the amygdala, orbitofrontal cortex, and portions of the basal ganglia.

limited capacity Concept that the stages of information processing have a finite processing capability, which leads to the need for the system to select high-priority information for access to these stages of analysis.

localization The belief that individual behaviors and perceptions are controlled by distinct regions of the brain.

long-term memory The persistence of information over the long term, from hours to days and years.

looking time Measure of the time infants look at presented items, which can be used to quantify their habituation to and dishabituation from presented stimuli.

magnetic resonance imaging (MRI) Imaging technique that exploits the magnetic properties of organic tissue. Certain atoms are especially sensitized to magnetic forces given the number of protons and neutrons in their nuclei. The orientation of these atoms can be altered by the presence of a strong magnetic field. When the magnetic field is removed, the atoms will gradually return to a randomly distributed orientation, and in the process of this transition will generate a small magnetic field that can be measured by sensitive detectors. Structural MRI studies usually measure variations in the density of hydrogen. Functional MRI measures changes over time in the distribution of the targeted atom.

magnetoencephalography (MEG) A measure of the magnetic signals generated by the brain. The electrical activity of neurons also produces small magnetic fields, which can be measured by sensitive magnetic detectors placed along the scalp, similar to the way EEG measures the surface electrical activity. MEG can be used in an event-related manner similar to ERPs, with similar temporal resolution. The spatial resolution, in theory, can be superior since magnetic signals are minimally distorted by organic tissue such as the brain or skull.

magnocellular (M) system The M system is composed of visually-responsive cells that are characterized by relatively large axons. The magno cells of the optic tract terminate in the bottom two layers of the LGN, and from there project to the cerebral cortex, primarily contributing to the dorsal pathway.

masking stimulus A stimulus immediately following a briefly presented stimulus, to prevent further conscious processing of the original stimulus.

materialism A major philosophical approach to describing consciousness, based on the theory that the mind and brain are both physical mediums. Variations include: philosophical behaviorism, reductive materialism, and functionalism.

medulla Also called myelencephalon, the medulla is the brainstem's most caudal portion. It is continuous with the spinal cord and contains the prominent, dorsally positioned nuclear groups known as the gracile and cuneate nuclei, which relay somatosensory information from the spinal cord to the brain, and the ventral pyramidal tracts, containing descending projection axons from the brain to the spinal cord. Various sensory and motor nuclei are found in the medulla.

memory The persistence of learning in a state that can be revealed at a later time.

mental lexicon A mental store of information about words that includes semantic information (the word meaning), syntactic information (rules for using the words), and the details of word forms (how they are spelled and what their sound pattern is).

mentalistic Related to the belief that one has a subjective understanding of what they and others see, and can act on this understanding.

microstimulation Injection of electrical current in the vicinity of a group of neurons of interest, in order to induce neural activity.

midbrain Consists of the tectum (meaning "roof," and representing the dorsal portion of the mesencephalon), tegmentum (the main portion of the midbrain), and ventral regions occupied by large fiber tracts (crus cerebri) from the forebrain to the spinal cord (corticospinal tract), cerebellum, and brainstem (cortico-bulbar tract). The midbrain contains neurons that participate in visuomotor functions (e.g., superior colliculus, oculomotor nucleus, trochlear nucleus), visual reflexes (e.g., pretectal region), auditory relays (inferior colliculus), and the mesencephalic tegmental nuclei involved in motor coordination (red nucleus). It is bordered anteriorly by the diencephalon, and caudally by the pons.

monitoring Refers to the executive function associated with evaluating whether current representations and/or actions are conducive to the achievement of current goals. Errors can be avoided or corrected by a monitoring system. One of the hypothesized operations of a supervisory attentional system.

mutation A sudden change in the genetic structure of an organism, which affects the development of the organism.

myelin A fatty substance that surrounds the axons of many neurons and increases the effective membrane resistance.

natural selection Individuals with characteristics that best suit an environment are more likely to survive and reproduce.

neglect syndrome Behavioral pattern exhibited by neurological patients with lesions to the forebrain, in which they fail or are slowed in acknowledging that objects or events exist in the hemispace opposite their lesion. Most closely associated with right parietal cortex damage.

neocortex Typically contains six main cortical layers (with sublayers) with a high degree of specialization of neuronal organization. The neocortex is composed of areas like the primary sensory and motor cortex and association cortex, and as its name suggests, is the most modern (evolved) type of cortex.

neural plat The dorsal surface of developing embryo early in embryogenesis, formed by ectoderm cells and from which the developing nervous system will arise.

neuroethology The study of the neurobiology of animal behavior.

neurology Branch of medicine focused on the function and dysfunction of the nervous system. Clinical neurologists will treat patients with a wide range of problems such as headache, trauma, stroke, and changes in mental ability. They may also study the patients as a way of understanding the operation of the nervous system in normal function.

neuron Information processing cell of the nervous system.

neuron doctrine The belief that brain functions are carried out through the synchronized activity of independent neurons.

neuronal determination and differentiation The specification of neuronal cell type during development. Cells produced from nondifferentiated precursor cells must take on the phenotype appropriate for their ultimate role in the nervous system.

neuronal migration The movement of new neurons to their final and appropriate location in the developing brain. Involves molecular cues and the assistance of special glial cells (in the cortex) that provide a superstructure (radial glia) for neurons to climb along en route to their position in the brain (e.g., specific layers of cortex).

neuronal proliferation Cell division in the developing embryo and fetus occurring over time, which populates the nervous system with the complete complement of neurons needed for the organism.

neurophysiology Study of the physiological processes of the nervous system. Neural activity is characterized by physiological changes that can be described both electrically and chemically. The changes can be observed at many different levels ranging from the gross changes recorded with EEG, to the firing of individual neurons, to the molecular changes that occur at the synapse.

neurotransmitter A chemical substance that transmits the signal between neurons at chemical synapses.

nodes of Ranvier The locations at which myelin is interrupted between successive patches of axon, and where action potentials can be generated.

nondeclarative memory Knowledge we typically have no conscious access to, such as motor and cognitive skills (procedural knowledge). For example, the ability to ride a bicycle is a nondeclarative form of knowledge. Although we can describe the action itself, the actual information one needs to ride a bicycle is not easily described.

object-based attention The view that attention processes act on object representations to achieve selection.

object constancy The ability to recognize invariant properties of an object across a wide range of contexts. For example, although the size of the retinal image changes dramatically when a car recedes in the distance, our percept is that the car remains the same size. Similarly, we are able to recognize that an object is the same when seen from different perspectives.

occipital lobe Cortical lobe located at the posterior of the cerebral cortex that contains primarily neurons involved in visual information processing.

olivocochlear bundle Neural projection from the neurons in the brainstem that project to the hair cells of the cochlea.

optic ataxia A neurological syndrome in which the patient has great difficulty in using visual information to guide her actions even though she is unimpaired in her ability to recognize objects. Associated with lesions of the parietal lobe.

orbitofrontal cortex A brain region whose primary functions are related to the processing of emotion.

out-group comparison A comparison that defines phylogenetic relationships among animals.

parietal lobe Located posterior to the central sulcus, anterior to the occipital lobe, and superior to the posterior temporal cortex. This cortical region contains a variety of neurons including the somatosensory cortex, gustatory cortex, and parietal association cortex, including regions involved in visuomotor orienting, attention, and representation of space.

Parkinson's disease Degenerative disorder of the basal ganglia in which the pathology results from the loss of dopaminergic cells in the substantia nigra. Primary symptoms include difficulty in initiating movement, slowness of movement, poorly articulated speech, and in some cases, resting tremor.

parsing The immediate assignment of syntactic structure to incoming words.

parvocellular (P) system The P system is composed of visually-responsive cells that are characterized by relatively small axons. The parvo cells of the optic tract terminate in the top four layers of the LGN and from there, project to the cerebral cortex, primarily contributing to the ventral pathway.

perceptual representation system A form of nondeclarative memory that acts within the perceptual system in which the structure and form of objects and words can be primed by prior experience and can be revealed later using implicit memory tests.

peripheral nervous system (PNS) A courier network that delivers sensory information to the CNS and then conducts the motor commands of the CNS to control muscles of the body; anything outside the brain and spinal cord.

permeability The extent to which ions can cross a neuronal membrane.

perseveration Tendency to produce a particular response on successive trials, even when the context has changed so that the response is no longer appropriate. Commonly observed in patients with prefrontal damage, and thought to reflect a loss of inhibitory control.

phenotype An observable trait or set of characteristics of an organism.

photoreceptors Specialized cells in the retina that transduce light energy into changes in membrane potential. The photoreceptors are the interface for the visual system between the external world and the nervous system.

phrenology The study of the physical shape of the human head, based on the belief that variations in the skull's surface can reveal specific intellectual and personality traits. Today, phrenology is understood to be without validity.

Piagetian theory Ideas of Jean Piaget that include a biological view of development and thus differ from prior, behaviorally-based theories. On the basis of his observations he characterized the cognitive development of humans in a four-stage model, which is currently undergoing rapid modification.

planum temporale Surface area of the temporal lobe that includes Wernicke's area; long believed to be larger in the left hemisphere due to the lateralization of language function, although this theory is currently controversial.

plasticity The ability to change, as in the malleability of plastic material. In the nervous system, plasticity is involved during development, recovery from injury, and in everyday learning.

plesiomorphic A trait that is a general feature of some group or lineage.

pons Includes the pontine tegmental regions on the floor of the fourth ventricle, and the pons itself, a vast system of fiber tracts interspersed with pontine nuclei. The fibers are continuations of the cortical projections to the spinal cord, brainstem, and cerebellar regions. Also includes the primary sensory nuclear groups for auditory and vestibular inputs, and somatosensory inputs from, and motor nuclei projecting to, the face and mouth. Neurons of the reticular formation can also be found in the anterior regions of the pons.

pop-out Phenomenon where target objects in a cluttered scene can be detected rapidly when they are defined by a unique feature (e.g., color) that distinguish them from distracters in the scene. This produces a flat search function, which plots reaction time to target detection as a function of the number of distracter stimuli. Implies that preattentive mechanisms can achieve target detection.

population vector Representation formed by considering the summed activity of a population of neurons. Motivated by the idea that a stronger correlation between neural signals and behavior can be obtained by considering the aggregate activity across a region of the brain rather than by considering the activity of individual neurons. For example, the population vector calculated from neurons in the motor cortex can predict the direction of a limb movement.

positron emission tomography (PET) PET scanning measures metabolic activity in the brain by monitoring the distribution of a radioactive tracer. The PET scanner measures the photons that are produced during the decay of the tracer. A popular tracer is O^{15} since the decay time is rapid and the distribution of oxygen increases to neural regions that are active.

postsynaptic The neuron located after the synapse with respect to information flow.

preattentive mechanism Mechanism based on perceptual processes that can signal the presence of a difference in the visual scene, rather than the action of focused attention.

precursor cells Undifferentiated cells from which all neurons are produced by cell division. They are located in the ventricular zone.

prefrontal cortex Takes part in the higher aspects of motor control and the planning and execution of behavior, perhaps especially tasks that require the integration of information over time requiring the involvement of working memory mechanisms. The prefrontal cortex has three or more main areas that are commonly referred to in descriptions of the gross anatomy of the frontal lobe: the dorsolateral prefrontal cortex, the anterior cingulate and medial frontal regions, and the orbitofrontal cortex.

premotor cortex Secondary motor area that includes the lateral aspect of Brodmann's area 6, just anterior to the primary motor cortex. While some neurons in premotor cortex project to the cortico-spinal tract, many terminate on neurons in the primary motor cortex and help shape the forthcoming movement.

presynaptic The neuron located before the synapse with respect to information flow.

primary motor cortex Region of the cerebral cortex that lies along the anterior bank of the central sulcus and precentral gyrus forming Brodmann's area 4. Some axons originating in the primary motor cortex form the majority of the cortico-spinal tract, while others project to cortical and subcortical regions involved in motor control. The primary motor cortex contains a prominent somatotopic representation of the body.

primary visual cortex Medial portion of the occipital lobe that is the primary receiving region in the cortex for axons originating in the LGN of the thalamus. Various terms are used to refer to this area, including Brodmann's area 17, V1, and striate cortex. The latter comes from the fact that the cytoarchitecture is regular and stippled.

principle of parsimony The idea that the best hypothesis is one that requires the fewest number of transformations to explain a phenomenon; the simplest of competing theories is preferred.

procedural memory One form of nondeclarative memory that involves the learning of a variety of motor (e.g., knowledge of how to ride a bike) and cognitive skills (e.g., the acquisition of reading skills).

propagation The movement of electrical signals down the axon during neuronal signaling.

prosopagnosia Neurological syndrome characterized by a deficit in the ability to recognize faces. Some patients will show a selective deficit in face perception, a type of category-specific deficit. In others, the prosopagnosia is one part of a more general agnosia. Frequently associated with bilateral lesions in the ventral pathway, although it can also occur with unilateral lesions of the right hemisphere.

pulvinar nucleus Large group of nuclei in the thalamus, located at the posterior end.

qualia Generally, the ways things seem to us. Specifically, the particular, personal, and subjective experiences we have while interacting with the world.

radial glia Special glial cells that are anchored on one end at the ventricular zone and at the other end at the cortical surface. Migrating neurons "crawl" along these glial cells, which act as scaffolding during corticogenesis.

rationalism The idea that through right thinking and rejection of unsupportable or superstitious beliefs, true beliefs can be determined.

recency memory Memory for the temporal order of previous events. Recency memory is a form of episodic memory in that it involves remembering when a specific event took place. Patients with prefrontal lesions do poorly on tests of recency memory despite the fact that their long-term memory is relatively intact.

receptive field The area of external space within a stimulus that must be presented in order to activate a cell. For example, cells in the visual cortex respond to stimuli that appear within a restricted region of space. In addition to spatial position, the cells may be selective to other stimulus features such as color or shape. Cells in the auditory cortex also have receptive fields. The cell's firing rate will increase when the sound comes from the region of space that defines its receptive field.

receptor There are two uses of the word receptor in the nervous system. One refers to sensory receptors such as the photoreceptors in the eye that transduce physical signals in the environment (light energy) to neuron signals. The other use refers to transmembrane proteins on the postsynaptic side of a synapse that bind neurotransmitters and lead to changes in the membrane potential of the postsynaptic neuron.

receptor potentials Electrical potentials generated by active processes (the opening of ion channels) in a sensory receptor such as the photoreceptors of the eye.

recombination A change in the array of alleles of existing genes.

reflexive attention The orienting of attention induced by bottom-up, or stimulus-driven effects, such as when a flash of light in the periphery capture one's attention. Also called automatic or exogenous attention.

refractory period Short period of time following an action potential when the neuron may not be able to generate action potentials or may only be able to do so with larger-than-normal depolarizing currents.

regional cerebral blood flow (rCBF) Distribution of the brain's blood supply, which can be measured with various imaging techniques. In PET scanning, rCBF is used as a measure of metabolic changes following increased neural activity in restricted regions of the brain.

reinforcement A method of influencing future behavior by rewarding or punishing a particular behavioral response to a stimulus.

response conflict Situation in which more than one response is activated, usually because of some ambiguity that is present in the stimulus information. It has been hypothesized that the anterior cingulate monitors the level of response conflict and modulates processing in active systems when conflict is high.

resting membrane potential The difference in voltage across the neuronal membrane at rest, when the neuron is not signaling.

reticular nucleus Thin sheet of neurons surrounding the thalamus and containing local interneurons, which receive input from the cortex and make projections onto thalamic relay neurons (also nucleus reticularis thalami).

retina Layer of neurons along the back surface of the eye. It contains a variety of cells including photoreceptors (the cells that respond to light) and ganglion cells (the cells whose axons form the optic nerve).

retinotopic Topographic map of visual space across a restricted region of the brain. Activation across the retina is determined by the reflectance of light from the environment. A retinotopic map in the brain is a representation in which some sort of orderly spatial relationship is maintained. Multiple retinotopic maps have been identified in the cortex and subcortex.

retrieval To utilize stored information to create a conscious representation, or to execute a learned behavior like a motor act.

saltatory conduction Mode of conduction in myelinated neurons where the action potentials are generated down the axon only at the locations of the nodes of Ranvier. Measurement of the propagation of the action potential gives it the appearance of jumping from node to node, hence the term saltatory, which means to jump.

scotoma A region in external space in which a person or animal fails to perceive a stimulus following neural damage. Scotoma occurs following lesions of primary visual cortex. The size and location of the scotoma vary depending on the extent and location of the lesion.

second messenger A system where chemical signals carry information from the postsynaptic receptor to ion channels or to other intracellular entities in the postsynaptic neuron (e.g., the nucleus).

selection of task-relevant information Refers to the interaction of stimulus information and current goals. At any point in time, the environment affords a number of different response options. The activation and maintenance of goals provides a means to ensure that task-relevant information is favored, underscoring the attentional operations of prefrontal cortex.

selective attention The ability to focus one's concentration upon a subset of sensory inputs, trains of thought, or actions, while simultaneously ignoring others. Selective attention can be distinguished from nonselective attention, which includes simple behavioral arousal (i.e., being generally more versus less attentive).

self-knowledge The ability of a being to access accurate information about its current and past states.

semantic paraphasias Patients with Wernicke's aphasia often produce semantic paraphasias; they produce words related in meaning to the intended word (e.g., horse for cow).

sensitive periods Time periods during development where the organization of the nervous system exhibits relatively high degrees of plasticity, and neuronal reorganization can occur during learning or under external pressures (e.g., damage or disease).

sensory memory The short-lived retention of sensory information, measurable in milliseconds to seconds, as when we recover what was said to us a moment before when we were not paying close attention to the speaker.

sentience What it is like to be or do something. Subjective experience, phenomenal awareness, and raw feelings.

serial position effect We are better at recalling items at the beginning and end of a list, known as primacy and recency effects, respectively.

short-term memory Refers to the retention of information over seconds to minutes.

simulation Method used in computer modeling to mimic some behavior or process. Simulations require a program that explicitly specifies the manner in which information is represented and processed. The resulting model can be tested to see if its output matches the simulated behavior or process. The program can then be used to generate new predictions.

single dissociation Method used to develop functional models of mental and/or neural processes. Evidence of a single dissociation requires a minimum of two groups and two tasks. A single dissociation is present when the groups differ in their performance on one task but not the other. Single dissociations provide weak evidence of functional specialization since it is possible the two tasks differ in terms of their sensitivity to detect group differences.

single-cell recording Neurophysiological method used to monitor the activity of individual neurons. The procedure requires positioning a small recording electrode either inside a cell, or more typically, near the outer membrane of a neuron. The electrode measures changes in the membrane potential and can be used to determine the conditions that cause the cell to respond.

sociobiology The systematic study of the biological basis of all social behavior.

soma The cell body of a neuron.

somatic marker A mechanism that helps sort through options. It provides a common metric for evaluating options with respect to their potential benefit.

somatotopy Point-for-point representation of the body surface in the nervous system. In the somatosensory cortex, regions of the body near one another (e.g., the index and middle fingers) are represented by neurons located near one another. Regions that are farther apart on the body surface (e.g., the nose and the big toe) are coded by neurons located farther apart in the somatosensory cortex.

source memory Memory for the context in which something was learned; a form of episodic memory. Patients with lateral prefrontal lesions may do well in remembering a list of facts, but they have difficulty in identifying the place and time at which the facts were learned.

spatial attention Selective attention based on location in the sensory world. For example, visuospatial attention is attention to one location while simultaneously ignoring another location.

spatial frequency hypothesis Hypothesis that the two hemispheres differ in how they process complex information, with the left hemisphere biased toward processing high-frequency information and the right hemisphere biased toward processing low-frequency information.

spike triggering zone The location at the juncture of the soma and axon of a neuron, where summation of currents from synaptic inputs on the soma and distant dendrites occurs and voltage-gate Na+ channels are located that can be triggered to generate action potentials that can propagate down the axon.

spine Little knobs attached by small necks to the surface of the dendrites. Where synapses are located.

splenum The posterior area of the corpus callosum that interconnects the occipital lobe.

split-brain research Studies involving patients who have had the corpus callosum severed, typically as a radical treatment for intractable epilepsy.

stimulus driven Behavior is said to be stimulus driven if it is dictated by the environmental context and fails to incorporate the animal's or person's goals. For example, a person with a lesion of prefrontal cortex might drink from a glass placed in front of him even if he isn't thirsty.

storage The result of acquisition and consolidation of information, which creates and maintains a permanent record.

subjectivity The phenomenal awareness of experiences unique to oneself that are tied together through time.

subliminal perception A technique used to study unconscious processing. Stimuli are rapidly presented so that participants are unaware of viewing the stimuli, but their subsequent actions are influenced by the nature of the stimuli.

substantia nigra One of the nuclei that form the basal ganglia. The nigra is composed of two parts. The axons of the substantia nigra pars compacta provide the primary source of the neurotransmitter dopamine and terminate in the striatum (caudate and putamen). The substantia nigra pars reticulata is one of the output nuclei from the basal ganglia.

subthreshold Stimuli and events that are only available to unconscious processing; they are perceived, maintained, and manipulated below the level of conscious awareness.

sulci The invaginated regions that appear as lines and creases of the surface of the cerebral cortex.

superior colliculus Subcortical visual structure located in the midbrain. It receives input from the retinal system and is interconnected with the subcortical and cortical systems. Plays a key role in visuomotor processes, and may be involved in the inhibitory component of reflexive attentional orienting.

supervisory attentional system (SAS) Psychological model used to explain how response selection is achieved in a flexible manner. Without the SAS, behavior is dictated by the context, with the selected action being the one that has been produced most often in the current context. The SAS allows for flexible behavior by biasing certain actions based on current goals or helping determine actions in unfamiliar situations.

supplementary motor area Secondary motor area that includes the medial aspect of Brodmann's area 6, just anterior to the primary motor cortex. The supplementary motor area plays an important role in the production of sequential movements, especially those that have been well learned.

Sylvian fissure A large fissure (sulcus) on the lateral surface of the cerebral cortex first described by the anatomist Sylvius. It separates the frontal cortex from the temporal lobe below.

synapse The locations on neurons where they come in contact with other neurons to transmit information. Synapses include both presynaptic (e.g., synaptic vesicles with neurotransmitter) and postsynaptic (e.g., receptors) specializations in the neurons that are involved in chemical transmission. There are also electrical synapses that involve special structures called gap junctions that make direct cytoplasmic connections between neurons.

synapse elimination Elimination of some synaptic contacts between neurons during development, including postnatally.

synaptogenesis Formation of synaptic connections between neurons in the developing nervous system.

syncytium A continuous mass of tissue that shares a common cytoplasm.

task-switching Experimental method used to study the processes involved in changing from one goal to another, an important feature of flexible behavior. In these studies, participants alternate between two or more tasks, allowing the experimenter to observe the processes involved in retrieving a new goal, discarding an old goal, and establishing a new set of stimulus-response conditions.

temporal lobe Lateral ventral portions of the cerebral cortex bounded superiorly by the Sylvian fissure and posteriorly by the anterior edge of the occipital lobe and ventral portion of the parietal lobe. The ventral medial portions contain the hippocampal complex and amygdala. The lateral neocortical regions are involved in higher-order vision (object analysis), the representation of conceptual information about the visual world, and linguistic representations. The superior portions within the depths of the Sylvian fissure contain auditory cortex.

thalamus Group of nuclei, primarily major sensory relay nuclei for somatosensory, gustatory, auditory, visual, and vestibular inputs to the cerebral cortex. Also contains nuclei involved in basal ganglia-cortical loops, and other specialized nuclear groups. It is a part of the diencephalon, a subcortical region, located in the center of the mass of the forebrain. There is one thalamus in each hemisphere, and these are connected at the midline in most humans by the massa intermedia.

theory of mind The ability to self-reflect and think about the mental states of others. This allows predictions of what others can understand, and how they will interact and behave in a given situation. This trait is considered unique to the human species.

threshold The value of the membrane potential to which the membrane must be depolarized to initiate an action potential.

topographic representation Systematic relationship between some property of the external world and the neural representation of that property. Examples include retinotopic maps in the visual cortex, tonotopic maps in the auditory cortex, and somatosensory maps in the motor and sensory cortices.

tract A bundle of axons in the CNS.

trait A distinguishing quality or attribute.

transcortical Communication between the left and right hemispheres, either via the corpus callosum or subcortical connections.

transcranial magnetic stimulation (TMS) Noninvasive method used to stimulate the cerebral cortex or motor neurons. A strong electrical current is rapidly generated in a coil placed over the targeted region. This current generates a magnetic field that causes the neurons in the underlying region to discharge. TMS is used in clinical settings to evaluate motor function by direct stimulation of motor cortex. Experimentally, the procedure is used to transiently disrupt neural processing, thus creating brief, reversible lesions.

unconscious processes Brain-mediated processing of the features of a given stimulus that presumably precede and give rise to conscious awareness of that stimulus.

universal grammar Innate predisposition to develop certain grammatical structures across all human languages.

ventral pathway One of the two main processing streams for visual information in the cerebral cortex. The ventral pathway extends from occipital lobe into inferior and middle temporal lobe. Functional hypotheses have centered on the idea that this pathway is important for the recognition of objects, including faces.

vesicle Small intracellular organelle located in the presynaptic terminals at synapses that contain neurotransmitter.

view-dependent theories Based on the idea that perception involves recognizing an object from a certain viewpoint. These theories assume that visual memory is based on previous experiences with objects in specific orientations and that the recognition of an object in a novel orientation involves an approximation process to the stored representations of specific perspectives.

view-invariant frame of reference Based on the idea that perception involves recognizing certain properties of an object that remain invariant, or constant across different perspectives. In this approach, these properties form the basis of visual memory and recognition entails matching the perceived properties to this knowledge base.

visual agnosia Failures of perception where the disorder is limited to the visual modality. In visual agnosia, the patient is relatively good in perceiving properties such as color, shape, or motion yet still fails to be able to recognize objects or identify their uses.

visual search The process of searching a cluttered visual scene for a relevant item using covert attention.

visual search task Behavioral task used to examine the processes involved in visual perception and attention. Rapid detection of targets from distracters has been used to identify the building blocks of perception.

voluntary attention The volitional or intentional focusing of attention upon a source of input, train of thought, or action.

Wernicke's aphasia A language deficit usually caused by brain lesions in the posterior parts of the left hemisphere, and result in comprehension deficits.

white matter Composed of millions of individual axons, each surrounded by myelin. It is the myelin that gives the fibers their whitish color, hence the name white matter.

Wisconsin Card Sorting Task Neuropsychological test that assesses various aspects of executive control. While the test has traditionally been viewed as an excellent diagnostic of prefrontal function, impaired performance is frequently observed in other patient groups, likely reflecting the complexity of this multidimensional sorting task in which rules are learned through trial-and-error.

working memory Transient representations of task-relevant information. These representations may be related to information that has just been activated from the distant past, linked to something in the current environment, or based on something that has recently been experienced. Representations in working memory guide behavior in the present, constituting what has been called, "the blackboard of the mind."

REFERENCES

ABRAMS, R.A., AND LANDGRAF, J.Z. (1990). Differential use of distance and location information for spatial localization. *Percept. Psychophys.* 47:349–359.

ABRAMS, R.A., VAN DILLEN, L., AND STEMMONS, V. (1994). Multiple Sources of Spatial Information for Aimed Limb Movements. In C. Umilta and M. Moscovitch (Eds.), *Attention and Performance.* Vol. 15: *Conscious and Nonconscious Information Processing* (pp. 267–290). Cambridge, MA: MIT Press.

ACREDOLO, L., AND GOODWYN, S. (1996). *Baby Signs: How To Talk with Your Baby Before Your Baby Can Talk.* Chicago: Contemporary Books.

ADOLPHS, R., TRANEL, D., AND DAMASIO, A.R. (1998). The human amygdala in social judgement. *Nature* 393:470–474.

ADOLPHS, R., TRANEL, D., AND DENBURG, N. (2000). Impaired emotional declarative memory following unilateral amygdala damage. *Learn. Mem.* 7:180–186.

ADOLPHS, R., TRANEL, D., HAMANN, S., YOUNG, A.W., CALDER, A.J., PHELPS, E.A., ANDERSON, A., LEE, G.P., AND DAMASIO, A.R. (1999). Recognition of facial emotion in nine individuals with bilateral amygdala damage. *Neuropsychologia* 37:1111–1117.

ADOLPHS, R.A., TRANEL, D., DAMASIO, H., AND DAMASIO, A.R. (1994). Impaired recognition of emotion in facial expressions following bilateral amygdala damage to the human amygdala. *Nature* 372:669–672.

AGGLETON, J.P. (Ed.). (1992). *The Amygdala: Neurobiological Aspects of Emotion, Memory and Mental Dysfunction.* New York: Wiley-Liss.

AITCHISON, J. (1987). *Words in the Mind: An Introduction to the Mental Lexicon.* Oxford, UK: Blackwell.

AKELAITIS, A.J. (1941). Studies on the corpus callosum: Higher visual functions in each homonymous visual field following complete section of corpus callosum. *Arch. Neurol. Psychiatry* 45:788.

ALCOCK, J. (1979). *Animal Behavior: An Evolutionary Approach,* 2nd edition. Sunderland, MA: Sinauer Assoc.

ALEXANDER, G.E., AND CRUTCHER, M.D. (1990). Preparation for movement: Neural representations of intended direction in three motor areas of the monkey. *J. Neurophysiol.* 64:133–150.

ALLEN, M. (1983). Models of hemisphere specialization. *Psychol. Bull.* 93:73–104.

ANDERSON, A.K., AND PHELPS, E.A. (2000). Expression without recognition: Contributions of the human amygdala to emotional communication. *Psychol. Sci.* 11:106–111.

ANDERSON, A.K., AND PHELPS, E.A. (2001). Lesions of the human amygdala impair enhanced perception of emotionally salient events. *Nature* 411:305–309.

ANDERSON, A.K., AND PHELPS, E.A. (In Press). Is the human amygdala critical for the subjective experience of emotion? Evidence of intact dispositional affect in patients with amygdala lesions. *J. Cogn. Neurosci.*

ANDERSON, A.K., LABAR, K.S., AND PHELPS, E.A. (1996). Facial affect processing abilities following unilateral temporal lobectomy. *Soc. Neurosci. Abstr.* 22:1866.

ANDERSON, A.K., SPENCER, D.D., FULBRIGHT, R.K., AND PHELPS, E.A. (2000). Contribution of the anteromedial temporal lobes to the evaluation of facial emotion. *Neuropsychology* 14:526–536.

ANDERSON, B.J., LI, X., ALCANTARA, A.A., ISAACS, K.R., BLACK, J.E., AND GREENOUGH, W.T. (1994). Glial hypertrophy is associated with synaptogenesis following motor-skill learning, but not with angiogenesis following exercise. *Glia* 11:73–80.

ANGRILLI, A., MARUI, A., PALOMBA, D., FLOR, H., BIRBAUMER, N., SARTORI, G., AND DI PAOLA, F. (1996). Startle reflex and emotion modulation impairment after a right amygdala lesion. *Brain* 119:1991–2000.

ARVANITAKI, A. (1939). Recherches sur la réponse oscillatoire locale de l'axone géant isolé de 'Sepia.' *Arch. Int. Physiol.* 49:209–256.

ASHE, J., TAIRA, M., SMYRNIS, N., PELLIZZER, G., GERORAKOPOULOS, T., LURITO, J.T., AND GEORGOPOULOS, A.P. (1993). Motor cortical activity preceding a memorized movement trajectory with an orthogonal bend. *Exp. Brain Res.* 95:118–130.

ATKINSON, R.C., AND SHIFFRIN, R.M. (1968). Human Memory: A Proposed System and Its Control Processes. In K.W. Spence and J.T. Spence (Eds.), *The Psychology of Learning and Motivation,* Vol. 2 (pp. 89–195). New York: Academic Press.

BADDELEY, A. (1986). *Working Memory.* Oxford, UK: Clarendon Press/Oxford University Press.

BADDELEY, A. (1995). Working Memory. In M.S. Gazzaniga (Ed.), *The Cognitive Neurosciences* (pp. 755–764). Cambridge, MA: MIT Press.

BADDELEY, A., AND HITCH, G. (1974). Working Memory. In G.H. Bower (Ed.), *The Psychology of Learning and Motivation*, Vol. 8 (pp. 47–89). New York: Academic Press.

BAILLARGEON, R. (1991). Reasoning about the height and location of a hidden object in 4½ and 6½ month old infants. *Cognition* 38:13–42.

BALDWIN, D.A., AND MOSES, L. (1994). Early Understanding of Referential Intent and Attentional Focus: Evidence from Language and Emotion. In C. Lewis and P. Mitchell (Eds.), *Children's Early Understanding of Mind: Origins and Development* (pp. 133–156). London: Lawrence Erlbaum Associates.

BANNERMAN, D.M., GOOD, M.A., BUTCHER, S.P., RAMSAY, M., AND MORRIS, R.G.M. (1995). Distinct components of spatial learning revealed by prior training and NMDA receptor blockade. *Nature* 378:182–186.

BARBUR, J.L., WATSON, J.D.G., FRACKOWIAK, R.S.J., AND ZEKI, S. (1993). Conscious visual perception without V1. *Brain* 116:1293–1302.

BARNEA, A. AND NOTTEBOHM, F. (1994). Seasonal recruitment of hippocampal neurons in adult free-ranging black-capped chickadees. *Proc. Natl. Acad. Sci. U.S.A.* 91:11217–11221.

BARRETT, A.M., CRUCIAN, G.P., RAYMER, A.M., AND HEILMAN, K.M. (1997). Spared comprehension of emotional prosody in a patient with global aphasia. *J. Int. Neuropsychol. Soc.* 3:57.

BARSALOU, L.W. (1991). Flexibility, Structure and Linguistic Vagary in Concepts: Manifestations of a Compositional System of Perceptual Symbols. In A.F. Collins, S.E. Gathercole, M.A. Conway, and P.A. Morris (Eds.), *Theories of Memory* pp. 22–101. Hove: Erlbaum.

BARTHOLOMEUS, B. (1974). Effects of task requirements on ear superiority for sung speech. *Cortex* 10:215–223.

BATISTA, A.P., BUNEO, C. A., SNYDER, L.H., AND ANDERSON, R.A. (1999). Reach plans in eye-centered coordinates. *Science* 285:257–260.

BAXTER, M.G., PARKER, A., LINDNER, C.C., IZQUIERDO, A.D., AND MURRAY, E.A. (2000). Control of response selection by reinforcer value requires interaction of amygdala and orbital prefrontal cortex. *J. Neurosci.* 20:4311–4319.

BAYLIS, G.C., ROLLS, E.T., AND LEONARD, C.M. (1985). Selectivity between faces in the responses of a population of neurons in the cortex in the superior temporal sulcus of the monkey. *Brain Res.* 342:91–102.

BEAR, M.F., CONNORS, B.W., AND PARADISO, M.A. (1996). *Neuroscience: Exploring the Brain*. Baltimore: Williams & Wilkins.

BECHARA, A., TRANEL, D., DAMASIO, H., ADOLPHS, R., ROCKLAND, C., AND DAMASIO, A.R. (1995). Double dissociation of conditioning and declarative knowledge relative to the amygdala and hippocampus in human. *Science* 269:1115–1118.

BECK, J. (1982). Textural Segmentation. In J. Beck (Ed.), *Organization and Representation in Perception* (pp. 285–317). Hillsdale, NJ: Lawrence Erlbaum Associates.

BECKERS, G., AND ZEKI, S. (1995). The consequences of inactivating areas V1 and V5 on visual motion perception. *Brain* 118:49–60.

BEHRMANN, M., AND TIPPER, S.P. (1994). Object-Based Attentional Mechanisms: Evidence from Patients with Unilateral Neglect. In C. Umilta and M. Moscovitch (Eds.), *Attention and Performance 15: Conscious and Nonconscious Information Processing* (pp. 351–375). Attention and performance series. Cambridge, MA: MIT Press.

BEHRMANN, M., MOSCOVITCH, M., AND WINOCUR, G. (1994). Intact visual imagery and impaired visual perception in a patient with visual agnosia. *J. Exp. Psychol.* 20:1068–1087.

BERENT, S., GIORDANI, B., LEHTINEN, S., MARKEL, D., PENNEY, J.B., BUCHTEL, H.A., STAROSTA-RUBENSTEIN, S., HICHWA, R., AND YOUNG, A.B. (1988). Positron emission tomographic scan investigation of Huntington's disease: Cerebral metabolic correlates of cognitive function. *Ann. Neurol.* 23:541–546.

BERGMAN, H., WICHMANN, T., AND DELONG, M.R. (1990). Reversal of experimental parkinsonism by lesions of the subthalamic nucleus. *Science* 249:1436–1438.

BERNS, G.S., AND SEJNOWSKI, T. (1996). How the Basal Ganglia Makes Decisions. In A. Damasio, H. Damasio, and Y. Christen (Eds.), *The Neurobiology of Decision Making* (pp. 101–113). Cambridge, MA: MIT Press.

BERRIDGE, K.C., AND WHISHAW, I.Q. (1992). Cortex, striatum and cerebellum: Control of serial order in a grooming sequence. *Exp. Brain Res.* 61:275–290.

BIANCHI, L. (1922). *The Mechanism of the Brain* (J.H. MacDonald, Trans.). Edinburgh: E. & S. Livingstone.

BIEDERMAN, I. (1990). Higher-Level Vision. In D.N. Osherson, S.M. Kosslyn, and J.M. Hollberbach (Eds.), *Visual Cognition and Action: An Invitation to Cognitive Science*, Vol. 2 (pp. 41–63). Cambridge, MA: MIT Press.

BINDER, J., AND PRICE, C. J. (2001). Functional Neuroimaging of Language Processes. In R. Cabeza and A. Kingstone (Eds.), *Handbook of Functional Neuroimaging of Cognition* (pp. 187–251). Cambridge, MA: MIT Press.

BINDER, J.R., FROST, J.A., HAMMEKE, T.A., BELLGOWAN, P.S.F., RAO, S.M., AND COX, J.A. (1999). Conceptual processing during the conscious resting state: A functional MRI study. *J. Cogn. Neurosci.* 11:80–93.

BINDER, J.R., FROST, J.A., HAMMEKE, T.A., BELLGOWAN, P.S.F., SPRINGER, J.A., KAUFMAN, J.N., AND POSSING, E.T. (2000). Human temporal lobe activation by speech and non-speech sounds. *Cereb. Cortex* 10:512–528.

BISIACH, E., AND LUZZATTI, C. (1978). Unilateral neglect of representational space. *Cortex* 14:129–133.

BIZZI, E., ACCORNERO, N., CHAPPLEL, W., AND HOGAN, N. (1984). Posture control and trajectory formation during arm movement. *J. Neurosci.* 4:2738–2744.

BLAIR, R.J.R., AND CIPOLOTTI, L. (2000). Impaired social response reversal: A case of 'acquired sociopathy.' *Brain* 123:1122–1141.

BLISS, T.V.P., AND LØMO, T. (1973). Long-lasting potentiation of synaptic transmission in the dentate area of the anaesthetized rabbit following stimulation of the perforant pathway. *J. Physiol.* 232:331–356.

BLONDERS, L.X., BOWERS, D., AND HEILMAN, K.M. (1991). The role of the right hemisphere in emotional communication. *Brain* 114:1115–1127.

BLOOM, F., NELSON, C.A., AND LAZERSON, A. (2001). *Brain, Mind and Behavior,* 3rd edition. New York: Worth Publishers.

BLOOM, R.L., BOROD, J.C., OBLER, L.K. AND GERSTMAN, L.J. (1993). A preliminary characterization of emotional expression in right and left brain-damaged patients. *Int. J. Neurosci.* 55:71–80.

BOROD, J.C., KOFF, E., PERLMAN LORCH, M. AND NICHOLAS, M. (1986). The expression and perception of facial emotion in brain damaged patients. *Neuropsychologia* 24:169–180.

BOTVINICK, M., NYSTROM, L.E., FISSELL, K., CARTER, C.S., AND COHEN, J.D. (1999). Conflict monitoring versus selection-for-action in anterior cingulate cortex. *Nature* 402:179–181.

BOWER, J.M. (1997). Control of sensory data acquisition. *Int. Rev. Neurobiol.* 41:489–513.

BOWERS, D., BAUER, R.M. AND HEILMAN, K. (1993). The nonverbal affect lexicon: Theoretical perspectives from neuropsychological studies of affect perception. *Neuropsychology* 7:433–444.

BRADSHAW, J., AND ROGERS, L. (1993). *The Evolution of Lateral Asymmetries, Language, Tool Use, and Intellect.* San Diego: Academic Press.

BRADSHAW, J.L., AND NETTLETON, N.C. (1981). The nature of hemispheric specialization in man. *Behav. Brain Sci.* 4:51–91.

BRAITENBERG, V. (1984). *Vehicles: Experiments in Synthetic Psychology.* Cambridge, MA: MIT Press.

BRANNON, E.M., AND TERRACE, H.S. (1998). Ordering of the numerosities 1-9 by monkeys. *Science* 282:746–749.

BRANNON, E.M., AND VAN DE WALLE, G. (2001). Ordinal numerical knowledge in young children. *Cogn. Psychol.* 43:53–81.

BREITER, H.C., ETCOFF, H.L., WHALAN, P.J., KENNEDY, W.A., RAUCH, S.L., BUCKNER, R.L., STRAUSS, M.M., HYMAN, S., AND ROSEN, B. (1996). Response and habituation of the human amygdala during visual processing of facial expression. *Neuron* 17:875–887.

BREWER, J., ZHAO, Z., DESMOND, J.E., GLOVER, G.H., AND GABRIELI, J.D.E. (1998). Making memories: Brain activity that predicts how well visual experience will be remembered. *Science* 281:1185–1187.

BRITTEN, K.H., SHALDEN, M.N., NEWSOME, W.T., AND MOVSHON, J.A. (1992). The analysis of visual motion: A comparison of neuronal and psychophysical performance. *J. Neurosci.* 12:4745–4765.

BROADBENT, D.A. (1958). *Perception and Communication.* New York: Pergamon.

BROADBENT, D.A. (1970). Stimulus Set and Response Set: Two Kinds of Selective Attention. In D.I. Motofsky (Ed.), *Attention: Contemporary Theory and Analysis* (pp. 51–60). New York: Appleton-Century-Crofts.

BRODAL, A. (1982). *Neurological Anatomy.* New York: Oxford University Press.

BRODMANN, K. (1909). *Vergleichende Lokalisationslehre der Grosshirnrinde in ihren Prinzipien dargestellt auf Grund des Zellenbaues.* Leipzig: J.A. Barth. In G. von Bonin, *Some Papers on the Cerebral Cortex (pp. 201–230).* Translated as, *"On the Comparative Localization of the Cortex."* Springfield, IL: Charles C. Thomas, 1960.

BROTCHIE, P., IANSEK, R., AND HORNE, M.K. (1991). Motor function of the monkey globus pallidus. *Brain* 114:1685–1702.

BROWN, C.M., AND HAGOORT, P. (1999). *The Neurocognition of Language.* Oxford, UK: Oxford University Press.

BROWN, R., AND KULIK, J. (1977). Flashbulb memories. *Cognition* 5:73–99.

BROWN, T. (1911). The intrinsic factors in the act of progression in the mammal. *Proc. R. Soc. Lond.,* Series B 84:308–319.

BRYDEN, M.P. (1982). *Laterality: Functional Asymmetry in the Intact Human Brain.* New York: Academic Press.

BUCHEL, C., MORRIS, J., DOLAN, R.J., AND FRISTON, K.J. (1998). Brain systems mediating aversive conditioning: An event-related fMRI study. *Neuron* 20:947–957.

BULLOCK, T.H. (1984a) Comparative neuroscience holds promise for a quiet revolution. *Science* 225:473–478.

BULLOCK, T.H., (1984b). Understanding brains by comparing taxa. *Perspect. Biol. Med.* 27:510–524.

BULLOCK, T.H. (1993). How are more complex brains different? One view and an agenda for comparative neurobiology. *Brain Behav. Evol.* 41:88–96.

BUSH, G., LUU, P., AND POSNER, M.I. (2000). Cognitive and emotional influences in anterior cingulate cortex. *Trends Cogn. Sci.* 4:215–222.

CABEZA, R., AND NYBERG, L. (2000). Imaging cognition II: An empirical review of 275 PET and fMRI studies. *J. Cogn. Neurol.* 12:1–47.

CABEZA, R., RAO, S.M., WAGNER, A.D., MAYER, A.R., AND SCHACTER, D.L. (2001). Can medial temporal lobe regions distinguish true from false? An event-related fMRI study of veridical and illusory recognition memory. *Proc. Natl. Acad. Sci. U.S.A.*

CAHILL, L., BABINSKY, R., MARKOWITSCH, H.J., AND MCGAUGH, J.L. (1995). The amygdala and emotional memory. *Science* 377:295–296.

CAHILL, L, HAIER, R.J., FALLON, J., ALKIRE, M.T., TANG, C., KEATOR, D., WU, J., AND MCGAUGH, J.L. (1996). Amygdala activity at encoding correlated with long-term, free recall of emotional information. *Proc. Natl. Acad. Sci. U.S.A.* 93:8016–8021.

CAJAL, RAMÓN Y. (1913). *Degeneration and Regeneration of the Nervous System* (R.M. Day, Transl., 1928). London: Oxford University Press.

CAMINITI, R., JOHNSON, P.B., GALLI, C., FERRAINA, S., AND BURNOD, Y. (1991). Making arm movements within different parts of space: The premotor and motor cortical representation of a coordinate system for reaching visual targets. *J. Neurosci.* 11:1182–1197.

CAPLAN, D. (1992). *Language: Structure, Processing, and Disorders.* Cambridge, MA: MIT Press.

CAPLAN, D. (1994). Language and the Brain. In M.A. Gernsbacher (Ed.), *Handbook of Psycholinguistics* (pp. 1023–1053). San Diego: Academic Press.

CAPLAN, D., ALPERT, N., WATERS, G., AND OLIVIERI, A. (2000). Activation of Broca's area by syntactic processing under conditions of concurrent articulation. *Hum. Brain Map.* 9:65–71.

CARAMAZZA, A. (1992). Is cognitive neuropsychology possible? *J. Cogn. Neurosci.* 4:80–95.

CARAMAZZA, A. (1996). The brain's dictionary. *Nature* 380:485–486.

CARAMAZZA, A., AND SHELTON, J. (1998). Domain-specific knowledge systems in the brain: Animate-inanimate distinction. *J. Cogn. Neurosci.* 10:1–34.

CARAMAZZA, A., HILLIS, A.E., RAPP, B.C., AND ROMANI, C. (1990). The multiple semantics hypothesis: Multiple confusions? *Cogn. Neuropsychol.* 7:161–189.

CARPENTER, M. (1976). *Human Neuroanatomy,* 7th edition. Baltimore: Williams & Wilkins.

CARTER, C.S., BRAVER, T.S., BARCH, D.M., BOTVINICK, M.M., NOLL, D.C., AND, COHEN, J.D. (1998). Anterior cingulate cortex, error detection and the on-line monitoring of performance. *Science* 280:747–749.

CASTIELLO, U., PAULIGNAN, Y., AND JEANNEROD, M. (1991). Temporal dissociation of motor responses and subjective awareness. *Brain* 114:2639–2655.

CATANIA, K.C., AND KAAS, J.H. (1995). Organization of the somatosensory cortex of the star-nosed mole. *J. Comp. Neurol.* 351:549–567.

CATANIA, K.C., NORTHCUTT, R.G., KAAS, J.H., AND BECK, P.D. (1993). Nose stars and brain stripes. *Nature* 364:493.

CAVINESS, V.S., JR., AND RAKIC, P. (1978). Mechanisms of cortical development: A view from mutations in mice. *Annu. Rev. Neurosci.* 1:297–326.

CHABRIS, C.F., AND HAMILTON, S.E. (1992). Hemispheric specialization for skilled perceptual organization by chess masters. *Neuropsychologia* 30:47–57.

CHAO, L.L., AND KNIGHT, R.T. (1995). Human prefrontal lesions increase distractibility to irrelevant sensory inputs. *Neuroreport: Int. J. Rapid Commun. Res. Neurosci.* 6:1605–1610.

CHAO, L.L., AND MARTIN, A. (1999). Cortical regions associated with perceiving, naming, and knowing about colors. *J. Cogn. Neurosci.* 11:25–35.

CHAO, L.L., HAXBY, J.V., AND MARTIN, A. (1999). Attribute-based neural substrates in temporal cortex for perceiving and knowing about objects. *Nat. Neurosci.* 2:913–919.

CHAPIN, J.K., MOXON, K.A., MARKOWITZ, R.S., AND NICOLELIS, M.A. (1999). Real-time control of a robot arm using simultaneously recorded neurons in the motor cortex. *Nat. Neurosci.* 2:664–670.

CHAWLA, D., LUMER, E.D., AND FRISTON, K.J. (1999a). The relationship between synchronization among neuronal populations and their mean activity levels. *Neural Comput.* 11:1389–1411.

CHAWLA, D., REES, G. AND FRISTON, K.J. (1999b). The physiological basis of attentional modulation in extrastriate visual areas. *Nat. Neurosci.* 7:671–676.

CHERRY, E.C. (1953). Some experiments on the recognition of speech, with one and two ears. *J. Acoustic Soc. Am.* 25:975–979.

CHIARELLO, C. (1991). Interpretation of Word Meanings by the Cerebral Hemispheres: One Is Not Enough. In P.J. Schwanenflugel (Ed.), *The Psychology of Word Meanings.* Hillsdale, NJ: Lawrence Erlbaum Associates.

CHOMSKY, N. (1957). *Syntactic Structures.* The Hague: Mouton.

CHRISTIANSON, S.A. (1992). *The Handbook of Emotion and Memory: Research and Theory.* Hillsdale, NJ: Lawrence Erlbaum Associates.

CHRISTMAN, S., AND KITTERLE, F.L. (1991). Hemispheric asymmetry in the processing of absolute versus relative spatial frequency. *Brain Cogn.* 16:62–73.

CHUGANI, H.T. (1998). A critical period of brain development: studies of cerebral glucose utilization with PET. *Prev. Med.* 27:184–188.

CHUGANI, H.T., MULLER, R.A., AND CHUGANI, D.C. (1996). Functional brain reorganization in children. *Brain Dev.* 18:347–356.

CHURCHILL, WINSTON, SIR. (1930). *My Early Life: A Roving Commission.* New York: Charles Scribner's Sons.

CHURCHLAND, P. (1988). *Matter and Consciousness.* Cambridge, MA: MIT Press.

CHURCHLAND, P. (1993). Book review: Consciousness explained. *J. Philos.* 90:181–193.

CHURCHLAND, P.S. (1986). *Neurophilosophy: Towards a Unified Science of the Mind/Brain.* Cambridge, MA: MIT Press.

COHEN, J.D., BOTVINICK, M., AND CARTER, C.S. (2000). Anterior cingulate and prefrontal cortex: Who's in control? *Nat. Neurosci.* 3:421–423.

COHEN, J.D., ROMERO, R.D., SERVAN-SCHREIBER, D., AND FARAH, M.J. (1994). Mechanisms of spatial attention: The relation of macrostructure to microstructure in parietal neglect. *J. Cogn. Neurosci.* 6:377–387.

COHEN, N.J., AND EICHENBAUM, H. (1993). *Memory, Amnesia and the Hippocampal System.* Cambridge, MA: MIT Press.

COHEN, R.M., SEMPLE, W.E., GROSS, M., AND NORDHAL, T.E. (1988). From syndrome to illness: Delineating the pathophysiology of schizophrenia with PET. *Schizophrenia Bull.* 14:169–176.

COLLINS, A.M., AND LOFTUS, E.F. (1975). A spreading-activation theory of semantic processing. *Psychol. Rev.* 82:407–428.

COLTHEART, M., CURTIS, B., ATKINS, P., AND HALLER, M. (1993). Models of reading aloud: Dual route and parallel-distributed-processing approaches. *Psychol. Rev.* 100:589–608.

COLTHEART, M., DAVELAAR, E., JONASSON, T.V., AND BESNER, D. (1977). Access to the Internal Lexicon. In D. Stanislav (Ed.), *Attention and Performance*, Vol. 6, (pp. 532–555). Hillsdale, NJ: Lawrence Erlbaum Associates.

CONEL, J.L. (1939–1967). *The Postnatal Development of the Human Cerebral Cortex.* Cambridge, MA: Harvard University Press.

CORBALLIS, M.C. (1991). *The Lopsided Ape: Evolution of the Generative Mind.* New York: Oxford University Press.

CORBETTA, M., KINCADE, J.M., OLLINGER, J.M., MCAVOY, M.P., AND SHULMAN, G.L. (2000). Voluntary orienting is dissociated from target detection in human posterior parietal cortex. *Nat. Neurosci.* 3:292–297.

CORBETTA, M., MIEZIN, F.M., DOBMEYER, S., SHULMAN, G.L., AND PETERSEN, S.E. (1991). Selective and divided attention during visual discriminations of shape, color and speed: Functional anatomy by positron emission tomography. *J. Neurosci.* 11:2383–2402.

CORBETTA, M., MIEZIN, F.M., SHULMAN, G.L., AND PETERSEN, S.E. (1993). A PET study of visuospatial attention. *J. Neurosci.* 13:1202–1226.

CORBETTA, M., SHULMAN, G., MIEZIN, F., AND PETERSEN, S. (1995). Superior parietal cortex activation during spatial attention shifts and visual feature conjunction. *Science* 270:802–805.

COREN, S., WARD, L.M., AND ENNS, J.T. (1994). *Sensation and Perception,* 4th edition. Ft. Worth, TX: Harcourt Brace College Publishers.

CORKIN, S. (1984). Lasting consequences of bilateral medial temporal lobectomy: Clinical course and experimental findings in HM. *Semin. Neurol.* 4:249–259.

CORKIN, S., AMARAL, D., GONZALEZ, R., JOHNSON, K., et al. (1997). H.M.'s medial temporal lobe lesion: Findings from magnetic resonance imaging. *J. Neurosci.* 17:3964–3979.

CORTHOUT, E., UTTL, B., ZIEMANN, U., COWEY, A., HALLETT, M. (1999). Two periods of processing in the (circum)striate visual cortex as revealed by transcranial magnetic stimulation. *Neuropsychologia* 37:137–145.

COSMIDES, L. (1984). The logic of social exchange: Has natural selection shaped how humans reason? Studies with the Wason Selection Task. *Cognition* 31:187–276.

COSMIDES, L., AND TOOBY, J. (1992). Cognitive Adaptations for Social Exchange. In J.H. Barkow, L. Cosmides, and J. Tooby (Eds.), *The Adapted Mind.* New York: Oxford University Press.

COTTERILL, R.M.J. (1997). On the neural correlates of consciousness. *Jpn. J. Cogn. Sci.* 4:31–34.

COURTNEY, S.M., UNGERLEIDER, L.G., KEIL, K., AND HAXBY, J.V. (1997). Transient and sustained activity in a disturbed neural system for human working memory. *Nature* 386:608–611.

COWEY, A., AND STOERIG, P. (1991). The neurobiology of blindsight. *Trends Neurosci.* 14:140–145.

CRAIK, F.I.M., AND LOCKHART, R.S. (1972). Levels of processing: A framework for memory research. *J. Verbal Learn. Verbal Behav.* 11:671–684.

CRICK, F. (1992). Function of the Thalamic Reticular Complex: The Searchlight Hypothesis. In S.M. Kosslyn and R.A. Andersen (Eds.), *Frontiers in Cognitive Neuroscience* (pp. 366–372). Cambridge, MA: MIT Press.

CRITCHLEY, H.D., ELLIOTT, R., MATHIAS, C.J., AND DOLAN, R.J. (2000). Neural activity relating to generation and representation of galvanic skin conductance responses: A functional magnetic resonance imaging study. *J. Neurosci.* 20:3033–3040.

CRONER, L.J., AND ALBRIGHT, T.D. (1999). Seeing the big picture: integration of image cues in the primate visual system. *Neuron* 24:777–789.

CROWLEY, J., AND KATZ, L. (1999). Development of ocular dominance columns in the absence of retinal input. *Nat. Neurosci.* 2:1125–1130.

CUTLER, A., AND NORRIS, D.G. (1988). The role of strong syllables in segmentation for lexical access. *J. Exp. Psychol. Hum. Percept. Perform.* 14:113–121.

DAMASIO, A., AND DAMASIO, H. (1983). The anatomic basis of pure dyslexia. *Neurology* 33:1573–1583.

DAMASIO, A.R. (1990). Category-related recognition defects as a clue to the neural substrates of knowledge. *Trends Neurosci.* 13:95–98.

DAMASIO, A.R. (1994). *Descartes' Error: Emotion, Reason, and the Human Brain.* New York: G.P. Putnam.

DAMASIO, A.R. (1999). *The Feeling of What Happens: Body and Emotion in the Making of Consciousness.* New York: Harcourt Brace & Company.

DAMASIO, A.R., GRABOWSKI, T.J., BECHARA, A., DAMASIO, H., PONTO, L.L.B., PARVIZI, J., AND HICHWA, R.D. (2000). Subcortical and cortical brain activity during the feeling of self-generated emotions. *Nat. Neurosci.* 3:1049–1056.

DAMASIO, H., GRABOWSKI, T., FRANK, R., GALABURDA, A.M., AND DAMASIO, A.R. (1994). The return of Phineas Gage: The skull of a famous patient yields clues about the brain. *Science* 264:1102–1105.

DAMASIO, H., GRABOWSKI, T.J., TRANEL, D., HICHWA, R.D., AND DAMASIO, A.R. (1996). A neural basis for lexical retrieval. *Nature* 380:499–505.

DARWIN, C. (1859). *On the Origin of Species.* London: J. Murray; reprinted. Cambridge, MA: Harvard University Press.

DARWIN, C. (1871). *The Descent of Man and Selection in Relation to Sex.* London: J. Murray.

DAVIDSON, R.J. (1995). Cerebral Asymmetry, Emotion, and Affective Style. In R.J. Davidson and K. Hugdahl (Eds.), *Brain Asymmetry* (pp. 361–387). Cambridge, MA: MIT Press.

DAVIDSON R.J. (2000). The Neuroscience of Affective Style. In R.D. Lane and L. Nadel (Eds.) *Cognitive Neuroscience of Emotion* (pp. 371–188). New York: Oxford University Press.

DAVIDSON, R.J., EKMAN, P., SARON, C., SENULIS, J., AND FRIESEN, W.V. (1990). Approach/withdrawal and cerebral asymmetry: Emotional expression and brain physiology. *J. Pers. Soc. Psychol.* 38L:330–341.

DAVIDSON, R.J., JACKSON, D.C., AND KALIN, N.H. (2000). Emotion, plasticity, context, and regulation: Perspectives from affective neuroscience. *Psychol. Bull.* 126:890–909.

DAVIDSON, R.J., MEDNICK, D., MOSS, E., SARON, C., AND SCHAFFER, C.E. (1987). Ratings of emotions in faces are influenced by the visual field to which stimuli are presented. *Brain Cogn.* 6:403–411.

DAVIS, M. (1992). The Role of the Amygdala in Conditioned Fear. In J.P. Aggleton (Ed.), *The Amygdala: Neurobiological Aspects of Emotion, Memory and Mental Dysfunction* (pp. 255–306). New York: Wiley-Liss.

DAWKINS, R. (1976). *The Selfish Gene.* New York: Oxford University Press.

DAWSON, J.L.M., CHEUNG, Y.M., AND LAU, R.T.S. (1973). Effects of neonatal sex hormones on sex-based cognitive abilities in the white rat. *Psychologia* 16:17–24.

DEARMOND, S., FUSCO, M., AND DEWEY, M. (1976). *A Photographic Atlas: Structure of the Human Brain,* 2nd edition. New York: Oxford University Press.

DEARMOND, S.J., FUSCO, M.M., AND DEWEY, M.M. (1989). *The Structure of the Human Brain: A Photographic Atlas,* 3rd edition. New York: Oxford University Press.

DEESE, J. (1959). On the prediction of occurrence of particular verbal intrusions in immediate recall. *J. Exp. Psychol.* 58:17–22.

DEHAENE, S. (1996). The organization of brain activations in number comparison: Event-related potentials and the additive-factors method. *J. Cogn. Neurosci.* 8:47–68.

DEHAENE, S., POSNER, M.I., AND TUCKER, D.M. (1994). Localization of a neural system for error detection and compensation. *Psychol. Sci.* 5:303–305.

DEIBERT, E., KRAUT, M., KREMEN, S., AND HART, J., JR. (1999). Neural pathways in tactile object recognition. *Neurology* 52:1413–1417.

DEJONG, R.N. (1979). *The Neurologic Examination,* 4th edition. New York: Harper & Row.

DELBRÜCK, M. (1986). *Mind from Matter?* London: Blackwell Scientific.

DELIS, D., ROBERTSON, L., AND EFRON, R. (1986). Hemispheric specialization of memory for visual hierarchical stimuli. *Neuropsychologia* 24:205–214.

DELL, G.S. (1986). A spreading activation theory of retrieval in sentence production. *Psychol. Rev.* 93:283–321.

DELONG, M.R. (1990). Primate models of movement disorders of basal ganglia origin. *Trends Neurosci.* 13:281–285.

DÉMONET, J.F., CHOLLET, F., RAMSAY, S., CERDEBAT, D., NESPOULES, J.D., WISE, R., et al. (1992). The anatomy of phonological and semantic processing in normal subjects. *Brain* 115:1753–1768.

DÉMONET, J.F., PRICE, C.J., WISE, R., AND FRACKOWIAK, R.S.J. (1994). Differential activation of right and left posterior sylvian regions by semantic and phonological tasks: A positron emission tomography study. *Neurosci. Lett.* 182:25–28.

DENNETT, D. (1995). *Darwin's Dangerous Idea.* New York: Simon and Schuster.

DENNETT, D. C. (1981). *Consciousness Explained.* Boston: Little, Brown.

DE RENZI, E., PERANI, D., CARLESIMO, G.A., SILVERI, M.C., and FAZIO, F. (1994). Prosopagnosia can be associated with damage confined to the right hemisphere—An MRI and PET study and a review of the literature. *Neuropsychologia* 32:893–902.

DESIMONE, R. (1991). Face-selective cells in the temporal cortex of monkeys. *J. Cogn. Neurosci.* 3:1–8.

DESIMONE, R. (1996). Neural mechanisms for visual memory and their role in attention. *Proc. Nat. Acad. Sci. U.S.A.* 93:13494–13499.

DESIMONE, R., ALBRIGHT, T.D., GROSS, C.G., AND BRUCE, C. (1984). Stimulus-selective properties of inferior temporal neurons in the macaque. *J. Neurosci.* 4:2051–2062.

DESIMONE, R., WESSINGER, M., THOMAS, L., AND SCHNEIDER, W. (1990). Attentional control of visual perception: Cortical and subcortical mechanisms. *Cold Spring Harb. Symp. Quant. Biol.* 55:963–971.

DESMOND, J.E., AND MOORE, J.W. (1991). Altering the synchrony of stimulus trace processes: Tests of a neural-network model. *Biol. Cybern.* 65:161–169.

DESMURGET, M., EPSTEIN, C.M., TURNER, R.S., PRABLANC, C., ALEXANDER, G.E., AND GRAFTON, S.T. (1999). Role of the posterior parietal cortex in updating reaching movements to a visual target. *Nat. Neurosci.* 2:563–567.

D'ESPOSITO, M., AGUIRRE, G.K., ZARAHN, E., BALLARD, D., SHIN, R.K., AND LEASE, J. (1998). Functional MRI studies of spatial and nonspatial working memory. *Cogn. Brain Res.* 7:1–13.

D'ESPOSITO, M., ZARAHN, E., AND AGUIRRE, G.K. (1999). Event-related functional MRI: Implications for cognitive psychology. *Psychol. Bull.* 125:155–164.

DEUTSCH, D. (1975). Musical illusions. *Sci. Am.* 233:92–104.

DEUTSCH, D. (1985). Dichotic listening to melodic patterns and its relationship to hemispheric specialization of function. *Music Percept.* 3:127–154.

DIAMOND, A. (1990). The Development and Neural Bases of Memory Functions as Indexed by the A(not)B and Delayed Response Tasks in Human Infants and Infant Monkeys. In A. Diamond (Ed.), *The Development and Neural Bases of Higher Cognitive Functions* (pp. 267–317). New York: New York Academy of Sciences.

DIAMOND, A. (1991). Neuropsychological Insights into the Meaning of Object Concept Development. In S. Carey and R. Gelman (Eds.), *The Epigenesis of Mind: Essays on Biology and Cognition*. Hillsdale, NJ: Erlbaum.

DOYLE, D., CABRAL, J., PFUETZNER, R., KUO, A., GULBIS, J., COHEN, S., CHAIT, B., AND MACKINNON, R. (1998). The structure of the potassium channel: Molecular basis of K+ conduction and selectivity; *Science* 280(5360): 69–77.

DRACHMAN, D.A., AND ARBIT, J. (1966). Memory and the hippocampal complex. II. Is memory a multiple process? *Arch. Neurol.* 15:52–61.

DREVETS, W.C. (1998). Functional neuroimaging studies of depression: The anatomy of melancholia. *Annu. Rev. Med.* 49:341–361.

DREVETS, W.C., AND RAICHLE, M.E. (1995). Positron Emission Tomographic Imaging Studies of Human Emotional Disorders. In M.S. Gazzaniga (Ed.), *The Cognitive Neurosciences* (pp. 1153–1164). Cambridge, MA: MIT Press.

DREVETS, W.C., VIDEN, T.O., PRICE, J.L., PRESKORN, S.H., CARMICHAEL, S.T., AND RAICHLE, M.E. (1992). A functional anatomical study of unipolar depression. *J. Neurosci.* 12:3628–3641.

DRONKERS, N. (1996). A new brain region for coordinating speech articulation. *Nature* 384:159–161.

DRONKERS, N.F., AND PINKER, S. (In Press). Language and the Aphasias. In E.R. Kandel, J. Schwartz, and T. Jessel (Eds.), *Principles in Neural Science*, 4th edition. New York: Elsevier.

DRONKERS, N.F., WILKINS, D.P., VAN VALIN, R.D., REDFERN, B.B., AND JAEGER, J.J. (1994). A reconsideration of the brain areas involved in the disruption of morphosyntactic comprehension. *Brain Lang.* 47:461–462.

DUNCAN, J. (1984). Selective attention and the organization of visual information. *J. Exp. Psychol. Gen* 113:501–517.

DUNCAN, J. (1995). Attention, Intelligence, and the Frontal Lobes. In M.S. Gazzaniga (Ed.), *The Cognitive Neurosciences* (pp. 721–733). Cambridge, MA: MIT Press.

DUONG, T.Q., KIM, D.S., UGURBIL, K., AND KIM, S.G. (2000). Spatiotemporal dynamics of the BOLD fMRI signals: Toward mapping submillimeter cortical columns using the early negative response. *Magn. Reson. Med.* 44:231–242.

EASON, R., HARTER, M., AND WHITE, C. (1969). Effects of attention and arousal on visually evoked cortical potentials and reaction time in man. *Physiol. Behav.* 4:283–289.

EFRON, R. (1990). *The Decline and Fall of Hemispheric Specialization*. Hillsdale, NJ: Lawrence Erlbaum Associates.

EGLIN, M., ROBERTSON, L.C., AND KNIGHT, R.T. (1989). Visual search performance in the neglect syndrome. *J. Cogn. Neurosci.* 1:372–385.

EGLY, R., DRIVER, J., RAFAL, R.D. (1994). Shifting visual attention between objects and locations—evidence from normal and parietal lesion subjects. *J. Exp. Psychol Gen.* 123:161–177.

EICHENBAUM, H. (2000). A cortical-hippocampal system for declarative memory. *Nat. Rev. Neurosci.* 1:41–50.

EISENBERG, J.F. (1981). *The Mammalian Radiations. An Analysis of Trends in Evolution, Adaptation and Behavior*. Chicago: University of Chicago Press.

EKMAN, P. (1971). Universals and Cultural Differences in Facial Expression. In J.K. Cole (Ed.), *Nebraska Symposium and Motivation* (pp. 207–284). Lincoln, NE: University of Nebraska Press.

EKMAN, P., (1973). Cross-Cultural Studies in Facial Expression. In P. Ekman (Ed.), *Darwin and Facial Expressions: A Century of Research in Review*. New York: Academic Press.

EKMAN, P. (1984). Expression and the Nature of Emotion. In P. Ekman and K. Scherer (Eds.), *Approaches to Emotion* (pp. 319–343). Hillsdale, NJ: Lawrence Erlbaum Associates.

ELBERT, T., PANTEV, C., WIENBRUCH, C., ROCKSTROH, B., AND TAUB, E. (1995). Increased cortical representation of the fingers of the left hand in string players. *Science* 270:305–307.

ENGEL, A.K., KREITER, A.K., KONIG, P., AND SINGER, W. (1991). Synchronization of oscillatory neuronal responses between striate and extrastriate visual cortical areas of the cat. *Proc. Natl. Acad. Sci. U.S.A.* 88:6048–6052.

ENGEL, S.A., RUMELHART, D.E., WANDELL, B.A., LEE, A.T., GLOVER G.H., CHICHILNISKY, E.J., AND SHADLEN, M.N. (1995). fMRI of human visual cortex. *Nature* 370:106.

ENNS, J.T., AND RENSINK, R.A. (1990). Sensitivity to three-dimensional orientation in visual search. *Psychol. Sci.* 1:323–326.

ERIKSEN, B., AND ERIKSEN, C. (1974). Effects of noise letters upon the identification of a target letter in a non search task. *Percept. Psychophys.* 16:143–149.

ERIKSSON, P.S., PERFILIEVA, E., BJÖRK-ERIKSSON, T., ALBORN, A., NORDBORG, C., PETERSON, D., AND GAGE, F. (1998). Neurogenesis in the adult humanhippocampus. *Nat. Med.* 4:1313–1317.

FARAH, M.J. (1988). Is visual imagery really visual? Overlooked evidence from neuropsychology. *Psychol. Rev.* 95:307–317.

FARAH, M.J. (1990). *Visual Agnosia: Disorders of Object Recognition and What They Tell Us about Normal Vision*. Cambridge, MA: MIT Press.

FARAH, M.J. (1994). Specialization Within Visual Object Recognition: Clues from Prosopagnosia and Alexia. In M.J. Farah and G. Ratcliff (Eds.), *The Neuropsychology of High-Level Vision: Collected Tutorial Essays* (pp. 133–146). Hillsdale, NJ: Lawrence Erlbaum Associates.

FARAH, M.J., AND MCCLELLAND, J.L. (1991). A computational model of semantic memory impairment: Modality specificity and emergent category specificity. *J. Exp. Psychol. Gen.* 120:339–357.

FENDRICH, R., WESSINGER, C.M., AND GAZZANIGA, M.S. (1992). Residual vision in a scotoma: Implications for blindsight. *Science* 258:1489–1491.

FERNÁNDEZ, G., EFFERN, A., GRUNWALD, T., PEZER, N., LEHNERTZ, K., DÜMPELMANN, M., VAN ROOST, D., AND ELGER, C.E. (1999) Real-time tracking of memory formation in the human rhinal cortex and hippocampus. *Science* 285:1582–1585.

FERRY, B., ROOZENDAAL, B., AND MCGAUGH, J.L. (1999). Basolateral amygdala noradrenergic influences on memory storage are mediated by an interaction between beta- and alpha1-adrenoceptors. *J. Neurosci.* 19:5119–5123.

FIEZ, J.A., RAIFE, E.A., BALOTA, D.A., SCHWARZ, J.P., RAICHLE, M.E., AND PETERSEN, S.E. (1996). A positron emission tomography study of short-term maintenance of verbal information. *J. Neurosci.* 16:808–822.

FINGER, S. (1994). *Origins of Neuroscience*. New York: Oxford University Press.

FLORENCE, S.L., AND KAAS, J.H. (1995). Large-scale reorganization at multiple levels of the somatosensory pathway follows therapeutic amputation of the hand in monkeys. *J. Neurosci.* 15:8083–8095.

FLOURENS, M.-J.P. (1824). *Recherches Expérimentales sur les proprieties et les functiones du Systeme Nerveux dans le Animaux Vertébrés*. Paris: J.B. Ballière.

FODOR, J.A. (1983). *The Modularity of Mind*. Cambridge, MA: MIT Press.

FOX, P.T., MIEZIN, F.M., ALLMAN, J.M., VAN ESSEN, D.C., AND RAICHLE, M.E. (1987). Retinotopic organization of human visual cortex mapped with positron-emission tomography. *J. Neurosci.* 7:913–922.

FRANZ, E., ELIASSEN, J., IVRY, R., AND GAZZANIGA, M. (1996). Dissociation of spatial and temporal coupling in the bimanual movements of callosotomy patients. *Psychol. Sci.* 7:306–310.

FREEDMAN, D.J., RIESENHUBER, M., POGGIO, T., AND MILLER, E.K. (2001). Categorical representations of visual stimuli in the primate prefontal cortex. *Science* 291:312–316.

FRIEDERICI, A., PFEIFER, E., AND HAHNE, A. (1993). Event-related brain potentials during natural speech processing: Effects of semantic morphological, and syntactic violations. *Cogn. Brain Res.* 1:183–192.

FRIEDMAN, H.R., AND GOLDMAN-RAKIC, P.S. (1994). Coactivation of prefrontal cortex and inferior parietal cortex in working memory tasks revealed by 2DG functional mapping in the rhesus monkey. *J. Neurosci.* 14:2775–2788.

FRITH, C.D., FRISTON, K., LIDDLE, P.F., AND FRACKOWIAK, R.S.J. (1991). Willed action and the prefrontal cortex in man: A study with PET. *Proc. R. Soc. Lond., Biol. Sci.* 244:241–246.

FUNAYAMA, E.S., GRILLON, C.G., DAVIS, M. AND PHELPS, E.A. (In Press). A double dissociation in the affective modulation of startle in humans: Effects of unilateral temporal lobectomy. *J. Cogn. Neurosci.*

FUSTER, J.M. (1989). *The Prefrontal Cortex: Anatomy, Physiology, and Neuropsychology of the Frontal Lobe,* 2nd edition. New York: Raven Press.

GABRIELI, J., FLEISCHMAN, D., KEANE, M., REMINGER, S., AND MORELL, F. (1995). Double dissociation between memory systems underlying explicit and implicit memory in the human brain. *Psychol. Sci.* 6:76–82.

GABRIELI, J.D., POLDRACK, R.A., AND DESMOND, J.E. (1998). The role of left prefrontal cortex in language and memory. *Proc. Natl. Acad. Sci. U.S.A.* 95:906–913.

GAFFAN, D., AND HARRISON, S. (1987). Amygdalectomy and disconnection in visual learning for auditory secondary reinforcement by monkeys. *J. Neurosci.* 7:2285–2292.

GAFFAN, D., AND HEYWOOD, C.A. (1993). A spurious category-specific visual agnosia for living things in normal human and nonhuman primates. *J. Cogn. Neurosci.* 5:118–128.

GALL, F.J., AND SPURZHEIM, J. (1810–1819). *Anatomie et Physiologie du Systè me Nerveux en Gèneral, et der Cerveauen Particulier.* Paris: F. Schoell.

GALLAGHER, M., AND HOLLAND, P.C. (1992). Understanding the Function of the Central Nucleus: Is Simple Conditioning Enough? In: J.P. Aggleton (Ed.), *The Amygdala: Neurobiological Aspects of Emotion, Memory, and Mental Dysfunction* (pp. 307–321). New York: Wiley-Liss.

GALLANT, J.L., SHOUP, R.E., AND MAZER, J.A. (2000). A human extrastriate area functionally homologous to macaque V4. *Neuron* 27:227–235.

GALLISTEL, C.R. (1995). The Replacement of General-Purpose Theories with Adaptive Specializations. In M.S. Gazzaniga (Ed.), *The Cognitive Neurosciences* (pp. 1255–1267). Cambridge, MA: MIT Press.

GALUSKE, R.A., SCHLOTE, W., BRATZKE, H., AND SINGER, W. (2000). Interhemispheric asymmetries of the modular structure in human temporal cortex. *Science* 289:1946–1949.

GAN , W.B. GRUTZENDLER, J., WONG, W.T., WONG, R.O., AND LICHTMAN, J.W. (2000). Multicolor "diolistic" labeling of the nervous system using lipophilic dye combinations. *Neuron* 27:219–225.

GASH, D.M., ZHANG, Z., OVADIA, A., CASS, W.A., YI, A., SIMMERMAN, L., RUSSELL, D., MARTIN, D., LAPCHAK, P.A., AND COLLINS, F. (1996). Functional recovery in parkinsonian monkeys treated with GDNF. *Nature* 380:252–255.

GAULIN, S.J.C. (1995). Does Evolutionary Theory Predict Sex Differences in the Brain? In M.S. Gazzaniga (Ed.), *The Cognitive Neurosciences* (pp. 1211–1225). Cambridge, MA: MIT Press.

GAULIN, S.J.C., AND FITZGERALD, R.W. (1989). Sexual selection for spatial-learning ability. *Anim. Behav.* 37:322–331.

GAUTHIER, I., SKUDLARSKI, P., GORE, J.C., AND ANDERSON, A.W. (2000). Expertise for cars and birds recruits brain areas involved in face recognition. *Nat. Neurosci.* 3:191–197.

GAUTHIER, I.L., BEHRMANN, M., AND TARR, M.J. (1999). Can face recognition really be dissociated from object recognition? *J. Cogn. Neurosci.* 11:349–370.

GAZZANIGA, M.S. (1983). Right hemisphere language following brain bisection: A twenty year perspective. *Am. Psychol.* 38:525–547.

GAZZANIGA, M.S. (1992). *Nature's Mind: The Biological Roots of Thinking, Emotions, Sexuality, Language, and Intelligence.* New York: Basic Books.

GAZZANIGA, M.S. (1995a). Principles of human brain organization derived from split-brain studies. *Neuron* 14:217–228.

GAZZANIGA, M.S. (Ed.). (1995b). *The Cognitive Neurosciences.* Cambridge, MA: MIT Press.

GAZZANIGA, M.S. (1995c). Consciousness and the Cerebral Hemispheres. In M.S. Gazzaniga (Ed.), *The Cognitive Neurosciences* (pp. 1391–1399). Cambridge, MA: MIT Press.

GAZZANIGA, M.S. (1998). *The Mind's Past.* Berkeley, CA: University of California Press.

GAZZANIGA, M.S. (2000). Cerebral specialization and interhemispheric communication: Does the corpus callosum enable the human condition? *Brain* 123:1293–1326.

GAZZANIGA, M.S., AND LE DOUX, J.E. (1978). *The Integrated Mind.* New York: Plenum Press

GAZZANIGA, M.S., AND SMYLIE, C.S. (1983). Facial recognition and brain asymmetries: Clues to underlying mechanisms. *Ann. Neurol.* 13:536–540.

GAZZANIGA, M.S., AND SMYLIE, C.S. (1990). Hemispheric mechanisms controlling voluntary and spontaneous facial expressions. *J. Cog. Neurosci.* 2:239–245.

GAZZANIGA, M.S., AND SPERRY, R.W. (1967). Language after section of the cerebral commissures. *Brain* 90:131–148.

GAZZANIGA, M.S., HOLTZMAN, J.D., AND SMYLIE, C.S. (1987). Speech without conscious awareness. *Neurology* 35:682–685.

GEARY, D.C. (1995). Reflections of evolution and culture in children's cognition. Implications for mathematical development and instruction. *Am. Psychol.* 50:24–37.

GEHRING, W.J., AND KNIGHT, R.T. (2000). Prefrontal-cingulate interactions in action monitoring. *Nat. Neurosci.* 3:516–520.

GEHRING, W.J., GOSS, B., COLES, M.G.H., MEYER, D.E., AND DONCHIN, E. (1993). A neural system for error detection and compensation. *Psychol. Sci.* 4:385–390.

GEORGOPOULOS, A.P. (1990). Neurophysiology of Reaching. In M. Jeannerod (Ed.), *Attention and Performance XIII: Motor Representation and Control* (pp. 227–263). Hillsdale, NJ: Lawrence Erlbaum Associates.

GEORGOPOULOS, A.P. (1995). Motor Cortex and Cognitive Processing. In M.S. Gazzaniga (Ed.), *The Cognitive Neurosciences* (pp. 507–517). Cambridge, MA: MIT Press.

GERAERTS, W.P.M., SMIT, A.B., AND LI, K.W. (1994). Constraints and Innovations in the Molecular Evolution of Neuronal Signaling: Implications for Behavior. In R.J. Greenspan and C.P. Kyriacou (Eds.), *Flexibility and Constraint in Behavioral Systems* (pp. 207–235). New York: John Wiley and Sons.

GERLOFF, C., CORWELL, B., CHEN, R., HALLETT, M., AND COHEN, L.G. (1997) Stimulation over the human supplementary motor area interferes with the organization of future elements in complex motor sequences. *Brain* 120:1587–1602.

GESCHWIND, N. (1967). The varieties of naming errors. *Cortex* 3:97–112.

GESCHWIND, N., AND GALABURDA, A.M. (1987). *Cerebral Lateralization: Biological Mechanisms, Associations, and Pathology.* Cambridge, MA: MIT Press.

GESCHWIND, N., AND LEVITSKY, W. (1968). Human brain: Left-right asymmetries in temporal speech region. *Science* 161:186–187.

GIARD, M.H., FORT, A., MOUCHETANT-ROSTAING, Y., AND PERNIER, J. (2000). Neurophysiological mechanisms of auditory selective attention in humans. *Frontiers Biosci.* 5:D84–D94.

GIBSON, J., BEIERLEIN, M., AND CONNORS, B. (1999). Two networks of electrically coupled inhibitory neurons in neocortex. *Nature* 402(6757):75–79.

GIEDD, J.N., BLUMENTHAL, J., JEFFRIES, N.O., CASTELLANOS, F.X., LIU, H., ZIJDENBOS, A., PAUS, T., EVANS, A.C., and RAPOPORT, J.L. (1999). Brain development during childhood and adolescense: A longitudinal MRI study. *Nat. Neurosci.* 2:861–863.

GILBERT, C.D., AND WIESEL, T.N. (1990). The influence of contextual stimuli on the orientation selectivity of cells in primary visual cortex. *Vision Res.* 30:1689–1701.

GLISKY, E.L., POLSTER, M.R., AND ROUTHUIEAUX, B.C. (1995). Double dissociation between item and source memory. *Neuropsychology* 9:229–235.

GOEL, V., GRAFMAN, J., TAJIK, J., GANA, S., AND DANTO, D. (1997). A study of the performance of patients with frontal lobe lesions in a financial planning task. *Brain* 120:1805–1822.

GOLDBERG, G. (1985). Supplementary motor area structure and function: Review and hypothesis. *Behav. Brain Sci.* 8:567–616.

GOLDMAN-RAKIC, P.S. (1992). Working memory and the mind. *Sci. Am.* 267:111–117.

GOLDMAN-RAKIC, P.S. (1995). Architecture of the Prefrontal Cortex and the Central Executive. In J. Grafman, K.J. Holyoak, and F. Boller (Eds.), *Structure and Functions of the Human Prefrontal Cortex* (pp. 71–83). New York: The New York Academy of Sciences.

GOLDSBY, R.A. (1976). *Basic Biology. (pp. 282–297).* New York: Harper and Row Publishers.

GOODALE, M.A., AND MILNER, A.D. (1992). Separate visual pathways for perception and action. *Trends Neurosci.* 15:22–25.

GOODMAN, G., QUAS, J., BATTERMAN-FAUNEE, J., RIDDLESBERGER, M., et al. (1994). Predictors of accurate and inaccurate memories of traumatic events experienced in childhood. *Conscious. Cogn.* 3:269–294.

GORDON, J., GHILARDI, M.F., AND GHEZ, C. (1995). Impairments of reaching movements in patients without proprioception. I. Spatial errors. *J. Neurophysiol.* 73:347–360.

GOULD, E., REEVES, A.J., GRAZIANO, M.S.A., AND GROSS, C.G. (1999). Neurogenesis in the neocortex of adult primates. *Science* 286:548–552.

GOULD, S.J., AND VRBA, E. (1981). Exaptation: A missing term in the science of form. *Paleobiology* 8:4–15.

GRABOWECKY, M., ROBERTSON, L.C., AND TREISMAN, A. (1993). Preattentive processes guide visual search. *J. Cogn. Neurosci.* 5:288–302.

GRAFTON, S., HAZELTINE, E., AND IVRY, R. (1995). Functional mapping of sequence learning in normal humans. *J. Cogn. Neurosci.* 7:497–510.

GRAHAM, J., CARLSON, G.R., AND GERARD, R.W. (1942). Membrane and injury potentials of single muscle fibers. *Fed. Proc.* 1:31.

GRANT, S., LONDON, E.D., NEWLIN, D.B., VILLEMAGNE, V.L., LIU, X., CONTOREGGI, C., PHILLIPS, R.L., KIMES, A.S., AND MARGOLIN, A. (1996). Activation of memory circuits during cue-elicited cocaine craving. *Proc. Natl. Acad. Sci. U.S.A.* 93:12040–12045.

GRATTON, G., AND FABIANI, M. (1998). Dynamic brain imaging: Event-related optical signal (EROS) measures of the time course and localization of cognitive-related activity. *Psycho. Bull. Rev.* 5:535–563.

GRAY, C.M., KONIG, P., ENGEL, A.K., AND SINGER, W. (1989). Oscillatory responses in cat visual cortex exhibit inter-columnar synchronization which reflects global stimulus properties. *Nature* 338:334–337.

GRAZIANO, M.S.A., AND GROSS, C.G. (1994). Mapping space with neurons. *Curr. Direct. Psychol. Sci.* 3:164–167.

GREENBERG, J.O. (1995). *Neuroimaging: A Companion to Adams and Victor's Principles of Neurology.* New York: McGraw-Hill.

GREENBLATT, S.H. (1984). The multiple roles of Broca's discovery in the development of modern neurosciences. *Brain Cogn.* 3:249–258.

GREENFIELD, P.M. (1991). Language, tools and brain: The ontogeny and phylogeny of hierarchically organized sequential behavior. *Behav. Brain Sci.* 14:531–595.

GROSS, C. (2000). Neurogenesis in the adult brain: Death of a dogma. *Nat. Rev. Neurosci.* 1:67–73.

GUR, R.C., SKOLNICK, B.E., AND GUR, R.E. (1994). Effects of emotional discrimination tasks on cerebral blood flow: Regional activation and its relation to performance. *Brain Cogn.* 25:271–286.

HAALAND, K.Y., HARRINGTON, M.F., AND KNIGHT, R.T. (1999). Spatial deficits in ideomotor limb apraxia. A kinematic analysis of aiming movements. *Brain* 122:1169–1182.

HADJIKHANI, N., LIU, A.K., DALE, A.M., CAVANAGH, P., and TOOTELL, R.B. (1998). Retinotopy and color sensitivity in human visual cortical area V8. *Nat. Neurosci.* 1:235–241.

HAEUSSER, M. (2000). The Hodgkin-Huxley theory of the action potential. *Nat. Rev. Neurosci.* 3:1165.

HAGOORT, P., BROWN, C., AND GROOTHUSEN, J. (1993). The syntactic positive shift (SPS) as an ERP measure of syntactic processing. Special Issue: Event-related brain potentials in the study of language. *Lang. Cogn. Processes.* 8:439–483.

HAGOORT, P., BROWN, C., AND SWAAB, T. (1996). Lexical semantic event-related potential effects in patients with left hemisphere lesions and aphasia, and patients with right hemisphere lesions without aphasia. *Brain* 119:627–649.

HAGOORT, P., INDEFREY, P., BROWN, C., HERZOG, H., STEINMETZ, H., AND SEITZ, R.J. (1999). The neural circuitry involved in the reading of German words and pseudowords. *J. Cogn. Neurosci.* 11:383–398.

HAMANN, S.B., ELY, T.D., GRAFTON, S.T., AND KILTS, C.D. (1999). Amygdala activity related to enhanced memory for pleasant and aversive stimuli. *Nat. Neurosci.* 2:289–293.

HAMANN, S.B., STEFANACCI, L., SQUIRE, L.R., ADOLPHS, R., TRANEL, D., DAMASIO, H., AND DAMASIO, A.R. (1996). Recognizing facial emotion. *Nature* 379:497.

HARKNESS, R.D., AND MAROUDAS, N.G. (1985). Central place foraging by an ant (*Cataglyphis bicolor* Fab.): A model of searching. *Anim. Behav.* 33:916–928.

HART, A.J., WHALEN, P.J., SHIN, L.M., McINEMY, S.C., FISCHER, H., AND RAUCH, S.L. (2000). Differential response in the human amygdala to racial outgroup vs. ingroup face stimuli. *Neuroreport* 11:231–2356.

HART, J., BERNDT, R.S., AND CARAMAZZA, A. (1985). Category-specific naming deficit following cerebral infarction. *Nature* 316:439–440.

HARTLINE, H.K. (1938). The response of single optic nerve fibers of the vertebrate eye to illumination of the retina. *Am. J. Physiol.* 121:400–415.

HAXBY, J., UNGERLEIDER, L., HORWITZ, B., MAISOG, J., ROPOPORT, S., AND GRADY, C. (1996). Face encoding and recognition in the human brain. *Proc. Natl. Acad. Sci. U.S.A.* 93:922–927.

HAXBY J.V., HORWITZ B., UNGERLEIDER, L.G., MAISOG, J.M., PIETRINI, P., AND GRADY, C.L. (1994). The functional organization of human extrastriate cortex: A PET-rCBF study of selective attention to faces and locations. *J. Neurosci.* 14:6336–6353.

HAYES, A., DAVIDSON, M., KEELE, S.W., AND RAFAL, R. (1998). Toward a functional analysis of the basal ganglia. *J. Cogn. Neurosci.* 10:178–198.

HAZELTINE, E., POLDRACK, R., AND GABRIELI, J.D.E. (2000). Neural activation during response competition. *J. Cogn. Neurosci.* 12:118–129.

HEBB, D. (1949). *The Organization of Behavior: A Neuropsychological Theory.* New York: John Wiley and Sons.

HECAEN, H., AND ALBERT, M.L. (1978). *Human Neuropsychology.* New York: Wiley.

HEILMAN, K.M., ROTHI, L.J., AND VALENSTEIN, E. (1982). Two forms of ideomotor apraxia. *Neurology* 32:342–346.

HEILMAN, K.M., SCHOLES, R., AND WATSON, R.T. (1975). Auditory affective agnosia: Disturbed comprehension of affective speech. *J. Neuro. Neurosurg. Psychiatry* 38:69–72.

HEINZE, H.J., MANGUN, G.R., BURCHERT, W., HINRICHS, H., SCHOLZ, M., MÜNTE, T.F., GÖS, A., SCHERG, M., JOHANNES, S., HUNDESHAGEN, H., GAZZANIGA, M.S., AND HILLYARD, S.A. (1994). Combined spatial and temporal imaging of brain activity during visual selective attention in humans. *Nature* 372:543–546.

HEITLER, W., AND EDWARDS, D. (1998). Effect of temperature on a voltage-sensitive electrical synapse in crayfish. *J. Exp. Biol.,* 201:503–513.

HELLIGE, J.B. (1993). *Hemispheric Asymmetry: What's Right and What's Left.* Cambridge, MA: Harvard University Press.

HENDERSON, L. (1982). *Orthography and Word Recognition in Reading.* London: Academic Press.

HERNANDEZ-PEON, R., SCHERRER, H., AND JOUVET, M. (1956). Modification of electrical activity in cochlear nucleus during attention in unanesthetized cats. *Science* 123:331–332.

HERRICK, C.J. (1963). *Brains of Rats and Men.* New York: Hafner Publishing.

HEYWOOD, C.A., WILSON, B., AND COWEY, A. (1987). A case study of cortical colour "blindness" with relatively intact achromatic discrimination. *J. Neurol. Neurosurg. Psychiatry* 50:22–29.

HILLYARD, S.A. (1993). Electrical and magnetic brain recordings: Contributions to cognitive neuroscience. *Curr. Opin. Neurobiol.* 3:710–717.

HILLYARD, S.A., HINK, R.F., SCHWENT, V.L., AND PICTON, T.W. (1973). Electrical signs of selective attention in the human brain. *Science* 182:177–180.

HILTS, P.J. (1995). *Memory's Ghost: The Strange Tale of Mr. M. and the Nature of Memory.* New York: Simon and Schuster.

HODGES, J., PATTERSON, K., OXBURY, S., AND FUNNELL, E. (1992). Semantic dementia: Progressive fluent aphasia with temporal lobe atrophy. *Brain* 115:1783–1806

HODOS, W., AND CAMPBELL, C.B.G. (1969), *Scala naturae*: Why there is no theory in comparative psychology. *Psychol. Rev.* 76:337–350.

HOLBOURN, A.H.S. (1943). Mechanics of head injury. *Lancet* 2:438–441.

HOLMES, G. (1919). Disturbances of visual orientation. *Bri. J. Ophthalmol.* 2:449–468.

HOLTZMAN, J.D. (1984). Interactions between cortical and subcortical visual areas: Evidence from human commissurotomy patients. *Vision Res.* 24:801–813.

HOLTZMAN, J.D., AND GAZZANIGA, M.S. (1982). Dual task interactions due exclusively to limits in processing resources. *Science* 218:1325–1327.

HOPFINGER, J., AND MANGUN, G.R. (1998). Reflexive attention modulates visual processing in human extrastriate cortex. *Psychol. Sci.* 9:441–447.

HOPFINGER, J.B., BUONOCORE, M.H. AND MANGUN, G.R. (2000). The neural mechanisms of top-down attentional control. *Nat. Neurosci.,* 3:284–291.

HORE, J., WILD, B., AND DIENER, H. (1991). Cerebellar dysmetria at the elbow, wrist, and fingers. *J. Neurophysiol.* 65:563–571.

HUBEL, D., AND WIESEL, T. (1968). Receptive fields and functional architecture of monkey striate cortex. *J. Physiol. (Lond.)* 195:215–243.

HUBEL, D.H., AND WIESEL, T.N. (1977). The Ferrier Lecture: Functional architecture of macaque monkey visual cortex. *Proc. R. Acad. Lond.,* Series B 198:1–59.

HUMPHREYS, G., AND RIDDOCH, J. (2001) Detection by action: Neuropsychological evidence for action-defined templates in search. *Nat. Neurosci.* 4:84–88.

HUMPHREYS, G.W., AND RIDDOCH, M.J. (1992). Interactions Between Objects and Space-Vision Revealed Through Neuropsychology. In D.E. Meyers and S. Kornblum (Eds.), *Attention and Performance XIV* (pp. 143–162). Hillsdale, NJ: Lawrence Erlbaum Associates.

HUMPHREYS, G.W., RIDDOCH, M.J., DONNELLY, N., FREEMAN, T., BOUCART, M., AND MULLER, H.M. (1994). Intermediate Visual Processing and Visual Agnosia. In M.J. Farah and G. Ratcliff (Eds.), *The Neuropsychology of High-Level Vision: Collected Tutorial Essays* (pp. 63–102). Hillsdale, NJ: Lawrence Erlbaum Associates.

HUTTENLOCHER, P.R. (1990) Morphometric study of human cerebral cortex development. *Neuropsychologia* 28:517–527.

HUTTENLOCHER, P.R., AND DABHOLKAR, A.S. (1997) Regional differences in synaptogenesis in human cerebral cortex. *J. Comp. Neurol.* 387:167–178.

HUTTENLOCHER, P.R., AND DE COURTEN, C. (1987). The development of synapses in striate cortex of man. *Hum. Neurobiol.* 6:1–9.

HUTTENLOCHER, P.R., DE COURTEN, C., GAREY, L.J., AND VAN DER LOOS, H. (1982) Synaptogenesis in human visual cortex—evidence for synapse elimination during normal development. *Neurosci. Lett.* 33:247–252.

HYDE, I.H. (1921). A micro-electrode and unicellular stimulation. *Biol. Bull.* 40:130–133.

IMAMIZU, H., MIYAUCHI, S., TAMADA, T., SASAKI, Y., TAKINO, R., PUTZ, B., YOSHIOKA, T., AND KAWATO, M. (2000). Human cerebellar activity reflecting an acquired internal model of a new tool. *Nature* 403:192–195.

INGLIS, J., AND LAWSON, J.S. (1982). A meta-analysis of sex differences in the effects of unilateral brain damage on intelligence test results. *Can. J. Psychol.* 36:670–683.

ITO, M., TAMURA, H., FUJITA, I., AND TANAKA, K. (1995). Size and position invariance of neuronal responses in monkey inferotemporal cortex. *J. Neurophysiol.* 73:218–226.

IVRY, R. (1993). Cerebellar involvement in the explicit representation of temporal information. *Ann. N.Y. Acad. Sci.* 682: 214–230.

IVRY, R.B., AND COHEN, A. (1992). Asymmetry in visual search for targets defined by differences in movement speed. *J. Exp. Psychol. Hum. Percept. Perform.* 18:1045–1057.

IVRY, R.B., AND LEBBY, P.C. (1993). Hemispheric differences in auditory perception are similar to those found in visual perception. *Psychol. Sci.* 4:41–45.

IVRY, R.B., AND ROBERTSON, L.C. (1998). *The Two Sides of Perception.* Cambridge, MA: MIT Press.

JACKENDOFF, R. (1987a). *Consciousness and the Computational Mind.* Cambridge, MA: MIT Press.

JACKENDOFF, R. (1987b). On beyond zebra: The relation of linguistic and visual information. *Cognition* 26:89–114.

JACOB, F. (1977) Evolution and tinkering. *Science* 196:1161–1166.

JACOBS, L.F., GAULIN, S.J.C., SHERRY, D.F., AND HOFFMAN, G.E. (1990). Evolution of spatial cognition: Sex-specific patterns of spatial behavior predict hippocampal size. *Proc. Natl. Acad. Sci. U.S.A.* 87:6349–6352.

JAMES, W. (1890). *Principles of Psychology.* New York: H. Holt.

JANER, K.W., AND PARDO, J.V. (1991). Deficits in selective attention following bilateral anterior cingulotomy. *J. Cogn. Neurosci.* 3:231–241.

JANOWSKY, J.S., OVIATT, S.K., AND ORWOLL, E.S. (1994). Testosterone influences spatial cognition in older men. *Behav. Neurosci.* 108:325–332.

JANOWSKY, J.S., SHIMAMURA, A.P., AND SQUIRE, L.R. (1989). Source memory impairment in patients with frontal lobe lesions. *Neuropsychologia* 27:1043–1056.

JASPER, H.H. (1995). A historical perspective: The rise and fall of prefrontal lobotomy. *Adv. Neurol.* 66:97–114.

JEANNEROD, M. (1997). *The Cognitive Neuroscience of Action.* Malden, MA: Blackwell Science.

JENKINS, I.H., BROOKS, D.J., NIXON, P.D., FRACKOWIAK, R.S.J., AND PASSINGHAM, R.E. (1994). Motor sequence learning: A study with positron emission tomography. *J. Neurosci.* 14:3775–3790.

JERNE, N. (1968). Antibodies and Learning: Selection Versus Instruction. In G. Quarton, T. Melnechuck, and F.O. Schmidt (Eds.), *The Neurosciences: A Study Program*, Vol. 1. New York: Rockefeller University Press.

JOHNSON, M.H. (Ed.) (1993). *Brain Development and Cognition: A Reader.* Cambridge, MA: Blackwell Science.

JOHNSON, M.H. (1997). *Developmental Cognitive Neuroscience.* Cambridge, MA: Blackwell Science.

JOHNSON, S.P., AND ASLIN, R.N. (1995). Perception of object unity in 2-month-old infants. *Dev. Psychol.* 31:739–745.

JOHNSRUDE, I.S., OWEN, A.M., WHITE, N.M., ZHAO, W.V., AND BOHBOT, V. (2000). Impaired preference conditioning after anterior temporal lobe resection in humans. *J. Neurosci.* 20:2649–2656.

JONES-GOTMAN, M., ZATORRE, R.J., CENDES, F., OLIVIER, A., ANDERMANN, F., MCMACKIN, D., STAUNTON, H., SIEGEL, A.M., AND WIESER, H.G. (1997). Contribution of medial versus lateral temporal-lobe structures to human odour identification. *Brain* 120:1845–1856.

JUST, M., CARPENTER, P., KELLER, T., EDDY, W., AND THULBORN, K. (1996). Brain activation modulated by sentence comprehension. *Science* 274:114–116.

KAAS, J. (1995). The Reorganization of Sensory and Motor Maps in Adult Mammals. In M.S. Gazzaniga (Ed.), *The Cognitive Neurosciences* (pp. 51–71). Cambridge, MA: MIT Press.

KAAS, J.H. (1997). What comparative studies of neocortex tell us about the human brain. *Rev. Brasil. Biol.* 56:315–322.

KAAS, J.H. (1999). The transformation of association cortex into sensory cortex. *Brain Res. Bull.* 50:425.

KAAS, J.H., NELSON, R.J., SUR, M., DYKES, R.W., AND MERZENICH, M.M. (1984). The somatotopic organization of the ventroposterior thalamus of the squirrel monkey, *Saimiri sciureus. J. Comp. Neurol.* 226:111–140.

KAKEI, S., HOFFMAN, D.S., AND STRICK, P.L. (1999). Muscle and movement representations in the primary motor cortex. *Science* 285:2136–2139.

KANDEL, E., SCHWARTZ, J., AND JESSELL, T. (Eds.) (1991). *Principles of Neural Science, 3rd edition.* New York: Elsevier.

KANDEL, E.R., SCHWARTZ, J.H., AND JESSELL, T.M. (1995). *Essentials of Neural Science and Behavior.* Norwalk, CT: Appleton and Lange.

KANWISHER, N. (2000). Domain specificity in face perception. *Nat. Neurosci.* 3:759–763.

KANWISHER, N. AND WOJCIULIK, E. (2000). Visual attention: Insights from brain imaging. *Nat. Rev. Neurosci.* 1:91–100.

KANWISHER, N., WOODS, R., IACOBONI, M., AND MAZZIOTTA, J.C. (1997). A locus in human extrastriate cortex for visual shape analysis. *J. Cogn. Neurosci.* 9:133–142.

KAPP, B.S., PASCOE, J.P., AND BIXLER, M.A. (1984). The Amygdala: A Neuroanatomical Systems Approach to Its Contributions to Aversive Conditioning. In N. Butters and L.R. Squire (Eds.), *Neuropsychology of Memory* (pp. 473–488). New York: Guilford Press.

KAPP, B.S., SUPPLE, W.F., AND WHALEN, P.J. (1994). Stimulation of the amygdaloid central nucleus produces EEG arousal. *Behav. Neurosci.* 108:81–93.

KAPP, B.S., WILSON, A., PASCOE, J.P., SUPPLE, W., AND WHALEN, P.J. (1990). A Neuroanatomical Systems Analysis of Conditioned Bradycardia in the Rabbit. In M. Gabriel and J. Moore (Eds.), *Learning and Computational Neuroscience: Foundations of Adaptive Networks* (pp. 53–90). Cambridge, MA: MIT Press.

KAPUR, S., CRAIK, F.I., TULVING, E., WILSON, A., HOULE, S., AND BROWN, G. (1994). Neuroanatomical correlates of encoding in episodic memory: Levels of processing effect. *Proc. Natl. Acad. Sci. U.S.A.* 91:2008–2011.

KARNI, A., MEYER, G., JEZZARD, P., ADAMS, M.M., TURNER, R., AND UNGERLEIDER, L.G. (1995). Functional MRI evidence for adult motor cortex plasticity during motor skill learning. *Nature* 377:155–158.

KARNI, A., MEYER, G., REY-HIPOLITO, C., JEZZARD, P., ADAMS, M., TURNER, R., AND UNGERLEIDER, L. (1998). The acquisition of skilled motor performance: Fast and slow experience-driven changes in primary motor cortex. *Proc. Natl. Acad. Sci. U.S.A.* 95:861–868.

KASS-SIMON, G., AND FARNES, P. (1990). *Women of Science: Righting the Record.* Bloomington, IN: Indiana University Press.

KASTNER, S. AND UNGERLEIDER, L. (2000). Mechanisms of visual attention in the human cortex. *Annu. Rev. Neurosci.* 23:315–341.

KASTNER, S., DEWEERD, P., DESIMONE, R., AND UNGERLEIDER, L.C. (1998). Mechanisms of directed attention in the human extrastriate cortex as revealed by functional MRI. *Science* 282:108–111.

KATZ, B. (1966). *Nerve, Muscle and Synapse.* New York: McGraw-Hill.

KEELE, S. (1986). Motor Control. In J.K. Boff, L. Kaufman, and J.P. Thomas (Eds.), *Handbook of Human Perception and Performance*, Vol. II (pp. 1–60). New York: John Wiley and Sons.

KELLMAN, P.J., AND SPELKE, E.S. (1983). Perception of partly occluded objects in infancy. *Cogn. Psychol.* 15:483–524.

KELLOGG, R.T. (1995). *Cognitive Psychology.* Thousand Oaks, CA: Sage.

KERETSZ, A., AND HOOPER, P. (1982). Praxis and language: The extent and variety of apraxia in aphasia. *Neuropsychologia* 20:275–286.

KESTENBAUM, R., TERMINE, N., AND SPELKE, E. S. (1987). Perception of objects and object boundaries by three-month-old infants. *Brit. J. Dev. Psychol.* 5:367–383.

KIMURA, D. (1973). The asymmetry of the human brain. *Sci. Am.* 228:70–78.

KINGSTONE, A., AND GAZZANIGA, M.S. (1995). Subcortical transfer of higher order information: More illusory than real? *Neuropsychology* 9:321–328.

KINGSTONE, A., AND KLEIN, R.M. (1993). Visual offsets facilitate saccadic latency: Does predisengagement of visuospatial attention mediate this gap effect? *J. Exp. Psychol. Hum. Percept. Perform.* 19:1251–1265.

KINGSTONE, A., ENNS, J., MANGUN, G.R., AND GAZZANIGA, M.S. (1995). Guided visual search is a left hemisphere process in split-brain patients. *Psychol. Sci.*, 6:118–121.

KINGSTONE, A., FRIESEN, C.K., AND GAZZANIGA, M.S. (2000). Reflexive joint attention depends on lateralized cortical connections. *Psychol. Sci.* 11:159–166.

KINSBOURNE, M. (1982). Hemispheric specialization and the growth of human understanding. *Am. Psychol.* 37:411–420.

KITTERLE, F., CHRISTMAN, S., AND HELLIGE, J. (1990). Hemispheric differences are found in identification, but not detection of low versus high spatial frequencies. *Percept. Psychophys.* 48:297–306.

KLATT, D.H. (1989). Review of Selected Models of Speech Perception. In W. Marslen-Wilson (Ed.), *Lexical Representation and Process* (pp. 169–226). Cambridge, MA: MIT Press.

KLEIM, J.A., SWAIN, R.A., ARMSTRONG, K.A., NAPPER, R.M., JONES, T.A., AND GREENOUGH, W.T. (1998). Selective synaptic plasticity within the cerebellar cortex following complex motor skill learning. *Neurobiol. Learn. Mem.* 69:274–289.

KLEINSMITH, L.J., AND KAPLAN, S. (1963). Paired-associate learning as a function of arousal and interpolated interval. *J. Exp. Psychol.* 65:190–193.

KLING, A.S., AND BROTHERS, L.A. (1992). The Amygdala and Social Behavior. In J.P. Aggleton (Ed.), *The Amygdala: Neurobiological Aspects of Emotion, Memory, and Mental Dysfunction* (pp. 353–377). New York: Wiley-Liss.

KLÜVER, H., AND BUCY, P.C. (1939). Preliminary analysis of functions of the temporal lobes in monkeys. *Arch. Neurol. Psychiatry Chicago* 42:979–1000.

KNIGHT, R., SCABINI, D., WOODS, D., AND CLAYWORTH, C. (1989). Contributions of temporal-parietal junction to the human auditory P3. *Brain Res.* 502:109–116.

KNIGHT, R.T., AND GRABOWECKY, M. (1995). Escape from Linear Time: Prefrontal Cortex and Conscious Experience. In M.S. Gazzaniga (Ed.), *The Cognitive Neurosciences* (pp. 1357–1371). Cambridge, MA: MIT Press.

KOHLER, S., KAPUR, S., MOSCOVITCH, M., WINOCUR, G., AND HOULE, S. (1995). Dissociation of pathways for object and spatial vision: A PET study in humans. *Neuroreport* 6:1865–1868.

KOLB, B., AND WHISHAW, I.Q. (1996). *Fundamentals of Human Neuropsychology,* 4th edition. New York: W.H. Freeman and Co.

KONISHI, M. (1993). Listening with two ears. *Sci. Am.* 2681:66–73.

KONISHI, S., NAKAJIMA, K., UCHIDA, I., KAMEYAMA, M., NAKAHARA, K., SEKIHARA, K., AND MIYASHITA, Y. (1998). Transient activation of inferior prefrontal cortex during cognitive set shifting. *Nat. Neurosci.* 1:80–84.

KOOB, G.F., AND LEMOAL, M. (2001). Drug addiction, dysregulation of reward, and allostasis. *Neuropsychopharmacology* 24:97–129.

KORDOWER, J.H., FREEMAN, T.B., SNOW, B.J., VINGERHOETS, F.J., MUFSON, E.J., SANBERG, P.R., HAUSER, R.A., SMITH, D.A., NAUERT, G.M., PERL, D.P., ET AL. (1995). Neuropathological evidence of graft survival and striatal reinnervation after the transplantation of fetal mesencephalic tissue in a patient with Parkinson's disease. *New Engl. J. Med.* 332:1118–1124.

KOSSLYN, S., AND ANDERSEN, R. (Eds.) (1992). *Frontiers in Cognitive Neuroscience.* Cambridge, MA: MIT Press.

KOSSLYN, S.M. (1987). Seeing and imagining in the cerebral hemispheres: A computational approach. *Psychol. Rev.* 94:148–175.

KOSSLYN, S.M. (1988). Aspects of cognitive neuroscience of mental imagery. *Science* 240:1621–1626.

KOSSLYN, S.M., ALPERT, N.M., THOMPSON, W.L., MALJKOVIK, V., WEISE, S.B., CHABRIS, C.F., HAMILTON, S.E., RAUCH, S.L., AND BUONANNO, F.S. (1993). Visual mental imagery activates topographically organized visual cortex: PET investigations. *J. Cogn. Neurosci.* 5:263–287.

KOSSLYN, S.M., KOENIG, O., BARRET, A., CAVE, C.B., TANG, J., AND GABRIELI, J.D.E. (1989). Evidence for two types of spatial representations: Hemispheric specialization for categorical and coordinate relations. *J. Exp. Psychol. Hum. Percept. Perform.* 15:723–735.

KOSSLYN, S.M., PASCUAL-LEONE, A., FELICIAN, O., CAMPOSANO, S., KEENAN, J.P., THOMPSON, W.L., GANIS, G., SUKEL, K.E., AND ALPERT, N.M. (1999). The role of area 17 in visual imagery: Convergent evidence from PET and rTMS. *Science* 284:167–170.

KOSSLYN, S.M., SHIN, L.M., THOMPSON, W.L., MCNALLY, P.J., RAUCH, S.L., PITMAN, R.K., AND ALPERT, N.M. (1996). Neural effects of visualizing and perceiving aversive stimuli: A PET investigation. *Neuroreport* 7:1569–1576.

KOTTER, R., AND MEYER, N. (1992). The limbic system: A review of its empirical foundation. *Behav. Brain Res.* 52:105–127.

KOURTZI, Z., AND KANWISHER, N. (2000). Activation in human MT/MST by static images with implied motion. *J. Cogn. Neurosci.* 12:48–55.

KRACK, P., POLLAK, P., LIMOUSIN, P., HOFFMANN, D., XIE, J., BENAZZOUZ, A., AND BENABID, A.L. (1998). Subthalamic nucleus or internal pallidal stimulation in young onset Parkinson's disease. *Brain* 121:451–457.

KRAUSE, M.A., AND FOUTS, R.S. (1997). Chimpanzee (*Pan troglodytes*) pointing: Hand shapes, accuracy, and the role of eye gaze. *J. Comp. Psychol.* 111:330–336.

KRUBITZER, L. (1998). What can monotremes tell us about brain evolution? *Philos. Trans. R. Soc. Lond. Biol. Sci.* 353:1127–1146.

KRUBITZER, L., MANGER, P., PETTIGREW, J., AND CALFORD, M. (1995). The organization of somatosensory cortex in monotremes: In search of the prototypical plan. *J. Comp. Neurol.* 351:261–306.

KUFFLER, S., AND NICHOLLS, J. (1976). *From Neuron to Brain.* Sunderland, MA: Sinauer Associates.

KUPFERMANN, I. (1991). Localization of Higher Cognitive and Affective Functions: The Association Cortices. In E.R. Kandel, J.H. Schwartz, and T.M. Jessell, (Eds.), *Principles of Neural Science* (pp. 821–838). Norwalk, CT: Appleton and Lange.

KUSHCH, A., GROSS-GLENN, K., JALLAD, B., LUBS, H., RABIN, M., FELDMAN, E., AND DUARA, R. (1993). Temporal lobe surface area measurements on MRI in normal and dyslexic readers. *Neuropsychologia* 31:811–821.

KUTAS, M., AND FEDERMEIER, K.D. (2000). Electrophysiology reveals semantic memory use in language comprehension. *Trends Cogn. Sci.* 4:463–470.

KUTAS, M., AND HILLYARD, S.A. (1980). Reading senseless sentences: Brain potentials reflect semantic incongruity. *Science* 207:203–205.

LABAR, K.S., AND PHELPS, E.A. (1998). Role of the human amygdala in arousal mediated memory consolidation. *Psychol. Sci.* 9:490–493.

LABAR, K.S., LE DOUX, J.E., SPENCER, D.D., AND PHELPS, E.A. (1995). Impaired fear conditioning following unilateral temporal lobectomy in humans. *J. Neurosci.* 15:6846–6855.

LABERGE, D. (1990). Thalamic and cortical mechanisms of attention suggested by recent positron emission tomographic experiments. *J. Cogn. Neurosci.* 2:358–372.

LACHICA, E.A., BECK, P.D., AND CASAGRANDE, V.A. (1993). Intrinsic connections of layer III of striate cortex in squirrel monkey and bush baby: Correlations with patterns of cytochrome oxidase. *J. Comp. Neurol.* 328:163–187.

LAENG, B. (1994). Lateralization of categorical and coordinate spatial functions: A study of unilateral stroke patients. *J. Cogn. Neurosci.* 6:189–203.

LANDAU, B., SMITH, L.B., AND JONES, S. (1992). Syntactic context and the shape bias in children's and adults' lexical learning. *J. Mem. Lang.* 31:807–825.

LANG, P.J., BRADLEY, M.M., AND CUTHBERT, B.N. (1995) *International Affective Picture System (IAPS): Technical Manual and Affective Ratings*. Bethesda, MD: NIMH Center for the Study of Emotion and Attention.

LANGSTON, W.J. (1984). I. MPTP neurotoxicity: An overview and characterization of phases of toxicity. *Life Sci.* 36:201–206.

LARKMAN, A.U., AND JACK, J.B. (1995). Synaptic plasticity: Hippocampal LTP. *Curr. Opin. Neurobiol.* 5:324–334.

LAZARUS, R.S. (1981). A cognitivist's reply to Zajonc on emotion and cognition. *Am. Psychol.* 36:222–223.

LAZARUS, R.S. (1984). On the primacy of cognition. *Am. Psychol.* 39:124–129.

LE DOUX, J.E. (1991). Emotion and the limbic system concept. *Concepts Neurosci.* 2:169–199.

LE DOUX, J.E. (1992). Emotion and the Amygdala. In J.P. Aggleton (Ed.), *The Amygdala: Neurobiological Aspects of Emotion, Memory, and Mental Dysfunction* (pp. 339–351). New York: Wiley-Liss.

LE DOUX, J.E. (1994). Emotion, memory, and the brain. *Sci. Am.* 270:50–57.

LE DOUX, J.E. (1995). In Search of an Emotional System in the Brain: Leaping from Fear to Emotion and Consciousness. In M.S. Gazzaniga (Ed.), *The Cognitive Neurosciences* (pp. 1047–1061). Cambridge, MA: MIT Press.

LE DOUX, J.E. (1996). *The Emotional Brain: The Mysterious Underpinnings of Emotional Life*. New York: Simon and Schuster.

LEE, K.M., CHANG, M.F., AND ROH, J.K. (1999). Subregions within the supplementary motor area activated at different stages of movement preparation and execution. *Neuroimage* 9:117–123.

LETTVIN, J.Y., MATURANA, H.R., MCCULLOCH, W.S., AND PITTS, W.H. (1959). What the frog's eye tells the frog's brain. *Proc. Inst. Radio Engineers* 47:1940–1951.

LEVELT, W.J.M. (1989). *Speaking: From Intention to Articulation*. Cambridge, MA: MIT Press.

LEVELT, W.J.M. (1993). The Architecture of Normal Spoken Language Use. In G. Blanken, J. Dittman, H. Grimm, J.C. Marshall, and C-W. Wallesh (Eds.), *Linguistic Disorders and Pathologies: An International Handbook*. Berlin: Walter de Gruyter.

LEVELT, W.J.M. (1994). The Skill of Speaking. In P. Bertelson, P. Eelen, and G. d'Ydewalle (Eds.), *International Perspectives on Psychological Science*. Vol. 1: *Leading Themes* (pp. 89–103). Hove, England: Lawrence Erlbaum Associates.

LEVELT, W.J.M. (1999). Models of word production. *Trends Cogn. Sci.* 3:223–232.

LEVELT, W.J.M., ROELOFS, A., AND MEYER, A.S. (1999). A theory of lexical access in speech production. *Behav. Brain Sci.* 22:1–75.

LEWIS, D.A., AND LUND, J.S. (1990). Heterogeneity of chandelier neurons in monkey neocortex: Corticotropin-releasing factor and parvalbumin-immunoreactive populations. *J. Comp. Neurol.* 293:599–615.

LEY, R.G., AND BRYDEN, M.P. (1982). A dissociation of right and left hemispheric effects for recognizing emotional tone and verbal content. *Brain Cogn.* 1:3–9.

LHERMITTE, F. (1983). "Utilization behaviour" and its relation to lesions of the frontal lobes. *Brain* 106:237–255.

LHERMITTE, F., PILLON, B., AND SERDARU, M. (1986). Human autonomy and the frontal lobes. Part I: Imitation and utilization behavior: A neuropsychological study of 75 patients. *Ann. Neurol.* 19:326–334.

LIBET, B. (1996). Neural Processes in the Production of Conscious Experience. In M. Velmans (Ed.), *The Science of Consciousness* (pp. 96–117). London: Routledge.

LINDVALL, O., BRUNDIN, P., WIDNER, H., REHNCRONA, S., GUSTAVII, B., FRACKOWIAK, R., LEENDERS, K.L., SAWLE, G., ROTHWELL, J.C., MARSDEN, C.D., AND BJORKLUND, A. (1990). Grafts of fetal dopamine neurons survive and improve motor function in Parkinson's disease. *Science* 247:574–577.

LINDZEY, G. (Ed.). (1936). *History of Psychology in Autobiography*, Vol. III. Worcester, MA: Clark University Press.

LINEBARGER, M., SCHWARTZ, M., AND SAFFRAN, E. (1983). Sensitivity to grammatical structure in so-called agrammatic aphasics. *Cognition* 13:361–392.

LISSAUER, H. (1890). Ein fall von seelenblindheit nebst einem Beitrage zur Theori derselben. *Archiv. Psychiatrie Nerv.* 21:222–270.

LIVINGSTONE, M., AND HUBEL, D. (1988). Segregation of form, color, movement, and depth: Anatomy, physiology, and perception. *Science* 240:740–749.

LIVINGSTONE, M.S., AND HUBEL, D.H. (1984). Anatomy and physiology of a color system in the primate visual cortex. *J. Neurosci.* 4:309–356.

LOFTUS, E., AND GREENE, E. (1980). Warning: Even memory for faces may be contagious. *Law Hum. Behav.* 4:323–334.

LOFTUS, E., MILLER, D., AND BURNS, H. (1978). Semantic integration of verbal information into visual memory. *J. Exp. Psychol. Hum. Learn. Mem.* 4:19–31.

LOFTUS, W.C., TRAMO, M.J., THOMAS, C.E., GREEN, R.L., NORDGREN, R.A., AND GAZZANIGA, M.S. (1993). Three-dimensional quantitative analysis of hemispheric asymmetry in the human superior temporal region. *Cereb. Cortex* 3:348–355.

LONDON, E.D., ERNST, M., GRANT, S., BONSON, K., AND WEINSTEIN, A. (2000). Orbitofrontal cortex and human drug abuse: Functional imaging. *Cereb. Cortex* 10:334–342.

LONDON, E.D., GRANT, S., MORGAN, M.J., AND ZUKIN, S.R. (1996). Neurobiology of Drug Abuse. In B.S. Fogel (Ed.), *Neuropsychiatry: A Comprehensive Textbook* (pp. 635–678). Baltimore, MD. Williams and Wilkins.

LOOMIS, J.M., FUJITA, N., DA SILVA, J.A., AND FUKUSIMA, S.S. (1992). Visual space perception and visually directed action. *J. Exp. Psychol.* 18:906–921.

LUCK, S.J., AND HILLYARD, S.A. (1994). Spatial filtering during visual search: Evidence from human electrophysiology. *J. Exp. Psychol. Hum. Percept. Perform.* 20:1000–1014.

LUCK, S.J., FAN, S., AND HILLYARD, S.A. (1993). Attention-related modulation of sensory-evoked brain activity in a visual search task. *J. Cogn. Neurosci.* 5:188–195.

LUCK, S.J., CHELAZZI, L., HILLYARD, S.A., AND DESIMONE, R. (1997). Neural mechanisms of spatial selective attention in areas V1, V2 and V4 of macaque visual cortex. *J. Neurophysiol.* 77:24–42.

LURIA, A.R. (1968). *The Mind of a Mnemonist: A Little Book About a Vast Memory*. New York: Basic Books.

MACCOBY, E., AND JACKLIN, C. (1974). *The Psychology of Sex Differences*. Stanford, CA: Stanford University Press.

MACDONALD, A.W., COHEN, J.D., STENGER, V.A., AND CARTER, C.S. (2000). Dissociating the role of the dorsolateral prefrontal and anterior cingulate cortex in cognitive control. *Science* 288:1835–1838.

MACKAY, D.G. (1987). *The Organization of Perception and Action: A Theory for Language and Other Cognitive Skills*. New York: Springer.

MACLEAN, P.D. (1949). Psychosomatic disease and the "visceral brain": Recent developments bearing on the Papez theory of emotion. *Psychosom. Med.* 11:338–353.

MacLean, P.D. (1952). Some psychiatric implications of physiological studies on frontotemporal portion of limbic system (visceral brain). *Electroencephalogr. Clin. Neurophysiol.* 4:407–418.

MacLeod, C. (1991). Half a century of research on the Stroop effect: An integrative review. *Psychol. Bull.* 109:163–203.

Magee, J.C. and Cook, E.P. (2000). Somatic EPSP amplitude is independent of synapse location in hippocampal pyramidal neurons. *Nat. Neurosci.* 3:895–903.

Malmo, R. (1942). Interference factors in delayed response in monkeys after removal of frontal lobes. *J. Neurophysiol.* 5:295–308.

Mangun, G.R., and Hillyard, S.A. (1991). Modulations of sensory-evoked brain potentials indicate changes in perceptual processing during visual-spatial priming. *J. Exp. Psychol. Hum. Percept. Perform.* 17:1057–1074.

Mangun, G.R., Hillyard, S., and Luck, S. (1993). Electrocortical Substrates of Visual Selective Attention. In D.E. Meyer and S. Kornblum (Eds.), *Attention and Performance XIV: Synergies in Experimental Psychology, Artificial Intelligence, and Cognitive Neuroscience* (pp. 219–243). Cambridge, MA: MIT Press.

Mangun, G.R., Hopfinger, J., Kussmaul, C., Fletcher E., and Heinze, H.J. (1997). Covariations in PET and ERP measures of spatial selective attention in human extrastriate visual cortex. *Hum. Brain Map.* 5:273–279.

Marcel, A. (1983a). Conscious and unconscious perception: Experiments on visual masking and word recognition. *Cogn. Psychol.* 15:197–237.

Marcel, A. (1983b). Conscious and unconscious perception: An approach to the relations between phenomenal experience and perceptual processes. *Cogn. Psychol.* 15:238–300.

Markowitsch, H.J. (1997). The functional neuroanatomy of episodic memory retrieval. *Trends Neurosci.* 20:557–558.

Markowitsch, H.J., Kalbe, E., Kessler, J., von Stockhausen, H.M., Ghaemi, M., and Heiss, W.D. (1999). Short-term memory deficit after focal parietal damage. *J. Clin. Exp. Neuropsychol.* 21:784–797.

Marler, P. (1991). Song-learning behavior: The interface with neuroethology. *Trends Neurosci.* 14:199–206.

Marr, D. (1982). *Vision: A Computational Investigation into the Human Representation and Processing of Visual Information.* San Francisco: Freeman.

Marr, D., and Nishihara, H.K. (1992). Visual Information Processing: Artificial Intelligence and the Sensorium of Sight. In S.M. Kosslyn and R.A. Andersen (Eds.), *Frontiers in Cognitive Neuroscience* (pp. 165–186). Cambridge, MA: MIT Press.

Marslen-Wilson, W., and Tyler, L.K. (1980). The temporal structure of spoken language understanding. *Cognition* 8:1–71.

Marsolek, C., Kosslyn, S., and Squire, L. (1992). Form-specific visual priming in the right cerebral hemisphere. *J. Exp. Psychol. Learn. Mem. Cogn.* 18:492–508.

Marsolek, C.J. (1995). Abstract visual-form representations in the left cerebral hemisphere. *J. Exp. Psychol. Hum. Percept. Perform.* 21:375–386.

Martin, A., Haxby, J.V., Lalonde, F.M., Wiggs, C.L., and Ungerleider, L.G. (1995). Discrete cortical regions associated with knowledge of color and knowledge of action. *Science* 270:102–105.

Martin, A., Wiggs, C.L., Ungerleider, L.G., and Haxby, J.V. (1996). Neural correlates of category specific behavior. *Nature* 379:649–652.

Mather, M., Henkel, L.A., and Johnson, M.K. (1997). Evaluating characteristics of false memories: Remember/know judgments and memory characteristics questionnaire compared. *Mem. Cogn.* 25:826–837.

Maunsell, J.H.R., and Van Essen, D.C. (1983). Functional properties of neurons in middle temporal visual area of the macaque monkey. I. Selectivity for stimulus direction, speed, and orientation. *J. Neurophysiol.* 49:1127–1147.

Mazoyer, B., Tzourio, N., Frak, V., Syrota, A., Murayama, N., Levier, O., et al. (1993). The cortical representation of speech. *J. Cogn. Neurosci.* 5:467–479.

McAdams, C.J., and Maunsell, J.H.R. (1999). Effects of attention on orientation-tuning functions of single neurons in macaque cortical area V4. *J. Neurosci.* 19:431–441.

McCarthy, G., Blamire, A.M., Puce, A., Nobe, A.C., Bloch, G., Hyder, F., Goldman-Rakic, P., and Shulman, R.G. (1994). Functional magnetic resonance imaging of human prefrontal cortex activation during a spatial working memory task. *Proc. Natl. Acad. Sci. U.S.A.* 91:8690–8694.

McCarthy, G., Puce, A., Gore, J.C., and Allison, T. (1997). Face-specific processing in the human fusiform gyrus. *J. Cogn. Neurosci.* 9:605–610.

McCarthy, R., and Warrington, E.K. (1986). Visual associative agnosia: A clinico-anatomical study of a single case. *J. Neurol. Neurosurg. Psychiatry* 49:1233–1240.

McClelland, J.L. (2000). Connectionist Models of Memory. In E. Tulving and F.I.M. Craik (Eds.), *The Oxford Handbook of Memory* (pp. 583–596). New York: Oxford University Press.

McClelland, J.L., and Rumelhart, D.E. (1981). An interactive activation model of context effects in letter perception: Part 1. An account of the basic findings. *Psychol. Rev.* 88:375–407.

McClelland, J.L., and Rumelhart, D.E. (1986). *Parallel Distributed Processing: Explorations in the Microstructure of Cognition.* Vol. 2: *Psychological and Biological Models.* Cambridge, MA: MIT Press.

McClelland, J., St. John, M., and Taraban, R. (1989). Sentence comprehension: A parallel distributed processing approach. *Lang. Cogn. Processes* 4:287–335.

McEwen, B.S. (1995). Stressful Experience, Brain, and Emotions: Developmental, Genetic, and Hormonal Influences. In M.S. Gazzaniga (Ed.), *The Cognitive Neurosciences* (pp. 1117–1135). Cambridge, MA: MIT Press.

McEwen, B.S., and Sapolsky, R.M. (1995). Stress and cognitive function. *Curr. Opin. Neurobiol.* 5:205–216.

McGaugh, J.L., Introini-Collision, I.B., Cahill, L., Munsoo, K., and Liang, K.C. (1992). Involvement of the Amygdala in Neuromodulatory Influences on Memory Storage. In J.P. Aggleton (Ed.), *The Amygdala: Neurobiological Aspects of Emotion, Memory, and Mental Dysfunction* (pp. 431–451). New York: Wiley-Liss.

McGinn, C. (1991). *The Problem of Consciousness: Essays Towards a Resolution.* Cambridge, MA: Blackwell Science.

McGuire, M.T., Raleigh, M.J., and Brammer, G.L. (1984). Adaptation, selection, and benefit-cost balances: Implications of behavioral-physiological studies of social dominance in male vervet monkeys. Symposium of the IXth International Congress of Primatology: The study of the adaptiveness of aggressive, dominance, and conflict resolution strategies in humans and nonhuman primates (1982, Atlanta, GA). *Ethnol. Sociobiol.* 5:269–277.

McNeil, J.E., and Warrington, E.K. (1993). Prosopagnosia: A face-specific disorder. *Q. J. Exp. Psychol. A* 46:1–10.

Meadows, J.C. (1974). Disturbed perception of colours associated with localized cerebral lesions. *Brain* 97:615–632.

Merzenich, M., and Jenkins, W.M. (1995). Cortical Plasticity, Learning and Learning Dysfunction. In B. Julesz and I. Kovacs (Eds.), *Maturational Windows and Adult Cortical Plasticity* (pp. 1–24). Reading, MA: Addison-Wesley.

MERZENICH, M.M., AND KAAS, J.H. (1980). *Principles of Organization of Sensory-Perceptual Systems of Mammals.* New York: Academic Press.

MERZENICH, M.M., KAAS, J.H., SUR, M., AND LIN, C.S. (1978). Double representation of the body surface within cytoarchitectonic areas 3b and 1 in "SI" in the owl monkey (*Aotus trivirgatus*). *J. Comp. Neurol.* 181:41–73.

MERZENICH, M., RECANZONE, G., JENKINS, W., ALLARD, T., AND NUDO, R. (1988). Cortical Representational Plasticity. In P. Rakic and W. Singer (Eds.), *Neurobiology of Neocortex* (pp. 41–67). New York: John Wiley and Sons.

MESULAM, M.M. (2000). *Principals of Behavioral and Cognitive Neurology.* New York: Oxford University Press.

METCALFE, J., FUNNELL, M., AND GAZZANIGA, M.S. (1995). Right hemisphere superiority: Studies of a split-brain patient. *Psychol. Sci.* 6:157–164.

METTER, E.J. (1995). PET in Aphasia and Language. In H.S. Kirsner (Ed.), *Handbook of Neurological Speech and Language Disorders. Neurological Disease and Therapy.* Vol. 33. New York: Marcel Dekker.

MIDDLETON, F.A., AND STRICK, P.L. (1994). Anatomical evidence for cerebellar and basal ganglia involvement in higher cognitive function. *Science* 266:458–461.

MIDDLETON, F.A., AND STRICK, P.L. (2000). Basal ganglia and cerebellar loops: Motor and cognitive circuits. *Brain Res. Rev.* 31:236–250.

MILBERG, W., AND BLUMSTEIN, S.E. (1981). Lexical decision and aphasia: Evidence for semantic processing. *Brain Lang.* 14:371–385.

MILLER, G. (1951). *Language and Communication.* New York: McGraw-Hill.

MILLER, G. (1956). The magical number seven, plus-or-minus two: Some limits on our capacity for processing information. *Psychol. Rev.* 101:343–352.

MILLER, G. (1962). *Psychology, the Science of Mental Life.* New York: Harper and Row.

MILLER, M.W. (1993). Migration of cortical neurons is altered by gestational exposure to ethanol. *Alcohol. Clin. Exp. Res.* 17:304–314.

MILLER, M.W. (1997). Effects of prenatal exposure to ethanol on callosal projection neurons in rat somatosensory cortex. *Brain Res.* 766:121–128.

MILLS, D.L., COFFEY-CORINA, S.A., AND NEVILLE, H.J. (1993). Language acquisition and cerebral specialization in 20-month-old infants. *J. Cogn. Neurosci.* 5:317–334.

MILLS, D.L., COFFEY-CORINA, S.A., AND NEVILLE, H.J. (1997). Language comprehension and cerebral specialization from 13 to 20 months. *Dev. Neuropsychol.* 13:397–445.

MILNER, A.D. AND GOODALE, M.A. (1995). *The Visual Brain in Action.* New York: Oxford University Press.

MILNER, B. (1995). Aspects of human frontal lobe function. *Adv. Neurol.* 66:67–84.

MILNER, B., CORKIN, S., AND TEUBER, H. (1968). Further analysis of the hippocampal amnesic syndrome: 14-year follow-up study of HM. *Neuropsychologia* 6:215–234.

MILNER, B., CORSI, P., AND LEONARD, G. (1991). Frontal-lobe contributions to recency judgements. *Neuropsychologia* 29:601–618.

MISHKIN, M. (1978). Memory in monkeys severely impaired by combined but not by separate removal of amygdala and hippocampus. *Nature* 273:297–298.

MOLFESE, D.L., WETZEL, W.F., AND GILL, L.A. (1993). Known versus unknown word discriminations in 12-month-old human infants: Electrophysiological correlates. *Dev. Neuropsychol.* 9:241–258.

MORAN, J., AND DESIMONE, R. (1985). Selective attention gates visual processing in extrastriate cortex. *Science* 229:782–784.

MORAY, N. (1959) Attention in dichotic listening: Effective cues and the influence of instructions. *Q. J. Exp. Psychol.* 9:56–60.

MORGAN, M., AND LE DOUX, J.E. (1995). Differential contribution of dorsal and ventral medial prefrontal cortex to the acquisition and extinction of conditioned fear. *Behav. Neurosci.* 109:681–688.

MORGAN, M.A., AND LE DOUX, J.E. (1999). Contribution of ventrolateral prefrontal cortex to the acquisition and extinction of conditioned fear in rats. *Neurobio. Learn. Mem.* 72:244–251.

MORRIS, J.S., FRISTON, K.J., BUCHEL, C., FRITH, C.D., YOUNG, A.W., CALDER, A.J., AND DOLAN, R.J. (1998). A neuromodulatory role for the human amygdala in processing emotional facial expressions. *Brain* 121:47–57.

MORRIS, R.G.M., ANDERSON, E., LYNCH, G., AND BAUDRY, M. (1986). Selective impairment of learning and blockade of long-term potentiation by an *N*-methyl-D-aspartate receptor antagonist, AP5. *Nature* 319:774–776.

MORUZZI, G., AND MAGOUN, H.W. (1949). Brainstem reticular formation and activation of the EEG. *Electroencephalogr. Clin. Neurophysiol.* 1:455–473.

MOSCOVITCH, M., WINOCUR, G., AND BEHRMANN, M. (1997). What is special about face recognition? Nineteen experiments on a person with visual object agnosia and dyslexia but normal face recognition. *J. Cogn. Neurosci.* 9:555–604.

MOTTER, B.C. (1993). Focal attention produces spatially selective processing in visual cortical areas V1, V2 and V4 in the presence of competing stimuli. *J. Neurophysiol.* 70:909–919.

MOUNTCASTLE, V. (Ed.) (1980). *Medical Physiology,* 14th edition. St. Louis: Mosby.

MOUNTCASTLE, V.B. (1976).The world around us: Neural command functions for selective attention. *Neurosci. Res. Prog. Bull.* 14(suppl):1–47.

MUKERJEE, M. (1997). Trends in animal research. *Sci. Am.* 276:86–93.

MÜNTE, T.F., HEINZE, H.-J., AND MANGUN, G.R. (1993). Dissociation of brain activity related to semantic and syntactic aspects of language. *J. Cogn. Neurosci.* 5:335–344.

MÜNTE, T.F., SCHILZ, K., AND KUTAS, M. (1998). When temporal terms belie conceptual order. *Nature* 395:71–73.

MUSHIAKE, H., MASAHIKO, I., AND TANJI, J. (1991). Neuronal activity in the primate premotor, supplementary, and precentral motor cortex during visually guided and internally determined sequential movements. *J. Neurophysiol.* 66:705–718.

MYERS, J.J., AND SPERRY, R.W. (1985). Interhemispheric communication after section of the forebrain commissures. *Cortex* 21:249–260.

MYERS, R.E., SWETT, C., AND MILLER, M. (1973). Loss of social group affinity following prefrontal lesions in free-ranging macaques. *Brain Res.* 64:257–269.

NAKAYAMA, K., AND SILVERMAN, G.H. (1986). Serial and parallel processing of visual feature conjunctions. *Nature* 320:264–265.

NASS, R.D., AND GAZZANIGA, M.S. (1987). Cerebral Lateralization and Specialization in Human Central Nervous System. In V.B. Mountcastle, F. Plum, and S.R. Geiger (Eds.), *Handbook of Physiology,* Vol. 5 (pp. 701–761). Bethesda: American Physiological Society.

NAVON, D. (1977). Forest before trees: The precedence of global features in visual perception. *Cogn. Psychol.* 9:353–383.

NEEDHAM, A. (1999). The role of shape in 4-month-old infants, segregation of adjacent objects. *Infant Behav. Dev.* 22:161–178.

NEEDHAM, A. (2000). Improvements in object exploration skills may facilitate the development of object segregation in early infancy. *J. Cogn. Dev.* 1:131–156.

NEEDHAM, A. (2001). Object recognition and object segregation in 4.5-month-old infants. *J. Exp. Child Psychol.* 78:3–24.

NEEDHAM, A., AND BAILLARGEON, R. (1997). Object segregation in 8-month-old infants. *Cognition* 62:121–149.

NEEDHAM, A., AND BAILLARGEON, R. (1998). Effects of prior experience in 4.5-month-old infants' object segregation. *Infant Behav. Dev.* 21:1–24.

NEISSER, U. (1982). Snapshots or Benchmarks? In U. Neisser (Ed.), *Memory Observed: Remembering in Natural Contexts.* New York: W.H. Freeman.

NEISSER, U., AND HARSCH, N. (1992). Phantom Flashbulbs: False Recollections of Hearing the News about Challenger. In E. Winograd and U. Neisser (Eds.), *Affect and Accuracy in Recall: Studies of "Flashbulb" Memories* (pp. 9–31). New York: Cambridge University Press.

NEISSER, V. (1967). *Cognitive Psychology.* New York: Appleton, Century, Crofts.

NESSE, R.M., AND WILLIAMS, G.C. (1996). *Why We Get Sick: The New Science of Darwinian Medicine.* New York: Vintage Books.

NETTER, F.H. (1983). *The CIBA Collection of Medical Illustrations.* Vol I: *Nervous System, Part 1: Anatomy and Physiology.* Summit, NJ: CIBA Pharmaceutical.

NEVILLE, H. (1995). Developmental Specificity in Neurocognitive Development in Humans. In M.S. Gazzaniga (Ed.), *The Cognitive Neurosciences* (pp. 219–234). Cambridge, MA: MIT Press.

NEWSOME, W.T., AND PARE, E.B. (1988). A selective impairment of motion perception following lesions of the middle temporal visual area (MT). *J. Neurosci.* 8:2201–2211.

NEWSOME, W.T., SHADLEN, M.N., ZOHARY, E., BRITTEN, K.H., AND MOVSHON, J.A. (1995). Visual Motion: Linking Neuronal Activity to Psychophysical Performance. In M.S. Gazzaniga (Ed.), *The Cognitive Neurosciences* (pp. 401–414). Cambridge, MA: MIT Press.

NISBETT, R.E., AND ROSS, L. (1980). *Human Inference: Strategies and Shortcomings of Social Judgment.* Englewood Cliffs, NJ: Prentice-Hall.

NISSEN, M.J., KNOPMAN, D.S., AND SCHACTER, D.L. (1987). Neurochemical dissociation of memory systems. *Neurology* 37:789–794.

NOBRE, A.C., ALLISON, T., AND MCCARTHY, G. (1994). Word recognition in the human inferior temporal lobe. *Nature* 372:260–263.

NORMAN, D., AND SHALLICE, T. (1980). *Attention to Action: Willed and Automatic Control of Behavior.* Center for Human Information Processing Report 99. La Jolla, CA: University of California, San Diego.

NORMAN, D.A., AND SHALLICE, T. (1986). Attention to Action: Willed and Automatic Control of Behavior. In R.J. Davidson, G.E. Schwartz, and D. Shapiro (Eds.), *Consciousness and Self-Regulation,* Vol. 4 (pp. 1–18). New York: Plenum Press.

NORMAN, K.A., AND SCHACTER, D.L. (1997). False recognition in younger and older adults: Exploring the characteristics of illusory memories. *Mem. Cogn.* 25:838–848.

NORRIS, D., MCQUEEN, J., AND CUTLER, A. (1995). Competition and segmentation in spoken word recognition. *J. Exp. Psychol. Learn. Mem. Cogn.* 21:1209–1228.

NOTTEBOHM, F. (1980). Brain pathways for vocal learning in birds: A review of the first 10 years. *Prog. Psychobiol. Physiol. Psychol.* 9:85–124.

NYBERG, L., CABEZA, R., AND TULVING, E. (1996). PET studies of encoding and retrieval: The HERA model. *Psychonomic Bull. Rev.* 3:134–147.

NYBERG, L., CABEZA, R., AND TULVING, E. (1998). Asymmetric frontal activation during episodic memory: What kind of specifity? *Trends Cogn. Sci.* 2:419–420.

NYBERG, L., MCINTOSH, A., CABEZA, R., HABIB, R., HOULE, S., AND TULVING, E. (1996). General and specific brain regions involved in encoding and retrieval of events: What, where, and when. *Proc. Natl. Acad. Sci. U.S.A.* 93:11280–11285.

O'CRAVEN, K.M., DOWNING, P.E., AND KANWISHER, N. (1999). fMRI evidence for objects as the units of attentional selection. *Nature* 401:584–587.

OJEMANN, G.A. (1983). Brain organization for language from the perspective of electrical stimulation mapping. *Behavi. Brain Sci.* 6:189–230.

OJEMANN, G., OJEMANN, J., LETTICH, E., AND BERGER, M. (1989). Cortical language localization in left, dominant hemisphere. *J. Neurosurg.* 71:316–326.

OSGOOD, C.E., SUCI, G.J., AND TANNENGAUM, P.H. (1957). *The Measurement of Meaning.* Urbana, IL: University of Illinois Press.

OSTERHOUT, L., AND HOLCOMB, P.J. (1992). Event-related brain potentials elicited by syntactic anomaly. *J. Mem. Lang.* 31:785–806.

OSTERHOUT, L., AND HOLCOMB, P.J. (1995). Event Related Potentials and Language Comprehension. In M.D. Rugg and M.G.H. Coles (Eds.), *Electrophysiology of mind: Event-related Brain Potentials and Cognition.* (pp. 171–215). Oxford Psychology Series, No. 25. Oxford, UK: Oxford University Press.

OWEN, A.M., ROBERTS, A.C., HODGES, J.R., SUMMERS, B.A., POLKEY, C.E., AND ROBBINS, T.W. (1993). Contrasting mechanisms of impaired attentional set-shifting in patients with frontal lobe damage or Parkinson's disease. *Brain* 116:1159–1175.

OZER, E., SARIOGLU, S., AND GURE, A. (2000). Effects of prenatal ethanol exposure on neuronal migration, neuronogenesis and brain myelination in the mice brain. *Clin. Neuropathol.* 19:21–25.

PAIVIO, A. (1991). Dual coding theory: Retrospect and current status. *Can. J. Psychol.* 45:255–287.

PALLER, K., KUTAS, M., AND MCISAAC, H. (1995). Monitoring conscious recollection via the electrical activity of the brain. *Psychol. Sci.* 6:107–111.

PALLIS, C.A. (1955). Impaired identification of faces and places with agnosia for colors. *J. Neurol. Neurosurg. Psychiatry* 18:218–224.

PALMER, S.E. (1999). *Vision Science: Photons to Phenomenology.* Cambridge, MA: MIT Press.

PAPEZ, J.W. (1937). A proposed mechanism of emotion. *Arch. Neurol. Psychiatry* 79:217–224.

PARR, L.A., DOVE, T., AND HOPKINS, W.D. (1998). Why faces may be special: Evidence of the inversion effect in chimpanzees. *J. Cogn. Neurosci.* 10:615–622.

PASCUAL-LEONE, A., BARTRES-FAZ, D., AND KEENAN, J.P. (1999). Transcranial magnetic stimulation: studying the brain-behaviour relationship by induction of 'virtual lesions.' *Philos. Trans. R. Soc. Lond. Biol. Sci.* 354:1229–1238.

PASCUAL-LEONE, A., GOMEZ-TORTOSA, E., GRAFMAN, J., ALWAY, D.P.N., AND HALLETT, M. (1994). Induction of visual extinction by rapid-rate transcranial magnetic stimulation of parietal lobe. *Neurology* 44:494–498.

PASSINGHAM, R. (1993). *The Frontal Lobes and Voluntary Action.* New York: Oxford University Press.

PASSINGHAM, R.E. (1982). *The Human Primate.* Oxford, UK: W.H. Freeman.

PAULESU, E., FRITH, D.D., AND FRACKOWIAK, R.S.J. (1993). The neural correlates of the verbal component of working memory. *Nature* 362:342–345.

PENFIELD, W., AND JASPER, H. (1954). *Epilepsy and the Functional Anatomy of the Human Brain.* Boston: Little, Brown.

PERANI, D., DEHAENE, S., GRASS, F., COHEN, L., CAPP, S.F., DUPOUX, E., FAZIO, F., AND MEHLER, J. (1996). Brain processes of native and foreign languages. *Neuroreport* 7:2439–2444.

PERETZ, I., KOLINSKY, R., TRAMO, M., LABRECQUE, R., HUBLET, C., DEMEURISSE, G., AND BELLEVILLE, S. (1994). Functional dissociations following bilateral lesions of auditory cortex. *Brain* 117:1283–1301.

PERFETTI, C.A. (1999). Comprehending Written Language: A Blueprint of the reader. In C. Brown and P. Hagoort (Eds.), *The Neurocognition of Language* (pp. 167–208). New York: Oxford University Press.

PERRETT, D.I., ORAM, M.W., HIETANEN, J.K., AND BENSON, P.J. (1994). Issues of Representations in Object Vision. In M.J. Farah and G. Ratcliff (Eds.), *The Neuropsychology of High-Level Vision: Collected Tutorial Essays* (pp. 33–62). Hillsdale, NJ: Lawrence Erlbaum Associates.

PERRETT, S., RUIZ, B., AND MAUK, M. (1993). Cerebellar cortex lesions disrupt learning-dependent timing of conditioned eyelid responses. *J. Neurosci.* 13:1708–1718.

PETERSEN, L.R., AND PETERSEN, M.R. (1959). Short-term retention of individual verbal items. *J. Exp. Psychol.* 58:193–198.

PETERSEN, S.E., AND FIEZ, J.A. (1993). The processing of single words studied with positron emission tomography. *Annu. Rev. Neurosci.* 16:509–530.

PETERSEN, S.E., FIEZ, J.A., AND CORBETTA, M. (1992). Neuroimaging. *Curr. Opin. Neurobiol.* 2:217–222.

PETERSEN, S.E., ROBINSON, D.L., AND MORRIS, J.D. (1987). Contributions of the pulvinar to visual spatial attention. *Neuropsychologia* 25:97–105.

PETERSEN, S.E., FOX, P.T., POSNER, M.I., MINTUN, M., AND RAICHLE, M. (1988). Positron emission tomographic studies of the cortical anatomy of single-word processing. *Nature* 331:585–589.

PETERSEN, S.E., FOX, P.T., SNYDER, A.Z., AND RAICHLE, M.E. (1990). Activation of extrastriate and frontal cortical areas by visual words and word-like stimuli. *Science* 249:1041–1044.

PETERSEN, S.E., VAN MIER, H., FIEZ, J.A., AND RAICHLE, M.E. (1998). The effects of practice on the functional anatomy of task performance. *Proc. Natl. Acad. Sci. U.S.A.* 95:853–860.

PETRIDES, M. (1994). Frontal lobes and behaviour. *Curr. Opin. Neurobiol.* 4:207–211.

PETRIDES, M. (2000). Middorsolateral and Midventrolateral Prefrontal Cortex: Two Levels of Executive Control for the Processing of Mnemonic Information. In S. Monsell and J. Driver (Eds.), *Control of Cognitive Processes. Attention and Performance XVIII* (pp. 535–548). Cambridge, MA: MIT Press.

PHELPS, E.A., AND ANDERSON, A.K. (1997). Emotional memory: What does the amygdala do? *Curr. Biol.* 7: 311–314.

PHELPS, E.A., AND GAZZANIGA, M.S. (1992). Hemispheric differences in mnemonic processing: The effects of left hemisphere interpretation. *Neuropsychologia* 30:293–297.

PHELPS, E.A., LABAR, D.S., ANDERSON, A.K., O'CONNOR, K.J., FULBRIGHT, R.K., AND SPENCER, D.S. (1998). Specifying the contributions of the human amygdala to emotional memory: A case study. *Neurocase* 4:527–540.

PHELPS, E.A., O'CONNOR, K.J., CUNNINGHAM, W.A., FUNAYMA, E.S., GATENBY, J.C., GORE, J.C., AND BANAJI, M.R. (2000). Performance on indirect measures of race evaluation predicts amygdala activity. *J. Cogn. Neurosci.* 12:729–738.

PHELPS, E.A., O'CONNOR, K.J., GATENBY, J.C., GRILLON, C., GORE, J.C., AND DAVIS, M. (2001). Activation of the human amygdala to a cognitive representation of fear. *Nature Neurosci.* 4:437–441.

PINKER, S. (1987). The Bootstrapping Problem in Language Acquisition. In B. MacWhinney (Ed.), *Mechanisms of Language Acquisition* (pp. 399–441). Hillsdale, NJ: Lawrence Erlbaum Associates.

PINKER, S. (1994). *The Language Instinct (pp. 370–403)*. New York: W. Morrow.

PINKER, S. (1995). Facts About Human Language Relevant to Its Evolution. In J. Changeux and J. Chavaillon (Eds.), *Origins of the Human Brain. Symposia of the Fyssen Foundation* (pp. 262–285). Oxford, UK: Clarendon Press/Oxford University Press.

PINKER, S. (1997a). *How the Mind Works.* New York: W.W. Norton.

PINKER, S. (1997b). Interview in M.S. Gazzaniga's *Conversations in Cognitive Neuroscience.* Cambridge, MA: MIT Press.

PISELLA, L, GRÉA, H., TILIKETE, C., VIGHETTO, A., DESMURGET, M., RODE, G., BOISSON, D., AND ROSETTI, Y. (2000). An 'automatic pilot' for the hand in human posterior parietal cortex: Toward reinterpreting optic ataxia. *Nat. Neurosci.* 3:729–736.

PLANT, G.T., LAXER, K.D., BARBARO, N.M., SCHIFFMAN, J.S., AND NAKAYAMA, K. (1993). Impaired visual motion perception in the contralateral hemifield following unilateral posterior cerebral lesions in humans. *Brain* 116:1303–1335.

PLAUT, D.C., MCCLELLAND, J.L., SEIDENBERG, M.S., AND PATTERSON, K. (1996). Understanding normal and impaired word reading: Computational principles in quasi regular domains. *Psychol. Rev.* 103:56–115.

PLOMIN, R., CORLEY, R., DEFRIES, J.C., AND FULKER, D.W. (1990). Individual differences in television viewing in early childhood: Nature as well as nurture. *Psychol. Sci.* 1:371–377.

PODGORNY, P., AND SHEPARD, R. (1978). Functional representations common to visual perception and imagination. *J. Exp. Psychol. Hum. Percept. Perform.* 4:21–35.

POHL, W. (1973). Dissociation of spatial discrimination deficits following frontal and parietal lesions in monkeys. *J. Comp. Physiol. Psychol.* 82:227–239.

POSNER, M. (1994). Attention: The mechanisms of consciousness. *Proc. Natl. Acad. Sci. U.S.A.* 91:7398–7403.

POSNER, M.I. (1986). *Chronometric Explorations of Mind.* New York: Oxford University Press.

POSNER, M.I., AND RAICHLE, M.E. (1994). *Images of Mind.* New York: W.H. Freeman.

POSNER, M.I., AND ROTHBART, M. (1980). The Development of Attentional Mechanisms. In J.H. Flowers (Ed.), *Nebraska Symposium on Motivation* (Volume 28). Lincoln, NE: University of Nebraska Press.

POSNER, M.I., SNYDER, C.R.R., AND DAVIDSON, J. (1980). Attention and the detection of signals. *J. Exp. Psychol. Gen.* 109:160–174.

POSNER, M.I., WALKER, J.A., FRIEDRICH, F.J., AND RAFAL, B.D. (1984). Effects of parietal injury on covert orienting of attention. *J. Neurosci.* 4:1863–1874.

POVEL, D.J., AND COLLARD, R. (1982). Structural factors in patterned finger tapping. *Acta Psychol.* 52:107–123.

POVINELLI, D.J., AND EDDY, T.J. (1996). What small chimpanzees know about seeing. *Monogr. Soc. Res. Child Dev.* Serial No. 247, Vol. 61, No. 3.

POVINELLI, D.J., NELSON, K.E., AND BOYSEN, S.T. (1990). Inferences about guessing and knowing by chimpanzees (*Pan troglodytes*). *J. Comp. Psychol.* 104:203–210.

POVINELLI, D.J., RULF, A.B., AND BIERSCHWALE, D.T. (1994). Absence of knowledge attribution and self-recognition in young chimpanzees. *J. Comp. Psychol.* 108:74–80.

PRABHAKARAN, V., NARAYANAN, K., ZHAO, Z., AND GABRIELI, J.D. (2000). Integration of diverse information in working memory within the frontal lobe. *Nat. Neurosci.* 3:85–90.

PREMACK, D., AND WOODRUFF, G. (1978). Does the chimpanzee have a theory of mind? *Behav. Brain Sci.* 1:515–526.

PREUSS, T.M. (1995). The Argument from Animals to Humans on Cognitive Neuroscience. In M.S. Gazzaniga (Ed.), *The Cognitive Neurosciences* (pp. 1227–1241). Cambridge, MA: MIT Press.

PREVIC, F.H. (1991). A general theory concerning the prenatal origins of cerebral lateralization in humans. *Psychol. Rev.* 98:299–334.

PRICE, C.J. (1998). The functional anatomy of word comprehension and production. *Trends Cogn. Sci.* 2:281–288.

PTITO, A., LEPORE, F., PTITO, M., AND LASSONDE, M. (1991). Target detection and movement discrimination in the blind field of hemispherectomized patients. *Brain* 114:497–512.

PUCE, A., ALLISON, T., ASGARI, M., GORE, J.C., AND MCCARTHY, G. (1996). Differential sensitivity of human visual cortex to faces, letterstrings, and textures: A functional magnetic resonance imaging study. *J. Neurosci.* 16:5205–5215.

PURVES, D., AUGUSTINE, G., AND FITZPATRICK, D. (2001). *Neuroscience*, 2nd edition. Sunderland, MA: Sinaur Associates.

RAFAL, R., AND POSNER, M.I. (1987). Deficits in human visual spatial field following thalamic lesions. *Proc. Natl. Acad. Sci. U.S.A.* 84:7349–7353.

RAFAL, R., SMITH, J., KRANTZ, J., COHEN, A., AND BRENNAN, C. (1990). Extrageniculate vision in hemianopic humans: Saccade inhibition by signals in the blind field. *Science* 250:118–121.

RAICHLE, M.E. (1994). Visualizing the mind. *Sci. Am.* 270:58–64.

RAICHLE, M.E. (1998). Behind the scenes of functional brain imaging: A historical and physiological perspective. *Proc. Natl. Acad. Sci. U.S.A.* 95:765–772.

RAICHLE, M.E., FIEZ, J.A., VIDEEN, T.O., MACLEOD, A.K., PARDO, J.V., FOX, P.T., AND PETERSEN, S.E. (1994). Practice-related changes in human brain functional anatomy during nonmotor learning. *Cereb. Cortex* 4:8–26.

RAINE, A., MELOY, J.R., BIHRLE, S., STODDARD, J., LACASSE, L., AND BUCHSBAUM, M.S. (1998). Reduced prefrontal and increased subcortical brain functioning assessed using positron emission tomography in predatory and affective murderers. *Behav. Sci. Law* 16:319–332.

RAKIC, P. (1995a). Corticogenesis in Human and Nonhuman Primates. In M.S. Gazzaniga (Ed.), *The Cognitive Neurosciences* (pp. 127–146). Cambridge, MA: MIT Press.

RAKIC, P. (1995b). A small step for the cell, a giant leap for mankind: A hypothesis of neocortical expansion during evolution. *Trends Neurosci.* 18:383–388.

RAKIC, P. (1995c). Setting the Stage for Cognition: Genesis of the Primate Cerebral Cortex. In M. Gazzaniga (Ed.), *The New Cognitive Neurosciences*, 2nd edition (pp. 7–22). Cambridge, MA: MIT Press.

RALEIGH, M.J. (1987). Differential behavioral effects of tryptophan and 5-hydroxytryptophan in vervet monkeys: Influence of catecholaminergic systems. *Psychopharmacology* 93:44–50.

RAMACHANDRAN, V.S. (1988). Perceiving shape from shading. *Sci. Am.* 259:76–83.

RAMACHANDRAN V.S. (1993). Behavioral and magnetoencephalographic correlates of plasticity in the adult human brain. *Proc. Natl. Acad. Sci. U.S.A.* 90:10413–10420.

RAMPON, C., TANG, Y.P., GOODHOUSE, J., SHIMIZU, E., KYIN, M., AND TSIEN, J.Z. (2000). Enrichment induces structural changes and recovery from nonspatial memory deficits in CA1 NMDAR1-knockout mice. *Nat. Neurosci.* 3:238–244.

RAMOA A.S., CAMPBELL G., AND SHATZ, C.J. (1988). Dendritic growth and remodeling of cat retinal ganglion cells during fetal and postnatal development. *J. Neurosci.* 8:4239–4261.

RAO, S.C., RAINER, G., AND MILLER, E.K. (1997) Integration of what and where in the primate prefrontal cortex. *Science* 276:821–824.

RAPP, B. (2001). *The Handbook of Cognitive Neuropsychology: What Deficits Reveal About the Human Mind.* Philadelphia, PA: Psychology Press.

RECANZONE, G., SCHREINER, C.E., AND MERZENICH, M. (1993). Plasticity in the frequency representation of primary auditory cortex following discrimination training in adult owl monkeys. *J. Neurosci.* 13:87–103.

REICHER, G.M. (1969). Perceptual recognition as a function of meaningfulness of stimulus material. *J. Exp. Psychol.* 81:275–280.

REMPEL-CLOWER, N., ZOLA, S., SQUIRE, L., AND AMARAL, D. (1996). Three cases of enduring memory impairment after bilateral damage limited to the hippocampal formation. *J. Neurosci.* 16:5233–5255.

RENTSCHLER, I., TREUTWEIN, B., AND LANDIS, T. (1994). Dissociation of local and global processing in visual agnosia. *Vision Res.* 34:963–971.

RIDDOCH, G. (1917). Dissociation of visual perceptions due to occipital injuries, with especial reference to appreciation of movement. *Brain* 40:15–47.

RIDDOCH, M.J. AND HUMPHREYS, G.W. (2001). Object Recognition. In B. Repp (Ed.), *The Handbook of Cognitive Neuropsychology: What Deficits Reveal About the Human Mind* (pp. 45–74). Philadelphia, PA: Psychology Press.

RIDDOCH, M.J., HUMPHREYS, G.W., GANNON, T., BOTT, W., AND JONES, V. (1999). Memories are made of this: The effects of time on stored visual knowledge in a case of visual agnosia. *Brain* 122:537–559.

RINGO, J.L., DOTY, R.W., DEMETER, S., AND SIMARD, P.Y. (1994). Time is of the essence: A conjecture that hemispheric specialization arises from interhemispheric conduction delays. *Cereb. Cortex* 4: 331–343.

RIZZOLATTI, G., FOGASSI, L., AND GALLESE, V. (2000). Cortical Mechanisms Subserving Object Grasping and Action Recognition: A New View on the Cortical Motor Functions. In M. Gazzaniga (Ed.), *The New Cognitive Neurosciences*, 2nd edition (pp. 539–552). Cambridge, MA: MIT Press.

RIZZOLATTI, G., GENTILUCCI, M., FOGASSI, L., LUPPINO, G., MATELLI, M., AND CAMARDA, R. (1988). Functional organization of inferior area 6 in the macaque monkey. *Exp. Brain Res.* 71:465–490.

ROBERTS, T.P.L., POEPPEL, D., AND ROWLEY, H.A. (1998). Magnetoencephalography and magnetic source imaging. *Neuropsychiatry Neuropsych. Behav. Neurol.* 11:49–64.

ROBERTSON, L.C., KNIGHT, R.T., RAFAL, R., AND SHIMAMURA, A.P. (1993). Cognitive neuropsychology is more than single-case studies. *J. Exp. Psychol. Learn. Mem. Cogn.* 19:710–717.

ROBERTSON, L.C., LAMB, M.R., AND KNIGHT, R.T. (1988). Effects of lesions of temporal-parietal junction on perceptual and attentional processing in humans. *J. Neurosci.* 8:3757–3769.

ROBERTSON, L.C., LAMB, M.R., AND ZAIDEL, E. (1993). Interhemispheric relations in processing hierarchical patterns: Evidence from normal and commissurotomized subjects. *Neuropsychology* 7:325–342.

ROBINSON, D.L., AND PETERSEN, S. (1992). The pulvinar and visual salience. *Trends Neurosci.* 15:127–132.

ROBINSON, D.L., GOLDBERG, M.E., AND STANTON, G.B. (1978). Parietal association cortex in the primate: Sensory mechanisms and behavioral modulation. *J. Neurophysiol.* 41:910–932.

ROCK, I. (1995). *Perception.* New York: W.H. Freeman.

ROEDIGER, H.L., AND MCDERMOTT, K.B. (1995). Creating false memories: Remembering words not presented in lists. *J. Exp. Psychol. Learn., Mem., Cogn.* 21:803–814.

ROGERS, R.D., SAHAKIAN, R.A., HODGES, J.R., POLKEY, C.E., KEN-NARD, C., AND ROBBINS, T.W. (1998). Dissociating executive mechanisms of task control following frontal lobe damage and Parkinson's disease. *Brain* 121:815–842.

ROLAND, P.E. (1993). *Brain Activation.* New York: Wiley-Liss.

ROLLS, E.T. (1992). Neurophysiological mechanisms underlying face processing within and beyond the temporal cortical visual areas. *Philos. Trans. R. Soc. Lond. Biol. Sci.* 335:11–20.

ROLLS, E.T. (1999). *The Brain and Emotion.* Oxford, UK: Oxford University Press.

ROLLS, E.T., HORNAK, J., WADE, D., AND MCGRATH, J. (1994). Emotion-related learning in patients with social and emotional changes associated with frontal lobe damage. *J. Neurol., Neurosurg., Psychiatry* 57:1518–1524.

ROSE, J.E., HIND, J.E., ANDERSON, D.J., AND BRUGGE, J.F. (1971). Some effects of stimulus intensity on response of auditory nerve fibers in the squirrel monkey. *J. Neurophysiol.* 24:685–699.

ROSENBAUM, D.A. (1991). *Human Motor Control.* San Diego: Academic Press.

ROSENBAUM, D.A., SLOTTA, J.D., VAUGHAN, J., AND PLAMONDON, R. (1991). Optimal movement selection. *Psychol. Sci.* 2:86–91.

ROSENZWEIG, M.R., LEIMAN, A.L., AND BREEDLOVE, S.M. (1996). *Biological Psychology.* Sunderland, MA: Sinaur Associates.

ROSS, E.D., (1993). Nonverbal aspects of language. *Neurol. Clin.* 11:9–23.

ROTHWELL, J.C., TRAUB, M.M., DAY, B.L., OBESO, J.A., THOMAS, P.K., AND MARSDEN, C.D. (1982). Manual motor performance in a deafferented man. *Brain* 105:515–542.

ROVET, J., AND NETELY, C. (1982). Processing deficits in Turner's syndrome. *Dev. Psychol.* 18:77–94.

ROWAN, A.N., AND ROLLIN, B.E. (1983). Animal research—For and against: A philosophical, social, and historical perspective. *Perspect. Biol. Med.* 27:1–17.

RUMELHART, D.E., MCCLELLAND, J.L., and the PDP Research Group. (1986). *Parallel Distributed Processing: Explorations in the Microstructure of Cognition.* Vol. 1: *Foundations.* Cambridge, MA: MIT Press.

RUSSEL, J.A. (1979). Affective space is bipolar. *J. Pers. Soc. Psychol.* 37:345–356.

SACKS, O.W. (1995). *An Anthropologist on Mars: Seven Paradoxical Tales.* New York: Knopf.

SADATO, N., PASCUAL-LEONE, A., GRAFMAN, J., IBANEZ, V., DEIBER, M-P., DOLD, G., AND HALLETT, M. (1996). Activation of the primary visual cortex by Braille reading in blind subjects. *Nature* 380:526–528.

SAMII, A., TURNBULL, I.M., KISHORE, A., SCHULZER, M., MAK, E., YARDLEY, S., AND CALNE, D.B. (1999). Reassessment of unilateral pallidotomy in Parkinson's disease. A 2-year follow-up study. *Brain* 122:417–425.

SAMS, M., HARI, R., RIF, J., AND KNUUTILA, J. (1993). The human auditory sensory memory trace persists about 10 sec—Neuromagnetic evidence. *J. Cogn. Neurosci.* 5:363–370.

SAMUEL, M., CEBALLOS-BAUMANN, A.O., BLIN, J., UEMA, T., BOECKER, H., PASSINGHAM, R.E., AND BROOKS, D.J. (1997a). Evidence for lateral premotor and parietal overactivity in Parkinson's disease during sequential and bimanual movements. A PET study. *Brain* 120:963–976.

SAMUEL, M., CEBALLOS-BAUMANN, A.O., TURJANSKI, N., BOECKER, H., GOROSPE, A., LINAZASORO, G., HOLMES, A.P., DE LONG, M.R., VITEK, J.L., AND THOMAS, D.G. (1997b). Pallidotomy in Parkinson's disease increases supplementary motor area and prefrontal activation during performance of volitional movements on H^2($_{15}$)O PET Study. *Brain* 120:1301–1313.

SAMUELSON, L.K., AND SMITH, L.B. (1999) Early noun vocabularies: Do ontology, category structure and syntax correspond? *Cognition* 73:1–33.

SAPIR, A., SOROKER, N., BERGER, A., AND HENIK, A. (1999). Inhibition of return in spatial attention: Direct evidence for collicular generation. *Nat. Neurosci.* 2:1053–1054.

SAPOLSKY, R.M. (1992). *Stress, the Aging Brain, and the Mechanisms of Neuron Death.* Cambridge, MA: MIT Press.

SATORI, G., AND JOB, R. (1988). The oyster with four legs: A neuropsychological study on the interaction of visual and semantic information. *Cogn. Neuropsychol.* 5:105–132.

SAUCIER, D., AND CAIN, D.P. (1995). Spatial learning without NMDA receptor-dependent long-term potentiation. *Nature* 378:186–189.

SAYWITZ, K., GOODMAN, G., NICHOLAS, E., AND MOAN, S. (1991). Children's memories of a physical examination involving genital touch: Implications for reports of child sexual abuse. *J. Consult. Clin. Psychol.* 59:682–691.

SCHACTER, D., ALPERT, N., SAVAGE, C., RAUCH, S., et al. (1996). Conscious recollection and the human hippocampal formation—Evidence from positron emission tomography. *Proc. Natl. Acad. Sci. U.S.A.* 93:321–325.

SCHACTER, D., COOPER, L., AND DELANEY, S. (1990). Implicit memory for unfamiliar objects depends on access to structural descriptions. *J. Exp. Psychol. Gen.* 119:5–24.

SCHACTER, D.L. (1987). Implicit memory: History and current status. *J. Exp. Psychol. Learn. Mem. Cogn.* 113:501–518.

SCHACTER, D.L., AND WAGNER, A.D. (1999). Remembrance of things past. *Science* 285:1503–1504.

SCHACTER, S., AND SINGER, J. (1962). Cognitive, social and physiological determinants of emotional state. *Psychol. Rev.* 69:379–399.

SCHILLER, P., AND LOGOTHETIS, N. (1990). The color-opponent and broad-band channels of the primate visual system. *Trends Neurosci.* 13:392–398.

SCHMIDT, R.A. (1987). The Acquisition of Skill: Some Modifications to the Perception-Action Relationship Through Practice. In H. Heuer and A.F. Sanders (Eds.), *Perspectives on Perception and Action* (pp. 77–103). Hillsdale, NJ: Lawrence Erlbaum Associates.

SCHNIEDER, G.E. (1969). Two visual systems. *Science* 163:895–902.

SCHWARTZ, M.F., MARIN, O.S.M., AND SAFFRAN, E.M. (1979). Dissociation of language function in dementia: A case study. *Brain Lang.* 7:277–306.

SCOTT, S.K., YOUNG, A.W., CALDER, A.J., HELLAWELL, D.J., AGGLETON, J.P., AND JOHNSON, M. (1997). Impaired auditory recognition of fear and anger following bilateral amygdala lesions. *Nature* 385:254–257.

SCOVILLE, W.B. (1954). The limbic lobe in man. *J. Neurosurg.* 11:64–66.

SCOVILLE, W.B., AND MILNER, B. (1957). Loss of recent memory after bilateral hippocampal lesions. *J. Neurol. Neurosurg. Psychiatry* 20:11–21.

SEARLE, J. (1992). *The Rediscovery of Mind.* Cambridge, MA: MIT Press.

SEARLE, J.R. (2000). Consciousness. *Annu. Rev. Neurosci.* 23:557–578.

SEIDENBERG, M.S., AND MCCLELLAND, J.L. (1989). A distributed, developmental model of word recognition and naming. *Psychol. Rev.* 96:523–568.

SEJNOWSKI, T.J., AND CHURCHLAND, P.S. (1989). Brain and Cognition. In M.I. Posner (Ed.), *Foundations of Cognitive Science* (pp. 301–356). Cambridge, MA: MIT Press.

SEKULER, R., AND BLAKE, R. (1990). *Perception,* 2nd edition. New York: McGraw-Hill.

SELFRIDGE, O.G. (1959). Pandemonium: A Paradigm for Learning. In *Proceedings of a Symposium on the Mechanisation of Thought Processes* (pp. 511–526). London: H.M. Stationary Office.

SERENO, M., DALE, A., REPPAS, J., KWONG, K., BELLIVEAU, J., BRADY, B., AND TOOTELL, R. (1995). Borders of multiple visual areas in humans revealed by functional MRI. *Science* 268:889–893.

SERGENT, J. (1982). The cerebral balance of power: Confrontation or cooperation: *J. Exp. Psychol. Hum. Percept. Perform.* 8:253–272.

SERGENT, J. (1985). Influence of task and input factors on hemispheric involvement in face processing. *J. Exp. Psychol. Hum. Percept. Perform.* 11:846–861.

SERVOS, P., ENGEL, S.A., GATI, J., AND MENON, R. (1999). fMRI evidence for an inverted face representation in human somatosensory cortex. *Neuroreport* 10:1393–1395.

SHADMEHR, R., AND HOLCOMB, H.H. (1997). Neural correlates of motor memory consolidation. *Science* 277:821–825.

SHALLICE, T. (1988a). Specialization within the semantic system. *Cogn. Neuropsychol.* 5:133–142.

SHALLICE, T. (1988b). *From Neuropsychology to Mental Structure.* Cambridge, UK: Cambridge University Press.

SHALLICE, T., AND BURGESS, W. (1991). Deficits in strategy application following frontal lobe damage in man. *Brain* 114:727–741.

SHALLICE, T., AND WARRINGTON, E. (1969). Independent functioning of verbal memory stores: A neuropsychological study. *Q. J. Exp. Psychol.* 22:261–273.

SHALLICE, T., BURGESS, P.W., SCHON, F., AND BAXTER, D.M. (1989). The origins of utilization behaviour. *Brain* 112:1587–1598.

SHEINBERG, D.L., AND LOGOTHETIS, N.K. (1997). The role of temporal cortical areas in perceptual organization. *Proc. Nat. Acad. Sci. U.S.A.* 94:3408–3413.

SHEPHERD, G.M. (1988). *Neurobiology,* 2nd edition. New York: Oxford University Press.

SHEPHERD, G.M. (1992). *Foundations of the Neuron Doctrine.* New York: Oxford University Press.

SHEPHERD, G.M., WOOLF, T.B., AND CARNEVALE, N.T.L. (1989). Comparisons between active properties and distal dendritic branches and spines: Implications for neuronal computations. *J. Cogn. Neurosci.* 1:273–286.

SHERRINGTON, C. (1947). *The Integrative Action of the Nervous System,* 2nd edition. New Haven: Yale University Press.

SHERRINGTON, C.S. (1935). Santiago Ramón y Cajal 1852–1934. *Obit Not. R. Soc.* 4:425–441.

SHERRY, D.F., JACOBS, L.F., AND GAULIN, S.J. (1992). Spatial memory and adaptive specialization of the hippocampus. *Trends Neurosci.* 15:298–303.

SHIMAMURA, A.P. (1995). Memory and Frontal Lobe Function. In M.S. Gazzaniga (Ed.), *The Cognitive Neurosciences* (pp. 803–813). Cambridge, MA: MIT Press.

SHIMAMURA, A.P. (2000). The role of the prefrontal cortex in dynamic filtering. *Psychobiology* 28:207–218.

SHUREN, J.E., BROTT, T.G., SCHEFFT, B.K., AND HOUSTON, W. (1996). Preserved color imagery in an achromatopsic. *Neuropsychologia* 34:485–489.

SIDTIS, J.J., VOLPE, B.T., HOLTZMAN, J.D., WILSON, D.H., AND GAZZANIGA, M.S. (1981). Cognitive interaction after staged callosal section: Evidence for transfer of semantic activation. *Science* 212(4492):344–346.

SIGNORET, J–L., CASTAIGNE, P., LEHERMITTE, F., ABELANET, R., AND LAVOREL, P. (1984). Rediscovery of Legorgre's brain: Anatomical description with CT scan. *Brain Lang.* 22:303–319.

SKINNER, J.E., AND YINGLING, C.D. (1976). Regulation of slow potential shifts in nucleus reticularis thalami by the mesencephalic reticular formation and the frontal granular cortex. *Electroencephalogr. Clin. Neurophysiol.* 40:288–296.

SLATER, A., MORISON, V., SOMERS, M., MATTOCK, A., BROWN, A., AND TAYLOR, D. (1990). Newborn and older infants' perception of partly occluded objects. *Infant Behav. Dev.* 13:33–49.

SMITH, E.E., AND JONIDES, J. (1994). Working Memory in Humans: Neuropsychological Evidence. In M.S. Gazzaniga (Ed.), *The Cognitive Neurosciences* (pp. 1009–1020). Cambridge, MA: MIT Press.

SMITH, E.E., AND JONIDES, J. (1999). Storage and executive processes in the frontal lobes. *Science* 283:1657–1661.

SNODGRASS, J.G., AND VANDERWART, M. (1980). A standardized set of 260 pictures: Norms for name agreement, image agreement, familiarity, and visual complexity. *J. Exp. Psychol. Hum. Learn. Mem.* 6:174–215.

SNYDER, A.Z., ABDULLAEV, Y.G., POSNER, M.I., AND RAICHLE, M.E. (1995). Scalp electrical potentials reflect regional cerebral blood flow responses during processing of written words. *Proc. Natl. Acad. Sci. U.S.A.* 92:1689–1693.

SOBEL, N., PRABHAKARAN, V., DESMOND, J.E., GLOVER, G.H., GOODE, R.L., SULLIVAN, E.V., AND GABRIELI, J.D. (1998). Sniffing and smelling: Separate subsystems in the human olfactory cortex. *Nature* 392:282–286.

SOBEL, N., PRABHAKARAN, V., ZHAO, Z., DESMOND, J.E., GLOVER, G.H., SULLIVAN, E.V., AND GABRIELI, J.D. (2000). Time course of odorant-induced activation in the human primary olfactory cortex. *J. Neurophys.* 83:537–551.

SPELKE, E., HIRST, W., AND NEISSER, U. (1976). Skills of divided attention. *Cognition* 4:215–230.

SPERLING, G. (1960). The information available in brief visual presentations. *Psychol. Monogr. Gen. Appl.* 74:1–29.

SPERRY, R.W., GAZZANIGA, M.S., AND BOGEN, J.E. (1969). Interhemispheric Relationships: The Neocortical Commissures; Syndromes of Hemisphere Disconnection. In P.J. Vinken and G.W. Bruyn (Eds.), *Handbook of Clinical Neurology,* Vol. 4 (pp. 273–290). Amsterdam: North-Holland Publishing Company; New York: John Wiley and Sons.

SQUIRE, L.R. (1987). *Memory and Brain.* New York: Oxford University Press.

SQUIRE, L.R., AND KNOWLTON, B.J. (1995). Memory, Hippocampus, and Brain Systems. In M.S. Gazzaniga (Ed.), *The Cognitive Neurosciences* (pp. 825–837). Cambridge, MA: MIT Press.

SQUIRE, L.R., AND SLATER, P. (1983). Electroconvulsive therapy and complaints of memory dysfunction: A prospective three-year follow-up study. *Br. J. Psychiatry* 142:1–8.

SQUIRE, L.R. AND ZOLA-MORGAN, S. (1991). The medial temporal lobe memory system. *Science* 253:1380–1386.

SQUIRE, L.R., OJEMANN, J.G., MIEZIN, F.M., PETERSEN, S.E., VIDEEN, T.O., AND RAICHLE, M.E. (1992). Activation of the hippocampus in normal humans: A functional anatomical study of memory. *Proc. Natl. Acad. Sci. U.S.A.* 89:1837–1841.

STERNBERG, S. (1966). High speed scanning in human memory. *Science* 153:652–654.

STERNBERG, S. (1975). Memory scanning: New findings and current controversies. *Q. J. Exp. Psychol.* 27:1–32.

STROOP, J. (1935). Studies of interference in serial verbal reaction. *J. Exp. Psychol.* 18:643–662.

STUSS, D.T., AND BENSON, D.F. (1986). *The Frontal Lobes.* New York: Raven Press.

SWAAB, T.Y., BROWN, C.M., AND HAGOORT, P. (1997). Spoken sentence comprehension in aphasia: Event-related potential evidence for a lexical integration deficit. *J. Cogn. Neurosci.* 9:39–66.

SWANSON, L.W. (1983). The Hippocampus and the Concept of the Limbic System. In W. Seifert (Ed.), *Neurobiology of the Hippocampus* (pp. 3–19). London: Academic Press.

TANAKA, J.W., AND FARAH, M.J. (1993). Parts and wholes in face recognition. *Q. J. Exp. Psychol. Hum. Exp. Psychol.* 46:225–245.

TARR, M.J., AND GAUTHIER, I. (2000). FFA: A flexible fusiform area for subordinate-level visual processing automatized by expertise. *Nat. Neurosci.* 3:764–769.

TARR, M.J., BUELTHOFF, H.H., ZABINSKI, M., AND BLANZ, V. (1997). To what extent do unique parts influence recognition across changes in viewpoint? *Psychol. Sci.* 8:282–289.

TAUB, E., AND BERMAN, A.J. (1968). Movement and Learning in the Absence of Sensory Feedback. In S.J. Freedman (Ed.), *The Neuropsychology of Spatially Oriented Behavior* (pp. 173–191). Homewood, IL: Dorsey.

TEATHER, L.A., PACKARD, M.G., AND BAZAN, N.G. (1998). Effects of posttraining intrahippocampal injections of platelet-activating factor and PAF antagonists on memory. *Neurobiol. Learn. Mem.* 70:349–363.

THACH, W.T. (1975). Timing of activity in cerebellar dentate nucleus and cerebral motor cortex during prompt volitional movements. *Brain Res.* 88:233–241.

THENIUS, E. (1988). From *Grzimek's Encyclopedia of Mammals*, vol. 2. New York: McGraw-Hill.

THOMPSON, C., COWAN, T., AND FRIEMAN, J. (1993). *Memory Search by a Memorist.* Hillsdale, NJ: Lawrence Erlbaum Associates.

THOMPSON-SCHILL, S.L., D'ESPOSITO, M., AGUIRRE, G.K., AND FARAH, M.J. (1997). Role of left inferior prefrontal cortex in retrieval of semantic knowledge: A reevaluation. *Proc. Natl. Acad. Sci. U.S.A.* 94:14792–14797.

THOMPSON-SCHILL, S.L., D'ESPOSITO, M., AND KAN, I.P. (1999). Effects of repetition and competition on activity in left prefrontal cortex during word generation. *Neuron* 23:513–522.

THOMPSON-SCHILL, S.L., SWICK, D., FARAH, M.J., D'ESPOSITO, M., KAN, I.P., AND KNIGHT, R.T. (1998). Verb generation in patients with focal frontal lesions: A nueropsychological test of neuroimaging findings. *Proc. Natl. Acad. Sci. U.S.A.* 95:15855–15860.

THORNDIKE, E. (1911). *Animal Intelligence: An Experimental Study of the Associative Processes in Animals.* New York: MacMillian.

TIFFANY, S.T. (1990). A cognitive model of drug urges and drug-use behavior: Role of automatic and nonautomatic processes. *Psychol. Rev.* 97:147–168.

TINBERGEN, N. (1951). *The Study of Instinct.* London: Oxford University Press.

TOMARKEN, A.J., DAVIDSON, R.J., WHEELER, R.E., AND DOSS, R.C. (1992). Individual differences in anterior brain asymmetry and fundamental dimensions of emotion. *J. Pers. Soc. Psychol.* 62:676–682.

TOOBY, J., AND COSMIDES, L. (1995). Mapping the Evolved Functional Organization of Mind and Brain. In M.S. Gazzaniga (Ed.), *The Cognitive Neurosciences.* Cambridge, MA: MIT Press.

TOOTELL, R.B., HADJIKHANI, N., HALL, E.K., MARRETT, S., VANDUFFEL, W., VAUGHAN, J.T., AND DALE, A.M. (1998). The retinotopy of visual spatial attention. *Neuron* 21:1409–1422.

TOOTELL, R.B., REPPA, J.B., KWONG, K.K., MALACH, R., BORN, R.T., BRADY, T.J., ROSEN, B.R., AND BELLIVEAU, J.W. (1995). Functional analysis of human MT and related visual cortical areas during magnetic resonance imaging. *J. Neurosci.* 15:3215–3230.

TOOTELL, R.B., SILVERMAN, M.S., SWITKES, E., AND DEVALOIS, R.L. (1982). Deoxyglucose analysis of retinotopic organization in primate striate cortex. *Science* 218:902–904.

TREISMAN, A. (1988). Features and objects: The Fourteenth Bartlett Memorial Lecture. *Q. J. Exp. Psychol. A* 40:201–237.

TREISMAN, A. (1991). Search, similarity, and integration of features between and within dimensions. *J. Exp. Psychol. Human Percep. Perform.* 17:652–676.

TREISMAN, A., AND GELADE, G. (1980). A feature-integration theory of attention. *Cogn. Psychol.* 12:97–136.

TREISMAN, A.M. (1969). Strategies and models of selective attention. *Psychol. Rev.* 76:282–299.

TRIVERS, R.L. (1971). The evolution of reciprocal altruism. *Q. Rev. Biol.* 46:35–57.

TULVING, E. (1995). Organization of Memory: Quo Vadis? In M.S. Gazzaniga (Ed.), *The Cognitive Neurosciences* (pp. 839–847). Cambridge, MA: MIT Press.

TULVING, E., AND SHACTER, D.L. (1990). Priming and human memory systems. *Science* 247:301–306.

TULVING, E., GORDON HAYMAN, C.A., AND MACDONALD, C.A. (1991). Long-lasting perceptual priming and semantic learning in amnesia. A case experiment. *J. Exp. Psychol.* 17:595–617.

TULVING, E., KAPUR, S., CRAIK, F.I.M., MOSCOVITCH, M., AND HOULE, S. (1994). Hemispheric encoding/retrieval asymmetry in episodic memory: Positron emission tomography findings. *Proc. Natl. Acad. Sci. U.S.A.* 91:2016–2020.

TVERSKY, A., AND KAHNEMAN, D. (1988). Rational Choice and the Framing of Decisions. In E.D. Bell, H. Raiffa, et al. (Eds.), *Decision Making: Descriptive, Normative, and Prescriptive Interactions.* (pp. 167–192). Cambridge, UK: Cambridge University Press.

UNGERLEIDER, L.G., AND HAXBY, J.V. (1994). "What" and "where" in the human brain. *Curr. Opin. Neurobiol.* 4:157–165.

UNGERLEIDER, L.G., AND MISHKIN, M. (1982). Two Cortical Visual Systems. In D.J. Engle, M.A. Goodale, and R.J. Mansfield (Eds.), *Analysis of Visual Behavior* (pp. 549–586). Cambridge, MA: MIT Press.

UPDIKE, J. (1989). *Self-consciousness: Memoirs.* New York: Knopf.

VALENSTEIN, E.S. (1986). *Great and Desperate Cures: The Rise and Decline of Psychosurgery and Other Radical Treatments for Mental Illness.* New York: Basic Books.

VAN ESSEN, D.C., AND DEYOE, E.A. (1995). Concurrent Processing in the Primate Visual Cortex. In M.S. Gazzaniga (Ed.), *The Cognitive Neurosciences* (pp. 383–400). Cambridge, MA: MIT Press.

VAN TURENNOUT, M., HAGOORT, P., AND BROWN, C.M. (1999). Brain activity during speaking: From syntax to phonology in 40 milliseconds. *Science* 280:572–574.

VOLFOVSKY, N., PARNAS, H., SEGAL, M., AND KORKOTIAN, E. (1999). Geometry of dendritic spines affects calcium dynamics in hippocampal neurons: Theory and experiments. *J. Neurophysiol.* 82:450–462.

VOLPE, B.T., LE DOUX, J.E., AND GAZZANIGA, M.S. (1979). Information processing of visual field stimuli in an "extinguished" field. *Nature* 282:722–724.

VON HELMHOLTZ, H. (1894). Handbuch der Physiologischen Optik. Leipzig: L. Vos., Hamburg, Germany. (See Van der Heijden, A.C.H. (1992). *Selective Attention in Vision* (pp. 32–33). London: Routledge.)

VON MELCHNER, L., PALLAS, S.L., AND SUR, M. (2000). Visual behaviour mediated by retinal projections directed to the auditory pathway. *Nature* 404:871–876.

WAGNER A.D., SCHACTER, D.L., ROTTE, M. KOUTSTAAL, W., MARIL, A., DALE, A.M., ROSEN, B.R., AND BUCKNER, R.L., (1998). Building memories: Remembering and forgetting of verbal experiences as predicted by brain activity. *Science* 281:1188–1191.

WALKER, E.P. (1983). *Walker's Mammals of the World*, 4th edition, vol. 1. Baltimore: Johns Hopkins University Press.

WALSH, V., AND COWEY, A. (2000). Transcranial magnetic stimulation and cognitive neuroscience. *Nat. Rev.*, 1:73–79.

WANG, X.T. (1996). Domain-specific rationality in human choices: Violations of utility axioms and social contexts. *Cognition* 60:31–63.

WAPNER, W., JUDD, T., AND GARDNER, H. (1978). Visual agnosia in an artist. *Cortex* 14:343–364.

WARRINGTON, E., AND SHALLICE, T. (1969). The selective impairment of auditory verbal short-term memory. *Brain* 92:885–896.

WARRINGTON, E.K. (1975). The selective impairment of semantic memory. *Q. J. Exp. Psychol.* 27:635–657.

WARRINGTON, E.K. (1982). Neuropsychological studies of object recognition. *Philos. Transact. R. Soc. Lond., Section B* 298:13–33.

WARRINGTON, E.K. (1985). Agnosia: The Impairment of Object Recognition. In P.J. Vinken, G.W. Bruyn, and H.L. Klawans (Eds.), *Handbook of Clinical Neurology* (pp. 333–349). New York: Elsevier Science.

WARRINGTON, E.K., AND MCCARTHY, R. (1983). Category specific access dysphasia. *Brain* 106:859–878.

WARRINGTON, E.K., AND MCCARTHY, R. (1987). Categories of knowledge: Further fractionation and an attempted integration. *Brain* 110:1273–1296.

WARRINGTON, E.K., AND MCCARTHY, R.A. (1994). Multiple meaning systems in the brain: A case for visual semantics. *Neuropsychologia* 32:1465–1473.

WARRINGTON, E.K., AND SHALLICE, T. (1984). Category specific semantic impairments. *Brain* 107:829–854.

WARRINGTON, E.K., AND TAYLOR, A.M. (1978). Two categorical stages of object recognition. *Perception* 7:695–705.

WARRINGTON, E.K., AND WHITELEY, A.M. (1977). Prosopagnosia: A clinical, psychological, and anatomical study of three patients. *J. Neurol. Neurosurg. Psychiatry* 40:395–403.

WATSON, J.D., AND CRICK, F.H.C. (1953). Genetic implications of the structure of deoxyribonucleic acid. *Nature* 171:964–967.

WAUGH, N.C., AND NORMAN, D.A. (1965). Primary memory. *Psychol. Rev.* 72:89–104.

WEINBERGER, D.R. (1988). Schizophrenia and the frontal lobes. *Trends Neurosci.* 11:367–370.

WEISKRANTZ, L. (1956). Behavioral changes associated with ablation of the amygdaloid complex in monkeys. *J. Comp. Physiol. Psychol.* 49:381–391.

WEISKRANTZ, L. (1986). *Blindsight: A Case Study and Implications.* Oxford, UK: Oxford University Press.

WEISKRANTZ, L., WARRINGTON, E.K., SANDERS M.D., AND MARSHALL, J. (1974). Visual capacity in the hemianopic field following a restricted occipital ablation. *Brain* 97:709–728.

WELLER, W.L. (1993). SmI cortical barrels in an Australian marsupial, *Trichosurus vulpecula* (brush-tailed possum): Structural organization, patterned distribution, and somatotopic relationships. *J. Comp. Neurol.* 337:471–492.

WENKE, R.J. (1980). *Patterns in Prehistory: Mankinds First Three Million Years.* New York: Oxford University Press.

WERTHEIMER, M. (1961). Psycho-motor coordination of auditory-visual space at birth. *Science* 134:1692.

WESSINGER, C.M., BUONOCORE, M.H., KUSSMAUL, C.L., AND MANGUN, G.R. (1997). Tonotopy in human auditory cortex examined with functional magnetic: resonance imaging. *Hum. Brain Map.* 5:18–25.

WESSINGER, C.M., FENDRICH, R., AND GAZZANIGA, M.S. (1997). Islands of residual vision in hemianopic patients. *J. Cogn. Neurosci.* 9:203–221.

WHALEN, P.J. (1998). Fear, vigilance, and ambiguity: Initial neuroimaging studies of the human amygdala. *Curr. Direct. Psychol. Sci.* 7:177–188.

WHALEN, P.J., RAUCH, S.L., ETCOFF, N.L., MCINERNEY, S.C., LEE, M.B., AND JENIKE, M.A. (1998). Masked presentations of emotional facial expressions modulate amygdala activity without explicit knowledge. *J. Neurosci.* 18:411–418.

WICHMANN, T., AND DELONG, M.R. (1996). Functional and pathophysiological models of the basal ganglia. *Curr. Opin. Neurobiol.* 6:751–758.

WIDNER, H., TZTRND, J., REHNERONA, S., SNOW, B.J., BRUMDIN, P., BJORKLURD, A., LINDVALL, O., AND LANGSTON, J. (1993). Fifteen months' follow-up on bilateral embryonic mesencephalic grafts in two cases of severe MPTP-induced parkinsonism. *Adv. Neurol.* 60:729–733.

WIESENDANGER, M., ROUILLER, E.M., KAZENNIKOV, O., AND PERRIG, S. (1996). Is the supplementary motor area a bilaterally organized system? *Adv. Neurol.* 70:85–93.

WILKERSON, I. (August 21, 1987). Apparent lapse on wing flaps shocks experts. *New York Times* 136:A8.

WILLIAMS, G.C. (1966). *Adaptation and Natural Selection.* Princeton, NJ: Princeton University Press.

WILLIAMSON, A., SPENCER, D.D., AND SHEPHERD, G.M. (1993). Comparisons between the membrane and synaptic properties of human and rodent dentate granule cells. *Brain Res.* 622:194–202.

WILSON, E.O. (1975). *Sociobiology, the New Synthesis.* Cambridge, MA: Belknap Press of Harvard University Press.

WILSON, E.O. (1994). *Naturalist.* Washington, DC: Shearwater Books/Island Press.

WILSON, F.A., SCALAIDHE, S.P., AND GOLDMAN-RAKIC, P.S. (1993). Dissociation of object and spatial processing domains in primate prefrontal cortex. *Science* 260:1955–1958.

WILSON, M.A., AND MCNAUGHTON, B.L. (1994). Reactivation of hippocampal ensemble memories during sleep. *Science* 265:676–679.

WILSON, M.A., AND TONEGAWA, S. (1997). Synaptic plasticity, place cells, and spatial memory: Study with second generation knockouts. *Trends Neurosci.* 20:102–106.

WISE, S.P., DI PELLEGRINO, G., AND BOUSSAOUD, D. (1996). The premotor cortex and nonstandard sensorimotor mapping. *Can. J. Physiol. Pharmacol.* 74:469–482.

WITELSON, S.F., KIGAR, D.L., AND HARVEY, T. (1999). The exceptional brain of Albert Einstein. *Lancet* 353:2149–2153.

WITHERS, G.S., GEORGE, J.M., BANKER, G.A., AND CLAYTON, D.F. (1997). Delayed localization of synelfin (synuclein, NACP) to presynaptic terminals in cultured rat hippocampal neurons. *Brain Res. Dev. Brain Res.* 99:87–94.

WOLDORFF, M.G., AND HILLYARD, S.A. (1991). Modulation of early auditory processing during selective listening to rapidly presented tones. *Electroencephalogr. Clin. Neurophysiol.* 79:170–191.

WOLDORFF, M.G., GALLEN, C.C., HAMPSON, S.A., HILLYARD, S.A., PANTEV, C., SOBEL, D., AND BLOOM, F.E. (1993). Modulation of early sensory processing in human auditory cortex during auditory selective attention. *Proc. Natl. Acad. Sci. U.S.A.* 90:8722–8726.

WOLFE, J.M., ALVAREZ, G.A., AND HOROWITZ, T.S. (2000). Attention is fast but volition is slow. *Nature* 406:691.

WOLFORD, G., MILLER, M.B., AND GAZZANIGA, M. (2000). The left hemisphere's role in hypothesis formation. *J. Neurosci.* (Online) 20(6):RC64.

WOODARD, J.S. (1973). *Histologic Neuropathology: A Color Slide Set.* Orange, CA: California Medical Publications.

WOODMAN, G., AND LUCK, S. (1999). Electrophysiological measurement of rapid shifts of attention during visual search. *Nature* 400:867.

WOOLSEY, C.N. (1952). Pattern of Localization in Sensory and Motor Areas of the Cerebral Cortex. In *The Biology of Mental Health*, (pp. 193–225). New York: Hoeber.

WOOLSEY, C.N. (1958). Organization of Somatic Sensory and Motor Areas of the Cerebral Cortex. University of Wisconsin Press.

WOOLSEY, T.A., WELKER, C., AND SCHWARTZ, R.H. (1975). Comparative anatomical studies of the SmI face cortex with special reference to the occurence of "barrels" in layer IV. *J. Comp. Neurol.* 164:79–94.

WURTZ, R.H., GOLDBERG, M.E., AND ROBINSON, D.L. (1982). Brain mechanisms of visual attention. *Sci. Am.* 246:124–135.

YINGLING, C.D., AND SKINNER, J.E. (1976). Selective regulation of thalamic sensory relay nuclei by nucleus reticularis thalami. *Electroencephalogr. Clin. Neurophysiol.* 41:476–482.

YONELINAS, A., KROLL, N., DOBBINS, I., LAZZARA, M., AND KNIGHT, R.T. (1998). Recollection and familiarity deficits in amnesia: Convergence of remember-know, process dissociation, and receiver operating characteristic data. *Neuropsychology* 12:323–339.

YOUDIM, M.B.H., AND RIEDERER, P. (1997). Understanding Parkinson's disease. *Sci. Am.* 276:52–59.

ZAJONC, R.B. (1980). Feeling and thinking: Preferences need no inferences. *Amer. Psychol.* 35:151–175.

ZAJONC, R.B. (1984). On the primacy of affect. *Amer. Psychol.* 39:117–123.

ZANGALADZE, A., EPSTEIN, C.M., GRAFTON, S.T., AND SATHIAN, K. (1999). Involvement of visual cortex in tactile discrimination of orientation. *Nature* 401:587–590.

ZEKI, S. (1993). *A Vision of the Brain.* Oxford, UK: Blackwell Scientific.

ZIHL, J., VON CRAMON, D., AND MAI, N. (1983). Selective disturbance of movement vision after bilateral brain damage. *Brain* 106:313–340.

ZOLA-MORGAN, S., SQUIRE, L.R., CLOWER, R.P., AND REMPEL, N.L. (1993). Damage to the perirhinal cortex exacerbates memory impairment following lesions to the hippocampal formation. *J. Neurosci.* 13:251–265.

ZWITSERLOOD, P., SCHRIEFERS, H., LAHIRI, A., AND DONSELAAR, W. (1993). The role of syllables in the perception of spoken Dutch. *J. Exp. Psychol. Learni. Mem. Cogn.* 19:260–271.

ACKNOWLEDGMENTS AND CREDITS

About the cover Courtesy of Dr. Kevin Wilson, Laboratory for Attention and Cognition, Center for Cognitive Neuroscience, Duke University.

PHOTOS

Chapter 1: **1.1** Corbis-Bettmann. **1.2** General Research Division, New York Public Library, Astor, Lenox and Tilden Foundations. **1.3A** Mary Evans Picture Library. **1.3B** From Luciani, Luigi. *Fisiologia del Homo.* Le Monnier, Firenze, 1901–1911. **1.4** Mary Evans Picture Library/Sigmund Freud Copyrights. **1.5A** New York Academy of Medicine. **1.6A** World Health Organization, Geneva, Switzerland. **1.6B** World Health Organization, Geneva, Switzerland. **1.6C** From *Arch. Anat. Physiol. Wiss. Medezin* (1852):199–216. **1.8B** From Golgi, Camillo, *Untersuchungen über den Flineren Bau des Centralen und Peripherischen Nerven Systems.* Fisher, Jena, 1894. **1.9A** Corbis-Bettmann. **1.9B** From *Histologie du Système Nerveaux de l'Homme et de Vertébrés* by Santiago Ramón y Cajal. Maloine, Paris. 1909–1911. **1.11A** Science Photo Library/Photo Researchers, Inc. **1.12A** Courtesy of Mary Brazier. **1.12B** From Sigmund Freud's *Über den Bau der Nervenfasern und Nervenzellen beim Flusskrebs, Sitz. Akad. Wiss.* 85 (1882): 9–14. Wien, Austria. **1.13A** McHenry, L.C., Jr., *Garrison's History of Neurology.* Springfield, Illinois: Charles C. Thomas, Publisher, 1969, p. 206. © 1969 by Charles C. Thomas, Publisher. Courtesy of Professor O.L. Zangwill, Cambridge. **1.13B** From Budge, Julius, *Lehrbuch der Speciallen Physiologie des Menschen,* vol. 8 (1862). **1.14A** Courtesy National Library of Medicine, Bethesda, Maryland. **1.14B and C** Courtesy of Mary Brazier.

1.15A Underwood & Underwood/Corbis-Bettmann. **1.15B** Buckley, K.W., Mechanical Man: John Broadus *Watson and the Beginnings of Behaviorism.* New York: Guilford Press, 1989, figure 9. Courtesy of Ben Harris. **1.16** UPI/Corbis-Bettmann. **1.17A** Underwood & Underwood/Corbis-Bettmann. **1.18** Courtesy of George A. Miller. **1.19** UPI/Corbis-Bettmann. **p. 10, Fig. A, left** © Copyright Museum Boerhaave, Leiden. From Science Photo Library/Photo Researchers, Inc. **p. 10, Fig A, right** Corbis-Bettmann. **p. 10, Fig B** Courtesy National Library of Medicine, Bethesda, Maryland. **p. 15, left** Reprinted from *The Physiologist* (1981) with permission. **p. 15, right** From *Biological Bulletin* 40 (1921).

Chapter 2: **2.1** Courtesy of Michael Silverman and Gary Banker, Oregon Health Sciences University. **2.5** Withers, G.S., George, J.M., Banker, G.A., and Clayton, D.F., Delayed localization of synelfin (synuclein, NACP) to presynaptic terminals in cultured rat hippocampal neurons, *Brain Research: Developmental Brain Research,* 99 (1997): 87–94. © 1997 with kind permission of Elsevier Science-NL **2.22** Copyright Dennis Kunkel, used with permission.

Chapter 3: **3.1A, B** DeArmond, S.J., Fusco, M.M., and Dewey, M.M., *The Structure of the Human Brain: A Photographic Atlas,* 2nd edition. New York: Oxford University Press, 1976. Copyright © 1976 by Oxford University Press, Inc. Reprinted with permission. **3.2A** Courtesy of Cindy Jordan, University of California, Berkeley. **3.2B, C and 3.3A, B** Courtesy of Carla J. Shatz, University of California, Berkeley. **3.4** Woodard, J.S., *Histologic Neuropathology: A Color Slide Set.* Orange, CA: California Medical Publications, 1973. **3.6** Adapted from McClelland, J.L., and Rummelhart, D.E. (1986). *Parallel Distributed Processing: Explorations in the Microstructure of Cognition. Vol. 2: Psychological and Biological Models.* Cambridge, MA: MIT Press. **3.8B** Courtesy of Allen Song, Duke University. **3.16** Drury, H.A., Van Essen, D.C., Anderson, C.H., Lee, C.W., Coogan, T.A., and Lewis, J.W., Computerized mappings of the cerebral cortex: a multiresolution flattening method and a surface-based coordinate system, *Journal of Cognitive Neuroscience* 8, no. 1 (1996): p. 13. © 1996 by the Massachusetts Institute of Technology. Reprinted by permission of MIT Press. **3.17B** Wessinger, C.M., Buonocore, M.H., Kussmaul, C.L., and Mangun, G.R., Tonotopy in human auditory cortex examined with functional magnetic resonance imaging, *Human Brain Mapping* 5 (1997): 18–25. New York: John Wiley & Sons, Inc., 1997. Reprinted by permission of the authors. **3.21B** Courtesy of Allen Song, Duke University. **3.24** Courtesy of David Amaral. **3.28B** Courtesy of Allen Song, Duke University. **p. 65A,B** Courtesy of John Walker, University of California, San Francisco.

Chapter 4: **4.9A** Reprinted with permission from Tootell, R.B., Silverman, M.S., Switkes, E., and De Valois, R.L., Deoxyglucose analysis of retinotopic organization in primate striate cortex, *Science* 218 (1982): 902–904. Copyright 1982 American Association for the Advancement of Science. **4.10** Rampon, C., Tang, Y.P., Goodhouse, J., Shimuzu, E., Kyin, M., Tsien, J.Z., Enrichment induces structural changes and recovery from nonspatial memory deficits in CA1 NMDAR1-knockout mice, *Nature Neuroscience* 3 (2000): 238–244, used with permission. **4.11B** Fig. 1–1 (A), Greenberg, J.O., and Adams, R.D. (Eds.), *Neuroimaging: A Companion to Adams and Victor's Principles of Neurology.* New York: McGraw-Hill, Inc., 1995. Reprinted by permission of McGraw-Hill, Inc. **4.12B** Fig. 1–1 (B), (C), and (D), Greenberg, J.O., and Adams, R.D. (Eds.), *Neuroimaging: A Companion to Adams and Victor's Principles of Neurology.* New York: McGraw-Hill, Inc., 1995. Reprinted by permission of McGraw-Hill, Inc. **4.13** DeArmond, S.J., Fusco, M.M., and Dewey, M.M., *The Structure of the Human Brain: A Photographic Atlas*, 2nd edition. New York: Oxford University Press, 1976. Copyright © 1976 by Oxford University Press, Inc. Reprinted with permission. **4.14, 4.15, and 4.16A** Woodard, J.S., *Histologic Neuropathology: A Color Slide Set.* Orange, CA: California Medical Publications, 1973. **4.16B** Fig. 8.24 (B) and (C), Greenberg, J.O., and Adams, R.D. (Eds.), *Neuroimaging: A Companion to Adams and Victor's Principles of Neurology.* New York: McGraw-Hill, Inc., 1995. Reprinted by permission of McGraw-Hill, Inc. **4.17A** Woodard, J.S., Histologic Neuropathology: A Color Slide Set. Orange, CA: California Medical Publications, 1973. **4.17B** Holbourn, A.H.S., Mechanics of head injury, *The Lancet* 2: 177–180, © by The Lancet 1943. **4.19** These cartoons appeared in *Life*, 3 March 1947. Used by permission of the estate of Mina Turner. **4.30** Courtesy of Marcus Raichle, M.D., School of Medicine, Washington University in St. Louis. **4.31** Figure 4, Fox et al., "Retinotopic Organization of Human Visual Cortex Mapped with Positron-emission Tomography," *The Journal of Neuroscience* 7 (3): 918, (1987). Reprinted with

permission of The Society for Neuroscience. **4.34** Wagner, A.D., Schacter, D.L., Rotte, M. Koutstaal, W., Maril, A., Dale, A.M., Rosen, B.R., and Buckner, R.L., Building memories: Remembering and forgetting of verbal experiences as predicted by brain activity, *Science* 281 (1998): 1188–1191. **4.35A** Deibert, E., Kraut, M., Kremen, S., and Hart, J., Jr., Neural pathways in tactile object recognition, *Neurology* 52 (1999): 1413–1417. **p. 109** Courtesy of Foundation for Biomedical Research, Washington, D.C.

Chapter 5: **5.1A** Claude Monet, Le Déjeuner sur l'Herbe, detail. Paris, Musée d'Orsay. Photo: Giraudon/Art Resource, New York. **5.1B** Pablo Picasso, Weeping Woman, 1937. Copyright © 1998 Estate of Pablo Picasso/Artists Rights Society (ARS), New York. **5.6 and 5.9** Woodard, J.S., *Histologic Neuropathology: A Color Slide Set.* Orange, CA: California Medical Publications, 1973. **5.25** Gallant, J.L., Shoup, R.E., and Mazer, J.A., A human extrastriate area functionally homologous to macaque V4, *Neuron* 27 (2000): 227–235. © 2000, reprinted with permission from Elsevier Science. **5.34** Wapner, W., Judd, T., and Gardner, H., Visual agnosia in an artist, *Cortex* 14 (1978): 343–364. Copyright 1978 by the Assoicazione per lo Sviluppo delle Recerche Neuropsicologiche. **5.35** Wessinger, C.M., Buonocore, M.H., Kussmaul, C.L., and Mangun, G.R., Tonotopy in human auditory cortex examined with functional magnetic: resonance imaging, *Human Brain Mapping* 5 (1997): 18–25. **p. 179** From Sadato, N., Pascual-Leone, A., Grafman, Jordan, Ibañez, V., Deiber, M., Doid, G. and Hallett, M., Activation of the primary visual cortex by Braille reading in blind subjects, *Nature* 380, 11 April, 1996. **p. 187A** Courtesy of N. Sobel. **p. 187B** Sobel, N., Prabhakaran, V., Desmond, J.E., Glover, G.H., Goode, R.L., Sullivan, E.V., and Gabrieli, J.D. (1998). Sniffing and smelling: Separate subsystems in the human olfactory cortex. *Nature* 392:282–286.

Chapter 6: **6.1** Corbis. **6.9B** Kanwisher, N., Woods, R., Iacobonie, M., and Mazziotta, J.C., A locus in human extrastriate cortex for visual shape analysis, *Journal of Cognitive Neuroscience* 9 (1997): 133–142. © 1997 by the Massachusetts Institute of Technology, reprinted with permission. **6.17** Sekuler, R., and Blake, R., *Perception,* 2nd edition. New York: McGraw-Hill, Inc., 1990. © 1990 by McGraw-Hill, Inc. Reprinted by permission of McGraw-Hill, Inc. **6.25** McCarthy, G., and Warrington, E.K., Visual associative agnosia: A Clinico-anatomical study of a single case, *Journal of Neurology, Neurosurgery and Psychiatry* 49 (1986): 1233–1240. **6.30** Wapner, W., Judd, T., and Gardner, H., Visual agnosia in an artist, *Cortex* 14 (1978): 343–364. Copyright 1978 by the Assoicazione per lo Sviluppo delle Recerche Neuropsicologiche. **6.36** Reprinted from *Brain Research* 342, Baylis, G.C., Rolls, E.T., and Leonard, C.M., Selectivity between faces in the responses of a population of neurons in the cortex in the superior temporal sulcus of the monkey, pp. 91–102, © 1985 with kind permission of Elsevier Science-NL, Sara Burgerhartstr 25, 1055 KV Amsterdam, The Netherlands. **6.37** McCarthy, G., Puce, A., Gore, J.C., and Allison, T., Face-specific processing in the human fusiform gyrus, *Journal of Cognitive Neuroscience* 9 (1997): 605-610. © 1997 by the Massachusetts Institute of Technology, used with permission. **6.38** Moscovitch, M., Winocur, G., and Behrmann, M., What is special about face recognition? Nineteen experiments on a person with visual object agnosia and dyslexia but normal face recognition, *Journal of Cognitive Neuroscience* 9 (1997): 555–604. © 1997 by the Massachusetts Institute of Technology, used with permission. **6.39** McNeil, J.E., and Warrington, E.K., "Prosopagnosia: A Face Specific Disorder" in *The Quarterly Journal of Experimental Psychology* 46A (1993): 1–10. Copyright 1993. Reprinted by permission of The Experimental Psychology Society. **6.40** Block, J.R., and Yuker, H.E., *Can You Believe Your Eyes: Over 250 Illusions and Other Visual Oddities.* Mattituck, New York: Amereon Press, 1992. **6.44** Behrmann, M., Moscovitch, M., and Winocur, G., Intact visual imagery and impaired visual perception in a patient with visual agnosia, *Journal of Experimental Psychology: Human Perception and Performance* 20 (1994): 1068–1087. Copyright © 1994 by the American Psychological Association. Reprinted with permission. **6.45** Sacks, O.W., *An Anthropologist on Mars: Seven Paradoxical Tales.* New York: Knopf, 1995. Reprinted with permission. **6.46** Riddoch, M.J., Humphreys, G.W., Gannon, T., Bott, W., and Jones, V., Memories are made of this: The effects of time on stored visual knowledge in a case of visual agnosia, *Brain* 122 (1999): 537–559. Reprinted with permission.

Chapter 7: **7.2** Courtesy National Library of Medicine, Bethesda, Maryland. **7.3 left** Corbis-Bettmann. **7.23** Adapted from Mangun, G.R. Hopfinger, J., Kussmaul, C., Fletcher E., and Heinze, H.J., Covariations in PET and ERP measures of spatial selective attention in human extrastriate visual cortex, *Human Brain Mapping* 5 (1997): 273–279. **7.24** Tootell, R.B., Hadjikhani, N., Hall, E.K., Marrett, S., Vanduffel, W., Vaughan, J.T., and Dale, A.M., The retinocopy of visual spatial attention, *Neuron* 21 (1998): 1409–1422. **7.25** O'Craven, K.M., Downing, P.E., and Kanwisher, N., fMRI evidence for objects as the units of attentional selection, *Nature* 401 (1999): 584–587. **7.26** Kastner, S., DeWeerd, P., Desimone, R., and Ungerleider, L.C., Mechanisms of directed attention in the human extrastriate cortex as revealed by functional MRI, *Science* 282 (1998): 108–111. **7.29** Hopfinger, J.B., Buonocore, M.H., and Mangun, G.R., The neural mechanism of top-down attentional control, *Nature Neuroscience* 3 (2000): 284–291. **7.30** Hopfinger, J.B., Buonocore, M.H., and Mangun, G.R., The neural mechanism of top-down attentional control, *Nature Neuroscience* 3 (2000): 284–291. **7.31** Hopfinger, J.B., Buonocore, M.H., and Mangun, G.R., The neural mechanism of top-down attentional control, *Nature Neuroscience* 3 (2000): 284–291. **7.41** Reproduced from *Scientific American* 246 (1982), p. 134. © 1998 Artists Rights Society (ARS), New York/VG Bild-Kunst, Bonn.

Chapter 8: **8.6** Markowitsch, H.J., Kalbe, E., Kessler, J., Von Stockhausen, H.M., Ghaemi, M., and Heiss, W.D., Short-term memory deficit after focal parietal damage, *Journal of Clinical & Experimental Neuropsychology* 21 (1999): 784–797. **8.12** Fig. 6.1, Corkin et al., "H.M.'s medial temporal lobe lesion: Findings from magnetic resonance imaging," *The Journal of Neuroscience* 17: 3964–3979, (1997). Adapted with permission of The Society for Neuroscience. **8.14A, B, and C** Courtesy of David Amaral. **8.22** M.C. Escher, "Waterfall," © 1998 Cordon Art B.V. -Baarn-Holland. All rights reserved. **8.31** Haxby, J., Ungerleider, L., Horowitzm B., Maisog, J., Ropoport, S., and Grady, C., Face encoding and recognition in the human brain, *Proc. Natl. Acad. Sci. U.S.A.* 93 (1996): 922–927. **8.33** Wagner, A.D., Schacter, D.L., Rotte, M. Koutstaal, W., Maril, A., Dale, A.M., Rosen, B.R., and Buckner, R.L., Building memories: Remembering and forgetting of verbal experiences as predicted by brain activity, *Science* 281 (1998):1188–1191. **8.34** Brewer, J., Zhao, Z., Desmond, J.E., Glover, G.H., and Gabrielli J.D.E., Making memories: Brain activity that predicts how well visual experience will be remembered, *Science* 281 (1998): 1185–1187. **8.36** Adapted from Schacter, D., Alpert, N., Savage, C., Rauch, S., et al., Conscious recollection and the human hippocampal formation—Evidence from positron emission tomography, *Proc. Natl. Acad. Sci. U.S.A.* 93 (1996): 321–325. Courtesy of Dr. Daniel Schacter. **p. 341** McClelland, J.L., Connectionist Models of Memory. In: Tulving, E. and Craik, F.I.M. (Eds.), *The Oxford Handbook of Memory*, pp. 583–596. Oxford University Press: New York, 2000.

Chapter 9: **9.10** Binder, J.R., Frost, J.A., Hammeke, T.A., Bellgowan, P.S.F., Springer, J.A., Kaufman, J.N., and Possing, E.T., Human temportal lobe activation by speech and non-speech sounds, *Cerebral Cortex* 10 (2000): 512–528. **9.13** Puce, A., Allison, T., Asgari, M., Gore, J.C., and McCarthy, G., Differential sensitivity of human visual cortex to faces, letterstrings, and textures: A functional magnetic resonance imaging study, *Journal of Neuroscience* 16 (1996): 5205–5215. **9.15** Binder, J.R., Frost, J.A., Hammeke, T.A., Bellgowan, P.S.F., Springer, J.A., Kaufman, J.N., and Possing, E.T., Human temporal lobe activation by speech and non-speech sounds. *Cerebral Cortex* 10 (2000): 512–528. **9.19** Caplan, D., Alpert, N., Waters, G., and Olivieri, A., Activiation of Broca's area by syntactic processing under conditions of cuncurrent articulation, *Human Brain Mapping* 9 (2000): 65–71. **9.20** Courtesy of Nina Dronkers. **9.24A** Courtesy Musée Depuytren, Paris.

Chapter 10: **10.4** Courtesy of Michael Gazzaniga. **10.5** Courtesy of Michael Gazzaniga. **10.29** Sergent, J., Influence of task and input factors on hemispheric involvement in face processing, *Journal of Experimental Psychology: Human Perception and Performance* 11 (1985): 846–861.

Chapter 11: **11.1 top** From Lewis P. Rowland (Ed.), *Merritt's Textbook of Neurology*, 8th edition. Philadelphia: Lea & Febiger, 1989, p. 661. Copyright © 1989 by Lea & Febiger. **11.1 bottom** From Jon Palfreman and J. William Langston, *The Case of the Frozen Addicts.* New York: Pantheon Books, 1995. Photograph by Russ Lee, © Pantheon Books, 1995. Used by permission. **11.9** From Taub, E., and Berman, A.J., Movement and Learning in the Absence of Sensory Feedback. In S.J. Freedman (Ed.), *The Neuropsychology of Spatially Oriented Behavior.* Homewood, IL: Dorsey, 1968, Figs. 2 and 8. Reprinted with permission. **11.10** Rothwell, J.C., Traub, M.M., Day, B.L., Obeso, J.A., Thomas, P.K., and Marsden, C.D., Manual motor performance in a deafferented man, *Brain* 105 (1982): 515–542. **11.23** Shadmehr, R., and Holcomb, H.H., Neural correlates of motor memory consolidation, *Science* 277 (1997): 821–825. **11.25** Lee, K.M, Chang, M.F., and Roh, J.K., Subregions within the supplementary motor area activated at different stages of movement preperation and execution, *Neuroimage* 9 (1999): 117–123. **11.44** Color plates 2 and 4 from Greenberg, J.O., and Adams, R.D. (Eds.), *Neuroimaging: A Companion to Adams and Victor's Principles of Neurology.* New York: McGraw-Hill, Inc., 1995. Reprinted by permission of McGraw-Hill, Inc.

Chapter 12: **12.5** McCarthy, G., Blamire, A.M., Puce, A., Nobe, A.C., Bloch, G., Hyder, F., Goldman-Rakic, P., and Shulman, R.G., Functional magnetic resonance imaging of human prefrontal cortex activation during a spatial working memory task, *Proc. Natl. Acad. Sci. U.S.A.* 91 (1994): 8690–8694. **12.9** Figure 5, Friedman et al., "Coactivation of Prefrontal Cortex and Inferior Parietal Cortex in Working Memory Tasks Revealed by 2DG Functional Mapping in the Rhesus Monkey," *The Journal of Neuroscience* 14 (5): 2782, (1994). Reprinted with permission of The Society for Neuroscience. **12.17B** Thompson-Schill, S.L., D'Esposito, M., Aquirre, G.K., and Farah, M.J., Role of left interior prefrontal cortex in retrieval of semantic knowledge: A reevaluation, *Proc. Natl. Acad. Sci. U.S.A.* 94 (1997): 14792–14797. **12.29** Botvinick, M., Nystrom, L.E., Fissell, K., Carter, C.S., and Cohen, J.D., Conflict monitoring versus selection-for-action in anterior cingulate cortex, *Nature* 402 (1999): 179–181.

Chapter 13: **13.1** Damasio, A.R., *Descartes' Error: Emotion, Reason, and the Human Brain.* New York: G.P. Putnam, 1994. Courtesy of Hanna Damasio. Reprinted by permission of The Putnam Publishing Group. Copyright © 1994 by Antonio R. Damasio, M.D. **13.2** Ekman, P., Cross-Cultural Studies in Facial Expression, in P. Ekman (Ed.), *Darwin and Facial Expressions: A Century of Research in Review.* New York: Academic Press, 1973. **13.3** Lang, P.J., Bradley, M.M., and Cuthbert, B.N., *International Affective Picture System (IAPS): Technical Manual and Affective Ratings.* Bethesda, MD: NIMH Center for the Study of Emotion and Attention, 1995. **13.7A, B, C, and D** Lhermitte, F., Pillon, B., and Serdaru, M., Human Autonomy and the Frontal Lobes. Part I: Imitation and Utilization Behavior: A Neuropsychological Study of 75 Patients, *Annals of Neurology* 19 (1986): 326–334. **13.7E, F, and G** From Lhermitte, F., Utilization behavior and its relation to lesions of the frontal lobes, *Brain* 106 (1983): 237–255. Oxford Journals, Oxford University Press. **13.8** Courtesy of David Tranel. **13.15** Ekman, P. (1971), Universals and cultural differences in facial expressions of emotions. In J.K. Cole (Ed.), *Nebraska Symposium on Motivation,* 207–283. Lincoln, NE: University of Nebraska

Press. © Paul Ekman, 1972. Reprinted by permission of the author. **13.18** Haerer, A.F. (revised by), DeJong's *The Neurologic Examination,* 5th edition. Philadelphia, Pennsylvania: J. B. Lippincott Company, 1992. Copyright © 1992 by J.B. Lippincott Company. **p. 567, top** Cohen, R.M., Semple, W.E., Gross, M., and Nordhal, T.E., From syndrome to illness: Delineating the pathophysiology of schizophrenia with PET, *Schizophrenia Bulletin* 14 (1988): 169–176. U.S. Public Health Service, National Institute of Mental Health. **p. 567, bottom** From Drevets, W.C., Videen, T.O., Price, J.L., Preskorn, S.H., Carmichael, S.T., and Raichle, M.E., "A functional anatomical study of unipolar depression," *The Journal of Neuroscience* 12: 3628–3641, (1992). Reprinted with permission of The Society for Neuroscience. **p. 569, top** From *Great and Desperate Cures* by Elliot S. Valenstein. Reprinted by permission of the author. **p. 569, bottom** From W. Freeman and J.W. Watts, *Psychosurgery: In the Treatment of Mental Disorders and Intractable Pain,* 2nd ed., Springfield, Illinois: Charles C. Thomas, Publisher, 1950. Courtesy of Charles C. Thomas, Publisher.

Chapter 14: **14.5** Weller, W.L., SmI cortical barrels in an Australian marsupial, *Trichosurus vulpecula* (brush-tailed possum): Structural organization, patterned distribution, and somatotopic relationships, *Journal of Comparative Neurology* 337 (1993): 471–492. **p. 587** Corbis.

Chapter 15: **15.4** UPI/Corbis-Bettmann. **15.9** Photograph © Garvis Kerimian/Peter Arnold, Inc. **15.11B** Photo: Lennart Nilsson/Bonnier Alba AB. **15.19 and 15.20** Eriksson, P.S., Perfilieva, E., Björk-Eriksson, T., Alborn, A., Nordborg, C., Peterson, D., and Cage, F., Neurogenesis in the adult hippocampus, *Nature Medicine* 4 (1998): 1313–1317. **15.30** Karni, A., Meyer, G., Rey-Hipolito, C., Jezzard, P., Adams, M. Turner, R., and Ungerleider, L., The acquisition of skilled motor performance: Fast and slow experience-driven changes in primary motor cortex, *Proc. Natl. Acad. Sci. U.S.A.* 95 (1998): 861–868. **p. 614** Crowley, J., and Katz, L., Development of ocular dominance columns in the absence of retinal input, *Nature Neuroscience* 2 (1999): 1125–1130. **p. 622** Winston Fraser, Fraser Photos. **p. 624** Courtesy of Linda Acredolo.

Chapter 16: **16.10** Petersen, S.E., van Mier, H., Fiez, J.A., and Raichle, M.E., The effects of practice on the functional anatomy of task performance, *Proc. Natl. Acad. Sci. U.S.A.* 95 (1998): 853–860.

FIGURES

Chapter 1: **1.7** Brodmann, K., *Vergleichende Lokalisationslehre der Grosshirnrinde in ihren Prinzipien dargestellt auf Grund des Zellenbaues.* Leipzig: J.A. Barth, 1909.

Chapter 2: **2.3** Carpenter, M., *Human Neuroanatomy,* 7th edition. Baltimore, Maryland: William & Wilkins, 1976. © 1976 by Williams & Wilkins. Adapted by permission of the publisher. **2.6** Adapted from Kandel, E.R. Schwartz, J.H., and Jessell, T.M. (Eds.), *Principles of Neural Science*, 3rd edition. Norwalk, Connecticut: Appleton & Lange, 1991. Copyright © 1991 by Appleton & Lange. **2.8 and 2.14** Netter, F.H., *The CIBA Collection of Medical Illustrations. Vol I: Nervous System, Part 1: Anatomy and Physiology*. Summit, NJ: CIBA Pharmaceutical Company, 1983. Adapted with permission of Novartis, formerly CIBA. **2.16A** Adapted from Kandel, E.R. Schwartz, J.H., and Jessell, T.M. (Eds.), *Principles of Neural Science,* 3rd edition. Norwalk, Connecticut: Appleton & Lange, 1991. Copyright © 1991 by Appleton & Lange. **2.16B** Adapted from Kuffler, S., and Nicholls, J., From Neuron to Brain. Sunderland, MA: Sinauer Associates, 1976. **2.19** Adapted from Kandel, E.R. Schwartz, J.H., and Jessell, T.M. (Eds.), *Principles of Neural Science,* 3rd edition. Norwalk, Connecticut: Appleton & Lange, 1991. Copyright © 1991 by Appleton & Lange. **2.21** Adapted from Kuffler, S., and Nicholls, J., *From Neuron to Brain.* Sunderland, MA: Sinauer Associates, 1976. **2.25 and 2.26** Doyle, D., Cabral, J., Pfuetzner, R., Kuo, A., Gulbis, J., Cohen, S., Chait, B., and MacKinnon, R. (1998), The structure of the potassium channel; Molecular basis of K⁺ conduction and selectivity, *Science* 280 (5360): 69-77. **2.29** Adapted from Kandel, E.R. Schwartz, J.H., and Jessell, T.M. (Eds.), *Principles of Neural Science,* 3rd edition. Norwalk, Connecticut: Appleton & Lange, 1991. Copyright © 1991 by Appleton & Lange. **2.31** Adapted from Purves, D., Augustine, G., and Fitzpatrick, D., *Neuroscience*, 2nd edition. Sunderland, MA: Sinaur Associates, 2001. **2.33** Adapted from Gibson, J., Beierlein, M., and Connors, B. (1999), Two networks of electrically coupled inhibitory neurons in neocortex, *Nature* 402 (6757): 75-79. **2.35** Adapted from Purves, D., Augustine, G., and Fitzpatrick, D., *Neuroscience*, 2nd edition. Sunderland, MA: Sinaur Associates, 2001. **p. 37** Netter, F.H., *The CIBA Collection of Medical Illustrations. Vol I: Nervous System, Part 1: Anatomy and Physiology.* Summit, NJ: CIBA Pharmaceutical Company, 1983. Adapted with permission of Novartis, formerly CIBA. **p. 41** Magee, J.C. and Cook, E.P., Somatic EPSP amplitude is independent of synapse location in hipocampal pyramidal neurons, *Nature Neuroscience* 3 (2000):895–903. **p. 42** From Haeusser, M., The Hodgkin-Huxley theory of the action potential, *Nature Reviews Neuroscience* 3 (2000): 1165.

Chapter 3: **3.8** DeArmond, S.J., Fusco, M.M., and Dewey, M.M., *The Structure of the Human Brain: A Photographic Atlas,* 2nd edition. New York: Oxford University Press, 1976. Copyright © 1976 by Oxford University Press, Inc. Adapted with permission. Finger, S., *Origins of Neuroscience.* Oxford University Press, 1994. Copyright © 1994 by Oxford University Press, Inc. Reprinted with permission. **3.11** Mesulam, M.M. (2000). *Principals of Behavioral and Cognitive Neurology.* New York: Oxford University Press. **3.12C** Bear, M.F., Connors, B.W., and Paradiso, M.A., *Neuroscience: Exploring the Brain.* Baltimore, Maryland: William & Wilkins, 1996. © 1996 by Williams & Wilkins. Adapted by permission of the publisher. **3.20** Adapted from Kandel, E.R., Schwartz, J.H., and Jessell, T.M. (Eds.), *Principles of Neural Science*, 3rd edition. Norwalk, Con-

necticut: Appleton & Lange, 1991. Copyright © 1991 by Appleton & Lange. **3.21A** Carpenter, M., *Human Neuroanatomy*, 7th edition. Baltimore, Maryland: William & Wilkins, 1976. © 1976 by Williams & Wilkins. Adapted by permission of the publisher. **3.22** Adapted from Kandel, E.R., Schwartz, J.H., and Jessell, T.M. (Eds.), *Principles of Neural Science*, 3rd edition. Norwalk, Connecticut: Appleton & Lange, 1991. Copyright © 1991 by Appleton & Lange.**3.26 and 3.27** Netter, F.H., *The CIBA Collection of Medical Illustrations. Vol I: Nervous System, Part 1: Anatomy and Physiology.* Summit, NJ: CIBA Pharmaceutical Company, 1983. Adapted with permission of Novartis, formerly CIBA. **3.29, 3.30, and 3.31** Carpenter, M., *Human Neuroanatomy*, 7th edition. Baltimore, Maryland: William & Wilkins, 1976. © 1976 by Williams & Wilkins. Adapted by permission of the publisher. **p. 65C, top** Kolb, B. and Whishaw, I.Q., *Fundamentals of Human Neuropsychology*, 4th edition. New York: W.H. Freeman and Company, 1996. © 1996 by W.H. Freeman and Company. Adapted by permission of W.H. Freeman and Company. **p. 65C, bottom** Adapted from Ramachandran, V.S., Behavioral and magnetoencephalographic correlates of plasticity in the adult human brain, *Proc. Natl. Acad. Sci. U.S.A.* 90 (1993): 10413–10420.

Chapter 4: **4.23** Corthout, E., Uttl, B., Ziemann, U., Cowey, A., Hallett, M., Two periods of processing in the circum(striate) visual cortex as revealed by transcranial magnetic stimulation, *Neuropsychologia* 37 (1999): 137–145. **4.29** Roberts, T.P.L., Poeppel, D., and Rowley, H.A., Magnetoencephalography and magnetic source imaging, *Neuropsychiatry, Neuropsychology and Behavioral Neurology* 11 (1998): 49–64. **4.30** Posner, M.I., and Raichle, M.E., *Images of Mind.* New York: W.H. Freeman and Company, 1994. © 1994 by W.H. Freeman and Company. Adapted by permission of the publisher. **4.32** Duong, T.Q., Kim, D.S., Ugurbil, K., and Kim, S.G., Spatiotemporal dynamics of the BOLD fMRI signals: Toward mapping submillimeter cortical columns using the early negative response, *Magnetic Resonance in Medicine* 44 (2000): 231–242. **4.34** Wagner, A.D., Schacter, D.L., Rotte, M. Koutstaal, W., Maril, A., Dale, A.M., Rosen, B.R., and Buckner, R.L., Building memories: Remembering and forgetting of verbal experiences as predicted by brain activity, *Science* 281 (1998): 1188–1191. **4.35B** Zangaladze, A., Epstein, C.M., Grafton, S.T., and Sathian, K., Involvement of visual cortex in tactile discrimination of orientation, *Nature* 401 (1999): 587–590.

Chapter 5: **5.2** Sekuler, R., and Blake, R., *Perception*, 2nd edition. New York: McGraw-Hill, Inc., 1990. © 1990 by McGraw-Hill, Inc. Adapted by permission of McGraw-Hill, Inc. **5.11** Zeki, S., *A Vision of the Brain.* Oxford, UK: Blackwell Scientific, 1993. **5.12** Adapted from Maunsell, J.H.R., and Van Essen, D.C., Functional properties of neurons in middle temporal visual area of the macaque monkey. I. Selectivity for stimulus direction, speed, and orientation, *Journal of Neurophysiology* 49 (1983): 1127–1147. **5.14 and 5.15** Zeki, S., A Vision of the Brain. Oxford, UK: Blackwell Scientific, 1993. **5.17 and 5.18** Hadjikhani, N., Liu, A.K., Dale, A.M., Cavanagh, P., and Tootell, R.B., Retinotopy and color sensitivity in human visual cortical area V8, *Nature Neuroscience* 1 (1998): 235–241. **5.20A** Adapted from Enns, J.T., and Rensink, R.A., Sensitivity to three-dimensional orientation in visual search, *Psychological Science* 1 (1990): 323–326. **5.24** Adapted from Heywood, C.A., Wilson, B., and Cowey, A., A case study of cortical colour blindness with relatively intact achromatic discrimination, *Journal of Neurology, Neurosurgery and Psychiatry* 50 (1987): 22–29. **5.26** Adapted from Gallant, J.L., Shoup, R.E., and Mazer, J.A., A human extrastriate area functionally homologous to macaque V4, *Neuron* 27 (2000): 227–235. **5.28** Beckers, G., and Zeki, S., The consequences of inactivating areas V1 and V5 on visual motion perception, *Brain* 118 (1995): 49–60. **p. 157C, top** Adapted from Hubel, D., and Wiesel, T., Receptive fields and functional architecture of monkey striate cortex, *Journal of Physiology* 195 (1968): 215–243. Published by the Cambridge University Press for the Physiological Society, London. **p. 157C, bottom** Bear, M.F., Connors, B.W., and Paradiso, M.A., *Neuroscience: Exploring the Brain.* Baltimore, Maryland: William & Wilkins, 1996. © 1996 by Williams & Wilkins. Adapted by permission of the publisher.

Chapter 6: **6.4** Pohl, W., Dissociation of spatial discrimination deficits following frontal and parietal lesions in monkeys, *Journal of Comparative and Physiological Psychology* 82 (1973): 227–239. Copyright © 1973 by the American Psychological Association. Adapted with permission. **6.6** Adapted from Robinson, D.L., and Petersen, S., The pulvinar and visual salience, *Trends in Neuroscience* 15 (1992): 127–132. **6.8** Kohler, S., Kapur, S., Moscovitch, M., Winocur, G., and Houle, S., Dissociation of pathways for object and spatial vision: A PET study in humans, *Neuroreport* 6 (1995): 1865–1868. **6.23** Adapted from Warrington, E.K., Agnosia: The Impairment of Object Recognition, in Vinken, P.J., Bruyn, G.W., and Klawans, H.L. (Eds.), *Handbook of Clinical Neurology.* New York: Elsevier Science, 1985, pp. 333–349. **6.24** Adapted from Warrington, E.K., Neuropsychological studies of object recognition, *Philosophical Transactions of the Royal Society, London, Section B* 298 (1982): 13–33. **6.26, top** Adapted from Warrington, E.K., Neuropsychological studies of object recognition, *Philosophical Transactions of the Royal Society, London, Section B* 298 (1982): 13–33. **6.26, bottom** Warrington, E.K., and Taylor, A.M., Two categorical stages of object recognition, *Perception* 7 (1978): 695–705. **6.28** Adapted from Humphreys, G.W., Riddoch, M.J., Donnelly, N., Freeman, T., Boucart, M., and Muller, H.M., Intermediate Visual Processing and Visual Agnosia, in Farah, M.J., and Ratcliff, G. (Eds.), *The Neuropsychology of High-Level Vision: Collected Tutorial Essays.* Hillsdale, NJ: Lawrence Erlbaum Associates, 1994, pp. 63–102. **6.29** Behrmann, M., Moscovitch, M., and Winocur, G., Intact visual imagery and impaired visual perception in a patient with visual agnosia, *Journal of Experimental Psychology: Human Perception and Performance* 20 (1994): 1068–1087. Copyright © 1994 by the American Psychological Association. Adapted with permission. **6.31** Adapted from Warrington, E.K., and McCarthy, R.A., Multiple meaning systems in the brain: A case

for visual semantics, *Neuropsychologia* 32 (1994): 1465–1473. **6.35** Farah, M.J., and McClelland, J.L., A computational model of semantic memory impairment: Modality specificity and emergent category specificity, *Journal of Experimental Psychology: General* 120 (1991): 339–357. Copyright © 1991 by the American Psychological Association. Adapted with permission. **6.36** Adapted from Brain Research 342, Baylis, G.C., Rolls, E.T., and Leonard, C.M., Selectivity between faces in the responses of a population of neurons in the cortex in the superior temporal sulcus of the monkey, pp. 91–102, © 1985 with kind permission of Elsevier Science-NL, Sara Burgerhartstr 25, 1055 KV Amsterdam, The Netherlands. **6.41** Adapted from Farah, M.J., Specialization Within Visual object Recognition: Clues from Prosopagnosia and Alexia, in Farah, M.J., and Ratcliff, G. (Eds.), *The Neuropsychology of High-Level Vision: Collected Tutorial Essays.* Hillsdale, NJ: Lawrence Erlbaum Associates, 1994, pp. 133–146. **6.43** Podgorny, P., and Shepard, R., Functional representations common to visual perception and imagination, *Journal of Experimental Psychology: Human Perception and Performance* 4 (1978): 21–35. Copyright © 1978 by the American Psychological Association. Adapted with permission.

Chapter 7: **7.6** Adapted from fig. 2, Broadbent, D.A., *Perception and Communication.* New York: Pergamon, 1958. **7.9** Posner, M.I., Snyder, C.R.R., and Davidson, J., Attention and the detection of signals, *Journal of Experimental Psychology: General* 109 (1980): 160–174. Copyright © 1980 by the American Psychological Association. Adapted with permission. **7.11** Wolfe, J.M., Alvarez, G.A., and Horowitz, T.S., Attention is fast but volition is slow. *Nature* 406 (2000): 691. **7.13** Adapted with permission from Hillyard, S.A., Hink, R.F., Schwent, V.L., and Picton, T.W., Electrical signs of selective attention in the human brain, *Science* 182 (1973): 177–180. Copyright 1973 American Association for the Advancement of Science. **7.14** Adapted from Woldorff, M.G., Gallen, C.C., Hampson, S.A., Hillyard, S.A., Pantev, C., Sobel, D., and Bloom, F.E., Modulation of early sensory processing in human auditory cortex during auditory selective attention, *Proc. Natl. Acad. Sci. U.S.A.* 90 (1993): 8722–8726. **7.19** Adapted from Hopfinger, J., and Mangun, G.R., Reflexive attention modulates visual processing in human extrastriate cortex. *Psychological Science* 9 (1998): 441–447. **7.21** Woodman, G., and Luck, S., Electrophysiological measurements of rapid shifts of attention during visual search, *Nature* 400 (1999): 867. **7.22** Figs. 4, 5, and 6, Corbetta et al., Selective and divided attention during visual discriminations of shape, color and speed: Functional anatomy by positron emission tomography, *The Journal of Neuroscience* 11: 2383–2402, (1991) Adapted with permission of The Society for Neuroscience. Adapted from Heinze, H.J., Mangun, G.R., Burchert, W., Hinrichs, H., Scholz, M., Münte, T.G., Gös, A., Scherg, M., Johannes, S., Hundeshagen, H., Gazzaniga, M.S. and Hillyard, S.A., Combined spatial and temporal imaging of brain activity during visual selective attention in humans, *Nature* 372 (1994): 543–546. **7.27** Figs. 1 and 4, Corbetta et al., "A PET study of visuospatial attention," *The*

Journal of Neuroscience 13: 1202–1226, (1993). Adapted with permission of The Society for Neuroscience. **7.32** Adapted with permission from Moran, J., and Desimone, R., Selective attention gates visual processing in extrastriate cortex, *Science* 229 (1985): 782–784. Copyright 1985 American Association for the Advancement of Science. **7.34** Adapted from Wurtz, R.H., Goldberg, M.E., and Robinson, D.L., Brain mechanisms of visual attention, *Scientific American* 246 (1982): 124–135. Copyright © 1982 by Scientific American, Inc. All rights reserved. **7.35** Adapted from Desimone, R., Wessinger, M., Thomas, L., and Schneider, W., Attentional control of visual perception: Cortical and subcortical mechanisms, Cold Spring Harbor Laboratory. *Symposia on Quantitative Biology* 55 (1990): 963–971. **7.37** Adapted from Robinson, D.L., and Petersen, S., The pulvinar and visual salience, *Trends in Neuroscience* 15 (1992): 127–132. **7.39** Adapted from Knight, R., Scabini, D., Woods, D., and Clayworth, C., Contributions of temporal-parietal junction to the human auditory P3, *Brain Research* 502 (1989): 109–116. © 1989 with kind permission of Elsevier Science-NL. **7.42** Fig. 2, Posner et al., "Effects of parietal injury on covert orienting of attention," *The Journal of Neuroscience* 4: 1863–1874, (1984). Adapted with permission of The Society for Neuroscience. **7.44** Cohen, J.D., Romero, R.D., Servan-Schreiber, D., and Farah, M.J., Mechanisms of spatial attention: The relation of macrostructure to microstructure in parietal neglect, *Journal of Cognitive Neuroscience* 6 (1994): 377–387. © 1994 by the Massachusetts Institute of Technology. Adapted with permission of MIT Press. **7.45** Eglin, M. Robertson, L.C., and Knight, R.T., Visual search performance in the neglect syndrome, *Journal of Cognitive Neuroscience* 1 (1989): 372–385. © 1989 by the Massachusetts Institute of Technology. Adapted with permission of MIT Press. **7.46** Grabowecky, M., Robertson, L.C., and Treisman, A., Preattentive processes guide visual search, *Journal of Cognitive Neuroscience* 5 (1993): 288–302. © 1993 by the Massachusetts Institute of Technology. Adapted with permission of MIT Press. **7.48** Adapted from Bisiach, E., and Luzzatti, C., Unilateral neglect of representational space, *Cortex* 14 (1978): 129–133. **7.49** Behrmann, M., and Tipper, S.P., Object-Based Attentional Mechanisms: Evidence from Patients with Unilateral Neglect, in Umilta, C., and Moscovitch, M. (Eds.), *Attention and Performance 15: Conscious and Nonconscious Information Processing*. Cambridge, MA: MIT Press, 1994, pp. 351–375. Adapted with permission of MIT Press. **7.51** Adapted from Humphreys, G.W., and Riddoch, M.J., Interactions Between Objects and Space-Vision Revealed Through Neuropsychology, in Meyers, D.E., and Kornblum, S. (Eds.), *Attention and Performance XIV*. Hillsdale, NJ: Lawrence Erlbaum Associates, 1992, pp. 143–162.

Chapter 8: **8.3** Petersen, L.R., and Petersen, M.R., Short-term retention of individual verbal items, *Journal of Experimental Psychology* 58 (1959): 193–198. Copyright © 1959 by the American Psychological Association. Adapted with permission. **8.5** Atkinson, R.C., and Shiffrin, R.M., Human Memory: A Proposed System and Its Control Processes, in Spence, K.W., and Spence, J.T. (Eds.), *The Psychology of Learning and Motivation*, Vol. 2. New York: Academic Press, 1968, pp. 89–195. Adapted by permission of the publisher. **8.7** Baddeley, A., and Hitch, G., Working Memory, in Bower, G.H. (Ed.), *The Psychology of Learning and Motivation*, Vol. 8. New York, Academic Press, 1974, pp. 47–89. Adapted by permission of the publisher. **8.8B** Adapted from Coren, S., Ward, L.M., and Enns, J.T., *Sensation and Perception*, 4th edition. Ft. Worth, TX: Harcourt Brace College Publishers, 1994. **8.10** Drachman, D.A., and Arbit, J., Memory and the hippocampal complex. II. Is memory a multiple process?, *Archives of Neurology* 15 (1966): 52–61. **8.11 and 8.13** Fig. 6.1, Corkin et al., "H.M.'s medial temporal lobe lesion: Findings from magnetic resonance imaging," *The Journal of Neuroscience* 17: 3964–3979, (1997). Adapted with permission of The Society for Neuroscience. **8.15** Adapted from Squire, L.R., and Slater, P., Electroconvulsive therapy and complaints of memory dysfunction: A prospective three-year follow-up study, *British Journal of Psychiatry* 142 (1983): 1–8. **8.17** Tulving, E., Gordon Hayman, C.A., and MacDonald, C.A., Long-lasting perceptual priming and semantic learning in amnesia. A case experiment, *Journal of Experimental Psychology* 17 (1991): 595–617. Copyright © 1991 by the American Psychological Association. Adapted with permission. **8.23** Adapted from Gabrieli, J., Fleischman, D., Keane, M., Reminger, S., and Morell, F., Double dissociation between memory systems underlying explicit and implicit memory in the human brain, *Psychological Science* 6 (1995): 76–82. **8.27** Figs. 2, 4, and 6, Zola-Morgan et al, Damage to the perirhinal cortex exacerbates memory impairment following lesions to the hippocampal formation, *The Journal of Neuroscience* 13: 251–265, (1993). Adapted with permission of The Society for Neuroscience. **8.32** Courtesy of Roberto Cabeza. **8.35** Grafton, S., Hazelttine, E., and Ivry, R., Functional mapping of sequence learning in normal humans, *Journal of Cognitive Neuroscience* 7 (1995): 497–510. © 1995 by the Massachusetts Institute of Technology. Adapted with permission of MIT Press. **p. 325** McLelland, J.L., Connectionist Models of Memory, in E. Tulving and F.I.M. Craik (Eds.), *The Oxford Handbook of Memory*, pp. 583–596. New York: Oxford University Press, 2000.

Chapter 9: **9.7 and 9.8** Courtesy of Tamara Swaab. © 1997 by the Massachusetts Institute of Technology. **9.12** McClelland, J.L., and Rumelhart, D.E., *Parallel Distributed Processing: Explorations in the Microstructure of Cognition*. Vol. 2: *Psychological and Biological Models*. Cambridge, MA: MIT Press, 1986. **9.18** Münte, T.F., Schilz, K., and Kutas, M., When temporal terms belie conceptual order, *Nature* 395 (1998): 71–73. **9.21** Adapted from Levelt, W.J.M., The Architecture of Normal Spoken Language Use, in Blanken, G., Dittman, J., Grimm, H., Marshall, J.C., and Wallesh, C-W. (Eds.), *Linguistic Disorders and Pathologies: An International Handbook*. Berlin: Walter de Gruyter, 1993. **9.22 and 9.23** Adapted from van Turennout, M., Hagoort, P., and Brown, C.M., Brain activity during speaking: From syntax to phonology in 40 milliseconds, *Science* 280 (1999): 572–574. **9.31** Kutas, M., and Federmeier,

K.D., Electrophysiology reveals semantic memory use in language comprehension, *Trends in Cognitive Sciences* 4 (2000): 463–470. **9.32** Adapted from Hagoort, P., Brown, C., and Groothusen, J., The syntactic positive shift (SPS) as an ERP measure of syntactic processing. Special issue: Event-related brain potentials in the study of language, *Language and Cognitive Processes* 8 (1993):439–483. **9.33** Münte, T.F., Heinze, H.-J., and Mangun, G.R., Dissociation of brain activity related to semantic and syntactic aspects of language, *Journal of Cognitive Neuroscience* 5 (1993): 335–344. © 1993 by the Massachusetts Institute of Technology. Adapted with permission of MIT Press.

Chapter 10: **10.19** Adapted from Delis, D., Robertson, L., and Efron, R., Hemispheric specialization of memory for visual hierarchical stimuli, *Neuropsychologia* 24 (1986): 205–214. **10.20 top** Adapted from Kimura, D., The Asymmetry of the human brain, *Scientific American* 228 (1973): 70–78. Copyright 1973 by Scientific American, Inc. All rights reserved. **10.21** Adapted from Bartholomeus, B., Effects of task requirements on ear superiority for sung speech, *Cortex* 10 (1974): 215–223. **10.23** Sergent, J., The cerebral balance of power: Confrontation or cooperation, *Journal of Experimental Psychology: Human Perception and Performance* 8 (1982): 253–272. Copyright © 1982 by the American Psychological Association. Adapted with permission. **10.28** Adapted from Kitterle, F., Christman, S., and Hellige, J., Hemispheric differences are found in identification, but not detection of low versus high spatial frequencies, *Perception and Psychophysics* 48 (1990), 297–306. **10.31** Kosslyn, S.M., Koenig, O., Barret, A., Cave, C.B., Tang, J., and Gabrieli, J.D.E., Evidence for two types of spatial representations: Hemispheric specialization for categorical and coordinate relations, *Journal of Experimental Psychology: Human Perception and Performance* 15 (1989): 723–735. Copyright © 1989 by the American Psychological Association. Adapted with permission. **10.35** Marsolek, C.J., Abstract visual-form representations in the left cerebral hemisphere, *Journal of Experimental Psychology: Human Perception and Performance* 21 (1995): 375–386. Copyright © 1995 by the American Psychological Association. Adapted with permission. **10.36** Gazzaniga, M.S., Cerebral specialization and interhemispheric communication: Does the corpus callosum enable the human condition? *Brain* 123 (2000): 1293–1326. **10.38** Previc, F.H., A general theory concerning the prenatal origins of cerebral lateralization in humans, *Psychological Review* 98 (1991): 299–334. Copyright © 1991 by the American Psychological Association. Adapted with permission.

Chapter 11: **11.8** Adapted from Kandel, E.R. Schwartz, J.H., and Jessell, T.M. (Eds.), *Principles of Neural Science*, 3rd edition. Norwalk, Connecticut: Appleton & Lange, 1991. Copyright © 1991 by Appleton & Lange. **11.11 and 11.12** Bizzi, E., Accornero, N. Chapplel, W., and Hogan, N., Posture control and trajectory formation during arm movement, *The Journal of Neuroscience* 4: 2738–2744, (1984). Reprinted with permission of The Society for Neuroscience. **11.14** Adapted from

Abrams, R.A., and Landgraf, J.Z., Differential use of distance and location information for spatial localization, *Perception and Psychophysics* 47 (1990): 349–359. **11.17, 11.18, and 11.19** Adapted from Georgopoulos, A.P., Neurophysiology of Reaching, in Jeannerod, M. (Ed.), *Attention and Performance XIII: Motor Representation and Control.* Hillsdale, NJ: Lawrence Erlbaum Associates, 1990, pp. 227–263. **11.20** Adapted from Ashe, J., Taira, M., Smyrnis, N., Pellizzer, G., Gerorakopoulos, T., Lurito, J.T., and Georgopoulos, A.P., Motor cortical activity preceding a memorized movement trajectory with an orthogonal bend, *Experimental Brain Research* 95 (1993): 118–130. **11.21** Adapted from Graziano, M.S.A., and Gross, C.G., Mapping space with neurons, Current Directions in *Psychological Science* 3 (1994): 164–167. **11.22** Desmurget, M., Epstein, C.M., Turner, R.S., Prablanc, C., Alexander, G.E., and Grafton, S.T., Role of the posterior parietal cortex in updating reaching movements to a visual target, *Nature Neuroscience* 2 (1999): 563–567. **11.24** Adapted from Roland, P.E., *Brain Activation.* New York: Wiley-Liss, 1993. **11.27** Adapted from Mushiake, H., Masahiko, I., and Tanji, J., Neuronal activity in the primate premotor, supplementary, and precentral motor cortex during visually guided and internally determined sequential movements, *Journal of Neurophysiology* 66 (1991): 705–718. **11.29** Karni, A., Meyer, G., Jezzard, P., Adams, M.M., Turner, R., and Ungerleider, L.G., Functional MRI evidence for adult motor cortex plasticity during motor skill learning, *Nature* 337 (1995): 155–158. **11.33** Adapted from Heilman, K.M., Rothi, L.J., and Valenstein, E., Two forms of ideomotor apraxia, *Neurology* 32 (1982): 342–346. **11.35** Adapted from Wise, S.P., Di Pellegrino, G., and Boussaoud, D., The premotor cortex and nonstandard sensorimotor mapping, *Canadian Journal of Physiology and Pharmacology* 74 (1996): 469–482. **11.40** Adapted from Wichmann, T., and DeLong, M.R., Functional and pathophysiological models of the basal ganglia, *Current Opinion in Neurobiology* 6 (1996): 751–758. **11.42** Berns, G.S., and Sejnowski, T., How the Basal Ganglia Makes Decisions, in A. Damasio, H. Damasio, and Y. Christen (Eds.), *The Neurobiology of Decision Making* (pp. 101-113). Cambridge, MA: MIT Press, 1996. **11.43** Adapted from Wichmann, T., and DeLong, M.R., Functional and pathophysiological models of the basal ganglia, *Current Opinion in Neurobiology* 6 (1996): 751–758. **p. 483** Adapted from Franz, E., Eliassen, J., Ivry, R., and Gazzaniga, M., Dissociation of spatial and temporal coupling in the bimanual movements of callosotomy patients, *Psychological Science* 7 (1996): 306–310.

Chapter 12: **12.2 and 12.6** Adapted from Fuster, J.M., *The Prefrontal Cortex: Anatomy, Physiology, and Neuropsychology of the Frontal Lobe*, 2nd edition. New York: Raven Press, 1989. **12.7** Rao, S.C., Rainer, G., and Miller, E.K., Integration of what and where in the primate prefrontal lobe cortex, *Science* 276 (1997): 821–824. **12.10** Adapted from Milner, B., Corsi, P., and Leonard, G., Frontal-lobe contributions to recency judgements, *Neuropsychologia* 29 (1991): 601–618. **12.12** Glisky, E.L., Polster, M.R., and Routhuieaux, B.C., Double dissociation between item and source memory, *Neuropsychol-*

ogy 9 (1995): 229–235. Copyright © 1995 by the American Psychological Association. Adapted with permission. **12.14** D'Esposito, M., Zarahn, E., Balard, D., Shin, R.K., and Lease, J., Functional MRI studies of spatial and nonspatial working memory, *Cognitive Brain Research* 7 (1998): 1–13. **12.17C** Thompson-Schill, S.L., Swick, D., Farah, M.J., D'Esposito, M., Kan, I.P., and Knight, R.T., Verb generation in patients with focal frontal lesions: A neuropsychological test of neuroimaging findings, *Proc. Natl. Acad. Sci. U.S.A.* 95 (1998): 15855–15860. **12.20** Chao, L.L., and Knight, R.T., Human prefrontal lesions increase distractability to irrelevant sensory inputs, *Neuroreport: Int. J. Rapid Commun. Res. Neurosci.* 6 (1995): 1605–1610 **12.22** Rogers, R.D., Shahakian, R.A., Hodges, J.R., Polkey, C.E., Kennard, C., and Robbins, T.W., Dissociating executive mechanisms of task control following frontal lobe damage and Parkinson's disease, *Brain* 121 (1998): 815–842. **12.23B** Konishi, S., Nakajima, K., Uchida, I., Kameyama, M., Nakahara, K., Sekihara, K., and Miyashita, Y., Transient activation of inferior prefrontal cortex during cognitive set shifting, *Nature Neuroscience* 1 (1998): 80–84. **12.24** Adapted from Shallice, T., Burgess, P.W., Schon, F., and Baxter, D.M., The origins of utilization behaviour, *Brain* 112 (1989): 1587–1598. **12.26** Adapted from Snyder, A.Z., Abdullaev, Y.G., Posner, M.I., and Raichle, M.E., Scalp electrical potentials reflect regional cerebral blood flow responses during processing of written words, *Proc. Natl. Acad. Sci. U.S.A.* 92 (1995): 1689–1693. **12.28** Adapted from Gehring, W.J., Goss, B., Coles, M.G.H., Meyer, D.E., and Donchin, E., A neural system for error detection and compensation, *Psychological Science* 43 (1993): 385–390.

Chapter 13: 13.4 MacLean, P.D., Psychosomatic disease and the "visceral brain": Recent developments bearing on the Papez theory of emotion, *Psychosomatic Medicine* 11 (1949): 338–353 **13.5** Davidson, R.J., Jackson, D.C., and Kalin, N.H., Emotion, plasticity, context, and regulation: Perspectives from affective neuroscience, *Psychological Bulletin* 126 (2000): 890–909. **13.10** Davis, M., The Role of the Amygdala in Conditioned Fear, in J.P. Aggleton (Ed.), *The Amygdala: Neurobiological Aspects of Emotion, Memory and Mental Dysfunction* (pp. 255-306). New York: Wiley-Liss, 1992. **13.16** Anderson, A.K., and Phelps, E.A., Lesions of the juman amygdala impair enhanced perception of emotionally salient events, *Nature* 411 (2001): 305–309. **13.19** Tomarken, A.J., Davidson, R.J., Wheeler, R.E., and Doss, R.C., Individual differences in anterior brain asymmetry and fundamental dimensions of emotion, *Journal of Personality and Social Psychology* 62 (1992): 676–682.

Chapter 14: 14.4B Adapted from Catania, K.C., Northcutt, R.G., Kaas, J.H., and Beck, P.D., Nose stars and brain stripes, *Nature* 364 (1993): 493 and Catania, K.C., Northcutt, R.G., Kaas, J.H., Organization of the somosensory cortex of the star-nosed mole, *Journal of Comparative Neurology* 351 (1995): 549–567. **14.10B** Florence, S.L., and Kaas, J.H., Large-scale reorganization at multiple levels of the somatosensory pathway follows therapeutic amputation of the hand in monkeys, *Journal of Neuroscience* 15 (1995): 8083–8095. **14.10C** Kaas, J.H., Nelson, R.J., Sur. M., Dykes, R.W., and Merzenich, M.M. (1984). The somatotopic organization of the ventroposterior thalamus of the squirrel monkey, Saimiri sciureus. *J. Comp. Neurol.* 226:111–140. **14.10D** Merzenich, M.M., Kaas, J.H., Sur, M., and Lin, C.S., Double representation of the body surface within cytoarchitectonic areas 3b and 1 in "SI" in the owl monkey (*Aotus trivigatus*). *Journal of Comparative Neurology* 181 (1978): 47–73. **14.13 and 14.14** Adapted from Gaulin, S.J.C., and Fitzgerald, R.W., Sexual selection for spatial-learning ability, *Animal Behavior* 37 (1989): 322–331. **14.15** Adapted from Jacobs, L.F., Gaulin, S.J.C., Sherry, D.F., and Hoffman, G.E., Evolution of spatial cognition: Sex-specific patterns of spatial behavior predict hippocampal size, *Proc. Natl. Acad. Sci. U.S.A.* 87 (1990): 6349–6352. **14.19** Gallistel, C.R., The Replacement of General-Purpose Theories with Adaptive Specializations, in Gazzaniga, M.S. (Ed.), *The Cognitive Neurosciences*. Cambridge, MA: MIT Press, 1995, pp. 1255–1267. © 1995 Massachusetts Institute of Technology. Adapted with permission of MIT Press. **14.20** Adapted from Harkness, R.D., and Maroudas, N.G., Central place foraging by an ant (Cataglyphis bicolor Fab.): A model of searching, *Animal Behaviour* 33 (1985): 916–928. Gallistel, C.R., The Replacement of General-Purpose Theories with Adaptive Specializations, in Gazzaniga, M.S. (Ed.), *The Cognitive Neurosciences*. Cambridge, MA: MIT Press, 1995, pp. 1255–1267. © 1995 Massachusetts Institute of Technology. Adapted with permission of MIT Press. **p. 599** Adapted from Lettvin, J.Y., Maturana, H.R., McCulloch, W.S., and Pitts, W.H., What the frog's eye tells the frog's brain, *Proc. Insti. Radio Engineers* 47 (1959): 1940–1951.

Chapter 15: 15.3 Adapted from Kandel, E.R. Schwartz, J.H., and Jessell, T.M. (Eds.), *Principles of Neural Science*, 3rd edition. Norwalk, Connecticut: Appleton & Lange, 1991. Copyright © 1991 by Appleton & Lange. **15.7** Courtesy of Amy Needham. **15.11A** Carpenter, M., *Human Neuroanatomy*, 7th edition. Baltimore, Maryland: William & Wilkins, 1976. © 1976 by Williams & Wilkins. Adapted by permission of the publisher. **15.17** Rakic, P., Corticogenesis in Human and Non-human Primates, in Gazzaniga, M.S. (Ed.), *The Cognitive Neurosciences*. Cambridge, MA: MIT Press, 1995, pp. 127–146. © 1995 Massachusetts Institute of Technology. Adapted with permission of MIT Press. **15.21** Huttenlocher, P.R., and Dabholkar, A.S., Regional differences in synaptogenesis in human cerebral cortex, *Journal of Comparative Neurology* 387 (1997): 167–178. **15.23** Von Melcher, L., Pallas, S.L., and Sur, M., Visual behavior mediated by retinal projections directed to the auditory pathway. *Nature* 404 (2000): 871–876. **15.24** Adapted from Kandel, E.R. Schwartz, J.H., and Jessell, T.M. (Eds.), *Principles of Neural Science*, 3rd edition. Norwalk, Connecticut: Appleton & Lange, 1991. Copyright © 1991 by Appleton & Lange. **15.29** Elbert, T., Pantev, C., Wienbruch, C., Rockstroh, B., and Taub, E., Increased cortical representation of the fingers of the left hand in string players, *Science* 270 (1995): 305–307. **p. 637** Adapted from Caviness, V.S., Jr., and Rakic, P.,

Mechanisms of cortical development: A view from mutations in mice, *Annual Review of Neuroscience* 1 (1978): 297–326.

Chapter 16: **16.1** Adapted from Weiskrantz, L., Warrington, E.K., Sanders, M.D., and Marshall, J., Visual capacity in the hemianopic field following a restricted occipital ablation, *Brain* 97 (1974): 709–728. **6.8** Adapted from Marcel, A., Conscious and unconscious perception: Experiments on visual masking and word recognition, *Cognitive Psychology* 15 (1983): 197–237. Adapted from Marcel, A., Conscious and unconscious perception: An approach to the relations between phenomenal experience and perceptual processes, *Cognitive Psychology* 15 (1983): 238–300. **16.11** Cohen Newsome, W.T., and Pare, E.B., A selective impairment of motion perception following lesions of the middle temporal visual area (MT), *The Journal of Neuroscience* 8 (1988): 2201–2211. Adapted with permission of The Society for Neuroscience. **16.12** Cotterill, R.M.J., On the neural correlates of consciousness, *Japanese Journal of Cognitive Science* 4 (1997): 31–34. **16.13** Cohen, J.D., Botvinick, M., and Carter, C.S., Anterior cingulate and prefrontal cortex: Who's in control? *Nature Neuroscience* 3 (2000): 421–423.

TABLES

6.1 Adapted from Farah, M.J., *Visual Agnosia: Disorders of Object Recognition and What They Tell Us about Normal Vision.* Cambridge, MA: MIT Press, 1990. Adapted with permission of MIT Press.

INDEX